## Channel Codes

Channel coding lies at the heart of digital communication and data storage. Fully updated to include current innovations in the field, including a new chapter on polar codes, this detailed introduction describes the core theory of channel coding, decoding algorithms, implementation details, and performance analyses. This edition includes over 50 new end-of-chapter problems to challenge students and numerous new figures and examples throughout.

The authors emphasize a practical approach and clearly present information on modern channel codes, including polar, turbo, and low-density parity-check (LDPC) codes, as well as detailed coverage of BCH codes, Reed–Solomon codes, convolutional codes, finite geometry codes, and product codes for error correction, providing a one-stop resource for both classical and modern coding techniques.

Assuming no prior knowledge in the field of channel coding, the opening chapters begin with basic theory to introduce newcomers to the subject. Later chapters then begin with classical codes, continue with modern codes, and extend to advanced topics such as code ensemble performance analyses and algebraic LDPC code design.

- 300 varied and stimulating end-of-chapter problems test and enhance learning, making this an essential resource for students and practitioners alike.
- Provides a one-stop resource for both classical and modern coding techniques.
- Starts with the basic theory before moving on to advanced topics, making it perfect for newcomers to the field of channel coding.
- 180 worked examples guide students through the practical application of the theory.

**William E. Ryan** is a Senior Associate at Zeta Associates, a Lockheed Martin company, where he designs communication waveforms and their receivers for difficult channels. Dr. Ryan has published approximately 150 journal and conference papers on channel codes and channel equalization for data communication and magnetic storage. He has also co-authored several book chapters and a graduate-level textbook, including the first edition of *Channel Codes: Classical and Modern*. He was an Associate Editor for *IEEE Transactions on Communications* from 1998 through 2005. He is a Fellow of the IEEE for his work in coding for data communication and storage.

**Shu Lin** is an Adjunct Professor in the Department of Electrical and Computer Engineering, University of California, Davis, and has been on the faculties of the University of Hawaii and Texas A&M University. He has authored and co-authored numerous technical papers and several books, including the successful *Error Control Coding* with Daniel J. Costello. He is an IEEE Life Fellow and has received several awards, including the Alexander von Humboldt Research Prize for U.S. Senior Scientists (1996) and the IEEE Third-Millennium Medal (2000). He holds a NASA service award and has won the 2020 IEEE Leon Kirchmayer Graduate Teaching Award.

**Stephen G. Wilson** is Professor Emeritus of Electrical and Computer Engineering at the University of Virginia, where he was on the faculty from 1976 to 2019. He taught courses and conducted research in digital communication systems, especially as relates to error control coding. He is the author of the textbook *Digital Modulation and Coding*, and has presented approximately 150 technical papers in this field. He served as Area Editor for Coding Techniques for *IEEE Transactions on Communications*. He consults to several industrial organizations in the areas of communication system analysis and digital signal processing.

# Channel Codes

## Classical and Modern

Second Edition

WILLIAM E. RYAN
*Zeta Associates*

SHU LIN
*University of California, Davis*

STEPHEN G. WILSON
*University of Virginia*

Shaftesbury Road, Cambridge CB2 8EA, United Kingdom

One Liberty Plaza, 20th Floor, New York, NY 10006, USA

477 Williamstown Road, Port Melbourne, VIC 3207, Australia

314–321, 3rd Floor, Plot 3, Splendor Forum, Jasola District Centre,
New Delhi – 110025, India

103 Penang Road, #05–06/07, Visioncrest Commercial, Singapore 238467

Cambridge University Press is part of Cambridge University Press & Assessment,
a department of the University of Cambridge.

We share the University's mission to contribute to society through the pursuit of
education, learning and research at the highest international levels of excellence.

www.cambridge.org
Information on this title: www.cambridge.org/9781009335904

DOI: 10.1017/9781009335928



When citing this work, please include a reference to the DOI 10.1017/9781009335928

First published 2009
Second edition 2025

*A catalogue record for this publication is available from the British Library*

*Library of Congress Cataloging-in-Publication Data*
Names: Ryan, William E., 1958– author. | Lin, Shu, 1937– author. |
Wilson, Stephen G., 1945–, author.
Title: Channel codes : classical and modern / William E. Ryan, Shu Lin, Stephen G. Wilson.
Description: Second edition. | New York, NY, USA : Cambridge University
Press, 2025. | Includes bibliographical references and index.
Identifiers: LCCN 2024009947 | ISBN 9781009335904 (hardback) |
ISBN 9781009335928 (ebook)
Subjects: LCSH: Coding theory. | Digital communications.
Classification: LCC TK5102.92 .R93 2025 | DDC 621.382–dc23/eng/20240405
LC record available at https://lccn.loc.gov/2024009947

ISBN 978-1-009-33590-4 Hardback

Additional resources for this publication at www.cambridge.org/channel-codes-2ed

**W. Ryan dedicates this book to Stephanie, Faith, Luke, and Grant**

**S. Lin gives special thanks to his wife, children, and grandchildren for their continuing love and support throughout this project**

**S. Wilson dedicates this work to Lynanne, Eric, and Elisa**

# Contents

# Detailed Contents

# Preface to the Second Edition

It has been approximately 15 years since the appearance of the first edition of *Channel Codes: Classical and Modern*. During that time, the field has progressed to the point where we felt an update was necessary. Most noteworthy among the advances in channel codes is the advent of polar codes, a topic to which we have devoted a whole new chapter. We have also extended four of our previous chapters to include some of the many advances in algebraic LDPC code design that have occurred. Another area that has received a good deal of attention in the coding community is code design and performance limits for short messages. We have added selected results on these topics in this second edition. Also contained in this new edition are several older topics that, in hindsight, we feel improve the first edition.

We list here the additions and modifications made to the first edition of *Channel Codes*. New problems were added to almost every chapter. The errata list from the first edition has been incorporated throughout. We acknowledge Professor Erik Agrell and his students for their contributions to this list.

In Chapter 1, the coverage was shortened and focused. A discussion on finite-blocklength code limits was added, as were a few problems. In Chapter 2, two problems were added.

In Chapter 3, subsections on the following topics were incorporated, along with several problems: error probability of syndrome decoders; the Meggitt decoder; CRC codes; soft-decision decoding via the Chase algorithm; errors-only RS decoding via the Berlekamp–Massey algorithm; errors-and-erasures RS decoding; and RS code performance calculations.

In Chapter 4, the new subsections cover the following topics: the WAVA decoder for tail-biting convolutional codes; the list Viterbi algorithm and a maximum-likelihood decoding algorithm for tail-biting convolutional codes; the Fano decoding algorithm; the list-BCJR algorithm; and a list of optimum rate-1/2 convolutional codes.

In Chapter 5, discussions of the less important sum–product algorithm implementations were removed and the coverage of a few topics was improved. A subsection on the single-scan min-sum algorithm has been added.

A section on protograph-based Raptor-like LDPC codes was added to Chapter 6. For Chapter 7, we improved the treatment of the turbo product code iterative decoding algorithm and appended a section on nonbinary turbo codes for short messages. Chapter 8 is unchanged from the first edition.

A subsection on the reciprocal-channel approximation algorithm was appended to Chapter 9. Chapter 10 is an entirely new chapter on polar codes.

Chapter 11 is Chapter 10 from the first edition with subsections on the following topics added: construction of QC-LDPC codes based on two-dimensional Euclidean geometries and quasi-cyclic codes based on partial geometries.

Chapter 12 is Chapter 11 from the first edition with subsections on the following topics added: construction of QC-LDPC codes based on two arbitrary subsets of a finite field and QC-LDPC codes constructed based on cyclic codes of prime length.

Chapter 13 is Chapter 12 from the first edition with a new subsection on globally coupled LDPC codes inserted. Chapter 14 is Chapter 13 from the first edition, unchanged.

Chapter 15 is Chapter 14 from the first edition with subsections on the following topics incorporated: decoding of nonbinary LDPC codes (this section has been fully rewritten for the second edition); construction of nonbinary LDPC codes based on RS codes; binary-to-NB replacement construction of nonbinary QC-LDPC codes. *Chapter 15 from the first edition has been removed for this second edition.*

# Acknowledgments (Second Edition)

We would first like to acknowledge Dr. Juane Li for running many of the simulations and creating many of the new figures that appear in Chapters 11 to 15. She designed many of the codes in Chapter 15. The authors also thank Phil Meyler and Anna Littleword, our first-edition editors, and Jane Adams, our second-edition editor, all of Cambridge University Press.

W. Ryan would like to acknowledge his home institutions over the past three decades: Zeta Associates Inc., The University of Arizona, and New Mexico State University. Each institution was supportive of the research and development of many of the techniques that appear in this book. He would also like to thank Dick Ryan for his LaTeX expertise, including installation and nuances.

# Preface to the First Edition

The title of this book, *Channel Codes: Classical and Modern*, was selected to reflect the fact that this book does indeed cover both classical and modern channel codes. It includes BCH codes, Reed–Solomon codes, convolutional codes, finite-geometry codes, turbo codes, low-density parity-check (LDPC) codes, and product codes. However, the title has a second interpretation. While the majority of this book is on LDPC codes, these can rightly be considered to be both classical (having first been discovered in 1961) and modern (having been rediscovered c. 1996). This is exemplified by David Forney's statement at his August 1999 IMA talk on codes on graphs: "It feels like the early days...." As another example of the classical/modern duality, finite-geometry codes were studied in the 1960s and thus are classical codes. However, they were rediscovered by Shu Lin *et al.* circa 2000 as a class of LDPC codes with very appealing features and are thus modern codes as well. The classical and modern incarnations of finite-geometry codes are distinguished by their decoders: one-step hard-decision decoding (classical) versus iterative soft-decision decoding (modern).

It has been 60 years since the publication in 1948 of Claude Shannon's celebrated *A Mathematical Theory of Communication*, which founded the fields of channel coding, source coding, and information theory. Shannon proved the existence of channel codes that ensure reliable communication provided the information rate for a given code did not exceed the so-called capacity of the channel. In the first 45 years that followed Shannon's publication, a large number of very clever and very effective coding systems were devised. However, none of these had been demonstrated, in a practical setting, to closely approach Shannon's theoretical limit. The first breakthrough came in 1993 with the discovery of turbo codes, the first class of codes shown to operate near Shannon's capacity limit. A second breakthrough came circa 1996 with the rediscovery of LDPC codes, which were also shown to have near-capacity performance. (LDPC codes were first invented in 1961 and mostly ignored thereafter. The computing power at that time made it difficult to demonstrate their practical capabilities.) Because it has been over a decade since the discovery of turbo and LDPC codes, the knowledge base for these codes is now quite mature and the time is ripe for a new book on channel codes.

This book was written for graduate students in engineering and computer science, as well as research and development engineers in industry and academia. We felt compelled to collect all of this information in one source as it has been scattered across many journal and conference papers. With this book, those entering the field of channel coding, and those wishing to advance their knowledge, have a convenient single resource for learning about both classical and modern channel codes. Further, whereas the archival literature is written for experts, this textbook is appropriate for both the novice (earlier chapters) and the expert (later chapters). The book begins slowly and does not presuppose prior knowledge in the field of channel coding. It then extends to frontiers of the channel coding field, as is evident from the table of contents.

The topics selected for this book of course reflect the experiences and interests of the authors, but they were also selected for their importance in the study of channel codes – not to mention the fact that additional chapters would make the book physically unwieldy. Thus, the emphasis of this book is on codes for binary-input channels, including the binary-input additive white Gaussian noise channel, the binary symmetric channel, and the binary erasure channel. One notable area of omission is coding for wireless channels, such as MIMO channels. While not covered, this book is useful for students and researchers in that area as well because many of the techniques applied to the additive white Gaussian noise channel, our main emphasis, can be extended to wireless channels. Another notable omission is soft-decision decoding of Reed–Solomon codes. While extremely important, this topic is not as mature as those in this book.

Several different course outlines are possible with this book. The most obvious for a first graduate course on channel codes includes selected topics from Chapters 1 to 5, and 7. Such a course introduces the student to capacity limits for several common channels (Chapter 1). It then provides the student with an introduction to just enough algebra (Chapter 2) to understand BCH and Reed–Solomon codes and their decoders (Chapter 3). Next, this course introduces the student to convolutional codes and their decoders (Chapter 4). This course next provides the student with an introduction to LDPC codes and iterative decoding (Chapter 5). Finally, with the knowledge gained from Chapters 4 and 5 in place, the student is ready to tackle turbo codes and turbo decoding (Chapter 7). The material contained in Chapters 1 to 5, and 7 is too much for a single-semester course and the instructor will have to select a preferred subset of that material.

Another possible course outline, for a more advanced course on LDPC code design, includes selected topics from Chapters 10 to 14. This course would first introduce the student to LDPC code design using Euclidean geometries and projective geometries (Chapter 10). Then the student would learn about LDPC code design using finite fields (Chapter 11) and combinatorics and graphs (Chapter 12). Next, the student would apply some of the techniques from these earlier chapters to design codes for the binary erasure channel (Chapter 13). Lastly, the student would learn design techniques for nonbinary LDPC codes (Chapter 14).

As a final example of a course outline, a course could be centered on computer-based design of LDPC codes. Such a course would include Chapters 5, 6, 8, and 9. This course would be for those who have had a course on classical channel codes, but who are interested in LDPC codes. The course would begin with an introduction to LDPC codes and various LDPC decoders (Chapter 5). Then the student would learn about various computer-based code design approaches, including Gallager codes, MacKay codes, codes based on protographs, and codes based on accumulators (Chapter 6). Next, the student would learn about assessing the performance of LDPC code ensembles from a weight distribution perspective (Chapter 8). Lastly, the student would learn about assessing the performance of (long) LDPC codes from a decoding threshold perspective via the use of density evolution and EXIT charts (Chapter 9).

All of the chapters contain a good number of problems. The problems are of various types, including those that require routine calculations or derivations, those that require computer solution or computer simulation, and those that might be characterized as a semester project. The authors have selected the problems to strengthen the student's knowledge of the material in each chapter (e.g., by requiring a computer simulation of a decoder) and to extend that knowledge (e.g., by requiring the proof of a result not contained in the chapter).

# Acknowledgments (First Edition)

We wish to thank, first of all, Professor Ian Blake, who read an early version of the entire manuscript and provided many important suggestions that led to a much improved book.

We also wish to thank the many graduate students who have been a tremendous help in the preparation of this book. They have not only helped with typesetting, computer simulations, proofreading, and figures, many of their research results can be found in this book. The students (former and current) who have contributed to W. Ryan's portion of the book are: Dr. Yang Han, Dr. Yifei Zhang, Dr. Michael (Sizhen) Yang, Dr. Yan Li, Dr. Gianluigi Liva, Dr. Fei Peng, Shadi Abu-Surra, Kristin Jagiello (who proofread eight chapters), and Matt Viens. Gratitude is also due to Li Zhang (S. Lin's student) who provided valuable feedback on Chapters 6 and 9. Finally, W. Ryan acknowledges Sara Sandberg of Luleå Institute of Technology for helpful feedback on an early version of Chapter 5. The students who have contributed to S. Lin's portion of the book are: Dr. Bo Zhou, Qin Huang, Dr. Ying Y. Tai, Dr. Lan Lan, Dr. Lingqi Zeng, Jingyu Kang and Li Zhang. Dr. Bo Zhou and Qin Huang deserve special appreciation for typing S. Lin's chapters and overseeing the preparation of the final version of his chapters.

Lastly, we thank Dr. Marc Fossorier who provided comments on Chapter 14 and Professor Ali Ghrayeb who provided comments on Chapter 7.

We are grateful to the National Science Foundation, the National Aeronautics and Space Administration, and the Information Storage Industry Consortium for their many years of funding support in the area of channel coding. Without their support, many of the results in this book would not have been possible. We also thank the University of Arizona and the University of California, Davis for their support in the writing of this book.

Finally, we would like to give special thanks to our wives, children, and grandchildren for their continuing love and affection throughout this project.

# 1 Coding and Capacity

## 1.1 Digital Data Communication and Storage

Digital communication systems are ubiquitous in our daily lives. The most obvious examples include cell phones, digital television via satellite or cable, digital radio, wireless internet connections via Wi-Fi, and wired internet connection via cable modem. Additional examples include digital data-storage devices, including magnetic ("hard") disk drives, magnetic tape drives, optical disk drives (e.g., CD, DVD, blu-ray), and flash drives. In the case of data storage, information is communicated from one point in time to another rather than one point in space to another. Each of these examples, while widely different in implementation details, generally fits into a common digital communication framework first established by C. Shannon in his 1948 seminal paper [1]. This framework is depicted in Figure 1.1, whose various components are described as follows.

*Source* and *user* (or *sink*). The information source may originally be in analog form (e.g., speech or music) and then later digitized, or it may originally be in digital form (e.g., computer files). We generally think of its output as a sequence of bits, which follow a probabilistic model. The user of the information may be a person, a computer, or some other electronic device.

*Source encoder* and *source decoder*. The encoder is a processor that converts the information source bit sequence into an alternative bit sequence with a more efficient representation of the information, that is, with fewer bits. Hence, this operation is often called *compression*. Depending on the source, the compression can be *lossless* (e.g., for computer data files) or *lossy* (e.g., for video, still images, or music, where the loss can be made to be imperceptible or acceptable). The source decoder is the encoder's counterpart that recovers the source sequence exactly, in the case of lossless compression, or approximately, in the case of lossy compression, from the encoder's output sequence.

*Channel encoder* and *channel decoder*. The role of the channel encoder is to protect the bits to be transmitted over a channel subject to noise, distortion, and interference. It does so by converting its input into an alternate sequence possessing redundancy, whose role is to provide immunity from the various channel impairments. The ratio of the number of bits that enter the channel encoder to the number that depart from it is called the *code rate*, denoted by $R$, with $0 < R < 1$. For example, if a 1000-bit codeword is assigned to each 500-bit information word, $R = 1/2$, and there are 500 redundant bits in each codeword.

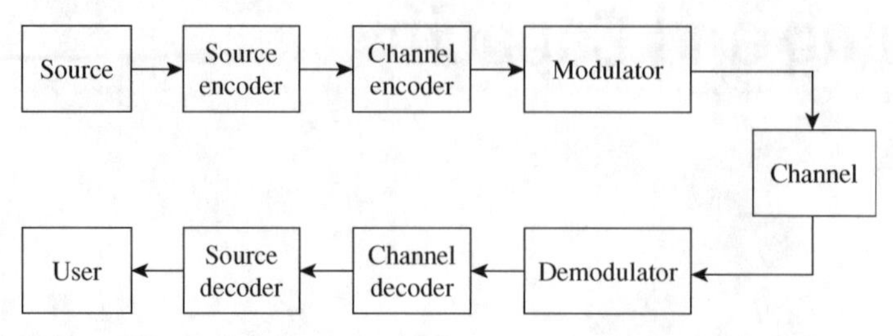


**Figure 1.1** Basic digital communication (or storage)-system block diagram due to Shannon.

The function of the channel decoder is to recover from the channel output the input to the channel encoder (i.e., the compressed sequence) in spite of the presence of noise, distortion, and interference in the received word.

*Modulator* and *demodulator*. The modulator converts the channel-encoder output bit stream into a form that is appropriate for the channel. For example, for a wireless communication channel, the bit stream must be represented by a high-frequency signal to facilitate transmission with an antenna of reasonable size. Another example is a so-called modulation code used in data storage. The modulation encoder output might be a sequence that satisfies a certain runlength constraint (runs of like symbols, for example) or a certain spectral constraint (the output contains a null at DC, for example). The demodulator is the modulator's counterpart that recovers the modulator input sequence from the modulator output sequence.

*Channel*. The channel is the physical medium through which the modulator output is conveyed, or by which it is stored. Our experience teaches us that the channel adds noise and often interference from other signals, on top of the signal distortion that is ever-present, albeit sometimes to a minor degree. For our purposes, the channel is modeled as a probabilistic device, and examples will be presented below. Physically, the channel can include antennas, amplifiers, and filters, both at the transmitter and at the receiver at the ends of the system. For a hard-disk drive, the channel would include the write head, the magnetic medium, the read head, the read amplifier and filter, and so on.

Following Shannon's model, Figure 1.1 does not include such blocks as encryption/decryption, symbol-timing recovery, and scrambling. The first of these is optional and the other two are assumed to be ideal and accounted for in the probabilistic channel models. On the basis of such a model, Shannon showed that a channel can be characterized by a parameter, $C$, called the *channel capacity*, which is a measure of how much information the channel can convey, much like the capacity of a plumbing system to convey water. Although $C$ can be represented in several different units, in the context of the channel code rate $R$, which has the unit of information bits per channel bit, Shannon showed that codes exist that provide arbitrarily reliable communication provided that the code rate satisfies $R < C$. He further showed that, conversely, if $R > C$, there exists no code that provides reliable communication.

Later in this chapter, we review the capacity formulas for a number of commonly studied channels for reference in subsequent chapters. Prior to that, however, we give an overview of various channel-coding approaches for error avoidance in data-transmission and data-storage scenarios. We then introduce the first channel code invented, the $(7,4)$ Hamming code, by which we mean a code that assigns to each 4-bit information word a 7-bit codeword according to a recipe specified by R. Hamming in 1950 [2]. This will introduce to the novice some of the elements of channel codes and will serve as a launching point for subsequent chapters. After the introduction to the $(7,4)$ Hamming code, we present code and decoder design criteria and code performance measures, all of which are used throughout this book.

## 1.2 Channel Coding Overview

The large number of coding techniques for error prevention may be partitioned into the set of automatic request-for-repeat (ARQ) schemes and the set of forward-error-correction (FEC) schemes. In ARQ schemes, the role of the code is simply to reliably detect whether or not the received word (e.g., received packet) contains one or more errors. In the event a received word does contain one or more errors, a request for retransmission of the same word is sent out from the receiver back to the transmitter. The codes in this case are said to be *error-detection codes*. In FEC schemes, the code is endowed with characteristics that permit error correction through an appropriately devised decoding algorithm; there is no receiver feedback. The codes for this approach are said to be *error-correction codes*, or sometimes *error-control codes*. There also exist *hybrid FEC/ARQ schemes* in which a request for retransmission occurs if the decoder fails to correct the errors incurred over the channel and detects this fact. Note that this is a natural approach for data-storage systems: if the FEC decoder fails, an attempt to reread the data is made. The codes in this case are said to be *error-detection-and-correction codes*.

The basic ARQ schemes can broadly be subdivided into the following protocols. First is the *stop-and-wait ARQ* scheme in which the transmitter sends a codeword (or encoded packet) and remains idle until the acknowledgment status signal is returned from the receiver. If a positive acknowledgment (ACK) is returned, a new packet is sent; otherwise, if a negative acknowledgment (NAK) is returned, the current packet, which was stored in a buffer, is retransmitted. The stop-and-wait method is inherently inefficient due to the idle time spent waiting for confirmation.

In *go-back-N ARQ*, the idle time is eliminated by continually sending packets while waiting for confirmations. If a packet is negatively acknowledged, that packet and the $N-1$ subsequent packets sent during the round-trip delay are retransmitted. Note that this preserves the ordering of packets at the receiver.

In *selective-repeat ARQ*, packets are continually transmitted as in go-back-$N$ ARQ, except only the packet corresponding to the NAK message is retransmitted. (The packets have "headers," which effectively number the information block for identification.)

Observe that, because only one packet is retransmitted rather than $N$, the throughput of accepted packets is increased with selective-repeat ARQ relative to go-back-$N$ ARQ. However, there is the added requirement of ample buffer space at the receiver to allow reordering of the blocks.

In *incremental-redundancy ARQ*, upon receiving a NAK message for a given packet, the transmitter transmits additional redundancy to the receiver. This additional redundancy is used by the decoder together with the originally received packet in a second attempt to recover the original data. This sequence of steps – NAK, additional redundancy, re-decode – can be repeated a number of times until the data are recovered or the packet is declared lost.

While ARQ schemes are very important, this book deals exclusively with FEC schemes. However, although the emphasis is on FEC, each of the FEC codes introduced can be used in a hybrid FEC/ARQ scheme where the code is used for both correction and detection. There exist many FEC schemes, employing both linear and nonlinear codes, although virtually all codes used in practice can be characterized as linear or linear at their core. Although the concept will be elucidated in Chapter 3, a *linear code* is one for which any sum of codewords is another codeword in the code. Linear codes are traditionally partitioned into the set of block codes and convolutional, or trellis-based, codes, although the turbo codes of Chapter 7 can be seen to be a hybrid of the two. Among the linear block codes are the cyclic and quasi-cyclic codes (defined in Chapter 3), both of which have more algebraic structure than do standard linear block codes. Also, we have been tacitly assuming binary codes, that is, codes whose code symbols are either 0 or 1. However, codes whose symbols are taken from a larger alphabet (e.g., 8-bit ASCII characters or 1000-bit packets) are possible, as described in Chapters 3 and 15.

This book will provide many examples of each of these code types, including nonbinary codes, and their decoders. For now, we introduce the first FEC code, due to Hamming [2], which provides a good introduction to the field of channel codes.

## 1.3 Channel Code Archetype: The (7, 4) Hamming Code

The (7, 4) Hamming code serves as an excellent channel-code prototype since it contains most of the properties of more practical codes. As indicated by the notation (7, 4), the codeword length is $n = 7$ and the data word length is $k = 4$, so the code rate is $R = 4/7$. As shown by R. McEliece, the Hamming code is easily described by the simple Venn diagram in Figure 1.2. In the diagram, the information word is represented by the vector $\mathbf{u} = (u_0, u_1, u_2, u_3)$ and the redundant bits (called *parity bits*) are represented by the parity vector $\mathbf{p} = (p_0, p_1, p_2)$. The codeword (also, code vector) is then given by the concatenation of $\mathbf{u}$ and $\mathbf{p}$:

$$\mathbf{v} = (\mathbf{u}\ \mathbf{p}) = (u_0, u_1, u_2, u_3, p_0, p_1, p_2) = (v_0, v_1, v_2, v_3, v_4, v_5, v_6).$$

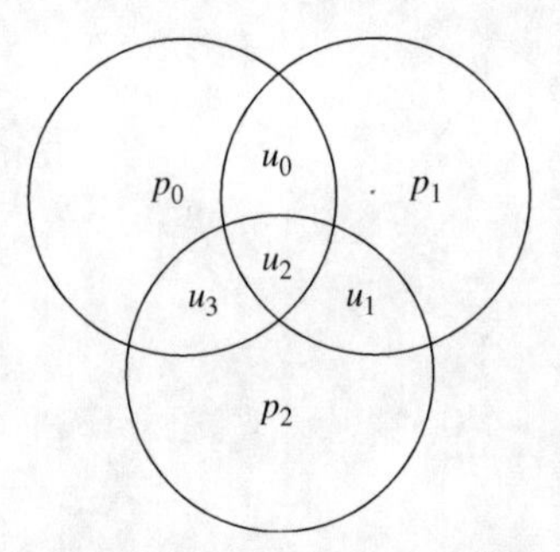


**Figure 1.2** Venn-diagram representation of (7, 4) Hamming-code encoding and decoding rules.

The encoding rule is trivial: the parity bits are chosen so that each circle has an even number of 1s, that is, the sum of bits in each circle is 0 modulo 2. From this encoding rule, we may write (where additions are modulo 2)

$$\begin{aligned} p_0 &= u_0 + u_2 + u_3, \\ p_1 &= u_0 + u_1 + u_2, \\ p_2 &= u_1 + u_2 + u_3, \end{aligned} \tag{1.1}$$

from which the 16 codewords are easily found:

| | | |
|---|---|---|
| 0000 000 | 1000 110 | 0010 111 |
| 1111 111 | 0100 011 | 1001 011 |
| | 1010 001 | 1100 101 |
| | 1101 000 | 1110 010 |
| | 0110 100 | 0111 001 |
| | 0011 010 | 1011 100 |
| | 0001 101 | 0101 110 |

As an example encoding, consider the third codeword in the middle column, (1010 001), for which the data word is $\mathbf{u} = (u_0, u_1, u_2, u_3) = (1, 0, 1, 0)$. Then

$$\begin{aligned} p_0 &= 1 + 1 + 0 = 0, \\ p_1 &= 1 + 0 + 1 = 0, \\ p_2 &= 0 + 1 + 0 = 1, \end{aligned}$$

yielding $\mathbf{v} = (\mathbf{u}\ \mathbf{p}) = (1010\,001)$. Observe that this code is *linear* because the sum of any two codewords yields a codeword. Note also that this code is *cyclic*: a cyclic shift of any codeword, rightward or leftward, gives another codeword.

Suppose now that $\mathbf{v} = (1010\,001)$ is transmitted, but $\mathbf{r} = (1011\,001)$ is received. That is, the channel has converted the 0 in code bit $v_3$ into a 1. The Venn diagram of Figure 1.3 can be used to decode $\mathbf{r}$ and correct the error. Note that Circle 2 in the figure has an even number of 1s in it, but Circles 1 and 3 do not. Thus, because the code rules are not satisfied by the bits in $\mathbf{r}$, we know that $\mathbf{r}$ contains one or more errors. Because a single error is more likely than two or more errors for most practical channels, we assume that $\mathbf{r}$ contains a single error. Then the error must be in the intersection of

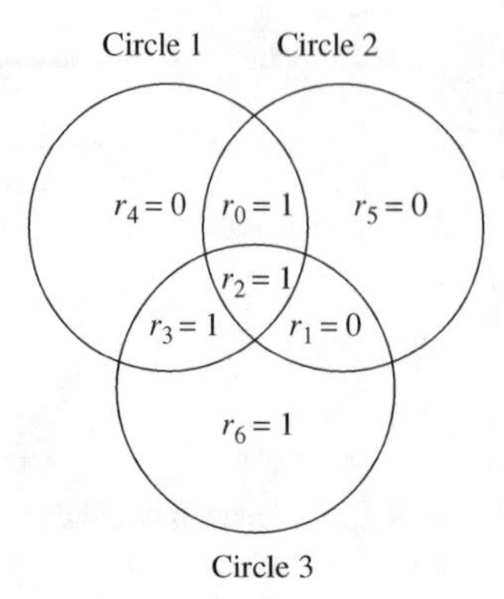


**Figure 1.3** Venn-diagram setup for the Hamming decoding example.

Circles 1 and 3. However, $r_2 = 1$ in the intersection cannot be in error because it is in Circle 2 whose rule is satisfied. Thus, it must be $r_3 = 1$ in the intersection that is in error. Thus, $v_3$ must be 0 rather than the 1 shown in Figure 1.3 for $r_3$. In conclusion, the decoded codeword is $\hat{\mathbf{v}} = (1010\,001)$, from which the decoded data $\hat{\mathbf{u}} = (1010)$ may be recovered.

It can be shown (see Chapter 3) that this particular single error is not special and that, independently of the codeword transmitted, all seven single errors are correctable and no error patterns with more than one error are correctable. The novice might ask what characteristic of these 16 codewords endows them with the ability to correct all single errors. This is easily explained using the concept of the *Hamming distance* $d_{\mathrm{H}}(\mathbf{x}, \mathbf{x}')$ between two length-$n$ words $\mathbf{x}$ and $\mathbf{x}'$, which is the number of locations in which they disagree. Thus, $d_{\mathrm{H}}(1000\;110, 0010\;111) = 3$. It can be shown, either exhaustively or using the principles developed in Chapter 3, that $d_{\mathrm{H}}(\mathbf{v}, \mathbf{v}') \geq 3$ for any two different codewords $\mathbf{v}$ and $\mathbf{v}'$ in the Hamming code. We say that the code's *minimum distance* is therefore $d_{\min} = 3$. Because $d_{\min} = 3$, a single error in some transmitted codeword $\mathbf{v}$ yields a received vector $\mathbf{r}$ that is closer to $\mathbf{v}$, in the sense of Hamming distance, than any other codeword. It is for this reason that all single errors are correctable.

Generalizations of the Venn diagram code description for the more complex codes used in applications are presented in Chapter 3 and subsequent chapters. In the chapters to follow, we will revisit the Hamming code a number of times, particularly in the problems. We will see how to reformulate encoding so that it employs a so-called generator matrix or, better, a simple shift-register circuit. We will also see how to reformulate decoding so that it employs a so-called parity-check matrix, and we will see many different decoding algorithms. Further, we will see applications of codes to a variety of channels, particularly ones introduced in the next section. Finally, we will see that a "good code" generally has the following properties: it is easy to encode, it is easy to decode, it has a large $d_{\min}$, and/or the number of codewords at distance $d_{\min}$ from any other codeword is small. We will see many examples of good codes in this book, and of their construction, their encoding, and their decoding.

## 1.4 Design Criteria and Performance Measures

Although there exist many channel models, it is usual to start with the two most frequently encountered *memoryless* channels, by which we mean a channel whose current output depends only on its current input. These most common memoryless channels are the binary symmetric channel (BSC) and the binary-input additive white Gaussian noise channel (BI-AWGNC). Examination of the BSC and BI-AWGNC illuminates many of the salient features of code and decoder design and code performance. For the sake of uniformity, for both channels we denote the $i$th channel input by $x_i$ and the $i$th channel output by $y_i$. As in the previous section, length-$n$ codewords to be sent over the channel will be represented by

$$\mathbf{v} = [v_0, v_1, \ldots, v_{n-1}]$$

whose binary code symbols $v_i$ will be mapped to channel inputs $x_i$ as follows. For the BSC, $x_i = v_i \in \{0, 1\}$; for the BI-AWGNC, $x_i = (-1)^{v_i} \in \{\pm 1\}$ so that $x_i = +1$ when $v_i = 0$.

Given channel input $x_i = v_i \in \{0, 1\}$ and channel output $y_i \in \{0, 1\}$, the BSC is completely characterized by the channel transition probabilities $P(y_i \mid x_i)$ given by

$$P(y_i = 1 \mid x_i = 0) = P(y_i = 0 \mid x_i = 1) = \varepsilon,$$
$$P(y_i = 1 \mid x_i = 1) = P(y_i = 0 \mid x_i = 0) = 1 - \varepsilon,$$

where $\varepsilon$ is called the *crossover probability*. The BI-AWGNC is completely characterized by the channel transition probability density function (pdf) $p(y_i \mid x_i)$ given by

$$p(y_i \mid x_i) = \frac{1}{\sqrt{2\pi}\sigma} \exp\left[-(y_i - x_i)^2 / (2\sigma^2)\right],$$

where $\sigma^2$ is the variance of the zero-mean Gaussian noise sample $n_i$ that the channel adds to the transmitted value $x_i$ (so that $y_i = x_i + n_i$). As a consequence of its memorylessness, we have for the BSC

$$P(\mathbf{y} \mid \mathbf{x}) = \prod_i P(y_i \mid x_i), \tag{1.2}$$

where $\mathbf{y} = [y_1, y_2, y_3, \ldots]$ and $\mathbf{x} = [x_1, x_2, x_3, \ldots]$. A similar expression exists for the BI-AWGNC with $P(\cdot)$ replaced by $p(\cdot)$.

The most obvious design criterion applicable to the design of a decoder is the minimum-probability-of-error criterion. When the design criterion is to minimize the probability that the decoder fails to decode to the correct codeword, that is, to minimize the probability of a *codeword error*, it can be shown that this is equivalent to maximizing the *a posteriori* probability $P(\mathbf{x} \mid \mathbf{y})$ (or $p(\mathbf{x} \mid \mathbf{y})$ for the BI-AWGNC). Recalling the one-to-one correspondence between $\mathbf{v}$ and $\mathbf{x}$, the optimal decision for the BSC is then given by

$$\hat{\mathbf{v}} = \arg\max_{\mathbf{v}} P(\mathbf{x} \mid \mathbf{y}) = \arg\max_{\mathbf{v}} \frac{P(\mathbf{y} \mid \mathbf{x})P(\mathbf{x})}{P(\mathbf{y})}, \tag{1.3}$$

where $\arg\max_{\mathbf{v}} f(\mathbf{v})$ equals the argument $\mathbf{v}$ that maximizes the function $f(\mathbf{v})$. Frequently, the channel-input words are equally likely and, hence, $P(\mathbf{x})$ is independent of $\mathbf{x}$ (hence, $\mathbf{v}$). Because $P(\mathbf{y})$ also does not change with $\mathbf{v}$, the maximum *a posteriori* (MAP) rule (1.3) can be replaced by the *maximum-likelihood* (ML) rule

$$\hat{\mathbf{v}} = \arg\max_{\mathbf{v}} P(\mathbf{y} \mid \mathbf{x}).$$

Using (1.2) and the monotonicity of the log function, we have for the BSC

$$\begin{aligned}
\hat{\mathbf{v}} &= \arg\max_{\mathbf{v}} \log \prod_i P(y_i \mid x_i) \\
&= \arg\max_{\mathbf{v}} \sum_i \log P(y_i \mid x_i) \\
&= \arg\max_{\mathbf{v}} \left[ d_{\mathrm{H}}(\mathbf{y}, \mathbf{x}) \log(\varepsilon) + (n - d_{\mathrm{H}}(\mathbf{y}, \mathbf{x})) \log(1 - \varepsilon) \right] \\
&= \arg\max_{\mathbf{v}} \left[ d_{\mathrm{H}}(\mathbf{y}, \mathbf{x}) \log\left( \frac{\varepsilon}{1 - \varepsilon} \right) + n \log(1 - \varepsilon) \right] \\
&= \arg\min_{\mathbf{v}} d_{\mathrm{H}}(\mathbf{y}, \mathbf{x}),
\end{aligned}$$

where $n$ is the codeword length and the last line follows since $\log[\varepsilon/(1-\varepsilon)] < 0$ and $n \log(1 - \varepsilon)$ is not a function of $\mathbf{v}$.

For the BI-AWGNC, the ML decision is that because the pdf $p(\mathbf{y} \mid \mathbf{x})$ is used:

$$\hat{\mathbf{v}} = \arg\max_{\mathbf{v}} p(\mathbf{y} \mid \mathbf{x}),$$

keeping in mind the mapping $\mathbf{x} = (-1)^{\mathbf{v}}$. Following a similar set of steps (and dropping irrelevant terms), we have

$$\begin{aligned}
\hat{\mathbf{v}} &= \arg\max_{\mathbf{v}} \sum_i \log\left( \frac{1}{\sqrt{2\pi}\sigma} \exp\left[ -(y_i - x_i)^2/(2\sigma^2) \right] \right) \\
&= \arg\min_{\mathbf{v}} \sum_i (y_i - x_i)^2 \\
&= \arg\min_{\mathbf{v}} d_{\mathrm{E}}(\mathbf{y}, \mathbf{x}),
\end{aligned}$$

where

$$d_{\mathrm{E}}(\mathbf{y}, \mathbf{x}) = \sqrt{\sum_i (y_i - x_i)^2}$$

is the *Euclidean distance* between $\mathbf{y}$ and $\mathbf{x}$, and on the last line we replaced $d_{\mathrm{E}}^2(\cdot)$ by $d_{\mathrm{E}}(\cdot)$ due to the monotonicity of the square-root function for non-negative arguments. Note that, once a decision is made on the codeword, the decoded data word $\hat{\mathbf{u}}$ may easily be recovered from $\hat{\mathbf{v}}$, particularly when the codeword is in the *systematic* form $\mathbf{v} = [\mathbf{u}\ \mathbf{p}]$.

In summary, for the BSC, the ML decoder chooses the codeword $\mathbf{v}$ that is closest to the channel output $\mathbf{y}$ in a Hamming-distance sense; for the BI-AWGNC, the ML decoder chooses the code sequence $\mathbf{x} = (-1)^{\mathbf{v}}$ that is closest to the channel output $\mathbf{y}$ in a Euclidean-distance sense. The implication for code design on the BSC is that the code should be designed to maximize the minimum Hamming distance between two codewords (and to minimize the number of codeword pairs at that distance). Similarly, the implication for code design on the BI-AWGNC is that the code should be designed to maximize the minimum Euclidean distance between any two code sequences on the channel (and to minimize the number of code-sequence pairs at that distance).

Finding the codeword $\mathbf{v}$ that minimizes the Hamming (or Euclidean) distance in a brute-force, exhaustive fashion is very complex, except for very simple codes such as the (7,4) Hamming code. Thus, ML decoding algorithms have been developed that exploit code structure, vastly reducing complexity. Such algorithms are presented in subsequent chapters. Suboptimal but less complex algorithms, which perform slightly worse than the ML decoder, will also be presented in subsequent chapters. These include so-called bounded-distance decoders, list decoders, and iterative decoders involving component sub-decoders. Often, these component decoders are based on the *bit-wise MAP criterion* that minimizes the probability of *bit error* rather than the probability of *codeword* error. This bit-wise MAP criterion is

$$\hat{v}_i = \arg\max_{v_i} P(x_i \mid \mathbf{y}) = \arg\max_{v_i} \frac{P(\mathbf{y} \mid x_i)P(x_i)}{P(\mathbf{y})},$$

where the *a priori* probability $P(x_i)$ is constant (and ignored together with $P(\mathbf{y})$) if the decoder is operating in isolation, but is supplied by a companion decoder if the decoder is part of an iterative decoding scheme. This topic will also be discussed in subsequent chapters.

The most commonly used performance measure is the *bit-error probability*, $P_b$, defined as the probability that the decoder output decision $\hat{u}_i$ does not equal the encoder input bit $u_i$,

$$P_b \triangleq \Pr\{\hat{u}_i \neq u_i\}.$$

Strictly speaking, we should average over all $i$ to obtain $P_b$. However, $\Pr\{\hat{u}_i \neq u_i\}$ is frequently independent of $i$, although, if it is not, the averaging is understood. $P_b$ is often called the *bit-error rate* (BER). Another commonly used performance measure is the *codeword-error probability*, $P_{cw}$, defined as the probability that the decoder output decision $\hat{\mathbf{v}}$ does not equal the encoder output codeword $\mathbf{v}$,

$$P_{cw} \triangleq \Pr\{\hat{\mathbf{v}} \neq \mathbf{v}\}.$$

In the coding literature, various alternative terms are used for $P_{cw}$, including *word-error rate* (WER) and *frame-error rate* (FER). A closely related error probability is the probability $P_{uw} \triangleq \Pr\{\hat{\mathbf{u}} \neq \mathbf{u}\}$, which can be useful for some applications, but we shall not emphasize this probability, particularly since $P_{uw} \approx P_{cw}$ for many coding systems. Lastly, for nonbinary codes, the *symbol-error probability* $P_s$ is pertinent. It is defined as

$$P_s \triangleq \Pr\{\hat{u}_i \neq u_i\},$$

where in this case the encoder input symbols $u_i$ and the decoder output symbols $\hat{u}_i$ are nonbinary. $P_s$ is also called the *symbol-error rate* (SER). We shall use the notation introduced in this paragraph throughout the book.

## 1.5 Channel Capacity Formulas for Common Channel Models

Recall from Section 1.1 that *channel capacity* is a theoretical limit on the (reliable) information rate that is possible through a channel. This is often summarized by the requirement $R < C$, where $R$ is the information rate and $C$ is the channel capacity in equivalent units (with examples given later).

From the time of Shannon's seminal work in 1948 until the early 1990s, it was thought that the only codes capable of operating near capacity are long, impractical codes, that is, unstructured codes that are essentially impossible to encode and decode in practice. However, the invention of turbo codes and low-density parity-check (LDPC) codes in the 1990s demonstrated that near-capacity performance was possible in practice. (As explained in Chapter 5, LDPC codes were first invented c. 1960 by R. Gallager and later independently reinvented by MacKay and others c. 1995. Their capacity-approaching properties with practical encoders/decoders could not be demonstrated with 1960s technology, so they were mostly ignored for about 35 years.) Because of the advent of these capacity-approaching codes, knowledge of information theory and channel capacity has become increasingly important for both the researcher and the practicing engineer. In this section we catalog capacity formulas for a variety of commonly studied channel models. We point out that these formulas correspond to infinite-length codes. However, we will see numerous examples in this book where finite-length codes operate very close to capacity, although this is possible only with long codes ($n > 5000$, say).

No derivations are given for the various capacity formulas. For such information, see [3–9]. However, it is useful to highlight the general formula for the mutual information between the channel output represented by $Y$ and the channel input represented by $X$. When the input and output take values from a discrete set, then the *mutual information* may be written as

$$I(X;Y) = H(Y) - H(Y \mid X), \tag{1.4}$$

where $H(Y)$ is the *entropy* of the channel output,

$$\begin{aligned} H(Y) &= -\mathbb{E}\{\log_2(\Pr(Y))\} \\ &= -\sum_y \Pr(y)\log_2(\Pr(y)) \end{aligned}$$

and $H(Y \mid X)$ is the *conditional entropy* of $Y$ given $X$,

$$\begin{aligned} H(Y \mid X) &= -\mathbb{E}\{\log_2(\Pr(Y \mid X))\} \\ &= -\sum_x\sum_y \Pr(x,y)\log_2(\Pr(y \mid x)) \\ &= -\sum_x\sum_y \Pr(x)\Pr(y \mid x)\log_2(\Pr(y \mid x)). \end{aligned}$$

In these expressions, $\mathbb{E}\{\cdot\}$ represents probabilistic expectation. The form (1.4) is most commonly used, although the alternative form $I(X;Y) = H(X) - H(X \mid Y)$ is sometimes useful. The capacity of a channel is then defined as

$$C = \max_{\{\Pr(x)\}} I(X;Y), \tag{1.5}$$

that is, the capacity is the maximum mutual information, where the maximization is over the channel input probability distribution $\{\Pr(x)\}$. As a practical matter, most channel models are *symmetric* (see [4]), in which case the optimal input distribution is uniform so that the capacity is given by

$$I(X;Y)|_{\text{uniform }\{\Pr(x)\}} = [H(Y) - H(Y \mid X)]_{\text{uniform }\{\Pr(x)\}}. \tag{1.6}$$

For cases in which the channel is not symmetric, the Blahut–Arimoto algorithm [3, 6] can be used to perform the optimization of $I(X;Y)$. Alternatively, the uniform-input information rate of (1.6) can be used as an approximation of capacity, as will be seen below for the Z channel. For a continuous channel output $Y$, the entropies in (1.4) are replaced by *differential entropies* $h(Y)$ and $h(Y \mid X)$, which are defined analogously to $H(Y)$ and $H(Y \mid X)$, with the probability mass functions replaced by probability density functions and the sums replaced by integrals.

Consistent with the code rate defined earlier, $C$ and $I(X;Y)$ are in units of information bits/code bit. Unless indicated otherwise, the capacities presented in the remainder of this chapter have these units, although we will see that it is occasionally useful to convert to alternative units. Also, all code rates $R$ for which $R < C$ are said to be *achievable rates* in the sense that reliable (low error rate) communication is achievable at these rates.

### 1.5.1 Capacity for Binary-Input Memoryless Channels

#### The BEC and the BSC

The binary erasure channel (BEC) and the binary symmetric channel (BSC) are illustrated in Figures 1.4 and 1.5. For the BEC, $p$ is the probability of a bit erasure, which is represented by the symbol $e$, or sometimes by ?, to indicate the fact that nothing is known about the bit that was erased. For the BSC, $\varepsilon$ is the probability of a bit error. While simple, both models are useful for practical applications and academic research. Each model is an abstraction of a physical waveform channel that can take many forms. For example, the binary erasure channel is a model for transmitting binary packets on the internet, where packets are either received correctly or are "lost" (erased). The BSC is a model of a noisy channel with binary inputs and binary outputs created by making early (called *hard*) decisions on the values of the bits at the receiver. The channel transition probability must be symmetric (must not favor either bit value) in order for the discrete channel model to be symmetric (i.e., to have equal conditional error probabilities).

The capacity of the BEC is easily shown from (1.6) to be

$$C_{\text{BEC}} = 1 - p. \tag{1.7}$$

It can similarly be shown that the capacity of the BSC is given by

$$C_{\text{BSC}} = 1 - \mathcal{H}(\varepsilon), \tag{1.8}$$

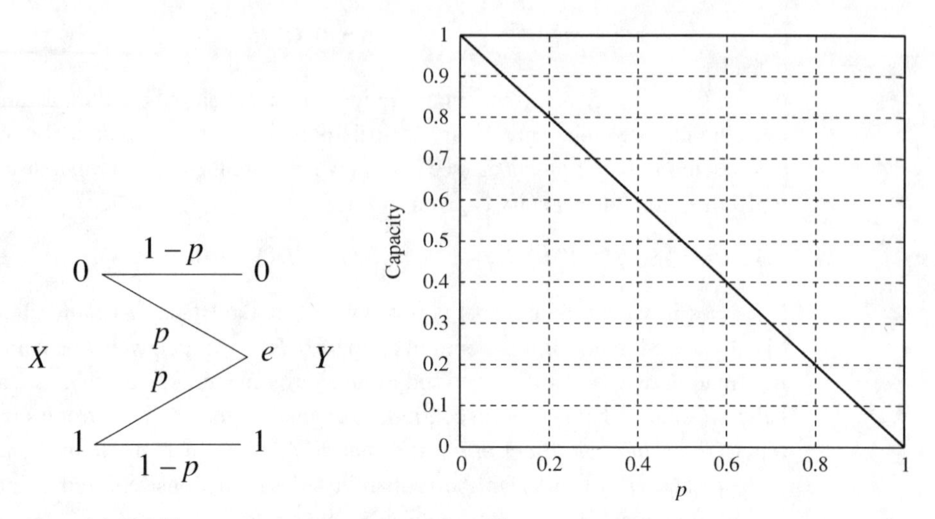


**Figure 1.4** The binary erasure channel and a plot of its capacity.

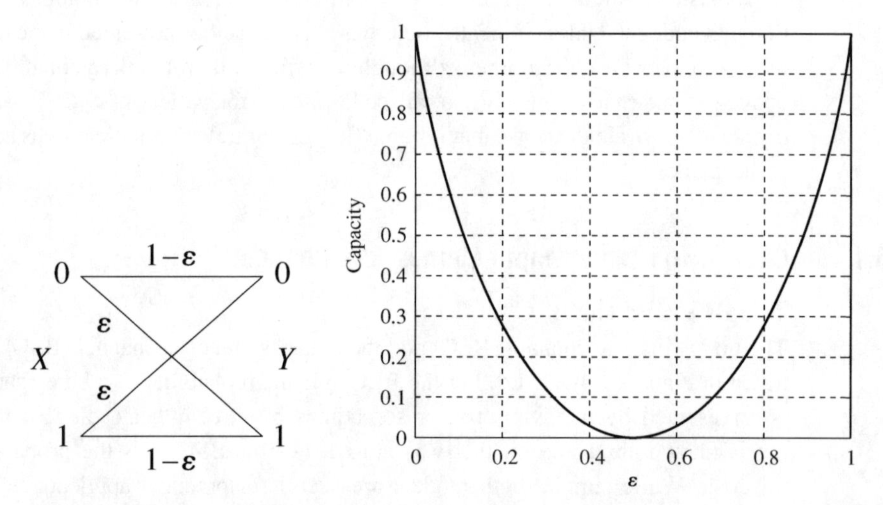


**Figure 1.5** The binary symmetric channel and a plot of its capacity.

where $\mathcal{H}(\varepsilon)$ is the *binary entropy function* given by

$$\mathcal{H}(\varepsilon) = -\varepsilon \log_2(\varepsilon) - (1-\varepsilon)\log_2(1-\varepsilon).$$

The derivations of these capacity formulas from (1.6) are considered in Problem 1.3. $C_{\text{BEC}}$ is plotted as a function of $p$ in Figure 1.4 and $C_{\text{BSC}}$ is plotted as a function of $\varepsilon$ in Figure 1.5.

### The Z Channel

The Z channel, depicted in Figure 1.6, is an idealized model of a free-space optical communication channel. It is an extreme case of an asymmetric binary-input/binary-output channel and is sometimes used to model solid-state memories. As indicated in the figure, the probability of error when transmitting a 0 is zero and the probability

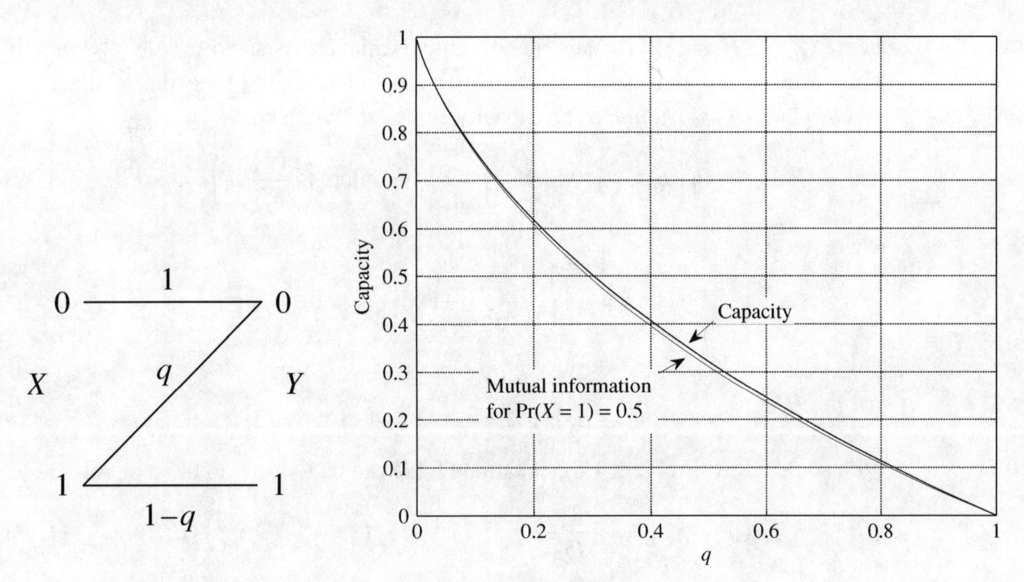


**Figure 1.6** The Z channel and a plot of its capacity.

of error when transmitting a 1 is $q$. Let $u$ equal the probability of transmitting a 1, $u = \Pr(X = 1)$. Then the capacity is given [10] by

$$C_Z = \max_u \{\mathcal{H}(up) - u\mathcal{H}(q)\} \tag{1.9}$$
$$= \mathcal{H}(u'p) - u'\mathcal{H}(q),$$

where $p = 1 - q$ and $u'$ is the maximizing value of $u$, given by

$$u' = \frac{q^{q/(1-q)}}{1 + (1-q)q^{q/(1-q)}}. \tag{1.10}$$

Our intuition tells us that it would be vastly advantageous to design error-correction codes that favor sending 0s, that is, whose codewords have more 0s than 1s, so that $u < 0.5$. However, consider the following example from [11]. Suppose that $q = 0.1$. Then $u' = 0.4563$ and $C_Z = 0.7628$. For comparison, suppose that we use a code for which the 0s and 1s occur equiprobably, that is, $u = 0.5$. In this case, the mutual information $I(X;Y) = \mathcal{H}(up) - u\mathcal{H}(q) = 0.7583$, so that little is lost by using such a code in lieu of an optimal code with $u' = 0.4563$. We have plotted both $C_Z$ and $I(X;Y)$ against $q$ in Figure 1.6, where it is seen that for all $q \in [0, 1]$ there is little difference between $C_Z$ and $I(X;Y)$. Thus, it appears that there is little to be gained by trying to invent codes with non-uniform symbols for the Z channel and similar asymmetric channels. This is fortuitous, for the design of codes with non-uniform symbols is difficult.

## The Binary-Input AWGN Channel

Consider the discrete-time channel model

$$y_\ell = x_\ell + z_\ell, \tag{1.11}$$

where $x_\ell \in \{\pm 1\}$ and $z_\ell$ is a real-valued additive white Gaussian noise (AWGN) sample with variance $\sigma^2$, that is, $z_\ell \sim \mathcal{N}(0, \sigma^2)$. This channel is called the *binary-input AWGN (BI-AWGN) channel.* The capacity can be shown to be

$$C_{\text{BI-AWGN}} = 0.5 \sum_{x=\pm 1} \int_{-\infty}^{+\infty} p(y \mid x) \log_2 \left( \frac{p(y \mid x)}{p(y)} \right) dy, \tag{1.12}$$

where

$$p(y \mid x = \pm 1) = \frac{1}{\sqrt{2\pi}\sigma} \exp\left[ -(y \mp 1)^2 / (2\sigma^2) \right]$$

and

$$p(y) = \frac{1}{2} \left[ p(y \mid x = +1) + p(y \mid x = -1) \right].$$

With some effort (Problem 1.6), equation (1.12) can be reduced to the integral

$$C_{\text{BI-AWGN}} = \int_{-\infty}^{+\infty} \frac{1}{\sqrt{2\pi}} e^{-z^2/2} \left( 1 - \log_2 \left( 1 + e^{-2\rho + 2\sqrt{\rho} z} \right) \right) dz, \tag{1.13}$$

where $\rho = E_s/\sigma^2$ and $E_s$ is the channel symbol energy and is equal to 1.0 for our model. An alternative formula, which follows from $C = h(Y) - h(Y \mid X) = h(Y) - h(Z)$, is

$$C_{\text{BI-AWGN}} = - \int_{-\infty}^{+\infty} p(y) \log_2 (p(y))\, dy - 0.5 \log_2 (2\pi e \sigma^2), \tag{1.14}$$

where we used $h(Z) = 0.5 \log_2 (2\pi e \sigma^2)$, which is shown in one of the problems.

Both forms, (1.13) and (1.14), require numerical integration to compute, something that is easily done with modern mathematical software. If that is not available, Monte-Carlo integration is a straightforward alternative. For example, the integral in (1.14) is simply the expectation $\mathbb{E}\{-\log_2 (p(y))\}$, which may be estimated as

$$\mathbb{E}\{-\log_2 (p(y))\} \simeq -\frac{1}{L} \sum_{\ell=1}^{L} \log_2 (p(y_\ell)), \tag{1.15}$$

where $\{y_\ell : \ell = 1, \ldots, L\}$ is a large number of realizations of the random variable $Y$.

In Figure 1.7, $C_{\text{BI-AWGN}}$ (labeled "soft") is plotted against the commonly used signal-to-noise-ratio (SNR) measure $E_b/N_0$, where $E_b$ is the average energy per information bit and $N_0/2 = \sigma^2$ is the two-sided power spectral density of the AWGN process, $z_\ell$. The $E_b/N_0$ SNR measure is usually favored because $E_b/N_0 = (P/N_0)/R_b$, where $P/N_0$ is received power SNR that fully characterizes the AWGN channel and $R_b$ is the data rate in bits per second. By comparing coding and modulation schemes on the basis of $E_b/N_0$, we are doing it for the *same channel and same data rate*.

On the other hand, as pointed out below (1.13), $C_{\text{BI-AWGN}}$ is a function of $\rho = E_s/\sigma^2 = 2E_s/N_0$. While $E_b = P/R_b$, $E_s = P/R_s$, where $R_s$ is the channel symbol rate. We then have that $E_b = (R_s/R_b)E_s = E_s/R$, where $R$ is the code rate (data bits per channel bit). Thus, to plot $C_{\text{BI-AWGN}}$ versus $E_b/N_0$ as in Figure 1.7, the $E_s/N_0$ values are converted to $E_b/N_0$ values via the relationship $E_b = E_s/R$. The value of $R$ used in this translation of SNR definitions is $R = C_{\text{BI-AWGN}}$, because $R$ is assumed to be just less than $C_{\text{BI-AWGN}}$ for the most efficient communication ($R = C_{\text{BI-AWGN}} - \delta$, where

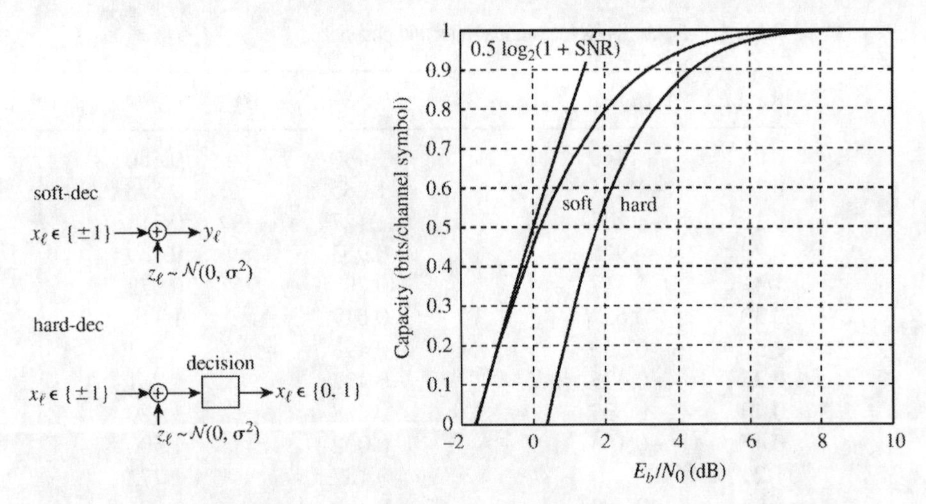


**Figure 1.7** Capacity curves for the soft-decision and hard-decision binary-input AWGN channel together with the unconstrained-input AWGN channel-capacity curve.

$\delta > 0$ is arbitrarily small). Each abscissa value in Figure 1.7 can be interpreted as the minimum $E_b/N_0$ consistent with reliably communicating at the rate $C_{\text{BI-AWGN}}$ that corresponds to that $E_b/N_0$ value (which equals $(E_s/N_0)/R$).

Also shown in Figure 1.7 is the capacity curve for a *hard decision* BI-AWGN channel (labeled "hard"). For this channel, so-called hard-decisions $\hat{x}_\ell$ are obtained from the *soft decisions* $y_\ell$ of (1.11) according to

$$\hat{x}_\ell = \begin{cases} 1, & \text{if } y_\ell \leq 0, \\ 0, & \text{if } y_\ell > 0. \end{cases}$$

The soft-decision and hard-decision models are also included in Figure 1.7. Note that the hard decisions convert the BI-AWGN channel into a BSC with error probability $\varepsilon = Q(1/\sigma) = Q\left(\sqrt{2RE_b/N_0}\right)$, where

$$Q(a) \triangleq \int_a^\infty \frac{1}{\sqrt{2\pi}} \exp\left(-\beta^2/2\right) d\beta.$$

Thus, the hard-decision curve in Figure 1.7 is plotted using $C_{\text{BSC}} = 1 - \mathcal{H}(\varepsilon)$. It is seen in the figure that the conversion to bits (i.e., hard decisions) prior to decoding results in a loss of 2 dB at low rates $C_{\text{BI-AWGN}}$ and decreases with increasing rate.

Finally, also included in Figure 1.7 is the capacity of the *unconstrained-input* AWGN channel discussed in Section 1.5.2. Often called the *Shannon capacity*, this capacity, $C = 0.5\log_2(1 + 2RE_b/N_0)$, gives the upper limit over all one-dimensional signaling schemes, including nonbinary ones, and is discussed in the next section. Observe that the capacity of the soft-decision binary-input AWGN channel is close to that of the unconstrained-input AWGN channel for low code rates. Observe also that reliable communication is not possible for $E_b/N_0 < -1.59$ dB $= 10\log_{10}(\ln(2))$ dB, as shown mathematically in the next section.

Table 1.1 lists the $E_b/N_0$ values required to achieve selected code rates for these three channels. Modern codes such as turbo and LDPC codes are capable of operating

**Table 1.1** $E_b/N_0$ limits for various rates and channels

| Rate $R$ | $(E_b/N_0)_{\text{Shannon}}$ (dB) | $(E_b/N_0)_{\text{soft}}$ (dB) | $(E_b/N_0)_{\text{hard}}$ (dB) |
|---|---|---|---|
| 0.05 | −1.440 | −1.440 | 0.480 |
| 0.10 | −1.284 | −1.285 | 0.596 |
| 0.15 | −1.133 | −1.127 | 0.713 |
| 0.20 | −0.976 | −0.963 | 0.839 |
| 1/4 | −0.817 | −0.794 | 0.972 |
| 0.30 | −0.657 | −0.617 | 1.112 |
| 1/3 | −0.550 | −0.495 | 1.211 |
| 0.35 | −0.495 | −0.432 | 1.261 |
| 0.40 | −0.333 | −0.238 | 1.420 |
| 0.45 | −0.166 | −0.032 | 1.590 |
| 1/2 | 0 | 0.187 | 1.772 |
| 0.55 | 0.169 | 0.423 | 1.971 |
| 0.60 | 0.339 | 0.679 | 2.188 |
| 0.65 | 0.511 | 0.960 | 2.428 |
| 2/3 | 0.569 | 1.060 | 2.514 |
| 0.70 | 0.686 | 1.272 | 2.698 |
| 3/4 | 0.860 | 1.627 | 3.007 |
| 4/5 | 1.037 | 2.040 | 3.370 |
| 0.85 | 1.215 | 2.543 | 3.815 |
| 9/10 | 1.396 | 3.198 | 4.399 |
| 0.95 | 1.577 | 4.192 | 5.295 |

very close to these $E_b/N_0$ "limits," within 0.5 dB for long codes. The implication of these $E_b/N_0$ limits is that, for the given code rate $R$, error-free communication is possible in principle via channel coding if $E_b/N_0$ just exceeds its limit; otherwise, if the SNR is less than the limit, reliable communication is not possible. While the table indicates that the lower code rates permit operation at smaller signal-to-ratio values $E_b/N_0$, it is at the expense of a wider signaling bandwidth occupancy and smaller $E_s/N_0 = RE_b/N_0$, making receiver synchronization more challenging. The values in Table 1.1 were computed in much the same way as the curves in Figure 1.7.

## 1.5.2 Coding Limits for *M*-ary-Input Memoryless Channels

### The Unconstrained-Input AWGN Channel

Consider the discrete-time channel model

$$y_\ell = x_\ell + z_\ell, \tag{1.16}$$

where $x_\ell$ is a real-valued signal whose power is constrained as $\mathbb{E}\left[x_\ell^2\right] \le P$ and $z_\ell$ is a real-valued AWGN sample with variance $\sigma^2$, that is, $z_\ell \sim \mathcal{N}\left(0, \sigma^2\right)$. For this model, since the channel symbol $x_\ell$ is not binary, the code rate $R$ would be defined in terms of information bits per channel symbol and it is not constrained to be less than unity. For convenience, we assume that the encoder input is binary, but the encoder output selects a sequence of nonbinary symbols $x_\ell$ that are transmitted over the AWGN channel. As an example, if $x_\ell$ draws from an alphabet of 16 symbols, then code rates of up to 4 bits per channel symbol are possible. The capacity of the channel in (1.16) is given by

$$C_{\text{Shannon}} = \frac{1}{2}\log_2\left(1 + \frac{P}{\sigma^2}\right) \text{ bits/channel symbol,} \tag{1.17}$$

and this expression is often called the *Shannon capacity*, since it is the upper limit among all signal sets (alphabets) from which $x_\ell$ may draw its values. As shown in Problem 1.12, the capacity is achieved when $x_\ell$ is drawn from a Gaussian distribution, $\mathcal{N}(0, P)$, which has an uncountably infinite alphabet size. In practice, it is possible to closely approach $C_{\text{Shannon}}$ with a finite (but large) alphabet that is Gaussian-like.

As before, reliable communication is possible on this channel only if $R < C$. Note that, for large values of SNR, $P/\sigma^2$, the capacity grows logarithmically. Thus, each doubling of SNR (increase by 3 dB) corresponds to a capacity increase of about 1 bit/channel symbol. We also point out that, because we assume a real-valued model, the units of this capacity formula might also be bits/channel symbol/dimension. For a complex-valued model, which requires two dimensions, the formula would increase by a factor of two and the power divided by two because it would be shared across the two dimensions: $C_{\text{Shannon}}^{\text{2-D}} = \log_2\left(1 + \frac{P}{2\sigma^2}\right)$. In general, for $N$ dimensions, it is

$$C_{\text{Shannon}}^{N\text{-D}} = \frac{N}{2}\log_2\left(1 + \frac{P}{N\sigma^2}\right) = \frac{N}{2}\log_2\left(1 + \frac{2E_s}{NN_0}\right) \text{ bits/channel use,} \tag{1.18}$$

where the second equality follows from $E_s = P/R_s$ and $\sigma^2 = R_s N_0/2$, with $R_s$ denoting the channel symbol rate.

Frequently, one is interested in a channel capacity in units of bits per second rather than bits per channel symbol. Such a channel capacity is easily obtainable via multiplication of $C_{\text{Shannon}}$ in (1.17) by the channel symbol rate $R_s$ (symbols per second):

$$C'_{\text{Shannon}} = R_s C_{\text{Shannon}} = \frac{R_s}{2}\log_2\left(1 + \frac{P}{\sigma^2}\right) \text{ bits/s.}$$

This formula still corresponds to the discrete-time model (1.16), but it leads us to the capacity formula for the continuous-time (baseband) AWGN channel bandlimited to $W$ Hz, with power spectral density $N_0/2$. Recall from Nyquist theory that the maximum distortion-free symbol rate in a bandwidth $W$ is $R_{s,\max} = 2W$. Substitution of this into the above equation gives

$$\begin{aligned} C'_{\text{Shannon}} &= W\log_2\left(1 + \frac{P}{\sigma^2}\right) \text{ bits/s} \\ &= W\log_2\left(1 + \frac{R_b E_b}{W N_0}\right) \text{ bits/s,} \end{aligned} \tag{1.19}$$

where in the second line we have used $P = R_b E_b$ and $\sigma^2 = W N_0$, where $R_b$ is the information bit rate in bits per second and $E_b$ is the average energy per information bit. Reliable communication is possible provided that $R_b < C'_{\text{Shannon}}$. Equation (1.19) is the classic result of Shannon for the bandlimited AWGN channel.

Letting $R_b = C'_{\text{Shannon}} = C'$ in (1.19) results in an inequality that, upon rearranging, is

$$\frac{C'}{W} < \log_2\left(1 + \frac{C'E_b}{WN_0}\right)$$

and, after isolating $E_b/N_0$, we have

$$E_b/N_0 > \frac{2^{C'/W} - 1}{C'/W}.$$

It is reasonable to assume that $C'$ increases with $W$ (and it can be proven mathematically from (1.21)). Further, this lower bound on $E_b/N_0$ reaches its minimum as $W \to \infty$ and this minimum value is $10\log_{10}(\ln(2)) = -1.59$ dB, a value pointed out earlier when discussing Figure 1.7.

### *M*-ary AWGN Channel

Consider an $M$-ary signal set $\{s_m\}_{m=1}^{M}$ existing in two dimensions so that each $s_m$ signal is representable as a complex number. The capacity formula for two-dimensional $M$-ary modulation in AWGN can be written as a straightforward generalization of the binary case. (We consider one-dimensional $M$-ary modulation in AWGN in the problems.) As is commonly done, we approximate capacity $C$ by $I(X;Y)$ with equiprobable inputs, often called the *symmetric information rate*. Thus, we begin with $C_{M\text{-ary}} \approx h(Y) - h(Y \mid X) = h(Y) - h(Z)$, from which we may write

$$C_{M\text{-ary}} \approx -\int_{-\infty}^{+\infty} p(y)\log_2(p(y))dy - \log_2(2\pi e\sigma^2), \tag{1.20}$$

where $h(Z) = \log_2(2\pi e\sigma^2)$ because the noise is now two-dimensional (or complex) with a variance of $2\sigma^2 = N_0$, or a variance of $\sigma^2 = N_0/2$ in each dimension. In this expression, $y$ has two components so that the integral is a double integral, and $p(y)$ is determined via

$$p(y) = \frac{1}{M}\sum_{m=1}^{M} p(y \mid s_m),$$

where $p(y \mid s_m)$ is the complex Gaussian pdf with mean $s_m$ and variance $2\sigma^2 = N_0$. The first term in (1.20) may be computed as in the binary case using (1.15). Note that an $M$-ary symbol is transmitted during each symbol epoch and so $C_{M\text{-ary}}$ has the unit of information bits per channel symbol. Because each $M$-ary symbol conveys $\log_2(M)$ code bits, $C_{M\text{-ary}}$ may be converted to information bits per code bit by dividing it by $\log_2(M)$.

For QPSK, 8-PSK, 16-PSK, and 16-QAM, $C_{M\text{-ary}}$ is plotted in Figure 1.8 against $E_b/N_0$, where $E_b$ is related to the average signal energy $E_s = \mathbb{E}[|s_m|^2]$ via the channel rate (channel capacity) as $E_b = E_s/C_{M\text{-ary}}$. Also included in Figure 1.8 is the capacity of the unconstrained-input, two-dimensional AWGN channel, $C^{2\text{-}D}_{\text{Shannon}} = \log_2(1 + E_s/N_0)$, described in the previous subsection. Recall that the Shannon

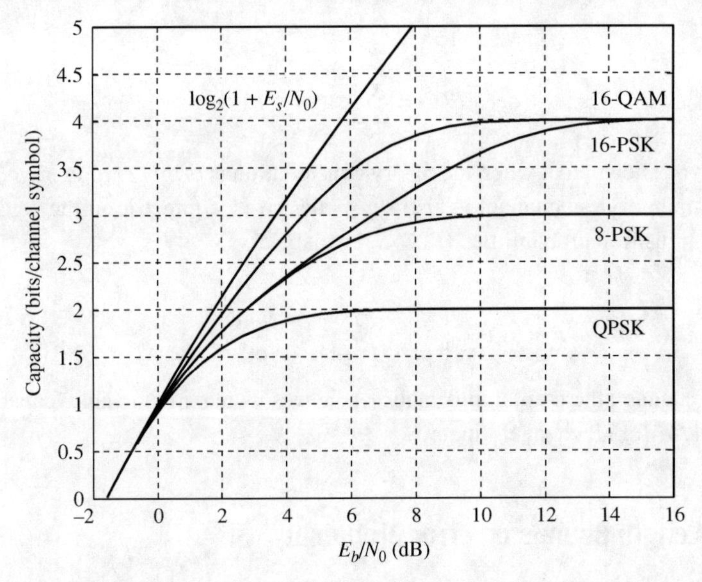


**Figure 1.8** Capacity versus $E_b/N_0$ curves for selected two-dimensional modulation schemes.

capacity gives the upper limit over all signaling schemes and, hence, its curve lies above all of the other curves. Observe, however, that, at the lower SNR values, the capacity curves for the standard modulation schemes are very close to that of the Shannon capacity curve. Next, observe that, again, reliable communication is not possible for $E_b/N_0 < -1.59$ dB.

One example of the use of Figure 1.8 is the observation that 16-PSK capacity is inferior to that of 16-QAM over the large range of SNR values that might be used in practice. This is understandable, given the smaller minimum inter-signal distance that exists in the 16-PSK signal set (for the same $E_s$). As another example, it is known that *uncoded* BPSK (not in the figure) provides a bit-error probability of $10^{-5}$ at $E_b/N_0$ near 9.6 dB at a throughput of 1 bit/channel use. Alternatively, the QPSK curve in Figure 1.8 indicates that codes exist that permit low error-rate coded-QPSK operation at 1 bit/use and at SNR values of $E_b/N_0 = 0.2$ dB or greater. This is a *coding gain* of about 9.4 dB. Lastly, QPSK can deliver 2 bits/use reliably at $E_b/N_0 = 9.6$ dB, whereas the 16-QAM curve indicates that coded 16-QAM can do the same at $E_b/N_0 = 2.1$ dB.

### *M*-ary Symmetric Channel

It was seen in Section 1.5.1 that a natural path to the binary-symmetric channel was via hard decisions at the output of a binary-input AWGN channel. When one applies hard decisions to the output of an $M$-ary-input AWGN channel, the resulting channel is called a discrete memoryless $M$-ary channel. The mutual information for the discrete memoryless $M$-ary channel is given by

$$I(X;Y) = \sum_{x=0}^{M-1} \sum_{y=0}^{M-1} P(y \mid x)P(x) \log_2\left(\frac{P(y \mid x)}{P(y)}\right), \tag{1.21}$$

where

$$P(y) = \sum_{x=0}^{M-1} P(y \mid x)P(x).$$

It can be shown that, when the $M$-ary constellation is *orthogonal* or *simplex* [8], the discrete memoryless channel is also *symmetric*. It is symmetric in the sense that the channel transition probabilities $\{P(y \mid x)\}$ satisfy

$$P(y \mid x) = \begin{cases} 1 - \varepsilon, & \text{when } y = x, \\ \varepsilon/(M-1), & \text{otherwise.} \end{cases}$$

For this channel, the mutual information achieves its maximum value (capacity) when $P(x) = 1/M$ for all channel inputs $x$.

## 1.6 Finite-Length Bounds on Error Probability

Shannon's capacity places sharp limits on what rate $R$ can, or cannot, be achieved for reliable communication over noisy channels, assuming the blocklength of the channel code becomes arbitrarily large (and implementation constraints are ignored!). Even when $R < C$, though, we cannot expect error probability to be arbitrarily small for typical code blocklengths. Practitioners naturally began to ask questions such as "For rate $R < C$ and blocklength $n$ used on a given physical channel, what block-error probability is guaranteed achievable?" Or "What error probability can never be beaten?" For a given codebook, decoding generally assumes a maximum likelihood assumption to optimize performance, even though ML processing may be unrealistic.

Initial work in this direction focused upon the achievability question, and more specifically upon the exponential dependence on blocklength, producing a variety of exponents that guarantee the existence of codes with error probability achieving (see [4])

$$P_e < e^{-nE(R)}, \tag{1.22}$$

where $E(R)$ is a so-called random-coding exponent that depends on the code rate $R$, as well as the physical channel of interest. (Similar exponential lower bounds were also produced.) This is the *Gallager random coding bound* on the codeword error probability $P_e$ and the *Gallager error exponent* is given by

$$E(R) = \max_{\{\Pr(x)\}} \max_{0 \le \rho \le 1} \left[E_0(\rho, \{\Pr(x)\}) - \rho R\right],$$

with

$$E_0(\rho, \{\Pr(x)\}) = -\log_2\left[\int_{-\infty}^{\infty}\left[\sum_{x\in\{\pm 1\}} \Pr(x)p(y \mid x)^{1/(1+\rho)}\right]^{1+\rho} dy\right].$$

Because the BI-AWGN channel is symmetric, the maximizing distribution on the channel input is $\Pr(x = +1) = \Pr(x = -1) = 1/2$. Also, for the BI-AWGN channel,

$p(y \mid x) = \left[1/\left(\sqrt{2\pi}\sigma\right)\right] \exp\left[-(y-x)/\left(2\sigma^2\right)\right]$ so that $E_0(\rho, \{\Pr(x)\})$ becomes, after some manipulation,

$$E_0(\rho, \{\Pr(x)\}) = -\log_2\left[\int_{-\infty}^{\infty} \frac{1}{\sqrt{2\pi}\sigma} \exp\left(-\frac{y^2+1}{2\sigma^2}\right)\left[\cosh\left(\frac{y}{\sigma^2(1+\rho)}\right)\right]^{1+\rho} dy\right].$$

It can be shown that $E(R) > 0$ if and only if $R < C$, thus proving from (1.22) that, as $n \to \infty$, arbitrarily reliable communication is possible provided that $R < C$.

While demonstrating the general importance of long codes for operation near the capacity limit, as well as exponential improvement with long blocklength, the bounds are not tight for moderate-length codes of growing interest in short-message, latency-limited digital communication. There we need a non-asymptotic relationship between block-error probability $P_e$, blocklength $n$, code rate $R$, which expresses the amount of redundancy in the code, and channel quality (e.g., crossover probability for a BSC or SNR for a Gausian channel model). Bounding will fix three of these and bound the fourth, often bounding $P_e$ for a given length, rate, and channel model.

Why settle for bounds? We avoid the difficult questions of finding the best code, and then of evaluating its performance. Both tasks are prohibitively difficult even for moderate-blocklength codes. The bounds discussed here are code-independent, that is, they hold for any coding system of a given blocklength and rate $R$, for a given channel model and channel quality. This section will give a review of old and more recent bounding results for the BI-AWGN model, in keeping with much of the coding applications in this text. Our main interest is in lower bounds to provide a benchmark for assessing the performance of real codes.

### 1.6.1 Shannon's Sphere-Packing (or Cone) Bound

A code for the Gaussian channel is a set of real-valued $n$-tuples, which can be seen as points in $\mathbb{R}^n$. Shannon [12] provided an elegant geometric argument for codes under one of three design constraints:

- Each codeword has equal energy, that is, all $2^{nR}$ codewords lie on a sphere in $n$ dimensions, with the common energy taken as $nE_s = nRE_b$.
- The *maximum* energy of a codeword is this same value.
- The *average* codeword energy matches the constraint.

The latter two allow codewords to be inside and/or outside the sphere of radius $r = (nE_s)^{1/2}$. For large $n$, all three constraints yield the same exponential dependence, and lead to the same channel capacity $C$, namely

$$C_{\text{AWGN}} = \frac{1}{2}\log_2\left(1 + \frac{2E_s}{N_0}\right) \text{ bits/channel use.} \tag{1.23}$$

However, for moderate-length codes, coding bounds differ somewhat under the constraints above, since they impose different geometric constraints on a code. To see why, consider simple two-dimensional constellations with eight points ($n = 2$ codes with rate $R = \log_2 3/2$ bits/channel use). Codes corresponding to 8-PSK, 7-around 1, and a "box" constellation rank differently if they are compared based on maximum, or peak, energy versus average energy [9]. Though coding with QAM modulation, for example, may produce codewords with unequal energy, our primary interest there is the equal-energy case appropriate to binary coding built on $\pm 1$ modulation coordinates.

Consider the equal-energy constraint. Then $M = 2^{nR}$ codewords may be viewed as points on the surface of a sphere, with sphere radius $r = \sqrt{nE_s}$, where $E_s$ is the energy per coordinate at the decoder. Ultimately, performance comparisons are made by relating $E_s$ to the energy per information bit, that is, $E_s = kE_b/n = RE_b$. Assuming independent Gaussian noise with variance $\sigma^2 = N_0/2$ is added in each coordinate, an ML decoder will choose that codeword closest in Euclidean distance to the received vector $\mathbf{y} \in \mathbb{R}^n$. The corresponding decision zones, or Voronoi zones, are pyramidal polyhedra in $\mathbb{R}^n$, bounded by at most $M - 1$ hyperplanes bisecting line segments connecting a given codeword and each remaining codeword. These decision zones intercept some amount of surface area on the $n$-sphere, and the ratio of this area to the radius is the solid angle of the given zone. Shannon first argues that the probabillity that noise carries $\mathbf{y}$ outside a given decision zone is lower bounded by the probability that noise carries $\mathbf{y}$ outside a right-circular cone with vertex at the origin and whose solid angle matches that of the actual decision zone. Then, to minimize the overall error probability, Shannon argues that decision zones should share the total solid angle of the sphere equally among $M$ codewords.

Using results for the area of a sphere in $n$ dimensions, as well as the area of a circular cone in $n$ dimensions that divides the total solid angle $M$ ways, [12] establishes that the half-cone angle $\theta_c$ satisfies

$$\int_0^{\theta_c} (\sin(\phi))^{n-2} d\phi = \frac{\Gamma((n-1)/2)\pi^{1/2}}{\Gamma(n/2)2^{nR}}. \tag{1.24}$$

As $n \to \infty$, it can be shown [13] that $\theta_c$ approaches $2^{-R}$, or $\pi/4$ for $R = 1/2$ codes.

Given the angle of this cone, Shannon's lower bound becomes

$$P_e \geq \Pr\{\text{noise moves } \mathbf{y} \text{ outside cone}\}, \tag{1.25}$$

where the mean of $\mathbf{y}$ is $((nE_s)^{1/2}, 0, 0, \ldots, 0)$ and the noise variance in each dimension is $\sigma^2 = N_0/2$. Computation of this probability is expressible in terms of the cumulative distribution function (cdf) of a non-central chi-squared random variable [12].

For even moderate blocklength $n$, numerical difficulties arise, and work by Valembois and Fossorier [13] and Wiechman and Sason [14] massages Shannon's bound into tractable computing recipes for any realistic $n$. (Both papers also study an improved sphere-packing bound for the BI-AWGN channel growing out of [15], but case studies show the original Shannon bound is tighter for most cases of interest.)

### 1.6.2 Hypothesis-Testing Bounds

More recently, Polyanskiy *et al.* [16] have developed a general lower bound, called the *meta-converse* (MC), as it reduces to known lower bounds or converse coding theorems in asymptotic cases. The decoder's task is, of course, a hypothesis- testing problem among $M = 2^{nR}$ alternatives. But the MC bound involves deciding between two substitute hypotheses, $H_0$ and $H_1$, where under $H_0$ the data obey the actual channel distribution $P_x W_{y|x}$ and under $H_1$, the data have the distribution $P_x Q_y$, for some auxilary distribution $Q_y$, with **y** independent of the input. Formally, the MC bound is given by a Neyman–Pearson formulation [17], fixing the false negative (or type II error probability) at $\beta = 1/M$, and asking for the minimum type I error probability (or false positive) $\alpha$. The MC bound then allows optimization over both $P_x$ and $Q_y$, and states

$$P_e \geq \inf_{P_x} \sup_{Q_y} \alpha_{\{\beta_{1/M}\}}[P_x W_{y|x}, P_x Q_y], \tag{1.26}$$

where the supremum is over all distributions $Q_y$ and the infimum is over all input distributions $P_x$.

Exact computation of this min–max problem is difficult, but has been made feasible by adopting the output distribution as independently and identically distributed (i.i.d.) Gaussian, with mean zero and variance subject to optimization. Vazquez-Vilar [18] and Erseghe [19] have provided a computational approach using saddle-point methods for integration, and software can be found in [20].

## Normal Approximation

Evolving from the work of Polyanskiy *et al.* was both the meta-converse (lower bound) and an upper bound called the RCU (random coding union bound). The latter defines a range of parameter space for which codes are guaranteed to exist. These upper and lower bounds merge as $n$ grows, for example, $P_e$ bounds plotted for a BI-AWGN mode for increasing $n$ have a shrinking difference on the SNR axis. Though neither an upper nor a lower bound, a simple approximation that can serve as a benchmark for assessing coding performance is the so-called *normal approximation* (NA). (Some presentations call this the *dispersion "bound."*) This approximation emerges from the bounding expressions as $n$ increases, and is also seen by eye-balling the family of upper and lower bounds. The NA holds for arbitrary channel models and one form of it is

$$P_{e_{\mathrm{NA}}} \simeq Q\left[\frac{n(C-R) + \frac{1}{2n}\log_2(n)}{\sqrt{V/n}}\right], \tag{1.27}$$

where $C$ is Shannon's capacity for the channel, $V$ is a channel parameter called the *channel dispersion*, and $Q(x)$ is the Gaussian tail integral function, thus the name "normal approximation." $V$ is a measure of channel variability, and is defined as the variance of the *information density*, where the (random) information density is

$$i(x; y) = \log_2 \frac{P(y|x)}{P(y)}. \tag{1.28}$$

(Note that the mean of the information density is the channel mutual information.) For the BSC, it may be shown that

$$V = p(1-p)\log_2^2\left(\frac{1-p}{p}\right), \tag{1.29}$$

and $C$ is given by (1.8). For the BI-AWGN model, numerical evaluation of the variance of information density is required, using

$$V = \int_{-\infty}^{+\infty} \frac{1}{\sqrt{2\pi}} e^{-z^2/2} \left(1 - \log_2\left(1 + e^{-2\rho + 2\sqrt{\rho} z}\right) - C\right)^2 dz, \tag{1.30}$$

where $\rho = (2E_s/N_0)^{1/2}$ and where $C$ is given by (1.13). The NA is generally accepted to be within a fraction of a decibel of both lower (converse) and upper (achievable) bounds on the BI-AWGN provided $n > 200$, and the rate is close to neither 0 or 1. Given its simpler computation, the NA serves as a convenient finite-blocklength standard for assessing code performance. For shorter blocklength applications, however, it is wise to use a true lower (or perhaps upper) bound for assessing performance.

### 1.6.3 Results

This finite-length bounding discussion is illustrated for two $R = 1/2$ coding applications of relatively short blocklength. Figure 1.9 shows these lower bounds, together with the normal approximation, for the case $n = 24, k = 12$. Also shown are simulation results for ML decoding of the extended Golay (24, 12) code, surely the best binary code with these parameters. We observe that the meta-converse lower bound is slightly tighter than Shannon's cone bound, though only at lower SNR. The simulation results are to the right of the meta-converse by about 0.15 dB as well. (The extended Golay code is surely the best binary code with these parameters, and the gap in the bounds is attributable to (1) bounding of performance and (2) the code employing binary-valued coordinates, something not required in Shannon's assumption.) Also note that the NA

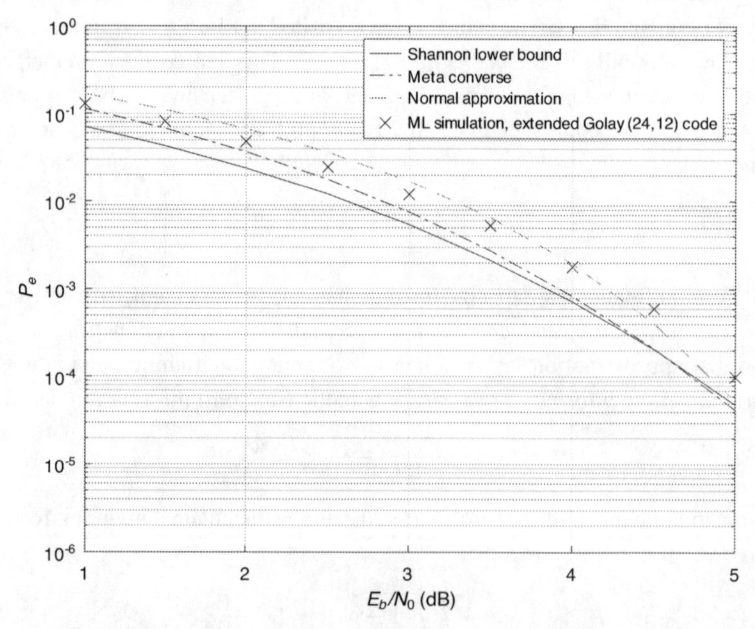


**Figure 1.9** Lower bounds and simulated performance for binary $(n, k) = (24, 12)$ codes.

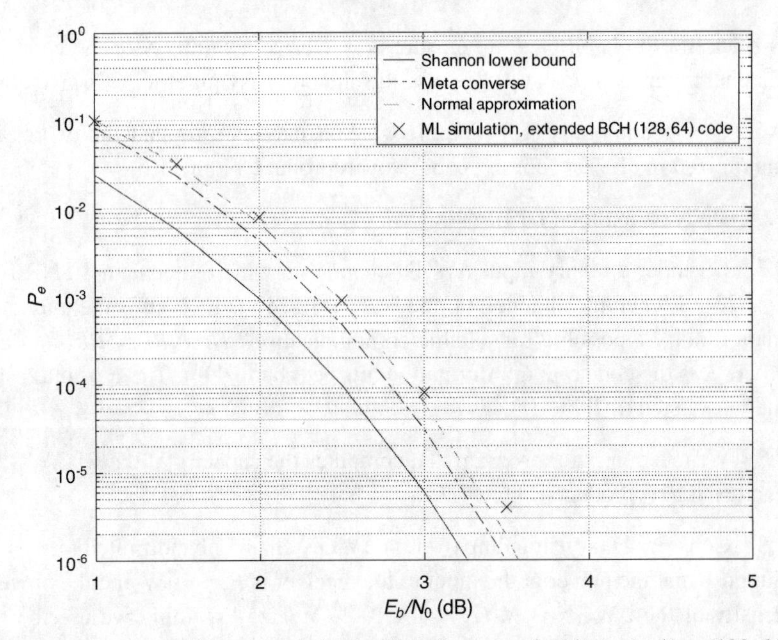


**Figure 1.10** Lower bounds and simulated performance for binary $(n, k) = (128, 64)$ codes.

fits the simulation points rather well, but this is coincidental – the NA does not claim to approximate the performance of any specific code.

Figure 1.10 repeats this for another $R = 1/2$ case, with $(n, k) = (128, 64)$. The extended BCH code with these parameters has the performance shown (data is taken from [13]). Similar conclusions hold about the two lower bounds relative to the performance of a very good binary code.

Further reading dealing with bounds and the performance of standard codes can be found in Coskun *et al*. [21] and Yang *et al*. [22].

## Problems

**1.1** A single error has been added (modulo 2) to a transmitted $(7, 4)$ Hamming codeword, resulting in the received word $\mathbf{r} = (0011\ 100)$. Using the decoding algorithm described in this chapter, find the error.

**1.2** For linear codes, the list $S_d(\mathbf{c})$ of Hamming distances from codeword $\mathbf{c}$ to all of the other codewords is the same, no matter what the choice of $\mathbf{c}$ is. (See Chapter 3.) $S_d(\mathbf{c})$ is called the *conditional distance spectrum* for the codeword $\mathbf{c}$. Find $S_d(0000\ 000)$ and $S_d(0111\ 001)$ for the $(7, 4)$ Hamming code and check that they are the same.

**1.3** Using (1.6), derive (1.7) and (1.8). Note that the capacity in each case is given by (1.6) because these channels are symmetric.

**1.4** Consider a hybrid of the BSC and the BEC in which there are two input symbols (0 and 1) and three output symbols (0, $e$, and 1). Further, the probability of an error is $\varepsilon$ and the probability of an erasure is $p$. Derive a formula for the capacity of this channel in terms of $\varepsilon$ and $p$. Because the channel is symmetric, $C$ is given by (1.6).

**1.5** Derive equations (1.9) and (1.10). You may assume that the expression being maximized in (1.9) is "convex cap." Now reproduce Figure 1.6.

**1.6** Derive equation (1.13).

**1.7** Consider a binary-input AWGN channel in which the channel SNR is $E_s/N_0 = -2.823$ dB when the channel bit rate is $R_s = 10$ Mbits/s. What is the maximum information rate $R_b$ possible for reliable communication? *Hint*: $R_b = RR_s$ and $E_s = RE_b$, where $R$ is the code rate in information bits per channel bit. The $R$ of interest is one of the rates listed in Table 1.1.

**1.8** Write a computer program that computes the capacity of the BI-AWGN channel and use it to reproduce Figure 1.7.

**1.9** Consider binary signaling over an AWGN channel bandlimited to $W = 5000$ Hz. Suppose that the power at the input to the receiver is $P = 1\ \mu\text{W}$ and the power spectral density of the AWGN is $N_0/2 = 5 \times 10^{-11}$ W/Hz. Assuming we use the maximum possible Nyquist signaling rate $R_s$ over the channel, find the maximum information rate $R_b$ possible for reliable communication through this channel. *Hint*: To start, you will have to find $E_s/N_0$ and then compute $C_{\text{BI-AWGN}}$ for the $E_s/N_0$ that you find. You will also need $R_b = RR_s$, $E_s = RE_b$, and $E_s = P/R_s$, where $R$ is the code rate in information bits per channel bit.

**1.10** Consider a bandlimited AWGN channel with bandwidth $W = 10\,000$ Hz and SNR $E_b/N_0 = 10$ dB. Assuming an unconstrained signaling alphabet, is reliable communication possible at $R_b = 50\,000$ bits/s? What about at $60\,000$ bits/s?

**1.11** From equation (1.19), we may write $\xi < \log_2(1 + \xi E_b/N_0)$, where $\xi = R_b/W$ is the spectral efficiency, a measure of the information rate per unit bandwidth in units of bits per second per hertz. Show that, when $\xi = 1$ bits/s/Hz, we must have $E_b/N_0 > 0$ dB; when $\xi = 4$ bits/s/Hz, we must have $E_b/N_0 > 5.74$ dB; and as $\xi \to 0$ bits/s/Hz, we must have $E_b/N_0 \to \ln(2) = -1.6$ dB.

**1.12**

(a) Let $G$ be a Gaussian random variable, $G \sim \mathcal{N}(\mu, \sigma^2)$. Show that its differential entropy is given by $h(G) \triangleq -\mathbb{E}[\log_2(p(G))] = \frac{1}{2}\log_2\left(2\pi e\sigma^2\right)$ by performing the integral implied by $-\mathbb{E}[\log_2(p(G))]$.

(b) We are given the channel model $Y = X + Z$, where $X \sim \mathcal{N}(0, P)$ and $Z \sim \mathcal{N}(0, \sigma^2)$ are independent. Show that the mutual information is given by $I(X;Y) = h(Y) - h(Y \mid X) = h(Y) - h(Z) = \frac{1}{2}\log_2\left(1 + P/\sigma^2\right) = C$. Thus, a Gaussian-distributed signal achieves capacity on the arbitrary-input AWGN channel.

**1.13** Let $G$ be a zero-mean complex Gaussian random variable with variance $\sigma^2$ on each component. Perform the integral implied by $-\mathbb{E}[\log_2(p(G))]$ to show that its differential entropy is given by $h(G) \triangleq -\mathbb{E}[\log_2(p(G))] = \log_2\left(2\pi e\sigma^2\right)$.

**1.14** Write a computer program to reproduce Figure 1.8.

**1.15** Analogous to (1.20), derive capacity formulas for one-dimensional $M$-ary modulation and plot the capacity of $M$-ary pulse amplitude modulation versus $E_b/N_0$ for $M = 2, 4, 8,$ and 16. Add the appropriate Shannon capacity curve to your plot for comparison.

**1.16** As mentioned in the chapter, the mutual information may alternatively be written as $I(X;Y) = H(X) - H(X \mid Y)$. Use this expression to obtain a formula for the capacity of the $M$-ary symmetric channel that is an alternative to (1.21).

**1.17** Consider an independent *Rayleigh fading channel* with channel output $y_k = a_k x_k + n_k$, where $n_k \sim \mathcal{N}(0, N_0/2)$ is an AWGN sample, $x_k \in \{\pm 1\}$ equally likely, and $a_k$ is a Rayleigh random variable with expectation $\mathbb{E}\{a_k^2\} = 1$. Derive a capacity formula for this channel and plot its capacity against $\bar{E}_b/N_0$, where $\bar{E}_b$ is the average information bit energy. *Hint*: Derive a capacity formula for a specific realization (value) for $a_k$ and then take the expected value of that formula. *Comment*: Independent fading is achieved by interleaving on a slowly fading channel. The resulting capacity computed here is known as the *ergodic capacity*.

**1.18**

(a) For a BSC with crossover probability $\varepsilon = 0.05$, calculate $C$ and the channel dispersion $V$ using (1.29).

(b) Use the normal approximation to estimate the performance of the famous $(n, k) = (23, 12)$ triple-error-correcting Golay code, and compare it with the probability that four or more errors occur.

(c) Repeat for a $(n, k) = (127, 64)$ BCH code (see Chapter 3) that has $d_{\min} = 21$ and assume the decoder only corrects error patterns with $t = 10$ or fewer errors.

**1.19** Consider $R = 1/2$ codes for the BI-AWGN channel with $E_b/N_0 = 3.01$ dB. This means $E_s/N_0 = 0$ dB.

(a) Show by numerical computation that $C = 0.721$ bits/channel use (and thus $R < C$) and that $V = 0.533$.

(b) Determine the normal approximation estimate of error probability for a (24, 12) extended Golay code with ML decoding and compare it with a simulation result that yields $P_e = 1.3 \times 10^{-2}$.

(c) Repeat for an extended BCH (128, 64) code and compare it with a simulation result of [13] giving $P_e = 10^{-5}$ at the same SNR.

## References

[1] C. Shannon, “A mathematical theory of communication,” *Bell System Technical Journal*, pp. 379–423 (Part I), pp. 623–656 (Part II), July 1948.

[2] R. Hamming, “Error detecting and correcting codes,” *Bell System Technical Journal*, pp. 147–160, April 1950.

[3] T. Cover and J. Thomas, *Elements of Information Theory*, 2nd ed., New York, Wiley, 2006.

[4] R. Gallager, *Information Theory and Reliable Communication*, New York, Wiley, 1968.

[5] R. McEliece, *The Theory of Information and Coding: Student Edition*, Cambridge, Cambridge University Press, 2004.
[6] R. Blahut, *Principles and Practice of Information Theory*, Reading, MA, Addison-Wesley, 1987.
[7] A. Viterbi and J. Omura, *Principles of Digital Communication and Coding*, New York, McGraw-Hill, 1979.
[8] J. Proakis, *Digital Communications*, 4th ed., New York, McGraw-Hill, 2000.
[9] S. Wilson, *Digital Modulation and Coding*, Englewood Cliffs, NJ, Prentice-Hall, 1996.
[10] S. Golomb, "The limiting behavior of the Z-channel," *IEEE Transactions on Information Theory*, vol. 26, no. 5, p. 372, May 1980.
[11] R. J. McEliece, "Are turbo-like codes effective on non-standard channels?," *IEEE Transactions onInformation Theory Society Newsletter,* vol. 51, no. 4, pp. 1–8, December 2001.
[12] C. E. Shannon, "Probability of error for optimal codes in a Gaussian channel," *Bell System Technical Journal*, vol. 38, pp. 611–656, 1959.
[13] A. Valembois and M. P. C. Fossorier, "Sphere-packing bounds revisited for moderate block lengths," *IEEE Transactions on Information Theory*, vol. 50, pp. 2998–3014, 2004.
[14] G. Wiechman and I. Sason, "An improved sphere-packing bound for finite-length codes on symmetric memoryless channels," *IEEE Transactions on Information Theory*, vol. 54, no. 5, pp. 1962–1990, May 2008.
[15] C. E. Shannon, R. Gallager, and E. Berlekamp, "Lower bounds to error probability for decoding on discrete memoryless channels, Part I," *Information and Control*, vol. 10, pp. 65–103, 1967.
[16] Y. Polayanskiy, H. V. Poor, and S. Verdu, "Channel coding rate in the finite blocklength regime," *IEEE Transactions on Information Theory*, vol. 56, pp. 2307–2593, 2010.
[17] J. Neyman and E. S. Pearson, "Sufficient statistics and uniformly most powerful tests of statistical hypotheses," in *Joint Statistical Papers*, Berkeley, CA, University of California Press, 1967, pp. 240–264.
[18] G. Vazquez-Vilar, "Error probability bounds for Gaussian channels under maximal and average power constraints," *IEEE Transactions on Information Theory*, vol. 67, pp. 3965–3985, 2021.
[19] T. Erseghe, "Coding in the finite-blocklength regime: Bounds based on Laplace integrals and their asymptotic approximations," arXiv:1511.04629,2015.
[20] G. Durisi, https://github.com/gdurisi/fbl-notes.
[21] M. C. Coskun, G. Durisi, T. Jerkovits, G. Liva, W. Ryan, B. Stein, F. Steiner, "Efficient error-correcting codes in the short blocklength regime," *Physical Communications*, vol. 34, pp. 66–79, 2019.
[22] H. Yang, E. Lang, M. Pan, and R. D. Wesel, "CRC-aided list decoding of convolutional codes in the short blocklength regime," *IEEE Transactions on Information Theory*, vol. 68, pp. 3744–3766, 2022.

# 2 Finite Fields, Vector Spaces, Finite Geometries, and Graphs

This chapter presents some important elements of modern algebra and combinatorial mathematics, namely, finite fields, vector spaces, finite geometries, and graphs, that are needed in the presentation of the fundamentals of classical channel codes and various constructions of modern channel codes in the following chapters. It will be seen that the algebraic structure imposed on code designs results in simplified encoders and decoders. The treatment of these mathematical elements is by no means complete or rigorous, and coverage is kept at an elementary level. There are many good textbooks on modern algebra, combinatorial mathematics, and graph theory that provide rigorous treatment and in-depth coverage of finite fields, vector spaces, finite geometries, and graphs. Some of these texts are listed at the end of this chapter.

## 2.1 Sets and Binary Operations

A set is a *collection* of certain objects, commonly called the *elements* of the set. A set and its elements will often be denoted by letters of an alphabet. Commonly, a set is represented by a capital letter and its elements are represented by lower-case letters (with or without subscripts). For example, $X = \{x_1, x_2, x_3, x_4, x_5\}$ is a set with five elements, $x_1$, $x_2$, $x_3$, $x_4$, and $x_5$. A set $S$ with a finite number of elements is called a *finite set*; otherwise, it is called an *infinite set*. In error-control coding theory, we mostly deal with finite sets. The number of elements in a set $S$ is called the *cardinality* of the set and is denoted by $|S|$. The cardinality of the above set $X$ is five. A part of a set $S$ is of course itself a set. Such a part of $S$ is called a *subset* of $S$. A subset of $S$ with cardinality smaller than that of $S$ is called a *proper subset* of $S$. For example, $Y = \{x_1, x_3, x_4\}$ is a proper subset of the above set $X$.

A *binary operation* on a set $S$ is a *rule* that assigns to any pair of elements in $S$, taken in a *definite order*, another element in the same set. For the present, we denote such a binary operation by $*$. Thus, if $a$ and $b$ are two elements of the set $S$, then the pair $(a, b)$ is an element of $S$, namely $a * b$. The operation also assigns to the pair $(b, a)$ an element in $S$, namely $b * a$. Note that $a * b$ and $b * a$ do not necessarily give the same element in $S$. When a binary operation $*$ is defined on $S$, we say that $S$ is *closed* under $*$. For example, let $S$ be the set of all integers and let the binary operation on $S$ be real (or ordinary) addition $+$. We all know that, for any two integers $i$ and $j$ in $S$, $i + j$ is another integer in $S$. Hence, the set of integers is closed under real (or ordinary) addition. Another well-known binary operation on the set of integers is real (or ordinary) multiplication, denoted by "$\cdot$."

The adjective *binary* is used because the operation produces from a pair of elements in $S$ (taken in a *specific order*) an element in $S$. If we take elements $a$, $b$, and $c$ of $S$ (not necessarily different), there are various ways in which we can combine them by the operation $*$. For example, we first combine $a$ and $b$ with $*$ to obtain an element $a * b$ of $S$ and then combine $a * b$ and $c$ with $*$ to produce an element $(a * b) * c$ of $S$. We can also combine the pair $a$ and $b * c$ to obtain an element $a * (b * c)$ in $S$. The parentheses are necessary to indicate just how the elements are to be grouped into pairs for combination by the operation $*$.

**Definition 2.1** Let $S$ be a set with a binary operation $*$. The operation is said to be *associative* if, given any elements $a$, $b$, and $c$ in $S$, the following equality holds:

$$(a * b) * c = a * (b * c). \tag{2.1}$$

The equality sign "=" is taken to mean "the same as." We also say that the binary operation $*$ satisfies the *associative law*, if the equality (2.1) holds.

The associative law simply implies that, when we combine elements of $S$ with repeated applications of the binary operation $*$, the result does not depend upon the *grouping* of the elements involved. Hence, we can write $(a*b)*c = a*(b*c) = a*b*c$. The associative law is a very important law, and most of the binary operations that we shall encounter will satisfy it. Addition $+$ and multiplication $\cdot$ on the set of integers (or real numbers) are associative binary operations.

**Definition 2.2** Let $S$ be a set with a binary operation $*$. The operation is said to be *commutative* if, given any two elements, $a$ and $b$, in $S$, the following equality holds:

$$a * b = b * a. \tag{2.2}$$

We also say that $*$ satisfies the *commutative law*, if (2.2) holds.

Note that the definition of the commutative law is independent of the associative law. The commutative law simply implies that, when we combine two elements of $S$ by the $*$ operation, the result does not depend on the order of combination of the two elements. Addition $+$ and multiplication $\cdot$ on the set of integers (or real numbers) are commutative binary operations. Therefore, addition and multiplication on the set of integers (or real numbers) are both associative and commutative.

Let $S$ be a set with an associative and commutative binary operation $*$. Then, if any combination of elements of $S$ is formed by means of repeated applications of $*$, the result depends neither on the grouping of the elements involved nor on their order. For example, let $a$, $b$, and $c$ be elements of $S$. Then

$$\begin{aligned}(a * b) * c &= a * (b * c) = a * (c * b) = (a * c) * b \\ &= c * (a * b) = (c * a) * b = (c * b) * a.\end{aligned}$$

Given two sets, $X$ and $Y$, we can form a set $X \cup Y$ that consists of the elements in either $X$ or $Y$, or both. This set $X \cup Y$ is called the *union* of $X$ and $Y$. We can also form a set $X \cap Y$ that consists of the set of all elements in both $X$ and $Y$. The set $X \cap Y$ is called the *intersection* of $X$ and $Y$. A set with no element is called an *empty set*, denoted by $\emptyset$. Two sets are said to be *disjoint* if their intersection is empty, that is, they have no element in common. If $Y$ is a subset of $X$, the set that contains the elements in $X$ but not in $Y$ is called the *difference* between $X$ and $Y$ and is denoted by $X \backslash Y$.

## 2.2 Groups

In this section, we introduce a simple algebraic system. By an algebraic system, we mean a set with one or more binary operations defined on it.

### 2.2.1 Basic Concepts of Groups

A group is an algebraic system with one binary operation defined on it.

**Definition 2.3** A *group* is a set $G$ together with a binary operation $*$ defined on $G$ such that the following axioms (conditions) are satisfied.

1. The binary operation $*$ is associative.
2. $G$ contains an element $e$ such that, for any element $a$ of $G$,

$$a * e = e * a = a.$$

   This element $e$ is called an *identity element* of $G$ with respect to the operation $*$.
3. For any element $a$ in $G$, there exists an element $a'$ in $G$ such that

$$a * a' = a' * a = e.$$

   The element $a'$ is called an *inverse* of $a$, and vice versa, with respect to the operation $*$.

A group $G$ is said to be commutative if the binary operation $*$ defined on it is also commutative. A commutative group is also called an *abelian* group.

**Theorem 2.4** *The identity element $e$ of a group $G$ is* unique.

*Proof* Suppose both $e$ and $e'$ are identity elements of $G$. Then $e*e' = e$ and $e*e' = e'$. This implies that $e$ and $e'$ are identical. Therefore, the identity of a group is unique. □

**Theorem 2.5** *The inverse of any element in a group $G$ is* unique.

*Proof* Let $a$ be an element of $G$. Suppose $a$ has two inverses, say $a'$ and $a''$. Then

$$a'' = e * a'' = (a' * a) * a'' = a' * (a * a'') = a' * e = a'.$$

This implies that $a'$ and $a''$ are identical and hence $a$ has a unique inverse. □

The set of all rational numbers with real addition $+$ forms a commutative group. We all know that real addition $+$ is both associative and commutative. The integer 0 is the identity element of the group. The rational number $-a/b$ ($b$ is a nonzero integer) is the additive inverse of the rational number $a/b$ and vice versa. The set of all rational numbers, *excluding* zero, forms a commutative group under real multiplication $\cdot$. The integer 1 is the identity element with respect to real multiplication, and the rational number $b/a$ (both $a$ and $b$ are nonzero integers) is the inverse of the rational number $a/b$.

The two groups given above have infinite numbers of elements. However, groups with a finite number of elements do exist, as will be seen later in this section.

**Definition 2.6** A group $G$ is called *finite* (or *infinite*) if it contains a finite (or an infinite) number of elements. The number of elements in a finite group is called the *order* of the group.

It is clear that the order of a finite group $G$ is simply its cardinality $|G|$. There is a convenient way of presenting a finite group $G$ with operation $*$. A table displaying the group operation $*$ is constructed by labeling the rows and the columns of the table by the group elements. The element appearing in the row labeled by $a$ and the column labeled by $b$ is then taken to be $a * b$. Such a table is called a *Cayley table*.

---

**Example 2.1** This example gives a simple group with two elements. We start with the set of two integers, $G = \{0, 1\}$. Define a binary operation, denoted by $\oplus$, on $G$ as shown by the Cayley table below:

| $\oplus$ | 0 | 1 |
|---|---|---|
| 0 | 0 | 1 |
| 1 | 1 | 0 |

From the table, we readily see that

$$0 \oplus 0 = 0, \quad 0 \oplus 1 = 1, \quad 1 \oplus 0 = 1, \quad 1 \oplus 1 = 0.$$

It is clear that the set $G = \{0, 1\}$ is closed under the operation $\oplus$ and $\oplus$ is commutative. To prove that $\oplus$ is associative, we simply determine $(a \oplus b) \oplus c$ and $a \oplus (b \oplus c)$ for eight possible combinations of $a$, $b$, and $c$ with $a$, $b$, and $c$ in $G = \{0, 1\}$, and show that

$$(a \oplus b) \oplus c = a \oplus (b \oplus c),$$

for each combination of $a$, $b$, and $c$. This kind of proof is known as *perfect induction*. From the table, we see that 0 is the identity element, the inverse of 0 is itself, and the inverse of 1 is also itself. Thus, $G = \{0, 1\}$ with operation $\oplus$ is a commutative group of order 2. The binary operation $\oplus$ defined on the set $G = \{0, 1\}$ by the above Cayley table is called *modulo*-2 *addition*.

---

## 2.2.2 Finite Groups

Finite groups are of practical importance in many areas, especially in the area of coding theory. In the following, we will present two classes of finite groups, one with a binary operation very similar to real addition and the other with a binary operation very similar to real multiplication.

Let $m$ be a positive integer. Consider the set of integers $G = \{0, 1, \ldots, m-1\}$. Let $+$ denote real addition. Define a binary operation, denoted by $\boxplus$. For any two integers $i$ and $j$ in $G$:

$$i \boxplus j = r, \tag{2.3}$$

where $r$ is the remainder resulting from dividing the sum $i + j$ by $m$. By Euclid's division algorithm, $r$ is a non-negative integer between 0 and $m - 1$, and is therefore an element in $G$. Thus, $G$ is closed under the binary operation $\boxplus$, which is called *modulo-m addition*.

The set $G = \{0, 1, \ldots, m-1\}$ with modulo-$m$ addition forms a commutative group. To prove this, we need to show that $G$ satisfies all the three axioms given in Definition 2.3. For any integer $i$ in $G$, since $0 \leq i < m$, it follows from the definition of $\boxplus$ that $0 \boxplus i = i$ and $i \boxplus 0 = i$. Hence, 0 is the identity element. For any nonzero integer $i$ in $G$, $m - i$ is also in $G$. Since $i + (m - i) = m$ and $(m - i) + i = m$, it follows from the definition of $\boxplus$ that $i \boxplus (m - i) = 0$ and $(m - i) \boxplus i = 0$. Therefore, $m - i$ is the inverse of $i$ with respect to $\boxplus$, and vice versa. Note that the inverse of 0 is itself. So every element in $G$ has an inverse with respect to $\boxplus$. For any two integers $i$ and $j$ in $G$, since real addition $+$ is commutative, $i + j = j + i$. On dividing $i + j$ and $j + i$ by $m$, both give the same remainder $r$ with $r$ in $G$. Hence, $i \boxplus j = j \boxplus i$ and $\boxplus$ is commutative. Next, we prove that $\boxplus$ is also associative. The proof is based on the fact that real addition is associative. For $i$, $j$, and $k$ in $G$,

$$i + j + k = (i + j) + k = i + (j + k).$$

On dividing the sum $i + j + k$ by $m$, we obtain

$$i + j + k = q \cdot m + r,$$

where $q$ and $r$ are the quotient and remainder, respectively, and $0 \leq r < m$. On dividing $i + j$ by $m$, we have

$$i + j = q_1 \cdot m + r_1, \tag{2.4}$$

where $0 \leq r_1 < m$. It follows from the definition of $\boxplus$ that $i \boxplus j = r_1$. Dividing $r_1 + k$ by $m$ results in

$$r_1 + k = q_2 \cdot m + r_2, \tag{2.5}$$

where $0 \leq r_2 < m$. Hence, $r_1 \boxplus k = r_2$ and

$$(i \boxplus j) \boxplus k = r_2.$$

On combining (2.4) and (2.5), we have $i + j + k = (q_1 + q_2) \cdot m + r_2$. This implies that $r_2$ is also the remainder when we divide $i + j + k$ by $m$. Since the remainder is unique when we divide an integer by another integer, we must have $r_2 = r$. Consequently, we have

$$(i \boxplus j) \boxplus k = r.$$

Similarly, we can prove that

$$i \boxplus (j \boxplus k) = r.$$

Therefore, $(i \boxplus j) \boxplus k = i \boxplus (j \boxplus k)$ and the binary operation $\boxplus$ is associative. This concludes our proof that the set $G = \{0, 1, \ldots, m-1\}$ with modulo-$m$ addition $\boxplus$ forms a commutative group of order $m$. For $m = 2$, $G = \{0, 1\}$ with modulo-2 addition gives the binary group constructed in Example 2.1. The above construction gives an infinite class of finite groups under modulo-$m$ addition with $m = 2, 3, 4, \ldots$. The groups in this class are called *additive groups*.

---

**Example 2.2** Let $m = 7$. Table 2.1 displays the additive group $G = \{0, 1, 2, 3, 4, 5, 6\}$ under modulo-7 addition.

---

Next we present a class of finite groups under a binary operation similar to real multiplication. Let $p$ be a prime. Consider the set of $p - 1$ positive integers less than $p$, $G = \{1, 2, \ldots, p - 1\}$. Let "$\cdot$" denote real multiplication. Every integer $i$ in $G$ is relatively prime to $p$. Define a binary operation, denoted by $\boxdot$, on $G$ as follows. For any two integers, $i$ and $j$, in $G$:

$$i \boxdot j = r, \tag{2.6}$$

where $r$ is the remainder resulting from dividing $i \cdot j$ by $p$. Since $i$ and $j$ are both relatively prime to $p$, $i \cdot j$ is also relatively prime to $p$. Hence, $i \cdot j$ is not divisible by $p$ and $r$ cannot be zero. As a result, $1 \leq r < p$ and $r$ is an element in $G$. Therefore, the set $G = \{1, 2, \ldots, p - 1\}$ is closed under the operation $\boxdot$, which is called *modulo-p multiplication*. $G$ with modulo-$p$ multiplication forms a commutative group of order $p-1$. It is easy to prove that modulo-$p$ multiplication $\boxdot$ is both associative and commutative, and that 1 is the identity element of $G$. The only thing left to be proved is that every element in $G$ has an inverse with respect to $\boxdot$. Let $i$ be an integer in $G$. Since $i$ and $p$ are relatively prime, there exist two integers $a$ and $b$ such that

$$a \cdot i + b \cdot p = 1, \tag{2.7}$$

**Table 2.1** The additive group with modulo-7 addition

| $\boxplus$ | 0 | 1 | 2 | 3 | 4 | 5 | 6 |
|---|---|---|---|---|---|---|---|
| 0 | 0 | 1 | 2 | 3 | 4 | 5 | 6 |
| 1 | 1 | 2 | 3 | 4 | 5 | 6 | 0 |
| 2 | 2 | 3 | 4 | 5 | 6 | 0 | 1 |
| 3 | 3 | 4 | 5 | 6 | 0 | 1 | 2 |
| 4 | 4 | 5 | 6 | 0 | 1 | 2 | 3 |
| 5 | 5 | 6 | 0 | 1 | 2 | 3 | 4 |
| 6 | 6 | 0 | 1 | 2 | 3 | 4 | 5 |

and $a$ and $p$ are relatively prime (Euclid's theorem). Rearrange (2.7) as follows:

$$a \cdot i = -b \cdot p + 1. \tag{2.8}$$

This says that, when we divide $a \cdot i$ (or $i \cdot a$) by $p$, the remainder is 1. If $0 < a < p$, $a$ is an integer in $G$. Then it follows from the definition of $\boxdot$ and (2.8) that

$$a \boxdot i = i \boxdot a = 1.$$

Therefore, $a$ is the inverse of $i$ with respect to operation $\boxdot$, and vice versa. If $a$ is not an integer in $G$, we divide $a$ by $p$, which gives

$$a = q \cdot p + r. \tag{2.9}$$

Since $a$ and $p$ are relatively prime, $r$ must be an integer between 1 and $p-1$ and hence is an element in $G$. On combining (2.8) and (2.9), we obtain

$$r \cdot i = -(b + q \cdot i) \cdot p + 1. \tag{2.10}$$

Since $r \cdot i = i \cdot r$, it follows from the definition of modulo-$p$ multiplication $\boxdot$ and (2.10) that $r \boxdot i = i \boxdot r = 1$ and $r$ is the inverse of $i$ with respect to $\boxdot$. Hence, every element $i$ in $G$ has an inverse with respect to modulo-$p$ multiplication $\boxdot$. This completes the proof that the set $G = \{1, 2, \ldots, p-1\}$ under modulo-$p$ multiplication $\boxdot$ is a commutative group. The above construction gives a class of finite groups under modulo-$p$ multiplication. The groups in this class are called *multiplicative groups*. Note that if $p$ is not a prime, the set $G = \{1, 2, \ldots, p-1\}$ does not form a group under modulo-$p$ multiplication (proof of this is left as an exercise).

**Example 2.3** Let $p = 7$, which is a prime. Table 2.2 gives the multiplicative group $G = \{1, 2, 3, 4, 5, 6\}$ under modulo-7 multiplication.

Consider the multiplicative group $G = \{1, 2, \ldots, p-1\}$ under modulo-$p$ multiplication. Let $a$ be an element in $G$. We define powers of $a$ as follows:

$$a^1 = a, \quad a^2 = a \boxdot a, \quad a^3 = a \boxdot a \boxdot a, \quad \ldots, \quad a^i = \underbrace{a \boxdot a \boxdot \cdots \boxdot a}_{i \text{ factors}}, \ldots.$$

Clearly, these powers of $a$ are elements of $G$.

**Table 2.2** The multiplicative group with modulo-7 multiplication

| $\boxdot$ | 1 | 2 | 3 | 4 | 5 | 6 |
|---|---|---|---|---|---|---|
| 1 | 1 | 2 | 3 | 4 | 5 | 6 |
| 2 | 2 | 4 | 6 | 1 | 3 | 5 |
| 3 | 3 | 6 | 2 | 5 | 1 | 4 |
| 4 | 4 | 1 | 5 | 2 | 6 | 3 |
| 5 | 5 | 3 | 1 | 6 | 4 | 2 |
| 6 | 6 | 5 | 4 | 3 | 2 | 1 |

**Definition 2.7** A multiplicative group $G$ is said to be *cyclic* if there exists an element $a$ in $G$ such that, for any $b$ in $G$, there is some integer $i$ with $b = a^i$. Such an element $a$ is called a *generator* of the cyclic group and we write $G = \langle a \rangle$.

A cyclic group may have more than one generator. This is illustrated by the next example.

---

**Example 2.4** Consider the multiplicative group $G = \{1, 2, 3, 4, 5, 6\}$ under modulo-7 multiplication given in Example 2.2. Taking the powers of 3 and using Table 2.2, we obtain

$$3^1 = 3,\ \ 3^2 = 3 \boxdot 3 = 2,\ \ 3^3 = 3^2 \boxdot 3 = 2 \boxdot 3 = 6,\ \ 3^4 = 3^3 \boxdot 3 = 6 \boxdot 3 = 4,$$
$$3^5 = 3^4 \boxdot 3 = 4 \boxdot 3 = 5,\ \ 3^6 = 3^5 \boxdot 3 = 5 \boxdot 3 = 1.$$

We see that the above six powers of 3 give all the six elements of $G$. Hence, $G$ is a cyclic group and 3 is a generator of $G$. The element 5 is also a generator of $G$, as shown below:

$$5^1 = 5,\ \ 5 \boxdot 5 = 4,\ \ 5^3 = 5^2 \boxdot 5 = 4 \boxdot 5 = 6,\ \ 5^4 = 5^3 \boxdot 5 = 6 \boxdot 5 = 2,$$
$$5^5 = 5^4 \boxdot 5 = 2 \boxdot 5 = 3,\ \ 5^6 = 5^5 \boxdot 5 = 3 \boxdot 5 = 1.$$

---

It can be proved that, for every prime $p$, the multiplicative group $G = \{1, 2, \ldots, p - 1\}$ under modulo-$p$ multiplication is cyclic.

### 2.2.3 Subgroups and Cosets

A group $G$ contains certain subsets that form groups in their own right under the operation of $G$. Such subsets are called *subgroups* of $G$.

**Definition 2.8** A non-empty subset $H$ of a group $G$ with an operation $*$ is called a *subgroup* of $G$ if $H$ is itself a group with respect to the operation $*$ of $G$, that is, $H$ satisfies all the axioms of a group under the same operation $*$ of $G$.

To determine whether a subset $H$ of a group $G$ with an operation $*$ is a subgroup, we need not verify all the axioms. The following axioms are sufficient.

(S1) For any two elements $a$ and $b$ in $H$, $a * b$ is also an element of $H$.
(S2) For any element $a$ in $H$, its inverse is also in $H$.

The axiom (S1) implies that $H$ is closed under the operation $*$, that is, $*$ is a binary operation on $H$. Axioms (S1) and (S2) together imply that $H$ contains the identity element $e$ of $G$. $H$ satisfies the associative law by virtue of the fact that every element of $H$ is also in $G$ and $*$ is an operation on $H$. Hence, $H$ is a group and a subgroup of $G$. The subset $\{e\}$ that contains the identity element $e$ of $G$ alone forms a subgroup of $G$. $G$ may be considered as a subgroup of itself. These two subgroups are trivial subgroups of $G$.

The set of all integers forms a subgroup of the group of all rational numbers under real addition $+$.

**Example 2.5** Table 2.3 gives an additive group $G = \{0, 1, 2, 3, 4, 5, 6, 7\}$ under modulo-8 addition. The subset $H = \{0, 2, 4, 6\}$ forms a subgroup of $G$ under modulo-8 addition. Table 2.4 shows the group structure of $H$.

Suppose that $m$ is not a prime. Let $k$ be a factor of $m$ and $m = ck$. Let $G = \{0, 1, \ldots, m-1\}$ be the additive group under modulo-$m$ addition. Then the subset $H = \{0, c, 2c, \ldots, (k-1)c\}$ of $G = \{0, 1, \ldots, m-1\}$ forms a subgroup of $G$ under modulo-$m$ addition. It can easily be checked that, for $1 \leq i < k$, $(k-i)c$ is the additive inverse of $ic$, and vice versa.

**Example 2.6** Consider the multiplicative group $G = \{1, 2, 3, 4, 5, 6\}$ under modulo-7 multiplication given by Table 2.2 (see Example 2.3). Table 2.5 shows that the subset $H = \{1, 2, 4\}$ of $G$ forms a subgroup of $G$ under modulo-7 multiplication.

**Definition 2.9** Let $H$ be a subgroup of a group $G$ with binary operation $*$. Let $a$ be an element of $G$. Define two subsets of elements of $G$ as follows: $a * H = \{a * h : h \in H\}$ and $H * a = \{h * a : h \in H\}$. These two subsets, $a * H$ and $H * a$, are called *left* and *right cosets* of $H$.

**Table 2.3** The additive group under modulo-8 addition

| ⊞ | 0 | 1 | 2 | 3 | 4 | 5 | 6 | 7 |
|---|---|---|---|---|---|---|---|---|
| 0 | 0 | 1 | 2 | 3 | 4 | 5 | 6 | 7 |
| 1 | 1 | 2 | 3 | 4 | 5 | 6 | 7 | 0 |
| 2 | 2 | 3 | 4 | 5 | 6 | 7 | 0 | 1 |
| 3 | 3 | 4 | 5 | 6 | 7 | 0 | 1 | 2 |
| 4 | 4 | 5 | 6 | 7 | 0 | 1 | 2 | 3 |
| 5 | 5 | 6 | 7 | 0 | 1 | 2 | 3 | 4 |
| 6 | 6 | 7 | 0 | 1 | 2 | 3 | 4 | 5 |
| 7 | 7 | 0 | 1 | 2 | 3 | 4 | 5 | 6 |

**Table 2.4** A subgroup of the additive group under modulo-8 addition

| ⊞ | 0 | 2 | 4 | 6 |
|---|---|---|---|---|
| 0 | 0 | 2 | 4 | 6 |
| 2 | 2 | 4 | 6 | 0 |
| 4 | 4 | 6 | 0 | 2 |
| 6 | 6 | 0 | 2 | 4 |

**Table 2.5** A subgroup of the multiplicative group $G = \{1, 2, 3, 4, 5, 6\}$ under modulo-7 multiplication

| $\boxdot$ | 1 | 2 | 4 |
|---|---|---|---|
| 1 | 1 | 2 | 4 |
| 2 | 2 | 4 | 1 |
| 4 | 4 | 1 | 2 |

It is clear that, if $a = e$ is the identity element of $G$, then $e * H = H * e = H$, and so $H$ is also considered as a coset of itself. If $G$ is a commutative group, then the left coset $a * H$ and the right coset $H * a$ are identical, that is, $a * H = H * a$ for any $a \in G$. In this text, we deal only with commutative groups. In this case, we do not differentiate between left and right cosets, and we simply call them cosets.

---

**Example 2.7** Consider the additive group $G = \{0, 1, 2, 3, 4, 5, 6, 7\}$ under modulo-8 addition given by Table 2.3 (see Example 2.5). We have shown that $H = \{0, 2, 4, 6\}$ is a subgroup of $G$ under modulo-8 addition. The coset $3 \boxplus H$ consists of the following elements:

$$\begin{aligned} 3 \boxplus H &= \{3 \boxplus 0, 3 \boxplus 2, 3 \boxplus 4, 3 \boxplus 6\} \\ &= \{3, 5, 7, 1\}. \end{aligned}$$

We see that $H$ and the coset $3 \boxplus H$ are disjoint, and that they together form the group $G = \{0, 1, 2, 3, 4, 5, 6, 7\}$.

Next we consider the multiplicative group $G = \{1, 2, 3, 4, 5, 6\}$ under modulo-7 multiplication given by Table 2.2 (see Example 2.3). We have shown in Example 2.6 that $H = \{1, 2, 4\}$ forms a subgroup of $G$ under modulo-7 multiplication. The coset $3 \boxdot H$ consists of the following elements:

$$\begin{aligned} 3 \boxdot H &= \{3 \boxdot 1, 3 \boxdot 2, 3 \boxdot 4\} \\ &= \{3, 6, 5\}. \end{aligned}$$

We see that $H$ and $3 \boxdot H$ are disjoint, and that they together form the group $G = \{1, 2, 3, 4, 5, 6\}$.

---

The cosets of a subgroup $H$ of a group $G$ have the following structural properties.

1. No two elements of a coset of $H$ are identical.
2. No two elements from two different cosets of $H$ are identical.
3. Every element of $G$ appears in one and only one coset of $H$.
4. All distinct cosets of $H$ are disjoint.
5. The union of all the distinct cosets of $H$ forms the group $G$.

The proofs of the above structural properties are not given here, but are left as an exercise. All the distinct cosets of a subgroup $H$ of a group $G$ are said to form a *partition* of $G$, denoted by $G/H$. It follows from the above structural properties of cosets of a subgroup $H$ of a group $G$ that we have the next theorem.

**Theorem 2.10** (Lagrange's theorem) *Let $G$ be a group of order $n$, and let $H$ be a subgroup of order $m$. Then $m$ divides $n$, and the partition $G/H$ consists of $n/m$ cosets of $H$.*

## 2.3 Fields

In this section, we introduce an algebraic system with two binary operations, called a *field*. Fields with finite numbers of elements, called *finite fields*, play an important role in developing algebraic coding theory and constructing error-correction codes that can be efficiently encoded and decoded. Well-known and widely used error-correction codes in communication and storage systems based on finite fields are *BCH* (Bose–Chaudhuri–Hocquenghem) codes and *RS* (Reed–Solomon) codes, which were introduced at the very beginning of the 1960s. Most recently, finite fields have been used successfully for constructing *low-density parity-check* (LDPC) codes that perform close to the Shannon limit with iterative decoding. In this section, we first introduce the basic concepts and fundamental properties of a field, and then give the construction of a class of finite fields that serve as *ground fields* from which *extension fields* are constructed in a later section. Some important properties of finite fields are also presented in this section.

### 2.3.1 Definitions and Basic Concepts

**Definition 2.11** Let $F$ be a set of elements on which two binary operations, called *addition* "+" and *multiplication* "·," are defined. $F$ is a *field* under these two operations, addition and multiplication, if the following conditions (or axioms) are satisfied.

1. $F$ is a commutative group under addition +. The identity element with respect to addition + is called the *zero element* (or *additive identity*) of $F$ and is denoted by 0.
2. The set $F\backslash\{0\}$ of nonzero elements of $F$ forms a commutative group under multiplication ·. The identity element with respect to multiplication · is called the *unit element* (or *multiplicative identity*) of $F$ and is denoted by 1.
3. For any three elements $a$, $b$, and $c$ in $F$,

$$a \cdot (b + c) = a \cdot b + a \cdot c. \tag{2.11}$$

   The equality of (2.11) is called the *distributive law*, that is, multiplication is distributive over addition.

From the definition, we see that a field consists of two groups with respect to two operations, addition and multiplication. The group with respect to addition is called the *additive group* of the field, and the group with respect to multiplication is called the *multiplicative group* of the field. Since each group must contain an identity element, a field must contain at least two elements. Later, we will show that a field with two elements does exist. In a field, the additive inverse of an element $a$ is denoted by "$-a$," and the multiplicative inverse of $a$ is denoted by "$a^{-1}$," provided that $a \neq 0$. From the additive and multiplicative inverses of elements of a field, two other operations, namely *subtraction* "$-$" and *division* "$\div$," can be defined. Subtracting a field element $b$ from a field element $a$ is defined as adding the additive inverse, $-b$, of $b$ to $a$, that is

$$a - b \triangleq a + (-b).$$

It is clear that $a - a = 0$. Dividing a field element $a$ by a nonzero element $b$ is defined as multiplying $a$ by the multiplicative inverse, $b^{-1}$, of $b$, that is

$$a \div b \triangleq a \cdot (b^{-1}).$$

It is clear that $a \div a = 1$ provided $a \neq 0$.

A field is simply an algebraic system in which we can perform addition, subtraction, multiplication, and division without leaving the field.

**Definition 2.12** The number of elements in a field is called the *order of the field*. A field with finite order is called a *finite field*.

Rational numbers under real addition and multiplication form a field with an infinite number of elements. All of the rational numbers under real addition form the additive group of the field, and all the nonzero rational numbers under real multiplication form the multiplicative group of the field. It is known that multiplication is distributive over addition in the system of rational numbers. Another commonly known field is the field of all real numbers under real addition and multiplication. An *extension* of the real-number field is the field of complex numbers under complex-number addition and multiplication. The complex-number field is constructed from the real-number field and a *root*, denoted by $\sqrt{-1}$, of the irreducible polynomial $X^2 + 1$ over the real-number field.

**Definition 2.13** Let $F$ be a field. A subset $K$ of $F$ that is itself a field under the operations of $F$ is called a *subfield* of $F$. In this context, $F$ is called an *extension field* of $K$. If $K \neq F$, we say that $K$ is a *proper subfield* of $F$. A field containing no proper subfields is called a *prime field*.

The rational-number field is a proper subfield of the real-number field, and the real-number field is a proper subfield of the complex-number field. The rational-number field is a prime field. The real- and complex-number fields are extension fields of the rational-number field.

Since the unit element 1 is an element of a field $F$, it follows from the closure property of the addition operation of $F$ that the sums

$$1, \quad 1+1, \quad 1+1+1, \quad \ldots, \quad \underbrace{1+1+\cdots+1}_{\text{sum of } k \text{ unit elements}}, \quad \ldots$$

are elements of $F$.

**Definition 2.14** Let $F$ be a field and 1 be its unit element (or multiplicative identity). The *characteristic* of $F$ is defined as the smallest positive integer $\lambda$ such that

$$\sum_{i=1}^{\lambda} 1 = \underbrace{1 + 1 + \cdots + 1}_{\lambda} = 0,$$

where the summation represents repeated applications of addition + of the field. If no such $\lambda$ exists, $F$ is said to have *zero characteristic*, that is, $\lambda = 0$, and $F$ is an infinite field.

The rational-, real-, and complex-number fields all have zero characteristics.

In the following, we state some fundamental properties of a field $F$ without proofs (left as exercises).

1. For every element $a$ in $F$, $a \cdot 0 = 0 \cdot a = 0$.
2. For any two nonzero elements $a$ and $b$ in $F$, $a \cdot b \neq 0$.
3. For two elements $a$ and $b$ in $F$, $a \cdot b = 0$ implies that either $a = 0$ or $b = 0$.
4. For any two elements $a$ and $b$ in $F$,

$$-(a \cdot b) = (-a) \cdot b = a \cdot (-b).$$

5. For $a \neq 0$, $a \cdot b = a \cdot c$ implies that $b = c$ (called the *cancellation law*).

**Theorem 2.15** *The characteristic $\lambda$ of a finite field is a prime.*

*Proof* Suppose that $\lambda$ is not a prime. Then $\lambda$ can be factored as a product of two smaller positive integers, say $k$ and $l$, with $1 < k, l < \lambda$, that is, $\lambda = kl$. It follows from the distributive law that

$$\sum_{i=1}^{\lambda} 1 = \left(\sum_{i=1}^{k} 1\right) \cdot \left(\sum_{i=1}^{l} 1\right) = 0.$$

It follows from property 3 given above that either $\sum_{i=1}^{k} 1 = 0$ or $\sum_{i=1}^{l} 1 = 0$. Since $1 < k, l < \lambda$, either $\sum_{i=1}^{k} 1 = 0$ or $\sum_{i=1}^{l} 1 = 0$ contradicts the definition that $\lambda$ is the smallest positive integer such that $\sum_{i=1}^{\lambda} 1 = 0$. Therefore, $\lambda$ must be a prime. □

### 2.3.2 Finite Fields

The rational-, real- and complex-number fields are infinite fields. In error-control coding, both theory and practice, we mainly use finite fields. In the following, we first present a class of finite fields that are constructed from prime numbers and then introduce some fundamental properties of a finite field. Finite fields are commonly called *Galois fields*, in honor of their discoverer, Évariste Galois.

Let $p$ be a prime number. In Section 2.2.2, we have shown that the set of integers, $\{0, 1, \ldots, p-1\}$, forms a commutative group under modulo-$p$ addition $\boxplus$. We have also shown that the set of nonzero elements, $\{1, 2, \ldots, p-1\}$, forms a commutative group under modulo-$p$ multiplication $\boxdot$. Following the definitions of modulo-$p$ addition and multiplication and the fact that real-number multiplication is distributive over real-number addition, we can readily show that the modulo-$p$ multiplication is distributive over modulo-$p$ addition. Therefore, the set of integers $\{0, 1, \ldots, p-1\}$ forms a finite field of order $p$ under modulo-$p$ addition and multiplication, where 0 and 1 are the zero and unit elements of the field, respectively. This finite field will be denoted by GF($p$). The following sums of the unit element 1 give all the elements of the field:

$$1, \quad \sum_{i=1}^{2} 1 = 2, \quad \sum_{i=1}^{3} 1 = 1 \boxplus 1 \boxplus 1 = 3, \quad \ldots, \quad \sum_{i=1}^{p-1} 1 = p-1, \quad \sum_{i=1}^{p} 1 = 0.$$

Since $p$ is the smallest positive integer such that $\sum_{i=1}^{p} 1 = 0$, the characteristic of the field is $p$. The field contains no proper subfield and hence is a prime field.

---

**Example 2.8** Consider the special case for which $p = 2$. For this case, the set $\{0, 1\}$ under modulo-2 addition and multiplication, as given by Tables 2.6 and 2.7, forms a field of two elements, denoted by GF(2).

Note that $\{1\}$ forms the multiplicative group of GF(2), a group with only one element. The two elements of GF(2) are simply the additive and multiplicative identities of the two groups of GF(2). The additive inverses of 0 and 1 are themselves. This implies that $1 - 1 = 1 + 1$. Hence, over GF(2), subtraction and addition are the same. The multiplicative inverse of 1 is itself. GF(2) is commonly called a *binary field*, the simplest finite field. GF(2) plays an important role in coding theory and is most commonly used as the alphabet of code symbols for error-correction codes.

---

**Example 2.9** Let $p = 5$. The set $\{0, 1, 2, 3, 4\}$ of integers under modulo-5 addition and multiplication, given by Tables 2.8 and 2.9, forms a field GF(5) of five elements.

The addition table is also used for subtraction, denoted by $\boxminus$. For example, suppose that 4 is subtracted from 2. First we use the addition table to find the additive inverse of 4, which is 1. Then we add 1 to 2 with modulo-5 addition. This gives the element 3. The entire subtraction process is expressed as follows:

$$2 \boxminus 4 = 2 \boxplus (-4) = 2 \boxplus 1 = 3.$$

For division, denoted by $\div$, we use the multiplication table. Suppose that we divide 3 by 2. From the multiplication table, we find the multiplicative inverse of 2, which is 3. We then multipy 3 by 3 with modulo-5 multiplication. The result is 4. The entire division process is expressed as follows:

$$3 \div 2 = 3 \boxdot (2^{-1}) = 3 \boxdot 3 = 4.$$

Let $p = 7$. With the modulo-7 addition and multiplication given by Tables 2.1 and 2.2, the set $\{0, 1, 2, 3, 4, 5, 6\}$ of integers forms a field, GF(7), of seven elements.

---

We have shown that, for every prime number $p$, there exists a prime field GF($p$). For any positive integer $m$, it is possible to construct a Galois field GF($p^m$) with $p^m$ elements based on the prime field GF($p$) and a root of a *special irreducible polynomial* with coefficients from GF($p$). This will be shown in a later section. The Galois

**Table 2.6** Modulo-2 addition

| $\boxplus$ | 0 | 1 |
|---|---|---|
| 0 | 0 | 1 |
| 1 | 1 | 0 |

**Table 2.7** Modulo-2 multiplication

| $\boxdot$ | 0 | 1 |
|---|---|---|
| 0 | 0 | 0 |
| 1 | 0 | 1 |

**Table 2.8** Modulo-5 addition

| $\boxplus$ | 0 | 1 | 2 | 3 | 4 |
|---|---|---|---|---|---|
| 0 | 0 | 1 | 2 | 3 | 4 |
| 1 | 1 | 2 | 3 | 4 | 0 |
| 2 | 2 | 3 | 4 | 0 | 1 |
| 3 | 3 | 4 | 0 | 1 | 2 |
| 4 | 4 | 0 | 1 | 2 | 3 |

**Table 2.9** Modulo-5 multiplication

| $\boxplus$ | 0 | 1 | 2 | 3 | 4 |
|---|---|---|---|---|---|
| 0 | 0 | 0 | 0 | 0 | 0 |
| 1 | 0 | 1 | 2 | 3 | 4 |
| 2 | 0 | 2 | 4 | 1 | 3 |
| 3 | 0 | 3 | 1 | 4 | 2 |
| 4 | 0 | 4 | 3 | 2 | 1 |

field GF($p^m$) contains GF($p$) as a subfield and is an extension field of GF($p$). A very important feature of any finite field is that its order must be a power of a prime.

In algebraic coding theory, Galois fields are used for code construction as well as decoding, such as BCH and RS codes and their decodings. In Chapters 11 to 15, we will use Galois fields to construct structured LDPC codes.

Hereafter, we use GF($q$) to mean a Galois (or finite) field with $q$ elements, where $q$ is a power of a prime. Before we consider the construction of extension fields of prime fields, we continue to develop some important properties of a finite field that will be used in later chapters for code construction.

Let $a$ be a nonzero element of GF($q$). The powers

$$a^1 = a, \quad a^2 = a \cdot a, \quad a^3 = a \cdot a \cdot a, \quad \ldots$$

are also elements of GF($q$). Since GF($q$) has only a finite number of elements, the above powers of $a$ cannot be all distinct. Hence, at some point of the sequence of powers of $a$, there must be a repetition, say $a^m = a^k$, with $m > k$. We express $a^m = a^k$ as $a^k \cdot a^{m-k} = a^k \cdot 1$. Using the cancellation law, we obtain the equality

$$a^{m-k} = 1.$$

This implies that, for any nonzero element $a$ in GF($q$), there exists at least one positive integer $n$ such that $a^n = 1$. This gives us the following definition.

**Definition 2.16** Let $a$ be a nonzero element of a finite field GF($q$). The smallest positive integer $n$ such that $a^n = 1$ is called the *order* of the nonzero field element $a$.

**Theorem 2.17** *Let a be a nonzero element of order n in a finite field GF(q). Then the powers of a,*

$$a^n = 1, \quad a, \quad a^2, \quad \ldots, \quad a^{n-1},$$

*form a cyclic subgroup of the multiplicative group of GF(q).*

*Proof* Let $G = \{a^n = 1, a, \ldots, a^{n-1}\}$. Since $n$ is the smallest positive integer such that $a^n = 1$, all the elements in $G$ are distinct nonzero elements in the multiplicative group of GF($q$), denoted by GF($q$)$\backslash\{0\}$. Hence, $G$ is a subset of GF($q$)$\backslash\{0\}$. For $1 \le i, j \le n$, consider $a^i \cdot a^j$. If $i + j \le n$, $a^i \cdot a^j = a^{i+j}$, which is an element in $G$. If $i + j > n$, then $i + j = n + k$, with $1 \le k < n$. In this case, $a^i \cdot a^j = a^k$, which is an element in $G$. Therefore, $G$ is closed under the multiplication operation of the field. $G$ contains the identity element 1 of the multiplicative group GF($q$)$\backslash\{0\}$ of GF($q$). For $1 \le i < n$, $a^{n-i}$ is the multiplicative inverse of the element $a^i$. The multiplicative inverse of the identity element 1 is itself. So every element in $G$ has a multiplicative inverse. Therefore, $G$ is a subgroup of GF($q$). It follows from Definition 2.5 that $G$ is a cyclic subgroup of GF($q$)$\backslash\{0\}$ and $a$ is a generator of $G$. □

**Theorem 2.18** *For any nonzero element a in a finite field GF(q),* $a^{q-1} = 1$.

*Proof* Let $b_1, b_2, \ldots, b_{q-1}$ be the $q - 1$ nonzero elements of GF($q$). Consider the $q - 1$ elements $a \cdot b_1, a \cdot b_2, \ldots, a \cdot b_{q-1}$, which are nonzero. These $q - 1$ elements must be distinct, otherwise, for some $i \ne j$, $a \cdot b_i = a \cdot b_j$, which implies that $b_i = b_j$ (use the cancellation law). Therefore, $a \cdot b_1, a \cdot b_2, \ldots, a \cdot b_{q-1}$ also form the $q - 1$ nonzero elements of GF($q$). Hence

$$(a \cdot b_1) \cdot (a \cdot b_2) \cdot \cdots \cdot (a \cdot b_{q-1}) = b_1 \cdot b_2 \cdot \cdots \cdot b_{q-1}.$$

Since the multiplication operation in GF($q$) is associative and commutative, the above equality can be put in the following form:

$$a^{q-1}(b_1 \cdot b_2 \cdot \cdots \cdot b_{q-1}) = b_1 \cdot b_2 \cdot \cdots \cdot b_{q-1}.$$

Since the product of nonzero elements is nonzero, it follows from the cancellation law that $a^{q-1} = 1$. This proves the theorem. □

**Theorem 2.19** *Let $n$ be the order of a nonzero element $a$ in GF($q$). Then $n$ divides $q - 1$.*

*Proof* Suppose $q - 1$ is not divisible by $n$. On dividing $q - 1$ by $n$, we have $q - 1 = kn + r$, where $1 \leq r < n$. Then $a^{q-1} = a^{kn+r} = a^{kn} \cdot a^r$. Since $a^{q-1} = 1$ and $a^n = 1$, we must have $a^r = 1$. This contradicts the fact that $n$ is the smallest positive integer such that $a^n = 1$. Therefore, $n$ must divide $q - 1$. □

**Definition 2.20** A nonzero element $a$ of a finite field GF($q$) is called a *primitive element* if its order is $q - 1$. That is, $a^{q-1} = 1, a, a^2, \ldots, a^{q-2}$ form all the nonzero elements of GF($q$).

Every finite field has at least one primitive element (see Problem 2.10). Consider the prime field GF(5) under modulo-5 addition and multiplication given in Example 2.9. If we take powers of 2 in GF(5) using the multiplication given by Table 2.9, we obtain all the nonzero elements of GF(5),

$$2^1 = 2, \quad 2^2 = 2 \boxdot 2 = 4, \quad 2^3 = 2^2 \boxdot 2 = 4 \boxdot 2 = 3,$$
$$2^4 = 2^3 \boxdot 2 = 3 \boxdot 2 = 1.$$

Hence, 2 is a primitive element of GF(5). The integer 3 is also a primitive element of GF(5), as shown below:

$$3^1 = 3, \quad 3^2 = 3 \boxdot 3 = 4, \quad 3^3 = 3^2 \boxdot 3 = 4 \boxdot 3 = 2,$$
$$3^4 = 3^3 \boxdot 3 = 2 \boxdot 3 = 1.$$

However, the integer 4 is not a primitive element and its order is 2:

$$4^1 = 4, \quad 4^2 = 4 \boxdot 4 = 1.$$

## 2.4 Vector Spaces

In this section, we present another algebraic system, called a *vector space*. A special type of vector space plays an important role in error-control coding. The main ingredients of a vector space are a field $F$, a group $V$ that is commutative under an addition, and a multiplication operation that combines the elements of $F$ and the elements of $V$. Therefore, a vector space may be considered as a super-algebraic system with a total of four binary operations: addition and multiplication on $F$, addition on $V$, and the multiplication operation that combines the elements of $F$ and the elements of $V$.

### 2.4.1 Basic Definitions and Properties

**Definition 2.21** Let $F$ be a field. Let $V$ be a set of elements on which a binary operation called addition + is defined. A multiplication operation, denoted by "·," is defined between the elements of $F$ and the elements of $V$. The set $V$ is called a *vector space* over the field $F$ if it satisfies the following axioms.

1. $V$ is a commutative group under addition + defined on $V$.
2. For any element $a$ in $F$ and any element $\mathbf{v}$ in $V$, $a \cdot \mathbf{v}$ is an element in $V$.
3. For any elements $a$ and $b$ in $F$ and any element $\mathbf{v}$ in $V$, the following associative law is satisfied:

$$(a \cdot b) \cdot \mathbf{v} = a \cdot (b \cdot \mathbf{v}).$$

4. For any element $a$ in $F$ and any elements $\mathbf{u}$ and $\mathbf{v}$ in $V$, the following distributive law is satisfied:

$$a \cdot (\mathbf{u} + \mathbf{v}) = a \cdot \mathbf{u} + a \cdot \mathbf{v}.$$

5. For any two elements $a$ and $b$ in $F$ and any element $\mathbf{v}$ in $V$, the following distributive law is satisfied:

$$(a + b) \cdot \mathbf{v} = a \cdot \mathbf{v} + b \cdot \mathbf{v}.$$

6. Let 1 be the unit element of $F$. Then, for any element $\mathbf{v}$ in $V$, $1 \cdot \mathbf{v} = \mathbf{v}$.

The elements of $V$ are called *vectors* and denoted by boldface lower case letters $\mathbf{u}$, $\mathbf{v}$, $\mathbf{w}$, and so on. The elements of $F$ are called *scalars* and denoted by lower case italic letters $a$, $b$, $c$, and so on. The addition + on $V$ is called a *vector addition*; the multiplication $\cdot$ that combines a scalar $a$ in $F$ and a vector $\mathbf{v}$ in $V$ into a vector $a \cdot \mathbf{v}$ in $V$ is referred to as *scalar multiplication*, and the vector $a \cdot \mathbf{v}$ is called the *product* of $a$ and $\mathbf{v}$. The additive identity of $V$ is denoted by boldface $\mathbf{0}$, called the *zero vector* of $V$. Note that we use + for additions on both $V$ and $F$. It should be clear that, when we combine two vectors in $V$, + means vector addition; when we combine two scalars in $F$, + means the addition defined on $F$. We also use $\cdot$ for both scalar multiplication and multiplication defined on $F$. When a scalar in $F$ and a vector in $V$ are combined, "$\cdot$" means scalar multiplication; when two scalars in $F$ are combined, "$\cdot$" means multiplication on $F$.

Some basic properties of a vector space $V$ over a field $F$ are given by the theorems below without proofs (left as exercises).

**Theorem 2.22** *Let 0 be the zero element of the field $F$. For any vector $\mathbf{v}$ in $V$, $0 \cdot \mathbf{v} = \mathbf{0}$.*

**Theorem 2.23** *For any element $a$ in $F$, $a \cdot \mathbf{0} = \mathbf{0}$.*

**Theorem 2.24** *For any element $a$ in $F$ and any vector $\mathbf{v}$ in $V$, $(-a) \cdot \mathbf{v} = a \cdot (-\mathbf{v}) = -(a \cdot \mathbf{v})$, that is, either $(-a) \cdot \mathbf{v}$ or $a \cdot (-\mathbf{v})$ is the additive inverse of the vector $a \cdot \mathbf{v}$.*

Let $R$ and $C$ be the real- and complex-number fields, respectively. $C$, with its usual operation of addition (but with multiplication ignored) is a vector space over $R$ if scalar multiplication is defined by the rule $c \cdot (a + bi) = (c \cdot a) + (c \cdot b)i$, where $a + bi$ is any complex number and $c$ is any real number.

A vector space $V$ over a field $F$ contains certain subsets that form vector spaces over $F$ in their own right. Such subsets are called *subspaces* of $V$.

**Definition 2.25** Let $S$ be a non-empty subset of a vector space $V$ over a field $F$. $S$ is called a *subspace* of $V$ if it satisfies the axioms for a vector space given by Definition 2.14.

To determine whether a subset $S$ of a vector space $V$ over a field $F$ is a subspace, it is not necessary to verify all the axioms given by Definition 2.14. The following axioms are sufficient.

(S1) For any two vectors $\mathbf{u}$ and $\mathbf{v}$ in $S$, $\mathbf{u} + \mathbf{v}$ is also a vector in $S$.
(S2) For any element $a$ in $F$ and any vector $\mathbf{u}$ in $S$, $a \cdot \mathbf{u}$ is also in $S$.

Axioms (S1) and (S2) simply say that $S$ is closed under vector addition and scalar multiplication of $V$. Axiom (S2) ensures that, for any vector $\mathbf{u}$ in $S$, its additive inverse $(-1) \cdot \mathbf{u}$ is also a vector in $S$. Consequently, the zero vector $\mathbf{0} = \mathbf{u} + (-1) \cdot \mathbf{u}$ is also in $S$. Therefore, $S$ is a subgroup of $V$ under vector addition. Since the vectors of $S$ are also vectors of $V$, the associative and distributive laws must hold for $S$. Furthermore, axiom (6) given by Definition 2.14 is obviously satisfied. Hence, $S$ is a vector space over $F$, a subspace of $V$.

### 2.4.2 Linear Independence and Dimension

Hereafter, when we take the product $a \cdot \mathbf{v}$ of a scalar $a$ and a vector $\mathbf{v}$ of a vector space $V$ over a field $F$, we drop the scalar multiplication and simply write the product $a \cdot \mathbf{v}$ as $a\mathbf{v}$.

Let $\mathbf{v}_1, \mathbf{v}_2, \ldots, \mathbf{v}_k$ be $k$ vectors in $V$. Let $a_1, a_2, \ldots, a_k$ be $k$ arbitrary elements of $F$. Then $a_1\mathbf{v}_1, a_2\mathbf{v}_2, \ldots, a_k\mathbf{v}_k$ and the sum

$$a_1\mathbf{v}_1 + a_2\mathbf{v}_2 + \cdots + a_k\mathbf{v}_k$$

are vectors in $V$. The sum is called a *linear combination* of $\mathbf{v}_1, \mathbf{v}_2, \ldots, \mathbf{v}_k$. It is clear that the sum of two linear combinations of $\mathbf{v}_1, \mathbf{v}_2, \ldots, \mathbf{v}_k$, namely

$$\begin{aligned}(a_1\mathbf{v}_1 + a_2\mathbf{v}_2 + \cdots + a_k\mathbf{v}_k) + (b_1\mathbf{v}_1 + b_2\mathbf{v}_2 + \cdots + b_k\mathbf{v}_k)\\ = (a_1 + b_1)\mathbf{v}_1 + (a_2 + b_2)\mathbf{v}_2 + \cdots + (a_k + b_k)\mathbf{v}_k,\end{aligned}$$

is also a linear combination of $\mathbf{v}_1, \mathbf{v}_2, \ldots, \mathbf{v}_k$, where $a_i + b_i$ is carried out with the addition of the field $F$. It is also clear that the product of an element $c$ in $F$ and a linear combination of $\mathbf{v}_1, \mathbf{v}_2, \ldots, \mathbf{v}_k$,

$$c(a_1\mathbf{v}_1 + a_2\mathbf{v}_2 + \cdots + a_k\mathbf{v}_k) = (c \cdot a_1)\mathbf{v}_1 + (c \cdot a_2)\mathbf{v}_2 + \cdots + (c \cdot a_k)\mathbf{v}_k,$$

is a linear combination of $\mathbf{v}_1, \mathbf{v}_2, \ldots, \mathbf{v}_k$, where $c \cdot a_i$ is carried out with the multiplication of the field $F$. Let $S$ be the set of all linear combinations of $\mathbf{v}_1, \mathbf{v}_2, \ldots, \mathbf{v}_k$. Then $S$ satisfies axioms (S1) and (S2) for a subspace of a vector space. Hence, $S$ is a subspace of the vector space $V$ over $F$.

**Definition 2.26** A set of vectors $\mathbf{v}_1, \mathbf{v}_2, \ldots, \mathbf{v}_k$ in a vector space $V$ over a field $F$ is said to be *linearly independent* over $F$ if and only if, for all $k$ scalars $a_i$ in $F$,

$$a_1\mathbf{v}_1 + a_2\mathbf{v}_2 + \cdots + a_k\mathbf{v}_k = \mathbf{0}$$

implies

$$a_1 = a_2 = \cdots = a_k = 0.$$

Vectors that are not linearly independent are said to be *linearly dependent.*

It follows from Definition 2.16 that, if $\mathbf{v}_1, \mathbf{v}_2, \ldots, \mathbf{v}_k$ are linearly dependent, then there exist $k$ scalars $a_1, a_2, \ldots, a_k$ in $F$, *not all zero*, such that

$$a_1\mathbf{v}_1 + a_2\mathbf{v}_2 + \cdots + a_k\mathbf{v}_k = \mathbf{0}.$$

Since there is at least one scalar, say $a_k$, not equal to zero, the above equality can be written as

$$\begin{aligned}\mathbf{v}_k &= -a_k^{-1}(a_1\mathbf{v}_1 + a_2\mathbf{v}_2 + \cdots + a_{k-1}\mathbf{v}_{k-1}) \\ &= c_1\mathbf{v}_1 + c_2\mathbf{v}_2 + \cdots + c_{k-1}\mathbf{v}_{k-1},\end{aligned}$$

where $c_i = -a_k^{-1}a_i$ for $1 \leq i < k$ and $a_k^{-1}$ is the multiplicative inverse of the field element $a_k$. In this case, we say that $\mathbf{v}_k$ is linearly dependent on $\mathbf{v}_1, \mathbf{v}_2, \ldots, \mathbf{v}_{k-1}$.

**Definition 2.27** Let $\mathbf{u}, \mathbf{v}_1, \mathbf{v}_2, \ldots, \mathbf{v}_k$ be vectors in a vector space $V$ over a field $F$. Then $\mathbf{u}$ is said to be *linearly dependent* on $\mathbf{v}_1, \mathbf{v}_2, \ldots, \mathbf{v}_k$ if it can be expressed as a linear combination of $\mathbf{v}_1, \mathbf{v}_2, \ldots, \mathbf{v}_k$ as follows:

$$\mathbf{u} = a_1\mathbf{v}_1 + a_2\mathbf{v}_2 + \cdots + a_k\mathbf{v}_k,$$

with $a_i$ in $F$ for $1 \leq i \leq k$.

**Definition 2.28** A set of vectors is said to *span* a vector space $V$ over a field $F$ if every vector in the vector space $V$ is equal to a linear combination of the vectors in the set.

In any vector space (or subspace), there exists at least one set $B$ of linearly independent vectors that spans the space. This set $B$ is called a *basis* (or *base*) of the vector space. Two different bases of a vector space have the *same number* of linearly independent vectors (the proof of this fact is left as an exercise). The number of linearly independent vectors in a basis of a vector space is called the *dimension* of the vector space. In the case that a basis of a vector space has a finite number, say $n$, of linearly independent vectors, we write $\dim(V) = n$. If a basis contains infinitely many linearly independent vectors, then we write $\dim(V) = \infty$. Let $V$ be a vector space with dimension $n$. For $0 \leq k \leq n$, a set of $k$ linearly independent vectors in $V$ spans a *$k$-dimensional* subspace of $V$.

### 2.4.3 Finite Vector Spaces over Finite Fields

In this subsection, we present a very useful vector space over a finite field GF($q$), which plays a central role in error-control coding theory. Let $n$ be a positive integer. Consider an *ordered sequence* of $n$ components,

$$\mathbf{v} = (v_0, v_1, \ldots, v_{n-1}),$$

where each component $v_i$, $0 \leq i < n$, is an element of the finite field GF($q$). This ordered sequence is called an *n-tuple* over GF($q$). Since each component $v_i$ can be any of the $q$ elements in GF($q$), there are $q^n$ distinct $n$-tuples over GF($q$). Let $V_n$ denote this set of $q^n$ distinct $n$-tuples. We define addition + of two $n$-tuples over GF($q$), $\mathbf{u} = (u_0, u_1, \ldots, u_{n-1})$ and $\mathbf{v} = (v_0, v_1, \ldots, v_{n-1})$, as follows:

$$\begin{aligned}\mathbf{u} + \mathbf{v} &= (u_0, u_1, \ldots, u_{n-1}) + (v_0, v_1, \ldots, v_{n-1}) \\ &= (u_0 + v_0, u_1 + v_1, \ldots, u_{n-1} + v_{n-1}),\end{aligned} \tag{2.12}$$

where $u_i + v_i$ is carried out with the addition of GF($q$) and is hence an element of GF($q$). Therefore, addition of two $n$-tuples over GF($q$) is also an $n$-tuple over GF($q$). Hence, $V_n$ is closed under the addition + defined by (2.12). We can readily verify that $V_n$ under the addition + defined by (2.12) is a commutative group. Since the addition operation on GF($q$) is associative and commutative, we can easily check that the addition defined by (2.12) is also associative and commutative. Let 0 be the zero element of GF($q$). The *all-zero* $n$-tuple $\mathbf{0} = (0, 0, \ldots, 0)$ is the additive identity of $V_n$, since

$$\begin{aligned}(0, 0, \ldots, 0) + (v_0, v_1, \ldots, v_{n-1}) &= (0 + v_0, 0 + v_1, \ldots, 0 + v_{n-1}) \\ &= (v_0, v_1, \ldots, v_{n-1})\end{aligned}$$

and

$$\begin{aligned}(v_0, v_1, \ldots, v_{n-1}) + (0, 0, \ldots, 0) &= (v_0 + 0, v_1 + 0, \ldots, v_{n-1} + 0) \\ &= (v_0, v_1, \ldots, v_{n-1}).\end{aligned}$$

Consider an $n$-tuple $\mathbf{v} = (v_0, v_1, \ldots, v_{n-1})$ in $V_n$. For $0 \leq i < n$, let $-v_i$ be the additive inverse of the field element $v_i$ of GF($q$). Then the $n$-tuple $-\mathbf{v} = (-v_0, -v_1, \ldots, -v_{n-1})$ in $V_n$ is the additive inverse of the $n$-tuple $\mathbf{v} = (v_0, v_1, \ldots, v_{n-1})$, since $\mathbf{v} + (-\mathbf{v}) = (-\mathbf{v}) + \mathbf{v} = \mathbf{0}$. Hence, every $n$-tuple in $V_n$ has an additive inverse with respect to the addition of two $n$-tuples over GF($q$). Therefore, $V_n$ under the addition defined by (2.12) satisfies all the axioms of a commutative group and is thus a commutative group.

Next, we define scalar multiplication of an $n$-tuple $\mathbf{v} = (v_0, v_1, \ldots, v_{n-1})$ by a field element $c$ as follows:

$$c \cdot \mathbf{v} = c \cdot (v_0, v_1, \ldots, v_{n-1}) = (c \cdot v_0, c \cdot v_1, \ldots, c \cdot v_{n-1}), \tag{2.13}$$

where $c \cdot v_i$, $0 \leq i < n$, is carried out with the multiplication of GF($q$). Since every component $c \cdot v_i$ of the ordered sequence $c \cdot \mathbf{v} = (c \cdot v_0, c \cdot v_1, \ldots, c \cdot v_{n-1})$ given by (2.13) is an element of GF($q$), $c \cdot \mathbf{v} = (c \cdot v_0, c \cdot v_1, \ldots, c \cdot v_{n-1})$ is an $n$-tuple in $V_n$. If $c = 1$, then $1 \cdot \mathbf{v} = \mathbf{v}$. We can easily verify that the addition of two $n$-tuples over GF($q$) and the scalar multiplication of an $n$-tuple by an element in GF($q$) defined by (2.12) and (2.13), respectively, satisfy the associative and distributive laws. Therefore, $V_n$ is a vector space over GF($q$) with all the $n$-tuples over GF($q$) as vectors.

Consider the following $n$ $n$-tuples over GF($q$):

$$\begin{aligned} \mathbf{e}_0 &= (1, 0, 0, \ldots, 0), \\ \mathbf{e}_1 &= (0, 1, 0, \ldots, 0), \\ &\vdots \\ \mathbf{e}_{n-1} &= (0, 0, 0, \ldots, 1), \end{aligned} \tag{2.14}$$

where the $n$-tuple $\mathbf{e}_i$ has a single nonzero component at position $i$, which is the unit element of GF($q$). Every $n$-tuple $\mathbf{v} = (v_0, v_1, \ldots, v_{n-1})$ over GF($q$) can be expressed as a linear combination of $\mathbf{e}_0, \mathbf{e}_1, \ldots, \mathbf{e}_{n-1}$ as follows:

$$\begin{aligned} \mathbf{v} &= (v_0, v_1, \ldots, v_{n-1}) \\ &= v_0\mathbf{e}_0 + v_1\mathbf{e}_1 + \cdots + v_{n-1}\mathbf{e}_{n-1}. \end{aligned} \tag{2.15}$$

Therefore, $\mathbf{e}_0, \mathbf{e}_1, \ldots, \mathbf{e}_{n-1}$ span the vector space $V_n$ of all the $q^n$ $n$-tuples over GF($q$). From (2.15), we readily see that the linear combination given by (2.15) is equal to the $\mathbf{0}$ vector if and only if $v_0, v_1, \ldots, v_{n-1}$ are all equal to the zero element of GF($q$). Hence, $\mathbf{e}_0, \mathbf{e}_1, \ldots, \mathbf{e}_{n-1}$ are linearly independent and they form a basis of $V_n$. The dimension of $V_n$ is then equal to $n$. Any basis of $V_n$ consists of $n$ linearly independent $n$-tuples over GF($q$).

For $0 \leq k \leq n$, let $\mathbf{v}_1, \mathbf{v}_2, \ldots, \mathbf{v}_k$ be $k$ linearly independent $n$-tuples in $V_n$. Then the set $S$ of $q^k$ linear combinations of $\mathbf{v}_1, \mathbf{v}_2, \ldots, \mathbf{v}_k$ of the form

$$\mathbf{w} = c_1\mathbf{v}_1 + c_2\mathbf{v}_2 + \cdots + c_k\mathbf{v}_k$$

forms a *k-dimensional subspace* of $V_n$.

The most commonly used vector space in error-control coding theory is the vector space $V_n$ of all the $2^n$ tuples over the binary field GF(2). In this case, the $n$-tuples over GF(2) are commonly called binary $n$-tuples. Adding two binary $n$-tuples requires adding their corresponding binary components using modulo-2 addition. In GF(2), the unit element 1 is its own inverse, that is, $1 + 1 = 0$. Therefore, the additive inverse of a binary $n$-tuple $\mathbf{v} = (v_0, v_1, \ldots, v_{n-1})$ is simply itself, since

$$\mathbf{v} + \mathbf{v} = (v_0 + v_0, v_1 + v_1, \ldots, v_{n-1} + v_{n-1}) = (0, 0, \ldots, 0).$$

---

**Example 2.10** Let $n = 4$. The vector space $V_4$ of all the 4-tuples over GF(2) consists of the following 16 binary 4-tuples:

(0000), (0001), (0010), (0011), (0100), (0101), (0110), (0111),
(1000), (1001), (1010), (1011), (1100), (1101), (1110), (1111).

The vector sum of (0101) and (1011) is

$$(0111) + (1011) = (0 + 1, 1 + 0, 1 + 1, 1 + 1) = (1100).$$

The four vectors (1000), (1100), (1110), and (1111) are linearly independent and they form a basis of $V_4$. The four vectors (0001), (0010), (0111), and (1110) are also linearly independent and they form another basis of $V_4$. The linear combinations of

(1000), (1110), and (1111) give a three-dimensional subspace of $V_4$ with the following eight vectors:

$$\begin{array}{cccc}(0000), & (1000), & (1110), & (1111), \\ (0110), & (0111), & (0001), & (1001).\end{array}$$

### 2.4.4 Inner Products and Dual Spaces

Let $\mathbf{u} = (u_0, u_1, \ldots, u_{n-1})$ and $\mathbf{v} = (v_0, v_1, \ldots, v_{n-1})$ be two $n$-tuples over GF($q$). We define the *inner product* of $\mathbf{u}$ and $\mathbf{v}$ as the following sum:

$$\mathbf{u} \cdot \mathbf{v} = u_0 \cdot v_0 + u_1 \cdot v_1 + \cdots + u_{n-1} \cdot v_{n-1},$$

where the multiplications and additions in the sum are carried out with multiplication and addition of GF($q$). The inner product $\mathbf{u} \cdot \mathbf{v}$ can also be written as $\mathbf{u}\mathbf{v}^{\mathrm{T}}$, where superscript T indicates transpose. The inner product of two $n$-tuples over GF($q$) is an element of GF($q$), that is, a scalar. If $\mathbf{u} \cdot \mathbf{v} = 0$, we say that $\mathbf{u}$ and $\mathbf{v}$ are *orthogonal* to each other. Often, the notation $\langle \mathbf{u}, \mathbf{v} \rangle$ is also used to denote the inner product of $\mathbf{u}$ and $\mathbf{v}$. The inner product has the following properties.

1. $\mathbf{u} \cdot \mathbf{v} = \mathbf{v} \cdot \mathbf{u}$ (commutative law).
2. $\mathbf{u} \cdot (\mathbf{v} + \mathbf{w}) = \mathbf{u} \cdot \mathbf{v} + \mathbf{u} \cdot \mathbf{w}$ (distributive law).
3. For any element $c$ in GF($q$), $(c\mathbf{u}) \cdot \mathbf{v} = c(\mathbf{u} \cdot \mathbf{v})$ (associative law).
4. $\mathbf{0} \cdot \mathbf{u} = 0$.

**Example 2.11** Consider the two 4-tuples (1011) and (1101) given in Example 2.10. The inner product of these two binary 4-tuples is

$$(1011) \cdot (1101) = 1 \cdot 1 + 0 \cdot 1 + 1 \cdot 0 + 1 \cdot 1 = 1 + 0 + 0 + 1 = 0.$$

Since their inner product is 0, they are orthogonal to each other.

For $0 \leq k < n$, let $S$ be a $k$-dimensional subspace of the vector space $V_n$ of all the $n$-tuples over GF($q$). Let $S_d$ be the set of $n$-tuples in $V_n$ such that, for any $\mathbf{u}$ in $S$ and $\mathbf{v}$ in $S_d$,

$$\mathbf{u} \cdot \mathbf{v} = 0,$$

that is

$$S_d = \{\mathbf{v} \in V_n : \mathbf{u} \cdot \mathbf{v} = 0, \mathbf{u} \in S\}. \tag{2.16}$$

Since $\mathbf{0} \cdot \mathbf{u} = 0$ for any $\mathbf{u} \in S$, $S_d$ contains at least the all-zero $n$-tuple of $V_n$ and hence is non-empty. Let $\mathbf{v}$ and $\mathbf{w}$ be any two tuples in $S_d$ and $\mathbf{u}$ be any $n$-tuple in $S$. It follows from the distributive law of the inner product that

$$\mathbf{u} \cdot (\mathbf{v} + \mathbf{w}) = \mathbf{u} \cdot \mathbf{v} + \mathbf{u} \cdot \mathbf{w} = 0 + 0 = 0.$$

Hence, $\mathbf{u}+\mathbf{w}$ is also in $S_d$. It follows from the associative law of the inner product that, for any scalar $a$ in GF($q$), any $n$-tuple $\mathbf{v}$ in $S_d$, and any $n$-tuple $\mathbf{u}$ in $S$,

$$(a \cdot \mathbf{v}) \cdot \mathbf{u} = a \cdot (\mathbf{u} \cdot \mathbf{v}) = a \cdot 0 = 0.$$

Thus, $a \cdot \mathbf{v}$ is also an $n$-tuple in $S_d$. Hence, $S_d$ satisfies the two axioms for a subspace of a vector space over a finite field. Consequently, $S_d$ is a subspace of the vector space $V_n$ of all the $n$-tuples over GF($q$). $S_d$ is called the *dual* (or *null*) space of $S$, and vice versa. The dimension of $S_d$ is given by the following theorem, whose proof is omitted here.

**Theorem 2.29** *For $0 \leq k \leq n$, let $S$ be a $k$-dimensional subspace of the vector space $V_n$ of all $n$-tuples over GF($q$). The dimension of its dual space $S_d$ is $n - k$. In other words,* $\dim(S) + \dim(S_d) = n$.

## 2.5 Polynomials over Finite Fields

In elementary algebra, one regards a polynomial as an expression of the following form: $f(X) = f_0 + f_1X + \cdots + f_nX^n$. The $f_i$s are called *coefficients* and are usually *real* or *complex* numbers, and $X$ is viewed as a *variable* such that, on substituting an arbitrary number $c$ for $X$, a well-defined number $f(c) = f_0 + f_1c + \cdots + f_nc^n$ is obtained. The arithmetic of polynomials is governed by familiar rules, such as addition, subtraction, multiplication, and division.

In this section, we consider polynomials with coefficients from a finite field GF($q$), which are used in error-control coding theory. A polynomial with one variable (or *indeterminate*) $X$ over GF($q$) is an expression of the following form:

$$a(X) = a_0 + a_1X + \cdots + a_nX^n,$$

where $n$ is a non-negative integer and the coefficients $a_i$, $0 \leq i \leq n$, are elements of GF($q$). The degree of a polynomial is defined as the *largest power* of $X$ with a nonzero coefficient. For the polynomial $a(X)$ above, if $a_n \neq 0$, its degree is $n$; if $a_n = 0$, its degree is less than $n$. The degree of a polynomial $a(X) = a_0$ with only the constant term is zero. We use the notation $\deg(a(X))$ to denote the degree of polynomial $a(X)$. A polynomial is called *monic* if the coefficient of the highest power of $X$ is 1 (the unit element of GF($q$)). The polynomial whose coefficients are all equal to zero is called the *zero polynomial*, denoted by 0. We adopt the convention that a term $a_iX^i$ with $a_i = 0$ need not be written down. The above polynomial $a(X)$ can also be written in the following *equivalent* form:

$$a(X) = a_0 + a_1X + \cdots + a_nX^n + 0X^{n+1} + \cdots + 0X^{n+k},$$

where $k$ is any non-negative integer. With this equivalent form, when comparing two polynomials over GF($q$), we can assume that they both involve the same powers of $X$.

Two polynomials over GF($q$) can be added and multiplied in the usual way. Let

$$a(X) = a_0 + a_1X + \cdots + a_nX^n, \qquad b(X) = b_0 + b_1X + \cdots + b_mX^m$$

be two polynomials over GF($q$) with degrees $n$ and $m$, respectively. Without loss of generality, we assume that $m \leq n$. To add $a(X)$ and $b(X)$, we simply add the coefficients of the same power of $X$ in $a(X)$ and $b(X)$ as follows:

$$\begin{aligned} a(X) + b(X) = (a_0 + b_0) + (a_1 + b_1)X + \cdots + (a_m + b_m)X^m \\ + (a_{m+1} + 0)X^{m+1} + \cdots + (a_n + 0)X^n, \end{aligned} \tag{2.17}$$

where $a_i + b_i$ is carried out with the addition of GF($q$). Multiplication (or the product) of $a(X)$ and $b(X)$ is defined as follows:

$$a(X) \cdot b(X) = c_0 + c_1X + \cdots + c_{n+m}X^{n+m}, \tag{2.18}$$

where, for $0 \leq k \leq n + m$,

$$c_k = \sum_{i+j=k} a_i \cdot b_j, \tag{2.19}$$

where $0 \leq i \leq n$ and $0 \leq j \leq m$. It is clear from (2.18) and (2.19) that, if $b(X) = 0$ (the zero polynomial), then

$$a(X) \cdot 0 = 0. \tag{2.20}$$

On the basis of the definitions of addition and multiplication of two polynomials and the facts that addition and multiplication of the field GF($q$) satisfy the associative, commutative, and distributive laws, we can readily verify that the polynomials over GF($q$) satisfy the following conditions.

1. *Associative laws*. For any three polynomials $a(X)$, $b(X)$, and $c(X)$ over GF($q$),

$$a(X) + [b(X) + c(X)] = [a(X) + b(X)] + c(X),$$
$$a(X) \cdot [b(X) \cdot c(X)] = [a(X) \cdot b(X)] \cdot c(X).$$

2. *Commutative laws*. For any two polynomials $a(X)$ and $b(X)$ over GF($q$),

$$a(X) + b(X) = b(X) + a(X),$$
$$a(X) \cdot b(X) = b(X) \cdot a(X).$$

3. *Distributive law*. For any three polynomials $a(X)$, $b(X)$, and $c(X)$ over GF($q$),

$$a(X) \cdot [b(X) + c(X)] = a(X) \cdot b(X) + a(X) \cdot c(X).$$

In algebra, the set of polynomials over GF($q$) under polynomial addition and multiplication defined above is called a *polynomial ring* over GF($q$).

Subtraction of $b(X)$ from $a(X)$ is carried out as follows:

$$\begin{aligned} a(X) - b(X) = (a_0 - b_0) + (a_1 - b_1)X + \cdots + (a_m - b_m)X^m \\ + (a_{m+1} - 0)X^{m+1} + \cdots + (a_n - 0)X^n, \end{aligned} \tag{2.21}$$

where $a_i - b_i = a_i + (-b_i)$ is carried out in GF($q$) and $-b_i$ is the additive inverse of $b_i$. These observations lead us to the following definition.

**Definition 2.30** A set $\mathcal{R}$ with two binary operations, called addition + and multiplication ·, is called a *ring* if the following axioms are satisfied.

1. $\mathcal{R}$ is a commutative group with respect to +.
2. Multiplication "·" is associative, that is, $(a \cdot b) \cdot c = a \cdot (b \cdot c)$ for all $a, b, c \in \mathcal{R}$.
3. The distributive laws hold: for $a, b, c \in \mathcal{R}$, $a \cdot (b + c) = a \cdot b + a \cdot c$ and $(b + c) \cdot a = b \cdot a + c \cdot a$.

---

**Example 2.12** Consider the ring of polynomials over the prime field GF(5) given by Example 2.9 (see Tables 2.8 and 2.9). Let $a(X) = 2 + X + 3X^2 + 4X^4$ and $b(X) = 1 + 3X + 2X^2 + 4X^3 + X^4$. On subtracting $b(X)$ from $a(X)$, we obtain

$$\begin{aligned} a(X) - b(X) &= (2-1) + (1-3)X + (3-2)X^2 + (0-4)X^3 + (4-1)X^4 \\ &= (2 + (-1)) + (1 + (-3))X + (3 + (-2))X^2 + (0 + (-4))X^3 \\ &\quad + (4 + (-1))X^4. \end{aligned}$$

Using Table 2.8 we find that $-1 = 4$, $-3 = 2$, $-2 = 3$, and $-4 = 1$. Then

$$a(X) - b(X) = (2+4) + (1+2)X + (3+3)X^2 + (0+1)X^3 + (4+4)X^4.$$

Using Table 2.8 for addition of the coefficients in the above expression, we have the following polynomial:

$$a(X) - b(X) = 1 + 3X + X^2 + X^3 + 3X^4,$$

which is a polynomial of degree 4 over GF(5).

---

The set of all polynomials over GF($q$) defined above satisfies all three axioms of a ring. Therefore, it is a ring of polynomials. The set of all integers under real addition and multiplication is a ring. For any positive integer $m \geq 2$, the set $\mathcal{R} = \{0, 1, \ldots, m-1\}$ under modulo-$m$ addition and multiplication forms a ring (the proof is left as an exercise). In Section 2.3, it was shown that, if $m$ is a prime, $\mathcal{R} = \{0, 1, \ldots, m-1\}$ is a prime field under modulo-$m$ addition and multiplication.

Let $a(X)$ and $b(X)$ be two polynomials over GF($q$). Suppose that $b(X)$ is not the zero polynomial, that is, $b(X) \neq 0$. When $a(X)$ is divided by $b(X)$, we obtain a *unique pair* of polynomials over GF($q$), $q(X)$ and $r(X)$, such that

$$a(X) = q(X) \cdot b(X) + r(X), \tag{2.22}$$

where $0 \leq \deg(r(X)) < \deg(b(X))$ and the polynomials $q(X)$ and $r(X)$ are called the *quotient* and *remainder*, respectively. This is known as *Euclid's division algorithm*. The quotient $q(x)$ and remainder $r(X)$ can be obtained by ordinary *long division* of polynomials. If $r(X) = 0$, we say that $a(X)$ is divisible by $b(X)$ (or $b(X)$ divides $a(X)$).

---

**Example 2.13** Let $a(x) = 3 + 4X + X^4 + 2X^5$ and $b(X) = 1 + 3X^2$ be two polynomials over the prime field GF(5) given by Example 2.9 (see Tables 2.8 and 2.9). Suppose we divide $a(X)$ by $b(X)$. Using long division, and Tables 2.8 and 2.9, we obtain

$$
\begin{array}{r|l}
 & 4X^3 + 2X^2 + 2X + 1 \\
\hline
3X^2 \qquad + 1 & 2X^5 + X^4 \qquad\qquad\qquad + 4X + 3 \\
 & -2X^5 \qquad - 4X^3 \\
 & \qquad X^4 + X^3 \qquad\quad + 4X + 3 \\
 & \qquad -X^4 \qquad - 2X^2 \\
 & \qquad\qquad X^3 + 3X^2 + 4X + 3 \\
 & \qquad\qquad -X^3 \qquad - 2X \\
 & \qquad\qquad\qquad 3X^2 + 2X + 3 \\
 & \qquad\qquad\qquad -3X^2 \qquad - 1 \\
 & \qquad\qquad\qquad\qquad 2X + 2.
\end{array}
$$

Thus, the quotient $q(X) = 4X^3 + 2X^2 + 2X + 1$ and the remainder $r(X) = 2X + 2$.

---

If a polynomial $c(X)$ is equal to a product of two polynomials $a(X)$ and $b(X)$, that is, $c(X) = a(X) \cdot b(X)$, then we say that $c(X)$ is divisible by $a(X)$ (or $b(X)$) or that $a(X)$ (or $b(X)$) divides $c(X)$, and that $a(X)$ and $b(X)$ are factors of $c(X)$.

**Definition 2.31** A polynomial $p(X)$ of degree $m$ over GF($q$) is said to be *irreducible* over GF($q$) if it is not divisible by any polynomial over GF($q$) that has a degree less than $m$ but greater than zero.

A polynomial over GF($q$) of positive degree that is not irreducible over GF($q$) is said to be reducible over GF($q$). The irreducibility of a given polynomial depends heavily on the field under consideration. An irreducible polynomial over GF($q$) may become reducible over a larger field that contains GF($q$) as a subfield. An important property of irreducible polynomials is stated in the following theorem without proof.

**Theorem 2.32** *Any irreducible polynomial $p(X)$ over GF($q$) of degree $m$ divides $X^{q^m-1} - 1$.*

**Definition 2.33** A monic irreducible polynomial $p(X)$ over GF($q$) of degree $m$ is said to be *primitive* if the smallest positive integer $n$ for which $p(X)$ divides $X^n - 1$ is $n = q^m - 1$.

Primitive polynomials are of fundamental importance for constructing Galois fields, as will be seen in the next section. Lists of primitive polynomials over various prime fields are given in Appendix A at the end of this chapter. Long lists of irreducible and primitive polynomials over various prime fields can be found in [1].

Let $n$ be a positive integer. Let $\mathcal{A}_n$ be the set of $q^n$ polynomials over GF($q$) with degree $n-1$ or less. Let $a(X)$ and $b(X)$ be two polynomials in $\mathcal{A}_n$. Take the product of $a(X)$ and $b(X)$ using the rules of (2.18) and (2.19). Dividing the product $a(X) \cdot b(X)$ by $X^n - 1$ gives a remainder $r(X)$ whose degree is $n - 1$ or less. It is clear that $r(X)$ is also a polynomial in $\mathcal{A}_n$. Now we define a multiplication "$\cdot$" on $\mathcal{A}_n$ as follows. For any two polynomials $a(X)$ and $b(X)$ in $\mathcal{A}_n$, assign $a(X) \cdot b(X)$ to $r(X)$, which is the remainder obtained from dividing $a(X) \cdot b(X)$ by $X^n - 1$. This multiplication is called *multiplication modulo*-$(X^n - 1)$ (or modulo-$(X^n - 1)$ multiplication). Mathematically, this multiplication is written as follows:

$$a(X) \cdot b(X) = r(X) \text{ modulo-}(X^n - 1). \tag{2.23}$$

$\mathcal{A}_n$ is closed under modulo-$(X^n - 1)$ multiplication. This multiplication satisfies the associative, commutative, and distributive laws. Note that addition of two polynomials in $\mathcal{A}_n$ is carried out as defined by (2.17). The sum of two polynomials in $\mathcal{A}_n$ is a polynomial with degree $n-1$ or less and is hence a polynomial in $\mathcal{A}_n$. The polynomials in $\mathcal{A}_n$ under addition and modulo-$(X^n - 1)$ multiplication form an *algebra* over GF($q$). Let

$$a(X) = a_0 + a_1X + \cdots + a_{n-1}X^{n-1}$$

be a polynomial in $\mathcal{A}_n$. On cyclically shifting the coefficients of $a(X)$ one place to the right, we obtain the following polynomial in $\mathcal{A}_n$:

$$a^{(1)}(X) = a_{n-1} + a_0X + \cdots + a_{n-2}X^{n-1}.$$

Polynomial $a^{(1)}(X)$ is called a *right cyclic shift* of $a(X)$. Actually, $\mathcal{A}_n$ is also a vector space over GF($q$). Each polynomial $a(X) = a_0 + a_1X + \cdots + a_{n-1}X^{n-1}$ in $\mathcal{A}_n$ can be represented by an $n$-tuple over GF($q$) using its $n$ coefficients as the components as follows:

$$\mathbf{a} = (a_0, a_1, \ldots, a_{n-1}).$$

With this representation, $\mathcal{A}_n$ is simply equivalent to the vector space of all the $n$-tuples over GF($q$).

## 2.6 Construction and Properties of Galois Fields

In Section 2.3 we have shown that, for any prime number $p$, there exists a prime field with $p$ elements under modulo-$p$ addition and multiplication. Construction of prime fields is very simple. In this section, we consider construction of *extension fields* of prime fields and give some important structural properties of these extension fields.

### 2.6.1 Construction of Galois Fields

Construction of an extension field of a prime field GF($p$)= $\{0, 1, \ldots, p-1\}$ begins with a primitive polynomial of degree $m$ over GF($p$),

$$p(X) = p_0 + p_1X + \cdots + p_{m-1}X^{m-1} + X^m. \tag{2.24}$$

Since the degree of $p(X)$ is $m$, it has $m$ roots. Since $p(X)$ is irreducible over GF($p$), its roots cannot be in GF($p$) and they must be in a larger field that contains GF($p$) as a *subfield*.

Let $\alpha$ be a root of $p(X)$. Let 0 and 1 be the zero and unit elements of GF($p$). We define a multiplication "$\cdot$" and introduce a sequence of powers of $\alpha$ as follows:

$$\begin{aligned}
&0 \cdot 0 = 0,\\
&0 \cdot 1 = 1 \cdot 0 = 0,\\
&0 \cdot \alpha = \alpha \cdot 0 = 0,\\
&1 \cdot 1 = 1,\\
&1 \cdot \alpha = \alpha \cdot 1 = \alpha,\\
&\alpha^2 = \alpha \cdot \alpha,\\
&\alpha^3 = \alpha \cdot \alpha \cdot \alpha,\\
&\vdots\\
&\alpha^j = \alpha \cdot \alpha \cdot \dots \cdot \alpha \ (j \text{ times}),\\
&\vdots
\end{aligned} \tag{2.25}$$

It follows from the definition of multiplication above that

$$\begin{aligned}
&0 \cdot \alpha^j = \alpha^j \cdot 0 = 0,\\
&1 \cdot \alpha^j = \alpha^j \cdot 1 = \alpha^j,\\
&\alpha^i \cdot \alpha^j = \alpha^j \cdot \alpha^i = \alpha^{i+j}.
\end{aligned}$$

Since $\alpha$ is a root of $p(X)$,

$$p(\alpha) = p_0 + p_1\alpha + \cdots + p_{m-1}\alpha^{m-1} + \alpha^m = 0. \tag{2.26}$$

Since $p(X)$ divides $X^{p^m-1} - 1$ (Theorem 2.9), we have

$$X^{p^m-1} - 1 = q(X) \cdot p(X). \tag{2.27}$$

On replacing $X$ by $\alpha$ in (2.27), we obtain

$$\alpha^{p^m-1} - 1 = q(\alpha) \cdot p(\alpha) = q(\alpha) \cdot 0. \tag{2.28}$$

If we regard $q(\alpha)$ as a polynomial of $\alpha$ over GF($p$), it follows from (2.20) that $q(\alpha) \cdot 0 = 0$. Consequently,

$$\alpha^{p^m-1} - 1 = 0. \tag{2.29}$$

This says that $\alpha$ is also a root of $X^{p^m-1} - 1$. On adding the unit element 1 of GF($p$) to both sides of (2.29) (use modulo-$p$ addition), we have the following identity:

$$\alpha^{p^m-1} = 1. \tag{2.30}$$

From (2.30), we see that the sequence of powers of $\alpha$ given by (2.25) (excluding the 0 element) repeats itself at $\alpha^{k(p^m-1)}$ for $k = 1, 2, \ldots$. Therefore, the sequence contains at most $p^m$ distinct elements including the 0 element, that is, $0, 1, \alpha, \ldots, \alpha^{p^m-2}$. Let

$$\mathcal{F} = \{0, 1, \alpha, \ldots, \alpha^{p^m-2}\}. \tag{2.31}$$

Later, we will show that the $p^m$ elements of $\mathcal{F}$ are distinct. The nonzero elements of $\mathcal{F}$ are closed under the multiplication operation "·" defined by (2.25). To see this, let $i$ and $j$ be two integers such that $0 \leq i,j < p^m - 1$. If $i + j < p^m - 1$, then $\alpha^i \cdot \alpha^j = \alpha^{i+j}$ is a nonzero element in $\mathcal{F}$. If $i + j \geq p^m - 1$, we can express $i + j$ as follows:

$$i + j = (p^m - 1) + r, \tag{2.32}$$

with $0 \leq r < p^m - 1$. Then

$$\alpha^i \cdot \alpha^j = \alpha^{i+j} = \alpha^{(p^m-1)+r} = \alpha^{p^m-1} \cdot \alpha^r = 1 \cdot \alpha^r = \alpha^r, \tag{2.33}$$

which is also a nonzero element of $\mathcal{F}$. Hence, the nonzero elements of $\mathcal{F}$ are closed under the multiplication "·" defined by (2.25). We readily see that this multiplication is associative and commutative. Since $1 \cdot \alpha^j = \alpha^j \cdot 1 = \alpha^j$, 1 is the identity element with respect to this multiplication. For $0 \leq j < p^m - 1$, $\alpha^{p^m-1-j}$ is the multiplicative inverse of $\alpha^j$. Note that, for $j = 0$, the unit element 1 is its own multiplicative inverse. Therefore, the nonzero elements of $\mathcal{F}$ form a commutative group of order $p^m - 1$ under the multiplication operation "·" defined by (2.25).

For $0 \leq i < p^m - 1$, on dividing $X^i$ by $p(X)$ we obtain

$$X^i = q_i(X) \cdot p(X) + a_i(X), \tag{2.34}$$

where $q_i(X)$ and $a_i(X)$ are the quotient and remainder, respectively. The remainder $a_i(X)$ is a polynomial over GF($p$) with degree $m - 1$ or less and is of the following form:

$$a_i(X) = a_{i,0} + a_{i,1}X + \cdots + a_{i,m-1}X^{m-1}, \tag{2.35}$$

where the $a_{i,j}$s are elements of the prime field GF($p$). Since $p(X)$ is an irreducible polynomial over GF($p$) and $X$ and $p(X)$ are relatively prime, $X^i$ is not divisible by $p(X)$. Hence, for $0 \leq i < p^m - 1$,

$$a_i(X) \neq 0. \tag{2.36}$$

To prove that all the nonzero elements of $\mathcal{F}$ are distinct, we need the following theorem.

**Theorem 2.34** *For $0 \leq i,j < p^m - 1$, and $i \neq j$,*

$$a_i(X) \neq a_j(X). \tag{2.37}$$

*Proof* Assume that $j > i$. Suppose $a_i(X) = a_j(X)$. Then it follows from (2.34) that

$$X^j - X^i = (q_j(X) - q_i(X)) \cdot p(X).$$

This says that $X^i(X^{j-i} - 1)$ is divisible by $p(X)$. Since $X^i$ and $p(X)$ are relatively prime, $p(X)$ must divide $X^{j-i} - 1$, where $j - i < p^m - 1$. However, this is not possible since $p(X)$ is a primitive polynomial of degree $m$ and the smallest $n$ for which $p(X)$ divides $X^n - 1$ is $n = p^m - 1$ (see Definition 2.21). Therefore, the hypothesis that $a_i(X) = a_j(X)$ is invalid. Consequently, $a_i(X) \neq a_j(X)$ for $i \neq j$. □

It follows from Theorem 2.10 that dividing $X^i$ by $p(X)$ with $i = 0, 1, \ldots, p^m - 2$ gives $p^m - 1$ distinct nonzero-remainder polynomials over GF($p$), $a_0(X), a_1(X), \ldots,$ $a_{p^m-2}(X)$. On replacing $X$ by $\alpha$ in (2.34) and (2.35), we obtain

$$\alpha^i = q(\alpha) \cdot p(\alpha) + a_{i,0} + a_{i,1}\alpha + \cdots + a_{i,m-1}\alpha^{m-1}. \tag{2.38}$$

Since $\alpha$ is a root of $p(X)$, $p(\alpha) = 0$ and $q(\alpha) \cdot 0 = 0$. Consequently, we have

$$\alpha^i = a_{i,0} + a_{i,1}\alpha + \cdots + a_{i,m-1}\alpha^{m-1}, \tag{2.39}$$

where the $a_{i,j}$s are elements of GF($p$). It follows from Theorem 2.10 and (2.39) that the $p^m - 1$ nonzero elements in $\mathcal{F}$ are distinct and represented by $p^m - 1$ distinct nonzero polynomials of $\alpha$ over GF($p$) with degree $m - 1$ or less. The 0 element in $\mathcal{F}$ may be represented by the zero polynomial. Note that there are exactly $p^m$ polynomials of $\alpha$ over GF($p$) with degree $m - 1$ or less. Each of the $p^m$ elements in $\mathcal{F}$ is represented by one and only one of these polynomials, and each of these polynomials represents one and only one element in $\mathcal{F}$. Among these polynomials of $\alpha$ over GF($p$) with degree $m - 1$ or less, there are $p$ polynomials of zero degree, which are $0, 1, 2, \ldots, p - 1$, the elements of GF($p$). Therefore, $\mathcal{F}$ contains the elements of GF($p$) as a subset.

Next, we define an addition "+" on $\mathcal{F}$ using polynomial representations of the elements in $\mathcal{F}$ as follows. For any two elements $\alpha^i$ and $\alpha^j$,

$$\begin{aligned} \alpha^i + \alpha^j &= (a_{i,0} + a_{i,1}\alpha + \cdots + a_{i,m-1}\alpha^{m-1}) + (a_{j,0} + a_{j,1}\alpha + \cdots + a_{j,m-1}\alpha^{m-1}) \\ &= (a_{i,0} + a_{j,0}) + (a_{i,1} + a_{j,1})\alpha + \cdots + (a_{i,m-1} + a_{j,m-1})\alpha^{m-1}, \end{aligned} \tag{2.40}$$

where $a_{i,l} + a_{j,l}$ is carried out over the prime field GF($p$) using modulo-$p$ addition. Hence, the polynomial given by (2.40) is a polynomial of $\alpha$ over GF($p$) and thus represents an element $\alpha^k$ in $\mathcal{F}$. Therefore, $\mathcal{F}$ is closed under the addition defined by (2.40). Since the addition of GF($p$) is associative and commutative, we can readily verify that the addition defined by (2.40) is also associative and commutative. From (2.40), we can easily verify that $0 + \alpha^j = \alpha^j + 0$. Hence, 0 is the additive identity with respect to the addition defined by (2.40). The additive inverse of $\alpha^i = a_{i,0} + a_{i,1}\alpha + \cdots + a_{i,m-1}\alpha^{m-1}$ is

$$-\alpha^i = (-a_{i,0}) + (-a_{i,1})\alpha + \cdots + (-a_{i,m-1})\alpha^{m-1},$$

where $-a_{i,l}$ is the additive inverse of the field element $a_{i,l}$ of GF($p$). Therefore, $\mathcal{F}$ satisfies all the axioms of a commutative group under the addition defined by (2.40).

So far, we have shown that all the elements of $\mathcal{F}$ form a commutative group under the addition defined by (2.40) and all the nonzero elements of $\mathcal{F}$ form a commutative group under the multiplication defined by (2.33). Using the polynomial representations for the elements of $\mathcal{F}$ and the fact that multiplication of polynomials over a field satisfies the distributive law, we can prove that the multiplication defined by (2.33) satisfies the distributive law over the addition defined by (2.40). Hence, the set $\mathcal{F} = \{0, 1, \alpha, \alpha^2, \ldots, \alpha^{p^m-2}\}$ forms a field with $p^m$ elements, denoted by GF($p^m$), under the addition defined by (2.40) and the multiplication defined by (2.33). GF($p^m$) contains GF($p$) as a subfield. Since GF($p^m$) is constructed from GF($p$) and a monic primitive polynomial over GF($p$), we call GF($p^m$) an *extension field* of GF($p$).

The prime field GF($p$) is called the *ground field* of GF($p^m$). Since the characteristic of GF($p$) is $p$ and GF($p$) is a subfield of GF($p^m$), the characteristic of GF($p^m$) is also $p$.

In the process of constructing the Galois field GF($p^m$) from the ground field GF($p$), we have developed two representations for each element of GF($p^m$), namely the power and polynomial representations. However, there is another useful representation for a field element of GF($p^m$). Let $a_{i,0} + a_{i,1}\alpha + \cdots + a_{i,m-1}\alpha^{m-1}$ be the polynomial representation of element $\alpha^i$. We can represent $\alpha^i$ in vector form by the following $m$-tuple:

$$(a_{i,0}, a_{i,1}, \ldots, a_{i,m-1}),$$

where the $m$ components are simply the coefficients of the polynomial representation of $\alpha^i$.

With vector representation, to add two elements $\alpha^i$ and $\alpha^j$ we simply add their corresponding vector representations $(a_{i,0}, a_{i,1}, \ldots, a_{i,m-1})$ and $(a_{j,0}, a_{j,1}, \ldots, a_{j,m-1})$ component-wise as follows:

$$\begin{aligned}&(a_{i,0}, a_{i,1}, \ldots, a_{i,m-1}) + (a_{j,0}, a_{j,1}, \ldots, a_{j,m-1})\\ &= (a_{i,0} + a_{j,0}, a_{i,1} + a_{j,1}, \ldots, a_{i,m-1} + a_{j,m-1}),\end{aligned}$$

where addition $a_{i,l} + a_{j,l}$ is carried out over the ground field GF($p$). The vector sum above is an $m$-tuple over GF($p$) and represents an element $\alpha^k$ in the extension field GF($p^m$). Now we have three representations for each element in GF($p^m$). The power representation is convenient for multiplication, the polynomial and vector representations are convenient for addition. With the power representation, the 0 element is represented by $\alpha^{-\infty}$.

The extension field GF($p^m$) may be viewed as an $m$-dimensional vector space over the ground field GF($p$). The elements of GF($p^m$) are regarded as the "vectors" and the elements of the ground field GF($p$) are regarded as the "scalars." To see this, we first note that all the elements of GF($p^m$) form a commutative group under the addition defined by (2.40). Since GF($p$) is a subfield of GF($p^m$), each element $\alpha^i \in$ GF($p^m$), $i = -\infty, 0, 1, \ldots, p^{m-2}$, can be multiplied by an element $\beta \in$ GF($p$) such that $\beta\alpha^i$ is also an element (or vector) in GF($p^m$). This multiplication is viewed as a "scalar" multiplication, with $\beta$ as a scalar and $\alpha^i$ as a vector. It is obvious that this scalar multiplication satisfies the associative and commutative laws given in Definition 2.14. Since every element $\alpha^i$ in GF($p^m$) is a linear combination of $\alpha^0 = 1, \alpha, \ldots, \alpha^{m-1}$,

$$\alpha^i = a_{i,0}\alpha^0 + a_{i,1}\alpha + \cdots + a_{i,m-1}\alpha^{m-1},$$

and $\alpha^i = 0$ if and only if all the coefficients of the $a_{i,j}$s are zero, the $m$ elements (or vectors) $\alpha^0, \alpha, \ldots, \alpha^{m-1}$ are linearly independent and they form a basis of GF($p^m$), viewed as an $m$-dimensional vector space over GF($p$). This basis is commonly referred to as the *polynomial basis* of GF($p^m$). GF($p^m$) is also called an *extension* of GF($p$) of degree $m$.

---

**Example 2.14** Let the binary field GF(2) be the ground field. Suppose we want to construct an extension field GF($2^5$) of GF(2) of degree $m = 5$. First, we choose a primitive polynomial of degree 5 over GF(2), say $p(X) = 1 + X^2 + X^5$. (This polynomial is the *reciprocal polynomial* $X^5q(X^{-1})$ of the primitive polynomial $q(X) = 1 + X^3 + X^5$ from the table given in Appendix A. See Problem (2.18).) Let $\alpha$ be a root of $p(X) = 1 + X^2 + X^5$. Then $p(\alpha) = 1 + \alpha^2 + \alpha^5 = 0$ and $\alpha^5 = 1 + \alpha^2$. Furthermore, it follows from (2.30) that $\alpha^{2^5-1} = \alpha^{31} = 1$. Using the equalities $\alpha^5 = 1 + \alpha^2$ and $\alpha^{31} = 1$, we can construct GF($2^5$). The elements of GF($2^5$) with three different representations are given in Table 2.10. In the construction of Table 2.10, we use the identity $\alpha^5 = 1 + \alpha^2$ repeatedly to form the polynomial representations for the elements of GF($2^5$). For example,

$$\begin{aligned}
\alpha^6 &= \alpha \cdot \alpha^5 = \alpha \cdot (1 + \alpha^2) = \alpha + \alpha^3, \\
\alpha^7 &= \alpha \cdot \alpha^6 = \alpha \cdot (\alpha + \alpha^3) = \alpha^2 + \alpha^4, \\
\alpha^8 &= \alpha \cdot \alpha^7 = \alpha \cdot (\alpha^2 + \alpha^4) = \alpha^3 + \alpha^5 = \alpha^3 + 1 + \alpha^2 = 1 + \alpha^2 + \alpha^3.
\end{aligned}$$

To multiply two elements $\alpha^i$ and $\alpha^j$, we simply add their exponents and use the identity $\alpha^{31} = 1$. For example, $\alpha^{13} \cdot \alpha^{25} = \alpha^{38} = \alpha^{31} \cdot \alpha^7 = 1 \cdot \alpha^7 = \alpha^7$. The multiplicative inverse of $\alpha^i$ is $\alpha^{31-i}$. For example, the multiplicative inverse of $\alpha^9$ is $\alpha^{31-9} = \alpha^{22}$. To divide $\alpha^j$ by $\alpha^i$, we simply multiply $\alpha^j$ by the multiplicative inverse $\alpha^{31-i}$ of $\alpha^i$. For example, $\alpha^7 \div \alpha^9 = \alpha^7 \cdot \alpha^{31-9} = \alpha^7 \cdot \alpha^{22} = \alpha^{29}$. To add two elements $\alpha^i$ and $\alpha^j$, we may use either their polynomial representations or vector representations. Suppose we want to add $\alpha^{11}$ and $\alpha^{17}$ using their vector representations. From Table 2.10, we find that the vector representations of $\alpha^{11}$ and $\alpha^{17}$ are (11100) and (11001), respectively. On adding these two vectors, we have

$$(11100) + (11001) = (00101).$$

From Table 2.10, we find that the vector (00101) represents the element $\alpha^7$. In polynomial form, it is $\alpha^2 + \alpha^4$.

Besides $\alpha$, $p(X)$ has four other roots. These four other roots can be found by substituting $X$ of $p(X) = 1 + X^2 + X^5$ by the elements of GF($2^5$) in turn. By doing so, we find that the other four roots are $\alpha^2$, $\alpha^4$, $\alpha^8$, and $\alpha^{16}$. For example, on replacing $X$ of $p(X) = 1 + X^2 + X^5$ by $\alpha^8$, we have

$$p(\alpha^8) = 1 + \alpha^{16} + \alpha^{40} = 1 + \alpha^{16} + \alpha^9.$$

Using polynomial representations of the elements in the above sum, we have

$$\begin{aligned}
p(\alpha^8) &= 1 + (1 + \alpha + \alpha^3 + \alpha^4) + (\alpha + \alpha^3 + \alpha^4) \\
&= (1 + 1) + (1 + 1)\alpha + (1 + 1)\alpha^3 + (1 + 1)\alpha^4 \\
&= 0 + 0 \cdot \alpha + 0 \cdot \alpha^3 + 0 \cdot \alpha^4 \\
&= 0,
\end{aligned}$$

where addition of two coefficients is modulo-2 addition. Hence, $\alpha^8$ is a root of $p(X) = 1+X^2+X^5$. Similarly, we can prove that $\alpha^2$, $\alpha^4$, and $\alpha^{16}$ are also roots of $p(X) = 1 + X^2 + X^5$. Later we will show how to find the roots of $p(X)$ in a much easier way.

---

**Table 2.10** $GF(2^5)$ generated by the primitive polynomial $p(X) = 1 + X^2 + X^5$ over GF(2)

| Power representation | Polynomial representation | Vector representation |
|---|---|---|
| $0$ | $0$ | (00000) |
| $1$ | $1$ | (10000) |
| $\alpha$ | $\alpha$ | (01000) |
| $\alpha^2$ | $\alpha^2$ | 00100) |
| $\alpha^3$ | $\alpha^3$ | (00010) |
| $\alpha^4$ | $\alpha^4$ | (00001) |
| $\alpha^5$ | $1 + \alpha^2$ | (10100) |
| $\alpha^6$ | $\alpha + \alpha^3$ | (01010) |
| $\alpha^7$ | $\alpha^2 + \alpha^4$ | (00101) |
| $\alpha^8$ | $1 + \alpha^2 + \alpha^3$ | (10110) |
| $\alpha^9$ | $\alpha + \alpha^3 + \alpha^4$ | (01011) |
| $\alpha^{10}$ | $1 + \alpha^4$ | (10001) |
| $\alpha^{11}$ | $1 + \alpha + \alpha^2$ | (11100) |
| $\alpha^{12}$ | $\alpha + \alpha^2 + \alpha^3$ | (01110) |
| $\alpha^{13}$ | $\alpha^2 + \alpha^3 + \alpha^4$ | (00111) |
| $\alpha^{14}$ | $1 + \alpha^2 + \alpha^3 + \alpha^4$ | (10111) |
| $\alpha^{15}$ | $1 + \alpha + \alpha^2 + \alpha^3 + \alpha^4$ | (11111) |
| $\alpha^{16}$ | $1 + \alpha + \alpha^3 + \alpha^4$ | (11011) |
| $\alpha^{17}$ | $1 + \alpha + \alpha^4$ | (11001) |
| $\alpha^{18}$ | $1 + \alpha$ | (11000) |
| $\alpha^{19}$ | $\alpha + \alpha^2$ | (01100) |
| $\alpha^{20}$ | $\alpha^2 + \alpha^3$ | (00110) |
| $\alpha^{21}$ | $\alpha^3 + \alpha^4$ | (00011) |
| $\alpha^{22}$ | $1 + \alpha^2 + \alpha^4$ | (10101) |
| $\alpha^{23}$ | $1 + \alpha + \alpha^2 + \alpha^3$ | (11110) |
| $\alpha^{24}$ | $\alpha + \alpha^2 + \alpha^3 + \alpha^4$ | (01111) |
| $\alpha^{25}$ | $1 + \alpha^3 + \alpha^4$ | (10011) |
| $\alpha^{26}$ | $1 + \alpha + \alpha^2 + \alpha^4$ | (11101) |
| $\alpha^{27}$ | $1 + \alpha + \alpha^3$ | (11010) |
| $\alpha^{28}$ | $\alpha + \alpha^2 + \alpha^4$ | (01101) |
| $\alpha^{29}$ | $1 + \alpha^3$ | (10010) |
| $\alpha^{30}$ | $\alpha + \alpha^4$ | (01001) |
| $\alpha^{31} = 1$ | | |

**Example 2.15** In this example, we construct an extension field $GF(3^2)$ of the prime field GF(3) of degree 2 using the primitive polynomial $p(X) = 2 + X + X^2$ over GF(3) (taken from the table given in Appendix A). The modulo-3 addition and multiplication for GF(3) are given in Tables 2.11 and 2.12. Let $\alpha$ be a root of $p(X)$. Then $p(\alpha) = 2 + \alpha + \alpha^2 = 0$ and $\alpha^2 = -2 - \alpha$. From Table 2.11, we see that the additive inverse $-2$ of 2 is 1 and the additive inverse $-1$ of 1 is 2. Consequently, $\alpha^2 = -2 - \alpha$ can be written as $\alpha^2 = 1 + 2\alpha$. Using the identities $\alpha^2 = 1 + 2\alpha$ and

$\alpha^8 = 1$, we can construct the extension field GF($3^2$). The elements of GF($3^2$) in three forms are given in Table 2.13. Since $p(X)$ has degree 2, it has two roots. Besides $\alpha$, the other root of $p(X)$ is $\alpha^3$. To see this, we replace $X$ by $\alpha^3$. This results in $p(\alpha^3) = 2 + \alpha^3 + \alpha^6$. On replacing $\alpha^3$ and $\alpha^6$ by their polynomial representations, we have $p(\alpha^3) = 2 + 2 + 2\alpha + 2 + \alpha = 0$. From Table 2.13, we also see that GF(3) is a subfield of GF($3^2$).

---

As we have shown, to construct an extension field GF($p^m$) of degree $m$ of the prime field GF($p$), we need a primitive polynomial $p(X)$ of degree $m$ over GF($p$). If a different primitive polynomial $p^*(X)$ of degree $m$ over GF($p$) is used, then the construction results in an extension field GF$^*$($p^m$) of degree $m$ of the prime field GF($p$) that has the *same set of elements* as that of GF($p^m$). There is a *one-to-one correspondence* between GF($p^m$) and GF$^*$($p^m$) such that, if $a \leftrightarrow a^*$, $b \leftrightarrow b^*$, and $c \leftrightarrow c^*$, then

**Table 2.11** Modulo-3 addition

| + | 0 | 1 | 2 |
|---|---|---|---|
| 0 | 0 | 1 | 2 |
| 1 | 1 | 2 | 0 |
| 2 | 2 | 0 | 1 |

**Table 2.12** Modulo-3 multiplication

| · | 0 | 1 | 2 |
|---|---|---|---|
| 0 | 0 | 0 | 0 |
| 1 | 0 | 1 | 2 |
| 2 | 0 | 2 | 1 |

**Table 2.13** GF($3^2$) generated by the primitive polynomial $p(X) = 2 + X + X^2$ over GF(3)

| Power representation | Polynomial representation | Vector representation |
|---|---|---|
| 0 | 0 | (00) |
| 1 | 1 | (10) |
| $\alpha$ | $\alpha$ | (01) |
| $\alpha^2$ | $1 + 2\alpha$ | (12) |
| $\alpha^3$ | $2 + 2\alpha$ | (22) |
| $\alpha^4$ | 2 | (20) |
| $\alpha^5$ | $2\alpha$ | (02) |
| $\alpha^6$ | $2 + \alpha$ | (21) |
| $\alpha^7$ | $1 + \alpha$ | (11) |
| $\alpha^8 = 1$ | | |

$$a + b \leftrightarrow a^* + b^*,$$
$$a \cdot b \leftrightarrow a^* \cdot b^*,$$
$$(a + b) + c \leftrightarrow a^* + (b^* + c^*),$$
$$(a \cdot b) \cdot c \leftrightarrow a^* \cdot (b^* \cdot c^*),$$
$$a \cdot (b + c) \leftrightarrow a^* \cdot (b^* + c^*).$$

GF($p^m$) and GF$^*$($p^m$) are said to be *isomorphic*. That is, GF($p^m$) and GF$^*$($p^m$) are structurally identical. In this sense, we may say that any primitive polynomial $p(X)$ of degree $m$ over the prime field GF($p$) gives the same extension field GF($p^m$). This implies uniqueness of the extension field GF($p^m$).

The above construction of an extension field GF($p^m$) of a prime field GF($p$) can be generalized. Let GF($q$) be a finite field with $q$ elements, where $q$ is a power of a prime, say $q = p^s$. It is known in modern algebra that, *for any positive integer $m$, there exists a primitive polynomial $p(X)$ of degree $m$ over GF($q$)*. Let $\alpha$ be a root of $p(X)$. Then the construction of the extension field GF($q^m$) of GF($q$) of degree $m$ is exactly the same as the construction of the extension field GF($p^m$) of the prime field GF($p$) of degree $m$, as given above.

## 2.6.2 Some Fundamental Properties of Finite Fields

As shown in the previous subsection, for any prime field GF($p$) and any positive integer $m$, it is possible to construct an extension field GF($p^m$) of degree $m$ with GF($p$) as the ground field. In fact, for any finite field GF($q$) with $q$ elements, $q$ must be a power of a prime $p$ and GF($q$) contains the prime field GF($p$) as a subfield. Proof of this can be found in any text on modern algebra and the theory of finite fields. The extension field GF($p^m$) has the same characteristic as that of the prime ground field GF($p$), that is, $p$. It has also been shown at the end of the last subsection that, for any positive integer $m$, it is possible to construct an extension field GF($q^m$) of degree $m$ with GF($q$) as the ground field. Since GF($q$) contains GF($p$) as a subfield, GF($q^m$) contains both GF($q$) and GF($p$) as subfields. GF($q^m$) may be viewed as an extension field of the prime field GF($p$) and has characteristic $p$.

In Section 2.3, some fundamental properties of finite fields have been presented. In the following, we will present some additional fundamental properties of finite fields that are important for constructing error-correcting codes. First, we start with an important property of the ring of polynomials over GF($q$) without proof. Let $f(X) = f_0 + f_1X + \cdots + f_kX^k$ be a polynomial over GF($q$). Then, for any non-negative integer $t$, the following equality holds:

$$[f(X)]^{q^t} = f\left(X^{q^t}\right). \tag{2.41}$$

**Theorem 2.35** *Let $f(X)$ be a polynomial over GF($q$). Let $\beta$ be an element of the extension field GF($q^m$) of GF($q$). If $\beta$ is a root of $f(X)$, then, for any non-negative integer $t$, $\beta^{q^t}$ is also a root of $f(X)$.*

*Proof* The proof of this theorem follows directly from the equality of (2.41). On substituting $X$ by $\beta$ in (2.41), we have

$$[f(\beta)]^{q^l} = f\left(\beta^{q^l}\right).$$

Since $\beta$ is a root of $f(X)$, $f(\beta) = 0$. Then it follows from the equality above that $f(\beta^{q^l}) = 0$. Hence $\beta^{q^l}$ is a root of $f(X)$. □

The element $\beta^{q^l}$ is called a *conjugate* of $\beta$. Theorem 2.11 says that if $\beta$, an element of GF($q^m$), is a root of a polynomial $f(X)$ over GF($q$), then all its conjugates $\beta, \beta^q, \beta^{q^2}, \ldots$ are also roots of $f(X)$.

**Theorem 2.36** *If $\beta$ is an element of order $n$ in GF($q^m$), then all its conjugates have the same order $n$. If $\beta$ is a primitive element of GF($q^m$), then all its conjugates are primitive elements. (The proof is left as an exercise.)*

---

**Example 2.16** Consider the polynomial $f(X) = 1 + X^3 + X^5$ over GF(2). This polynomial has $\alpha^{15}$, an element of GF($2^5$) given by Table 2.10, as a root. To verify this, we use Table 2.10 and the identity $\alpha^{31} = 1$. On substituting the variable $X$ of $f(X)$ by $\alpha^{15}$, we obtain

$$\begin{aligned} f(\alpha^{15}) &= 1 + (\alpha^{15})^3 + (\alpha^{15})^5 = 1 + \alpha^{45} + \alpha^{75} \\ &= 1 + \alpha^{14} + \alpha^{13} \\ &= 1 + (1 + \alpha^2 + \alpha^3 + \alpha^4) + (\alpha^2 + \alpha^3 + \alpha^4) \\ &= (1 + 1) + (1 + 1) \cdot \alpha^2 + (1 + 1) \cdot \alpha^3 + (1 + 1) \cdot \alpha^4 \\ &= 0 + 0 \cdot \alpha^2 + 0 \cdot \alpha^3 + 0 \cdot \alpha^4 = 0 + 0 + 0 + 0 = 0. \end{aligned}$$

The conjugates of $\alpha^{15}$ are

$$(\alpha^{15})^2 = \alpha^{30}, \quad (\alpha^{15})^{2^2} = \alpha^{60} = \alpha^{29},$$
$$(\alpha^{15})^{2^3} = \alpha^{120} = \alpha^{27}, \quad (\alpha^{15})^{2^4} = \alpha^{240} = \alpha^{23}.$$

Note that $(\alpha^{15})^{2^5} = \alpha^{480} = \alpha^{15}$. It follows from Theorem 2.11 that $\alpha^{30}$, $\alpha^{29}$, $\alpha^{27}$, and $\alpha^{23}$ are also roots of $f(X) = 1 + X^3 + X^5$. Consider the conjugate $\alpha^{23}$. On replacing $X$ by $\alpha^{23}$ in $f(X)$, we have

$$\begin{aligned} f(\alpha^{23}) &= 1 + (\alpha^{23})^3 + (\alpha^{23})^5 = 1 + \alpha^{69} + \alpha^{115} \\ &= 1 + \alpha^7 + \alpha^{22} = 1 + (\alpha^2 + \alpha^4) + (1 + \alpha^2 + \alpha^4) \\ &= 0. \end{aligned}$$

Hence, $\alpha^{23}$ is a root of $f(X)$. Since the degree of $f(X)$ is 5, it must have five roots. The elements $\alpha^{15}$, $\alpha^{23}$, $\alpha^{27}$, $\alpha^{29}$, and $\alpha^{30}$ give all five roots of $f(X) = 1 + X^3 + X^5$. So all the roots of $f(X)$ are in GF($2^5$). With some computational effort, we can check that $\alpha^{15}$ and its conjugates have order 31, so they are all primitive.

---

**Example 2.17** Consider the polynomial $f(X) = 2 + 2X + X^2$ over GF(3) given in Example 2.15 (see Tables 2.11 and 2.12). Using modulo-3 addition and multiplication, the extension field GF($3^2$) of GF(3) given by Table 2.13, and the fact

$\alpha^8 = 1$, we find that $\alpha^5$ and its conjugate $(\alpha^5)^3 = \alpha^{15} = \alpha^7$ are roots of $f(X) = 2 + 2X + X^2$. Suppose we take the powers of $\alpha^5$,

$$\alpha^5, \quad (\alpha^5)^2 = \alpha^{10} = \alpha^2, \quad (\alpha^5)^3 = \alpha^{15} = \alpha^7, \quad (\alpha^5)^4 = \alpha^{20} = a^4,$$
$$(\alpha^5)^5 = \alpha^{25} = \alpha, \quad (\alpha^5)^6 = \alpha^{30} = \alpha^6, \quad (\alpha^5)^7 = \alpha^{35} = a^3,$$
$$(\alpha^5)^8 = \alpha^{40} = 1.$$

We see that the powers of $\alpha^5$ give all the elements of GF($3^2$). Therefore, $\alpha^5$ is a primitive element. Similarly, we can show that the powers of the conjugate $\alpha^7$ of $\alpha^5$ also give all the elements of GF($3^2$). Hence, $\alpha^7$ is also a primitive element of GF($3^2$).

---

**Theorem 2.37** *Let GF($q^m$) be an extension field of GF($q$). The $q^m - 1$ nonzero elements of GF($q^m$) form all the roots of the polynomial $X^{q^m-1} - 1$ over GF($q$).*

*Proof* Let $\beta$ be a nonzero element of GF($q^m$). On substituting $X$ of $X^{q^m-1} - 1$ by $\beta$, we have $\beta^{q^m-1} - 1$. However, it follows from Theorem 2.6 that $\beta^{q^m-1} = 1$. Hence, $\beta^{q^m-1} - 1 = 1 - 1 = 0$. This says that $\beta$ is a root of $X^{q^m-1} - 1$. Therefore, every nonzero element of GF($q^m$) is a root of $X^{q^m-1} - 1$. Since the degree of $X^{q^m-1} - 1$ is $q^m - 1$, the $q^m - 1$ nonzero elements of GF($q^m$) form all the roots of $X^{q^m-1}$. □

Since the zero element 0 of GF($q^m$) is the root of $X$, Theorem 2.13 has the following corollary.

**Corollary 2.1** *The $q^m$ elements of GF($q^m$) form all the roots of $X^{q^m} - X$.*

It follows from Corollary 2.1 that

$$X^{q^m} - X = \prod_{\beta \in \mathrm{GF}(q^m)} (X - \beta).$$

Since any element $\beta$ in GF($q^m$) is a root of the polynomial $X^{q^m} - X$ over GF($q$), $\beta$ may be a root of a polynomial over GF($q$) with degree less than $q^m$.

**Definition 2.38** Let $\beta$ be an element of GF($q^m$), an extension field of GF($q$). Let $\phi(X)$ be the monic polynomial of smallest degree over GF($q$) that has $\beta$ as a root, that is, $\phi(\beta) = 0$. This polynomial $\phi(X)$ is called the *minimal polynomial* of $\beta$.

It is clear that the minimal polynomial of the 0 element of GF($q^m$) is $X$. In the following, we present a number of properties of minimal polynomials of elements of GF($q^m$).

**Theorem 2.39** *The minimal polynomial $\phi(X)$ of a field element $\beta$ is irreducible.*

*Proof* Suppose that $\phi(X)$ is not irreducible. Then $\phi(X)$ can be expressed as a product of two factors, $\phi(X) = \phi_1(X)\phi_2(X)$, where both $\phi_1(X)$ and $\phi_2(X)$ are monic and have degrees greater than 0 but less than that of $\phi(X)$. Since $\phi(\beta) = \phi_1(\beta)\phi_2(\beta) = 0$, either $\phi_1(\beta) = 0$ or $\phi_2(\beta) = 0$. This contradicts the hypothesis that $\phi(X)$ is a polynomial of smallest degree over GF($q$) that has $\beta$ as a root. Hence, $\phi(X)$ must be irreducible. □

**Theorem 2.40** *The minimal polynomial $\phi(X)$ of a field element $\beta$ is unique.*

*Proof* Suppose that $\beta$ has two different minimal polynomials, $\phi(X)$ and $\phi'(X)$. Then these two minimal polynomials are monic and have the same degree, say $k$. Then $m(X) = \phi(X) - \phi'(X)$ is a polynomial of degree $k-1$ or less over GF($q$). If $m(X)$ is not monic, it can be made monic by multiplying it by the inverse of the coefficient of the highest degree of $m(X)$. Since both $\phi(X)$ and $\phi'(X)$ have $\beta$ as a root, $m(\beta) = \phi(\beta) - \phi'(\beta) = 0 - 0 = 0$. If $m(X) \neq 0$, it is a polynomial over GF($q$) with degree less than that of $\phi(X)$ and $\phi'(X)$ that has $\beta$ as a root. This contradicts the hypothesis that $\phi(X)$ and $\phi'(X)$ are the smallest-degree polynomials over GF($q$) with $\beta$ as a root. Therefore, $m(X)$ must be the zero polynomial and $\phi(X) = \phi'(X)$. This proves the theorem. □

**Theorem 2.41** *Let $f(X)$ be a polynomial over GF($q$). Let $\phi(X)$ be the minimal polynomial of an element $\beta$ in GF($q^m$). If $\beta$ is a root of $f(X)$, then $f(X)$ is divisible by $\phi(X)$.*

*Proof* On dividing $f(X)$ by $\phi(X)$ we obtain

$$f(X) = a(X)\phi(X) + r(X),$$

where $r(X)$ is the remainder with degree less than that of $\phi(X)$. On substituting $X$ of the equation above with $\beta$, we have $f(\beta) = a(\beta)\phi(\beta) + r(\beta)$. Since both $f(X)$ and $\phi(X)$ have $\beta$ as a root, we have $r(\beta) = 0$. This says that $\beta$ is also a root of $r(X)$. If $r(X) \neq 0$, then $r(X)$ is a polynomial over GF($q$) with degree lower than the degree of $\phi(X)$ that has $\beta$ as a root. This contradicts the fact that $\phi(X)$ is the polynomial over GF($q$) of smallest degree that has $\beta$ as a root. Hence, we must have $r(X) = 0$ and $f(X)$ must be divisible by $\phi(X)$. □

A direct consequence of the above theorem is the following corollary.

**Corollary 2.2** *Let $p(X)$ be a monic irreducible polynomial over GF($q$). Let $\beta$ be an element of GF($q^m$) and $\phi(X)$ be its minimal polynomial. If $\beta$ is a root of $p(X)$, then $p(X) = \phi(X)$.*

Corollary 2.2 simply says that, if a monic irreducible polynomial has $\beta$ as a root, it is the minimal polynomial $\phi(X)$ of $\beta$. It follows from Theorem 2.11 that $\beta$ and its conjugates $\beta^q, \ldots, \beta^{q^t}, \ldots$ are roots of $\phi(X)$. Let $e$ be the smallest positive integer such that $\beta^{q^e} = \beta$. Then $\beta, \beta^q, \ldots, \beta^{q^{e-1}}$ are all the distinct conjugates of $\beta$. Since $\beta^{q^m} = \beta$ and $e$ is the smallest positive integer such that $\beta^{q^e} = \beta$, we must have $e \leq m$. In fact, $e$ divides $m$ (the proof is left as an exercise).

**Theorem 2.42** *Let $\phi(X)$ be the minimal polynomial of an element $\beta$ of GF($q^m$). Then $\phi(X)$ divides $X^{q^m} - X$.*

*Proof* The proof of this theorem follows from Corollary 2.1 and Theorem 2.16. □

It follows from Theorem 2.11, Theorem 2.14, and Corollary 2.2 that we have the following theorem.

**Theorem 2.43** *Let $\beta$ be an element of GF($q^m$). Then all its conjugates $\beta, \beta^q, \ldots, \beta^{q^{e-1}}$ have the same minimal polynomial.*

A direct consequence of Theorems 2.17 and 2.18 is that $X^{q^m} - X$ is equal to the product of the distinct minimal polynomials of the elements of GF($q^m$).

The next theorem tells us how to determine the minimal polynomials of elements of the extension field GF($q^m$) of degree $m$ of GF($q$). The proof of this theorem is left as an exercise.

**Theorem 2.44** *Let $\phi(X)$ be the minimal polynomial of an element $\beta$ of GF($q^m$). Let $e$ be the smallest positive integer such that $\beta^{q^e} = \beta$. Then*

$$\phi(X) = \prod_{i=0}^{e-1} \left(X - \beta^{q^i}\right). \tag{2.42}$$

Since $e \leq m$, the degree of the minimal polynomial of any element $\beta$ of GF($q^m$) is $m$ or less.

---

**Example 2.18** Consider the extension field GF($2^5$) of the binary field GF(2) given by Table 2.10. Let $\beta = \alpha^3$. The conjugates of $\beta$ are

$$\beta^2 = (\alpha^3)^2 = \alpha^6, \quad \beta^{2^2} = (\alpha^3)^{2^2} = \alpha^{12},$$
$$\beta^{2^3} = (\alpha^3)^{2^3} = \alpha^{24}, \quad \beta^{2^4} = (\alpha^3)^{2^4} = \alpha^{48} = \alpha^{17}.$$

Note that $\beta^{2^5} = (\alpha^3)^{2^5} = \alpha^{96} = \alpha^3$. Hence, $e = 5$. The minimal polynomial of $\alpha^3$ is then

$$\phi(X) = (X - \alpha^3)(X - \alpha^6)(X - \alpha^{12})(X - \alpha^{17})(X - \alpha^{24}).$$

On multiplying out the right-hand side of the equation above with the aid of Table 2.10 and the modulo-2 addition of coefficients, we obtain

$$\phi(X) = 1 + X^2 + X^3 + X^4 + X^5.$$

---

**Example 2.19** Consider the extension field GF($3^2$) of degree 2 of the prime field GF(3) given by Table 2.13. Let $\beta = \alpha^5$. The only conjugate of $\beta$ is $\alpha^7$. Then the minimal polynomial of $\alpha^3$ and $\alpha^7$ is given by

$$\phi(X) = (X - \alpha^5)(X - \alpha^7)$$
$$= \alpha^{12} - (\alpha^5 + \alpha^7)X + X^2.$$

With the aid of Table 2.13, we find that

$$\phi(X) = \alpha^4 - (2\alpha + 1 + \alpha)X + X^2$$
$$= 2 - X + X^2 = 2 + 2X + X^2.$$

---

In the construction of the extension field GF($q^m$) of degree $m$ of GF($q$), we use a primitive polynomial $p(X)$ of degree $m$ over GF($q$) and require that the element $\alpha$ be a root of $p(X)$. Since the powers of $\alpha$ generate all the nonzero elements of GF($q^m$), $\alpha$ is a primitive element. It follows from Theorem 2.12 that all its conjugates are primitive elements. There is a simple rule to determine whether a nonzero element of $GF(q^m)$ is primitive. For $0 < j < q^m - 1$, a nonzero element $\alpha^j$ of $GF(q^m)$ is a primitive element if and only if $j$ and $q^m - 1$ are relatively prime. Consequently, the number of primitive elements in $GF(q^m)$ is given by $K_p = \phi(q^m - 1)$, which is known as *Euler's formula*. To determine $K_p$, we first factor $q^m - 1$ as a product of powers of primes,

$$q^m - 1 = p_1^{k_1} p_2^{k_2} \cdots p_t^{k_t}.$$

Then

$$K_p = (q^m - 1) \prod_{i=1}^{t} (1 - 1/p_i). \tag{2.43}$$

If $q^m - 1$ is a prime, then $K_p = q^m - 2$. In this case, except for the unit element 1, every nonzero element of $GF(q^m)$ is a primitive element.

---

**Example 2.20** Consider the extension field $GF(2^6)$ of degree 6 of the binary field GF(2). The number $2^6 - 1 = 63$ can be factored as the product $63 = 3^2 \times 7$. It follows from (2.43) that $GF(2^6)$ consists of

$$K_p = 63(1 - 1/3)(1 - 1/7) = 63 \times (2/3)(6/7) = 36$$

primitive elements.

---

### 2.6.3 Additive and Cyclic Subgroups

Consider the extension field $GF(q^m)$ of degree $m$ with $GF(q)$ as the ground field. Let $\alpha$ be a primitive element of $GF(q^m)$. Then $\alpha^0 = 1, \alpha, \ldots, \alpha^{m-1}$ form a polynomial basis of $GF(q^m)$ over $GF(q)$. For $i = -\infty, 0, 1, \ldots, q^m - 2$, every element $\alpha^i$ can be expressed as a linear sum (or combination) of $\alpha^0, \alpha, \ldots, \alpha^{m-1}$ as follows:

$$\alpha^i = a_{i,0}\alpha^0 + a_{i,1}\alpha + \cdots + a_{i,m-1}\alpha^{m-1}, \tag{2.44}$$

with $a_{i,j} \in GF(q)$. For $0 \le t < m - 1$, let $j_0, j_1, \ldots, j_{t-1}$ be $t$ integers such that $0 \le j_0 < j_1 < \cdots < j_{t-1} \le m - 1$. Consider the $q^t$ elements of $GF(q^m)$ of the following form:

$$\alpha^k = c_{k,0}\alpha^{j_0} + c_{k,1}\alpha^{j_1} + \cdots + c_{k,t-1}\alpha^{j_{t-1}}, \tag{2.45}$$

where $c_{k,j} \in GF(q)$. Let $S$ denote the set of these $q^t$ elements. $S$ forms a subgroup of the additive group of $GF(q^m)$ and is called an additive subgroup of $GF(q^m)$.

---

**Example 2.21** Consider the extension field $GF(2^5)$ of the binary field GF(2) given by Table 2.10 (see Example 2.14). The elements $\alpha^0, \alpha, \alpha^2, \alpha^3, \alpha^4$ form a polynomial basis. Let $j_0 = 1, j_1 = 3$, and $j_2 = 4$. Consider the following eight linear sums of $\alpha$, $\alpha^3$, and $\alpha^4$:

$$\alpha^k = c_{k,0}\alpha + c_{k,1}\alpha^3 + c_{k,2}\alpha^4,$$

with $c_{k,j} \in \mathrm{GF}(2)$. Using Table 2.10, we find that the linear sums of $\alpha$, $\alpha^3$, and $\alpha^4$ give the following eight elements:

$$\alpha^{-\infty} = 0,\ \alpha,\ \alpha^3,\ \alpha^4,\ \alpha^6,\ \alpha^9,\ \alpha^{21},\ \alpha^{30}.$$

These eight elements give a subgroup of the additive group of $\mathrm{GF}(2^5)$.

---

Suppose $q^m - 1$ is not a prime. It can be factored as a product of two positive integers greater than 1, say $q^m - 1 = kn$. Let $\alpha$ be a primitive element of $\mathrm{GF}(q^m)$ and $\beta = \alpha^k$. Then $\beta$ is a nonzero element of order $n$, that is, $n$ is the smallest positive integer such that $\beta^n = 1$ (see Section 2.3). The powers $\beta^0 = 1, \beta, \ldots, \beta^{n-1}$ form a subgroup of the multiplicative group of $\mathrm{GF}(q^m)$. Such a subgroup is called a *cyclic subgroup* of $\mathrm{GF}(q^m)$ with $\beta = \alpha^k$ as the generator.

---

**Example 2.22** Consider the extension field $\mathrm{GF}(3^2)$ of the prime field GF(3) given by Table 2.13. The integer $3^2 - 1 = 8$ can be factored as a product of $k = 2$ and $n = 4$. Let $\beta = \alpha^2$. The order of $\beta$ is 4. The four distinct powers of $\beta$ are $\beta^0 = 1, \beta = \alpha^2, \beta^2 = \alpha^4$, and $\beta^3 = \alpha^6$. These four elements form a cyclic subgroup of $\mathrm{GF}(3^2)$.

---

## 2.7 Finite Geometries

In contrast to ordinary geometry, a finite geometry has finite numbers of points, lines, and flats. (A *flat* is a generalization of a plane.) However, finite and ordinary geometries also have some fundamental structures in common, such as the following: (1) two points are connected by a line; (2) two lines are either disjoint (i.e., have no point in common) or they intersect at one and only one point (i.e., they have one point in common); and (3) if two lines have two points in common, they are the same line. In this section, we present two families of finite geometries over finite fields, namely Euclidean and projective geometries. These geometries will be used to construct LDPC codes for iterative decoding in Chapters 11 and 12.

### 2.7.1 Euclidean Geometries

Let $m$ be a positive integer greater than unity and $\mathrm{GF}(q)$ be a finite field with $q$ elements, where $q$ is a power of prime $p$, say $p^s$, with $s \geq 1$. The *m-dimensional Euclidean geometry over* $\mathrm{GF}(q)$, denoted by $\mathrm{EG}(m, q)$, consists of points, lines, and flats [1–4]. Each point is an $m$-tuple over $\mathrm{GF}(q)$. Therefore, there are

$$n = q^m \tag{2.46}$$

points in $\mathrm{EG}(m, q)$, which actually form the vector space of all the $m$-tuples over $\mathrm{GF}(q)$, denoted by $V_m$. The all-zero $m$-tuple $(0, 0, \ldots, 0)$ is referred to as the *origin* of the geometry.

A *line* in EG$(m, q)$ is either a one-dimensional subspace or a coset of a one-dimensional subspace of the vector space $V_m$ of all the $m$-tuples over GF$(q)$. There are

$$J_{\mathrm{EG}}(m, 1) = q^{m-1}(q^m - 1)/(q - 1) \tag{2.47}$$

lines in EG$(m, q)$ and each line consists of $q$ points. Any two points are connected by one and only one line. Any two lines are either disjoint (i.e., have no point in common) or they intersect at one and only one point (i.e., have one point in common). Lines that correspond to two cosets of the same one-dimensional subspace of $V_m$ do not have any point in common and they are said to be *parallel*. Each one-dimensional subspace of $V_m$ has $q^{m-1}$ cosets. Lines in EG$(m, q)$ that correspond to these $q^{m-1}$ cosets are parallel to each other and they form a *parallel bundle*. A parallel bundle of lines contains all the points of EG$(m, q)$; each point appears once and only once on only one line in the parallel bundle. It follows from (2.47) and the fact that each parallel bundle of lines consists of $q^{m-1}$ parallel lines that the lines of EG$(m, q)$ can be partitioned into

$$K_{\mathrm{EG}}(m, 1) = (q^m - 1)/(q - 1) \tag{2.48}$$

parallel bundles.

For any point $\mathbf{a}$ in EG$(m, q)$, there are exactly

$$g_{\mathrm{EG}}(m, 1, 0) = (q^m - 1)/(q - 1) \tag{2.49}$$

lines in EG$(m, q)$ that intersect at $\mathbf{a}$ (this will be shown later). Note that we have $K_{\mathrm{EG}}(m, 1) = g_{\mathrm{EG}}(m, 1, 0)$. The $g_{\mathrm{EG}}(m, 1, 0)$ lines in EG$(m, q)$ that intersect at a given point $\mathbf{a}$ are said to form an *intersecting bundle of lines* at $\mathbf{a}$. A line that contains a point $\mathbf{a}$ is said to pass through $\mathbf{a}$.

For $0 \le \mu \le m$, a $\mu$-*flat* in EG$(m, q)$ is either a $\mu$-dimensional subspace or a coset of a $\mu$-dimensional subspace of the vector space $V_m$ over GF$(q)$. For $\mu = 0$, a 0-flat is simply a point of EG$(m, q)$. For $\mu = 1$, a 1-flat is simply a line of EG$(m, q)$. Two flats that correspond to two cosets of the same $\mu$-dimensional subspace of $V_m$ are said to be parallel, and they do not have any point in common. For each $\mu$-dimensional subspace of $V_m$, there are $q^{m-\mu}$ parallel $\mu$-flats in EG$(m, q)$ and they form a parallel bundle of $\mu$-flats. These parallel $\mu$-flats are mutually disjoint and they contain all the points of EG$(m, q)$, each point appearing once and only once on only one of these parallel flats. In EG$(m, q)$, there are

$$J_{\mathrm{EG}}(m, \mu) = q^{m-\mu} \prod_{i=1}^{\mu} \frac{q^{m-i+1} - 1}{q^{\mu-i+1} - 1} \tag{2.50}$$

$\mu$-flats. These $\mu$-flats can be partitioned into

$$K_{\mathrm{EG}}(m, \mu) = \prod_{i=1}^{\mu} \frac{q^{m-i+1} - 1}{q^{\mu-i+1} - 1} \tag{2.51}$$

parallel bundles of $\mu$-flats.

For $1 \le \mu < m$, if two $\mu$-flats are not disjoint, they intersect on a flat of smaller dimension. The largest flat that two $\mu$-flats can intersect on is a $(\mu - 1)$-flat. If two

$\mu$-flats in EG$(m, q)$ intersect on a $(\mu - 1)$-flat, then they can intersect on one and only one $(\mu-1)$-flat. The number of $\mu$-flats in EG$(m, q)$ that intersect on a given $(\mu-1)$-flat is given by

$$g_{\mathrm{EG}}(m, \mu, \mu - 1) = \frac{q^{m-\mu+1} - 1}{q - 1}. \tag{2.52}$$

As shown in Section 2.6.1, GF$(q^m)$ as an extension field of GF$(q)$ can be viewed as an $m$-dimensional vector space $V_m$ of all $m$-tuples over GF$(q)$; each element in GF$(q^m)$ can be represented by an $m$-tuple over GF$(q)$. Therefore, GF$(q^m)$ as an extension field of GF$(q)$ is a realization of EG$(m, q)$. Let $\alpha$ be a primitive element of GF$(q^m)$. Then the elements $\alpha^{-\infty} = 0, \alpha^0 = 1, \alpha, \ldots, \alpha^{q^m-2}$ of GF$(q^m)$ represent the $n = q^m$ points of EG$(m, q)$, and the 0 element represents the origin of the geometry.

Using GF$(q^m)$ as a realization of the $m$-dimensional Euclidean geometry EG$(m, q)$ over GF$(q)$, it is much easier to develop the fundamental properties of the geometry. Hereafter, we refer to the powers of $\alpha$ as points in EG$(m, q)$. Let $\alpha^{j_1}$ be a non-origin point in EG$(m, q)$. The set of $q$ points

$$\{\beta\alpha^{j_1} : \beta \in \mathrm{GF}(q)\} \tag{2.53}$$

constitutes a line (or 1-flat) of EG$(m, q)$. For $\beta = 0$, $0 \cdot \alpha^{j_1} = 0$ is the origin point of EG$(m, q)$. Therefore, this line passes through the origin of EG$(m, q)$. Viewing the $q$ points on this line as vectors in the vector space $V_m$ of all the $m$-tuples over GF$(q)$, this line forms a one-dimensional subspace of $V_m$. For convenience, we use the notation $\{\beta\alpha^{j_1}\}$ to denote this line. Let $\alpha^{j_0}$ and $\alpha^{j_1}$ be two independent points in EG$(m, q)$. Then the collection of $q$ points

$$\{\alpha^{j_0} + \beta\alpha^{j_1}\}, \tag{2.54}$$

with $\beta \in$ GF$(q)$, forms a line passing through the point $\alpha^{j_0}$. We may view this as a line connecting the points $\alpha^{j_0}$ and $\alpha^{j_1}$. This line is simply a coset of the one-dimensional subspace $\{\beta\alpha^{j_1}\}$. The $q^{m-1}$ cosets of $\{\beta\alpha^{j_1}\}$ form a parallel bundle of lines in EG$(m, q)$. Since there are $(q^m-1)/(q-1)$ different one-dimensional subspaces in GF$(q^m)$ (viewed as an $m$-dimensional vector space $V_m$ over GF$(q)$) and each one-dimensional subspace of GF$(q^m)$ has $q^{m-1}$ cosets, we readily see that the number of lines in EG$(m, q)$ is given by (2.47).

Let $\alpha^{j_0}$, $\alpha^{j_1}$, and $\alpha^{j_2}$ be three points in EG$(m, q)$ that are pair-wise linearly independent. Then $\{\alpha^{j_0} + \beta\alpha^{j_1}\}$ and $\{\alpha^{j_0} + \beta\alpha^{j_2}\}$ form two different lines in EG$(m, q)$ that intersect at the point $\alpha^{j_0}$. On the basis of this fact, we readily determine that the number of lines in EG$(m, q)$ that intersect at any point in EG$(m, q)$ is given by (2.49).

---

**Example 2.23** Consider the two-dimensional Euclidean geometry EG$(2, 2^2)$ over GF$(2^2)$. This geometry consists of $2^{2\times 2} = 16$ points, each point is a 2-tuple over GF$(2^2)$. Each line in EG$(2, 2^2)$ consists of four points. Consider the Galois field GF$(2^4)$ generated by the primitive polynomial $1 + X + X^4$ over GF(2) given by

Table 2.14. Let $\alpha$ be a primitive element of GF($2^4$) and $\beta = \alpha^5$. The order of $\beta$ is 3. The set

$$\text{GF}(2^2) = \{0, \beta^0 = 1, \beta = \alpha^5, \beta^2 = \alpha^{10}\}$$

forms the subfield GF($2^2$) of GF($2^4$). Every element $\alpha^i$ in GF($2^4$) can be expressed as the following polynomial form:

$$\alpha^i = \beta_0\alpha^0 + \beta_1\alpha,$$

with $\beta_0, \beta_1 \in \text{GF}(2^2)$. In vector form, $\alpha^i$ is represented by the 2-tuple $(\beta_0, \beta_1)$ over GF($2^2$). GF($2^4$) as an extension field of GF($2^2$) is given by Table 2.15. Regard GF($2^4$) as a realization of the two-dimensional Euclidean geometry EG($2, 2^2$) over GF($2^2$). Then

$$\mathcal{L}_0 = \{\beta_1\alpha\} = \{0, \alpha, \alpha^6, \alpha^{11}\}$$

with $\beta_1 \in \text{GF}(2^2)$ forms a line passing through the origin of GF($2^4$). The three lines parallel to $\mathcal{L}_0$ are (using Table 2.14 or Table 2.15)

$$\begin{aligned}
\mathcal{L}_1 &= \{1 + \beta_1\alpha\} = \{1, \alpha^4, \alpha^{12}, \alpha^{13}\},\\
\mathcal{L}_2 &= \{\alpha^5 + \beta_1\alpha\} = \{\alpha^2, \alpha^3, \alpha^5, \alpha^9\},\\
\mathcal{L}_3 &= \{\alpha^{10} + \beta_1\alpha\} = \{\alpha^8, \alpha^7, \alpha^{10}, \alpha^{14}\}.
\end{aligned}$$

$\mathcal{L}_0$, $\mathcal{L}_1$, $\mathcal{L}_2$, and $\mathcal{L}_3$ together form a parallel bundle of lines in EG($2, 2^2$).

For every point in EG($2, 2^4$), there are $(2^{2\times 2} - 1)/(2^2 - 1) = 5$ lines intersecting. For example, consider the point $\alpha$. The lines that intersect at $\alpha$ are

$$\begin{aligned}
\mathcal{L}'_0 &= \{\alpha + \beta_1\alpha^2\} = \{\alpha, \alpha^5, \alpha^{13}, \alpha^{14}\},\\
\mathcal{L}'_1 &= \{\alpha + \beta_1\alpha^3\} = \{\alpha, \alpha^9, \alpha^{10}, \alpha^{12}\},\\
\mathcal{L}'_2 &= \{\alpha + \beta_1\alpha^4\} = \{1, \alpha, \alpha^3, \alpha^7\},\\
\mathcal{L}'_3 &= \{\alpha + \beta_1\alpha^5\} = \{\alpha, \alpha^2, \alpha^4, \alpha^8\},\\
\mathcal{L}'_4 &= \{\alpha + \beta_1\alpha^6\} = \{0, \alpha, \alpha^{11}, \alpha^6\}.
\end{aligned}$$

These five lines form an intersecting bundle of lines at point $\alpha$.

---

Besides the parallel and intersecting structures, the lines of EG($m, q$) also have *cyclic structure*. Let EG*($m, q$) be the subgeometry obtained from EG($m, q$) by removing the origin and all the lines passing through the origin. This subgeometry consists of $q^m - 1$ non-origin points and

$$J_{0,\text{EG}}(m, 1) = (q^{m-1} - 1)(q^m - 1)/(q - 1) \tag{2.55}$$

lines not passing through the origin. Let $\mathcal{L} = \{\alpha^{j_0} + \beta\alpha^{j_1}\}$ be a line in EG*($m, q$). For $0 \le i < q^m - 1$, let

$$\alpha^i\mathcal{L} \triangleq \{\alpha^{j_0+i} + \beta\alpha^{j_1+i}\}.$$

Since $\alpha^{j_0}$ and $\alpha^{j_1}$ are linearly independent points, it is clear that $\alpha^{j_0+i}$ and $\alpha^{j_1+i}$ are also linearly independent points. Therefore, $\alpha^i\mathcal{L}$ is also a line in EG*($m, q$) that passes

**Table 2.14** GF($2^4$) generated by the primitive polynomial $p(X) = 1 + X + X^4$

| Power representation | Polynomial representation | | | | | | | Vector representation |
|---|---|---|---|---|---|---|---|---|
| 0 | 0 | | | | | | | (0 0 0 0) |
| 1 | 1 | | | | | | | (1 0 0 0) |
| $\alpha$ | | | $\alpha$ | | | | | (0 1 0 0) |
| $\alpha^2$ | | | | | $\alpha^2$ | | | (0 0 1 0) |
| $\alpha^3$ | | | | | | | $\alpha^3$ | (0 0 0 1) |
| $\alpha^4$ | 1 | + | $\alpha$ | | | | | (1 1 0 0) |
| $\alpha^5$ | | | $\alpha$ | + | $\alpha^2$ | | | (0 1 1 0) |
| $\alpha^6$ | | | | | $\alpha^2$ | + | $\alpha^3$ | (0 0 1 1) |
| $\alpha^7$ | 1 | + | $\alpha$ | | | + | $\alpha^3$ | (1 1 0 1) |
| $\alpha^8$ | 1 | | | + | $\alpha^2$ | | | (1 0 1 0) |
| $\alpha^9$ | | | $\alpha$ | | | + | $\alpha^3$ | (0 1 0 1) |
| $\alpha^{10}$ | 1 | + | $\alpha$ | + | $\alpha^2$ | | | (1 1 1 0) |
| $\alpha^{11}$ | | | $\alpha$ | + | $\alpha^2$ | + | $\alpha^3$ | (0 1 1 1) |
| $\alpha^{12}$ | 1 | + | $\alpha$ | + | $\alpha^2$ | + | $\alpha^3$ | (1 1 1 1) |
| $\alpha^{13}$ | 1 | | | + | $\alpha^2$ | + | $\alpha^3$ | (1 0 1 1) |
| $\alpha^{14}$ | 1 | | | | | + | $\alpha^3$ | (1 0 0 1) |

**Table 2.15** GF($2^4$) as an extension field GF($2^2$)= $\{0, 1, \beta, \beta^2\}$ with $\beta = \alpha^5$

| Power representation | Polynomial representation | | | Vector representation |
|---|---|---|---|---|
| 0 | 0 | | | $(0, 0)$ |
| 1 | 1 | | | $(1, 0)$ |
| $\alpha$ | | | $\alpha$ | $(0, 1)$ |
| $\alpha^2$ | $\beta$ | + | $\alpha$ | $(\beta, 1)$ |
| $\alpha^3$ | $\beta$ | + | $\beta^2\alpha$ | $(\beta, \beta^2)$ |
| $\alpha^4$ | 1 | + | $\alpha$ | $(1, 1)$ |
| $\alpha^5$ | $\beta$ | | | $(\beta, 0)$ |
| $\alpha^6$ | | | $\beta\alpha$ | $(0, \beta)$ |
| $\alpha^7$ | $\beta^2$ | + | $\beta\alpha$ | $(\beta^2, \beta)$ |
| $\alpha^8$ | $\beta^2$ | + | $\alpha$ | $(\beta^2, 1)$ |
| $\alpha^9$ | $\beta$ | + | $\beta\alpha$ | $(\beta, \beta)$ |
| $\alpha^{10}$ | $\beta^2$ | | | $(\beta^2, 0)$ |
| $\alpha^{11}$ | | | $\beta^2\alpha$ | $(0, \beta^2)$ |
| $\alpha^{12}$ | 1 | + | $\beta^2\alpha$ | $(1, \beta^2)$ |
| $\alpha^{13}$ | 1 | + | $\beta\alpha$ | $(1, \beta)$ |
| $\alpha^{14}$ | $\beta^2$ | + | $\beta^2\alpha$ | $(\beta^2, \beta^2)$ |

through the non-origin point $\alpha^{j_0+i}$. It can be proved for $0 \leq i, k < q^m - 1$, and $i \neq k$ that $\alpha^i \mathcal{L} \neq \alpha^k \mathcal{L}$ (this is left as an exercise). A line with this property is said to be *primitive*. Every line in EG$(m, q)$ is primitive [5]. Then $\mathcal{L}, \alpha\mathcal{L}, \ldots, \alpha^{q^m-2}\mathcal{L}$ form $q^m - 1$ different lines in EG$^*(m, q)$. Since $\alpha^{q^m-1} = 1$, $\alpha^{q^m-1}\mathcal{L} = \mathcal{L}$. The line $\alpha^i\mathcal{L}$ is called the *i*th *cyclic shift* of $\mathcal{L}$. The set $\{\mathcal{L}, \alpha\mathcal{L}, \ldots, \alpha^{q^m-2}\mathcal{L}\}$ of $q^m - 1$ lines is said to form a *cyclic class* of lines in EG$^*(m, q)$ of size $q^m - 1$. It follows from (2.55) that the lines in EG$^*(m, q)$ can be partitioned into

$$K_{\text{c,EG}}(m, 1) = (q^{m-1} - 1)/(q - 1) \tag{2.56}$$

cyclic classes of size $q^m - 1$.

The mathematical formulation of lines in the $m$-dimensional Euclidean geometry EG$(m, q)$ over GF$(q)$ in terms of the extension field GF$(q^m)$ of GF$(q)$ can be extended to flats of EG$(m, q)$. For $1 \leq \mu < m$, let $\alpha^{j_1}, \alpha^{j_2}, \ldots, \alpha^{j_\mu}$ be $\mu$ linearly independent points of EG$(m, q)$. The set of $q^\mu$ points

$$\mathcal{F}^{(\mu)} = \{\beta_1\alpha^{j_1} + \beta_2\alpha^{j_2} + \cdots + \beta_\mu\alpha^{j_\mu} : \beta_i \in \text{GF}(q), 1 \leq i \leq \mu\} \tag{2.57}$$

forms a $\mu$-flat in EG$(m, q)$ passing through the origin (a $\mu$-dimensional subspace of the $m$-dimensional vector space $V_m$ over GF$(q)$). For simplicity, we use $\{\beta_1\alpha^{j_1} + \beta_2\alpha^{j_2} + \cdots + \beta_\mu\alpha^{j_\mu}\}$ to denote the $\mu$-flat given by (2.57). Let $\alpha^{j_0}$ be a point not on $\mathcal{F}^{(\mu)}$. Then

$$\alpha^{j_0} + \mathcal{F}^{(\mu)} = \{\alpha^{j_0} + \beta_1\alpha^{j_1} + \cdots + \beta_\mu\alpha^{j_\mu}\} \tag{2.58}$$

is a $\mu$-flat passing through the point $\alpha^{j_0}$ and parallel to the $\mu$-flat $\mathcal{F}^{(\mu)}$. It follows from (2.57) and (2.58) that, for each $\mu$-flat of EG$(m, q)$ passing through the origin, there are $q^{m-\mu} - 1$ $\mu$-flats parallel to it. $\mathcal{F}^{(\mu)}$ and the $q^{m-\mu} - 1$ $\mu$-flats parallel to it form a parallel bundle of $\mu$-flats. The $\mu$-flats of EG$(m, q)$ can be partitioned into $K_{\text{EG}}(m, \mu)$ parallel bundles of $\mu$-flats, where $K_{\text{EG}}(m, \mu)$ is given by (2.51).

If $\alpha^{j_{\mu+1}}$ is not a point on the $\mu$-flat $\{\alpha^{j_0} + \beta_1\alpha^{j_0} + \cdots + \beta_\mu\alpha^{j_\mu}\}$, then the $(\mu + 1)$-flat

$$\{\alpha^{j_0} + \beta_1\alpha^{j_1} + \cdots + \beta_\mu\alpha^{j_\mu} + \beta_{\mu+1}\alpha^{j_{\mu+1}}\}$$

contains the $\mu$-flat $\{\alpha^{j_0} + \beta_1\alpha^{j_1} + \cdots + \beta_\mu\alpha^{j_\mu}\}$. Let $\alpha^{j'_{\mu+1}}$ be a point not on the $(\mu+1)$-flat

$$\{\alpha^{j_0} + \beta_1\alpha^{j_1} + \cdots + \beta_{\mu+1}\alpha^{j_{\mu+1}}\},$$

then the two $(\mu + 1)$-flats $\{\alpha^{j_0} + \beta_1\alpha^{j_0} + \cdots + \beta_\mu\alpha^{j_\mu} + \beta_{\mu+1}\alpha^{j_{\mu+1}}\}$ and $\{\alpha^{j_0} + \beta_1\alpha^{j_1} + \cdots + \beta_\mu\alpha^{j_\mu} + \beta_{\mu+1}\alpha^{j'_{\mu+1}}\}$ intersect on the $\mu$-flat $\{\alpha^{j_0} + \beta_1\alpha^{j_1} + \cdots + \beta_\mu\alpha^{j_\mu}\}$. For a given $\mu$-flat $\mathcal{F}^{(\mu)}$, the number of $(\mu + 1)$-flats in EG$(m, q)$ that intersect on $\mathcal{F}^{(\mu)}$ is given by

$$g(m, \mu + 1, \mu) = \frac{q^{m-\mu} - 1}{q - 1}, \tag{2.59}$$

which is obtained from (2.52) with $\mu$ replaced by $\mu + 1$.

Let $\mathcal{F}^{(\mu)} = \{\alpha^{j_0} + \beta_1\alpha^{j_1} + \cdots + \beta_\mu\alpha^{j_\mu}\}$ be a $\mu$-flat not passing through the origin of EG$(m, q)$ (i.e., $\alpha^{j_0} \neq 0$). For $0 \leq i < q^m - 1$, the set of $q^\mu$ points

$$\alpha^i\mathcal{F}^{(\mu)} = \{\alpha^{j_0+i} + \beta_1\alpha^{j_1+i} + \cdots + \beta_\mu\alpha^{j_\mu+i}\}$$

also forms a $\mu$-flat of EG$(m, q)$ (or the subgeometry EG$^*(m, q)$) not passing through the origin. It can also be proved that $\mathcal{F}^{(\mu)}, \alpha\mathcal{F}^{(\mu)}, \ldots, \alpha^{q^m-2}\mathcal{F}^{(\mu)}$ are $q^m - 1$ different $\mu$-flats in EG$(m, q)$ not passing through the origin [5]. This is to say that every $\mu$-flat not passing through the origin of EG$(m, q)$ is primitive. A $\mu$-flat not passing through the origin of EG$(m, q)$ and its $q^m - 2$ cyclic shifts are said to form a cyclic class of $\mu$-flats. The $\mu$-flats in EG$^*(m, q)$ not passing through the origin can be partitioned into cyclic classes of $\mu$-flats of size $q^m - 1$.

In Chapters 11 and 12, LDPC codes will be constructed from the parallel, intersecting, and cyclic structures of the lines and flats of finite Euclidean geometries.

### 2.7.2 Projective Geometries

Projective geometries form another family of geometries over finite fields, consisting of finite numbers of points, lines, and flats [1–4]. For any positive integer $m$ and any finite field GF$(q)$, there exists an $m$-dimensional projective geometry over GF$(q)$, denoted by PG$(m, q)$. A point of PG$(m, q)$ is a *nonzero* $(m + 1)$-tuple over GF$(q)$. Let $\beta$ be a primitive element of GF$(q)$. For any point $(a_0, a_1, \ldots, a_m)$ in PG$(m, q)$ and $0 \leq j < q - 1$, the $(m + 1)$-tuple $(\beta^j a_0, \beta^j a_1, \ldots, \beta^j a_m)$ represents the same point $(a_0, a_1, \ldots, a_m)$. That is to say that the $q - 1$ nonzero $(m + 1)$-tuples in the set

$$\{(\beta^j a_0, \beta^j a_1, \ldots, \beta^j a_m) \colon \beta^j \in \mathrm{GF}(q), 0 \leq j < q - 2\}$$

represent the same point $(a_0, a_1, \ldots, a_m)$. Therefore, no point in PG$(m, q)$ is a *multiple* (or scalar product) of another point in PG$(m, q)$. Since all the points of PG$(m, q)$ are nonzero $(m + 1)$-tuples over GF$(q)$, PG$(m, q)$ does not have an origin.

The $m$-dimensional projective geometry PG$(m, q)$ over GF$(q)$ can be constructed from the extension field GF$(q^{m+1})$ of GF$(q)$. We view GF$(q^{m+1})$ as the $(m + 1)$-dimensional vector space $V_{m+1}$ over GF$(q)$. Each element in GF$(q^{m+1})$ can be represented by an $(m+1)$-tuple over GF$(q)$. Let $\alpha$ be a primitive element of GF$(q^{m+1})$. Then $\alpha^0, \alpha, \ldots, \alpha^{q^{m+1}-2}$ give all the nonzero elements of GF$(q^{m+1})$. Let

$$n = \frac{q^{m+1} - 1}{q - 1} \tag{2.60}$$

and let $\beta = \alpha^n$. The order of $\beta$ is $q - 1$. The $q$ elements $0, \beta^0 = 1, \beta, \ldots, \beta^{q-2}$ form the ground field GF$(q)$ of GF$(q^{m+1})$.

Partition the nonzero elements of GF$(q^m)$ into $n$ disjoint subsets as follows:

$$\begin{aligned} &\{\alpha^0, \beta\alpha^0, \ldots, \beta^{q-2}\alpha^0\}, \\ &\{\alpha^1, \beta\alpha^1, \ldots, \beta^{q-2}\alpha^1\}, \\ &\{\alpha^2, \beta\alpha^2, \ldots, \beta^{q-2}\alpha^2\}, \\ &\vdots \\ &\{\alpha^{n-1}, \beta\alpha^{n-1}, \ldots, \beta^{q-2}\alpha^{n-1}\}. \end{aligned} \tag{2.61}$$

Each set consists of $q-1$ nonzero elements of GF($q^{m+1}$) and each element in the set is a multiple of the first element. We can represent each set by its first element as follows:

$$(\alpha^i) = \{\alpha^i, \beta\alpha^i, \ldots, \beta^{q-2}\alpha^i\}, \quad (2.62)$$

with $0 \le i < n$. For any element $\alpha^k$ in GF($q^{m+1}$), if $\alpha^k = \beta^j\alpha^i$ with $0 \le i < n$, then $\alpha^k$ is represented by $\alpha^i$. If each element in GF($q^{m+1}$) is represented by an $(m+1)$-tuple over GF($q$), then $(\alpha^i)$ consists of $q-1$ nonzero $(m+1)$-tuples over GF($q$), and each is a multiple of the $(m+1)$-tuple representation of $\alpha^i$. It follows from the definition of a point of PG($m, q$) given above that $(\alpha^i)$ is a point in PG($m, q$). Therefore

$$(\alpha^0), (\alpha^1), \ldots, (\alpha^{n-1})$$

form all the points of PG($m, q$) and the number of points in PG($m, q$) is $n$ given by (2.60).

Note that, if the 0 element of GF($q^m$) is added to the set $(\alpha^i)$, we obtain a set $\{0, \alpha^i, \beta\alpha^i, \ldots, \beta^{q-2}\alpha^i\}$ of $q$ elements. This set of $q$ elements, viewed as $(m+1)$-tuples over GF($q$), is simply a one-dimensional subspace of the vector space $V_{m+1}$ of all the $(m+1)$-tuples over GF($q$), and hence it is a line in the $(m+1)$-dimensional Euclidean geometry EG($m+1,q$) over GF($q$) that passes through the origin of EG($m+1, q$). Therefore, we may regard a point $(\alpha^i)$ of PG($m, q$) as a *projection* of a line of EG($m+1,q$) passing through the origin of EG($m+1, q$).

Let $(\alpha^i)$ and $(\alpha^j)$ be two distinct points in PG($q^{m+1}$). The line that connects $(\alpha^i)$ and $(\alpha^j)$ consists of points of the following form:

$$(\delta_1\alpha^i + \delta_2\alpha^j), \quad (2.63)$$

where $\delta_1$ and $\delta_2$ are scalars from GF($q$) and are *not both equal to zero*. We denote this line by

$$\{(\delta_1\alpha^i + \delta_2\alpha^j)\}. \quad (2.64)$$

There are $q^2-1$ choices of $\delta_1$ and $\delta_2$ from GF($q$) (excluding $\delta_1 = \delta_2 = 0$). However, for each choice of $(\delta_1, \delta_2)$, there are $q-2$ multiples of $(\delta_1, \delta_2)$. Consequently, $(\delta_1, \delta_2)$ and its $q-2$ multiples result in the same points of the form given by (2.63). Therefore, the line connecting points $(\alpha^i)$ and $(\alpha^j)$ consists of

$$\frac{q^2-1}{q-1} = q+1 \quad (2.65)$$

points. For any two points in PG($m, q$), there is one and only one line connecting them. From the number of points and the definition of a line in PG($m, q$), we can enumerate the number of lines in PG($m, q$), which is given by

$$J_{\mathrm{PG}}(m, 1) = \frac{(q^{m+1}-1)(q^m-1)}{(q^2-1)(q-1)}. \quad (2.66)$$

Let $(\alpha^k)$ be a point not on the line that connects the points $(\alpha^i)$ and $(\alpha^j)$. Then the line $\{(\delta_1\alpha^i+\delta_2\alpha^j)\}$ and the line $\{(\delta_1\alpha^i+\delta_2\alpha^k)\}$ intersect at the point $(\alpha^i)$ (with $\delta_1 = 1$

and $\delta_2 = 0$). The number of lines in PG$(m, q)$ that intersect at a point in PG$(m, q)$ is given by

$$g_{\text{PG}}(m, 1, 0) = \frac{q^m - 1}{q - 1}. \tag{2.67}$$

The lines that intersect at a point of PG$(m, q)$ are said to form an intersecting bundle of lines at the point.

If we take all the $q^2$ linear combinations of $\alpha^i$ and $\alpha^j$ in $\{(\delta_1\alpha^i + \delta_2\alpha^j)\}$ (including $\delta_1 = \delta_2 = 0$), these $q^2$ linear combinations give $q^2$ vectors over GF$(q)$ that form a two-dimensional subspace of the vector space $V_{m+1}$ over GF$(q)$. This two-dimensional space of $V_{m+1}$ is a 2-flat of EG$(m+1, q)$ that passes through the origin of EG$(m+1, q)$. Thus, a line of PG$(m, q)$ may be regarded as a projection of a 2-flat (a two-dimensional plane) of EG$(m + 1, q)$.

For $0 \leq \mu < m$, let $(\alpha^{j_0}), (\alpha^{j_1}), \ldots, (\alpha^{j_\mu})$ be $\mu + 1$ linearly independent points. Then a $\mu$-flat in PG$(m, q)$ consists of points of the following form:

$$(\delta_0\alpha^{j_0} + \delta_1\alpha^{j_1} + \cdots + \delta_\mu\alpha^{j_\mu}), \tag{2.68}$$

where $\delta_k \in$ GF$(q)$ with $0 \leq k \leq \mu$ and not all of $\delta_0, \delta_1, \ldots, \delta_\mu$ are zero. We denote this $\mu$-flat by

$$\{(\delta_0\alpha^{j_0} + \delta_1\alpha^{j_1} + \cdots + \delta_\mu\alpha^{j_\mu})\}. \tag{2.69}$$

It can be shown that the number of points on a $\mu$-flat is

$$\frac{q^{\mu+1} - 1}{q - 1}. \tag{2.70}$$

There are

$$J_{\text{PG}}(m, \mu) = \prod_{i=0}^{\mu} \frac{q^{m-i+1} - 1}{q^{\mu-i+1} - 1} \tag{2.71}$$

$\mu$-flats in PG$(m, q)$.

If we allow $\delta_i$ in (2.69) to be any element in GF$(q)$ without restriction, this results in $q^{\mu+1}$ points in the $(m + 1)$-dimensional Euclidean geometry EG$(m + 1, q)$ over GF$(q)$, which form a $(\mu + 1)$-flat in EG$(m + 1, q)$. Therefore, we may regard a $\mu$-flat in the $m$-dimensional projective geometry PG$(m, q)$ over GF$(q)$ as a projection of a $(\mu + 1)$-flat of the $(m + 1)$-dimensional Euclidean geometry EG$(m + 1, q)$ over GF$(q)$.

For $1 \leq \mu < m$, let $\alpha^{j'_\mu}$ be a point not on the $\mu$-flat $\{(\delta_0\alpha^{j_0} + \delta_1\alpha^{j_1} + \cdots + \delta_\mu\alpha^{j_\mu})\}$. Then the $\mu$-flat

$$\{(\delta_0\alpha^{j_0} + \delta_1\alpha^{j_1} + \cdots + \delta_{\mu-1}\alpha^{j_{\mu-1}} + \delta_\mu\alpha^{j_\mu})\}$$

and the $\mu$-flat

$$\{(\delta_0\alpha^{j_0} + \delta_1\alpha^{j_1} + \cdots + \delta_{\mu-1}\alpha^{j_{\mu-1}} + \delta_\mu\alpha^{j'_\mu})\}$$

intersect on the $(\mu - 1)$-flat $\{(\delta_0\alpha^{j_0} + \delta_1\alpha^{j_1} + \cdots + \delta_{\mu-1}\alpha^{j_{\mu-1}})\}$. For $1 \le \mu < m$, the number of $\mu$-flats that intersect on a given $(\mu - 1)$-flat is given by

$$g_{\mathrm{PG}}(m, \mu, \mu - 1) = \frac{q^{m-\mu+1} - 1}{q - 1}. \tag{2.72}$$

The $\mu$-flats in PG$(q, m)$ that intersect on a $(\mu - 1)$-flat $\mathcal{F}^{(\mu-1)}$ are called an intersecting bundle of $\mu$-flats on $\mathcal{F}^{(\mu-1)}$.

Next we consider the cyclic structure of lines in an $m$-dimensional projective geometry PG$(m, q)$ over GF$(q)$. Let $\mathcal{L}_0 = \{(\delta_0\alpha^{j_0} + \delta_1\alpha^{j_1})\}$ be a line in PG$(m, q)$. For $0 \le i < n = (q^{m+1} - 1)/(q - 1)$,

$$\alpha^i\mathcal{L}_0 = (\delta_0\alpha^{j_0+i} + \delta_1\alpha^{j_1+i})$$

is also a line in PG$(m, q)$, which is called the $i$th cyclic shift of $\mathcal{L}_0$. If $m$ is even, every line is primitive [5]. In this case, $\mathcal{L}_0, \alpha\mathcal{L}_0, \ldots, \alpha^{n-1}\mathcal{L}_0$ are all different lines and they form a cyclic class. Then the lines of PG$(m, q)$ can be partitioned into

$$K^{(\mathrm{e})}_{\mathrm{c,PG}}(m, 1) = \frac{q^m - 1}{q^2 - 1} \tag{2.73}$$

cyclic classes of size $(q^{m+1} - 1)/(q - 1)$. If $m$ is odd, there are $l_0 = (q^{m+1} - 1)/(q^2 - 1)$ nonprimitive lines, which can be represented by $\alpha^i\mathcal{L}_0$ for a certain line $\mathcal{L}_0$ in PG$(m, q)$, where $0 \le i < l_0$ [5]. These $l_0$ lines form a cyclic class of size $l_0$. All the other lines in PG$(m, q)$ are primitive and they can be partitioned into

$$K^{(\mathrm{o})}_{\mathrm{c,PG}}(m, 1) = \frac{q(q^{m-1} - 1)}{q^2 - 1} \tag{2.74}$$

cyclic classes of size $(q^{m+1} - 1)/(q - 1)$.

For $1 < \mu < m$, a $\mu$-flat in PG$(m, q)$ is in general not primitive, but it is primitive in the following cases [6]: (1) when $\mu = m - 1$; and (2) when the number of points in a $\mu$-flat and the number of points in PG$(m, q)$ are relatively prime. For the first case, there are $(q^{m+1} - 1)/(q - 1)$ $(m - 1)$-flats and they together form a single cyclic class of size $(q^{m+1} - 1)/(q - 1)$. For the second case, the $\mu$-flats in PG$(m, q)$ can be partitioned into

$$K_{\mathrm{c,PG}}(m, \mu) = \prod_{i=0}^{\mu-1} \frac{q^{m-i} - 1}{q^{\mu-i+1} - 1} \tag{2.75}$$

cyclic classes.

Since the $\mu$-flats of PG$(m, q)$ are obtained from the $(\mu + 1)$-flats of EG$(m + 1, q)$ not passing through the origin, the $\mu$-flats in PG$(m, q)$ do not have parallel structure.

## 2.8 Graphs

The theory of graphs is an important part of combinatorial mathematics. A graph is a simple geometrical figure consisting of *vertices* (or *nodes*) and *edges* (*lines* or *branches*), which connect some of the vertices. Because of their diagrammatic representation, graphs have been found to be extremely useful in modeling systems arising in physical sciences, engineering, social sciences, economics, computer science, and so on. In error-control coding, graphs are used to construct codes and to represent codes for specific ways of decoding. This will be seen in later chapters. In this section, we first introduce some basic concepts and fundamental properties of graphs and then present some special graphs that are useful for designing codes and decoding algorithms.

### 2.8.1 Basic Concepts

As with every mathematical discussion, we have to begin with a list of definitions.

**Definition 2.45** A *graph*, denoted by $\mathcal{G} = (\mathcal{V}, \mathcal{E})$, consists of a non-empty set $\mathcal{V} = \{v_1, v_2, \ldots\}$ of elements called *vertices* and a prescribed set $\mathcal{E} = \{(v_i, v_j)\}$ of unordered pairs of vertices called *edges*. The set $\mathcal{V}$ is called the *vertex set* and the set $\mathcal{E}$ is called the *edge set*. The vertices $v_i$ and $v_j$ associated with an edge $(v_i, v_j)$ are called the *end vertices* of the edge $(v_i, v_j)$. The set $\mathcal{E}$ may be a *multi-set* in the most general case; that is, it may have element repetitions (parallel edges).

If $(v_i, v_j)$ is an edge in a graph $\mathcal{G} = (\mathcal{V}, \mathcal{E})$, the edge $(v_i, v_j)$ is said to be *incident with* the end vertices $v_i$ and $v_j$, and $v_i$ and $v_j$ are said to be *adjacent*. Two adjacent vertices $v_i$ and $v_j$ are said to be connected by edge $(v_i, v_j)$. If $(v_i, v_j)$ and $(v_j, v_k)$ are two edges in $\mathcal{G} = (\mathcal{V}, \mathcal{E})$, these two edges are incident with the common vertex $v_j$ and they are said to be adjacent edges. A vertex is said to be an *isolated vertex* if there is no edge incident with it. A vertex is called a *pendant* if it is incident with only one edge. The edge $(v_i, v_i)$ (if it exists) is called a *self-loop*. Multiple (two or more) edges are said to be parallel if they are incident with the same two vertices.

The most common and useful representation of a graph is by means of a diagram in which the vertices are represented as points on a plane and an edge $(v_i, v_j)$ is represented by a line joining vertices $v_i$ and $v_j$. For example, consider the graph $\mathcal{G} = (\mathcal{V}, \mathcal{E})$ with

$$\mathcal{V} = \{v_1, v_2, v_3, v_4, v_5, v_6\},$$
$$\mathcal{E} = \{(v_1, v_1), (v_1, v_2), (v_1, v_3), (v_1, v_4), (v_2, v_3), (v_3, v_4), (v_4, v_3), (v_4, v_5)\}.$$

The geometrical representation of this graph is shown in Figure 2.1. It is clear that $(v_1, v_1)$ forms a self-loop, vertex $v_6$ is an isolated vertex, and $v_5$ is a pendant.

**Definition 2.46** A graph with finite numbers of vertices as well as a finite number of edges is called a *finite graph*; otherwise, it is called an *infinite graph*.

**Definition 2.47** In a graph $\mathcal{G} = (\mathcal{V}, \mathcal{E})$, the *degree* of a vertex, denoted by $d(v)$, is defined as the number of edges that are incident with $v$.

It is clear that the degree of an isolated vertex is zero and the degree of a pendant is 1. Consider the graph shown in Figure 2.1. The degrees of the vertices are $d(v_1) = 5$, $d(v_2) = 2$, $d(v_3) = 4$, $d(v_4) = 4$, $d(v_5) = 1$, and $d(v_6) = 0$.

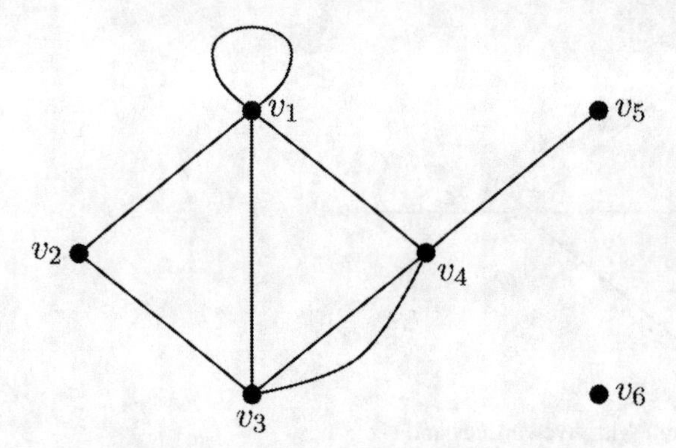


**Figure 2.1** A graph with six vertices and eight edges.

**Theorem 2.48** *For a graph $\mathcal{G} = (\mathcal{V}, \mathcal{E})$ with m edges and n vertices, $v_1, v_2, \ldots, v_n$, the sum of degrees of vertices is equal to 2m, that is*

$$\sum_{i=1}^{n} d(v_i) = 2m.$$

*Proof* The proof follows from the fact that each edge contributes two to the sum of degrees. □

A direct consequence of Theorem 2.20 is the following corollary.

**Corollary 2.3** *In a graph $\mathcal{G} = (\mathcal{V}, \mathcal{E})$, the number of vertices with odd degree is even.*

**Definition 2.49** A graph that has neither self-loops nor parallel edges is called a *simple graph*. A graph is said to be *regular* of degree $k$ if the degree of each vertex is $k$.

The graph shown in Figure 2.2 is a simple graph, while the graph shown in Figure 2.3 is a simple regular graph of degree 3.

**Definition 2.50** A simple graph is said to be *complete* if there exists an edge for every pair of vertices.

It is clear that, for a complete graph with $n$ vertices, each vertex has degree $n - 1$ and hence it is a regular graph of degree $n - 1$. Figure 2.4 shows a simple complete graph.

**Definition 2.51** A graph $\mathcal{G}_s = (\mathcal{V}_s, \mathcal{E}_s)$ is called a *subgraph* of graph $\mathcal{G} = (\mathcal{V}, \mathcal{E})$ if $\mathcal{V}_s$ is a subset of $\mathcal{V}$ and $\mathcal{E}_s$ is a subset of $\mathcal{E}$.

In many applications, self-loops and parallel multiple edges do not arise. For our purpose in this book, we mostly restrict our attention to simple graphs.

A simple finite graph $\mathcal{G} = (\mathcal{V}, \mathcal{E})$ with $n$ vertices and $m$ edges can be represented by two matrices. Let $\mathcal{V} = \{v_1, v_2, \ldots, v_n\}$ and $\mathcal{E} = \{e_1, e_2, \ldots, e_m\}$. The first matrix is an $n \times n$ matrix, denoted by $\mathbf{A}(\mathcal{G}) = [a_{i,j}]$, in which the rows and columns correspond to

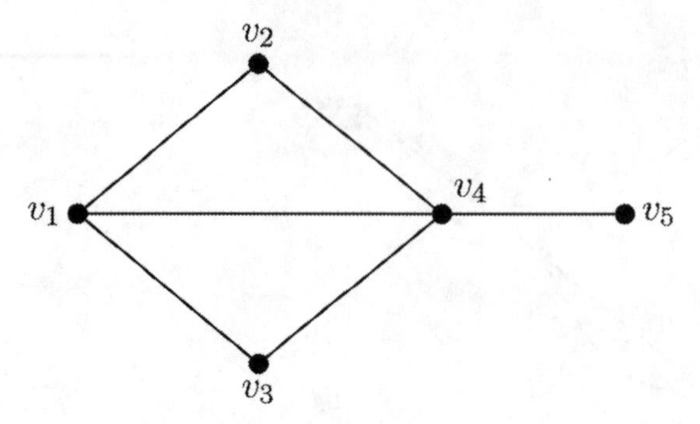


**Figure 2.2** A simple graph with five vertices and six edges.

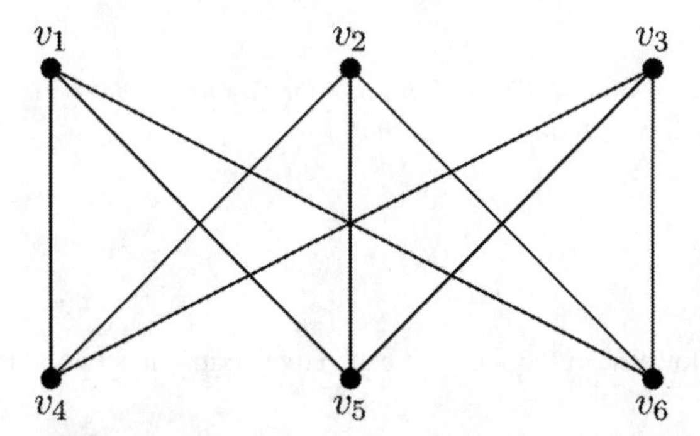


**Figure 2.3** A regular graph of degree 3.

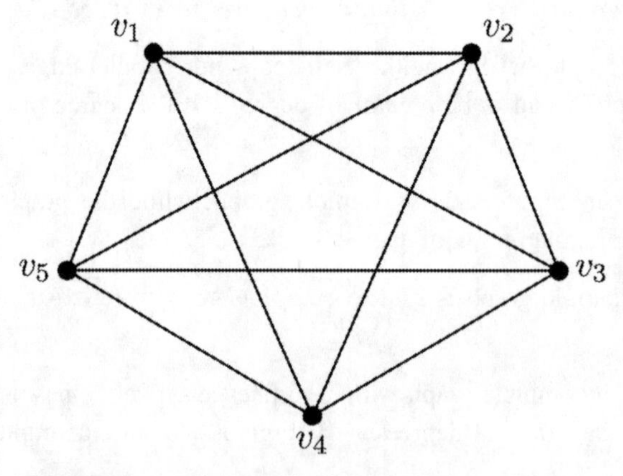


**Figure 2.4** A complete graph with five vertices.

the $n$ vertices of $\mathcal{G}$ and the entry $a_{i,j} = 1$ if $(v_i, v_j)$ is an edge in $\mathcal{E}$, otherwise $a_{i,j} = 0$. $\mathbf{A}(\mathcal{G})$ is called the *adjacency matrix* of $\mathcal{G}$. The second matrix is an $m \times n$ matrix, denoted by $\mathbf{M}(\mathcal{G}) = [m_{i,j}]$, in which the rows correspond to the $m$ edges of $\mathcal{G}$, the columns correspond to the $n$ vertices of $\mathcal{G}$, and the entry $m_{i,j} = 1$ if the $j$th vertex $v_j$ is an end vertex of the $i$th edge $e_i$, otherwise $m_{i,j} = 0$. $\mathbf{M}(\mathcal{G})$ is called the *incidence matrix* of $\mathcal{G}$.

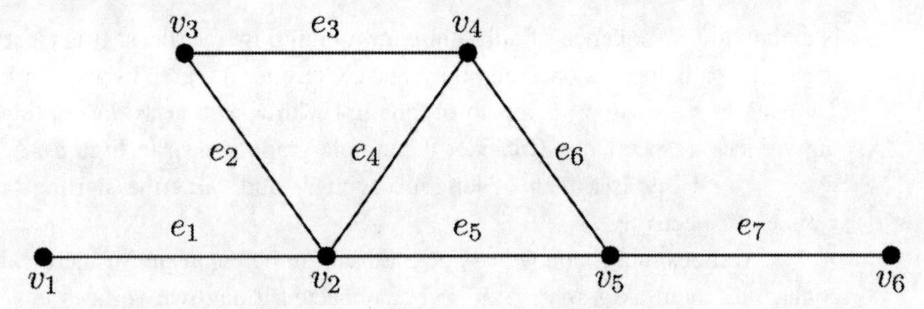


**Figure 2.5** A simple graph with six vertices and seven edges.

Consider the graph $\mathcal{G} = (\mathcal{V}, \mathcal{E})$ shown in Figure 2.5. The adjacency and incidence matrices of this graph are given below:

$$\mathbf{A}(\mathcal{G}) = \begin{array}{c} \\ v_1 \\ v_2 \\ v_3 \\ v_4 \\ v_5 \\ v_6 \end{array} \begin{array}{c} \begin{array}{cccccc} v_1 & v_2 & v_3 & v_4 & v_5 & v_6 \end{array} \\ \begin{bmatrix} 0 & 1 & 0 & 0 & 0 & 0 \\ 1 & 0 & 1 & 1 & 1 & 0 \\ 0 & 1 & 0 & 1 & 0 & 0 \\ 0 & 1 & 1 & 0 & 1 & 0 \\ 0 & 1 & 0 & 1 & 0 & 1 \\ 0 & 0 & 0 & 0 & 1 & 0 \end{bmatrix} \end{array},$$

$$\mathbf{M}(\mathcal{G}) = \begin{array}{c} \\ e_1 \\ e_2 \\ e_3 \\ e_4 \\ e_5 \\ e_6 \\ e_7 \end{array} \begin{array}{c} \begin{array}{cccccc} v_1 & v_2 & v_3 & v_4 & v_5 & v_6 \end{array} \\ \begin{bmatrix} 1 & 1 & 0 & 0 & 0 & 0 \\ 0 & 1 & 1 & 0 & 0 & 0 \\ 0 & 0 & 1 & 1 & 0 & 0 \\ 0 & 1 & 0 & 1 & 0 & 0 \\ 0 & 1 & 0 & 0 & 1 & 0 \\ 0 & 0 & 0 & 1 & 1 & 0 \\ 0 & 0 & 0 & 0 & 1 & 1 \end{bmatrix} \end{array}.$$

### 2.8.2 Paths and Cycles

A sequence of connected edges that form a continuous route plays an important role in graph theory and has many practical applications.

**Definition 2.52** A *path* in a graph $\mathcal{G} = (\mathcal{V}, \mathcal{E})$ is defined as an alternating sequence of vertices and edges, such that each edge is incident with the vertices preceding and following it, and no vertex appears more than once. The number of edges on a path is called the *length* of the path.

A simple way to represent a path is by arranging the vertices on the path in consecutive order as follows:

$$v_{i_0}, v_{i_1}, \ldots, v_{i_{k-1}}, v_{i_k},$$

in which any two consecutive vertices are adjacent (or connected by an edge), where $v_{i_0}$ and $v_{i_k}$ are called the *starting* and *ending* vertices of the path. Basically, a path is a continuous sequence of adjacent edges joining two vertices. It is clear that each edge on a path appears once and only once. Consider the graph shown in Figure 2.1. The path $v_1, v_2, v_3, v_4, v_5$ is a path of length 4 with $v_1$ and $v_5$ as the starting and ending vertices, respectively. Consider the simple graph shown in Figure 2.5. The path $v_1, v_2, v_3, v_4, v_5, v_6$ is a path of length 5 with $v_1$ and $v_6$ as the starting and ending vertices, respectively.

Two vertices in a graph $\mathcal{G} = (\mathcal{V}, \mathcal{E})$ are said to be *connected* if there exists a path connecting them. A graph is said to be connected if any two vertices in $\mathcal{G} = (\mathcal{V}, \mathcal{E})$ are connected by at least one path; otherwise, it is said to be disconnected. The graphs shown in Figures 2.3 to 2.5 are simple and connected graphs. In a connected graph $\mathcal{G} = (\mathcal{V}, \mathcal{E})$, the *distance* between two of its vertices $v_i$ and $v_j$, denoted by $d(v_i, v_j)$, is defined as the length of the shortest path (i.e., the number of edges on the shortest path) between $v_i$ and $v_j$. Consider the connected graph shown in Figure 2.5. There are three paths connecting $v_1$ and $v_6$: (1) $v_1, v_2, v_5, v_6$; (2) $v_1, v_2, v_4, v_5, v_6$; and (3) $v_1, v_2, v_3, v_4, v_5, v_6$. They have lengths 3, 4, and 5, respectively. The shortest path between $v_1$ and $v_6$ is $v_1, v_2, v_5, v_6$, which has length 3, and hence the distance between $v_1$ and $v_6$ is 3.

**Definition 2.53** The *eccentricity* $E(v_i)$ of a vertex $v_i$ in a graph $\mathcal{G} = (\mathcal{V}, \mathcal{E})$ is defined as the distance from $v_i$ to the vertex farthest from $v_i$ in $\mathcal{G} = (\mathcal{V}, \mathcal{E})$, that is

$$E(v_i) = \max\{d(v_i, v_j): v_j \in \mathcal{V}\}.$$

A vertex with minimum eccentricity in graph $\mathcal{G} = (\mathcal{V}, \mathcal{E})$ is called a *center*, denoted by $c(\mathcal{G})$, of $\mathcal{G} = (\mathcal{V}, \mathcal{E})$.

It is clear that a graph may have more than one center. Consider the graph shown in Figure 2.6. The graph is a simple connected graph with 14 vertices. The eccentricity of each vertex is shown next to the vertex. This graph has seven vertices with eccentricities of 6, three vertices with eccentricities of 5, two vertices with eccentricities of 4, and one vertex (the center) with eccentricity of 3.

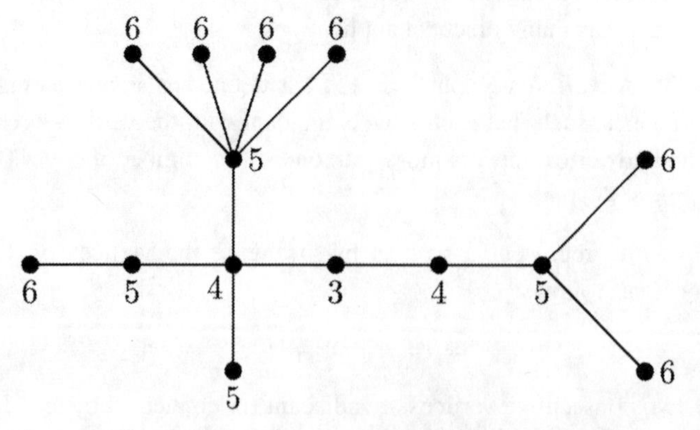


**Figure 2.6** Eccentricities of the vertices and center of a graph.

**Definition 2.54** The *diameter* of a connected graph $\mathcal{G} = (\mathcal{V}, \mathcal{E})$, denoted by dia $(\mathcal{G})$, is defined as the largest distance between two vertices in the graph, that is

$$\text{dia}(\mathcal{G}) = \max_{v_i, v_j \in \mathcal{V}} d(v_i, v_j).$$

The *radius* of a connected graph $\mathcal{G} = (\mathcal{V}, \mathcal{E})$, denoted by $r(\mathcal{G})$, is defined as the eccentricity of a center of the graph, that is

$$r(\mathcal{G}) = \min\{E(v_i): v_i \in \mathcal{V}\}.$$

It follows from the definitions of eccentricity of a vertex and diameter of a graph that the diameter of a connected graph equals the maximum of the vertex eccentricities, that is

$$\text{dia}(\mathcal{G}) = \max\{E(v_i): v_i \in \mathcal{V}\}.$$

Consider the connected graph shown in Figure 2.6. The diameter and radius are 6 and 3, respectively.

If two vertices $v_i$ and $v_j$ in a graph $\mathcal{G} = (\mathcal{V}, \mathcal{E})$ are connected, there may be more than one path between them. Among the paths connecting $v_i$ and $v_j$, the one (or ones) with the shortest length is (or are) called the *shortest path* (or paths) between $v_i$ and $v_j$. Finding shortest paths between connected vertices has many applications in communication networks, operations research, and other areas. There are several known algorithms for finding shortest paths between two vertices, which can be found in many texts on graphs.

**Definition 2.55** A closed path in a graph $\mathcal{G} = (\mathcal{V}, \mathcal{E})$ that begins and ends at the same vertex is called a *cycle*. The number of edges on the cycle is called the *length* of the cycle. The *girth* of a graph with cycles is defined as the length of its shortest cycle. A graph with no cycles has infinite girth.

Consider the graph shown in Figure 2.7. It has many cycles. The girth of this graph is 5.

**Definition 2.56** A graph with no cycle is said to be *acyclic*. A connected acyclic graph is called a *tree*. An edge in a tree is called a *branch*. A *pendant* in a tree is called an *end vertex*.

The graph shown in Figure 2.6 contains no cycle and hence is a tree. There are eight end vertices.

### 2.8.3 Bipartite Graphs

In this section we present a special type of graph, known as a *bipartite graph*, which is particularly useful in the design (or construction) of a class of Shannon-limit-approaching codes and provides a pictorial explanation of a suboptimal decoding algorithm for this class of codes.

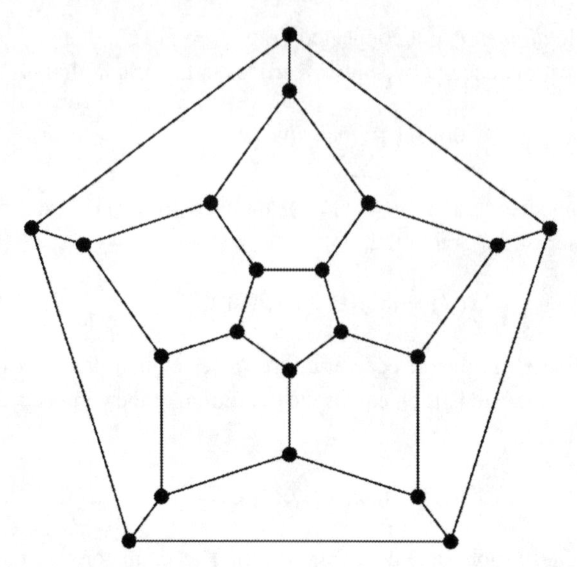

**Figure 2.7** A graph with girth 5.

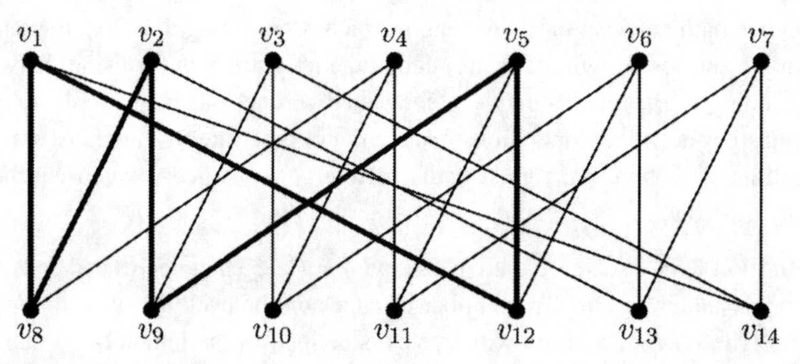


**Figure 2.8** A bipartite graph.

**Definition 2.57** A graph $\mathcal{G} = (\mathcal{V}, \mathcal{E})$ is called a *bipartite graph* if its vertex set $\mathcal{V}$ can be decomposed into two disjoint subsets $\mathcal{V}_1$ and $\mathcal{V}_2$ such that every edge in the edge set $\mathcal{E}$ joins a vertex in $\mathcal{V}_1$ and a vertex in $\mathcal{V}_2$, and no two vertices in either $\mathcal{V}_1$ or $\mathcal{V}_2$ are connected.

It is obvious that a bipartite graph contains no self-loop. Figure 2.8 shows a bipartite graph with $\mathcal{V}_1 = \{v_1, v_2, v_3, v_4, v_5, v_6, v_7\}$ and $\mathcal{V}_2 = \{v_8, v_9, v_{10}, v_{11}, v_{12}, v_{13}, v_{14}\}$. It is a simple connected graph. Since all the vertices of the graph have the same degree, 3, it is also a regular graph.

**Theorem 2.58** *If a bipartite graph $\mathcal{G} = (\mathcal{V}, \mathcal{E})$ contains cycles, then all cycles have even lengths.*

*Proof* Suppose we trace a cycle in $\mathcal{G}$ beginning from a vertex $v_i$ in $\mathcal{V}_1$. The first edge of this cycle must connect vertex $v_i$ to a vertex $v_j$ in $\mathcal{V}_2$. The second edge of the cycle must connect $v_j$ to a vertex $v_k$ in $\mathcal{V}_1$. If $v_k$ is identical to the starting vertex $v_i$, then we have completed the tracing of a cycle of length 2. If $v_k$ is not identical to the starting vertex $v_i$, we must trace the third edge of the cycle that connects $v_k$ to a vertex $v_l \in \mathcal{V}_2$. Then we must take the fourth edge of the cycle to return to a vertex in $\mathcal{V}_1$. If this vertex is not identical to the starting vertex $v_i$, we need to continue tracing the next edge of the cycle. This back and forth tracing process continues until the last edge of the cycle terminates at the starting vertex $v_i$. Leaving a vertex in $\mathcal{V}_1$ and returning to a vertex in $\mathcal{V}_1$ requires two edges each time. Therefore, the length of the cycle must be a multiple of 2, an even number. □

If a bipartite graph $\mathcal{G} = (\mathcal{V}, \mathcal{E})$ is a simple graph (i.e., there exist no multiple edges between a vertex in $\mathcal{V}_1$ and a vertex in $\mathcal{V}_2$), then it has no cycle of length 2. In this case, the shortest cycle in $\mathcal{G}$ has a length of at least 4. For a simple bipartite graph $\mathcal{G} = (\mathcal{V}, \mathcal{E})$, if no two vertices in $\mathcal{V}_1$ are jointly connected (or jointly adjacent) to two vertices in $\mathcal{V}_2$, or vice versa, then $\mathcal{G}$ contains no cycle of length 4. In this case, the shortest cycle in $\mathcal{G}$ has a length of at least 6. That is to say, the girth of $\mathcal{G}$ is at least 6. Consider the bipartite graph shown in Figure 2.8. Since it is a simple graph, it has no cycle of length 2. Since no two vertices in $\mathcal{V}_1$ are jointly connected to two vertices in $\mathcal{V}_2$, the graph has no cycle of length 4. It does have cycles of length 6; one such cycle is marked by heavy lines. Therefore, the girth of this bipartite graph is 6. How to construct bipartite graphs with large girth is a very challenging problem.

Consider a simple bipartite graph $\mathcal{G} = (\mathcal{V}, \mathcal{E})$ with disjoint subsets of vertices, $\mathcal{V}_1$ and $\mathcal{V}_2$. Let $m$ and $n$ be the numbers of vertices in $\mathcal{V}_1$ and $\mathcal{V}_2$, respectively. This bipartite graph can then be represented by an $m \times n$ adjacency matrix $\mathbf{A}(\mathcal{G}) = [a_{i,j}]$ whose rows correspond to the vertices of $\mathcal{V}_1$ and whose columns correspond to the vertices of $\mathcal{V}_2$, where $a_{i,j} = 1$ if and only if the $i$th vertex of $\mathcal{V}_1$ is adjacent to the $j$th vertex of $\mathcal{V}_2$, otherwise $a_{i,j} = 0$. The adjacency matrix of the bipartite graph $\mathcal{G}$ shown in Figure 2.8 is given below (top vertices are $\mathcal{V}_2$ and bottom vertices are $\mathcal{V}_1$):

$$\mathbf{A}(\mathcal{G}) = \begin{bmatrix} 1 & 1 & 0 & 1 & 0 & 0 & 0 \\ 0 & 1 & 1 & 0 & 1 & 0 & 0 \\ 0 & 0 & 1 & 1 & 0 & 1 & 0 \\ 0 & 0 & 0 & 1 & 1 & 0 & 1 \\ 1 & 0 & 0 & 0 & 1 & 1 & 0 \\ 0 & 1 & 0 & 0 & 0 & 1 & 1 \\ 1 & 0 & 1 & 0 & 0 & 0 & 1 \end{bmatrix}.$$

For a simple bipartite graph $\mathcal{G} = (\mathcal{V}, \mathcal{E})$, if no two vertices in $\mathcal{V}_1$ are jointly connected to two vertices in $\mathcal{V}_2$, then there are not four ones at the four corners of a rectangle in its adjacency matrix $\mathbf{A}(\mathcal{G})$. Conversely, if there are not four ones at the four corners of a rectangle in the adjacency matrix $\mathbf{A}(\mathcal{G})$ of a simple bipartite graph $\mathcal{G} = (\mathcal{V}, \mathcal{E})$, then no two vertices in $\mathcal{V}_1$ are jointly connected to two vertices in $\mathcal{V}_2$. This implies that, if there is no rectangle in the adjacency matrix of a bipartite graph with four ones at its four corners, then the graph has no cycle of length 4.

For additional reading on the topics in this chapter, the reader is referred to [7–15].

## Problems

**2.1** Construct the group under modulo-11 addition.

**2.2** Construct the group under modulo-11 multiplication.

**2.3** Let $m$ be a positive integer. If $m$ is not a prime, prove that the set $\{1, 2, \ldots, m-1\}$ does not form a group under modulo-$m$ multiplication.

**2.4** Find all the generators of the multiplicative group constructed in Problem 2.2.

**2.5** Starting with the subgroup of the additive group constructed in Example 2.5, give its cosets.

**2.6** Find a cyclic subgroup of the multiplicative group constructed using the prime integer 13 and give its cosets.

**2.7** Prove the following properties of a field $F$.

1. For every element $a$ in $F$, $a \cdot 0 = 0 \cdot a = 0$.
2. For any two nonzero elements $a$ and $b$ in $F$, $a \cdot b \neq 0$.
3. For two elements $a$ and $b$ in $F$, $a \cdot b = 0$ implies that either $a = 0$ or $b = 0$.
4. For $a \neq 0$, $a \cdot b = a \cdot c$ implies that $b = c$.
5. For any two elements $a$ and $b$ in $F$, $-(a \cdot b) = (-a) \cdot b = a \cdot (-b)$.

**2.8** Let $m$ be a positive integer. If $m$ is not a prime, prove that the set $\{0, 1, \ldots, m-1\}$ does not form a field under modulo-$m$ addition and multiplication.

**2.9** Consider the integer group $G = \{0, 1, 2, \ldots, 31\}$ under modulo-32 addition. Show that $H = \{0, 4, 8, 12, 16, 20, 24, 28\}$ forms a subgroup of $G$. Decompose $G$ into cosets with respect to $H$ (or modulo $H$).

**2.10** Prove that every finite field has at least one primitive element.

**2.11** Prove the following properties of a vector space $V$ over a field $F$.

1. Let 0 be the zero element of $F$. For any vector $\mathbf{v}$ in $V$, $0 \cdot \mathbf{v} = \mathbf{0}$.
2. For any element $a$ in $F$, $a \cdot \mathbf{0} = \mathbf{0}$.
3. For any element $a$ in $F$ and any vector $\mathbf{v}$ in $V$, $(-a) \cdot \mathbf{v} = a \cdot (-\mathbf{v}) = -(a \cdot \mathbf{v})$.

**2.12** Prove that two different bases of a vector space over a field have the same number of linearly independent vectors.

**2.13** Prove that the inner product of two $n$-tuples over GF($q$) has the following properties.

1. $\mathbf{u} \cdot \mathbf{v} = \mathbf{v} \cdot \mathbf{u}$ (commutative law).
2. $\mathbf{u} \cdot (\mathbf{v} + \mathbf{w}) = \mathbf{u} \cdot \mathbf{v} + \mathbf{u} \cdot \mathbf{w}$ (distributive law).
3. For any element $c$ in GF($q$), $(c\mathbf{u}) \cdot \mathbf{v} = c(\mathbf{u} \cdot \mathbf{v})$ (associative law).
4. $\mathbf{0} \cdot \mathbf{u} = \mathbf{0}$.

**2.14** Prove that the addition and multiplication of two polynomials over GF($q$) defined by (2.17), (2.18), and (2.19) satisfy the associative, commutative, and distributive laws.

**2.15** Let $m$ be a positive integer. Prove that the set $\{0, 1, \ldots, m - 1\}$ forms a ring under modulo-$m$ addition and multiplication.

**2.16** Let $a(X) = 3X^2 + 1$ and $b(X) = X^6 + 3X + 2$ be two polynomials over the prime field GF(5). Divide $b(X)$ by $a(X)$ and find the quotient and remainders of the division.

**2.17** Let $a(X) = 3X^3 + X + 2$ be a polynomial over the prime field GF(5). Determine whether or not $a(X)$ is irreducible over GF(5).

**2.18** Let $f(X)$ be a polynomial of degree $n$ over GF($q$). The reciprocal of $f(X)$ is defined as $f^*(X) = X^n f(1/X)$.

(a) Prove that $f^*(X)$ is irreducible over GF($q$) if and only if $f(X)$ is irreducible over GF($q$).

(b) Prove that $f^*(X)$ is primitive if and only if $f(X)$ is primitive.

**2.19** Prove that the extension field GF($p^m$) of the prime field GF($p$) is an $m$-dimensional vector space over GF($p$).

**2.20** Prove Theorem 2.12.

**2.21** Let $\beta$ be a nonzero element of GF($q^m$). Let $e$ be the smallest non-negative integer such that $\beta^{q^e} = \beta$. Prove that $e$ divides $m$.

**2.22** Prove Theorem 2.19.

**2.23** Consider the Galois field GF($2^5$) given by Table 2.10. Find the minimum polynomials of $\alpha^5$ and $\alpha^7$.

**2.24** If $q - 1$ is a prime, prove that every nonzero element of GF($q$) except the unit element 1 is primitive.

**2.25** Consider the Galois field GF($2^4$) given by Table 2.14. Let $\beta = \alpha^5$. Then the set $\{0, 1, \beta^1, \beta^2\}$ forms a subfield GF($2^2$) of GF($2^4$). Regard GF($2^4$) as the two-dimensional Euclidean geometry EG($2, 2^2$) over GF($2, 2^2$). Find the lines that intersect at point $\alpha^3$.

**2.26** Construct the Galois field GF($2^6$) based on the primitive polynomial $p(X) = 1 + X + X^6$ over GF(2). Let $\alpha$ be a primitive element of GF($2^6$) and $\beta = \alpha^{21}$. Then $\{0, 1, \beta, \beta^2\}$ forms a subfield GF($2^2$) of GF($2^6$). Regard GF($2^6$) as the three-dimensional Euclidean geometry EG($3, 2^2$) over GF($2^2$). Find all the 2-flats that intersect on the 1-flat, $\{\alpha^{63} + \delta\alpha\}$, where $\delta \in$ GF($2^2$).

**2.27** Let $\mathcal{L}$ be a line in the $m$-dimensional Euclidean geometry EG($m, q$) over GF($q$), not passing through the origin. Let $\alpha$ be a primitive element of GF($q^m$). Prove that $\mathcal{L}, \alpha\mathcal{L}, \ldots, \alpha^{q^m-2}\mathcal{L}$ form $q^m - 1$ different lines in EG($m, q$), not passing through the origin.

**2.28** Find the cyclic classes of lines in EG($3, 2^2$) constructed in Problem 2.26.

**2.29** Find the points of the two-dimensional projective geometry PG($2, 2^2$) over GF($2^2$). Choose a point in PG($2, 2^2$) and give the lines that intersect at this point.

**2.30** For odd $m$, prove that there are $(q^{m+1} - 1)/(q^2 - 1)$ nonprimitive lines.

**2.31** Construct the finite field GF(8) from the primitive polynomial $X^3 + X + 1$, providing the addition and multiplication tables in terms of powers of primitive element $\alpha$ (and 0). Use those tables to solve this system of linear equations:

$$\alpha^6 X + \alpha^4 Y = \alpha^4,$$
$$\alpha^2 X + \alpha Y = \alpha^3.$$

**2.32** Consider the primitive polynomial $p(X) = X^5 + X^2 + 1$.

(a) Show that the vector representations of the nonzero elements of the field GF(32) generated by $p(X)$ can be computed from the recursion $\mathbf{v}_{n+1} = \mathbf{A}\,\mathbf{v}_n$ with initialization $\mathbf{v}_1 = [0\,0\,0\,0\,1]^{\mathrm{T}}$, where

$$\mathbf{A} = \begin{bmatrix} 0 & 1 & 0 & 0 & 0 \\ 0 & 0 & 1 & 0 & 0 \\ 1 & 0 & 0 & 1 & 0 \\ 0 & 0 & 0 & 0 & 1 \\ 1 & 0 & 0 & 0 & 0 \end{bmatrix}$$

and all operations are over GF(2).

(b) Draw the feedback shift register that implements this recursion.

(c) Draw the feedback shift register that implements the recursion $\mathbf{v}_{n+1} = \mathbf{A}^{\mathrm{T}}\mathbf{v}_n$. (Note that $\mathbf{A}$ is transposed here, unlike above.) Observe that this is the standard implementation of the pseudo-random sequence (also called the *maximum-length sequence*) generated by the primitive polynomial $p(X)$.

(d) Discuss why this generalizes to any primitive polynomial $p(X) = p_m X^m + \cdots + p_2 X^2 + p_1 X + p_0$ and give the general form for the matrix $\mathbf{A}$.

**2.33** (See the previous problem.) Write a computer program that checks if a binary polynomial of degree $m$ is primitive, for any $m \leq 20$. One way to do this is to check the length of period of the sequence of nonzero states in the linear feedback shift register implied by the polynomial in question. (The polynomial is not primitive if this length is less than $2^m - 1$.) Observe that this program can be extended to generate all primitive polynomials of degree $m$.

## References

[1] R. Lidl and H. Niederreiter, *Introduction to Finite Fields and Their Applications*, revised ed., Cambridge, Cambridge University Press, 1994.

[2] I. F. Blake and R. C. Mullin, *The Mathematical Theory of Coding*, New York, Academic Press, 1975.

[3] R. D. Carmichael, *Introduction to the Theory of Groups of Finite Orders*, Boston, MA, Ginn & Co., 1937.

[4] H. B. Mann, *Analysis and Design of Experiments*, New York, Dover, 1949.

[5] H. Tang, J. Xu, S. Lin, and K. Abdel-Ghaffar, "Codes on finite geometries," *IEEE Transactions on Information Theory*, vol. 51, no. 2, pp. 572–596, February 2005.

[6] J. E. Maxfield and M. W. Maxfield, *Abstract Algebra and Solution by Radicals*, New York, Dover, 1992.
[7] G. Birkhoff and S. MacLane, *A Survey of Modern Algebra*, New York, Macmillan, 1953.
[8] A. Clark, *Elements of Abstract Algebra*, New York, Dover, 1984.
[9] R. A. Dean, *Classical Abstract Algebra*, New York, Harper & Row, 1990.
[10] N. Deo, *Graph Theory with Applications to Engineering and Computer Science*, Englewood Cliffs, NJ, Prentice-Hall, 1974.
[11] D. S. Dummit and R. M. Foote, *Abstract Algebra*, 3rd ed., New York, Wiley, 2004.
[12] T. W. Hungerford, *Abstract Algebra: An Introduction*, 2nd ed., New York, Saunders College Publishing, 1997.
[13] S. Lang, *Algebra*, 2nd ed., Reading, MA, Addison-Wesley, 1984.
[14] B. L. Van der Waerden, *Modern Algebra*, vols. 1 and 2, New York, Ungar, 1949.
[15] D. B. West, *Introduction to Graph Theory*, 2nd ed., Upper Saddle River, NJ, Prentice-Hall, 2001.

## Appendix A: Primitive Polynomials

The reciprocal polynomial of each of the primitive polynomials in the following tables is also primitive (see Problem 2.19). For example, the reciprocal polynomial $X^4p(X^{-1}) = X^4 + X + 1$ of the primitive polynomial $p(X) = X^4 + X^3 + 1$ is primitive. In these tables, $m$ represents the polynomial degree.

Characteristic 2

| $m$ | Polynomial | $m$ | Polynomial |
|---|---|---|---|
| 2 | $X^2 + X + 1$ | 14 | $X^{14} + X^{13} + X^{12} + X^2 + 1$ |
| 3 | $X^3 + X^2 + 1$ | 15 | $X^{15} + X^{14} + 1$ |
| 4 | $X^4 + X^3 + 1$ | 16 | $X^{16} + X^{15} + X^{13} + X^4 + 1$ |
| 5 | $X^5 + X^3 + 1$ | 17 | $X^{17} + X^{14} + 1$ |
| 6 | $X^6 + X^5 + 1$ | 18 | $X^{18} + X^{11} + 1$ |
| 7 | $X^7 + X^6 + 1$ | 19 | $X^{19} + X^{18} + X^{17} + X^{14} + 1$ |
| 8 | $X^8 + X^7 + X^6 + X + 1$ | 20 | $X^{20} + X^{17} + 1$ |
| 9 | $X^9 + X^5 + 1$ | 21 | $X^{21} + X^{19} + 1$ |
| 10 | $X^{10} + X^7 + 1$ | 22 | $X^{22} + X^{21} + 1$ |
| 11 | $X^{11} + X^9 + 1$ | 23 | $X^{23} + X^{18} + 1$ |
| 12 | $X^{12} + X^{11} + X^{10} + X^4 + 1$ | 24 | $X^{24} + X^{23} + X^{22} + X^{17} + 1$ |
| 13 | $X^{13} + X^{12} + X^{11} + X^8 + 1$ | 25 | $X^{25} + X^3 + 1$ |

Characteristic 3

| $m$ | Polynomial | $m$ | Polynomial |
|---|---|---|---|
| 2 | $X^2 + X + 2$ | 7 | $X^7 + X^6 + X^4 + 1$ |
| 3 | $X^3 + 2X^2 + 1$ | 8 | $X^8 + X^5 + 2$ |
| 4 | $X^4 + X^3 + 2$ | 9 | $X^9 + X^7 + X^5 + 1$ |
| 5 | $X^5 + X^4 + X^2 + 1$ | 10 | $X^{10} + X^9 + X^7 + 2$ |
| 6 | $X^6 + X^5 + 2$ | 11 | $X^{11} + X^{10} + X^4 + 1$ |

Characteristic 5

| $m$ | Polynomial | $m$ | Polynomial |
|---|---|---|---|
| 2 | $X^2 + X + 2$ | 6 | $X^6 + X^5 + 2$ |
| 3 | $X^3 + X^2 + 2$ | 7 | $X^7 + X^6 + 2$ |
| 4 | $X^4 + X^3 + X + 3$ | 8 | $X^8 + X^5 + X^3 + 3$ |
| 5 | $X^5 + X^2 + 2$ | 9 | $X^9 + X^7 + X^6 + 3$ |

Characteristic 7

| $m$ | Polynomial | $m$ | Polynomial |
|---|---|---|---|
| 2 | $X^2 + X + 3$ | 5 | $X^5 + X^4 + 4$ |
| 3 | $X^3 + X^2 + X + 2$ | 6 | $X^6 + X^5 + X^4 + 3$ |
| 4 | $X^4 + X^3 + X^2 + 3$ | 7 | $X^7 + X^5 + 4$ |

Characteristic 11

| $m$ | Polynomial | $m$ | Polynomial |
|---|---|---|---|
| 2 | $X^2 + X + 7$ | 4 | $X^4 + X^3 + 8$ |
| 3 | $X^3 + X^2 + 3$ | 5 | $X^5 + X^4 + X^3 + 3$ |

# 3 Linear Block Codes

There are two structurally different types of codes, *block* and *convolutional* codes. Both types of codes have been widely used for error control in communication and/or storage systems. Block codes can be divided into two categories, linear and nonlinear block codes. Nonlinear block codes are almost never used in practical applications and not investigated much. This chapter gives an introduction to linear block codes. The coverage of this chapter includes (1) fundamental concepts and structures of linear block codes; (2) specifications of these codes in terms of their generator and parity-check matrices; (3) their error-correction capabilities; (4) several important classes of linear block codes; and (5) encoding of two special classes of linear block codes, namely cyclic and quasi-cyclic codes. In our presentation, proofs and derivations are often not provided; mostly descriptions and constructions of codes are given. We will begin our introduction with linear block codes with symbols from the binary field GF(2). Linear block codes over nonbinary fields will be given at the end of the chapter.

There are many excellent texts on the subject of error-control coding theory [1–15], which have extensive coverage of linear block codes. For in-depth study of linear block codes, readers are referred to these texts.

## 3.1 Introduction to Linear Block Codes

We assume that the output of an information source is a continuous sequence of binary symbols over GF(2), called an *information sequence*. The binary symbols in an information sequence are commonly called information bits, where a "bit" stands for a binary digit. In block coding, an information sequence is segmented into message blocks of fixed length; each message block consists of $k$ information bits. There are $2^k$ distinct messages. At the channel encoder, each input message $\mathbf{u} = (u_0, u_1, \ldots, u_{k-1})$ of $k$ information bits is encoded into a longer binary sequence $\mathbf{v} = (v_0, v_1, \ldots, v_{n-1})$ of $n$ binary digits with $n > k$, according to certain encoding rules. This longer binary sequence $\mathbf{v}$ is called the *codeword* of the message $\mathbf{u}$. The binary digits in a codeword are called code bits. Since there are $2^k$ distinct messages, there are $2^k$ codewords, one for each distinct message. This set of $2^k$ codewords is said to form an $(n, k)$ block code. For a block code to be useful, the $2^k$ codewords for the $2^k$ distinct messages must be distinct. The $n - k$ bits added to each input message by the channel encoder are called *redundant bits*. These redundant bits carry no new information and their main function is to provide the code with the capability of *detecting* and *correcting* transmission errors caused by the channel noise or interferences. How to form these redundant bits such that an $(n, k)$ block has good error-correcting capability is a major concern in designing the channel encoder. The ratio $R = k/n$ is called the *code rate*, which is interpreted as the average number of information bits carried by each code bit.

For a block code of length $n$ with $2^k$ codewords, unless it has certain special structural properties, the encoding and decoding apparatus would be prohibitively complex for large $k$ since the encoder has to store $2^k$ codewords of length $n$ in a dictionary and the decoder has to perform a table (with $2^n$ entries) look-up to determine the transmitted codeword. Therefore, we must restrict our attention to block codes that can be implemented in a practical manner. A desirable structure for a block code is *linearity*.

**Definition 3.1** A binary block code of length $n$ with $2^k$ codewords is called an $(n, k)$ *linear block code* if and only if its $2^k$ codewords form a $k$-dimensional subspace of the vector space $V$ of $n$-tuples over GF(2).

### 3.1.1 Generator and Parity-Check Matrices

Since a binary $(n, k)$ linear block code $\mathcal{C}$ is a $k$-dimensional subspace of the vector space of all the $n$-tuples over GF(2), there exist $k$ *linearly independent codewords* $\mathbf{g}_0, \mathbf{g}_1, \ldots, \mathbf{g}_{k-1}$ in $\mathcal{C}$ such that every codeword $\mathbf{v}$ in $\mathcal{C}$ is a *linear combination* of these $k$ linearly independent codewords. These $k$ linearly independent codewords in $\mathcal{C}$ form a *basis* $\mathcal{B}_c$ of $\mathcal{C}$. Using this basis, encoding can be done as follows. Let $\mathbf{u} = (u_0, u_1, \ldots, u_{k-1})$ be the message to be encoded. The codeword $\mathbf{v} = (v_0, v_1, \ldots, v_{n-1})$ for this message is given by the following linear combination of $\mathbf{g}_0, \mathbf{g}_1, \ldots, \mathbf{g}_{k-1}$, with the $k$ message bits of $\mathbf{u}$ as the coefficients:

$$\mathbf{v} = u_0\mathbf{g}_0 + u_1\mathbf{g}_1 + \cdots + u_{k-1}\mathbf{g}_{k-1}. \tag{3.1}$$

We may arrange the $k$ linearly independent codewords $\mathbf{g}_0, \mathbf{g}_1, \ldots, \mathbf{g}_{k-1}$ of $\mathcal{C}$ as rows of a $k \times n$ matrix over GF(2) as follows (where our vectors are rows, following convention in the coding field):

$$\mathbf{G} = \begin{bmatrix} \mathbf{g}_0 \\ \mathbf{g}_1 \\ \vdots \\ \mathbf{g}_{k-1} \end{bmatrix} = \begin{bmatrix} g_{0,0} & g_{0,1} & \cdots & g_{0,n-1} \\ g_{1,0} & g_{1,1} & \cdots & g_{1,n-1} \\ \vdots & \vdots & \ddots & \vdots \\ g_{k-1,0} & g_{k-1,1} & \cdots & g_{k-1,n-1} \end{bmatrix}. \tag{3.2}$$

Then the codeword $\mathbf{v} = (v_0, v_1, \ldots, v_{n-1})$ for message $\mathbf{u} = (u_0, u_1, \ldots, u_{k-1})$ given by (3.1) can be expressed as the matrix product of $\mathbf{u}$ and $\mathbf{G}$ as follows:

$$\mathbf{v} = \mathbf{u} \cdot \mathbf{G}. \tag{3.3}$$

Therefore, the codeword $\mathbf{v}$ for a message $\mathbf{u}$ is simply a linear combination of the rows of matrix $\mathbf{G}$ with the information bits in the message $\mathbf{u}$ as the coefficients. $\mathbf{G}$ is called a *generator matrix* of the $(n, k)$ linear block code $\mathcal{C}$. Since $\mathcal{C}$ is spanned by the rows of $\mathbf{G}$, it is called the *row space* of $\mathbf{G}$. In general, an $(n, k)$ linear block code has more than one basis. Consequently, a generator matrix of a given $(n, k)$ linear block code is not unique. Any choice of a basis of $\mathcal{C}$ gives a generator matrix of $\mathcal{C}$. Obviously, the rank of a generator matrix of a linear block code $\mathcal{C}$ is equal to the dimension of $\mathcal{C}$.

Since a binary $(n, k)$ linear block code $\mathcal{C}$ is a $k$-dimensional subspace of the vector space $V$ of all the $n$-tuples over GF(2), its *null* (or *dual*) space, denoted by $\mathcal{C}_d$, is an $(n-k)$-dimensional subspace of $V$ that is given by the following set of $n$-tuples in $V$:

$$\mathcal{C}_d = \{\mathbf{w} \in V{:}\langle \mathbf{w}, \mathbf{v} \rangle = 0 \text{ for all } \mathbf{v} \in \mathcal{C}\}, \tag{3.4}$$

where $\langle \mathbf{w}, \mathbf{v} \rangle$ denotes the inner product of $\mathbf{w}$ and $\mathbf{v}$ (see Section 2.4). $\mathcal{C}_d$ may be regarded as a binary $(n, n-k)$ linear block code and is called the *dual code* of $\mathcal{C}$. Let $\mathcal{B}_d$ be a basis of $\mathcal{C}_d$. Then $\mathcal{B}_d$ consists of $n-k$ linearly independent codewords in $\mathcal{C}_d$. Let $\mathbf{h}_0, \mathbf{h}_1, \ldots, \mathbf{h}_{n-k-1}$ be the $n-k$ linearly independent codewords in $\mathcal{B}_d$. Then every codeword in $\mathcal{C}_d$ is a linear combination of these $n-k$ linearly independent codewords in $\mathcal{B}_d$. Form the following $(n-k) \times n$ matrix over GF(2):

$$\mathbf{H} = \begin{bmatrix} \mathbf{h}_0 \\ \mathbf{h}_1 \\ \vdots \\ \mathbf{h}_{n-k-1} \end{bmatrix} = \begin{bmatrix} h_{0,0} & h_{0,1} & \cdots & h_{0,n-1} \\ h_{1,0} & h_{1,1} & \cdots & h_{1,n-1} \\ \vdots & \vdots & \ddots & \vdots \\ h_{n-k-1,0} & h_{n-k-1,1} & \cdots & h_{n-k-1,n-1} \end{bmatrix}. \tag{3.5}$$

Then $\mathbf{H}$ is a generator matrix of the dual code $\mathcal{C}_d$ of the binary $(n, k)$ linear block code $\mathcal{C}$. It follows from (3.2), (3.4), and (3.5) that $\mathbf{G} \cdot \mathbf{H}^{\mathrm{T}} = \mathbf{O}$, where $\mathbf{O}$ is a $k \times (n-k)$ zero matrix. Furthermore, $\mathcal{C}$ is also uniquely specified by the $\mathbf{H}$ matrix as follows: a binary $n$-tuple $\mathbf{v} \in V$ is a codeword in $\mathcal{C}$ if and only if $\mathbf{v} \cdot \mathbf{H}^{\mathrm{T}} = \mathbf{0}$ (the all-zero $(n-k)$-tuple), that is

$$\mathcal{C} = \{\mathbf{v} \in V{:}\, \mathbf{v} \cdot \mathbf{H}^{\mathrm{T}} = \mathbf{0}\}. \tag{3.6}$$

$\mathbf{H}$ is called a *parity-check* matrix of $\mathcal{C}$ and $\mathcal{C}$ is said to be the *null space* of $\mathbf{H}$. Therefore, a linear block code is uniquely specified by two matrices, a generator matrix and a parity-check matrix. Conventionally, encoding of a linear block code is based on a generator matrix of the code using (3.3) and decoding is based on a parity-check matrix of the code. Many classes of well-known linear block codes are constructed in terms of their parity-check matrices, as will be seen later in this chapter and other chapters. A parity-check matrix $\mathbf{H}$ of a linear block code is said to be a *full-rank* matrix if its rank is equal to the number of rows of $\mathbf{H}$. However, in many cases a parity-check matrix of an $(n, k)$ linear block code is not given as a full-rank matrix, that is, the number of its rows is greater than its row rank, $n-k$. In this case, some rows of the given parity-check matrix $\mathbf{H}$ are linear combinations of a set of $n-k$ linearly independent rows. These extra rows are called *redundant rows*. The LDPC codes constructed in Chapters 11 to 15 are specified by parity-check matrices that are in general not full-rank matrices.

It is also desirable to require a codeword to have the format shown in Figure 3.1, where a codeword is divided into parts, namely the message part and the redundant check part. The message part consists of the $k$ unaltered information digits and the redundant check part consists of $n-k$ parity-check digits. A linear block code with

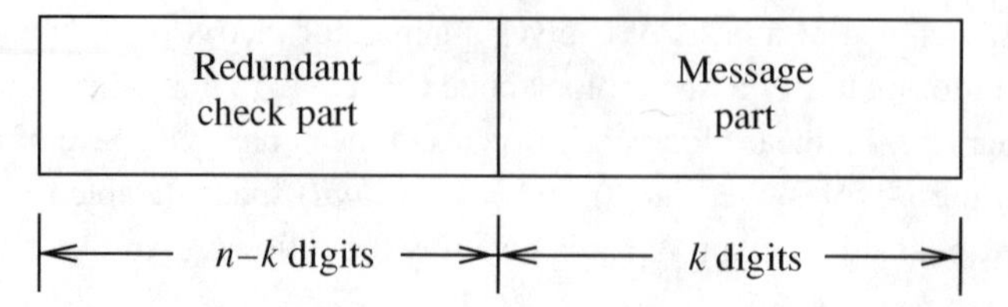


**Figure 3.1** Systematic format of a codeword. The bit ordering goes as [check bits | message bits] $= [b_0, b_1, \ldots, b_{n-k-1}, m_0, m_1, \ldots, m_{k-1}]$.

this structure is referred to as a *linear systematic code*. A linear systematic $(n, k)$ block code is completely specified by a $k \times n$ generator matrix of the following form:

$$\mathbf{G} = \begin{bmatrix} \mathbf{g}_0 \\ \mathbf{g}_1 \\ \vdots \\ \mathbf{g}_{k-1} \end{bmatrix} = \Bigg[ \underbrace{\begin{matrix} p_{0,0} & p_{0,1} & \cdots & p_{0,n-k-1} \\ p_{1,0} & p_{1,1} & \cdots & p_{1,n-k-1} \\ \vdots & \vdots & \ddots & \vdots \\ p_{k-1,0} & p_{k-1,1} & \cdots & p_{k-1,n-k-1} \end{matrix}}_{\mathbf{P}\text{ matrix}} \quad \underbrace{\begin{matrix} 1 & 0 & \cdots & 0 \\ 0 & 1 & \cdots & 0 \\ \vdots & \vdots & \ddots & \vdots \\ 0 & 0 & \cdots & 1 \end{matrix}}_{k \times k \text{ identity matrix } \mathbf{I}_k} \Bigg]. \tag{3.7}$$

No performance loss is suffered by putting the code in systematic form because doing so involves elementary row (and possibly column) operations. (See also the concept of combinatorial equivalence below.)

The generator matrix **G** consists of two submatrices, a $k \times (n-k)$ submatrix **P** on the left with entries over GF(2) and a $k \times k$ identity matrix $\mathbf{I}_k$ on the right. For simplicity, we write $\mathbf{G} = [\mathbf{P} \quad \mathbf{I}_k]$. Let $\mathbf{u} = (u_0, u_1, \ldots, u_{k-1})$ be the message to be encoded. Taking the linear combination of the rows of **G** given by (3.7) with the information bits in **u** as coefficients, we obtain the following corresponding codeword:

$$\begin{aligned} \mathbf{v} &= (v_0, v_1, \ldots, v_{n-k-1}, v_{n-k}, \ldots, v_{n-1}) \\ &= (u_0, u_1, \ldots, u_{k-1}) \cdot \mathbf{G}, \end{aligned} \tag{3.8}$$

where, for $l = 0, 1, \ldots, k-1$, the $n$ code bits are given by

$$v_{n-k+l} = u_l, \tag{3.9}$$

and, for $j = 0, 1, \ldots, n-k-1$,

$$v_j = u_0 p_{0,j} + u_1 p_{1,j} + \cdots + u_{k-1} p_{k-1,j}. \tag{3.10}$$

Expression (3.9) shows that the rightmost $k$ code bits of codeword **v** are identical to $k$ information bits $u_0, u_1, \ldots, u_{k-1}$ to be encoded, and (3.10) shows that the leftmost $n-k$ code bits of **v** are linear sums of information bits. These $n-k$ code bits given by linear sums of information bits are called *parity-check bits* (or simply parity bits) and they are completely specified by the $n-k$ columns of the **P** submatrix of **G** given by (3.7). The $n-k$ equations given by (3.10) are called *parity-check equations* of the code and the **P** submatrix of **G** is called the parity submatrix of **G**. A generator matrix in the form given by (3.7) is said to be in systematic form.

Given a $k \times n$ generator matrix $\mathbf{G}'$ of an $(n, k)$ linear block code $C'$ not in systematic form, a generator matrix **G** in the systematic form of (3.7) can always be obtained by performing elementary operations on the rows of $\mathbf{G}'$ and then possibly taking column permutations. The $k \times n$ matrix $\mathbf{G}$ is called a *combinatorially equivalent matrix* of $\mathbf{G}'$. The systematic $(n, k)$ linear block code $\mathcal{C}$ generated by $\mathbf{G}$ is called a *combinatorially equivalent code* of $\mathcal{C}'$. $\mathcal{C}$ and $\mathcal{C}'$ are possibly different in the arrangement (or order) of code bits in their codewords, that is, a codeword in $\mathcal{C}$ is obtained by a fixed permutation of the positions of the code bits in a codeword of $\mathcal{C}'$, and vice versa. Two combinatorially equivalent $(n, k)$ linear block codes give the same error performance.

If a generator matrix of an $(n, k)$ linear block code $\mathcal{C}$ is given in the systematic form of (3.7), then its corresponding parity-check matrix in systematic form is given below:

$$\begin{aligned}\mathbf{H} &= \begin{bmatrix}\mathbf{I}_{n-k} & \mathbf{P}^{\mathrm{T}}\end{bmatrix} \\ &= \begin{bmatrix} 1 & 0 & \cdots & 0 & p_{0,0} & p_{1,0} & \cdots & p_{k-1,0} \\ 0 & 1 & \cdots & 0 & p_{0,1} & p_{1,1} & \cdots & p_{k-1,1} \\ \vdots & \vdots & \ddots & \vdots & \vdots & \vdots & \ddots & \vdots \\ 0 & 0 & \cdots & 1 & p_{0,n-k-1} & p_{1,n-k-1} & \cdots & p_{k-1,n-k-1} \end{bmatrix}.\end{aligned} \tag{3.11}$$

Observe that $\mathbf{G} \cdot \mathbf{H}^{\mathrm{T}} = \mathbf{O}$, as follows from $\mathbf{G} = [\mathbf{P} \;\; \mathbf{I}_k]$ and $\mathbf{H} = \begin{bmatrix}\mathbf{I}_{n-k} & \mathbf{P}^{\mathrm{T}}\end{bmatrix}$.

### 3.1.2 Error Detection with Linear Block Codes

Consider an $(n, k)$ linear block code $\mathcal{C}$ with an $(n - k) \times n$ parity-check matrix $\mathbf{H}$. Suppose a codeword $\mathbf{v} = (v_0, v_1, \ldots, v_{n-1})$ in $\mathcal{C}$ is transmitted over a BSC (or any binary-input, binary-output channel). Let $\mathbf{r} = (r_0, r_1, \ldots, r_{n-1})$ be the corresponding hard-decision received vector ($n$-tuple or sequence) at the input of the channel decoder. Because of the channel noise, the received vector $\mathbf{r}$ and the transmitted codeword $\mathbf{v}$ may differ in some places. Define the following vector sum of $\mathbf{r}$ and $\mathbf{v}$:

$$\begin{aligned}\mathbf{e} &= \mathbf{r} + \mathbf{v} \\ &= (e_0, e_1, \ldots, e_{n-1}) \\ &= (r_0 + v_0, r_1 + v_1, \ldots, r_{n-1} + v_{n-1}),\end{aligned} \tag{3.12}$$

where $e_j = r_j + v_j$ for $0 \leq j < n$ and the addition $+$ is modulo-2 addition. Then $e_j = 1$ for $r_j \neq v_j$ and $e_j = 0$ for $r_j = v_j$. Therefore, the positions of the 1-components in the $n$-tuple $\mathbf{e}$ are the places where $\mathbf{r}$ and $\mathbf{v}$ differ. At these places, transmission errors have occurred. Since $\mathbf{e}$ displays the pattern of transmission errors in $\mathbf{r}$, we call $\mathbf{e}$ the *error pattern* (or *vector*) during the transmission of the codeword $\mathbf{v}$. The 1-components in $\mathbf{e}$ are called *transmission errors* caused by the channel noise. For a BSC, an error can occur at any place with the same probability $p$ over the span of $n$ places (the length of the code). There are $2^n - 1$ possible nonzero error patterns.

It follows from (3.12) that we can express the received vector $\mathbf{r}$ as

$$\mathbf{r} = \mathbf{v} + \mathbf{e}. \tag{3.13}$$

At the receiver, neither the transmitted codeword $\mathbf{v}$ nor the error pattern $\mathbf{e}$ is known. Upon receiving $\mathbf{r}$, the decoder must first determine whether there are transmission errors in $\mathbf{r}$. If the presence of errors is detected, then the decoder must estimate the error pattern **e** on the basis of the code $\mathcal{C}$ and the channel information provided to the decoder. Let $\mathbf{e}^*$ denote the estimated error pattern. Then the estimated transmitted codeword is given by $\mathbf{v}^* = \mathbf{r} + \mathbf{e}^*$, noting that this subtracts the error estimate from the received word (because addition and subtraction are identical, modulo 2).

To check whether a received vector **r** contains transmission errors, we compute the following $(n-k)$-tuple over GF(2):

$$\begin{aligned}\mathbf{s} &= (s_0, s_1, \ldots, s_{n-k-1}) \\ &= \mathbf{r} \cdot \mathbf{H}^{\mathrm{T}}, \end{aligned} \quad (3.14)$$

called a *syndrome* or syndrome vector, for it represents the collection of bit errors that "ails" the received vector. Note that **r** is an $n$-tuple in the vector space $V$ of all the $n$-tuples over GF(2). Recall that an $n$-tuple **r** in $V$ is a codeword in $\mathcal{C}$ if and only if $\mathbf{r} \cdot \mathbf{H}^{\mathrm{T}} = 0$. Therefore, if $\mathbf{s} \neq \mathbf{0}$, **r** is not a codeword in $\mathcal{C}$. In this case, the transmitter transmitted a codeword but the receiver received a vector that is not a codeword. Hence, the presence of transmission errors is detected. If $\mathbf{s} = \mathbf{0}$, then **r** is a codeword in $\mathcal{C}$. In this case, the channel decoder assumes that **r** is error-free and accepts **r** as the transmitted codeword. However, in the event that **r** is a codeword in $\mathcal{C}$ but differs from the transmitted codeword **v**, on accepting **r** as the transmitted codeword, the decoder commits a decoding error. This occurs when the error pattern **e** caused by the noise changes the transmitted codeword **v** into another codeword in $\mathcal{C}$. This happens when and only when the error pattern **e** is identical to a nonzero codeword in $\mathcal{C}$, and so there are $2^k - 1$ such *undetectable error patterns*. Since the $(n-k)$-tuple **s** over GF(2) is used for detecting whether the received vector **r** contains transmission errors, it is called the *syndrome* of **r**.

### 3.1.3 Weight Distribution and Minimum Hamming Distance of a Linear Block Code

Let $\mathbf{v} = (v_0, v_1, \ldots, v_{n-1})$ be an $n$-tuple over GF(2). The *Hamming weight* (or simply weight) of **v**, denoted by $w(\mathbf{v})$, is defined as the number of nonzero components in **v**. Consider an $(n, k)$ linear block code $\mathcal{C}$ with code symbols from GF(2). For $0 \leq i \leq n$, let $A_i$ be the number of codewords in $\mathcal{C}$ with Hamming weight $i$. Then the numbers $A_0, A_1, \ldots, A_n$ are called the *weight distribution* of $\mathcal{C}$. It is clear that $A_0 + A_1 + \cdots + A_n = 2^k$. Since there is one and only one all-zero codeword in a linear block code, $A_0 = 1$. The *smallest weight* of the nonzero codewords in $\mathcal{C}$, denoted by $w_{\min}(\mathcal{C})$, is called the *minimum weight* of $\mathcal{C}$. Mathematically, the minimum weight of $\mathcal{C}$ is given as follows:

$$w_{\min}(\mathcal{C}) = \min\{w(\mathbf{v}) : \mathbf{v} \in \mathcal{C}, \mathbf{v} \neq 0\}. \quad (3.15)$$

Suppose $\mathcal{C}$ is used for error control over a BSC with transition probability $p$. Recall that an undetectable error pattern is an error pattern that is identical to a nonzero codeword in $\mathcal{C}$. When such an error pattern occurs, the decoder will not be able to detect the presence of transmission errors and hence will commit a decoding error. There are $A_i$ undetectable error patterns of weight $i$; each occurs with probability

$p^i(1-p)^{n-i}$. Hence, the total probability that an undetectable error pattern with $i$ errors occurs is $A_i p^i(1-p)^{n-i}$. Then the probability that the decoder fails to detect the presence of transmission errors, called the *probability of an undetected error*, is equal to

$$P_u(E) = \sum_{i=1}^{n} A_i p^i (1-p)^{n-i}. \tag{3.16}$$

Therefore, the weight distribution of a linear block code completely determines its probability of an undetected error. It has been proved that in the *ensemble* of $(n, k)$ linear block codes over GF(2), there exists at least one code with the probability of an undetected error, $P_u(E)$, upper bounded by $2^{-(n-k)}$, that is

$$P_u(E) \leq 2^{-(n-k)}. \tag{3.17}$$

Note that the bound is independent of $p$. A code that satisfies this upper bound is said to be a good error-detection code.

Let $\mathbf{v}$ and $\mathbf{w}$ be two $n$-tuples over GF(2). The Hamming distance (or simply distance) between $\mathbf{v}$ and $\mathbf{w}$, denoted by $d(\mathbf{v}, \mathbf{w})$, is defined as the number of places where $\mathbf{v}$ and $\mathbf{w}$ differ. The Hamming distance is a metric function that satisfies the *triangle inequality*. Let $\mathbf{v}$, $\mathbf{w}$, and $\mathbf{x}$ be three $n$-tuples over GF(2). Then

$$d(\mathbf{v}, \mathbf{w}) + d(\mathbf{w}, \mathbf{x}) \geq d(\mathbf{v}, \mathbf{x}). \tag{3.18}$$

It follows from the definition of the Hamming distance between two $n$-tuples and the Hamming weight of an $n$-tuple that the Hamming distance between $\mathbf{v}$ and $\mathbf{w}$ is equal to the Hamming weight of the vector sum of $\mathbf{v}$ and $\mathbf{w}$, that is, $d(\mathbf{v}, \mathbf{w}) = w(\mathbf{v} + \mathbf{w})$. Further, because the code is linear, $\mathbf{v} + \mathbf{w}$ is in the code whenever $\mathbf{v}$ and $\mathbf{w}$ are.

The minimum distance of an $(n, k)$ linear block code $\mathcal{C}$, denoted by $d_{\min}(\mathcal{C})$, is defined as the *smallest* Hamming distance between two different codewords in $\mathcal{C}$, that is

$$d_{\min}(\mathcal{C}) = \min\{d(\mathbf{v}, \mathbf{w}) : \mathbf{v}, \mathbf{w} \in \mathcal{C}, \mathbf{v} \neq \mathbf{w}\}. \tag{3.19}$$

Using the fact that $d(\mathbf{v}, \mathbf{w}) = w(\mathbf{v} + \mathbf{w})$, we can easily prove that the minimum distance $d_{\min}(\mathcal{C})$ of $\mathcal{C}$ is equal to the minimum weight $w_{\min}(\mathcal{C})$ of $\mathcal{C}$. This follows from (3.19),

$$\begin{aligned} d_{\min}(\mathcal{C}) &= \min\{d(\mathbf{v}, \mathbf{w}) : \mathbf{v}, \mathbf{w} \in \mathcal{C}, \mathbf{v} \neq \mathbf{w}\} \\ &= \min\{w(\mathbf{v} + \mathbf{w}) : \mathbf{v}, \mathbf{w} \in \mathcal{C}, \mathbf{v} \neq \mathbf{w}\} \\ &= \min\{w(\mathbf{x}) : \mathbf{x} \in \mathcal{C}, \mathbf{x} \neq \mathbf{0}\} \\ &= w_{\min}(\mathcal{C}). \end{aligned} \tag{3.20}$$

Therefore, for a linear block code, determining its minimum distance is equivalent to determining its minimum weight. The weight structure (or weight distribution) of a linear block code $\mathcal{C}$ is related to a parity-check matrix $\mathbf{H}$ of the code, as given by the following three theorems. We state these theorems without proofs.

**Theorem 3.2** *Let $\mathcal{C}$ be an $(n, k)$ linear block code with a parity-check matrix* $\mathbf{H}$. *For each codeword in $\mathcal{C}$ with weight $i$, there exist $i$ columns in* $\mathbf{H}$ *whose vector sum gives a* *zero (column) vector. Conversely, if there are i columns in* **H** *whose vector sum results in a zero (column) vector, there is a codeword in* $C$ *with weight i.*

**Theorem 3.3** *The minimum weight (or minimum distance) of an* $(n, k)$ *linear block code* $C$ *with a parity-check matrix* **H** *is equal to the smallest number of columns in* **H** *whose vector sum is a zero vector.*

**Theorem 3.4** *For an* $(n, k)$ *linear block code* $C$ *given by the null space of a parity-check matrix* **H**, *if there are no* $d - 1$ *or fewer columns in* **H** *that sum to a zero vector, the minimum distance (or weight) of* $C$ *is at least* $d$.

Theorem 3.3 gives a *lower bound* on the minimum distance (or weight) of a linear block code. In general, it is very hard to determine the exact minimum distance (or weight) of a linear block code; however, it is much easier to give a lower bound on its minimum distance (or weight). This will be seen later.

The weight distribution of a linear block code $C$ actually gives the *distance distribution* of the nonzero codewords with respect to the all-zero codeword $\mathbf{0}$. For $1 \leq i \leq n$, the number $A_i$ of codewords in $C$ with weight $i$ is simply equal to the number of codewords that are at distance $i$ from the all-zero codeword $\mathbf{0}$. Owing to the linear structure of $C$, $A_i$ also gives the number of codewords in $C$ that are at a distance $i$ from *any* fixed codeword $\mathbf{v}$ in $C$. Therefore, the weight distribution $\{A_0, A_1, \ldots, A_n\}$ of $C$ is also the distance distribution of $C$ with respect to any codeword in $C$.

The ability of a linear block code $C$ to detect and correct random errors depends heavily on its distance (or weight) distribution for both hard-decision (BSC) decoding and soft-decision (binary-input AWGN channel) decoding, with $d_{\min}(C)$ the most important parameter in both cases.

For an $(n, k)$ linear block code $C$ with minimum distance $d_{\min}(C)$, no error pattern with $d_{\min}(C) - 1$ or fewer errors can change a transmitted codeword into another codeword in $C$. Therefore, any error pattern with $d_{\min}(C) - 1$ or fewer errors will result in a received vector that is not a codeword in $C$ and hence its syndrome will not be equal to zero. Therefore, all the error patterns with $d_{\min}(C) - 1$ or fewer errors are detectable by the channel decoder. However, if a codeword $\mathbf{v}$ is transmitted and an error pattern with $d_{\min}(C)$ errors that happens to be a codeword in $C$ at a distance $d_{\min}(C)$ from $\mathbf{v}$ occurs, then the received vector $\mathbf{r}$ is a codeword and its syndrome is zero. Such an error pattern is an undetectable error. This is to say that all the error patterns with $d_{\min}(C) - 1$ or fewer errors are guaranteed to be detectable; however, detection is not guaranteed for error patterns with $d_{\min}(C)$ or more errors. For this reason, $d_{\min}(C) - 1$ is called the *error-detecting capability* of the code $C$. The number of guaranteed detectable error patterns is

$$\binom{n}{1} + \binom{n}{2} + \cdots + \binom{n}{d_{\min}(C) - 1},$$

where $\binom{n}{i}$ is a binomial coefficient. For large $n$, this number of guaranteed detectable error patterns is only a small fraction of the $2^n - 2^k + 1$ detectable error patterns.

So far, we have focused on error detection with a linear block code. Decoding and the error-correcting capability of a linear block code will be discussed in the next section. We will show that, with hard-decision decoding, a linear block code $\mathcal{C}$ with minimum distance $d_{\min}(\mathcal{C})$ is capable of correcting $\lfloor (d_{\min}(\mathcal{C})-1)/2 \rfloor$ or fewer random errors over a span of $n$ transmitted code digits.

### 3.1.4 Decoding of Linear Block Codes

Consider an $(n, k)$ linear block code $\mathcal{C}$ with a parity-check matrix $\mathbf{H}$ and minimum distance $d_{\min}(\mathcal{C})$. Suppose a codeword in $\mathcal{C}$ is transmitted and $\mathbf{r}$ is the received vector. For maximum-likelihood decoding (MLD) as described in Chapter 1, $\mathbf{r}$ is decoded into a codeword $\mathbf{v}$ such that the conditional probability $P(\mathbf{r}|\mathbf{v})$ is maximized. For a BSC, this is equivalent to decoding $\mathbf{r}$ into a codeword $\mathbf{v}$ such that the Hamming distance $d(\mathbf{r}, \mathbf{v})$ between $\mathbf{r}$ and $\mathbf{v}$ is minimized. This is called *minimum-distance* (or *nearest-neighbor*) decoding. With minimum-distance decoding, the decoder has to compute the distance between $\mathbf{r}$ and every codeword in $\mathcal{C}$ and then choose a codeword $\mathbf{v}$ (not necessarily unique) that is closest to $\mathbf{r}$ (i.e., $d(\mathbf{r}, \mathbf{v})$ is the smallest) as the decoded codeword. This decoding is called a *complete error-correction decoding* and apparently requires a total of $2^k$ computations of the distances between $\mathbf{r}$ and the $2^k$ codewords in $\mathcal{C}$. For large $k$, implementation of this complete decoder is practically impossible. However, for many linear block codes, efficient algorithms have been developed for incomplete error-correction decoding to achieve good error performance with vastly reduced decoding complexity.

No matter which codeword in $\mathcal{C}$ is transmitted over a noisy channel, the received vector $\mathbf{r}$ is one of the $2^n$ $n$-tuples over GF(2). Let $\mathbf{v}_0 = \mathbf{0}, \mathbf{v}_1, \ldots, \mathbf{v}_{2^k-1}$ be the codewords in $\mathcal{C}$. Any decoding scheme used at the decoder is a rule to partition the vector space $V$ of all the $n$-tuples over GF(2) into $2^k$ regions; each region contains one and only one codeword in $\mathcal{C}$, as shown in Figure 3.2. Decoding is done to find the region

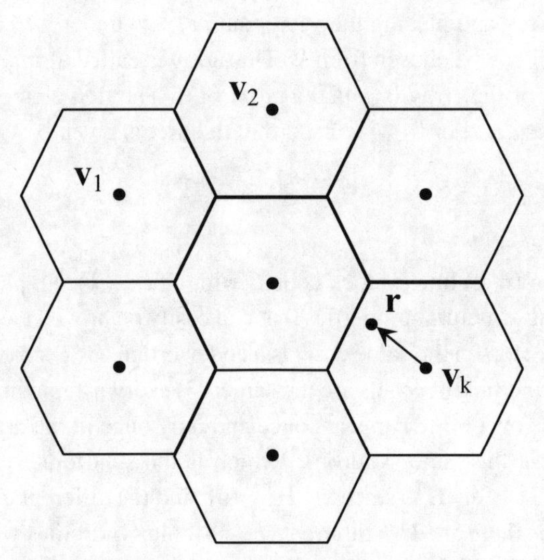


**Figure 3.2** Decoding regions. (Hexagons provide only a schematic representation of the decoding regions in $n$-dimensional space.) The space of received vectors is discrete for hard decisions and continuous for soft decisions.

**Table 3.1** A standard array for an $(n, k)$ linear block code

| Cosets | Coset leaders | | | | | |
|---|---|---|---|---|---|---|
| $\mathcal{C}$ | $\mathbf{e}_0 = \mathbf{v}_0 = \mathbf{0}$ | $\mathbf{v}_1$ | $\cdots$ | $\mathbf{v}_i$ | $\cdots$ | $\mathbf{v}_{2^k-1}$ |
| $\mathbf{e}_1 + \mathcal{C}$ | $\mathbf{e}_1$ | $\mathbf{e}_1 + \mathbf{v}_1$ | $\cdots$ | $\mathbf{e}_1 + \mathbf{v}_i$ | $\cdots$ | $\mathbf{e}_1 + \mathbf{v}_{2^k-1}$ |
| $\mathbf{e}_2 + \mathcal{C}$ | $\mathbf{e}_2$ | $\mathbf{e}_2 + \mathbf{v}_1$ | $\cdots$ | $\mathbf{e}_2 + \mathbf{v}_i$ | $\cdots$ | $\mathbf{e}_2 + \mathbf{v}_{2^k-1}$ |
| $\vdots$ | $\vdots$ | $\vdots$ | $\cdots$ | $\vdots$ | $\cdots$ | $\vdots$ |
| $\mathbf{e}_{2^{n-k}-1} + \mathcal{C}$ | $\mathbf{e}_{2^{n-k}-1}$ | $\mathbf{e}_{2^{n-k}-1} + \mathbf{v}_1$ | $\cdots$ | $\mathbf{e}_{2^{n-k}-1} + \mathbf{v}_i$ | $\cdots$ | $\mathbf{e}_{2^{n-k}-1} + \mathbf{v}_{2^k-1}$ |

that contains the received vector $\mathbf{r}$. Then decode $\mathbf{r}$ into the codeword $\mathbf{v}$ that is contained in the region. These regions are called *decoding regions*. For MLD, the decoding regions for the $2^k$ codewords are, for $0 \leq i < 2^k$,

$$D(\mathbf{v}_i) = \{\mathbf{r} \in V: P(\mathbf{r}|\mathbf{v}_i) \geq P(\mathbf{r}|\mathbf{v}_j), j \neq i\}.$$

For minimum-distance decoding (MLD for the BSC), the decoding regions for the $2^k$ codewords are, for $0 \leq i < 2^k$,

$$D(\mathbf{v}_i) = \{\mathbf{r} \in V: d(\mathbf{r}, \mathbf{v}_i) \leq d(\mathbf{r}, \mathbf{v}_j), j \neq i\}.$$

An algebraic method to partition the $2^n$ possible received vectors into $2^k$ decoding regions can be implemented as follows. First, we arrange the $2^k$ codewords in $\mathcal{C}$ as the top row (or 0th row) of a $2^{n-k} \times 2^k$ array with the all-zero codeword $\mathbf{v}_0 = \mathbf{0}$ as the first entry, as shown in Table 3.1. Then we form the rest of the rows of the array one at a time. Suppose we have formed the $(j-1)$th row of the array $1 \leq j \leq 2^{n-k}$. To form the $j$th row of the array, we choose a vector $\mathbf{e}_j$ in $V$ that is not contained in the previous $j-1$ rows of the array. Form the $j$th row of the array by adding $\mathbf{e}_j$ to each codeword $\mathbf{v}_i$ in the top row of the array and placing the vector sum $\mathbf{e}_j + \mathbf{v}_i$ under $\mathbf{v}_i$. The array is completed when no vector can be chosen from $V$. This array is called a *standard array* for the code $\mathcal{C}$. Each row of the array is called a *coset* of $\mathcal{C}$. The first element of each coset is called the *coset leader*. For $0 \leq j < 2^{n-k}$, the $j$th coset is given by

$$\mathbf{e}_j + \mathcal{C} = \{\mathbf{e}_j + \mathbf{v}_i: \mathbf{v}_i \in \mathcal{C}, 0 \leq i < 2^k\}, \tag{3.21}$$

where $\mathbf{e}_j$ is the coset leader.

A standard array for an $(n, k)$ linear block code $\mathcal{C}$ with an $(n-k) \times n$ parity-check matrix $\mathbf{H}$ has the following structural properties that can easily be proved (see Problem 3.9): (1) the sum of two vectors in the same coset is a codeword in $\mathcal{C}$; (2) no two vectors in the same coset or in two different cosets are the same; (3) every $n$-tuple in the vector space $V$ of all the $n$-tuples over GF(2) appears once and only once in the array; (4) all the vectors in a coset have the same syndrome, which is the syndrome of the coset leader, that is, $(\mathbf{e}_j + \mathbf{v}_i) \cdot \mathbf{H}^{\mathrm{T}} = \mathbf{e}_j \cdot \mathbf{H}^{\mathrm{T}}$ (since $\mathbf{v}_i \cdot \mathbf{H}^{\mathrm{T}} = \mathbf{0}$); and (5) different cosets have different syndromes. Since there are $2^{n-k}$ different $(n-k)$-tuple syndromes with respect to the parity-check matrix $\mathbf{H}$ of the code $\mathcal{C}$ and there are $2^{n-k}$ cosets in a standard array for $\mathcal{C}$, it follows from properties (4) and (5) that there is a *one-to-one correspondence* between a coset leader and an $(n-k)$-tuple syndrome.

A standard array for an $(n, k)$ linear block code $\mathcal{C}$ consists of $2^k$ columns; each column contains one and only one codeword at the top of the column. For $0 \leq i < 2^k$, the $i$th column consists of the following set of $2^{n-k}$ $n$-tuples:

$$D(\mathbf{v}_i) = \{\mathbf{v}_i, \mathbf{e}_1 + \mathbf{v}_i, \mathbf{e}_2 + \mathbf{v}_i, \ldots, \mathbf{e}_{2^{n-k}-1} + \mathbf{v}_i\}, \tag{3.22}$$

where each element is the vector sum of the $i$th codeword $\mathbf{v}_i$ and a coset leader $\mathbf{e}_j$ (note that $\mathbf{e}_0 = \mathbf{0}$) with $0 \leq j < 2^{n-k}$. We see that the $i$th codeword $\mathbf{v}_i$ is the only codeword in $D(\mathbf{v}_i)$. The $2^k$ columns of a standard array for $\mathcal{C}$ form a partition of the vector space $V$ of all the $n$-tuples over GF(2). These $2^k$ columns of the array can be used as the regions for decoding $\mathcal{C}$. If the received vector $\mathbf{r}$ is found in the $i$th column $D(\mathbf{v}_i)$, we decode $\mathbf{r}$ into $\mathbf{v}_i$. From the structure of a standard array for $\mathcal{C}$, we can easily check that (1) if the $i$th codeword $\mathbf{v}_i$ is transmitted and the error pattern caused by the channel noise is a coset leader $\mathbf{e}_j$, then the received vector $\mathbf{r} = \mathbf{v}_i + \mathbf{e}_j$ is in the column $D(\mathbf{v}_i)$ that contains $\mathbf{v}_i$; and (2) if $\mathbf{v}_i$ is transmitted but the error pattern is not a coset leader, then the received vector $\mathbf{r}$ is not in column $D(\mathbf{v}_i)$ (see Problem 3.10). Therefore, using the columns of a standard array of an $(n, k)$ linear block code $\mathcal{C}$ as decoding regions, decoding is correct (i.e., $\mathbf{r}$ is decoded into the transmitted codeword) if and only if a codeword $\mathbf{v}_i$ is transmitted and the error pattern caused by the channel noise is identical to a coset leader. This is to say that the $2^{n-k} - 1$ nonzero coset leaders of a standard array are all the correctable error patterns (i.e., they result in correct decoding). To minimize the probability of a decoding error, the error patterns that are *most likely* to occur for a given channel should be chosen as the coset leaders.

For a BSC, an error pattern of smaller weight (or smaller number of errors) is more probable than an error pattern with larger weight (or larger number of errors). So, when a standard array for a linear block code $\mathcal{C}$ is formed, each coset leader should be chosen to be an $n$-tuple of least weight from the remaining available $n$-tuples in $V$. Choosing coset leaders in this manner, each coset leader has the minimum weight in each coset. In this case, decoding based on a standard array for $\mathcal{C}$ is minimum-distance decoding (or MLD for a BSC). A standard array formed in this way is called an *optimal standard array* for $\mathcal{C}$.

The minimum distance $d_{\min}(\mathcal{C})$ of a linear block code $\mathcal{C}$ is either odd or even. Let $t = \lfloor (d_{\min}(\mathcal{C}) - 1)/2 \rfloor$, where $\lfloor x \rfloor$ denotes the integer part of $x$ (or the largest integer equal to or less than $x$). Then $2t + 1 \leq d_{\min}(\mathcal{C}) \leq 2t + 2$. It can be shown that all the $n$-tuples over GF(2) of weight $t$ or less can be used as coset leaders in an optimal standard array for $\mathcal{C}$ (see Problem 3.11). Also, it can be shown that there is at least one $n$-tuple of weight $t + 1$ that cannot be used as a coset leader (see Problem 3.11). When this error pattern occurs, the received vector $\mathbf{r}$ will be decoded incorrectly. Therefore, for a linear code $\mathcal{C}$ with minimum distance $d_{\min}(\mathcal{C})$, any error pattern with $t = \lfloor (d_{\min}(\mathcal{C}) - 1)/2 \rfloor$ or fewer errors is guaranteed correctable (i.e., results in correct decoding), but not all the error patterns with $t + 1$ or more errors. The parameter $t = \lfloor (d_{\min}(\mathcal{C}) - 1)/2 \rfloor$ is called the *error-correction capability* of $\mathcal{C}$. We say that $\mathcal{C}$ is capable of correcting $t$ or fewer random errors and $\mathcal{C}$ is called a $t$-error-correcting code.

**Table 3.2** A look-up decoding table

| Syndromes | Correctable error patterns |
|---|---|
| $\mathbf{0}$ | $\mathbf{e}_0 = \mathbf{0}$ |
| $\mathbf{s}_1$ | $\mathbf{e}_1$ |
| $\mathbf{s}_2$ | $\mathbf{e}_2$ |
| ... | ... |
| $\mathbf{s}_{2^{n-k}-1}$ | $\mathbf{e}_{2^{n-k}-1}$ |

Decoding of an $(n, k)$ linear block code $\mathcal{C}$ based on an optimal standard array for the code requires sufficient memory to store $2^n$ $n$-tuples. For large $n$, the size of the memory will be prohibitively large and implementation of standard-array-based decoding becomes practically impossible. However, the decoding can be significantly simplified by using the following facts: (1) the coset leaders form all the correctable error patterns; and (2) there is a one-to-one correspondence between an $(n - k)$-tuple syndrome and a coset leader. From these two facts, we form a table with only two columns that consists of $2^{n-k}$ coset leaders (correctable error patterns) in one column and their corresponding syndromes in another column, as shown in Table 3.2. Decoding of a received vector $\mathbf{r}$ is carried out in three steps:

1. Compute the syndrome of $\mathbf{r}$, $\mathbf{s} = \mathbf{r} \cdot \mathbf{H}^T$.
2. Find the coset leader $\mathbf{e}$ in the table whose syndrome is equal to $\mathbf{s}$. Then $\mathbf{e}$ is assumed to be the error pattern caused by the channel noise.
3. Decode $\mathbf{r}$ into the codeword $\mathbf{v} = \mathbf{r} + \mathbf{e}$.

The above decoding is called *syndrome decoding* or *table-look-up decoding*. With this decoding, the decoder complexity is drastically reduced compared with standard-array-based decoding.

For a long code with large $n - k$, a complete table-look-up decoder is still very complex, requiring a very large memory to store the look-up table. If we limit ourself to correcting only the error patterns guaranteed by the error-correcting capability $t = \lfloor (d_{\min}(\mathcal{C}) - 1)/2 \rfloor$ of the code, then the size of the look-up table can be further reduced. The new table consists of only

$$N_t = \binom{n}{0} + \binom{n}{1} + \cdots + \binom{n}{t}$$

correctable error patterns guaranteed by the minimum distance $d_{\min}(\mathcal{C})$ of the code. With this new table, decoding of a received vector $\mathbf{r}$ consists of the following four steps:

1. Compute the syndrome $\mathbf{s}$ of $\mathbf{r}$.
2. If the syndrome $\mathbf{s}$ corresponds to a correctable error pattern $\mathbf{e}$ listed in the table, go to Step 3; otherwise, go to Step 4.
3. Decode $\mathbf{r}$ into the codeword $\mathbf{v} = \mathbf{r} + \mathbf{e}$. Stop.
4. Declare a *decoding failure*. In this case, the presence of errors is detected but the decoder fails to correct the errors.

With the above partial table-look-up decoding, the number of errors to be corrected is bounded by the error-correction capability $t = \lfloor (d_{\min}(\mathcal{C}) - 1)/2 \rfloor$ of the code. This is called *bounded-distance decoding*.

Many classes of linear block codes with good error-correction capabilities have been constructed over the years. Efficient algorithms to carry out the bounded-distance decoding of these classes of codes have been devised. Two classes of these codes will be presented in later sections of this chapter.

### 3.1.5 Error Probability of Syndrome Decoders

We are interested in the performance improvement that coding with syndrome decoding provides relative to no coding. For convenience, we will assume a binary symmetric channel (BSC) with bit error probability $p$, which may be the result of applying hard decisions to a binary-input AWGN channel. As a baseline, for a message of size $k$ bits with no coding, the message block received from the BSC will contain at least one error with probability

$$P_e(\text{uncoded}) = \Pr(\text{block error}) = \Pr(\text{at least 1 error in } k \text{ bits}).$$

But this is also

$$P_e(\text{uncoded}) = 1 - \Pr(0 \text{ errors in } k \text{ bits}) = 1 - (1-p)^k,$$

where the last expression follows from the fact that each of $k$ bits will not be in error with probability $1 - p$, and these probabilities multiply due to the fact that the bit transmissions are independent.

Consider now the case for an $(n, k)$ linear code with syndrome decoding:

$$\begin{aligned} P_e(\text{coded}) &= \Pr(\text{error pattern is } \textit{not} \text{ in the syndrome-error table}) \\ &= 1 - \Pr(\text{error pattern is in the syndrome-error table}) \\ &= 1 - \sum_{w=0}^{n} N_w p^w (1-p)^{n-w}, \end{aligned}$$

where $N_w, w = 0, 1, \ldots, n$, is the number of syndromes in the error-syndrome table corresponding to weight-$w$ error patterns and $p^w(1-p)^{n-w}$ is the probability of occurrence of a specific weight-$w$ error pattern on the BSC. Note that, for the case of $w = 0$ errors, $N_0 = 1$. Finally, observe that the summation above is the probability of correct decoding so that its complement gives $P_e$(coded), the probability of decoding error, on the left-hand side.

---

**Example 3.1** Consider the $(7, 4)$ Hamming code whose syndrome decoder can correct all single-error patterns and no others. Then $N_0 = 1, N_1 = 7$, and $N_w = 0$ for $w > 1$. Suppose $p = 10^{-3}$. In this case, $P_e(\text{uncoded}) = 1 - (1-p)^k \approx kp = 4 \times 10^{-3}$ and $P_e(\text{coded}) = 1 - N_0(1-p)^7 - N_1 p^1 (1-p)^6 = 2.1 \times 10^{-5}$.

---

## 3.2 Cyclic Codes

Let $\mathbf{v} = (v_0, v_1, v_2, \ldots, v_{n-1})$ be an $n$-tuple over GF(2). If we shift every component of $\mathbf{v}$ cyclically one place to the right, we obtain the following $n$-tuple:

$$\mathbf{v}^{(1)} = (v_{n-1}, v_0, v_1, \ldots, v_{n-2}), \tag{3.23}$$

which is called the *right cyclic shift* (or simply cyclic shift) of $\mathbf{v}$.

**Definition 3.5** An $(n, k)$ linear block code $\mathcal{C}$ is said to be *cyclic* if the cyclic shift of each codeword in $\mathcal{C}$ is also a codeword in $\mathcal{C}$.

Cyclic codes form a very special type of linear block code. They have encoding advantage over many other types of linear block codes. Encoding of this type of code can be implemented with simple shift registers with feedback connections. Many classes of cyclic codes with large minimum distances have been constructed; and, furthermore, efficient algebraic hard-decision decoding algorithms for some of these classes of cyclic codes have been developed. This section gives an introduction to cyclic codes.

To analyze the structural properties of a cyclic code, a codeword $\mathbf{v} = (v_0, v_1, \ldots, v_{n-1})$ is represented by a polynomial over GF(2) of degree $n - 1$ or less with the components of $\mathbf{v}$ as coefficients, as follows:

$$\mathbf{v}(X) = v_0 + v_1 X + \cdots + v_{n-1} X^{n-1}. \tag{3.24}$$

This polynomial is called a *code polynomial*. In polynomial form, an $(n, k)$ cyclic code $\mathcal{C}$ consists of $2^k$ code polynomials. The code polynomial corresponding to the all-zero codeword is the zero polynomial. All the other $2^k - 1$ code polynomials corresponding to the $2^k - 1$ nonzero codewords in $\mathcal{C}$ are nonzero polynomials.

Some important structural properties of cyclic codes are presented in the following without proofs. References [1–6, 11] contain good and extensive coverage of the structure and construction of cyclic codes.

In an $(n, k)$ cyclic code $\mathcal{C}$, every nonzero code polynomial has degree *at least* $n - k$ but not greater than $n - 1$. There exists one and only one code polynomial $\mathbf{g}(X)$ of degree $n - k$ of the following form:

$$\mathbf{g}(X) = 1 + g_1 X + g_2 X^2 + \cdots + g_{n-k-1} X^{n-k-1} + X^{n-k}. \tag{3.25}$$

Therefore, $\mathbf{g}(X)$ is a nonzero code polynomial of *minimum degree* and is *unique*. Every code polynomial $\mathbf{v}(X)$ in $\mathcal{C}$ is divisible by $\mathbf{g}(X)$, that is, is a *multiple* of $\mathbf{g}(X)$. Moreover, every polynomial over GF(2) of degree $n - 1$ or less that is divisible by $\mathbf{g}(X)$ is a code polynomial in $\mathcal{C}$. Therefore, an $(n, k)$ cyclic code $\mathcal{C}$ is completely specified by the unique polynomial $\mathbf{g}(X)$ of degree $n - k$ given by (3.25). This unique nonzero code polynomial $\mathbf{g}(X)$ of minimum degree in $\mathcal{C}$ is called the *generator polynomial* of the $(n, k)$ cyclic code $\mathcal{C}$. The degree of $\mathbf{g}(X)$ is simply the number of parity-check bits of the code.

Since each code polynomial $\mathbf{v}(X)$ in $\mathcal{C}$ is a multiple of $\mathbf{g}(X)$ (including the zero code polynomial), it can be expressed as the following product:

$$\mathbf{v}(X) = \mathbf{m}(X)\mathbf{g}(X), \tag{3.26}$$

where $\mathbf{m}(X) = m_0 + m_1X + \cdots + m_{k-1}X^{k-1}$ is a polynomial over GF(2) of degree $k-1$ or less. If $\mathbf{m} = (m_0, m_1, \ldots, m_{k-1})$ is the message to be encoded, then $\mathbf{m}(X)$ is the polynomial representation of $\mathbf{m}$ (called a *message polynomial*) and $\mathbf{v}(X)$ is the corresponding code polynomial of the message polynomial $\mathbf{m}(X)$. With this encoding, the corresponding $k\times n$ generator matrix of the $(n, k)$ cyclic code $\mathcal{C}$ is given as follows:

$$\mathbf{G} = \begin{bmatrix} 1 & g_1 & g_2 & \cdots & g_{n-k-1} & 1 & 0 & 0 & \cdots & 0 & 0 \\ 0 & 1 & g_1 & \cdots & g_{n-k-2} & g_{n-k-1} & 1 & 0 & \cdots & 0 & 0 \\ \vdots & \vdots & \vdots & \ddots & \vdots & \vdots & \vdots & \vdots & \ddots & \vdots & \vdots \\ 0 & 0 & 0 & \cdots & 1 & g_1 & g_2 & g_3 & \cdots & g_{n-k-1} & 1 \end{bmatrix}. \tag{3.27}$$

Note that $\mathbf{G}$ is simply obtained by using the $n$-tuple representation of the generator polynomial $\mathbf{g}(X)$ as the first row and its $k-1$ right cyclic shifts as the other $k-1$ rows. $\mathbf{G}$ is not in systematic form but can be put into systematic form by elementary row operations without column permutations.

A very important property of the generator polynomial $\mathbf{g}(X)$ of an $(n, k)$ cyclic code $\mathcal{C}$ over GF(2) is that $\mathbf{g}(X)$ divides $X^n + 1$. Consequently, $X^n + 1$ can be expressed as the following product:

$$X^n + 1 = \mathbf{g}(X)\mathbf{f}(X), \tag{3.28}$$

where $\mathbf{f}(X) = 1 + f_1X + \cdots + f_{k-1}X^{k-1} + X^k$ is a polynomial of degree $k$ over GF(2). Let

$$\begin{aligned} \mathbf{h}(X) &= X^k\mathbf{f}(X^{-1}) \\ &= 1 + h_1X + \cdots + h_{k-1}X^{k-1} + X^k \end{aligned} \tag{3.29}$$

be the *reciprocal polynomial* of $\mathbf{f}(X)$. It is easy to prove that $\mathbf{h}(X)$ also divides $X^n + 1$. Form the following $(n-k) \times n$ matrix over GF(2) with the $n$-tuple representation $\mathbf{h} = (1, h_1, \ldots, h_{k-1}, 1, 0, \ldots, 0)$ of $\mathbf{h}(X)$ as the first row and its $n-k-1$ cyclic shifts as the other $n-k-1$ rows:

$$\mathbf{H} = \begin{bmatrix} 1 & h_1 & h_2 & \cdots & h_{k-1} & 1 & 0 & 0 & \cdots & 0 & 0 \\ 0 & 1 & h_1 & \cdots & h_{k-2} & h_{k-1} & 1 & 0 & \cdots & 0 & 0 \\ \vdots & \vdots & \vdots & \ddots & \vdots & \vdots & \vdots & \vdots & \ddots & \vdots & \vdots \\ 0 & 0 & 0 & \cdots & 1 & h_1 & h_2 & h_3 & \cdots & h_{k-1} & 1 \end{bmatrix}. \tag{3.30}$$

Then $\mathbf{H}$ is a parity-check matrix of the $(n, k)$ cyclic code $\mathcal{C}$ corresponding to $\mathbf{G}$ given by (3.27). The polynomial $\mathbf{h}(X)$ given by (3.29) is called the *parity-check polynomial* of $\mathcal{C}$. In fact, $\mathbf{h}(X)$ is the generator polynomial of the dual code, an $(n,n-k)$ cyclic code, of the $(n, k)$ cyclic code $\mathcal{C}$.

Systematic encoding of an $(n, k)$ cyclic code with generator polynomial $\mathbf{g}(X)$ can be accomplished easily. Suppose $\mathbf{m} = (m_0, m_1, \ldots, m_{k-1})$ is the message to be encoded.

On multiplying the message polynomial $\mathbf{m}(X) = m_0 + m_1X + \cdots + m_{k-1}X^{k-1}$ by $X^{n-k}$, we obtain

$$X^{n-k}\mathbf{m}(X) = m_0X^{n-k} + m_1X^{n-k+1} + \cdots + m_{k-1}X^{n-1}, \tag{3.31}$$

which is a polynomial of degree $n - 1$ or less. On dividing $X^{n-k}\mathbf{m}(X)$ by the generator polynomial $\mathbf{g}(X)$, we have

$$X^{n-k}\mathbf{m}(X) = \mathbf{a}(X)\mathbf{g}(X) + \mathbf{b}(X), \tag{3.32}$$

where $\mathbf{a}(X)$ and $\mathbf{b}(X)$ are the quotient and remainder, respectively. Since the degree of $\mathbf{g}(X)$ is $n - k$, the degree of the remainder $\mathbf{b}(X)$ must be $n - k - 1$ or less. Then $\mathbf{b}(X)$ must be of the following form:

$$\mathbf{b}(X) = b_0 + b_1X + \cdots + b_{n-k-1}X^{n-k-1}. \tag{3.33}$$

We rearrange the expression in (3.32) as follows:

$$\mathbf{b}(X) + X^{n-k}\mathbf{m}(X) = \mathbf{a}(X)\mathbf{g}(X). \tag{3.34}$$

Expression (3.34) shows that $\mathbf{b}(X) + X^{n-k}\mathbf{m}(X)$ is divisible by $\mathbf{g}(X)$. Since the degree of $\mathbf{b}(X)+X^{n-k}\mathbf{m}(X)$ is $n-1$ or less, $\mathbf{b}(X)+X^{n-k}\mathbf{m}(X)$ is hence a code polynomial of the $(n, k)$ cyclic code $\mathcal{C}$ with $\mathbf{g}(X)$ as its generator polynomial. The $n$-tuple representation of the code polynomial $\mathbf{b}(X) + X^{n-k}\mathbf{m}(X)$ is

$$(b_0, b_1, \ldots, b_{n-k-1}, m_0, m_1, \ldots, m_{k-1}),$$

which is in systematic form, where the $n - k$ parity-check bits $b_0, b_1, \ldots, b_{n-k-1}$ are simply the coefficients of the remainder $\mathbf{b}(X)$ given by (3.33).

For $0 \leq i < k$, let $\mathbf{m}_i(X) = X^i$ be the message polynomial with a single nonzero information bit at the $i$th position of the message $\mathbf{m}_i$ to be encoded. Dividing $X^{n-k}\mathbf{m}_i(X) = X^{n-k+i}$ by $\mathbf{g}(X)$, we obtain

$$X^{n-k+i} = \mathbf{a}_i(X)\mathbf{g}(X) + \mathbf{b}_i(X), \tag{3.35}$$

where the remainder $\mathbf{b}_i(X)$ is of the following form:

$$\mathbf{b}_i(X) = b_{i,0} + b_{i,1}X + \cdots + b_{i,n-k-1}X^{n-k-1}. \tag{3.36}$$

Since $\mathbf{b}_i + X^{n-k+i}$ is divisible by $\mathbf{g}(X)$, it is a code polynomial in $\mathcal{C}$. On arranging the $n$-tuple representations of the $n - k$ code polynomials, $\mathbf{b}_i + X^{n-k+i}$ for $0 \leq i < k$, as the rows of a $k \times n$ matrix over GF(2), we obtain

$$\mathbf{G}_{\text{c,sys}} = \begin{bmatrix} b_{0,0} & b_{0,1} & \cdots & b_{0,n-k-1} & 1 & 0 & 0 & \cdots & 0 \\ b_{1,0} & b_{1,1} & \cdots & b_{1,n-k-1} & 0 & 1 & 0 & \cdots & 0 \\ \vdots & \vdots & \ddots & \vdots & \vdots & \vdots & \vdots & \ddots & \vdots \\ b_{k-1,0} & b_{k-1,1} & \cdots & b_{k-1,n-k-1} & 0 & 0 & 0 & \cdots & 1 \end{bmatrix}, \tag{3.37}$$

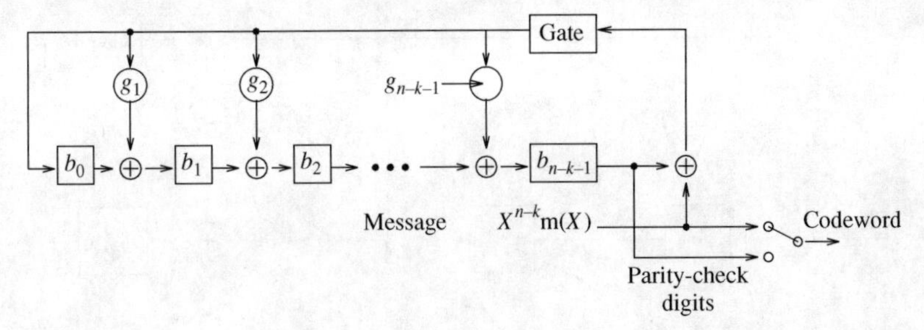


**Figure 3.3** An encoding circuit for an $(n, k)$ cyclic code with generator polynomial $\mathbf{g}(X) = 1 + g_1X + \cdots + g_{n-k-1}X^{n-k-1} + X^{n-k}$. After the $k$ message symbols are shifted into the circuit, the switch at bottom right shifts downward and the gate opens so that zeros enter the adders.

which is the generator matrix of the $(n, k)$ cyclic code $\mathcal{C}$ in systematic form. The corresponding parity-check matrix of $\mathcal{C}$ in systematic form is

$$\mathbf{H}_{\text{c,sys}} = \begin{bmatrix} 1 & 0 & 0 & \cdots & 0 & b_{0,0} & b_{1,0} & \cdots & b_{k-1,0} \\ 0 & 1 & 0 & \cdots & 0 & b_{0,1} & b_{1,1} & \cdots & b_{k-1,1} \\ \vdots & \vdots & \vdots & \ddots & \vdots & \vdots & \vdots & \ddots & \vdots \\ 0 & 0 & 0 & \cdots & 1 & b_{0,n-k-1} & b_{1,n-k-1} & \cdots & b_{k-1,n-k-1} \end{bmatrix}. \tag{3.38}$$

From (3.32) and (3.34), we see that encoding of an $(n, k)$ cyclic code $\mathcal{C}$ can be achieved with a division circuit that divides the message polynomial $X^{n-k}\mathbf{m}(X)$ by its generator polynomial $\mathbf{g}(X)$ and takes the remainder as the parity part of the codeword. This division circuit can be implemented with an $(n-k)$-stage shift register with feedback connections based on the coefficients of the generator polynomial, as shown in Figure 3.3. The pre-multiplication of the message polynomial $\mathbf{m}(X)$ by $X^{n-k}$ is accomplished by inserting the message into the right end of the length-$(n-k)$ feedback shift register instead of the left end, as shown in Figure 3.3.

Suppose an $(n, k)$ cyclic code $\mathcal{C}$ with a generator polynomial $\mathbf{g}(X)$ is used for error control over a noisy channel. Let $\mathbf{r}(X) = r_0 + r_1X + \cdots + r_{n-1}X^{n-1}$ be the received polynomial. Then $\mathbf{r}(X)$ is the sum of a transmitted polynomial $\mathbf{v}(X)$ and an error polynomial $\mathbf{e}(X)$, that is, $\mathbf{r}(X) = \mathbf{v}(X) + \mathbf{e}(X)$. The first step in decoding $\mathbf{r}(X)$ is to compute its syndrome. The syndrome of $\mathbf{r}(X)$, denoted by $\mathbf{s}(X)$, is given by the remainder obtained from dividing $\mathbf{r}(X)$ by the generator polynomial $\mathbf{g}(X)$ of code $\mathcal{C}$. We use the shorthand $\mathbf{s}(X) = R_{\mathbf{g}}[\mathbf{r}(X)]$ to signify this fact. If $\mathbf{s}(X) = 0$, then $\mathbf{r}(X)$ is a code polynomial and is accepted by the receiver as the transmitted code polynomial. If $\mathbf{s}(X) \neq 0$, then $\mathbf{r}(X)$ is not a code polynomial and the presence of transmission errors is detected. Computation of the syndrome $\mathbf{s}(X)$ of $\mathbf{r}(X)$ can again be accomplished with a division circuit, as shown in Figure 3.4.

Similar to the decoding of a linear block code described in Section 3.1.4, decoding of a cyclic code involves associating the computed syndrome $\mathbf{s}(X)$ of a received polynomial $\mathbf{r}(X)$ with a specific correctable error pattern $\mathbf{e}(X)$. Then the estimated transmitted code polynomial $\hat{\mathbf{v}}(X)$ is obtained by removing that specific error pattern $\mathbf{e}(X)$ from the received polynomial $\mathbf{r}(X)$, that is, $\hat{\mathbf{v}}(X) = \mathbf{r}(X) - \mathbf{e}(X)$. For syndrome

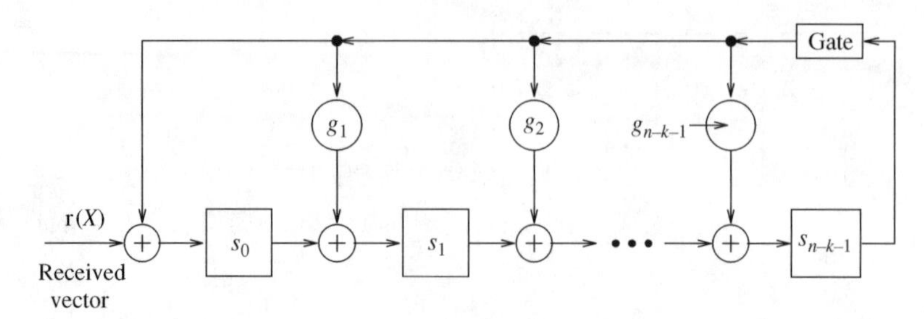


**Figure 3.4** A syndrome-computation circuit for an $(n, k)$ cyclic code with generator polynomial $\mathbf{g}(X) = 1 + g_1X + \cdots + g_{n-k-1}X^{n-k-1} + X^{n-k}$. The gate closes after $n - k$ symbols are shifted in.

and bounded-distance decoding, the architecture and complexity of the decoding circuit very much depends on the decoding algorithm devised for a specific class of codes. For more on decoding of cyclic codes, readers are referred to the next subsection, the BCH code section, and [1, 5, 11].

Consider a systematic $(n, k)$ cyclic code $\mathcal{C}$. Let $l$ be a non-negative integer less than $k$. Consider the set of code polynomials whose $l$ leading high-order information digits $v_{n-l}, \ldots, v_{n-2}, v_{n-1}$ are zeros. There are $2^{k-l}$ such code polynomials. If the $l$ zero-information digits are deleted from each of these code polynomials, we obtain a set of $2^{k-l}$ polynomials over GF(2) with degree $n - l - 1$ or less. These $2^{k-l}$ shortened polynomials form an $(n - l, k - l)$ linear block code. This code is called a *shortened cyclic code* (or *polynomial code*) and it is not cyclic. A shortened cyclic code has at least the same error-correction capability as the code from which it is shortened. The encoding and syndrome computation for a shortened cyclic code can be accomplished by the same circuits as employed by the original cyclic code. This is so because the deleted $l$ leading zero-information digits do not affect the parity-check and syndrome computation. The decoding circuit for the original cyclic code can be used for decoding the shortened code simply by prefixing each received vector with $l$ zeros. This prefixing can be eliminated by modifying the decoding circuit.

So far, we have discussed only the structure, encoding, and decoding of cyclic codes, saying nothing about the existence of cyclic codes. Recall that the generator polynomial $\mathbf{g}(X)$ of an $(n, k)$ cyclic code divides $X^n + 1$. With this fact, it can be proved that any factor $\mathbf{g}(X)$ over GF(2) of $X^n + 1$ with degree $n - k$ can be used as a generator polynomial of an $(n, k)$ cyclic code. In general, any factor of $X^n + 1$ over GF(2) generates a cyclic code.

### 3.2.1 The Meggitt Decoder

Syndrome decoding can be very complex for more powerful codes because the size of the syndrome-error table is $2^{n-k}$. For cyclic codes and, as we shall see, shortened cyclic codes, syndrome decoding can be vastly simplified by exploiting the cyclic nature of the code. The syndrome decoder that does this is called a *Meggitt decoder*, named after its inventor.

For cyclic codes, codeword positions $X^0$, $X^j$, and $X^{n-1}$ are in some sense on the same footing because each cyclic shift of a codeword is another codeword. Because all code bit positions have the same status, the Meggitt decoder strives to correct only errors in position $X^{n-1}$. While this may appear insufficient, note that the received word may be cyclically shifted once so that there is a new code bit in position $X^{n-1}$ and so that the Meggitt decoder then attempts to correct an error, if it exists, in the new position $X^{n-1}$. With this strategy, instead of having to store $2^{n-k}$ syndromes in a table, the designer need only store syndromes corresponding to error patterns having one of the errors in position $X^{n-1}$. For example, if the cyclic code is a three-error corrector, the syndrome-error table need only store syndromes for one weight-one error pattern (the error pattern with an error in position $X^{n-1}$), for $n - 1$ weight-two error patterns (each having an error in position $X^{n-1}$), and for $(n-1)(n-2)/2$ weight-three error patterns. For some three-error-correcting codes, $n - k = 24$ so that the standard syndrome table would be of size $2^{24} \approx 16.7$ million, a much larger table than a Meggitt syndrome table.

Additional complexity savings can be realized by a simplified update to the syndrome calculation for each cyclic shift $\mathbf{r}^{(1)}(X)$ (by one place) of the received word $\mathbf{r}(X)$. This simplified update relies on the following result.

**Theorem 3.6** *Let* $\mathbf{g}(X) = (X^n+1)/\mathbf{h}(X)$ *be a generator of a cyclic code and define the syndrome of the received word* $\mathbf{r}(X)$ *by* $\mathbf{s}(X) = R_{\mathbf{g}}[\mathbf{r}(X)]$, *the remainder upon dividing* $\mathbf{r}(X)$ *by* $\mathbf{g}(X)$. *Then the syndrome for the cyclic shift of* $\mathbf{r}(X)$, $\mathbf{r}^{(1)}(X) = X\mathbf{r}(X)$ *mod* $(X^n + 1)$, *is given by* $R_{\mathbf{g}}[X\mathbf{s}(X)]$.

*Proof* The definition $\mathbf{s}(X) = R_{\mathbf{g}}[\mathbf{r}(X)]$ implies

$$\begin{aligned}
\mathbf{r}(X) &= \mathbf{g}(X)Q(X) + \mathbf{s}(X),\\
X\mathbf{r}(X) &= X\mathbf{g}(X)Q(X) + X\mathbf{s}(X),\\
X\mathbf{r}(X) \bmod (X^n+1) &= [X\mathbf{g}(X)Q(X) + X\mathbf{s}(X)] \bmod (X^n+1),\\
X\mathbf{r}(X) \bmod (X^n+1) &= [X\mathbf{g}(X)Q(X)] \bmod (X^n+1) + X\mathbf{s}(X),\\
R_{\mathbf{g}}[X\mathbf{r}(X) \bmod (X^n+1)] &= R_{\mathbf{g}}[X\mathbf{g}(X)Q(X) \bmod (X^n+1)] + R_{\mathbf{g}}[X\mathbf{s}(X)],\\
R_{\mathbf{g}}[X\mathbf{r}(X) \bmod (X^n+1)] &= R_{\mathbf{g}}[X\mathbf{s}(X)].
\end{aligned}$$

The fourth line holds because $X\mathbf{s}(X)$ has degree less than $n$ for nontrivial codes, so that $X\mathbf{s}(X) \bmod (X^n + 1) = X\mathbf{s}(X)$. The last line holds because in the previous line $R_{\mathbf{g}}[X\mathbf{g}(X)Q(X) \bmod (X^n+1)] = 0$ as a consequence of the fact that, from the division algorithm, the remainder $X\mathbf{g}(X)Q(X) \bmod (X^n + 1)$ can be written as $X\mathbf{g}(X)Q(X) - (X^n + 1)Q_0(X)$, which is divisible by $\mathbf{g}(X)$. □

It is assumed that the Meggitt decoder we now present possesses a *syndrome list* that contains all of the syndromes corresponding to correctable error patterns with an error in position $X^{n-1}$. Let the first entry in this list be $\mathbf{s}_1(X) = R_{\mathbf{g}}[X^{n-1}]$, the syndrome corresponding to an error in position $X^{n-1}$ (and no other errors). The Meggitt decoding algorithm is as follows.

1. Compute the syndrome $\mathbf{s}(X) = R_{\mathbf{g}}[\mathbf{r}(X)]$. If the syndrome is zero, then declare no errors exist, decoding is complete.
2. If the nonzero syndrome $\mathbf{s}(X)$ is in the syndrome list:
   (a) Invert (i.e., correct) the bit in position $X^{n-1}$.
   (b) Remove from $\mathbf{s}(X)$ the syndrome contribution corresponding to the bit error just removed: $\mathbf{s}(X) \leftarrow \mathbf{s}(X) + \mathbf{s}_1(X)$.
   (c) If updated syndrome $\mathbf{s}(X)$ is zero, cease decoding.
3. If all bit positions are exhausted, cease decoding; else, cyclically shift the received word, $\mathbf{r}(X) \leftarrow X\mathbf{r}(X) \bmod (X^n + 1)$, and update the syndrome so that it corresponds to the shifted received word: $\mathbf{s}(X) \leftarrow R_{\mathbf{g}}[X\mathbf{s}(X)]$. Return to Step 2.

Observe that if $\mathbf{s}(X)$ is not in the syndrome list, nothing is done in Step 2 and the decoder proceeds to Step 3. Observe also that cyclic shifting in Step 3 is unnecessary and instead a pointer to the bit under consideration can be moved to the next bit. Further, a shortened cyclic code – which is not cyclic – can be decoded using this algorithm. Lastly, there is nothing special about position $X^{n-1}$ and the syndrome table can be designed for all correctable error patterns with an error in position $X^0$, for example.

What we have presented here is a specific type of Meggitt decoder. Variations on, and simplifications of, this decoder may be found in [2, 5].

### 3.2.2 CRC Codes

Cyclic codes, usually shortened, used for error detection are called *cyclic redundancy check* (CRC) codes. The error detection feature of possibly shortened cyclic codes derives from the fact that every code polynomial is a multiple of the code's generator polynomial, $\mathbf{g}(X)$. With the guarantee that every transmitted code polynomial has the form $\mathbf{v}(X) = \mathbf{a}(X)\mathbf{g}(X)$, the presence of errors in the received polynomial $\mathbf{r}(X)$ can be discerned by checking if $\mathbf{r}(X)$ is a multiple of $\mathbf{g}(X)$, that is, if the remainder upon dividing $\mathbf{r}(X)$ by $\mathbf{g}(X)$ is the zero polynomial. We denote this remainder by $\mathrm{R}_{\mathbf{g}}[\mathbf{r}(X)]$. With $\mathbf{r}(X)$ of the form $\mathbf{r}(X) = \mathbf{v}(X) + \mathbf{e}(X)$, where $\mathbf{e}(X)$ is a (possibly zero) degree-$(n-1)$ error polynomial, we have

$$\begin{aligned} \mathrm{R}_{\mathbf{g}}[\mathbf{r}(X)] &= \mathrm{R}_{\mathbf{g}}[\mathbf{v}(X) + \mathbf{e}(X)] \\ &= \mathrm{R}_{\mathbf{g}}[\mathbf{v}(X)] + \mathrm{R}_{\mathbf{g}}[\mathbf{e}(X)] \\ &= \mathrm{R}_{\mathbf{g}}[\mathbf{e}(X)] \\ &\triangleq \mathbf{s}(X). \end{aligned}$$

Here, $\mathbf{s}(X)$ is called the *syndrome polynomial* and an error is said to be detected whenever $\mathbf{s}(X)$ is nonzero. $\mathbf{r}(X)$ can be claimed to be error-free if $\mathbf{s}(X)$ is the zero polynomial.

If $\mathbf{s}(X) \neq 0$, the decoder's conclusion that an error is present will always be correct. If $\mathbf{s}(X) = 0$, the decoder's conclusion that no errors are present can be wrong, but typically with a small probability, as we will now explain.

When $\mathbf{s}(X) = 0$, this means $R_{\mathbf{g}}[\mathbf{e}(X)] = 0$, implying that $\mathbf{e}(X)$ is zero or is a nonzero multiple of $\mathbf{g}(X)$, that is, $\mathbf{e}(X)$ is a code polynomial. The probability that $\mathbf{e}(X)$ is a code polynomial depends on the channel in place and equation (3.16) gives this probability for the binary symmetric channel assuming the weight distribution of the code is known. In the event that the weight distribution is not known, we can still make some useful statements independently of the channel specifics. Because $\mathbf{e}(X)$ has $n$ coefficients, there are $2^n$ possible error polynomials. Because there are only $2^k$ code polynomials, the fraction of error polynomials that are code polynomials (hence, are undetectable) is $2^k/2^n = 1/2^{n-k}$. It is for this reason that the number of CRC parity bits, $r = n - k$, is usually chosen to be at least 16.

Another useful fact is that all error patterns of length $r = n-k$ or less are detectable. To see this, first recall that the decoder is only fooled when $\mathbf{e}(X)$ is a nonzero multiple of $\mathbf{g}(X)$, which has degree $n - k$, so that the decoder can always detect $\mathbf{e}(X)$ if the first and last nonzero coefficients in $\mathbf{e}(X)$ are contained in a span of $n - k$ or fewer bits. This is because error polynomials of the form $X^j\mathbf{g}(X)$ and $\mathbf{b}(X)\mathbf{g}(X)$ (for some polynomial $\mathbf{b}(X)$) cannot be contained in such a span since the coefficients $g_0$ and $g_{n-k}$ in $\mathbf{g}(X)$ are both nonzero. In the case when the CRC is a (not shortened) cyclic code, these statements need to be refined to include "wrap-around" error patterns. In this case, we would consider error patterns of the form $X^j\mathbf{g}(X) \bmod (X^n + 1)$ and $\mathbf{b}(X)\mathbf{g}(X) \bmod (X^n + 1)$.

Because they are shortened cyclic codes, CRC codes are encoded using the cyclic code encoder in Figure 3.3. Interestingly, the CRC code encoder can also be used as the core of the decoder, that is, as a syndrome computer. Recall that the encoder takes an input, multiplies it by $X^{n-k}$ (in effect), and then divides the result by $\mathbf{g}(X)$, leaving the remainder in the shift register. If the input is the received word $\mathbf{r}(X)$ and all $n$ bits of this word are shifted into the encoder, the remainder $R_{\mathbf{g}}[X^{n-k}\mathbf{r}(X)]$ in the encoder shift register will be zero if and only if $\mathbf{r}(X)$ is a multiple of $\mathbf{g}(X)$. Thus, the encoder remainder serves as a syndrome that is equally as effective as $\mathbf{s}(X) = R_{\mathbf{g}}[\mathbf{r}(X)]$.

CRC code design can be based on the BCH codes to be described in the next section, or they can involve computer-aided designs as in [16–18]. Frequently, a CRC generator is chosen to be a primitive polynomial (so $d_{\min} = 3$) times $X + 1$ (so $d_{\min} = 4$). Some commonly used CRC generator polynomials are listed in the table below, along with their origins or associations in parentheses. The polynomials are represented as hex strings and the convention is that the least significant bit corresponds to the $X^0$ coefficient. For example, the entry 1A7 (hex) corresponds to the polynomial $X^8 + X^7 + X^5 + X^2 + X + 1$. It should be pointed out that, if the coefficients were to be reversed, the CRC code would have identical performance in virtually all situations.

| Degree $r$ | $\mathbf{g}(X)$ | $\mathbf{g}(X)$ |
|---|---|---|
| 8 | 1A7 (Bluetooth) | 107 (CCITT) |
| 16 | 18005 (IBM) | 11021 (CCITT) |
| 24 | 131FF19 ([17]) | 15BC4F5 ([17]) |
| 32 | 104C110B7 (IEEE-802) | 11EDC6F41 ([17]) |

Our discussion to this point on CRC codes has assumed that CRC error detection is the last function performed before data is delivered to the user, usually with a flag indicating whether or not the data is to be trusted (i.e., whether or not it has passed the CRC check). We will see in later chapters that CRC codes can also aid the error-correction process for convolutional codes and polar codes (and likely other codes). As an example, a CRC outer code is concatenated with a convolutional inner code and a convolutional *list decoder* produces a metric-ordered list of candidates that is sent to the CRC decoder. The first candidate in the list that passes the CRC check is the ultimate decoder output.

## 3.3 BCH Codes

BCH (Bose–Chaudhuri–Hocquenghem) codes form a large class of cyclic codes for correcting multiple random errors. This class of codes was first discovered by Hocquenghem in 1959 [19] and then independently by Bose and Chaudhuri in 1960 [20]. The first algorithm for decoding of binary BCH codes was devised by Peterson [10]. Since then, the Peterson decoding algorithm has been improved and generalized by others [1, 21]. In this section, we introduce a subclass of binary BCH codes that is the most important subclass from the standpoint of both theory and practice.

### 3.3.1 Code Construction

BCH codes are specified in terms of the roots of their generator polynomials in finite fields. For any positive integers $m$ and $t$ with $m \geq 3$ and $t < 2^{m-1}$, there exists a binary BCH code of length $2^m - 1$ with minimum distance at least $2t + 1$ and at most $mt$ parity-check digits. This BCH code is capable of correcting $t$ or fewer random errors over a span of $2^m - 1$ transmitted code bits. It is called a *t-error-correcting* BCH code.

Construction of a $t$-error-correcting BCH code begins with a Galois field GF($2^m$). Let $\alpha$ be a primitive element in GF($2^m$). The generator polynomial $\mathbf{g}(X)$ of the $t$-error-correcting binary BCH code of length $2^m - 1$ is the *smallest-degree* (or *minimum-degree*) polynomial over GF(2) that has the following $2t$ consecutive powers of $\alpha$:

$$\alpha, \alpha^2, \alpha^3, \ldots, \alpha^{2t} \tag{3.39}$$

as *roots*. It follows from Theorem 2.11 that $\mathbf{g}(X)$ has $\alpha, \alpha^2, \ldots, \alpha^{2t}$ and their conjugates as all of its roots. Since all the nonzero elements of GF($2^m$) are roots of $X^{2^m-1}+1$ and all the roots of $\mathbf{g}(X)$ are elements in GF($2^m$), $\mathbf{g}(X)$ divides $X^{2^m-1} + 1$. Since $\mathbf{g}(X)$ is a factor of $X^{2^m-1} + 1$, it can be used to generate a cyclic code of length $2^m - 1$.

For $1 \leq i \leq 2t$, let $\phi_i(X)$ be the minimal polynomial of $\alpha^i$. Then the generator polynomial of the $t$-error-correcting binary BCH code of length $2^m - 1$ is given by the *least common multiple* (*LCM*) of $\phi_1(X), \phi_2(X), \ldots, \phi_{2t}(X)$, that is

$$\mathbf{g}(X) = \text{LCM}\{\phi_1(X), \phi_2(X), \ldots, \phi_{2t}(X)\}. \tag{3.40}$$

If $i$ is an even integer, it can be expressed as a product of an odd integer $i'$ and a power of 2 as follows:

$$i = i'2^l.$$

Then $\alpha^i$ is a conjugate of $\alpha^{i'}$, and hence $\alpha^i$ and $\alpha^{i'}$ have the same minimal polynomial, that is

$$\phi_i(X) = \phi_{i'}(X).$$

Therefore, every even power of $\alpha$ in the sequence of (3.39) has the same minimal polynomial as some previous odd power of $\alpha$ in the sequence. As a result, the generator polynomial of the $t$-error-correcting binary BCH code of length $2^m - 1$ given by (3.40) can be simplified as follows:

$$\mathbf{g}(X) = \text{LCM}\{\phi_1(X), \phi_3(X), \ldots, \phi_{2t-1}(X)\}. \tag{3.41}$$

There are only $t$ minimal polynomials in the LCM expression of (3.41). Since the degree of a minimal polynomial of an element in GF($2^m$) is *at most m*, the degree of $\mathbf{g}(X)$ is then at most $mt$. Therefore, the $t$-error-correcting binary BCH code generated by $\mathbf{g}(X)$ has at most $mt$ parity-check bits and hence its dimension is at least $2^m - mt - 1$. For a given field GF($2^m$) with $m \geq 3$, the above construction gives a family of binary BCH codes of length $2^m - 1$ with various rates and error-correction capabilities. The BCH codes defined above are called *primitive* (or *narrow-sense*) BCH codes since they are defined by consecutive powers of a primitive element $\alpha$ in GF($2^m$).

Consider a primitive $t$-error-correcting BCH code $\mathcal{C}$ constructed using GF($2^m$), whose generator polynomial $\mathbf{g}(X)$ has $\alpha, \alpha^2, \ldots, \alpha^{2t}$ and their conjugates as roots. Then, for $1 \leq i \leq 2t$, $\mathbf{g}(\alpha^i) = 0$. Let $\mathbf{v}(X) = v_0 + v_1X + \cdots + v_{2^m-2}X^{2^m-2}$ be a code polynomial in $\mathcal{C}$. Since $\mathbf{v}(X)$ is divisible by $\mathbf{g}(X)$, a root of $\mathbf{g}(X)$ is also a root of $\mathbf{v}(X)$. Hence, for $1 \leq i \leq 2t$,

$$\mathbf{v}(\alpha^i) = v_0 + v_1\alpha^i + v_2\alpha^{2i} + \cdots + v_{2^m-2}\alpha^{(2^m-2)i} = 0. \tag{3.42}$$

The above equality can be written as the following matrix product:

$$(v_0, v_1, \ldots, v_{2^m-2}) \cdot \begin{pmatrix} 1 \\ \alpha^i \\ \alpha^{2i} \\ \vdots \\ \alpha^{(2^m-2)i} \end{pmatrix} = 0. \tag{3.43}$$

Equation (3.43) simply says that the inner product of the codeword $\mathbf{v} = (v_0, v_1, \ldots, v_{2^m-1})$ and $(1, \alpha^i, \ldots, a^{(2^m-2)i})$ over GF($2^m$) is equal to zero. Form the following $2t \times (2^m - 1)$ matrix over GF($2^m$):

$$\mathbf{H} = \begin{bmatrix} 1 & \alpha & \alpha^2 & \cdots & \alpha^{2^m-2} \\ 1 & \alpha^2 & (\alpha^2)^2 & \cdots & (\alpha^2)^{2^m-2} \\ 1 & \alpha^3 & (\alpha^3)^2 & \cdots & (\alpha^3)^{2^m-2} \\ \vdots & \vdots & \vdots & \ddots & \vdots \\ 1 & \alpha^{2t} & (\alpha^{2t})^2 & \cdots & (\alpha^{2t})^{2^m-2} \end{bmatrix}. \tag{3.44}$$

It follows from (3.43) that, for every codeword $\mathbf{v} = (v_0, v_1, \ldots, v_{2^m-2})$ in the $t$-error-correcting BCH code $\mathcal{C}$ of length $2^m - 1$ generated by $\mathbf{g}(X)$, the following condition holds:

$$\mathbf{v} \cdot \mathbf{H}^{\mathrm{T}} = \mathbf{0}. \tag{3.45}$$

On the other hand, if a $(2^m - 1)$-tuple $\mathbf{v} = (v_0, v_1, \ldots, v_{2^m-1})$ over GF(2) satisfies the condition given by (3.45), then it follows from (3.43) and (3.42) that its corresponding polynomial $\mathbf{v}(X) = v_0 + v_1X + \cdots + v_{2^m-2}X^{2^m-2}$ has $\alpha, \alpha^2, \ldots, \alpha^{2t}$ and their conjugates as roots. As a result, $\mathbf{v}(X)$ is divisible by the generator polynomial $\mathbf{g}(X)$ of the $t$-error-correcting BCH code of length $2^m - 1$, and hence is a code polynomial. Therefore, the $t$-error-correcting primitive BCH code of length $2^m - 1$ generated by the generator polynomial $\mathbf{g}(X)$ given by (3.41) is the null space over GF(2) of $\mathbf{H}$; and $\mathbf{H}$ is a parity-check matrix of the code.

For $d \leq 2t$, any $d \times d$ submatrix of the parity-check matrix $\mathbf{H}$ is a generalized *Vandermonde matrix* whose determinant is nonzero. This implies that any $d$ columns of $\mathbf{H}$ are linearly independent and hence cannot be summed to a zero column vector. It follows from Theorem 3.3 that the BCH code given by the null space of $\mathbf{H}$ has minimum distance at least $2t + 1$. Hence, the BCH code can correct any error pattern with $t$ or fewer random errors during the transmission of a codeword over the BSC. The parameter $2t + 1$ is called the *designed minimum distance* of the $t$-error-correcting BCH code. The true minimum distance of the code may be greater than its designed minimum distance $2t + 1$. So, $2t + 1$ is a lower bound on the minimum distance of the BCH code. The proof of this bound is based on the fact that the generator polynomial of the code has $2t$ consecutive powers of a primitive element $\alpha$ in GF($2^m$) as roots. This bound is referred to as the *BCH bound*.

---

**Example 3.2** Let GF($2^6$) be the code construction field that is generated by the primitive polynomial $p(X) = 1 + X + X^6$. The elements of this field in three forms are given in Table 3.3. Let $\alpha$ be a primitive element of GF($2^6$). Set $t = 3$. Then the generator polynomial $\mathbf{g}(X)$ of the triple-error-correcting primitive BCH code of length 63 has

$$\alpha, \alpha^2, \alpha^3, \alpha^4, \alpha^5, \alpha^6$$

and their conjugates as roots. Note that $\alpha$, $\alpha^2$, and $\alpha^4$ are conjugate elements and have the same minimal polynomial, which is $1 + X + X^6$. The elements $\alpha^3$ and $\alpha^6$ are conjugates and their minimal polynomial is $1 + X + X^2 + X^4 + X^6$. The minimal polynomial of $\alpha^5$ is $1 + X + X^2 + X^5 + X^6$. It follows from (3.41) that the generator polynomial of the triple-error-correction primitive BCH code of length 63 is given by

$$\begin{aligned}\mathbf{g}(X) &= (1 + X + X^6)(1 + X + X^2 + X^4 + X^6)(1 + X + X^2 + X^5 + X^6)\\ &= 1 + X + X^2 + X^3 + X^6 + X^7 + X^9 + X^{15} + X^{16} + X^{17} + X^{18}.\end{aligned}$$

Since the degree of $\mathbf{g}(X)$ is 18, the BCH code generated by $\mathbf{g}(X)$ is a $(63, 45)$ code with designed minimum distance 7. However, it can be proved that the true minimum distance is also 7.

---

**Table 3.3** GF($2^6$) generated by the primitive polynomial $p(X) = 1 + X + X^6$ over GF(2)

| Power representation | Vector representation | Power representation | Vector representation |
|---|---|---|---|
| 0 | (0 0 0 0 0 0) | $\alpha^{32}$ | (1 0 0 1 0 0) |
| 1 | (1 0 0 0 0 0) | $\alpha^{33}$ | (0 1 0 0 1 0) |
| $\alpha$ | (0 1 0 0 0 0) | $\alpha^{34}$ | (0 0 1 0 0 1) |
| $\alpha^2$ | (0 0 1 0 0 0) | $\alpha^{35}$ | (1 1 0 1 0 0) |
| $\alpha^3$ | (0 0 0 1 0 0) | $\alpha^{36}$ | (0 1 1 0 1 0) |
| $\alpha^4$ | (0 0 0 0 1 0) | $\alpha^{37}$ | (0 0 1 1 0 1) |
| $\alpha^5$ | (0 0 0 0 0 1) | $\alpha^{38}$ | (1 1 0 1 1 0) |
| $\alpha^6$ | (1 1 0 0 0 0) | $\alpha^{39}$ | (0 1 1 0 1 1) |
| $\alpha^7$ | (0 1 1 0 0 0) | $\alpha^{40}$ | (1 1 1 1 0 1) |
| $\alpha^8$ | (0 0 1 1 0 0) | $\alpha^{41}$ | (1 0 1 1 1 0) |
| $\alpha^9$ | (0 0 0 1 1 0) | $\alpha^{42}$ | (0 1 0 1 1 1) |
| $\alpha^{10}$ | (0 0 0 0 1 1) | $\alpha^{43}$ | (1 1 1 0 1 1) |
| $\alpha^{11}$ | (1 1 0 0 0 1) | $\alpha^{44}$ | (1 0 1 1 0 1) |
| $\alpha^{12}$ | (1 0 1 0 0 0) | $\alpha^{45}$ | (1 0 0 1 1 0) |
| $\alpha^{13}$ | (0 1 0 1 0 0) | $\alpha^{46}$ | (0 1 0 0 1 1) |
| $\alpha^{14}$ | (0 0 1 0 1 0) | $\alpha^{47}$ | (1 1 1 0 0 1) |
| $\alpha^{15}$ | (0 0 0 1 0 1) | $\alpha^{48}$ | (1 0 1 1 0 0) |
| $\alpha^{16}$ | (1 1 0 0 1 0) | $\alpha^{49}$ | (0 1 0 1 1 0) |
| $\alpha^{17}$ | (0 1 1 0 0 1) | $\alpha^{50}$ | (0 0 1 0 1 1) |
| $\alpha^{18}$ | (1 1 1 1 0 0) | $\alpha^{51}$ | (1 1 0 1 0 1) |
| $\alpha^{19}$ | (0 1 1 1 1 0) | $\alpha^{52}$ | (1 0 1 0 1 0) |
| $\alpha^{20}$ | (0 0 1 1 1 1) | $\alpha^{53}$ | (0 1 0 1 0 1) |
| $\alpha^{21}$ | (1 1 0 1 1 1) | $\alpha^{54}$ | (1 1 1 0 1 0) |
| $\alpha^{22}$ | (1 0 1 0 1 1) | $\alpha^{55}$ | (0 1 1 1 0 1) |
| $\alpha^{23}$ | (1 0 0 1 0 1) | $\alpha^{56}$ | (1 1 1 1 1 0) |
| $\alpha^{24}$ | (1 0 0 0 1 0) | $\alpha^{57}$ | (0 1 1 1 1 1) |
| $\alpha^{25}$ | (0 1 0 0 0 1) | $\alpha^{58}$ | (1 1 1 1 1 1) |
| $\alpha^{26}$ | (1 1 1 0 0 0) | $\alpha^{59}$ | (1 0 1 1 1 1) |
| $\alpha^{27}$ | (0 1 1 1 0 0) | $\alpha^{60}$ | (1 0 0 1 1 1) |
| $\alpha^{28}$ | (0 0 1 1 1 0) | $\alpha^{61}$ | (1 0 0 0 1 1) |
| $\alpha^{29}$ | (0 0 0 1 1 1) | $\alpha^{62}$ | (1 0 0 0 0 1) |
| $\alpha^{30}$ | (1 1 0 0 1 1) | $\alpha^{63} = 1$ | |
| $\alpha^{31}$ | (1 0 1 0 0 1) | | |

### 3.3.2 Bounded-Distance Decoding

Suppose a code polynomial $\mathbf{v}(X)$ of a $t$-error-correcting primitive BCH code is transmitted over the BSC channel and $\mathbf{r}(X)$ is the corresponding received polynomial. The syndrome of $\mathbf{r}(X)$ is given by

$$\mathbf{S} = (S_1, S_2, \ldots, S_{2t}) = \mathbf{r} \cdot \mathbf{H}^{\mathrm{T}}, \tag{3.46}$$

which consists of $2t$ syndrome components. It follows from (3.44), (3.45), and (3.46) that, for $1 \leq i \leq 2t$, the $i$th syndrome component is given by

$$S_i = \mathbf{r}(\alpha^i) = r_0 + r_1\alpha^i + \cdots + r_{2^m-2}\alpha^{(2^m-2)i}, \tag{3.47}$$

which is an element in GF($2^m$). An effective algebraic hard-decision decoding algorithm, called the *Berlekamp–Massey* algorithm [1, 21], has been devised assuming as input the syndrome $\mathbf{S}$ (also see [2, 5, 11] for details).

Let $\mathbf{e}(X) = e_0 + e_1X + \cdots + e_{2^m-2}X^{2^m-2}$ be the error pattern induced by the channel noise during the transmission of the code polynomial $\mathbf{v}(X)$, then the received polynomial $\mathbf{r}(X) = \mathbf{v}(X) + \mathbf{e}(X)$. It follows from (3.47) that, for $1 \leq i \leq 2t$,

$$S_i = \mathbf{r}(\alpha^i) = \mathbf{v}(\alpha^i) + \mathbf{e}(\alpha^i). \tag{3.48}$$

Since $\mathbf{v}(\alpha^i) = 0$, we have

$$S_i = \mathbf{e}(\alpha^i), \tag{3.49}$$

for $1 \leq i \leq 2t$. Suppose $\mathbf{e}(X)$ contains $\nu$ errors at the locations $j_1, j_2, \ldots, j_\nu$, where $0 \leq j_1 < j_2 < \cdots < j_\nu < n$. Then

$$\mathbf{e}(X) = X^{j_1} + X^{j_2} + \cdots + X^{j_\nu}. \tag{3.50}$$

From (3.49) and (3.50), we obtain the following $2t$ equalities that relate the error locations $j_1, j_2, \ldots, j_\nu$ of $\mathbf{e}(X)$ and the $2t$ syndrome components $S_1, S_2, \ldots, S_{2t}$:

$$\begin{aligned} S_1 &= \mathbf{e}(\alpha) = \alpha^{j_1} + \alpha^{j_2} + \cdots + \alpha^{j_\nu}, \\ S_2 &= \mathbf{e}(\alpha^2) = (\alpha^{j_1})^2 + (\alpha^{j_2})^2 + \cdots + (\alpha^{j_\nu})^2, \\ &\vdots \\ S_{2t} &= \mathbf{e}(\alpha^{2t}) = (\alpha^{j_1})^{2t} + (\alpha^{j_2})^{2t} + \cdots + (\alpha^{j_\nu})^{2t}. \end{aligned} \tag{3.51}$$

Note that the $2t$ syndrome components $S_1, S_2, \ldots, S_{2t}$ are known, but the error locations $j_1, j_2, \ldots, j_\nu$ are unknown. If we can solve the above $2t$ equations, we can determine $\alpha^{j_1}, \alpha^{j_2}, \ldots, \alpha^{j_\nu}$, whose exponents then give the locations of errors in the error pattern $\mathbf{e}(X)$. Since the elements $\alpha^{j_1}, \alpha^{j_2}, \ldots, \alpha^{j_\nu}$ in GF($2^m$) give the locations of the errors in $\mathbf{e}(X)$, they are referred to as the *error-location numbers*.

For $1 \leq l \leq \nu$, define

$$\beta_l = \alpha^{j_l}. \tag{3.52}$$

Then the $2t$ equations given by (3.51) can be expressed in a simpler form as follows:

$$\begin{aligned} S_1 &= \beta_1 + \beta_2 + \cdots + \beta_\nu, \\ S_2 &= \beta_1^2 + \beta_2^2 + \cdots + \beta_\nu^2, \\ &\vdots \\ S_{2t} &= \beta_1^{2t} + \beta_2^{2t} + \cdots + \beta_\nu^{2t}. \end{aligned} \tag{3.53}$$

The $2t$ equations given by (3.53) are called the *power-sum symmetric functions*. Since they are nonlinear equations, solving them directly would be very difficult, if not impossible. In the following, an indirect method for solving these equations is provided.

Define the following polynomial of degree $v$ over GF($2^m$):

$$\begin{aligned}\sigma(X) &= (1+\beta_1 X)(1+\beta_2 X)\cdots(1+\beta_v X)\\ &= \sigma_0 + \sigma_1 X + \cdots + \sigma_v X^v,\end{aligned} \tag{3.54}$$

so that

$$\begin{aligned}&\sigma_0 = 1,\\ &\sigma_1 = \textstyle\sum_{i=1}^{v} \beta_i = \beta_1 + \beta_2 + \cdots + \beta_v,\\ &\sigma_2 = \textstyle\sum_{1\le i<j\le v} \beta_i\beta_j = \beta_1\beta_2 + \beta_1\beta_3 + \cdots + \beta_{v-1}\beta_v,\\ &\sigma_3 = \textstyle\sum_{1\le i<j<k\le v} \beta_i\beta_j\beta_k = \beta_1\beta_2\beta_3 + \beta_1\beta_2\beta_4 + \cdots + \beta_{v-1}\beta_{v-2}\beta_{v-3},\\ &\vdots\\ &\sigma_v = \beta_1\beta_2\cdots\beta_v.\end{aligned} \tag{3.55}$$

The polynomial $\sigma(X)$ has $\beta_1^{-1}, \beta_2^{-1}, \ldots, \beta_v^{-1}$ (the *inverses* of the location numbers $\beta_1, \beta_2, \ldots, \beta_v$) as roots. This polynomial is called the *error-location polynomial*. If this polynomial can be found, then the inverses of its roots give the error-location numbers. Using (3.52), we can determine the error locations of the error pattern $\mathbf{e}(X)$. The $v$ equalities given by (3.55) are called the *elementary-symmetric functions* that relate the coefficients of the error-location polynomial $\sigma(X)$ to the error-location numbers $\beta_1, \beta_2, \ldots, \beta_v$.

From (3.53) and (3.55), we see that the syndrome components $S_1, S_2, \ldots, S_{2t}$ and the coefficients of the error-location polynomial $\sigma(X)$ are related through the error-location numbers $\beta_1, \beta_2, \ldots, \beta_v$. From these two sets of equations, it is possible to derive the following equations that relate the coefficients of the error-location polynomial $\sigma(X)$ to the syndrome components computed from the received polynomial $\mathbf{r}(X)$ [1, 10, 21]:

$$\begin{aligned}&S_1 + \sigma_1 = 0,\\ &S_2 + \sigma_1 S_1 + 2\sigma_2 = 0,\\ &S_3 + \sigma_1 S_2 + \sigma_2 S_1 + 3\sigma_3 = 0,\\ &\vdots\\ &S_v + \sigma_1 S_{v-1} + \sigma_2 S_{v-2} + \cdots + \sigma_{v-1} S_1 + v\sigma_v = 0,\\ &S_{v+1} + \sigma_1 S_v + \sigma_2 S_{v-1} + \cdots + \sigma_{v-1} S_2 + \sigma_v S_1 = 0,\\ &\vdots\end{aligned} \tag{3.56}$$

Since $1 + 1 = 0$ under modulo-2 addition, we have $i\sigma_i = \sigma_i$ for odd $i$ and $i\sigma_i = 0$ for even $i$. The identities given by (3.56) are called the *Newton identities*.

If we can determine the coefficients $\sigma_1, \sigma_2, \ldots, \sigma_v$ of the error-location polynomial $\sigma(X)$ from the Newton identities, then we can determine $\sigma(X)$. Once $\sigma(X)$ has been determined, we find its roots. By taking the inverses of the roots of $\sigma(X)$, we obtain the error-location numbers $\beta_1, \beta_2, \ldots, \beta_v$ of the error pattern $\mathbf{e}(X)$. From

(3.52) and (3.50), we can determine the error pattern $\mathbf{e}(X)$. On removing $\mathbf{e}(X)$ from the received polynomial $\mathbf{r}(X)$ (i.e., adding $\mathbf{e}(X)$ to $\mathbf{r}(X)$ using modulo-2 addition of their corresponding coefficients), we obtain the decoded codeword $\mathbf{v}(X)$. From the description given above, a procedure for decoding a BCH code is as follows.

1. Compute the syndrome $\mathbf{S} = (S_1, S_2, \ldots, S_{2t})$ of the received polynomial $\mathbf{r}(X)$.
2. Determine the error-location polynomial $\sigma(X)$ from the Newton identities.
3. Determine the roots $\beta_1^{-1}, \beta_2^{-1}, \ldots, \beta_\nu^{-1}$ of $\sigma(X)$ in GF($2^m$). Take the inverses of these roots to obtain the error-location numbers $\beta_1 = \alpha^{j_1}, \beta_2 = \alpha^{j_2}, \ldots, \beta_\nu = \alpha^{j_\nu}$. Form the error pattern

$$\mathbf{e}(X) = X^{j_1} + X^{j_2} + \cdots + X^{j_\nu}.$$

4. Perform the error correction by adding $\mathbf{e}(X)$ to $\mathbf{r}(X)$. This gives the decoded code polynomial $\mathbf{v}(X) = \mathbf{r}(X) + \mathbf{e}(X)$.

Steps 1, 3, and 4 can be carried out easily. However, Step 2 involves solving the Newton identities. There will in general be more than one error pattern for which the coefficients of its error-location polynomial satisfy the Newton identities. To minimize the probability of a decoding error, we need to find the *most probable* error pattern for error correction. For the BSC, finding the most probable error pattern means determining the error-location polynomial of minimum degree whose coefficients satisfy the Newton identities given by (3.56). This can be achieved iteratively by using an algorithm first devised by Berlekamp [1] and then improved by Massey [21]. Such an algorithm is commonly referred to as the Berlekamp–Massey (BM) algorithm.

The BM algorithm is used to find the error-location polynomial $\sigma(X)$ iteratively in $2t$ steps. For $1 \leq k \leq 2t$, the algorithm at the $k$th step gives an error-location polynomial of minimum degree,

$$\sigma^{(k)}(X) = \sigma_0^{(k)} + \sigma_1^{(k)} X + \cdots + \sigma_{l_k}^{(k)} X^{l_k}, \tag{3.57}$$

whose coefficients satisfy the *first* $k$ Newton identities. At the $(k+1)$th step, the algorithm is used to find the next minimum-degree error-location polynomial $\sigma^{(k+1)}(X)$ whose coefficients satisfy the first $k+1$ Newton identities based on $\sigma^{(k)}(X)$. First, we check whether the coefficients of $\sigma^{(k)}(X)$ also satisfy the $(k+1)$th Newton identity. If yes, then

$$\sigma^{(k+1)}(X) = \sigma^{(k)}(X) \tag{3.58}$$

is the minimum-degree error-location polynomial whose coefficients satisfy the first $k+1$ Newton identities. If no, a correction term is added to $\sigma^{(k)}(X)$ to form a minimum-degree error-location polynomial,

$$\sigma^{(k+1)}(X) = \sigma_0^{(k+1)} + \sigma_1^{(k+1)} X + \cdots + \sigma_{l_{k+1}}^{(k+1)} X^{l_{k+1}}, \tag{3.59}$$

whose coefficients satisfy the first $k+1$ Newton identities. To test whether the coefficients of $\sigma^{(k)}(X)$ satisfy the $(k+1)$th Newton identity, we compute

$$d_k = S_{k+1} + \sigma_1^{(k)} S_k + \sigma_2^{(k)} S_{k-1} + \cdots + \sigma_{l_k}^{(k)} S_{k+1-l_k}. \tag{3.60}$$

Note that the sum on the right-hand side of the equality (3.60) is actually equal to the left-hand side of the $(k+1)$th Newton identity given by (3.56). If $d_k = 0$, then the coefficients of $\sigma^{(k)}(X)$ satisfy the $(k+1)$th Newton identity. If $d_k \neq 0$, then the coefficients of $\sigma^{(k)}(X)$ do not satisfy the $(k+1)$th Newton identity. In this case, $\sigma^{(k)}(X)$ needs to be adjusted to obtain a new minimum-degree error-location polynomial $\sigma^{(k+1)}(X)$ whose coefficients satisfy the first $k+1$ Newton identities. The quantity $d_k$ is called the $k$th *discrepancy*.

If $d_k \neq 0$, a *correction term* is added to $\sigma^{(k)}(X)$ to obtain the next minimum-degree error-location polynomial $\sigma^{(k+1)}(X)$. First, we go back to the steps prior to the $k$th step and determine a step $i$ at which the minimum-degree error-location polynomial is $\sigma^{(i)}(X)$ such that the $i$th discrepancy $d_i \neq 0$ and $i - l_i$ has the largest value, where $l_i$ is the degree of $\sigma^{(i)}(X)$. Then

$$\sigma^{(k+1)}(X) = \sigma^{(k)}(X) + d_k d_i^{-1} X^{k-i} \sigma^{(i)}(X), \tag{3.61}$$

where

$$d_k d_i^{-1} X^{k-i} \sigma^{(i)}(X) \tag{3.62}$$

is the correction term. Since $i$ is chosen to maximize $i - l_i$, this choice of $i$ is equivalent to minimizing the degree $k-(i-l_i)$ of the correction term. We repeat the above testing and correction process until we reach the $2t$th step. Then

$$\sigma(X) = \sigma^{(2t)}(X). \tag{3.63}$$

To begin using the above BM iterative algorithm to find $\sigma(X)$, two sets of initial conditions are needed: (1) for $k = -1$, set $\sigma^{(-1)}(X) = 1$, $d_{-1} = 1$, $l_{-1} = 0$, and $-1 - l_{-1} = -1$; and (2) for $k = 0$, set $\sigma^{(0)}(X) = 1$, $d_0 = S_1$, $l_0 = 0$, and $0 - l_0 = 0$. Then the BM algorithm for finding the error-location polynomial $\sigma(X)$ can be formulated as Algorithm 3.1.

---

**Algorithm 3.1** The Berlekamp–Massey Algorithm for Finding the Error-Location Polynomial of a BCH Code

---

**Initialization.** For $k = -1$, set $\sigma^{(-1)}(X) = 1$, $d_{-1} = 1$, $l_{-1} = 0$, and $-1 - l_{-1} = -1$. For $k = 0$, set $\sigma^{(0)}(X) = 1$, $d_0 = S_1$, $l_0 = 0$ and $0 - l_0 = 0$.

1. If $k = 2t$, output $\sigma^{(k)}(X)$ as the error-location polynomial $\sigma(X)$; otherwise, go to Step 2.
2. Compute $d_k$ and go to Step 3.
3. If $d_k = 0$, set $\sigma^{(k+1)}(X) = \sigma^{(k)}(X)$; otherwise, set $\sigma^{(k+1)}(X) = \sigma^{(k)}(X) + d_k d_i^{-1} X^{k-i} \sigma^{(i)}(X)$. Go to Step 4.
4. $k \leftarrow k+1$. Go to Step 1.

---

If the number $\nu$ of errors induced by the channel noise during the transmission of a codeword in a $t$-error-correcting primitive BCH is $t$ or fewer (i.e., $\nu \leq t$), the BM algorithm guarantees to produce a *unique* error-location polynomial $\sigma(X)$ of degree $\nu$ that has $\nu$ roots in GF($2^m$). The inverses of the roots of $\sigma(X)$ give the correct locations of the errors in the received sequence $\mathbf{r}(X)$. Flipping the received bits at these locations will result in a correct decoding. The BM algorithm can be executed by setting up and filling in the following table.

| Step $k$ | Partial solution $\sigma^{(k)}(X)$ | Discrepancy $d_k$ | Degree $l_k$ | Step/degree difference $k - l_k$ |
|---|---|---|---|---|
| $-1$ | 1 | 1 | 0 | $-1$ |
| $-0$ | 1 | $S_1$ | 0 | 0 |
| 1 | $\sigma^{(1)}(X)$ | $d_1$ | $l_1$ | $1 - l_1$ |
| 2 | $\sigma^{(2)}(X)$ | $d_2$ | $l_2$ | $2 - l_2$ |
| ⋮ | | | | |
| $2t$ | $\sigma^{(2t)}(X)$ | — | — | — |

**Example 3.3** Let GF($2^4$) be a field constructed using the primitive polynomial $p(X) = 1 + X + X^4$ over GF(2) from Table 2.14. Let $\alpha$ be a primitive element of GF($2^4$). Suppose we want to construct the triple-error-correction BCH code of length $2^4 - 1 = 15$. The generator polynomial $\mathbf{g}(X)$ of this code is the minimum-degree polynomial over GF(2) that has $\alpha, \alpha^2, \alpha^3, \alpha^4, \alpha^5, \alpha^6$ and their conjugates as roots. The minimal polynomials of $\alpha$, $\alpha^3$, and $\alpha^5$ are $1 + X + X^4$, $1 + X + X^2 + X^3 + X^4$, and $1 + X + X^2$, respectively. Then it follows from (3.41) that the generator polynomial $\mathbf{g}(X)$ of the triple-error-correction BCH code is

$$\begin{aligned}\mathbf{g}(X) &= (1 + X + X^4)(1 + X + X^2 + X^3 + X^4)(1 + X + X^2)\\ &= 1 + X + X^2 + X^4 + X^5 + X^8 + X^{10}.\end{aligned}$$

Since the degree of $\mathbf{g}(X)$ is 10, the triple-error-correction BCH code generated by $\mathbf{g}(X)$ is a (15, 5) code.

Suppose the zero code polynomial is transmitted and $\mathbf{r}(X) = X^3 + X^5 + X^{12}$ is received. In this case, the error polynomial $\mathbf{e}(X)$ is identical to the received polynomial. To decode $\mathbf{r}(X)$, we compute its syndrome $\mathbf{S} = (S_1, S_2, S_3, S_4, S_5, S_6)$, where

$$\begin{aligned}S_1 &= \mathbf{r}(\alpha) = \alpha^3 + \alpha^5 + \alpha^{12} = 1, & S_2 &= \mathbf{r}(\alpha^2) = \alpha^6 + \alpha^{10} + \alpha^{24} = 1,\\ S_3 &= \mathbf{r}(\alpha^3) = \alpha^9 + \alpha^{15} + \alpha^{36} = \alpha^{10}, & S_4 &= \mathbf{r}(\alpha^4) = \alpha^{12} + \alpha^{20} + \alpha^{48} = 1,\\ S_5 &= \mathbf{r}(\alpha^5) = \alpha^{15} + \alpha^{25} + \alpha^{60} = \alpha^{10}, & S_6 &= \mathbf{r}(\alpha^6) = \alpha^{18} + \alpha^{30} + \alpha^{72} = \alpha^5.\end{aligned}$$

With the above syndrome values, executing the BM algorithm results in Table 3.4. From this table, we find that the error-location polynomial is $\sigma(X) = \sigma^{(6)}(X) = 1 + X + \alpha^5 X^3$. On substituting for the indeterminate $X$ of $\sigma(X)$ the nonzero elements of GF($2^4$), we find that the roots of $\sigma(X)$ are $\alpha^3$, $\alpha^{10}$, and $\alpha^{12}$. The inverses of these roots are $\alpha^{-3} = \alpha^{12}$, $\alpha^{-10} = \alpha^5$, and $\alpha^{-12} = \alpha^3$, respectively. Therefore, the locations of errors of the error pattern $\mathbf{e}(X)$ are $X^3$, $X^5$, and $X^{12}$, and

**Table 3.4** The decoding table for the (15, 5) BCH code given in Example 3.3

| Step $k$ | Partial solution $\sigma^{(k)}(X)$ | Discrepancy $d_k$ | Degree $l_k$ | Step/degree difference $k - l_k$ |
|---|---|---|---|---|
| −1 | 1 | 1 | 0 | −1 |
| 0 | 1 | 1 | 0 | 0 |
| 1 | $1 + X$ | 0 | 1 | 0 |
| 2 | $1 + X$ | $\alpha^5$ | 1 | 1 |
| 3 | $1 + X + \alpha^5 X^2$ | 0 | 2 | 1 |
| 4 | $1 + X + \alpha^5 X^2$ | $\alpha^{10}$ | 2 | 2 |
| 5 | $1 + X + \alpha^5 X^3$ | 0 | 3 | 2 |
| 6 | $1 + X + \alpha^5 X^3$ | — | — | — |

the error pattern is $\mathbf{e}(X) = X^3 + X^5 + X^{12}$. On adding $\mathbf{e}(X)$ to $\mathbf{r}(x)$, we obtain the zero code polynomial that is the transmitted code polynomial. This completes the decoding process.

---

Finding the roots of the error-location polynomial $\sigma(X)$, determining the error-location numbers, and the error correction can be carried out systematically all together. The received polynomial $\mathbf{r}(X) = r_0 + r_1 X + \cdots + r_{2^m-1-i} X^{2^m-1-i} + \cdots + r_{2^m-2} X^{2^m-2}$ is read out from a buffer register one bit at a time from the high-order end (i.e., $r_{2^m-2}$ first). At the same time, the indeterminate $X$ of $\sigma(X)$ is replaced by the nonzero elements of GF($2^m$), one at a time, in this order: $\alpha, \alpha^2, \ldots, \alpha^{2^m-2}, \alpha^{2^m-1}$. For $1 \leq i \leq 2^m - 1$, if $\alpha^i$ is a root of $\sigma(X)$ (i.e., $\sigma(\alpha^i) = 0$), then $\alpha^{-i} = \alpha^{2^m-1-i}$ is an error-location number. In this case, the received bit $r_{2^m-1-i}$ is changed either from 0 to 1 or from 1 to 0 as it is shifted out of the buffer register.

The BM algorithm can be used to find the error-location polynomial for either a binary BCH code or a $q$-ary BCH code over GF($q$) with $q$ as a power of a prime (see the next section). For application to a binary BCH code, it is possible to prove that, if the first, third, ..., $(2t - 1)$th Newton identities are satisfied, then the second, fourth, ..., $2t$th Newton identities are also satisfied. This implies that, with the BM algorithm for finding the error-location polynomial $\sigma(X)$, the solution $\sigma^{(2k-1)}(X)$ at the $(2k - 1)$th step of the BM algorithm is also the solution $\sigma^{(2k)}(X)$ at the $2t$th step of the BM algorithm, that is

$$\sigma^{(2k)}(X) = \sigma^{(2k-1)}(X), \tag{3.64}$$

for $1 \leq k \leq t$. This is seen in the second column of Table 3.4.

Given the above fact, in executing the BM algorithm the $(2k - 1)$th and $2k$th steps can be combined into one step. Consequently, for decoding a binary BCH code, the BM algorithm can be simplified as described in Algorithm 3.2. The simplified BM algorithm can be executed by setting up and filling in the table immediately below Algorithm 3.2.

**Algorithm 3.2** The Simplified BM Algorithm for Finding the Error-Location Polynomial of a Binary BCH Code

**Initialization.** For $k = -1/2$, set $\sigma^{(-1/2)}(X) = 1$, $d_{-1/2} = 1$, $l_{-1/2} = 0$, and $2(-1/2) - l_{-1/2} = -1$. For $k = 0$, set $\sigma^{(0)}(X) = 1$, $d_0 = S_1$, $l_0 = 0$, and $0 - l_0 = 0$.

1 If $k = t$, output $\sigma^{(k)}(X)$ as the error-location polynomial $\sigma(X)$; otherwise, go to Step 2.
2 Compute $d_k = S_{2k+1} + \sigma_1^{(k)} S_{2k} + \sigma_2^{(k)} S_{2k-1} + \cdots + \sigma_{l_k}^{(k)} S_{2k+1-l_k}$ and go to Step 3.
3 If $d_k = 0$, set $\sigma^{(k+1)}(X) = \sigma^{(k)}(X)$; otherwise, set

$$\sigma^{(k+1)}(X) = \sigma^{(k)}(X) + d_k d_i^{-1} X^{2(k-i)} \sigma^{(i)}(X), \tag{3.65}$$

where $i$ is a step prior to the $k$th step at which the discrepancy $d_i \neq 0$ and $2i - l_i$ is the largest. Go to Step 4.
4 $k \leftarrow k + 1$. Go to Step 1.

| Step $k$ | Partial solution $\sigma^{(k)}(X)$ | Discrepancy $d_k$ | Degree $l_k$ | Step/degree difference $2k - l_k$ |
|---|---|---|---|---|
| $-1/2$ | 1 | 1 | 0 | $-1$ |
| 0 | 1 | $S_1$ | 0 | 0 |
| 1 | $\sigma^{(1)}(X)$ | $d_1$ | $l_1$ | $2 - l_1$ |
| 2 | $\sigma^{(2)}(X)$ | $d_2$ | $l_2$ | $4 - l_2$ |
| $\vdots$ | | | | |
| $t$ | $\sigma^{(t)}(X)$ | — | — | — |

**Example 3.4** Consider the binary $(15, 5)$ triple-error-correction BCH code given in Example 3.2. Using the simplified BM algorithm for binary BCH code, Table 3.4 is reduced to Table 3.5.

### 3.3.3 Soft-Decision Decoding: The Chase Algorithm

The *Chase algorithm* [22] is a pragmatic, low-complexity soft-decision decoder that employs a hard-decision syndrome decoder (e.g., Meggitt decoder) at its core. While there are variations on the algorithm, we present only one Chase algorithm (see Algorithm 3.3). We assume a binary linear code with minimum distance $d_{\min}$ whose codewords are transmitted over the binary-input AWGN channel. The soft-decision channel output word is $\mathbf{y} = (y_0, y_1, \ldots, y_i, \ldots, y_{n-1})$ and the hard-decision word is $\mathbf{h} = (h_0, h_1, \ldots, h_i, \ldots, h_{n-1})$ so that $y_i$ is a real number and $h_i \in \{0, 1\}$. Now choose Chase decoder parameter $\rho = d_{\min}$ and define reliabilities $r_i = |y_i|$, for $i = 0, 1, 2, \ldots, n - 1$. The Chase algorithm is displayed as Algorithm 3.3.

**Table 3.5** A simplified decoding table for the (15, 5) BCH code given in Example 3.3

| Step $k$ | Partial solution $\sigma^{(k)}(X)$ | Discrepancy $d_k$ | Degree $l_k$ | Step/degree difference $2k - l_k$ |
|---|---|---|---|---|
| $-1$ | 1 | 1 | 0 | $-1$ |
| 0 | 1 | 1 | 0 | 0 |
| 1 | $1 + X$ | $\alpha^5$ | 1 | 1 |
| 2 | $1 + X + \alpha^5 X^2$ | $\alpha^{10}$ | 2 | 2 |
| 3 | $1 + X + \alpha^5 X^3$ | 0 | 3 | 3 |

**Algorithm 3.3** The Chase Soft-Decoding Algorithm

1. Identify the $\rho$ least reliable positions in the received word. These are the $\rho$ positions with the smallest $r_i$ values.
2. Starting with the hard-decision word **h**, create $2^\rho$ candidate words by placing all possible $2^\rho$ binary *test patterns* into the $\rho$ least reliable positions of **h**.
3. Send the $2^\rho$ candidate words through a syndrome decoder and record all of the decoder output words of the candidates that successfully decode.
4. The recorded output word that is closest to $\mathbf{y} = (y_0, y_1, \ldots, y_{n-1})$ in the Euclidean distance sense is the Chase decoder output word.

The Euclidean distance calculation in Step 4 is as follows. Let $\mathbf{c} = (c_0, c_1, \ldots, c_{n-1})$ be one of the successful decoder outputs, where $c_i \in \{0, 1\}$. For each $i$, define $x_i = 2c_i - 1$ so that $x_i \in \{\pm 1\}$. The *squared* Euclidean distance between **y** and **c** (or $\mathbf{x} = (x_0, x_1, \ldots, x_{n-1})$) is then given by $d_E^2(\mathbf{y}, \mathbf{c}) = \sum_{i=0}^{n-1} (y_i - x_i)^2$.

**Example 3.5** We revisit the (63, 45) BCH code of Example 3.2 with generator polynomial $\mathbf{g}(X) = (X^6 + X + 1)(X^6 + X^4 + X^2 + X + 1)(X^6 + X^5 + X^2 + X + 1)$ and minimum distance $d_{\min} = 7$. We compare a bounded-distance Meggitt decoder with a Chase decoder; so, a hard-decision decoder versus a soft-decision decoder. The Meggitt decoder syndrome table has syndrome/error pairs for all single-, double-, and triple-error patterns with a 1 in position $X^{63}$. It thus corrects any error pattern containing $t = (d_{\min} - 1)/2 = 3$ or fewer errors. The $\rho = d_{\min} = 7$ Chase decoder has the same Meggitt decoder at its core, but has the extra test pattern and Euclidean distance calculation steps. The performance of each of these two decoders for this code is presented in Figure 3.5, where we see that the Chase decoder gains around 2 dB relative to the hard-decision decoder. Further, the Chase decoder has performance nearly identical to the maximum-likelihood decoder discussed in [23], while requiring a much lower complexity.

This example is not the final answer on the Chase algorithm for soft-decoding of cyclic and shortened cyclic codes. For example, for longer codes with larger $d_{\min}$, the Chase decoder complexity will grow as a result of a much larger syndrome/error table

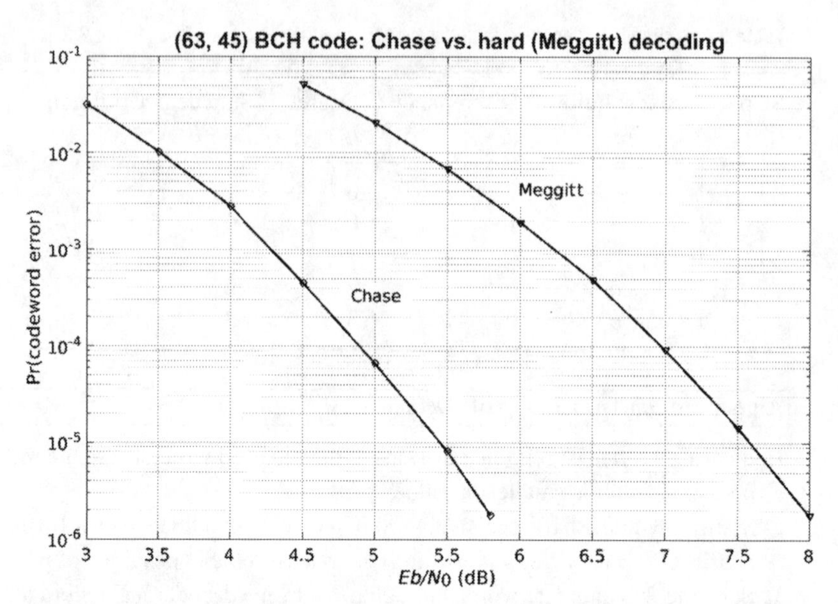


**Figure 3.5** Performance comparison between the Chase algorithm and the (bounded-distance) Meggitt decoder for (63, 45) BCH code with generator polynomial $\mathbf{g}(X) = (X^6 + X + 1)(X^6 + X^4 + X^2 + X + 1)(X^6 + X^5 + X^2 + X + 1)$.

and the exponentially larger number of test patterns to consider. The large number of test patterns will lead to a performance degradation relative to a maximum-likelihood decoder.

## 3.4 Nonbinary Linear Block Codes and Reed–Solomon Codes

So far, we have considered only block codes with symbols from the binary field GF(2). Block codes with symbols from nonbinary fields can be constructed in exactly the same manner as for binary block codes. Block codes with code symbols from GF($q$), where $q$ is a power of a prime, are called $q$-ary block codes (or block codes over GF($q$)). A $q$-ary $(n, k)$ block code has length $n$ and $q^k$ codewords. A message for a $q$-ary $(n, k)$ block code consists of $k$ information symbols from GF($q$).

**Definition 3.7** A $q$-ary $(n, k)$ block code of length $n$ with $q^k$ codewords is called a *$q$-ary $(n, k)$ linear block code* if and only if its $q^k$ codewords form a $k$-dimensional subspace of the vector space of all the $q^n$ $n$-tuples over GF($q$).

All the fundamental concepts and structural properties developed for binary linear block codes (including cyclic codes) in the previous three sections apply to $q$-ary linear block codes with few modifications. We simply replace GF(2) by GF($q$). A $q$-ary $(n, k)$ linear block code is specified by either a $k \times n$ generator matrix $\mathbf{G}$ or an $(n - k) \times n$ parity-check matrix $\mathbf{H}$ over GF($q$). Generator and parity-check matrices of a $q$-ary $(n, k)$ linear block code in systematic form have exactly the same forms as given by (3.7) and (3.11), except that their entries are from GF($q$). (The parameter $q$ is usually a power of 2. If $q$ is a power of some other prime, then $\mathbf{H} = [\mathbf{I_{n-k}} - \mathbf{P^T}]$.) Encoding and decoding of $q$-ary linear block codes are the same as for binary codes, except that operations and computations are performed over GF($q$).

A $q$-ary $(n, k)$ cyclic code $C$ is generated by a monic polynomial of degree $n - k$ over GF($q$),

$$\mathbf{g}(X) = g_0 + g_1 X + \cdots + g_{n-k-1} X^{n-k-1} + X^{n-k}, \tag{3.66}$$

where $g_0 \neq 0$ and $g_i \in \text{GF}(q)$. This generator polynomial $\mathbf{g}(X)$ is a factor of $X^n - 1$. A polynomial $\mathbf{v}(X)$ of degree $n-1$ or less over GF($q$) is a code polynomial if and only if $\mathbf{v}(X)$ is divisible by the generator polynomial of the code.

Let GF($q^m$) be an extension field of GF($q$) and $\alpha$ be a primitive element of GF($q^m$). A $t$-symbol-error-correcting primitive BCH code of length $q^m - 1$ over GF($q$) is a cyclic code generated by the lowest-degree polynomial $\mathbf{g}(X)$ over GF($q$) that has $\alpha, \alpha^2, \ldots, \alpha^{2t}$ and their conjugates as roots. For $1 \leq i \leq 2t$, let $\phi_i(X)$ be the (monic) minimal polynomial of $\alpha^i$ over GF($q$). Then

$$\mathbf{g}(X) = \text{LCM}\{\phi_1(X), \phi_2(X), \ldots, \phi_{2t}(X)\}. \tag{3.67}$$

Since the degree of a minimal polynomial of an element in GF($q^m$) is at most $m$, the degree of $\mathbf{g}(X)$ is at most $2mt$ and $\mathbf{g}(X)$ divides $X^{q^m-1} - 1$. The $q$-ary $t$-symbol-correcting primitive BCH code has length $q^m - 1$ with dimension at least $q^m - 2mt - 1$. Its parity-check matrix $\mathbf{H}$ given in terms of its roots has exactly the same form as that given by (3.44). In the same manner, we can prove that no $2t$ or fewer columns of $\mathbf{H}$ can be added to a zero column vector. Hence, the code has minimum distance at least $2t + 1$ (BCH bound) and is capable of correcting $t$ or fewer random symbol errors over a span of $q^m - 1$ symbol positions. For a given field GF($q^m$), a family of $q$-ary BCH codes can be constructed. The BM algorithm given in the previous section can be used for decoding both binary and $q$-ary BCH codes.

The most important and most widely used class of $q$-ary codes is the class of Reed–Solomon (RS) codes, which was discovered in 1960 [24], in the same year as the discovery of binary BCH codes by Bose and Chaudhuri [20]. For an RS code, the symbol field and construction field are the same. Such codes can be put into either cyclic or non-cyclic form. We define RS codes in cyclic form. Let $\alpha$ be a primitive element of GF($q$). For a positive integer $t$ such that $2t < q$, the generator polynomial of a $t$-symbol-error-correcting cyclic RS code over GF($q$) of length $q - 1$ is given by

$$\begin{aligned}\mathbf{g}(X) &= (X - \alpha)(X - \alpha^2) \cdots (X - \alpha^{2t}) \\ &= g_0 + g_1 X + \cdots + g_{2t-1} X^{2t-1} + X^{2t},\end{aligned} \tag{3.68}$$

with $g_i \in \text{GF}(q)$. Since $\alpha, \alpha^2, \ldots, \alpha^{2t}$ are roots of $X^{q-1} - 1$, $\mathbf{g}(X)$ divides $X^{q-1} - 1$. Therefore, $\mathbf{g}(X)$ generates a cyclic RS code of length $q-1$ with exactly $2t$ parity-check symbols. Actually, RS codes form a special subclass of $q$-ary BCH codes with $m = 1$. However, since they were discovered independently and before $q$-ary BCH codes and are much more important than other $q$-ary BCH codes in practical applications, we consider them as an independent class of $q$-ary codes. Since the BCH bound gives a minimum distance (or minimum weight) of at least $2t+1$ and $\mathbf{g}(X)$ is a code polynomial with $2t+1$ nonzero terms (weight $2t+1$), its minimum distance is exactly $2t+1$. Note that the minimum distance $2t+1$ of an RS code is 1 greater than the number $2t$ of its parity-check symbols. A code with minimum distance that is 1 greater than the number of its parity-check symbols is called a *maximum-distance-separable* (*MDS*) code. In summary, a $t$-symbol-error-correction RS code over GF($q$) has the following parameters:

| | |
|---|---|
| Length | $q-1$ |
| Number of parity-check symbols | $2t$ |
| Dimension | $q-2t-1$ |
| Minimum distance | $2t+1$ |

Such an RS code is commonly called a $(q-1, q-2t-1, 2t+1)$ RS code over GF($q$) with the length, dimension, and minimum distance displayed.

In all practical applications of RS codes in digital communication or data-storage systems, $q$ is commonly chosen as a power of 2, say $q = 2^s$, and the code symbols are from GF($2^s$). If each code symbol is represented by an $s$-tuple over GF(2), then an RS code can be transmitted using binary signaling, such as BPSK. In decoding, every $s$ received bits are grouped into a received symbol over GF($2^s$). This results in a received sequence of $2^s - 1$ symbols over GF($2^s$). Then decoding is performed on this received symbol sequence. Since RS codes are special $q$-ary BCH codes, they can be decoded with the BM algorithm. The Euclidean algorithm is also commonly used for decoding RS codes [1, 2, 5, 25].

---

**Example 3.6** A RS code that has widely been used in optical communications, data-storage systems, and hard-disk drives is the (255, 239, 17) RS code over GF($2^8$), which is capable of correcting eight or fewer symbol errors. Another important RS code is the (255, 223, 33) RS code over GF($2^8$), which is a NASA standard code for deep-space and satellite communications. This code is capable of correcting 16 or fewer random symbol errors. The symbol and block-error performances of these two codes for the binary-input AWGN channel are shown in Figure 3.6. Eight-bit bytes are code symbols and eight consecutive hard decisions at the channel output lead to a code symbol. In Problem 3.22 we examine how these curves were produced.

---

The $(q-1, q-2t-1, 2t+1)$ cyclic RS code $\mathcal{C}$ over GF($q$) generated by $\mathbf{g}(X) = (X-a)(X-\alpha^2)\cdots(X-\alpha^{2t})$ can be extended by adding an extra code symbol to each codeword in $\mathcal{C}$. Let $\mathbf{v}(X) = v_0 + v_1X + \cdots + v_{q-2}X^{q-2}$ be a nonzero code polynomial in $\mathcal{C}$. Its corresponding codeword is $\mathbf{v} = (v_0, v_1, \ldots, v_{q-2})$. Extend this codeword by adding the following code symbol (replacing $X$ of $\mathbf{v}(X)$ by 1):.

$$\begin{aligned} v_\infty &= \mathbf{v}(1) \\ &= v_0 + v_1 + \cdots + v_{q-2}. \end{aligned} \tag{3.69}$$

This added code symbol is called the *overall parity-check symbol* of $\mathbf{v}$. The extended codeword is then $\mathbf{v}_{\text{ext}} = (v_\infty, v_0, v_1, \ldots, v_{q-2})$. The extension results in a

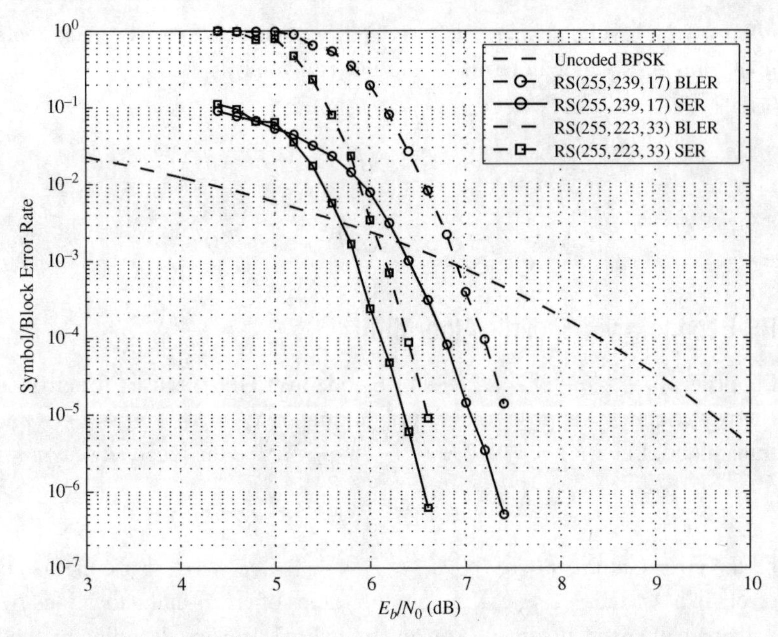


**Figure 3.6** Symbol- and block-error performances of the (255, 239, 17) and (255, 223, 33) RS codes over GF($2^8$). The BPSK curve is bit-error rate.

$(q, q-2t-1, 2t+2)$ code $\mathcal{C}_{\text{ext}}$, called an *extended* RS code. To determine the minimum distance of the extended RS code, there are two cases to be considered. The first case is that $v_\infty = \mathbf{v}(1) = 0$. This implies that 1 is a root of $\mathbf{v}(X)$ and hence $\mathbf{v}(X)$ is divisible by $\mathbf{g}^*(X) = (X-1)\mathbf{g}(X)$. Consequently, $\mathbf{v}(X)$ is a nonzero codeword in the $(q-1, q-2t-2, 2t+2)$ RS code generated by $\mathbf{g}^*(X) = (X-1)\mathbf{g}(X) = (X-1)(X-\alpha)\cdots(X-\alpha^{2t})$ and the weight of $\mathbf{v}(X)$ must be at least $2t+2$. Therefore, the extended codeword $(v_\infty = 0, v_0, \ldots, v_{q-2})$ has weight at least $2t+2$. The second case is that $v_\infty = \mathbf{v}(1) \neq 0$. In this case, adding the overall parity-check symbol $v_\infty$ to $\mathbf{v} = (v_0, v_1, \ldots, v_{q-2})$ increases the weight of $\mathbf{v}$ by 1. This implies that the weight of the extended codeword $\mathbf{v}_{\text{ext}} = (v_\infty, v_1, \ldots, v_{q-2})$ is at least $2t+2$. Note that $\mathbf{g}(X)$ does not have 1 as a root. Hence the overall parity-check symbol for the codeword corresponding to $\mathbf{g}(X)$ is nonzero. Adding the overall parity-check symbol to the codeword corresponding to $\mathbf{g}(X)$ results in an extended codeword of weight exactly $2t+2$. It follows from the above analysis that the minimum weight of the extended $(q, q-2t-1, 2t+2)$ RS code $\mathcal{C}_{\text{ext}}$ is exactly $2t+2$. This implies that the minimum distance of the extended RS code $\mathcal{C}_{\text{ext}}$ is exactly $2t+2$.

Suppose $q-1$ is not a prime and can be factored as $q-1 = cn$. Let $\alpha$ be a primitive element in GF($q$). Let $\beta = \alpha^c$. Then the order $\beta$ is $n$. The $n$ elements $\beta^0 = 1, \beta, \beta^2, \ldots, \beta^{n-1}$ form a cyclic subgroup of the multiplicative group of GF($q$). They also form all the roots of $X^n - 1$. For a positive integer $t$ such that $2t < n$, let $\mathbf{g}(X) = (X-\beta)(X-\beta^2)\cdots(X-\beta^{2t})$. It is clear that $\mathbf{g}(X)$ divides $X^n - 1$. Then all the polynomials of degree $n-1$ or less over GF($q$) that are divisible by $\mathbf{g}(X)$ give a *nonprimitive* cyclic $(n, n-2t, 2t+1)$ RS code. The parity-check matrix of this nonprimitive RS code in terms of its roots is given by

$$\mathbf{H} = \begin{bmatrix} 1 & \beta & \beta^2 & \cdots & \beta^{n-1} \\ 1 & \beta^2 & (\beta^2)^2 & \cdots & (\beta^2)^{n-1} \\ \vdots & \vdots & \vdots & \ddots & \vdots \\ 1 & \beta^{2t} & (\beta^{2t})^2 & \cdots & (\beta^{2t})^{n-1} \end{bmatrix}. \tag{3.70}$$

## 3.4.1 RS Decoding via Euclid's Algorithm

Suppose a $(q-1, q-2t-1, 2t+1)$ RS code over GF($q$) is used for error control over a noisy channel. Suppose a code polynomial $\mathbf{v}(X) = v_0 + v_1X + \cdots + v_{q-2}X^{q-2}$ is transmitted. Let $\mathbf{r}(X) = r_0 + r_1X + \cdots + r_{q-2}X^{q-2}$ be the received polynomial and let

$$\mathbf{e}(X) = e_{j_1}X^{j_1} + e_{j_2}X^{j_2} + \cdots + e_{j_\nu}X^{j_\nu} \tag{3.71}$$

be the error pattern, where $0 \leq j_1 < j_2 < \cdots < j_\nu < q-1$ are the locations of the errors in $\mathbf{e}(X)$ and $e_{j_1}, e_{j_2}, \ldots, e_{j_\nu}$ are the values of errors at the locations $j_1, j_2, \ldots, j_\nu$, which are elements in GF($q$). Any algebraic hard-decision decoding method must find the locations and values of errors in $\mathbf{e}(X)$ given the $2t$ syndrome components $S_1 = \mathbf{r}(\alpha), S_2 = \mathbf{r}(\alpha^2), \ldots, S_{2t} = \mathbf{r}(\alpha^{2t})$. Two well-known and efficient algorithms for decoding RS codes (or $q$-ary BCH codes) are the BM algorithm and Euclid's algorithm [25]. The computational complexities of these two algorithms are about the same; however, Euclid's algorithm is much easier to understand so we present it first in this subsection. In the next subsection, we present the BM algorithm for error correction in RS codes and extend it in the subsection after that to solving for error and erasure values (where erasures will also be defined).

In the following, we present the decoding procedure for Euclid's algorithm. For $1 \leq i \leq \nu$, define

$$\beta_i \triangleq \alpha^{j_i}. \tag{3.72}$$

The exponents of $\alpha^{j_1}, \alpha^{j_2}, \ldots, \alpha^{j_\nu}$ are simply the locations of the errors in $\mathbf{e}(X)$. We call $\beta_1, \beta_2, \ldots, \beta_\nu$ the *error-location numbers*. Define the following polynomial:

$$\begin{aligned} \sigma(X) &= (1-\beta_1X)(1-\beta_2X)\cdots(1-\beta_\nu X) \\ &= \sigma_0 + \sigma_1X + \sigma_2X^2 + \cdots + \sigma_\nu X^\nu, \end{aligned} \tag{3.73}$$

where $\sigma_0 = 1$. Note that $\sigma(X)$ has $\beta_1^{-1}, \beta_2^{-1}, \ldots, \beta_\nu^{-1}$ (the reciprocals (or inverses) of the error-location numbers) as roots. If we can determine $\sigma(X)$ from the $2t$ syndrome components $S_1, S_2, \ldots, S_{2t}$, then the roots of $\sigma(X)$ give us the reciprocals of the error-location numbers. From the error-location numbers, we can determine the error locations $j_1, j_2, \ldots, j_\nu$. The polynomial $\sigma(X)$ is called the *error-location polynomial*.

Define

$$\mathbf{S}(X) = S_1 + S_2X + \cdots + S_{2t}X^{2t-1} \tag{3.74}$$

and

$$\mathbf{Z}_0(X) = S_1 + (S_2 + \sigma_1 S_1)X + (S_3 + \sigma_1 S_2 + \sigma_2 S_1)X^2 + \cdots$$
$$+ (S_0 + \sigma_1 S_{\nu-1} + \cdots + \sigma_{\nu-1} S_1)X^{\nu-1}. \tag{3.75}$$

$\mathbf{S}(X)$ and $\mathbf{Z}_0(X)$ are called the *syndrome polynomial* and *error-value evaluator*, respectively. Once $\sigma(X)$ and $\mathbf{Z}_0(X)$ have been found, the locations and values of errors in the error pattern $\mathbf{e}(X)$ can be determined. Note that the degree of $\mathbf{Z}_0(X)$ is at least one less than the degree of the error-location polynomial $\sigma(X)$. The three polynomials $\sigma(X)$, $\mathbf{S}(X)$, and $\mathbf{Z}_0(X)$ are related by the following equation [1]:

$$\sigma(X)\mathbf{S}(X) \equiv \mathbf{Z}_0(X) \bmod X^{2t} \tag{3.76}$$

(i.e., $\sigma(X)S(X) - \mathbf{Z}_0$ is divisible by $X^{2t}$), which is called the *key equation*. Any method of solving this key equation to find $\sigma(X)$ and $\mathbf{Z}_0(X)$ is a decoding method for RS codes (or $q$-ary BCH codes). If the number of errors in $\mathbf{e}(X)$ is less than or equal to $t$ (the error-correction capability of the code), then the key equation has a unique pair of solutions ($\sigma(X)$ and $\mathbf{Z}_0(X)$), with

$$\deg[\mathbf{Z}_0(X)] < \deg[\sigma(X)] \leq t. \tag{3.77}$$

The key equation of (3.76) can be solved by using *Euclid's iterative division algorithm* to find the *greatest common divisor* (GCD) of $X^{2t}$ and $\mathbf{S}(X)$. The first step is to divide $X^{2t}$ by $\mathbf{S}(X)$. This results in the following expression:

$$X^{2t} = \mathbf{q}_1(X)\mathbf{S}(X) + \mathbf{Z}_0^{(1)}(X), \tag{3.78}$$

where $\mathbf{q}_1(X)$ and $\mathbf{Z}_0^{(1)}(X)$ are the quotient and remainder, respectively. Then we divide $\mathbf{S}(X)$ by $\mathbf{Z}_0^{(1)}(X)$. Let $\mathbf{q}_2(X)$ and $\mathbf{Z}^{(2)}(X)$ be the resultant quotient and remainder, respectively. Then we divide $\mathbf{Z}_0^{(1)}(X)$ by $\mathbf{Z}_0^{(2)}(X)$. We repeat the above division process. Let $\mathbf{Z}_0^{(i-2)}(X)$ and $\mathbf{Z}_0^{(i-1)}(X)$ be the remainders at the $(i-2)$th and $(i-1)$th division steps, respectively. Euclid's algorithm for finding $\sigma(X)$ and $\mathbf{Z}_0(X)$ carries out the following two computations iteratively.

1. At the $i$th division step with $i = 1, 2, \ldots$, divide $\mathbf{Z}_0^{(i-2)}(X)$ by $\mathbf{Z}_0^{(i-1)}(X)$ to obtain the quotient $\mathbf{q}_i(X)$ and remainder $\mathbf{Z}_0^{(i)}(X)$.
2. Find $\sigma^{(i)}(X)$ from

$$\sigma^{(i)}(X) = \sigma^{(i-2)}(X) - \mathbf{q}_i(X)\sigma^{(i-1)}(X). \tag{3.79}$$

Iteration begins with the following initial conditions:

$$\begin{aligned} \mathbf{Z}_0^{(-1)}(X) &= X^{2t}, \quad \mathbf{Z}_0^{(0)}(X) = \mathbf{S}(X), \\ \sigma^{(-1)}(X) &= 0, \quad \sigma^{(0)}(X) = 1. \end{aligned} \tag{3.80}$$

Iteration stops when a step $\rho$ is reached for which

$$\deg[\mathbf{Z}_0^{(\rho)}(X)] < \deg[\sigma^{(\rho)}(X)] \leq t. \tag{3.81}$$

Then the error-location polynomial $\sigma(X)$ and error-value evaluator are given by

$$\sigma(X) = \sigma^{(\rho)}(X), \tag{3.82}$$

$$\mathbf{Z}_0(X) = \mathbf{Z}_0^{(\rho)}(X). \tag{3.83}$$

If the number of errors in $\mathbf{e}(X)$ is $t$ or less, there always exists a step $\rho \leq 2t$ for which the condition given by (3.81) holds.

Once $\sigma(X)$ and the error-value evaluator $\mathbf{Z}_0(X)$ have been found, the locations and values of errors in the error pattern $\mathbf{e}(X)$ can be determined as follows.

1. Find the roots of $\sigma(X)$ and take the reciprocal of each of the roots, which gives the error-location numbers $\alpha^{j_1}, \alpha^{j_2}, \ldots, \alpha^{j_v}$. Then the exponents of $\alpha$, $j_1, j_2, \ldots, j_v$, give the locations of errors in the error pattern $\mathbf{e}(X)$.
2. Let $\sigma'(X)$ be the derivative of $\sigma(X)$. Then the error value at location $j_i$ is given by

$$e_{j_i} = \frac{-\mathbf{Z}_0(\alpha^{-j_i})}{\sigma'(\alpha^{-j_i})}. \tag{3.84}$$

The above two steps completely determine the error pattern $\mathbf{e}(X)$. Then the estimated transmitted code polynomial is given by $\mathbf{v}^*(X) = \mathbf{r}(X) - \mathbf{e}(X)$. The iteration process for finding $\sigma(X)$ and $\mathbf{Z}_0(X)$ can be carried out by setting up and filling in a table as shown below.

| Iteration step $i$ | $\mathbf{Z}_0^{(i)}(X)$ | $\mathbf{q}_i(X)$ | $\sigma^{(i)}(X)$ |
|---|---|---|---|
| $-1$ | $X^{2t}$ | — | 0 |
| 0 | $\mathbf{S}(X)$ | — | 1 |
| 1 | | | |
| 2 | | | |
| ⋮ | | | |
| $i$ | | | |
| ⋮ | | | |

---

**Example 3.7** Let GF($2^4$) be the field constructed from the primitive polynomial $p(X) = 1 + X + X^4$ (see Table 2.14). Let $\alpha$ be a primitive element of GF($2^4$). Consider the triple-error-correction $(15, 9, 7)$ RS code over GF($2^4$) generated by

$$\mathbf{g}(X) = (X + \alpha)(X + \alpha^2)(X + \alpha^3)(X + \alpha^4)(X + \alpha^5)(X + \alpha^6).$$

Suppose a code polynomial $\mathbf{v}(X)$ is transmitted and the received polynomial is $\mathbf{r}(X) = \alpha^7 X^3 + \alpha^{11} X^{10}$. The syndrome components of $\mathbf{r}(X)$ are

$$S_1 = \mathbf{r}(\alpha) = \alpha^{10} + \alpha^{21} = \alpha^7,$$
$$S_2 = \mathbf{r}(\alpha^2) = \alpha^{13} + \alpha^{31} = \alpha^{12},$$
$$S_3 = \mathbf{r}(\alpha^3) = \alpha^{16} + \alpha^{41} = \alpha^6,$$

**Table 3.6** The steps of Euclid's algorithm for finding the error-location polynomial and the error-value evaluator of the (15, 9, 7) RS code given in Example 3.5

| $i$ | $\mathbf{Z}_0^{(i)}(X)$ | $\mathbf{q}_i(X)$ | $\sigma^{(i)}(X)$ |
|---|---|---|---|
| $-1$ | $X^6$ | — | 0 |
| 0 | $\mathbf{S}(X)$ | — | 1 |
| 1 | $\alpha^8 + \alpha^3 X + \alpha^5 X^2 + \alpha^5 X^3 + \alpha^6 X^4$ | $\alpha + \alpha X$ | $\alpha + \alpha X$ |
| 2 | $\alpha^3 + \alpha^2 X$ | $\alpha^{11} + \alpha^8 X$ | $\alpha^{11} + \alpha^8 X + \alpha^9 X^2$ |

$$
\begin{aligned}
S_4 &= \mathbf{r}(\alpha^4) = \alpha^{19} + \alpha^{51} = \alpha^{12},\\
S_5 &= \mathbf{r}(\alpha^5) = \alpha^7 + \alpha = \alpha^{14},\\
S_6 &= \mathbf{r}(\alpha^6) = \alpha^{10} + \alpha^{11} = \alpha^{14}.
\end{aligned}
$$

The syndrome polynomial is

$$\mathbf{S}(X) = \alpha^7 + \alpha^{12}X + \alpha^6 X^2 + \alpha^{12}X^3 + \alpha^{14}X^4 + \alpha^{14}X^5.$$

On executing Euclid's algorithm, we obtain Table 3.6.

We find that at Step 2, the condition given by (3.81) holds. Hence

$$
\begin{aligned}
\sigma(X) &= \sigma^{(2)}(X)\\
&= \alpha^{11} + \alpha^8 X + \alpha^9 X^2 = \alpha^{11}(1 + \alpha^{10}X)(1 + \alpha^3 X),\\
\mathbf{Z}_0(X) &= \alpha^3 + \alpha^2 X.
\end{aligned}
$$

The roots of $\sigma(X)$ are $\alpha^5$ and $\alpha^{12}$. Their reciprocals are $\alpha^{10}$ and $\alpha^3$ and hence there are two errors in the estimated error pattern $\mathbf{e}(X)$ at the locations 3 and 10. The error values at the locations 3 and 10 are

$$
\begin{aligned}
e_3 &= \frac{-\mathbf{Z}_0(\alpha^{-3})}{\sigma'(\alpha^{-3})} = \frac{\alpha^3 + \alpha^2\alpha^{-3}}{\alpha^{11}\alpha^3(1 + \alpha^{10}\alpha^{-3})} = \frac{1}{\alpha^8} = \alpha^7,\\
e_{10} &= \frac{-\mathbf{Z}_0(\alpha^{-10})}{\sigma'(\alpha^{-10})} = \frac{\alpha^3 + \alpha^2\alpha^{-10}}{\alpha^{11}\alpha^{10}(1 + \alpha^3\alpha^{-10})} = \frac{\alpha^4}{\alpha^8} = \alpha^{11}.
\end{aligned}
$$

Therefore, the estimated error pattern is $\mathbf{e}(X) = \alpha^7 X^3 + \alpha^{11}X^{10}$ and the decoded codeword $\mathbf{v}^*(X) = \mathbf{r}(X) - \mathbf{e}(X)$ is the all-zero codeword.

### 3.4.2 Errors-Only RS Decoding via the Berlekamp–Massey Algorithm

We present in this subsection the *Berlekamp–Massey algorithm* for the correction of symbols errors in Reed–Solomon codes and we extend it in the next subsection to the correction of errors and erasures. The development borrows from [26, 27]. We allow a more general definition of the RS code generator polynomial,

$$\mathbf{g}(X) = \prod_{i=1}^{2t}(X + \alpha^{b+i-1}),$$

where $\alpha$ is primitive in GF($q$) and $q$ is a power of 2 and where $b = 1$ in our earlier definition of $\mathbf{g}(X)$. Because $\{\alpha^{b+i-1}\}$ are roots of $\mathbf{g}(X)$ and code polynomials are of the form $\mathbf{v}(X) = \mathbf{a}(X)\mathbf{g}(X)$, we have that $\mathbf{v}(\alpha^{b+i-1}) = 0$ for $i = 1, 2, \ldots, 2t$. This leads to the more general parity-check matrix

$$\mathbf{H} = \begin{bmatrix} 1 & \alpha^b & \alpha^{2b} & \cdots & \alpha^{(n-1)b} \\ 1 & \alpha^{b+1} & \alpha^{2(b+1)} & \cdots & \alpha^{(n-1)(b+1)} \\ \vdots & \vdots & \vdots & \vdots & \vdots \\ 1 & \alpha^{b+2t-1} & \alpha^{2(b+2t-1)} & \cdots & \alpha^{(n-1)(b+2t-1)} \end{bmatrix},$$

so that $\mathbf{v}\mathbf{H}^{\mathrm{T}} = \mathbf{0}$ for any code vector $\mathbf{v}$.

Given a received polynomial $\mathbf{r}(X) = \mathbf{v}(X) + \mathbf{e}(X)$, the syndromes are given by

$$S_i = \mathbf{r}(\alpha^{b+i-1}) = \mathbf{e}(\alpha^{b+i-1}), \quad \text{for } i = 1, 2, \ldots, 2t.$$

As before, we assume there are $\nu$ errors and write the error polynomial as $\mathbf{e}(X) = e_{j_1}X^{j_1} + e_{j_2}X^{j_2} + \cdots + e_{j_\nu}X^{j_\nu}$ where, for $m = 1, 2, \ldots, \nu$, $\{e_{j_m}\}$ are the error values and $\{j_m\}$ are the error locations. From $S_i = \mathbf{e}(\alpha^{b+i-1})$, we may write

$$S_i = \sum_{m=1}^{\nu} e_{j_m} \beta_m^{b+i-1}, \tag{3.85}$$

where $\beta_m = \alpha^{j_m}$ is the error-location number as before (also called the *error locator*).

Substitution of root $\beta_m^{-1}$ into $\sigma(X) = \prod_{m=1}^{\nu}(1 + \beta_m X) = \sigma_0 + \sigma_1 X + \cdots + \sigma_\nu X^\nu$, the error-location polynomial introduced in the previous subsection, gives

$$\sigma_0 + \sigma_1\beta_m^{-1} + \sigma_2\beta_m^{-2} + \cdots + \sigma_\nu\beta_m^{-\nu} = 0.$$

If we multiply this equation by $e_{j_m}\beta_m^{b+i-1}$ and sum over all $m = 1, 2, \ldots, \nu$, the result is

$$\sigma_0 S_i + \sigma_1 S_{i-1} + \sigma_2 S_{i-2} + \cdots + \sigma_\nu S_{i-\nu} = 0.$$

With $i = \nu+1, \nu+2, \ldots, 2t$, we obtain $2t - \nu$ linear equations in $\nu$ unknowns and these are solvable provided $2\nu \leq 2t = d_{\min} - 1$.

In Massey's version of Berlekamp's original decoding algorithm, Massey regards the previous equation (where $\sigma_0 = 1$) to be that of a linear feedback shift register (LFSR),

$$S_i = \sigma_1 S_{i-1} + \sigma_2 S_{i-2} + \cdots + \sigma_\nu S_{i-\nu},$$

but with an unknown register length $\nu$. Looking for the most likely error pattern to produce the received codeword amounts to finding the shortest shift register (smallest $\nu$) for which the above recursion holds, for $i = \nu + 1, \nu + 2, \ldots, 2t$. The BM algorithm, presented below, finds this shortest LFSR.

---

**Example 3.8** Let us consider a (15, 9) RS code over GF(16). This code can correct up to $t = (15 - 9)/2 = 3$ errors and has $d_{\min} = 2t + 1 = 7$. With $\alpha$ a primitive element of $X^4 + X + 1$, let the generator polynomial be $\mathbf{g}(X) = \prod_{i=1}^{6}(X + \alpha^{5+i-1})$ so that $b = 5$. Using the GF(16) representations give in Table 2.14, we find that $\mathbf{g}(X) = X^6 + \alpha^{14}X^5 + \alpha^7X^4 + \alpha X^3 + \alpha^7X^2 + \alpha^{14}X + 1$, a reversible polynomial

**Algorithm 3.4** Berlekamp–Massey Algorithm for RS Codes

DEFINITIONS

1. $i$ = syndrome counter
2. $L$ = current register length (eventually, $L = \nu$)
3. $\delta$ = current discrepancy
4. $D(X)$ = correction term

ALGORITHM

1. Initialize algorithm variables: $\sigma(X) = 1,\ L = 0,\ D(X) = X,\ i = 1$.
2. Compute discrepancy: $\delta = S_i + \sum_{l=1}^{L} \sigma_i S_{i-l}$.
3. Test discrepancy: If $\delta = 0$, go to Step 8; else, go to Step 4.
4. Modify $\sigma(X)$: $\sigma^*(X) = \sigma(X) - \delta D(X)$.
5. Test register length $L$: If $2L \geq i$, go to Step 7 (i.e., do no extend register). Otherwise, go to Step 6.
6. Update $L$ and $D(X)$: $L \leftarrow i - L$ and $D(X) = \sigma(X)/\delta$.
7. Update $\sigma(X)$ and $D(X)$: $\sigma(X) = \sigma^*(X)$ and $D(X) \leftarrow XD(X)$.
8. Increment and test $i$: $i \leftarrow i + 1$. If $i \leq 2t$, go to Step 2; else, stop.

(or palindrome) so that multipliers may be shared in a hardware implementation. Suppose the all-zero codeword was sent but $\mathbf{r}(X) = \alpha^3 + \alpha^9 X^5 + \alpha^2 X^{10}$ is received. Note that $\nu = 3$ (three errors) and $j_1 = 0, j_2 = 5, j_3 = 10$. So when we find $\sigma(X)$, it should have as roots $\alpha^{-0} = 1, \alpha^{-5} = \alpha^{10}$, and $\alpha^{-10} = \alpha^5$, so that $\sigma(X) = (1 + X)(1 + \alpha^{10}X)(1 + \alpha^5 X) = 1 + X^3$ See Algorithm 3.4.

The first step in the decoding algorithm is to compute the syndromes that, using Table 2.14, simplify as $S_1 = \mathbf{r}(\alpha^5) = 0,\ S_2 = \mathbf{r}(\alpha^6) = \alpha^5,\ S_3 = \mathbf{r}(\alpha^7) = \alpha^{11}$, $S_4 = \mathbf{r}(\alpha^8) = 0,\ S_5 = \mathbf{r}(\alpha^9) = \alpha^5,\ S_6 = \mathbf{r}(\alpha^{10}) = \alpha^{11}$. The BM algorithm that uses these syndrome values as input is presented in Table 3.7, where we see the final result is indeed $\sigma(X) = 1 + X^3$. The error locations are found by first substituting all nonzero elements of GF(16), $1, \alpha, \alpha^2, \ldots, \alpha^{14}$, into $\sigma(X)$ to determine its roots. The roots are found to be $\alpha^0 = \alpha^{-0}$, $\alpha^5 = \alpha^{-10}$, and $\alpha^{10} = \alpha^{-5}$ so that the error locations are 0, 10, and 5.

**Table 3.7** Berlekamp–Massey decoding steps for Example 3.8

| $i$ | $L$ | $\delta$ | $D(X)$ | $\sigma(X)$ | $\sigma^*(X)$ |
|---|---|---|---|---|---|
| 1 | 0 | 0 | $X$ | 1 | |
| 2 | | $\alpha^5$ | $X^2$ | | $1 + \alpha^5 X^2$ |
| 3 | 2 | $\alpha^{11}$ | $\alpha^{10}, \alpha^{10}X$ | $1 + \alpha^5 X^2$ | $1 + \alpha^6 X + \alpha^5 X^2$ |
| 4 | | $\alpha^4$ | $\alpha^{10}X^2$ | $1 + \alpha^6 X + \alpha^5 X^2$ | $1 + \alpha^6 X + \alpha^{12} X^2$ |
| 5 | 3 | $\alpha^4$ | $\alpha^{10}X^3$ | $1 + \alpha^6 X + \alpha^{12} X^2$ | $1 + \alpha^6 X + \alpha^{12} X^2 + \alpha^{14} X^3$ |
| 6 | | $\alpha^{10}$ | $\alpha^{11} + \alpha^2 X + \alpha^8 X^2$, $\alpha^{11}X + \alpha^2 X^2 + \alpha^8 X^3$ | $1 + \alpha^6 X + \alpha^{12} X^2$ $+\alpha^{14} X^3$ | $1 + X^3$ |
| 7 | | | $XD(X)$ | $1 + X^3$ | |

Unlike the binary code case for which all error values are 1, for nonbinary codes such as RS codes, we need to determine the $q$-ary error values $\{e_{j_m}\}$ once the error locations are known. One way to do this is through the syndrome equations $S_i = \sum_{m=1}^{\nu} e_{j_m}\beta_m^{b+i-1}$, but a more efficient way will now be developed.

We combine the syndrome polynomial $\mathbf{S}(X)$ defined in equation (3.74) with the error-location polynomial $\sigma(X)$ to define the *evaluator polynomial*

$$\Omega(X) = \mathbf{S}(X)\sigma(X) \bmod X^{2t}.$$

Using the definition of the coefficients of $\mathbf{S}(X)$, we may write

$$\mathbf{S}(X) = \sum_{i=1}^{2t}\left[\sum_{m=1}^{\nu} e_{j_m}\beta_m^{b+i-1}\right]X^{i-1}$$

so that $\Omega(X)$ may be expanded as follows:

$$\begin{aligned}\Omega(X) &= \left[\sum_{i=1}^{2t}\sum_{m=1}^{\nu} e_{j_m}\beta_m^{b+i-1}X^{i-1}\right]\left[\prod_{s=1}^{\nu}(1+\beta_s X)\right] \bmod X^{2t}\\ &= \sum_{m=1}^{\nu} e_{j_m}\beta_m^{b}\left[(1+\beta_m X)\sum_{i=1}^{2t}(\beta_m X)^{i-1}\right]\prod_{\substack{s=1\\ s\neq m}}^{\nu}(1+\beta_s X) \bmod X^{2t}.\end{aligned}$$

The bracketed term is a factorization of $\left(1+\beta_m^{2t}X^{2t}\right)$ and therefore

$$\Omega(X) = \sum_{m=1}^{\nu} e_{j_m}\beta_m^{b}\prod_{\substack{s=1\\ s\neq m}}^{\nu}(1+\beta_s X). \tag{3.86}$$

The computation of the error values $\{e_{j_m}\}$ is now easy. From our new expression for $\Omega(X)$, we may write, for $k = 1, 2, \ldots, \nu$,

$$\Omega\left(\beta_k^{-1}\right) = e_{j_k}\beta_k^{b}\prod_{\substack{s=1\\ s\neq k}}^{\nu}\left(1+\beta_s\beta_k^{-1}\right)$$

so that the error values are given by

$$e_{j_k} = \frac{\Omega\left(\beta_k^{-1}\right)}{\beta_k^{b}\prod_{k\neq s=1}^{\nu}\left(1+\beta_s\beta_k^{-1}\right)}. \tag{3.87}$$

---

**Example 3.9** We continue the previous example where we found $S_1 = 0,\ S_2 = \alpha^5,\ S_3 = \alpha^{11},\ S_4 = 0,\ S_5 = \alpha^5,\ S_6 = \alpha^{11};\ \beta_1 = 1,\ \beta_2 = \alpha^5,\ \beta_3 = \alpha^{10};$ and $\sigma(X) = 1 + X^3$. From this, the syndrome polynomial is $\mathbf{S}(X) = \alpha^5X + \alpha^{11}X^2 + \alpha^5X^4 + \alpha^{11}X^5$, from which we may write

$$\begin{aligned}\Omega(X) &= \mathbf{S}(X)\sigma(X) \bmod X^6\\ &= (\alpha^5X + \alpha^{11}X^2 + \alpha^5X^4 + \alpha^{11}X^5)(1+X^3) \bmod X^6\\ &= \alpha^5X + \alpha^{11}X^2.\end{aligned}$$

The error values can now be found using the error-location numbers $\beta_1 = 1$, $\beta_2 = \alpha^5$, and $\beta_3 = \alpha^{10}$ and $b = 5, \nu = 3$:

$$e_{j_1} = \frac{\Omega\left(1^{-1}\right)}{1^5 \prod_{s\neq 1}\left(1+\beta_s 1^{-1}\right)} = \frac{\alpha^5\cdot 1 + \alpha^{11}\cdot 1^2}{(1+\alpha^5)(1+\alpha^{10})} = \alpha^3,$$

$$e_{j_2} = \frac{\Omega\left(\alpha^{-5}\right)}{(\alpha^5)^5 \prod_{s\neq 2}\left(1+\beta_s \alpha^{-5}\right)} = \frac{\alpha^5\cdot\alpha^{-5} + \alpha^{11}\cdot\alpha^{-10}}{\alpha^{10}(1+1\cdot\alpha^{-5})(1+\alpha^{10}\cdot\alpha^{-5})} = \alpha^9,$$

$$e_{j_3} = \frac{\Omega\left(\alpha^{-10}\right)}{(\alpha^{10})^5 \prod_{s\neq 3}\left(1+\beta_s \alpha^{-10}\right)} = \frac{\alpha^5\cdot\alpha^{-10} + \alpha^{11}\cdot\alpha^{-20}}{\alpha^{5}(1+1\cdot\alpha^{-10})(1+\alpha^{5}\cdot\alpha^{-10})} = \alpha^2.$$

Thus, the decoder found the error polynomial $\mathbf{e}(X) = \alpha^3 + \alpha^9 X^5 + \alpha^2 X^{10}$ so that the decoded codeword is $\mathbf{c}(X) = \mathbf{r}(X) + \mathbf{e}(X) = 0$, the all-zeros codeword.

---

### 3.4.3 Errors-and-Erasures RS Decoding

We have seen in Chapter 1 and this chapter the advantage soft-decision decoding has over hard-decision decoding. There exist a number of soft-decision decoding algorithms for RS codes, but they are beyond the scope of this book. A very important and useful step toward soft-decision decoding of RS codes is errors-and-erasures decoding. An *erasure* is a receive code symbol that has been flagged by an independent mechanism as erroneous, or at least suspect, for the benefit of the decoder. The flagging mechanism can be a fade detector, a sync-loss detector, or an independent detection/correction code at work. These flagged symbols, erasures, may be interpreted as errors whose locations are known but whose values are not. Thus, the decoder has half the work when resolving erasures: only values need be determined, locations are already known. That errors cost twice as much as erasures to clean up is summarized in the result that says that any pattern of $\nu$ errors and $\mu$ erasures can be corrected whenever

$$2\nu + \mu \leq d_{\min} - 1. \tag{3.88}$$

This fact will be made evident in the following development of the errors-and-erasures RS decoding algorithm.

Suppose that there are $\nu$ errors in positions $j_1, j_2, \ldots, j_\nu$ and $\mu$ erasures in positions $i_1, i_2, \ldots, i_\mu$. Then if $\mathbf{v} = [v_0, v_1, \ldots, v_{n-1}]$ is the transmitted codeword, the erasure values are $d_{i_\ell} = r_{i_\ell} - v_{i_\ell}$ for $\ell = 1, 2, \ldots, \mu$, where $\{r_{i_\ell}\}$ are generally assigned arbitrary values prior to decoding. We write the received polynomial $\mathbf{r}(X)$ as

$$\mathbf{r}(X) = \mathbf{v}(X) + \mathbf{e}(X) + \mathbf{d}(X)$$

so that the syndromes are in this case

$$S_i = \mathbf{r}(\alpha^{b+i-1}) = \mathbf{e}(\alpha^{b+i-1}) + \mathbf{d}(\alpha^{b+i-1}), \quad \text{for } i = 1, 2, \ldots, 2t.$$

This then expands as

$$S_i = \sum_{m=1}^{\nu} e_{j_m} \beta_m^{b+i-1} + \sum_{\ell=1}^{\mu} d_{i_\ell} \lambda_\ell^{b+i-1}, \tag{3.89}$$

where $\beta_m = \alpha^{j_m}$ are error locators as before and $\lambda_\ell = \alpha^{i_\ell}$ are *erasure locators*. Any method for solving these equations for $\{j_m\}$, $\{e_{j_m}\}$, and $\{d_{i_\ell}\}$ is an error-and-erasures decoding procedure. We now present a decoding procedure due to Forney [29].

Define the *erasure-location polynomial*

$$\sigma'(X) = \prod_{\ell=1}^{\mu} (1 + \lambda_\ell X) = \sigma'_0 + \sigma'_1 X + \cdots + \sigma'_\mu X^\mu$$

for which the roots are clearly the inverses of the erasure locators $\lambda_\ell$: $\sigma'\left(\lambda_\ell^{-1}\right) = 0$ for $\ell = 1, 2, \ldots, \mu$. Next form Forney's modified syndromes

$$T_k = \sum_{\ell=0}^{\mu} \sigma'_\ell S_{k+\mu+1-\ell} \quad \text{for } k = 0, 1, \ldots, 2t - \mu - 1. \tag{3.90}$$

In the case of no erasures, $T_k = S_{k+1}$ since in this case $\mu = 0$ and $\sigma'(X) = \sigma'_0 = 1$. Placing (3.89) into (3.90) gives, for $k = 0, 1, \ldots, 2t - \mu - 1$,

$$\begin{aligned}
T_k &= \sum_{l=0}^{\mu} \sigma'_l \left[ \sum_{m=1}^{\nu} e_{j_m} \beta_m^{b+k+\mu-l} + \sum_{\ell=1}^{\mu} d_{i_\ell} \lambda_\ell^{b+k+\mu-\ell} \right] \\
&= \sum_{m=1}^{\nu} e_{j_m} \beta_m^{b+\mu} \beta_m^k \left[ \sum_{l=0}^{\mu} \sigma'_l \beta_m^{-l} \right] + \sum_{\ell=1}^{\mu} d_{i_\ell} \lambda_\ell^{b+k+\mu} \left[ \sum_{l=0}^{\mu} \sigma'_l \lambda_\ell^{-l} \right] \\
&= \sum_{m=1}^{\nu} e_{j_m} \beta_m^{b+\mu} \beta_m^k \sigma'\left(\beta_m^{-1}\right) \\
&= \sum_{m=1}^{\nu} E_m \beta_m^k,
\end{aligned}$$

where we have used the fact that on the second line, the rightmost bracketed expression is $\sigma'\left(\lambda_\ell^{-1}\right) = 0$ and where we have defined $E_m = e_{j_m} \beta_m^{b+\mu} \sigma'\left(\beta_m^{-1}\right)$.

Repeating steps analogous to those that followed (3.85), we obtain an analogous LFSR equation

$$T_k = \sigma_1 T_{k-1} + \sigma_2 T_{k-2} + \cdots + \sigma_\nu T_{k-\nu}, \quad \text{for } k = 0, 1, \ldots, 2t - \mu - 1, \tag{3.91}$$

giving us $2t - \nu - \mu$ linear equations in $\nu$ unknowns ($\{\sigma_l\}$). These are solvable provided $2\nu \leq 2t - \mu = d_{\min} - \mu - 1$, which is equivalent to (3.88). (For the more general case where $d_{\min}$ may be equal to $2t + 1$ or $2t + 2$, we everywhere replace $2t$ by $d_{\min} - 1$ and still arrive at (3.88).) The BM algorithm applied to this new shift-register equation will yield the error-location polynomial $\sigma(X)$, which will in turn give us the error locations $\{j_m\}$ via the roots $\beta_m^{-1} = \alpha^{-j_m}$ of $\sigma(X)$, found via exhaustive search.

The error values $e_{j_m}$ and erasure values $d_{i_\ell}$ may be obtained from the syndromes $\{S_i\}$ using the same procedure as in the errors-only case. The details are as follows.

Once the error locations $j_m$ have been determined, the errors may be treated as erasures. In this light, let us define, for $\ell = \mu+1, \mu+2, \ldots, \mu+\nu$,

$$\begin{aligned} i_\ell &= j_{\ell-\mu}, \\ \lambda_\ell &= \beta_{\ell-\mu} = \alpha^{i_{\ell-\mu}}, \\ d_{i_\ell} &= e_{j_{\ell-\mu}}, \end{aligned}$$

so that the syndromes may be written as

$$S_i = \sum_{\ell=1}^{\nu+\mu} d_{i_\ell} \lambda_\ell^{b+i-1}, \quad \text{for } i = 1, 2, \ldots, 2t,$$

which is a rewriting of (3.89).

Now the *errata* (i.e., error and erasure) values $d_{i_\ell}$ may be shown to be a function of the $\{S_i\}$ and $\lambda_\ell$ (which are known), by mimicking our earlier procedure in the derivation of (3.87). This begins with a simple generalization of the evaluator polynomial $\Omega(X)$ to contain both error locators and erasure locators:

$$\Omega(X) = \mathbf{S}(X)\sigma(X)\sigma'(X) \bmod X^{2t} = \mathbf{S}(X) \prod_{\ell=1}^{\nu+\mu} (1 + \lambda_\ell X) \bmod X^{2t},$$

where $\mathbf{S}(X)$ is defined in (3.74). Using steps analogous to those used to derive (3.87), it can be shown (see Problem 3.26) that, for $\ell = 1, 2, \ldots, \mu+\nu$,

$$d_{i_\ell} = \frac{\Omega\left(\lambda_\ell^{-1}\right)}{\lambda_\ell^b \prod_{\ell \neq s=1}^{\nu+\mu} \left(1 + \lambda_s \lambda_\ell^{-1}\right)}. \tag{3.92}$$

---

**Example 3.10** We revisit the example above where now we allow the decoder to correct erasures. Recall that we considered a (15, 9) RS code over GF(16) generated by $\mathbf{g}(X) = \prod_{i=1}^{6}(X + \alpha^{4+i})$ so that $b = 5$ and $d_{\min} = 7$. Suppose the zero codeword was sent and that $\mathbf{e}(X) = \alpha^5 X^2$ and $\mathbf{d}(X) = \alpha^2 X^6 + \alpha^7 X^7$; that is, there is an error in location 2 with value $\alpha^5$ and there are two erasures in locations 6 and 7 with values $\alpha^2$ and $\alpha^7$, respectively.

Following the errors-and-erasures decoding algorithm, we first compute the syndromes $S_i = \mathbf{r}(\alpha^{4+i})$, $i = 1, 2, \ldots, 6$, where in this case $\mathbf{r}(X) = 0 + \mathbf{e}(X) + \mathbf{d}(X) = \alpha^5 X^2 + \alpha^2 X^6 + \alpha^7 X^7$. The syndromes are then

$$\begin{aligned} S_1 &= \mathbf{r}(\alpha^5) = \alpha^5\alpha^{10} + \alpha^2\alpha^{30} + \alpha^7\alpha^{35} = 1 + \alpha^2 + \alpha^{12} = \alpha^9, \\ S_2 &= \mathbf{r}(\alpha^6) = \alpha, \\ S_3 &= \mathbf{r}(\alpha^7) = \alpha^2, \\ S_4 &= \mathbf{r}(\alpha^8) = \alpha, \\ S_5 &= \mathbf{r}(\alpha^9) = \alpha^6, \\ S_6 &= \mathbf{r}(\alpha^{10}) = \alpha^{10}. \end{aligned}$$

**Table 3.8** Berlekamp–Massey decoding steps for Example 3.10

| $i$ | $L$ | $\delta$ | $D(X)$ | $\sigma(X)$ | $\sigma^*(X)$ |
|---|---|---|---|---|---|
| 1 | 0 | 1 | $X$ | 1 | 1+X |
| 2 | 1 | $\alpha^8$ | $X$ | 1+X | $1+\alpha^2 X$ |
| 3 | 1 | 0 | $X^2$ | $1+\alpha^2 X$ | — |
| 4 | 1 | 0 | $X^3$ | $1+\alpha^2 X$ | — |

The next step is to compute the erasure-location polynomial $\sigma'(X)$,

$$\sigma'(X) = (1+\alpha^6 X)(1+\alpha^7 X) = 1 + \alpha^{10}X + \alpha^{13}X^2,$$

and the modified syndromes $T_k, k = 0, 1, \ldots, d_{\min} - 1 - \mu$ (where $d_{\min} = 7$ and $\mu = 2$ here) are, using (3.90),

$$T_0 = \sum_{\ell=0}^{2} \sigma'_\ell S_{3-\ell} = 1 \cdot \alpha^2 + \alpha^{10}\alpha + \alpha^{13}\alpha^9 = 1,$$

$$T_1 = \sum_{\ell=0}^{2} \sigma'_\ell S_{4-\ell} = \alpha^2,$$

$$T_2 = \sum_{\ell=0}^{2} \sigma'_\ell S_{5-\ell} = \alpha^4,$$

$$T_3 = \sum_{\ell=0}^{2} \sigma'_\ell S_{6-\ell} = \alpha^6.$$

Recalling the LFSR relationship (3.90), we use the BM algorithm to find the error-location polynomial $\sigma(X)$ as shown in Table 3.8, where we see that $\sigma(X) = 1 + \alpha^2 X$. From this, $\alpha^2$ is the error locator and thus there is an error in location 2 ($j_1 = 2$).

Now form the errata locator $\Omega(X)$,

$$\begin{aligned}\Omega(X) &= \mathbf{S}(X)\sigma(X)\sigma'(X) \bmod X^6 \\ &= \alpha^9 + \alpha^{12}X + \alpha^8 X^2,\end{aligned}$$

and compute the errata values $d_{i_\ell}, \ell = 1, 2, 3$, using (3.92) and

$$\begin{aligned}\nu + \mu &= 1 + 2 = 3, \\ i_1 &= 6 \text{ (known erasure location)} \implies \lambda_1 = \alpha^6, \\ i_2 &= 7 \text{ (known erasure location)} \implies \lambda_2 = \alpha^7, \\ i_3 &= j_1 = 2 \text{ (error location from } \sigma(X)) \implies \lambda_3 = \alpha^2.\end{aligned}$$

From (3.92),

$$\begin{aligned} d_{i_1} &= \frac{\Omega\left(\lambda_1^{-1}\right)}{\lambda_1^5 \prod_{1\neq s=1}^{3}\left(1+\lambda_s\lambda_1^{-1}\right)} \\ &= \frac{\alpha^9+\alpha^{12}\alpha^{-6}+\alpha^8\alpha^{-12}}{\alpha^{30}(1+\alpha^7\alpha^{-6})(1+\alpha^2\alpha^{-6})} = \alpha^2, \\ d_{i_2} &= \frac{\Omega\left(\lambda_2^{-1}\right)}{\lambda_2^5 \prod_{2\neq s=1}^{3}\left(1+\lambda_s\lambda_2^{-1}\right)} = \alpha^7, \\ d_{i_3} &= \frac{\Omega\left(\lambda_3^{-1}\right)}{\lambda_3^5 \prod_{3\neq s=1}^{3}\left(1+\lambda_s\lambda_3^{-1}\right)} = \alpha^5. \end{aligned}$$

Thus, $\mathbf{e}(X) = \alpha^5 X^2$ and $\mathbf{d}(X) = \alpha^2 X^6 + \alpha^7 X^7$ and the corrected codeword is $\mathbf{r}(X) + \mathbf{e}(X) + \mathbf{d}(X) = 0$.

### 3.4.4 RS Code Performance Calculations

The nominal code length for a RS code over GF($q$) is $q - 1$. In practice, the code length is often shortened to some length $n < q - 1$. Further, while $\nu$ errors and $\mu$ erasures are correctable whenever $2\nu + \mu \leq d_{\min} - 1$, in practice the correction capability is often limited to some amount less than $d_{\min} - 1$, say $d' - 1$, where $d' < d_{\min}$. Doing so improves the detection capability of the code system so that most of the uncorrectable received words are detected as such, rather than decoded to some incorrect codeword. This is explained intuitively in Figure 3.7 where we see that smaller

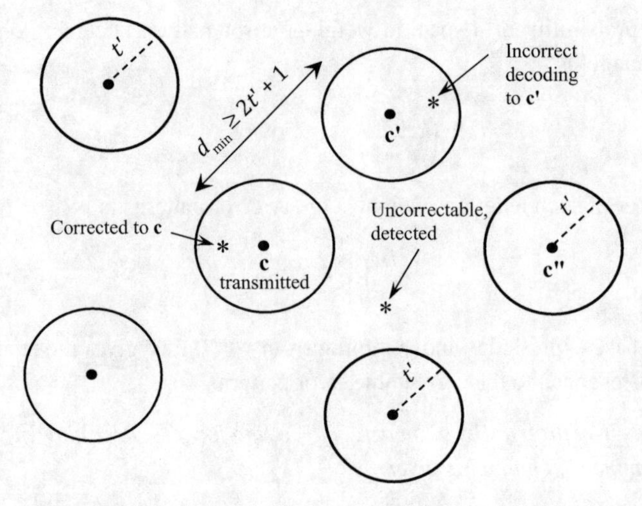


**Figure 3.7** Illustration of possible decoding events at the RS decoder for decoding spheres with radius $t'$, where $t' \leq t = \lfloor (d_{\min} - 1)/2 \rfloor$. The black dots represent codewords and the asterisks represent possible received words.

*decoding spheres* reduce the correction capability, but improve the uncorrectable-error detection capability. That is, the probability of decoding to an incorrect codeword is reduced.

From Figure 3.7, three events may occur at the decoder.

1. The decoder may choose the *correct codeword* (with probability $P_c$).
2. The decoder may *detect* an uncorrectable error (with probability $P_d$).
3. Or the decoder may choose an *incorrect codeword* (with probability $P_{ic}$).

Another way of describing the third event is that an uncorrectable error is also undetectable. Observe that these probabilities must sum to one, $P_c + P_d + P_{ic} = 1$, and that the probability of codeword error is $P_e = 1 - P_c = P_d + P_{ic}$. Notice from this that two types of errors contribute to the overall error probability, $P_e$.

We now provide mathematical expressions for the performance charactistics of RS code systems based on the *operational distance* $d' \leq d_{\min}$ and the *operational correction capability* $t' = \lfloor \frac{d'-1}{2} \rfloor$, which may be less than the true values, $d_{\min}$ and $t = \lfloor \frac{d_{\min}-1}{2} \rfloor$. Because the probabilities sum to one, we need only provide expressions for two of the three probabilities. We will assume a $q$-ary symmetric channel (which may have an underlying binary channel) for which the probability of error will be denoted by $P$ and the probability of a specific error (value) is $P/(q-1)$. Errors-only decoding is assumed first; modifications for errors-and-erasures decoding will come next. These results follow [30], with a few differences.

**Theorem 3.8** *The probability of correct decoding for a $t'$-correcting $(n, k)$ RS code on a $q$-ary symmetric channel is given by*

$$P_c = \sum_{\nu=0}^{t'} \binom{n}{\nu} P^{\nu}(1-P)^{n-\nu}.$$

*Proof* The probability of a specific weight-$\nu$ error pattern occurring on the $q$-ary symmetric channel is

$$\left(\frac{P}{q-1}\right)^{\nu} (1-P)^{n-\nu}$$

and the number of such length-$n$, weight-$\nu$, $q$-ary error patterns is

$$\binom{n}{\nu}(q-1)^{\nu}.$$

Multiplying these expressions and summing from $\nu = 0$ to $t'$ gives the result because $0 \leq \nu \leq t'$ corresponds to the correctable error patterns. □

**Theorem 3.9** *The probability of incorrect decoding for a $t'$-correcting $(n, k)$ RS code on a $q$-ary symmetric channel is given by*

$$P_{ic} = \sum_{\nu=0}^{n} \left(\frac{P}{q-1}\right)^{\nu} (1-P)^{n-\nu} \sum_{d=0}^{t'} \sum_{\ell=d_{\min}}^{n} A_{\ell} N(\ell, d; \nu),$$

*where $A_\ell$ is the number of weight-$\ell$ codewords and $N(\ell, d;v)$ is the number of weight-$v$ error patterns that are at a distance $d$ from a weight-$\ell$ codeword.*

*Proof* Because the code is linear, we may assume that the zero codeword was transmitted. We are then interested in the probability that the decoder decodes to a nonzero codeword. For $\ell = d_{\min}, \ldots, n$, there are $A_\ell$ codewords of weight $\ell$ and, for each distance $d = 0, 1, \ldots, t'$, there are $N(\ell, d;v)$ error patterns of weight $v$ that will decode to a weight-$\ell$ codeword. Thus, the total number of weight-$v$ error patterns that will be decoded to a nonzero (hence, wrong) codeword is $\sum_{d=0}^{t'} \sum_{\ell=d_{\min}}^{n} A_\ell N(\ell, d;v)$, each occurring with probability $\left(\frac{P}{q-1}\right)^v (1-P)^{n-v}$. Now sum over all possible error-pattern weights $v$. □

**Corollary 3.1** *For an $(n, k)$ RS code and for $t'$ satisfying $2t' + 1 \le d_{\min} - 1$, the probability of incorrect decoding given the received word contains a weight-$v$ error pattern, $P_{ic|v}$, is equal to the fraction of weight-$v$ error patterns within distance $t'$ of a codeword:*

$$P_{ic|v} = \frac{\sum_{d=0}^{t'} \sum_{\ell=d_{\min}}^{n} A_\ell N(\ell, d;v)}{\binom{n}{v}(q-1)^v}.$$

*Proof* The numerator was derived in the theorem proof and the denominator is simply the number of weight-$v$ $q$-ary error patterns of length $n$. □

In order to make use of the expressions for $P_{ic}$ and $P_{ic|v}$, we need formulas for $A_\ell$ and $N(\ell, d;v)$. We provide the formula $A_\ell$ without proof (see [31]).

**Theorem 3.10** *The weight spectrum $\{A_\ell\}_{\ell=0}^{n}$ of an $(n, k)$ RS code over GF($q$) is given by $A_0 = 1$, $A_\ell = 0$ for $\ell = 1, 2, \ldots, d_{\min} - 1$, and*

$$A_\ell = \binom{n}{\ell} \sum_{j=0}^{\ell - d_{\min}} (-1)^j \binom{\ell}{j} \left(q^{\ell - d_{\min} + 1 - j} - 1\right)$$

*for $d_{\min} \le \ell \le n$.*

**Theorem 3.11** *The number of error patterns of weight $v$ that are at a distance $d$ from a particular codeword of weight $\ell$ is given by*

$$N(\ell, d;v) = \sum_{\substack{0 \le i,j,k \le n \\ i+j+k=v \\ \ell+k-j=d}} \left[\binom{n-\ell}{k}(q-1)^k\right] \left[\binom{\ell}{i}(q-2)^i\right] \left[\binom{\ell-i}{j}\right].$$

*Proof* Figure 3.8 should be helpful in the following descriptions. Consider a weight-$\ell$ codeword. The first bracketed factor in the sum is the number of words obtained by changing any $k$ of the $n-\ell$ zero components of the codeword to any of the $q-1$ nonzero elements. The second bracketed factor is the number of words obtained by changing any $i$ of the $\ell$ nonzero components of the codeword to any of the other $q-2$ nonzero elements. The third bracketed factor is the number of words obtained by changing any $j$ of the remaining $\ell - i$ nonzero components to zeros. The constraint $\ell + k - j = d$

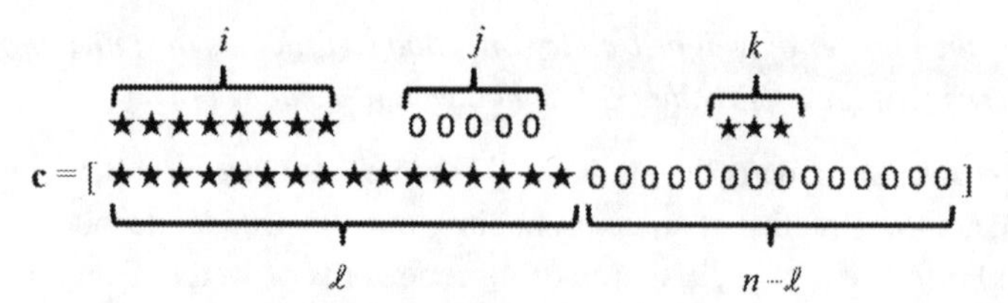


**Figure 3.8** Illustration of the various variables used in Theorem 3.11. The ★ symbols represent nonzero components of a weight-$\ell$ codeword, which are grouped together for convenience in the illustration. The row above the codeword represents the changes in the values of the codeword caused by an error pattern of weight $\nu = i + j + k$.

ensures that the resulting word is at a distance $d$ from the codeword and the constraint $i + j + k = \nu$ ensures that the error word has weight $\nu$. □

As mentioned, the expressions given above for $P_c$, $P_{ic}$, and $P_{ic|\nu}$ assume an errors-only RS decoder. For errors-and-erasures decoding, we condition on the presence of $\mu$ erasures, in which case $n$ in the expressions should be replaced by $n - \mu$ and $d_{\min}$ should be replaced by $d_{\min} - \mu$. If the probability distribution on $\mu$ is known, it can be used to compute the expected value of the expressions conditioned on $\mu$. If not, the conditional expressions are useful in themselves.

---

**Example 3.11** We consider a shortened (60, 50) RS code over GF(64) for which $d_{\min} = 11$ and $t = 5$. For various combinations of $\nu$ and $\mu$, Table 3.9 illustrates the impact of an operational $d'_{\min} < d_{\min}$ (and an operational $t' < t$) and of a large number of erasures on the probability of incorrect decoding, $P_{ic|\nu}$. The leftmost four columns in Table 3.9 correspond to $d'_{\min} = d_{\min} = 11$ and $t = t' = 5$. The rightmost four columns correspond to $d'_{\min} = 9$ and $t' = 4$; that is, the decoder limits its correction capability to improve $P_{ic|\nu}$.

We see in the $t' = 5$ case that when $\mu \geq 8$, $P_{ic|\nu}$ is utterly unacceptable when errors are present in addition to the erasures; the decoder (almost) always chooses the wrong codeword. When the operational correction limit is reduced to $t' = 4$, fewer errors and erasures are correctable, but the decoder rarely chooses the wrong codeword.

---

*Decoder Implementation*. The $P_{ic|\nu}$ values reported in Table 3.9 do not hold for just any RS errors-and-erasures decoder implementation. An effort must be made in the decoder design to detect uncorrectable errors. Recall that $L$ is the degree of the error-location polynomial $\sigma(X)$ found by the BM algorithm. Then the decoder must declare the *detection of an uncorrectable error* (or decoding failure) if any of the following checks fail.

1. If $\sigma(X)$ has degree $L$, then $L$ roots of $\sigma(X)$ must be found. (When $2\nu + \mu$ exceeds $d'_{\min} - 1$, occasionally fewer than $L$ roots from GF($q$) are found; some roots are in a higher field. This is a clear sign of too many errors.)
2. For a shortened RS code (i.e., $n < q - 1$), the error locations must be in the set $\{0, 1, \ldots, n-1\}$. (As an example, for the (60, 50) RS code in the example above, if an error location is found in position 62, there is clearly a problem.)
3. $2L + \mu$ cannot exceed $d'_{\min} - 1$.

**Table 3.9** Calculations of $P_{ic|\nu}$ for Example 3.11

| $t'$ | $\nu$ | $\mu$ | $P_{ic\|\nu}$ | $t'$ | $\nu$ | $\mu$ | $P_{ic\|\nu}$ |
|---|---|---|---|---|---|---|---|
| 5 | 1–5 | 0 | 0 | 4 | 1–6 | 0 | 0 |
| 5 | 6 | 0 | $3.2 \times 10^{-3}$ | 4 | 7 | 0 | $4.7 \times 10^{-6}$ |
| 5 | 7 | 0 | $4.5 \times 10^{-3}$ | 4 | 8 | 0 | $6.4 \times 10^{-6}$ |
| 5 | 8–60 | 0 | $4.7 \times 10^{-3}$ | 4 | 9–60 | 0 | $6.6 \times 10^{-6}$ |
| 5 | 0–1 | 8 | 0 | 4 | 1–3 | 6 | 0 |
| 5 | 2 | 8 | 0.79 | 4 | 4 | 6 | $2.0 \times 10^{-4}$ |
| 5 | 3–52 | 8 | 0.80 | 4 | 5–54 | 6 | $2.03 \times 10^{-4}$ |
| 5 | 0 | 10 | 0 | 4 | 1–2 | 8 | 0 |
| 5 | 1–50 | 10 | 1 | 4 | 3–52 | 8 | $\sim 2.5 \times 10^{-4}$ |

*Adaptive Currency Decoder.* It is helpful to think of the innate correction capability of an RS code, $d_{\min} - 1 = n - k$, as *currency*, where it costs two units of currency to correct an error and one unit to correct an erasure, as follows from the requirement $2\nu + \mu \leq n - k$. In this section, we have discussed a *fixed* currency $d' - 1$, which is made less than $d_{\min} - 1 = n - k$ for improved undetected error (incorrect decoding) performance, as seen in Example 3.11. As pointed out by Gronholz [29], for a given requirement on $P_{ic|\nu}$, an RS decoder can be made more effective by making the currency *adaptive*.

To explain the adaptive currency decoding algorithm, denote the currency variable by $c$, where $c \leq n - k$, and the worst-case $P_{ic|\nu}$ requirement by $P_{ic}^{req}$ (for all possible $\nu \geq 1$). Then, once $\mu \leq n - k$ erasures have been received by the decoder together with the received codeword, $P_{ic|\nu}(c)$ is computed for $c = n - k,\ n - k - 1,\ n - k - 2, \ldots$ until $P_{ic|\nu}(c) \leq P_{ic}^{req}$ is satisfied. The largest currency $c$ satisfying this requirement is the one used to decode the received codeword. This is done for each received codeword and set of erasures. The computation of $P_{ic|\nu}(c)$ for each candidate $c$ is quick and need only be performed for $\nu = n - k - \mu + 1$, a value we found empirically to give the largest $P_{ic|\nu}(c)$. Note that $P_{ic|\nu}(c)$ is a function of $n' \triangleq n - \mu$, $d'_{\min} \triangleq d_{\min} - \mu$, and $c \triangleq d' - 1$.

## 3.5 Product, Interleaved, and Concatenated Codes

In this section, we present three well-known and widely used coding techniques to construct long, powerful codes from short component codes. These coding techniques are called *product* [28], *interleaving*, and *concatenation* [31], which are effective in correcting mixed types of errors, such as combinations of random errors and bursts of errors (or random errors and erasures) [5].

### 3.5.1 Product Codes

Let $\mathcal{C}_1$ be a binary $(n_1, k_1)$ linear block code and $\mathcal{C}_2$ be a binary $(n_2, k_2)$ linear block code. A code with $n_1 n_2$ symbols can be constructed by making a *rectangular array* of

$$
\left[\begin{array}{cccc|ccc}
v_{0,0} & v_{0,1} & \cdots & v_{0,n_1-k_1-1} & v_{0,n_1-k_1} & \cdots & v_{0,n_1-1} \\
v_{1,0} & v_{1,1} & \cdots & v_{1,n_1-k_1-1} & v_{1,n_1-k_1} & \cdots & v_{1,n_1-1} \\
\vdots & \vdots & \ddots & \vdots & \vdots & & \vdots \\
v_{k_2-1,0} & v_{k_2-1,1} & \cdots & v_{k_2-1,n_1-k_1-1} & v_{k_2-1,n_1-k_1} & \cdots & v_{k_2-1,n_1-1} \\
\hline
v_{k_2,0} & v_{k_2,1} & \cdots & v_{k_2,n_1-k_1-1} & v_{k_2,n_1-k_1} & \cdots & v_{k_2,n_1-1} \\
\vdots & \vdots & \ddots & \vdots & \vdots & & \vdots \\
v_{n_2-1,0} & v_{n_2-1,1} & \cdots & v_{n_2-1,n_1-k_1-1} & v_{n_2-1,n_1-k_1-1} & \cdots & v_{n_2-1,n_1-1}
\end{array}\right]
$$

**Figure 3.9** A code array in a two-dimensional product code.

$n_1$ columns and $n_2$ rows in which every row is a codeword in $C_1$ and every column is a codeword in $C_2$, as shown in Figure 3.9. The $k_1k_2$ symbols in the *upper-right quadrant* of the array are information symbols. The $k_2(n_1 - k_1)$ symbols in the *upper-left quadrant* of the array are parity-check symbols formed from the parity-check rules for code $C_1$, and the $k_1(n_2-k_2)$ symbols in the *lower-right quadrant* are parity-check symbols formed from the parity-check rules for $C_2$. The $(n_1 - k_1)(n_2 - k_2)$ parity-check symbols in the *lower-left quadrant* can be formed by using either the parity-check rules for $C_2$ on columns or the parity-check rules for $C_1$ on rows. The rectangular array shown in Figure 3.9 is called a *code array* that consists of $k_1k_2$ information symbols and $n_1n_2 - k_1k_2$ parity-check symbols. There are $2^{k_1k_2}$ such code arrays. The sum of two code arrays is an array obtained by adding either their corresponding rows or their corresponding columns. Since the rows (or columns) are codewords in $C_1$ (or $C_2$), the sum of two corresponding rows (or columns) in two code arrays is a codeword in $C_1$ (or in $C_2$). Consequently, the sum of two code arrays is another code array. Hence, the $2^{k_1k_2}$ code arrays form a two-dimensional $(n_1n_2, k_1k_2)$ linear block code, denoted by $C_1 \times C_2$, which is called the *direct product* (or simply product) of $C_1$ and $C_2$.

Encoding can be accomplished in two stages. A sequence of $k_1k_2$ information symbols is first arranged as a $k_2 \times k_1$ array **A**, the upper-right quadrant of the code array given in Figure 3.9. At the first stage of encoding, each row of **A** is encoded into a codeword in $C_1$. This results in a $k_2 \times n_1$ array **B**. The first stage of encoding is referred to as the *row encoding*. At the second stage of encoding, each column of **B** is encoded into a codeword in $C_2$. After the completion of the second stage of encoding, we obtain an $n_2 \times n_1$ code array as shown in Figure 3.9. The second stage of encoding is referred to as *column encoding*. The above two-stage encoding of the information array **A** is referred to as *row/column* encoding. Clearly, two-stage encoding can be accomplished by first performing column encoding of the information array **A** and then row encoding. This two-stage encoding is referred to as *column/row* encoding. With row/column encoding, a code array is transmitted column by column. A column codeword in $C_2$ is transmitted as soon as it is formed. With column/row encoding, a code array is transmitted row by row. A row codeword in $C_1$ is transmitted as soon as it is formed. With two-stage encoding, a transmitter buffer is needed to store the code array.

If the minimum weights (or minimum distances) of $C_1$ and $C_2$ are $d_1$ and $d_2$, respectively, then the minimum weight (or distance) of the product code $C_1 \times C_2$ is $d_1d_2$.

A minimum-weight code array in the product code can be formed as follows: (1) choose a minimum-weight codeword $\mathbf{v}_1$ in $C_1$ and a minimum-weight codeword $\mathbf{v}_2$ in $C_2$; and (2) form a code array in which all columns corresponding to the zero-components of $\mathbf{v}_1$ are zero columns and all columns corresponding to the 1-components of $\mathbf{v}_1$ are $\mathbf{v}_2$.

Decoding of a product code can be accomplished in two stages. Suppose a code array is transmitted, either column by column or row by row. At the receiving end, the received $n_1 n_2$ symbols are arranged back into an $n_2 \times n_1$ array, called a *received array*. Then decoding is carried out first on columns and then on rows (or first on rows and then on columns), which is referred to as column/row decoding (or row/column decoding). Residue errors that are not corrected at the first stage will be, with high probability, corrected at the second stage. Decoding performance can be improved by carrying out the column/row (row/column) decoding iteratively [5, 24]. Residue errors that are not corrected in one iteration will be, with high probability, corrected in the next iteration. This type of decoding is referred to as *iterative decoding*, which will be discussed in detail in Chapter 7. The complexity of two-stage decoding is roughly the sum of the complexities of the two component code decodings.

A product code over GF($q$) can be constructed by using two component codes over GF($q$).

### 3.5.2 Interleaved Codes

Let $C$ be an $(n, k)$ linear block code and $\lambda$ a positive integer. We can construct a $(\lambda n, \lambda k)$ linear block code $C(\lambda)$ by arranging $\lambda$ codewords from $C$ as rows of a $\lambda \times n$ array as shown in Figure 3.10 and then transmitting the array, column by column. This process basically interleaves $\lambda$ codewords in $C$ such that two neighboring bits of a codeword are separated by $\lambda - 1$ bits from $\lambda - 1$ other codewords. This code-construction technique is called *block interleaving* and the parameter $\lambda$ is referred to as the *interleaving depth* (or *degree*). $C(\lambda)$ is called an interleaved code with interleaving depth $\lambda$.

At the receiving end, before decoding, every $\lambda n$ received symbols must be rearranged column by column back to a $\lambda \times n$ array, called a received array. This process is referred to as *de-interleaving*. Then each row of the received array is decoded on the basis of $C$ and a decoding algorithm. Interleaving is a very effective technique for correcting *multiple bursts of errors*. Let $C$ be a $t$-error-correcting code. Suppose a codeword in the interleaved code $C(\lambda)$ is transmitted and multiple bursts of errors occur during its transmission. After de-interleaving, if the multiple bursts of errors

$$\begin{bmatrix} v_{0,0} & v_{0,1} & \cdots & v_{0,n-1} \\ v_{1,0} & v_{1,1} & \cdots & v_{1,n-1} \\ \vdots & \vdots & \ddots & \vdots \\ v_{\lambda-1,0} & v_{\lambda-1,1} & \cdots & v_{\lambda-1,n-1} \end{bmatrix}$$

**Figure 3.10** A block-interleaved array.

in the received sequence do not cause more than $t$ errors in each row of the received array, then row decoding based on $\mathcal{C}$ will correct the errors in each row and hence corrects the multiple bursts of errors in the received sequence. The de-interleaving simply makes the multiple bursts of errors in the received sequence appear as random errors in each row of the received array.

### 3.5.3 Concatenated Codes

Let GF($2^m$) be an extension field of GF(2). Then each element in GF($2^m$) can be represented by an $m$-tuple over GF(2). A simple concatenated code is formed by two component codes, an $(n_1, k_1)$ code $\mathcal{C}_1$ over GF($2^m$) and a binary $(n_2, m)$ code $\mathcal{C}_2$. Let $\mathbf{u} = (u_0, u_1, \ldots, u_{n_1-1})$ be a codeword in $\mathcal{C}_1$. On expanding each code symbol of $\mathbf{u}$ into an $m$-tuple of GF(2), we obtain an $mn_1$-tuple $\mathbf{w} = (w_0, w_1, \ldots, w_{mn_1-1})$ over GF(2). This $mn_1$-tuple $\mathbf{w}$ is called the *binary image sequence* of $\mathbf{u}$. Encode every $m$ consecutive binary symbols of $\mathbf{w}$ into a codeword in $\mathcal{C}_2$. This results in a sequence of $n_1n_2$ binary symbols, which consists of a sequence of $n_1$ binary codewords in $\mathcal{C}_2$. This concatenated sequence of $n_1n_2$ binary symbols contains $k_1m$ information bits. Since there are $2^{k_1m}$ codewords in $\mathcal{C}_1$, there are $2^{k_1m}$ such concatenated $n_1n_2$-bit sequences, which form a binary $(n_1n_2, k_1m)$ linear block code, called a *concatenated code* [31].

Encoding of a concatenated code consists of two stages, as shown in Figure 3.11. This particular configuration is sometimes called a *serial concatenated code*. First a binary information sequence of $k_1m$ bits is divided into $k_1$ bytes of $m$ information bits each. Each $m$-bit byte is regarded as an element in GF($2^m$). At the first stage of encoding, the $k_1$ bytes, regarded as $k_1$ information symbols over GF($2^m$), are encoded into an $n_1$-byte codeword $\mathbf{u}$ in $\mathcal{C}_1$. The first stage of encoding results in a coded sequence $\mathbf{w}$ of $mn_1$ bits (or $n_1$ $m$-bit bytes). At the second stage of encoding, every group of $m$ consecutive bits of $\mathbf{w}$ is encoded into an $n_2$-bit codeword in $\mathcal{C}_2$, resulting in a string of $n_1$ codewords in $\mathcal{C}_2$. This string of $n_1$ codewords in $\mathcal{C}_2$ is then transmitted, one $\mathcal{C}_2$ codeword at a time, in succession. Since $\mathcal{C}_1$ and $\mathcal{C}_2$ are used at *outer* and *inner* encoding stages, respectively, as shown in Figure 3.11, they are called outer and inner codes, respectively. If the minimum distances of $\mathcal{C}_1$ and $\mathcal{C}_2$ are $d_1$ and $d_2$, then the minimum distance of the concatenation of $\mathcal{C}_1$ and $\mathcal{C}_2$ is at least $d_1d_2$.

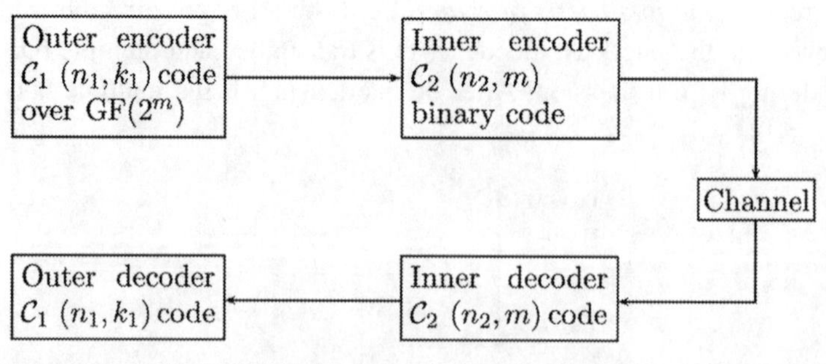


**Figure 3.11** A classical (serial) concatenated coding system.

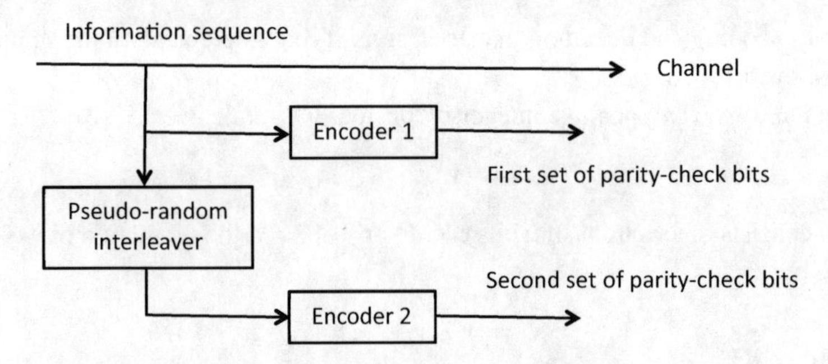


**Figure 3.12** A parallel-concatenated coding scheme.

Decoding of a concatenated code is also done in two stages. First, decoding is carried out for each inner $n_2$-bit received word as it arrives, based on a decoding method for the inner code $C_2$, and the parity-check bits are then removed, leaving a sequence of $n_1$ $m$-bit bytes. This stage of decoding is called the *inner* decoding. The $n_1$ decoded bytes at the end of the inner decoding are then decoded based on the outer code $C_1$ using a certain decoding method. This second decoding stage, called the *outer decoding*, results in $k_1$ decoded information bytes (or $k_1m$ decoded information bits).

In concatenated coding systems, RS codes over extension fields of GF(2) are commonly used as outer codes for their ability to correct error bursts at the output of an inner code decoder. In the above presentation, a binary linear block code is used as the inner code. However, the convolutional codes to be presented in the next chapter can also be used as inner codes. Usually simple codes are used as inner codes and they are decoded with soft-decision decoding to achieve good error performance.

Binary codes can also be used as outer codes. A concatenated code with both binary outer and binary inner codes can be put into a *parallel concatenation*, in which a single information sequence is encoded by two encoders independently using a pseudo-random interleaver as shown in Figure 3.12. The encoding generates two independent (or uncorrelated) sets of parity-check bits for the same information sequence. At the receiving end, iterative decoding is performed using two decoders based on the two independent sets of parity-check bits, with messages passing between the two decoders. To achieve good error performance, both decoders use soft-decision decoding. Parallel concatenation is also known as *turbo coding*, which will be discussed in Chapter 7.

Chapter 7 also introduces a serial concatenated code (or *serial turbo code*) that uses a binary convolutional code as the outer code in place of the nonbinary outer code discussed earlier (Figure 3.11).

## 3.6 Quasi-cyclic Codes

Quasi-cyclic codes are a generalization of cyclic codes. As such, they form a large class of codes, simplifying the search for good codes. Their utility lies in the structure of their encoding and decoding matrices, simplifying implementation. We now define a quasi-cyclic code.

Let $t$ and $b$ be two positive integers. Consider a $tb$-tuple over GF(2),

$$\mathbf{v} = (\mathbf{v}_0, \mathbf{v}_1, \ldots, \mathbf{v}_{t-1}), \tag{3.93}$$

which consists of $t$ sections of $b$ bits each. For $0 \leq j < t$, the $j$th section of $\mathbf{v}$ is a $b$-tuple over GF(2),

$$\mathbf{v}_j = (v_{j,0}, v_{j,1}, \ldots, v_{j,b-1}). \tag{3.94}$$

Let $\mathbf{v}_j^{(1)}$ be the $b$-tuple over GF(2) obtained by cyclically shifting each component of $\mathbf{v}_j$ one place to the right. We call $\mathbf{v}_j^{(1)}$ the (right) cyclic shift of $\mathbf{v}_j$. Let

$$\mathbf{v}^{(1)} = (\mathbf{v}_0^{(1)}, \mathbf{v}_1^{(1)}, \ldots, \mathbf{v}_{t-1}^{(1)}). \tag{3.95}$$

The $tb$-tuple $\mathbf{v}^{(1)}$ is called the *t-sectionized* cyclic shift of $\mathbf{v}$.

**Definition 3.12** Let $b$, $k$, and $t$ be positive integers such that $k < tb$. A $(tb, k)$ linear block code $\mathcal{C}_{qc}$ over GF(2) is called a *quasi-cyclic* (QC) code if the following conditions hold: (1) each codeword in $\mathcal{C}_{qc}$ consists of $t$ sections of $b$ bits each; and (2) every $t$-sectionized cyclic shift of a codeword in $\mathcal{C}_{qc}$ is also a codeword in $\mathcal{C}_{qc}$. Such a QC code is called a *t-section QC code.*

If $t = 1$, $\mathcal{C}_{qc}$ is a cyclic code of length $b$. Therefore, cyclic codes form a subclass of QC codes. For $k = cb$ with $1 \leq c < t$, the generator matrix $\mathbf{G}_{qc}$ of a binary $(tb, cb)$ QC code consists of a $c \times t$ *array* of $b \times b$ circulants over GF(2). A $b \times b$ circulant is a $b \times b$ matrix for which each row is a right cyclic shift of the row above it and the first row is the right cyclic shift of the last row. For a circulant, each column is a downward cyclic shift of the column on its left and the first column is the downward cyclic shift of the last column. The top row (or the leftmost column) of a circulant is called the *generator* of the circulant. The set of columns of a circulant, read from bottom to top (or from top to bottom), is the same as the set of rows of the circulant, read from left to right (or from right to left). Therefore, the rows and columns of a circulant have the same weight. A $b \times b$ zero matrix may be regarded as a (trivial) circulant.

The generator matrix of a $(tb, cb)$ QC code in systematic form is given as

$$\mathbf{G}_{qc,\text{sys}} = \begin{bmatrix} \mathbf{G}_0 \\ \mathbf{G}_1 \\ \vdots \\ \mathbf{G}_{c-1} \end{bmatrix} = \Bigg[ \underbrace{\begin{matrix} \mathbf{G}_{0,0} & \mathbf{G}_{0,1} & \cdots & \mathbf{G}_{0,t-c-1} \\ \mathbf{G}_{1,0} & \mathbf{G}_{1,1} & \cdots & \mathbf{G}_{1,t-c-1} \\ \vdots & \vdots & \ddots & \vdots \\ \mathbf{G}_{c-1,0} & \mathbf{G}_{c-1,1} & \cdots & \mathbf{G}_{c-1,t-c-1} \end{matrix}}_{\mathbf{P}} \quad \underbrace{\begin{matrix} \mathbf{I} & \mathbf{O} & \cdots & \mathbf{O} \\ \mathbf{O} & \mathbf{I} & \cdots & \mathbf{O} \\ \vdots & \vdots & \ddots & \vdots \\ \mathbf{O} & \mathbf{O} & \cdots & \mathbf{I} \end{matrix}}_{\mathbf{I}_{cb}} \Bigg], \tag{3.96}$$

where $\mathbf{I}$ is a $b \times b$ identity matrix, $\mathbf{O}$ a $b \times b$ zero matrix, and $\mathbf{G}_{i,j}$ a $b \times b$ circulant with $0 \leq i < c$ and $0 \leq j < t - c$. The generator matrix of the form given by (3.96) is said to be in *systematic circular form.* We see that the parity submatrix $\mathbf{P}$ on the left-hand side of $\mathbf{G}_{qc,\text{sys}}$ is a $c \times (t-c)$ array of $b \times b$ circulants. $\mathbf{G}_{qc,\text{sys}}$ is a $cb \times tb$ matrix over GF(2). Let

$$\begin{aligned}\mathbf{u} &= (\mathbf{u}_0, \mathbf{u}_1, \ldots, \mathbf{u}_{c-1}) \\ &= (u_0, u_1, \ldots, u_{cb-1})\end{aligned}$$

be an information sequence of $cb$ bits that consists of $c$ sections, of $b$ bits each. The codeword for $\mathbf{u}$ in systematic form is given by

$$\mathbf{v} = \mathbf{u}\mathbf{G} = \mathbf{u}_0\mathbf{G}_0 + \mathbf{u}_1\mathbf{G}_1 + \cdots + \mathbf{u}_{c-1}\mathbf{G}_{c-1}. \tag{3.97}$$

---

**Example 3.12** Consider the four-section (20, 10) systematic QC code generated by the following generator matrix in systematic circular form:

$$\mathbf{G}_{qc,\text{sys}} =$$

$$\left[\begin{array}{ccccc|ccccc|ccccc|ccccc}
1&0&1&0&0&1&0&0&0&1&1&0&0&0&0&0&0&0&0&0\\
0&1&0&1&0&1&1&0&0&0&0&1&0&0&0&0&0&0&0&0\\
0&0&1&0&1&0&1&1&0&0&0&0&1&0&0&0&0&0&0&0\\
1&0&0&1&0&0&0&1&1&0&0&0&0&1&0&0&0&0&0&0\\
0&1&0&0&1&0&0&0&1&1&0&0&0&0&1&0&0&0&0&0\\
1&1&0&0&0&0&1&0&1&0&0&0&0&0&0&1&0&0&0&0\\
0&1&1&0&0&0&0&1&0&1&0&0&0&0&0&0&1&0&0&0\\
0&0&1&1&0&1&0&0&1&0&0&0&0&0&0&0&0&1&0&0\\
0&0&0&1&1&0&1&0&0&1&0&0&0&0&0&0&0&0&1&0\\
1&0&0&0&1&1&0&1&0&0&0&0&0&0&0&0&0&0&0&1
\end{array}\right].$$

Suppose the information sequence to be encoded is $\mathbf{u} = (\mathbf{u}_0, \mathbf{u}_1) = (10000, 00011)$, which consists of two sections, 5 bits each. The corresponding codeword for $\mathbf{u}$ is

$$\begin{aligned}\mathbf{v} &= (\mathbf{v}_0, \mathbf{v}_1, \mathbf{v}_2, \mathbf{v}_3) \\ &= (00110, 01100, 10000, 000011),\end{aligned}$$

which consists of four sections, 5 bits each. On cyclically shifting each section of $\mathbf{v}$ one place to the right, we obtain the following vector:

$$\begin{aligned}\mathbf{v}^{(1)} &= \left(\mathbf{v}_0^{(1)}, \mathbf{v}_1^{(1)}, \mathbf{v}_2^{(1)}, \mathbf{v}_3^{(1)}\right) \\ &= (00011, 00110, 01000, 10001),\end{aligned}$$

which is the codeword for the information sequence $\mathbf{u}' = (01000, 10001)$.

---

Encoding of a QC code in systematic circular form can be implemented using simple shift registers with linear complexity [16]. For $0 \le i < c$ and $0 \le j < t-c$, let $\mathbf{g}_{i,j}$ be the generator of the circulant $\mathbf{G}_{i,j}$ in the $\mathbf{P}$-matrix of the generator matrix $\mathbf{G}_{qc,\text{sys}}$ given by (3.96). For $0 \le l < b$, let $\mathbf{g}_{i,j}^{(l)}$ be the $b$-tuple obtained by cyclically shifting every component of $\mathbf{g}_{i,j}$ $l$ places to the right. This $b$-tuple $\mathbf{g}_{i,j}^{(l)}$ is called the $l$th right-cyclic shift of $\mathbf{g}_{i,j}$. It is clear that $\mathbf{g}_{i,j}^{(0)} = \mathbf{g}_{i,j}^{(b)} = \mathbf{g}_{i,j}$. Let $\mathbf{u} = (\mathbf{u}_0, \mathbf{u}_1, \ldots, \mathbf{u}_{c-1}) = (u_0, u_1, \ldots, u_{cb-1})$

be the information sequence of $cb$ bits to be encoded. Divide this sequence into $c$ sections of $b$ bits each, where the $i$th section of $\mathbf{u}$ is $\mathbf{u}_i = (u_{ib}, u_{ib+1}, \ldots, u_{(i+1)b-1})$ for $0 \leq i < c$. Then the codeword for $\mathbf{u}$ is $\mathbf{v} = \mathbf{u}\mathbf{G}_{qc,\text{sys}}$, which has the following systematic form:

$$\mathbf{v} = (\mathbf{p}_0, \mathbf{p}_1, \ldots, \mathbf{p}_{t-c}, \mathbf{u}_0, \mathbf{u}_1, \ldots, \mathbf{u}_{c-1}), \tag{3.98}$$

where, for $0 \leq j < t - c$, $\mathbf{p}_j = (p_{j,0}, p_{j,1}, \ldots, p_{j,b-1})$ is a section of $b$ parity-check bits. It follows from $\mathbf{v} = \mathbf{u}\mathbf{G}_{qc,\text{sys}}$ that, for $0 \leq j < t - c$,

$$\mathbf{p}_j = \mathbf{u}_0\mathbf{G}_{0,j} + \mathbf{u}_1\mathbf{G}_{1,j} + \cdots + \mathbf{u}_{c-1}\mathbf{G}_{c-1,j}, \tag{3.99}$$

where, for $0 \leq i < c$,

$$\mathbf{u}_i\mathbf{G}_{i,j} = u_{ib}\mathbf{g}_{i,j}^{(0)} + u_{ib+1}\mathbf{g}_{i,j}^{(1)} + \cdots + u_{(i+1)b-1}\mathbf{g}_{i,j}^{(b-1)}. \tag{3.100}$$

It follows from (3.98) and (3.99) that the $j$th parity-check section $\mathbf{p}_j$ can be computed, step by step, as the information sequence $\mathbf{u}$ is shifted into the encoder. The information sequence $\mathbf{u} = (\mathbf{u}_0, \mathbf{u}_1, \ldots, \mathbf{u}_{c-1})$ is shifted into the encoder in the order from $\mathbf{u}_{c-1}$ to $\mathbf{u}_0$, that is, the section $\mathbf{u}_{c-1}$ is shifted into the encoder first and $\mathbf{u}_0$ last. For $1 \leq l < c$, at the $l$th step, the accumulated sum

$$\mathbf{s}_{l,j} = \mathbf{u}_{c-1}\mathbf{G}_{c-1,j} + \mathbf{u}_{c-2}\mathbf{G}_{c-2,j} + \cdots + \mathbf{u}_{c-l}\mathbf{G}_{c-l,j} \tag{3.101}$$

is formed and stored in an accumulator. At the $(l + 1)$th step, the partial sum $\mathbf{u}_{c-l-1}\mathbf{G}_{c-l-1,j}$ is computed from (3.100) and added to $\mathbf{s}_{l,j}$ to form the accumulated sum $\mathbf{s}_{l+1,j}$. At the $c$th step, the accumulated sum $\mathbf{s}_{c,j}$ gives the $j$th parity-check section $\mathbf{p}_j$.

By application of the above encoding process and (3.100), the $j$th parity-check section $\mathbf{p}_j$ can be formed with a *cyclic shift register–adder–accumulator* (CSRAA) circuit as shown in Figure 3.13. At the beginning of the first step, $\mathbf{g}_{c-1,j}^{(b-1)}$, the $(b - 1)$th right cyclic shift of the generator $\mathbf{g}_{c-1,j}$ of the circulant $\mathbf{G}_{c-1,j}$ is stored in the feedback shift register B, and the content of register A is set to zero. When the information bit $u_{cb-1}$ is shifted into the encoder and the channel, the product $u_{cb-1}\mathbf{g}_{c-1,j}^{(b-1)}$ is formed at the output of AND gates, and is added to the content stored in register A (zero at this time). The sum is then stored back into register A. The feedback shift register B is shifted once to the left. The new content in A is $\mathbf{g}_{c-1,j}^{(b-2)}$. When the next information bit $u_{cb-2}$ is shifted into the encoder and the channel, the product $u_{cb-2}\mathbf{g}_{c-1,j}^{(b-2)}$ is formed at the output of the AND gates. This product is then added to the sum $u_{cb-1}\mathbf{g}_{c-1,j}^{(b-1)}$ in the accumulator register A. The sum $u_{cb-2}\mathbf{g}_{c-1,j}^{(b-2)} + u_{cb-1}\mathbf{g}_{c-1,j}^{(b-1)}$ is then stored back into A. The above *shift–add–store* process continues. When the last information bit $u_{(c-1)b}$ of information section $\mathbf{u}_{c-1}$ has been shifted into the encoder, register A stores the partial sum $\mathbf{u}_{c-1}\mathbf{G}_{c-1,j}$, which is the contribution to the parity-check section $\mathbf{p}_j$ from the information section $\mathbf{u}_{c-1}$. At this time, $\mathbf{g}_{c-2,j}^{(b-1)}$, the $(b - 1)$th right cyclic shift of the generator $\mathbf{g}_{c-2,j}$ of circulant $\mathbf{G}_{c-2,j}$, is loaded into register B. The shift–add–store process repeats. When the information section $\mathbf{u}_{c-2}$ has been completely shifted into the encoder, register A stores the accumulated partial sum $\mathbf{u}_{c-2}\mathbf{G}_{c-2,j} + \mathbf{u}_{c-1}\mathbf{G}_{c-1,j}$, that is the contribution to the parity-check section $\mathbf{p}_j$ from the information sections $\mathbf{u}_{c-1}$ and

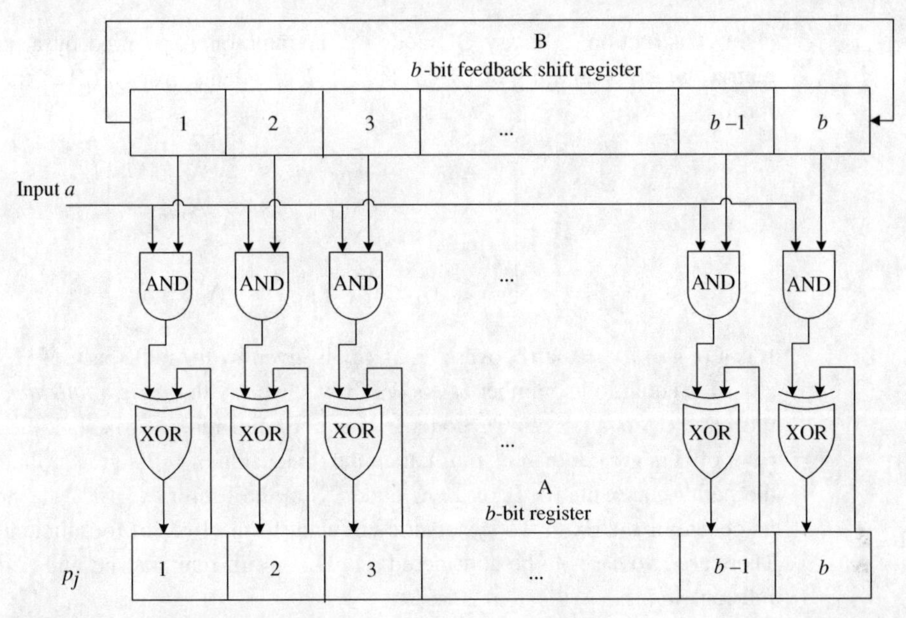


**Figure 3.13** A CSRAA encoder circuit.

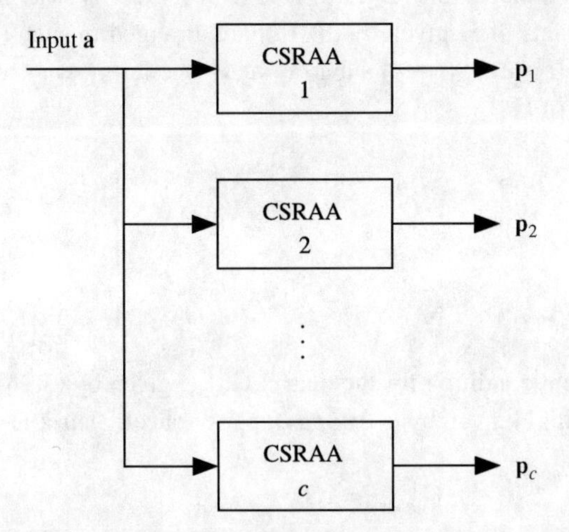


**Figure 3.14** A CSRAA-based QC-LDPC code encoder.

$\mathbf{u}_{c-2}$. The above process repeats until the last information section $\mathbf{u}_0$ has been shifted into the encoder. At this time, the accumulator register A contains the $j$th parity-check section $\mathbf{p}_j$. To form $t-c$ parity-check sections, $t-c$ CSRAA circuits are needed, one for computing each parity-check section. A block diagram for the entire encoder is shown in Figure 3.14. All the parity-check sections are formed at the same time in parallel, and they are then shifted into the channel serially. The encoding circuit consists of $t-c$ CSRAA circuits with a total of $2(t-c)b$ flip-flops, $(t-c)b$ AND gates, and $(t-c)b$ two-input XOR gates (modulo-2 adders). The encoding is accomplished in linear time with complexity linearly proportional to the number of parity-check bits $(t-c)b$.

In construction, a binary QC code $\mathcal{C}_{qc}$ is commonly specified by a parity-check matrix, which is an array of $(t-c) \times t$ $b \times b$ circulants over GF(2) of the following form:

$$\mathbf{H} = \begin{bmatrix} \mathbf{A}_{0,0} & \mathbf{A}_{0,1} & \cdots & \mathbf{A}_{0,t-1} \\ \mathbf{A}_{1,0} & \mathbf{A}_{1,1} & \cdots & \mathbf{A}_{1,t-1} \\ \vdots & \vdots & \ddots & \vdots \\ \mathbf{A}_{t-c-1,0} & \mathbf{A}_{t-c-1,1} & \cdots & \mathbf{A}_{t-c-1,t-1} \end{bmatrix}. \tag{3.102}$$

It is a $(t-c)b \times tb$ matrix over GF(2). $\mathcal{C}_{qc}$ is given by the null space of $\mathbf{H}$. If the rank $r$ of $\mathbf{H}$ is equal to its number $(t-c)b$ of rows, we say that $\mathbf{H}$ is a *full-rank* matrix. In this case, $\mathcal{C}_{qc}$ is a $(tb, cb)$ QC code. If $\mathbf{H}$ is not a full-rank matrix (i.e., the number of rows of $\mathbf{H}$ is greater than its rank), then the dimension of $\mathcal{C}_{qc}$ is greater than $cb$. Given the parity-check matrix $\mathbf{H}$ of a QC code $\mathcal{C}_{qc}$ in the form of (3.102), we need to find its generator matrix in the systematic circulant form of (3.96) for efficient encoding. There are two cases to be considered: (1) $\mathbf{H}$ is a full-rank matrix; and (2) $\mathbf{H}$ is not a full-rank matrix.

We first consider the case that the parity-check matrix $\mathbf{H}$ of a QC code given by (3.102) is a full-rank matrix, that is, the rank $r$ of $\mathbf{H}$ is $(t-c)b$. In this case, we assume that the columns of circulants of $\mathbf{H}$ given by (3.102) are arranged in such a way that the rank of the following $(t-c) \times (t-c)$ subarray of $\mathbf{H}$ (the leftmost $t-c$ columns of $\mathbf{H}$) is $(t-c)b$, the rank of $\mathbf{H}$:

$$\mathbf{D} = \begin{bmatrix} \mathbf{A}_{0,0} & \mathbf{A}_{0,1} & \cdots & \mathbf{A}_{0,t-c-1} \\ \mathbf{A}_{1,0} & \mathbf{A}_{1,1} & \cdots & \mathbf{A}_{1,t-c-1} \\ \vdots & \vdots & \ddots & \vdots \\ \mathbf{A}_{t-c-1,0} & \mathbf{A}_{t-c-1,1} & \cdots & \mathbf{A}_{t-c-1,t-c-1} \end{bmatrix}. \tag{3.103}$$

The necessary and sufficient condition for the matrix $\mathbf{G}_{qc,\text{sys}}$ given by (3.96) to be the generator matrix of $\mathcal{C}_{qc}$ with $\mathbf{H}$ given by (3.102) as a parity-check matrix is

$$\mathbf{H}\mathbf{G}_{qc,\text{sys}}^{\text{T}} = [\mathbf{O}], \tag{3.104}$$

where $[\mathbf{O}]$ is a $(t-c)b \times cb$ zero matrix.

For $0 \le i < c$ and $0 \le j < t-c$, let $\mathbf{g}_{i,j}$ be the generator of the circulant $\mathbf{G}_{i,j}$ in the $\mathbf{P}$-matrix of $\mathbf{G}_{qc,\text{sys}}$. Once we have found the $\mathbf{g}_{i,j}$s from (3.104), we can form all the circulants $\mathbf{G}_{i,j}$ of $\mathbf{G}_{qc,\text{sys}}$. Let $\mathbf{u} = (1, 0, \ldots, 0)$ be the unit $b$-tuple with a "1" in the first position, and $\mathbf{0} = (0, 0, \ldots, 0)$ be the all-zero $b$-tuple. For $0 \le i < c$, the first row of the submatrix $\mathbf{G}_i$ of $\mathbf{G}_{qc,\text{sys}}$ given by (3.96) is

$$\mathbf{g}_i = (\mathbf{g}_{i,0}\ \mathbf{g}_{i,1} \cdots \mathbf{g}_{i,t-c-1}\ \mathbf{0} \cdots \mathbf{0}\ \mathbf{u}\ \mathbf{0} \cdots \mathbf{0}), \tag{3.105}$$

where the unit $b$-tuple $\mathbf{u}$ is at the $(t-c+i)$th position of $\mathbf{g}_i$ (section-wise). The row $\mathbf{g}_i$ consists of $t$ sections, each consisting of $b$ bits. The first $t-c$ sections are simply the generators of the $t-c$ circulants $\mathbf{G}_{i,0}, \mathbf{G}_{i,1}, \ldots, \mathbf{G}_{i,t-c-1}$ of the $i$th submatrix $\mathbf{G}_i$ of $\mathbf{G}_{qc,\text{sys}}$. If follows from (3.104) that $\mathbf{H}\mathbf{g}_i^{\mathrm{T}} = \mathbf{0}$ (the all-zero $(t-c)b$-tuple) for $0 \le i < c$. Define

$$\mathbf{z}_i = (\mathbf{g}_{i,0}\mathbf{g}_{i,1} \cdots \mathbf{g}_{i,t-c-1}) \tag{3.106}$$

and

$$\mathbf{M}_{t-c+i} = \begin{bmatrix} \mathbf{A}_{0,t-c+i} \\ \mathbf{A}_{1,t-c+i} \\ \vdots \\ \mathbf{A}_{t-c-1,t-c+i} \end{bmatrix} \tag{3.107}$$

with $0 \le i < c$. $\mathbf{M}_{t-c+i}$ is simply the $(t-c+i)$th column of circulants of $\mathbf{H}$. Then $\mathbf{H}\mathbf{g}_i^{\mathrm{T}} = \mathbf{0}$ gives the following equality:

$$\mathbf{D}\mathbf{z}_i^{\mathrm{T}} + \mathbf{M}_{t-c+i}\mathbf{u}^{\mathrm{T}} = \mathbf{0}. \tag{3.108}$$

Since $\mathbf{D}$ is a $(t-c)b \times (t-c)b$ square matrix and has full rank $(t-c)b$, it is nonsingular and has an inverse $D^{-1}$. Since all the matrices and vectors are over GF(2), it follows from (3.108) that

$$\mathbf{z}_i^{\mathrm{T}} = \mathbf{D}^{-1}\mathbf{M}_{t-c+i}\mathbf{u}^{\mathrm{T}}. \tag{3.109}$$

On solving (3.109) for $0 \le i < c$, we obtain $\mathbf{z}_0, \mathbf{z}_1, \ldots, \mathbf{z}_{c-1}$. From $\mathbf{z}_0, \mathbf{z}_1, \ldots, \mathbf{z}_{c-1}$ and (3.106), we obtain all the generators $\mathbf{g}_{i,j}$ of the circulants $\mathbf{G}_{i,j}$ of $\mathbf{G}_{qc,\text{sys}}$. Then $\mathbf{G}_{qc,\text{sys}}$ can be constructed readily.

Now we consider the case for which the rank $r$ of the parity-check matrix $\mathbf{H}$ given by (3.102) is either less than the number $(t-c)b$ of rows of $\mathbf{H}$, or given by $r = (t-c)b$, but there does not exist a $(t-c)\times(t-c)$ subarray $\mathbf{D}$ in $\mathbf{H}$ with rank $r$. For this case, the generator matrix of the QC code $\mathcal{C}_{qc}$ given by the null space of $\mathbf{H}$ cannot be put exactly into the systematic circulant form given by (3.96), however, it can be put into a *semi-systematic circulant form* that allows encoding with simple shift registers, as shown in Figure 3.13. To construct the generator matrix in semi-systematic circulant form, we first find the *least number* of columns of circulants in $\mathbf{H}$, say $l$ with $t-c \le l \le t$, such that these $l$ columns of circulants form a $(t-c)\times l$ subarray $\mathbf{D}^*$ whose rank is equal to the rank $r$ of $\mathbf{H}$. We permute the columns of circulants of $\mathbf{H}$ to form a new $(t-c)\times t$ array $\mathbf{H}^*$ of circulants, such that the first (or the leftmost) $l$ columns of $\mathbf{H}^*$ form the array $\mathbf{D}^*$. Note that the null space of $\mathbf{H}^*$ gives a code that is combinatorially equivalent to the code given by the null space of $\mathbf{H}$, that is, they give the same error performance with the same decoding.

Let

$$\mathbf{D}^* = \begin{bmatrix} \mathbf{A}_{0,0} & \mathbf{A}_{0,1} & \cdots & \mathbf{A}_{0,l-1} \\ \mathbf{A}_{1,0} & \mathbf{A}_{1,1} & \cdots & \mathbf{A}_{1,l-1} \\ \vdots & \vdots & \ddots & \vdots \\ \mathbf{A}_{t-c,0} & \mathbf{A}_{t-c,1} & \cdots & \mathbf{A}_{t-c,l-1} \end{bmatrix}. \tag{3.110}$$

Then, given $\mathbf{D}^*$, the generator matrix of the QC code $C_{qc}$ given by the null space of $\mathbf{H}^*$ can be put into the following *semi-systematic circulant form*:

$$\mathbf{G}_{qc,\text{semi-sys}} = \begin{bmatrix} \mathbf{Q} \\ \mathbf{G}^*_{qc,\text{sys}} \end{bmatrix}, \tag{3.111}$$

which consists of two submatrices $\mathbf{G}^*_{qc,\text{sys}}$ and $\mathbf{Q}$. The submatrix $\mathbf{G}^*_{qc,\text{sys}}$ is a $(t-l)b \times tb$ matrix in systematic circulant form,

$$\mathbf{G}^*_{qc,\text{sys}} = \begin{bmatrix} \mathbf{G}_{0,0} & \mathbf{G}_{0,1} & \cdots & \mathbf{G}_{0,l-1} & \mathbf{I} & \mathbf{O} & \cdots & \mathbf{O} \\ \mathbf{G}_{1,0} & \mathbf{G}_{1,1} & \cdots & \mathbf{G}_{1,l-1} & \mathbf{O} & \mathbf{I} & \cdots & \mathbf{O} \\ \vdots & \vdots & \ddots & \vdots & \vdots & \vdots & \ddots & \vdots \\ \mathbf{G}_{t-l,0} & \mathbf{G}_{t-l,1} & \cdots & \mathbf{G}_{t-l,l-1} & \mathbf{O} & \mathbf{O} & \cdots & \mathbf{I} \end{bmatrix}, \tag{3.112}$$

where each $\mathbf{G}_{i,j}$ is a $b \times b$ circulant, $\mathbf{I}$ is a $b \times b$ identity matrix, and $\mathbf{O}$ is a $b \times b$ zero matrix. Then the generators $\mathbf{g}_{i,j}$ of the $b \times b$ circulants $\mathbf{G}_{i,j}$ in $\mathbf{G}^*_{qc,\text{sys}}$ with $0 \leq i < t-l$ and $0 \leq j < l$ can be obtained by solving equation (3.108) with $\mathbf{D}$ replaced by $\mathbf{D}^*$ and setting the $bl - r$ linearly dependent elements in $\mathbf{z}_i = (\mathbf{g}_{i,0}\mathbf{g}_{i,1} \cdots \mathbf{g}_{i,l-1})$ to zeros. The linearly dependent elements in $\mathbf{z}_i$ correspond to the linearly dependent columns of $\mathbf{D}^*$.

The submatrix $\mathbf{Q}$ of $\mathbf{G}_{qc,\text{semi-sys}}$ is an $(lb - r) \times tb$ matrix whose rows are linearly independent, and also linearly independent of the rows of the submatrix $\mathbf{G}^*_{qc,\text{sys}}$. For $\mathbf{G}_{qc,\text{semi-sys}}$ to be a generator matrix of the null space of $\mathbf{H}^*$, $\mathbf{Q}$ must satisfy the condition $\mathbf{H}^*\mathbf{Q}^T = \mathbf{O}$, where $\mathbf{O}$ is a $(t-c)b \times (lb-r)$ zero matrix. To obtain $\mathbf{Q}$, let $d_0, d_1, \ldots, d_{l-1}$ be the number of linearly dependent columns in the 0th, 1st, ..., $(l-1)$th columns of circulants in $\mathbf{D}^*$, respectively, such that

$$d_0 + d_1 + \cdots + d_{l-1} = lb - r. \tag{3.113}$$

For a $b \times b$ circulant, if its rank is $b$, then all the columns (or rows) are linearly independent. If its rank $\lambda$ is less than $b$, then any $\lambda$ consecutive columns (or rows) of the circulant are linearly independent and the other $b - \lambda$ columns (or rows) are linearly dependent. Starting from this structure, we take the last $b - d_j$ columns of the $j$th column of circulants of $\mathbf{D}^*$ as the linearly independent columns. Then the first $d_0, d_1, \ldots, d_{l-1}$ columns of the 0th, 1st, ..., $(l-1)$th columns of circulants of $\mathbf{D}^*$ are linearly dependent columns. $\mathbf{Q}$ can be put into the following circulant form:

$$\mathbf{Q} = \begin{bmatrix} \mathbf{Q}_0 \\ \mathbf{Q}_1 \\ \vdots \\ \mathbf{Q}_{l-1} \end{bmatrix} = \begin{bmatrix} \mathbf{Q}_{0,0} & \mathbf{Q}_{0,1} & \cdots & \mathbf{Q}_{0,l-1} & \mathbf{O}_{0,0} & \mathbf{O}_{0,1} & \cdots & \mathbf{O}_{0,t-l-1} \\ \mathbf{Q}_{1,0} & \mathbf{Q}_{1,1} & \cdots & \mathbf{Q}_{1,l-1} & \mathbf{O}_{1,0} & \mathbf{O}_{1,1} & \cdots & \mathbf{O}_{1,t-l-1} \\ \vdots & \vdots & \ddots & \vdots & \vdots & \vdots & \ddots & \vdots \\ \mathbf{Q}_{l-1,0} & \mathbf{Q}_{l-1,1} & \cdots & \mathbf{Q}_{l-1,l-1} & \mathbf{O}_{l-1,0} & \mathbf{O}_{l-1,1} & \cdots & \mathbf{Q}_{l-1,t-l-1} \end{bmatrix}, \tag{3.114}$$

where (1) $\mathbf{O}_{i,j}$ is a $d_i \times b$ zero matrix with $0 \leq i < l$ and $0 \leq j < t-l$; and (2) $\mathbf{Q}_{i,j}$ is a $d_i \times b$ *partial circulant* obtained by cyclically shifting its first row $d_i - 1$ times to form the other $d_i - 1$ rows with $0 \leq i,j < l$.

For $0 \leq i < l$, let

$$\begin{aligned} \mathbf{q}_i &= (\mathbf{w}_i, \mathbf{0}) \\ &= (q_{i,0}, q_{i,1}, \ldots, q_{i,lb-1}, 0, 0, \ldots, 0) \end{aligned} \tag{3.115}$$

be the first row of the $i$th submatrix $\mathbf{Q}_i$ of $\mathbf{Q}$, which consists of two parts, the right part $\mathbf{0}$ and the left part $\mathbf{w}_i$. The right part $\mathbf{0} = (0, 0, \ldots, 0)$ consists of $(t-l)b$ zeros. The left part $\mathbf{w}_i = (q_{i,0}, q_{i,1}, \ldots, q_{i,lb-1})$ consists of $lb$ bits, which correspond to the $lb$ columns of $\mathbf{D}^*$. The $lb - r$ bits of $\mathbf{w}_i$ that correspond to the linearly dependent columns of $\mathbf{D}^*$, called dependent bits, have the following form:

$$(\mathbf{0}_0, \ldots, \mathbf{0}_{i-1}, \mathbf{u}_i, \mathbf{0}_{i+1}, \ldots, \mathbf{0}_{l-1}), \tag{3.116}$$

where, for $e \neq i$, $\mathbf{0}_e$ is a zero $d_e$-tuple and $\mathbf{u}_i = (1, 0, \ldots, 0)$ is a unit $d_i$-tuple. From the structure of $\mathbf{w}_i$, the number of unknown components in $\mathbf{w}_i$ is $r$, the rank of $\mathbf{D}^*$ (or $\mathbf{H}^*$).

The condition $\mathbf{H}^*\mathbf{Q}^{\mathrm{T}} = \mathbf{O}$ gives the following equation for $0 \leq i < l$:

$$\mathbf{D}^*\mathbf{w}_i^{\mathrm{T}} = \begin{bmatrix} \mathbf{A}_{0,0} & \mathbf{A}_{0,1} & \cdots & \mathbf{A}_{0,l-1} \\ \mathbf{A}_{1,0} & \mathbf{A}_{1,1} & \cdots & \mathbf{A}_{1,l-1} \\ \vdots & \vdots & \ddots & \vdots \\ \mathbf{A}_{t-c-1,0} & \mathbf{A}_{t-c-1,1} & \cdots & \mathbf{A}_{t-c-1,l-1} \end{bmatrix} \begin{bmatrix} q_{i,0} \\ q_{i,1} \\ \vdots \\ q_{i,lb-1} \end{bmatrix} = \mathbf{0}. \tag{3.117}$$

On solving (3.117), we find $\mathbf{w}_i = (q_{i,0}, q_{i,1}, \ldots, q_{i,lb-1})$ for $0 \leq i < l$. Divide $\mathbf{w}_i$ into $l$ sections, denoted by $\mathbf{w}_{i,0}, \mathbf{w}_{i,1}, \ldots, \mathbf{w}_{i,l-1}$, each consisting of $b$ consecutive components of $\mathbf{w}_i$. For $0 \leq i,j < l$, the partial circulant $\mathbf{Q}_{i,j}$ of $\mathbf{Q}$ is obtained by using $\mathbf{w}_{i,j}$ as the first row, and then cyclically shifting it $d_i - 1$ times to form the other $d_i - 1$ rows. From the $\mathbf{Q}_{i,j}$s with $0 \leq i,j < l$, we form the $\mathbf{Q}$ submatrix of $\mathbf{G}_{qc,\text{semi-sys}}$. By combining $\mathbf{Q}$ and $\mathbf{G}^*_{qc,\text{sys}}$ into the form of (3.116), we obtain the generator matrix $\mathbf{G}_{qc,\text{semi-sys}}$ of the QC code $\mathcal{C}_{qc}$ given by the null space of a non-full-rank array $\mathbf{H}$ of circulants.

Given $\mathbf{G}_{qc,\text{semi-sys}}$ as described in (3.111), an encoder with $l$ CSRAA circuits of the form given by Figures 3.13 and 3.14 can be implemented. Encoding consists of two phases. An information sequence $\mathbf{a} = (a_0, a_1, \ldots, a_{tb-r-1})$ of $tb - r$ bits is divided into two parts, $\mathbf{a}^{(1)} = (a_{lb-r}, a_{lb-r+1}, \ldots, a_{tb-r-1})$ and $\mathbf{a}^{(2)} = (a_0, a_1, \ldots, a_{lb-r-1})$. Then $\mathbf{a} = (\mathbf{a}^{(2)}, \mathbf{a}^{(1)})$. The information bits are shifted into the encoder serially in the order of bit $a_{tb-r-1}$ first and bit $a_0$ last. The first part $\mathbf{a}^{(1)}$ of $\mathbf{a}$, consisting of $(t-l)b$ bits, is first shifted into the encoder and is encoded into a codeword $\mathbf{v}^{(1)}$ in the subcode $\mathcal{C}_{qc}^{(1)}$ generated by $\mathbf{G}^*_{qc,\text{sys}}$. The $l$ parity sections are generated at the same time when $\mathbf{a}^{(1)}$ has been completely shifted into the encoder, as described in the first case. After encoding of $\mathbf{a}^{(1)}$, the second part $\mathbf{a}^{(2)}$ of $\mathbf{a}$ is then shifted into the encoder and is encoded into a codeword $\mathbf{v}^{(2)}$ in the subcode $\mathcal{C}_{qc}^{(2)}$ generated by $\mathbf{Q}$. By adding $\mathbf{v}^{(1)}$ and $\mathbf{v}^{(2)}$, we obtain the codeword $\mathbf{v} = \mathbf{v}^{(1)} + \mathbf{v}^{(2)}$ for the information sequence $\mathbf{a} = (\mathbf{a}^{(2)}, \mathbf{a}^{(1)})$.

To encode $\mathbf{a}^{(2)}$ using $\mathbf{Q}$, we divide $\mathbf{a}^{(2)}$ into $l$ sections, $\mathbf{a}_0^{(2)}, \mathbf{a}_1^{(2)}, \ldots, \mathbf{a}_{l-1}^{(2)}$, with $d_0, d_1, \ldots, d_{l-1}$ bits, respectively. Then the codeword for $\mathbf{a}^{(2)}$ is of the form

$$\mathbf{v}^{(2)} = \left(\mathbf{v}_0^{(2)}, \mathbf{v}_1^{(2)}, \ldots, \mathbf{v}_{l-1}^{(2)}, \mathbf{0}, \mathbf{0}, \ldots, \mathbf{0}\right), \tag{3.118}$$

which consists of $t - l$ zero sections and $l$ nonzero sections, $\mathbf{v}_0^{(2)}, \mathbf{v}_1^{(2)}, \ldots, \mathbf{v}_{l-1}^{(2)}$, each section, zero or nonzero, consisting of $b$ bits. For $0 \leq j < l$,

$$\mathbf{v}_j^{(2)} = \mathbf{a}_0^{(2)}\mathbf{Q}_{0,j} + \mathbf{a}_1^{(2)}\mathbf{Q}_{1,j} + \cdots + \mathbf{a}_{l-1}^{(2)}\mathbf{Q}_{l-1,j}. \tag{3.119}$$

Since each $\mathbf{Q}_{i,j}$ in $\mathbf{Q}$ with $0 \leq i,j < l$ is a partial circulant with $d_i$ rows, encoding of $\mathbf{a}^{(2)}$ can be accomplished with the same $l$ CSRAA circuits as used for encoding $\mathbf{a}^{(1)}$. For $0 \leq j < l$, at the end of encoding $\mathbf{a}^{(1)}$, the accumulator A of the $j$th CSRAA stores the $j$th parity-check section $\mathbf{v}_j^{(1)}$ of $\mathbf{v}^{(1)}$. In the second phase of encoding, $\mathbf{a}_0^{(2)}, \mathbf{a}_1^{(2)}, \ldots, \mathbf{a}_{l-1}^{(2)}$ are shifted into the encoder one at a time and the generators $\mathbf{w}_{0,j}, \mathbf{w}_{1,j}, \ldots, \mathbf{w}_{l-1,j}$ of the partial circulants $\mathbf{Q}_{0,j}, \mathbf{Q}_{1,j}, \ldots, \mathbf{Q}_{l-1,j}$ are stored in register B of the $j$th CSRAA in turn. Then the CSRAA circuit cyclically shifts register B, $d_0, d_1, \ldots, d_{l-1}$ times, respectively. At the end of $d_0 + d_1 + \cdots + d_{l-1} = lb - r$ shifts, the $j$th parity section, $\mathbf{v}_j^{(1)} + \mathbf{v}_j^{(2)}$, is stored in the accumulator register A of the $j$th CSRAA circuit.

Note that the codeword $\mathbf{v} = (v_0, v_1, \ldots, v_{tb-1})$ for the information sequence $\mathbf{a} = \left(\mathbf{a}^{(2)}, \mathbf{a}^{(1)}\right)$ is not completely systematic. Only the rightmost $(t - l)b$ bits of $\mathbf{v}$ are identical to the information bits in $\mathbf{a}^{(1)}$, that is, $(v_{lb}, v_{lb+1}, \ldots, v_{tb-1}) = \mathbf{a}^{(1)}$. The next $lb - r$ bits, $v_r, v_{r+1}, \ldots, v_{lb-1}$, of $\mathbf{v}$ are not identical to the information bits in $\mathbf{a}^{(2)}$. The rightmost bits $v_0, v_1, \ldots, v_{r-1}$ of $\mathbf{v}$ are parity-check bits.

## 3.7 Repetition and Single-Parity-Check Codes

Repetition and single-parity-check codes are two very simple types of linear block codes. A repetition code $C_{\text{rep}}$ over GF(2) of length $n$ is a binary $(n, 1)$ linear code with a single information bit. The code is simply obtained by repeating a single information bit $n$ times. Therefore, it consists of only two codewords, namely the all-zero codeword $(00 \ldots 0)$ and the all-one codeword $(11 \ldots 1)$. Obviously, its generator matrix is

$$\mathbf{G}_{\text{rep}} = [1\ 1 \ldots 1].$$

A single-parity-check (SPC) code $C_{\text{spc}}$ over GF(2) of length $n$ is a binary $(n, n-1)$ linear code for which each codeword consists of $n - 1$ information bits and a single parity-check bit. Let $\mathbf{u} = (u_0, u_1, \ldots, u_{n-2})$ be the message to be encoded. Then a single parity-check bit $c$ is added to it to form an $n$-bit codeword $(c, u_0, u_1, \ldots, u_{n-2})$. This single parity-check bit $c$ is simply the modulo-2 sum of the $n - 1$ information bits of the message $\mathbf{u}$, that is

$$c = u_0 + u_1 + \cdots + u_{n-2}.$$

Hence, every codeword in $C_{\text{spc}}$ has even weight. $C_{\text{spc}}$ simply consists of all the $n$-tuples over GF(2) with even weight and hence its minimum distance is 2. Any error pattern with an odd number of errors will change a codeword in $C_{\text{spc}}$ into a non-codeword and any error pattern with a nonzero even number of errors will change a codeword in $C_{\text{spc}}$ into another codeword. This implies that $C_{\text{spc}}$ is capable of detecting any error pattern containing an odd number of errors but not any error pattern containing a nonzero even number of errors. Such SPC codes are commonly used in communication and storage systems for simple error detection.

The generator matrix of an $(n, n-1)$ SPC code in systematic form is given as follows:

$$\mathbf{G}_{\text{spc}} = \begin{bmatrix} 1 & \\ 1 & \\ 1 & \mathbf{I}_{n-1} \\ \vdots & \\ 1 & \end{bmatrix} = \begin{bmatrix} 1 & 1 & 0 & 0 & 0\cdots 0 \\ 1 & 0 & 1 & 0 & 0\cdots 0 \\ 1 & 0 & 0 & 1 & 0\cdots 0 \\ \vdots & \vdots & & \ddots & \vdots \\ 1 & 0 & 0 & 0 & 0\cdots 1 \end{bmatrix},$$

where $\mathbf{I}_{n-1}$ is an $(n-1) \times (n-1)$ identity matrix. It is easy to check that the inner product of the single row of the generator matrix $\mathbf{G}_{\text{rep}}$ of the $(n, 1)$ repetition code and any row of the generator matrix $\mathbf{G}_{\text{spc}}$ of the $(n, n-1)$ SPC code is zero, that is, $\mathbf{G}_{\text{rep}}\mathbf{G}_{\text{spc}}^{\text{T}} = 0$ (an $n-1$ zero-tuple). Therefore, the $(n, 1)$ repetition code and the $(n, n-1)$ SPC code are dual codes.

Repetition and SPC codes are extremely simple codes; however, they are quite useful in many applications, as will be seen in later chapters.

## Problems

**3.1** There are constraints on the code bits in the $(7, 4)$ Hamming code example in Section 1.3 of Chapter 1. In equation (3.6), there are also constraints on the code bits imposed by a parity-check matrix for a linear code. Connect the two by finding the parity-check matrix for the $(7, 4)$ Hamming code.

**3.2** Consider a binary linear block code with the following matrix as generator matrix:

$$\mathbf{G} = \begin{bmatrix} 1 & 1 & 1 & 0 & 1 & 0 & 0 & 0 \\ 1 & 0 & 0 & 1 & 1 & 1 & 0 & 0 \\ 1 & 1 & 0 & 0 & 0 & 1 & 1 & 0 \\ 0 & 1 & 1 & 0 & 0 & 0 & 1 & 1 \end{bmatrix}.$$

(a) Put the given generator matrix into systematic form and find all of the systematic-form codewords.
(b) Find the parity-check matrix of the code in systematic form. Determine the parity-check equations.
(c) What is the minimum distance of the code?
(d) Determine the weight distribution of the code.

**3.3** Consider the code given in Problem 3.2 in systematic form. Suppose a codeword is transmitted over the BSC with transition probability $p = 0.01$ and $\mathbf{r} = (01110110)$ is the corresponding received vector.
(a) Compute the syndrome of the received vector.
(b) Find all of the error patterns that have the syndrome you have computed.
(c) Compute the probabilities of the error patterns you have found and determine the most probable error pattern.

(d) Assuming that the code is used solely for error detection, compute the probability of an undetected error.

(e) Assuming that the code is used solely for error correction, find $\{N_w\}$, the number of weight-$w$ error patterns in the syndrome table, and compute the probability of decoding error, $P_e$(coded), for $p = 0.01$.

**3.4** Prove that the Hamming distance satisfies the triangle inequality given by (3.18).

**3.5** Prove Theorem 3.1.

**3.6** Prove Theorem 3.2.

**3.7** Prove Theorem 3.3.

**3.8** Construct a standard array of the code given in Problem 3.2 that realizes the maximum-likelihood decoding for a BSC with transition probability $p < 1/2$. Compute the probability of a decoding error based on the standard array that you have constructed.

**3.9** In a standard array for an $(n, k)$ linear block code $\mathcal{C}$, prove that (a) the sum of two vectors in the same coset is a codeword in $\mathcal{C}$; (b) no two vectors in the same coset or two different cosets are the same; (c) all the vectors in a coset have the same syndrome; and (d) different cosets have different syndromes.

**3.10** Using a standard array of a linear block code $\mathcal{C}$ for decoding, prove that, if an error pattern is not a coset leader, then the decoding of a received vector is incorrect.

**3.11** Let $\mathcal{C}$ be a binary $(n, k)$ linear block code with minimum distance $d_{\min}$. Let $t = \lfloor (d_{\min} - 1)/2 \rfloor$. Show that all the $n$-tuples over GF(2) with weights $t$ or less can be used as coset leaders in a standard array of $\mathcal{C}$. Show that at least one $n$-tuple over GF(2) with weight $t + 1$ cannot be used as a coset leader.

**3.12** Determine the code given by the null space of the following matrix over GF(2):

$$\mathbf{H} = \begin{bmatrix} 1 & 0 & 1 & 1 & 0 & 0 & 0 \\ 0 & 1 & 0 & 1 & 1 & 0 & 0 \\ 0 & 0 & 1 & 0 & 1 & 1 & 0 \\ 0 & 0 & 0 & 1 & 0 & 1 & 1 \\ 1 & 0 & 0 & 0 & 1 & 0 & 1 \\ 1 & 1 & 0 & 0 & 0 & 1 & 0 \\ 0 & 1 & 1 & 0 & 0 & 0 & 1 \end{bmatrix}.$$

If we regard matrix $\mathbf{H}$ as the adjacency matrix of a bipartite graph, draw this bipartite graph. What is the girth of this bipartite graph?

**3.13** Let $\mathbf{H} = [1 \quad 1 \quad \cdots \quad 1]$ be a $1 \times n$ matrix over GF(2) where all the entries of the single row are the 1 element of GF(2). Determine the code given by the null space of $\mathbf{H}$ and its weight distribution. This code is called a *single-parity-check* (SPC) code. Draw the bipartite graph with $\mathbf{H}$ as the incidence matrix.

**3.14** Let $\mathbf{g}(X) = 1 + X + X^2 + X^4 + X^5 + X^8 + X^{10}$ be the generator polynomial of a $(15, 5)$ cyclic code over GF(2).

(a) Construct the generator matrix of the code in systematic form.

(b) Find the parity-check polynomial of the code.

**3.15** Consider the $(15, 7)$ cyclic code with generator $\mathbf{g}(X) = X^8 + X^7 + X^6 + X^4 + 1$ and $d_{\min} = 5$. This code is capable of correcting all error patterns containing two or fewer errors, and additional error patterns with more than two errors.

(a) Design and simulate a Meggitt decoder that corrects only error patterns of two or fewer errors. (There are 15 Meggitt-type error patterns.)

(b) Simulate the performance of the Meggitt decoder on the BSC and plot $P_e$(coded) against the BSC parameter $p$ on a log–log plot.

(c) Derive the $P_e$(coded) performance of this code on the BSC channel and plot it with your simulation results.

**3.16** For the code in Problem 3.15, notice that three cyclic shifts of the error pattern (00000 00000 01001) gives another two-error pattern (00100 00000 00001) with a 1 in position $X^{n-1}$. Show that the 15 syndromes mentioned in the first step of Problem 3.15 can be reduced to 8 syndromes using the fact that these two-error patterns come in pairs as in this example. That is, the syndrome for the first error pattern is unnecessary because the second error pattern will arise after three cyclic shifts. (For shortened codes, the full set of syndromes is needed.)

**3.17** We return to Problem 3.15, for which we used bounded-distance decoding: we limited the decoder correction capability to at most two errors and no others.

(a) Show that with a bounded-distance decoder, 120 error patterns are correctable.

(b) Show that for a full syndrome decoder, 255 nonzero error patterns are correctable.

(c) Derive and (log–log) plot the theoretical $P_e$(coded) BSC performances of the bounded-distance decoder and the full syndrome decoder against the BSC parameter $p$. Comment.

**3.18** Consider an 8-bit CRC code with message length $k = 12$ and generator polynomial $\mathbf{g}(X) = X^8 + X^4 + X^3 + X^2 + 1$.

(a) Show that the number error pattern spanning $n - k = 8$ bits is $2^6 \cdot 13$.

(b) Check via computer simulation that all $2^6 \cdot 13$ of these error patterns are detectable.

(c) Check via computer simulation that $2^k = 2^{12}$ of the $2^n = 2^{20}$ error polynomials are not detectable.

**3.19** Let $\alpha$ be a primitive element of the Galois field GF($2^5$) generated by the primitive polynomial $\mathbf{p}(X) = 1 + X^2 + X^5$ (see Table 2.10). Find the generator polynomial of the triple-error-correcting primitive BCH code of length 31.

**3.20** Consider the primitive triple-error-correction BCH code over GF(2) constructed in Problem 3.19. Suppose a code polynomial $\mathbf{v}(X)$ is transmitted and $\mathbf{r}(X) = 1 + X^5 + X^{21}$ is the received polynomial. Decode $\mathbf{r}(X)$ using the BM algorithm.

**3.21** Consider the primitive triple-error-correcting (63, 45) BCH code over GF(2) constructed in Example 3.2. Suppose a code polynomial $\mathbf{v}(X)$ is transmitted and $\mathbf{r}(X) = 1+X^{61}+X^{62}$ is the received polynomial. Decode $\mathbf{r}(X)$ using the BM algorithm.

**3.22** First argue that the block-error rate for a $t$-error correcting $(n, k)$ RS code over GF($2^m$) on a memoryless channel is given by

$$P_{blk} = \sum_{l=t+1}^{n} \binom{n}{l} P_s^l (1 - P_s)^{n-l},$$

where $P_s$ is the channel symbol-error probability. Now argue that when the underlying channel is the binary-input AWGN channel with hard decisions, $P_s$ is given by $P_s = 1 - (1 - p)^m$, where $p$ is the error probability of the binary symmetric channel induced by the hard decsions and is given by $p = Q\left(\sqrt{2RE_b/N_0}\right)$, where $R$ is the code rate. Finally, argue that the decoded symbol rate is given by

$$P_{sym} = \sum_{l=t+1}^{n} \frac{l}{n} \binom{n}{l} P_s^l (1 - P_s)^{n-l}.$$

Using these results, reproduce the curves in Figure 3.6.

**3.23** Using the Galois field GF($2^5$) generated by the primitive polynomial $p(X) = 1+X^2+X^5$ (see Table 2.10), find the generator polynomial of the triple-error-correcting RS code $\mathcal{C}$ over GF($2^5$) of length 31. Let $\alpha$ be a primitive element of GF($2^5$). Suppose a code polynomial $\mathbf{v}(X)$ of $\mathcal{C}$ is transmitted and $\mathbf{r}(X) = \alpha^2 + \alpha^{21}X^{12} + \alpha^7X^{20}$ is the corresponding received polynomial. Decode $\mathbf{r}(X)$ using Euclid's algorithm.

**3.24** Decode the received polynomial $\mathbf{r}(X) = \alpha^2 + \alpha^{21}X^{12} + \alpha^7X^{20}$ given in Problem 3.23 using the Berlekamp–Massey algorithm.

**3.25** (Berlekamp) This problem involves the design of a RS code with a low-complexity encoder, a characteristic of interest for deep-space applications.

(a) Write a program module that produces a table for GF(256) generated by the primitive polynomial $p(x) = x^8 + x^4 + x^3 + x^2 + 1$. The table you create should allow you to multiply two elements of GF(256) by adding the exponents of the elements modulo 255 (the special case of the zero element, which has no exponent, is easily taken care of). The table should also enable you to add any two elements by adding their vector representations component-wise over $\mathbb{F}_2$.

(b) Find the generator polynomial $g(x)$ for a 16-error-correcting RS code with the characteristic that $\mathbf{g}(X) = \mathbf{g}^*(X)$, called a palindrome, where $\mathbf{g}^*(X) = X^{n-k}\mathbf{g}(X^{-1})$ is the reciprocal polynomial of $\mathbf{g}(X)$. Show that it is indeed a palindrome by multiplying out the factors of $\mathbf{g}(X)$. *Hint*: How are the roots of $\mathbf{g}(X)$ related to the roots of $\mathbf{g}^*(X)$? What must be true of the roots of a polynomial satisfying $\mathbf{g}(X) = \mathbf{g}^*(X)$?

(c) Write a program module for the systematic encoder corresponding to this generator polynomial. Such a generator polynomial was first proposed by E. Berlekamp, who noticed that the number of hardware Galois field multipliers can be halved.

**3.26** Derive equation (3.92) by mimicking the derivation of equation (3.87).

**3.27** For the (15, 9) RS code over GF(16) in Example 3.8, suppose the received polynomial is $\mathbf{r}(X) = \alpha^{11}X^4 + \alpha^3X^9 + \alpha^{13}X^{12}$. Use the Berlekamp–Massey algorithm to decode it.

**3.28** For the (15, 9) RS code over GF(16) in Example 3.8 shortened to a rate-1/2 (12, 6) RS code, suppose the received polynomial is $\mathbf{r}(X) = \alpha X^2 + \alpha^{10}X^5 + \alpha^6X^8 + \alpha^{12}X^{11}$. Use the Berlekamp–Massey algorithm to decode it. Comment on your findings.

**3.29** For the (15, 9) RS code over GF(16) in Example 3.8 shortened to a rate-1/2 (12, 6) RS code, suppose the received polynomial is $\mathbf{r}(X) = \alpha X^2 + \alpha^{10}X^5$ and there are two erasures such that $\mathbf{d}(X) = \alpha^5X^3 + \alpha^7X^{10}$. Use the Berlekamp–Massey algorithm to decode it.

**3.30** Referring to Theorem 3.8, show that when $P < 10^{-3}$, say, the probability of codeword error $P_e = 1 - P_c$ can be closely approximated as $P_e \approx \binom{n}{t'+1}P^{t'+1}(1-P)^{n-(t'+1)}$, a simpler formula to use.

**3.31** A *perfect code* over GF($q$), $q \geq 2$, with parameters $(n, k)$ is one for which the combined "volume" of the decoding spheres in Figure 3.7 with $t' = t = (d_{\min} - 1)/2$ is equal to the total number of length-$n$ words.

(a) Show that the volume of a radius-$t$ sphere is given by

$$V_t = \sum_{j=0}^{t} \binom{n}{j}(q-1)^j.$$

This is the number of length-$n$ $q$-ary words within distance $t$ from a codeword.

(b) The total number of codewords is $q^k$ and so the combined volume of the $q^k$ decoding spheres is $q^kV_t$, so that a perfect code must satisfy

$$q^kV_t = q^n.$$

Show that a (63, 50) RS code over GF(64) is not perfect.

(c) Quantify the imperfectness of this code, defined to be the fraction of $q$-ary $n$-symbol words occupied by the decoding spheres in $q$-ary $n$-space:

$$D = q^kV_t/q^n.$$

This is also the *sphere-packing density*. Comment on the implications of your finding on the probability of incorrect decoding. Repeat for a (255, 223) RS code over GF(256).

(d) Check if a (7, 4) Hamming code is perfect. Repeat for the (23, 12) binary Golay code for which $t = 3$.

**3.32** (Berlekamp) Show that the RS code sphere-packing density $D$ defined in the previous problem is approximately $1/t!$ when $q \gg t \geq 1$ and $n = q - 1$.

**3.33** For $i = 1$ and 2, let $\mathcal{C}_i$ be an $(n_i, k_i)$ linear block code over GF(2) with generator and parity-check matrices $\mathbf{G}_i$ and $\mathbf{H}_i$, respectively.
(a) Find a generator matrix of the product code $\mathcal{C} = \mathcal{C}_1 \times \mathcal{C}_2$ in terms of $\mathbf{G}_1$ and $\mathbf{G}_2$.
(b) Find a parity-check matrix of the product code $\mathcal{C} = \mathcal{C}_1 \times \mathcal{C}_2$ in terms of $\mathbf{H}_1$ and $\mathbf{H}_2$.

**3.34** Let $\mathbf{g}(X)$ be the generator polynomial of an $(n, k)$ cyclic code $\mathcal{C}$. Suppose we interleave $\mathcal{C}$ by a depth of $\lambda$. Prove that the interleaved code $\mathcal{C}_\lambda$ is also a cyclic code with the generator polynomial $\mathbf{g}(X^\lambda)$.

**3.35** Referring to Figure 3.11, let $\mathcal{C}_1$ and $\mathcal{C}_2$ be the outer and inner codes of a serial concatenated code $\mathcal{C}$ with minimum distances $d_1$ and $d_2$, respectively. Prove that the minimum distance of the concatenated code $\mathcal{C}$ is at least $d_1 d_2$.

## References


[1] E. R. Berlekamp, *Algebraic Coding Theory*, revised ed., Laguna Hills, CA, Aegean Park Press, 1984.
[2] R. E. Blahut, *Algebraic Codes for Data Transmission*, Cambridge, Cambridge University Press, 2003.
[3] I. F. Blake and R. C. Mullin, *The Mathematical Theory of Coding*, New York, Academic Press, 1975.
[4] G. Clark and J. Cain, *Error-Correcting Codes for Digital Communications*, New York, Plenum Press, 1981.
[5] S. Lin and D. J. Costello, Jr., *Error Control Coding: Fundamentals and Applications*, 2nd ed., Upper Saddle River, NJ, Prentice-Hall, 2004.
[6] F. J. MacWilliams and N. J. A. Sloane, *The Theory of Error-Correcting Codes*, Amsterdam, North-Holland, 1977.
[7] R. J. McEliece, *The Theory of Information Theory and Coding*, Reading, MA, Addison-Wesley, 1977.
[8] A. M. Michaelson and A. J. Levesque, *Error Control Coding Techniques for Digital Communications*, New York, Wiley, 1985.
[9] J. C. Moreira and P. G. Farrell, *Essentials of Error-Control Coding*, Chichester, Wiley, 2006.
[10] W. W. Peterson, "Encoding and error-correction procedures for the Bose–Chaudhuri codes," *IRE Transactions on Information Theory*, vol. 6, no. 5, pp. 459–470, September 1960.
[11] W. W. Peterson and E. J. Weldon, Jr., *Error-Correcting Codes*, 2nd ed., Cambridge, MA, MIT Press, 1972.
[12] V. Pless, *Introduction to the Theory of Error Correcting Codes*, 3rd ed., New York, Wiley, 1998.
[13] A. Poli and L. Huguet, *Error Correcting Codes, Theory and Applications*, Hemel Hempstead, Prentice-Hall, 1992.
[14] R. M. Roth, *Introduction to Coding Theory*, Cambridge, Cambridge University Press, 2006.
[15] S. B. Wicker, *Error Control Systems for Digital Communication and Storage*, Englewood Cliffs, NJ, Prentice-Hall, 1995.

[16] G. Castagnoli, S. Brauer, and M. Herrmann, "Optimization of cyclic redundancy-check codes with 24 and 32 parity bits," *IEEE Transactions on Communications*, vol. 41, no. 6, pp. 883–892, June 1993.

[17] P. Koopman and T. Chackravary, "Cyclic redundancy code (CRC) polynomial selection for embedded networks," *Proceedings of the 2004 International Conference on Dependable Systems and Networks*, July 2004.

[18] P. Koopman, "Best CRC polynomials," https://users.ece.cmu.edu/koopman/crc/.

[19] A. Hocquenghem, "Codes correcteurs d'erreurs," *Chiffres*, vol. 2, pp. 147–156, 1959.

[20] R. C. Bose and D. K. Ray-Chaudhuri, "On a class of error correcting binary group codes," *Information Control*, vol. 3, pp. 68–79, March 1960.

[21] J. L. Massey, "Shift-register synthesis and BCH decoding," *IEEE Transactions on Information Theory*, vol. 15, no. 1, pp. 122–127, January 1969.

[22] D. Chase, "Class of algorithms for decoding block codes with channel measurement information," *IEEE Transactions on Information Theory*, vol. 18, no. 1, pp. 170–182, January 1972.

[23] H. Yang, H. Yao, A. Vardy, D. Divsalar, and R. Wesel, "A list-decoding approach to low-complexity soft maximum-likelihood decoding of cyclic codes," *Proceedings of the 2019 IEEE GlobeCom Conference*, December 2019.

[24] I. S. Reed and G. Solomon, "Polynomial codes over certain finite fields," *Journal of the Society for Industrial and Applied Mathematics,* vol. 8, pp. 300–304, June 1960.

[25] Y. Sugiyama, M. Kasahara, and T. Namekawa, "A method for solving key equation for decoding Goppa codes," *Information Control*, vol. 27, pp. 87–99, January 1975.

[26] A. Michelson and A. Levesque, *Error-Control Techniques for Digital Communication*, New York, Wiley, 1985.

[27] G. D. Forney, "On Decoding BCH Codes," *IEEE Transactions on Information Theory*, vol. 11, no. 4, pp. 549–557, October 1965.

[28] G. D. Forney, Jr., *Concatenated Codes*, Cambridge, MA, MIT Press, 1966.

[29] B. Gronholz, Zeta Associates Inc., Personal Communication, May 2023.

[30] P. Elias, "Error-free coding," *IRE Transactions on Information Theory*, vol. 4, no. 5, pp. 29–37, September 1954.

[31] Z. Li, L. Chen, L. Zeng, S. Lin, and W. Fong, "Efficient encoding of quasi-cyclic low-density parity-check codes," *IEEE Transactions on Communications*, vol. 54, no. 1, pp. 71–81, January 2006.

# 4 Convolutional Codes

The class of convolutional codes was invented by Elias in 1955 [1] and has been widely in use for wireless, space, and broadcast communications since about 1970. Their popularity stems from the relative ease with which the maximum-likelihood sequence decoder may be implemented and from their effectiveness when concatenated with a Reed–Solomon code. Since the early 1990s, they have enjoyed new respect and popularity due to the efficacy of concatenations of multiple convolutional codes in turbo codes.

In this chapter, we first present algebraic descriptions of convolutional codes, which were pioneered by Forney [2] and extended by Johannesson and Zigangirov [3]. We also discuss various encoder realizations and matrix representations of convolutional codes. We then discuss a graphical (trellis) representation, which aids two optimal decoding algorithms, the Viterbi algorithm [4, 5] and the BCJR algorithm [6]. Also discussed are the list Viterbi algorithm [7] for CRC-convolutional concatenations and the wrap-around Viterbi algorithm [8] for tail-biting codes. The near-optimal Fano decoding algorithm [9], which is based on a tree representation, is then discussed. Finally, we present techniques for bounding the performance of convolutional codes based on the technique of Viterbi [10]. Our discussion in this chapter will include only binary convolutional codes. Extensions of the various details to the nonbinary case are straightforward.

## 4.1 The Convolutional Code Archetype

Convolutional codes are linear codes with a very distinct algebraic structure. While they can be utilized in block-oriented (packet-based) situations, their encoders are frequently described as stream-oriented. That is, in contrast to a block code, whose encoder assigns an $n$-bit codeword to each block of $k$ data bits, a convolutional encoder assigns code bits to an incoming information bit sequence continuously, in a stream-oriented fashion.

The convolutional code archetype is a four-state, rate-1/2 code whose encoder is depicted in Figure 4.1. Observe that two code bits are produced for each data bit that enters the encoder, so the rate is 1/2. The state of the encoder is defined to be the contents of the two binary memory cells; hence, the encoder is a four-state device and is initialized to the zero state. As with binary block codes, all operations are over the binary field GF(2), so the two adders in Figure 4.1 are modulo-2 adders.

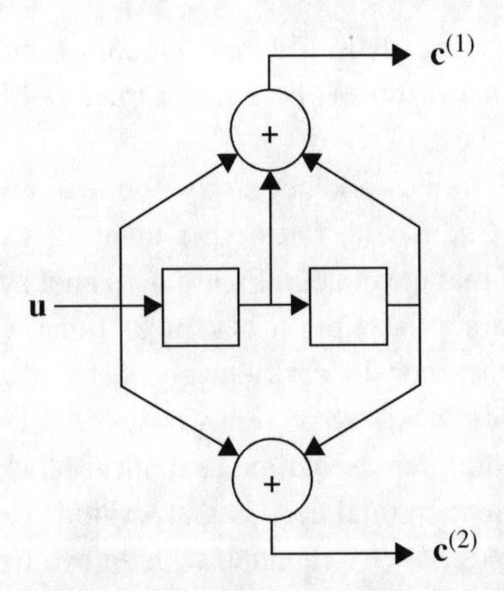


**Figure 4.1** A four-state, rate-1/2 convolutional code encoder.

Observe that the top and bottom parts of the encoder act as two discrete-time finite-impulse-response filters with operations in GF(2). The top filter has impulse response $\mathbf{g}^{(1)} = [1\ 1\ 1]$ and the bottom filter has impulse response $\mathbf{g}^{(2)} = [1\ 0\ 1]$. It is from this perspective that the nomenclature "convolutional" originates: encoder output $\mathbf{c}^{(1)}$ is the convolution of the input $\mathbf{u}$ and the impulse response $\mathbf{g}^{(1)}$, and similarly for encoder output $\mathbf{c}^{(2)}$. Thus, we may write

$$\mathbf{c}^{(j)} = \mathbf{u} \circledast \mathbf{g}^{(j)}$$

for $j = 1, 2$. Moreover, since convolution in the time domain corresponds to multiplication in the transform domain, this equation may be rewritten as

$$c^{(j)}(D) = u(D)g^{(j)}(D),$$

where the coefficients of the polynomials in $D$ are the elements of the corresponding vectors, so that $g^{(1)}(D) = 1 + D + D^2$ and $g^{(2)}(D) = 1 + D^2$. Note that $D$ is equivalently the discrete-time delay operator $z^{-1}$, although $D$ is generally used in the coding literature.

We may also model the two encoding operations more compactly via the following matrix equation:

$$\begin{aligned}\left[c^{(1)}(D) \quad c^{(2)}(D)\right] &= u(D)\left[g^{(1)}(D) \quad g^{(2)}(D)\right] \\ &= u(D)\mathbf{G}(D).\end{aligned}$$

The $1 \times 2$ matrix $\mathbf{G}(D) = \left[g^{(1)}(D) \quad g^{(2)}(D)\right]$ is the code's *generator matrix* and the polynomials $g^{(j)}(D)$ are called *generator polynomials*.

The codeword for such a convolutional encoder is generally taken to be the word formed by multiplexing the bits corresponding to the polynomials $c^{(1)}(D)$ and $c^{(2)}(D)$. For an input word length of $L$ (i.e., the degree of $u(D)$ is $L - 1$), the output word length is $2L + 4$. The added four bits are a consequence of the two extra 0 inputs used to return the encoder to the zero state, resulting in four extra code bits. Thus, in practice the code rate is $L/(2L + 4)$, which approaches $1/2$ for large $L$. We remark that, in addition to low-complexity encoding (as in Figure 4.1), convolutional codes allow low-complexity decoding. This is a result of the highly structured but simple relationship between the data word and the codeword.

The following sections provide the framework necessary for a more thorough understanding of convolutional codes. Codes with rates other than 1/2 will be considered, as will systematic codes. (Note that the code in Figure 4.1 is not systematic.) We first consider a more general algebraic description of convolutional codes than the one in the foregoing example. We then consider various encoder realizations for the various classes of convolutional codes. Next, we present various alternative representations of convolutional codes, including representation as a linear block code and graphical representations. Next, we present optimal decoding algorithms for convolutional codes operating over a binary-input AWGN channel, with an eye toward their use in iterative decoders, to be discussed in subsequent chapters. Finally, we present techniques for estimating the performance of these codes on the AWGN channel.

## 4.2 Algebraic Description of Convolutional Codes

In order to present an algebraic description of convolutional codes, the following notation for various algebraic structures is necessary. We denote by $\mathbb{F}_2[D]$ the *ring of polynomials over* $\mathbb{F}_2$ in the indeterminate $D$, where $\mathbb{F}_2$ is a shorthand for the finite field GF(2). Thus, an element of $\mathbb{F}_2[D]$ may be represented by $\sum_{i=0}^{s} a_i D^i$ for some integer $s \geq 0$, where $a_i \in \mathbb{F}_2$. We denote by $\mathbb{F}_2(D)$ the *field of rational functions over* $\mathbb{F}_2$ in the indeterminate $D$. Thus, an element of $\mathbb{F}_2(D)$ may be represented as a ratio of polynomials $a(D)/b(D)$ for some $a(D)$ and $b(D) \neq 0$ in $\mathbb{F}_2[D]$. Finally, we denote by $\mathbb{F}_2\langle D\rangle$ the *field of Laurent series over* $\mathbb{F}_2$ in the indeterminate $D$. Thus, an element of $\mathbb{F}_2\langle D\rangle$ may be represented by $\sum_{i=r}^{\infty} s_i D^i$ for some integer $r$, where $s_i \in \mathbb{F}_2$. The following relationships are obvious: $\mathbb{F}_2[D] \subset \mathbb{F}_2(D) \subset \mathbb{F}_2\langle D\rangle$. Note that $n$-tuples of elements of $\mathbb{F}_2\langle D\rangle$, denoted by $\mathbb{F}_2^n\langle D\rangle$, form a vector space with $\mathbb{F}_2\langle D\rangle$ as the scalar field.

---

**Example 4.1** The polynomials $1 + D$, $1 + D^2$, and $D + D^2 + D^3$ are all elements of the ring $\mathbb{F}_2[D]$ (and, hence, are in $\mathbb{F}_2(D)$ and $\mathbb{F}_2\langle D\rangle$ as well). The rational functions

$$\frac{1+D}{D+D^2+D^3}, \quad \frac{1+D}{1+D+D^2}, \quad \text{and} \quad \frac{1+D^2}{1+D}$$

are all elements of the field $\mathbb{F}_2(D)$. Note that the numerator of the third rational function factors as $1 + D^2 = (1 + D)(1 + D)$ and so the third rational function reduces to $1 + D$, an element of $\mathbb{F}_2[D]$. The first two are not elements of $\mathbb{F}_2[D]$, however. Application of long division to the second rational function yields

$$\frac{1+D}{1+D+D^2} = 1 + D^2 + D^3 + D^5 + D^6 + D^8 + D^9 + \cdots,$$

that is, a power series whose coefficients are 1011011011 .... Thus,

$$(1 + D)/(1 + D + D^2)$$

is clearly in $\mathbb{F}_2\langle D\rangle$ as well as in $\mathbb{F}_2(D)$. The elements of $\mathbb{F}_2(D)$ must have power-series coefficients that are periodic after some initial transient, as seen in the foregoing example. To see this, first note that an element $a(D)/b(D) \in \mathbb{F}_2(D)$ satisfies $a(D) = b(D) \cdot \sum_{i=0}^{\infty} s_i D^i = b(D) \cdot s(D)$. Now note that the only way that the product between the power series $s(D)$ and the polynomial $b(D)$ can be a polynomial $a(D)$ (which has finite length, by definition) is if $s(D)$ is periodic. Thus, an example of an element of a series that is in $\mathbb{F}_2\langle D\rangle$ but not in $\mathbb{F}_2(D)$ is one whose coefficients are not periodic, for example, are randomly chosen.

---

Given these definitions, we may now proceed to define a binary convolutional code. Whereas a linear binary block code of length $n$ is a subspace of the vector space $\mathbb{F}_2^n$ of $n$-tuples over the field $\mathbb{F}_2$, a linear binary convolutional code of length $n$ is a subspace of the vector space $\mathbb{F}_2^n\langle D\rangle$ of $n$-tuples over the field $\mathbb{F}_2\langle D\rangle$. Thus, given a convolutional code $\mathcal{C}$, a codeword $\mathbf{c}(D) \in \mathcal{C} \subset \mathbb{F}_2^n\langle D\rangle$ has the form

$$\mathbf{c}(D) = [c^{(1)}(D) \quad c^{(2)}(D) \ \dots \ c^{(n)}(D)],$$

where $c^{(j)}(D) = \sum_{i=r}^{\infty} c_i^{(j)} D^i$ is an element of $\mathbb{F}_2\langle D\rangle$. Further, since $\mathcal{C}$ is a subspace of $\mathbb{F}_2^n\langle D\rangle$, each $\mathbf{c}(D) \in \mathcal{C}$ may be written as a linear combination of basis vectors $\{\mathbf{g}_i(D)\}_{i=1}^k$ from that subspace,

$$\mathbf{c}(D) = \sum_{i=1}^{k} u^{(i)}(D)\mathbf{g}_i(D), \tag{4.1}$$

where $u^{(i)}(D) \in \mathbb{F}_2\langle D\rangle$ and $k$ is the dimension of $\mathcal{C}$. Analogously to linear block codes, we may rewrite (4.1) as

$$\mathbf{c}(D) = \mathbf{u}(D)\mathbf{G}(D), \tag{4.2}$$

where $\mathbf{u}(D) = [u^{(1)}(D) \quad u^{(2)}(D) \quad \dots \quad u^{(k)}(D)]$ and the basis vectors $\mathbf{g}_i(D) = \left[g_i^{(1)}(D) \quad g_i^{(2)}(D) \quad \dots \quad g_i^{(n)}(D)\right]$ form the rows of the $k \times n$ *generator matrix* $\mathbf{G}(D)$, so that the element in row $i$ and column $j$ of $\mathbf{G}(D)$ is $g_i^{(j)}(D)$.

In practice, the entries of $\mathbf{G}(D)$ are confined to the field of rational functions $\mathbb{F}_2(D)$, so that encoding and decoding are realizable. (Additional necessary constraints on $\mathbf{G}(D)$ are given in the next section.) Further, while $\left\{u^{(i)}(D)\right\}$ and $\left\{c^{(j)}(D)\right\}$ are still taken from $\mathbb{F}_2\langle D\rangle$, they are restricted to be delay-free, that is, $r = 0$. Observe that, if we multiply a matrix $\mathbf{G}(D)$ with entries in $\mathbb{F}_2(D)$ by the least common multiple of the denominators of its entries $g_i^{(j)}(D)$, we obtain a generator matrix for the same code whose entries are confined to the ring of polynomials $\mathbb{F}_2[D]$. We denote this *polynomial form* of a generator matrix by $\mathbf{G}_{\text{poly}}(D)$ and emphasize that there exists a polynomial-form generator matrix for any realizable convolutional code. Hereafter, we will let $\mathbf{G}(D)$ represent a generic generator matrix, so that it may be in either systematic or non-systematic form, and its entries may be from $\mathbb{F}_2(D)$ or $\mathbb{F}_2[D]$.

There exists an $(n-k) \times n$ *parity-check matrix* $\mathbf{H}(D)$ whose null space is $\mathcal{C}$. This may be restated as

$$\mathbf{c}(D)\mathbf{H}^{\mathrm{T}}(D) = \mathbf{0}$$

for any $\mathbf{c}(D) \in \mathcal{C}$, or as

$$\mathbf{G}(D)\mathbf{H}^{\mathrm{T}}(D) = \mathbf{0}.$$

In the previous two equations, $\mathbf{0}$ represents a $1 \times (n-k)$ zero vector and a $k \times (n-k)$ zero matrix, respectively. As for linear block codes, $\mathbf{G}(D)$ and $\mathbf{H}(D)$ each have a *systematic form*, $\mathbf{G}_{\mathrm{sys}}(D) = [\mathbf{I}\ \mathbf{P}(D)]$ and $\mathbf{H}_{\mathrm{sys}}(D) = [\mathbf{P}^{\mathrm{T}}(D)\ \mathbf{I}]$. Also, as for linear block codes, $\mathbf{G}_{\mathrm{sys}}(D)$ is obtained by row reduction of $\mathbf{G}(D)$ combined with possible column permutations to obtain the $[\mathbf{I}\ \mathbf{P}(D)]$ format. As we will see in the following example, there also exists a polynomial form of the parity-check matrix, denoted by $\mathbf{H}_{\mathrm{poly}}(D)$.

---

**Example 4.2** Consider the generator matrix for a rate-2/3 convolutional code given by

$$\mathbf{G}(D) = \begin{bmatrix} \dfrac{1+D}{1+D+D^2} & 0 & 1+D \\ 1 & \dfrac{1}{1+D} & \dfrac{D}{1+D^2} \end{bmatrix}.$$

If we (a) multiply the first row by $(1+D+D^2)/(1+D)$, (b) add the new first row to the second row, and (c) multiply the new second row by $1+D$, with all operations over $\mathbb{F}_2(D)$, the result is the systematic form

$$\mathbf{G}_{\mathrm{sys}}(D) = \begin{bmatrix} 1 & 0 & 1+D+D^2 \\ 0 & 1 & \dfrac{1+D^3+D^4}{1+D} \end{bmatrix}.$$

Noting that the submatrix $\mathbf{P}(D)$ is the rightmost column of $\mathbf{G}_{\mathrm{sys}}(D)$, we may immediately write

$$\mathbf{H}_{\mathrm{sys}}(D) = \begin{bmatrix} 1+D+D^2 & \dfrac{1+D^3+D^4}{1+D} & 1 \end{bmatrix}$$

because we know that $\mathbf{H}_{\mathrm{sys}}(D) = [\mathbf{P}^{\mathrm{T}}(D)\ \mathbf{I}]$. The least common multiple of the denominators of the entries of $\mathbf{G}(D)$ is $(1+D+D^2)(1+D^2)$, from which

$$\mathbf{G}_{\mathrm{poly}}(D) = \begin{bmatrix} (1+D)^3 & 0 & (1+D+D^2)(1+D)^3 \\ (1+D+D^2)(1+D^2) & (1+D+D^2)(1+D) & D(1+D+D^2) \end{bmatrix}.$$

$(1+D)^3$ can be factored out of the first row and $1+D+D^2$ can be factored out of the second row to obtain a reduced-complexity form of $\mathbf{G}_{\mathrm{poly}}(D)$. Lastly, by multiplying each of the entries of $\mathbf{H}_{\mathrm{sys}}(D)$ by $1+D$, we obtain

$$\mathbf{H}_{\mathrm{poly}}(D) = \begin{bmatrix} 1+D^3 & 1+D^3+D^4 & 1+D \end{bmatrix}.$$

---

We emphasize that $\mathbf{G}(D)$, $\mathbf{G}_{\text{sys}}(D)$, and $\mathbf{G}_{\text{poly}}(D)$ all generate the same code, that is, their rows span the same subspace of $\mathbb{F}_2^n\langle D\rangle$. In (4.2), $\mathbf{u}(D)$ represents the data word to be encoded, and $\mathbf{u}(D)\mathbf{G}(D)$ represents the codeword corresponding to that data word. Thus, while the three generator matrices yield the same code, that is, the same list of codewords, they correspond to different $\mathbf{u}(D) \mapsto \mathbf{c}(D)$ mappings, that is, to different encoders. From this point forward, we use "encoder" and "generator matrix" interchangeably because they both describe the same $\mathbf{u}(D) \mapsto \mathbf{c}(D)$ mapping. In the next section we divide the different encoder possibilities for a given code into four classes and describe two realizations for any given encoder, that is, for any given $\mathbf{G}(D)$.

## 4.3 Encoder Realizations and Classifications

When considering encoder realizations, it is useful to let $D$ represent the unit-delay operator, which is equivalent to $z^{-1}$ in the discrete-time signal-processing literature. In the encoding operation represented by $\mathbf{c}(D) = \mathbf{u}(D)\mathbf{G}(D)$, we note that there is a discrete-time transfer function $g_i^{(j)}(D)$ between input $u^{(i)}(D)$ and output $c^{(j)}(D)$. Further, when an element $g_i^{(j)}(D)$ of $\mathbf{G}(D)$ is a polynomial from $\mathbb{F}_2[D]$, this transfer function may be implemented as if it were a finite-impulse-response (FIR) filter, but with arithmetic over $\mathbb{F}_2$. When $g_i^{(j)}(D)$ is a rational function from $\mathbb{F}_2(D)$, it may be implemented as if it were an infinite-impulse-response (IIR) filter, but with arithmetic over $\mathbb{F}_2$.

When $g_j^{(i)}(D)$ is a rational function, that is, $g_i^{(j)}(D) = a(D)/b(D)$, we assume that it has the form

$$g_i^{(j)}(D) = \frac{a_0 + a_1 D + \cdots + a_m D^m}{1 + b_1 D + \cdots + b_m D^m}, \tag{4.3}$$

so that $b(D)|_{D=0} = b_0 = 1$. This assumption implies that $g_i^{(j)}(D)$ is a causal transfer function and, hence, is realizable. To see this, note that, by definition, the transfer function $g_i^{(j)}(D)$ implies that

$$c^{(j)}(D) = u^{(i)}(D)g_i^{(j)}(D) = u^{(i)}(D)\frac{a(D)}{b(D)}$$

or, in the time domain with discrete-time parameter $t$,

$$\begin{aligned} b_0 c_t^{(j)} &= a_0 u_t^{(i)} + a_1 u_{t-1}^{(i)} + \cdots + a_m u_{t-m}^{(i)} \\ &\quad - b_1 c_{t-1}^{(j)} - b_2 c_{t-2}^{(j)} - \cdots - b_m c_{t-m}^{(j)}. \end{aligned}$$

Observe that $b_0 = 0$ makes the situation untenable, that is, $c_t^{(j)}$ cannot be determined from the inputs $u_t^{(i)}, u_{t-1}^{(i)}, \ldots, u_{t-m}^{(i)}$ and the outputs $c_{t-1}^{(j)}, c_{t-2}^{(j)}, \ldots, c_{t-m}^{(j)}$.

Figure 4.2 depicts the *Type I IIR filter realization* of the transfer function $g_i^{(j)}(D) = a(D)/b(D)$ and Figure 4.3 presents the *Type II IIR filter realization*. The Type I form is also called the *controller canonical form* or the *direct canonical form*. The Type II form is also called the *observer canonical form* or the *transposed canonical form*.

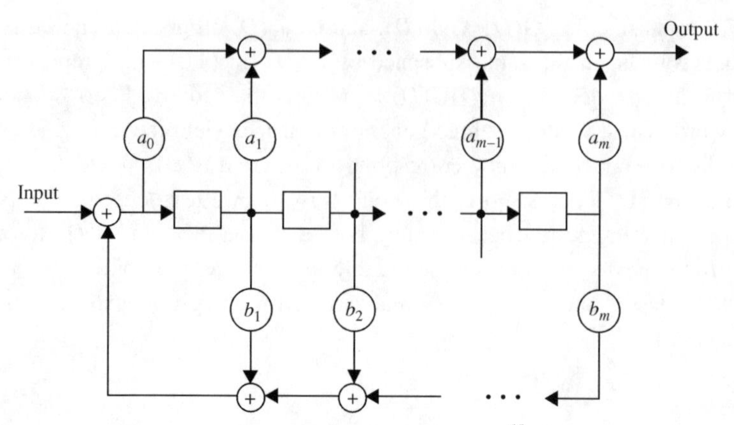


**Figure 4.2** Type I realization of the transfer function $g_i^{(j)}(D) = a(D)/b(D)$, with $b_0 = 1$. The input is $u^{(i)}(D)$ and the output is $c^{(j)}(D)$.

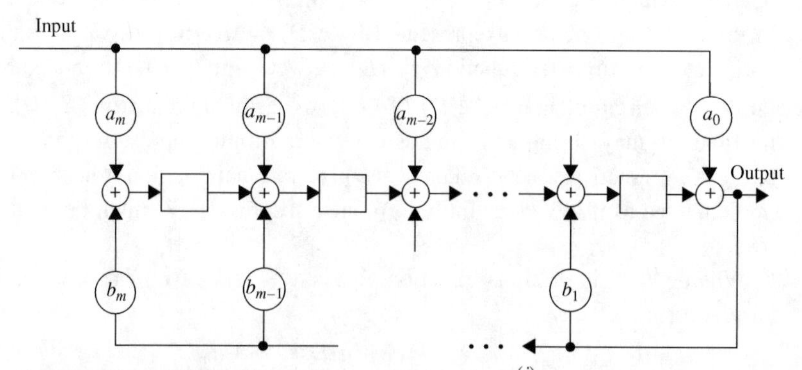


**Figure 4.3** Type II realization of the transfer function $g_i^{(j)}(D) = a(D)/b(D)$, with $b_0 = 1$. The input is $u^{(i)}(D)$ and the output is $c^{(j)}(D)$.

When $g_i^{(j)}(D)$ is a polynomial, that is, $g_i^{(j)}(D) = a(D)$, this is a special case of a rational function with $b(D) = 1$. Thus, the implementation of $g_i^{(j)}(D)$ in this case is also depicted in Figure 4.2 or Figure 4.3, but with $b_1 = b_2 = \cdots = b_m = 0$, that is, without the feedback. The derivations of both types of realizations are explored in Problems 4.4 and 4.5. See also Sections 6.3 and 6.4 of [11], which discuss implementations of IIR systems.

Throughout the rest of this chapter, we shall require that $\mathbf{G}(D)$ be *delay-free*, that is, that $\mathbf{G}(D)$ correspond to a delay-free encoder. This implies that $\mathbf{G}(0)$ is not the zero matrix or, equivalently, that $D^l$, $l > 0$, cannot be factored out of $\mathbf{G}(D)$. We shall also require that $\mathbf{G}(D)$ be *realizable*, meaning that $\mathbf{G}(D)$ corresponds to a realizable encoder. When $\mathbf{G}(D)$ consists of rational functions, this means that the denominators of all entries of $\mathbf{G}(D)$ have nonzero constant terms. (Consider the impact of a zero constant term, $b_0 = 0$, in either Figure 4.2 or Figure 4.3.) $\mathbf{G}(D)$ in polynomial form is always realizable.

**Example 4.3** The Type I encoder realization of the rate-$1/2$ convolutional code with generator matrix $\mathbf{G}(D) = [1 + D + D^2 \quad 1 + D^2]$, originally depicted in Figure 4.1, is presented in Figure 4.4(a). The Type I encoder for the systematic form of this code, for which the generator matrix is

$$\mathbf{G}_{\text{sys}}(D) = \left[1 \quad \frac{1 + D^2}{1 + D + D^2}\right],$$

is presented in Figure 4.4(b). The Type II encoder for $\mathbf{G}_{\text{sys}}(D)$ is presented in Figure 4.4(c). The two outputs in each case are multiplexed in practice.

In the table below we give an example input/output pair for each of the encoders in Figures 4.4(a) and (b). The examples were specifically selected to demonstrate how different inputs lead to the same codeword in each case. The (semi-infinite) vector notation is used in the table. In this notation a polynomial (power series) is represented by a vector (semi-infinite vector) of its coefficients. Thus, the vector representation for $u(D) = u_0 + u_1D + u_2D^2 + \cdots$ is $\mathbf{u} = [u_0 \quad u_1 \quad u_2 \quad \ldots]$ and the vector representation for $c^{(j)}(D) = c_0^{(j)} + c_1^{(j)}D + c_2^{(j)}D^2 + \cdots$ is $\mathbf{c}^{(j)} = \left[c_0^{(j)} \quad c_1^{(j)} \quad c_2^{(j)} \quad \ldots\right]$, $j = 1, 2$. As indicated in Figures 4.4(a) and (b), in the table the vector $\mathbf{c}$ is the codeword corresponding to multiplexing the bits of $\mathbf{c}^{(1)}$ and $\mathbf{c}^{(2)}$, as is done in practice.

(a)

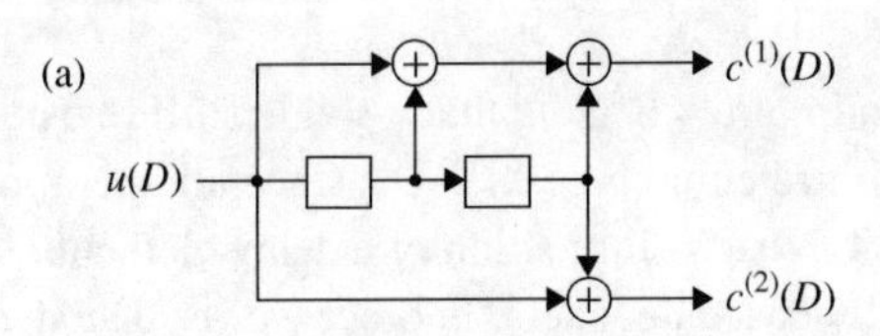


(b)

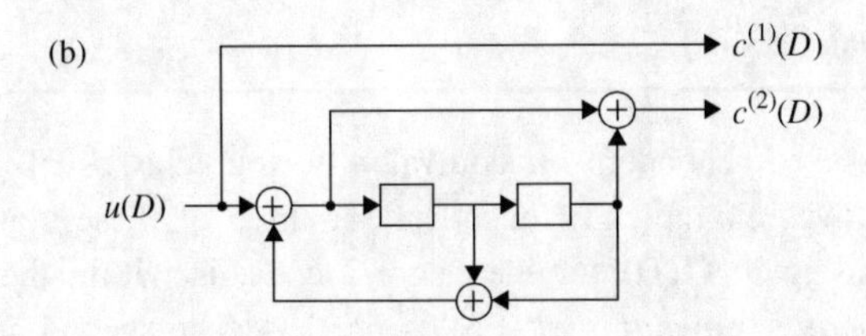


(c)

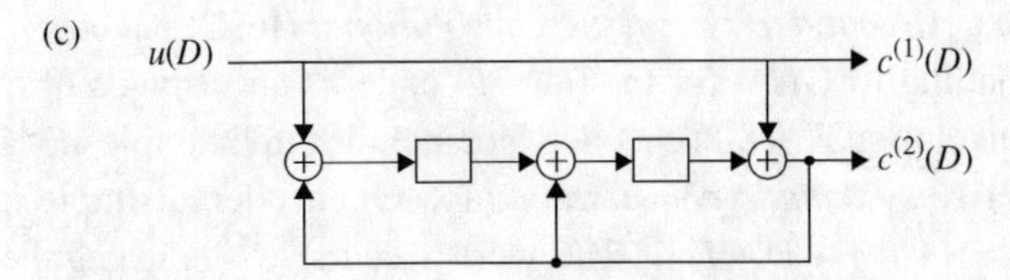


**Figure 4.4** Encoder realizations of the non-systematic and systematic versions of the rate-$1/2$ convolutional code given in Example 4.3 with generator matrix $\mathbf{G}(D) = [1 + D + D^2 \quad 1 + D^2]$. (a) Type I realization of a non-systematic encoder. (b) Type I realization of a systematic encoder. (c) Type II realization of a systematic encoder.

| | Figure 4.4(a) encoder | Figure 4.4(b) encoder |
|---|---|---|
| $u(D)$ | 1 | $1 + D + D^2$ |
| $\mathbf{u}$ | 10... | 1110... |
| $c^{(1)}(D)$ | $1 + D + D^2$ | $1 + D + D^2$ |
| $c^{(2)}(D)$ | $1 + D^2$ | $1 + D^2$ |
| $\mathbf{c}^{(1)}$ | 1110... | 1110... |
| $\mathbf{c}^{(2)}$ | 1010... | 1010... |
| $\mathbf{c}$ | 111011000... | 111011000... |

Figure 4.5 presents the Type I encoder realization corresponding to the generator matrix $\mathbf{G}(D)$ of the rate-2/3 convolutional code given in Example 4.2,

$$\mathbf{G}(D) = \begin{bmatrix} \frac{1+D}{1+D+D^2} & 0 & 1+D \\ 1 & \frac{1}{1+D} & \frac{D}{1+D^2} \end{bmatrix}.$$

Note that, in an attempt to minimize the number of delay elements, $1 + D$ was implemented as $(1 + D^3)/(1 + D + D^2)$ and $1/(1 + D)$ was implemented as $(1 + D)/(1 + D^2)$. However, we can reduce the number of encoder delay elements further for this code by implementing the encoder for

$$\mathbf{G}_{\text{sys}}(D) = \begin{bmatrix} 1 & 0 & \frac{1+D^3}{1+D} \\ 0 & 1 & \frac{1+D^3+D^4}{1+D} \end{bmatrix}$$

in the Type II form as depicted in Figure 4.6. Note that a single shift register may be shared in the computation of the third column of $\mathbf{G}_{\text{sys}}(D)$, since all operations are linear. As we will see in Section 4.5, each elimination of a delay element results in a halving of the complexity of trellis-based decoders. In other words, decoder complexity is proportional to $2^{\mu}$, where $\mu$ is the number of encoder delay elements; $\mu$ is called the encoder *memory*.

---

There exist four classes of encoders or, equivalently, four classes of generator matrices. Example 4.3 gives examples of encoders for three of these classes. The first example, corresponding to $\mathbf{G}(D)$ for the rate-1/2 code, is within the class of *non-recursive non-systematic convolutional* ($\bar{\text{R}}\bar{\text{S}}$C) encoders. The second and fourth examples, corresponding to $\mathbf{G}_{\text{sys}}(D)$ for the rate-1/2 and rate-2/3 codes, respectively, are instances of *recursive systematic convolutional* (RSC) encoders. The third example, corresponding to $\mathbf{G}(D)$ for the rate-2/3 code, is an instance of a *recursive non-systematic* convolutional (R$\bar{\text{S}}$C) encoder. Not included in Example 4.3 is an example of a *non-recursive systematic convolutional* ($\bar{\text{R}}$SC) encoder, a simple instance of which has $\mathbf{G}(D) = \begin{bmatrix} 1 & 1 + D + D^2 \end{bmatrix}$. The word "recursive" indicates the presence of at least one rational function in $\mathbf{G}(D)$ or, equivalently, the presence of feedback in the encoder realization. Conversely, a "non-recursive" encoder indicates the absence of any rational functions in $\mathbf{G}(D)$, that is, $\mathbf{G}(D)$ is in polynomial form, so that the encoder realization is devoid of any feedback. In fact, in the literature a recursive systematic

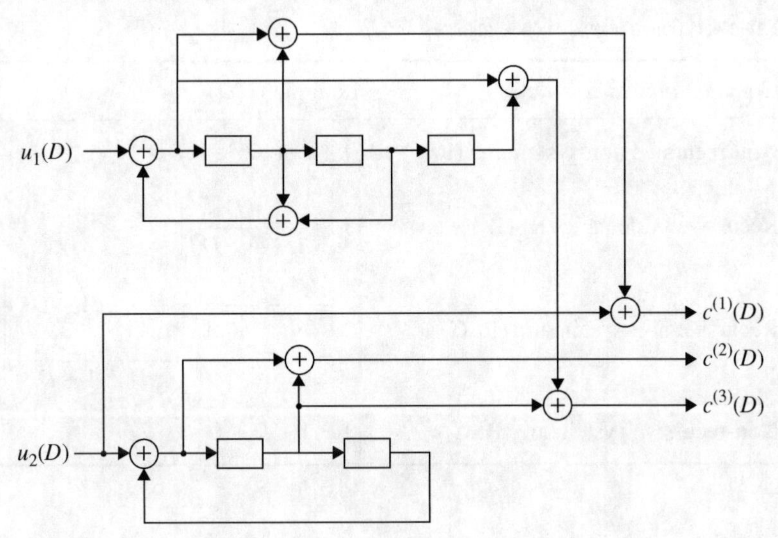


**Figure 4.5** Type I encoder realization of $\mathbf{G}(D)$ for the rate-2/3 encoder of the convolutional code given in Example 4.2 and discussed further in Example 4.3.

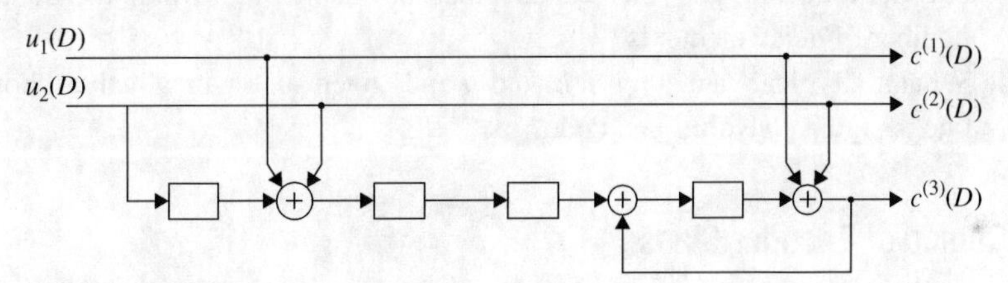


**Figure 4.6** Type II encoder realization of $\mathbf{G}_{\text{sys}}(D)$ for the rate-2/3 convolutional code given in Examples 4.2 and 4.3.

encoder is often called a *systematic encoder with feedback*, and similar nomenclature exists for the other three classes.

The four convolutional encoder classes are summarized in Table 4.1, together with their abbreviations and example generator matrices. A generator matrix $\mathbf{G}(D)$ may be realized as either a Type I or a Type II encoder for all four encoder classes listed. It is important to observe that, while we use $\mathbf{G}(D)$ and "encoder" interchangeably, an "encoder" (the $\mathbf{u}(D) \mapsto \mathbf{c}(D)$ mapping) is distinct from the "encoder realization" (Type I or Type II circuit).

Within a given encoder class there exist many encoders (generator matrices) for a given convolutional code. For example, scaling each entry of a generator matrix $\mathbf{G}(D)$ by the same arbitrary polynomial changes the encoder, but does not change the code. As another example, replacing a given row in $\mathbf{G}(D)$ by the sum of that row and another row does not change the code. The notion that many encoders exist for a single code and, on top of that, two encoder realizations exist, leads us to ask the following questions, each of which will be addressed below.

**Table 4.1** Convolutional encoder classes

| Convolutional encoder class | Example $\mathbf{G}(D)$ |
|---|---|
| Non-recursive non-systematic ($\bar{\text{R}}\bar{\text{S}}$C) | $\begin{bmatrix}1+D+D^2 & 1+D^2\end{bmatrix}$ |
| Recursive systematic (RSC) | $\begin{bmatrix}1 & \dfrac{1+D^2}{1+D+D^2}\end{bmatrix}$ |
| Recursive non-systematic (R$\bar{\text{S}}$C) | $\begin{bmatrix}\dfrac{1+D}{1+D+D^2} & 0 & 1+D \\ 1 & \dfrac{1}{1+D} & \dfrac{D}{1+D^2}\end{bmatrix}$ |
| Non-recursive systematic ($\bar{\text{R}}$SC) | $\begin{bmatrix}1 & 1+D+D^2\end{bmatrix}$ |

1. Is one encoder class superior to the others?
2. Are there any encoders to be avoided?
3. Since decoder complexity is proportional to $2^{\mu}$, can an encoder class and an encoder realization be selected such that the encoder realization has the minimum number of delay elements?
4. What are typical values for $n$, $k$, and $\mu$ and, given such typical values, how does one design a convolutional code?

### 4.3.1 Choice of Encoder Class

The answer to the first question depends on the application. For example, many applications require that the encoder be systematic so that the message bits are readily observable in the code stream. As another example, parallel turbo codes (Chapter 7) require that the constituent convolutional codes be of the recursive type, and usually systematic. For many applications the encoder class matters very little, particularly applications in which the convolutional code is not a constituent code of a turbo(-like) code. As seen in Example 4.3 and as we shall see in the discussion below of the third question, the choice of encoder class can affect the encoder memory $\mu$.

### 4.3.2 Catastrophic Encoders

The answer to the second question is that catastrophic encoders are to be avoided, a topic we develop via the following example. Consider the rate-1/2 code with $\mathbf{G}(D) = \begin{bmatrix}1+D^2 & 1+D\end{bmatrix}$. Note that the entries of $\mathbf{G}(D)$ have the common polynomial factor $1+D$. Now set the encoder input $u(D) = 1/(1+D)$ so that the resulting codeword is $\mathbf{c}(D) = [1+D \quad 1]$. Observe that the input $u(D) = 1/(1+D)$ expands into the power series $1+D+D^2+D^3+\cdots$ corresponding to the binary input sequence of coefficients $\mathbf{u} = 1111\ldots$. Thus, the input has an infinite Hamming weight. On the other hand, the two output sequences are $\mathbf{c} = [\mathbf{c}^{(1)} \quad \mathbf{c}^{(2)}] = [11000\ldots \quad 1000\ldots]$, which has a Hamming weight of three. In practice, these two sequences would be multiplexed, yielding the code sequence $\mathbf{c}' = 111000\ldots$.

Now consider that any decoder implementation must have a finite memory "window." Even in the absence of noise, this decoder, which initially observes the three ones from the encoder output, will eventually see only zeros in its memory window. Because the all-zeros input word maps to the all-zeros codeword, any decoder will eventually produce zeros at its output, whereas the correct encoder input sequence was all ones. Such a phenomenon is called *catastrophic error propagation* and this encoder is called a *catastrophic encoder*. Formally, an encoder is a catastrophic encoder if there exists an infinite-weight input that yields a finite-weight output. Clearly, a systematic encoder can never be catastrophic. (The literature often loosely uses the term "catastrophic code" when in fact it is only the encoder that can be catastrophic.)

For rate-$1/n$ convolutional codes, the generalization of the above example is as follows. Let $\mathbf{G}(D) = \left[g^{(1)}(D) \quad g^{(2)}(D) \quad \ldots \quad g^{(n)}(D)\right]$, where only one subscript is necessary for the entries $g^{(j)}(D)$ since there is only one row. We let the $g^{(j)}(D)$ be polynomials without loss of generality since, if $\mathbf{G}(D)$ contains rational functions, it is a simple matter to put it into polynomial form. Suppose now that the $\{g^{(j)}(D)\}$ have a common polynomial factor $a(D)$, so that $g^{(j)}(D) = a(D)f^{(j)}(D)$. Now let the encoder input be $u(D) = 1/a(D)$, which is clearly not a polynomial, but a rational function of infinite length when expanded into a power series. The codeword corresponding to $u(D)$ would then be $\mathbf{c}(D) = [f^{(1)}(D) \quad f^{(2)}(D) \quad \ldots \quad f^{(n)}(D)]$, which has finite weight since $\{f^{(j)}(D)\}$ are polynomials. Thus, a rate-$1/n$ encoder is catastrophic if (and only if) $\gcd\{g^{(j)}(D)\}$ is a non-monomial (a polynomial with more than one term).

To generalize the catastrophic encoder for rate-$k/n$ convolutional codes, we first define $\{\Delta^{(i)}(D)\}$ to be the determinants of the set of $\binom{n}{k}$ possible $k \times k$ submatrices of the generator matrix $\mathbf{G}(D)$. Then the encoder is catastrophic if $\gcd\{\Delta^{(i)}(D),\ i = 1, 2, \ldots, \binom{n}{k}\}$ is a non-monomial. See [3] or [12] for a proof of this fact.

## 4.3.3 Minimal Encoders

We now address the third question: encoders that achieve the minimum encoder memory are called *minimal encoders*. Consider first a rate-$1/n$ code represented by $\mathbf{G}(D) = \left[g^{(1)}(D) \quad g^{(2)}(D) \quad \ldots \quad g^{(n)}(D)\right]$ in polynomial form. Clearly, if this code is catastrophic, the entries $g^{(j)}(D)$ of $\mathbf{G}(D)$ have a common (non-monomial) polynomial factor and this representation of the encoder is not minimal. However, if the greatest common factor of the $\{g^{(j)}(D)\}$ is factored out, then the resulting generator matrix (encoder) will be minimal. Alternatively, one may convert this generator matrix into the RSC form $\mathbf{G}_{\text{sys}}(D) = \left[1 \quad g^{(2)}(D)/g^{(1)}(D) \quad \ldots \quad g^{(n)}(D)/g^{(1)}(D)\right]$, which automatically removes any common factor after the rational functions $g^{(j)}(D)/g^{(1)}(D)$ are reduced to their lowest terms. Note that a rate-$1/n$ minimal encoder is non-catastrophic (and this fact extends to all code rates).

A close inspection reveals that the Type I realization is naturally suited to rate-$1/n$ codes in the RSC format. This is because each of the $n-1$ parity bits may be computed by "tapping off" the same length-$\mu$ shift register in the appropriate locations and summing (over $\mathbb{F}_2$). In this way, the Type I circuit of Figure 4.2 may be considered to be a single-input/multiple-output device, as is necessary when computing $n-1$ parity bits for each input bit. On the other hand, the Type II realization is particularly ill-suited as an encoder realization of $\mathbf{G}_{sys}(D)$ given above because it has rate $1/n$. In summary, a minimal encoder for a rate-$1/n$ convolutional code is obtained by applying the Type I encoder realization to $\mathbf{G}_{sys}(D)$, or to $\mathbf{G}(D)$ with the gcd factored out of its entries.

---

**Example 4.4** For the rate-$1/2$ code of Example 4.3, the Type I encoder realization for $\mathbf{G}(D)$ and the Type I and Type II encoder realizations for $\mathbf{G}_{sys}(D)$ are all minimal with $\mu = 2$, whereas the Type II realization of $\mathbf{G}(D)$ has $\mu = 4$.

Consider the rate-$1/3$ code with $\mathbf{G}(D) = [1 + D^3 \quad 1 + D + D^2 + D^3 \quad 1 + D^2]$. If we divide out the gcd, $1 + D$, from each of the entries of $\mathbf{G}(D)$, we obtain $\mathbf{G}(D) = [1 + D + D^2 \quad 1 + D^2 \quad 1 + D]$. The Type I and Type II realizations for this generator matrix are presented in Figures 4.7(a) and (b), which we observe have memories $\mu = 2$ and $\mu = 5$, respectively. Noting that

$$\mathbf{G}_{sys}(D) = \left[1 \quad \frac{1 + D^2}{1 + D + D^2} \quad \frac{1 + D}{1 + D + D^2}\right],$$

the Type I and II realizations for this systematic generator matrix are presented in Figures 4.7(c) and (d), which we observe have memories $\mu = 2$ and $\mu = 4$, respectively. Note that the Type I realization applied to $\mathbf{G}_{sys}(D)$ yields a minimal encoder as claimed above. Note also that the Type I realization applied to $\mathbf{G}(D)$, after we have divided out the gcd from its entries, also yields a minimal encoder.

---

Next we consider a rate-$k/(k + 1)$ code whose encoder is represented in polynomial form by

$$\mathbf{G}_{poly}(D) = \left[g_i^{(j)}(D)\right],$$

or in systematic form by

$$\mathbf{G}_{sys}(D) = \left[\begin{array}{cc} & h^{(k)}(D)/h^{(0)}(D) \\ \mathbf{I}_k & h^{(k-1)}(D)/h^{(0)}(D) \\ & \vdots \\ & h^{(1)}(D)/h^{(0)}(D) \end{array}\right], \tag{4.4}$$

where $h^{(j)}(D)\big|_{j\neq 0}$ and $h^{(0)}(D)$ are relatively prime polynomials. We may immediately write, from (4.4),

$$\mathbf{H}_{sys}(D) = \left[h^{(k)}(D)/h^{(0)}(D) \quad h^{(k-1)}(D)/h^{(0)}(D) \ \ldots \ h^{(1)}(D)/h^{(0)}(D) \quad 1\right] \tag{4.5}$$

and

$$\mathbf{H}_{poly}(D) = \left[h^{(k)}(D) \quad h^{(k-1)}(D) \ \ldots \ h^{(1)}(D) \quad h^{(0)}(D)\right]. \tag{4.6}$$

We now argue that the Type II realization of $\mathbf{G}_{\text{sys}}(D)$ as given by (4.4) is a minimal encoder, since the rational functions in $\mathbf{G}_{\text{sys}}(D)$ are in lowest terms and the Type II realization is naturally suited to rate $k/(k+1)$. Note that the encoder realization is completely specified by the $k+1$ *parity-check polynomials* $\left\{h^{(j)}(D)\right\}_{j=0}^{k}$ and $\gcd\left\{h^{(0)}(D), h^{(1)}(D), \ldots, h^{(k)}(D)\right\} = 1$ under the relatively prime assumption and the delay-free encoder assumption. The Type II realization is naturally suited to rate-$k/(k+1)$ codes in the systematic format in (4.4) because the single parity bit may be computed by "tapping into" the same length-$\mu$ shift register at the appropriate adders. Thus, the Type II circuit of Figure 4.3 may be considered to be a multiple-input/single-output device, as is necessary when computing a single parity bit from $k$ input bits. On the other hand, the Type I realization is particularly ill-suited as an encoder realization of $\mathbf{G}_{\text{sys}}(D)$. In summary, a minimal encoder for a rate-$k/(k+1)$ convolutional code is obtained by applying the Type II encoder realization to $\mathbf{G}_{\text{sys}}(D)$ or, equivalently, to $\mathbf{H}_{\text{poly}}(D)$ with the gcd factored out of its entries.

---

**Example 4.5** Returning to the rate-2/3 code of Example 4.2, we have

$$\mathbf{G}_{\text{poly}}(D) = \begin{bmatrix} (1+D)^3 & 0 & (1+D+D^2)(1+D)^3 \\ (1+D+D^2)(1+D^2) & (1+D+D^2)(1+D) & D(1+D+D^2) \end{bmatrix},$$

$$\mathbf{G}_{\text{sys}}(D) = \begin{bmatrix} 1 & 0 & \dfrac{1+D^3}{1+D} \\ 0 & 1 & \dfrac{1+D^3+D^4}{1+D} \end{bmatrix},$$

$$\mathbf{H}_{\text{sys}}(D) = \begin{bmatrix} \dfrac{1+D^3}{1+D} & \dfrac{1+D^3+D^4}{1+D} & 1 \end{bmatrix},$$

$$\mathbf{H}_{\text{poly}}(D) = \begin{bmatrix} 1+D^3 & 1+D^3+D^4 & 1+D \end{bmatrix}.$$

Note that the element $1+D+D^2$ in $\mathbf{G}_{\text{sys}}(D)$ and $\mathbf{H}_{\text{sys}}(D)$ in Example 4.2 has been adjusted to $(1+D^3)/(1+D)$, so that the expressions above conform to the forms given in (4.4), (4.5), and (6.3). Clearly, $h^{(0)}(D) = 1 + D$. The Type II implementation of $\mathbf{G}_{\text{sys}}(D)$, which gives the minimal encoder, is depicted in Figure 4.6, where a memory size of $\mu = 4$ is evident. The Type I implementation of $\mathbf{G}_{\text{poly}}(D)$ requires $\mu = 9$ when a length-5 shift register is used to implement the top row of $\mathbf{G}_{\text{poly}}(D)$ and a length-4 shift register is used to implement the bottom row of $\mathbf{G}_{\text{poly}}(D)$. The shift registers in each case are shared by the elements in a row, with appropriate taps connecting a binary adder to the shift register. The Type II implementation of $\mathbf{G}_{\text{poly}}(D)$ requires $\mu = 9$ when circuit sharing is utilized: one circuit is used to implement the factor $(1+D)^3$ in the first row of $\mathbf{G}_{\text{poly}}(D)$ and one circuit is used to implement the factor $(1+D+D^2)$ in the second row of $\mathbf{G}_{\text{poly}}(D)$. The Type II implementation of $\mathbf{G}_{\text{sys}}(D)$ requires $\mu = 6$, which is achieved when the element $(1+D^3)/(1+D)$ in $\mathbf{G}_{\text{sys}}(D)$ is implemented as $1+D+D^2$.

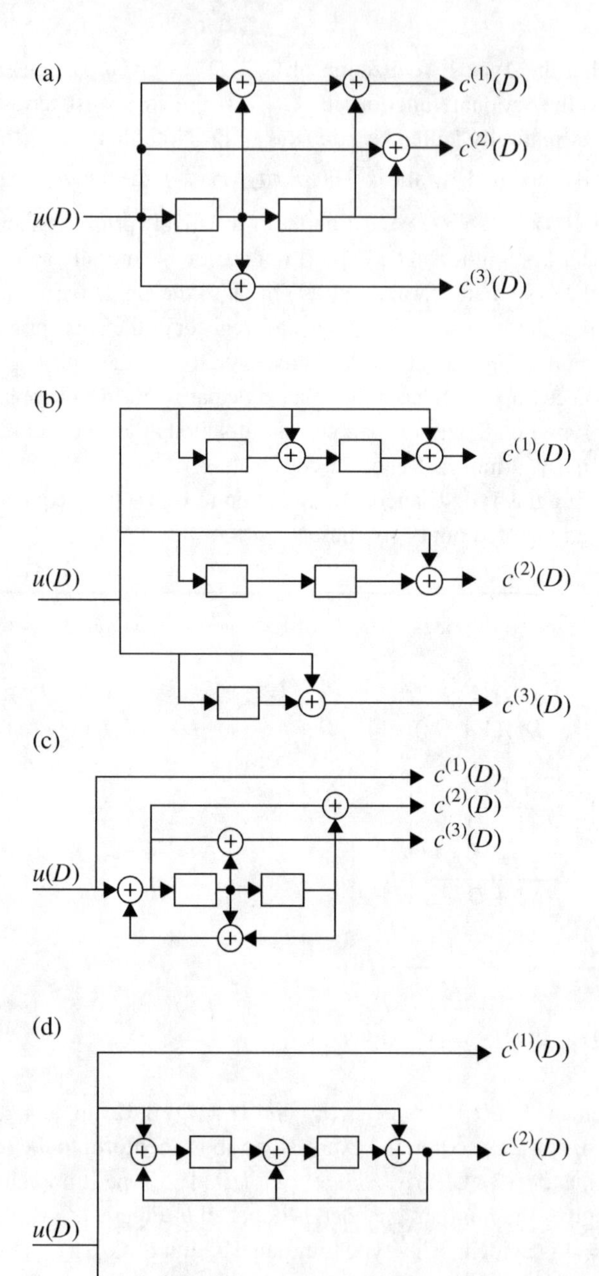


**Figure 4.7** Encoder realizations for the rate-1/3 code of Example 4.4. (a) Type I $\bar{\text{R}}\bar{\text{S}}$C encoder. (b) Type II $\bar{\text{R}}\bar{\text{S}}$C encoder. (c) Type I RSC encoder. (d) Type II RSC encoder.

In the foregoing discussion of minimal encoder realizations for rate-$1/n$ and rate-$k/(k+1)$ convolutional codes, we saw that the first step was to put the generator matrix in systematic form. The minimal realization for the rate-$1/n$ case is then achieved by the Type I form and that for the rate-$k/(k+1)$ case is achieved by the Type II form. For the general rate-$k/n$ convolutional code, where $k/n$ is neither $1/n$ nor $k/(k+1)$, the first step is again to put the generator matrix in systematic form. However, the minimal realization of this generator matrix may be neither Type I nor Type II. This topic is beyond our scope (see [3] for a comprehensive treatment), but we have the following remarks: (1) in practice, rate-$1/n$ and rate-$k/(k+1)$ are generally used, and tables of these codes appear in minimal form; (2) in the rare event that a rate-$k/n$ convolutional code is required, one may use standard sequential-circuit minimization techniques to achieve the minimal encoder realization.

### 4.3.4 Design of Convolutional Codes

Regarding the fourth question listed in Section 4.3, for reasons affecting decoder complexity, typical values of $n$ and $k$ are very small: $k = 1, 2, 3$ and $n = 2, 3, 4$. Typical code rates are 1/2, 1/3, 1/4, 2/3, and 3/4, with 1/2 by far the most common rate. Higher code rates are generally achieved by *puncturing* a lower-rate code, which means that code bits at the encoder output are periodically deleted and hence not transmitted. For example, to achieve a rate-2/3 convolutional code by puncturing a rate-1/2 encoder, one need only puncture every fourth code bit at the encoder output. Since each group of four code bits is produced by two information bits, but only three of the four will be sent, rate 2/3 has been achieved. Convolutional code rates close to unity (e.g., 0.95) are not viable in practice, with or without puncturing, because of the implications for encoder and decoder complexity as well as performance. Tables of puncturing patterns may be found in [3, 12, 13].

As mentioned briefly in Example 4.3, the complexity of the (optimum) convolutional decoder (see Section 4.5) is proportional to $2^\mu$, where $\mu$ is the encoder memory. Thus, typical values of $\mu$ are less than 10, with $\mu = 6$ corresponding to the widely used rate-1/2 convolutional code, which has generator polynomials $g^{(1)}(D) = 1+D+D^2+D^3+D^6$ and $g^{(2)}(D) = 1+D^2+D^3+D^5+D^6$. It is conventional to represent these polynomials as octal numbers, where the coefficients $g_0^{(1)}$ and $g_0^{(2)}$ of $D^0$ correspond to the least-significant bit of the octal numbers. Thus, in octal notation, these generator polynomials are $\mathbf{g}_{\text{octal}}^{(1)} = 117$ and $\mathbf{g}_{\text{octal}}^{(2)} = 155$.

While convolutional codes possess very nice algebraic structure, no satisfactory algebraic design algorithm exists for designing convolutional codes. Code design has been performed in the past via computer search. In the computer-aided-design approach, one usually specifies $n$, $k$, and $\mu$, and chooses a (recursive) systematic encoder realization to ensure that the code is non-catastrophic. Then the computer algorithm runs through an exhaustive list of encoder connections (polynomials), each time examining the resulting code for its distance spectrum. The distance spectra ($d_{\min}$ and beyond, plus multiplicities) are determined via a Viterbi-like algorithm (see Section 4.5 for a discussion of the Viterbi algorithm).

## 4.4 Alternative Convolutional Code Representations

In the previous two sections we presented an algebraic description of convolutional codes, which we have seen is very useful in terms of encoder realization and classification. In this section, we consider alternative representations of convolutional codes that are useful for many other aspects of convolutional codes, such as decoding, code design, and analysis.

### 4.4.1 Convolutional Codes as Semi-infinite Linear Codes

In our algebraic description of a convolutional code, the information word $\mathbf{u}(D) = [u^{(1)}(D) \quad u^{(2)}(D) \quad \ldots \quad u^{(k)}(D)]$ was mapped to a codeword

$$\mathbf{c}(D) = [c^{(1)}(D) \quad c^{(2)}(D) \quad \ldots \quad c^{(n)}(D)]$$

via a generator matrix (or encoder) $\mathbf{G}(D)$, where $u^{(i)}(D)$ and $c^{(j)}(D)$ are elements of the field $\mathbb{F}_2\langle D\rangle$ and the entries $g_i^{(j)}(D)$ of $\mathbf{G}(D)$ are elements of the field $\mathbb{F}_2(D)$ or the ring $\mathbb{F}_2[D]$. It is often useful to take a more system-theoretic view and represent $u^{(i)}(D)$, $c^{(j)}(D)$, and $g_i^{(j)}(D)$ by (generally semi-infinite) vectors of their coefficients, denoted by $\mathbf{u}^{(i)}$, $\mathbf{c}^{(j)}$, and $\mathbf{g}_i^{(j)}$, respectively. Thus, the vector representation of $u^{(i)}(D) = u_0^{(i)} + u_1^{(i)}D + u_2^{(i)}D^2 + \cdots$, is $\mathbf{u}^{(i)} = \left[u_0^{(i)} \quad u_1^{(i)} \quad u_2^{(i)} \quad \ldots\right]$, and similarly for $\mathbf{c}^{(j)}$ and $\mathbf{g}_i^{(j)}$. Under this vector notation, we may write $\mathbf{u} = [\mathbf{u}^{(1)} \quad \mathbf{u}^{(2)} \quad \ldots \quad \mathbf{u}^{(k)}]$ and $\mathbf{c} = [\mathbf{c}^{(1)} \quad \mathbf{c}^{(2)} \quad \ldots \quad \mathbf{c}^{(n)}]$. We would like to determine a generator matrix that is a function of $\left\{\mathbf{g}_i^{(j)}\right\}$.

Since multiplication in the $D$ domain is convolution in the coefficient domain, we may replace $c^{(j)}(D) = \sum_{i=1}^{k} u^{(i)}(D)g_i^{(j)}(D)$ by the convolution

$$\mathbf{c}^{(j)} = \sum_{i=1}^{k} \mathbf{u}^{(i)} \circledast \mathbf{g}_i^{(j)}. \tag{4.7}$$

Given this, one may loosely write $\mathbf{c} = \mathbf{u} \circledast \mathbf{G}$ for some generator matrix $\mathbf{G}$ depending on $\left\{\mathbf{g}_i^{(j)}\right\}$. However, this notation is awkward and not too useful. We can replace the convolution operation in (4.7) with a multiplication operation by employing a matrix that explicitly performs the convolution. Specifically, we have

$$\begin{aligned}\mathbf{u}^{(i)} \circledast \mathbf{g}_i^{(j)} &= \left[u_0^{(i)} \quad u_1^{(i)} \quad u_2^{(i)} \quad \ldots\right] \begin{bmatrix} g_{i,0}^{(j)} & g_{i,1}^{(j)} & g_{i,2}^{(j)} & \cdots & & \\ & g_{i,0}^{(j)} & g_{i,1}^{(j)} & g_{i,2}^{(j)} & \cdots & \\ & & g_{i,0}^{(j)} & g_{i,1}^{(j)} & g_{i,2}^{(j)} & \cdots \\ & & & \ddots & \ddots & \ddots \end{bmatrix} \\ &= \mathbf{u}^{(i)}\mathbf{G}_i^{(j)},\end{aligned}$$

where $\mathbf{u}^{(i)} = \begin{bmatrix} u_0^{(i)} & u_1^{(i)} & u_2^{(i)} & \ldots \end{bmatrix}$, $\mathbf{g}_i^{(j)} = \begin{bmatrix} g_{i,0}^{(j)} & g_{i,1}^{(j)} & g_{i,2}^{(j)} & \ldots \end{bmatrix}$, and

$$\mathbf{G}_i^{(j)} = \begin{bmatrix} g_{i,0}^{(j)} & g_{i,1}^{(j)} & g_{i,2}^{(j)} & \cdots & & \\ & g_{i,0}^{(j)} & g_{i,1}^{(j)} & g_{i,2}^{(j)} & \cdots & \\ & & g_{i,0}^{(j)} & g_{i,1}^{(j)} & g_{i,2}^{(j)} & \cdots \\ & & & \ddots & \ddots & \ddots \end{bmatrix}.$$

We may then replace (4.7) by

$$\begin{aligned} \mathbf{c}^{(j)} &= \sum_{i=1}^{k} \mathbf{u}^{(i)} \mathbf{G}_i^{(j)} \\ &= \mathbf{u}\mathbf{G}^{(j)}, \end{aligned} \tag{4.8}$$

where

$$\mathbf{G}^{(j)} = \begin{bmatrix} \mathbf{G}_1^{(j)} \\ \mathbf{G}_2^{(j)} \\ \vdots \\ \mathbf{G}_k^{(j)} \end{bmatrix}.$$

Finally, we have that encoding in the coefficient domain may be represented by the product

$$\mathbf{c} = \mathbf{u}\mathbf{G}, \tag{4.9}$$

where

$$\mathbf{G} = \begin{bmatrix} \mathbf{G}_1^{(1)} & \mathbf{G}_1^{(2)} & \cdots & \mathbf{G}_1^{(n)} \\ \mathbf{G}_2^{(1)} & \mathbf{G}_2^{(2)} & \cdots & \mathbf{G}_2^{(n)} \\ \vdots & \vdots & \cdots & \vdots \\ \mathbf{G}_k^{(1)} & \mathbf{G}_k^{(2)} & \cdots & \mathbf{G}_k^{(n)} \end{bmatrix}. \tag{4.10}$$

Encoding via (4.9) puts the codeword into the unmultiplexed format $\mathbf{c} = [\mathbf{c}^{(1)} \quad \mathbf{c}^{(2)} \quad \ldots \quad \mathbf{c}^{(n)}]$, analogous to $\mathbf{c}(D) = [c^{(1)}(D) \quad c^{(2)}(D) \quad \ldots \quad c^{(n)}(D)]$. However, in practice, the convolutional encoder outputs are multiplexed in a round-robin fashion, so that one bit is taken from $c^{(1)}(D)$, then $c^{(2)}(D)$, …, then $c^{(n)}(D)$, after which a second bit is taken from $c^{(1)}(D)$, and so on. This leads to the multiplexed codeword format

$$\mathbf{c}' = \begin{bmatrix} c_0^{(1)} & c_0^{(2)} & \ldots & c_0^{(n)} & c_1^{(1)} \, c_1^{(2)} & \ldots & c_1^{(n)} & c_2^{(1)} & c_2^{(2)} & \ldots & c_2^{(n)} & \cdots \end{bmatrix}. \tag{4.11}$$

It will be convenient to also have the multiplexed information word format

$$\mathbf{u}' = \begin{bmatrix} u_0^{(1)} & u_0^{(2)} & \ldots & u_0^{(k)} & u_1^{(1)} & u_1^{(2)} & \ldots & u_1^{(k)} & u_2^{(1)} & u_2^{(2)} & \ldots & u_2^{(k)} & \cdots \end{bmatrix}. \tag{4.12}$$

We may then write

$$\mathbf{c}' = \mathbf{u}'\mathbf{G}',$$

where $\mathbf{G}'$, also called a *generator matrix*, is given by

$$\mathbf{G}' = \begin{bmatrix} g_{1,0}^{(1)} & \cdots & g_{1,0}^{(n)} & g_{1,1}^{(1)} & \cdots & g_{1,1}^{(n)} & g_{1,2}^{(1)} & \cdots & g_{1,2}^{(n)} & \\ \vdots & \ddots & \vdots & \vdots & \ddots & \vdots & \vdots & \ddots & \vdots & \cdots \\ g_{k,0}^{(1)} & \cdots & g_{k,0}^{(n)} & g_{k,1}^{(1)} & \cdots & g_{k,1}^{(n)} & g_{k,2}^{(1)} & \cdots & g_{k,2}^{(n)} & \\ & & & g_{1,0}^{(1)} & \cdots & g_{1,0}^{(n)} & g_{1,1}^{(1)} & \cdots & g_{1,1}^{(n)} & \\ & & & \vdots & \ddots & \vdots & \vdots & \ddots & \vdots & \cdots \\ & & & g_{k,0}^{(1)} & \cdots & g_{k,0}^{(n)} & g_{k,1}^{(1)} & \cdots & g_{k,1}^{(n)} & \\ & & & & & & g_{1,0}^{(1)} & \cdots & g_{1,0}^{(n)} & \\ & & & & & & \vdots & \ddots & \vdots & \cdots \\ & & & & & & g_{k,0}^{(1)} & \cdots & g_{k,0}^{(n)} & \\ & & & & & & & & & \ddots \end{bmatrix} \qquad (4.13)$$

or

$$\mathbf{G}' = \begin{bmatrix} \mathbf{G}_0' & \mathbf{G}_1' & \mathbf{G}_2' & \mathbf{G}_3' & \cdots \\ & \mathbf{G}_0' & \mathbf{G}_1' & \mathbf{G}_2' & \cdots \\ & & \mathbf{G}_0' & \mathbf{G}_1' & \cdots \\ & & & \mathbf{G}_0' & \cdots \\ & & & & \ddots \end{bmatrix}, \qquad (4.14)$$

where

$$\mathbf{G}_l' = \begin{bmatrix} g_{1,l}^{(1)} & \cdots & g_{1,l}^{(n)} \\ \vdots & \ddots & \vdots \\ g_{k,l}^{(1)} & \cdots & g_{k,l}^{(n)} \end{bmatrix}.$$

---

**Example 4.6** We are given the rate-$1/2$ convolutional code with generator matrix $\mathbf{G}(D) = [g^{(1)}(D) \quad g^{(2)}(D)] = [1 + D + D^2 \quad 1 + D^2]$, so that therefore $\mathbf{g}^{(1)} = \left[g_0^{(1)} \quad g_1^{(1)} \quad g_2^{(1)}\right] = [1\ 1\ 1]$ and $\mathbf{g}_2 = \left[g_2^{(0)} \quad g_2^{(1)} \quad g_2^{(2)}\right] = [1\ 0\ 1]$. The generator matrix corresponding to the multiplexed codeword is given by

$$\mathbf{G}' = \begin{bmatrix} g_0^{(1)}g_0^{(2)} & g_1^{(1)}g_1^{(2)} & g_2^{(1)}g_2^{(2)} & & \\ & g_0^{(1)}g_0^{(2)} & g_1^{(1)}g_1^{(2)} & g_2^{(1)}g_2^{(2)} & \\ & & g_0^{(1)}g_0^{(2)} & g_1^{(1)}g_1^{(2)} & g_2^{(1)}g_2^{(2)} \\ & & \ddots & \ddots & \ddots \end{bmatrix}$$

$$= \begin{bmatrix} 11 & 10 & 11 & & \\ & 11 & 10 & 11 & \\ & & 11 & 10 & 11 \\ & & \ddots & \ddots & \ddots \end{bmatrix}.$$

Consider a rate-2/3 convolutional code with

$$\mathbf{G}(D) = \begin{bmatrix} 1+D & 0 & 1 \\ 1+D^2 & 1+D & 1+D+D^2 \end{bmatrix}.$$

From this, we have $\mathbf{g}_1^{(1)} = \begin{bmatrix} g_{1,0}^{(1)} & g_{1,1}^{(1)} & g_{1,2}^{(1)} \end{bmatrix} = [1\ 1\ 0]$,
$\mathbf{g}_2^{(1)} = \begin{bmatrix} g_{2,0}^{(1)} & g_{2,1}^{(1)} & g_{2,2}^{(1)} \end{bmatrix} = [1\ 0\ 1]$, $\mathbf{g}_1^{(2)} = [0\ 0\ 0]$, and so on, so that from (4.13)

$$\mathbf{G}' = \left[\begin{array}{ccccccccccccc}
1 & 0 & 1 & 1 & 0 & 0 & 0 & 0 & 0 & & & & \\
1 & 1 & 1 & 0 & 1 & 1 & 1 & 0 & 1 & & & & \\
 & & & 1 & 0 & 1 & 1 & 0 & 0 & 0 & 0 & 0 & \\
 & & & 1 & 1 & 1 & 0 & 1 & 1 & 1 & 0 & 1 & \\
 & & & & & & 1 & 0 & 1 & 1 & 0 & 0 & \ddots \\
 & & & & & & 1 & 1 & 1 & 0 & 1 & 1 & \\
 & & & & & & & & & 1 & 0 & 1 & \ddots \\
 & & & & & & & & & 1 & 1 & 1 & \\
 & & & & & & & & & & & & \ddots
\end{array}\right].$$

It is even possible for the multiplexed generator-matrix representation to accommodate the recursive systematic encoder

$$\mathbf{G}_{\text{sys}}(D) = \begin{bmatrix} 1 & 0 & \dfrac{1}{1+D} \\ 0 & 1 & \dfrac{D^2}{1+D} \end{bmatrix}$$

of this code. In this case, the following is obvious: $\mathbf{g}_1^{(1)} = [1\ 0\ 0\ \ldots]$,
$\mathbf{g}_2^{(1)} = [0\ 0\ 0\ \ldots]$, $\mathbf{g}_1^{(2)} = [0\ 0\ 0\ \ldots]$, and $\mathbf{g}_2^{(2)} = [1\ 0\ 0\ \ldots]$. Since
$g_{13}(D) = 1/(1+D)$ has the sequence representation 111 ... and
$g_{23}(D) = D^2/(1+D)$ has the sequence representation 00111 ..., it follows that
$\mathbf{g}_1^{(3)} = [1\ 1\ 1\ \ldots]$ and $\mathbf{g}_2^{(3)} = [0\ 0\ 1\ 1\ 1\ \ldots]$. We then have

$$\mathbf{G}' = \left[\begin{array}{ccccccccccccc}
1 & 0 & 1 & 0 & 0 & 1 & 0 & 0 & 1 & 0 & 0 & 1 & \cdots \\
0 & 1 & 0 & 0 & 0 & 0 & 0 & 0 & 1 & 0 & 0 & 1 & \\
 & & & 1 & 0 & 1 & 0 & 0 & 1 & 0 & 0 & 1 & \ddots \\
 & & & 0 & 1 & 0 & 0 & 0 & 0 & 0 & 0 & 1 & \\
 & & & & & & 1 & 0 & 1 & 0 & 0 & 1 & \ddots \\
 & & & & & & 0 & 1 & 0 & 0 & 0 & 0 & \\
 & & & & & & & & & 1 & 0 & 1 & \ddots \\
 & & & & & & & & & 0 & 1 & 0 & \\
 & & & & & & & & & & & & \ddots
\end{array}\right].$$

In Example 4.6, we gave the form of $\mathbf{G}'$ for a specific rate-2/3 RSC encoder. The form of $\mathbf{G}'$ for the general systematic encoder of the form

$$\mathbf{G}_{\text{sys}}(D) = \begin{bmatrix} & g_1^{(n-k+1)}(D) & \cdots & g_1^{(n)}(D) \\ \mathbf{I} & \vdots & \ddots & \vdots \\ & g_k^{(n-k+1)}(D) & \cdots & g_k^{(n)}(D) \end{bmatrix},$$

where each $g_i^{(j)}(D)$ is a rational function in general, is given by

$$\mathbf{G}'_{\text{sys}} = \begin{bmatrix} \mathbf{I}\,\mathbf{P}_0 & \mathbf{O}\,\mathbf{P}_1 & \mathbf{O}\,\mathbf{P}_2 & \cdots & & \\ & \mathbf{I}\,\mathbf{P}_0 & \mathbf{O}\,\mathbf{P}_1 & \mathbf{O}\,\mathbf{P}_2 & \cdots & \\ & & \mathbf{I}\,\mathbf{P}_0 & \mathbf{O}\,\mathbf{P}_1 & \mathbf{O}\,\mathbf{P}_2 & \cdots \\ & & \ddots & \ddots & \ddots & \end{bmatrix},$$

where $\mathbf{I}$ is the $k \times k$ identity matrix, $\mathbf{O}$ is the $k \times k$ all-zero matrix, and $\mathbf{P}_l$ is the $k \times (n-k)$ matrix given by

$$\mathbf{P}_l = \begin{bmatrix} g_{1,l}^{(n-k+1)} & \cdots & g_{1,l}^{(n)} \\ \vdots & \ddots & \vdots \\ g_{k,l}^{(n-k+1)} & \cdots & g_{k,l}^{(n)} \end{bmatrix}.$$

Here, $g_{i,l}^{(j)}$ is the coefficient of $D^l$ in the series expansion of $g_i^{(j)}(D)$.

It is easy to check that the corresponding parity-check matrix $\mathbf{H}'_{\text{sys}}$ is given by

$$\mathbf{H}'_{\text{sys}} = \begin{bmatrix} \mathbf{P}_0^{\mathrm{T}}\,\mathbf{I} & & & & \\ \mathbf{P}_1^{\mathrm{T}}\,\mathbf{O} & \mathbf{P}_0^{\mathrm{T}}\,\mathbf{I} & & & \\ \mathbf{P}_2^{\mathrm{T}}\,\mathbf{O} & \mathbf{P}_1^{\mathrm{T}}\,\mathbf{O} & \mathbf{P}_0^{\mathrm{T}}\,\mathbf{I} & & \\ \mathbf{P}_3^{\mathrm{T}}\,\mathbf{O} & \mathbf{P}_2^{\mathrm{T}}\,\mathbf{O} & \mathbf{P}_1^{\mathrm{T}}\,\mathbf{O} & \mathbf{P}_0^{\mathrm{T}}\,\mathbf{I} & \\ \vdots & \ddots & \ddots & \ddots & \ddots \end{bmatrix}. \tag{4.15}$$

Unless the code rate is 1/2, finding $\mathbf{H}'$ for the non-systematic case is non-trivial. (See [4], Section 2.9, and Problem 4.29.) For the rate-1/2 non-systematic case with $\mathbf{G}(D) = [g^{(1)}(D) \quad g^{(2)}(D)]$, clearly $\mathbf{H}(D) = [g^{(2)}(D) \quad g^{(1)}(D)]$ since $\mathbf{G}(D)\,\mathbf{H}^{\mathrm{T}}(D) = 0$ in this case. The non-multiplexed parity-check matrix is then

$$\mathbf{H} = \left[\mathbf{G}^{(2)} \ \mathbf{G}^{(1)}\right]$$

(so that $\mathbf{G}\mathbf{H}^{\mathrm{T}} = \mathbf{0}$), where, for $j = 1, 2$,

$$\mathbf{G}^{(j)} = \begin{bmatrix} g_0^{(j)} & g_1^{(j)} & g_2^{(j)} & \cdots & & \\ & g_0^{(j)} & g_1^{(j)} & g_2^{(j)} & \cdots & \\ & & g_0^{(j)} & g_1^{(j)} & g_2^{(j)} & \cdots \\ & & \ddots & \ddots & \ddots & \ddots \end{bmatrix}.$$

The multiplexed version of the parity-check matrix is then

$$\mathbf{H}' = \begin{bmatrix} \mathbf{H}'_0 & \mathbf{H}'_1 & \cdots & \mathbf{H}'_m & & \\ & \mathbf{H}'_0 & \mathbf{H}'_1 & \cdots & \mathbf{H}'_m & \\ & & \ddots & & & \ddots \end{bmatrix}, \tag{4.16}$$

where

$$\mathbf{H}'_l = \begin{bmatrix} g_l^{(2)} & g_l^{(1)} \end{bmatrix}$$

and $m$ is the maximum degree of $g^{(1)}(D)$ and $g^{(2)}(D)$. For the general non-systematic case, that is, for code rates other than 1/2, one usually puts $\mathbf{G}(D)$, and hence $\mathbf{G}$, in systematic form, from which the parity-check matrix $\mathbf{H}'_{\text{sys}}$ given by (4.15) follows.

---

**Example 4.7** Consider again the rate-1/2 convolutional code with generator matrix $\mathbf{G}(D) = [1 + D + D^2 \quad 1 + D^2]$. We have $\mathbf{g}^{(1)} = \begin{bmatrix} g_0^{(1)} & g_1^{(1)} & g_2^{(1)} \end{bmatrix} = [1\ 1\ 1]$ and $\mathbf{g}^{(2)} = \begin{bmatrix} g_0^{(2)} & g_1^{(2)} & g_2^{(2)} \end{bmatrix} = [1\ 0\ 1]$, so that

$$\mathbf{H} = \begin{bmatrix} 101 & & & 111 & & \\ & 101 & & & 111 & \\ & & 101 & & & 111 \\ & & & \ddots & & & \ddots \end{bmatrix}$$

and

$$\mathbf{H}' = \begin{bmatrix} 11 & 01 & 11 & & \\ & 11 & 01 & 11 & \\ & & 11 & 01 & 11 \\ & & & \ddots & \ddots & \ddots \end{bmatrix}.$$

---

### 4.4.2 Graphical Representations for Convolutional Code Encoders

There exist several graphical representations for the encoders of convolutional codes, with each of these representations playing different roles. It is sufficient to describe the graphical models for our archetypal encoder, specifically, the rate-1/2 encoder described by $\mathbf{G}(D) = [1 + D + D^2 \quad 1 + D^2]$. The procedure for deriving graphical models for other convolutional codes is identical.

We start with the *finite-state transition-diagram* (FSTD), or *state-diagram*, graphical model for $\mathbf{G}(D)$. The encoder realization for $\mathbf{G}(D)$ was presented earlier in Figure 4.4(a), and the *encoder state* is defined to be the contents of the two memory elements in the encoder circuit (read from left to right). Observe that the number of states is $2^{\mu}$, where $\mu$ is the memory of the encoder. From that figure and the state definition, we may produce the state-transition table below, from which the code's FSTD may easily be drawn, as depicted in Figure 4.8(a). The state diagram is useful

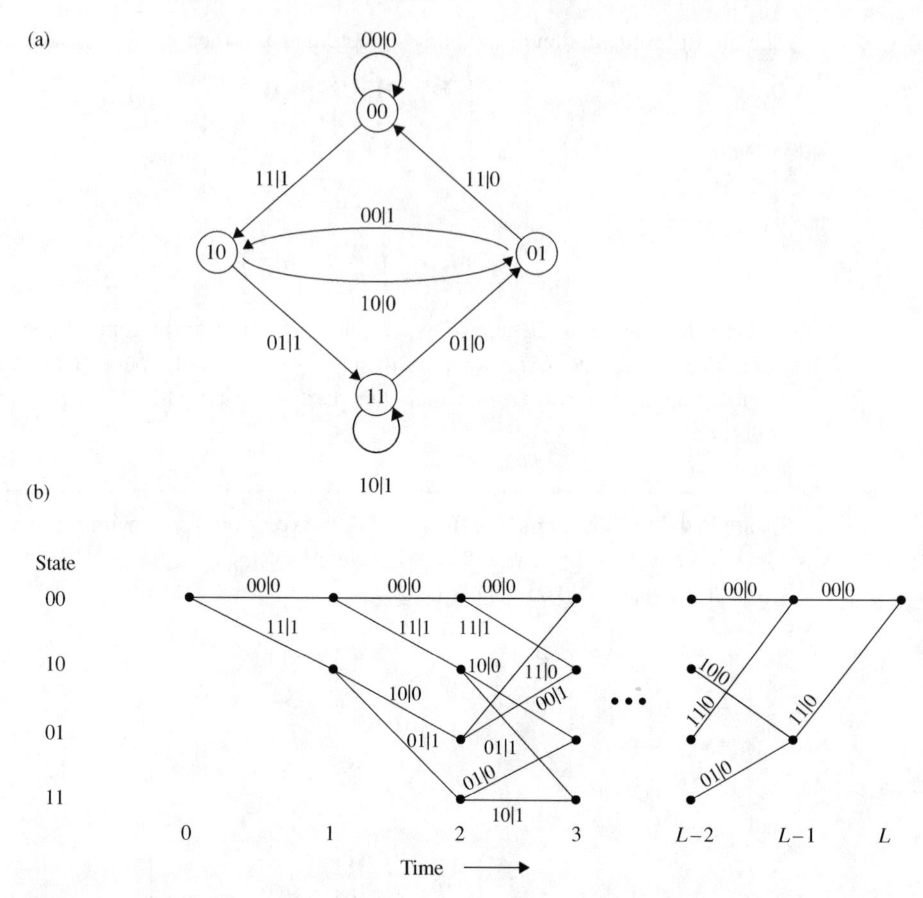


**Figure 4.8** (a) FSTD for the encoder of Figure 4.4(a). The edge labels are $c_1(i)c_2(i)|u(i)$ or output|input. (b) A trellis diagram for the encoder of Figure 4.4(a) for input length $L$ and input containing two terminating zeros.

for analytically determining the distance spectrum of a convolutional code and it also leads directly to the code's *trellis diagram*, which is simply the state diagram with the dimension of discrete time added in the horizontal direction.

| Input $u_i$ | Current state $u_{i-1}u_{i-2}$ | Next state $u_iu_{i-1}$ | Output $c_i^{(1)}c_i^{(2)}$ |
|---|---|---|---|
| 0 | 00 | 00 | 00 |
| 1 | 00 | 10 | 11 |
| 0 | 01 | 00 | 11 |
| 1 | 01 | 10 | 00 |
| 0 | 10 | 01 | 10 |
| 1 | 10 | 11 | 01 |
| 0 | 11 | 01 | 01 |
| 1 | 11 | 11 | 10 |

The trellis diagram for our example code is presented in Figure 4.8(b); it has been drawn under the assumption that the encoder starts in the 00 state, as is usually the case in practice. Note that the trellis stages between times $i$ and $i + 1$, $i \geq 2$, are replicas of the state diagram. We remark that a list of code sequences of any given length may be obtained from the trellis by tracing through all possible trellis paths corresponding to that length, along the way picking off the code symbols that label the *branches* (or *edges*) within each path.

When we consider trellis-based decoding in the next subsection, we will see that, unless the final trellis state is known, the last several decoded bits (about $5\mu$ bits) will be somewhat unreliable. To counter this circumstance, the system designer will frequently *terminate* the trellis to a known state. Clearly this necessitates the appending of $\mu$ bits to the information sequence, resulting in an increase in coding overhead (or, equivalently, a decrease in code rate). These $\mu$ appended bits are usually zeros so that the encoder starts and ends in the zero state. These codes are called *zero-terminated convolutional codes* (ZTCC). For cases in which the message is long relative to $\mu$, the apparent wastefulness of the appended zeros is negligible.

A clever way to circumvent this short-message inefficiency while still maintaining the advantage of trellis termination is via *tail-biting*. The encoder of a tail-biting convolutional code (TBCC) assumes the $\bar{\text{R}}$SC or $\bar{\text{R}}\bar{\text{S}}$C forms and initializes the encoder state with the final $\mu$ bits in the information sequence so that the initial and the final encoder states are identical. Note that this has the effect of rendering both the initial and the final encoder state unknown (albeit identical). However, with a little increase in decoder complexity, the decoder can be made to be near optimal and, in particular, the reliability of all of the decoded bits is made more uniform. A decoder for a tail-biting convolutional code will be discussed in the next section after the development of the decoder for ZTCCs.

## 4.5 Decoders for Convolutional Codes

We discuss in this section several optimum and suboptimum decoders. Foremost among them is the Viterbi algorithm (VA) decoder, which is a maximum-likelihood *sequence* decoder (MLSD) for trellis-based codes. Also very important is the BCJR algorithm, which is an efficient *bit-wise* maximum *a posteriori* (MAP) decoder for trellis-based codes. The BCJR algorithm is a necessary ingredient for turbo codes and other iteratively decoded concatenated codes. Also contained here is an adaptation of the VA to tail-biting convolutional codes called the wrap-around Viterbi algorithm (WAVA). Another important decoder is the list Viterbi algorithm (LVA), which extends the power of a convolutional code through the use of an outer error-detection code code, typically a CRC code. While the above decoders all rely on a trellis representation of a convolutional code, we present here the Fano algorithm decoder, which is an extremely low-complexity near-optimum decoder that is based on a tree represention. Lastly, this section presents the differential Viterbi algorithm, which is a lower-complexity alternative to the VA that is applicable to a subset of convolutional codes in common use.

### 4.5.1 MLSD and the Viterbi Algorithm

Our focus is the binary symmetric channel (BSC) and the binary-input AWGN channel (BI-AWGNC). For the sake of uniformity, for both channels we denote the $i$th channel input by $x_i$ and the $i$th channel output by $y_i$. Given channel input $x_i = c_i \in \{0, 1\}$ and channel output $y_i \in \{0, 1\}$, the BSC has channel transition probabilities $P(y_i \mid x_i)$ given by

$$P(y_i \neq x \mid x_i = x) = \varepsilon,$$
$$P(y_i = x \mid x_i = x) = 1 - \varepsilon,$$

where $\varepsilon$ is the crossover probability. (For simplicity of notation, we renumber the code bits in (4.11) so that the superscripts may be dropped.) For the BI-AWGNC, the code bits are mapped to the channel inputs as $x_i = (-1)^{c_i} \in \{\pm 1\}$. The BI-AWGNC has the channel transition probability density function (pdf) $p(y_i \mid x_i)$ given by

$$p(y_i \mid x_i) = \frac{1}{\sqrt{2\pi}\sigma} \exp\left[-(y_i - x_i)^2 / (2\sigma^2)\right],$$

where $\sigma^2$ is the variance of the zero-mean Gaussian-noise sample $n_i$ that the channel adds to the transmitted value $x_i$ (so that $y_i = x_i + n_i$). Considering first the BSC, the maximum-likelihood (ML) decision is

$$\hat{\mathbf{c}} = \arg\max_{\mathbf{c}} P(\mathbf{y} \mid \mathbf{x}),$$

which was shown in Chapter 1 to reduce to

$$\hat{\mathbf{c}} = \arg\min_{\mathbf{c}} d_{\mathrm{H}}(\mathbf{y}, \mathbf{x}),$$

where $d_{\mathrm{H}}(\mathbf{y}, \mathbf{x})$ represents the Hamming distance between $\mathbf{y}$ and $\mathbf{x}$ (note that $\mathbf{x} = \mathbf{c}$). For the BI-AWGNC, the ML decision is

$$\hat{\mathbf{c}} = \arg\max_{\mathbf{c}} p(\mathbf{y} \mid \mathbf{x}),$$

which was shown in Chapter 1 to reduce to

$$\hat{\mathbf{c}} = \arg\min_{\mathbf{c}} d_{\mathrm{E}}(\mathbf{y}, \mathbf{x}) = \arg\min_{\mathbf{c}} d_{\mathrm{E}}^2(\mathbf{y}, \mathbf{x}),$$

where $d_{\mathrm{E}}(\mathbf{y}, \mathbf{x})$ represents the Euclidean distance between $\mathbf{y}$ and $\mathbf{x}$ (note that $\mathbf{x} = (-1)^{\mathbf{c}}$). Owing to monotonicity, $d_{\mathrm{E}}^2(\mathbf{y}, \mathbf{x})$ may be used for lower complexity.

Thus, for the BSC (BI-AWGNC), the MLSD chooses the code sequence $\mathbf{c}$ that is closest to the channel output $\mathbf{y}$ in a Hamming (Euclidean) distance sense. In principle, one could use the trellis to exhaustively search for the sequence that is closest to $\mathbf{y}$, since the trellis enumerates all of the code sequences. The computation for $L$ trellis branches would be

$$\Gamma_L = \sum_{l=1}^{L} \lambda_l, \tag{4.17}$$

where $\lambda_l$ is the $l$th *branch metric*, given by

$$\lambda_l = \sum_{j=1}^{n} d_{\mathrm{H}}\left(y_l^{(j)}, x_l^{(j)}\right) = \sum_{j=1}^{n} y_l^{(j)} \oplus x_l^{(j)} \quad \text{for the BSC,} \tag{4.18}$$

$$\lambda_l = \sum_{j=1}^{n} d_{\mathrm{E}}^2\left(y_l^{(j)}, x_l^{(j)}\right) = \sum_{j=1}^{n} \left(y_l^{(j)} - x_l^{(j)}\right)^2 \quad \text{for the BI-AWGNC,} \tag{4.19}$$

where we have returned to the notation of (4.11) so that $x_l^{(j)} = c_l^{(j)}$ for the BSC and $x_l^{(j)} = (-1)^{c_l^{(j)}}$ for the BI-AWGNC. The metric in (4.18) is called a *Hamming metric* and the metric in (4.19) is called a *(squared) Euclidean metric*.

Clearly, the number of computations implied by both $\arg\min_{\mathbf{c}} d_{\mathrm{H}}(\mathbf{y}, \mathbf{x})$ and $\arg\min_{\mathbf{c}} d_{\mathrm{E}}(\mathbf{y}, \mathbf{x})$ is enormous, since $L$ trellis stages imply $2^{k(L-\mu)}$ code sequences for a rate-$k/(k+1)$ ZTCC code. However, Viterbi [4] noticed that the complexity can be reduced to $O(2^{\mu})$ with no loss in optimality under the following observation. Consider the two code sequences in Figure 4.8(b) that begin with [00 00 00 . . . ] and [11 10 11 . . . ], corresponding to input sequences [0 0 0 . . . ] and [1 0 0 . . . ]. Observe that the trellis paths corresponding to these two sequences diverge at state 0/time 0 and remerge at state 0 after three branches (equivalently, after three trellis stages). The significance of this remerging after three stages is that thereafter the two paths have identical extensions into the latter trellis stages. Thus, if one of these two paths possesses the superior *cumulative metric* $\Gamma_l$ at time $l = 3$, then that path plus its extensions will also possess superior cumulative metrics. Consequently, we may remove the inferior path from further consideration at time $l = 3$. The path that is maintained is called the *survivor*. This argument applies to all merging paths in the trellis and to trellises with more than two merging paths per trellis node, and it leads us to the *Viterbi-algorithm* implementation of the MLSD.

## Decision Stage

There are several ways to choose the ML code sequence (hence, ML information sequence) using the Viterbi decoder path metrics. We list a few here assuming ZTCCs. See Algorithm 4.2 in Section 4.5.3 for a detailed description of the Viterbi decoder at the *traceback decision stage*.

*Block-oriented approach.* Because the code is assumed to be zero terminated, the Viterbi decoder is designed to account for this fact (that is, with the knowledge that the last $\mu$ encoder inputs are zero) so that there will be a single survivor at stage $L$ in the trellis and that survivor will be in the zero state. The branch labels (in particular, the encoder input labels) along this surviving path give the decoder output. Note that it is assumed that $L$ is small enough for the encoder memory to accommodate all of the information associated with the $2^{\mu}$ survivors in a typical trellis stage.

**Algorithm 4.1** The Viterbi Algorithm: Add–Compare–Select Iterations

DEFINITIONS

1. $\lambda_l(s', s)$ = branch metric from state $s'$ at time $l-1$ to state $s$ at time $l$, where $s', s \in \{0, 1, \ldots, 2^\mu - 1\}$.
2. $\Gamma_{l-1}(s')$ = cumulative metric for the survivor state $s'$ at time $l-1$; the sum of branch metrics for the surviving path.
3. $\Gamma_l(s', s)$ = tentative cumulative metric for the paths extending from state $s'$ at time $l-1$ to state $s$ at time $l$; $\Gamma_l(s', s) = \Gamma_{l-1}(s') + \lambda_l(s', s)$.

ADD–COMPARE–SELECT ITERATION

**Initialization**

Set $\Gamma_0(0) = 0$ and $\Gamma_0(s') = -\infty$ for all $s' \in \{1, \ldots, 2^\mu - 1\}$. (The encoder, and hence the trellis, is initialized to state 0.)

**for** $l = 1$ to $L$ **do**

Compute the possible branch metrics $\lambda_l(s', s)$.

For each state $s'$ at time $l-1$ and all possible states $s$ at time $l$ that may be reached from $s'$, compute the tentative cumulative metrics $\Gamma_l(s', s) = \Gamma_{l-1}(s') + \lambda_l(s', s)$ for the paths extending from state $s'$ to state $s$.

For each state $s$ at time $l$, select and store the path possessing the minimum among the metrics $\Gamma_l(s', s)$. The cumulative metric for state $s$ will be $\Gamma_l(s) = \min_{s'}\{\Gamma_l(s', s)\}$. (See Figure 4.9 for a depiction of these steps.)

**end for**

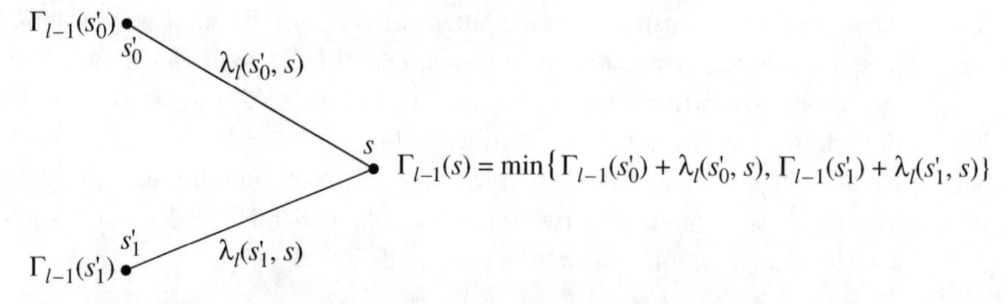


**Figure 4.9** Depiction of Steps 2 and 3 of the ACS operation of the Viterbi algorithm (Algorithm 4.1) for a rate-$1/n$ code and a distance metric. For a rate-$k/(k+1)$ code, $2^k$ branches merge into each state. For the correlation metric discussed below, "min" becomes "max."

*Stream-oriented approach 1*. This approach exploits the fact that survivors tend to have a common "tail," as demonstrated in Figure 4.10. After the $l$th ACS iteration, the decoder traces back $\delta$ branches along an *arbitrary surviving path* and produces as its output decision the information bit(s) that label the branch at the $(l-\delta)$th trellis stage. This is effectively a *sliding-window decoder* that stores and processes information within $\delta$ trellis stages as it slides down the length of the trellis. The value of the *decoding depth* $\delta$ (also called *decoding delay*) used is usually quoted to be on the order of $5\mu$. However, it is prudent to employ computer simulations to determine $\delta$ since the appropriate value is code dependent. Alternatively, one can run a computer

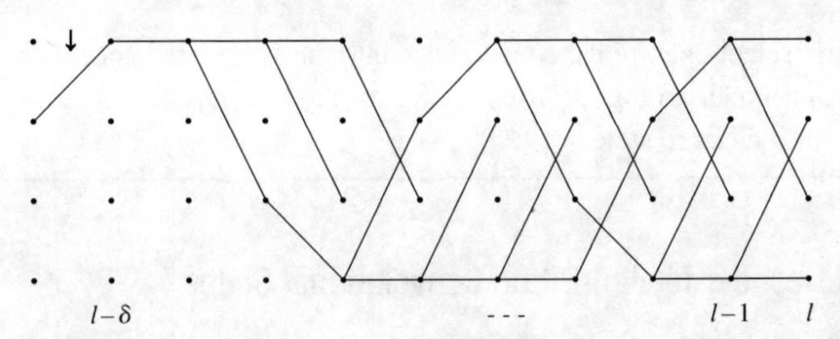


**Figure 4.10** An example of survivors sharing a common tail in a four-state trellis. Because of this tail, a decision may be made for the $(l - \delta)$th trellis stage, in accordance with the label on the trellis branch indicated by the arrow.

search algorithm over the code's trellis to determine the maximum path length among the nonzero code sequences that diverge from the all-zeros path at state 0/time 0 and remerge later, and whose weights are less than some threshold. The decoding delay $\delta$ would be set to a value greater than this maximum path length.

*Stream-oriented approach 2*. After the $l$th ACS iteration, rather than choosing an arbitrary path, choose the trellis path with the *best cumulative metric* and *trace back* $\delta$ branches. The corresponding output decision is selected as the bit(s) that label the branch at the $(l - \delta)$th trellis stage.

The Euclidean distance metric given in (4.19) for the BI-AWGNC can be simplified as follows. Observe that

$$\begin{aligned}
\arg\min_{\mathbf{c}} d_{\mathrm{E}}^2(\mathbf{y},\mathbf{x}) &= \arg\min_{\mathbf{c}} \sum_{l=1}^{L}\sum_{j=1}^{n} \left(y_l^{(j)} - x_l^{(j)}\right)^2 \\
&= \arg\min_{\mathbf{c}} \sum_{l=1}^{L}\sum_{j=1}^{n} \left[\left(y_l^{(j)}\right)^2 + \left(x_l^{(j)}\right)^2 - 2y_l^{(j)}x_l^{(j)}\right] \\
&= \arg\max_{\mathbf{c}} \sum_{l=1}^{L}\sum_{j=1}^{n} y_l^{(j)}x_l^{(j)},
\end{aligned}$$

where the last line follows since the squared terms on the second line are independent of $\mathbf{c}$. Thus, the AWGN branch metric in (4.19) can be replaced by the *correlation metric*

$$\lambda_l = \sum_{j=1}^{n} y_l^{(j)}x_l^{(j)} \qquad (4.20)$$

under the BI-AWGNC assumption. Further, the min operation is replaced by the max operation in the ACS iteration.

---

**Example 4.8** Figure 4.11 presents a Viterbi decoding example on the BSC using the block-oriented decision approach and the Hamming distance metric. Figure 4.12 presents a Viterbi decoding example on the BI-AWGNC using the block-oriented decision approach and the correlation metric. Both examples are for a four-state ZTCC whose initial state is the zero state and both only demonstrate the add–compare–select operations; convergence into the zero state is not shown. An actual Viterbi decoder arranges for the metrics of the last $\mu$ trellis stages to "force" the survivor into the zero state.

---

### 4.5.2 The WAVA Decoder for Tail-Biting Convolutional Codes

The TBCC decoding algorithm we present here is the *wrap-around Viterbi algorithm* (WAVA) introduced in [8]. Recall, for tail-biting convolutional codes, the initial and

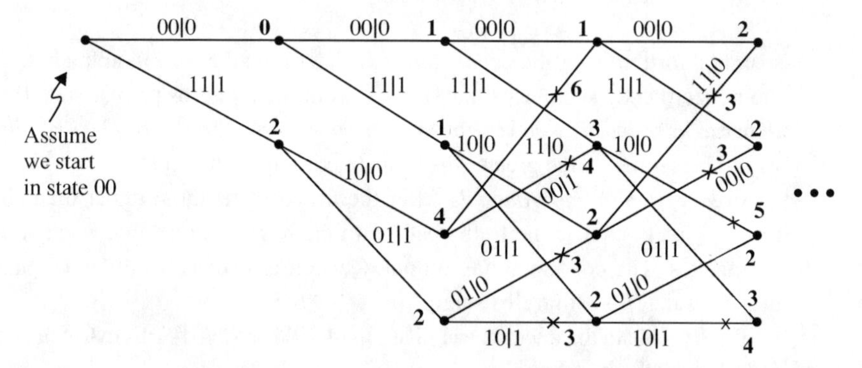


**Figure 4.11** A Viterbi decoding example for the code of Figure 4.8 on the BSC. Here $\mathbf{y} = [00, 01, 00, 01]$. Non-surviving paths are indicated by an "X" and cumulative metrics are written in bold near merging branches. For longer codewords, the algorithm continues in this manner until the $L$th stage is reached, at which time the ML path is the lone survivor in state 0.

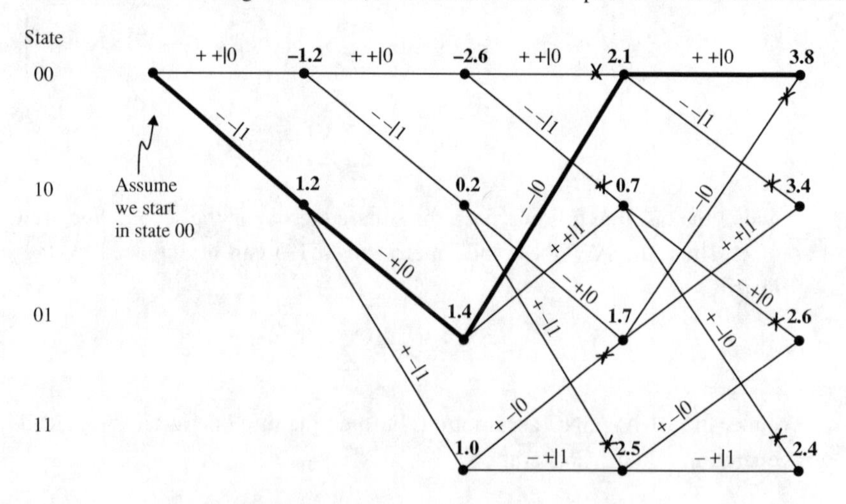


**Figure 4.12** A Viterbi decoding example for the code of Figure 4.8 on the BI-AWGNC using the correlation metric. The received word is $\mathbf{y} = [(-0.7, -0.5), (-0.8, -0.6), (-1.1, +0.4), (+0.9, +0.8)]$. Non-surviving paths are indicated by an "X" and cumulative metrics are written in bold near merging branches. If forced to make a decision at this point, the ML path is the one with the cumulative metric of 3.8.

final trellis states are identical but unknown. Consequently, the TBCC decoding algorithm must obtain an estimate of the proper state-metric initialization. The WAVA algorithm forces multiple passes through the trellis in an attempt to obtain an appropriate state metric initialization and to simultaneously find the tail-biting trellis path with the best path metric (if one exists). To do this, for each trellis pass, the WAVA uses as initial state metrics the final cumulative metrics from the last trellis pass, except on the first pass where all of the state metrics are initialized to zero (or an arbitrary constant). At the end of each pass, the algorithm looks for survivors that are tail-biting and chooses the tail-biting path with the largest cumulative metric. Observe that it is possible after a pass for no tail-biting paths to exist, that is, it is possible that no final state equals the path's initial state. If so, the algorithm proceeds to the next pass, again using as initial state metrics the final state metrics from the just completed pass. A maximum of three passes through the trellis is usually sufficient to obtain near maximum-likelihood performance, although the algorithm usually completes in one or two passes.

Observe that, in each trellis pass, the WAVA in effect computes a probability mass function (pmf) on the final (and therefore, initial) trellis states and uses that probabilistic information to initialize the trellis states for the next pass. For comparison, for ZTCCs, the initialization pmf is a unit delta function situated at state 0 so that the probabilities for all nonzero states are all zero.

To summarize, the WAVA decoding algorithm is as follows.

1. Initialize all state metrics to zero and run the standard Viterbi algorithm on the full trellis starting from all states and ending in all states.
2. Find the surviving path at the end of the trellis with the best metric. If this path is also tail-biting, this is the decoder output. (Note this requires that each path record its initial state.)
3. After the first iteration, run the standard Viterbi algorithm on the full trellis as in Step 1, but start from (continue from) the final state metrics from the previous trellis pass. Find the tail-biting paths among the $2^\mu$ survivors and choose as the decoder output the tail-biting path with the largest cumulative metric. If no tail-biting path exists, repeat this step, unless the preset maximum number of iterations has been reached.

---

**Example 4.9** As discussed, tail-biting convolutional codes are more efficient than zero-terminated convolutional codes because they avoid terminating the trellis with $\mu$ dummy zeros. Here, we consider a memory $\mu = 8$, rate-1/2 convolutional code with binary encoder taps given by the pair of octal numbers $(561, 753)_8$, with the generator matrix $\mathbf{G}(D) = [1 + D^2 + D^3 + D^4 + D^8 \quad 1 + D + D^2 + D^3 + D^5 + D^7 + D^8]$. Consider the message length $K = 100$ bits. The codeword length for the ZTCC will be $N = 2(100 + 8) = 216$, whereas the codeword length for the TBCC will be $N = 2(100) = 200$, giving code rates $R_{ZT} = 0.463$ and $R_{TB} = 1/2$, respectively. We will learn later in this chapter how to estimate the performance difference between the maximum-likelihood decoders for these codes and that it is given by

$10\log_{10}(R_{TB}/R_{ZT}) = 0.33$ dB. Indeed, our simulations have shown a block-error rate of 0.0011 at $E_b/N_0 = 3.3$ dB for the ZTCC with a Viterbi decoder and at $E_b/N_0 = 3$ dB for the TBCC with a WAVA decoder. This agrees with the code rate loss value, 0.33 dB. As another example, for a message length of $K = 500$ bits, the *code rate loss* would be $10\log_{10}(R_{TB}/R_{ZT}) = 10\log_{10}(1016/1000) = 0.069$ dB.

### 4.5.3 The List Viterbi Algorithm

The weight spectrum, hence error-rate performance, of convolutional codes improves with increasing encoder memory size $\mu$. However, the goal of attaining improved performance using this strategy is problematic if the decoder is required to be a standard Viterbi decoder whose complexity goes as $2^{\mu}$. There are two ways around this issue of long-memory convolutional codes and an exponentially growing number of states. The first is to use the Fano algorithm described in Section 4.5.5. The Fano route works well if the message lengths are not too short (Fano only works with ZTCCs) and not too long (periodically inserted pilot bits are necessary for long messages). The second is to use the list Viterbi algorithm on convolutional codes having generator matrices of the form

$$\mathbf{G}_{cat}(D) = \left[g(D)g_1(D) \;\; g(D)g_2(D)\right].$$

An explanation is required here because in Section 4.3.2 we advised against using generator matrices of this form to avoid catastrophic error propagation. There are two situations in which this form is not a problem. The first is if the message length $K$ is short enough, say less than 1000 bits, a Viterbi decoder can be designed to store entire path histories of length $K+\mu$, that is, for the block-oriented Viterbi decoding algorithm presented earlier. This will work for ZTCCs because the trellis termination eliminates catastrophic error sequences that otherwise might have survived in a stream-oriented Viterbi decoder with an insufficient decoding depth. It will not work for TBCCs, short message or not (as far as we know).

The second situation in which the generator matrix form $\mathbf{G}_{cat}(D)$ is not a problem is when the *list Viterbi algorithm* (LVA) is employed by the decoder. As we will detail shortly, the LVA consists of an augmented Viterbi decoder that produces a likelihood-ordered list of candidate codewords and an outer CRC code that goes down the ordered list and chooses as the LVA decoder output the first candidate that passes the CRC check.[1] In the context of $\mathbf{G}_{cat}(D)$, the LVA is designed for the trellis of the code with generator $\mathbf{G}(D) = [g_1(D) \;\; g_2(D)]$ and the polynomial $g(D)$ is used as the generator for the CRC code. Thus, after CRC encoding, the words entering the $\mathbf{G}(D)$ convolutional encoder are multiples of $g(D)$ so that the two streams at the output of the $\mathbf{G}(D)$ encoder are multiples of $\mathbf{G}_{cat}(D) = \left[g(D)g_1(D) \;\; g(D)g_2(D)\right]$. Thus, the LVA is

[1] The original LVA of [7] did not mention a CRC code, but all of the more recent publications ([14]–[19]) on the LVA include a CRC code, so that will be our assumption throughout.

a decoder for a concatenated coding system with a CRC code as the outer code and a convolutional code as the inner code.

Observe that the list decoder just described will only choose as its output words that are a multiple of both $g(D)$ and $\mathbf{G}(D) = [g_1(D)\ g_2(D)]$, that is, words that are a multiple of $\mathbf{G}_{cat}(D)$. Thus, assuming the list is sufficiently long, we can expect the LVA performance with the setup described to have the performance of the large-memory code $\mathbf{G}_{cat}(D)$ rather than the small-memory code $\mathbf{G}(D)$. Note also that the trellis involved is that of the small-memory code, thus reducing complexity. The complexity is reduced even though the trellis decoder is now a list decoder, and this is particularly true because the average list size is typically less than 2 for the adaptive list decoder that we will discuss.

The list Viterbi algorithm was first introduced in [7], where the authors proposed a *parallel LVA* and a *serial LVA*. Here, we present only the parallel LVA because it is much simpler to describe and implement, and because it can form the core of a hybrid parallel–serial LVA, described below, that has near-serial LVA average list size. The *hybrid LVA* has also been called an *adaptive LVA*.

The parallel LVA is a straightforward extension of the standard Viterbi algorithm, the difference being that, rather than a single survivor at each state in the trellis there will be $\mathcal{L}$ survivors at each state. Because of the close connection between the VA and the parallel LVA, we review here the VA, following the description in [7]. The discussion will be for a rate-1/2 ZTCC; extension to other code rates is straightforward. An extension to TBCCs will be presented later.

We denote by $\mathcal{S}$ the set of trellis states, $\mathcal{S} = \{0, 1, \ldots, 2^\mu - 1\}$, and by $\mathcal{S}(s)$ the left-neighborhood of state $s$ (trellis states leading to state $s$), where $|\mathcal{S}(s)| = 2$ for rate-1/2 convolutional codes. We denote by $L$ the length of the trellis in stages so that the discrete-time variable $l$ takes on values in the set $\{0, 1, 2, \ldots, L\}$. As before, $\lambda_l(s', s)$ is the branch metric from state $s'$ at time $l-1$ to state $s$ at time $l$ and $\Gamma_l(s)$ is the cumulative metric for state $s$ at time $l$. We assume the correlation metric so that $\lambda_l(s', s) = -\infty$ if $s'$ and $s$ are not connected. Considering the surviving path for state $s$ at time $l$, we define the *state history* array $h_l(s)$ to be the state at time $l-1$ from which the survivor into state $s$ at time $l$ came. Then the Viterbi algorithm for ZTCCs (with *traceback*) is presented in Algorithm 4.2.

At each discrete time instant $l$, the standard VA chooses the best path between two competing paths entering each state. The parallel LVA with list size $\mathcal{L}$, at each discrete time instant $l$, chooses the $\mathcal{L}$ best paths among the $2\mathcal{L}$ competing paths entering each state. (See Figure 4.13.) For this reason, for the parallel LVA, two of the metric arrays above need an $\mathcal{L}$-fold expansion. Thus, let $\Gamma_l(s, r)$ be the $r$th-best cumulative metric among the surviving paths at state $s$ at time $l$, where the *ranking* variable $r$ takes on values in the set $\mathcal{R} = \{1, 2, \ldots, \mathcal{L}\}$. Similarly, let $h_l(s, r)$ hold the state at time $l-1$ of the $r$th-best path into state $s$ at time $l$. Lastly, $\rho_l(s, r)$ will be the ranking of that $r$th-best path into state $s$ at time $l$. The parallel LVA can now be presented as in Algorithm 4.3 below (where $\max^{(r)}$ returns the $r$th-largest value, so that the parallel LVA necessitates a sorting algorithm).

**Algorithm 4.2** The Viterbi Algorithm for ZTCCs with Traceback

1. Initialization: $l = 0, s = 0$

   $\Gamma_0(0) = 0$ and $\Gamma_0(s) = -\infty$ for all $s \in \mathcal{S} \setminus \{0\}$

   $h_0(s) = 0$ for all $s \in \mathcal{S}$.
2. ACS recursion: For $l = 1, 2, \ldots, L-1$, for all $s \in \mathcal{S}$

   $$\Gamma_l(s) = \max_{s' \in \mathcal{S}(s)} [\Gamma_{l-1}(s') + \lambda_l(s', s)]$$

   $$h_l(s) = \arg\max_{s' \in \mathcal{S}(s)} [\Gamma_{l-1}(s') + \lambda_l(s', s)].$$
3. Termination: $l = L, s = 0$

   $$\Gamma_L(0) = \max_{s' \in \mathcal{S}(s)} [\Gamma_{L-1}(s') + \lambda_L(s', 0)]$$

   $$h_L(0) = \arg\max_{s' \in \mathcal{S}(s)} [\Gamma_{L-1}(s') + \lambda_L(s', 0)].$$
4. Traceback:

   With $s_0 = 0$ and $s_L = 0$, the maximum-likelihood state sequence is $(0, s_1, s_2, \ldots, s_{L-1}, 0)$, where $s_l = h_{l+1}(s_{l+1})$ for $l = L-1, \ldots, 2, 1$.

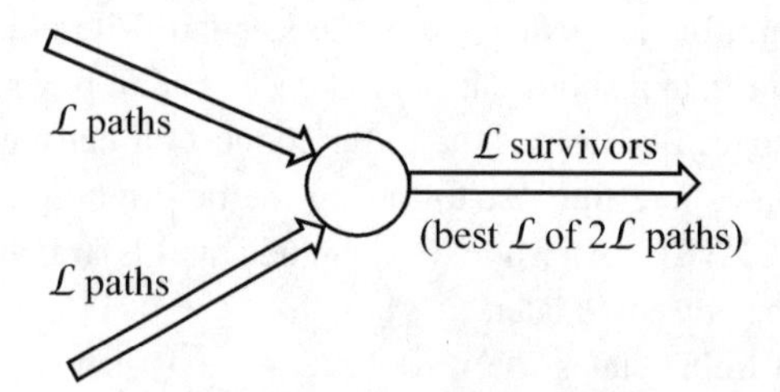


**Figure 4.13** Illustration of the $\mathcal{L}$-fold add–compare–select operation in a parallel list Viterbi algorithm. Compare to Figure 4.9. (Adapted from [7].)

We see in the traceback step the need for the ranking sequence array $\rho_l(s, r)$. Specifically, the $r$th-best path at time $l$ and state $s$ may have come from a path at time $l-1$ and state $h_l(s, r)$ whose ranking $\rho_l(s, r)$ is not equal to $r$. Thus, we need to store the rankings along with the states in order to properly recover the $\mathcal{L}$ surviving sequences at time $L$.

An outer error-detection code is necessary to complete the list Viterbi algorithm decoder. Thus, prior to the ZTCC encoder is a CRC encoder. After the parallel LVA produces its ordered list of $\mathcal{L}$ candidate messages (which typically contains the CRC codeword), the CRC decoder goes down the list, from most likely to least likely, until one of the decoded messages passes the CRC check or the list is exhausted. There are two failure modes: the correct CRC codeword never appears on the list or a valid but incorrect CRC codeword appears on the list. If the list size $\mathcal{L}$ is sufficiently large or the noise is sufficiently low, the correct message is very likely to be in the list. Further, if the CRC code is properly designed, an incorrect message is unlikely to be chosen before the correct one.

*Modification of the Parallel LVA for TBCCs*. Because the interest in the LVA is for message lengths of 1000 bits or less, it is natural to consider applying list decoding to tail-biting convolutional codes. Unsurprisingly, the modification of the parallel LVA

**Algorithm 4.3** The Parallel List Viterbi Algorithm for ZTCCs

1. Initialization: $l = 0, s = 0$

   $\Gamma_0(0, r) = 0$ and $\Gamma_0(s) = -\infty$ for all $s \in \mathcal{S} \setminus \{0\}$ and for all $r \in \mathcal{R}$

   $h_0(s, r) = 0$ for all $s \in \mathcal{S}$ and for all $r \in \mathcal{R}$.

2. ACS recursion: For $l = 1, 2, \ldots, L-1$, for all $s \in \mathcal{S}$

   $\Gamma_l(s, r) = \max^{(r)}_{s' \in \mathcal{S}(s), r \in \mathcal{R}} [\Gamma_{l-1}(s') + \lambda_l(s', s)]$

   $(s^*, r^*) = \arg\max^{(r)}_{s' \in \mathcal{S}(s), r \in \mathcal{R}} [\Gamma_{l-1}(s') + \lambda_l(s', s)]$

   $h_l(s, r) = s^*$

   $\rho_l(s, r) = r^*$.

3. Termination: $l = L, s = 0$

   $\Gamma_L(0, r) = \max^{(r)}_{s' \in \mathcal{S}(s), r \in \mathcal{R}} [\Gamma_{L-1}(s') + \lambda_L(s', 0)]$

   $(s^*, r^*) = \arg\max^{(r)}_{s' \in \mathcal{S}(s), r \in \mathcal{R}} [\Gamma_{L-1}(s') + \lambda_L(s', 0)]$

   $h_L(0, r) = s^*$

   $\rho_L(0, r) = r^*$.

4. Traceback:

   With $s_0 = 0$ and $s_L = 0$, for all rankings $r$, the $r$th-best sequence is $(0, s_1^*, s_2^*, \ldots, s_{L-1}^*, 0)$, where

   $s_{L-1}^* = h_L(0, r)$

   $r_{L-1}^* = \rho_L(0, r)$

   $s_l^* = h_{l+1}(s_{l+1}^*, r_{l+1}^*)$ for $l = L-2, \ldots, 2, 1$

   $r_l^* = \rho_{l+1}(s_{l+1}^*, r_{l+1}^*)$ for $l = L-2, \ldots, 2, 1$.

for TBCCs borrows from the WAVA algorithm, but there are only two runs through the trellis whereas the WAVA may allow three or four. See Algorithm 4.4 for details.

*The Adaptive Parallel LVA Decoder*. To realize the full power of the CRC/ZTCC or CRC/TBCC combination, a large list size, say $\mathcal{L} = 256$ or $\mathcal{L} = 1024$, is necessary to accommodate those occasions on which the correct message is not near the top of the list. However, as we will see by example, for normal operating conditions, for the vast majority of the received codewords, a list size of $\mathcal{L} = 2$ will suffice because the correct message is one of the top two messages in the list. This motivates the idea of an adaptive parallel LVA (AP-LVA) decoder, which invokes a parallel LVA decoder with increasing size until the CRC check passes or until that largest-size parallel LVA decoder has completed its work without a CRC pass. It is common to use list sizes that are powers of 2 ($\mathcal{L} = 1, 2, 4, \ldots, \mathcal{L}_{\max}$) or powers of 4 ($\mathcal{L} = 1, 4, 16, \ldots, \mathcal{L}_{\max}$). We will use the former in our algorithm description for both CRC/ZTCC codes and CRC/TBCC codes. The AP-LVA is presented in Algorithm 4.5.

**Example 4.10** We first consider a memory $\mu = 8$, rate-1/2 convolutional code treated as a ZTCC (decoded by the VA) and as a TBCC (decoded by the WAVA). The message size is $k = 150$ bits in both cases. The code rates for the two cases are

**Algorithm 4.4** The Parallel LVA for TBCCs

1. Assume a list size of $\mathcal{L} = 1$, that is, the standard Viterbi algorithm, and initialize all cumulative metrics to zero at time $l = 0$.
2. Run the standard Viterbi algorithm through the trellis, *from all states* at time $l = 0$ *to all states* at time $l = L$.
3. Instead of the metric initialization given for the parallel LVA for ZTCCs, at time $l = 0$ set the first-ranked metrics, $\{\Gamma_l(s, 1)\}$, in the TBCC LVA to the final-state metrics found from the standard VA run in the previous step. All of the metrics for the lower rankings can be set to 0.
4. Now run the parallel LVA as described above for ZTCCs, but do not terminate in state 0: allow there to be a list of $\mathcal{L}$ candidates at each of the $2^{\mu}$ states. Thus, there will be $2^{\mu} \cdot \mathcal{L}$ candidates after parallel LVA processing, each of which has a metric indicating its likelihood.
5. In order of likelihood metrics, go down the list of $2^{\mu} \cdot \mathcal{L}$ candidates and choose as the decoder output the first candidate that is both a tail-biting path and passes the CRC check. If no such candidate exists, declare decoding failure.

**Algorithm 4.5** The Adaptive Parallel LVA for ZTCCs and TBCCs

**for** $\mathcal{L} = 1, 2, 4, \ldots, \mathcal{L}_{max}$ **do**
  Process received word with ZTCC/TBCC parallel LVA with list size $\mathcal{L}$.
  If the parallel LVA with list size $\mathcal{L}$ produces an output, then break.
**end for**
If there was a parallel LVA decoder output for some list size $\mathcal{L}$, then output decoded codeword, else output decode-failure flag.

$R_{ZTCC} = 150/316$ and $R_{TBCC} = 150/300$. Problem 4.25 shows that the *code rate loss* is $10\log_{10}(R_{TBCC}/R_{ZTCC}) \approx 0.23$ dB, approximately the gap seen between the ZTCC and TBCC Viterbi decoder curves in Figure 4.14.

To examine the effect of AP-LVA decoding, a 16-bit CRC with generator polynomial coefficients 0x11021 (hex) was implemented. (The least-significant bit is the $X^{16}$ coefficient.) In this case, the code rates are $R_{ZTCC} = 150/348$ and $R_{TBCC} = 150/332$. The power-of-4 adaptive parallel LVA with $\mathcal{L}_{\max} = 256$ was applied to both the ZTCC and TBCC with the outer 16-bit CRC code. In Figure 4.14 we observe a gap of about 0.2 dB between the list decoding ZTCC and TBCC curves, consistent with the code rate loss of $10\log_{10}(R_{TBCC}/R_{ZTCC}) \approx 0.20$ dB.

Observe that the CRC-convolutional codes with LVA decoding provide a gain of about 1 dB over the conventional VA-decoded convolutional codes. Further, the presence of the CRC and a moderate $\mathcal{L}_{\max}$ endows the CRC-convolutional scheme with an error-detection capability not present in the conventional schemes. A much larger $\mathcal{L}_{\max}$ (say 10000) moves the curves to the left a few tenths of a decibel and removes any error-detection capability. A much smaller $\mathcal{L}_{\max}$ (say 16) moves the curves to the right a few tenths of a decibel and enhances the error-detection

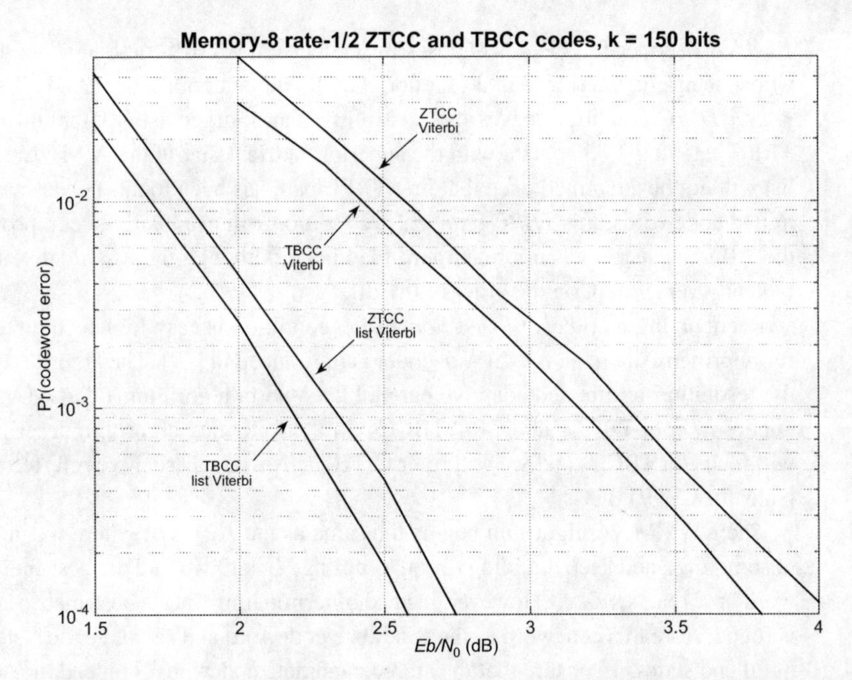


**Figure 4.14** Comparison of various coding and decoding schemes for a $k = 150$-bit message and a rate-1/2, $\mu = 8$ convolutional code. A 16-bit CRC was used for the cases with list Viterbi decoding.

capability of the LVA decoder. At an error rate of $2 \times 10^{-4}$ (about 2.7 dB for ZTCC and about 2.5 dB for TBCC), the power-of-4 adaptive LVA has an average list size of about 1.4 in both cases and the power-of-2 adaptive LVA has an average list size of about 1.3 in both cases.

---

In the example above, the CRC used was chosen arbitrarily. It is possible to choose CRC codes that are *distance-spectrum optimum* (DSO), that is, they optimize the distance spectrum of the combined CRC-convolutional code with generator $\mathbf{G}_{cat}(D)$ described earlier in this subsection. Methods for designing such DSO CRC codes given an optimal ZTCC may be found in [14]. Methods for designing DSO CRC codes given an optimal TBCC may be found in [15]. Methods for modifying a Viterbi decoder for error detection are described in [16].

The authors of [17] apply the AP-LVA algorithm to the convolutional code and CRC code specified in the CCSDS TM Synchronization and Channel Coding standard, including high-rate puncturings of the convolutional code. They show list decoding gains of around 3 dB at a frame-error rate of $10^{-6}$.

*Modification of the AP-LVA Decoder to Make it Maximum Likelihood*. In the context of AP-LVA decoding of CRC/TBCCs, a trellis path that is tail-biting and passes the CRC check will be called a TBPC path. Recall that an AP-LVA decoder chooses as its decision the TBPC path with the best decoding metric among the $2^{\mu} \cdot \mathcal{L}$ paths at the end of the trellis. It was noticed by Hulse and Wesel [18] that occasionally the

AP-LVA decoder for TBCCs chooses as its decision a TBPC path at some final state whose metric is inferior to that of a non-TBCP path at another final state. This *TBPC < non-TBPC* condition leaves the possibility that a larger list (LL) could include a TBPC path at the other state with the superior metric. (But, in the AP-LVA, the larger list will not be attempted once that first TBPC path has been found.) Over many transmitted codewords, this *TBPC path in LL* condition will eventually occur, proving that the AP-LVA decoder is not maximum likelihood, although the loss relative to ML is typically very small, on the order of 0.1 dB.

Recognizing that the *TBPC < non-TBPC* condition is easy to detect, an algorithm for improving upon the AP-LVA decoder is introduced in [19]. The algorithm is called the resolution-terminated adaptive parallel list Viterbi algorithm (RT-AP-LVA) and it applies to both TBCCs and CRC/TBCCs. In the following algorithm description, we will focus on TBCCs and we will refer to TB paths where earlier we referred to TBPC paths for CRC/TBCCs.

The RT-AP-LVA algorithm begins the same as the AP-LVA, where we initialize $\mathcal{L}$ to some $\mathcal{L}_{\min}$ and keep doubling (or quadrupling) $\mathcal{L}$ until we find the best-metric codeword or $\mathcal{L}$ reaches $\mathcal{L}_{\max}$. However, instead of terminating once a best-metric codeword is found, store this codeword as the candidate codeword and calculate the state metrics of all end states. To ensure that the stored candidate codeword is indeed the ML codeword, we first need to check that all other state metrics are worse than the candidate state metric. If there exists a state with a non-TB path having a better metric than the candidate, then it is possible that increasing the list size could find a better candidate codeword (TB path) at this end state.

Future trellis passes can never increase any state metric, since all codewords at a given end state are found in metric order. Thus, if the state metric of a given state is worse than the candidate state metric, we can deactivate this starting state since it will never find a codeword better than the candidate codeword. Such a state is said to be *resolved.* Any state for which it is still possible to find a codeword with a better metric than that of the candidate codeword is said to be *unresolved*.

Once a codeword (TB path) is found, we examine the path metrics of all states and resolve the candidate state and any states with a worse state metric than the candidate codeword. If every state is resolved, then the candidate codeword must be the ML codeword and we terminate. If there is at least one unresolved state, we run another pass through the trellis where every resolved starting state is deactivated. (The list size may or may not be doubled in a subsequent pass; more on this below.) If a better codeword than the candidate codeword is found, we replace the candidate codeword and candidate state with the new codeword and state. This continues until all states are resolved, at which point the decoder selects the candidate codeword and terminates.

When a starting state is resolved, future passes through the trellis will not find any codewords that start at this state. As long as at least one state was switched from unresolved to resolved following a trellis pass, it is possible for the remaining unresolved states to find new codewords without increasing $\mathcal{L}$. In [19], $\mathcal{L}$ was increased only when there was no change in the number of resolved states after a trellis pass.

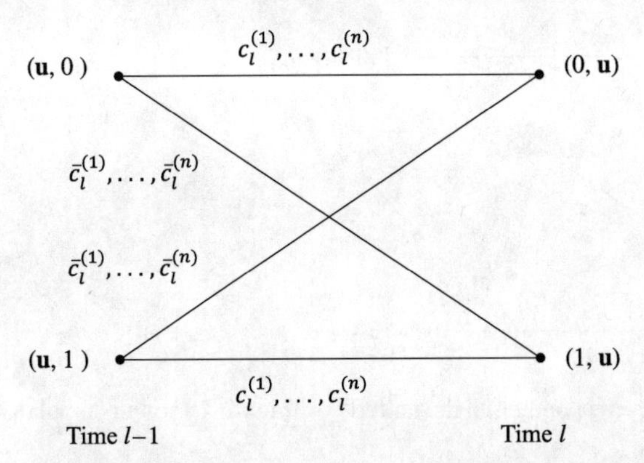


**Figure 4.15** A generic butterfly from a trellis of a rate-$1/n$ convolutional code whose generator polynomial coefficients are "1" in the first and last place.

## 4.5.4 Differential Viterbi Decoding

It is possible to employ a (lower-complexity) *differential Viterbi algorithm* for the decoding of certain rate-$1/n$ non-recursive, non-systematic convolutional codes on the BI-AWGNC [20]. In particular, the algorithm applies to such convolutional codes whose generator polynomials have the same degree $\mu$ and have first and last coefficients equal to "1": $g_0^{(1)} = \cdots = g_0^{(n)} = 1$ and $g_\mu^{(1)} = \cdots = g_\mu^{(n)} = 1$. This characteristic holds for most of the best rate-$1/n$ codes. For example, it applies to the industry-standard convolutional code for which $\mathbf{g}_{\text{octal}}^{(1)} = 117$ and $\mathbf{g}_{\text{octal}}^{(2)} = 155$, and it applies to the rate-1/2, memory-2 code with $\mathbf{g}_{\text{octal}}^{(1)} = 7$ and $\mathbf{g}_{\text{octal}}^{(2)} = 5$ that was examined in the previous section.

The differential algorithm derives from the fact that (after the initial $\mu$ stages) trellises for rate-$1/n$ $\bar{\text{R}}\bar{\text{S}}\text{C}$ codes are composed of multiple "butterflies" of the form shown in Figure 4.15. Note that the two states at time $l-1$ have the forms $(\mathbf{u}, 0)$ and $(\mathbf{u}, 1)$, where $\mathbf{u} = \left(u_{l-1}, u_{l-2}, \ldots, u_{l-\mu+1}\right)$ is the contents of the first $\mu-1$ memory elements in the memory-$\mu$ encoder. At time $l$, the two states have the forms $(0, \mathbf{u})$ and $(1, \mathbf{u})$. Note also that the labels on the two branches that diverge from a given state at time $l$ are complements of each other. Lastly, the two labels on one pair of diverging branches are equal to the two labels on the other pair. These facts allow us to display the various metrics for the butterfly as in Figure 4.16. As seen in this figure, the four branch metrics $\{\lambda_l(s', s)\}$ can take on only one of two values, $+\lambda$ or $-\lambda$, as a consequence of the fact that diverging branches are complements of each other (see also equation (4.20)).

In view of Figure 4.16, observe that the tentative cumulative metric difference at state $(0, \mathbf{u})$ is

$$D_l(0, \mathbf{u}) = \Delta\Gamma_{l-1} + 2\lambda,$$

where $\Delta\Gamma_{l-1} \triangleq \Gamma_{l-1}(\mathbf{u}, 0) - \Gamma_{l-1}(\mathbf{u}, 1)$. Similarly, the tentative cumulative metric difference at state $(1, \mathbf{u})$ is

$$D_l(1, \mathbf{u}) = \Delta\Gamma_{l-1} - 2\lambda.$$

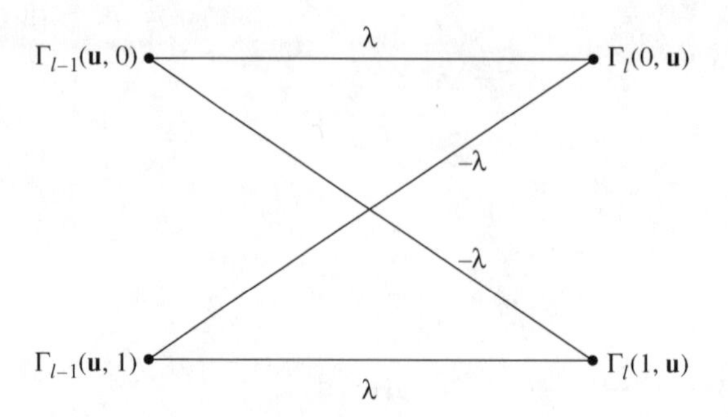


**Figure 4.16** The butterfly corresponding to the butterfly of Figure 4.15 with cumulative and branch metrics displayed.

This can be written compactly as

$$D_l(u_l) = \Delta\Gamma_{l-1} + (-1)^{u_l} 2\lambda,$$

where $u_l \in \{0, 1\}$ is the encoder input at time $l$.

From Figure 4.16, the following survivor-selection rules may be determined:

$$\begin{array}{l} \text{the survivor into state } (0, \mathbf{u}) \text{ comes from} \\ \qquad \text{state } (\mathbf{u}, 0), \quad \text{if } D_l(0, \mathbf{u}) > 0 \\ \qquad \text{state } (\mathbf{u}, 1), \quad \text{if } D_l(0, \mathbf{u}) < 0 \\ \text{the survivor into state } (1, \mathbf{u}) \text{ comes from} \\ \qquad \text{state } (\mathbf{u}, 0), \quad \text{if } D_l(1, \mathbf{u}) > 0 \\ \qquad \text{state } (\mathbf{u}, 1), \quad \text{if } D_l(1, \mathbf{u}) < 0 \end{array}$$

where an arbitrary decision may be made if $D_l = 0$. It is easily shown that these survivor-selection rules can be represented alternatively as in Figure 4.17. Observe from this that in order to determine the two survivors in each butterfly within the full trellis, one must first determine $\max\{|\Delta\Gamma_{l-1}|, |2\lambda|\}$ and then find the sign of $\Delta\Gamma_{l-1}$ if $|\Delta\Gamma_{l-1}|$ is maximum, or find the sign of $\lambda$ if $|2\lambda|$ is maximum. Once the survivors have been found, their cumulative metrics can be determined via

$$\Gamma_l(u_l, \mathbf{u}) = \Gamma_{l-1}(\mathbf{u}, u_{l-\mu}) + (-1)^{u_l}(-1)^{u_{l-\mu}}\lambda.$$

Note that the core operation is effectively *compare–select–add* (CSA) as opposed to add–compare–select. It can be shown that a reduction of approximately 33% in additions is achieved by employing the differential Viterbi decoder instead of the conventional one.

### 4.5.5 The Fano Algorithm

In 1963, several years prior to the publication of the Viterbi algorithm, R. Fano published what he described as a probabilistic decoding algorithm for decoding convolutional codes [9]. The *Fano algorithm* provides near maximum-likelihood decoding performance and has very low complexity. The Viterbi algorithm is much more complex, but has the advantage of being truly optimal and it has no issues with very long message lengths. As will be explained, the Fano algorithm requires periodically

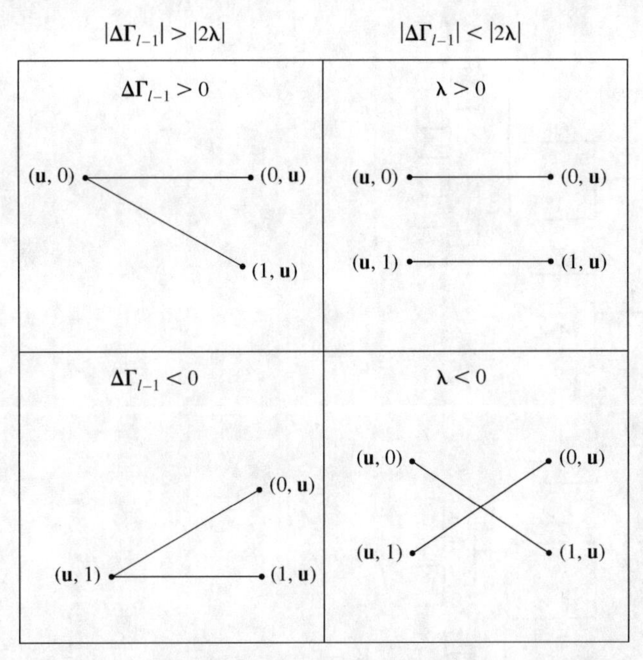


**Figure 4.17** Survivor-selection rules for the butterfly of Figure 4.16.

inserted pilot bits for long messages to avoid buffer overflows or erased messages. Still, the Fano algorithm is very important for its low complexity and, when the code memory is large, it can be at least 100 times faster than a Viterbi decoder with negligible performance loss. Additionally, it has resurfaced recently in the context of decoding short (modified) polar codes. We focus in this section only on rate-1/2 zero-terminated convolutional codes. Extension to other code rates is straightforward. Extension to a tail-biting convolutional code (in an effective way) is an open research problem.

Rather than traversing the code's trellis to determine the transmitted codeword, the Fano algorithm traverses a *tree* representation of the encoder. Figure 4.18 presents the tree for the memory-2 encoder of Figure 4.1, which has octal encoder taps $(7, 5)_8$. The tree root, or *root node*, on the left represents the encoder zero state prior to any encoder inputs. The labels of the horizontal *branches* give encoder outputs corresponding to the encoder inputs that led to that branch, and the branches themselves lead to the next tree *node*. More specifically, as indicated by the arrows in the diagram, each encoder input of 0 results in two encoder outputs found by going upward from the current tree node to the next branch rightward. Similarly, each encoder input of 1 results in the two encoder outputs that label the branch below and right of the current node.

Observe for message size $K + \mu$ (which includes $\mu$ terminating bits), the tree will possess $2^{K+\mu}$ terminal nodes (or leaves). This is clearly too much information to store for $K > 32$, say, so instead the decoder stores encoder state diagram information. That is, the decoder is equipped with one function that gives the encoder output bits and a second function that gives the next encoder state. Both functions (or look-up tables) take as inputs the current state and the current encoder input bit.

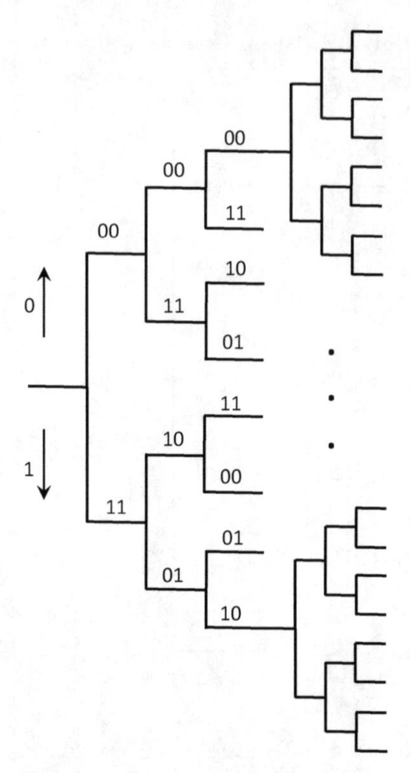


**Figure 4.18** Tree representation of the encoder of Figure 4.1.

The Fano decoder traverses a single path at a time in the code's tree (cf. Figure 4.18), moving rightward from a current node to an adjacent node, occasionally in a backward (leftward) direction when metric computations indicate that the decoder might be on the wrong path. Using the Fano branch metric to be described below, the decoder steps through the tree forward (rightward), and often backward, on the basis of cumulative metric comparisons with a dynamically changing *threshold*, $T$. If the *forward cumulative metric*, $M_F$, is greater than $T$, the decoder steps forward to the next branch; otherwise, the decoder looks back to the previous metric and moves backward if the *backward cumulative metric*, $M_B$, is greater than $T$. If the look-back metric $M_B$ is also less than $T$, then $T$ is decremented and the look-forward processing begins again, but this time an alternative path is taken. The threshold $T$ is continually adjusted ("tightened") by some multiple of an increment $\Delta$ so that $T$ is just less than the current path's cumulative metric. Fano's ingenius design arranges it so that each time a tree node is visited in the forward direction, the threshold is lower than it was on the previous visit, eliminating the possibility of an infinite loop. The path that reaches the end of the tree is the decoded path. This algorithm is summarized in Figure 4.19 using a flowchart that will be discussed in greater detail shortly.

Much like a Viterbi decoder, the Fano decoder takes channel output values and computes branch metrics that depend on branch labels and the corresponding (in time) channel output values. A significant difference is that, while the Viterbi decoder extends and metric-updates $2^{\mu}$ trellis paths at each decoding iteration, the Fano

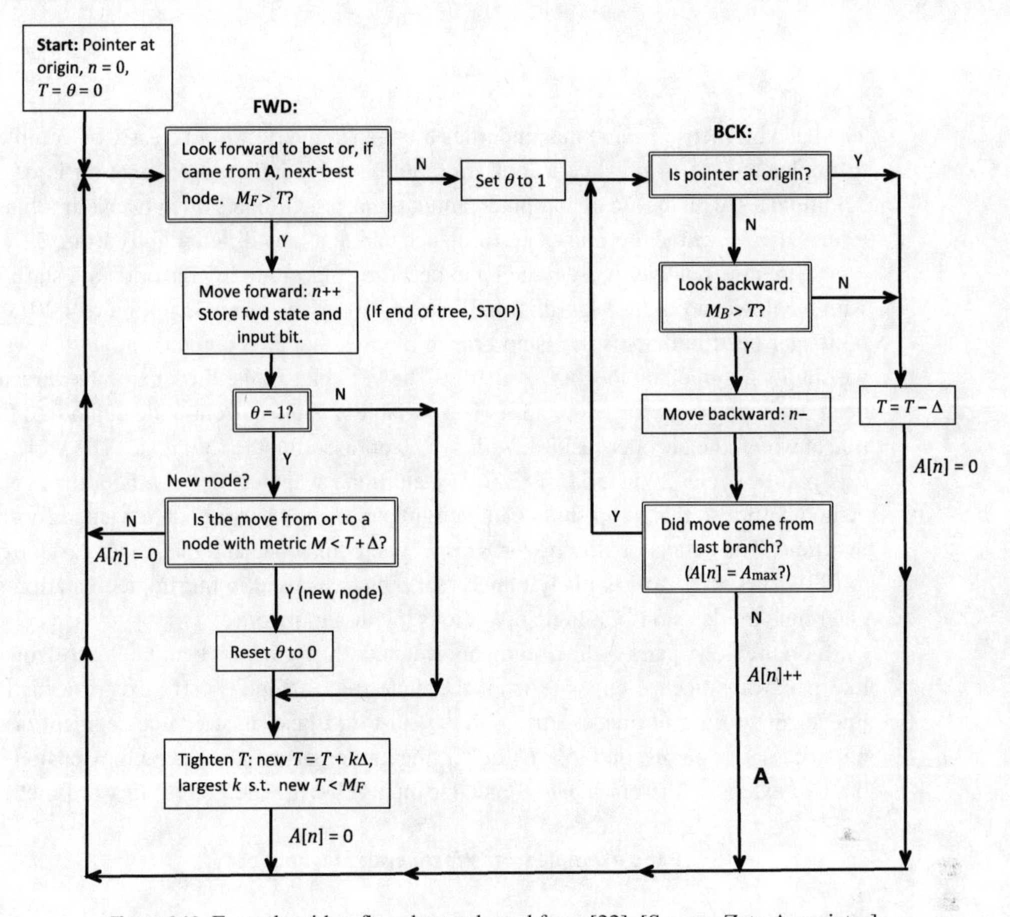


**Figure 4.19** Fano algorithm flowchart, adapted from [22]. [Source: Zeta Associates]

algorithm only extends and metric-updates a single tree path. The Viterbi decoder need only compare a path's cumulative metric to that of its competitors to assess its quality. By contrast, the Fano decoder compares its cumulative metric to a threshold. Clearly, the threshold should change (increase in magnitude) as the decoder moves deeper into the tree. This is aided by a metric, the *Fano metric*, that self-adjusts for the length of the tree path.

Denoting the code rate by $R$ (bits per channel symbol), the Fano metric [26] on the BI-AWGN channel for channel output $y_l$ and code bit $x_l \in \{\pm 1\}$ is given by

$$\begin{aligned} F(x, y) &= \log_2 \left[ p(y \mid x)/p(y) \right] - R \\ &= 1 - \log_2 \left[ 1 + \exp(-4yx/N_0) \right] - R, \end{aligned}$$

where $p(y \mid x)$ is the conditional density on the channel output $y$ given input $x$. It is the code rate $R$ in this expression that counters the effect of a growing cumulative metric that is to be compared to a threshold (albeit itself growing). Analogous to the correlation branch metric in (4.20), the Fano branch metric (for a rate-1/2 code) is given by

$$\lambda_l = \sum_{j=1}^{2} F\left(y_l^{(j)}, x_l^{(j)}\right). \tag{4.21}$$

The Fano bit metric $F(y, x)$ that underlies the branch metric would clearly be complex to implement, so in practice a look-up table or polynomial approximation is used. Additionally, with the aid of computer simulations, practitioners often tweak the "bias" term $R$ to a value that optimizes performance and that may differ slightly from $R$.

We are now ready to present the Fano decoding algorithm in greater detail, starting with the flowchart of the algorithm in Figure 4.19 and the example in Figure 4.20. We no longer need to discuss the Fano branch metrics that are computed along the way, we simply assume that they are available. The variables in the flowchart have already been defined, except for $n$, $\theta$, and $A$. The variable $n$ is the current node depth into the tree at which the decoder resides, with $n = 0$ representing the root node. The variable $\theta$ is part of a clever method of Fano to determine whether or not a node has been visited before so that a threshold can be tightened only when a node is visited for the first time. The details of why it works are beyond our scope and the reader is referred to [22]. The array $A[n]$ is likely unnecessary, but was used in our implementation to determine bit decisions via logic operations in our simulator.

In Figure 4.20, pairs of line segments emanate from each node in the figure, much like pairs of branches emanate from the code tree in Figure 4.18, except a higher line segment does not necessarily correspond to a 0 and a lower line segment does not necessarily correspond to a 1. Rather, the higher segment is the path extension of the two extensions (from 0 and 1 encoder inputs) having the greater forward metric.

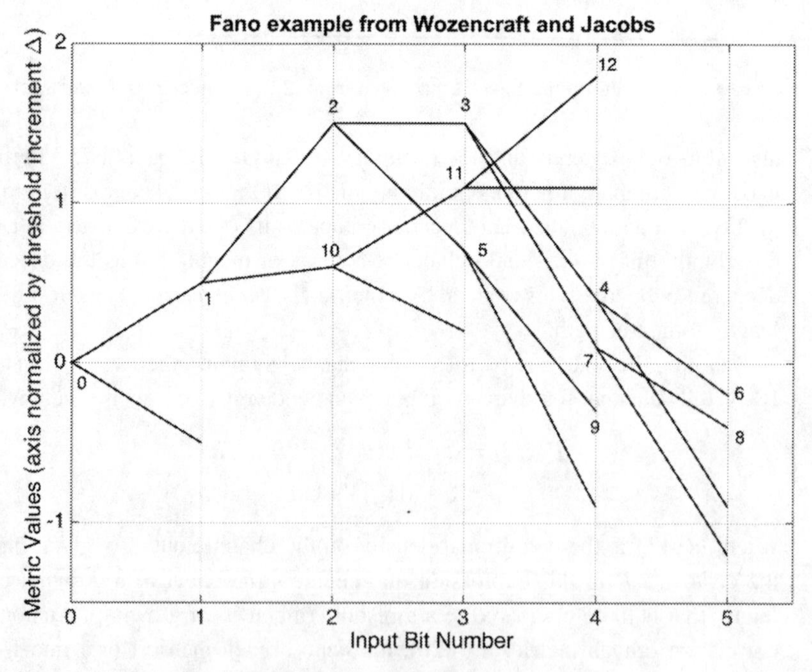


**Figure 4.20** Fano algorithm example taken from [22].

We imagine the decoder keeps track of which encoder input labels each branch (therefore, each segment), but we are only concerned at the moment with how to iteratively choose the sequence of segments in accordance with Fano's algorithm.

| Node | Threshold | $\theta$ | Actions ($\times$ means "set $\theta = 1$") | | |
|---|---|---|---|---|---|
| 0 | 0 | 0 | look at 1 | move to 1 | |
| 1 | 0 | 0 | look at 2 | move to 2 | set $T = \Delta$ |
| 2 | $\Delta$ | 0 | look at 3 | move to 3 | |
| 3 | $\Delta$ | 0 | look at 4 | $\times$ look at 2 | move to 2 |
| 2 | $\Delta$ | 1 | look at 5 | $\times$ look at 1 | set $T = 0$ |
| 2 | 0 | 1 | look at 3 | move to 3 | |
| 3 | 0 | 1 | look at 4 | move to 4 | set $\theta = 0$ |
| 4 | 0 | 0 | look at 6 | $\times$ look at 3 | move to 3 |
| 3 | 0 | 1 | look at 7 | move to 7 | set $\theta = 0$ |
| 7 | 0 | 0 | look at 8 | $\times$ look at 3 | move to 3 |
| 3 | 0 | 1 | look at 2 | move to 2 | |
| 2 | 0 | 1 | look at 5 | move to 5 | set $\theta = 0$ |
| 5 | 0 | 0 | look at 9 | $\times$ look at 2 | move to 2 |
| 2 | 0 | 1 | look at 1 | move to 1 | |
| 1 | 0 | 1 | look at 10 | move to 10 | set $\theta = 0$ |
| 10 | 0 | 0 | look at 11 | move to 11 | set $T = \Delta$ |
| 11 | $\Delta$ | 0 | look at 12 | move to 12 | |

The table above, adapted from [22], explains how the Fano algorithm of Figure 4.19 would proceed given the situation in Figure 4.20. Below, we provide additional details to the table and the reader is advised to closely follow Figure 4.19 while going through the table and the steps below. Each item below corresponds to a row in the table.

1. From the root node with threshold value $T = 0$ and current metric set to 0, both look-forward metrics are computed. The decoder compares the greater of the two to $T$ and, since it is greater than $T$, it moves to node 1. $\theta = 0$ and threshold $T$ cannot be tightened, for increasing it by $\Delta$ would make it greater than the node 1 metric, $M_F(1)$.
2. From node 1, the larger of the two forward metrics computed (at node 2, $M_F(2)$) is greater than $T = 0$ and so the decoder can move to node 2. $\theta = 0$ and threshold $T$ can be increased to $\Delta$ because $M_F(2)$ is greater than $\Delta$.
3. From node 2, node 3 has the larger of the two forward metrics and its metric, $M_F(3)$, is greater than $T = \Delta$ and so the decoder moves to node 3. $\theta = 0$ and threshold $T$ cannot be tightened because it would become greater than $M_F(3)$.
4. From node 3, the metric at node 4 is the greater of the two forward metrics, but it is less than $T = \Delta$ and so the decoder sets $\theta = 1$ and looks back at node 2. Because $M_B(2) > T$, the decoder moves back to node 2. (Note that there is no reason to consider node 7 because the node 4 metric violated the threshold, the node 7 metric would as well.)

5. The segment from node 2 to node 3 has already been attempted, and so the "next-best" node, node 5, will be attempted. Looking at node 5, $M_F(5) < T$, violating the threshold, so the decoder sets $\theta = 1$ and looks back at node 1, which also violates the threshold, so the threshold is reduced to $T = 0$.
6. Still at node 2, with $T = 0$, the decoder looks at node 3 and moves to node 3.
7. From node 3, the decoder looks at node 4 and moves to node 4, because its metric is greater than $T$. $\theta$ is reset to 0.
8. From node 4, the decoder looks at node 6 (which has the greater metric), and $M_F(6)$ violates the threshold of $T = 0$. So the decoder sets $\theta = 1$ and looks back to node 3, and moves to that node because $M_B(3) > T$.
9. From node 3, the decoder looks at the "next-best" node, that is, node 7, whose metric is greater than $T$, and so the decoder moves to node 7 and resets $\theta = 0$. The threshold $T$ clearly cannot be tightened (i.e., increased).
10. And so on, following the flowchart, until node 12 is reached.

---

**Example 4.11** Figure 4.21 compares the performances of the Viterbi and Fano decoding algorithms for a memory-14, rate-1/2 convolutional code with octal encoder taps $(75063, 56711)_8$ on a BI-AWGN channel. The common message length is 150 bits and in both cases the encoder was terminated with 14 zeros. As seen in the figure, the two algorithms have near-identical performance for the range of

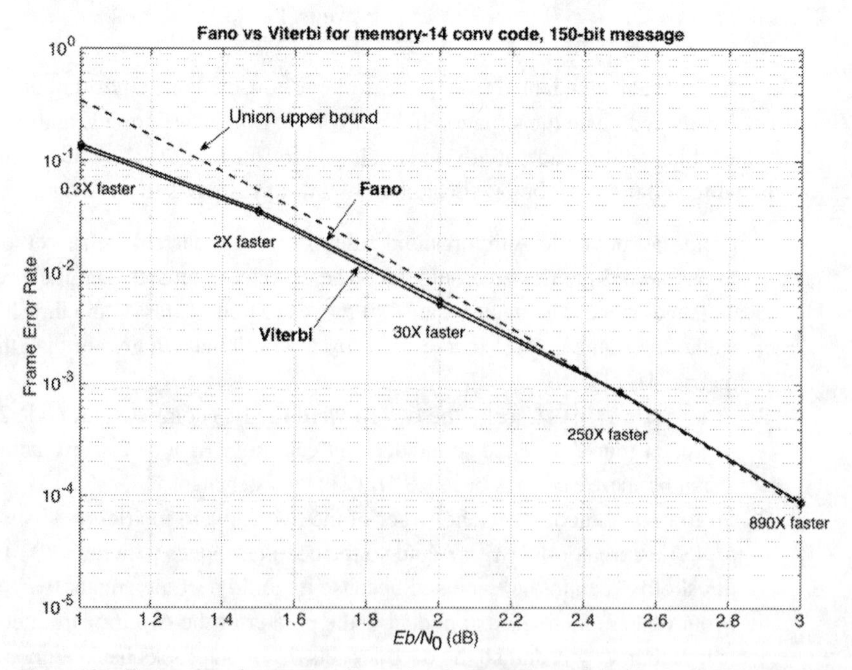


**Figure 4.21** Comparison between software implementations of the Fano algorithm and the Viterbi algorithm for a memory $\mu = 14$, rate-1/2, ZTCC with encoder taps $(75063, 56711)_8$ on a BI-AWGN channel. Labels on the simulation data points indicate the speed advantage of the Fano decoder.

signal-to-noise ratios and error rates simulated. There is no reason to expect a divergence in their performance at even lower error rates and so the decoders have essentially identical error-rate characteristics for this code. What is vastly different, as indicated in the figure, is that the Fano algorithm is orders of magnitude faster because it is much less complex as a result of the fact that the Viterbi decoder must provide add–compare–select computations for $2^{14} = 16,384$ states at each of the 164 trellis stages. Also shown in the figure is the union bound estimate on performance developed later in this chapter. Our Fano decoder used $\Delta = 0.9$ and the bit-wise Fano metric bias of $R = 150/328 = 0.4573$ was replaced by a bias of 0.55, which performed much better.

---

While the example above demonstrates an important advantage of the Fano decoder, the reader must be reminded that the message length was only 150 bits. For longer messages, say 10,000 bits, the Fano decoder will certainly lose its way before it reaches the end of the tree. That is, it is prone to eventually following extremely long paths, with millions of forward and backward moves, in the maze that is the code tree. The number of computations can be very large in this case and implementers of Fano decoders would discard the entire block in this event so that the decoder does not fall behind as newer codewords arrived. (In the simulator for the example above, we made no such effort because the message is short.) Discarded code blocks in this context are called *erasures* and this erasure phenomenon is often cited as one reason the Viterbi algorithm has been preferred to the Fano algorithm. One solution to this problem is to periodically insert a known pattern of bits (typically zeros) into the encoder, say every 1000 bits. The Fano decoder would exploit knowledge of these "pilot" bits, effectively resynchronizing. Another approach is to instead use a turbo code or an LDPC code for long blocklengths and only use a convolutional code for blocklengths less than about 1000 bits.

The Fano algorithm is one example of a type of decoder called a *sequential decoder*. Another well-known sequential decoding algorithm is the so-called *stack algorithm*. Using the Fano metric, this algorithm also extends one code path at a time, but puts the (best) path being extended (and its metric) on top of a stack that includes previously considered paths of differing lengths and their cumulative metrics. The path on top of the stack is repeatedly extended until its metric falls below that of at least one other candidate in the stack. In this event, the extended path is inserted into the stack and the stack is reordered in accordance with the cumulative metric values of the paths in the stack. After reordering, the top path is extended until it no longer has the best metric and has to be replaced, as described. When the top path reaches the end of the tree, it becomes the decoder output. We remark that reordering (also called sorting) is a very time-consuming process and care must be taken to sort sparingly so that the stack algorithm decoder can operate efficiently. There are many variations to the stack algorithm, but we have given the main ideas here.

Lastly, we mention that a good deal of work has gone into understanding the statistics of the number of computations performed by sequential decoders. Discussions of some of this work can be found in [3] and [12] and their references. We point out that the number of computations can be widely varying from codeword to codeword. This is in contrast to the Viterbi algorithm, which has a fixed number of computations for every codeword. This is another cited reason why Viterbi decoders have been preferred over Fano decoders. Still, the example above is compelling enough to show that the Fano algorithm can be advantageous in certain situations.

### 4.5.6 Bit-wise MAP Decoding and the BCJR Algorithm

Whereas the MLSD minimizes the probability of sequence error, the bit-wise MAP decoder minimizes the bit-error probability. While they have similar performance characteristics, both bit-wise and sequence-wise, they have different applications. For example, the MAP decoder is used in turbo decoders and turbo equalizers where soft decoder outputs are necessary.

The bit-wise MAP decoding criterion is given by

$$\hat{u}_l = \arg\max_{u_l} P(u_l \mid \mathbf{y}),$$

where $P(u_l \mid \mathbf{y})$ is the *a posteriori* probability (APP) of the information bit $u_l$ given the received word $\mathbf{y}$. For this discussion, we will favor $u_l \in \{+1, -1\}$ over $u_l \in \{0, 1\}$ under the mapping $\{0 \leftrightarrow +1, 1 \leftrightarrow -1\}$, and similarly for other binary variables. With $u_l \in \{+1, -1\}$, the bit-wise MAP rule simplifies to

$$\hat{u}_l = \text{sign}[L(u_l)], \tag{4.22}$$

where $L(u_l)$ is the logarithmic *a posteriori* probability (log-APP) ratio defined as

$$L(u_l) \triangleq \log\left[\frac{P(u_l = +1 \mid \mathbf{y})}{P(u_l = -1 \mid \mathbf{y})}\right].$$

The log-APP ratio is typically called the log-likelihood ratio (LLR) in the literature and we shall follow this convention here.

For convenience, we consider the BCJR-algorithm implementation of the MAP decoder applied to a rate-1/2 RSC code on a BI-AWGNC. Generalizing to other code rates is straightforward. The BCJR algorithm for the BSC will be considered in Problem 4.20. Further, the transmitted codeword $\mathbf{c}$ will have the form $\mathbf{c} = [c_1, c_2, \ldots, c_L] = [u_1, p_1, u_2, p_2, \ldots, u_L, p_L]$ with $c_l \triangleq [u_l, p_l]$, where $u_l$ signifies the systematic bit, $p_l$ signifies the parity bit, and $L$ is the length of the encoder input sequence including termination bits. The received word $\mathbf{y} = \mathbf{c} + \mathbf{n}$ will have the form $\mathbf{y} = [y_1, y_2, \ldots, y_L] = [y_1^u, y_1^p, y_2^u, y_2^p, \ldots, y_L^u, y_L^p]$, where $y_l \triangleq [y_l^u, y_l^p]$, and similarly for $\mathbf{n}$.

Our goal is the development of the BCJR algorithm for computing the LLR $L(u_l)$ given the received word $\mathbf{y}$. In order to incorporate the RSC code trellis into this computation, we rewrite $L(u_l)$ as

$$L(u_l) = \log\left[\frac{\sum_{U^+} p(s_{l-1} = s', s_l = s, \mathbf{y})}{\sum_{U^-} p(s_{l-1} = s', s_l = s, \mathbf{y})}\right], \tag{4.23}$$

where $s_l$ is the encoder state at time $l$, $U^+$ is the set of pairs $(s', s)$ for the state transitions $(s_{l-1} = s') \to (s_l = s)$ that correspond to the event $u_l = +1$, and $U^-$ is similarly defined for the event $u_l = -1$. To write (7.10) we used Bayes' rule, total probability, and then cancelled out $1/p(\mathbf{y})$ in the numerator and denominator. We see from (7.10) that we need only compute $p(s', s, \mathbf{y}) = p(s_{l-1} = s', s_l = s, \mathbf{y})$ for all state transitions and then sum over the appropriate transitions in the numerator and denominator. We now present the crucial results that facilitate the computation of $p(s', s, \mathbf{y})$.

**Lemma 4.1** *The pdf $p(s', s, \mathbf{y})$ may be factored as*

$$p(s', s, \mathbf{y}) = \alpha_{l-1}(s')\gamma_l(s', s)\beta_l(s), \tag{4.24}$$

*where*

$$\begin{aligned}
\alpha_l(s) &\triangleq p\left(s_l = s, \mathbf{y}_1^l\right), \\
\gamma_l(s', s) &\triangleq p(s_l = s, y_l \mid s_{l-1} = s'), \\
\beta_l(s) &\triangleq p\left(\mathbf{y}_{l+1}^L \mid s_l = s\right),
\end{aligned}$$

*and* $\mathbf{y}_a^b \triangleq [y_a, y_{a+1}, \ldots, y_b]$.

*Proof* By several applications of Bayes' rule, we have

$$\begin{aligned}
p(s', s, \mathbf{y}) &= p\left(s', s, \mathbf{y}_1^{l-1}, y_l, \mathbf{y}_{l+1}^L\right) \\
&= p\left(\mathbf{y}_{l+1}^L \mid s', s, \mathbf{y}_1^{l-1}, y_l\right)p\left(s', s, \mathbf{y}_1^{l-1}, y_l\right) \\
&= p\left(\mathbf{y}_{l+1}^L \mid s', s, \mathbf{y}_1^{l-1}, y_l\right)p\left(s, y_l \mid s', \mathbf{y}_1^{l-1}\right) \cdot p\left(s', \mathbf{y}_1^{l-1}\right) \\
&= p\left(\mathbf{y}_{l+1}^L \mid s\right)p(s, y_l \mid s')p\left(s', \mathbf{y}_1^{l-1}\right) \\
&= \beta_l(s)\gamma_l(s', s)\alpha_{l-1}(s'),
\end{aligned}$$

where the fourth line follows from the third because the variables omitted on the fourth line are conditionally independent. □

**Lemma 4.2** *The probability $\alpha_l(s)$ may be computed in a "forward recursion" via*

$$\alpha_l(s) = \sum_{s'} \gamma_l(s', s)\alpha_{l-1}(s'), \tag{4.25}$$

*where the sum is over all possible encoder states.*

*Proof* By several applications of Bayes' rule and the theorem on total probability, we have

$$\begin{aligned}
\alpha_l(s) &\triangleq p\left(s, \mathbf{y}_1^l\right) \\
&= \sum_{s'} p\left(s', s, \mathbf{y}_1^l\right) \\
&= \sum_{s'} p\left(s, y_l \mid s', \mathbf{y}_1^{l-1}\right)p\left(s', \mathbf{y}_1^{l-1}\right)
\end{aligned}$$

$$= \sum_{s'} p(s, y_l \mid s') p\left(s', \mathbf{y}_1^{l-1}\right)$$
$$= \sum_{s'} \gamma_l(s', s) \alpha_{l-1}(s'),$$

where the fourth line follows from the third due to conditional independence of $\mathbf{y}_1^{l-1}$. □

**Lemma 4.3** *The probability $\beta_l(s)$ may be computed in a "backward recursion" via*

$$\beta_{l-1}(s') = \sum_{s} \beta_l(s)\, \gamma_l(s', s). \tag{4.26}$$

*Proof* Applying Bayes' rule and the theorem on total probability, we have

$$\begin{aligned}
\beta_{l-1}(s') &\triangleq p\left(\mathbf{y}_l^L \mid s'\right) \\
&= \sum_{s} p\left(\mathbf{y}_l^L, s \mid s'\right) \\
&= \sum_{s} p\left(\mathbf{y}_{l+1}^L \mid s', s, y_l\right) p(s, y_l \mid s') \\
&= \sum_{s} p\left(\mathbf{y}_{l+1}^L \mid s\right) p(s, y_l \mid s') \\
&= \sum_{s} \beta_l(s) \gamma_l(s', s),
\end{aligned}$$

where conditional independence led to the omission of variables in the fourth line. □

The recursion for $\{\alpha_l(s)\}$ is initialized according to

$$\alpha_0(s) = \begin{cases} 1, & s = 0, \\ 0, & s \neq 0, \end{cases}$$

following from the reasonable assumption that the convolutional encoder is initialized to the zero state. The recursion for $\{\beta_l(s)\}$ is initialized according to

$$\beta_L(s) = \begin{cases} 1, & s = 0, \\ 0, & s \neq 0, \end{cases}$$

which assumes that "termination bits" have been appended at the end of the data word so that the convolutional encoder is again in state zero at time $L$.

All that remains at this point is the computation of $\gamma_l(s', s) = p(s, y_l \mid s')$. Observe that $\gamma_l(s', s)$ may be written as

$$\begin{aligned}
\gamma_l(s', s) &= \frac{P(s', s)}{P(s')} \cdot \frac{p(s', s, y_l)}{P(s', s)} \\
&= P(s \mid s') p(y_l \mid s', s) \\
&= P(u_l) p(y_l \mid u_l),
\end{aligned} \tag{4.27}$$

where the event $u_l$ corresponds to the event $s' \to s$. Note that $P(s \mid s') = \Pr(s' \to s) = 0$ if $s$ is not a valid state from state $s'$ and $\Pr(s' \to s) = 1/2$ otherwise (since we assume a binary-input encoder with equal *a priori* probabilities $P(u_l) = P(s \mid s'))$. Hence, $\gamma_l(s', s) = 0$ if $s' \to s$ is not valid and, otherwise,

$$\gamma_l(s', s) = \frac{P(u_l)}{2\pi\sigma^2} \exp\left[-\frac{\|y_l - c_l\|^2}{2\sigma^2}\right] \tag{4.28}$$

$$= \frac{1}{2(2\pi)\sigma^2} \exp\left[-\frac{(y_l^u - u_l)^2 + (y_l^p - p_l)^2}{2\sigma^2}\right], \tag{4.29}$$

where $\sigma^2 = N_0/2$.

In summary, we may compute $L(u_l)$ via (7.10) using (4.24), (7.8), (7.9), and (7.11). This "probability-domain" version of the BCJR algorithm is numerically unstable for long and even moderate codeword lengths, so we now present the stable "log-domain" version of it. (Note that, in the presentation of the Viterbi algorithm, it was easy to immediately go to the log domain, although there exists a probability-domain Viterbi algorithm.)

In the log-BCJR algorithm, $\alpha_l(s)$ is replaced by the *forward metric*

$$\begin{aligned}\tilde{\alpha}_l(s) &\triangleq \log(\alpha_l(s)) \\ &= \log \sum_{s'} \alpha_{l-1}(s')\gamma_l(s', s) \\ &= \log \sum_{s'} \exp(\tilde{\alpha}_{l-1}(s') + \tilde{\gamma}_l(s', s)),\end{aligned} \tag{4.30}$$

where the *branch metric* $\tilde{\gamma}_l(s', s)$ is given by

$$\begin{aligned}\tilde{\gamma}_l(s', s) &= \log \gamma_l(s', s) \\ &= -\log\left(4\pi\sigma^2\right) - \frac{\|y_l - c_l\|^2}{2\sigma^2}.\end{aligned} \tag{4.31}$$

We will see that the first term in (4.31) may be dropped. Note that (4.30) not only defines $\tilde{\alpha}_l(s)$, but also gives its recursion. These log-domain forward metrics are initialized as

$$\tilde{\alpha}_0(s) = \begin{cases} 0, & s = 0, \\ -\infty, & s \neq 0. \end{cases} \tag{4.32}$$

The probability $\beta_{l-1}(s')$ is replaced by the *backward metric*

$$\begin{aligned}\tilde{\beta}_{l-1}(s') &\triangleq \log(\beta_{l-1}(s')) \\ &= \log\left(\sum_{s} \exp\left(\tilde{\beta}_l(s) + \tilde{\gamma}_l(s', s)\right)\right)\end{aligned} \tag{4.33}$$

with initial conditions

$$\tilde{\beta}_L(s) = \begin{cases} 0, & s = 0, \\ -\infty, & s \neq 0, \end{cases} \tag{4.34}$$

under the assumption that the encoder has been terminated to the zero state.

As before, $L(u_l)$ is computed as

$$L(u_l) = \log\left[\frac{\sum_{U^+}\alpha_{l-1}(s')\,\gamma_l(s',s)\,\beta_l(s)}{\sum_{U^-}\alpha_{l-1}(s')\,\gamma_l(s',s)\beta_l(s)}\right]$$

$$= \log\left[\sum_{U^+}\exp\left(\tilde{\alpha}_{l-1}(s') + \tilde{\gamma}_l(s',s) + \tilde{\beta}_l(s)\right)\right]$$

$$- \log\left[\sum_{U^-}\exp\left(\tilde{\alpha}_{l-1}(s') + \tilde{\gamma}_l(s',s) + \tilde{\beta}_l(s)\right)\right]. \tag{4.35}$$

It is evident from (4.35) that the constant term in (4.31) may be ignored since it may be factored all the way out of both summations. At first glance, equations (4.30)–(4.35) do not look any simpler than the probability-domain algorithm, but we use the following results to attain the simplification.

It can be shown (Problem 4.19) that

$$\max(x,y) = \log\left(\frac{e^x + e^y}{1 + e^{-|x-y|}}\right). \tag{4.36}$$

Now define

$$\max{}^*(x,y) \triangleq \log\,(e^x + e^y) \tag{4.37}$$

so that, from (4.36),

$$\max{}^*(x,y) = \max(x,y) + \log\left(1 + e^{-|x-y|}\right). \tag{4.38}$$

This may be extended to more than two variables. For example,

$$\max{}^*(x,y,z) \triangleq \log(e^x + e^y + e^z),$$

which may be computed in pair-wise fashion according to

$$\max{}^*(x,y,z) = \max{}^*\left[\max{}^*(x,y), z\right].$$

Given the function $\max^*(\cdot)$, we may now rewrite (4.30), (4.33), and (4.35) as

$$\tilde{\alpha}_l(s) = \max_{s'}{}^*\left[\tilde{\alpha}_{l-1}(s') + \tilde{\gamma}_l(s',s)\right], \tag{4.39}$$

$$\tilde{\beta}_{l-1}(s') = \max_{s}{}^*\left[\tilde{\beta}_l(s) + \tilde{\gamma}_l(s',s)\right], \tag{4.40}$$

and

$$L(u_l) = \max_{U^+}{}^*\left[\tilde{\alpha}_{l-1}(s') + \tilde{\gamma}_l(s',s) + \tilde{\beta}_l(s)\right]$$

$$- \max_{U^-}{}^*\left[\tilde{\alpha}_{l-1}(s') + \tilde{\gamma}_l(s',s) + \tilde{\beta}_l(s)\right]. \tag{4.41}$$

**Algorithm 4.6** The Log-Domain BCJR Algorithm

ASSUMPTIONS

We assume as above a rate-1/2 RSC encoder, a data block **u** of length $L$, and that the encoder starts and terminates in the zero state (the last $\mu$ bits of **u** are so selected). In practice, the value $-\infty$ used in initialization is simply some large-magnitude negative number.

THE BCJR ALGORITHM

**Initialization**

Set $\tilde{\alpha}_0(s)$ and $\tilde{\beta}_L(s)$ according to (4.32) and (4.34).

**for** $l = 1$ to $L$ **do**

get $y_l = [y_l^u, y_l^p]$

compute $\tilde{\gamma}_l(s', s) = -\|y_l - c_l\|^2 / (2\sigma^2)$ for all allowable state transitions $s' \rightarrow s$ (note that $c_l = c_l(s', s)$ here)[2]

compute $\tilde{\alpha}_l(s)$ for all $s$ using the recursion (4.39)

**end for**

**for** $l = L$ to 2 step $-1$ **do**

compute $\tilde{\beta}_{l-1}(s')$ for all $s'$ using (4.40)

**end for**

**for** $l = 1$ to $L$ **do**

compute $L(u_l)$ using (4.41)

compute hard decisions via $\hat{u}_l = \text{sign}[L(u_l)]$

**end for**

The BCJR algorithm that uses these three equations at its core is presented in Algorithm 4.6 above. Figure 4.22 illustrates pictorially the trellis-based computations that these three equations represent. It is also illuminating to compare Figure 4.22 with Figure 4.9. We see from (4.39), (4.40), and (4.41) how the log-domain computation of $L(u_l)$ is vastly simplified relative to the probability-domain computation. From (4.38), implementation of the $\max^*(\cdot)$ function involves only a two-input $\max(\cdot)$ function plus a look-up table or some other approximation of the high-complexity "correction term" $\log\left(1 + e^{-|x-y|}\right)$. The size of the look-up table has been investigated in [23] for specific cases. When $\max^*(x, y)$ is replaced by $\max(\cdot)$ in (4.39) and (4.40), these recursions become forward and reverse Viterbi algorithms, respectively. The performance loss associated with this approximation in turbo decoding depends on the specific turbo code, but a loss of about 0.5 dB is typical [23]. See [24] for the BCJR algorithm for TBCCs.

## 4.5.7 The List-BCJR Algorithm

The list-BCJR algorithm can be characterized as a serial list decoder and operates as follows. As with the list-Viterbi algorithm, there is typically an outer CRC code that

[2] We may alternatively use the branch metric given in Problem 4.21.

(a)

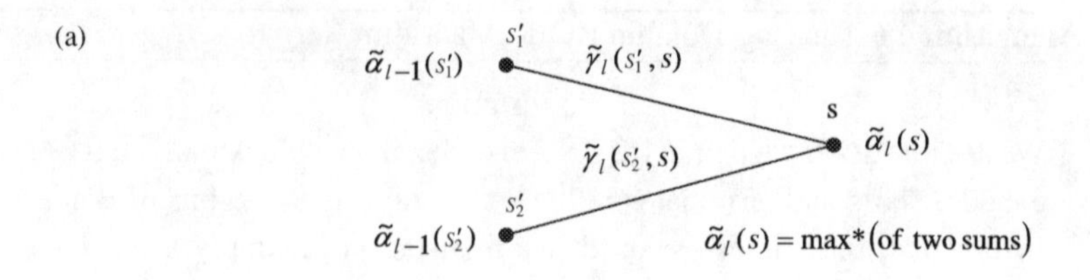


(b)

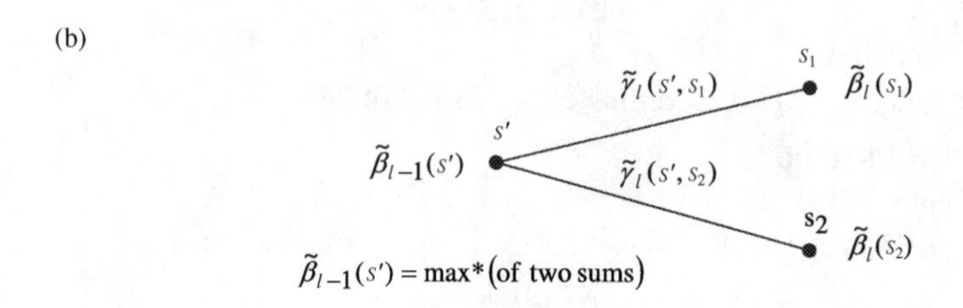


(c)

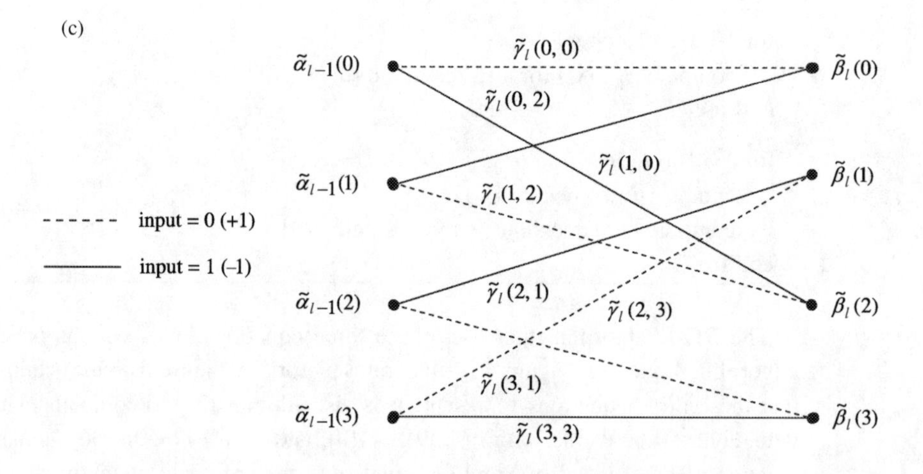


$$L(u_l) = \max^*\left\{\tilde{\alpha}_{l-1} + \tilde{\gamma}_l + \tilde{\beta}_l \text{ for dashed lines}\right\} - \max^*\left\{\tilde{\alpha}_{l-1} + \tilde{\gamma}_l + \tilde{\beta}_l \text{ for solid lines}\right\}$$

**Figure 4.22** An illustration of the BCJR algorithm. (a) The forward recursion in (4.39); (b) the backward recursion in (4.40); and (c) the computation of $L(u_l)$ via (4.41).

aids the list decoding process. If CRC checking fails after the BCJR decoder passes through the trellis the first time, the bit with the weakest bit metric is identified and BCJR decoding is repeated with a "strong 0" in that bit position; if CRC checking fails, BCJR decoding is repeated with a "strong 1" in that bit position. (By this, we mean that the channel values in those positions are replaced by large-magnitude negative or positive values, with the actual magnitude being unimportant.) If CRC checking fails again, then the two weakest positions are found in the first decoding output and strong "00," "01," "10," and "11" patterns are each tried in the two weak positions, with CRC checking each time. The algorithm continues until there is a CRC success or the maximum number of weak positions (usually 5, 6, or 7) have been tried without a CRC success. Observe that this algorithm is a type of iterative Chase algorithm. (The Chase algorithm is discussed in Chapter 3.) For TBCCs, a wrap-around BCJR (see [24]) is used, with only two trellis passes being necessary.

We have found, for both ZTCCs and TBCCs, that the list-BCJR and list-Viterbi algorithms have similar error-rate characteristics for similar list sizes. One advantage that the list-BCJR algorithm has over the (parallel) list-Viterbi algorithm described in Section 4.5.3 is that sorting of lists is avoided in the former case. More details on the list-BCJR algorithm can be found in [25].

## 4.6 Performance Estimates for Trellis-Based Decoders

### 4.6.1 ML Decoder Performance for Block Codes

Before we present formulas for estimating the performance of maximum-likelihood sequence decoding (Viterbi decoding) of convolutional codes on the BSC and BI-AWGN channels, we do this for linear binary block codes. This approach is taken because the block-code case is simpler conceptually, the block-code results are useful in their own right, and the block-code results provide a large step toward the convolutional-code results. We assume that the codeword length is $N$, the information word length is $K$, the code rate is $R$, and the minimum distance is $d_{\min}$.

To start, we observe that both channels are symmetric in the sense that for the BSC $\Pr(y = 1 \mid x = 1) = \Pr(y = 0 \mid x = 0)$ and for the BI-AWGN channel $p(y \mid x) = p(-y \mid -x)$, where $x$ and $y$ represent the channel input and output, respectively. Given these conditions together with code linearity, the probability of error given that some codeword $\mathbf{c}$ was transmitted is identical to the probability of error given that any other codeword $\mathbf{c}'$ was transmitted. Thus, it is convenient to assume that the all-zeros codeword $\mathbf{0}$ was transmitted in performance analyses. We can therefore bound the probability of codeword error $P_{cw}$ using the union bound as follows:

$$\begin{aligned} P_{cw} &= \sum_{\mathbf{c}} \Pr(\text{error} \mid \mathbf{c}) \Pr(\mathbf{c}) \\ &= \Pr(\text{error} \mid \mathbf{0}) \\ &= \Pr(\cup_{\hat{\mathbf{c}} \neq \mathbf{0}} \text{ decide } \hat{\mathbf{c}} \mid \mathbf{0}) \\ &\leq \sum_{\hat{\mathbf{c}} \neq \mathbf{0}} \Pr(\text{decide } \hat{\mathbf{c}} \mid \mathbf{0}), \end{aligned} \tag{4.42}$$

where $\hat{\mathbf{c}}$ denotes the decoder output. The probability $\Pr(\text{decide } \hat{\mathbf{c}} \mid \mathbf{0})$, which is the probability that the decoder chooses $\hat{\mathbf{c}}$ given that $\mathbf{0}$ was transmitted, is called the *two-codeword error probability* (or *pair-wise error probability*). It is derived under the assumption that only two codewords are involved: $\hat{\mathbf{c}}$ and $\mathbf{0}$.

As we will see, both for the BSC and for the AWGN channel, $\Pr(\text{decide } \hat{\mathbf{c}} \mid \mathbf{0})$ is a function of the Hamming distance between $\hat{\mathbf{c}}$ and $\mathbf{0}$, that is, the Hamming weight $w$ of $\hat{\mathbf{c}}$. Further, if $\hat{\mathbf{c}}$ and $\hat{\mathbf{c}}'$ both have weight $w$, then $\Pr(\text{decide } \hat{\mathbf{c}} \mid \mathbf{0}) = \Pr(\text{decide } \hat{\mathbf{c}}' \mid \mathbf{0})$ and we denote this common probability by $P_w$. In view of this, we may rewrite (4.42) as

$$P_{cw} \leq \sum_{w=d_{\min}}^{N} A_w P_w, \tag{4.43}$$

where $A_w$ is the number of codewords of weight $w$.

For the BSC, when $w$ is odd, $P_w \triangleq \Pr(\text{decide a codeword } \hat{\mathbf{c}} \text{ of weight } w \,|\mathbf{0})$ is the probability that the channel produces $\lceil w/2 \rceil$ or more errors that place the received word closer to $\hat{\mathbf{c}}$ than to $\mathbf{0}$ in a Hamming-distance sense. ($\lceil x \rceil$ is the integer greater than or equal to $x$.) Thus, we focus only on the $w$ nonzero positions of $\hat{\mathbf{c}}$ to write

$$P_w = \begin{cases} \sum\limits_{j=\lceil w/2 \rceil}^{w} \binom{w}{j,} \varepsilon^j (1-\varepsilon)^{w-j} & \text{for } w \text{ odd}, \\ \Pr(w/2) + \sum\limits_{j=w/2+1}^{w} \binom{w}{j} \varepsilon^j (1-\varepsilon)^{w-j}, & \text{for } w \text{ even}, \end{cases} \tag{4.44}$$

where the term

$$\Pr(w/2) \triangleq \frac{1}{2}\binom{w}{w/2} \varepsilon^{w/2} (1-\varepsilon)^{w/2}$$

accounts for fact that ties in the decoder are resolved arbitrarily when $w$ is even.

For the BI-AWGN channel, let $m(\mathbf{c})$ denote the channel representation of the codeword $\mathbf{c}$. For example, $m(\mathbf{0}) = \left[+\sqrt{E_c}, +\sqrt{E_c}, +\sqrt{E_c}, +\sqrt{E_c}, \ldots\right]$ and $m(\hat{\mathbf{c}}) = m([1\ 1\ 1\ 0\ \ldots]) = \left[-\sqrt{E_c}, -\sqrt{E_c}, -\sqrt{E_c}, +\sqrt{E_c}, \ldots\right]$, where $E_c$ is the average code-bit energy on the channel and is related to the code rate $R$ and the average data-bit energy $E_b$ as $E_c = RE_b$. Now, $P_w$ is the probability that the white-noise sequence results in a receiver output $\mathbf{r}$ that is closer to $m(\hat{\mathbf{c}})$ than it is to $m(\mathbf{0})$ in a Euclidean-distance sense. However, $m(\hat{\mathbf{c}})$ and $m(\mathbf{0})$ are separated by the Euclidean distance $d_{\mathrm{E}} = 2\sqrt{wE_c} = 2\sqrt{wRE_b}$. Thus, since the white noise has variance $\sigma^2 = N_0/2$ in all directions, $P_w$ is the probability that the noise in the direction of $\hat{\mathbf{c}}$ has magnitude greater than $d_{\mathrm{E}}/2 = \sqrt{wRE_b}$, that is

$$P_w = Q\left(\frac{d_{\mathrm{E}}}{2\sigma}\right) = Q\left(\sqrt{\frac{2wRE_b}{N_0}}\right). \tag{4.45}$$

The bit error probability $P_b$, another commonly used performance measure, can be obtained from the above results. We first define $A_{i,w}$ to be the number of weight-$w$ codewords produced by weight-$i$ encoder inputs. Then the probability of bit error can be bounded as

$$\begin{aligned} P_b &\leq \frac{1}{K} \sum_{w=d_{\min}}^{N} \sum_{i=1}^{K} i A_{i,w} P_w \\ &= \frac{1}{K} \sum_{w=d_{\min}}^{N} B_w P_w, \end{aligned} \tag{4.46}$$

where $B_w \triangleq \sum_{i=1}^{K} i A_{i,w}$ is the number of nonzero information bits corresponding to all of the weight-$w$ codewords. $P_w$ in the above expressions is given by (4.44) or (4.45), depending on whether the channel is BSC or BI-AWGN.

### 4.6.2 Weight Enumerators for Convolutional Codes

Because digital communication links are generally packet-based, convolutional codes are often used as block codes. That is, the inputs to a rate-$k/n$ convolutional encoder in this case are blocks of $K$ information bits, which produce blocks of $N = K(n/k)$ code bits. When a convolutional code is treated as a block code, all of the results of the preceding subsection apply. However, as we will see, it is more convenient to settle for approximations of the bounds given above. This is so because the computation of $A_w$ and $A_{i,w}$ (or $B_w$) can be demanding (although it is possible; see Chapter 8). Recall that $A_w$ is the number of weight-$w$ codewords. Or, in the context of the code's trellis, it is the number of weight-$w$ paths that diverge from the all-zeros trellis path *one or more times* in $L = K/k$ trellis stages. Finding $\{A_w\}$ is possible by computer for all practical codes, and so this is one possible avenue. Then the formulas of the previous section may be utilized.

Traditionally, however, one uses an alternative weight spectrum $\{A'_w\}$, where $A'_w$ is the number of weight-$w$ paths that diverge from the all-zeros trellis path *one time* in $L$ trellis stages. The attractiveness of $A'_w$ stems from the fact that the entire spectrum $\{A'_w\}$ is easily found analytically for small-memory codes, or by computer for any practical code. These comments apply as well to the alternative weight spectrum $\{A'_{i,w}\}$, where $A'_{i,w}$ is the number of weight-$w$ paths, corresponding to weight-$i$ encoder inputs, that diverge from the all-zeros trellis path *one time* in $L$ trellis stages.

We remark, incidentally, that a portion of a path that diverges from the correct path and remerges later is called an *error event*. Thus, as shown in Figure 4.23, we say that a path that diverges/remerges from/to the correct path multiple times has multiple error events; otherwise, it has a single error event. Further, in this context, the minimum weight among the error events is called the minimum *free distance* and is denoted by $d_{\text{free}}$.

Consider any convolutional code and its state diagram and trellis. It is possible to determine $\{A'_{i,w}\}$ via computer by having the algorithm traverse the $L$-stage trellis until all paths have been covered, updating the $A'_{i,w}$ tables along the way. Alternatively, the

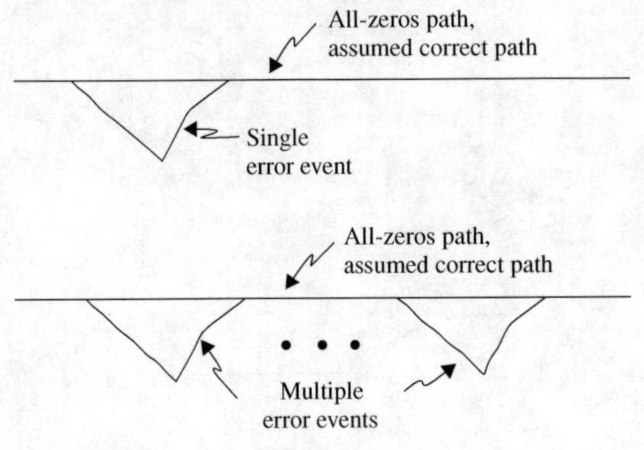


**Figure 4.23** An illustration of nonzero code paths comprising single error events and multiple error events.

state diagram can be utilized to analytically determine $\{A'_{i,w}\}$. To do this, recognize that diverging from the all-zeros trellis path one time (single error event) is equivalent to leaving state zero in the state diagram and returning to that state only once. Thus, we can count the cumulative information weight and codeword weight in either the trellis or the state diagram. However, signal flow-graph techniques allow us to perform this computation analytically using the state diagram. Note that $\{A'_w\}$ is obtainable from $\{A'_{i,w}\}$ since $A'_w = \sum_i A'_{i,w}$.

---

**Example 4.12** Consider again the rate-1/2 encoder described by $\mathbf{G}(D) = [1 + D + D^2 \quad 1 + D^2]$ whose state diagram and trellis are presented in Figure 4.8. We are interested in all of the paths that leave state $S_0 = 00$ and then return to it some time later. Thus, we split the state diagram at state $S_0$. Then, in order to exploit signal flow-graph techniques, we relabel the state-transition branches with bivariate monomials whose exponents contain the information weight and code weight. For example, the label 11|1 would be replaced by the label $I^1W^2 = IW^2$ and the label 01|0 would be replaced by the label $I^0W^1 = W$. The resulting *split state diagram* is shown in Figure 4.24.

The split state diagram is now regarded as a transform-domain linear system with input $S_0$, and output $S'_0$, whose subsystem gains are given by the new branch labels. If one obtains the transfer function $S'_0/S_0$ of this linear system, one has actually obtained the *input–output weight enumerator* (IO-WE) $A'(I, W) = \sum_{i,w} A'_{i,w} I^i W^w$. This is because taking the product of the gains of the form $I^i W^w$ that label a series of branches has the effect of adding exponents, thereby counting the information weight and code weight along the series of branches. The transfer-function computation also keeps track of the number of paths from $S_0$ to $S'_0$ for the weight profiles $(i, w)$; these are the coefficients $A'_{i,w}$.

The most obvious way to compute $A'(I, W) = S'_0/S_0$ is to write down a system of linear equations based on the split state diagram and then solve for $S'_0/S_0$. For example, from Figure 4.24 we may write

$$S_2 = IW^2 S_0 + IS_1,$$
$$S_1 = WS_2 + WS_3,$$

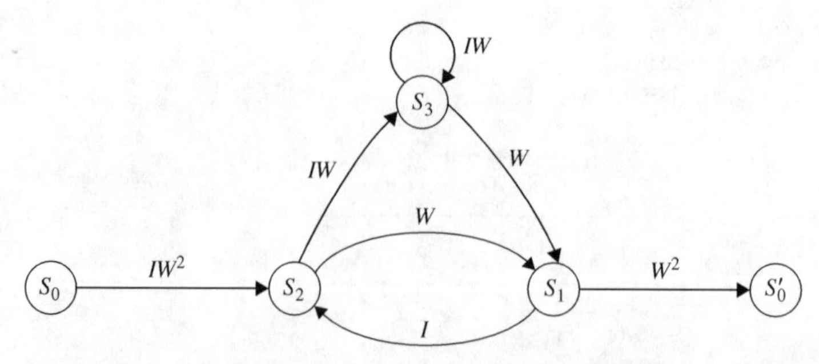


**Figure 4.24** The branch-labeled split state diagram for the $\bar{\text{R}}\bar{\text{S}}$C rate-1/2 convolutional encoder with octal generators $(7, 5)$.

$$S_3 = IWS_3 + IWS_2,$$

$$S_0' = W_2 S_1.$$

The solution is

$$A'(I, W) = S_0'/S_0 = \frac{IW^5}{1 - 2IW}$$

$$= IW^5 + 2I^2W^6 + 4I^3W^7 + 8I^4W^8 + \cdots.$$

Thus, $A'_{1,5} = 1$ (one weight-5 path created by a weight-1 encoder input), $A'_{2,6} = 2$ (two weight-6 paths created by weight-2 encoder inputs), $A'_{3,7} = 4$ (four weight-7 paths created by weight-3 encoder inputs), and so on. Note that the codeword weight enumerator $A'(W)$ may be obtained from $A'(I, W)$ as

$$A'(W) = A'(I, W)|_{I=1}$$

$$= \frac{W^5}{1 - 2W}$$

$$= W^5 + 2W^6 + 4W^7 + 8W^8 + \cdots.$$

Thus, $A'_5 = 1$ (one weight-5 path), $A'_6 = 2$ (two weight-6 paths), $A'_7 = 4$ (four weight-7 paths), and so on. Of course, $A'(W)$ may be obtained more directly by using branch labels of the form $W^w$ in the split state diagram (which effectively sets $I = 1$).

By analogy to the definition of $B_w$, we define $B'_w \triangleq \sum_{i=1}^{K} iA'_{i,w}$, the number of nonzero information bits corresponding to all of the weight-$w$ paths that diverge from state zero (at an arbitrary time) and remerge later. The values $B'_w$ may be obtained as coefficients of the enumerator

$$B'(W) = \left[\frac{d}{dI}A'(I, W)\right]\Big|_{I=1}$$

$$= \frac{W^5}{1 - 4W + 4W^2}$$

$$= W^5 + 4W^6 + 12W^7 + 32W^8 + \cdots.$$

---

**Example 4.13** An even simpler approach to arriving at $A'(I, W)$ avoids writing out a system of linear equations. The approach involves successively simplifying the split state diagram until one obtains a single branch with a single label that is equal to $A'(I, W)$. This is demonstrated in Figure 4.25 for the code of the previous example. Except for the last flow graph, whose derivation is obvious, explanations are given for the derivation of each flow graph from the one above it. For extremely large split state diagrams, one generally utilizes a computer program based on *Mason's gain rule*.

---

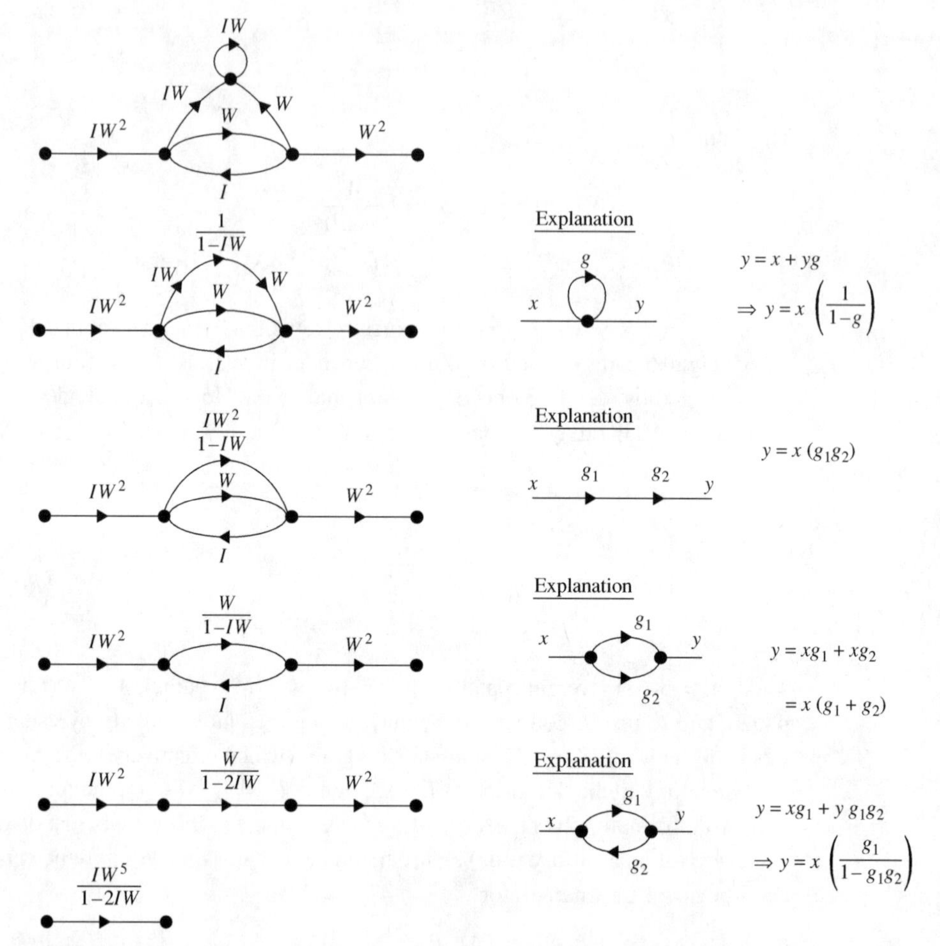


**Figure 4.25** Successive reduction of the split state diagram of Figure 4.24 to obtain the transfer function.

The techniques in the foregoing examples allow us to derive $A'(I,W)$ for information words and codewords of unlimited length. For finite-length codewords, the techniques may be simply modified to obtain an *augmented* weight enumerator $A'(I,W,B) = \sum_{i,w,b} A_{i,w,b} I^i W^w B^b$, where $A_{i,w,b}$ is the number of weight-$w$ paths spanning $b$ trellis branches, created by weight-$i$ encoder inputs. The split-state-diagram labels in this case are of the form $I^i W^w B$, where the exponent of $B$ is always unity since it labels and accounts for a single branch.

Following this procedure for our example code above results in

$$\begin{aligned} A'(I,W,B) &= \frac{IW^5B^3}{1 - IW(B + B^2)} \\ &= IW^5B^3 + I^2W^6B^4 + I^2W^6B^5 + \cdots. \end{aligned}$$

Thus, the weight-5 path consists of three trellis branches. Also, one weight-6 path consists of four branches and the other consists of five (both correspond to weight-2 inputs). Note that $A'(I,W) = A'(I,W,B)|_{B=1}$.

Suppose now we are interested in codewords in the above code corresponding to length-$K$ information words. Then we can find the weight enumerator for this situation by omitting the terms in $A'(I, W, B)$ whose $B$ exponents are greater than $K$.

### 4.6.3 ML Decoder Performance for Convolutional Codes

For packet-based communications, we make the assumption that the rate-$k/n$ convolutional encoder accepts $K$ information bits and produces $N$ code bits. As mentioned at the beginning of the previous section, traditional flow-graph techniques yield only weight-enumerator data for incorrect paths that have single error events (e.g., $A'_{i,w}$ or $A'_w$). On the other hand, our $P_{cw}$ and $P_b$ performance bounds for $(N, K)$ linear codes, equations (4.43) and (4.46), require enumerator data for all incorrect paths ($A_w$ and $A_{i,w}$). Chapter 8 shows how one can carefully obtain $A_w$ and $A_{i,w}$ from $A'_{i,w}$ or $A'_w$. Here, we make a couple of approximations that simplify our work and result in simple formulas. We will see that the approximate formulas are sufficiently accurate for most purposes.

The first approximation is that at most one error event occurs in each code block. This is clearly inaccurate for extremely long code blocks and low SNR values, but it is quite accurate otherwise, as we will see in the example below. The second approximation is that the number of different locations at which an error event may occur in a code block is $L = K/k$. The more accurate value is $L - \ell$ locations, where $\ell$ is the length of the error event measured in trellis stages, but frequently in practice $\ell \ll L$.

Given these two approximations, we may write

$$A_w \simeq LA'_w$$

and

$$A_{i,w} \simeq LA'_{i,w}.$$

Application of these relations to (4.43) and (4.46) yields

$$P_{cw} \lesssim \sum_{w=d_{\min}}^{N} LA'_w P_w = \frac{K}{k} \sum_{w=d_{\min}}^{N} A'_w P_w \tag{4.47}$$

and

$$\begin{aligned} P_b &\lesssim \frac{1}{K} \sum_{w=d_{\min}}^{N} \sum_{i=1}^{K} iLA'_{i,w} P_w \\ &= \frac{1}{K} \frac{K}{k} \sum_{w=d_{\min}}^{N} \left( \sum_{i=1}^{K} iA'_{i,w} \right) P_w \\ &= \frac{1}{k} \sum_{w=d_{\min}}^{N} B'_w P_w, \end{aligned} \tag{4.48}$$

where $B'_w$ is as defined above. As before, $P_w$ in the above expressions is given by (4.44) for the BSC and (4.45) for the BI-AWGN channel.

**Example 4.14** We demonstrate the accuracy of equations (4.47) and (4.48) in Figure 4.26, which displays these bounds together with simulation results for our running-example rate-1/2 code for which $\mu = 2$, $\mathbf{g}^{(1)}_{\text{octal}} = 7$, and $\mathbf{g}^{(2)}_{\text{octal}} = 5$. ($\{A'_w\}$ and $\{B'_w\}$ were found in Example 4.12.) The figure presents $P_{cw}$ curves for $K = 100$, 1000, and 5000. Observe that for all three $P_{cw}$ curves and the $P_b$ curve there is close agreement between the bounds (lines) and the simulation results (circles). Also included in the figure are the $P_b$ bound (4.48) and $P_b$ simulation results for the $\mu = 6$ industry-standard code for which $\mathbf{g}^{(1)}_{\text{octal}} = 117$ and $\mathbf{g}^{(2)}_{\text{octal}} = 155$ (see Section 4.3.4). It was found by computer search that, for this code, $[B'_{10}, B'_{11}, \ldots, B'_{21}] = [36, 0, 211, 0, 1404, 0, 11633, 0, 77433, 0, 502690, 0]$. Clearly, from the figure, this limited-weight-spectrum information is sufficient to obtain a tight bound.

## 4.6.4 Optimum Rate-1/2 Convolutional Codes

Historically, a convolutional code of a given rate and memory size has been called optimum if its information-weight spectrum $B'_w$ minimizes the union bound on $P_b$ given in equation (4.48), with $d_{\min}$ replaced by $d_{\text{free}}$. Thus, for that rate and memory

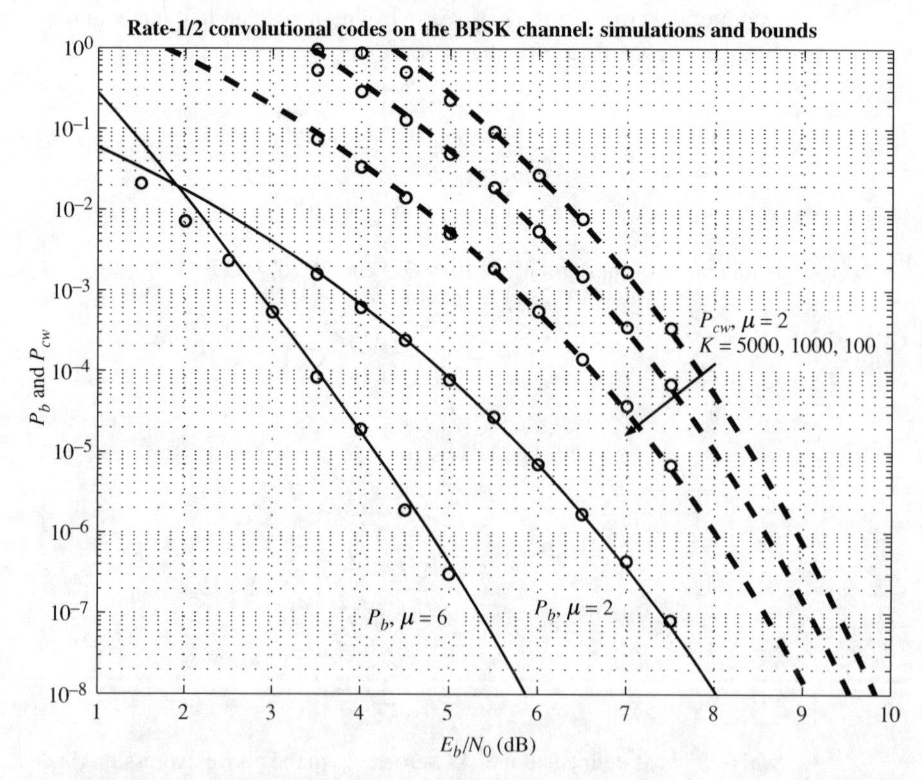


**Figure 4.26** Maximum-likelihood decoding performance bounds and simulation results for the rate-1/2 convolutional codes of Example 4.10 on the BI-AWGN channel.

**Table 4.2** Optimum rate-1/2 convolutional codes

| $\mu$ | $(g_1, g_2)$ | $d_{\text{free}}$ | $\{A'_w\}, \{B'_w\}$ for $w = d_{\text{free}}, \ldots, d_{\text{free}} + 7$ |
|---|---|---|---|
| 2 | (5, 7) | 5 | {1, 2, 4, 8, 16, 32, 64, 128} |
| | | | {1, 4, 12, 32, 80, 192, 448, 1024} |
| 3 | (15, 17) | 6 | {1, 3, 5, 11, 25, 55, 121, 267, 589} |
| | | | {2, 7, 18, 49, 130, 333, 836, 2069} |
| 4 | (23, 35) | 7 | {2, 3, 4, 16, 37, 68, 176, 432} |
| | | | {4, 12, 20, 72, 225, 500, 1324, 3680} |
| 5 | (53, 75) | 8 | {1, 8, 7, 12, 48, 95, 281, 605}, |
| | | | {2, 36, 32, 62, 332, 701, 2342, 5503} |
| 6 | (133, 171) | 10 | {11, 0, 38, 0, 193, 0, 1331, 0}, |
| | | | {36, 0, 211, 0, 1404, 0, 11 633, 0} |
| 7 | (247, 371) | 10 | {1, 6, 12, 26, 52, 132, 317, 730} |
| | | | {2, 22, 60, 148, 340, 1008, 2642, 6748} |
| 8 | (561, 753) | 12 | {11, 0, 50, 0, 286, 0, 1630, 0} |
| | | | {33, 0, 281, 0, 2179, 0, 15 035, 0} |
| 9 | (1151, 1753) | 12 | {1, 7, 19, 28, 69, 185, 411, 1010} |
| | | | {2, 21, 100, 186, 474, 1419, 3542, 9774} |
| 10 | (3345, 3613) | 14 | {14, 0, 92, 0, 426, 0, 2595, 0} |
| | | | {56, 0, 656, 0, 3708, 0, 27 503, 0} |
| 11 | (5261, 7173) | 15 | {14, 21, 34, 101, 249, 597, 1373, 3317} |
| | | | {66, 98, 220, 788, 2083, 5424, 13 771, 35 966} |
| 12 | (12 767, 16 461) | 16 | {14, 38, 35, 108, 342, 724, 1604, 4020} |
| | | | {60, 188, 288, 952, 2754, 6628, 16 606, 44 640} |
| 13 | (27 251, 37 363) | 16 | {1, 17, 38, 69, 158, 414, 944, 2210} |
| | | | {2, 99, 234, 513, 1316, 3890, 9642, 24 478 } |
| 14 | (63 057, 44 735) | 18 | {26, 0, 165, 0, 845, 0, 4844, 0} |
| | | | {133, 0, 1321, 0, 7901, 0, 54 864, 0} |

size, the optimum convolutional code possesses the largest free distance, $d_{\text{free}}$, and the smallest $B'_w$ for each $w \geq d_{\text{free}}$. An alternative definition of an optimum convolutional code would require that the code have the largest $d_{\text{free}}$ and the smallest $A'_w$ for each $w \geq d_{\text{free}}$. Often, a code that is optimum in the former sense is also optimum in the latter sense. Because rate 1/2 is most commonly used, we present in Table 4.2 optimum information-weight ($B'_w$) convolutional codes of rate 1/2 taken from [26]. Codes of different rates can be found in that reference and in [12, 3]. The generator polynomials are given in octal notation with the least-significant bit corresponding to the $D^0$ coefficient and the most-significant bit corresponding to the $D^\mu$ coefficient. For example, the entry for $\mu = 6$, (133, 171), corresponds to the polynomials $(D^6 + D^4 + D^3 + D + 1, D^6 + D^5 + D^4 + D^3 + 1)$. However, codes with this convention reversed are also optimal, as shown in Problem 10(b).

These codes were found by computer search. Specifically, for a given memory size $\mu$, the computer algorithm steps through the possible degree-$\mu$ polynomial pairs (with common-sense omissions) and traverses the trellis of each candidate pair, accumulating weight spectrum information up to a pre-specified maximum weight. (Recall that the weight spectrum is equivalent to the distance spectrum for linear codes.) See Problem 4.26 for a hint on the algorithm. This algorithm can be used to find codes for rates 1/3 and 1/4. For rates 2/3 and 3/4, puncturing of a rate-1/2 code is usually used, where a trellis search technique is used to find optimum puncturing patterns. For punctured codes, the rate-1/2 decoder places a channel value of zero in each of the punctured positions.

## Problems

**4.1** Find the power series $s(D)$ for the rational function

$$\frac{a(D)}{b(D)} = \frac{1 + D + D^2}{1 + D + D^3}$$

and note that, after a transient, its coefficients are periodic. Show that $a(D) = b(D)s(D)$.

**4.2** Show that, if the entries of the generator matrix $\mathbf{G}(D)$ are confined to the field of rational functions, $\mathbb{F}_2(D)$, then the entries of the minimal representation of the parity-check matrix $\mathbf{H}(D)$ must also be confined to $\mathbb{F}_2(D)$.

**4.3** Find $\mathbf{G}_{\text{sys}}(D)$, $\mathbf{H}_{\text{sys}}(D)$, $\mathbf{G}_{\text{poly}}(D)$, and $\mathbf{H}_{\text{poly}}(D)$ for the rate-2/3 convolutional-code generator matrix given by

$$\mathbf{G}(D) = \begin{bmatrix} \dfrac{1+D}{1+D+D^2} & 1+D & 0 \\ 1 & \dfrac{1}{1+D} & \dfrac{D}{1+D^2} \end{bmatrix}.$$

**4.4** Show that the transfer function

$$g_{ij}(D) = \frac{a_0 + a_1 D + \cdots + a_m D^m}{1 + b_1 D + \cdots + b_m D^m}$$

can be implemented as in Figure 4.2. To do so, write $c^{(j)}(D)$ as

$$c^{(j)}(D) = \left(u^{(i)}(D) \cdot \frac{1}{b(D)}\right) \cdot a(D) \tag{4.49}$$

then sketch the direct implementation of the leftmost "filtering" operation $v(D) = u^{(i)}(D) \cdot 1/b(D)$, which can be determined from the difference equation

$$v_t = u_t^{(i)} - b_1 v_{t-1} - b_2 v_{t-2} - \cdots - b_m v_{t-m}.$$

Next, sketch the direct implementation of the second filtering operation in (4.49), $c^{(j)}(D) = v(D)a(D)$, which can be determined from the difference equation

$$c_t^{(j)} = a_0 v_t^{(i)} + a_1 v_{t-1}^{(i)} + \cdots + a_m v_{t-m}^{(i)}.$$

Finally, sketch the two "filters" in cascade, the $1/b(D)$ filter followed by the $a(D)$ filter (going from left to right), and notice that the $m$ delay (memory) elements may be shared by the two filters.

**4.5** Show that Figures 4.2 and 4.3 implement equivalent transfer functions and that Figure 4.3 is the transposed form of Figure 4.2. That is, note that Figure 4.3 may be obtained from Figure 4.2 by (a) changing all nodes to adders; (b) changing all adders to nodes; (c) reversing the directions of the arrows; and (d) naming the input the output, and vice versa.

**4.6** Show that every non-systematic encoder is equivalent to a systematic encoder in the sense that they generate the same code. Show that for every non-recursive non-systematic encoder there exists an equivalent recursive systematic encoder in the sense that they generate the same code.

**4.7** Consider the rate-2/3 convolutional code in Example 4.2. Let the input for the encoder $\mathbf{G}(D)$ be $\mathbf{u}(D) = [1 \quad 1 + D]$ and find the corresponding codeword $\mathbf{c}(D) = [\mathbf{c}_1(D) \quad \mathbf{c}_2(D) \quad \mathbf{c}_3(D)]$. Find the input that yields the same codeword when the encoder is given by $\mathbf{G}_{\mathrm{sys}}(D)$. Repeat for $\mathbf{G}_{\mathrm{poly}}(D)$.

**4.8** Show that every rate-$1/n$ convolutional code has a catastrophic encoder. Repeat for rate $k/n$.

**4.9** Determine whether either of the following encoders is catastrophic:

$$\mathbf{G}(D) = \left[1 + D + D^2 + D^3 + D^4 \quad D + D^6\right],$$

$$\mathbf{G}(D) = \begin{bmatrix} (1 + D)^3 & 0 & (1 + D + D^2)(1 + D)^3 \\ (1 + D + D^2)(1 + D^2) & (1 + D + D^2)(1 + D) & D(1 + D + D^2) \end{bmatrix}.$$

**4.10** Generalize the catastrophic-code condition for rate-$1/n$ convolutional codes to the case in which the entries of $\mathbf{G}(D)$ are rational functions. Using this result, determine whether either of the following encoders is catastrophic:

$$\mathbf{G}(D) = \begin{bmatrix} \dfrac{1 + D}{1 + D + D^2} & 1 + D \end{bmatrix},$$

$$\mathbf{G}(D) = \begin{bmatrix} \dfrac{D}{1 + D + D^2} & \dfrac{1 + D^2}{1 + D^3} \end{bmatrix}.$$

**4.11** Consider the four generator matrices given in the previous two problems.

(a) Determine to which class of encoder each of these generator matrices belongs (RSC, $\bar{\text{R}}$SC, R$\bar{\text{S}}$C, or $\bar{\text{R}}\bar{\text{S}}$C). If an encoder is not systematic, find its systematic form.

(b) Sketch the Type I and Type II realizations of the systematic and non-systematic versions of each of these codes.

**4.12** Consider a rate-2/3 convolutional code with

$$\mathbf{G}_{\mathrm{poly}}(D) = \begin{bmatrix} 1 + D & 0 & 1 \\ 1 + D^2 & 1 + D & 1 + D + D^2 \end{bmatrix}.$$

Show that the memory required for the Type I and Type II realizations of $\mathbf{G}_{\text{poly}}(D)$ is $\mu = 3$ and $\mu = 6$, respectively. Show that

$$\mathbf{G}_{\text{sys}}(D) = \begin{bmatrix} 1 & 0 & \dfrac{1}{1+D} \\ 0 & 1 & \dfrac{D^2}{1+D} \end{bmatrix},$$

and that the memory required for its Type I realization is $\mu = 3$. Finally, show that the Type II realization of $\mathbf{G}_{\text{sys}}(D)$ requires only $\mu = 2$. Thus, the Type II realization of $\mathbf{G}_{\text{sys}}(D)$ is the minimal encoder as claimed in Section 4.3.3 for rate-$k/(k+1)$ convolutional codes.

**4.13** Sketch the minimal encoder realizations for the code corresponding to the following generator matrices:

(a) $\mathbf{G}(D) = \left[1+D^4 \quad 1+D^3 \quad 1+D+D^2+D^3\right]$,

(b) $\mathbf{G}(D) = \begin{bmatrix} 1+D^3 & 0 & (1+D+D^2)(1+D^3) \\ (1+D+D^2)(1+D^2) & 1+D^3 & 1+D+D^2+D^3 \end{bmatrix}.$

**4.14** Suppose a generator matrix $\mathbf{G}(D)$ is in polynomial form and the maximum degree of any polynomial in $\mathbf{G}(D)$ is two. Give the form of (4.13) specialized to this case.

**4.15** Find the multiplexed generator matrix $\mathbf{G}'$ and the multiplexed parity-check matrix $\mathbf{H}'$ for the rate-1/2 encoder matrix

$$\mathbf{G}(D) = \left[\ 1+D^2+D^3+D^4 \quad 1+D+D^4\ \right].$$

Find $\mathbf{G}_{\text{sys}}(D)$ corresponding to $\mathbf{G}(D)$ and repeat for the encoder matrix $\mathbf{G}_{\text{sys}}(D)$.

**4.16** Consider the rate-2/3 convolutional code with polynomial parity-check matrix

$$\begin{aligned}\mathbf{H}(D) &= \left[\ h_2(D) \quad h_1(D) \quad h_0(D)\ \right] \\ &= \left[\ 1+D \quad 1+D^2 \quad 1+D+D^2\ \right].\end{aligned}$$

Show that a generator matrix for this code is given by

$$\mathbf{G}(D) = \begin{bmatrix} h_1(D) & h_2(D) & 0 \\ 0 & h_0(D) & h_1(D) \end{bmatrix}.$$

Find the multiplexed generator matrix $\mathbf{G}'$. Find $\mathbf{G}_{\text{sys}}(D)$ and find $\mathbf{G}'$ for $\mathbf{G}_{\text{sys}}(D)$.

**4.17** Assuming the BI-AWGNC, simulate Viterbi decoding of the rate-1/2 convolutional code whose encoder matrix is given by

$$\mathbf{G}(D) = \left[\ 1+D^2+D^3+D^4 \quad 1+D+D^4\ \right].$$

Plot the bit-error rate from $P_b = 10^{-1}$ to $P_b = 10^{-6}$. Repeat for the BCJR decoder and comment on the performance of the two decoders.

**4.18** Assuming the BI-AWGNC, simulate Viterbi decoding of the rate-2/3 convolutional code whose parity-check matrix is given by (see Problem 4.16)

$$\begin{aligned}\mathbf{H}(D) &= \begin{bmatrix} h_2(D) & h_1(D) & h_0(D) \end{bmatrix} \\ &= \begin{bmatrix} 1+D & 1+D^2 & 1+D+D^2 \end{bmatrix}.\end{aligned}$$

Plot the bit-error rate from $P_b = 10^{-1}$ to $P_b = 10^{-6}$. Repeat for the BCJR decoder and comment on the performance of the two decoders.

**4.19** Show that

$$\max(x, y) = \log\left(\frac{e^x + e^y}{1 + e^{-|x-y|}}\right).$$

*Hint*: First suppose $x > y$.

**4.20** Derive the BCJR algorithm for the BSC. Initialization from the channel output values proceeds as follows. Let $y_i \in \{0, 1\}$ be the BSC output and $c_i \in \{0, 1\}$ be the BSC output at time $i$. Define $\varepsilon = \Pr(y_i = b^c \mid c_i = b)$ to be the error probability. Then

$$\Pr(c_i = b \mid y_i) = \begin{cases} 1 - \varepsilon, & \text{when } y_i = b, \\ \varepsilon, & \text{when } y_i = b^c, \end{cases}$$

and, from this, we have

$$L(c_i \mid y_i) = (-1)^{y_i} \log\left(\frac{1-\varepsilon}{\varepsilon}\right).$$

**4.21** Show that in the log-domain BCJR algorithm we may alternatively use the branch metric[3].

**4.22** Show that there is a BCJR algorithm equivalent to the one presented in this chapter, as follows. Define

$$\begin{aligned}\min{}^*(x, y) &\triangleq -\log(e^{-x} + e^{-y}) \\ &= \min(x, y) + \log\left(1 + e^{-|x-y|}\right).\end{aligned}$$

A similar definition exists when more than two variables are involved, for example, $\min^*(x, y, z) = -\log(e^{-x} + e^{-y} + e^{-z})$. Define also

$$\begin{aligned}\breve{\alpha}_l(s) &\triangleq -\log(\alpha_l(s)) \\ &= -\log \sum_{s'} \exp(-\breve{\alpha}_{l-1}(s') - \breve{\gamma}_l(s', s)) \\ &= \min{}^*_{s'}\{\breve{\alpha}_{l-1}(s') + \breve{\gamma}_l(s', s)\}, \\ \breve{\gamma}_l(s', s) &\triangleq -\log \gamma_l(s', s) \\ &= \frac{\|y_l - c_l\|^2}{2\sigma^2} \text{ (ignores a useless constant)},\end{aligned}$$

[3] $\tilde{\gamma}_l(s', s) = u_l y_l^u/\sigma^2 + p_l y_l^p/\sigma^2$

$$\begin{aligned}\breve{\beta}_{l-1}(s') &\triangleq -\log\left(\beta_{l-1}(s')\right) \\ &= -\log\left(\sum_s \exp\left(-\breve{\beta}_l(s) - \breve{\gamma}_l(s',s)\right)\right) \\ &= \min_s^*\left\{\breve{\beta}_l(s) + \breve{\gamma}_l(s',s)\right\}.\end{aligned}$$

Specify the initial conditions and show that $L(u_l)$ can be computed as

$$\begin{aligned}L(u_l) &= \log\left[\frac{\sum_{U^+}\alpha_{l-1}(s')\,\gamma_l(s',s)\,\beta_l(s)}{\sum_{U^-}\alpha_{l-1}(s')\,\gamma_l(s',s)\,\beta_l(s)}\right] \\ &= \min_{U^+}{}^*\left[\breve{\alpha}_{l-1}(s') + \breve{\gamma}_l(s',s) + \breve{\beta}_l(s)\right] \\ &\quad - \min_{U^-}{}^*\left[\breve{\alpha}_{l-1}(s') + \breve{\gamma}_l(s',s) + \breve{\beta}_l(s)\right].\end{aligned}$$

**4.23** Repeat Example 4.12 for the RSC code with

$$\mathbf{G}(D) = \begin{bmatrix} 1 & \dfrac{1+D^2}{1+D+D^2} \end{bmatrix}.$$

Find also the augmented weight enumerator $A'(I,W,B)$. You should find, after dividing the denominator of $A'(I,W,B)$ into its numerator, that the series begins as $I^3W^5B^3 + I^2W^6B^4 + \left(I^3W^7 + I^4W^6\right)B^5 + \cdots$ . Show in the state diagram of the code the four paths corresponding to these four terms.

**4.24** Repeat Example 4.13 and Figure 4.25 for the RSC code with

$$\mathbf{G}(D) = \begin{bmatrix} 1 & \dfrac{1+D^2}{1+D+D^2} \end{bmatrix}.$$

Compare your results with those of the $\bar{\text{R}}\bar{\text{S}}$C code of Figure 4.19 for which $\mathbf{G}(D) = \begin{bmatrix} 1+D+D^2 & 1+D \end{bmatrix}$.

**4.25** Consider a memory-8, rate-1/2 convolutional code having some $d_{\text{free}}$ and some $A_{d_{\text{free}}}$. Suppose it is used first as a zero-terminated convolutional code and then a tail-biting convolutional code and that $d_{\text{free}}$ does not change in the latter case (and $A_{d_{\text{free}}}$ does not change by much). Assuming a data word length of 150 bits, using a union bound expression, show that for maximum-likelihood decoding the TBCC is superior to the ZTCC asymptotically (large SNR) by $10\log_{10}(R_{TBCC}/R_{ZTCC}) = 10\log_{10}(158/150) \approx 0.23$ dB.

**4.26** Write a computer program to check that the first few terms of the weight spectrum of the optimum rate-1/2, memory-10 convolutional code are given by $A'_{14} = 14$, $A'_{16} = 92$, $A'_{18} = 426$, $A'_{20} = 2595$, with $A'_d = 0$ for $d$ odd and $d < 14$. (See Table 4.2.) The program should have a model of the code's trellis and should count the paths of weight less than 21 that leave the zero state and remerge some time later.

**4.27** Derive the expression given for the Fano metric on the BI-AWGN channel.

**4.28**

(a) Show that if you reverse the order of the taps on a rate-1/2 convolutional code, the weight spectrum and therefore the performance will be unchanged.

(b) Argue that a corollary to this fact is that, if the coefficients of the generators of an optimal rate-$1/n$ convolutional code are reversed, the resulting code will also be optimal. *Hint*: How are the input-word/output-word mappings of the two cases related? (Here, an output word is a binary word at one of the $n$ binary adder outputs.)

**4.29** We showed in Section 4.4.1 that, for rate-1/2 $\bar{\text{R}}\bar{\text{S}}\text{C}$ codes for which $\mathbf{G}(D) = \left[g^{(1)}(D)\ g^{(2)}(D)\right]$, we have $\mathbf{H}(D) = \left[g^{(2)}(D)\ g^{(1)}(D)\right]$. Show for rate-1/3 $\bar{\text{R}}\bar{\text{S}}\text{C}$ codes with $\mathbf{G}(D) = \left[g^{(1)}(D)\ g^{(2)}(D)\ g^{(3)}(D)\right]$, that

$$\mathbf{H}(D) = \begin{bmatrix} g^{(2)}(D) & g^{(1)}(D) & 0 \\ g^{(3)}(D) & 0 & g^{(1)}(D) \end{bmatrix}$$

is a valid parity-check matrix. Next, find $\mathbf{H}(D)$ for the rate-1/4 $\bar{\text{R}}\bar{\text{S}}\text{C}$ code case.

## References


[1] P. Elias, "Coding for noisy channels," *IRE Convention Record*, vol. 3, no. 4 pp. 37–47, 1955.

[2] G. D. Forney, Jr., "Convolutional codes I: Algebraic structure," *IEEE Transactions on Information Theory*, vol. 16, no. 11, pp. 720–738, November 1970.

[3] R. Johannesson and K. Zigangirov, *Fundamentals of Convolutional Coding*, New York, IEEE Press, 1999.

[4] A. J. Viterbi, "Error bounds for convolutional codes and an asymptotically optimum decoding algorithm," *IEEE Transactions on Information Theory*, vol. 13, no. 4, pp. 260–269, April 1967.

[5] G. D. Forney, Jr., "The Viterbi algorithm," *Proceedings of the IEEE*, pp. 268–278, March 1973.

[6] L. Bahl, J. Cocke, F. Jelinek, and J. Raviv, "Optimal decoding of linear codes for minimizing symbol error rate," *IEEE Transactions on Information Theory*, vol. 20, no. 3, pp. 284–287, March 1974.

[7] N. Seshadri and C-E. Sundberg, "List Viterbi decoding algorithms with applications," *IEEE Transactions on Communications*, pp. 313–323, February–April 1994.

[8] R. Shao, S. Lin, and M. Fossorier, "Two decoding algorithms for tailbiting codes," *IEEE Transactions on Communications*, vol. 51, no. 10, pp. 1658–1665, October 2003.

[9] R. M. Fano, "A heuristic discussion of probabilistic decoding," *IEEE Transactions on Information Theory*, pp. 64–74, April 1963.

[10] A. J. Viterbi, "Convolutional codes and their performance in communication systems," *IEEE Transactions on Communication Technology*, vol. 19, no. 5, pp. 751–772, October 1971.

[11] A. V. Oppenheim and R. W. Schafer, *Discrete-Time Signal Processing*, Englewood Cliffs, NJ, Prentice-Hall, 1989.

[12] S. Lin and D. J. Costello, Jr., *Error Control Coding*, 2nd ed., New Saddle River, NJ, Prentice-Hall, 2004.

[13] S. Wicker, *Error Control Systems for Digital Communication and Storage*, Englewood Cliffs, NJ, Prentice-Hall, 1995.

[14] C.-Y. Lou, B. Daneshrad, and R. D. Wesel, "Convolutional-code-specific CRC code design," *IEEE Transactions on Communications*, vol. 63, no. 10, pp. 3459–3470, October 2015.

[15] H. Yang, E. Liang, M. Pan, and R. D. Wesel, "CRC-aided list decoding of convolutional codes in the short blocklength regime," *IEEE Transactions on Information Theory,* vol. 68, no. 6, pp. 3744–3766, June 2022.

[16] A. Baldauf, A. Belhouchat, S. Kalantarmoradian, A. Sung-Miller, D. Song, N. Wong, and R. D. Wesel, "Efficient computation of Viterbi decoder reliability with an application to variable-length coding," *IEEE Transactions on Communications*, vol. 70, no. 9, pp. 5711–5723, September 2022.

[17] R. Schiavone, R. Garello, and G. Liva, "Performance improvement of space missions using convolutional codes by CRC-aided list Viterbi algorithms," *IEEE Access*, vol. 11, pp. 55925–55937, June 2023.

[18] C. Hulse, "FPGA implementation of decoders for CRC-aided tail-biting convolutional codes," Master's thesis, ECE Department, UCLA, 2022.

[19] J. King, W. E. Ryan, and R. Wesel, "Efficient maximum-likelihood decoding for TBCC and CRC-TBCC codes via parallel list Viterbi," *2023 International Symposium on Topics in Coding*.

[20] M. Fossorier and S. Lin, "Differential trellis decoding of convolutional codes," *IEEE Transactions on Information Theory*, vol. 46, no. 5, pp. 1046–1053, May 2000.

[21] J. L. Massey, "Variable-length codes and the Fano metric," *IEEE Transactions on Information Theory*, pp. 196–198, January 1972.

[22] J. M. Wozencraft and I. M. Jacobs, *Principles of Communication Engineering*, Chichester, Wiley, 1965.

[23] P. Robertson, E. Villebrun, and P. Hoeher, "A comparison of optimal and suboptimal MAP decoding algorithms operating in the log domain," *Proceedings of the 1995 International Conference on Communications*, pp. 1009–1013.

[24] J. B. Anderson and S. M. Hladik, "Tailbiting MAP decoders," *IEEE Journal on Selected Areas in Communications*, vol. 16, no. 2, pp. 297–302, February 1998.

[25] M. Coşkun, G. Durisi, T. Jerkovits, G. Liva, W. Ryan, B. Stein, and F. Steiner, "Efficient error-correcting codes in the short blocklength regime," *Physical Communication*, vol. 34, pp. 66–79, June 2019.

[26] P. Robertson, E. Villebrun, and P. Hoeher, "A comparison of optimal and suboptimal MAP decoding algorithms operating in the log domain," *Proceedings of the 1995 International Conference on Communications*, November 1999, pp. 317–319.

# 5 Low-Density Parity-Check Codes

Low-density parity-check (LDPC) codes are a class of linear block codes with implementable decoders that provide near-capacity performance on a large set of data-transmission and data-storage channels. LDPC codes were invented by Gallager in his 1960 doctoral dissertation [1] and were mostly ignored during the 35 years that followed. One notable exception is the important work of Tanner in 1981 [2], in which Tanner generalized LDPC codes and introduced a graphical representation of LDPC codes, now called a Tanner graph. The study of LDPC codes was resurrected in the mid-1990s with the work of MacKay, Luby, and others [3–6], who noticed, apparently independently of Gallager's work, the advantages of linear block codes with sparse (low-density) parity-check matrices.

This chapter introduces LDPC codes and creates a foundation for further study of LDPC codes in later chapters. We start with the fundamental representations of LDPC codes via parity-check matrices and Tanner graphs. We then learn about the decoding advantages of linear codes that possess sparse parity-check matrices. We will see that this sparseness characteristic makes the code amenable to various iterative decoding algorithms, which in many instances provide near-optimal performance. Gallager [1] of course recognized the decoding advantages of such low-density parity-check codes and he proposed a decoding algorithm for the BI-AWGNC and a few others for the BSC. In this chapter, we present these decoding algorithms together with several others, most of which are related to Gallager's original algorithms. We point out that some of the algorithms were independently obtained by other coding researchers (e.g., MacKay and Luby [4, 5]), who were unaware of Gallager's work at the time, as well as by researchers working on graph-based problems unrelated to coding [7].

## 5.1 Representations of LDPC Codes

### 5.1.1 Matrix Representation

We shall consider only binary LDPC codes for the sake of simplicity, although LDPC codes can be generalized to nonbinary alphabets as is done in Chapter 15. A *low-density parity-check code* is a linear block code given by the null space $\{\mathbf{v}:\mathbf{v}\mathbf{H}^{\mathrm{T}} = \mathbf{0}\}$ of an $m \times n$ parity-check matrix $\mathbf{H}$ that has a low density of 1s. A *regular LDPC code* is a linear block code whose parity-check matrix $\mathbf{H}$ has column weight $g$ and row weight $r$, where $r = g(n/m)$ and $g \ll m$. If $\mathbf{H}$ is low density, but its row and column weight are not both constant, then the code is an *irregular LDPC code*. For irregular LDPC codes, the various row and column weights are determined by one of the code-design procedures discussed in subsequent chapters. For reasons that will become apparent later, almost all LDPC code constructions impose the following additional structural property on $\mathbf{H}$: no two rows (or two columns) have more than one position in common that contains a nonzero element. This property is called the *row–column constraint*, or simply the *RC constraint*.

The descriptor "low density" is unavoidably vague and cannot be precisely quantified, although a density of 0.01 or lower can be called low density (1% or fewer of the entries of $\mathbf{H}$ are 1s). As will be seen later in this chapter, the density need only be sufficiently low to permit effective iterative decoding. This is in fact the key innovation behind the invention of LDPC codes. As is well known, optimum (e.g., maximum-likelihood) decoding of the general linear block code that is useful for applications is not possible due to the vast complexity involved. The low-density aspect of LDPC codes accommodates iterative decoding, which typically has near-maximum-likelihood performance at error rates of interest for many applications. These ideas will be made clearer as the reader progresses through this chapter.

As will be seen below and in later chapters, the construction of LDPC codes usually involves the construction of $\mathbf{H}$, which need not be full rank. In this case, the code rate $R$ for a regular LDPC code is bounded as

$$R \geq 1 - \frac{m}{n} = 1 - \frac{g}{r},$$

with equality when $\mathbf{H}$ is full rank.

### 5.1.2 Graphical Representation

The *Tanner graph* of an LDPC code is analogous to the trellis of a convolutional code in that it provides a complete representation of the code and it aids in the description of decoding algorithms. A Tanner graph is a *bipartite graph* (introduced in Chapter 2), that is, a graph whose nodes may be separated into two types, with edges connecting only nodes of different types. The two types of nodes in a Tanner graph are the *variable nodes* (or *code-bit nodes*) and the *check nodes* (or *constraint nodes*), which we denote by VN and CN, respectively. The Tanner graph of a code is drawn as follows: CN $i$ is connected to VN $j$ whenever element $h_{ij}$ in $\mathbf{H}$ is a 1. Observe from this rule that there are $m$ CNs in a Tanner graph, one for each check equation, and $n$ VNs, one for each code bit. Further, the $m$ rows of $\mathbf{H}$ specify the $m$ CN connections, and the $n$ columns of $\mathbf{H}$ specify the $n$ VN connections. Accordingly, the allowable $n$-bit words represented by the $n$ VNs are precisely the codewords in the code. Throughout this book, we shall use both the notation CN $i$ and VN $j$ and the notation CN $c_i$ and VN $v_j$, where $v_j$ is the $j$th variable node or the $j$th code bit, depending on context.

**Example 5.1** Consider a (10, 5) linear block code with $w_c = 2$ and $w_r = 4$, with the following **H** matrix:

$$\mathbf{H} = \begin{bmatrix} 1 & 1 & 1 & 1 & 0 & 0 & 0 & 0 & 0 & 0 \\ 1 & 0 & 0 & 0 & 1 & 1 & 1 & 0 & 0 & 0 \\ 0 & 1 & 0 & 0 & 1 & 0 & 0 & 1 & 1 & 0 \\ 0 & 0 & 1 & 0 & 0 & 1 & 0 & 1 & 0 & 1 \\ 0 & 0 & 0 & 1 & 0 & 0 & 1 & 0 & 1 & 1 \end{bmatrix}.$$

The Tanner graph corresponding to **H** is depicted in Figure 5.1. Observe that VNs 0, 1, 2, and 3 are connected to CN 0 in accordance with the fact that, in the zeroth row of **H**, $h_{00} = h_{01} = h_{02} = h_{03} = 1$ (all others are zero). Observe that analogous situations hold for CNs 1, 2, 3, and 4, which correspond to rows 1, 2, 3, and 4 of **H**, respectively. Note, as follows from $\mathbf{vH}^{\mathrm{T}} = \mathbf{0}$, that the bit values connected to the same check node must sum to zero (mod 2). We may also proceed along columns to construct the Tanner graph. For example, note that VN 0 is connected to CNs 0 and 1 in accordance with the fact that, in the zeroth column of **H**, $h_{00} = h_{10} = 1$. The sum of rows of **H** is the all-zero vector, so **H** is not full rank and $R \geq 1 - 5/10$. It is easily seen that the first row is dependent (it is the sum of the other rows). From this, we have rank(**H**) = 4 and $R = 1 - 4/10 = 3/5$.

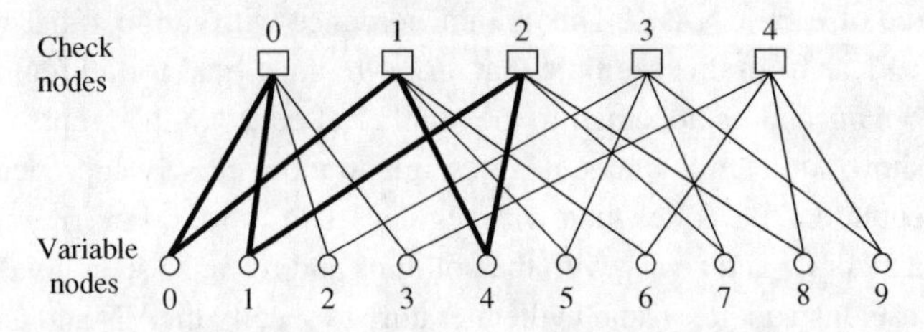


**Figure 5.1** The Tanner graph for the code given in Example 5.1.

The Tanner graph of an LDPC code acts as a blueprint for the iterative decoder in the following way. Each of the nodes acts as a locally operating processor and each edge acts as a bus that conveys information from a given node to each of its neighbors. The information conveyed is typically probabilistic information, for example, log-likelihood ratios (LLRs), pertaining to the values of the bits assigned to the variable nodes. The LDPC decoder is initiated by $n$ LLRs from the channel, which are received by the $n$ VN processors. At the beginning of each half-iteration in the basic iterative decoding algorithm, each VN processor takes inputs from the channel and each of its neighboring CNs, and from these computes outputs for each one of its neighboring CN processors. In the next half-iteration, each CN processor takes inputs from each of its neighboring VNs, and from these computes outputs for each one of its neighboring VN processors. The VN $\leftrightarrow$ CN iterations continue until a codeword is found or until the preset maximum number of iterations has been reached.

The effectiveness of the iterative decoder depends on a number of structural properties of the Tanner graph on which the decoder is based. Observe the six thickened edges in Figure 5.1. A sequence of edges such as these, which form a closed path, is called a *cycle* (see Chapter 2). We are interested in cycles because short cycles degrade the performance of the iterative decoding algorithms employed by LDPC codes. This fact will be made evident in the discussion of the decoding algorithms later in this chapter, but it can also be seen from the brief algorithm description in the previous paragraph. It should be clear from the description that cycles force the decoder to operate locally in some portions of the graph (e.g., continually around a short cycle) so that a globally optimum solution is impossible. Observe also from the decoder description the necessity of a low-density matrix **H**: at high densities (about half of the entries are 1s), many short cycles will exist, thus precluding the use of an iterative decoder.

The *length of a cycle* is equal to the number of edges that form the cycle, so the length of the cycle in Figure 5.1 is 6. A cycle of length $l$ is often called an $l$-cycle. The minimum cycle length in a given bipartite graph is called the graph's *girth*. The girth of the Tanner graph for the example code is clearly 6. The shortest possible cycle in a bipartite graph is clearly a length-4 cycle, and such cycles manifest themselves in the **H** matrix as four 1s that lie on the four corners of a rectangular submatrix of **H**. Observe that the RC constraint eliminates length-4 cycles.

The Tanner graph in the above example is regular: each VN has two edge connections and each CN has four edge connections. We say that the *degree* of each VN is 2 and the degree of each CN is 4. This is in accordance with the fact that $g = 2$ and $r = 4$. It is also clear from this example that $mr = ng$ must hold for all regular LDPC codes, since both $mr$ and $ng$ are equal to the number of edges in the graph.

As will be shown in later chapters, it is possible to more closely approach capacity limits with irregular LDPC codes than with regular LDPC codes. For irregular LDPC codes, the parameters $g$ and $r$ vary with the columns and rows, so such notation is not useful in this case. Instead, it is usual in the literature to specify the VN and CN *degree-distribution polynomials*, denoted by $\lambda(X)$ and $\rho(X)$, respectively. In the polynomial

$$\lambda(X) = \sum_{d=1}^{d_v} \lambda_d X^{d-1}, \tag{5.1}$$

$\lambda_d$ denotes the fraction of all edges connected to degree-$d$ VNs and $d_v$ denotes the maximum VN degree. Similarly, in the polynomial

$$\rho(X) = \sum_{d=1}^{d_c} \rho_d X^{d-1}, \tag{5.2}$$

$\rho_d$ denotes the fraction of all edges connected to degree-$d$ CNs and $d_c$ denotes the maximum CN degree. Note that, for the regular code above, for which $g = 2$ and $r = 4$, we have $\lambda(X) = X$ and $\rho(X) = X^3$.

Let us denote the number of VNs of degree $d$ by $N_v(d)$ and the number of CNs of degree $d$ by $N_c(d)$. Let us further denote by $E$ the number of edges in the graph. Then it can be shown (Problem 5.3) that

$$E = \frac{n}{\int_0^1 \lambda(X)dX} = \frac{m}{\int_0^1 \rho(X)dX}, \tag{5.3}$$

$$N_v(d) = E\lambda_d/d = \frac{n\lambda_d/d}{\int_0^1 \lambda(X)dX}, \tag{5.4}$$

$$N_c(d) = E\rho_d/d = \frac{m\rho_d/d}{\int_0^1 \rho(X)dX}. \tag{5.5}$$

From the two expressions for $E$, we may easily conclude that the code rate is bounded as

$$R \geq 1 - \frac{m}{n} = 1 - \frac{\int_0^1 \rho(X)dX}{\int_0^1 \lambda(X)dX}. \tag{5.6}$$

The polynomials $\lambda(X)$ and $\rho(X)$ represent a Tanner graph's degree distributions from an "edge perspective." The degree distributions may also be represented from a "node perspective" using the notation $\tilde{\lambda}(X)$ and $\tilde{\rho}(X)$, where the coefficient $\tilde{\lambda}_d$ is the fraction of all VNs that have degree $d$ and $\tilde{\rho}_d$ is the fraction of CNs that have degree $d$. It is easily shown (Problem 5.4) that

$$\tilde{\lambda}_d = \frac{\lambda_d/d}{\int_0^1 \lambda(X)dX}, \tag{5.7}$$

$$\tilde{\rho}_d = \frac{\rho_d/d}{\int_0^1 \rho(X)dX}. \tag{5.8}$$

## 5.2 Classifications of LDPC Codes

As described in the next chapter, the original LDPC codes are random in the sense that their parity-check matrices possess little structure. This is problematic in that both encoding and decoding become quite complex when the code possesses no structure beyond being a linear code. More recently, LDPC codes with structure have been constructed.

The most obvious type of structure is that of a cyclic code. As discussed in Chapter 3, the encoder of a cyclic code consists of a single length-$(n-k)$ shift register, some binary adders, and a gate. The nominal parity-check matrix $\mathbf{H}$ of a cyclic code is an $n \times n$ circulant; that is, each row is a cyclic shift of the one above it, with the first row a cyclic shift of the last row. The implication of a sparse circulant matrix $\mathbf{H}$ for LDPC decoder complexity is substantial: because each check equation is closely related to its predecessor and its successor, implementation can be vastly simplified compared with the case of random sparse $\mathbf{H}$ matrices for which wires are randomly routed. However, beside being regular, a drawback of cyclic LDPC codes is that the nominal $\mathbf{H}$ matrix is $n \times n$, independently of the code rate, implying a more complex decoder. Another drawback is that the known cyclic LDPC codes tend to have large row weights, which makes decoder implementation challenging. On the other hand, as discussed in Chapter 11, cyclic LDPC codes tend to have large minimum distances and very low iteratively decoded error-rate floors. (See a discussion of floors in Section 5.4.6.)

Quasi-cyclic (QC) codes also possess structure, leading to simplified encoder and decoder designs. Additionally, they permit more flexibility in code design, particularly irregularity, and, hence, lead to improved codes relative to cyclic LDPC codes. The **H** matrix of a QC code is generally represented as an array of circulants, for example

$$\mathbf{H} = \begin{bmatrix} A_{11} & \cdots & A_{1N} \\ \vdots & & \vdots \\ A_{M1} & \cdots & A_{MN} \end{bmatrix}, \tag{5.9}$$

where each matrix $A_{rc}$ is a $Q \times Q$ circulant. For LDPC codes, the circulants must be sparse, and in fact weight-1 circulants, which means that the weight of each row and column of the circulants is 1, are common. To effect irregularity, some of the circulants may be the all-zero $Q \times Q$ matrix using a technique called masking (see Chapter 11 and 12). Chapter 3 discusses how the QC characteristic leads to simplified encoders. Chapter 6 discusses how QC codes permit simplified, modular decoders. The second half of the book presents many code-design techniques that result in QC codes.

In addition to partitioning LDPC codes into three classes – cyclic, quasi-cyclic, and random (but linear) – the LDPC code-construction techniques can be partitioned as well. The first class of construction techniques can be described as algorithmic or computer-based. These techniques will be introduced in Chapter 6 and covered in greater detail in Chapters 8 and 9. The second class of construction techniques consists of those based on finite mathematics, including algebra, combinatorics, and graph theory. These techniques are covered in Chapters 11 to 15. We note that the computer-based construction techniques can lead to either random or structured LDPC codes. The mathematical construction techniques generally lead to structured LDPC codes, although exceptions exist.

### 5.2.1 Generalized LDPC Codes

It is useful in the study of iteratively decodable codes to consider generalizations of LDPC codes. Tanner was the first to generalize Gallager's low-density parity-check code idea in his pioneering work [2], the paper in which Tanner graphs were introduced. In the Tanner graph of a generalized LDPC (G-LDPC) code, depicted in Figure 5.2, constraint nodes are more general than single parity-check (SPC) constraints. In fact, the variable nodes may represent more complex codes than the repetition codes they represent in standard LDPC codes [8], but we shall not consider such "doubly generalized" LDPC codes here.

In Figure 5.2, the CNs $C_0, C_1, \ldots, C_{m_c-1}$ signify the $m_c$ code constraints placed on the code bits connected to the CNs. Note that, while a bipartite graph gives a complete description of an LDPC code, for a generic G-LDPC code the specifications of the

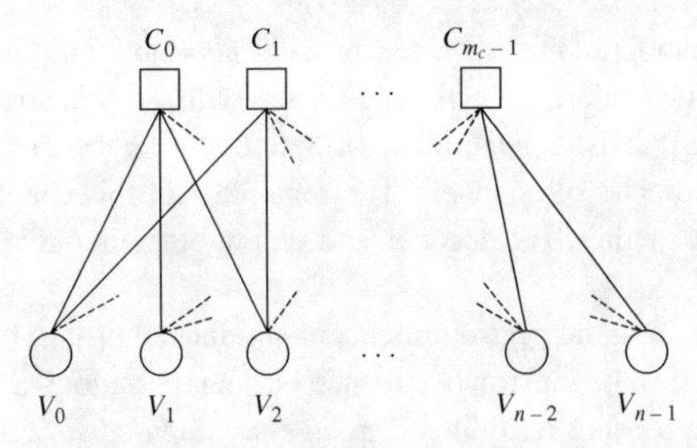


**Figure 5.2** A Tanner graph for a G-LDPC code. Circle nodes represent code bits as before, but the square nodes represent generalized constraints.

component (linear) block codes are also required. Let us define $m$ as the summation of all the redundancies introduced by the $m_c$ CNs, namely

$$m = \sum_{i=0}^{m_c-1} m_i,$$

where $m_i = n_i - k_i$ and $(n_i, k_i)$ represent the parameters of the component code for constraint node $C_i$, whose parity-check matrix is denoted by $\mathbf{H}_i$. Then the rate of a length-$n$ G-LDPC code satisfies

$$R \geq 1 - \frac{m}{n},$$

with equality only if all of the check equations derived from the CNs are linearly independent.

Let $V = \{V_j\}_{j=0}^{n-1}$ be the set of $n$ VNs and $C = \{C_i\}_{i=0}^{m_c-1}$ be the set of $m_c$ CNs in the bipartite graph of a G-LDPC code (Figure 5.2). The connection between the nodes in $V$ and $C$ can be summarized in an $m_c \times n$ adjacency matrix $\mathbf{\Gamma}$. We now introduce the relationship between the adjacency matrix $\mathbf{\Gamma}$ and the parity-check matrix $\mathbf{H}$ for a G-LDPC code.

While $\mathbf{\Gamma}$ for an LDPC code serves as its parity-check matrix, for a G-LDPC code, to obtain $\mathbf{H}$, one requires also knowledge of the parity-check matrices $\mathbf{H}_i$ of the CN component codes. The $n_i$ 1s in the $i$th binary row of $\mathbf{\Gamma}$ indicate which of the $n$ G-LDPC code bits are constrained by constraint node $C_i$. Because the parity checks corresponding to $C_i$ are represented by the rows of $\mathbf{H}_i$, $\mathbf{H}$ is easily obtained by replacing the $n_i$ 1s in row $i$ of $\mathbf{\Gamma}$ by the columns of $\mathbf{H}_i$, for all $i$. The zeros in row $i$ are replaced by a length-$m_i$ all-zero column vector. Note that this procedure allows for the case where $C_i$ is an SPC code, in which case $\mathbf{H}_i$ has only one row. Note also that $\mathbf{H}$ will be of size $m \times n$, where $m = \sum_{i=0}^{m_c-1} m_i$.

## 5.3 Message Passing and the Turbo Principle

The key innovation behind LDPC codes is the low-density nature of the parity-check matrix, which facilitates iterative decoding. Although the various iterative decoding algorithms are suboptimal, they often provide near-optimal performance, at least for the error rates of interest. In particular, we will see that the so-called *sum–product algorithm* (SPA) is a general algorithm that provides near-optimal performance across a broad class of channels. The remainder of this chapter is dedicated to the development of the SPA decoder and a few other iterative decoding algorithms.

Toward the goal of presenting the near-optimal sum–product algorithm for the iterative decoding of LDPC codes, it is instructive to step back and first look at the larger picture of generic message-passing decoding. *Message-passing decoding* refers to a collection of low-complexity decoders working in a distributed fashion to decode a received codeword in a concatenated coding scheme. Variants of the SPA and other message-passing algorithms have been invented independently in a number of different contexts, including *belief propagation* for inference in Bayesian networks and turbo decoding for parallel concatenated convolutional codes (which are the original "turbo codes"). Researchers in belief propagation liken the algorithm to the distributed counting of soldiers, and the inventors of turbo codes were inspired by turbo-charged engines in which the engine is supplied an increased amount of air by exploiting its own exhaust in a feedback fashion. In this section we develop an intuition for message-passing decoding and the so-called turbo principle through a few simple examples. This will be followed by a detailed presentation of the SPA message-passing decoder in the next section for four binary-input channels: the BEC, the BSC, the AWGNC, and the independent Rayleigh channel.

We can consider an LDPC code to be a generalized concatenation of many SPC codes. A message-passing decoder for an LDPC code employs an individual decoder for each SPC code and these decoders operate cooperatively in a distributed fashion to determine the correct code bit values. As indicated in the discussion above, Tanner graphs often serve as decoder models. In the context of a message-passing decoder, the check nodes represent SPC decoders that work cooperatively to decode the received word.

---

**Example 5.2** As our first decoding example, Figure 5.3 shows how an SPC product code may be represented as an LDPC code and iteratively decoded. Figure 5.3(a) gives the product code array and Figure 5.3(b) gives the parity-check matrix. The graph-based decoder model is depicted in Figure 5.3(c). The top check nodes in the graph represent the row SPC decoders and the bottom check nodes represent the column SPC decoders. Figure 5.3(d) presents an example decoding on the BEC, which may be taken to be standard iterative row/column decoding on the BEC, a technique that has been known for several decades. As an example decoding operation, because each row and column of the code array corresponds to an SPC codeword, we know that the middle erasure in the received word must equal 1, since $1 + e + 0 = 0 \pmod 2$. Note in Figure 5.3(d) how certain erasures may be resolved only by the row decoder and other erasures may be resolved only by the column decoder, but by working together all erasures are eventually resolved. Although it is not necessary for this simple example, one can equivalently view this decoding from the point of view of the Tanner graph in which the component SPC decoders iteratively solve for the erasures: first the top decoders and then the bottom decoders, and so on. This decoder is also equivalent to the iterative erasure-filling algorithm to be discussed in Section 5.7.1.

---

We can continue to expand this point of view of message passing with a crossword-puzzle analogy. In solving a crossword puzzle, as one iterates between "Across" clues and "Down" clues, the answers in one direction tend increasingly to help one to find the answers in the other direction. Certainly it is more difficult to first find all of the "Across" answers and then all of the "Down" answers than it is to iterate back and forth, receiving clues (messages) from intersecting answers.

We consider a final illustrative example of the idea of message passing in a distributed-processor system. This example allows us to introduce the important notion of extrinsic information. It also makes clear (without proof) the fact that message-passing decoding for a collection of *constituent decoders* arranged in a graph is optimal provided that the graph contains no cycles, but it is not optimal for graphs with cycles. The example is the well-known soldier-counting problem in the Bayesian-inference literature [7].

Consider Figure 5.4(a), which depicts six soldiers in a linear formation. The goal is for each of the soldiers to learn the total number of soldiers present by counting

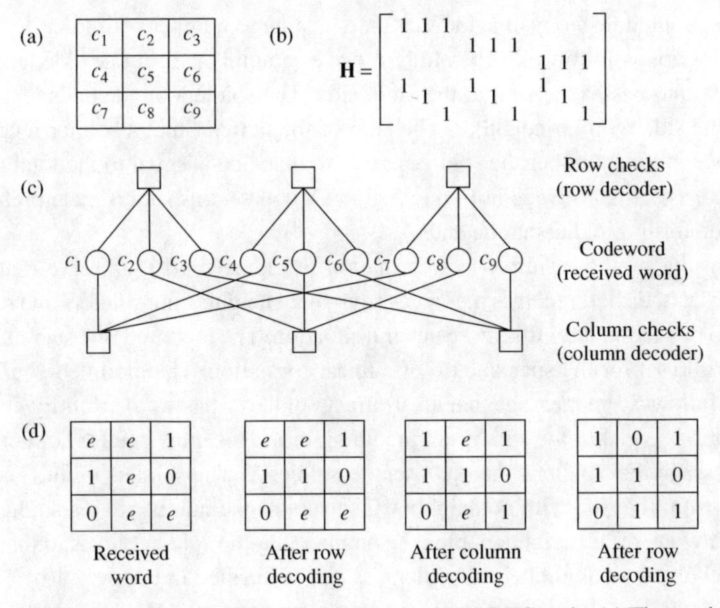


**Figure 5.3** $(3,2) \times (3,2)$ SPC product code as an LDPC code. (a) The product-code array. (b) The parity-check matrix **H**. (c) The Tanner graph; each CN represents an SPC erasure decoder. The six CNs correspond to the six rows of **H** and the rows and columns of the product code array. (d) An iterative decoding example on the BEC.

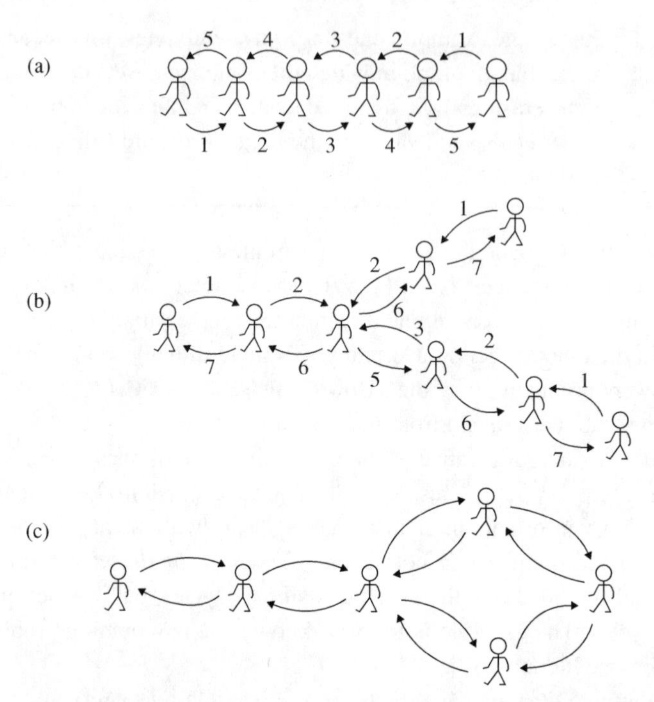


**Figure 5.4** Distributed soldier counting. (After Pearl [7].) (a) Soldiers in a line. (b) Soldiers in a tree formation. (c) Soldiers in a formation containing a cycle.

in a distributed fashion (imagine that there is a large number of soldiers). The counting rules are as follows. Each soldier receives a number from one side, adds one for himself, and passes the sum to the other side. The soldiers on the ends receive a zero from the side with no neighbor. The sum of the number that a soldier receives from one side and the number the soldier passes to that side is equal to the total number of soldiers, as can be verified in Figure 5.4(a). Of course, this sum is meaningful only on one side for the soldiers at the ends.

Consider now the simple tree formation in Figure 5.4(b). Observe that the third soldier from the left receives messages from three neighboring soldiers and so the rules need to be modified for this more general situation. The modified rule set reduces to the above rule set for the special case of a linear formation. The modified counting rules are as follows. The message that an arbitrary soldier $X$ passes to arbitrary neighboring soldier $Y$ is equal to the sum of all incoming messages, plus one for soldier $X$, minus the message that soldier $Y$ had just sent to soldier $X$. The soldiers on the ends receive a zero from the side with no neighbor. The sum of the number that a soldier receives from any one of his neighbors plus the number that the soldier passes to that neighbor is equal to the total number of soldiers, as can be verified in Figure 5.4(b).

This message-passing rule introduces the concept of *extrinsic information*. The idea is that a soldier does not pass to a neighboring soldier any information that the neighboring soldier already has, that is, only extrinsic information is passed. It is for this reason that soldier $Y$ receives the sum total of messages that soldier $X$ has available *minus* the information that soldier $Y$ already has. We say that soldier $X$ passes to soldier $Y$ only extrinsic information, which may be computed as

$$\begin{aligned} I_{X\to Y} &= \sum_{Z\in N(X)} I_{Z\to X} - I_{Y\to X} + I_X \\ &= \sum_{Z\in N(X)-\{Y\}} I_{Z\to X} + I_X, \end{aligned} \tag{5.10}$$

where $N(X)$ is the set of neighbors of soldier $X$, $I_{X\to Y}$ is the extrinsic information sent from soldier $X$ to soldier $Y$ (and similarly for $I_{Z\to X}$ and $I_{Y\to X}$), and $I_X$ is sometimes called the *intrinsic information*. $I_X$ is the "one" that a soldier counts for himself, so $I_X = 1$ for this example.

Finally, we consider the formation of Figure 5.4(c), which contains a cycle. The reader can easily verify that the situation is untenable: no matter what counting rule one may devise, the cycle represents a type of positive feedback, both in the clockwise and in the counter-clockwise direction, so that the messages passed within the cycle will increase without bound as the trips around the cycle continue. This example demonstrates that message passing on a graph cannot be claimed to be optimal if the graph contains one or more cycles. However, while most practical codes contain cycles, it is well known that message-passing decoding performs very well for properly designed codes for most error-rate ranges of interest. It is for this reason that we are interested in message-passing decoding.

The notion of extrinsic-information passing described above has been called the *turbo principle* in the context of the iterative decoding of concatenated codes in communication channels. A depiction of the turbo principle is contained in Figure 5.5. In the figure, $I_X^{\text{total}} = \sum_{Z\in N(X)} I_{Z\to X} + I_X$, and similarly for $I_Y^{\text{total}}$. $I_X$ and $I_Y$ represent intrinsic information received from the channel. Note that Figure 5.5 is simply a block diagram of equation (5.10), which was discussed in connection with Figure 5.4(b). We remark that addition and subtraction operations in (5.10) can be generalized operations, such as the "box-plus" ($\boxplus$) operation introduced later in this chapter.

For simplicity, our discussion of soldier counting omitted the details of the actual iterative counting procedure: the values given in Figure 5.4 represent the values after the procedure had already reached convergence. The top sequence of five numbers in Figure 5.4(a) would be initialized to all zeros and then be updated as $(1, 1, 1, 1, 1) \to (2, 2, 2, 2, 1) \to (3, 3, 3, 2, 1) \to (4, 4, 3, 2, 1) \to (5, 4, 3, 2, 1)$; updating stops when no numbers change. The sequence of five numbers at the bottom of Figure 5.4(a) would be computed in a similar fashion (the updated sequences are simply the reverse of those just given). The situation is nearly identical for iterative decoding schemes with multiple decoders. Thus, the message being passed from decoder $X$ to decoder $Y$ in Figure 5.5(a) will happen multiple times in practice, although the timing depends on the selected *scheduling scheme* that coordinates the message passing among the various decoders. The most commonly used flooding schedule will be discussed in the next section.

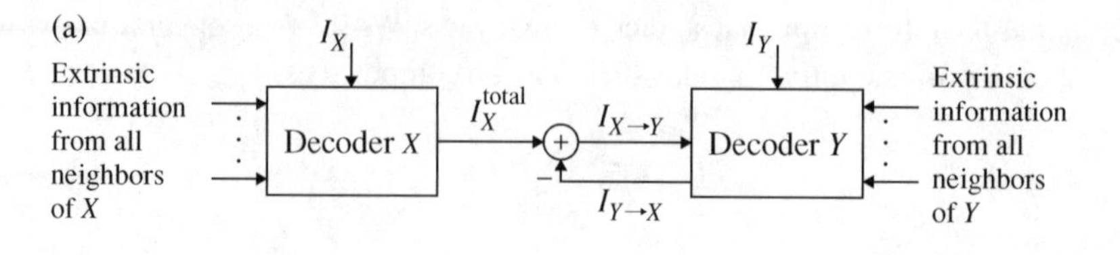


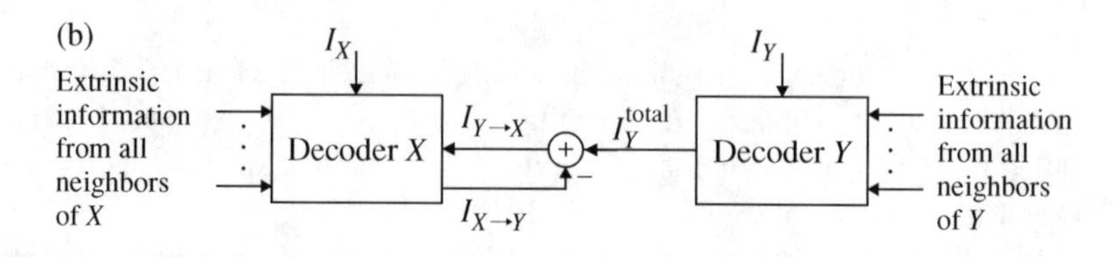


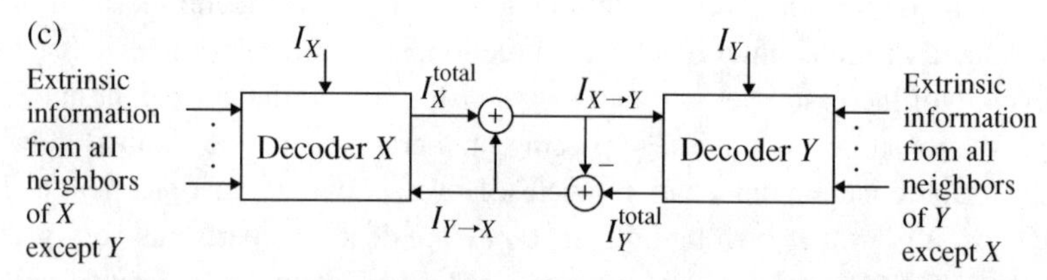


**Figure 5.5** A depiction of the turbo principle for concatenated coding schemes. (a) Extrinsic information $I_{X \to Y}$ from decoder $X$ to decoder $Y$. (b) Extrinsic information $I_{Y \to X}$ from decoder $Y$ to decoder $X$. (c) A compact symmetric combination of (a) and (b).

## 5.4 The Sum–Product Algorithm

In this section we derive the sum–product algorithm for general memoryless binary-input channels, applying the turbo principle in our development. As special cases, we consider the BEC, the BSC, the BI-AWGNC, and the independent Rayleigh fading channel. We first present an overview of the SPA, after which the so-called log-domain version of the SPA is developed.

### 5.4.1 Overview

In addition to introducing LDPC codes in his seminal work in 1960, Gallager also provided a near-optimal decoding algorithm that is now called the sum–product algorithm. This algorithm is also sometimes called the belief-propagation algorithm (BPA), a name taken from the Bayesian-inference literature, where the algorithm was derived independently [7]. The algorithm is identical for all memoryless channels, so our development will be general.

The optimality criterion underlying the development of the SPA decoder is symbol-wise maximum *a posteriori* (MAP). Much like optimal symbol-by-symbol decoding of trellis-based codes (i.e., the BCJR algorithm), we are interested in computing the *a posteriori* probability (APP) that a specific bit in the transmitted codeword $\mathbf{v} = [v_0\ v_1\ \ldots\ v_{n-1}]$ equals 1, given the received word $\mathbf{y} = [y_0\ y_1\ \ldots\ y_{n-1}]$. Without loss of generality, we focus on the decoding of bit $v_j$, so that we are interested in computing the APP,

$$\Pr(v_j = 1 \mid \mathbf{y}),$$

the APP ratio (also called the likelihood ratio, LR),

$$l(v_j \mid \mathbf{y}) \triangleq \frac{\Pr(v_j = 0 \mid \mathbf{y})}{\Pr(v_j = 1 \mid \mathbf{y})},$$

or the more numerically stable log-APP ratio, also called the log-likelihood ratio (LLR),

$$L(v_j \mid \mathbf{y}) \triangleq \log\left(\frac{\Pr(v_j = 0 \mid \mathbf{y})}{\Pr(v_j = 1 \mid \mathbf{y})}\right).$$

Here and in that which follows, the natural logarithm is assumed for LLRs.

The SPA for the computation of $\Pr(v_j = 1 \mid \mathbf{y})$, $l(v_j|\mathbf{y})$, or $L(v_j \mid \mathbf{y})$ is a distributed algorithm that is an application of the turbo principle to a code's Tanner graph. This is more easily seen in Figure 5.6, which presents a different rendition of the Tanner graph of Figure 5.1. Figure 5.6 makes more obvious the notion that an LDPC code can be deemed a collection of SPC codes concatenated through an interleaver $\Pi$ to a collection of repetition (REP) codes. Further, the SPC codes are treated as outer codes, that is, they are not connected to the channel. The edges dangling from the VNs in Figure 5.6 are connected to the channel. We can apply the turbo principle of Figure 5.5 to this concatenation of SPC and REP codes as follows.

Figure 5.7 depicts the REP (VN) decoder situation for a VN with degree greater than the degree-2 VNs of Figure 5.6. Note that the VN $j$ decoder receives LLR information both from the channel and from its neighbors. However, in the computation of the extrinsic information $L_{j\to i}$, VN $j$ need not receive $L_{i\to j}$ from CN $i$ since it would be subtracted out anyway in accordance with (5.10) or Figure 5.5. Figure 5.8 depicts the SPC (CN) decoder situation. Similar to the VN case, in the computation of $L_{i\to j}$ for some specific $j$, CN $i$ need not receive $L_{j\to i}$, for its impact would be subtracted out anyway.

The VN and CN decoders work cooperatively and iteratively to estimate $L(v_j \mid \mathbf{y})$ for $j = 0, 1, \ldots, n-1$. Throughout our development, we shall assume that the

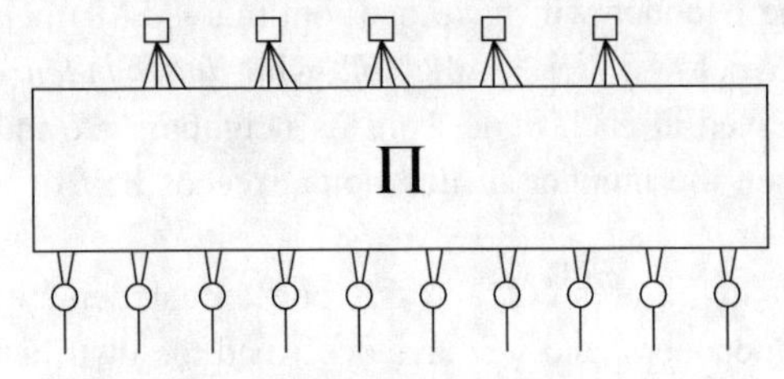


**Figure 5.6** Graphical representation of an LDPC code as a concatenation of SPC and REP codes. "Π" represents an interleaver.

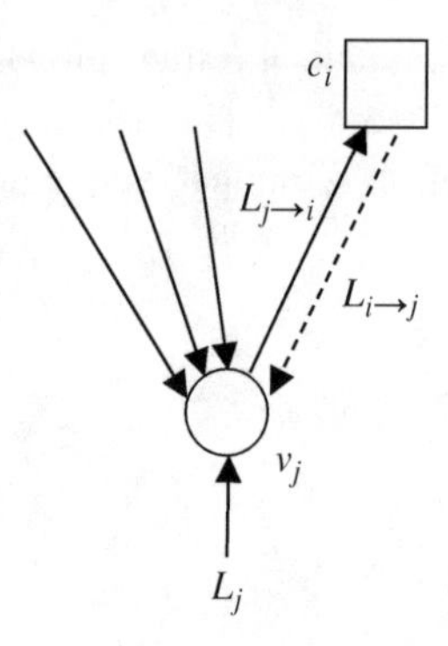


**Figure 5.7** A VN decoder (REP code decoder). VN $j$ receives LLR information from the channel and from all of its neighbors, excluding message $L_{i \to j}$ from CN $i$, from which VN $j$ composes message $L_{j \to i}$ that is sent to CN $i$.

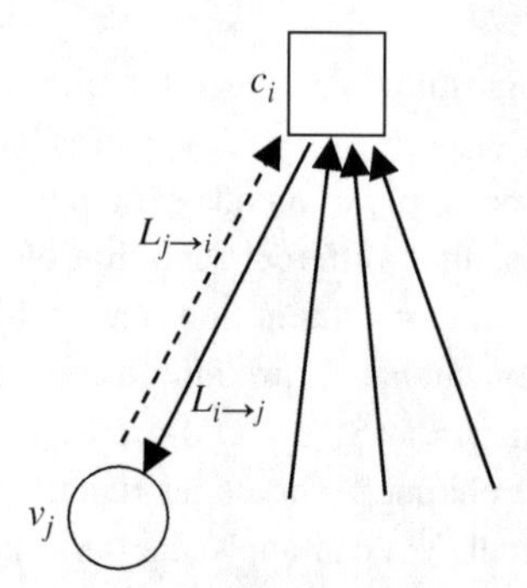


**Figure 5.8** A CN decoder (SPC decoder). CN $i$ receives LLR information from all of its neighbors, excluding message $L_{j \to i}$ from VN $j$, from which CN $i$ composes message $L_{i \to j}$ that is sent to VN $j$.

*flooding schedule* is employed. According to this schedule, all VNs process their inputs and pass extrinsic information up to their neighboring check nodes; the check nodes then process their inputs and pass extrinsic information down to their neighboring variable nodes; and the procedure repeats, starting with the variable nodes. After a preset maximum number of repetitions (or *iterations*) of this VN/CN decoding round, or after some stopping criterion has been met, the decoder computes the LLRs $L(v_j \mid \mathbf{y})$ from which decisions on the bits $v_j$ are made. When the cycles are large, the estimates will be very accurate and the decoder will have near-optimal (MAP) performance. The development of the SPA below relies on the following *independence assumption*: the LLR quantities received at each node from its neighbors are independent. Clearly this breaks down when the number of iterations exceeds half of the Tanner graph's girth.

So far, we have provided in Figures 5.7 and 5.8 conceptual depictions of the functions of the CN and VN decoders, and we have described the distributed decoding schedule in the context of Figure 5.6. At this point, we need to develop the detailed operations within each constituent (CN and VN) decoder. That is, we digress a bit to develop the MAP decoders and APP processors for REP and SPC codes. These decoders are integral to the overall LDPC decoder, which will be discussed subsequently.

### 5.4.2 Repetition Code MAP Decoder and APP Processor

Consider a REP code in which the binary code symbol $c \in \{0, 1\}$ is transmitted over a memoryless channel $d$ times so that the $d$-vector $\mathbf{r}$ is received. By definition, the MAP decoder computes the log-APP ratio

$$L(c \mid \mathbf{r}) = \log\left(\frac{\Pr(c = 0 \mid \mathbf{r})}{\Pr(c = 1 \mid \mathbf{r})}\right)$$

or, under an equally likely assumption for the value of $c$, $L(c|\mathbf{r})$ is the LLR

$$L(c \mid \mathbf{r}) = \log\left(\frac{\Pr(\mathbf{r} \mid c = 0)}{\Pr(\mathbf{r} \mid c = 1)}\right).$$

This simplifies as

$$\begin{aligned} L(c \mid \mathbf{r}) &= \log\left(\frac{\prod_{l=0}^{d-1} \Pr(r_l \mid c = 0)}{\prod_{l=0}^{d-1} \Pr(r_l \mid c = 1)}\right) \\ &= \sum_{l=0}^{d-1} \log\left(\frac{\Pr(r_l \mid c = 0)}{\Pr(r_l \mid c = 1)}\right) \\ &= \sum_{l=0}^{d-1} L(r_l \mid x), \end{aligned}$$

where $L(r_l \mid x)$ is obviously defined. Thus, the MAP receiver for a REP code computes the LLRs for each channel output $r_l$ and adds them. The MAP decision is $\hat{c} = 0$ if $L(c \mid \mathbf{r}) \geq 0$ and $\hat{c} = 1$ otherwise.

In the context of LDPC decoding, the above expression is adapted to compute the extrinsic information to be sent from VN $j$ to CN $i$ (cf. Figure 5.7),

$$L_{j \to i} = L_j + \sum_{i' \in N(j) - \{i\}} L_{i' \to j}. \tag{5.11}$$

The quantity $L_j$ in this expression is the LLR value computed from the channel sample $y_j$,

$$L_j = L(c_j \mid y_j) = \log\left(\frac{\Pr(c_j = 0 \mid y_j)}{\Pr(c_j = 1 \mid y_j)}\right).$$

Note that the update equation (5.11) follows the format of (5.10). In the context of LDPC decoding, we call the VN an *APP processor* instead of a MAP decoder. At the last iteration, VN $j$ produces a decision based on

$$L_j^{\text{total}} = L_j + \sum_{i \in N(j)} L_{i \to j}. \tag{5.12}$$

### 5.4.3 Single-Parity-Check Code MAP Decoder and APP Processor

To develop the MAP decoder for an SPC code, we first need the following result due to Gallager [1].

**Lemma 5.1** *Consider a vector of $d$ independent binary random variables* $\mathbf{a} = [a_0, a_1, \ldots, a_{d-1}]$ *in which* $\Pr(a_l = 1) = p_1^{(l)}$ *and* $\Pr(a_l = 0) = p_0^{(l)}$*. Then the probability that* **a** *contains an even number of 1s is*

$$\frac{1}{2} + \frac{1}{2}\prod_{l=0}^{d-1}\left(1 - 2p_1^{(l)}\right) \tag{5.13}$$

*and the probability that* **a** *contains an odd number of 1s is*

$$\frac{1}{2} - \frac{1}{2}\prod_{l=0}^{d-1}\left(1 - 2p_1^{(l)}\right). \tag{5.14}$$

*Proof* The proof is by induction on $d$. (Problem 5.9.) □

Armed with this result, we now derive the MAP decoder for SPC codes. Consider the transmission of a length-$d$ SPC codeword **c** over a memoryless channel whose output is **r**. The bits $c_l$ in the codeword **c** have a single constraint: there must be an even number of 1s in **c**. Without loss of generality, we focus on bit $c_0$, for which the MAP decision rule is

$$\hat{c}_0 = \arg\max_{b \in \{0,1\}} \Pr(c_0 = b \mid \mathbf{r},\ \text{SPC}),$$

where the conditioning on SPC is a reminder that there is an SPC constraint imposed on **c**. Consider now

$$\begin{aligned}\Pr\{c_0 = 0 \mid \mathbf{r},\ \text{SPC}\} &= \Pr\{c_1, c_2, \ldots, c_{d-1} \text{ has an even no. of 1s} \mid \mathbf{r}\}\\ &= \frac{1}{2} + \frac{1}{2}\prod_{l=1}^{d-1}(1 - 2\Pr(c_l = 1 \mid r_l)),\end{aligned}$$

where the second line follows from (5.13). Rearranging gives

$$1 - 2\Pr(c_0 = 1 \mid \mathbf{r},\ \text{SPC}) = \prod_{l=1}^{d-1}(1 - 2\Pr(c_l = 1 \mid r_l)), \tag{5.15}$$

where we used the fact that $p(c_0 = 0 \mid \mathbf{r},\ \text{SPC}) = 1 - p(c_0 = 1 \mid \mathbf{r},\ \text{SPC})$. We can change this to an LLR representation using the easily proven relation for a generic binary random variable with probabilities $p_1$ and $p_0$,

$$1 - 2p_1 = \tanh\left(\frac{1}{2}\log\left(\frac{p_0}{p_1}\right)\right) = \tanh\left(\frac{1}{2}\text{LLR}\right), \tag{5.16}$$

where here $\text{LLR} = \log(p_0/p_1)$. Applying this relation to (5.15) gives

$$\tanh\left(\frac{1}{2}L(c_0 \mid \mathbf{r},\ \text{SPC})\right) = \prod_{l=1}^{d-1}\tanh\left(\frac{1}{2}L(c_l \mid r_l)\right)$$

or

$$L(c_0 \mid \mathbf{r},\ \mathrm{SPC}) = 2\tanh^{-1}\left(\prod_{l=1}^{d-1} \tanh\left(\frac{1}{2}L(c_l \mid r_l)\right)\right). \tag{5.17}$$

Thus, the MAP decoder for bit $c_0$ in a length-$d$ SPC code makes the decision $\hat{c}_0 = 0$ if $L(c_0 \mid \mathbf{r},\ \mathrm{SPC}) \geq 0$ and $\hat{c}_0 = 1$ otherwise.

In the context of LDPC decoding, following (5.17), when the check nodes function as APP processors instead of MAP decoders, CN $i$ computes the extrinsic information

$$L_{i\to j} = 2\tanh^{-1}\left(\prod_{j' \in N(i)-\{j\}} \tanh\left(\frac{1}{2}L_{j'\to i}\right)\right) \tag{5.18}$$

and transmits it to VN $j$. Note that, because the product is over the set $N(i) - \{j\}$ (cf. Figure 5.8), the message $L_{j\to i}$ has in effect been subtracted out to obtain the extrinsic information $L_{i\to j}$. Ostensibly, this update equation does not follow the format of (5.10), but we will show in Section 5.4.5 that it can essentially be put in that format for an appropriately defined addition operation.

### 5.4.4 The Gallager SPA Decoder

Equations (5.11) and (5.18) form the core of the SPA decoder. The information $L_{j\to i}$ that VN $j$ sends to CN $i$ at each iteration is the best (extrinsic) estimate of the value of $v_j$ (the sign bit of $L_{j\to i}$) and the confidence or reliability level of that estimate (the magnitude of $L_{j\to i}$). This information is based on the REP constraint for VN $j$ and all inputs from the neighbors of VN $j$, excluding CN $i$. Similarly, the information $L_{i\to j}$ that CN $i$ sends to VN $j$ at each iteration is the best (extrinsic) estimate of the value of $v_j$ (sign bit of $L_{i\to j}$) and the confidence or reliability level of that estimate (magnitude of $L_{i\to j}$). This information is based on the SPC constraint for CN $i$ and all inputs from the neighbors of CN $i$, excluding VN $j$.

While (5.11) and (5.18) form the core of the SPA decoder, we still need an iteration-stopping criterion and an initialization step. A standard stopping criterion is to stop iterating when $\hat{\mathbf{v}}\mathbf{H}^{\mathrm{T}} = \mathbf{0}$, where $\hat{\mathbf{v}}$ is a tentatively decoded codeword. The decoder is initialized by setting all VN messages $L_{j\to i}$ equal to

$$L_j = L(v_j \mid y_j) = \log\left(\frac{\Pr(v_j = 0 \mid y_j)}{\Pr(v_j = 1 \mid y_j)}\right), \tag{5.19}$$

for all $j$, $i$ for which $h_{ij} = 1$. Here, $y_j$ represents the channel value that was actually received, that is, it is not a variable here. We consider the following special cases, each of which can be incorporated in the initialization step of Algorithm 5.1.

**BEC.** In this case, $y_j \in \{0, 1, e\}$ and we define $p = \Pr(y_j = e \mid v_j = b)$ to be the erasure probability, where $b \in \{0, 1\}$. Then it is easy to see that

$$\Pr\left(v_j = b \mid y_j\right) = \begin{cases} 1, & \text{when } y_j = b, \\ 0, & \text{when } y_j = b^c, \\ 1/2, & \text{when } y_j = e, \end{cases}$$

where $b^c$ represents the complement of $b$. From this, it follows that

$$L(v_j \mid y_j) = \begin{cases} +\infty, & y_j = 0, \\ -\infty, & y_j = 1, \\ 0, & y_j = e. \end{cases}$$

**BSC.** In this case, $y_j \in \{0, 1\}$ and we define $\varepsilon = \Pr(y_j = b^c \mid v_j = b)$ to be the error probability. Then it is obvious that

$$\Pr\left(v_j = b \mid y_j\right) = \begin{cases} 1 - \varepsilon, & \text{when } y_j = b, \\ \varepsilon, & \text{when } y_j = b^c. \end{cases}$$

From this, we have

$$L(v_j \mid y_j) = (-1)^{y_j} \log\left(\frac{1 - \varepsilon}{\varepsilon}\right).$$

Note that knowledge of $\varepsilon$ is necessary.

**BI-AWGNC.** We first let $x_j = (-1)^{v_j}$ be the $j$th transmitted binary value; note that $x_j = +1(-1)$ when $v_j = 0(1)$. We shall use $x_j$ and $v_j$ interchangeably hereafter. The $j$th received sample is $y_j = x_j + n_j$, where the $n_j$ are independent and normally distributed as $\mathcal{N}\left(0, \sigma^2\right)$. Then it is easy to show that

$$\Pr\left(x_j = x \mid y_j\right) = \left[1 + \exp\left(-2y_jx/\sigma^2\right)\right]^{-1},$$

where $x \in \{\pm 1\}$ and, from this, that

$$L(v_j \mid y_j) = 2y_j/\sigma^2.$$

In practice, an estimate of $\sigma^2$ is necessary.

**Rayleigh.** Here we assume an independent Rayleigh fading channel and that the decoder has perfect knowledge of the fading variables. The model is similar to that of the AWGNC: $y_j = \alpha_j x_j + n_j$, where $\{\alpha_j\}$ are independent Rayleigh random variables with unity variance. The channel transition probability in this case is

$$\Pr\left(x_j = x \mid y_j\right) = \left[1 + \exp\left(-2\alpha_j y_j x/\sigma^2\right)\right]^{-1}.$$

Then the variable nodes are initialized by

$$L(v_j \mid y_j) = 2\alpha_j y_j/\sigma^2.$$

Estimates of $\alpha_j$ and $\sigma^2$ are necessary in practice. This expression is applicable to other fading statistics as well, simply by making the random variables $\alpha_j$ have a distribution different from Rayleigh.

**Remarks on Algorithm 5.1**

1. The origin of the name "sum–product" is not evident from this "log-domain" version of the sum–product algorithm (log-SPA). We refer the reader to [9] for a more general discussion of this algorithm and its applications.
2. For most LDPC codes, this algorithm is able to detect an uncorrected codeword with near-unity probability (in Step 5).

**Algorithm 5.1** The Gallager Sum–Product Algorithm

1. **Initialization.** For all $j$, initialize $L_j$ according to (5.19) for the appropriate channel model. Then, for all $i, j$ for which $h_{ij} = 1$, set $L_{j\to i} = L_j$.
2. **CN update.** Compute outgoing CN messages $L_{i\to j}$ for each CN using (5.18),

$$L_{i\to j} = 2\tanh^{-1}\left(\prod_{j'\in N(i)-\{j\}} \tanh\left(\frac{1}{2}L_{j'\to i}\right)\right),$$

and then transmit to the VNs. (This step is shown diagrammatically in Figure 5.8.)
3. **VN update.** Compute outgoing VN messages $L_{j\to i}$ for each VN using Equation (5.11),

$$L_{j\to i} = L_j + \sum_{i'\in N(j)-\{i\}} L_{i'\to j},$$

and then transmit to the CNs. (This step is shown diagrammatically in Figure 5.7.)
4. **LLR total.** For $j = 0, 1, \ldots, n-1$ compute

$$L_j^{\text{total}} = L_j + \sum_{i\in N(j)} L_{i\to j}.$$

5. **Stopping criteria.** For $j = 0, 1, \ldots, n-1$, set

$$\hat{v}_j = \begin{cases} 1, & \text{if } L_j^{\text{total}} < 0, \\ 0, & \text{else}, \end{cases}$$

to obtain $\hat{\mathbf{v}}$. If $\hat{\mathbf{v}}\mathbf{H}^{\mathrm{T}} = \mathbf{0}$ or the number of iterations equals the maximum limit, stop; else, go to Step 2.

3. As mentioned, the SPA assumes that the messages passed are statistically independent throughout the decoding process. When the $y_j$ are independent, this independence assumption would hold true if the Tanner graph possessed no cycles. Further, the SPA would yield exact LLRs in this case. However, for a graph of girth $\gamma$, the independence assumption is true only up to the $(\gamma/2)$th iteration, after which messages start to loop back on themselves in the graph's various cycles.
4. This algorithm has been presented for pedagogical clarity, but may be adjusted to optimize the number of computations. For example, fewer operations are required if Step 4 is performed before Step 3, and Step 3 is modified as $L_{j\to i} = L_j^{\text{total}} - L_{i\to j}$. See also Section 5.5.5.
5. This algorithm can be simplified further for the BEC and BSC channels since the initial LLRs are ternary in the first case and binary in the second case. In fact, the algorithm for the BEC is precisely the iterative erasure filling algorithm of Example 5.2 and Section 5.7.1, as will be shown when discussing the min-sum decoder below.

The update equation (5.18) for Step 2 is numerically challenging due to the presence of the product and the tanh and $\tanh^{-1}$ functions. Following Gallager, we can improve the situation as follows. First, factor $L_{j\to i}$ into its sign and magnitude (or *bit value* and *bit reliability*):

$$L_{j\to i} = \alpha_{ji}\beta_{ji},$$
$$\alpha_{ji} = \text{sign}\left(L_{j\to i}\right),$$
$$\beta_{ji} = \left|L_{j\to i}\right|,$$

so that (5.18) may be rewritten as

$$\tanh\left(\frac{1}{2}L_{i\to j}\right) = \prod_{j'\in N(i)-\{j\}} \alpha_{j'i} \cdot \prod_{j'\in N(i)-\{j\}} \tanh\left(\frac{1}{2}\beta_{j'i}\right).$$

We then have

$$\begin{aligned} L_{i\to j} &= \prod_{j'} \alpha_{j'i} \cdot 2\tanh^{-1}\left(\prod_{j'} \tanh\left(\frac{1}{2}\beta_{j'i}\right)\right) \\ &= \prod_{j'} \alpha_{j'i} \cdot 2\tanh^{-1}\log^{-1}\log\left(\prod_{j'} \tanh\left(\frac{1}{2}\beta_{j'i}\right)\right) \\ &= \prod_{j'} \alpha_{j'i} \cdot 2\tanh^{-1}\log^{-1}\sum_{j'} \log\left(\tanh\left(\frac{1}{2}\beta_{j'i}\right)\right), \end{aligned}$$

which yields a new form for (5.18), Step 2,

$$\textbf{CN update: } L_{i\to j} = \prod_{j'\in N(i)-\{j\}} \alpha_{j'i} \cdot \phi\left(\sum_{j'\in N(i)-\{j\}} \phi\left(\beta_{j'i}\right)\right), \tag{5.20}$$

where we have defined

$$\phi(x) = -\log[\tanh(x/2)] = \log\left(\frac{e^x+1}{e^x-1}\right)$$

and used the fact that $\phi^{-1}(x) = \phi(x)$ when $x > 0$. Thus, (5.20) may be used instead of (5.18) in Step 2 of the SPA presented above. The function $\phi(x)$, shown in Figure 5.9, may be implemented by use of a look-up table. However, if a very low error rate is a requirement, a table-based decoder can suffer from an "error-rate floor" (see Section 5.4.6 for a discussion of "floors"). One possible alternative to a table is a piecewise linear approximation, as discussed in [10, 11]. Other alternatives are presented in the next subsection and in Section 5.5.

### 5.4.5 The Box-Plus SPA Decoder

Implementation of the $\phi(x)$ function by use of a look-up table is sufficient for some software simulations, but for most hardware applications it is not. The reason is that, due to dynamic-range issues, it is a difficult function to approximate, even with a large table. Thus, hardware implementations of the $\phi(x)$ function generally suffer from performance degradations, especially in the error-rate floor region. In this section, we

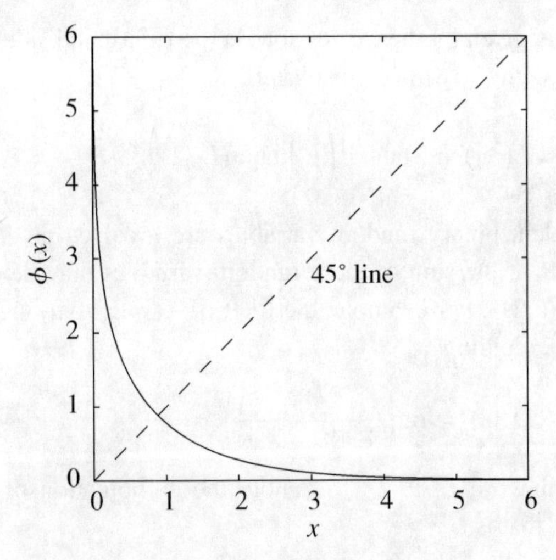


**Figure 5.9** A plot of the $\phi(x)$ function.

approach check-node processing from a slightly different perspective to yield an alternative SPA decoder. This alternative decoder is preferable in some applications, and it leads to certain low-complexity approximations as discussed later in the chapter. We first need the following general result.

**Lemma 5.2** *Consider two independent binary random variables $a_1$ and $a_2$ with probabilities $\Pr(a_l = b) = p_b^{(l)}$, $b \in \{0, 1\}$, and LLRs $L_l = L(a_l) = \log\left(p_0^{(l)}/p_1^{(l)}\right)$. The LLR of the binary sum $A_2 = a_1 \oplus a_2$, defined as*

$$L(A_2) = \log\left(\frac{\Pr(A_2 = 0)}{\Pr(A_2 = 1)}\right),$$

*is given by*

$$L(A_2) = \log\left(\frac{1 + e^{L_1 + L_2}}{e^{L_1} + e^{L_2}}\right). \tag{5.21}$$

*Proof* To see this, observe that

$$\begin{aligned}
L(A_2) &= \log\left(\frac{\Pr(a_1 \oplus a_2 = 0)}{\Pr(a_1 \oplus a_2 = 1)}\right) \\
&= \log\left(\frac{p_0^{(1)}p_0^{(2)} + p_1^{(1)}p_1^{(2)}}{p_0^{(1)}p_1^{(2)} + p_1^{(1)}p_0^{(2)}}\right) = \log\left(\frac{1 + \dfrac{p_0^{(1)}}{p_1^{(1)}}\dfrac{p_0^{(2)}}{p_1^{(2)}}}{\dfrac{p_0^{(1)}}{p_1^{(1)}} + \dfrac{p_0^{(2)}}{p_1^{(2)}}}\right) \\
&= \log\left(\frac{1 + e^{L_1 + L_2}}{e^{L_1} + e^{L_2}}\right).
\end{aligned}$$

□

This result may be applied directly to the computation of a CN output with $d_c - 1 = 2$ inputs because CN outputs are LLRs of mod-2 sums of binary r.v.s. Thus, a CN output

$L_{\text{out}}$ with inputs $L_1$ and $L_2$ is given by the expression on the right-hand side of (5.21). Moreover, to be consistent with (5.18), we must have

$$L_{\text{out}} = L(A_2) = 2\tanh^{-1}\left(\prod_{l=1}^{2} \tanh(L_l/2)\right).$$

If more than two independent binary random variables are involved, as is the case when $d_c > 3$, then the LLR of the sum of these random variables may be computed by repeated application of (5.21). For example, the LLR of $A_3 = a_1 \oplus a_2 \oplus a_3$ may be computed via $A_3 = A_2 \oplus a_3$, yielding

$$L(A_3) = \log\left(\frac{1 + e^{L(A_2)+L_3}}{e^{L(A_2)} + e^{L_3}}\right).$$

As a shorthand, we will write $L_1 \boxplus L_2$ to denote the computation of $L(A_2) = L(a_1 \oplus a_2)$ from $L_1$ and $L_2$; that is,

$$L_1 \boxplus L_2 = \log\left(\frac{1 + e^{L_1+L_2}}{e^{L_1} + e^{L_2}}\right). \tag{5.22}$$

Similarly, $L_1 \boxplus L_2 \boxplus L_3$ will denote the computation of $L(A_3) = L(a_1 \oplus a_2 \oplus a_3)$ from $L_1, L_2$, and $L_3$; and so on for more variables. $\boxplus$ is called the "box-plus" operator. The box-plus counterpart to (5.18) or (5.20) is

$$\textbf{CN update: } L_{i\to j} = \mathop{\boxplus}_{j' \in N(i)-\{j\}} L_{j'\to i}, \tag{5.23}$$

where $\boxplus$ represents the summation operator for the binary operator $\boxplus$. Observe that this is essentially in the form of (5.10), except that $I_X = 0$ in (5.10) because the CNs are not directly connected to the channel. Observe also the symmetry with the VN update equation (that also follows (5.10)), which we reproduce here to emphasize the point:

$$\textbf{VN update: } L_{j\to i} = L_j + \sum_{i' \in N(j)-\{i\}} L_{i'\to j}.$$

The expression (5.23) may be computed via repeated application of Lemma 5.2. For example, representing the binary operator $\boxplus$ as the two-input function

$$\mathcal{B}(a, b) = 2\tanh^{-1}(\tanh(a/2)\tanh(b/2)), \tag{5.24}$$

$L_1 \boxplus L_2 \boxplus L_3 \boxplus L_4$ would be computed as

$$\mathcal{B}(L_1, \mathcal{B}(L_2, \mathcal{B}(L_3, L_4))).$$

$\mathcal{B}(a, b)$ can be implemented by use of a two-input look-up table.

A degree-$d_c$ CN processor has $d_c$ inputs $\{L_\ell\}$ and $d_c$ outputs $\{L_{out,\ell}\}$, where, ostensibly, each output must be computed using $d_c - 1$ $\boxplus$ operations, for a total of $d_c(d_c - 1)$ $\boxplus$ operations. The $d_c$ outputs may be computed more efficiently with a total of only $3(d_c - 2)$ $\boxplus$ operations as follows. Let $F_1 = L_1$ and, for $\ell = 2$ to $d_c - 1$, let $F_\ell = F_{\ell-1} \boxplus L_\ell$. Next, let $B_{d_c} = L_{d_c}$ and, for $\ell = d_c - 1$ to 2, let $B_\ell = B_{\ell+1} \boxplus L_\ell$. Now the outputs can be computed from the forward-going and backward-going partial "sums" as follows: $L_{\text{out},1} = B_2$, $L_{\text{out},d_c} = F_{d_c-1}$, and for $\ell = 2$ to $d_c - 1$, $L_{\text{out},\ell} = F_{\ell-1} \boxplus B_{\ell+1}$.

### 5.4.6 Comments on the Performance of the SPA Decoder

In contrast with the error-rate curves for classical codes (e.g., Reed–Solomon codes with an algebraic decoder or convolutional codes with a Viterbi decoder), the error-rate curves for iteratively decoded codes generally have a region in which the slope *decreases* as the channel SNR increases (or, for a BSC, as $\varepsilon$ decreases). An example of such of an error-rate curve for the BI-AWGNC appears in Figure 5.11 in association with Example 5.4 below. The region of the curve just before the slope transition region is called the *waterfall region* of the error-rate curve and the region of the curve with the reduced slope is called the *error-rate floor region*, or simply the *floor region*.

A floor seen in the performance curve of an iteratively decoded code is occasionally attributable to a small minimum distance, particularly when the code is a parallel turbo code (see Chapter 7). However, it is possible to have a floor in the error-rate curve of an LDPC code (or serial turbo code) with a large minimum distance. In this case, the floor is attributable to so-called trapping sets. An $(\omega, \nu)$ *trapping set* is a set of $\omega$ VNs that induce a subgraph with $\nu$ odd-degree checks so that, when the $\omega$ bits are all in error, there will be $\nu$ failed parity checks. An essentially equivalent notion is a near-codeword. An $(\omega, \nu)$ *near-codeword* is a length-$n$ error pattern of weight $\omega$ that results in $\nu$ check failures (i.e., the syndrome weight is $\nu$), where $\omega$ and $\nu$ are "small." Near-codewords tend to lead to error situations from which the SPA decoder (and its approximations) cannot escape. The implication of a small $\omega$ is that the error pattern is more likely; the implication of a small $\nu$ is that only a few check equations are affected by the pattern, making it more likely to escape the notice of the iterative decoder. Iterative decoders are susceptible to trapping sets (near-codewords) since an iterative decoder works locally in a distributed-processing fashion, unlike an ML decoder, which finds the globally optimum solution. The floor seen in Example 5.4 is due to a trapping-set-induced graph, as discussed in that example.

## 5.5 Reduced-Complexity SPA Approximations

Problem 5.10 presents the so-called probability-domain decoder. It should be clear from the above that the log-SPA decoder has lower complexity and is more numerically stable than the probability-domain SPA decoder. We now present decoders of even lower complexity, which often suffer only a little in terms of performance. As should be clear, the focus of these reduced-complexity approximate-SPA decoders is on the complex check-node processor. The amount of degradation of a given algorithm is a function of the code and the channel, as demonstrated in the examples that follow.

### 5.5.1 The Min-Sum Decoder

Consider the update equation (5.20) for $L_{i \to j}$ in the log-domain SPA decoder. Note from the shape of $\phi(x)$ that the largest term in the sum corresponds to the smallest $\beta_{ji}$ so that, assuming this term dominates the sum,

$$\phi\left(\sum_{j'}\phi\left(\beta_{j'i}\right)\right) \simeq \phi\left(\phi\left(\min_{j'}\beta_{j'i}\right)\right)$$

$$= \min_{j' \in N(i)-\{j\}} \beta_{j'i}.$$

Thus, the min-sum algorithm is simply the log-domain SPA with Step 2 replaced by

$$\textbf{CN update: } L_{i\to j} = \prod_{j' \in N(i)-\{j\}} \alpha_{j'i} \cdot \min_{j' \in N(i)-\{j\}} \beta_{j'i}.$$

It can also be shown that, in the AWGNC case, the initialization $L_{j\to i} = 2y_j/\sigma^2$ may be replaced by $L_{j\to i} = y_j$ when the min-sum algorithm is employed (Problem 5.11). The advantage, of course, is that an estimate of the noise power $\sigma^2$ is unnecessary in this case.

---

**Example 5.3** Note that, for the BEC, $\phi\left(\sum_{j'}\phi\left(\beta_{j'i}\right)\right)$ is 0 or $\infty$ exactly when $\min_{j'}\beta_{j'i}$ is 0 or $\infty$, respectively. That is, the log-SPA and min-sum algorithms are identical on the BEC. Figure 5.10 repeats the $(3,2)\times(3,2)$ SPC product-code decoding example of Figure 5.3 using the min-sum decoder. Note that the result of the min operation is either 0 or $\infty$. Note also that this is equivalent to the iterative erasure-filling algorithm of Figure 5.3. This erasure decoding algorithm will be discussed in further generality in Section 5.7.1.

---

---

**Example 5.4** We consider a rate-0.9 quasi-cyclic IRA code with parameters (4544, 4096). The simulated bit-error-rate and frame-error-rate performance of this code with the SPA decoder and the min-sum decoder is depicted in Figure 5.11. We observe that, while the SPA decoder is superior in the waterfall region by about 0.3 dB, it suffers from an *error-rate floor*. This floor is attributable to trapping sets seen by the SPA decoder, which are apparently transparent to the min-sum decoder. Figure 5.12 displays the induced graphs for the dominant trapping sets for SPA decoding of this code. (Actually, the graphs represent equivalence classes of trapping sets. There are 64 trapping sets within each class because the parity-check matrix for this code is an array of $64\times 64$ circulant permutation matrices.) For some applications, the floor is not an issue. However, for applications requiring extremely low error rates (e.g., magnetic or optical storage), this floor is unacceptable. Decoding techniques for lowering floors, specifically for this code, have been studied in [12–15], where this code was studied.

---

We remark that the floor seen in the previous example is not uncommon. In fact it is a characteristic of iteratively decodable codes and was first seen with turbo codes. Moreover, if one sees an iterative-decoder error-rate curve without a floor, then, in all likelihood, the simulations have not been run at a high enough SNR. That is, the floor likely exists, but it requires more time and effort to find it, since it might be out of

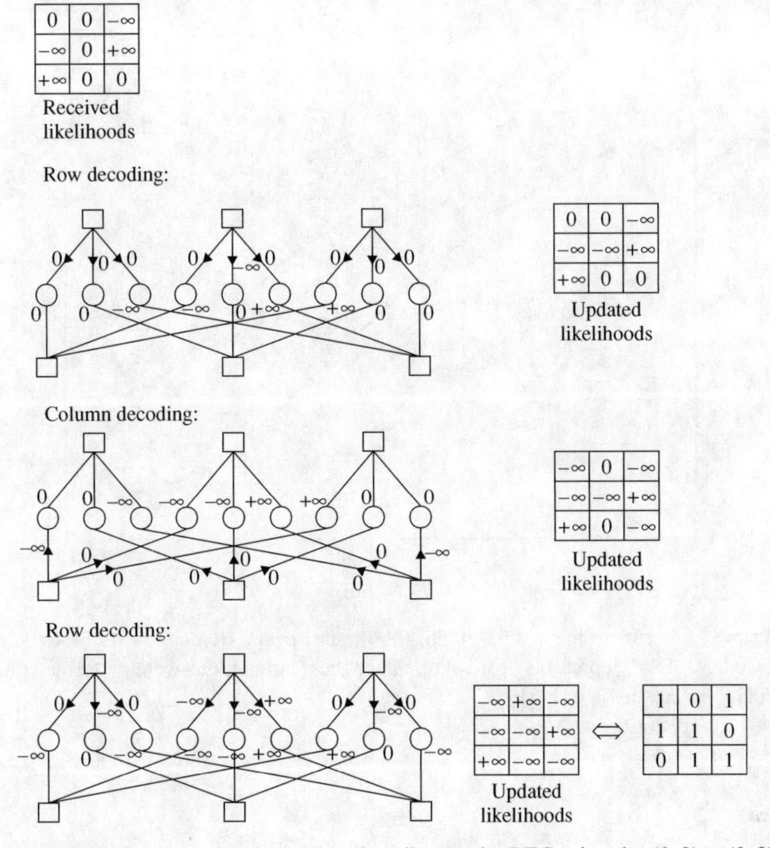


**Figure 5.10** An example of min-sum decoding on the BEC using the $(3, 2) \times (3, 2)$ SPC product code of Figure 5.3.

the reach of standard computer simulations. The floor can be due to trapping sets (as is usually the case for LDPC codes and serial turbo codes) or to a small minimum distance (as is usually the case for parallel turbo codes).

## 5.5.2 The Attenuated and Offset Min-Sum Decoders

As demonstrated in Figure 5.13 for the first iteration of a min-sum decoder, the extrinsic information passed from a CN to a VN for a min-sum decoder is on average larger in magnitude than the extrinsic information that would have been passed by an SPA decoder. That is, the min-sum decoder is generally too optimistic in assigning reliabilities. A very simple and effective adjustment that accounts for this fact is the attenuation of such extrinsic information before it is passed to a VN. The resulting update equation is

$$\textbf{CN update: } L_{i\to j} = \prod_{j' \in N(i)-\{j\}} \alpha_{j'i} \cdot A \cdot \min_{j' \in N(i)-\{j\}} \beta_{j'i},$$

where $0 < A < 1$ is the attenuation factor.

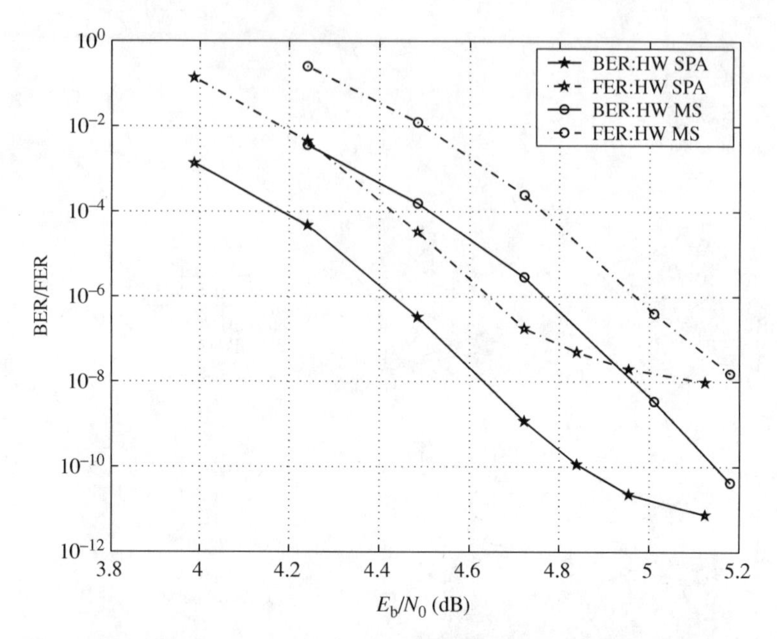


**Figure 5.11** Performance of a 0.9(4550, 4096) quasi-cyclic IRA code with SPA and min-sum decoders. "HW" represents "hardware," since these simulations were performed on an FPGA-based simulator [12].

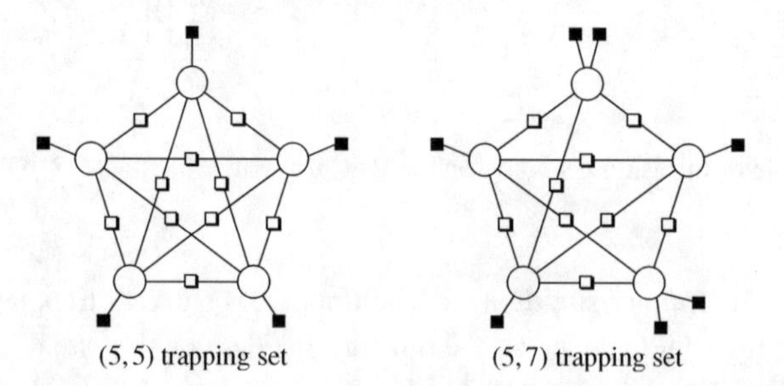


**Figure 5.12** Two classes of trapping sets (or graphs induced by trapping sets) for the 0.9(4544, 4096) code of Example 5.4. The black squares are unsatisfied check nodes and the white squares are "mis-satisfied" check nodes. The white circles represent erroneous bits (VNs).

A particularly convenient attenuation factor is $A = 0.5$, since it is implementable by a register shift. The impact of a 0.5 attenuator is presented in Figure 5.14, where it can be seen that, for the first iteration, the expected value of a min-sum reliability is about equal to that of an SPA reliability. The performance improvement realized by the attenuated min-sum decoder will be demonstrated by example later in the chapter. We point out that, depending on the code involved, it may be advantageous to allow the attenuation value to change with each iteration or to periodically turn off the attenuator [12, 13]. It may also be advantageous to attenuate messages

An alternative technique for mitigating the overly optimistic extrinsic information that occurs in a min-sum decoder is to subtract a constant offset $c_{\text{offset}} > 0$ from

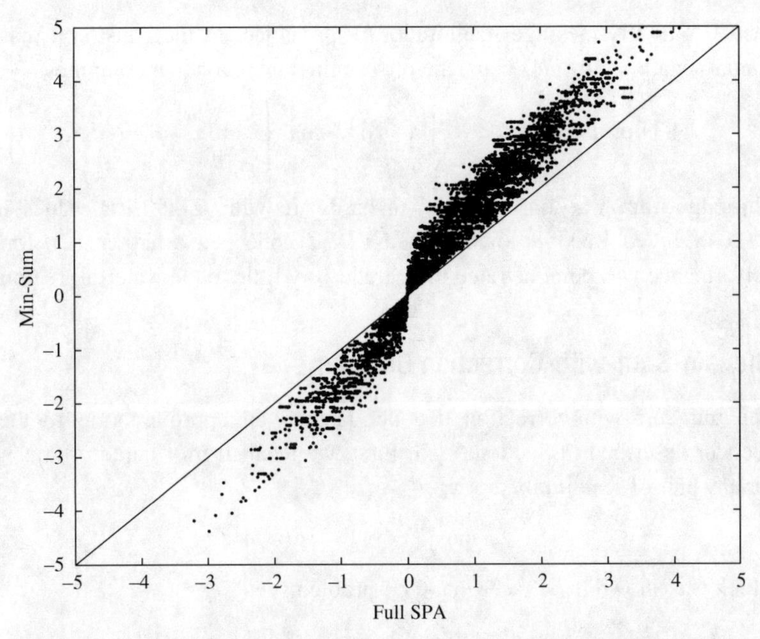


**Figure 5.13** Comparison between CN outputs for a min-sum decoder and an SPA decoder.

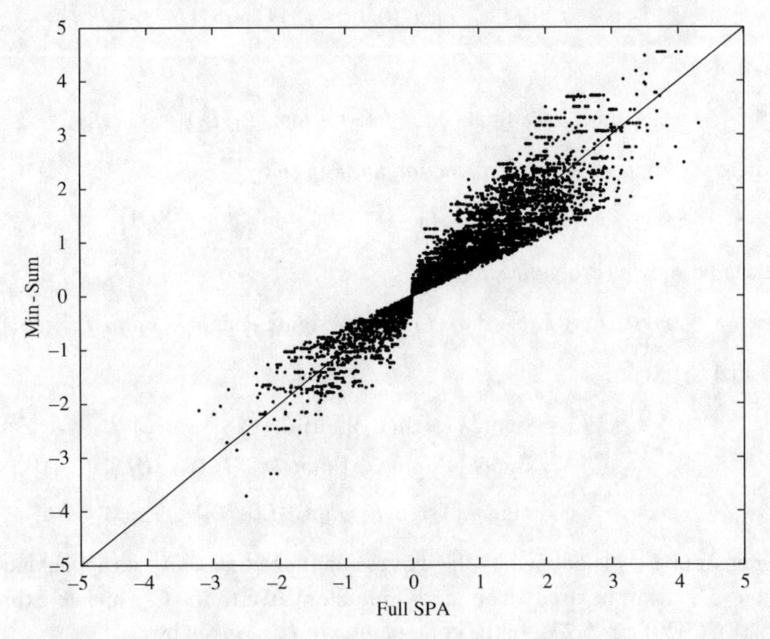


**Figure 5.14** Comparison between CN outputs for an attenuated min-sum decoder (with attenuator $A = 0.5$) and an SPA decoder.

each CN-to-VN message magnitude $\left|L_{i\to j}\right|$, subject to the constraint that the result is non-negative [16, 17]. Thus, the offset-min-sum algorithm computes

$$\textbf{CN update: } L_{i\to j} = \prod_{j'\in N(i)-\{j\}} \alpha_{j'i} \cdot \max\left\{ \min_{j'\in N(i)-\{j\}} \beta_{j'i} - c_{\text{offset}},\ 0 \right\}.$$

This algorithm was implemented in hardware with 4-bit LLR values in [18] for a (2048, 1723) Reed–Solomon-based LDPC code (see Chapter 12), for which its performance was demonstrated to degrade very little in the waterfall region.

### 5.5.3 The Min-Sum-with-Correction Decoder

The min-sum-with-correction decoder [19] is an approximation to the box-plus decoder described in Section 5.4.5. First, we recall from Chapter 4 that we defined, for any pair of real numbers $x, y$,

$$\max{}^*(x,\ y) \triangleq \log(e^x + e^y), \tag{5.25}$$

which was shown to be (see Chapter 4 problems)

$$\max{}^*(x, y) = \max(x, y) + \log\left(1 + e^{-|x-y|}\right). \tag{5.26}$$

Observe from (5.22) and (5.25) that we may write

$$L_1 \boxplus L_2 = \max{}^*(0, L_1 + L_2) - \max{}^*(L_1, L_2), \tag{5.27}$$

so that

$$L_1 \boxplus L_2 = \max(0, L_1 + L_2) - \max(L_1, L_2) + s(L_1, L_2), \tag{5.28}$$

where $s(x, y)$ is a so-called *correction term* given by

$$s(x, y) = \log\left(1 + e^{-|x+y|}\right) - \log\left(1 + e^{-|x-y|}\right).$$

It can be shown (Problem 5.21) that

$$\max(0, L_1 + L_2) - \max(L_1, L_2) = \operatorname{sign}(L_1)\operatorname{sign}(L_2)\min(|L_1|, |L_2|),$$

so that

$$\begin{aligned} L_1 \boxplus L_2 &= \operatorname{sign}(L_1)\operatorname{sign}(L_2)\min(|L_1|, |L_2|) + s(L_1, L_2) && (5.29) \\ &= \operatorname{sign}(L_1)\operatorname{sign}(L_2)\left[\min(|L_1|, |L_2|) + s(|L_1|, |L_2|)\right] \\ &\triangleq \operatorname{sign}(L_1)\operatorname{sign}(L_2)\min{}^*(|L_1|, |L_2|), && (5.30) \end{aligned}$$

where $\min^*(\cdot, \cdot)$ is defined by the expression in the line above its appearance. Expression (5.29) can be used as an algorithm alternative to the CN update expressions in (5.18), (5.20), or (5.23). In the computation of $L_{i\to j}$, given by

$$L_{i\to j} = \boxplus_{j'\in N(i)-\{j\}} L_{j'\to i},$$

in (5.23), one can repeatedly apply (5.29). This gives one implementation of the box-plus SPA decoder (with $s(x, y)$ possibly implemented by a look-up table).

From the definition of min*, $L_1 \boxplus L_2$ may be approximated as

$$L_1 \boxplus L_2 \simeq \text{sign}(L_1)\,\text{sign}(L_2)\min(|L_1|, |L_2|), \tag{5.31}$$

since $|s(x, y)| \leq 0.693$, which is simply the min-sum algorithm. An improvement to the min-sum algorithm replaces $s(x, y)$ in (5.29) by an approximation $\tilde{s}(x, y)$, where

$$\tilde{s}(x, y) = \begin{cases} c, & \text{if } |x + y| < 2 \text{ and } |x - y| > 2\,|x + y|, \\ -c, & \text{if } |x - y| < 2 \text{ and } |x + y| > 2\,|x - y|, \\ 0, & \text{otherwise,} \end{cases}$$

and where $c$ on the order of 0.5 is typical.

---

**Example 5.5** We consider a regular quasi-cyclic rate-0.875 (8176, 7156) LDPC code derived from a regular-Euclidean-geometry (EG) code (discussed in Chapter 11). The performance of this code is examined with three of the decoders discussed above: the (log-)SPA, the min-sum, and the min-sum with a correction factor (which we denote by min-sum-$c$, with $c$ set to 0.5). The $\mathbf{H}$ matrix for this code is $1022 \times 8176$ and has column weight 4 and row weight 32. It is a $2 \times 16$ array of circulants, each with column and row weight equal to 2. This code has been adopted for NASA near-Earth missions [20, 21]. The performance of this code for the three decoders on a BI-AWGNC is presented in Figure 5.15. With measurements taken at a BER of $10^{-5}$, the loss relative to the SPA decoder suffered by the min-sum decoder is 0.3 dB and the loss suffered by the min-sum-$c$ decoder is 0.01 dB. We remark that these losses are a strong function of the code, in particular, the row weight. For codes with a larger row weight, the losses can be more substantial.

---

### 5.5.4 The Approximate min* Decoder

We now consider another approximate SPA decoder, called the approximate min* (a-min*) decoder [10], which reduces the number of operations at each check node. For the sake of simplicity, we focus on the computation of values for $\{L_{i\to j}\}$ using (5.30), although this is not necessary. Note that, for a degree-$d_c$ CN, $d_c$ quantities are computed and $d_c - 1$ applications of the $\boxplus$ operator are required in order to compute each quantity. We can reduce the $d_c(d_c - 1)$ applications of the $\boxplus$ operator at a degree-$d_c$ CN with negligible performance loss as follows.

1. At CN $i$, find the incoming message of least reliability (i.e., having minimum $|L_{j\to i}|$) and label the source of that message VN $j_{\min}$.
2. The message sent to VN $j_{\min}$ is identically the one used in the standard log-SPA algorithm:

$$L_{i\to j_{\min}} = \boxplus_{j \in N(i) - \{j_{\min}\}} L_{j\to i}.$$

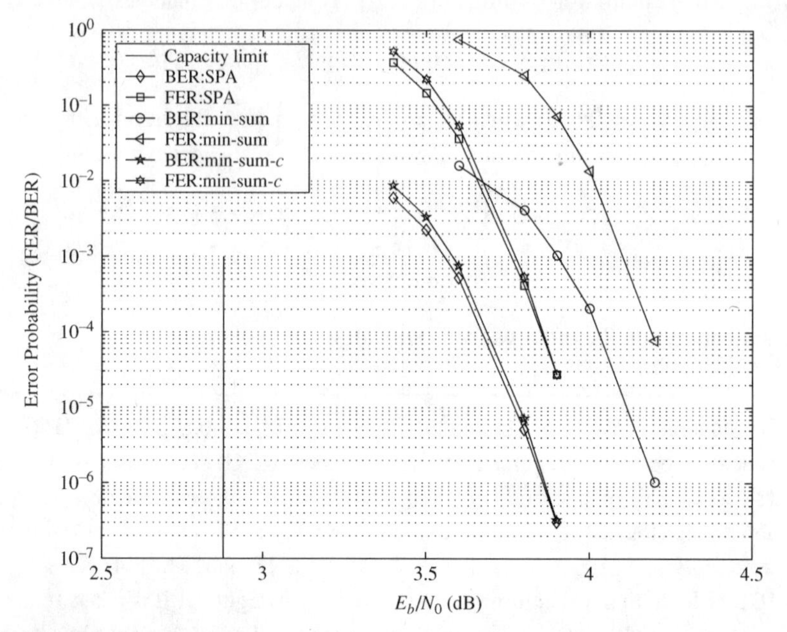


**Figure 5.15** Performance of a quasi-cyclic EG (8176, 7156) LDPC code on a binary-input AWGN channel and three decoding algorithms, with 50 decoding iterations.

3. For all $v$ in the exclusive neighborhood $N(i) - \{j_{\min}\}$, the message sent to VN $v$ is

$$L_{i\to v} = \left(\prod_{j\in N(i)-\{v\}} \alpha_{ji}\right) \cdot \left|L_{i\to j_{\min}} \boxplus L_{j_{\min}\to i}\right|$$
$$= \left(\prod_{j\in N(i)-\{v\}} \alpha_{ji}\right) \cdot \left|\boxplus_{j\in N(i)} L_{j\to i}\right|,$$

where $\alpha_{ji} = \operatorname{sign}\left(L_{j\to i}\right)$ as before.

Observe that, in this approximation of the SPA, only two magnitudes are computed at each check node, requiring only $d_c - 1$ applications of the $\boxplus$ operator to compute both. That transmitting only one of two magnitudes from a CN node results in negligible performance loss can be intuitively explained via the following consideration of the log-SPA, min-sum, and a-min* decoders. If two or more unreliable LLRs are received at a CN, then all of the outgoing LLRs will be unreliable for all three decoders (this is best seen from the perspective of the min-sum algorithm). If there is only one unreliable LLR received at a CN, then the outgoing likelihood to the unreliable VN will be reliable for all three decoders. However, in both of these situations, the LLR sent to the least reliable bit in the a-min* decoder is exactly that of the log-SPA decoder. This explains the performance improvement of the a-min* decoder over the min-sum decoder and its negligible loss relative to the log-SPA decoder.

### 5.5.5 Single-Scan Min-Sum Algorithm

It was remarked after the presentation of Algorithm 5.1 that the SPA algorithm can be rearranged for improved computational efficiency. We present here one such rearrangement, called the *single-scan min-sum algorithm* [22], applicable to the min-sum algorithm and its modifications. For simplicity, we focus only on the attenuated min-sum algorithm and leave other modifications for the reader to devise. Our experience is that software implementation of the single-scan algorithm is more than twice as fast as the conventional (attenuated) min-sum algorithm presented earlier.

As mentioned in Remark 4 after Algorithm 5.1, the VN-to-CN message $L_{j\to i}$ is simply the total LLR at VN $j$ minus the incoming CN-to-VN message $L_{i\to j}$, that is, $L_{j\to i} = L_j^{\text{total}} - L_{i\to j}$. The single-scan algorithm exploits this fact by maintaining the total LLRs for each of the VNs and subtracts the CN-to-VN messages $L_{i\to j}$ when needed.

In the computation of a CN-to-VN message $L_{i\to j}$, of the $d_c$ messages received by a CN $i$, one of them has the minimum magnitude $|L_{\min}|$ that was sent by VN $j_{\min}$. Of the other $d_c - 1$ messages received by CN $i$, one of them will possess the second-smallest magnitude, $|L_{\min}^{(2)}|$, of the $d_c$ messages into CN $i$. Variable node $j_{\min}$ will receive this second-smallest magnitude $|L_{\min}^{(2)}|$ and the $d_c - 1$ other VNs will receive $|L_{\min}|$. These two magnitudes will be scaled by an attenuation factor $A$ and by products of signs of the form $\prod_{j'\neq j} \alpha_{j'i}$, in accordance with the (attenuated) min-sum algorithm. (Recall $\alpha_{ji} = \text{sign}\left(L_{j\to i}\right)$.)

Rather than switching control back and forth between the variable nodes and the check nodes, the single-scan min-sum algorithm centers control at the check nodes, leveraging the facts pointed out in the previous two paragraphs to accomplish the message passing. At each iteration, while cycling ("scanning") through the check nodes and their edges, the single-scan algorithm computes the messages $L_{i\to j}$ and the total variable node LLRs $L_j^{\text{total}}$, after which the messages $L_{j\to i}$ are computed from $L_{j\to i} = L_j^{\text{total}} - L_{i\to j}$. The algorithm is presented below.

## 5.6 Iterative Decoders for Generalized LDPC Codes

The iterative decoder for a generalized LDPC code, introduced in Section 5.2.1, is a straightforward generalization of the iterative decoder for LDPC codes. Recall that, in deriving that decoder, we derived the MAP decoder for the check nodes (which represent SPC codes) and converted the MAP decoder into an APP soft-in/soft-out processor; we then repeated this for the variable nodes (which represent REP codes). For generalized LDPC codes, we need only concern ourselves with the generalized constraint nodes (G-CNs); the variable nodes remain as REP nodes. Thus, we require a MAP decoder (actually, an APP processor) for an arbitrary G-CN, which is typically a (low-complexity) binary linear code.

The most well-known MAP decoder for binary linear codes is the BCJR algorithm applied to a trellis that represents the linear code. A commonly used trellis, known for

**Algorithm 5.2** Single-Scan Min-Sum Algorithm

DEFINITIONS

1. $n$ = number of VNs, $m$ = number of CNs.
2. $N(i)$ = neighborhood of CN $i$ = set of VNs connected to CN $i$.
3. $y_j$ = the $j$th channel output value, $j = 0, 1, 2, \ldots, n-1$.
4. $L_{i\to j}(\text{old}) = L_{i\to j}$ from previous iteration.
5. $L_j^{\text{total}}(\text{old}) = L_j^{\text{total}}$ from previous iteration.
6. $A$ = attenuation factor.

INITIALIZATION

For all $j = 0, 1, 2, \ldots, n-1$, set $L_j^{\text{total}}(\text{old}) = L_{j\to i} = y_j$ (for all $i \in N(j)$).

ITERATIONS

For all $j = 0, 1, 2, \ldots, n-1$, set $L_j^{\text{total}} = y_j$.
**for** $i = 0$ to $m-1$ **do** ▹ Loop over the check nodes
  $S_i = 1$ ▹ to compute $L_{j\to i}$ and $L_{i\to j}$
  **for** next $j \in N(i)$ **do**
    $L_{j\to i} = L_j^{\text{total}}(\text{old}) - L_{i\to j}(\text{old})$
    $S_i = S_i \cdot \text{sign}(L_{j\to i})$
    Use $L_{j\to i}$ to update $|L_{\min}|$ and $|L_{\min}^{(2)}|$. Store $j_{\min}$.
  **end for**
  $L_{\min} = A \cdot S_i \cdot |L_{\min}|$
  $L_{\min}^{(2)} = A \cdot S_i \cdot |L_{\min}^{(2)}|$
  **for** next $j \in N(i)$ **do**
    **if** $j = j_{\min}$ **then** $L_{i\to j} = \text{sign}(L_{j\to i}) \cdot L_{\min}^{(2)}$
    **else** $L_{i\to j} = \text{sign}(L_{j\to i}) \cdot L_{\min}$
    $L_j^{\text{total}} \mathrel{+}= L_{i\to j}$
  **end for**
**end for**
**for** $i = 0$ to $m-1$ **do** ▹ Loop over the check nodes to check if
  $S_i = 1$ ▹ word estimate satisfies code contraints
  **for** next $j \in N(i)$ **do**
    $S_i = S_i \cdot \text{sign}(L_j^{\text{total}})$
  **end for**
  **if** $S_i < 0$ **then** break ▹ $S_i < 0$ if odd number of 1s at CN $i$,
**end for** ▹ odd number of 1s fails parity check
**if** $S_i > 0$ **then** found codeword, break from iterations
**else** For all $i$ and $j$, set $L_j^{\text{total}}(\text{old}) = L_j^{\text{total}}$, $L_{i\to j}(\text{old}) = L_{i\to j}$.

its simplicity and its optimality [23, 24], is the so-called *BCJR trellis*. To derive the BCJR trellis for an arbitrary binary linear code, we start with the equation $\mathbf{cH}^{\mathrm{T}} = \mathbf{0}$, where $\mathbf{H}$ is $m \times n$. If we let $\mathbf{h}_j$ represent the $j$th column of $\mathbf{H}$, we have

$$c_1\mathbf{h}_1 + c_2\mathbf{h}_2 + \cdots + c_n\mathbf{h}_n = \mathbf{0}.$$

This equation leads directly to a trellis, for its solutions yield the list of codewords just as the paths through a code's trellis gives the codeword list. Thus, the possible states at the $\ell$th trellis stage can be computed as

$$S_\ell = c_1\mathbf{h}_1 + c_2\mathbf{h}_2 + \cdots + c_\ell\mathbf{h}_\ell.$$

Further, since the $\mathbf{h}_j$ are $m$-vectors, there are at most $2^m$ states for trellis stages $\ell = 1, 2, \ldots, n-1$, and of course $S_0 = S_n = \mathbf{0}$.

---

**Example 5.6** Following [23], let

$$\mathbf{H} = \begin{bmatrix} 1 & 1 & 0 & 1 & 0 \\ 0 & 1 & 1 & 0 & 1 \end{bmatrix}.$$

Then at stage 1 there are two possible states derived from $S_1 = c_1\mathbf{h}_1$, namely

$$\begin{bmatrix} 0 \\ 0 \end{bmatrix} \text{ and } \begin{bmatrix} 1 \\ 0 \end{bmatrix},$$

corresponding to $c_1 = 0$ and $c_1 = 1$, respectively. At stage 2, there are four possible states derived from

$$S_2 = c_1\mathbf{h}_1 + c_2\mathbf{h}_2,$$

and so on. The resulting trellis is given in Figure 5.16.

---

In summary, the iterative decoder for the generalized LDPC codes is much like that of LDPC codes, except that the G-CNs require APP processors more general than those for standard SPC CNs. One example is the BCJR-based APP processor designed for the code's BCJR trellis. Because the BCJR algorithm naturally accepts LLRs and produces outputs that are LLRs, it serves well as an APP processor. Of course, alternative, possibly suboptimal, soft-in/soft-out algorithms may be employed in G-CN processors, but we mention the BCJR algorithm because it is optimal and well known.

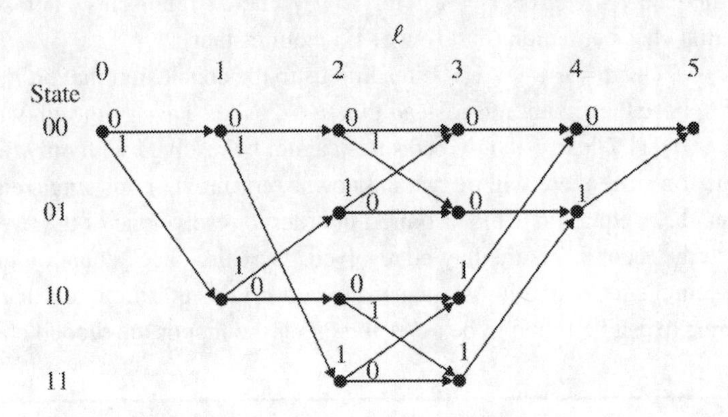


**Figure 5.16** The BCJR trellis for example code (after [23]).

Lastly, irrespective of the G-CN processor used, it is important that *extrinsic* LLR information is passed from the G-CN to each of its neighboring VNs. For example, following Figure 5.5, at the $s$th iteration, the message from G-CN $i$ to VN $j$ can be computed as

$$L_{i\to j}^{(s)} = \text{LLR}^{(s)} - L_{j\to i}^{(s-1)},$$

where $\text{LLR}^{(s)}$ is the LLR for the G-CN code bit corresponding to edge $j \leftrightarrow i$.

## 5.7 Decoding Algorithms for the BEC and the BSC

In this section, we consider various decoders for the BEC and the BSC. We consider only those decoders that appear frequently in the literature or will aid our discussion elsewhere in the book. For example, Chapter 14 considers the design of LDPC codes for the BEC.

### 5.7.1 Iterative Erasure Filling for the BEC

The iterative decoding algorithm for the BEC simply works to fill in the erased bits in such a way that all of the check equations contained in $\mathbf{H}$ are eventually satisfied. We have already seen one example of this algorithm in Example 5.2, but we now consider a simpler example to aid our development of the general iterative erasure-filling algorithm.

---

**Example 5.7** Consider an $\mathbf{H}$ matrix for the (7, 4) Hamming code,

$$\mathbf{H} = \begin{bmatrix} 1 & 0 & 1 & 1 & 1 & 0 & 0 \\ 1 & 1 & 1 & 0 & 0 & 1 & 0 \\ 0 & 1 & 1 & 1 & 0 & 0 & 1 \end{bmatrix}.$$

Suppose the received word is $\mathbf{r} = [e\ 1\ 1\ 1\ 1\ 1\ 1]$, where $e$ represents an erasure. Then the first check equation (first row of $\mathbf{H}$) requires that $0 = r_0 + r_2 + r_3 + r_4 = e + 1 + 1 + 1$, leading us to the conclusion that bit $r_0$ must equal 1. Suppose the situation is instead that two erasures have occurred, with $\mathbf{r} = [e\ 1\ e\ 1\ 1\ 1\ 1]$. Clearly, the two erasures cannot be resolved with only the first check equation since there will be two unknowns (erasures) in this single equation. A different check equation is first required in order to resolve one of the two erasures, after which the second erasure may be resolved. The third check equation involves only the second erasure and it may be used to convert this erasure into a 1. After this, the first erasure can be found to be a 1 using either the first or the second check equation.

---

In summary, the erasure decoding algorithm is as follows.

**Algorithm 5.3** Iterative Erasure Decoding Algorithm

1. Find all check equations (rows of **H**) involving a single erasure and solve for the values of bits corresponding to these erasures.
2. Repeat Step 1 until there exist no more erasures or until all equations are unsolvable.

Note that the above example and algorithm description tell us exactly which erasure patterns will fail to be resolved by this algorithm. Namely, if no parity-check equation that involves only one of the erasures in the erasure set can be found, then none of the equations will be solvable and decoding will stop. Code-bit sets that lead to unsolvable equations when they are erased are called stopping sets. Thus, a subset $S$ of code bits forms a *stopping set* if each equation that involves the bits in $S$ involves two or more of them. In the context of the Tanner graph, $S$ is a stopping set if all of its neighbors are connected to $S$ at least twice. Such a configuration is problematic for an iterative decoder because, if these bits are erased, all equations involved will contain at least two erasures. Notice that the trapping sets depicted in Figure 5.12 are also stopping sets.

**Example 5.8** Consider the (7, 4) Hamming code again, together with the received pattern $\mathbf{r} = [e\ e\ e\ 1\ 1\ 1\ 1]$. Note that each check equation involves at least two erasures and so a stopping set is $\{c_0, c_1, c_2\}$ (See Figure 1.2.).

## 5.7.2 ML Decoder for the BEC

The algorithm described in the previous subsection is not maximum-likelihood, but we can obtain an ML decoding algorithm by applying the above algorithm to a modified or "adapted" **H** matrix as described below. Following the description of this "adapted-**H** algorithm," we provide an example in which the stopping set in the previous example is resolvable.

Suppose we transmit any codeword in the code $C$ over a BEC and we receive the word $\mathbf{r}$ whose elements are taken from the set $\{0, 1, e\}$. Let $J$ represent the index set of unerased positions in $\mathbf{r}$ and $J'$ the set of erased positions. It is easy to show (Problem 5.24) that the ML decoder chooses a codeword $\mathbf{c} \in C$ satisfying $c_j = r_j$ for all $j \in J$ and, further, that $\mathbf{c}$ is a solution to the equation

$$\mathbf{c}_{J'}\mathbf{H}_{J'}^{\mathrm{T}} = \mathbf{c}_J\mathbf{H}_J^{\mathrm{T}}, \tag{5.32}$$

where $\mathbf{H}_J$ ($\mathbf{H}_{J'}$) is the submatrix of the code's parity-check matrix **H** obtained by taking only the columns of **H** corresponding to $J$ ($J'$), and similarly for $\mathbf{c}_J$ ($\mathbf{c}_{J'}$). Equation (5.32) follows from the fact that $\mathbf{0} = \mathbf{cH}^{\mathrm{T}} = \mathbf{c}_{J'}\mathbf{H}_{J'}^{\mathrm{T}} + \mathbf{c}_J\mathbf{H}_J^{\mathrm{T}}$. There is a unique solution if and only if the rows of $\mathbf{H}_{J'}^{\mathrm{T}}$ are linearly independent, in which case the elements of the unknown $\mathbf{c}_{J'}$ may be determined by Gaussian elimination. Because $\mathbf{H}_{J'}^{\mathrm{T}}$ has $n - k$ columns, its rows can be linearly independent only if $|J'| \leq n-k$, giving us a necessary condition for the uniqueness of the solution to (5.32). Also, because any $d_{\min} - 1$ rows of $\mathbf{H}^{\mathrm{T}}$ (and hence $\mathbf{H}_{J'}^{\mathrm{T}}$) are linearly independent, (5.32) is guaranteed to have a unique solution whenever $|J'| < d_{\min}$.

One may solve for the unknowns $\mathbf{c}_{J'}$ as follows. First apply Gaussian elimination to $\mathbf{H}$, targeting the columns in the index set $J'$ to produce a modified matrix $\tilde{\mathbf{H}}$, which we assume without loss of generality to be of the form $\tilde{\mathbf{H}} = \left[\tilde{\mathbf{H}}_J \ \tilde{\mathbf{H}}_{J'}\right]$. The submatrix $\tilde{\mathbf{H}}_{J'}$ will have the form

$$\tilde{\mathbf{H}}_{J'} = \begin{bmatrix} \mathbf{T} \\ \mathbf{M} \end{bmatrix},$$

where $\mathbf{T}$ is a $|J'| \times |J'|$ lower-triangular matrix with ones along the diagonal and $\mathbf{M}$ is an arbitrary binary matrix (which may be empty).

Thus, one may solve for the unknowns in $\mathbf{c}_{J'}$ by successively solving the $|J'|$ parity-check equations represented by the top $|J'|$ rows of $\tilde{\mathbf{H}}$. This is possible provided that none of the top $|J'|$ rows of $\tilde{\mathbf{H}}_J$ are all zeros. None of these rows can be all zeros since, for example, if row $p \leq |J'|$ of $\tilde{\mathbf{H}}_J$ is all zeros, then the corresponding row of $\tilde{\mathbf{H}}$ has Hamming weight 1. Further, any word with a one in the same position as the single one in this row of $\tilde{\mathbf{H}}$ will have $\mathbf{c}\tilde{\mathbf{H}}^{\mathrm{T}} \neq \mathbf{0}$ and thus is not a valid codeword. However, half of the codewords of any reasonable linear code have a one in any given position. (A "reasonable code" is one whose generator matrix $\mathbf{G}$ has no all-zero columns.) Thus, it follows that none of the top $|J'|$ rows of $\tilde{\mathbf{H}}_J$ are all zeros and the unknowns in $\mathbf{c}_{J'}$ may be determined.

---

**Example 5.9** We continue with the (7, 4) Hamming code and the received pattern $\mathbf{r} = [e\ e\ e\ 1\ 1\ 1\ 1]$, which contains a stopping set in the positions $J' = \{0, 1, 2\}$. A few elementary row operations on $\mathbf{H}$ yield an alternative $\mathbf{H}$ matrix adapted to the erasure positions,

$$\tilde{\mathbf{H}} = \begin{bmatrix} 1 & 0 & 0 & 1 & 0 & 1 & 1 \\ 0 & 1 & 0 & 1 & 1 & 1 & 0 \\ 0 & 1 & 1 & 1 & 0 & 0 & 1 \end{bmatrix},$$

which is in the form $\tilde{\mathbf{H}} = \left[\mathbf{T}\ \tilde{\mathbf{H}}_J\right]$. It is evident that, due to the presence of the triangular matrix, the three check equations may be solved in succession and all three erasures found to be 1.

---

### 5.7.3 Gallager's Algorithm A and Algorithm B for the BSC

Gallager presented in his seminal work [1] several decoding algorithms for LDPC codes on the BSC. These algorithms have been called Gallager Algorithm A, Gallager Algorithm B, and the bit-flipping algorithm. The first two algorithms will be discussed here, and the bit-flipping algorithm will be discussed in the next subsection.

Gallager Algorithms A and B are message-passing algorithms much like the SPA, so we focus on decoding Steps 2 and 3. Because LLRs are not involved, we use $M_{j \to i}$

instead of $L_{j\to i}$ to denote the message to be passed from VN $j$ to CN $i$. Similarly, we use $M_{i\to j}$ instead of $L_{i\to j}$. Additionally, $M_j$ will be used in lieu of $L_j$ for the (intrinsic) information coming from the channel. Note that $M_j$ is a binary quantity since we are dealing with the BSC and $M_{j\to i}$, $M_{i\to j}$ are binary messages.

The check-node update equation for Algorithm A is

$$M_{i\to j} = \bigoplus_{j' \in N(i)-\{j\}} M_{j'\to i},$$

where $\oplus$ represents modulo-2 summation. Note that this ensures that the sum of the incoming messages $\{M_{j'\to i}\}$ and the outgoing message $M_{i\to j}$ is zero, modulo 2. At the variable nodes, the outgoing message $M_{j\to i}$ is set to $M_j$ unless all of the incoming extrinsic messages $M_{i'\to j}$ disagree ($i' \neq i$), in which case, $M_{j\to i}$ is set to $M_j^{\mathrm{c}}$, the logical complement of $M_j$. Notationally, this is

$$M_{j\to i} = \begin{cases} M_j, & \text{if } \exists\, i' \in N(j) - \{i\} \text{ s.t. } M_{i'\to j} = M_j, \\ M_j^{\mathrm{c}}, & \text{otherwise.} \end{cases}$$

Gallager's decoding Algorithm B is a refinement of Algorithm A in that outgoing VN messages $M_{j\to i}$ do not require that all of the extrinsic messages $M_{i'\to j}$ disagree with $M_j$ in order for the outgoing message to be $M_j^{\mathrm{c}}$. That is, if there are at least $t$ such messages, then $M_{j\to i}$ will be set to $M_j^{\mathrm{c}}$, otherwise it will be set to $M_j$. Thus, the VN update rule for Algorithm B is

$$M_{j\to i} = \begin{cases} M_j^{\mathrm{c}}, & \text{if } \left|\left\{i' \in N(j) - \{i\}\colon M_{i'\to j} = M_j^{c}\right\}\right| \geq t, \\ M_j, & \text{otherwise.} \end{cases}$$

The final code-bit decision at the $j$th variable node is set to the binary value held by the majority of all incoming messages, $M_j$ and $M_{i\to j}$. Gallager showed that for $(d_v, d_c)$-regular LDPC codes, the optimum value of $t$ is the smallest integer $t$ satisfying

$$\frac{1-\varepsilon}{\varepsilon} \leq \left[\frac{1+(1-2p)^{d_c-1}}{1-(1-2p)^{d_c-1}}\right]^{2t-d_v+1},$$

where $\varepsilon$ and $p$ are the intrinsic and extrinsic message error rates, respectively. ($\varepsilon$ is, of course, the BSC transition probability.) Observe that, when $d_v = 3$, Algorithms A and B are identical.

The foregoing discussion focused on Steps 2 and 3 in the Gallager A and B decoding algorithms. The code bit decisions for these algorithms are majority-based, as follows. If the degree of a variable node is odd, then the decision is set to the majority of the incoming check-node messages. If the degree of a variable node is even, then the decision is set to the majority of the check-node messages and the channel message.

Ardakani and Kschischang [25] have shown that, among binary message-passing algorithms, Algorithm B is optimal for regular LDPC codes and that it is also optimal for irregular LDPC codes under very loose conditions.

### 5.7.4 The Bit-Flipping Algorithm for the BSC

Briefly, the bit-flipping algorithm first evaluates all parity-check equations in $\mathbf{H}$ and then "flips" (complements) any bits in the received word that are involved in more than some fixed number of failed parity checks. This step is then repeated with the modified received word until all parity checks are satisfied or until some maximum number of iterations has been executed.

Noting that failed parity-check equations are indicated by the elements of the syndrome, $\mathbf{s} = \mathbf{r}\mathbf{H}^{\mathrm{T}}$, and that the number of failed parity checks for each code bit is contained in the $n$-vector $\mathbf{f} = \mathbf{s}\mathbf{H}$ (operations over the integers), the algorithm is presented below.

---

**Algorithm 5.4** The Bit-Flipping Algorithm

---

1. Compute $\mathbf{s} = \mathbf{r}\mathbf{H}^{\mathrm{T}}$ (operations over $\mathbb{F}_2$). If $\mathbf{s} = \mathbf{0}$, stop, since $\mathbf{r}$ is a codeword.
2. Compute $\mathbf{f} = \mathbf{s}\mathbf{H}$ (operations over $\mathbb{Z}$).
3. Identify the elements of $\mathbf{f}$ greater than some preset threshold, and then flip all the bits in $\mathbf{r}$ corresponding to those elements.
4. If the maximum number of iterations has not been reached, go to Step 1 using the updated $\mathbf{r}$.

---

The threshold will depend on the channel conditions and Gallager has derived the optimum threshold for regular LDPC codes, which we do not repeat here. Instead, we point out a particularly convenient way to obtain a threshold that effectively adapts to the channel quality.

3. Identify the elements of $\mathbf{f}$ equal to $\max\left\{f_j\right\}$ and then flip all the bits in $\mathbf{r}$ corresponding to those elements.

Examples of the performance of this algorithm will be presented in subsequent chapters. Additional discussions of the bit-flipping algorithm, including weighted bit-flipping, may be found in Chapter 11.

## 5.8 Concluding Remarks

This chapter has presented a number of message-passing decoding algorithms for several important communication channels. As mentioned early on in the chapter, the flooding schedule was adopted in each case. However, it should be emphasized that other schedules are possible, most designed with an eye toward lowering the error floor. Examples can be found in [28–31]. We also mention that other variations and approximations of the sum–product decoding algorithm exist in the literature [28–31], too many to present here.

## Problems

**5.1** One version of the **H** matrix for the (7, 4) Hamming code can be created by listing as its columns all of the nonzero 3-vectors:

$$\mathbf{H}_1 = \begin{bmatrix} 1 & 0 & 1 & 0 & 1 & 0 & 1 \\ 0 & 1 & 1 & 0 & 0 & 1 & 1 \\ 0 & 0 & 0 & 1 & 1 & 1 & 1 \end{bmatrix}.$$

Draw the Tanner graph for this code and identify all cycles. Repeat for the cyclic version of the (7, 4) Hamming code whose generator matrix is given by $g(x) = (1 + x + x^3)$ and whose **H** matrix is

$$\mathbf{H}_2 = \begin{bmatrix} 1 & 0 & 1 & 1 & 1 & 0 & 0 \\ 0 & 1 & 0 & 1 & 1 & 1 & 0 \\ 0 & 0 & 1 & 0 & 1 & 1 & 1 \end{bmatrix}.$$

**5.2** Reordering the rows of an **H** matrix does not change the code. Reordering the columns of an **H** matrix changes the code in the sense that the code bits in each codeword are reordered, but the weight spectrum is not changed. Neither of these operations eliminate cycles in the code's Tanner graph. Check whether it is possible to eliminate the 4-cycles in $\mathbf{H}_1$ in the previous problem by replacing rows by a sum of rows. Repeat for $\mathbf{H}_2$.

**5.3** Derive equations (5.3)–(5.6).

**5.4** Derive equations (5.7) and (5.8).

**5.5** Via iterative erasure decoding applied to a (3, 2) × (3, 2) SPC product code, decode the following codeword received at the output of a BEC (see Example 5.2):

$$\begin{bmatrix} e & 1 & e \\ e & e & 0 \\ e & 0 & 1 \end{bmatrix}.$$

Also, find all minimal stopping sets for the graph in Figure 5.3(c).

**5.6** Derive $L(v_j \mid y_j)$ for the BEC, the BSC, the BI-AWGNC, and the Rayleigh channel starting from the channel transition probabilities in each case. Expressions for $L(v_j \mid y_j)$ together with the models and assumptions are given in Section 5.4.4, but put in all of the missing steps.

**5.7** (a) Suppose that, instead of the mapping $x_j = 1 - 2v_j$ defined in this chapter for the BI-AWGNC, we let $x_j = 2v_j - 1$. How does this change the initial messages $L(v_j \mid y_j)$? (b) Suppose we change the LLR definition so that the zero hypothesis is in the denominator:

$$L(v_j \mid \mathbf{y}) = \log\left(\frac{\Pr(v_j = 1 \mid \mathbf{y})}{\Pr(v_j = 0 \mid \mathbf{y})}\right).$$

Under this assumption, derive the updated equation for $L_{i\to j}$ that would replace (5.18). (c) Now combine parts (a) and (b) so that $x_j = 2v_j - 1$ and both $L(v_j \mid y_j)$ and $L(v_j \mid \mathbf{y})$ are defined with the zero hypothesis in the denominator. Find the initial messages $L(v_j \mid y_j)$ and the updated equation for $L_{i\to j}$.

**5.8** Consider an LDPC-coded $M$-ary pulse-position modulation (PPM) free-space optical system configured as follows: LDPC encoder → PPM modulator → PPM soft-out demodulator → LDPC decoder. Under this configuration, each group of $\log_2, M$ LDPC code bits, denoted by $\mathbf{d}$, is mapped to an $M$-bit channel word, denoted by $\mathbf{x}$, representing the PPM channel symbol. For example, with $M = 4$, $\mathbf{d} = 00 \to \mathbf{x} = 0001$, $\mathbf{d} = 01 \to \mathbf{x} = 0010$, $\mathbf{d} = 10 \to \mathbf{x} = 0100$, and $\mathbf{d} = 11 \to \mathbf{x} = 1000$. With a channel bit $x_j = 1$ representing a pulse of light and a channel bit $x_j = 0$ representing the absence of light, the channel with a photon-counting detector at the receiver's front end can be modeled as Poisson, so that

$$\Pr\left(y_j \mid x_j = 1\right) = \frac{(n_s + n_b)^{y_j} e^{-(n_s+n_b)}}{y_j!},$$

$$\Pr\left(y_j \mid x_j = 0\right) = \frac{(n_b)^{y_j} e^{-n_b}}{y_j!}.$$

In these expressions, $y_j$ is the number of received photons for the $j$th PPM slot (corresponding to $x_j$), $n_s$ is the mean number of signal photons in a PPM slot, and $n_b$ is the mean number of noise photons in a PPM slot. Show that the initial LLRs passed to the LDPC decoder from the PPM soft-out demodulator can be computed as

$$\begin{aligned} L(d_k \mid \mathbf{y}) &= \log\left[\frac{\Pr(d_k = 1 \mid \mathbf{y})}{\Pr(d_k = 0 \mid \mathbf{y})}\right] \\ &= \log\left[\frac{\sum_{\mathbf{x}\in X_1^k} \Pr(\mathbf{x} \mid \mathbf{y})}{\sum_{\mathbf{x}\in X_0^k} \Pr(\mathbf{x} \mid \mathbf{y})}\right] \\ &= \log\left[\frac{\sum_{\mathbf{x}\in X_1^k} (1 + n_s/n_b)^{y_{l(\mathbf{x})}}}{\sum_{\mathbf{x}\in X_0^k} (1 + n_s/n_b)^{y_{l(\mathbf{x})}}}\right], \end{aligned}$$

where $X_1^k \triangleq \{\mathbf{x}: d_k = 1\}$, $X_0^k \triangleq \{\mathbf{x}: d_k = 0\}$, and the subscript $l(\mathbf{x})$ is the index of the PPM symbol $\mathbf{x}$ at which the 1 is located. (As an example, $l(1000) = 3$, $l(0010) = 1$, and $l(0001) = 0$.) Further, show that this can be rewritten as

$$L(d_k|\mathbf{y}) = \max_{\mathbf{x}\in X_1^k}{}^{*}\{y_{l(\mathbf{x})} \cdot \log\left(1 + n_s/n_b\right)\} - \max_{\mathbf{x}\in X_0^k}{}^{*}\{y_{l(\mathbf{x})} \cdot \log\left(1 + n_s/n_b\right)\}.$$

**5.9** Prove (5.13) and (5.14) using induction.

**5.10** Derive the probability-domain SPA decoder presented below. (The $q$-ary version is presented in Chapter 15.) Define the messages to be passed, which are probabilities rather than LLRs, as follows:

- $p_{j\to i}(b)$ is the probabilistic message that VN $j$ sends to CN $i$ about the probability that code bit $j$ is $b \in \{0, 1\}$, given (i) the decoder input from the channel, (ii) the probabilistic messages, $\{p_{i'\to j}(b): i' \in N(j) - \{i\}\}$, received from the exclusive neighborhood, and (iii) the REP constraint for VN $j$;

- $p_{i \to j}(b)$ is the probabilistic message that CN $i$ sends to VN $j$ about the probability that code bit $j$ is $b \in \{0, 1\}$, given (i) the probabilistic messages, $\{p_{j' \to i}(b) : j' \in N(i) - \{j\}\}$, received from the exclusive neighborhood of $j$, and (ii) the SPC constraint for CN $i$;
- $P_j(b)$ is the probability that code bit $j$ is $b \in \{0, 1\}$, given (i) the decoder input from the channel, (ii) the probabilistic messages, $\{p_{i \to j}(b) : i \in N(j)\}$, received from the neighborhood of $j$, and (iii) the REP constraint for VN $j$.

**5.11** Show that the min-sum algorithm on the BI-AWGNC requires no knowledge of the noise sample variance $\sigma^2$. That is, show that the min-sum decoder will operate identically if the initial LLRs (and hence all subsequent LLRs) are scaled by $\sigma^2/2$, yielding initial LLRs of $y_j$.

**5.12** Show that (5.21) is equivalently

$$L(A_2) = 2\tanh^{-1}\left(\prod_{l=1}^{2} \tanh(L_l/2)\right).$$

*Hint*: There is more than one way to show this, but you might find this useful:

$$\tanh^{-1}(z) = \frac{1}{2}\ln\left(\frac{1+z}{1-z}\right).$$

---

**Probability-Domain SPA**

1. For all $i, j$ for which $h_{ij} = 1$, set $p_{j \to i}(0) = \Pr(c_j = 0 \mid y_j)$ and $p_{j \to i}(1) = \Pr(c_j = 1 | y_j)$, where $y_j$ is the $j$th received channel value.
2. For each $b \in \{0, 1\}$, update $\{p_{i \to j}(b)\}$ at each CN using

$$p_{i \to j}(b) = \frac{1}{2} + \frac{(-1)^b}{2} \prod_{j' \in N(i) - \{j\}} \left(1 - 2p_{j' \to i}(1)\right).$$

3. For each $b \in \{0, 1\}$, update $\{p_{j \to i}(b)\}$ for each VN using

$$p_{j \to i}(b) = K_{ji} \Pr(c_j = b \mid y_j) \prod_{i' \in N(j) - \{i\}} p_{i' \to j}(b),$$

   where the constants $K_{ji}$ are selected to ensure that $p_{j \to i}(0) + p_{j \to i}(1) = 1$.
4. For each $b \in \{0, 1\}$, and for each $j = 0, 1, \ldots, n-1$, compute

$$P_j(b) = K_j \Pr(c_j = b \mid y_j) \prod_{i \in N(j)} p_{i \to j}(b),$$

   where the constants $K_j$ are chosen to ensure that $P_j(0) + P_j(1) = 1$.
5. For $j = 0, 1, \ldots, n-1$, set

$$\hat{v}_j = \begin{cases} 1, & \text{if } P_j(1) > P_j(0), \\ 0, & \text{else.} \end{cases}$$

   If $\hat{\mathbf{v}}H^{\mathrm{T}} = \mathbf{0}$ or the number of iterations equals the maximum limit, stop; else, go to Step 2.

---

**5.13** Consider $d$ binary random variables $a_1, a_2, \ldots, a_d$ and denote by $L_l$ the LLR of $a_l$:

$$L_l = L(a_l) = \log\left(\frac{\Pr(a_l = 0)}{\Pr(a_l = 1)}\right).$$

Show, by induction, that

$$\boxplus_{l=1}^{d} L_l = \ln\left(\frac{\prod_{l=1}^{d}\left(e^{L_l}+1\right)+\prod_{l=1}^{d}\left(e^{L_l}-1\right)}{\prod_{l=1}^{d}\left(e^{L_l}+1\right)-\prod_{l=1}^{d}\left(e^{L_l}-1\right)}\right),$$

from which we may write

$$\boxplus_{l=1}^{d} L_l = \ln\left(\frac{1+\prod_{l=1}^{d}\tanh(L_l/2)}{1-\prod_{l=1}^{d}\tanh(L_l/2)}\right)$$

$$= 2\tanh^{-1}\left(\prod_{l=1}^{d}\tanh(L_l/2)\right).$$

Note that

$$\tanh^{-1}(z) = \frac{1}{2}\ln\left(\frac{1+z}{1-z}\right).$$

**5.14** Consider $d$ independent binary random variables $a_1, a_2, \ldots, a_d$ and denote by $L_l$ the LLR of $a_l$:

$$L_l = L(a_l) = \log\left(\frac{\Pr(a_l = 0)}{\Pr(a_l = 1)}\right).$$

Consider also a binary-valued constraint $\chi$ on these variables. As an example, for the single parity-check constraint, $\chi(\mathbf{a}) = 0$ when the Hamming weight of $\mathbf{a} = (a_1, a_2, \ldots, a_d)$ is even and $\chi(\mathbf{a}) = 1$ otherwise. Show that the LLR for $\chi(\mathbf{a})$ is given by

$$L(\chi(\mathbf{a})) = \log\left(\frac{\Pr(\chi(\mathbf{a}) = 0)}{\Pr(\chi(\mathbf{a}) = 1)}\right)$$

$$= \max_{\mathbf{a}:\chi(\mathbf{a})=0}{}^{*}\left(\sum_{l=1}^{d}\delta_{a_l}L_l\right) - \max_{\mathbf{a}:\chi(\mathbf{a})=1}{}^{*}\left(\sum_{l=1}^{d}\delta_{a_l}L_l\right),$$

where $\delta_k$ is the Kronecker delta function. Note that this may alternatively be written as

$$L(\chi(\mathbf{a})) = \max_{\mathbf{a}:\chi(\mathbf{a})=0}{}^{*}\left(\sum_{l=1}^{d}a_l^{c}L_l\right) - \max_{\mathbf{a}:\chi(\mathbf{a})=1}{}^{*}\left(\sum_{l=1}^{d}a_l^{c}L_l\right),$$

where $a_l^{c}$ is the logical complement of the binary number $a_l$. Note also, for the special case of $d = 2$ and an SPC constraint, that this becomes

$$L(\chi(\mathbf{a})) = L(a_1 \oplus a_2) = \max{}^{*}(0, L_1 + L_2) - \max{}^{*}(L_1, L_2).$$

**5.15** Consider the rate-1/2 product code depicted below, whose rows and columns form (3, 2) SPC codewords:

$$\begin{array}{ccc} c_0 & c_1 & c_2 \\ c_3 & c_4 & c_5 \\ c_6 & c_7 & \end{array}$$

(a) Program the probability-domain SPA decoder (see Problem 5.10) and decode the following received word assuming $\sigma^2 = 0.5$: $\mathbf{y} = (0.2, 0.2, -0.9, 0.6, 0.5, -1.1, -0.4, -1.2)$. You should find that the decoder converges to the codeword $\mathbf{c} = (1, 0, 1, 0, 1, 1, 1, 1)$ after seven iterations with the following cumulative probabilities at the eight VNs: $(0.740, 0.338, 0.969, 0.409, 0.787, 0.957, 0.775, 0.992)$.

(b) Program the log-domain SPA decoder and repeat part (a). You should find that the decoder converges to the codeword $\mathbf{c} = (1, 0, 1, 0, 1, 1, 1, 1)$ after seven iterations with the following cumulative LLRs at the eight VNs: $(-1.05, 0.672, -3.45, 0.370, -1.306, -3.095, -1.239, -4.863)$.

(c) Show that these results are in agreement.

**5.16** Simulate SPA decoding of the rate-1/2 product code of the previous problem on the binary-input AWGN channel and plot the bit error probability $P_b$ versus $E_b/N_0$ (dB). Simulate an exhaustive-search maximum-likelihood decoder (i.e., minimum-Euclidean-distance decoder) on the binary-input AWGN channel and compare its $P_b$ versus $E_b/N_0$ curve with that of the SPA decoder.

**5.17** Carry out the previous problem with the SPA decoder replaced by the min-sum decoder.

**5.18** Simulate SPA decoding of the (7, 4) Hamming code on the binary-input AWGN channel and plot the bit error probability $P_b$ versus $E_b/N_0$ (dB). Use the following $\mathbf{H}$ matrix for the (7, 4) Hamming code to design your SPA decoder:

$$\mathbf{H} = \begin{bmatrix} 1 & 0 & 1 & 0 & 1 & 0 & 1 \\ 0 & 1 & 1 & 0 & 0 & 1 & 1 \\ 0 & 0 & 0 & 1 & 1 & 1 & 1 \end{bmatrix}.$$

Simulate an exhaustive-search maximum-likelihood decoder (i.e., a minimum-Euclidean-distance decoder) on the binary-input AWGN channel and compare its $P_b$ versus $E_b/N_0$ curve with that of the SPA decoder.

**5.19** Carry out the previous problem with the SPA decoder replaced by the min-sum decoder.

**5.20** Let $a$ be a binary r.v. and let $p_1 = \Pr(a = 1)$. Show that

$$p_1 = \frac{1}{1 + \exp(L(a))}$$

and

$$p_0 = \frac{1}{2} + \frac{1}{2}\tanh\left(\frac{L(a)}{2}\right),$$

where $L(a) = \ln(p_0/p_1)$.

**5.21** Show that

$$\max(0, x + y) - \max(x, y) = \text{sign}(x)\text{sign}(y)\min(|x|, |y|)$$

for any pair of real numbers $x, y$.

**5.22** Assume LDPC-coded transmission on the BEC. Show that the iterative erasure-filling decoder is precisely the SPA decoder and, hence, is precisely the min-sum decoder. (Consult Section 5.5.1 and the example therein.)

**5.23** Assuming iterative erasure-filling decoding, find all the uncorrectable $x$-erasure patterns, $x = 1, 2, 3$, for the two Hamming code representations of Problem 5.1. Explain why the decoder is unable to resolve these erasure patterns (consider the stopping-set terminology).

**5.24** Show that, for the BEC, the ML decoder chooses a codeword $\mathbf{c} \in C$ satisfying $c_j = r_j$ for all $j \in J$ and, further, that this is a solution to the equation

$$\mathbf{c}_{J'}\mathbf{H}_{J'}^{\mathrm{T}} = \mathbf{c}_J\mathbf{H}_J^{\mathrm{T}},$$

where $\mathbf{H}_J$ ($\mathbf{H}_{J'}$) is the submatrix of the code's parity-check matrix $\mathbf{H}$ obtained by taking only the columns of $\mathbf{H}$ corresponding to $J$ ($J'$), and similarly for $\mathbf{c}_J$ ($\mathbf{c}_{J'}$). See (5.32) and surrounding discussion.

## References

[1] R. G. Gallager, *Low-Density Parity-Check Codes*, Cambridge, MA, MIT Press, 1963. (Also, R. G. Gallager, "Low density parity-check codes," *IRE Transactions on Information Theory*, vol. 8, no. 1, pp. 21–28, January 1962.)

[2] R. M. Tanner, "A recursive approach to low complexity codes," *IEEE Transactions on Information Theory*, vol. 27, no. 9, pp. 533–547, September 1981.

[3] D. MacKay and R. Neal, "Good codes based on very sparse matrices," *Cryptography and Coding, 5th IMA Conference,* C. Boyd, (Ed.), Berlin, Springer-Verlag, October 1995.

[4] D. MacKay, "Good error correcting codes based on very sparse matrices," *IEEE Transactions on Information Theory*, vol. 45, no. 3, pp. 399–431, March 1999.

[5] N. Alon and M. Luby, "A linear time erasure-resilient code with nearly optimal recovery," *IEEE Transactions on Information Theory*, vol. 42, no. 11, pp. 1732–1736, November 1996.

[6] J. Byers, M. Luby, M. Mitzenmacher, and A. Rege, "A digital fountain approach to reliable distribution of bulk data," *ACM SIGCOMM Computer Communication Review*, vol. 28, issue 4, October 1998.

[7] J. Pearl, *Probabilistic Reasoning in Intelligent Systems*, San Mateo, CA, Morgan Kaufmann, 1988.

[8] Y. Wang and M. Fossorier, "Doubly generalized LDPC codes," *2006 IEEE International Symposium on Information Theory*, July 2006, pp. 669–673.

[9] F. Kschischang, B. Frey, and H.-A. Loeliger, "Factor graphs and the sum-product algorithm," *IEEE Transactions on Information Theory*, vol. 47, no. 2, pp. 498–519, February 2001.

[10] C. Jones, E. Valles, M. Smith, and J. Villasenor, "Approximate-min* constraint node updating for LDPC code decoding," *IEEE Military Communications Conference*, October 2003, pp. 157–162.

[11] C. Jones, S. Dolinar, K. Andrews, D. Divsalar, Y. Zhang, and W. Ryan, "Functions and architectures for LDPC decoding," *2007 Information Theory Workshop*, pp. 577–583, September 2007.

[12] Y. Zhang, *Design of Low-Floor Quasi-Cyclic IRA Codes and their FPGA Decoders*, PhD thesis, ECE Department, University of Arizona, May 2007.

[13] Y. Zhang and W. E. Ryan, "Toward low LDPC-code floors: A case study," *IEEE Transactions on Communications*, pp. 1566–1573, June 2009.

[14] Y. Han and W. E. Ryan, "Low-floor decoders for LDPC codes," *IEEE Transactions on Communications*, pp. 1663–1673, June 2009.

[15] B. Butler and P. Siegel, "Error floor approximation for LDPC code in the AWGN channel," *IEEE Transactions on Information Theory*, vol. 60, no. 12, pp. 7416–7441, December 2014.

[16] J. Zhao, F. Zarkeshvari, and A. Banihashemi, "On implementation of min-sum algorithm and its modifications for decoding low-density parity-check (LDPC) codes," *IEEE Transactions on Communications*, vol. 53, no. 4, pp. 549–554, April 2005.

[17] J. Chen, A. Dholakia, E. Eleftheriou, M. Fossorier, and X.-Y. Hu, "Reduced-complexity decoding of LDPC codes," *IEEE Transactions on Communications*, vol. 53, no. 8, pp. 1288–1299, August 2005.

[18] Z. Zhang, L. Dolecek, B. Nikolic, V. Anantharam, and M. Wainwright, "Lowering LDPC error floors by postprocessing," *2008 IEEE GlobeCom Conference*, November–December 2008, pp. 1–6.

[19] X.-Y. Hu, E. Eleftheriou, D.-M. Arnold, and A. Dholakia, "Efficient implementation of the sum–product algorithm for decoding LDPC codes," *proceedings of the 2001 IEEE GlobeCom Conference*, November 2001, pp. 1036–1036E.

[20] Z.-W. Li, L. Chen, L.-Q. Zeng, S. Lin, and W. Fong, "Efficient encoding of quasi-cyclic low-density parity-check codes," *IEEE Transactions on Communications*, vol. 54, no. 1, pp. 71–81, January 2006.

[21] L. H. Miles, J. W. Gambles, G. K. Maki, W. E. Ryan, and S. R. Whitaker, "An 860-Mb/s (8158, 7136) low-density parity-check encoder," *IEEE Journal of Solid-State Circuits*, pp. 1686–1691, August 2006.

[22] X. Huang, "Single-scan min-sum algorithms for fast decoding of LDPC codes," *IEEE Information Theory Workshop*, Chengdu, China, September 2006.

[23] L. Bahl, J. Cocke, F. Jelinek, and J. Raviv, "Optimal decoding of linear codes for minimizing symbol error rate," *IEEE Transactions on Information Theory*, vol. 20, no. 3, pp. 284–287, March 1974.

[24] R. McEliece, "On the BCJR trellis for linear block codes," *IEEE Transactions on Information Theory*, vol. 42, no. 7, pp. 1072–1092, July 1996.

[25] M. Ardakani, *Efficient Analysis, Design and Decoding of Low-Density Parity-Check Codes*, PhD thesis, University of Toronto, 2004.

[26] H. Xiao and A. H. Banihashemi, "Graph-based message-passing schedules for decoding LDPC codes," *IEEE Transactions on Communications*, vol. 52, no. 12, pp. 2098–2105, December 2004.

[27] A. Vila Casado, M. Griot, and R. D. Wesel, "Informed scheduling for belief-propagation decoding of LDPC codes," *2007 International Conference on Communications*, June 2007, pp. 932–937.

[28] M. Mansour and N. Shanbhag, "High-throughput LDPC decoders," *IEEE Transactions on VLSI Systems*, pp. 976–996, December 2003.

[29] J. Chen and M. Fossorier, "Near optimum universal belief propagation based decoding of low-density parity check codes," *IEEE Transactions on Communications*, vol. 50, no. 3, pp. 406–414, March 2002.

[30] J. Chen, A. Dholakia, E. Eleftheriou, M. Fossorier, and X.-Y Hu. "Reduced-complexity decoding of LDPC codes," *IEEE Transactions on Communications*, vol. 53, no. 8, pp. 1288–1299, August 2005.

[31] M. Fossorier, "Iterative reliability-based decoding of low-density parity check codes," *IEEE Journal on Selected Areas in Communications*, vol. 19, no. 5, pp. 908–917, May 2001.

[32] Y. Kou, S. Lin, and M. Fossorier, "Low-density parity-check codes based on finite geometries: a rediscovery and new results," *IEEE Transactions on Information Theory*, vol. 47, no. 11, pp. 2711–2736, November 2001.

[33] S. Lin and D. J. Costello, Jr., *Error Control Coding* &, 2nd edn., New Saddle River, NJ, Prentice-Hall, 2004.

# 6 Computer-Based Design of LDPC Codes

Unlike BCH and Reed–Solomon codes, for which there exists essentially a single code-design procedure, for LDPC codes there are many code-design approaches. Many of these design approaches, including that for the original LDPC codes, are computer-based algorithms. Many others rely on finite mathematics such as the design techniques of Chapters 11 to 15. In the present chapter, we first present the original design techniques of Gallager [1] and MacKay [2]. We then provide an overview of two popular computer-based design algorithms: the progressive-edge-growth algorithm [3] and the approximate cycle extrinsic message degree algorithm [4]. Next, we present structured classes of LDPC codes, including protograph-based LDPC codes, accumulator-based LDPC codes, and generalized LDPC codes. The discussion includes selected code-design case studies to provide the student with some useful code-design approaches.

## 6.1 The Original LDPC Codes

As mentioned in the previous chapter, LDPC codes were originally invented by R. Gallager circa 1960 and were later reinvented by others, including D. MacKay, circa 1995. In this section, we will briefly overview Gallager's and MacKay's LDPC code-design approaches.

### 6.1.1 Gallager Codes

Gallager's original definition of a (regular) low-density parity-check code was identical to that given in Chapter 5: it is a linear code whose $m \times n$ parity-check matrix $\mathbf{H}$ has $g \ll m$ ones in each column and $r \ll n$ ones in each row. The LDPC code-construction technique he presented is as follows. The matrix $\mathbf{H}$ has the form

$$\mathbf{H} = \begin{bmatrix} \mathbf{H}_1 \\ \mathbf{H}_2 \\ \vdots \\ \mathbf{H}_g \end{bmatrix}, \tag{6.1}$$

where the submatrices $\mathbf{H}_a$, $a = 1, \ldots, g$, have the following structure. For any integers $\mu$ and $r$ greater than 1, each submatrix $\mathbf{H}_a$ is $\mu \times \mu r$ with row weight $r$ and column weight 1. The submatrix $\mathbf{H}_1$ has the following specific form: for $i = 0, 1, \ldots, \mu - 1$, the $i$th row contains all of its $r$ 1s in columns $ir$ to $(i+1)r-1$. The other submatrices are obtained by column permutations of $\mathbf{H}_1$. It is evident that $\mathbf{H}$ is regular, has dimension $\mu g \times \mu r$, and has row and column weights $r$ and $g$, respectively. The absence of length-4 cycles in $\mathbf{H}$ is not guaranteed, but they can be avoided via computer design of $\mathbf{H}$. Gallager showed that the ensemble of such codes has excellent distance properties, provided that $g \geq 3$. Further, Gallager pointed out that such codes have low-complexity encoders since parity bits can be solved for as a function of the user bits via the parity-check matrix.

---

**Example 6.1** The following $\mathbf{H}$ matrix is the first example given by Gallager [1]:

$$\mathbf{H} = \left[\begin{array}{cccccccccccccccccccc}
1 & 1 & 1 & 1 & 0 & 0 & 0 & 0 & 0 & 0 & 0 & 0 & 0 & 0 & 0 & 0 & 0 & 0 & 0 & 0 \\
0 & 0 & 0 & 0 & 1 & 1 & 1 & 1 & 0 & 0 & 0 & 0 & 0 & 0 & 0 & 0 & 0 & 0 & 0 & 0 \\
0 & 0 & 0 & 0 & 0 & 0 & 0 & 0 & 1 & 1 & 1 & 1 & 0 & 0 & 0 & 0 & 0 & 0 & 0 & 0 \\
0 & 0 & 0 & 0 & 0 & 0 & 0 & 0 & 0 & 0 & 0 & 0 & 1 & 1 & 1 & 1 & 0 & 0 & 0 & 0 \\
0 & 0 & 0 & 0 & 0 & 0 & 0 & 0 & 0 & 0 & 0 & 0 & 0 & 0 & 0 & 0 & 1 & 1 & 1 & 1 \\
\hline
1 & 0 & 0 & 0 & 1 & 0 & 0 & 0 & 1 & 0 & 0 & 0 & 1 & 0 & 0 & 0 & 0 & 0 & 0 & 0 \\
0 & 1 & 0 & 0 & 0 & 1 & 0 & 0 & 0 & 1 & 0 & 0 & 0 & 0 & 0 & 0 & 1 & 0 & 0 & 0 \\
0 & 0 & 1 & 0 & 0 & 0 & 1 & 0 & 0 & 0 & 0 & 0 & 0 & 1 & 0 & 0 & 0 & 1 & 0 & 0 \\
0 & 0 & 0 & 1 & 0 & 0 & 0 & 0 & 0 & 0 & 1 & 0 & 0 & 0 & 1 & 0 & 0 & 0 & 1 & 0 \\
0 & 0 & 0 & 0 & 0 & 0 & 0 & 1 & 0 & 0 & 0 & 1 & 0 & 0 & 0 & 1 & 0 & 0 & 0 & 1 \\
\hline
1 & 0 & 0 & 0 & 0 & 1 & 0 & 0 & 0 & 0 & 0 & 1 & 0 & 0 & 0 & 0 & 0 & 1 & 0 & 0 \\
0 & 1 & 0 & 0 & 0 & 0 & 1 & 0 & 0 & 0 & 1 & 0 & 0 & 0 & 0 & 1 & 0 & 0 & 0 & 0 \\
0 & 0 & 1 & 0 & 0 & 0 & 0 & 1 & 0 & 0 & 0 & 0 & 1 & 0 & 0 & 0 & 0 & 0 & 1 & 0 \\
0 & 0 & 0 & 1 & 0 & 0 & 0 & 0 & 1 & 0 & 0 & 0 & 0 & 1 & 0 & 0 & 1 & 0 & 0 & 0 \\
0 & 0 & 0 & 0 & 1 & 0 & 0 & 0 & 0 & 1 & 0 & 0 & 0 & 0 & 1 & 0 & 0 & 0 & 0 & 1
\end{array}\right].$$

It corresponds to a (20, 5) code with $g = 3$ and $r = 4$. Observe that this matrix has the form of (6.1) and $\mathbf{H}_1$ has the form described above, with $\mu = 5$.

---

### 6.1.2 MacKay Codes

Thirty-five years after Gallager had done so, MacKay, unaware of Gallager's work, independently discovered the benefits of designing binary codes with sparse $\mathbf{H}$ matrices and was the first to show by computer simulation the ability of these codes to perform near capacity limits on the BSC and BI-AWGNC. MacKay has archived on a web page [5] a large number of LDPC codes he has designed for application to data communication and storage. A few of the computer-based design algorithms suggested by MacKay are listed below in order of increasing algorithm complexity (but not necessarily improved performance).

1. $\mathbf{H}$ is created by randomly generating weight-$g$ columns and, as nearly as possible, uniform row weight.

2. $\mathbf{H}$ is created by randomly generating weight-$g$ columns, while ensuring weight-$r$ rows, and the ones-overlap of any two columns is at most one (i.e., ensuring the RC constraint to avoid length-4 cycles).
3. $\mathbf{H}$ is generated as in algorithm 2, plus additional short cycles are avoided.
4. $\mathbf{H}$ is generated as in algorithm 3, plus $\mathbf{H} = \begin{bmatrix}\mathbf{H}_1 & \mathbf{H}_2\end{bmatrix}$ is constrained so that $\mathbf{H}_2$ is invertible.

One drawback of MacKay codes is that they lack sufficient structure to enable low-complexity encoding. Encoding is performed by putting $\mathbf{H}$ in the form $\begin{bmatrix}\mathbf{P}^{\mathrm{T}} & \mathbf{I}\end{bmatrix}$ via Gauss–Jordan elimination (or by multiplying $\mathbf{H}$ by $\mathbf{H}_2^{-1}$ for algorithm 4), from which the generator matrix can be put into the systematic form $\mathbf{G} = \begin{bmatrix}\mathbf{I} & \mathbf{P}\end{bmatrix}$.[1] The problem with encoding via $\mathbf{G}$ is that the submatrix $\mathbf{P}$ is generally not sparse, so for codes of practical interest the encoding complexity is high. An efficient encoding technique based on the $\mathbf{H}$ matrix was proposed in [6] for arbitrary $\mathbf{H}$ matrices. Later in this chapter and in Chapters 11 to 15, we give various approaches for introducing structure into parity-check matrices so that encoding is facilitated, and with essentially no performance loss relative to less structured parity-check matrices.

## 6.2 The PEG and ACE Code-Design Algorithms

### 6.2.1 The PEG Algorithm

As indicated in the previous chapter, cycles in the Tanner graph of an LDPC code present problems for iterative decoders. Because of this, many papers dealing with the design of Tanner graphs with large girths have been published. While some design techniques have relied on mathematics for achieving large girths (see Chapters 11 to 15 and [7]), the progressive-edge-growth (PEG)algorithm [3] is very effective and has been used widely for computer-based code design. Although the PEG algorithm is described in detail in Chapter 13, we here provide an overview of this algorithm, for it is a popular computer-based design algorithm.

The PEG algorithm is initialized by the number of variable nodes, $n$, the number of check nodes, $m$, and a variable node-degree sequence $D_v$, which is the list of degrees for each of the $n$ variable nodes. Given these parameters, the goal of the algorithm is to build the graph one edge at a time, and each edge is added to the graph in a manner that maximizes the local girth. Thus, the PEG algorithm is a greedy algorithm for creating a Tanner graph with a large girth.

Because the low-degree VNs are the most susceptible to error (they receive the least amount of neighborly help), edge placement begins with the lowest-degree VNs and progresses to VNs of increasing (or non-decreasing) degree. The algorithm does not move to the next VN until all of the edges of the current VN have been attached. The first edge attached to a VN is connected to a lowest-degree CN under the current

[1] We note that Chapter 3 introduces the generator matrix form $\mathbf{G} = \begin{bmatrix}\mathbf{P} & \mathbf{I}\end{bmatrix}$, whereas here we use the form $\mathbf{G} = \begin{bmatrix}\mathbf{I} & \mathbf{P}\end{bmatrix}$. Both are commonly used in the literature and we use both in this book.

state of the graph. Subsequent attachments of edges to the VN are done in such a way that the (local) girth for that VN is maximum. Thus, if the current state of the graph is such that one or more CNs cannot be reached from the current VN by traversing the edges connected so far, then the edge should be connected to an unreachable CN so that no cycle is created. Otherwise, if all CNs are reachable from the current VN along some number of edges, the new edge should be connected to a CN of lowest degree that results in the largest girth seen by the current VN. This lowest-degree CN strategy will yield a fairly uniform CN degree distribution.

### 6.2.2 The ACE Algorithm

The ACE algorithm [4] (acronym defined below) is another computer-based code-design algorithm that accounts for the pitfalls of iterative decoding. Specifically, it accounts for the fact that the iterative decoder not only has difficulties with cycles, but also is hindered by the overlap of multiple cycles. The ACE algorithm was motivated by the following observations made in [4].

1. In a Tanner graph for which each VN degree is at least 2, every stopping set contains multiple cycles, except for the special case in which all VNs in the stopping set are degree-2 VNs, in which case the stopping set is a single cycle. (Recall from Chapter 5 that a stopping set $S$ is a set of VNs whose neighboring CNs are connected to $S$ at least twice. Stopping sets thwart iterative decoding on erasure channels.)
2. For a code with minimum distance $d_{\min}$, each set of $d_{\min}$ columns of $\mathbf{H}$ that sum to the zero vector corresponds to a set of VNs that form a stopping set.
3. The previous observation implies that preventing small stopping sets in the design of an LDPC code (i.e., the construction of $\mathbf{H}$) also prevents a small $d_{\min}$.

The upshot of these results is that a code that is well suited for iterative decoding should have a minimum stopping-set size that is as large as possible. One approach for doing this is of course maximizing the girth for the code's Tanner graph, using the PEG algorithm, for example. However, as noted in [4], this is not sufficient. Because short cycles are susceptible to iterative decoding errors, the ACE algorithm seeks to have high connectivity from the VNs within a short cycle to CNs outside of the cycle. In this way, information from outside the cycle flows into the short cycle, under the assumption that such *extrinsic* information is helpful on average.

The description of the ACE algorithm requires the following definitions. An *extrinsic CN* relative to the VNs in a cycle is a CN that is connected to the cycle only once. The *extrinsic message degree* (EMD) for a cycle is the number of such extrinsic CNs connected to it. The *approximate cycle EMD* (ACE) of a length-$2\delta$ cycle is the maximum possible EMD for the cycle. Because a degree-$d$ VN within a cycle can be connected to at most $d-2$ extrinsic CNs, the ACE for a length-$2\delta$ cycle is given by $\sum_{l=1}^{\delta}(d_l-2)$, where $d_l$ is the $l$th VN in the cycle.

The ACE code-design algorithm builds a Tanner graph ($\mathbf{H}$ matrix) in such a way that the short cycles have large EMD values. Note that this has the effect of increasing the size of the smallest stopping set because the EMD of a stopping set is zero. The ACE algorithm generates a $(\delta_{\text{ACE}}, \epsilon)$ LDPC code, defined as an LDPC code whose cycles of length $2\delta_{\text{ACE}}$ or less have ACE values of $\epsilon$ or more. The algorithm begins with the lowest-weight column of $\mathbf{H}$ and progresses to columns of increasing (non-decreasing) weight. Each (randomly generated) candidate column for $\mathbf{H}$ is added only if, in the event its addition creates a cycle of length $2\delta_{\text{ACE}}$ or less, the cycle's ACE value is $\epsilon$ or more.

Note that it is possible to combine the PEG and ACE algorithms to obtain a hybrid PEG/ACE algorithm. This was proposed in [8], where the PEG/ACE algorithm was shown to result in codes with good iterative decoding performance. An application of the PEG/ACE algorithm may be found in Section 6.5.2. See also Chapter 13 for a description of the PEG/ACE algorithm.

## 6.3 Protograph LDPC Codes

Soon after the reinvention of LDPC codes in the late 1990s, researchers sought improvements in performance and complexity. Performance improvements were achieved via the design of irregular LDPC codes with optimal and near-optimal degree distributions. Such degree distributions may be obtained using so-called density evolution and EXIT chart algorithms, which are discussed in Chapter 9. Notice that the LDPC codes discussed above possess very little structure, that is, their parity-check matrices are quite random, implying complex encoders and decoders. Improvements in complexity were achieved by considering classes of LDPC codes whose parity-check matrices contain additional structure that facilitates encoding and/or decoding. Some of the more common approaches, each of which may be used in conjunction with the PEG and ACE algorithms, are considered in this and subsequent sections.

Motivated by simplified design, implementation, and analysis, techniques that rely on the expansion of a smaller matrix or graph prototype (protograph) into a full matrix or graph are in common use. These are called *protograph-based codes*, or simply *protograph codes* [9, 10]. A *protograph* is a relatively small bipartite graph from which a larger graph can be obtained by a copy-and-permute procedure: the protograph is copied $Q$ times, and then the edges of the individual replicas are permuted among the $Q$ replicas (under restrictions described below) to obtain a single, large bipartite graph. When the protograph possesses $n_p$ variable nodes and $m_p$ constraint nodes, the derived graph will consist of $n = n_p Q$ variable nodes and $m_c = m_p Q$ constraint nodes. This process of expanding a smaller graph into a larger graph by interconnecting copies of the smaller graph is called *lifting*.

In order to preserve the node degree distributions of the original protograph, the edge permutations among the copies are not arbitrary. In particular, the nodes of the protograph are labeled so that, if variable node $j$ is connected to constraint node $i$ in the protograph, then a "type-$j$" variable node in a replica can connect only to one of the $Q$ replicated "type-$i$" constraint nodes. In addition to preserving the degree distributions of the protograph (see Chapters 8 and 9), this also permits the design

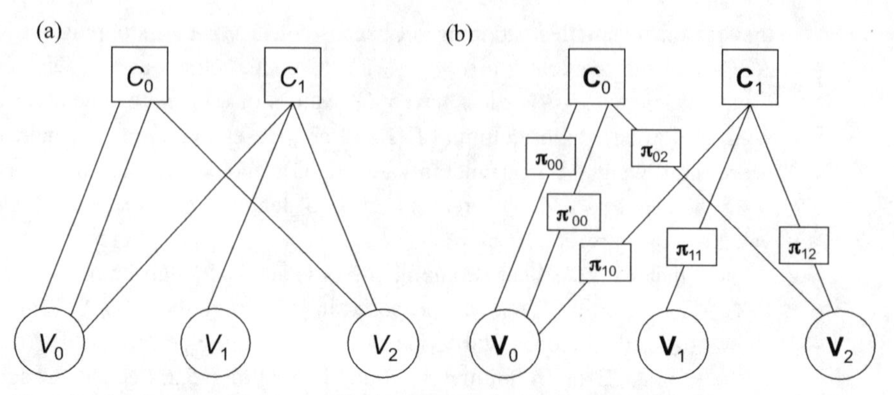


**Figure 6.1** (a) An example protograph. (b) The Tanner graph that results after the copy-and-permute procedure is applied to the protograph in (a). Each permutation matrix is $Q \times Q$, each node is interpreted to be a $Q$-bundle (or $Q$-vector) of nodes of the same type, and each edge represents a $Q$-bundle of edges.

of quasi-cyclic codes. In particular, if the edge permutations correspond to weight-1 circulants, which are also called *circulant permutation matrices*, then the resulting $\mathbf{H}$ matrix will be an array of such circulants, that is, the code will be quasi-cyclic.

Figure 6.1(a) presents a protograph and Figure 6.1(b) presents the Tanner graph derived from it by replacing each protograph node by a bundle (or vector) of $Q$ nodes and each protograph edge by a bundle of $Q$ edges. The connections between node bundles via the edge bundles are specified by the $Q \times Q$ matrices $\pi_{ij}$ seen in the figure. Thus, each variable node $j$ becomes a bundle of type-$j$ variable nodes and each check node $i$ becomes a bundle of type-$i$ check nodes. Notice that parallel edges are permissible in a protograph, but the permutation matrices are selected so that the parallel edges no longer exist in the expanded graph.

---

**Example 6.2** Consider the protograph in Figure 6.1(a). The adjacency matrix for this protograph, sometimes called a *base matrix*, is given by

$$\mathbf{B} = \begin{bmatrix} 2 & 0 & 1 \\ 1 & 1 & 1 \end{bmatrix},$$

where the “2” entry signifies the two parallel edges between nodes $C_0$ and $V_0$, and the “0” entry signifies the absence of an edge between nodes $C_0$ and $V_1$. With $Q = 3$, a possible $\mathbf{H}$ matrix derived from $\mathbf{B}$ is

$$\mathbf{H} = \left[\begin{array}{ccccccccc} 1 & 1 & 0 & 0 & 0 & 0 & 0 & 0 & 1 \\ 0 & 1 & 1 & 0 & 0 & 0 & 1 & 0 & 0 \\ 1 & 0 & 1 & 0 & 0 & 0 & 0 & 1 & 0 \\ \\ 0 & 0 & 1 & 0 & 1 & 0 & 1 & 0 & 0 \\ 1 & 0 & 0 & 0 & 0 & 1 & 0 & 1 & 0 \\ 0 & 1 & 0 & 1 & 0 & 0 & 0 & 0 & 1 \end{array}\right],$$

where the permutation matrices $\pi_{00}$ and $\pi'_{00}$ were chosen so that $\mathbf{H}$ is a meaningful parity-check matrix and its corresponding Tanner graph contains no parallel edges. Because there are parallel edges between $V_0$ and $C_0$, the sum of $\pi_{00}$ and $\pi'_{00}$ appears as a circulant submatrix in the upper-left corner of $\mathbf{H}$. Note that $\mathbf{H}$ is an array of circulants, so the code is quasi-cyclic.

---

In this example we saw how replacing the nonzero elements in the base matrix of a protograph by circulants gives rise to a quasi-cyclic code. Observe that, conversely, any quasi-cyclic code has a protograph representation. This is easily seen from the previous example because, once $\mathbf{H}$ is given, the base matrix $\mathbf{B}$ is obvious, from which the protograph directly follows. For this reason, we will often speak of QC codes and protograph codes interchangeably. We will see in Chapter 13 that the class of protograph codes is a special case of the class of *superposition codes*.

We remark also that, when the protograph is small and the desired code length is large, one is tempted to use large circulant permutation matrices. However, code performance can suffer when $Q$ is too large, for the code becomes too structured. Thus, what is typically done in practice is to replace each “1” element of $\mathbf{B}$ by a $Q \times Q$ matrix that is an array of $Q' \times Q'$ circulant permutation matrices, where $Q'$ divides $Q$. Note that such an array is still a permutation matrix. A “2” element in $\mathbf{B}$ would be replaced by a $Q \times Q$ matrix that is a sum of two arrays of $Q' \times Q'$ circulant permutation matrices (equivalently, an array of $Q' \times Q'$ weight-2 circulants); and so on for values larger than 2. As discussed later, Figure 6.13, in conjunction with Example 6.7, gives an illustrative example.

A brute-force approach to designing a protograph code is as follows. First pick a rate $R$ and the number of protograph VNs $n_p$. The number $m_p$ of protograph CNs follows from $n_p$ and $R$. Once $R$, $n_p$, and $m_p$ have been selected, the node degree distributions $\lambda(x)$ and $\rho(x)$ may be determined. Using the techniques of Chapter 9, for example, the degree distributions can be chosen (via computer search) to optimize error-rate performance in the waterfall region. Alternatively, using the techniques of Chapter 8, the degree distributions can be chosen to optimize the performance in the floor region. Section 6.6.2 presents a technique for obtaining protographs for code ensembles with exceptional performance both in the waterfall region and in the floor region. Once the degree distributions have been determined, the protograph can be expanded into a full Tanner graph via the choice of permutation matrices, one for each protograph edge. The idea is that, if the code ensemble specified by the protograph is good, an arbitrary member of the ensemble can be expected to be good. The selection of permutation matrices (i.e., converting $\mathbf{B}$ to $\mathbf{H}$) can be directed by the PEG and/or ACE algorithms. More systematic protograph code-design approaches involve accumulator-based protograph codes as discussed in [11, 12] and Section 6.6.2.

### 6.3.1 Decoding Architectures for Protograph Codes

As discussed in Chapter 5, a Tanner graph for a code represents a blueprint of its iterative decoder, with the nodes representing processors and the edges representing links between the processors. In principle, an LDPC decoder architecture could mimic the code's Tanner graph, but frequently such a fully parallel architecture exceeds the hardware specifications of an application. The protograph representation leads to two *partially parallel* alternatives [9, 13–15].

We call the first partially parallel architecture the *circulant-based architecture*. Note that, for a QC code, the set of VNs and the set of CNs are naturally partitioned into subsets of size $Q$ because the $\mathbf{H}$ matrix for QC codes is an $M \times N$ array of $Q \times Q$ circulants. The protograph corresponding to these partitions therefore has $M$ check nodes and $N$ variable nodes. In the circulant-based architecture, there are $M$ check-node processors (CNPs) and $N$ variable-node processors (VNPs). The decoder also requires $MN$ random-access memory (RAM) blocks to store the intermediate messages being passed between the node processors. In the first half-iteration, the $N$ VNPs operate in parallel, but the $Q$ variable nodes assigned to each VNP are updated serially. In the second half-iteration, the $M$ CNPs operate in parallel, but the $Q$ check nodes assigned to each CNP are updated serially. Another way of stating this is that, in the first half-iteration, the circulant-based decoder simultaneously updates all of the VNs within one copy of the protograph, and then does the same for each of the other copies, one copy at a time. Then, in the second half-iteration, the decoder simultaneously updates all of the CNs within one copy of the protograph, and then does so for each of the other copies, one copy at a time.

The second partially parallel architecture is called the *protograph-based architecture*. This decoder architecture uses the fact that the code's Tanner graph was created by $Q$ copies of the protograph. Thus, there are $Q$ "protograph decoders" that run simultaneously, but each protograph decoder is a serial decoder with one VNP and one CNP.

The throughput $T$ for both architectures can be related to other parameters as

$$T \text{ (bits/s)} = \frac{k \text{ (bits)} \cdot f_{\text{clock}} \text{ (cycles/s)}}{I_{\text{ave}} \text{ (iterations)} \cdot S_d \text{ (cycles/iteration)}},$$

where $k$ is the number of information bits per codeword, $f_{\text{clock}}$ is the clock frequency, $I_{\text{ave}}$ is the average number of decoding iterations, and $S_d$ is the number of clock cycles required per iteration. For typical node processor implementations, $S_d$ is proportional to $Q$ for the circulant-based architecture and is proportional to $n_e$ for the protograph-based architecture, where $n_e$ is the number of edges in the protograph. Thus, we would want $Q$ small for the circulant-based architecture. From the discussion in the previous subsection, $n_e$ increases as $Q$ decreases. Thus, we want $Q$ large in the case of the protograph-based architecture.

## 6.4 Multi-edge-Type LDPC Codes

*Multi-edge-type LDPC codes* [16] represent a generalization of protograph LDPC codes. Multi-edge-type (MET) codes were invented first, but we present them second because they are slightly more involved. Before we give a formal definition of MET

LDPC codes, we present an example of a generalized protograph that encapsulates the concepts behind this class of codes.

---

**Example 6.3** Figure 6.2 represents a generalized (expanded) protograph that is much like the graph in Figure 6.1(b). Observe that, in contrast with Figure 6.1(b), in Figure 6.2 multiple edges may be connected to a permutation matrix. Further, since there is a $Q$-fold copy-and-permute procedure associated with this graph, the matrix $\pi_1$ is the sum of two distinct $Q \times Q$ permutation matrices since two parallel edges are connected to it. The matrix $\pi_2$ is $Q \times 3Q$ since three skew (non-parallel) edges (actually, edge bundles) are connected to it. Finally, $\pi_3$ is a $Q \times Q$ permutation matrix. With $Q = 3$, an example $\mathbf{H}$ matrix is

$$\mathbf{H} = \left[\begin{array}{ccc|ccc|ccc} 0 & 1 & 1 & 0 & 0 & 0 & 0 & 1 & 0 \\ 1 & 0 & 1 & 0 & 0 & 0 & 0 & 0 & 1 \\ 1 & 1 & 0 & 0 & 0 & 0 & 1 & 0 & 0 \\ \hline 0 & 1 & 0 & 1 & 1 & 0 & 0 & 0 & 0 \\ 0 & 0 & 1 & 0 & 0 & 1 & 0 & 1 & 0 \\ 1 & 0 & 0 & 0 & 0 & 0 & 1 & 0 & 1 \end{array}\right].$$

In the graph of Figure 6.2, there are three *edge types*: the "type-1" edges connected through $\pi_1$, the "type-2" edges connected through $\pi_2$, and the "type-3" edges connected through $\pi_3$. One may see from Figure 6.2 that the MET class of codes contains the protograph codes as a special case. In Figure 6.1(b), there are six edge types.

---

MET LDPC codes may be formally defined as follows [16]. First, a node of degree $d$, whether VN or CN, possesses $d$ "sockets" of various types. Further, we associate with each node a *degree vector* $\mathbf{d}$, which enumerates the number of sockets of each type, and whose sum of elements is $d$. Thus, for example, in a Tanner graph containing three socket types, a degree-8 constraint node with degree vector $\mathbf{d} = [2, 5, 1]$ has two type-1 sockets, five type-2 sockets, and one type-3 socket. An edge of a given type then connects a VN socket to a CN socket, both of which are of the same type as

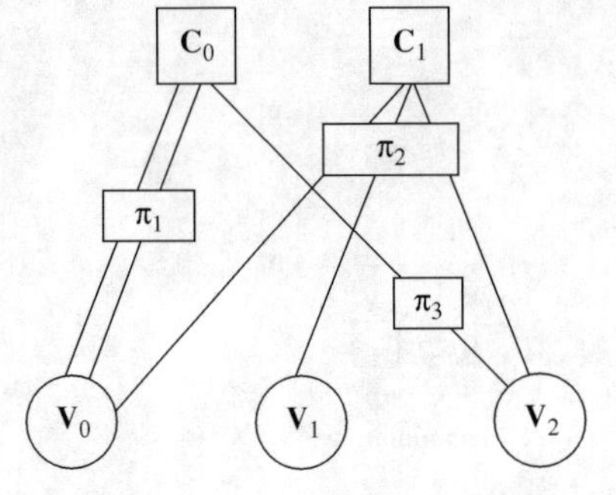


**Figure 6.2** A vectorized MET generalized protograph.

the edge. As in the previous example, the routing of a bundle of edges that connects node sockets of the same type is described by a permutation matrix. Design techniques for MET LDPC codes are covered in [16].

## 6.5 Single-Accumulator-Based LDPC Codes

The accumulator-based codes that were invented first are the so-called repeat-accumulate (RA) codes [17]. Despite their simple structure, they were shown to provide good performance and, more importantly, they paved a path toward the design of efficiently encodable LDPC codes. RA codes and other accumulator-based codes are LDPC codes that can be decoded as serial turbo codes, but are more commonly treated as LDPC codes. We discuss in this section only the most well known among the accumulator-based LDPC codes.

### 6.5.1 Repeat–Accumulate Codes

As shown in Figure 6.3, an RA code consists of a serial concatenation, through an interleaver, of a single rate-$1/q$ repetition code with an *accumulator* having transfer function $1/(1+D)$. Referring to Figures 4.2 and 4.3, the implementation of the transfer function $1/(1+D)$ is identical to that of an accumulator, although the accumulated value can be only 0 or 1 since operations are over $\mathbb{F}_2$. To ensure a large minimum Hamming distance, the interleaver should be designed so that consecutive 1s at its input are widely separated at its output. To see this, observe that two 1s separated by $s-1$ 0s at the interleaver output (and hence at the accumulator input) will yield a run of $s$ 1s at the accumulator output. Hence, we would like $s$ to be large in order to achieve large Hamming weight at the output of the accumulator.

RA codes can be either non-systematic or systematic. In the first case, the accumulator output, $\mathbf{p}$, is the codeword and the code rate is $1/q$. For systematic RA codes, the information word, $\mathbf{u}$, is combined with $\mathbf{p}$ to yield the codeword $\mathbf{c} = [\mathbf{u}\ \mathbf{p}]$, so that the code rate is $1/(1+q)$. The main limitations of RA codes are the code rate, which cannot be higher than $1/2$, and mediocre performance at short and medium lengths. They perform surprisingly well on the AWGN channel, but they are not capacity-approaching

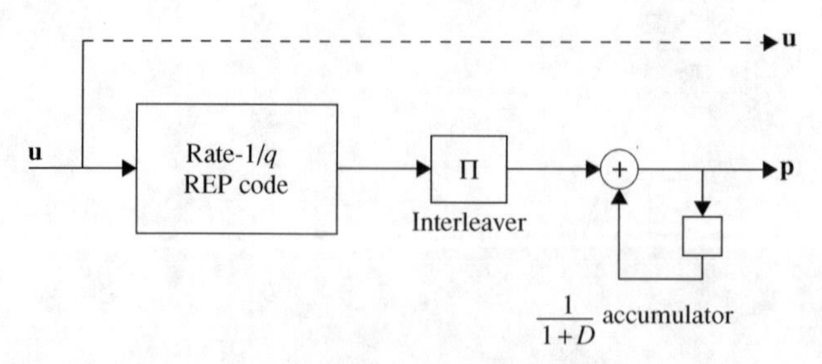


**Figure 6.3** A repeat–accumulate code block diagram.

codes. The following subsections will present a brief overview of the major enhancements to RA codes, which permit operation closer to capacity and at high code rates.

### 6.5.2 Irregular Repeat–Accumulate Codes

The irregular repeat–accumulate (IRA) codes [18, 19] generalize the RA codes in that the repetition rate may differ for each of the $k$ information bits and that linear combinations of the repeated bits are sent through the accumulator. Further, IRA codes are typically systematic. IRA codes provide two important advantages over RA codes. First, they allow flexibility in the choice of the repetition rate for each information bit so that high-rate codes may be designed. Second, their irregularity allows operation closer to the capacity limit.

The Tanner graph for IRA codes is presented in Figure 6.4(a) and the encoder structure is depicted in Figure 6.4(b) (the matrices $\mathbf{A}$ and $\mathbf{H}_u$ are described below). The variable repetition rate is accounted for in the graph by letting the variable node degrees $d_{b,j}$ vary with $j$. The accumulator is represented by the rightmost part of the graph, where the dashed edge is added to include the possibility of a *tail-biting trellis*, that is, a trellis whose first and last states are identical. Also, we see that $d_{c,i}$ interleaver output bits are added to produce the $i$th accumulator input. Figure 6.4 also includes the representation for RA codes. As indicated in the table in the figure, for an RA code, each information bit node connects to exactly $q$ check nodes ($d_{b,j} = q$) and each check node connects to exactly one information bit node ($d_{c,i} = 1$).

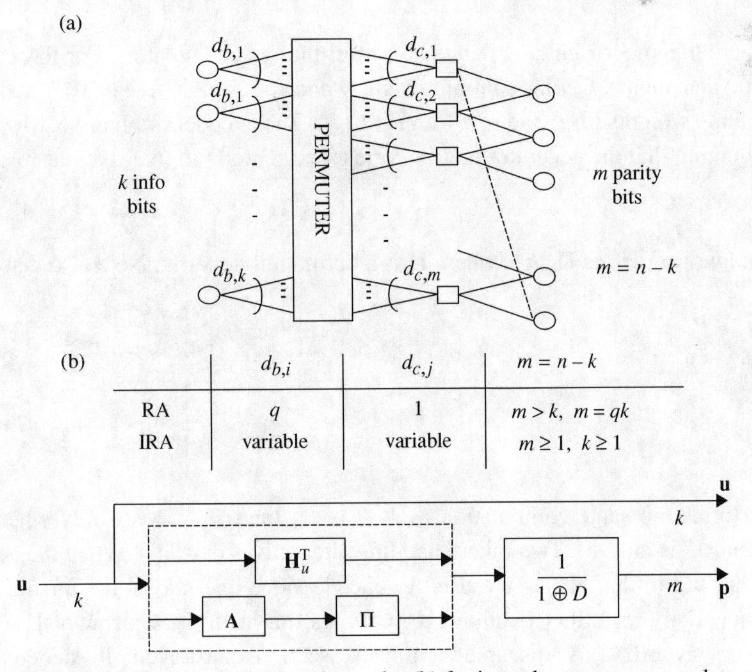


**Figure 6.4** A Tanner graph (a) and encoder (b) for irregular repeat–accumulate codes.

To determine the code rate for an IRA code, define $\overline{q}$ to be the average repetition rate of the information bits,

$$\overline{q} = \frac{1}{k}\sum_{j=1}^{k} d_{b,j},$$

and $\bar{d}_c$ as the average of the degrees $\{d_{c,i}\}$,

$$\bar{d}_c = \frac{1}{m}\sum_{i=1}^{m} d_{c,i}.$$

Then the code rate for systematic IRA codes is

$$R = \frac{1}{1 + \overline{q}/\bar{d}_c}. \tag{6.2}$$

For non-systematic IRA codes, $R = \bar{d}_c/\overline{q}$.

The parity-check matrix for systematic RA and IRA codes has the form

$$\mathbf{H} = \begin{bmatrix} \mathbf{H}_u & \mathbf{H}_p \end{bmatrix}, \tag{6.3}$$

where $\mathbf{H}_p$ is an $m \times m$ "dual-diagonal" matrix,

$$\mathbf{H}_p = \begin{bmatrix} 1 & & & & (1) \\ 1 & 1 & & & \\ & \ddots & \ddots & & \\ & & 1 & 1 & \\ & & & 1 & 1 \end{bmatrix}, \tag{6.4}$$

where the upper-right 1 is included for tail-biting accumulators. For RA codes, $\mathbf{H}_u$ is a regular matrix having column weight $q$ and row weight 1. For IRA codes, $\mathbf{H}_u$ has column weights $\{d_{b,j}\}$ and row weights $\{d_{c,i}\}$. The encoder of Figure 6.4(b) is obtained by noting that the generator matrix corresponding to $\mathbf{H}$ in (6.3) is

$$\mathbf{G} = \begin{bmatrix} \mathbf{I} & \mathbf{H}_u^{\mathrm{T}}\mathbf{H}_p^{-\mathrm{T}} \end{bmatrix}$$

and writing $\mathbf{H}_u$ as $\mathbf{\Pi}^{\mathrm{T}}\mathbf{A}^{\mathrm{T}}$, where $\mathbf{\Pi}$ is a permutation matrix. Note also that

$$\mathbf{H}_p^{-\mathrm{T}} = \begin{bmatrix} 1 & 1 & \cdots & & 1 \\ & 1 & 1 & \cdots & 1 \\ & & \ddots & & \vdots \\ & & & 1 & 1 \\ & & & & 1 \end{bmatrix}$$

performs the same computation as $1/(1 \oplus D)$ (and $\mathbf{H}_p^{-\mathrm{T}}$ exists only when the "tail-biting 1" is absent). Two other encoding alternatives exist. (1) When the accumulator is not tail-biting, $\mathbf{H}$ may be used to encode since one may solve for the parity bits sequentially from the equation $\mathbf{c}\mathbf{H}^{\mathrm{T}} = \mathbf{0}$, starting with the top row of $\mathbf{H}$ and moving on downward. (2) As discussed in the next section, quasi-cyclic IRA code designs are possible, in which case the QC encoding techniques of Chapter 3 may be used.

Given the code rate, length, and degree distributions, an IRA code is defined entirely by the matrix $\mathbf{H}_u$ (equivalently, by $\mathbf{A}$ and $\mathbf{\Pi}$). From the form of $\mathbf{G}$, a weight-1 encoder input simply selects a row of $\mathbf{H}_u^{\mathrm{T}}$ and sends it through the accumulator (modeled by $\mathbf{H}_p^{-\mathrm{T}}$). Thus, to maximize the weight of the accumulator output for weight-1 inputs, the 1s in the columns of $\mathbf{H}_u$ (rows of $\mathbf{H}_u^{\mathrm{T}}$) should be widely separated. Similarly, weight-2 encoder inputs send the sum of two columns of $\mathbf{H}_u$ through the accumulator, so the 1s in the sums of all pairs of columns of $\mathbf{H}_u$ should be widely separated. In principle, $\mathbf{H}_u$ could be designed so that even-larger-weight inputs yield sums with widely separated 1s, but the complexity of doing so eventually becomes unwieldy. Further, accounting for only weight-1 and weight-2 inputs generally results in good codes. As a final design guideline, as shown in [20] and Chapter 8, the column weight of $\mathbf{H}_u$ should be at least 4, otherwise there will be a high error floor due to a small minimum distance.

While an unstructured $\mathbf{H}_u$ would generally give good performance, it leads to high-complexity decoder implementations. This is because a substantial amount of memory would be required to store the connection information implicit in $\mathbf{H}_u$. In addition, although standard message-passing decoding algorithms for LDPC codes are inherently parallel, the physical interconnections required to realize a code's bipartite graph become an implementation hurdle and prohibit a fully parallel decoder. Using a structured $\mathbf{H}_u$ matrix mitigates these problems.

Tanner [21] was the first to consider structured RA codes, more specifically, quasi-cyclic RA codes, which require tail-biting in the accumulator. Simulation results in [21] demonstrate that the QC-RA codes compete well with unstructured RA codes and surpass their performance at high SNR values. We now generalize the result of [21] to IRA codes, following [20].

To attain structure in $\mathbf{H}$ for IRA codes, one cannot simply choose $\mathbf{H}_u$ to be an array of circulant permutation matrices. It is easy to show that doing so will produce a poor LDPC code in the sense of minimum distance. (Consider weight-2 encoder inputs with adjacent 1s assuming such an $\mathbf{H}_u$.) Instead, the following strategy has been used [20]. Let $\mathbf{P}$ be an $L \times J$ array of $Q \times Q$ circulant permutation matrices (for some convenient $Q$). Then set $\mathbf{A}^{\mathrm{T}} = \mathbf{P}$ so that $\mathbf{H}_u = \mathbf{\Pi}^{\mathrm{T}}\mathbf{P}$ and

$$\mathbf{H}_a = \left[\mathbf{\Pi}^{\mathrm{T}}\mathbf{P} \;\; \mathbf{H}_p\right], \tag{6.5}$$

where $\mathbf{H}_p$ represents the tail-biting accumulator. Note that $m = L \times Q$ and $k = J \times Q$.

We now choose $\mathbf{\Pi}$ to be a standard deterministic "row–column" interleaver so that row $lQ + q$ in $\mathbf{P}$ becomes row $qL + l$ in $\mathbf{\Pi}^{\mathrm{T}}\mathbf{P}$, for all $0 \le l < L$ and $0 \le q < Q$. Next, we permute the rows of $\mathbf{H}_a$ by $\mathbf{\Pi}^{-\mathrm{T}}$ to obtain

$$\mathbf{H}_b = \mathbf{\Pi}^{-\mathrm{T}}\mathbf{H} = \left[\mathbf{P} \;\; \mathbf{\Pi}\mathbf{H}_p\right], \tag{6.6}$$

where we have used the fact that $\mathbf{\Pi}^{-\mathrm{T}} = \mathbf{\Pi}$. Finally, we permute only the columns corresponding to the parity part of $\mathbf{H}_b$, which gives

$$\mathbf{H}_{\text{QC-IRA}} = \left[\mathbf{P} \;\; \mathbf{\Pi}\mathbf{H}_p\mathbf{\Pi}^{\mathrm{T}}\right] = \left[\mathbf{P} \;\; \mathbf{H}_{p,\text{QC}}\right]. \tag{6.7}$$

It is easily shown that the parity part of $\mathbf{H}_{\text{QC-IRA}}$, that is, $\mathbf{H}_{p,\text{QC}} \triangleq \mathbf{\Pi}\mathbf{H}_p\mathbf{\Pi}^{\text{T}}$, is in the quasi-cyclic form

$$\mathbf{H}_{p,\text{QC}} = \begin{bmatrix} I_0 & & & & I_1 \\ I_0 & I_0 & & & \\ & \ddots & \ddots & & \\ & & I_0 & I_0 & \\ & & & I_0 & I_0 \end{bmatrix}, \tag{6.8}$$

where $I_0$ is the $Q \times Q$ identity matrix and $I_1$ is obtained from $I_0$ by cyclically shifting all of its rows leftward once. Therefore, $\mathbf{H}_{\text{QC-IRA}}$ corresponds to a quasi-cyclic IRA code since $\mathbf{P}$ is also an array of $Q \times Q$ circulant permutation matrices. Observe that, except for a reordering of the parity bits, $\mathbf{H}_{\text{QC-IRA}}$ describes the same code as $\mathbf{H}_a$ and $\mathbf{H}_b$. If the upper-right "1" in (6.4) is absent, then the upper-right $I_1$ in (6.8) will be absent as well. In this case we refer to the code as simply "structured." For the structured (non-QC) case, encoding may be performed directly from the $\mathbf{H}$ matrix by solving for the parity bits given the data bits.

Given (6.7), fairly good QC-IRA codes are easily designed by choosing the permutation matrices in $\mathbf{P}$ such that 4-cycles are avoided. For enhanced performance, particularly in the error-floor region, additional considerations are necessary, such as incorporating the PEG and ACE algorithms into the design. These additional considerations are considered in the following subsection and [20].

### Quasi-cyclic IRA Code Design

Before presenting the design and performance of QC-IRA codes, we discuss the *potential* of these codes in an ensemble sense. An iterative *decoding threshold* is the theoretical performance limit for the iterative decoding of an ensemble of codes with a given degree distribution assuming infinite code length and an infinite number of decoder iterations. Chapter 9 presents several methods for numerically estimating such thresholds. Table 6.1 compares the binary-input AWGN decoding thresholds ($E_b/N_0$) for QC-IRA codes with those of regular QC-LDPC codes for selected code rates. The thresholds were calculated using the multidimensional EXIT algorithm presented in

**Table 6.1** Comparison of thresholds of QC-IRA codes and regular QC-LDPC codes

| Code rate | Capacity (dB) | QC-IRA threshold (dB) | Regular QC-LDPC threshold (dB) |
|---|---|---|---|
| 1/2 | 0.187 | 0.97 | 2.0 |
| 2/3 | 1.059 | 1.77 | 2.25 |
| 3/4 | 2.040 | 2.66 | 2.87 |
| 4/5 | 2.834 | 3.4 | 3.5 |
| 7/8 | 2.951 | 3.57 | 3.66 |
| 8/9 | 3.112 | 3.72 | 3.8 |

Chapter 9 (see also [22]). The QC-IRA codes are semi-regular, with column weight 5 for the systematic part and 2 for the parity part. The regular QC-LDPC codes have constant column weight 5. Observe in Table 6.1 that the QC-IRA codes have better thresholds for all rates, but the advantage decreases with increasing code rate. Also listed in Table 6.1 are the $E_b/N_0$ capacity limits for each code rate. We note that the gap to capacity for QC-IRA codes is about 0.8 dB for rate 1/2 and about 0.6 dB for rate 8/9. It is possible to achieve performance closer to capacity with non-constant systematic column weights (e.g., weight 3 and higher), but here we target finite code length, in which case a constant column weight of 5 has been shown to yield large minimum distance (see Chapter 8).

Recall that the parity-check matrix for a quasi-cyclic IRA (QC-IRA) code is given by $\mathbf{H} = \begin{bmatrix} \mathbf{P} & \mathbf{H}_{p,\mathrm{QC}} \end{bmatrix}$ per (6.7) and (6.8). Let us first define the submatrix $\mathbf{P}$ to be the following $L \times S$ array of $Q \times Q$ circulant permutation submatrices:

$$\mathbf{P} = \begin{bmatrix} b_{0,0} & b_{0,1} & \dots & b_{0,S-1} \\ b_{1,0} & b_{1,1} & \dots & b_{1,S-1} \\ & & \vdots & \\ b_{L-1,0} & b_{L-1,1} & \dots & b_{L-1,S-1} \end{bmatrix}, \tag{6.9}$$

where $b_{l,s} \in \{\infty, 0, 1, \dots, Q-1\}$ is the exponent of the circulant permutation matrix $\pi$ formed by cyclically shifting (mod $Q$) each row of a $Q \times Q$ identity matrix $\mathbf{I}_Q$ to the right once. Thus, (6.9) is a shorthand notation in which each entry $b_{l,s}$ should be replaced by $\pi^{b_{l,s}}$. We define $\pi^\infty$ to be the $Q \times Q$ zero matrix $\mathbf{O}_Q$ which corresponds to the "masked-out" entries of $\mathbf{P}$ (see Chapter 11 for a discussion of masking). Because of the structure of $\mathbf{P}$, rows and columns occur in groups of size $Q$. We will refer to *row group* $l$, $0 \le l \le L$, and *column group* $s$, $0 \le s \le S$.

We now present a design algorithm for finite-length QC-IRA codes and selected performance results for each. The design algorithm is the hybrid PEG/ACE algorithm (see Section 6.2) tailored to QC-IRA codes. It is also a modification for QC-IRA codes of the PEG-like algorithm in [23] proposed for unstructured IRA codes. Recall that the ACE algorithm attempts to maximize the minimum stopping-set size in the Tanner graph by ensuring that cycles of length $2\delta_{\mathrm{ACE}}$ or less have an ACE value no less than $\epsilon$. Clearly a length-$\ell$ cycle composed of only systematic nodes has a higher ACE value than does a length-$\ell$ cycle that includes also (degree-2) parity nodes. As in [23], we differentiate between these two cycle types and denote the girths for them by $g_{\mathrm{sys}}$ and $g_{\mathrm{all}}$.

By targeting a quasi-cyclic IRA code, the algorithm complexity and speed are improved by a factor of $Q$ compared with those for unstructured IRA codes. This is because for each $Q$-group of variable nodes (column group) only one variable node in the group need be condition tested during matrix construction. Further, in contrast with [23], when conducting PEG conditioning, rather than enforcing a single girth value across all columns, we assign an independent target girth value $g_{\mathrm{all}}$ to each column group. The advantage of this modification is that, if the girth value for one $Q$-group of nodes cannot be attained in the design process, only this girth value need be

**Algorithm 6.1** QC-IRA Code-Design Algorithm

**Initialization.** Initialize the parity part of **H** to the quasi-cyclic form of (6.8).
**Generation of P**
**while P incomplete do**

1. randomly select an unmasked position $(l, j)$ in matrix **P**, $l \in [0, L)$ and $j \in [0, J)$, for which $b_{l,j}$ is as yet undetermined
2. randomly generate an integer $x \in [0, Q)$ different from others already generated $b_{l',j}$
3. write the circulant $\pi^x$ in position $(l, j)$ in matrix **P**

**if** all conditions ($g_{\mathrm{sys}}$, $g_{\mathrm{all}}[j]$, and $(d_{\mathrm{ACE}}, \eta)$) are satisfied, then continue;
**else**
  **if** all $x$s in $[0, Q)$ have been tried, then decrease $g_{\mathrm{all}}[j]$ by 2 and go to Step 2
  **else** clear current circulant $\pi^x$, set $x = (x + 1) \bmod Q$, go to Step 3
**end (while)**

decremented, keeping the girth targets for the other column groups unchanged. Thus, at the end of the design process $g_{\mathrm{all}}$ will be a vector of length $S$. This modification produces a better cycle distribution in the code's graph than would be the case if a single value of $g_{\mathrm{all}}$ were used. Finally, in addition to this PEG-like girth conditioning, the ACE algorithm is also included in the condition testing.

**Example 6.4** Using the above algorithm, a rate-0.9 $(4544, 4096)$ QC-IRA code was designed with the parameters $Q = 64$, $L = 7$, $S = 64$, $g_{\mathrm{sys}} = 6$, initial $g_{\mathrm{all}}[s] = 10$ (for $0 \leq s < S$), and ACE parameters $(\delta_{\mathrm{ACE}}, \eta) = (4, 6)$. In Figure 6.5, software simulation results obtained using an SPA decoder on an AWGN channel are displayed for a maximum number of iterations $I_{\max}$ equal to 1, 2, 5, 10, and 50. Observe that the performance of this code at BER $= 10^{-6}$ and $I_{\max} = 50$ is about 1.3 dB from the Shannon limit. Notice also that the gap between $I_{\max} = 10$ and $I_{\max} = 50$ is only about 0.1 dB, so decoder convergence is reasonably fast and only 10 iterations (or fewer) would be necessary for most applications. Finally, we point out that the hardware decoder performance of this code was presented in Figure 5.11 of Chapter 5, where it was seen that there exists a floor below BER $= 10^{-10}$.

We now show two approaches for designing a family of QC-IRA codes of constant information word length but varying code rate. Such families have applications in communication systems that are subject to a range of operating SNRs, but require a single encoder/decoder structure. We consider an encoder/decoder-compatible family of codes comprising code rates 1/2, 2/3, and 4/5. In [24], a design method is presented that combines masking and "extension" techniques to design an S-IRA code family with fixed message length $k$. In this method, the lower-rate codes are constructed by

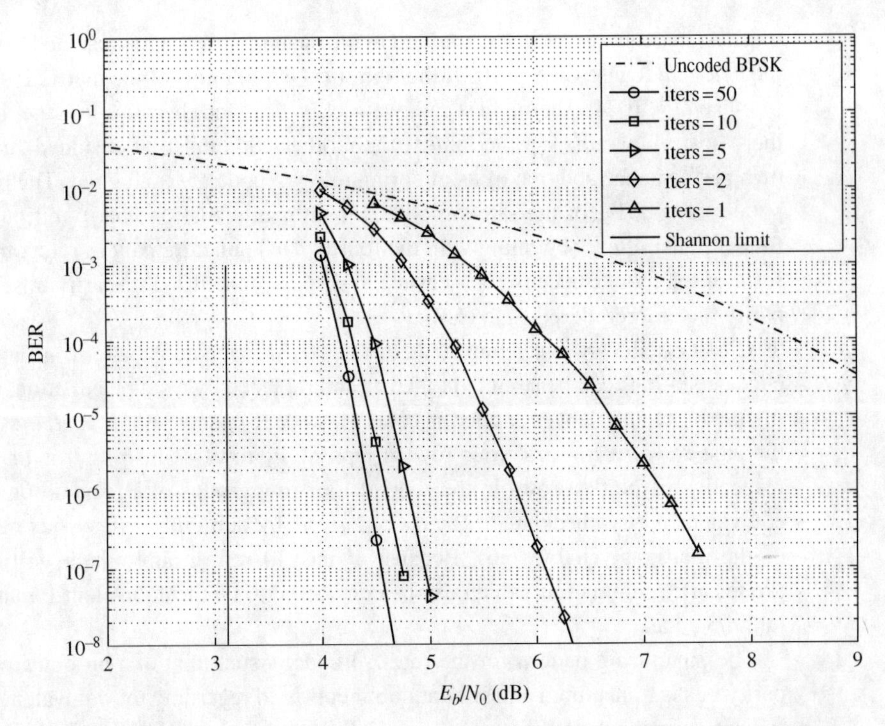


**Figure 6.5** The BER and convergence performance of a QC-IRA (4544, 4096) code on the AWGN channel.

adding rows to the **P** matrix of a higher-rate code while masking certain circulant permutation matrices of **P** to maintain a targeted column weight. Thus, the codes in the family share some circulants, therefore making the encoder/decoder implementation more efficient. The design of such codes can be enhanced by including girth conditioning (using the PEG algorithm). Also, in contrast with [24], the design order can go from low rate to high rate. Specifically, after an initial low rate design, higher-rate codes inherit circulants from lower-rate codes. The design technique just discussed will be called the "Method I" technique and the performance of such a family for $k = 1024$ will be presented below.

The design of a *rate-compatible* (RC) [25] family of QC-IRA codes is simple via the use of *puncturing* (i.e., deleting code bits to achieve a higher code rate), an approach we call "Method II." In a rate-compatible code family, the parity bits of higher-rate codes are embedded in those of lower-rate codes. The designer must be careful because, when the percentage of check nodes affected by multiple punctured bits is large, decoder message exchange becomes ineffective, leading to poor performance.

As an illustrative design example, we choose as the mother code a rate-1/2 (2044, 1024) S-IRA code with parameters $Q = 51$, $L = 20$, $S = 21$, $g = 5$, $g_{\text{sys}} = 6$, and initial $g_{\text{all}}[s] = 16$ (for $0 \leq s < Q$). ACE conditioning turns out to be unnecessary for this design example, so it was omitted in the design of the code presented in the example below. Because $Q \cdot S = 1071$, the rightmost 47 columns of the matrix $\mathbf{P}$ are deleted to make $k = 1024$. The highest rate we seek is 4/5, and, because the rate-4/5 code is the "most vulnerable" in the family, the mother code must be designed such that its own performance and that of its offspring rate-4/5 code are both good. The puncturing pattern is "0001" for the rate-4/5 code, which means that one parity bit out of every four is transmitted, beginning with the fourth. This puncture pattern refers to the (6.6) form of the matrix because it is easiest to discuss puncturing of parity bits before the parity bits are reordered as in (6.7).

It can be shown that, equivalent to the "0001" puncture pattern, groups of four check equations can be summed together and replaced by a single equation. Considering the block interleaver applied in (6.5), rows 0, 1, 2, and 3 of $\mathbf{\Pi}^{\mathrm{T}}\mathbf{P}$ are, respectively, the first rows in the first four row groups of $\mathbf{P}$; rows 4, 5, 6, and 7 of $\mathbf{\Pi}^{\mathrm{T}}\mathbf{P}$ are, respectively, the first rows in the second four row groups of $\mathbf{P}$; and so on. Thus, an equivalent parity-check matrix can be derived by summing every four row groups of the matrix given by (6.6). Because $\mathbf{P}$ has 20 row groups, there will be 5 row groups after summation, which can be considered to be the equivalent $\mathbf{P}$ matrix of the rate-4/5 code.

The puncturing pattern for the rate-2/3 code is such that one bit out of every two parity bits is transmitted and similar comments hold regarding the equivalent $\mathbf{P}$ matrix. In order to make the family rate-compatible, the unpunctured bit is selected so that the parity bits in the rate-4/5 code are embedded in those of the rate-2/3 code. A cycle test (CT) can also be applied to the equivalent parity-check matrix of the rate-4/5 code to guarantee that it is free of length-4 cycles. As shown in the example below, this additional CT rule improves the performance of the rate-4/5 code at the higher SNR values without impairing the performance of the mother code. The example also shows that the CT test is not necessary for the rate-2/3 code.

---

**Example 6.5** Figure 6.6 shows the frame-error-rate (FER) performance of the QC-IRA code families designed using Methods I and II (with CT), with $k = 1024$ and rates of 1/2, 2/3, and 4/5. They are simulated using the approximate-min* decoder with 50 iterations. The rate-2/3 and rate-4/5 codes in the Method I family are slightly better than those in the Method II family in the waterfall region. However, Method II is much more flexible: (1) the decoder is essentially the same for every member of the family; and (2) any rate between 1/2 and 4/5 is easy to obtain by puncturing only, with performance no worse than that of the rate-4/5 code. To verify the improvement by using the cycle test for the equivalent parity-check matrix of the rate-4/5 code, we also plot the curve for a rate-4/5 code obtained without using the CT rule. The results show that the CT rule produces a gain of 0.15 dB at FER $= 10^{-6}$. We did not include the results for the other two rates since the performances for these two rates are the same with and without the CT rule in the region simulated.

---

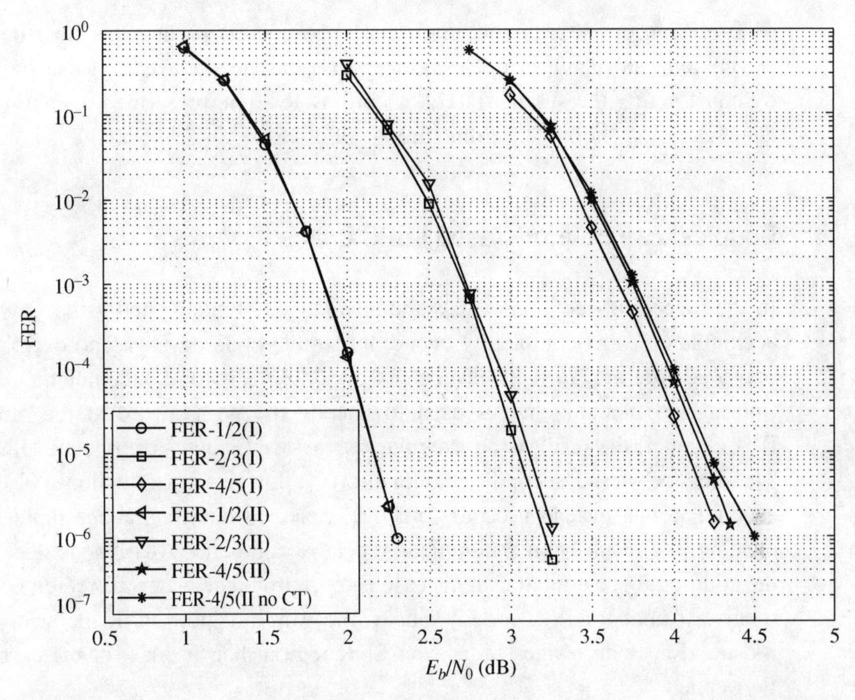


**Figure 6.6** The FER performance of QC-IRA code families with $k = 1024$ designed with Methods I and II.

### 6.5.3 Generalized Accumulator LDPC Codes

IRA codes based on *generalized accumulators* (IRGA codes) [26, 27] increase the flexibility in choosing degree distributions relative to IRA codes. The encoding algorithms for IRGA codes are efficient and similar to those of non-QC IRA codes. For IRGA codes, the accumulator $1/(1 \oplus D)$ in Figure 6.4(b) is replaced by a generalized accumulator with transfer function $1/g(D)$, where $g(D) = \sum_{l=0}^{t} g_l D^l$, $t > 1$, and $g_l \in \{0, 1\}$, except $g_0 = 1$. The systematic encoder therefore has the same generator-matrix format, $\mathbf{G} = \left[\mathbf{I} \;\; \mathbf{H}_u^{\mathrm{T}} \mathbf{H}_p^{-\mathrm{T}}\right]$, but now

$$
\mathbf{H}_p = \begin{bmatrix}
1 & & & & & & \\
g_1 & 1 & & & & & \\
g_2 & g_1 & \ddots & & & & \\
\vdots & g_2 & \ddots & \ddots & & & \\
g_t & \vdots & \ddots & \ddots & \ddots & & \\
 & g_t & \ddots & \ddots & \ddots & \ddots & \\
 & & \ddots & \ddots & \ddots & \ddots & \ddots \\
 & & & g_t & \cdots & g_2 & g_1 & 1
\end{bmatrix}.
$$

Further, the parity-check-matrix format is unchanged, $\mathbf{H} = [\mathbf{H}_u \;\; \mathbf{H}_p]$.

To design an IRGA code, one must choose $g(D)$ so that the bipartite graph for $\mathbf{H}_p$ contains no length-4 cycles. Once $g(D)$ has been chosen, $\mathbf{H}$ can be completed by constructing the submatrix $\mathbf{H}_u$, according to some prescribed degree distribution, again avoiding short cycles, this time in all of $\mathbf{H}$.

## 6.6 Double-Accumulator-Based LDPC Codes

In the preceding section, we saw the advantages of LDPC codes that involve single accumulators: low-complexity encoding, simple code designs, and excellent performance. There are, in fact, advantages to adding a second accumulator. If a second (interleaved) accumulator is used to encode the parity word at the output of an IRA encoder, the result is an irregular repeat–accumulate–accumulate (IRAA) code. The impact of the second accumulator is a lower error-rate floor. If the second accumulator is instead used to "precode" selected data bits at the input to an IRA encoder, the result is an accumulate–repeat–accumulate (ARA) code. The impact of the additional accumulator in this case is to improve the waterfall-region performance relative to that for IRA codes. That is, the waterfall portion of the error-rate curve for an ARA code resides in a lower-SNR region than it does for the corresponding IRA code.

### 6.6.1 Irregular Repeat–Accumulate–Accumulate Codes

We now consider IRAA codes that are obtained by concatenating the parity arm of the IRA encoder of Figure 6.4(b) with another accumulator, through an interleaver, as shown in Figure 6.7. The IRAA codeword can be either $\mathbf{c} = [\mathbf{u} \;\; \mathbf{p}]$ or $\mathbf{c} = [\mathbf{u} \;\; \mathbf{b} \;\; \mathbf{p}]$, depending on whether the intermediate parity bits $\mathbf{b}$ are punctured or not. The parity-check matrix of the general IRAA code corresponding to Figure 6.7 is

$$\mathbf{H}_{\text{IRAA}} = \begin{bmatrix} \mathbf{H}_u & \mathbf{H}_p & \mathbf{0} \\ \mathbf{0} & \mathbf{\Pi}_1^{\text{T}} & \mathbf{H}_p \end{bmatrix}, \tag{6.10}$$

where $\mathbf{\Pi}_1$ is the interleaver between the two accumulators. $\mathbf{H}_u$ for an IRAA code can be designed as for an IRA code. An IRAA code will typically have a lower floor but worse waterfall region than an IRA code of the same length, rate, and complexity, as shown in [14] and what follows.

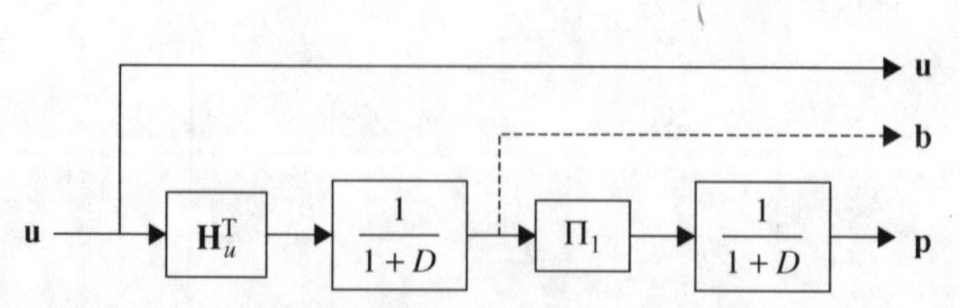


**Figure 6.7** An IRAA encoder.

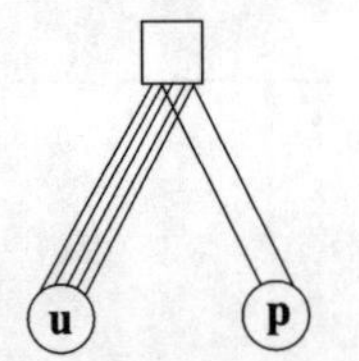

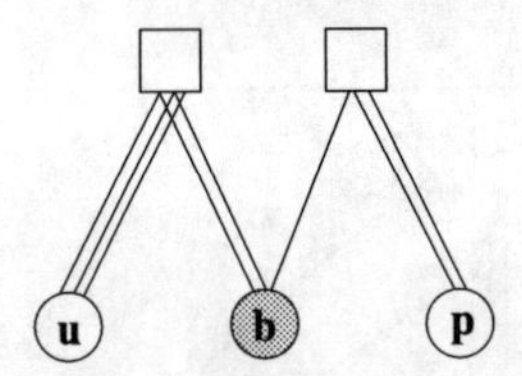


**Figure 6.8** Rate-1/2 IRA and IRAA protographs. The shaded node in the IRAA protograph represents punctured bits.

---

**Example 6.6** Example rate-1/2 protographs for IRA and IRAA codes are presented in Figure 6.8. For the IRA protograph, $d_{b,j} = 5$ for all $j$, and $d_{c,i} = 5$ for all $i$. For the IRAA protograph, $d_{b,j} = d_{c,i} = 3$ and the intermediate parity vector $\mathbf{b}$ is not transmitted in order to maintain the code rate at 1/2. Because the decoder complexity is proportional to the number of edges in a code's parity-check matrix, the complexity of the IRAA decoder is about 14% greater than that of the IRA decoder. To see this, note that the IRAA protograph has eight edges whereas the IRA protograph has seven edges.

---

Analogous to (6.7), the parity-check matrix for a quasi-cyclic IRAA code is given by

$$\mathbf{H}_{\text{QC-IRAA}} = \begin{bmatrix} \mathbf{P} & \mathbf{H}_{p,\text{QC}} & \mathbf{0} \\ \mathbf{0} & \mathbf{\Pi}_{\text{QC}}^{\text{T}} & \mathbf{H}_{p,\text{QC}} \end{bmatrix}, \tag{6.11}$$

where $\mathbf{P}$ and $\mathbf{H}_{p,\text{QC}}$ are as described in (6.9) respectively, and (6.8), and $\mathbf{\Pi}_{\text{QC}}$ is a permutation matrix consisting of $Q \times Q$ circulant permutation matrices and $Q \times Q$ zero matrices arranged to ensure that the row and column weights of $\mathbf{\Pi}_{\text{QC}}$ are both 1. The design algorithm for QC-IRAA codes is analogous to that for QC-IRA codes. Specifically, (1) the $\mathbf{H}_{p,\text{QC}}$ and $\mathbf{0}$ matrices in (6.11) are fixed components of $\mathbf{H}_{\text{QC-IRAA}}$ and (2) the $\mathbf{P}$ and $\mathbf{\Pi}_{\text{QC}}^{\text{T}}$ submatrices of $\mathbf{H}_{\text{QC-IRAA}}$ may be constructed using a PEG/ACE-like algorithm, much like Algorithm 6.1 for QC-IRA codes.

Let us now compare the performance of rate-1/2 (2048, 1024) QC-IRA and QC-IRAA codes. For the QC-IRA code $d_{b,j} = 5$ for all $j$, whereas for the QC-IRAA code $d_{b,j} = 3$ for all $j$. For the IRAA code, $\mathbf{c} = \begin{bmatrix} \mathbf{u} & \mathbf{p} \end{bmatrix}$, that is, the intermediate parity bits $\mathbf{b}$ are punctured. The QC-IRA code was designed using the algorithm of the previous section. The QC-IRAA code was designed using the algorithm given in the previous paragraph. We observe in Figure 6.9 that, for both codes, there are no error floors in the BER curves down to BER $= 5 \times 10^{-8}$ and in the FER curves down to FER$= 10^{-6}$. While the S-IRAA code is 0.2 dB inferior to the S-IRA code in the waterfall region, it is conjectured that it has a lower floor (which is difficult to measure), which would be due to the second accumulator, whose function is to improve the weight spectrum.

As an example of a situation in which the IRAA class of code is superior, consider the comparison of rate-1/3 (3072, 1024) QC-IRA and QC-IRAA codes, with $d_{b,j} = 4$ for the QC-IRA code and $d_{b,j} = 3$ for the QC-IRAA code. In this case, $\mathbf{c} = \begin{bmatrix} \mathbf{u} & \mathbf{b} & \mathbf{p} \end{bmatrix}$,

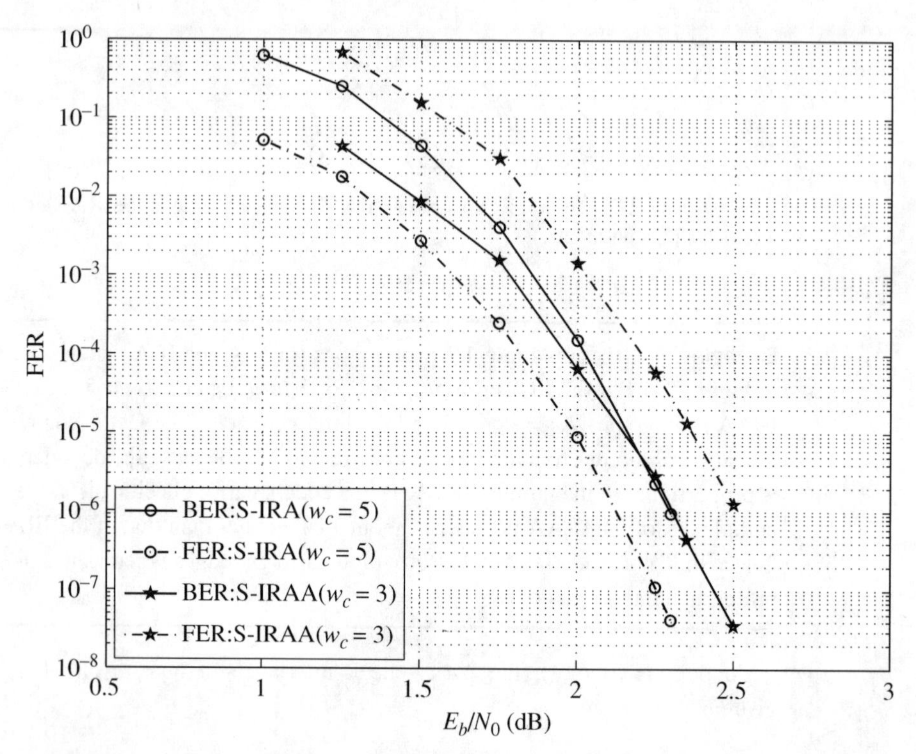


**Figure 6.9** Performance of an IRAA code and an IRA code with $n = 2048$ and $k = 1024$ on the AWGN channel ($I_{\max} = 50$). $w_c$ is the column weight of the submatrix $\mathbf{P}$.

that is, the intermediate parity bits $\mathbf{b}$ are not punctured. Also, the decoder complexities are the same. We see in Figure 6.10 that, in the low-SNR region, the performance of the IRA code is 0.4 dB better than that of the IRAA code. However, as is evident from Figure 6.10, the IRAA code will outperform the IRA code in the high-SNR region due to its lower error floor.

### 6.6.2 Accumulate–Repeat–Accumulate Codes

For ARA codes, which were introduced in [28, 29], an accumulator is added to precode a subset of the information bits of an IRA code. The primary role of this second accumulator is to improve the decoding threshold of a code (see Chapter 9), that is, to shift the error-rate curve leftward toward the capacity limit. Precoding is generally useful only for relatively low code rates because satisfactory decoding thresholds are easier to achieve for high-rate LDPC codes. Figure 6.11 presents a generic ARA Tanner graph in which punctured variable nodes are blacked. The enhanced performance provided by the precoding accumulator is achieved at the expense of these punctured variable nodes which act as auxiliary nodes that enlarge the $\mathbf{H}$ used by the decoder. The iterative graph-based ARA decoder thus has to deal with a redundant representation of the code, implying a larger $\mathbf{H}$ matrix than the nominal $(n-k) \times n$. Note that this is much like the case for IRAA codes.

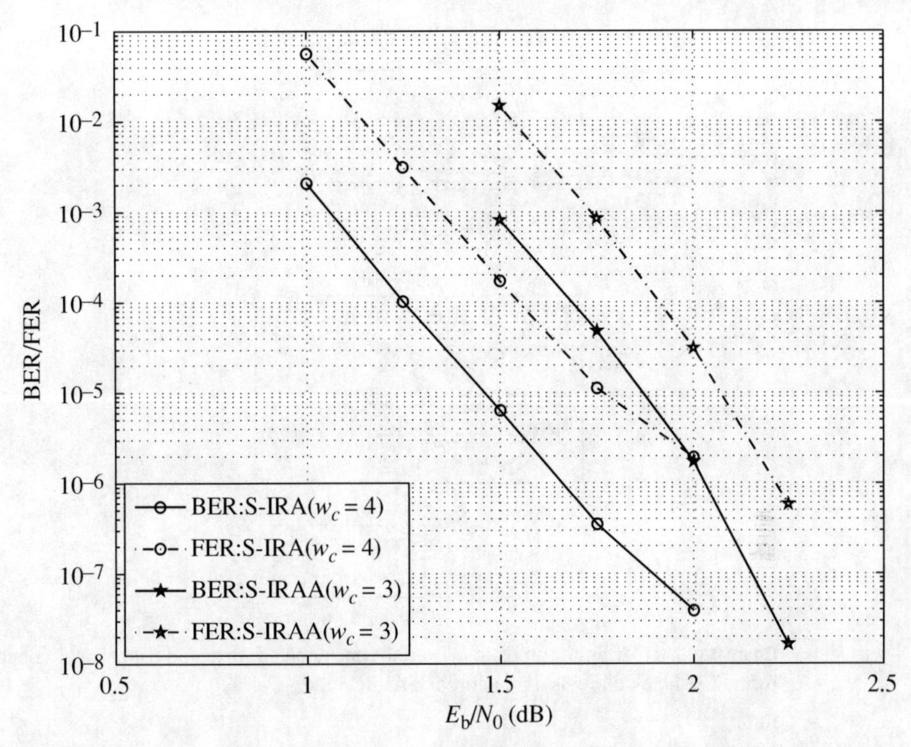


**Figure 6.10** Performance of an IRAA code and an IRA code with $n = 3072$ and $k = 1024$ on the AWGN channel ($I_{\max} = 50$). $w_c$ is the column weight of the submatrix $\mathbf{P}$.

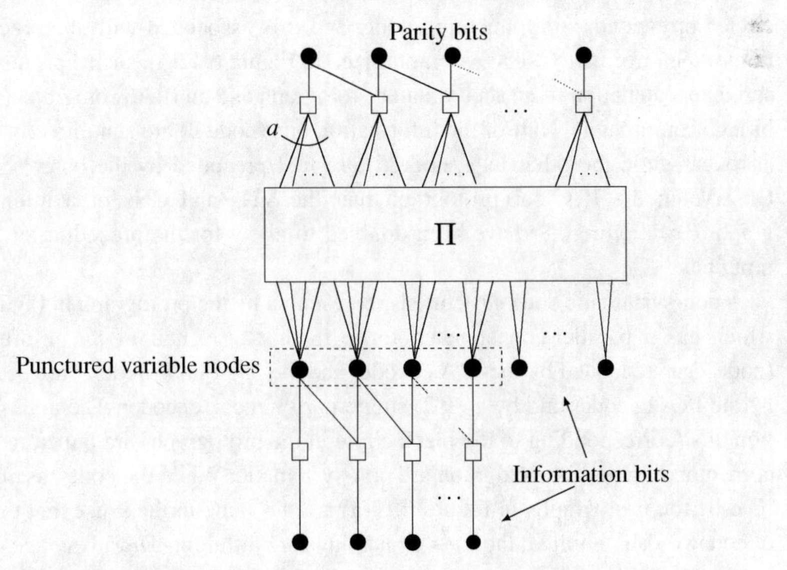


**Figure 6.11** A generic bipartite graph for ARA codes.

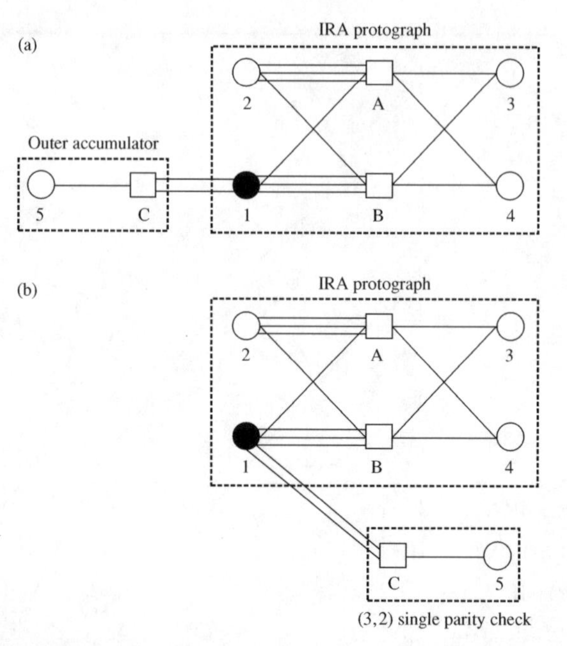


**Figure 6.12** AR4A protographs in (a) serial-concatenated form and (b) parallel-concatenated form. The black circle is a punctured variable node.

ARA codes typically rely on very simple protographs. The protograph of a rate-1/2 ARA code ensemble with repetition rate 4, denoted by AR4A, is depicted in Figure 6.12(a). This encoding procedure corresponds to a systematic code. The black circle corresponds to a punctured node, and it is associated with the precoded fraction of the information bits. As emphasized in Figure 6.12(a), such a protograph is the serial concatenation of an accumulator protograph and an IRA protograph (with a tail-biting accumulator). Half of the information bits (node 2) are sent directly to the IRA encoder, while the other half (node 5) are first precoded by the outer accumulator. Observe in the IRA sub-protograph that the VNs and CNs of a minimal protograph (e.g., Figure 6.8) have been doubled to allow for the precoding of half of the input bits.

A non-systematic code structure is represented by the protograph in Figure 6.12(b), which has a parallel-concatenated form. In this case, half of the information bits (node 2) are encoded by the IRA encoder and the other half (node 1) are encoded both by the IRA encoder and by a (3, 2) single-parity-check encoder. The node-1 information bits (corresponding to the black circle in the protograph) are punctured, so codes corresponding to this protograph are non-systematic. While the code ensembles specified by the protographs in Figure 6.12(a) are the same in the sense that the same set of codewords is implied, the $\mathbf{u} \to \mathbf{c}$ mappings are different. The advantage of the non-systematic protograph is that, although the node-1 information bits in Figure 6.12(b) are punctured, the node degree is 6, in contrast with the node-5 information bits in Figure 6.12(a), in which the node degree is only 1. In the iterative decoder, the bit log-likelihood values associated with the degree-6 node tend to converge faster and to

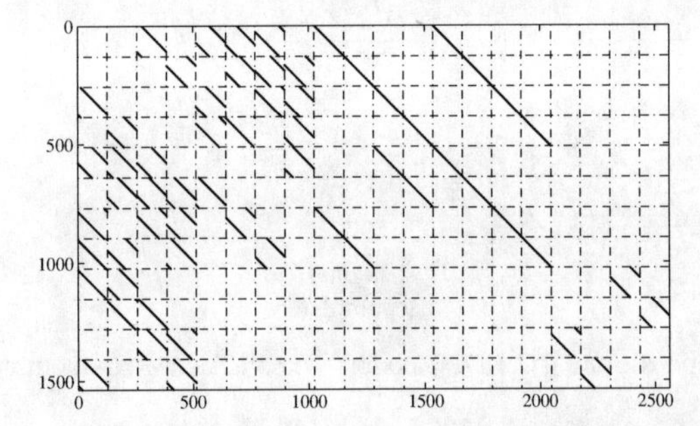


**Figure 6.13** The **H** matrix for the (2048, 1024) AR4A code.

a larger value than do those associated with the degree-1 node. Hence, these bits will be more reliably decoded.

The design of ARA codes is quite involved and requires the material presented in Chapters 8 and 9. An overview of the design of excellent ARA codes is presented in Section 6.6.2.

---

**Example 6.7** A pixelated image of the **H** matrix for a (2048, 1024) QC AR4A code is depicted in Figure 6.13. The first group of 512 columns (of weight 6) corresponds to variable-node type 1 of degree 6 (Figure 6.12), whose bits are punctured, and the subsequent four groups of 512 columns correspond, respectively, to node types 2, 3, 4, and 5. The first group of 512 rows (of weight 6) corresponds to check-node type A (of degree 6), and the two subsequent groups of rows correspond to node types B and C, respectively.

---

## Protograph-Based ARA Code Design

Section 6.3 describes a brute-force technique for designing protograph codes. Here we present two approaches for designing protograph-based ARA codes, following [11, 12]. The goal is to obtain a protograph for an ensemble that has a satisfactory decoding threshold and a minimum distance $d_{\min}$ that grows linearly with the codeword length $n$. This linear-distance-growth idea was investigated by Gallager, who showed that (1) regular LDPC code ensembles had linear $d_{\min}$ growth, provided that the VN degrees were greater than 2; and (2) LDPC code ensembles with a randomly constructed **H** matrix had linear $d_{\min}$ growth. As an example, for rate-1/2 randomly constructed LDPC codes, $d_{\min}$ grows as $0.11n$ for large $n$. As another example, for the rate-1/2 (3, 6)-regular LDPC code ensemble, $d_{\min}$ grows as $0.023n$.

The first approach to designing protograph-based ARA codes is the result of a bit of LDPC code-design experience and a bit of trial and error. As in the example presented in Figure 6.12, we start with a rate-1/2 IRA protograph and precode 50% of its encoder input bits to improve the decoding threshold. Then, noticing that linear $d_{\min}$ growth is thwarted by a large number of degree-2 VNs, one horizontal branch is added to the accumulator part of the IRA protograph to convert half of the degree-2 VNs to degree-3 VNs. This would result in the protograph in Figure 6.12(a), but with two edges

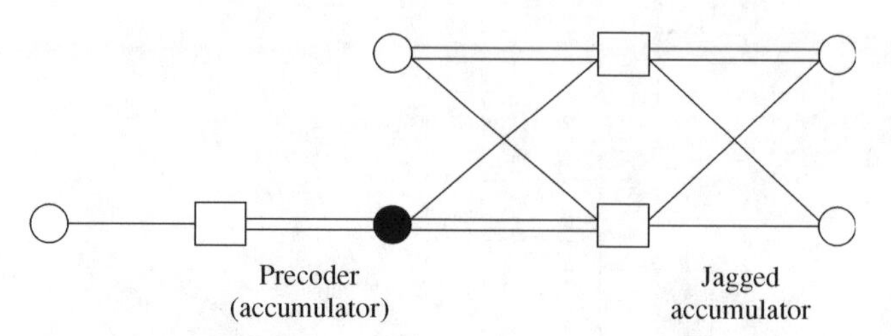


**Figure 6.14** An ARJA protograph for rate-1/2 LDPC codes. The black VN is punctured.

connecting node A to node 1. One could also experiment with the number of edges between nodes 2 and A and nodes 1 and B in that figure. It was found in [11, 12] that performance is improved if the number of edges connecting node 2 to node A is reduced from three to two. The final protograph, called an ARJA protograph for its "jagged" accumulator [11, 12], is shown in Figure 6.14. Given this protograph, one can construct an **H** matrix from the base matrix corresponding to the protograph, using a PEG/ACE-like algorithm.

The second approach to arriving at this particular ARJA protograph begins with a rate-2/3 protograph known to have linear $d_{\min}$ growth, that is, a protograph whose VN degrees are 3 or greater. This protograph is shown at the top of Figure 6.15 and the sequence of protographs that we will derive from it lies below that protograph. To lower the code rate to 1/2, an additional CN must be added. To do this, the CN in the top protograph is split into two CNs and the edges connected to the original CN are distributed between the two new CNs. Further, a degree-2 (punctured) VN is connected to the two CNs. Observe that this second protograph is equivalent to the first one: the set of 3-bit words satisfying the constraints of the first protograph is identical to that of the second. If we allow the recently added VN to correspond to transmitted bits instead of punctured bits, we arrive at the first rate-1/2 protograph in Figure 6.15. Further, because the underlying rate-2/3 ensemble has linear $d_{\min}$ growth, the derived rate-1/2 ensemble must have this property as well. Finally, the bottom protograph in Figure 6.15 is obtained with the addition of a precoder (accumulator).

Now, this rate-1/2 protograph has an AWGN channel decoding threshold of $(E_b/N_0)_{\text{thresh}} = 0.64$ dB. Also, its asymptotic $d_{\min}$ growth rate goes as $0.015n$. It is possible to further split CNs to obtain protographs with rates lower than 1/2. Alternatively, higher-rate protographs are obtained via the addition of VNs as in Figure 6.16 [11, 12]. Table 6.2 presents the ensemble decoding thresholds for this code family on the binary-input AWGN channel and compares these values with their corresponding capacity limits. Moderate-length codes designed from these protographs were presented in [11, 12], where is it shown that the codes have excellent decoding thresholds and very low error floors, both of which many other code-design techniques fail to achieve. The drawback, however, is that these codes suffer from slow decoder convergence due to punctured VNs and to degree-1 VNs (precoded bits).

*Encoding*. Encoding of ARJA-type codes is discussed in [30–33]. These techniques are generalizable to other quasi-cyclic protograph-based codes. See also Section 3.6 of Chapter 3.

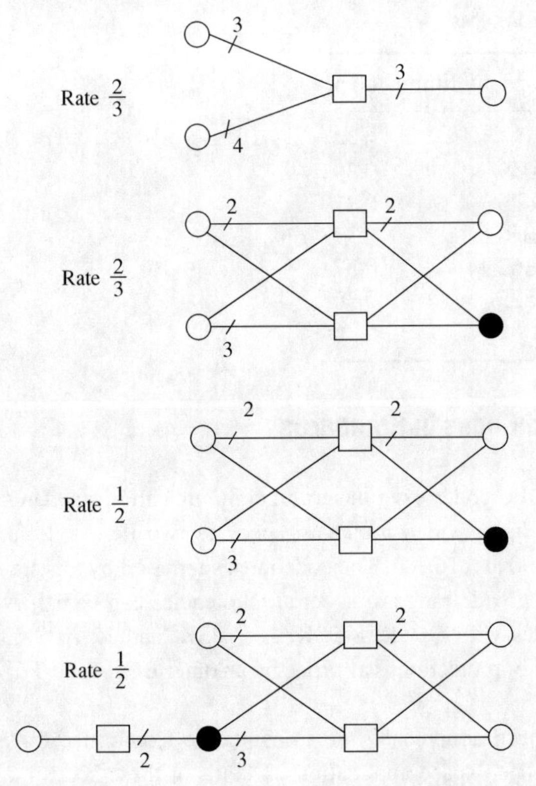


**Figure 6.15** Constructing a rate-1/2 ARJA LDPC code from a rate-2/3 protograph to preserve the linear $d_{\min}$ growth property and to achieve a good decoding threshold. Black circles are punctured VNs.

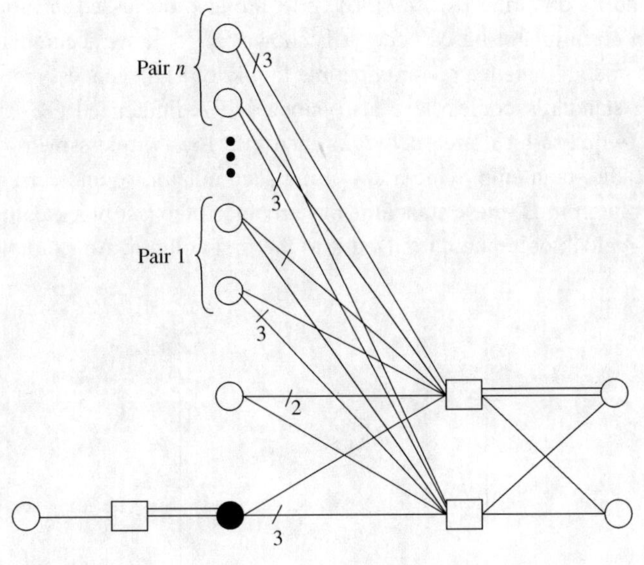


**Figure 6.16** An ARJA protograph for a family of LDPC codes with rates $(n + 1)/(n + 2)$, $n = 0, 1, 2, \ldots$.

**Table 6.2** Thresholds of ARJA codes

| Rate | Threshold | Capacity limit | Gap |
|---|---|---|---|
| 1/2 | 0.628 | 0.187 | 0.441 |
| 2/3 | 1.450 | 1.059 | 0.391 |
| 3/4 | 2.005 | 1.626 | 0.379 |
| 4/5 | 2.413 | 2.040 | 0.373 |
| 5/6 | 2.733 | 2.362 | 0.371 |
| 6/7 | 2.993 | 2.625 | 0.368 |
| 7/8 | 3.209 | 2.845 | 0.364 |

## 6.7 Accumulator-Based Codes in Standards

Accumulator-based LDPC codes exist in several communication standards. The ETSI DVB S2 standard for digital video broadcast specifies two IRA code families with code blocklengths 64 800 and 16 200. The code rates supported by this standard range from $1/4$ to $9/10$, and a wide range of spectral efficiencies can be achieved by coupling these LDPC codes with QPSK, 8-PSK, 16-APSK, and 32-APSK modulation formats. A further level of protection is afforded by an outer BCH code. For details, see [34, 35].

ARJA codes have been adopted by the Consultative Committee for Space Data Systems (CCSDS) for deep-space applications. Details of these codes may be found in [36]. Codes with rates $1/2$, $2/3$, and $4/5$ and data blocklengths of $k = 1024, 4096,$ and 16 384 are presented in that document. The $(8176, 7156)$ Euclidean-geometry code presented in Section 11.5 is part of the CCSDS standard for near-earth applications. This code, and its offspring $(8160, 7136)$ code, are also discussed in [36]. Although this is not an accumulator-based code, it is shown in [37] how accumulators can be added to this code to obtain a rate-compatible family of LDPC codes.

The IEEE standards bodies have also adopted IRA-influenced QC LDPC codes for 802.11n (wireless local-area networks) and 802.16e (wireless metropolitan-area networks). Rather than employing a tail-biting accumulator, or one corresponding to a weight-1 column in $\mathbf{H}$, these standards have replaced the last block-column in (6.8) by a weight-3 block-column and moved it to the first column. An example of such a format is

$$\begin{bmatrix} I_1 & I_0 & & & & & \\ & I_0 & I_0 & & & & \\ & & I_0 & \ddots & & & \\ I_0 & & & \ddots & \ddots & & \\ & & & & \ddots & I_0 & \\ & & & & & I_0 & I_0 \\ I_1 & & & & & & I_0 \end{bmatrix},$$

where $I_1$ is the matrix that results after cyclically shifting rightward all rows of the identity matrix $I_0$. Encoding is facilitated by this matrix since the sum of all block-rows gives the block-row $\begin{pmatrix} I_0 & 0 & \cdots & 0 \end{pmatrix}$, so that encoding is initialized by summing all of the block-rows of $\mathbf{H}$ and solving for the first $Q$ parity bits using the resulting block-row.

## 6.8 Protograph-Based Raptor-Like LDPC Codes

Protograph-based Raptor-like (PBRL) codes [38] are a class of LDPC codes that are designed via protographs leading to a family of codes of differing rates by extending a core high-rate code (HRC) instead of puncturing a parental low-rate code. The extension approach makes them Raptor-like [39] and simultaneously rate-compatible [25]. PBRL codes demonstrate excellent performance when compared to other LDPC code designs and to theoretical limits. The search for effective protographs is performed using the low-complexity reciprocal-channel approximation (RCA) approach instead of the high-complexity density evolution technique, both of which are discussed in Chapter 9. The RCA algorithm provides fast and accurate decoding thresholds for protographs. The PEG and ACE algorithms (Section 6.2) are the *lifting* algorithms used to expand a protograph into a full parity-check matrix [38]. (The lifting concept was introduced in Section 6.3.)

Recalling that protograph codes are conveniently represented by *adjacency* or base matrices, the base matrix for PBRL codes has the form

$$\mathbf{B} = \begin{bmatrix} \mathbf{B}_{HRC} & \mathbf{0} \\ \mathbf{B}_{IRC} & \mathbf{I} \end{bmatrix}, \tag{6.12}$$

where $\mathbf{B}_{HRC}$ is the base matrix for the core high-rate code, $\mathbf{B}_{IRC}$ is the base matrix for an incremental-redundancy code, $\mathbf{0}$ is a zero matrix, and $\mathbf{I}$ is an identity matrix. The size of $\mathbf{B}_{IRC}$ depends on the lowest code rate desired and the sizes of $\mathbf{I}$ and $\mathbf{0}$ are chosen to be compatible with the sizes of $\mathbf{B}_{HRC}$ and $\mathbf{B}_{IRC}$.

As an example, from [38], with

$$\mathbf{B}_{HRC} = \begin{bmatrix} 1 & 1 & 2 & 1 & 2 & 1 \\ 2 & 2 & 1 & 2 & 1 & 2 \end{bmatrix} \tag{6.13}$$

and

$$\mathbf{B}_{IRC} = \begin{bmatrix} 1 & 1 & 1 & 1 & 1 & 1 \\ 1 & 1 & 1 & 0 & 1 & 0 \\ 0 & 1 & 0 & 0 & 1 & 1 \\ 1 & 0 & 0 & 1 & 0 & 1 \\ 0 & 0 & 1 & 0 & 1 & 0 \\ 0 & 1 & 0 & 1 & 0 & 1 \\ 1 & 0 & 1 & 0 & 1 & 0 \end{bmatrix}, \tag{6.14}$$

the corresponding protograph is shown in Figure 6.17. It can be interpreted as a high-rate code followed by an incremental-redundancy code consisting of degree-1 variable

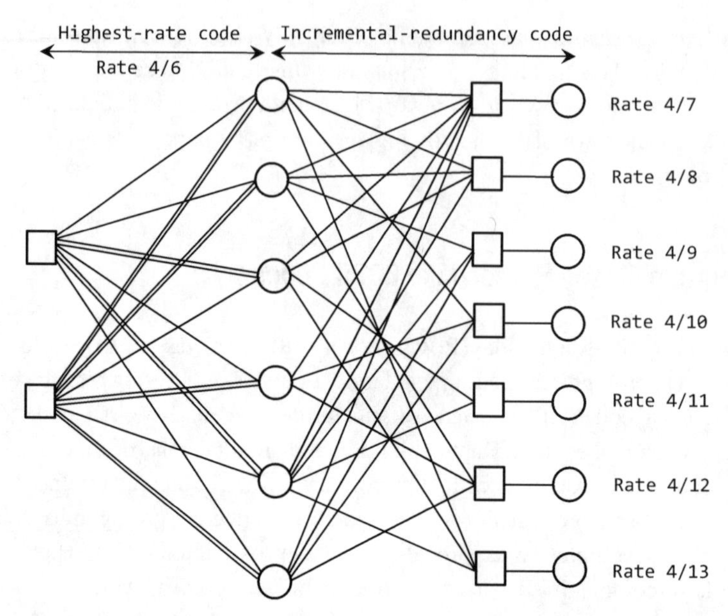


**Figure 6.17** Protograph for the PBRL specified by (6.12)–(6.14). (Adapted from [38].)

nodes. As indicated in the figure, the IRC portion of the figure corresponds to the family of lower-rate codes, where the next-lower rate is attained by including the next degree-1 VN (and the next row of $\mathbf{B}_{IRC}$). The degree-1 VNs correspond to the submatrix $\mathbf{I}$ in $\mathbf{B}$. The edges to the CNs connected to these degree-1 VNs correspond to $\mathbf{B}_{IRC}$.

Once a base matrix has been lifted into a full parity-check matrix $\mathbf{H}$ using circulant permutation matrices, encoding can occur in two stages: the HRC part ("precode" part) followed by the IRC part. The latter part is easily accomplished by applying exclusive-or operations to the HRC encoder output. The HRC encoding can be accomplished using the protograph code encoding algorithms referenced at the end of Section 6.6. Decoding is performed using the sum–product algorithm on the entire Tanner graph for $\mathbf{H}$, that is, the HRC and IRC portions of the graph are decoded jointly. In this section, following [38], we will discuss the design of good $\mathbf{B}_{HRC}$ and $\mathbf{B}_{IRC}$ matrices, leveraging the RCA algorithm.

*Design of the HRC protograph*. The design of the HRC protograph is discussed is greater detail in [38]. The design of the HRC protograph begins with the knowledge that a protograph with a minimum VN degree of 3 is guaranteed to have linear minimum distance ($d_{\min}$) growth with increasing code length. On the other hand, while a larger average VN-node degree tends to lower the error-rate floor, it also degrades the waterfall region of a code ensemble's error-rate curve. Thus, there must be a balance between low- and high-degree VN counts in a protograph. In fact, carefully adding several degree-1 and degree-2 VNs improves the ensemble decoding threshold, provided there are no graphical cycles involving only degree-2 nodes. Further, judiciously adding several degree-4 VNs improves the error-rate floor. Lastly, puncturing a node in the HRC generally improves the decoding threshold and so [38] recommends puncturing the first VN. The RCA algorithm is used to check the decoding threshold of candidate matrices in the search for a good base matrix (or protograph) $\mathbf{B}_{HRC}$.

A protograph is characterized by the triple $(n_v, n_c, n_p)$, where $n_v$ is the number of VNs, $n_c$ is the number of CNs, and $n_p$ is the number of punctured nodes ($n_p = 1$ here). Assuming that $\mathbf{B}_{HRC}$ is full rank, the set of possible code rates in the PBRL code family is then (for $i > 0$)

$$R_i = \frac{n_v - n_c}{n_v - n_p - i}$$

because we can indefinitely add degree-1 IRC variable nodes (as for Raptor codes).

*Design of the IRC protograph*. In searching for the IRC protograph, the designer must be mindful of (a) linear $d_{\min}$ growth, (b) the decoding threshold, and (c) the level of the error-rate floor. Further, two rules of thumb are followed in the design: (a) parallel edges improve the thresholds and (b) the punctured node should connect to (almost) all CNs in the IRC portion of the protograph. At the center of the design process is a greedy algorithm that leverages the RCA technique to optimize the threshold at each code rate given the protograph of the previous (higher) code rate. This greedy algorithm is as follows.

1. Add a CN and the degree-1 VN connected to it.
2. Use the RCA algorithm to find the next IRC row with the best decoding threshold under PEG and ACE constraints. The sum of the values in each IRC row must be at least two, that is, each IRC check node must have degree at least two.
3. If the next lower rate is desired, go to Step 1; else, stop.

Further details on the design of PBRL codes, for both short (or moderate-length) and long codes may be found in [38].

## 6.9 Generalized LDPC Codes

Generalized LDPC codes were introduced in Chapter 5. For these codes, the CNs are more general than SPC constraints. There are $m_c$ code constraints placed on the $n$ code bits connected to the CNs. Let $V = \{V_j\}_{j=0}^{n-1}$ be the set of $n$ VNs and $C = \{C_i\}_{i=0}^{m_c-1}$ be the set of $m_c$ CNs in the bipartite graph of a G-LDPC code. Recall that the connections between the nodes in $V$ and $C$ can be summarized in an $m_c \times n$ adjacency matrix $\Gamma$. We sometimes call $\mathbf{\Gamma}$ the code's graph because of this correspondence. Chapter 5 explains the relationship between the adjacency matrix $\mathbf{\Gamma}$ and the parity-check matrix $\mathbf{H}$ for a generic G-LDPC code. Here, following [40], we consider the derivation of $\mathbf{H}$ for quasi-cyclic G-LDPC codes and then present a simple QC G-LDPC code design.

To design a quasi-cyclic G-LDPC code, that is, to obtain a matrix $\mathbf{H}$ that is an array of circulant permutation matrices, we exploit the protograph viewpoint. Consider a G-LDPC protograph $\mathbf{\Gamma}_p$ with $m_p$ generalized constraint nodes and $n_p$ variable nodes. The adjacency matrix for the G-LDPC code graph, $\mathbf{\Gamma}$, is constructed by substituting

$Q \times Q$ circulant permutation matrices for the 1s in $\mathbf{\Gamma}_p$ in such a manner that short cycles are avoided and parallel edges are eliminated. The 0s in $\mathbf{\Gamma}_p$ are replaced by $Q \times Q$ zero matrices. As demonstrated in Figure 6.1, the substitution of circulant permutation matrices for 1s in $\mathbf{\Gamma}_p$ effectively applies the copy-and-permute procedure to the protograph corresponding to $\mathbf{\Gamma}_p$. The resulting adjacency matrix for the G-LDPC code will be an $m_p \times n_p$ array of circulant permutation matrices of the form

$$\mathbf{\Gamma} = \begin{bmatrix} \pi_{0,0} & \pi_{0,1} & \cdots & \pi_{0,n_p-1} \\ \pi_{1,0} & \pi_{1,1} & \cdots & \pi_{1,n_p-1} \\ & & \ddots & \\ \pi_{m_p-1,0} & \pi_{m_p-1,1} & \cdots & \pi_{m_p-1,n_p-1} \end{bmatrix},$$

where each $\pi_{\mu,\nu}$ is either a $Q \times Q$ circulant permutation matrix or a $Q \times Q$ zero matrix. Note that $\mathbf{\Gamma}$ is an $m_p Q \times n_p Q = m_c \times n$ binary matrix. Now $\mathbf{H}$ can be obtained from $\mathbf{\Gamma}$ and $\{\mathbf{H}_i\}$ as in the previous paragraph. To see that this procedure leads to a quasi-cyclic code, we consider some examples.

---

**Example 6.8** Consider a protograph with the $m_p \times n_p = 2 \times 3$ adjacency matrix

$$\mathbf{\Gamma}_p = \begin{bmatrix} 1 & 0 & 1 \\ 1 & 1 & 1 \end{bmatrix}.$$

Let $Q = 2$ and expand $\mathbf{\Gamma}_p$ by substituting $2 \times 2$ circulant permutation matrices for each 1 in $\mathbf{\Gamma}_p$. The result is

$$\mathbf{\Gamma} = \begin{bmatrix} 1 & 0 & 0 & 0 & 1 & 0 \\ 0 & 1 & 0 & 0 & 0 & 1 \\ 1 & 0 & 0 & 1 & 0 & 1 \\ 0 & 1 & 1 & 0 & 1 & 0 \end{bmatrix},$$

so that $m_c = m_p Q = 4$ and $n = n_p Q = 6$. Suppose now that the CN constraints are given by

$$\mathbf{H}_1 = \begin{bmatrix} 1 & 1 \end{bmatrix}$$

and

$$\mathbf{H}_2 = \begin{bmatrix} 1 & 1 & 0 \\ 1 & 0 & 1 \end{bmatrix}.$$

Then, upon replacing the 1s in $\mathbf{\Gamma}$ by the corresponding columns of $\mathbf{H}_1$ and $\mathbf{H}_2$, the $\mathbf{H}$ matrix for the G-LDPC code is found to be

$$\mathbf{H} = \begin{bmatrix} \mathbf{1} & 0 & 0 & 0 & \mathbf{1} & 0 \\ 0 & \mathbf{1} & 0 & 0 & 0 & \mathbf{1} \\ \mathbf{1} & 0 & 0 & \mathbf{1} & 0 & \mathbf{0} \\ \mathbf{1} & 0 & 0 & \mathbf{0} & 0 & \mathbf{1} \\ 0 & \mathbf{1} & \mathbf{1} & 0 & \mathbf{0} & 0 \\ 0 & \mathbf{1} & \mathbf{0} & 0 & \mathbf{1} & 0 \end{bmatrix},$$

where the inserted columns of $\mathbf{H}_1$ and $\mathbf{H}_2$ are highlighted in bold. Note that because the $\mu$th row of $\mathbf{\Gamma}_p$, $\mu = 0, 1, \ldots, m_p - 1$, corresponds to constraint $C_\mu$ (matrix $\mathbf{H}_\mu$), the $\mu$th block-row (the $\mu$th group of $Q$ binary rows) within $\mathbf{\Gamma}$ corresponds to constraint $C_\mu$ (matrix $\mathbf{H}_\mu$). In the form given above, it is not obvious that $\mathbf{H}$ corresponds to a quasi-cyclic code. However, if we permute the last four rows of $\mathbf{H}$, we obtain the array of circulants

$$\mathbf{H}' = \begin{bmatrix} 1 & 0 & 0 & 0 & 1 & 0 \\ 0 & 1 & 0 & 0 & 0 & 1 \\ & & & & & \\ 1 & 0 & 0 & 1 & 0 & 0 \\ 0 & 1 & 1 & 0 & 0 & 0 \\ & & & & & \\ 1 & 0 & 0 & 0 & 0 & 1 \\ 0 & 1 & 0 & 0 & 1 & 0 \end{bmatrix}.$$

The permutation used on the last four rows of $\mathbf{H}$ can be considered to be a mod-$m_2$ de-interleave of the rows, where $m_2 = 2$ is the number of rows in $\mathbf{H}_2$. No de-interleaving was necessary for the first two rows of $\mathbf{H}$ because $m_1 = 1$.

---

The fact that in general a code's $\mathbf{H}$ matrix will be an array of permutation matrices (after appropriate row permutation) whenever $\Gamma$ is an array of permutation matrices should be obvious from the construction of $\mathbf{H}$ from $\mathbf{\Gamma}$ in the above example. We now consider an example in which the protograph contains parallel edges.

---

**Example 6.9** Change the upper-left element of $\mathbf{\Gamma}_p$ of the previous example to "2" so that

$$\mathbf{\Gamma}_p = \begin{bmatrix} 2 & 0 & 1 \\ 1 & 1 & 1 \end{bmatrix}.$$

Thus, there are two edges connecting variable node $V_0$ and constraint node $C_0$. A possible $q$-fold expansion of $\mathbf{\Gamma}_p$ is

$$\mathbf{\Gamma} = \begin{bmatrix} \pi_0 & \pi_1 & 0 & 0 & 0 & \pi_2 \\ \pi_3 & \pi_4 & 0 & 0 & \pi_5 & 0 \\ & & & & & \\ \pi_6 & 0 & 0 & \pi_8 & \pi_{10} & 0 \\ 0 & \pi_7 & \pi_9 & 0 & 0 & \pi_{11} \end{bmatrix},$$

where the permutation matrices are $Q/2 \times Q/2$ and are selected so as to avoid short cycles. $\mathbf{H}$ is then obtained by replacing the 1s in the rows of $\mathbf{\Gamma}$ by the columns of $\mathbf{H}_1$ and $\mathbf{H}_2$; the first $Q$ rows of $\mathbf{\Gamma}$ correspond to $\mathbf{H}_1$ and the second $Q$ rows correspond to $\mathbf{H}_2$.

---

Note that a G-LDPC code has a parity-check matrix that is 4-cycle-free if its adjacency matrix is 4-cycle-free and the component codes possess parity-check matrices that are 4-cycle-free.

### 6.9.1 A Rate-1/2 G-LDPC Code

The design of G-LDPC codes is still a developing area. Some successful design approaches may be found in [40]. We now present one of these designs, which leads to an excellent quasi-cyclic rate-1/2 G-LDPC code.

The approach starts with a very simple protograph: 2 CNs, 15 VNs, with both CNs connected to each of the 15 VNs. Thus, the CNs have degree 15 and the VNs have degree 2. Both CNs correspond to the (15, 11) Hamming code constraint, but with different code-bit orders. Specifically, protograph CN $C_0$ is described by the parity-check matrix

$$\mathbf{H}_0 = [\mathbf{M}_1 \quad \mathbf{M}_2] = \begin{bmatrix} 10101010 & 1010101 \\ 01100110 & 0110011 \\ 00011110 & 0001111 \\ 00000001 & 1111111 \end{bmatrix} \tag{6.15}$$

and protograph CN $C_1$ is described by

$$\mathbf{H}_1 = [\mathbf{M}_2 \quad \mathbf{M}_1]. \tag{6.16}$$

Next, the protograph is replicated $Q = 146$ times, yielding a derived graph with 2190 VNs and 292 Hamming CNs, 146 of which are described by (6.15) and 146 of which are described by (6.16). Because the number of parity bits is $m = 292 \cdot (15 - 11) = 1168$, the resulting code has as parameters (2190, 1022). The connections between the VNs and the CNs are given by the adjacency matrix $\mathbf{\Gamma}$ at the top of Figure 6.18, which was chosen simply to avoid 4-cycles and 6-cycles in the Tanner graph corresponding to that matrix. Therefore, the girth of the Tanner graph corresponding to $\mathbf{\Gamma}$ is 8. The $\mathbf{H}$ matrix (with appropriately re-ordered rows) is given at the bottom of Figure 6.17. Observe that an alternative approach for obtaining the reordered matrix $\mathbf{H}$ is to replace each 1 in the rows of the matrix in (6.15) (the matrix in (6.16)) by the corresponding permutation matrices of the first (second) block-row of the adjacency matrix in Figure 6.18 and then stack the first resulting matrix on top of the second.

We may obtain a quasi-cyclic rate-1/2 (2044, 1022) G-LDPC code by puncturing the first 146 bits of each (2190, 1022) codeword. Observe that this corresponds to puncturing a single VN in the code's protograph and the first column of circulants of $\mathbf{\Gamma}$. The FER performance of this rate-1/2 code on the binary-input AWGN channel is depicted in Figure 6.19. For the simulations, the maximum number of iterations was set to $I_{\max} = 50$. The G-LDPC code does not display a floor down to FER $\simeq 5 \times 10^{-8}$. As shown in Figure 6.18, the code's performance is within 1 dB of Gallager's random-coding bound for (2044, 1022) block codes.

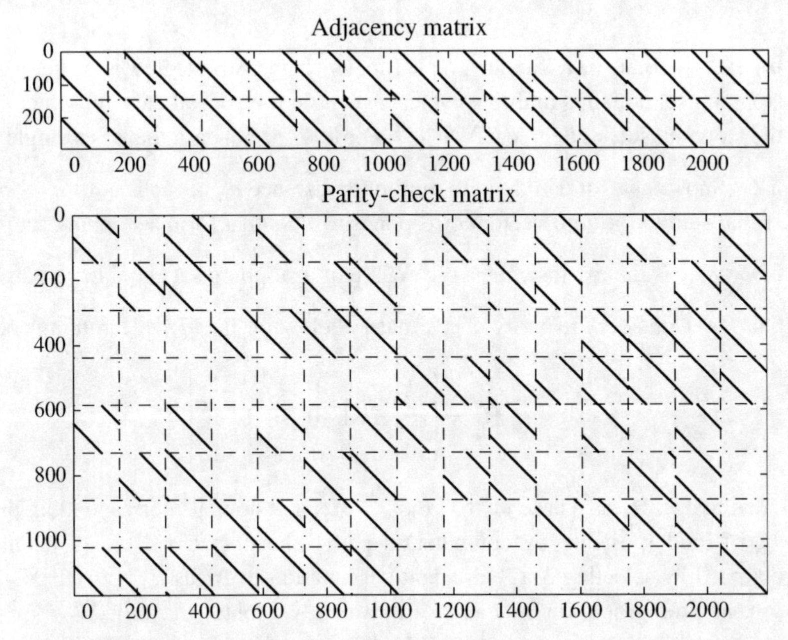


**Figure 6.18** The adjacency matrix of the (2190, 1022) G-LDPC code and its block-circulant parity-check matrix **H**.

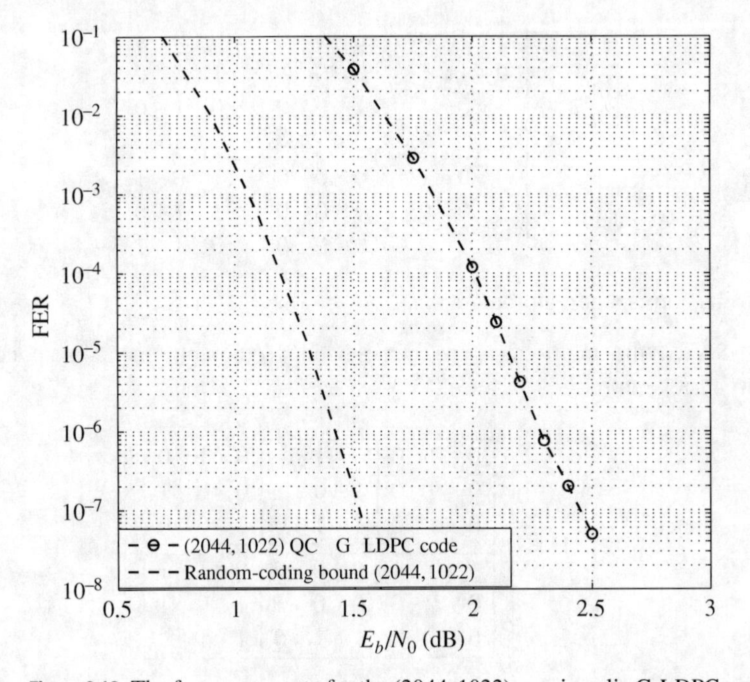


**Figure 6.19** The frame-error rate for the (2044, 1022) quasi-cyclic G-LDPC code, compared with the random-coding bound. $I_{\max}$ was set to 50.

## Problems

**6.1** Show that, in a Tanner graph for which the VN degree is at least two, every stopping set contains multiple cycles, except in the special case for which all VNs in the stopping set are degree-2 VNs [4]. For this special case there is a single cycle.

**6.2** Show that, for a code with minimum distance $d_{\min}$, each set of $d_{\min}$ columns of $\mathbf{H}$ that sum to the zero vector corresponds to VNs that form a stopping set [4].

**6.3** Why is the extrinsic message degree of a stopping set equal to zero?

**6.4** (a) Consider the parity-check matrix below for the $(7, 4)$ Hamming code:

$$\mathbf{H}_1 = \begin{bmatrix} 1 & 0 & 1 & 0 & 1 & 0 & 1 \\ 0 & 1 & 1 & 0 & 0 & 1 & 1 \\ 0 & 0 & 0 & 1 & 1 & 1 & 1 \end{bmatrix}.$$

Treat matrix $\mathbf{H}_1$ as a base matrix, $\mathbf{B}_{\text{base}} = \mathbf{H}_1$, and draw its corresponding protograph.

(b) Find (possibly by computer) a parity-check matrix $\mathbf{H}$ for a length-21 code obtained by selecting $3 \times 3$ circulant permutation matrices $\pi_{ij}$ between VN $j$ and CN $i$ in the protograph, for all $j$ and $i$. Equivalently, replace the 1s in $\mathbf{B}_{\text{base}}$ by the matrices $\pi_{ij}$. The permutation matrices should be selected to avoid 4-cycles to the extent possible.

**6.5** A quasi-cyclic code has the parity-check matrix

$$\mathbf{H} = \begin{bmatrix} 1 & 1 & 0 & 0 & 1 & 0 & 0 & 0 & 1 \\ 0 & 1 & 1 & 0 & 0 & 1 & 1 & 0 & 0 \\ 1 & 0 & 1 & 1 & 0 & 0 & 0 & 1 & 0 \\ & & & & & & & & \\ 0 & 0 & 1 & 0 & 0 & 0 & 1 & 0 & 1 \\ 1 & 0 & 0 & 0 & 0 & 0 & 1 & 1 & 0 \\ 0 & 1 & 0 & 0 & 0 & 0 & 0 & 1 & 1 \end{bmatrix}.$$

Draw its corresponding protograph.

**6.6** Sketch the multi-edge-type generalized protograph that corresponds to the parity-check matrix

$$\mathbf{H} = \left[ \begin{array}{ccc|ccc|ccc} 0 & 0 & 0 & 0 & 1 & 1 & 0 & 1 & 0 \\ 0 & 0 & 0 & 1 & 0 & 1 & 0 & 0 & 1 \\ 0 & 0 & 0 & 1 & 1 & 0 & 1 & 0 & 0 \\ \hline 1 & 0 & 1 & 0 & 0 & 0 & 0 & 1 & 1 \\ 0 & 1 & 0 & 0 & 0 & 1 & 1 & 0 & 1 \\ 0 & 0 & 0 & 1 & 1 & 0 & 1 & 1 & 0 \end{array} \right]$$

and indicate which permutation matrix corresponds to each edge (or pair of edges).

**6.7** You are given the following parity-check matrix for an IRA code:

$$\mathbf{H} = \begin{bmatrix} 1 & 1 & 0 & 0 & 1 & 1 & 0 & 0 & 0 & 0 \\ 0 & 1 & 1 & 1 & 0 & 1 & 1 & 0 & 0 & 0 \\ 1 & 0 & 0 & 1 & 1 & 0 & 1 & 1 & 0 & 0 \\ 0 & 1 & 1 & 0 & 1 & 0 & 0 & 1 & 1 & 0 \\ 0 & 0 & 1 & 1 & 0 & 0 & 0 & 0 & 1 & 1 \end{bmatrix}.$$

Use (6.2) to determine the code rate and compare your answer with the code rate that is determined from the dimensions of $\mathbf{H}$. Find the codeword corresponding to the data word [1 0 0 1 1].

**6.8** The $m \times n$ parity-check matrix for systematic RA and IRA codes has the form $\mathbf{H} = [\mathbf{H}_u \ \ \mathbf{H}_p]$, where $\mathbf{H}_p$ is $m \times m$ and is given by (6.4). Let each codeword have the form $\mathbf{c} = [\mathbf{u} \ \ \mathbf{p}]$, where $\mathbf{u}$ is the data word and $\mathbf{p}$ is the length-$m$ parity word. Show that, depending on the data word $\mathbf{u}$, there exist either two solutions or no solutions for $\mathbf{p}$ when the "1" in the upper-right position in $\mathbf{H}_p$ in (6.4) is present. This statement is true as well for the quasi-cyclic form in (6.8). The impediment is that $\text{rank}(\mathbf{H}_p) = m - 1$. Given this, it would be wise to choose one of the parity-bit positions to correspond to one of the columns of $\mathbf{H}_u$, assuming that $\mathbf{H}$ is full rank. (With acknowledgment of Y. Han.)

**6.9** Find the generator matrix $\mathbf{G}$ corresponding to the parity-check matrix $\mathbf{H}$ given in (6.10) assuming that no puncturing occurs. Figure 6.7 might be helpful. Show that $\mathbf{G}\mathbf{H}^{\mathrm{T}} = \mathbf{0}$. Repeat for the quasi-cyclic form of the IRAA parity-check matrix given in (6.11).

**6.10** Design a rate-1/2 (2048, 1024) LDPC code based on the ARJA protograph of Figure 6.14 by constructing an $\mathbf{H}$ matrix. It is not necessary to include PEG/ACE conditioning, but the design must be free of 4-cycles. Consider values of $Q$ from the set $\{16, 32, 64\}$. Present the $\mathbf{H}$ matrix both as a matrix of exponents of $\pi$ (the identity cyclically shifted rightward once) and as displayed in Figure 6.13. Simulate the designed code down to an error rate of $10^{-8}$ on the AWGN channel using a maximum of 100 decoder iterations.

**6.11** Repeat the previous problem for a rate-2/3 code of dimension $k = 1024$ using the ARJA protograph of Figure 6.16.

**6.12** Prove the following. In a given Tanner graph (equivalently, $\mathbf{H}$ matrix) for an $(n, k)$ LDPC code, the maximum number of degree-2 variable nodes possible before a cycle is created involving only these degree-2 nodes is $n - k - 1 = m - 1$. Furthermore, for codes free of such "degree-2 cycles" and possessing this maximum, the submatrix of $H$ composed of only its weight-2 columns is simply a permutation of the following $m \times (m - 1)$ parent matrix:

$$T = \begin{bmatrix} 1 & & & & & \\ 1 & 1 & & & & \\ & 1 & 1 & & & \\ & & & \cdots & & \\ & & & & 1 & 1 \\ & & & & & 1 \end{bmatrix}.$$

**6.13** Suppose we would like to design a rate-1/3 protograph-based RA code. Find the protograph that describes this collection of codes. Note that there will be one input VN that will be punctured, three output VNs that will be transmitted, and three CNs. Find a formula for the code rate as a function of $N_{v,p}$, the number of punctured VNs, $N_{v,t}$, the number of transmitted VNs, and $N_c$, the number of CNs. (With acknowledgment of D. Divsalar *et al.*)

**6.14** (a) Draw the protograph for the ensemble of rate-1/2 (6, 3) LDPC codes, that is, LDPC codes with degree-3 VNs and degree-6 CNs. (b) Suppose we would like to precode half of the codeword bits of a rate-1/2 (6, 3) LDPC code using an accumulator. Draw a protograph for this ensemble of rate-1/2 codes that maximizes symmetry. Since half of the code bits are precoded, the number of VNs and CNs in your protograph of part (a) will need to be doubled prior to adding the precoder. The protograph should have three degree-3 VNs, one degree-5 VN (that will be punctured), and one degree-1 VN (this one is precoded).

**6.15** Instead of an accumulate–repeat–accumulate code, consider an accumulate–repeat–generalized-accumulate code with generalized accumulator transfer function $1/(1 + D + D^2)$. Find the rate-1/2 protograph for such a code. Set the column weight for the systematic part of the parity-check matrix equal to 5 (i.e., $d_{b,j} = 5$ for all $j$). Construct an example parity-check matrix for a (200, 100) realization of such a code and display a pixelated version of it as in Figure 6.13.

**6.16** Consider a generalized protograph-based LDPC with protograph adjacency matrix

$$\mathbf{\Gamma}_p = \begin{bmatrix} 1 & 0 & 1 & 1 \\ 1 & 2 & 0 & 1 \end{bmatrix}.$$

(a) Sketch the protograph. (b) With $Q = 3$, find an adjacency matrix $\Gamma$ for the G-LDPC code. Use circulant permutation matrices. (c) Now suppose the first constraint in the protograph is the (3, 2) SPC code and the second constraint is the (4, 1) code with parity-check matrix

$$\begin{bmatrix} 1 & 0 & 0 & 1 \\ 0 & 1 & 0 & 1 \\ 0 & 0 & 1 & 1 \end{bmatrix}.$$

Find the parity-check matrix $\mathbf{H}$ for the G-LDPC code and from this give its quasi-cyclic form $\mathbf{H}'$.

**6.17** Consider a product code as a G-LDPC code with two constraints, the row constraint (row code) and the column constraint (column code). Let the row code have length $N_r$ and the column code have length $N_c$. (a) Show that the adjacency matrix is of the form

$$\Gamma = \begin{bmatrix} V & & & \\ & V & & \\ & & \ddots & \\ & & & V \\ I & I & \cdots & I \end{bmatrix},$$

where $V$ is the length-$N_r$ all-1s vector and $I$ is the $N_r \times N_r$ identity matrix. $V$ and $I$ each appear $N_c$ times in the above matrix. (b) Sketch the Tanner graph for this G-LDPC code. (c) Determine the parity-check matrix $\mathbf{H}$ for such a product code if both the row and the column code are the $(7, 4)$ Hamming code with parity-check matrix given by $\mathbf{H}_1$ in Problem 6.4.

## References


[1] R. G. Gallager, *Low-Density Parity-Check Codes*, Cambridge, MA, MIT Press, 1963. (Also, R. G. Gallager, "Low density parity-check codes," *IRE Transactions on Information Theory*, vol. 8, no. 1, pp. 21–28, January 1962.)

[2] D. MacKay, "Good error correcting codes based on very sparse matrices," *IEEE Transactions on Information Theory*, vol. 45, no. 3, pp. 399–431, March 1999.

[3] X.-Y. Hu, E. Eleftheriou, and D.-M. Arnold, "Progressive edge-growth Tanner graphs," *2001 IEEE Global Telecommunications Conference*, November 2001, pp. 995–1001.

[4] T. Tian, C. Jones, J. Villasenor, and R. Wesel, "Construction of irregular LDPC codes with low error floors," *2003 IEEE International Conference on Communications*, May 2003, pp. 3125–3129.

[5] www.inference.phy.cam.ac.uk/mackay/CodesFiles.html.

[6] T. J. Richardson and R. Urbanke, "Efficient encoding of low-density parity-check codes," *IEEE Transactions on Information Theory*, vol. 47, no. 2, pp. 638–656, February 2001.

[7] B. Vasic and O. Milenkovic, "Combinatorial constructions of low-density parity-check codes for iterative decoding," *IEEE Transactions on Information Theory*, vol. 50, no. 6, pp. 1156–1176, June 2004.

[8] H. Xiao and A. Banihashemi, "Improved progressive-edge-growth (PEG) construction of irregular LDPC codes," *2004 IEEE Global Telecommunications Conference*, November–December 2004, pp. 489–492.

[9] T. Richardson and V. Novichkov, "Methods and apparatus for decoding LDPC codes," U.S. Patent 6,633,856, October 14, 2003.

[10] J. Thorpe, *Low Density Parity Check (LDPC) Codes Constructed from Protographs*. JPL INP Progress Report 42-154, August 15, 2003.

[11] D. Divsalar, C. Jones, S. Dolinar, and J. Thorpe, "Protograph based LDPC codes with minimum distance linearly growing with block size," *2005 IEEE Global Telecomunications Conference*.

[12] D. Divsalar, S. Dolinar, and C. Jones, "Construction of protograph LDPC codes with linear minimum distance," *2006 International Symposium on Information Theory*.

[13] T. Richardson and V. Novichkov, "Method and apparatus for decoding LDPC codes," U.S. Patent 7,133,853, November 7, 2006.

[14] Y. Zhang, *Design of Low-Floor Quasi-Cyclic IRA Codes and their FPGA Decoders*, PhD thesis, ECE Department, University of Arizona, May 2007.

[15] C. Jones, S. Dolinar, K. Andrews, D. Divsalar, Y. Zhang, and W. Ryan, "Functions and architectures for LDPC decoding," *2007 IEEE Information Theory Workshop*, pp. 577–583, September 2–6, 2007.

[16] T. Richardson, "Multi-edge type LDPC codes," paper presented at the workshop honoring Professor R. McEliece on his 60th birthday, California Institute of Technology, Pasadena, CA, May 24–25, 2002.

[17] D. Divsalar, H. Jin, and R. McEliece, "Coding theorems for turbo-like codes," *Proceedings of the 36th Annual Allerton Conference on Communication, Control, and Computing*, September 1998, pp. 201–210.

[18] H. Jin, A. Khandekar, and R. McEliece, "Irregular repeat–accumulate codes," *Proceedings of the 2nd International Symposium on Turbo Codes and Related Topics*, Brest, France, September 2000, pp. 1–8.

[19] M. Yang, W. E. Ryan, and Y. Li, "Design of efficiently encodable moderate-length high-rate irregular LDPC codes," *IEEE Transactions on Communications*, vol. 52, no. 4, pp. 564–571, April 2004.

[20] Y. Zhang and W. E. Ryan, "Structured IRA codes: Performance analysis and construction," *IEEE Transactions on Communications*, vol. 55, no. 5, pp. 837–844, May 2007.

[21] R. M. Tanner, "On quasi-cyclic repeat–accumulate codes," *Proceedings of the 37th Allerton Conference on Communication, Control, and Computing*, September 1999.

[22] G. Liva and M. Chiani, "Protograph LDPC codes design based on EXIT analysis," *Proceedings of the 2007 IEEE GlobeCom Conference*, November 2007, pp. 3250–3254.

[23] L. Dinoi, F. Sottile, and S. Benedetto, "Design of variable-rate irregular LDPC codes with low error floor," *2005 IEEE International Conference on Communications*, May 2005, pp. 647–651.

[24] Y. Zhang, W. E. Ryan, and Y. Li, "Structured eIRA codes," *Proceedings of the 38th Asilomar Conference on Signals, Systems, and Computing*, Pacific Grove, CA, November 2004, pp. 7–10.

[25] J. Hagenauer, "Rate-compatible punctured convolutional codes and their applications," *IEEE Transactions on Communications*, vol. 36, no. 4, pp. 389–400, April 1988.

[26] G. Liva, E. Paolini, and M. Chiani, "Simple reconfigurable low-density parity-check codes," *IEEE Communications Letters*, vol. 9, no. 3, pp. 258–260, March, 2005.

[27] S. J. Johnson and S. R. Weller, "Constructions for irregular repeat–accumulate codes," *Proceedings of the IEEE International Symposium on Information Theory*, Adelaide, September 2005.

[28] A. Abbasfar, D. Divsalar, and K. Yao, "Accumulate repeat accumulate codes," *Proceedings of the 2004 IEEE GlobeCom Conference*, Dallas, TX, November 2004.

[29] A. Abbasfar, D. Divsalar, and K. Yao, "Accumulate–repeat–accumulate codes," *IEEE Transactions on Communications*, vol. 55, no. 4, pp. 692–702, April 2007.

[30] K. Andrews, S. Dolinar, D. Divsalar, and J. Thorpe, *Design of Low-Density Parity-Check (LDPC) Codes for Deep-Space Applications*, JPL IPN Progress Report 42–159, November 15, 2004.

[31] K. Andrews, S. Dolinar, and J. Thorpe, "Encoders for block-circulant LDPC codes," *Proceedings of the 2005 International Symposium on Information Theory*, September 2005.

[32] K. Andrews, S. Dolinar, D. Divsalar, and J. Thorpe, *Low-Density Parity-Check Code Design Techniques to Simplify Encoding*, JPL IPN Progress Report 42–171, November 15, 2007.

[33] CCSDS Report, *TM Synchronization and Channel Coding – Summary of Concept and Rationale*, Green Book, June 2020.

[34] ETSI Standard TR 102 376 V1.1.1: Digital Video Broadcasting (DVB) User Guidelines for the Second Generation System for Broadcasting, Interactive Services, News Gathering and Other Broadband Satellite Applications (DVB-S2). http://webapp.etsi.org/workprogram/Report_WorkItem.asp?WKI_ID=21402.

[35] M. C. Valenti, S. Cheng, and R. Iyer Seshadri, "Turbo and LDPC codes for digital video broadcasting," Chapter 12 of *Turbo Code Applications: A Journey from a Paper to Realization*, Berlin, Springer-Verlag, 2005.

[36] *Low Density Parity Check Codes for Use in Near-Earth and Deep Space*. Orange Book, Issue 2, September 2007, http://public.ccsds.org/publications/OrangeBooks.aspx.

[37] S. Dolinar, "A rate-compatible family of protograph-based LDPC codes built by expurgation and lengthening," *Proceedings of the 2005 International Symposium on Information Theory*, September 2005, pp. 1627–1631.

[38] T.-Y. Chen, K. Vakilinia, D. Divsilar, and R. Wesel, "Protograph-based Raptor-like codes," *IEEE Transactions on Communications*, pp. 1522–1532, May 2015.

[39] A. Shokrollahi, "Raptor codes," *IEEE Transactions on Information Inf. Theory*, pp. 2551–2567, June 2006.

[40] G. Liva, W. E. Ryan, and M. Chiani, "Quasi-cyclic generalized LDPC codes with low error floors," *IEEE Transactions on Communications*, pp. 49–57, January 2008.

# 7 Turbo Codes

Turbo codes, which were first presented to the coding community in 1993 [1, 2], represent one of the most important breakthroughs in coding since Ungerboeck introduced trellis codes in 1982 [3]. In fact, the invention of turbo codes and their iterative "turbo" decoders started a revolution in iteratively decodable codes and iterative receiver processing (such as "turbo equalization"). Most frequently, a turbo code refers to a concatenation of two (or more) convolutional encoders separated by interleavers. The turbo decoder consists of two (or more) soft-in/soft-out convolutional decoders, which iteratively feed probabilistic information back and forth to each other in a manner that is reminiscent of a turbo engine. In this chapter we introduce the most important classes of turbo codes, provide some heuristic justification as to why they should perform well, and present their iterative decoders. Our focus will be on parallel- and serial-concatenated convolutional codes (PCCCs and SCCCs) on the binary-input AWGN channel, but we also include the important class of turbo product codes. These codes involve block codes arranged as rows and columns in a rectangular array of bits. The decoder is similar to that for PCCCs and SCCCs, except that the constituent decoders are typically suboptimal soft-in/soft-out list decoders. The reader is advised to consult the leading papers in the field for additional information on the codes considered in this chapter [4–19]. Nonbinary turbo codes for the noncoherent reception of orthogonal signals are considered in [20].

## 7.1 Parallel-Concatenated Convolutional Codes

Figure 7.1 depicts the encoder for a standard parallel-concatenated convolutional code (PCCC), the first class of turbo code invented. As seen in the figure, for a $K$-bit message, the PCCC encoder consists of two binary rate-1/2 recursive systematic convolutional (RSC) encoders separated by a $K$-bit interleaver or permuter, together with an optional puncturing mechanism. Clearly, without the puncturer, the encoder is rate 1/3, mapping $K$ data bits to $3K$ code bits. We observe that the encoders are configured in a manner reminiscent of classical concatenated codes. However, instead of cascading the encoders in the usual *serial* fashion, the encoders are arranged in a so-called *parallel concatenation*. As will be argued below, recursive convolutional encoders are necessary in order to attain the exceptional performance for which turbo codes are known. Before describing further details of the PCCC encoder in its entirety, we shall first discuss its individual components.

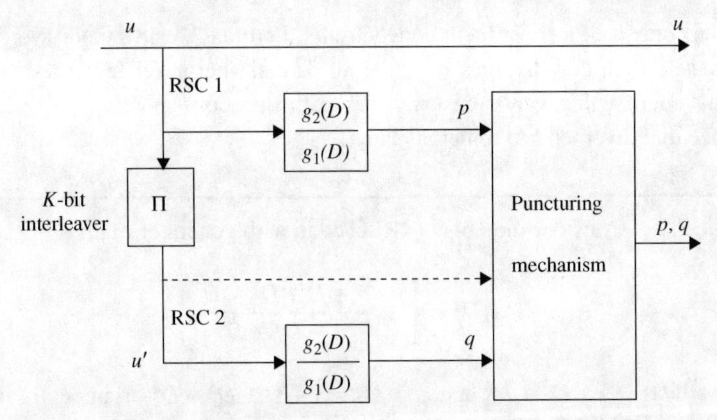


**Figure 7.1** The encoder for a standard parallel-concatenated convolutional code.

### 7.1.1 Critical Properties of RSC Codes

Without any essential loss of generality, we assume that the constituent RSC codes are identical, with generator matrix

$$\mathbf{G}(D) = \left[ 1 \quad \frac{g^{(2)}(D)}{g^{(1)}(D)} \right].$$

Observe that the code sequence $u(D)\mathbf{G}(D)$ will be of finite weight if and only if the input sequence is divisible by $g^{(1)}(D)$. We have from this fact the following immediate results, which we shall use later.

**Lemma 7.1** *A weight-1 input into an RSC encoder will produce an infinite-weight output, for such an input is never divisible by a (nontrivial) polynomial $g^{(1)}(D)$. (In practice, "infinite" should be replaced by "large" since the input length would be finite.)*

**Lemma 7.2** *For any nontrivial $g^{(1)}(D)$, there exists a family of weight-2 inputs into an RSC encoder of the form $D^j(1+D^P)$, $j \geq 0$, which produce finite-weight outputs, that is, which are divisible by $g^{(1)}(D)$. When $g^{(1)}(D)$ is a primitive polynomial of degree $m$, then $P = 2^m - 1$. More generally, $P$ is the period of the pseudo-random sequence generated by $g^{(1)}(D)$.*

*Proof* Because the encoder is linear, its output due to a weight-2 input $D^j(1 + D^t)$ is equal to the sum of its outputs due to $D^j$ and $D^jD^t$. The output due to $D^j$ will be periodic with period $P$, since the encoder is a finite-state machine: the state at time $j$ must be reached again in a finite number of steps $P$, after which the state sequence is repeated indefinitely with period $P$. Now, letting $t = P$, the output due to $D^jD^P$ is just the output due to $D^j$ shifted by $P$ bits. Thus, the output due to $D^j(1 + D^P)$ is the sum of the outputs due to $D^j$ and $D^jD^P$, which must be of finite length and weight since all but one period will cancel out in the sum. □

In the context of a constituent code's trellis, Lemma 7.1 says that a weight-1 input will create a path that diverges from the all-0s path, but never remerges. Lemma 7.2 says that there will always exist a trellis path that corresponds to a weight-2 data sequence that diverges and remerges later.

---

**Example 7.1** Consider the 16-state RSC code with generator matrix

$$\mathbf{G}(D) = \left[ 1 \quad \frac{1 + D + D^3 + D^4}{1 + D^3 + D^4} \right].$$

Thus, $g^{(1)}(D) = 1 + D^3 + D^4$ and $g^{(2)}(D) = 1 + D + D^3 + D^4$ or, in octal form, the generators are $(31, 33)$. Observe that $g^{(1)}(D)$ is primitive of order 15, so that, for example, $u(D) = 1 + D^{15}$ produces the finite-length code sequence $(1 + D^{15}, 1 + D + D^4 + D^5 + D^7 + D^9 + D^{10} + D^{11} + D^{12} + D^{15})$. Of course, any delayed version of this input, say $D^7(1 + D^{15})$, will simply produce a delayed version of this code sequence.

---

### 7.1.2 Critical Properties of the Interleaver

The function of the interleaver is to take each incoming block of bits and rearrange them in a pseudo-random fashion prior to encoding by the second RSC encoder. It is crucial that this interleaver permute the bits in a manner that ostensibly lacks any apparent order, although it should be tailored in a certain way for weight-2 and weight-3 inputs, as will be made clearer below. The $S$-random interleaver [17] is quite effective in this regard. This particular interleaver ensures that any two input bits whose positions are within $S$ of each other are separated by an amount greater than $S$ at the interleaver output. $S$ should be selected to be as large as possible for a given value of $K$. Observe that the implication of this design is that, if the input $u(D)$ is of the form $D^j(1 + D^P)$, where $P = 2^m - 1$ (see Lemma 7.2), then the interleaver output $u'(D)$ will be of the form $D^{j'}(1 + D^{P'})$, where $P' > 2^m - 1$ and is somewhat large. This means that the parity weight at the second encoder output is not likely to also be low. Lastly, as we shall see, performance increases with $K$, so $K \geq 1000$ is typical.

The $S$-random interleaver is created by an algorithm involving a pseudo-random computer search. It has the disadvantage that $K$ permutation values must be stored. An alternative is an algorithmic interleaver that allows the generation of the interleaver values via computation, omitting the storage requirement. See, for example, the papers by Takeshita and Ryu [18–22]. Most common are *quadratic polynomial* interleavers whose interleaver values are produced by a polynomial of the form $aX^2 + bX + c$ (modulo $K$), where $a, b, c$ are positive integers and $X$ ranges from 0 to $K - 1$. The parameter $c$ is usually set to zero, although a set of nonzero values for $c$ can give a family of permuters.

### 7.1.3 The Puncturer

The role of the turbo-code puncturer is identical to that of any other code, that is, to delete selected bits to reduce coding overhead. It is most convenient to delete only parity bits, otherwise the decoder (described in Section 7.2) will have to be substantially redesigned. For example, to achieve a rate of 1/2, one might delete all even-parity bits from the top encoder and all odd-parity bits from the bottom one. At the receiver, the punctured bits are assigned log-likelihood values of zero at the decoder input, indicating that no information is known about these bits. There is no guarantee that deleting only parity bits and no data bits will yield the maximum possible minimum codeword distance for the code rate and length of interest, but this is the most convenient. Approaches to puncturing may be found in the literature, such as [23].

### 7.1.4 Performance Estimate on the BI-AWGNC

A maximum-likelihood (ML) sequence decoder would be far too complex for a PCCC due to the presence of the interleaver. However, the suboptimum iterative decoding algorithm to be described in Section 7.2 offers near-ML performance. Hence, we shall now estimate the performance of an ML decoder for a PCCC code on the BI-AWGNC. A more careful analysis of turbo codes and iteratively decodable codes in general may be found in Chapter 8.

Armed with the above descriptions of the components of the PCCC encoder of Figure 7.1, it is easy to conclude that it is linear since its components are linear. (We ignore the nuisance issue of terminating the trellises of the two constituent encoders, which generally makes the turbo code nonlinear.) The constituent codes are certainly linear, and the permuter is linear since it may be modeled by a permutation matrix. Further, the puncturer does not affect linearity since all codewords share the same puncture locations. As usual, the importance of linearity is that, in considering the performance of a code, one may choose the all-zeros sequence as a reference. Thus, hereafter we shall assume that the all-zeros codeword was transmitted.

Now consider the all-zeros codeword (the 0th codeword) and the $k$th codeword, for some $k \in \{1, 2, \ldots, 2^K - 1\}$. The ML decoder will choose the $k$th codeword over the 0th codeword with probability $Q\left(\sqrt{2d_k RE_b/N_0}\right)$, where $R$ is the code rate and $d_k$ is the weight of the $k$th codeword. The bit-error rate for this two-codeword situation would then be

$$\begin{aligned} P_b(k|0) &= w_k \text{ (bit errors/cw error)} \\ &\quad \times \frac{1}{K} \text{ (cw/data bits)} \\ &\quad \times Q\left(\sqrt{2Rd_k E_b/N_0}\right) \text{ (cw errors/cw)} \\ &= \frac{w_k}{K} Q\left(\sqrt{\frac{2Rd_k E_b}{N_0}}\right) \text{ (bit errors/data bit),} \end{aligned}$$

where $w_k$ is the weight of the $k$th data word. Now, including all of the codewords and invoking the usual union bounding argument, we may write

$$
\begin{aligned}
P_b &= P_b\left(\text{choose any } k \in \{1, 2, \ldots, 2^K - 1\}|0\right) \\
&\leq \sum_{k=1}^{2^K-1} P_b(k|0) \\
&= \sum_{k=1}^{2^K-1} \frac{w_k}{K} Q\left(\sqrt{\frac{2Rd_kE_b}{N_0}}\right).
\end{aligned}
$$

Note that every nonzero codeword is included in the above summation. Let us now reorganize the summation as

$$
P_b \leq \sum_{w=1}^{K} \sum_{v=1}^{\binom{K}{w}} \frac{w}{K} Q\left(\sqrt{\frac{2Rd_{wv}E_b}{N_0}}\right), \tag{7.1}
$$

where the first sum is over the weight-$w$ inputs, the second sum is over the $\binom{K}{w}$ different weight-$w$ inputs, and $d_{wv}$ is the weight of the codeword produced by the $v$th weight-$w$ input.

Consideration of the first few terms in the outer summation of (7.1) leads to a certain characteristic of the code's weight spectrum, called *spectral thinning* [9], explained in the following.

$\underline{w = 1}$: From Lemma 7.1 and the associated discussion above, weight-1 inputs will produce only large-weight codewords at both constituent encoder outputs since the trellis paths created never remerge with the all-zeros path. Thus, each $d_{1v}$ is significantly greater than the minimum codeword weight so that the $w = 1$ terms in (7.1) will be negligible.

$\underline{w = 2}$: Of the $\binom{K}{2}$ weight-2 encoder inputs, only a fraction will be divisible by $g_1(D)$ (i.e., yield remergent paths) and, of these, only certain ones will yield the smallest weight, $d_{2,\min}^{\text{CC}}$, at a constituent encoder output (here, CC denotes "constituent code"). Further, with the permuter present, if an input $u(D)$ of weight 2 yields a weight-$d_{2,\min}^{\text{CC}}$ codeword at the first encoder's output, it is unlikely that the permuted input, $u'(D)$, seen by the second encoder will also correspond to a weight-$d_{2,\min}^{\text{CC}}$ codeword (much less be divisible by $g_1(D)$). We can be sure, however, that there will be some minimum-weight turbo codeword produced by a $w = 2$ input, and that this minimum weight can be bounded as

$$
d_{2,\min}^{\text{PCCC}} \geq 2d_{2,\min}^{\text{CC}} - 2,
$$

with equality when both of the constituent encoders produce weight-$d_{2,\min}^{\text{CC}}$ codewords (minus 2 for the bottom encoder). The exact value of $d_{2,\min}^{\text{PCCC}}$ is permuter-dependent. We will denote the number of weight-2 inputs that produce weight-$d_{2,\min}^{\text{PCCC}}$ turbo codewords by $n_2$ so that, for $w = 2$, the inner sum in (7.1) may be approximated as

$$\sum_{v=1}^{\binom{K}{2}} \frac{2}{K} Q\left(\sqrt{\frac{2Rd_{2v}E_b}{N_0}}\right) \simeq \frac{2n_2}{K} Q\left(\sqrt{\frac{2Rd_{2,\min}^{\text{PCCC}}E_b}{N_0}}\right). \tag{7.2}$$

From the foregoing discussion, we can expect $n_2$ to be small relative to $K$. This fact is the *spectral-thinning* effect brought about by the combination of the recursive encoder and the interleaver.

$\underline{w = 3}$: Following an argument similar to the $w = 2$ case, we can approximate the inner sum in (7.1) for $w = 3$ as

$$\sum_{v=1}^{\binom{K}{3}} \frac{3}{K} Q\left(\sqrt{\frac{2Rd_{3v}E_b}{N_0}}\right) \simeq \frac{3n_3}{K} Q\left(\sqrt{\frac{2Rd_{3,\min}^{\text{PCCC}}E_b}{N_0}}\right), \tag{7.3}$$

where $n_3$ and $d_{3,\min}^{\text{PCCC}}$ are obviously defined. While $n_3$ is clearly dependent on the interleaver, we can make some comments on its size relative to $n_2$ for a randomly generated interleaver. Although there are $(K-2)/3$ times as many $w = 3$ terms in the inner summation of (7.1) as there are $w = 2$ terms, we can expect the number of weight-3 terms divisible by $g_1(D)$ to be on the order of the number of weight-2 terms divisible by $g_1(D)$. Thus, most of the $\binom{K}{3}$ terms in (7.1) can be removed from consideration for this reason. Moreover, given a weight-3 encoder input $u(D)$ divisible by $g_1(D)$ (e.g., $g_1(D)$ itself in the above example), it becomes very unlikely that the permuted input $u'(D)$ seen by the second encoder will also be divisible by $g_1(D)$. For example, suppose $u(D) = g_1(D) = 1+D+D^4$. Then the permuter output will be a multiple of $g_1(D)$ if the three input 1s become the $j$th, $(j+1)$th, and $(j+4)$th bits out of the permuter, for some $j$. If we imagine that the permuter acts in a purely random fashion so that the probability that one of the 1s lands in a given position is $1/K$, the permuter output will be $D^j g_1(D) = D^j\left(1+D+D^4\right)$ with probability $3!/K^3$. This is not the only weight-3 pattern divisible by $g_1(D)$ ($g_1^2(D) = 1 + D^2 + D^8$ is another one, but this too has probability $3!/K^3$ of occurring). For comparison, for $w = 2$ inputs, a given permuter output pattern occurs with probability $2!/K^2$. Thus, we would expect the number of weight-3 inputs, $n_3$, resulting in remergent paths in both encoders to be much less than $n_2$, $n_3 \ll n_2$, with the result being that the inner sum in (7.1) for $w = 3$ is negligible relative to that for $w = 2$. Because our argument assumes a "purely random" permuter, the inequality $n_3 \ll n_2$ has to be interpreted probabilistically. Thus, it is more accurate to write $E\{n_3\} \ll E\{n_2\}$, where the expectation is over all interleavers. Alternatively, for the *average* interleaver, we would expect $n_3 \ll n_2$.

$\underline{w \geq 4}$: Again, we can approximate the inner sum in (7.1) for $w = 4$ in the same manner as in (7.2) and (7.3). Still, we would like to make some comments on its size for the "random" interleaver. A weight-4 input might appear to the first encoder as a weight-3 input concatenated some time later with a weight-1 input, leading to a non-remergent path in the trellis and, hence, a negligible term in the inner sum in (7.1). It might also appear as a concatenation of two weight-2 inputs, in which case the turbo codeword weight is at least $2d_{2,\min}^{\text{PCCC}}$, again leading to a negligible term in (7.1).

Finally, if it happens to be some other pattern divisible by $g_1(D)$ at the first encoder, with probability on the order of $1/K^3$, it will be simultaneously divisible by $g_1(D)$ at the second encoder. (The value of $1/K^3$ derives from the fact that ideally a particular divisible output pattern occurs with probability $4!/K^4$, but there will be approximately $K$ shifted versions of that pattern, each divisible by $g_1(D)$.) Thus, we may expect $n_4 \ll n_2$ so that the $w \geq 4$ terms are negligible in (7.1). The cases for $w > 4$ are argued similarly.

To summarize, the bound in (7.1) can be approximated as

$$P_b \simeq \sum_{w\geq 2} \frac{wn_w}{K} Q\left(\sqrt{\frac{2Rd_{w,\min}^{\text{PCCC}} E_b}{N_0}}\right)$$

$$\simeq \max_{w\geq 2}\left\{\frac{wn_w}{K} Q\left(\sqrt{\frac{2Rd_{w,\min}^{\text{PCCC}} E_b}{N_0}}\right)\right\}, \tag{7.4}$$

where $n_w$ and $d_{w,\min}^{\text{PCCC}}$ are functions of the particular interleaver employed. From our discussion above, we would expect that the $w = 2$ term dominates for a randomly generated interleaver, although it is easy to find interleavers for which this is not true. One example is the $S$-random interleaver mentioned earlier. In any case, we observe that $P_b$ decreases with $K$, so that the error rate can be reduced simply by increasing the interleaver length ($w$ and $n_w$ do not change appreciably with $K$). This effect is called *interleaver gain* and demonstrates the necessity of large interleavers. Finally, we emphasize that recursive encoders are crucial elements of a turbo code since, for non-recursive encoders, there is no division by $g_1(D)$ (non-remergent trellis paths), so that (7.4) would not hold with small values of $n_w$.

---

**Example 7.2** We consider the $P_b$ performance of a rate-1/2 (23, 33) parallel turbo code for two different interleavers of size $K = 1000$. We start first with an interleaver that was randomly generated. We found, for this particular interleaver, $n_2 = 0$ and $n_3 = 1$, with $d_{3,\min}^{\text{PC}} = 9$, so that the $w = 3$ term dominates in (7.4). The interleaver input corresponding to this dominant error event was $D^{168}(1 + D^5 + D^{10})$, which produces the interleaver output $D^{88}(1 + D^{15} + D^{848})$, where of course both polynomials are divisible by $g_1(D) = 1 + D + D^4$. Figure 7.2 gives the simulated $P_b$ performance of this code for 15 iterations of the iterative decoding algorithm detailed in the next section. Also included in Figure 7.2 is the estimate of (7.4) for the same interleaver, which is observed to be very close to the simulated values. The interleaver, was then modified to improve the weight spectrum of the code. It was a simple matter to attain $n_2 = 1$ with $d_{2,\min}^{\text{PC}} = 12$ and $n_3 = 4$ with $d_{3,\min}^{\text{PC}} = 15$ for this second interleaver, so that the $w = 2$ term now dominates in (7.4). The simulated and estimated performance curves for this second interleaver are also included in Figure 7.2.

---

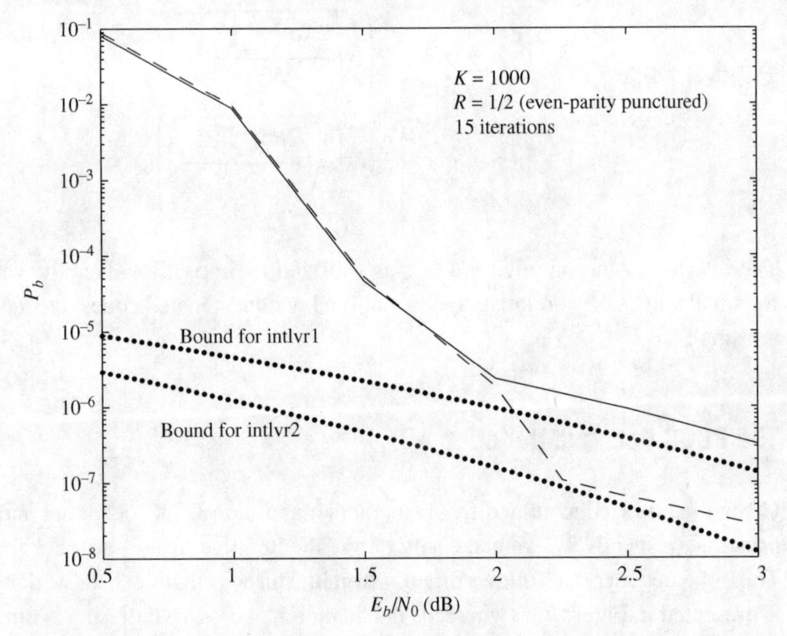


**Figure 7.2** Simulated performance of a rate-1/2 (23, 33) PCCC for two different interleavers (intlvr1 and intlvr2) ($K = 1000$) together with the asymptotic performance estimates of each given by (7.4).

In addition to illustrating the use of the estimate (7.4), this example helps explain the unusual shape of the error-rate curve: it may be interpreted as the usual $Q$-function shape for a signaling scheme in AWGN with a modest $d_{\min}$, "pushed down" by the interleaver gain $w^* n_{w^*}/K$, where $w^*$ is the maximizing value of $w$ in (7.4) and $w^* n_{w^*}$ is small relative to $K$.

The spectral-thinning effect that leads to a small $w^* n_{w^*}$ factor is clearly a consequence of the combination of RSC encoders and pseudo-random interleavers. "Thinning" refers to the small multiplicities of low-weight codewords relative to classical codes. For example, consider a cyclic code of length $N$. Because each cyclic shift is a codeword, the multiplicity $n_{d_{\min}}$ of the minimum-weight codewords will be a multiple of $N$. Similarly, consider a convolutional code (recursive or not) with length-$K$ inputs. If the input $u(D)$ creates a minimum-weight codeword, so does $D^j u(D)$ for $j = 1, 2, \ldots$. Thus, the multiplicity $n_{d_{\min}}$ will be a multiple of $K$. By contrast, for a turbo code, $D^j u(D)$ and $D^k u(D)$, $j \neq k$, will almost surely lead to codewords of different weights because the presence of the interleaver makes the turbo code not shift-invariant.

Occasionally, the codeword error rate, $P_{cw} = \Pr\{\text{decoder output contains one or more residual errors}\}$, is the preferred performance metric. (As mentioned in Chapter 1, $P_{cw}$ is also denoted by FER or WER in the literature, representing "frame" or "word" error rate, respectively.) An estimate of $P_{cw}$ may be derived using steps almost identical to those used to derive $P_b$. The resulting expressions are nearly identical, except that the factor $w/K$ is removed from the expressions. The result is

$$
\begin{aligned}
P_{cw} &\simeq \sum_{w \geq 2} n_w Q\left(\sqrt{\frac{2Rd_{w,\min}^{\text{PCCC}} E_b}{N_0}}\right) \\
&\simeq \max_{w \geq 2}\left\{n_w Q\left(\sqrt{\frac{2Rd_{w,\min}^{\text{PCCC}} E_b}{N_0}}\right)\right\}.
\end{aligned} \tag{7.5}
$$

The interleaver has an effect on $P_{cw}$ as well and its impact involves the values of $n_w$ for small values of $w$: relative to conventional unconcatenated codes, the values for $n_w$ are small.

## 7.2 The PCCC Iterative Decoder

Given the knowledge gained from Chapters 4 and 5 on BCJR decoding and the turbo principle, respectively, we can easily derive the iterative (turbo) decoder for a PCCC. The turbo decoder that follows directly from the turbo principle discussed in Chapter 5 is presented in Figure 7.3 (where, as in Chapter 5, necessary buffering is omitted). The two decoders are soft-in/soft-out (SISO) decoders, typically BCJR decoders, matched to the top and bottom RSC encoders of Figure 7.1. SISO decoder 1 (D1) receives noisy data ($y_k^u$) and parity ($y_k^p$) bits from the channel, corresponding to the bits $u_k$ and $p_k$ transmitted by RSC 1 in Figure 7.1. Decoder D2 receives only the noisy parity bits $y_k^q$, corresponding to the output $q_k$ of RSC 2 in Figure 7.1. As per the turbo principle, only extrinsic log-likelihood-ratio (LLR) information is sent from one decoder to the other, with appropriate interleaving/de-interleaving in accordance with the PCCC encoder. We will learn below how a BCJR decoder processes information from both the channel and from a companion decoder. After a certain number of iterations, either decoder can sum the LLRs from the channel, from its companion decoder, and from its own

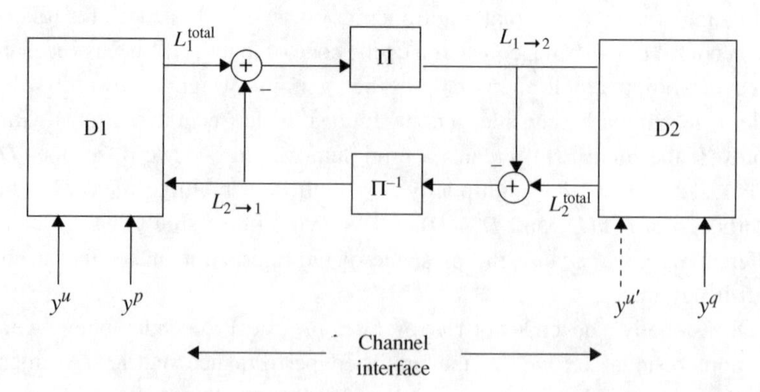


**Figure 7.3** A turbo decoder obtained directly from the turbo principle (Chapter 5). $L_1^{\text{total}}$ is the total log-likelihood information at the output of RSC decoder D1, and similarly for $L_2^{\text{total}}$. The log-likelihood quantity $L_{1\to 2}$ is the extrinsic information sent from RSC decoder 1 (D1) to RSC decoder 2 (D2), and similarly for $L_{2\to 1}$.

computations to produce the total LLR values for the data bits $u_k$. Decisions are made from the total LLR values: $\hat{u}_k = \text{sign}[LLR(u_k)]$.

An alternative interpretation of the decoder can be appreciated by considering the two graphical representations of a PCCC in Figure 7.4. The top graph follows straightforwardly from Figure 7.1 and the bottom graph follows straightforwardly from the top graph (where the bottom interleaver permutes only the edges between the second convolutional code constraint node (CC2) and the systematic bit nodes). Now observe that the bottom graph is essentially that of a generalized LDPC code whose decoder was discussed in Chapter 5. Thus, the decoder for a PCCC is very much like that of a generalized LDPC code.

In spite of the foregoing discussion, we still find it useful and instructive to present the details of the iterative PCCC decoder ("turbo decoder"). In particular, it is helpful to illuminate the computation of extrinsic information within the constituent BCJR decoders and how these BJCR decoders process incoming extrinsic information together with incoming channel information.

## 7.2.1 Overview of the Iterative Decoder

The goal of the iterative decoder is to iteratively estimate the *a posteriori* probabilities (APPs) $\Pr(u_k|\mathbf{y})$, where $u_k$ is the $k$th data bit, $k = 1, 2, \ldots, K$, and $\mathbf{y}$ is the received

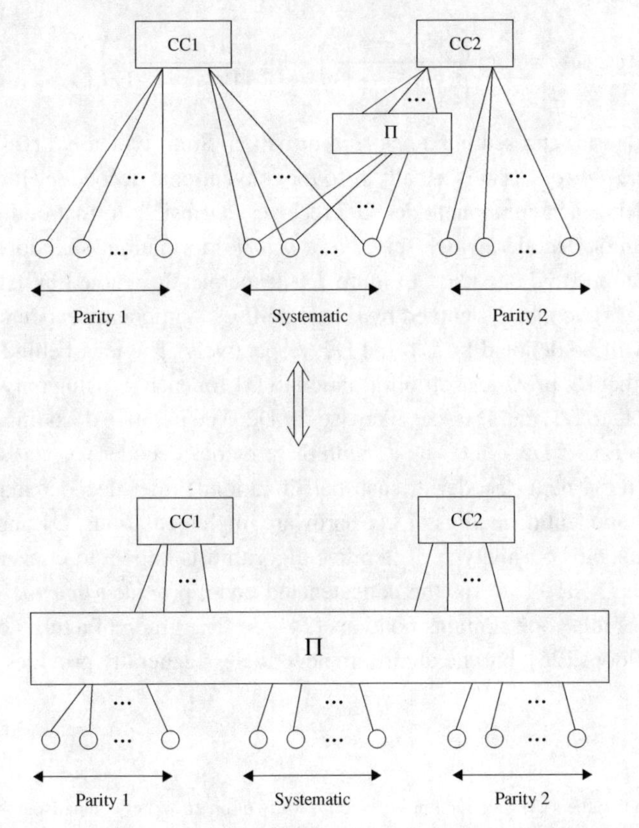


**Figure 7.4** Interpretation of a PCCC code's graph as the graph of a generalized LDPC code.

codeword in AWGN, $\mathbf{y} = \mathbf{c} + \mathbf{n}$. In this equation, we assume for convenience that the components of $\mathbf{c}$ take values in the set $\{\pm 1\}$ (and similarly for $\mathbf{u}$) and that $\mathbf{n}$ is a noise vector whose components are AWGN samples. Knowledge of the APPs allows one to make optimal decisions on the bits $u_k$ via the maximum *a posteriori* (MAP) rule[1]

$$\frac{\Pr(u_k = +1|\mathbf{y})}{\Pr(u_k = -1|\mathbf{y})} \underset{-1}{\overset{+1}{\gtrless}} 1$$

or, more conveniently,

$$\hat{u}_k = \text{sign}[L(u_k)], \tag{7.6}$$

where $L(u_k)$ is the log *a posteriori* probability (log-APP) ratio defined as

$$L(u_k) \triangleq \log\left(\frac{\Pr(u_k = +1|\mathbf{y})}{\Pr(u_k = -1|\mathbf{y})}\right).$$

We shall use the term log-likelihood ratio (LLR) in place of log-APP ratio for consistency with the literature.

From Bayes' rule, the LLR for an arbitrary SISO decoder can be written as

$$L(u_k) = \log\left(\frac{p(\mathbf{y}|u_k = +1)}{p(\mathbf{y}|u_k = -1)}\right) + \log\left(\frac{\Pr(u_k = +1)}{\Pr(u_k = -1)}\right), \tag{7.7}$$

with the second term representing *a priori* information. Since typically $\Pr(u_k = +1) = \Pr(u_k = -1)$, the *a priori* term is usually zero for conventional decoders. However, for *iterative* decoders, each component decoder receives extrinsic information for each $u_k$ from its companion decoder, which serves as *a priori* information. We adopt the convention that the top RSC encoder in Figure 7.1 is encoder 1, denoted by E1, and the bottom encoder is encoder 2, denoted by E2. The SISO component decoders matched to E1 and E2 will be denoted by D1 and D2, respectively. The idea behind extrinsic information is that D2 provides soft information to D1 for each $u_k$ using only information not available to D1, and D1 does likewise for D2. The iterative decoding proceeds as D1 $\to$ D2 $\to$ D1 $\to$ D2 $\to$ D1 $\to \ldots$, with the previous decoder passing soft information along to the next decoder at each half-iteration. Either decoder may initiate the chain of component decodings or, for hardware implementations, D1 and D2 may operate simultaneously. This type of iterative algorithm is known to converge to the true value of the LLR $L(u_k)$ for the concatenated code, provided that the graphical representation of this code contains no loops [24–25]. The graph of a turbo code does in fact contain loops [26], but the algorithm nevertheless generally provides near-ML performance.

[1] It is well known that the MAP rule minimizes the probability of bit error. For comparison, the ML rule, which maximizes the likelihoods $P(\mathbf{y}|\mathbf{c})$ over the codewords $\mathbf{c}$, minimizes the probability of codeword error. See Chapter 1.

### 7.2.2 Decoder Details

We assume no puncturing in the PCCC encoder of Figure 7.1, so that the overall code rate is 1/3. The transmitted codeword $\mathbf{c}$ will have the form $\mathbf{c} = [c_1, c_2, \ldots, c_K] = [u_1, p_1, q_1, \ldots, u_K, p_K, q_K]$, where $c_k \triangleq [u_k, p_k, q_k]$. The received word $\mathbf{y} = \mathbf{c} + \mathbf{n}$ will have the form $\mathbf{y} = [y_1, y_2, \ldots, y_K] = [y_1^u, y_1^p, y_1^q, \ldots, y_K^u, y_K^p, y_K^q]$, where $y_k \triangleq [y_k^u, y_k^p, y_k^q]$, and similarly for $\mathbf{n}$. We denote the codewords produced by E1 and E2 as, respectively, $\mathbf{c}_1 = [c_1^1, c_2^1, \ldots, c_K^1]$, where $c_k^1 \triangleq [u_k, p_k]$, and $\mathbf{c}_2 = [c_1^2, c_2^2, \ldots, c_K^2]$, where $c_k^2 \triangleq [u_k', q_k]$. Note that $\left\{u_k'\right\}$ is a permuted version of $\{u_k\}$ and is not actually transmitted (see Figure 7.1). We define the noisy received versions of $\mathbf{c}_1$ and $\mathbf{c}_2$ to be $\mathbf{y}_1$ and $\mathbf{y}_2$, respectively, having components $y_k^1 \triangleq [y_k^u, y_k^p]$ and $y_k^2 \triangleq [y_k^{u'}, y_k^q]$, respectively. Note that $\mathbf{y}_1$ and $\mathbf{y}_2$ can be assembled from $\mathbf{y}$ in an obvious fashion: using an interleaver to obtain $\left\{y_k^{u'}\right\}$ from $\left\{y_k^u\right\}$. For doing so, the component decoder inputs are the two vectors $\mathbf{y}_1$ and $\mathbf{y}_2$.

Each SISO decoder is essentially a BCJR decoder (developed in Chapter 4), except that each BCJR decoder is modified so that it may accept extrinsic information from a companion decoder and produce soft outputs. Thus, rather than using the LLRs to obtain hard decisions, the LLRs become the SISO BCJR decoder's outputs. We shall often use SISO and BCJR interchangeably. To initiate the discussion, we present the BCJR algorithm summary for a single RSC code, which has been adapted from Chapter 4. (For the time being, we will discuss a generic SISO decoder so that we may avoid using cumbersome superscripts for the two constituent decoders until it is necessary to do so.)

In the algorithm above, the branch metric $\tilde{\gamma}_k(s', s) = \log \gamma_k(s', s)$, where, from Chapter 4,

$$\gamma_k(s', s) = \frac{P(u_k)}{2\pi\sigma^2} \exp\left[-\frac{\|y_k - c_k\|^2}{2\sigma^2}\right]. \tag{7.11}$$

$P(u_k)$ is the probability that the data bit $u_k(s', s)$ corresponding to the transition $s' \to s$ is equal to the specific value $u_k \in \{\pm 1\}$, that is, $P(u_k) = \Pr(u_k(s', s) = u_k)$. In Chapter 4, assuming equiprobable encoder inputs, we promptly set $P(u_k) = 1/2$ for either value of $u_k$. As indicated earlier, the extrinsic information received from a companion decoder takes the role of *a priori* information in the iterative decoding algorithm (cf. (7.7) and surrounding discussion), so that

$$L^{\mathrm{e}}(u_k) \Leftrightarrow \log\left(\frac{\Pr(u_k = +1)}{\Pr(u_k = -1)}\right). \tag{7.12}$$

To incorporate extrinsic information into the BCJR algorithm, we consider the log-domain version of the branch metric,

$$\tilde{\gamma}_k(s', s) = \log(P(u_k)) - \log\left(2\pi\sigma^2\right) - \frac{\|y_k - c_k\|^2}{2\sigma^2}. \tag{7.13}$$

Now observe that we may write

$$\begin{aligned} P(u_k) &= \left(\frac{\exp[-L^{\mathrm{e}}(u_k)/2]}{1 + \exp[-L^{\mathrm{e}}(u_k)]}\right) \cdot \exp[u_k L^{\mathrm{e}}(u_k)/2] \\ &= A_k \exp[u_k L^{\mathrm{e}}(u_k)/2], \end{aligned} \tag{7.14}$$

**Algorithm 7.1** BCJR Algorithm Summary

**Initialize** $\tilde{\alpha}_0(s)$ and $\tilde{\beta}_K(s)$ according to

$$\tilde{\alpha}_0(s) = \begin{cases} 0, & s = 0 \\ -\infty, & s \neq 0 \end{cases}$$

$$\tilde{\beta}_K(s) = \begin{cases} 0, & s = 0 \\ -\infty, & s \neq 0 \end{cases}$$

**for** $k = 1$ to $K$ **do** (forward sweep)
  get $y_k = [y_k^u, y_k^p]$
  compute $\tilde{\gamma}_k(s', s)$ for all allowable state transitions $s' \to s$
    (note that $c_k = c_k(s', s)$ here)
  compute $\tilde{\alpha}_k(s)$ for all $s$ using the recursion

$$\tilde{\alpha}_k(s) = \max_{s'}{}^* \left[\tilde{\alpha}_{k-1}(s') + \tilde{\gamma}_k(s', s)\right] \tag{7.8}$$

**end for**

**for** $k = K$ to 2 step $-1$ **do** (backward sweep)
  compute $\tilde{\beta}_{k-1}(s')$ for all $s'$ using the recursion

$$\tilde{\beta}_{k-1}(s') = \max_{s}{}^* \left[\tilde{\beta}_k(s) + \tilde{\gamma}_k(s', s)\right] \tag{7.9}$$

**end for**

**for** $k = 1$ to $K$ **do** (decision stage)
  compute $L(u_k)$ using

$$L(u_k) = \max_{U^+}{}^* \left[\tilde{\alpha}_{k-1}(s') + \tilde{\gamma}_k(s', s) + \tilde{\beta}_k(s)\right] - \max_{U^-}{}^* \left[\tilde{\alpha}_{k-1}(s') + \tilde{\gamma}_k(s', s) + \tilde{\beta}_k(s)\right] \tag{7.10}$$

**end for**

where the first equality follows since the right-hand side equals

$$\left(\frac{\sqrt{P_-/P_+}}{1 + P_-/P_+}\right) \sqrt{P_+/P_-} = P_+$$

when $u_k = +1$ and

$$\left(\frac{\sqrt{P_-/P_+}}{1 + P_-/P_+}\right) \sqrt{P_-/P_+} = P_-$$

when $u_k = -1$, and we have defined $P_+ \triangleq \Pr(u_k = +1)$ and $P_- \triangleq \Pr(u_k = -1)$ for convenience. Substitution of (7.14) into (7.13) yields

$$\tilde{\gamma}_k(s', s) = \log\left[A_k/(2\pi\sigma^2)\right] + u_k L^{\mathrm{e}}(u_k)/2 - \frac{\|y_k - c_k\|^2}{2\sigma^2}, \tag{7.15}$$

where the first term may be ignored since it is independent of $u_k$ (equivalently, of $s' \to s$). (It may appear from the notation that $A_k$ is a function of $u_k$ since it is a function of $L^{\mathrm{e}}(u_k)$, but from (7.12) the latter quantity is a function of the probability mass function for $u_k$, not $u_k$.) In summary, the extrinsic information received from a companion decoder is included in the computation through the modified branch metric

$$\tilde{\gamma}_k(s', s) = u_k L^{\mathrm{e}}(u_k)/2 - \frac{\|\mathbf{y}_k - \mathbf{c}_k\|^2}{2\sigma^2}. \tag{7.16}$$

The rest of the SISO BCJR algorithm proceeds exactly as before.

Upon substitution of (7.16) into (7.10), we have

$$\begin{aligned} L(u_k) = \; & L^{\mathrm{e}}(u_k) + \max_{U^+}{}^{*}\left[\tilde{\alpha}_{k-1}(s') + u_k y_k^u/\sigma^2 + p_k y_k^p/\sigma^2 + \tilde{\beta}_k(s)\right] \\ & - \max_{U^-}{}^{*}\left[\tilde{\alpha}_{k-1}(s') + u_k y_k^u/\sigma^2 + p_k y_k^p/\sigma^2 + \tilde{\beta}_k(s)\right], \end{aligned} \tag{7.17}$$

where we have used the fact that

$$\begin{aligned} \|\mathbf{y}_k - \mathbf{c}_k\|^2 &= (y_k^u - u_k)^2 + (y_k^p - p_k)^2 \\ &= \left(y_k^u\right)^2 - 2u_k y_k^u + u_k^2 + \left(y_k^p\right)^2 - 2p_k y_k^p + p_k^2 \end{aligned}$$

and only the terms in this expression dependent on $U^+$ or $U^-$, namely $u_k y_k^u/\sigma^2$ and $p_k y_k^p/\sigma^2$, survive after the subtraction. Now note that $u_k y_k^u/\sigma^2 = y_k^u/\sigma^2$ under the first $\max^*(\cdot)$ operation in (7.17) ($U^+$ is the set of state transitions for which $u_k = +1$) and $u_k y_k^u/\sigma^2 = -y_k^u/\sigma^2$ under the second $\max^*(\cdot)$ operation. Using the definition for $\max^*(\cdot)$, it is easy to see that these terms may be brought out of the $\max^*(\cdot)$ functions so that

$$\begin{aligned} L(u_k) = 2y_k^u/\sigma^2 + L^{\mathrm{e}}(u_k) + & \max_{U^+}{}^{*}\left[\tilde{\alpha}_{k-1}(s') + p_k y_k^p/\sigma^2 + \tilde{\beta}_k(s)\right] \\ - & \max_{U^-}{}^{*}\left[\tilde{\alpha}_{k-1}(s') + p_k y_k^p/\sigma^2 + \tilde{\beta}_k(s)\right]. \end{aligned} \tag{7.18}$$

The interpretation of this new expression for $L(u_k)$ is that the first term is likelihood information received directly from the channel, the second term is extrinsic likelihood information received from a companion decoder, and the third "term" $(\max_{U^+}{}^{*} - \max_{U^-}{}^{*})$ is extrinsic likelihood information to be passed to a companion decoder. Note that this third term is likelihood information gleaned from received parity that is not available to the companion decoder. Thus, specializing to decoder D1, for example, on any given iteration, D1 computes

$$L_1(u_k) = 2y_k^u/\sigma^2 + L_{21}^{\mathrm{e}}(u_k) + L_{12}^{\mathrm{e}}(u_k), \tag{7.19}$$

where $L_{21}^{\mathrm{e}}(u_k)$ is extrinsic information received from D2, and $L_{12}^{\mathrm{e}}(u_k)$ is the $(\max_{U^+}{}^{*} - \max_{U^-}{}^{*})$ term in (7.18) that is to be used as extrinsic information from D1 to D2.

To recapitulate, the turbo decoder for a PCCC includes two SISO-BCJR decoders matched to the two constituent RSC encoders within the PCCC encoder. As summarized in (7.19) for D1, each SISO decoder accepts scaled soft channel information $(2y_k^u/\sigma^2)$ as well as soft extrinsic information from its counterpart $(L_{21}^{\mathrm{e}}(u_k))$. Each

SISO decoder also computes soft extrinsic information to send to its counterpart ($L_{12}^{\mathrm{e}}(u_k)$). The two SISO decoders would be configured in a turbo decoder as in Figure 7.5. Although this more symmetric decoder configuration appears to be different from the decoder configuration of Figure 7.3, the main difference being that $\{y_k^{u'}\}$ are not fed to D2 in Figure 7.3, they are essentially equivalent. To see this, suppose $L_{2\to1}$ in Figure 7.3 is equal to $L_{21}^{\mathrm{e}}(u_k)$ so that the inputs to D1 in that figure are $y_k^u$, $y_k^p$, and $L_{21}^{\mathrm{e}}(u_k)$. Then, from our development that led to (7.19), the output of D1 in Figure 7.3 is clearly $L_1^{\mathrm{total}} = L_1(u_k) = 2y_k^u/\sigma^2 + L_{21}^{\mathrm{e}}(u_k) + L_{12}^{\mathrm{e}}(u_k)$. From this, $L_{1\to2}$ must be $L_1^{\mathrm{total}} - L_{2\to1} = 2y_k^u/\sigma^2 + L_{12}^{\mathrm{e}}(u_k)$ so that $L_2^{\mathrm{total}} = L_2(u_k) = 2y_k^u/\sigma^2 + L_{12}^{\mathrm{e}}(u_k) + L_{21}^{\mathrm{e}}(u_k)$. Lastly, $L_{2\to1} = L_2^{\mathrm{total}} - L_{1\to2} = L_{21}^{\mathrm{e}}(u_k)$, which agrees with our initial assumption.

### 7.2.3 Summary of the PCCC Iterative Decoder

We will present in this section pseudo-code for the turbo decoder for PCCCs. The algorithm given below for the iterative decoding of a parallel turbo code follows directly from the development above. The constituent decoder order is D1, D2, D1, D2, and so on. The interleaver and de-interleavers are represented by the arrays $P[\cdot]$ and $P\mathrm{inv}[\cdot]$, respectively. For example, the permuted word $\mathbf{u}'$ is obtained from the original word $\mathbf{u}$ via the following pseudo-code statement: `for` $k = 1{:}K$, $u_k' = u_{P[k]}$, `end`. We point out that knowledge of the noise variance $\sigma^2 = N_0/2$ by each SISO BCJR decoder is necessary. Also, a simple way to obtain higher code rates via puncturing is, in the computation of $\gamma_k(s', s)$, to set to zero the received parity samples, $y_k^p$ or $y_k^q$, corresponding to the punctured parity bits, $p_k$ or $q_k$. (This will set to zero the term in the branch metric corresponding to the punctured bit.) Thus, puncturing need not be performed at the encoder for computer simulations.

When discussing the BCJR algorithm, it is usually assumed that the trellis starts in the zero state and terminates in the zero state. This is accomplished for a single convolutional code by appending $\mu$ appropriately chosen "termination bits" at the end of the data word ($\mu$ is the RSC memory size), and it is accomplished in the same way for E1 in the PCCC encoder. However, termination of encoder E2 to the zero state

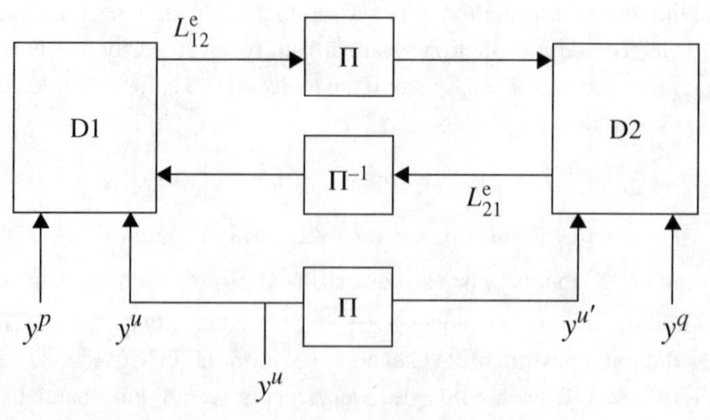


**Figure 7.5** The turbo decoder for a PCCC code.

can be problematic due to the presence of the interleaver. (Various solutions exist in the literature.) Fortunately, there is generally a small performance loss when E2 is not terminated. In this case, $\beta_K(s)$ for D2 may be set to $\alpha_K(s)$ for all $s$, or it may be set to a nonzero constant (e.g., $1/S_2$, where $S_2$ is the number of E2 states).

We remark that some sort of iteration-stopping criterion is necessary. The most straightforward criterion is to set a maximum number of iterations. However, this can be inefficient since the correct codeword is often found after only a few iterations. An efficient technique utilizes a carefully chosen outer error-detection code. After each iteration, a parity check is carried out and the iterations stop whenever no error is detected. Other stopping criteria are presented in the literature. Observe that it is not possible to use the parity-check matrix $\mathbf{H}$ for the PCCC together with the check equation $\mathbf{cH}^{\mathrm{T}} = \mathbf{0}$ because the decoder outputs are decisions on the systematic bits, not on the entire codeword.

We first present an outline of the turbo decoding algorithm assuming D1 decodes first. Then we present the algorithm in detail.

**Outline**

1. Initialize all state metrics appropriately and set all extrinsic information to zero.
2. D1 decoder: Run the SISO BJCR algorithm with inputs $y_k^1 = [y_k^u, y_k^p]$ and $L_{21}^{\mathrm{e}}(u_{P\mathrm{inv}[k]})$ to obtain $L_{12}^{\mathrm{e}}(u_k)$. Send extrinsic information $L_{12}^{\mathrm{e}}(u_{P[k]})$ to the D2 decoder.
3. D2 decoder: Run the SISO BJCR algorithm with inputs $y_k^2 = [y_{P[k]}^u, y_k^q]$ and $L_{12}^{\mathrm{e}}(u_{P[k]})$ to obtain $L_{21}^{e}(u_k)$. Send extrinsic information $L_{21}^{\mathrm{e}}(u_{P\mathrm{inv}[k]})$ to the D1 decoder.
4. Repeat Steps 2 and 3 until the preset maximum number of iterations is reached or some other stopping criterion is satisfied. Make decisions on bits according to $\mathrm{sign}[L_1(u_k)]$, where $L_1(u_k) = 2y_k^u/\sigma^2 + L_{21}^{\mathrm{e}}(u_{P\mathrm{inv}[k]}) + L_{12}^{\mathrm{e}}(u_k)$. (One can instead use $\mathrm{sign}[L_2(u_k)]$.)

---

**PCCC Iterative Decoder**

---

**Initialization**

**D1:**

$$\tilde{\alpha}_0^{(1)}(s) = \begin{cases} 0, & \text{for } s = 0 \\ -\infty, & \text{for } s \neq 0 \end{cases}$$

$$\tilde{\beta}_K^{(1)}(s) = \begin{cases} 0, & \text{for } s = 0 \\ -\infty, & \text{for } s \neq 0 \end{cases}$$

$L_{21}^{\mathrm{e}}(u_k) = 0$ for $k = 1, 2, \ldots, K$

**D2:**

$$\tilde{\alpha}_0^{(2)}(s) = \begin{cases} 0, & \text{for } s = 0 \\ -\infty, & \text{for } s \neq 0 \end{cases}$$

$\tilde{\beta}_K^{(2)}(s) = \tilde{\alpha}_K^{(2)}(s)$ for all $s$ $\left(\text{set once after computation of } \{\tilde{\alpha}_K^{(2)}(s)\} \text{ in the } \textit{first} \text{ iteration}\right)$

$L_{12}^{\mathrm{e}}(u_k)$ is to be determined from D1 after the first half-iteration and so need not be initialized

**The *n*th iteration**

**D1:**

**for** $k = 1$ to $K$ (forward sweep)

• get $\mathrm{y}_k^1 = [y_k^u, y_k^p]$
• compute $\tilde{\gamma}_k(s', s)$ for all allowable state transitions $s' \to s$ from (7.16) or the simplified form (see the discussion following (7.16))

$$\tilde{\gamma}_k(s', s) = u_k L_{21}^{\mathrm{e}}(u_{P\mathrm{inv}[k]})/2 + u_k y_k^u/\sigma^2 + p_k y_k^p/\sigma^2$$

($u_k$ ($p_k$) in this expression is set to the value of the encoder input (output) corresponding to the transition $s' \to s$)
• compute $\tilde{\alpha}_k^{(1)}(s)$ for all $s$ using (7.8)

**end**

**for** $k = K$ to 2 step $-1$ (backward sweep)

• compute $\tilde{\beta}_{k-1}^{(1)}(s)$ for all $s$ using (7.9)

**end**

**for** $k = 1$ to $K$ (extrinsic information computation)

• compute $L_{12}^{\mathrm{e}}(u_k)$ using

$$\begin{aligned} L_{12}^{\mathrm{e}}(u_k) = & \max_{U^+}{}^* \left[\tilde{\alpha}_{k-1}^{(1)}(s') + p_k y_k^p/\sigma^2 + \tilde{\beta}_k^{(1)}(s)\right] \\ & - \max_{U^-}{}^* \left[\tilde{\alpha}_{k-1}^{(1)}(s') + p_k y_k^p/\sigma^2 + \tilde{\beta}_k^{(1)}(s)\right] \end{aligned}$$

**end**

**D2:**

**for** $k = 1$ to $K$ (forward sweep)

• get $\mathrm{y}_k^2 = [y_{P[k]}^u, y_k^q]$
• compute $\tilde{\gamma}_k(s', s)$ for all allowable state transitions $s' \to s$ from

$$\tilde{\gamma}_k(s', s) = u_k L_{12}^{\mathrm{e}}(u_{P[k]})/2 + u_k y_{P[k]}^u/\sigma^2 + q_k y_k^q/\sigma^2$$

($u_k$ ($q_k$) in this expression is set to the value of the encoder input (output) corresponding to the transition $s' \to s$)
• compute $\tilde{\alpha}_k^{(2)}(s)$ for all $s$ using (7.8)

**end**

**for** $k = K$ to 2 step $-1$ (backward sweep)

• compute $\tilde{\beta}_{k-1}^{(2)}(s)$ for all $s$ using (7.9)

**end**

**for** $k = 1$ to $K$ (extrinsic information computation)

• compute $L_{21}^{\mathrm{e}}(u_k)$ using

$$L_{21}^{\mathrm{e}}(u_k) = \max_{U^+}{}^* \left[\tilde{\alpha}_{k-1}^{(2)}(s') + q_k y_k^q/\sigma^2 + \tilde{\beta}_k^{(2)}(s)\right] - \max_{U^-}{}^* \left[\tilde{\alpha}_{k-1}^{(2)}(s') + q_k y_k^q/\sigma^2 + \tilde{\beta}_k^{(2)}(s)\right]$$

**end**

**Decision after the last iteration**

**for** $k = 1$ to $K$

• compute: $L_1(u_k) = 2y_k^u/\sigma^2 + L_{21}^{\mathrm{e}}(u_{\mathrm{Pinv}[k]}) + L_{12}^{\mathrm{e}}(u_k)$
• $\hat{u}_k = \mathrm{sign}[L(u_k)]$

**end**

### 7.2.4 Lower-Complexity Approximation: The SOVA

The core algorithm within the turbo decoder, the BCJR algorithm, uses the max* function that (from Chapter 4) has the following representations:

$$\begin{aligned}\max{}^*(x, y) &\triangleq \log(e^x + e^y) \\ &= \max(x, y) + \log\left(1 + e^{-|x-y|}\right).\end{aligned} \tag{7.20}$$

From the second expression, we can see that this function may be implemented by a max function followed by use of a look-up table for the term $\log\left(1 + e^{-|x-y|}\right)$. It has been shown that such a table can be quite small (e.g., of size 8), but its size depends on the turbo code in question and the performance requirements. Alternatively, since $\log\left(1 + e^{-|x-y|}\right) \leq \ln(2) = 0.693$, this term can be dropped in (7.20), with the max function used in place of the max* function. This will, of course, be at the expense of some performance loss, which depends on turbo-code parameters. Note that, irrespective of whether it is max* or max that is used, when more than two quantities are involved, the computation may take place pair-wise. For example, for the three quantities $x$, $y$, and $z$, $\max^*(x, y, z) = \max^*[\max^*(x, y), z]$.

An alternative to the BCJR algorithm is the *soft-output Viterbi algorithm* (SOVA) [7, 12, 13], which is a modification of the Viterbi algorithm that produces soft (reliability) outputs for use in iterative decoders and also accepts extrinsic information. The BJCR algorithm is an algorithm that has two Viterbi-like algorithms, one forward and one backward. Thus, the complexity of the SOVA is about half that of the BCJR algorithm. Moreover, as we will see, unlike the BCJR decoder, the SOVA decoder does not require knowledge of the noise variance. Of course, the advantages of the SOVA are gained at the expense of a performance degradation. For most situations, this degradation can be made to be quite small by attenuating the SOVA output (its reliability estimates are too large, on average). We present here the original SOVA algorithm of [12]. Improvements may be found in [25–30].

We assume a rate-$1/n$ convolutional code so that the cumulative (squared Euclidean distance) metric $\Gamma_k(s)$ for state $s$ at time $k$ is updated according to

$$\Gamma_k(s) = \min\{\Gamma_{k-1}(s') + \lambda_k(s', s), \Gamma_{k-1}(s'') + \lambda_k(s'', s)\}, \tag{7.21}$$

where $s'$ and $s''$ are the two states leading to state $s$, $\lambda_k(s', s)$ is the branch metric for the transition from state $s'$ to state $s$, and the rest of the quantities are obviously defined. When the Viterbi decoder chooses a survivor according to (7.21), it chooses in favor of the smaller of the two metrics. The difference between these two metrics is a measure of the reliability of that decision and, hence, of the reliability of the corresponding bit decision. This is intuitively so, but we can demonstrate it as follows.

Define the reliability of a decision to be the LLR of the correct/error binary hypothesis:

$$\rho = \ln\left(\frac{1 - \Pr(\text{error})}{\Pr(\text{error})}\right).$$

To put this in terms of the Viterbi algorithm, consider the decision made at one node in a trellis according to the add–compare–select operation of (7.21). Owing to code linearity and channel symmetry, the Pr(error) derivation is independent of the correct path. Thus, let $S_1$ be the event that the correct path comes from state $s'$ and let $D_2$ be the event that the path from $s''$ is chosen. Define also

$$M'_k = \Gamma_{k-1}(s') + \lambda_k(s', s)$$

and

$$M''_k = \Gamma_{k-1}(s'') + \lambda_k(s'', s)$$

so that $M''_k \leq M'_k$ when event $D_2$ occurs. Then, with $\mathbf{y}$ the received codeword and $S_1$ the event that the correct path comes from $s''$, several applications of Bayes' rule and cancellation of equal probabilities yield the following:

$$\begin{aligned}
\Pr(\text{error}) &= \Pr(D_2|\mathbf{y}, S_1) \\
&= \frac{\Pr(D_2)}{\Pr(S_1)} \Pr(S_1|\mathbf{y}, D_2) \\
&= \Pr(S_1|\mathbf{y}) \\
&= \frac{p(\mathbf{y}|S_1)\Pr(S_1)}{p(\mathbf{y})} \\
&= \frac{p(\mathbf{y}|S_1)}{p(\mathbf{y}|S_1) + p(\mathbf{y}|S_2)}
\end{aligned}$$

$$= \frac{\exp\left[-M_k'/(2\sigma^2)\right]}{\exp\left[-M_k'/(2\sigma^2)\right] + \exp\left[-M_k''/(2\sigma^2)\right]}$$

$$= \frac{1}{1 + \exp\left[\left(M_k' - M_k''\right)/(2\sigma^2)\right]}$$

$$= \frac{1}{1 + \exp\left[\Delta_k/(2\sigma^2)\right]}, \tag{7.22}$$

where $\Delta_k$ is the tentative metric-difference magnitude at time $k$:

$$\Delta_k = \left|M_k' - M_k''\right|. \tag{7.23}$$

Rearranging (7.22) yields

$$\frac{\Delta_k}{2\sigma^2} = \ln\left(\frac{1 - \Pr(\text{error})}{\Pr(\text{error})}\right) = \rho, \tag{7.24}$$

which coincides with our earlier intuition that the metric difference is a measure of reliability.

Note that $\rho$ includes the noise variance $\sigma^2$, whereas the standard Viterbi algorithm does not, that is, the branch metrics $\lambda_k(.,.)$ in (7.21) are unscaled Euclidean distances. If we use the same branch metrics as used by the BCJR algorithm, then the noise variance will be automatically included in the metric-difference computation. That is, if we replace the squared Euclidean-distance branch metric $\lambda_k(s', s)$ by the branch metric (we now include the incoming extrinsic information term)

$$\tilde{\gamma}_k(s', s) = u_k L^{\mathrm{e}}(u_k)/2 + u_k y_k^u/\sigma^2 + p_k y_k^p/\sigma^2, \tag{7.25}$$

we have the alternative normalized metrics

$$\tilde{M}_k' = \Gamma_{k-1}(s') + \tilde{\gamma}_k(s', s) \tag{7.26}$$

and

$$\tilde{M}_k'' = \Gamma_{k-1}(s'') + \tilde{\gamma}_k(s'', s). \tag{7.27}$$

For the correlation metric (7.25), the cumulative metrics are computed as

$$\Gamma_k(s) = \max\{M_k', M_k''\}.$$

The metric difference $\Delta_k$ is now defined in terms of these normalized metrics as

$$\Delta_k = \left|\tilde{M}_k' - \tilde{M}_k''\right|/2$$

and, under this new definition, the metric reliability is given by $\rho = \Delta_k/2$.

Observe that $\Delta_k$ gives a path-wise reliability, whereas we need to compute bit-wise reliabilities to be used as soft outputs. To obtain the soft output for a given bit $u_k$, we first obtain the hard decision $\hat{u}_k$ after a delay $\delta$ (i.e., at time $k + \delta$), where $\delta$ is the decoding depth. At time $k + \delta$, we select the surviving path with the largest metric. We trace back the largest-metric path to obtain the hard decision $\hat{u}_k$. Along this path, there are $\delta + 1$ non-surviving paths that have been discarded (due to the add–compare–select algorithm), and each non-surviving path has a certain difference metric $\Delta_j$, where $k \leq j \leq k+\delta$. The *bit-wise reliability* for the decision $\hat{u}_k$ is defined as

$$\Delta_k^* = \min\{\Delta_k, \Delta_{k+1}, \ldots, \Delta_{k+\delta}\},$$

where the minimum is taken only over the non-surviving paths along the largest-metric path within the time window $[k \leq j \leq k+\delta]$ that would have led to a different decision for $\hat{u}_k$. We may now write that the soft output for bit $u_k$ is given by

$$L_{\text{SOVA}}(u_k) = \hat{u}_k \Delta_k^*.$$

Thus, the turbo decoder of Figure 7.3 would have as constituent decoders two SOVA decoders with $\{L_{\text{SOVA}}(u_k)\}$ acting as the total LLR outputs. We mention some additional details that are often incorporated in practice. First, the scaling by $\sigma^2$ is often dropped in practice, making knowledge of the noise variance unnecessary. For this version of the SOVA, the computed extrinsic information tends to be larger on average than that of a BCJR. Thus, attenuation of $L_{1\to 2}$ and $L_{2\to 1}$ in Figure 7.3 generally improves performance. See [29, 30] for additional details.

Next, since the two RSC encoders start at the zero state, and E1 is terminated at the zero state, the SOVA-based turbo decoder should exploit this as follows. For times $k = 1, 2, \ldots, K$, the SOVA decoders determine the survivors and their difference metrics, but no decisions are made until $k = K$. At this time, select for D1 the single surviving path that ends at state zero. For D2, select as the single surviving path at time $k = K$ the path with the highest metric. For each constituent decoder, the decisions may be obtained by tracing back along their respective sole survivors, computing $\{\hat{u}_k\}, \{\Delta_k\}$, and $\{\Delta_k^*\}$ along the way.

## 7.3 Serial-Concatenated Convolutional Codes

Figure 7.6 depicts the encoder for a standard serial-concatenated convolutional code (SCCC). An SCCC encoder consists of two binary convolutional encoders (E1 and E2) separated by a length-$(K/R_1)$ interleaver. The outer code has rate $R_1$, the inner code has rate $R_2$, with possible puncturing occurring at either encoder, and the overall code rate is $R = R_1 R_2$. As will be made evident below, an SCCC can achieve interleaving gain only if the inner code is a recursive convolutional code (usually systematic), although the outer code need not be recursive. Much like in the PCCC case, the interleaver is important in that it ensures that if the inner encoder produces a low-weight codeword it is unlikely that the outer code does so too.

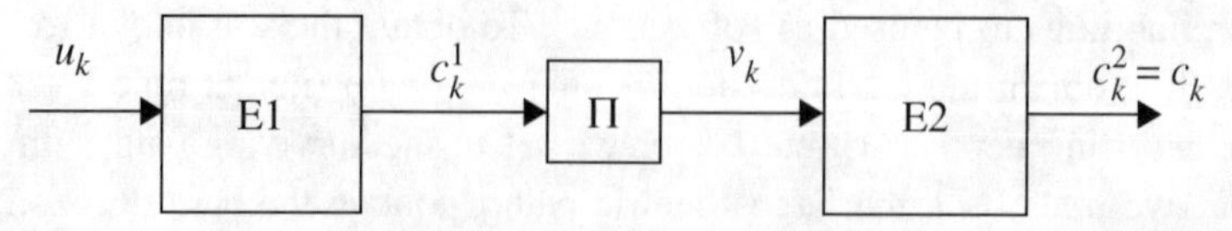


**Figure 7.6** The encoder for a standard serial-concatenated convolutional code.

**Example 7.3** Suppose the outer and inner codes have identical generator matrices

$$\mathbf{G}_1(D) = \mathbf{G}_2(D) = \begin{bmatrix} 1 & \dfrac{1+D^2}{1+D+D^2} \end{bmatrix}.$$

Now suppose the input is $u(D) = 1 + D + D^2$. Then the output of the outer encoder, E1, is $\begin{bmatrix} 1+D+D^2 & 1+D^2 \end{bmatrix}$ or, as a multiplexed binary sequence, the output is 111011000. . .. The output of the interleaver (e.g., an $S$-random interleaver) will be five scattered 1s. But, ignoring delays (or leading zeros), the response of encoder E2 to each of the five isolated 1s is

$$\begin{bmatrix} 1 & \dfrac{1+D^2}{1+D+D^2} \end{bmatrix} = \begin{bmatrix} 1 & 1+D+D^2+D^4+D^5+D^7+D^8+\cdots \end{bmatrix}$$

or, as a multiplexed binary sequence, 1101010001010001010. . .. Thus, the combined effect of the interleaver and the recursive inner code is the amplification of inner codeword weight so that the serial-concatenated code will have a large minimum distance (when the interleaver is properly designed).

### 7.3.1 Performance Estimate on the BI-AWGNC

Because of their similarity, and their linearity, the derivation of an ML performance estimate for SCCCs would be very similar to the development given in Section 7.1.4 for the PCCCs. Thus, from Section 7.1.4 we may immediately write

$$\begin{aligned} P_b &\le \sum_{k=1}^{2^K-1} \frac{w_k}{K} Q\left(\sqrt{\frac{2Rd_kE_b}{N_0}}\right) \\ &= \sum_{w=1}^{K} \sum_{v=1}^{\binom{K}{w}} \frac{w}{K} Q\left(\sqrt{\frac{2Rd_{wv}E_b}{N_0}}\right) \end{aligned}$$

and examine the impact of input weights $w = 1, 2, 3 \ldots$ on the SCCC encoder output weight. As before, weight-1 encoder inputs have negligible impact on performance because at least one of the encoders is recursive. Also, as before weight-2, weight-3, and weight-4 inputs can be problematic. Thus, we may write

$$\begin{aligned} P_b &\simeq \sum_{w \ge 2} \frac{wn_w}{K} Q\left(\sqrt{\frac{2Rd_{w,\min}^{\text{SCCC}}E_b}{N_0}}\right) \\ &\simeq \max_{w \ge 2} \left\{ \frac{wn_w}{K} Q\left(\sqrt{\frac{2Rd_{w,\min}^{\text{SCCC}}E_b}{N_0}}\right) \right\}, \end{aligned} \tag{7.28}$$

where $n_w$ and $d_{w,\min}^{\text{SCCC}}$ are functions of the particular interleaver employed. Analogous to $d_{w,\min}^{\text{PCCC}}$, $d_{w,\min}^{\text{SCCC}}$ is the minimum SCCC codeword weight among weight-$w$ inputs.

Again, there exists an interleaver gain because $wn_w/K$ tends to be much less than unity (for sufficiently large $K$). An interleaver gain of the form $wn_w/K^f$ for some integer $f > 1$ is often quoted for SCCC codes, but this form arises in the average performance analysis for an ensemble of SCCC codes. The ensemble result for large $K$ and $f > 1$ implies that, among the codes in the ensemble, a given SCCC is unlikely to have the worst-case performance (which depends on the worst-case $d_{w,\min}^{\text{SCCC}}$). This is discussed in detail in Chapter 8.

When codeword error rate, $P_{cw}$ = Pr{decoder output contains one or more residual errors}, is the preferred performance metric, the result is

$$P_{cw} \simeq \sum_{w\geq 2} n_w Q\left(\sqrt{\frac{2Rd_{w,\min}^{\text{SCCC}}E_b}{N_0}}\right)$$
$$\simeq \max_{w\geq 2}\left\{n_w Q\left(\sqrt{\frac{2Rd_{w,\min}^{\text{SCCC}}E_b}{N_0}}\right)\right\}. \tag{7.29}$$

---

**Example 7.4** We consider in this example a PCCC and an SCCC, both rate 8/9 with parameters $(N, K) = (1152, 1024)$. The PCCC encoder uses two identical four-state RSC encoders whose generator polynomials are $\left(\mathbf{g}_{\text{octal}}^{(1)}, \mathbf{g}_{\text{octal}}^{(2)}\right) = (7, 5)$. To achieve a code rate of 8/9, only one bit is saved in every 16-bit block of parity bits at each RSC encoder output. As for the SCCC encoder, the outer constituent encoder is the same

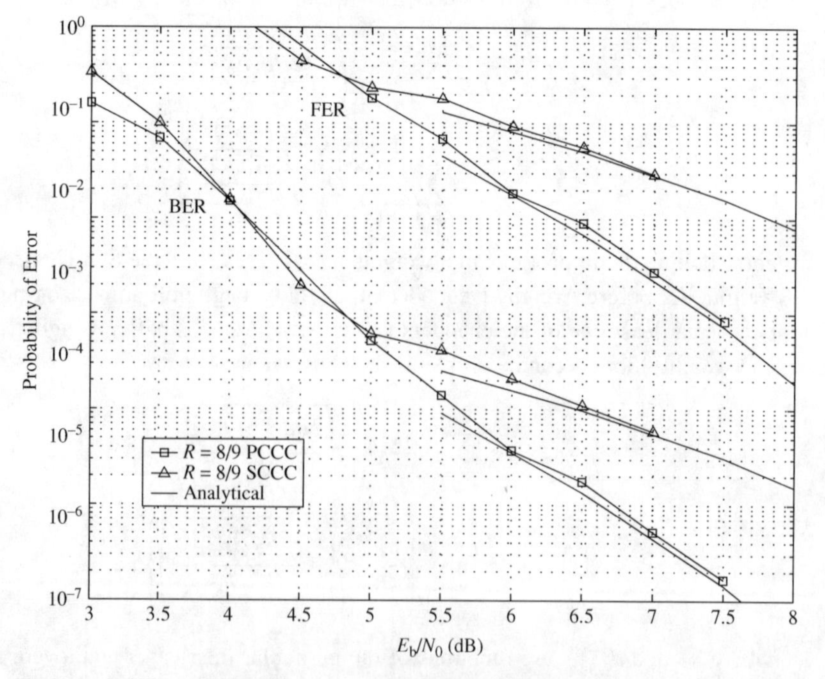


**Figure 7.7** PCCC and SCCC bit-error-rate (BER) and frame-error-rate (FER) simulation results for rate-8/9 (1152, 1024) codes together with analytical results in (7.4) and (7.28).

four-state RSC encoder, and the inner code is a rate-1 differential encoder (accumulator) with transfer function $1/(1 \oplus D)$. A rate of 8/9 is achieved in this case by saving one bit in every 8-bit block of parity bits at the RSC encoder output. The PCCC interleaver is a 1024-bit pseudo-random interleaver with no constraints added (e.g., no $S$-random constraint). The SCCC interleaver is a 1152-bit pseudo-random interleaver with no constraints added. Figure 7.7 presents simulated performance results for these codes using iterative decoders (the SCCC iterative decoder will be discussed in the next section). Simulation results both for the bit-error rate $P_b$ (BER in the figure) and for the frame or codeword-error rate $P_{cw}$ (FER in the figure) are presented. Also included in Figure 7.7 are analytic performance estimates for ML decoding using (7.4), (7.5), (7.28), and (7.29). We can see the close agreement between the analytical and simulated results in this figure.

---

We comment on the fact that the PCCC in Figure 7.7 is substantially better than the SCCC, whereas it is known that SCCCs generally have lower floors [11]. We attribute this to the fact that the outer RSC code in the SCCC has been punctured so severely that $d_{\min} = 2$ for this outer code (although $d_{\min}$ for the SCCC is certainly larger than 2). The RSC encoders for the PCCC are punctured only half as much, so $d_{\min} > 2$ for each of these encoders. We also attribute this to the fact that we have not used an optimized interleaver for this example. In support of these comments, we present the following example.

---

**Example 7.5** We have simulated rate-$1/2$ versions of the same code structure as in the previous example, but no puncturing occurs for the SCCC and much less occurs for the PCCC. In this case, $(N, K) = (2048, 1024)$ and $S$-random interleavers were used ($S = 16$ for PCCC and $S = 20$ for SCCC). The results are presented in Figure 7.8, where we observe that the SCCC has a much lower error-rate floor, particularly for the FER curves. Finally, we remark that the $w \geq 4$ terms in (7.28) and (7.29) are necessary for an accurate estimate of the floor level of the SCCC case in Figure 7.8.

---

### 7.3.2 The SCCC Iterative Decoder

We present in this section the iterative decoder for an SCCC consisting of two constituent rate-1/2 RSC encoders. We assume no puncturing so that the overall code rate is 1/4. Higher code rates are achievable via puncturing and/or by replacing the inner encoder by a rate-1 differential encoder with transfer function $1/(1 \oplus D)$. It is straightforward to derive the iterative decoding algorithm for other SCCC codes from the special case that we consider here.

A block diagram of the SCCC iterative decoder with component SISO decoders is presented in Figure 7.9. We denote by $\mathbf{c}_1 = [c_1^1, c_2^1, \ldots, c_{2K}^1] = [u_1, p_1, u_2, p_2, \ldots, u_K, p_K]$ the codeword produced by E1 whose input is $\mathbf{u} = [u_1, u_2, \ldots, u_K]$. We denote by $\mathbf{c}_2 = [c_1^2, c_2^2, \ldots, c_{2K}^2] = [v_1, q_1, v_2, q_2, \ldots, v_{2K}, q_{2K}]$ (with $c_k^2 \triangleq [v_k, q_k]$) the codeword produced by E2 whose input $\mathbf{v} = [v_1, v_2, \ldots, v_{2K}]$ is the interleaved version

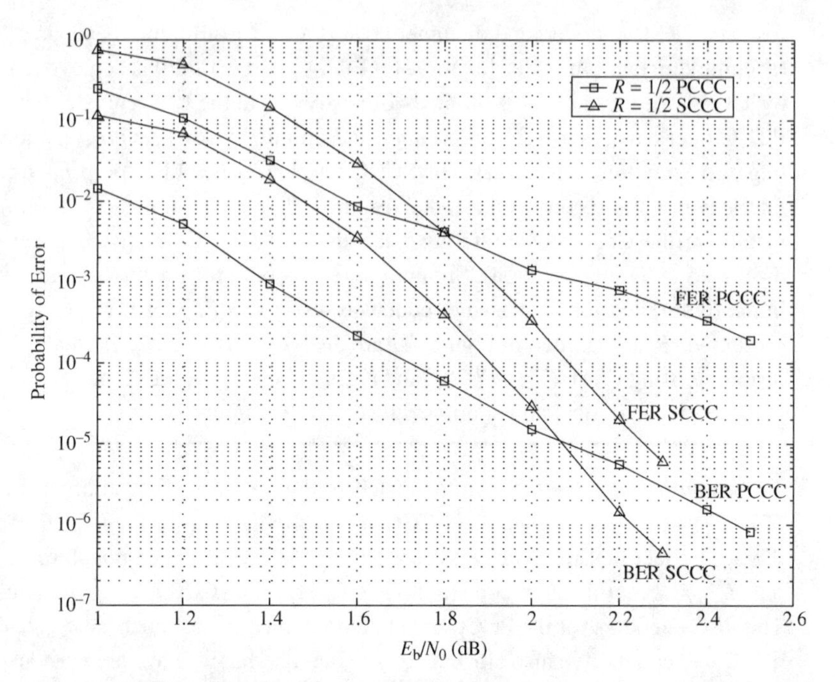


**Figure 7.8** PCCC and SCCC bit-error-rate (BER) and frame-error-rate (FER) simulation results for rate-1/2 (2048, 1024) codes.

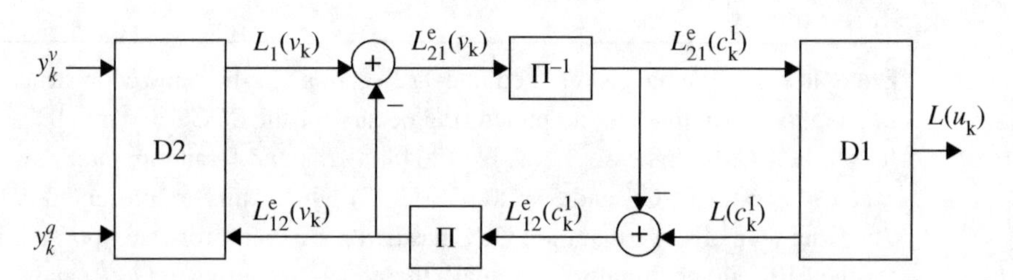


**Figure 7.9** The SCCC iterative decoder.

of $\mathbf{c}_1$, that is, $\mathbf{v} = \mathbf{c}_1'$. As indicated in Figure 7.6, the transmitted codeword $\mathbf{c}$ is the codeword $\mathbf{c}_2$. The received word $\mathbf{y} = \mathbf{c} + \mathbf{n}$ will have the form $\mathbf{y} = [y_1, y_2, \ldots, y_{2K}] = [y_1^v, y_1^q, \ldots, y_{2K}^v, y_{2K}^q]$, where $y_k \triangleq [y_k^v, y_k^q]$.

The iterative SCCC decoder in Figure 7.9 employs two SISO decoding modules. Note that, unlike the PCCC case, which focuses on the systematic bits, these SISO decoders share extrinsic information on the E1 code bits $\{c_k^1\}$ (equivalently, on the E2 input bits $\{v_k\}$) in accordance with the fact that these are the bits known to both encoders. A consequence of this is that D1 must provide likelihood information on E1 *output* bits, whereas D2 produces likelihood information on E2 *input* bits, as indicated in Figure 7.9. Further, because LLRs must be obtained on the original data bits $u_k$ so that final decisions may be made, D1 must also compute likelihood information on E1 input bits. Note also that, because E1 feeds no bits directly to the channel, D1 receives no samples directly from the channel. Instead, the only input to D1 is the extrinsic information it receives from D2.

In light of the foregoing discussion, the SISO module D1 requires two features that we have not discussed when discussing the BCJR algorithm or the PCCC decoder. The first feature is the requirement for likelihood information on a constituent encoder's input *and* output. In prior discussions, such LLR values were sought only for a constituent encoder's inputs (the $u_k$s). Since we assume the SCCC constituent codes are systematic, computing LLRs on an encoder's output bits gives LLRs for both input and output bits.

If we retrace the development of the BCJR algorithm, it will become clear that the LLR for an encoder's output $c_k$ can be computed by modifying the sets in (7.10) over which the max* operations are taken. For a given constituent code, let $C^+$ equal the set of trellis transitions at time $k$ for which $c_k = +1$ and let $C^-$ equal the set of trellis transitions at time $k$ for which $c_k = -1$. Then, with all other steps in the BCJR algorithm the same, the LLR $L(c_k)$ can be computed as

$$\begin{aligned} L(c_k) = &\max_{C^+}{}^* \left[\tilde{\alpha}_{k-1}(s') + \tilde{\gamma}_k(s', s) + \tilde{\beta}_k(s)\right] \\ &- \max_{C^-}{}^* \left[\tilde{\alpha}_{k-1}(s') + \tilde{\gamma}_k(s', s) + \tilde{\beta}_k(s)\right]. \end{aligned} \tag{7.30}$$

We note also that a trellis-based BCJR/SISO decoder is generally capable of decoding either the encoder's input or its output, irrespective of whether the code is systematic. This is evident since the trellis branches are labeled both by inputs and by outputs, and again one need only perform the max* operation over the appropriate sets of trellis transitions.

The second feature required by the SCCC iterative decoder that was not required by the PCCC decoder is that constituent decoder D1 has only extrinsic information as input. In this case the branch metric (7.16) is simply modified as

$$\tilde{\gamma}_k(s', s) = u_k L_{21}^{\mathrm{e}}(u_k)/2 + p_k L_{21}^{\mathrm{e}}(p_k)/2. \tag{7.31}$$

Other than these modifications, the iterative SCCC decoder proceeds much like the PCCC iterative decoder and as indicated in Figure 7.9.

### 7.3.3 Summary of the SCCC Iterative Decoder

Essentially all of the comments mentioned for the PCCC decoder hold also for the SCCC, so we do not repeat them. The only difference is that the decoding order is D2 $\to$ D1 $\to$ D2 $\to$ D1 $\to \cdots$. We first present an outline of the SCCC decoding algorithm, and then we present the algorithm in detail.

**Outline**

1. Initialize all state metrics appropriately and set all extrinsic information to zero.
2. D2 decoder: Run the SISO BJCR algorithm with inputs $y_k = [y_k^v, y_k^q]$ and $L_{12}^{\mathrm{e}}(v_k)$ to obtain $L_{21}^{\mathrm{e}}(v_k)$. Send extrinsic information $L_{21}^{\mathrm{e}}(v_k)$ to the D1 decoder.
3. D1 decoder: Run the SISO BJCR algorithm with inputs $L_{21}^{\mathrm{e}}(v_k)$ to obtain $L_{12}^{\mathrm{e}}(v_k)$. Send extrinsic information $L_{12}^{\mathrm{e}}(v_k)$ to the D2 decoder.

4. Repeat Steps 2 and 3 until the preset maximum number of iterations is reached (or some other stopping criterion is satisfied). Make decisions on bits according to sign$[L(u_k)]$, where $L(u_k)$ is computed by D1.

---

**SCCC Iterative Decoder**

---

**Initialization**

**D1:**

$$\tilde{\alpha}_0^{(1)}(s) = \begin{cases} 0, & \text{for } s = 0 \\ -\infty, & \text{for } s \neq 0 \end{cases}$$

$$\tilde{\beta}_K^{(1)}(s) = \begin{cases} 0, & \text{for } s = 0 \\ -\infty, & \text{for } s \neq 0 \end{cases}$$

$L_{12}^{\mathrm{e}}(u_k)$ is to be determined from D1 after the first half-iteration and so need not be initialized

**D2:**

$$\tilde{\alpha}_0^{(2)}(s) = \begin{cases} 0, & \text{for } s = 0 \\ -\infty, & \text{for } s \neq 0 \end{cases}$$

$$\tilde{\beta}_{2K}^{(2)}(s) = \tilde{\alpha}_{2K}^{(2)}(s) \text{forall} s \left( \text{setaftercomputationof} \left\{ \tilde{\alpha}_{2K}^{(2)}(s) \right\} \text{inthe} \textit{first} \text{iteration} \right)$$

$L_{12}^{\mathrm{e}}(u_k)$ is to be determined from D1 after the first half-iteration and so need not be initialized

$L_{12}^{\mathrm{e}}(v_k) = 0$ for $k = 1, 2, \ldots, 2K$

**The *n*th iteration**

**D2:**

**for** $k = 1$ to $2K$

• get $y_k = [y_k^v, y_k^q]$

• compute $\tilde{\gamma}_k(s', s)$ for all allowable state transitions $s' \to s$ from

$$\tilde{\gamma}_k(s', s) = v_k L_{12}^{\mathrm{e}}(v_k)/2 + v_k y_k^v/\sigma^2 + q_k y_k^q/\sigma^2$$

($v_k$ ($q_k$) in the above expression is set to the value of the encoder input (output) corresponding to the transition $s' \to s$; $L_{12}^{\mathrm{e}}(v_k)$ is $L_{12}^{\mathrm{e}}\left(c_{P[k]}^1\right)$, the interleaved extrinsic information from the previous D1 iteration)

• compute $\tilde{\alpha}_k^{(2)}(s)$ for all $s$ using (7.8)

**end**

**for** $k = 2K$ to 2 step $-1$

• compute $\tilde{\beta}_{k-1}^{(2)}(s)$ for all $s$ using (7.9)

**end**

**for** $k = 1$ to $2K$

• compute $L_{21}^{\mathrm{e}}(v_k)$ using

$$\begin{aligned}
L_{21}^{\mathrm{e}}(v_k) = &\max_{V^+}{}^*\left[\tilde{\alpha}_{k-1}^{(2)}(s') + \tilde{\gamma}_k(s',s) + \tilde{\beta}_k^{(2)}(s)\right] \\
&- \max_{V^-}{}^*\left[\tilde{\alpha}_{k-1}^{(2)}(s') + \tilde{\gamma}_k(s',s) + \tilde{\beta}_k^{(2)}(s)\right] - L_{12}^{\mathrm{e}}(v_k) \\
= &\max_{V^+}{}^*\left[\tilde{\alpha}_{k-1}^{(2)}(s') + v_k y_k^v/\sigma^2 + q_k y_k^q/\sigma^2 + \tilde{\beta}_k^{(2)}(s)\right] \\
&- \max_{V^-}{}^*\left[\tilde{\alpha}_{k-1}^{(2)}(s') + v_k y_k^v/\sigma^2 + q_k y_k^q/\sigma^2 + \tilde{\beta}_k^{(2)}(s)\right],
\end{aligned}$$

where $V^+$ is the set of state-transition pairs $(s', s)$ corresponding to the event $v_k = +1$, and $V^-$ is similarly defined

**end**

**D1:**
**for** $k = 1$ to $K$

• for all allowable state transitions $s' \to s$ set $\tilde{\gamma}_k(s', s)$ via

$$\begin{aligned}
\tilde{\gamma}_k(s',s) &= u_k L_{21}^{\mathrm{e}}(u_k)/2 + p_k L_{21}^{\mathrm{e}}(p_k)/2 \\
&= u_k L_{21}^{\mathrm{e}}(c_{2k-1}^1)/2 + p_k L_{21}^{\mathrm{e}}(c_{2k}^1)/2
\end{aligned}$$

$\big($$u_k$ ($p_k$) in the above expression is set to the value of the encoder input (output) corresponding to the transition $s' \to s$; $L_{21}^{\mathrm{e}}(c_{2k-1}^1)$ is $L_{21}^{\mathrm{e}}(v_{Pinv[2k-1]})$, the de-interleaved extrinsic information from the previous D2 iteration, and similarly for $L_{21}^{\mathrm{e}}(c_{2k}^1)$$\big)$
• compute $\tilde{\alpha}_k^{(1)}(s)$ for all $s$ using (7.8)

**end**

**for** $k = K$ to 2 step $-1$

• compute $\tilde{\beta}_{k-1}^{(1)}(s)$ for all $s$ using (7.9)

**end**

**for** $k = 1$ to $K$

• compute $L_{12}^{\mathrm{e}}(u_k) = L_{12}^{\mathrm{e}}(c_{2k-1}^1)$ using

$$\begin{aligned}
L_{12}^{\mathrm{e}}(u_k) = &\max_{U^+}{}^*\left[\tilde{\alpha}_{k-1}^{(1)}(s') + \tilde{\gamma}_k(s',s) + \tilde{\beta}_k^{(1)}(s)\right] \\
&- \max_{U^-}{}^*\left[\tilde{\alpha}_{k-1}^{(1)}(s') + \tilde{\gamma}_k(s',s) + \tilde{\beta}_k^{(1)}(s)\right] - L_{21}^{\mathrm{e}}(c_{2k-1}^1) \\
= &\max_{U^+}{}^*\left[\tilde{\alpha}_{k-1}^{(1)}(s') + p_k L_{21}^{\mathrm{e}}(p_k)/2 + \tilde{\beta}_k^{(1)}(s)\right] \\
&- \max_{U^-}{}^*\left[\tilde{\alpha}_{k-1}^{(1)}(s') + p_k L_{21}^{\mathrm{e}}(p_k)/2 + \tilde{\beta}_k^{(1)}(s)\right]
\end{aligned}$$

- compute $L_{12}^{e}(p_k) = L_{12}^{e}(c_{2k}^{1})$ using

$$
\begin{aligned}
L_{12}^{e}(p_k) &= \max_{P^+}{}^{*}\left[\tilde{\alpha}_{k-1}^{(1)}(s') + \tilde{\gamma}_k(s',s) + \tilde{\beta}_k^{(1)}(s)\right] \\
&\quad - \max_{P^-}{}^{*}\left[\tilde{\alpha}_{k-1}^{(1)}(s') + \tilde{\gamma}_k(s',s) + \tilde{\beta}_k^{(1)}(s)\right] - L_{21}^{e}(c_{2k}^{1}) \\
&= \max_{P^+}{}^{*}\left[\tilde{\alpha}_{k-1}^{(1)}(s') + u_k L_{21}^{e}(u_k)/2 + \tilde{\beta}_k^{(1)}(s)\right] \\
&\quad - \max_{P^-}{}^{*}\left[\tilde{\alpha}_{k-1}^{(1)}(s') + u_k L_{21}^{e}(u_k)/2 + \tilde{\beta}_k^{(1)}(s)\right]
\end{aligned}
$$

**end**

**After the last iteration**

**for** $k = 1$ to $K$

- for all allowable state transitions $s' \to s$ set $\tilde{\gamma}_k(s',s)$ via

$$\tilde{\gamma}_k(s',s) = u_k L_{21}^{e}(c_{2k-1}^{1})/2 + p_k L_{21}^{e}(c_{2k}^{1})/2$$

- compute $L(u_k)$ using

$$
\begin{aligned}
L(u_k) &= \max_{U^+}{}^{*}\left[\tilde{\alpha}_{k-1}^{(1)}(s') + \tilde{\gamma}_k(s',s) + \tilde{\beta}_k^{(1)}(s)\right] \\
&\quad - \max_{U^-}{}^{*}\left[\tilde{\alpha}_{k-1}^{(1)}(s') + \tilde{\gamma}_k(s',s) + \tilde{\beta}_k^{(1)}(s)\right]
\end{aligned}
$$

- $\hat{u}_k = \text{sign}[L(u_k)]$

**end**

## 7.4 Turbo Product Codes

A *turbo product code* (TPC), also called *block turbo code* (BTC) [15], is best explained via the diagram of Figure 7.10, which depicts a two-dimensional codeword. The codeword is an element of a product code (see Chapter 3), where the first constituent code has parameters $(n_1, k_1)$ and the second constituent code has parameters $(n_2, k_2)$. The $k_1 \times k_2$ submatrix in Figure 7.10 contains the $k_1 k_2$-bit data word. The columns of this submatrix are encoded by the "column code," after which the rows of the resulting $n_1 \times k_2$ matrix are encoded by the "row code." Alternatively, row encoding may occur first, followed by column encoding. Because the codes are linear, the resulting two-dimensional codeword is independent of the encoding order. In particular, the "checks-on-checks" submatrix will be unchanged. This is shown in Problem 7.15.

The aggregate code rate for this product code is $R = R_1 R_2 = (k_1 k_2)/(n_1 n_2)$, where $R_1$ and $R_2$ are the code rates of the individual codes. The minimum distance of the product code is $d_{\min} = d_{\min,1} d_{\min,2}$, where $d_{\min,1}$ and $d_{\min,2}$ are the minimum distances of the individual codes. (See Problem 7.16.) The constituent codes are typically

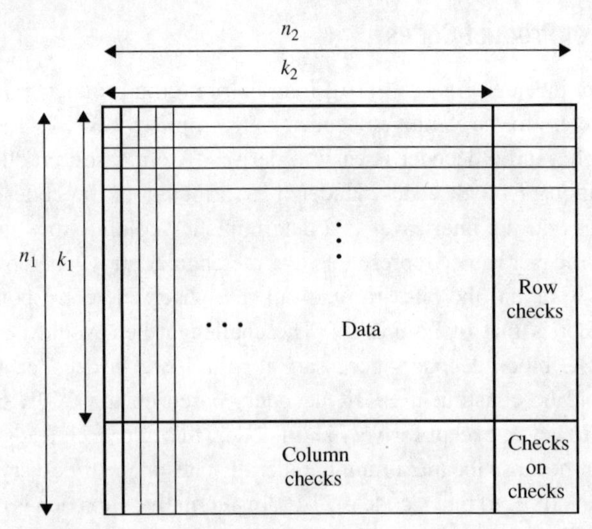


**Figure 7.10** The product code of two constituent block codes with parameters $(n_1, k_1)$ and $(n_2, k_2)$.

extended BCH codes (including extended Hamming codes). The extended BCH codes are particularly advantageous because the extension beyond the nominal BCH code increases the (design) minimum distance of each constituent code by one, at the expense of only one extra parity bit per constituent codeword, while achieving an increase in aggregate minimum distance by $d_{\min,1} + d_{\min,2} + 1$.

---

**Example 7.6** Let the constituent block codes be two $(15, 11)$ Hamming codes, for which $d_{\min,1} = d_{\min,2} = 3$. The minimum distance of the product code is then $d_{\min} = 9$. For comparison, let the constituent block codes be two $(16, 11)$ extended Hamming codes, for which $d_{\min,1} = d_{\min,2} = 4$. The minimum distance of the resulting TPC is then $d_{\min} = 16$. In the first case the product code is a rate-0.538 $(225, 121)$ code and in the second case the product code is a rate-0.473 $(256, 121)$ code.

---

We remark that the $P_b$ and $P_{cw}$ expressions for ML decoding performance of a TPC on the AWGN channel are no different from those of any other code:

$$P_b \sim \frac{w_{\min}}{k_1 k_2} Q\left(\sqrt{\frac{2Rd_{\min}E_b}{N_0}}\right),$$

$$P_{cw} \sim A_{\min} Q\left(\sqrt{\frac{2Rd_{\min}E_b}{N_0}}\right),$$

where $w_{\min}$ is the total information weight corresponding to all of the $A_{\min}$ TPC codewords at the minimum distance $d_{\min}$. We note that, unlike for PCCCs and SCCCs, $w_{\min}$ and $A_{\min}$ are quite large for TPCs. That is, there exists no spectral thinning because the interleaver is deterministic and the constituent codes are not recursive.

### 7.4.1 Turbo Decoding of Product Codes

Product codes were invented (in 1954 [16]) long before turbo product codes, but the qualifier "turbo" refers to the iterative decoder that comprises two SISO constituent decoders [15]. Such a turbo decoder is easy to derive if we recast a product code as a serial concatenation of block codes. Under this formulation, the codes in Figure 7.6 are block codes and the interleaver is a deterministic "column–row" interleaver. A column–row interleaver can be represented as a rectangular array whereby the array is written column-wise and the bits are read out row-wise. The corresponding iterative (turbo) decoder is that of Figure 7.9. The challenge then would be to design the constituent SISO block decoders necessary for the iterative decoder. One obvious approach would be constituent BCJR decoders based on the BCJR trellises of each block code. However, except for very short codes, this approach leads to a high-complexity decoder because the maximum number of states in the time-varying BCJR trellis (Chapter 5) of an $(n, k)$ block code is $2^{n-k}$. An alternative approach is to use constituent soft-output Viterbi decoders instead of BCJR decoders. Yet another involves a SISO Chase decoder [15], which has substantially lower implementation complexity, at the expense of some performance loss.

We shall now describe the turbo-product-code decoder based on the SISO Chase decoder. As before, we assume knowledge of the turbo principle and turbo decoding, as in Figure 7.9, so our main focus need only be on the constituent soft-output Chase decoders. We will put in details of the turbo product decoder later. Each *SISO Chase decoder* performs the following steps, which we will discuss in the following.

1. Produce a list of candidate codewords.
2. From that list, make a decision on which codeword was transmitted.
3. For each code bit in the selected codeword, compute soft outputs and identify extrinsic information.

#### Obtaining the List of Candidate Codewords

Let $\mathbf{y} = \mathbf{c} + \mathbf{n}$ be the received word, where the components of the transmitted codeword $\mathbf{c}$ take values in the set $\{\pm 1\}$ and $\mathbf{n}$ is a noise word whose components are AWGN samples. Next, make hard decisions $h_k$ on the elements $y_k$ of $\mathbf{y}$ according to $h_k = (1 - \text{sign}(y_k))/2$. Next, we rely on the fact that, at least for high-rate codes for which the SNR is relatively high, the transmitted word $\mathbf{c}$ is likely to be within a radius $\delta - 1$ of the hard-decision word $\mathbf{h}$, where $\delta$ is the constituent code's minimum distance. To find such a list of candidate words, we require the reliabilities of each sample $y_k$. We assign reliability $r_k$ to sample $y_k$ according to

$$r_k = \left| \ln\left( \frac{\Pr(y_k | c_k = +1)}{\Pr(y_k | c_k = -1)} \right) \right|,$$

which is equal to $2|y_k|/\sigma^2$ for the binary-input AWGN channel. The list $\mathcal{L}$ of candidate codewords is then constructed via these first steps of the Chase algorithm.

1. Determine the positions of the $p =$ least reliable elements of $\mathbf{y}$ (and hence of the decisions in $\mathbf{h}$). Typically, $p < \delta$, but we have found $p = \delta$ or greater can improve performance at the expense of complexity.
2. Form $2^p$ test words obtained from $\mathbf{h}$ by placing all $2^p$ $p$-bit patterns in the $p$ least reliable positions. (Alternatively, the $p$-bit patterns can be XOR'd with the bits residing in those positions in $\mathbf{h}$.)
3. Decode the test words using an algebraic (or syndrome) decoder (which have correction capability $\lfloor(\delta - 1)/2\rfloor$). The candidate list $\mathcal{L}$ is the set of codewords at the decoder output. Decoder outputs for failed decodings are omitted.

### The Codeword Decision

For an arbitrary code $\mathcal{C}$, the optimum (ML) decision on the binary-input AWGN channel is given by

$$\hat{\mathbf{c}}_{\mathrm{ML}} = \arg\min_{\mathbf{c}\in\mathcal{C}} \|\mathbf{y} - \mathbf{c}\|^2.$$

Unless the code is described by a relatively low-complexity trellis, performing this operation requires unacceptable complexity. The utility of the Chase decoder is that the minimization is applied over a smaller set of codewords, those in $\mathcal{L}$:

$$\hat{\mathbf{c}} = \arg\min_{\mathbf{c}\in\mathcal{L}} \|\mathbf{y} - \mathbf{c}\|^2.$$

This is the final step in the conventional Chase algorithm. We now discuss how the algorithm can be extended to produce soft outputs and extrinsic information.

### Computing Soft Outputs and Extrinsic Information

Given the codeword decision $\hat{\mathbf{c}}$, we now must obtain reliability values for each of the bits in $\hat{\mathbf{c}}$. To do this, we start with the standard log-APP ratio involving the transmitted bits $c_k$,

$$\ln\left(\frac{\Pr(c_k = +1|\mathbf{y})}{\Pr(c_k = -1|\mathbf{y})}\right),$$

and then we add conditioning on the neighborhood of $\hat{\mathbf{c}}$, to obtain the log-APP ratio:

$$L(c_k) = \ln\left(\frac{\Pr(c_k = +1|\mathbf{y}, \mathcal{L})}{\Pr(c_k = -1|\mathbf{y}, \mathcal{L})}\right). \tag{7.32}$$

Now let $\mathcal{L}_k^+$ represent the set of codewords in $\mathcal{L}$ for which $c_k = +1$ and let $\mathcal{L}_k^-$ represent the set of codewords in $\mathcal{L}$ for which $c_k = -1$ (recalling the correspondence $0 \leftrightarrow +1$

and $1 \leftrightarrow -1$). Then, after applying Bayes' rule to (7.32) (under a uniform-codeword-distribution assumption), we may write

$$L(c_k) = \ln\left(\frac{\sum_{\mathbf{c}\in\mathcal{L}_k^+} p(\mathbf{y}|\mathbf{c})}{\sum_{\mathbf{c}\in\mathcal{L}_k^-} p(\mathbf{y}|\mathbf{c})}\right), \tag{7.33}$$

where

$$p(\mathbf{y}|\mathbf{c}) = \left(\frac{1}{\sqrt{2\pi}\sigma}\right)^n \exp\left(-\frac{\|\mathbf{y}-\mathbf{c}\|^2}{2\sigma^2}\right) \tag{7.34}$$

and $\sigma^2$ is the variance of each AWGN sample. Now, ignoring the scale factor in (7.34), (7.33) contains two logarithms of sums of exponentials, a form we have found convenient to express using the max* function. Thus, (7.33) becomes

$$L(c_k) = \max_{\mathbf{c}\in\mathcal{L}_k^+} * \left(-\|\mathbf{y}-\mathbf{c}\|^2/(2\sigma^2)\right) - \max_{\mathbf{c}\in\mathcal{L}_k^-} * \left(-\|\mathbf{y}-\mathbf{c}\|^2/(2\sigma^2)\right). \tag{7.35}$$

The preceding expression may be used to compute the reliabilities $|L(c_k)|$, but a simplified, approximate expression is possible that typically leads to a small performance loss. This is done by replacing the max* functions in (7.35) by max functions so that

$$\begin{aligned}\tilde{L}(c_k) &= \frac{1}{2\sigma^2}\left(\left\|\mathbf{y}-\mathbf{c}_k^-\right\|^2 - \left\|\mathbf{y}-\mathbf{c}_k^+\right\|^2\right) \\ &= \frac{1}{\sigma^2}\mathbf{y}\cdot\left(\mathbf{c}_k^+ - \mathbf{c}_k^-\right),\end{aligned} \tag{7.36}$$

where $\mathbf{c}_k^+$ is the codeword in $\mathcal{L}_k^+$ that is closest to $\mathbf{y}$ and $\mathbf{c}_k^-$ is the codeword in $\mathcal{L}_k^-$ that is closest to $\mathbf{y}$. Clearly, the decision $\hat{\mathbf{c}}$ must either be $\mathbf{c}_k^+$ or $\mathbf{c}_k^-$, so we must find its counterpart, which we will denote by $\mathbf{c}'$. Given this, we may write (7.36) as

$$\tilde{L}(c_k) = \frac{1}{\sigma^2}\mathbf{y}\cdot(\hat{\mathbf{c}} - \mathbf{c}')\,\hat{c}_k \tag{7.37}$$

because $\hat{\mathbf{c}} = \mathbf{c}_k^+$ when $\hat{c}_k = +1$ and $\hat{\mathbf{c}} = \mathbf{c}_k^-$ when $\hat{c}_k = -1$ (compare (7.37) with (7.36)). By combining (7.37) with (7.36), the output LLR for the $k$th bit can be written as

$$\tilde{L}(c_k) = \frac{1}{2\sigma^2}\left|\|\mathbf{y}-\hat{\mathbf{c}}\|^2 - \|\mathbf{y}-\mathbf{c}'\|^2\right|\hat{c}_k.$$

At this point, we need to show how extrinsic information is computed. Recall from our earlier discussions in this chapter, extrinsic information can be computed by subtracting the bit LLR at the decoder's input from the (enhanced) bit LLR at the decoder's output. The input LLR for the $k$th bit is $2y_k/\sigma^2$. The extrinsic information is given then by the difference

$$L_e(c_k) = \tilde{L}(c_k) - 2y_k/\sigma^2 = \frac{1}{2\sigma^2}\left|\|\mathbf{y}-\hat{\mathbf{c}}\|^2 - \|\mathbf{y}-\mathbf{c}'\|^2\right|\hat{c}_k - 2y_k/\sigma^2.$$

It can be shown that, because we used the max function instead of the max* function, the noise variance $\sigma^2$ can be removed from this expression. Dividing the expression by $2/\sigma^2$, we obtained a simpler extrinsic information expression,

$$L_e(c_k) = \frac{1}{4}\left|\|\mathbf{y}-\hat{\mathbf{c}}\|^2 - \|\mathbf{y}-\mathbf{c}'\|^2\right|\hat{c}_k - y_k. \tag{7.38}$$

It is important to point out that $\mathbf{c}'$, the counterpart to $\hat{\mathbf{c}}$, might not exist, in which case, $\tilde{L}(c_k)$ does not exist so that $L_e(c_k)$ cannot be computed. Recall that $\hat{\mathbf{c}} \in \{\mathbf{c}_k^+, \mathbf{c}_k^-\}$ and $\mathbf{c}' = \{\mathbf{c}_k^+, \mathbf{c}_k^-\}\backslash\{\hat{\mathbf{c}}\}$ only if $\{\mathbf{c}_k^+, \mathbf{c}_k^-\}\backslash\{\hat{\mathbf{c}}\}$ exists in $\mathcal{L}$. If $\{\mathbf{c}_k^+, \mathbf{c}_k^-\}\backslash\{\hat{\mathbf{c}}\} \notin \mathcal{L}$, then one might increase the Chase algorithm parameter $p$ to increase the size of $\mathcal{L}$, but this would be at the expense of increased complexity. A very effective low-complexity alternative is to let the extrinsic information be

$$L_e(c_k) = \hat{c}_k \beta \tag{7.39}$$

for some constant $\beta \in (0, 1)$ that can be optimized experimentally. In the context of iterative decoding, $\beta$ can change with each iteration and, in fact, slightly better performance is usually achieved in this case.

## The Turbo Decoder

Given the above details on the SISO constituent Chase algorithm block decoders, we are now ready to present the turbo-product-code decoder. The decoder is depicted in Figure 7.11, which we note is very similar to the PCCC and SCCC decoders in Figures 7.3 and 7.9. With all extrinsic information initialized to zero, the row Chase decoder inputs at each iteration are $\{y_k + \alpha L_{cr,k}^{\mathrm{e}}\}$, that is, the sum of the input from the channel and the extrinsic information $L_{cr,k}^{\mathrm{e}}$ from the column decoder about bit $c_k$. Similarly, the column Chase decoder has inputs $\{y_k + \alpha L_{rc,k}^{\mathrm{e}}\}$ at each iteration, where $L_{rc,k}^{\mathrm{e}}$ is extrinsic information from the row decoder about $c_k$. The extrinsic information in each case is given by (7.38) or (7.39). The parameter $\alpha \in (0, 1)$ is a scale factor that advantageously changes with each decoding iteration.

1. Initialize $L_{cr,k}^{\mathrm{e}} = L_{rc,k}^{\mathrm{e}} = 0$ for all $k$.
2. *Row decoder*. Run the SISO Chase algorithm with inputs $y_k + \alpha L_{cr,k}^{\mathrm{e}}$ to obtain $\{\hat{c}_k\}$ and $\{L_{rc,k}^{\mathrm{e}}\}$. Send extrinsic information $\{L_{rc,k}^{\mathrm{e}}\}$ to the column decoder.
3. *Column decoder*. Run the SISO Chase algorithm with inputs $y_k + \alpha L_{rc,k}^{\mathrm{e}}$ to obtain $\{\hat{c}_k\}$ and $\{L_{cr,k}^{\mathrm{e}}\}$. Send extrinsic information $\{L_{cr,k}^{\mathrm{e}}\}$ to the row decoder.
4. Repeat Steps 2 and 3 until the preset maximum number of iterations is reached or some other stopping criterion is satisfied. A natural stopping criterion is that, for all rows and all columns, the hard-decision words $\mathbf{h}$ at the constituent decoder inputs all give zero syndromes.

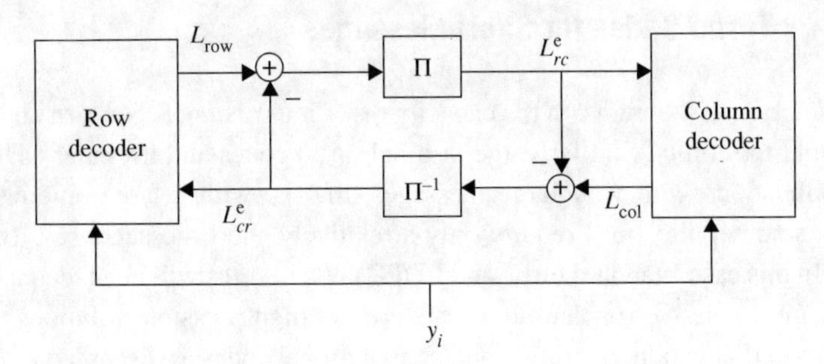


**Figure 7.11** The TPC iterative decoder.

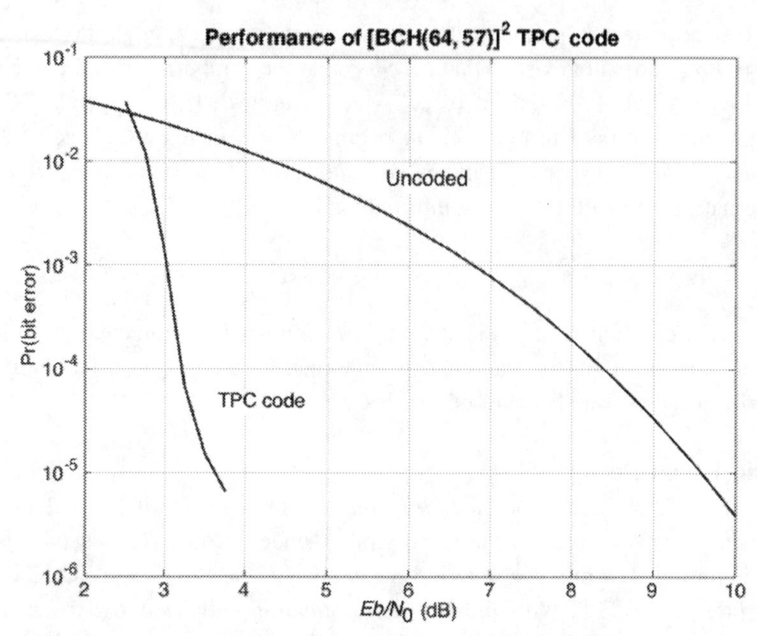


**Figure 7.12** Simulation results for the $[\mathrm{BCH}(64, 57)]^2$ TPC.

---

**Example 7.7** We consider as an example a product code with extended (64, 57) BCH codewords (for which $\delta = 4$) along rows and along columns, sometimes denoted by $[\mathrm{BCH}(64, 57)]^2$ (see Figure 7.12). The TPC iterative decoder was simulated on the binary-input AWGN channel with $p = 6$ and the scale factors *fixed* at $\alpha = 0.4$ and $\beta = 0.9$. The maximum number of decoder iterations was 40. The capacity limit for a code of this rate ($R = 0.7932$) is $E_b/N_0 = 2$ dB. So, while this code provides excellent bit-error-rate performance for such a high-rate code, there is considerable distance to the capacity limit due to the fact that the code is highly structured and the code length is not very large. Pyndiah [15] was able to get 0.25 dB better performance with only four iterations and scale factors $\alpha$ and $\beta$ that varied with each iteration.

---

## 7.5 Nonbinary Turbo Codes for Short Messages

Shannon's original theorems on the capacity of various channels concern codes whose lengths tend to infinity. Similarly, the original turbo codes and irregular LDPC codes target applications with very large messages, that is, with a large amount of data. However, some applications require only a relatively short message, say, 100 bits or 500 bits. In this case, standard turbo and LDPC code designs will be inadequate as they will typically possess error-rate floors that are too high. Possible solutions are CRC-aided list-Viterbi decoding of tail-biting convolutional codes as discussed in Chapter 4 or CRC-aided successive cancellation list decoding of polar codes as discussed in

Chapter 10. Other alternatives can be found in [31]. In this section we will provide a brief coverage of the non-binary turbo codes (over a finite field GF($q$) = $\mathbb{F}_q$) introduced in [32] that perform very close to the short-code theoretical limits discussed in Chapter 1.

Two classes of turbo codes were presented in [32]. The first is a parallel concatenation of two *memory-one, time-variant* recursive convolutional codes over $\mathbb{F}_q$. The second is a serial concatenation of a non-recursive convolutional code with a recursive convolutional code, both over $\mathbb{F}_q$. The component codes in each code type are separated by interleavers. As we will see, each code type can be represented as an LDPC code over $\mathbb{F}_q$ and can be decoded that way (see Section 15.2 for information on decoding an LDPC code over $\mathbb{F}_q$).

The encoder for the short *parallel turbo code over* $\mathbb{F}_q$ is as in Figure 7.1, except for the following differences: the input is a block of $K$ symbols from $\mathbb{F}_q$, not bits; the interleaver is a symbol interleaver, not a bit interleaver; the ratio $g^{(2)}(D)/g^{(1)}(D)$ for the top encoder in Figure 7.1 is given by

$$\frac{g_i^{(t)}}{1+f_i^{(t)}D}$$

and for the bottom encoder it is

$$\frac{g_i^{(b)}}{1+f_i^{(b)}D},$$

where "$t$" signifies "top" and "$b$" signifies "bottom." In these expressions, the parameters $g_i$ and $f_i$ are nonzero elements of $\mathbb{F}_q$ that change with the discrete-time index $i$, $i = 0, 1, \ldots, K-1$, making the component encoders time-variant. Observe that when these parameters are equal to 1, the rate-1/2 component codes have the generator

$$\left[1 \quad \frac{1}{1+D}\right], \tag{7.40}$$

so that the parity symbols are produced by the *accumulator* $\frac{1}{1+D}$. Thus, for the nonbinary parallel turbo code, the parity symbols are produced by *generalized accumulators*. We emphasize also that the design requires tail-biting accumulators so that the initial and final states are identical for each accumulator.

As shown in Problem 7.18, the parity-check matrix for a code with the generator in (7.40) *with tail-biting* is given by [**I P**], where **I** is the $K \times K$ identity matrix and

$$\mathbf{P} = \begin{bmatrix} 1 & 0 & 0 & \cdots & 0 & 0 & 1 \\ 1 & 1 & 0 & \cdots & 0 & 0 & 0 \\ 0 & 1 & 1 & \cdots & 0 & 0 & 0 \\ \vdots & \vdots & \vdots & \ddots & \vdots & \vdots & \vdots \\ 0 & 0 & 0 & \cdots & 1 & 1 & 0 \\ 0 & 0 & 0 & \cdots & 0 & 1 & 1 \end{bmatrix}, \tag{7.41}$$

a $K \times K$ circulant matrix. The parity-check matrix for the parallel turbo code whose component codes are given by (7.40) is then

$$\mathbf{H} = \begin{bmatrix} \mathbf{I} & \mathbf{P} & \mathbf{0} \\ \mathbf{\Pi} & \mathbf{0} & \mathbf{P} \end{bmatrix}, \tag{7.42}$$

where $\mathbf{0}$ is a $K \times K$ all-zeros matrix and $\mathbf{\Pi}$ is a $K \times K$ binary permutation matrix whose $K$ ones are in positions $(i, \pi(i))$, where $\pi(i)$ is a "random" bijective map on the set $\{0, 1, \ldots, K-1\}$.

For our $q$-ary turbo code that employs time-variant, $q$-ary accumulators, the parity-check matrix generalizes to

$$\mathbf{H}_q = \begin{bmatrix} \mathbf{I}_q & \mathbf{P}_q^{(t)} & \mathbf{0} \\ \mathbf{\Pi}_q & \mathbf{0} & \mathbf{P}_q^{(b)} \end{bmatrix}, \tag{7.43}$$

where

- $\mathbf{I}_q$ is a $K \times K$ diagonal matrix with nonzero elements $g_i^{(t)} \in \mathbb{F}_q$ down the diagonal, $i = 0, 1, \ldots, K-1$.
- $\mathbf{\Pi}_q = [\Pi_{i,j}]$ is a $K \times K$ generalized permutation matrix with nonzero entries $\Pi_{i,j} = \Pi_{i,\pi(i)} = g_i^{(b)} \in \mathbb{F}_q \backslash 0$, for $i = 0, 1, \ldots, K-1$. Observe that $\mathbf{\Pi}_q = \text{diag}\left(g_0^{(b)}, g_1^{(b)}, \ldots, g_{K-1}^{(b)}\right)$.
- $\mathbf{P}_q^{(t)}$ and $\mathbf{P}_q^{(b)}$ are both of the form

$$\mathbf{P}_q^{(z)} = \begin{bmatrix} 1 & 0 & 0 & \cdots & 0 & 0 & f_0^{(z)} \\ f_1^{(z)} & 1 & 0 & \cdots & 0 & 0 & 0 \\ 0 & f_2^{(z)} & 1 & \cdots & 0 & 0 & 0 \\ \vdots & \vdots & \vdots & \ddots & \vdots & \vdots & \vdots \\ 0 & 0 & 0 & \cdots & & 1 & 0 \\ 0 & 0 & 0 & \cdots & 0 & f_{K-1}^{(z)} & 1 \end{bmatrix}, \tag{7.44}$$

where $z \in \{t, b\}$ and $f_i^{(z)} \in \mathbb{F}_q \backslash 0$, for $i = 0, 1, \ldots, K-1$.

The four nonzero submatrices that make up $\mathbf{H}_q$ completely specify the nonbinary parallel turbo code. We refer the reader to [32] for details on the design of these submatrices. A sum–product algorithm decoder based on $\mathbf{H}_q$ can be used to decode this code as for any nonbinary LDPC code (see Section 15.2).

The *serial turbo code over* $\mathbb{F}_q$ of [32] consists of two memory-one encoders in series, but separated by a symbol interleaver. The outer encoder has rate 1/2 and has the generator

$$\left[1 \quad 1 + f_i^{(O)} D\right],$$

where "$O$" signifies "outer" and $f_i^{(O)} \in \mathbb{F}_q \backslash 0$, for $i = 0, 1, \ldots, K-1$. Only the $K$ parity symbols produced by the *generalized differentiator* $1 + f_i^{(O)} D$ enter the symbol interleaver represented by the matrix $(\mathbf{\Pi}_q^T)^{-1}$, following the convention of [32]. The $K$ outputs of the interleaver enter the inner encoder that has rate 1 and the generator

$$\frac{g_i^{(I)}}{1 + f_i^{(I)} D},$$

where "$I$" signifies "inner" and $g_i^{(I)}, f_i^{(I)} \in \mathbb{F}_q \backslash 0$, for $i = 0, 1, \ldots, K-1$.

Two overall code rates are possible for the nonbinary serial turbo code. For rate 1/2, the codeword has the form $\mathbf{c} = [\,\mathbf{u}\;\mathbf{p}\,]$, where $\mathbf{u}$ is the $K$-symbol data word at the input of the inner encoder and $\mathbf{p}$ is the $K$-symbol parity word at the output of the outer encoder. It can be shown [32] (and Problem 7.20) that the parity-check matrix in this case is

$$\mathbf{H}_q = \left[\, \mathbf{\Pi}_q^{-1} \mathbf{P}_q^{(O)} \quad \mathbf{I}_q^{-1} \mathbf{P}_q^{(I)} \right], \tag{7.45}$$

where $\mathbf{P}_q^{(O)}$ is analogous to $\mathbf{P}_q^{(z)}$ in (7.44) with $z$ replaced by $O$ and similarly for $\mathbf{P}_q^{(I)}$.

For the rate-1/3 case of the serial turbo code, in addition to the subvectors $\mathbf{u}$ and $\mathbf{p}$ being transmitted, the output of the symbol interleaver, $\mathbf{v}$, is transmitted. The codeword in this case has the form $\mathbf{c} = [\,\mathbf{v}\;\mathbf{p}\;\mathbf{u}\,]$ and the corresponding parity-check matrix is the one given in (7.43) with "$t$" replaced by "$I$" and "$b$" replaced by "$O$," as verified in Problem 7.19.

## Problems

**7.1** Consider an RSC encoder with octal generators $(7, 5)$. Find the output of the weight-1 input sequence 1000... and note that it is of infinite length. (The length is measured by the location of the last "1," or the degree-plus-1 of the polynomial representation of the sequence.) Find the output of the weight-2 input sequence 1001000... and note that it is of finite length. Why does the weight-2 sequence 10001000... produce an infinite-length sequence?

**7.2** The chapter describes the PCCC decoder for the BI-AWGNC. How would the decoder be modified for the binary erasure channel? How would it be modified for the binary symmetric channel? Provide all of the details necessary for one to be able to write decoder-simulation programs for these channels.

**7.3** Consider a rate-1/3 PCCC whose constituent encoders are both RSC codes with octal generators $(7, 5)$. Let the interleaver be specified by the linear congruential relationship $P[k] = (13P[k-1] + 7) \bmod 32$ for $k = 1, 2, \ldots, 31$ and $P[0] = 11$. Thus, $P[0], P[1], \ldots, P[31]$ is $11, 22, \ldots, 20$. This means that the first bit out of the interleaver will be the 12th bit in, and the last bit out of the interleaver will be the 21st bit in. Find the codeword corresponding to the input sequences 1001000...0 (32 bits) and 01001000...0 (32 bits). Notice that, unlike its constituent codes, the PCCC is not time-invariant. That is, the codeword for the second input sequence is not a shift of the codeword for the first input sequence.

**7.4** Consider the rate-1/3 PCCC of the previous problem, except now increase the interleaver length to 128 so that $P[k] = (13P[k-1] + 7) \bmod 128$ for $k = 1, 2, \ldots, 127$ and $P[0] = 11$. By computer search, find $d_{2,\min}^{\mathrm{PCCC}}$, the minimum codeword weight corresponding to weight-2 PCCC encoder inputs, and $n_2$, the number of such codewords.

Repeat for $d^{\text{PCCC}}_{3,\min}$, the minimum codeword weight corresponding to weight-3 PCCC encoder inputs, and $n_3$, the number of such codewords. Which is dominant, the codewords due to weight-2 inputs or those due to weight-3 inputs? Can you improve any of these four parameters ($d^{PCCC}_{2,\min}$, $n_2$, $d^{\text{PCCC}}_{3,\min}$, and $n_3$) with an improved interleaver design? Consider an $S$-random interleaver.

**7.5** Do the previous problem, except use constituent RSC encoders with octal generators $(5, 7)$. Note that, for the $(7, 5)$ RSC code, the generator polynomial $g^{(1)}(D) = 1 + D + D^2$ is primitive, whereas for the $(5, 7)$ RSC code $g^{(1)}(D) = 1 + D^2$ is not primitive. This affects $d^{\text{PCCC}}_{2,\min}$. See the discussion in Chapter 8 regarding these two PCCCs.

**7.6** Consider a *dicode* partial-response channel with transfer function $1 - D$ and AWGN. Thus, for inputs $u_k \in \{\pm 1\}$, the outputs of this channel are given by $r_k = c_k + n_k$, where $c_k = u_k - u_{k-1} \in \{0, \pm 2\}$ and the noise samples $n_k$ are distributed as $\eta\left(0, \sigma^2\right)$. (a) Draw the two-state trellis for this channel. (b) Suppose the channel is known to start and end in channel state $+1$. Perform the BCJR algorithm by hand to detect the received sequence $(r_1, r_2, r_3, r_4) = (0.5, -1.3, 1.5, 0.2)$. You should find that the detected bits are $(\hat{u}_1, \hat{u}_2, \hat{u}_3, \hat{u}_4) = (+1, -1, +1, -1)$.

**7.7** Simulate the rate-1/3 PCCC with RSC octal generators $(7, 5)$ on the binary-input AWGN channel. Use the length-128 interleaver given by $P[k] = (13P[k-1] + 7) \bmod 128$ for $k = 1, 2, \ldots, 127$ and $P[0] = 11$ (the first bit out is the 11th bit in). You should find that, for $E_b/N_0 = 3, 4$, and 5 dB, the bit-error rate $P_b$ is in the vicinity of $3 \times 10^{-5}$, $4 \times 10^{-6}$, and $5 \times 10^{-7}$, respectively. Repeat using an $S$-random interleaver in an attempt to lower the "floor" of the $P_b$ versus $E_b/N_0$ curve.

**7.8** Do the previous problem using the max function instead of the max* function in the constituent BCJR decoders.

**7.9** This chapter describes the SCCC decoder for the BI-AWGNC. How would the decoder be modified for the binary erasure channel? How would it be modified for the binary symmetric channel? Provide all of the details necessary for one to be able to write decoder-simulation programs for these channels.

**7.10** Consider a rate-1/2 SCCC whose outer code is the rate-1/2 $\bar{\text{R}}\bar{\text{S}}\text{C}$ code with octal generators $(7, 5)$ and whose inner code is the rate-1 accumulator with transfer function $1/(1 \oplus D)$. Let the interleaver be specified by the linear congruential relationship $P[k] = (13P[k-1] + 7) \bmod 32$ for $k = 1, 2, \ldots, 31$ and $P[0] = 11$. Thus, $P[0], P[1], \ldots, P[31]$ is $11, 22, \ldots, 20$. This means that the first bit out of the interleaver will be the 11th bit in, and the last bit out of the interleaver will be the 20th bit in. Find the codeword corresponding to the input sequences $100\ldots0$ (32 bits) and $0100\ldots0$ (32 bits). Notice that, unlike its constituent codes, the SCCC is not time-invariant. That is, the codeword for the second input sequence is not a shift of the codeword for the first input sequence.

**7.11** Consider the rate-1/2 SCCC of the previous problem, except now increase the interleaver length to 128 so that $P[k] = (13P[k-1] + 7) \bmod 128$ for $k = 1, 2, \ldots, 127$ and $P[0] = 11$. By computer search, find $d^{\text{SCCC}}_{1,\min}$, the minimum codeword weight corresponding to weight-1 PCCC encoder inputs, and $n_1$, the number of such codewords. Repeat for $d_{2,\min}^{\text{PCCC}}$, the minimum codeword weight corresponding to weight-2 PCCC encoder inputs, and $n_2$, the number of such codewords. Which is dominant, the codewords due to weight-1 inputs or those due to weight-2 inputs? Can you improve any of these four parameters ($d_{1,\min}^{\text{PCCC}}$, $n_1$, $d_{2,\min}^{\text{PCCC}}$, and $n_2$) with an improved interleaver design? Consider an $S$-random interleaver.

**7.12** Do the previous problem, except use an RSC encoder with octal generators $(7, 5)$ instead of an $\overline{\text{RSC}}$ encoder. Comment on your findings.

**7.13** Simulate the rate-1/2 SCCC whose outer code is the rate-1/2 $\overline{\text{RSC}}$ code with octal generators $(7, 5)$ and whose inner code is the rate-1 accumulator with transfer function $1/(1 \oplus D)$. Use the length-128 interleaver specified by the linear congruential relationship $P[k] = (13P[k-1] + 7) \bmod 128$ for $k = 1, 2, \ldots, 127$ and $P[0] = 11$. Repeat your simulation using an $S$-random interleaver in an attempt to lower the "floor" of the $P_b$ versus $E_b/N_0$ curve.

**7.14** Do the previous problem using the max function instead of the max* function in the constituent BCJR decoders.

**7.15** To show that we obtain the same two-dimensional product codeword (Figure 7.10) whether rows or columns are encoded first, first argue that encoding by rows and then by columns is mathematically represented by

$$\left[(\mathbf{U}\mathbf{G}_{\text{row}})^{\text{T}}\mathbf{G}_{\text{col}}\right]^{\text{T}},$$

where $\mathbf{U}$ is the $k_1 \times k_2$ data matrix. Now apply the matrix identity $(\mathbf{AB})^{\text{T}} = \mathbf{B}^{\text{T}}\mathbf{A}^{\text{T}}$ to this expression twice to arrive at an expression corresponding to encoding by columns and then by rows.

**7.16** Show that the minimum distance of a product code is given by $d_{\min} = d_{\min,1}d_{\min,2}$, where $d_{\min,1}$ and $d_{\min,2}$ are the minimum distances of the row and column codes.

**7.17** Would the turbo-product-code iterative-decoding algorithm described in this chapter work if the row code and the column code were both single parity-check codes? Explain your answer in appropriate detail. Can the sum–product algorithm used for LDPC codes be used to decode turbo product codes? Explain your answer in appropriate detail.

**7.18** Forgetting tail-biting, show that the binary generator matrix for the RSC code specified by (7.40) has the form $[\mathbf{I} \;\; \mathbf{P}_{acc}]$, where $\mathbf{P}_{acc}$ is upper triangular. Show that the corresponding parity-check matrix is given by $[\mathbf{P}' \;\; \mathbf{I}]$, where $\mathbf{P}'$ is given by (7.41) with the upper-right 1 set to 0 (since tail-biting is ignored here). Now show that $\mathbf{P}'$ is the transpose of the inverse of $\mathbf{P}_{acc}$. Finally, argue that (7.41) is the appropriate matrix for the tail-biting case.

**7.19** Consider the rate-1/3 nonbinary serial turbo code discussed in Section 7.5 whose codewords have the form $\mathbf{c} = [\,\mathbf{v}\;\mathbf{p}\;\mathbf{u}\,]$. Show that $\mathbf{v} = \mathbf{u} \cdot (\mathbf{P}^{(O)})^{\text{T}} \cdot \boldsymbol{\Pi}_q^{-\text{T}}$ and

$\mathbf{p} = \mathbf{v} \cdot (\mathbf{P}^{(I)})^{-\mathrm{T}}$, where superscript "–T" signifies "inverse transpose." From these expressions, verify that the parity-check matrix for this code is the one given in (7.43) with "$t$" replaced by "$I$" and "$b$" replaced by "$O$."

**7.20** Show that the matrix in (7.45) is indeed a parity-check matrix for the rate-1/2 nonbinary serial turbo code presented in Section 7.5. (Going through the previous problem first might be helpful.)

## References


[1] C. Berrou, A. Glavieux, and P. Thitimajshima, "Near Shannon limit error-correcting coding and decoding: Turbo codes," *Proceedings of the 1993 International Conference on Communications*, 1993, pp. 1064–1070.

[2] C. Berrou and A. Glavieux, "Near optimum error correcting coding and decoding: Turbo-codes," *IEEE Transactions on Communications*, pp. 1261–1271, October 1996.

[3] G. Ungerboeck, "Channel coding with multilevel/phase signals," *IEEE Transactions on Information Theory*, vol. 28, no. 1, pp. 55–67, January 1982.

[4] P. Robertson, "Illuminating the structure of code and decoder of parallel concatenated recursive systematic (turbo) codes," *Proceedings of GlobeCom 1994*, 1994, pp. 1298–1303.

[5] S. Benedetto and G. Montorsi, "Unveiling turbo codes: Some results on parallel concatenated coding schemes," *IEEE Transactions on Information Theory*, vol. 40, no. 3, pp. 409–428, March 1996.

[6] S. Benedetto and G. Montorsi, "Design of parallel concatenated codes," *IEEE Transactions on Communications*, pp. 591–600, May 1996.

[7] J. Hagenauer, E. Offer, and L. Papke, "Iterative decoding of binary block and convolutional codes," *IEEE Transactions on Information Theory*, vol. 42, no. 3, pp. 429–445, March 1996.

[8] D. Arnold and G. Meyerhans, *The Realization of the Turbo Coding System*, Semester Project Report, ETH Zürich, July 1995.

[9] L. Perez, J. Seghers, and D. Costello, "A distance spectrum interpretation of turbo codes," *IEEE Transactions on Information Theory*, vol. 42, no. 11, pp. 1698–1709, November 1996.

[10] L. Bahl, J. Cocke, F. Jelinek, and J. Raviv, "Optimal decoding of linear codes for minimizing symbol error rate," *IEEE Transactions on Information Theory*, vol. 20, no. 3, pp. 284–287, March 1974.

[11] S. Benedetto, D. Divsalar, G. Montorsi, and F. Pollara, "Serial concatenation of interleaved codes: Performance analysis, design, and iterative decoding," *IEEE Transactions on Information Theory*, vol. 44, no. 5, pp. 909–926, May 1998.

[12] J. Hagenauer and P. Hoeher, "A Viterbi algorithm with soft-decision outputs and its applications," *1989 IEEE Global Telecommunications Conference*, November 1989, pp. 1680–1686.

[13] P. Robertson, E. Villebrun, and P. Hoeher, "A comparison of optimal and suboptimal MAP decoding algorithms operating in the log domain," *Proceedings of the 1995 International Conference on Communications*, pp. 1009–1013.

[14] A. Viterbi, "An intuitive justification and a simplified implementation of the MAP decoder for convolutional codes," *IEEE Journal on Selected Areas in Communications*, vol. 18, no. 2, pp. 260–264, February 1998.

[15] R. Pyndiah, "Near-optimum decoding of product codes: Block turbo codes," *IEEE Transactions on Communications*, vol. 46, no. 8, pp. 1003–1010, August 1998.

[16] P. Elias, "Error-free coding," *IEEE Transactions on Information Theory*, vol. 1, no. 9, pp. 29–37, September 1954.

[17] D. Divsalar and F. Pollara, *Multiple Turbo Codes for Deep-Space Communications*. JPL TDA Progress Report 42-121, May 15, 1995.

[18] J. Sun and O. Takeshita, "Interleavers for turbo codes using permutation polynomials over integer rings," *IEEE Transactions on Information Theory*, pp. 101–119, January 2005.

[19] O. Takeshita, "On maximum contention-free interleavers and permutation polynomials over integer rings," *IEEE Transactions on Information Theory*, pp. 1249–1253, March 2006.

[20] M. C. Valenti and S. Cheng, "Iterative demodulation and decoding of turbo coded *M*-ary noncoherent orthogonal modulation," *IEEE Journal on Selected Areas in Communications (Special Issue on Differential and Noncoherent Wireless Communications)*, vol. 23, no. 9, pp. 1738–1747, September 2005.

[21] O. Takeshita, "Permutation polynomial interleavers: An algebraic-geometric perspective," *IEEE Transactions on Information Theory*, pp. 2116–2132, June 2007.

[22] J. Ryu, *Permutation Polynomial Based Interleavers for Turbo Codes over Integer Rings: Theory and Applications*, PhD thesis, ECE Department, Ohio State University, 2007.

[23] O. Acikel and W. Ryan, "Punctured turbo codes for BPSK/QPSK channels," *IEEE Transactions on Communications*, vol. 47, no. 9, pp. 1315–1323, September 1999.

[24] O. Acikel and W. Ryan, "Punctured high rate SCCCs for BPSK/QPSK channels," *Proceedings of the 2000 IEEE International Conference on Communications*, June 2000, pp. 434–439.

[25] G. Battail, "Ponderation des symboles decodes par l'algorithme de Viterbi," *Annals of Telecommunications*, pp. 31–38, January 1987.

[26] N. Wiberg, *Codes and Decoding on General Graphs*, PhD thesis, University of Linköping, Sweden, 1996.

[27] L. Lin and R. S. Cheng, "Improvements in SOVA-based decoding for turbo codes," *Proceedings of ICC*, Montreal, Canada, pp. 1473–1478, June 1997.

[28] M. Fossorier, F. Burkert, S. Lin, and J. Hagenauer, "On the equivalence between SOVA and Max-Log-MAP Decodings," *IEEE Communications Letters*, pp. 137–139, May 1998.

[29] C. X. Huang and A. Ghrayeb, "A simple remedy for the exaggerated extrinsic information produced by the SOVA algorithm," *IEEE Transactions on Wireless Communications*, vol. 5, no. 5, pp. 996–1002, May 2006.

[30] A. Ghrayeb and C. X. Huang, "Improvements in SOVA-based decoding for turbo-coded storage channels," *IEEE Transactions on Magnetics*, vol. 41, no. 12, pp. 4435–4442, December 2005.

[31] M. CoÅ§kun, G. Durisi, T. Jerkovits, G. Liva, W. Ryan, B. Stein, and F. Steiner, "Efficient error-correcting codes in the short blocklength regime," *Physical Communication*, vol. 34, pp. 66–79, June 2019.

[32] G. Liva, E. Paolini, B. Matuz, S. Scalise, and M. Chiani, "Short turbo codes over high order fields," *IEEE Transactions on Communications*, pp. 2201–2211, June 2013.

# 8 Ensemble Enumerators for Turbo and LDPC Codes

*Weight enumerators* or *weight-enumerating functions* are polynomials that represent in a compact way the input and/or output weight characteristics of the encoder for a code. The utility of weight enumerators is that they allow us to easily estimate, via the union bound, the performance of a maximum-likelihood (ML) decoder for the code. Given that turbo and LDPC codes employ suboptimal iterative decoders this may appear meaningless, but it is actually quite sensible for at least two reasons. One reason is that knowledge of ML-decoder performance bounds allows us to weed out weak codes. That is, if a code performs poorly for the ML decoder, we can expect it to perform poorly for an iterative decoder. Another reason for the ML-decoder approach is that the performance of an iterative decoder is generally approximately equal to that of its counterpart ML decoder, at least over a restricted range of SNRs. We saw this in Chapter 7 and we will see it again in Figure 8.5 in this chapter.

A drawback to the union-bound/ML-decoder approach is that the bound diverges in the low-SNR region, which is precisely the region of interest when one is attempting to design codes that perform very close to the capacity limit. Thus, when attempting to design codes that are simultaneously effective in the floor region (high SNRs) and the waterfall region (low SNRs), the techniques introduced in this chapter should be supplemented with the techniques in the next chapter which are applicable to the low-SNR region. We also introduce in this chapter trapping-set and stopping-set enumerators because trapping sets and stopping sets are also responsible for floors in the high-SNR region. The presentation in this chapter follows [1–12].

We emphasize that the weight enumerators derived in this chapter are for code ensembles rather than for a specific code. Thus, the performance estimates correspond to the *average* performance over a code ensemble rather than for a specific code, although the formulas can be applied to a specific code if its weight enumerator is known. The motivation for the ensemble approach is that finding the weight enumerator for a specific code is generally much more difficult than finding an ensemble enumerator. Further, if one finds an ensemble with excellent performance, then one can pick a code at random from the ensemble and expect excellent performance.

## 8.1 Notation

The weight distribution $\{A_0, A_1, \ldots, A_N\}$ of a length-$N$ linear code $\mathcal{C}$ was first introduced in Chapter 3. When the values $A_1, \ldots, A_N$ are used as coefficients of a degree-$N$

polynomial, the resulting polynomial is called a weight enumerator. Thus, we define the *weight enumerator* (WE) $A(W)$ for an $(N, K)$ linear block code to be

$$A(W) = \sum_{w=1}^{N} A_w W^w,$$

where $A_w$ is the number of codewords of weight $w$. Note that the weight-0 codeword is not included in the summation. Various weight enumerators were discussed in detail in Chapter 4 when discussing the performance of the ML decoder. In this chapter, we introduce various other enumerators necessary for evaluating the performance of turbo and LDPC codes.

To develop the notation for the various enumerating functions, we use as a running example the cyclic (7, 4) Hamming code generated in systematic form by the binary polynomial $g(x) = 1+x+x^3$. Below we list the codewords in three columns, where the first column lists the weight-0 and the weight-7 codewords, the second column lists the seven weight-3 codewords, and the third column lists the seven weight-4 codewords:

| | | |
|---|---|---|
| 0000 000 | 1000 110 | 0010 111 |
| 1111 111 | 0100 011 | 1001 011 |
| | 1010 001 | 1100 101 |
| | 1101 000 | 1110 010 |
| | 0110 100 | 0111 001 |
| | 0011 010 | 1011 100 |
| | 0001 101 | 0101 110 |

The WE for the (7, 4) Hamming code is then

$$A(W) = 7W^3 + 7W^4 + W^7.$$

We also define the *information–parity weight enumerator* (IP-WE) $A(I, P)$ for systematic encoders as

$$A(I, P) = \sum_{i=1}^{K} \sum_{p=0}^{N-K} A_{i,p} I^i P^p,$$

where $A_{i,p}$ is the number of codewords with information weight $i$ and parity weight $p$. (Throughout our discussions of enumerating functions, "transform variables" such as $W$, $I$, and $P$ are simply indeterminates.) Again, the zero codeword is excluded. The IP-WE for the Hamming code is

$$A(I, P) = I\left(3P^2 + P^3\right) + I^2\left(3P + 3P^2\right) + I^3(1 + 3P) + I^4P^3.$$

Thus, there are three codewords with information weight 1 and parity weight 2, there is one codeword with information weight 1 and parity-weight 3, and so on. We will also find it useful to define the *conditional parity weight enumerator* (C-PWE), written as

$$A_i(P) = \sum_{p=0}^{N-K} A_{i,p} P^p,$$

which enumerates the parity weight for a given information weight $i$. Thus, for the Hamming code,

$$A_1(P) = 3P^2 + P^3,$$
$$A_2(P) = 3P + 3P^2,$$
$$A_3(P) = 1 + 3P,$$
$$A_4(P) = P^3.$$

We next define the *input–output weight enumerator* (IO-WE) as follows:

$$A(I, W) = \sum_{i=1}^{K} \sum_{w=1}^{N} A_{i,w} I^i W^w,$$

where $A_{i,w}$ is the number of weight-$w$ codewords produced by weight-$i$ encoder inputs. The IO-WE for the $(7, 4)$ Hamming code is

$$A(I, W) = I\left(3W^3 + W^4\right) + I^2\left(3W^3 + 3W^4\right) + I^3\left(W^3 + 3W^4\right) + I^4 W^7.$$

The IO-WE gives rise to the *conditional output-weight enumerator* (C-OWE) expressed as

$$A_i(W) = \sum_{w=1}^{N} A_{i,w} W^w,$$

which enumerates the codeword weight for a given information weight $i$. For the Hamming code, the following is clear:

$$A_1(W) = 3W^3 + W^4,$$
$$A_2(W) = 3W^3 + 3W^4,$$
$$A_3(W) = W^3 + 3W^4,$$
$$A_4(W) = W^7.$$

Observe the following straightforward relationships:

$$A(W) = A(I, P)|_{I=P=W} = A(I, W)|_{I=1},$$
$$A(I, P) = \sum_{i=1}^{K} I^i A_i(P).$$

The above weight-enumerating functions are useful for estimating a code's codeword error rate $P_{cw}$ (sometimes called the frame-error rate, FER). When $P_b$, the bit-error rate (BER), is the metric of interest, bit-wise enumerators are appropriate. For information word length $K$, the *cumulative information-weight enumerator* (CI-WE) is given by

$$B(W) = \left. \sum_{i=1}^{K} i I^i A_i(P) \right|_{I=P=W},$$

the *scaled C-PWE* is given by

$$B_i(P) = iA_i(P) = i\sum_{p=0}^{N-K} A_{i,p}P^p,$$

and the *cumulative IP-WE* is given by

$$B(I,P) = \sum_{i=1}^{K} I^i B_i(P) = \sum_{i=1}^{K}\sum_{p=0}^{N-K} B_{i,p}I^iP^p,$$

where $B_{i,p} = iA_{i,p}$. Observe that we may write

$$B(W) = B(I,P)|_{I=P=W} = \sum_{w=1}^{N} B_w W^w,$$

where $B_w = \sum_{i,p:i+p=w} B_{i,p}$ is the total information weight for the weight-$w$ codewords. For the (7,4) Hamming code, it is easy to show that

$$B(I,P) = I\left(3P^2 + P^3\right) + 2I^2\left(3P + 3P^2\right) + 3I^3(1 + 3P) + 4I^4P^3$$

and

$$B(W) = B(I,P)|_{I=P=W} = 12W^3 + 16W^4 + 4W^7.$$

Given the WE $A(W)$ for a length-$N$ code, we may upper bound its FER on a binary-input AWGN channel with two-sided power-spectral density $N_0/2$ as

$$P_{cw} \le \sum_{w=1}^{N} A_w Q\left(\sqrt{2wRE_b/N_0}\right), \tag{8.1}$$

where

$$Q(\alpha) = \int_{\alpha}^{\infty} \frac{1}{\sqrt{2\pi}} \exp\left(-\lambda^2/2\right)d\lambda,$$

$R$ is the code rate, and $E_b/N_0$ is the well-known SNR measure. A more compact, albeit looser, bound is given by

$$P_{cw} < A(W)|_{W=\exp(-RE_b/N_0)}, \tag{8.2}$$

which follows from the bound $Q(\alpha) < \exp(-\alpha^2/2)$. Similarly, we can impose an upper bound on the BER as

$$P_b \le \frac{1}{K}\sum_{w=1}^{N} B_w Q\left(\sqrt{2wRE_b/N_0}\right), \tag{8.3}$$

and the corresponding looser bound is given by

$$P_b < \frac{1}{K}B(W)\bigg|_{W=\exp(-RE_b/N_0)} = \frac{1}{K}\sum_{i=1}^{K} iI^iA_i(P)\bigg|_{I=P=\exp(-RE_b/N_0)}. \tag{8.4}$$

This expression extends to other channels with an appropriate substitution for $I$ and $P$.

## 8.2 Ensemble Enumerators for Parallel-Concatenated Codes

### 8.2.1 Preliminaries

In the previous section we discussed enumerators for specific codes. While finding such enumerators is possible for relatively short codes, it is not possible for moderate-length or long codes, even with the help of a fast computer. For this reason, code designers typically study the average performance of ensembles of codes and then pick a code from a good ensemble. In this section, we consider enumerators for ensembles of parallel-concatenated codes (PCCs), configured as in Figure 8.1, where the constituent codes can be either block codes or convolutional codes. The ensembles we consider are obtained by fixing the constituent codes and then varying the length-$K$ interleaver over all $K!$ possibilities. It is possible to consider interleaving multiple input blocks using a length-$pK$ interleaver, $p > 1$, but we shall not do so here.

Prior to studying ensemble enumerators, it is helpful to first develop an understanding of how an enumerator for a PCC is related to the corresponding enumerators of its constituent codes. This also further motivates the ensemble approach. Thus, consider the simple PCC whose constituent codes, $C_1$ and $C_2$, are both $(7, 4)$ Hamming codes and fix the interleaver $\mathbf{\Pi}$ to one of its $4! = 24$ possibilities. Now consider weight-1 encoder inputs. Then the C-PWEs are $A_1^{C_1}(P) = A_1^{C_2}(P) = P^2 + P^2 + P^2 + P^3$, where we have intentionally written four separate terms for the four possible weight-1 inputs. That is, for the four possible weight-1 inputs into either encoder, E1 or E2, three will yield parity-weight-2 outputs and one will yield a parity-weight-3 output. As an example of a C-PWE computation for the PCC, suppose a particular weight-1 input to E1 produces parity weight 3, producing a $P^3$ term, and, after the input has passed through the interleaver, E2 produces parity weight 2, producing a $P^2$ term. The corresponding term for the PCC is clearly $P^3 \cdot P^2 = P^5$, since the PCC parity weight for that input will be 5. Note how each of the four terms in $A_1^{C_1}(P)$ pairs with each of the four terms in $A_1^{C_2}(P)$, depending on $\mathbf{\Pi}$, as in the $P^3 \cdot P^2$ calculation. Except for when $K$ is small,

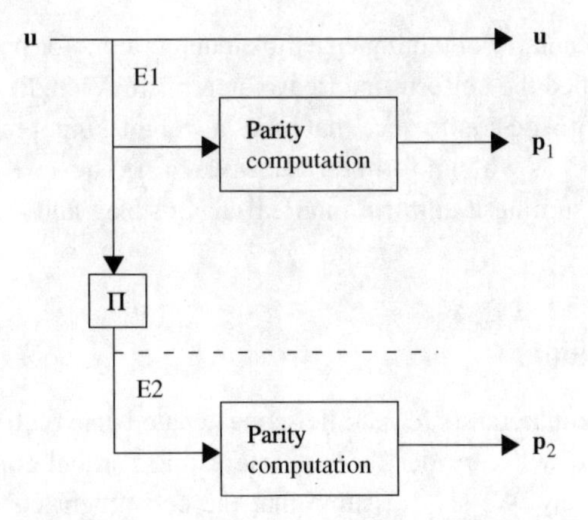


**Figure 8.1** An encoder for generic parallel-concatenated code.

keeping track of such pairings is impossible in general. This fact suggests the ensemble approach whereby we compute the average C-PWE for a PCC ensemble, where the average is over all $K!$ interleavers ($C_1$ and $C_2$ are fixed).

To see how this is done, note that the set of all length-$K$ interleavers permutes a given weight-$i$ input into all $\binom{K}{i}$ possible permutations with equal probability. Thus, in the present example for which $K = 4$, when averaged over the 4! interleavers, a given term in $A_1^{C_1}(P)$ will be paired with a given term in $A_1^{C_2}(P)$ with probability $1/\binom{4}{1} = 1/4$ in the computation of the C-PWE $A_1^{\text{PCC}}(P)$ for the PCC. Thus, averaged over the ensemble of such parallel-concatenated codes, the average C-PWE is given by

$$A_1^{\text{PCC}}(P) = \left(P^2 + P^2 + P^2 + P^3\right) \cdot \frac{1}{4} \cdot \left(P^2 + P^2 + P^2 + P^3\right)$$
$$= \frac{A_1^{C_1}(P) \cdot A_1^{C_2}(P)}{\binom{4}{1}}.$$

More generally, for arbitrary constituent codes, for a length-$K$ interleaver and weight-$i$ inputs, it can be shown that the ensemble enumerator is

$$A_i^{\text{PCC}}(P) = \frac{A_i^{C_1}(P) \cdot A_i^{C_2}(P)}{\binom{K}{i}}. \tag{8.5}$$

Note that, given the ensemble C-PWE in (8.5), one can determine any of the other ensemble enumerators, including the bit-wise enumerators. For example,

$$A^{\text{PCC}}(I, P) = \sum_{i=1}^{K} I^i A_i^{\text{PCC}}(P)$$

and

$$A^{\text{PCC}}(W) = A^{\text{PCC}}(I, P)\Big|_{I=P=W}.$$

We remark that this approach for obtaining ensemble enumerators for parallel concatenated codes has been called the uniform-interleaver approach. A length-$K$ *uniform interleaver* [1, 2] is a probabilistic contrivance that maps a weight-$i$ input binary word into each of its $\binom{K}{i}$ permutations with uniform probability $1/\binom{K}{i}$. Thus, we can repeat the development of (8.5) assuming a uniform interleaver is in play and arrive at the same result.

### 8.2.2 PCCC Ensemble Enumerators

We now consider ensemble enumerators for parallel-concatenated convolutional codes (PCCCs) in some detail. We will consider the less interesting parallel-concatenated block-code case along the way. We shall assume that the constituent convolutional codes are identical, terminated, and, for convenience, rate 1/2. Termination of $C_2$ is, of course, problematic in practice, but the results developed closely match practice when $K$ is large, $K \gtrsim 100$, which is generally the case. Our development will follow that of [1, 2, 4].

In view of (8.5), to derive the C-PWE for a PCCC, we require $A_i^{C_1}(P)$ and $A_i^{C_2}(P)$ for the constituent convolutional codes considered as block codes with length-$K$ inputs (including the termination bits). This statement is not as innocent as it might appear. Recall that the standard enumeration technique for a convolutional code (Chapter 4) considers all paths in a split-state graph that depart from state 0 at time zero and return to state 0 some time later. The enumerator is obtained by calculating the transfer function for that split-state graph. Because the all-zero code sequence is usually chosen as a reference in such analyses, the paths that leave and then return to state 0 are typically called *error events*. Such paths also correspond to nonzero codewords that are to be included in our enumeration. However, the transfer-function technique for computing enumerators *does not include paths that return to state 0 more than once*, that is, that contain multiple error events. On the other hand, for an accurate computation of the enumerators for the equivalent block codes of convolutional codes $C_1$ and $C_2$, we require *all* paths, including those that contain multiple error events. We need to consider such paths because they are included in the list of nonzero codewords, knowledge of which is necessary in order to compute performance bounds.

Although there is a clear distinction between a convolutional code (with semi-infinite codewords) and its equivalent block code, we point out that the weight enumerator for the former may be used to find the weight enumerator for the latter. This is so because the nonzero codewords of the equivalent block code may be thought of as concatenations of single error events of the convolutional code. Further, error-event enumerators for convolutional codes are easily obtained via the transfer-function technique or computer search. For now, assume that we know the parity-weight enumerator $A_{i,n,\ell}(P)$ conditioned on convolutional encoder input weight $i$, which results in $n$ concatenated error events whose lengths total $\ell$ trellis stages (or $\ell$ input bits). As stated, $A_{i,n,\ell}(P)$ can be derived from the convolutional-code single-error-event transfer-function-based enumerator, but the counting procedure includes only one form of concatenation of the $n$ convolutional-code error events. An example would be $n$ consecutive error events, the first of which starts at time zero, and their permutations. That is, the counting procedure that yields $A_{i,n,\ell}(P)$ does not include the number of locations within the $K$-block in which the $n$ error events may occur. Thus, we introduce the notation $N(K, n, \ell)$, which is the number of ways $n$ error events having total length $\ell$ may be situated in a $K$-block in a particular order. We may now write for the enumerator for the equivalent block code (EBC)

$$A_i^{\text{EBC}}(P) = \sum_{n=1}^{n_{\max}} \sum_{\ell=1}^{K} N(K, n, \ell) A_{i,n,\ell}(P). \tag{8.6}$$

It can be shown that

$$N(K, n, \ell) = \binom{K - \ell + n}{n},$$

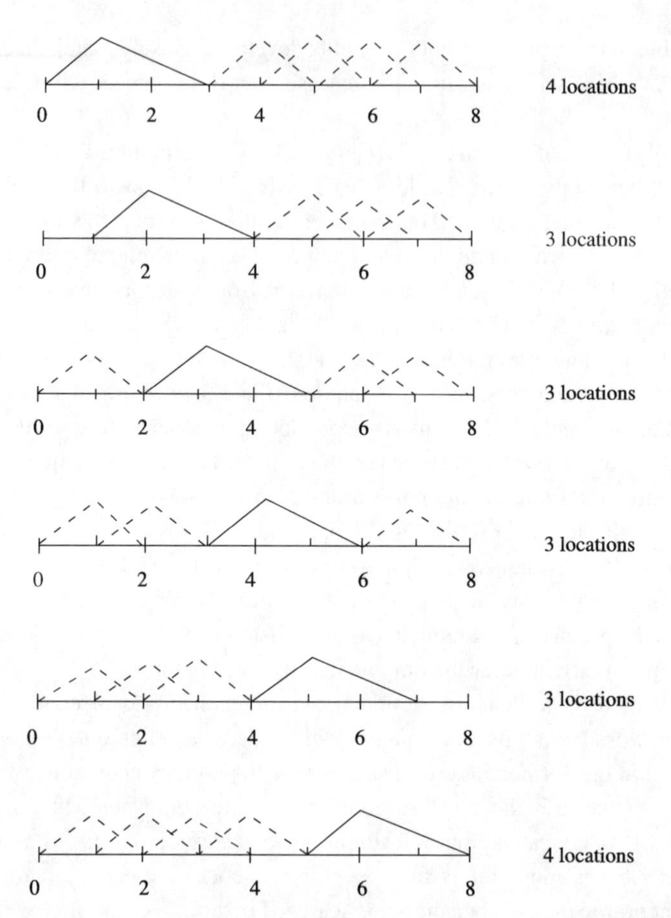


**Figure 8.2** An error-event location-counting example for $K = 8$, $n = 2$, and $\ell = 5$. The horizontal lines represent the all-zero path and the triangular figures represent error events.

independently of the individual error-event lengths. This follows since $N(K, n, \ell)$ is equivalent to the number of ways of choosing $n$ error event starting locations among $K - \ell$, partitioned into numbers that sum to $K - \ell + n$ slots. (There must be a total of $K - \ell$ zeros in the zero-strings that separate the $n$ error events.) Figure 8.2 provides an example showing that the total number of ways in which the $n = 2$ error events of total length $\ell = 5$ may be situated in a block of length $K = 8$ is

$$\frac{4 \cdot 3 + 4 \cdot 2}{2!} = \frac{5 \cdot 4}{2!} = \binom{8 - 5 + 2}{2} = N(8, 2, 5).$$

We divide by 2!, or by $n!$ in the general case, because $A_{i,n,\ell}(P)$ in (8.6) already accounts for the $n!$ different permutations of the $n$ error events. Thus, although 20 configurations are shown in the figure, $N(8, 2, 5) = 10$.

Under the assumptions $K \gg \ell$ and $K \gg n$, as is generally the case, we may write

$$N(K, n, \ell) \simeq \binom{K}{n},$$

so that (8.6) becomes

$$A_i^{\mathrm{EBC}}(P) \simeq \sum_{n=1}^{n_{\max}} \binom{K}{n} A_{i,n}(P), \tag{8.7}$$

where

$$A_{i,n}(P) = \sum_{\ell=1}^{K} A_{i,n,\ell}(P). \tag{8.8}$$

Substitution of (8.7) into (8.5) yields

$$A_i^{\mathrm{PCCC}}(P) \simeq \sum_{n_1=1}^{n_{\max}} \sum_{n_2=1}^{n_{\max}} \frac{\binom{K}{n_1}\binom{K}{n_2}}{\binom{K}{i}} A_{i,n_1}(P) A_{i,n_2}(P).$$

Three applications of the approximation

$$\binom{K}{s} \simeq \frac{K^s}{s!},$$

which assumes $s \ll K$, gives

$$A_i^{\mathrm{PCCC}}(P) \simeq \sum_{n_1=1}^{n_{\max}} \sum_{n_2=1}^{n_{\max}} \frac{i!}{n_1! n_2!} K^{n_1+n_2-i} A_{i,n_1}(P) A_{i,n_2}(P).$$

Also, since $K$ is large, this last expression can be approximated by the term in the double summation with the highest power of $K$ ($n_1 = n_2 = n_{\max}$), so that

$$A_i^{\mathrm{PCCC}}(P) \simeq \frac{i!}{(n_{\max}!)^2} K^{2n_{\max}-i} \left[A_{i,n_{\max}}(P)\right]^2.$$

We may now use this expression together with (8.4) to estimate PCCC ensemble BER performance on the binary-input AWGN channel as

$$\begin{aligned} P_b^{\mathrm{PCCC}} &< \left.\sum_{i=1}^{K} \frac{i}{K} I^i A_i^{\mathrm{PCCC}}(P)\right|_{I=P=\exp(-RE_b/N_0)} \\ &\simeq \left.\sum_{i=i_{\min}}^{K} i \frac{i!}{(n_{\max}!)^2} K^{2n_{\max}-i-1} I^i \left[A_{i,n_{\max}}(P)\right]^2\right|_{I=P=\exp(-RE_b/N_0)}, \end{aligned} \tag{8.9}$$

where $i_{\min}$ is the minimum information weight leading to non-negligible terms in (8.9). For example, when the constituent encoders for the PCCC are non-recursive $i_{\min} = 1$, and when they are recursive $i_{\min} = 2$. We now examine these two cases individually.

### PCCCs with Non-recursive Constituent Encoders

We will examine (8.9) more closely, assuming a non-recursive memory-$\mu$ constituent encoder, in which case $i_{\min} = 1$. Further, for a weight-$i$ input, the greatest number of error events possible is $n_{\max} = i$ (consider isolated 1s). In this case,

$$A_{i,n_{\max}}(P) = A_{i,i}(P) = \left[A_{1,1}(P)\right]^i,$$

that is, the error events due to the $i$ isolated 1s are identical and their weights simply add, since $A_{1,1}(P)$ has a single term. Applying this to (8.9) with $n_{\max} = i$ yields

$$P_b^{\mathrm{PCCC}} \simeq \sum_{i=1}^{K} \frac{K^{i-1}}{(i-1)!} I^i \left[A_{1,1}(P)\right]^{2i} \Big|_{I=P=\exp(-RE_b/N_0)}.$$

Observe that the dominant term ($i = 1$) is independent of the interleaver size $K$. Thus, as argued more superficially in Chapter 7, interleaver gain does not exist for non-recursive constituent encoders. This is also the case for block constituent codes, for which $i_{\min} = 1$ and $n_{\max} = i$ are also true. By contrast, we will see that interleaver gain is realized for PCCCs with recursive constituent encoders.

## PCCCs with Recursive Constituent Encoders

As we saw in Chapter 7, the minimum PCCC encoder input weight leading to non-negligible terms in the union bound on error probability is $i_{\min} = 2$. Further, for input weight $i$, the largest number of error events is $n_{\max} = \lfloor i/2 \rfloor$. Given this, consider now the $i$th term in (8.9) for $i$ odd:

$$i \frac{i!}{(\lfloor i/2 \rfloor !)^2} K^{2\lfloor i/2 \rfloor - i - 1} I^i \left[A_{i,\lfloor i/2 \rfloor}(P)\right]^2 = i \frac{i!}{(\lfloor i/2 \rfloor !)^2} K^{-2} I^i \left[A_{i,\lfloor i/2 \rfloor}(P)\right]^2.$$

When $i$ is even, $n_{\max} = i/2$ and the $i$th term in (8.9) is

$$i \frac{i!}{((i/2)!)^2} K^{-1} I^i \left[A_{i,i/2}(P)\right]^2. \tag{8.10}$$

Thus, odd terms go as $K^{-2}$ and even terms go as $K^{-1}$, so the odd terms may be ignored. Moreover, the factor $K^{-1}$ in (8.10) represents interleaver gain. This result, which indicates that interleaver gain is realized when the constituent convolutional encoders are recursive, was seen in Chapter 7, albeit at a more superficial level.

Keeping only the even terms in (8.9) that have the form (8.10), with $i = 2k$,

$$\begin{aligned} P_b^{\mathrm{PCCC}} &\simeq \sum_{k=1}^{\lfloor K/2 \rfloor} 2k \frac{(2k)!}{(k!)^2} K^{-1} I^{2k} \left[A_{2k,k}(P)\right]^2 \Big|_{I=P=\exp(-RE_b/N_0)} \\ &= \sum_{k=1}^{\lfloor K/2 \rfloor} 2k \binom{2k}{k} K^{-1} I^{2k} \left[A_{2,1}(P)\right]^{2k} \Big|_{I=P=\exp(-RE_b/N_0)}, \end{aligned} \tag{8.11}$$

where the second line follows since

$$\binom{2k}{k} = \frac{(2k)!}{(k!)^2}$$

and

$$A_{2k,k}(P) = \left[A_{2,1}(P)\right]^k. \tag{8.12}$$

Equation (8.12) holds because the codewords corresponding to $A_{2k,k}(P)$ must be the concatenation of $k$ error events, each of which is produced by weight-2 recursive convolutional-encoder inputs, each of which produces a single error event.

$A_{2,1}(P)$ has the simple form

$$\begin{aligned} A_{2,1}(P) &= P^{p_{\min}} + P^{2p_{\min}-2} + P^{3p_{\min}-4} + \cdots \\ &= \frac{P^{p_{\min}}}{1 - P^{p_{\min}-2}}. \end{aligned} \tag{8.13}$$

To show this, observe that the first 1 into the parity computation circuit with transfer function $g^{(2)}(D)/g^{(1)}(D)$ will yield the impulse response that consists of a transient $t(D)$ of length $\tau$ (and degree $\tau - 1$) followed by a periodic binary sequence having period $\tau$, where $\tau$ is dictated by $g^{(1)}(D)$. If the second 1 comes along at the start of the second period, it will in effect "squelch" the output so that the subsequent output of the parity circuit is all zeros. Thus, the response to the two 1s separated by $\tau - 2$ zeros is the transient followed by a 1, that is, $t(D) + D^{\tau}$, and the weight of this response will be denoted by $p_{\min}$. If the second 1 instead comes along at the start of the third period, by virtue of linearity the parity output will be $[t(D) + D^{\tau}] + D^{\tau}[t(D) + D^{\tau}]$, which has weight $2p_{\min} - 2$ since the 1 that trails the first transient will cancel out the 1 that starts the second transient. If the second 1 comes along at the start of the fourth period, the parity weight will be $3p_{\min} - 4$, and so on. From this, we conclude (8.13).

As an example, with $g^{(2)}(D)/g^{(1)}(D) = (1 + D^2)/(1 + D + D^2)$, the impulse response is 111 011 011 011 . . ., where $\tau = 3$ is obvious. The response to 10010. . . is clearly 111 011 011 011 . . . + 000 111 011 011 . . . = 11110. . ., and $p_{\min} = 4$. The response to 10000010. . . is 111 011 011 011,. . . + 000 000 111 011 011 . . . = 11101110. . ., which has weight $2p_{\min} - 2$.

Substitution of (8.13) into (8.11) yields

$$P_b^{\text{PCCC}} \simeq \sum_{k=1}^{\lfloor K/2 \rfloor} 2k \binom{2k}{k} K^{-1} I^{2k} \left[ \frac{P^{p_{\min}}}{1 - P^{p_{\min}-2}} \right]^{2k} \Bigg|_{I=P=\exp(-RE_b/N_0)}. \tag{8.14}$$

When $\exp(-RE_b/N_0)$ is relatively small, that is, the SNR is relatively large, the dominant ($k = 1$) term has the approximation

$$4K^{-1} \left[ I P^{p_{\min}} \right]^2 \Big|_{I=P=\exp(-RE_b/N_0)} = \frac{4}{K} \exp[-(2 + 2p_{\min}) RE_b/N_0], \tag{8.15}$$

from which we may make the following observations. First, the interleaver gain is evident from the factor $4/K$. Second, the exponent has the factor $2 + 2p_{\min}$, which is due to the worst-case inputs for both constituent encoders: two 1s separated by $\tau - 2$ zeros. Such a situation will yield information weight 2 and weight $p_{\min}$ from each parity circuit. Of course, a properly designed interleaver can avoid such a misfortune for a specific code, but our ensemble result must include such worst-case instances.

## PCCC Design Principles

We would now like to consider how the code polynomials, $g^{(1)}(D)$ and $g^{(2)}(D)$, employed at both constituent encoders should be selected to determine an ensemble with (near-)optimum performance. The optimization criterion is usually one of two criteria: the polynomials can be selected to minimize the level of the floor of the error-rate curve or they can be selected to minimize the SNR at which the waterfall region occurs. The latter topic will be considered in Chapter 9, where the waterfall region is shown to be connected to the idea of a "decoding threshold." We will now consider the former topic.

In view of the previous discussion, it is clear that the ensemble floor will be lowered (although not necessarily minimized) if $p_{\min}$ can be made as large as possible. This statement assumes a given complexity, that is, a given memory size $\mu$. However, $p_{\min}$ is the weight of $t(D) + D^{\tau}$, where $t(D)$ is the length-$\tau$ transient part of the impulse response of $g^{(2)}(D)/g^{(1)}(D)$. Assuming that about half of the bits in $t(D)$ are 1s, one obvious way to ensure that the weight of $t(D)$, and hence of $p_{\min}$, is (nearly) maximal is to choose $g^{(1)}(D)$ to be a primitive polynomial so that $t(D)$ is of maximal length, that is, length $2^{\mu} - 1$.

## Example Performance Bounds

We now consider the ensemble performance bounds for two PCCCs, one for which $g^{(1)}(D)$ is primitive and one for which $g^{(1)}(D)$ is not primitive. Specifically, for PCCC1, both constituent codes have the generator matrix

$$G_1(D) = \begin{bmatrix} 1 & \dfrac{1+D^2}{1+D+D^2} \end{bmatrix} \qquad \text{(PCCC1)}$$

and for PCCC2 both constituent codes have the generator matrix

$$G_2(D) = \begin{bmatrix} 1 & \dfrac{1+D+D^2}{1+D^2} \end{bmatrix} \qquad \text{(PCCC2).}$$

Note that $g^{(1)}(D) = 1 + D + D^2$ for $G_1(D)$ primitive, whereas $g^{(1)}(D) = 1 + D^2$ for $G_2(D)$ not primitive. Earlier we showed for $G_1(D)$ that the input $1+D^3$ yields the parity output $1+D+D^2+D^3$. so that $p_{\min} = 4$ for this code. Correspondingly, the worst-case PCCC1 codeword weight for weight-2 inputs is $d_{2,\min}^{\text{PCCC1}} = 2 + 4 + 4 = 10$. The overall minimum distance for the PCCC1 ensemble is produced by the input $1+D+D^2$, which yields the output $\begin{bmatrix} 1+D+D^2 & 1+D^2 & 1+D^2 \end{bmatrix}$, so that $d_{\min}^{\text{PCCC1}} = 3 + 2 + 2 = 7$. As for PCCC2, it is easy to see that the input $1 + D^2$ yields the parity output $1 + D + D^2$, so that $p_{\min} = 3$. The worst-case PCCC2 codeword weight for weight-2 inputs is then $d_{2,\min}^{\text{PCCC2}} = 2 + 3 + 3 = 8$. The overall minimum distance for the PCCC2 ensemble is also produced by the input $1 + D^2$, so that $d_{\min}^{\text{PCCC2}} = d_{2,\min}^{\text{PCCC2}} = 8$.

Thus, while PCCC1 has the smaller minimum distance $d_{\min}^{\text{PCCC}}$, it has the larger *effective minimum distance* $d_{2,\min}^{\text{PCCC}}$ and, thus, superior performance. The impact of its smaller minimum distance is negligible since it is due to an odd-weight encoder input (weight 3), as discussed in Section 8.2.2. This can also easily be seen by plotting (8.14) with $p_{\min} = 4$ for PCCC1 and $p_{\min} = 3$ for PCCC2. However, we shall work toward a tighter bound starting with (8.9) and using $Q\left(\sqrt{2iRE_b/N_0}\right)$ in place of its upper bound, $\exp(-iRE_b/N_0)$. The looser bound (8.14) is useful for code-ensemble performance comparisons, but we pursue the tighter result that has closer agreement with simulations.

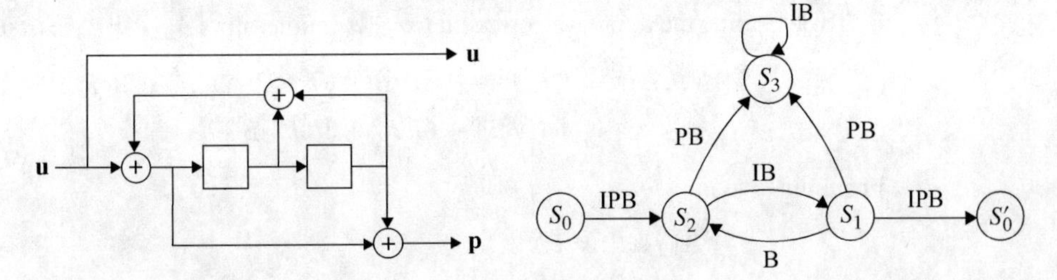


**Figure 8.3** The encoder and split-state diagram for a constituent encoder for PCCC1 with $G_1(D) = \left[1 \quad \dfrac{1+D^2}{1+D+D^2}\right]$.

To use (8.9), we require $\{A_{i,n_{\max}}(P)\}$ for the constituent codes for each PCCC. We find these enumerators first for $G_1(D)$, whose state diagram (split at state 0) is displayed in Figure 8.3. From this diagram, we may obtain the augmented IP-WE,

$$
\begin{aligned}
A(I,P,B) &= \frac{I^3P^2B^3 + I^2P^4B^4 - I^4P^2B^4}{1 - P^2B^3 - I(B+B^2) + I^2B^3} \\
&= I^3P^2B^3 + I^2P^4B^4 + \left(I^3P^4 + I^4P^2\right)B^5 \\
&\quad + \left(2I^3P^4 + I^4P^4\right)B^6 + \left(I^2P^6 + 2I^4P^4 + I^5P^2 + I^5P^4\right)B^7 \\
&\quad + \left(2I^3P^6 + 3I^4P^4 + 2I^5P^4 + I^6P^4\right)B^8 \\
&\quad + \left(3I^3P^6 + 3I^4P^6 + 3I^5P^4 + I^6P^2 + 2I^6P^4 + I^7P^4\right)B^9 + \cdots,
\end{aligned}
$$

where the exponent of the augmented indeterminate $B$ in a given term indicates the length of the error event(s) for that term. From this, we may identify the single-error-event ($n = 1$) enumerators $A_{i,1,\ell}(P)$ as follows (for $\ell \leq 9$):

$$
\begin{aligned}
&A_{2,1,4}(P) = P^4, \quad A_{2,1,7}(P) = P^6, \\
&A_{3,1,3}(P) = P^2, \quad A_{3,1,5}(P) = P^4, \quad A_{3,1,6}(P) = 2P^4, \quad A_{3,1,8}(P) = 2P^6, \\
&A_{3,1,9}(P) = 3P^6, \\
&A_{4,1,5}(P) = P^2, \quad A_{4,1,6}(P) = P^4, \quad A_{4,1,7}(P) = 2P^4, \quad A_{4,1,8}(P) = 3P^4, \\
&A_{4,1,9}(P) = 3P^6 \\
&\ldots
\end{aligned}
$$

Note that $A_{1,n,\ell}(P) = 0$ for all $n$ and $\ell$ because weight-1 inputs produce trellis paths that leave the zero path and never remerge. From the list above and (8.8), we can write the first few terms of $A_{i,1}(P)$:

$$
\begin{aligned}
A_{2,1}(P) &= P^4 + P^6 + \cdots, \\
A_{3,1}(P) &= P^2 + 3P^4 + 5P^6 + \cdots, \\
A_{4,1}(P) &= P^2 + 6P^4 + 3P^6 + \cdots, \\
&\ldots
\end{aligned}
$$

To determine the double-error-event ($n = 2$) enumerators $A_{i,2,\ell}(P)$, we first compute

$$[A(I,P,B)]^2 = I^6P^4B^6 + 2I^5P^6B^7 + (I^4P^8 + 2I^6P^6 + 2I^7P^4)B^8 + (2I^5P^8 + 6I^6P^6 + 2I^7P^6)B^9 + \cdots.$$

From this, we may write

$$\begin{aligned} &A_{4,2,8}(P) = P^8, \\ &A_{5,2,7}(P) = 2P^6, \quad A_{5,2,9}(P) = 2P^8, \\ &A_{6,2,6}(P) = P^4, \quad A_{6,2,8}(P) = 2P^6, \quad A_{6,2,9}(P) = 6P^6, \\ &\ldots. \end{aligned}$$

Again using (8.8), we can write the first few terms of $A_{i,2}(P)$:

$$\begin{aligned} &A_{4,2}(P) = P^8 + \cdots, \\ &A_{5,2}(P) = 2P^6 + 2P^8 + \cdots, \\ &A_{6,2}(P) = P^4 + 8P^6 + \cdots, \\ &\ldots. \end{aligned}$$

We consider only the first few terms of (8.9), which are dominant. For input weights $i = 2$ and 3, $n_{\max} = 1$, and for $i = 4$ and 5, $n_{\max} = 2$. Thus, with $I = P = \exp(-RE_b/N_0)$, we have for PCCC1

$$\begin{aligned} P_b^{\text{PCCC1}} &< \frac{4}{K}I^2[A_{2,1}(P)]^2 + \frac{18}{K^2}I^3[A_{3,1}(P)]^2 + \frac{24}{K}I^4[A_{4,2}(P)]^2 \\ &\quad + \frac{150}{K^2}I^5[A_{5,2}(P)]^2 + \cdots \\ &\approx \frac{4}{K}I^2\left[P^4 + P^6\right]^2 + \frac{18}{K^2}I^3\left[P^2 + 3P^4 + 5P^6\right]^2 + \frac{24}{K}I^4\left[P^8\right]^2 \\ &\quad + \frac{150}{K^2}I^5\left[P^4 + 8P^6\right]^2 + \cdots \\ &= \frac{18}{K^2}W^7 + \frac{108}{K^2}W^9 + \frac{4}{K}W^{10} + \frac{342}{K^2}W^{11} + \frac{8}{K}W^{12} + \frac{540}{K^2}W^{13} \\ &\quad + \frac{4}{K}W^{14} + \cdots, \end{aligned}$$

where in the last line we set $W = I = P$. The bit-error-rate estimate can be obtained by setting $W = \exp(-RE_b/N_0)$ in the above expression, but we prefer the tighter expression

$$\begin{aligned} P_b^{\text{PCCC1}} &< \frac{18}{K^2}f(7) + \frac{108}{K^2}f(9) + \frac{4}{K}f(10) + \frac{342}{K^2}f(11) + \frac{8}{K}f(12) + \frac{540}{K^2}f(13) \\ &\quad + \frac{4}{K}f(14) + \cdots, \end{aligned} \tag{8.16}$$

where $f(w) = Q\left(\sqrt{2wRE_b/N_0}\right)$ and $R = 1/3$. It is expression (8.16) that we plot, but we shall first develop the analogous expression for PCCC2.

To find the enumerators $\{A_{i,n_{\max}}(P)\}$ for $G_2(D)$, we start with the state diagram displayed in Figure 8.4. Upon comparison with the state diagram for $G_1(D)$ in Figure 8.3, we see that one may be obtained from the other by exchanging $I$ and $P$. In

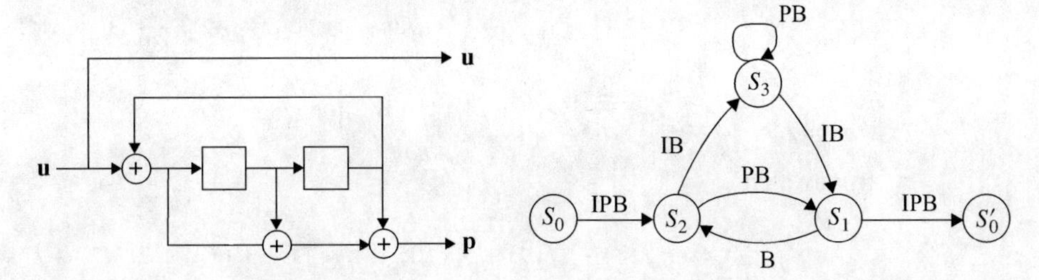


**Figure 8.4** The encoder and split-state diagram for a constituent encoder for PCCC2 with $G_2(D) = \left[1 \quad \frac{1+D+D^2}{1+D^2}\right]$.

fact, the enumerator $A_{G_2}(I,P,B)$ for $G_2(D)$ may be obtained by exchanging $I$ and $P$ in the enumerator $A_{G_1}(I,P,B)$ for $G_1(D)$. That is, $A_{G_2}(I,P,B) = A_{G_1}(P,I,B)$, yielding for $G_2(D)$

$$\begin{aligned}
A(I,P,B) &= \frac{I^2P^3B^3 + I^4P^2B^4 - I^2P^4B^4}{1 - I^2B^3 - P(B+B^2) + P^2B^3} \\
&= I^2P^3B^3 + I^4P^2B^4 + \left(I^4P^3 + I^2P^4\right)B^5 \\
&\quad + \left(2I^4P^3 + I^4P^4\right)B^6 + \left(I^6P^2 + 2I^4P^4 + I^2P^5 + I^4P^5\right)B^7 \\
&\quad + \left(2I^6P^3 + 3I^4P^4 + 2I^4P^5 + I^4P^6\right)B^8 \\
&\quad + \left(3I^6P^3 + 3I^6P^4 + 3I^4P^5 + I^2P^6 + 2I^4P^6 + I^4P^7\right)B^9 + \cdots.
\end{aligned}$$

By following steps similar to those in the $G_1(D)$ case, we may identify the single-error-event ($n = 1$) enumerators $A_{i,1,\ell}(P)$ (for $\ell \leq 9$) for $G_2(D)$, from which we obtain the enumerators $A_{i,1}(P)$,

$$\begin{aligned}
A_{2,1}(P) &= P^3 + P^4 + P^5 + P^6 + \cdots, \\
A_{3,1}(P) &= 0, \\
A_{4,1}(P) &= P^2 + 3P^3 + 6P^4 + 6P^5 + 3P^6 + P^7 + \cdots, \\
&\cdots.
\end{aligned}$$

Similarly, we can obtain the double-error-event ($n = 2$) enumerators $A_{i,2}(P)$,

$$\begin{aligned}
A_{4,2}(P) &= P^6 + P^7 + \cdots, \\
A_{5,2}(P) &= 0, \\
A_{6,2}(P) &= P^5 + 2P^6 + \cdots, \\
&\cdots.
\end{aligned}$$

Again, we consider only the first few terms of (8.9). For input weight $i = 2$, $n_{\max} = 1$; for $i = 4$, $n_{\max} = 2$; and for $i = 3$ and 5, $A_{i,1}(P) = 0$. Thus, we have for PCCC2

$$\begin{aligned}
P_b^{\text{PCCC2}} &< \frac{4}{K} I^2 \left[A_{2,1}(P)\right]^2 + \frac{24}{K} I^4 \left[A_{4,2}(P)\right]^2 + \cdots \\
&\approx \frac{4}{K} I^2 \left[P^3 + P^4 + P^5 + P^6\right]^2 + \frac{24}{K} I^4 \left[P^6 + P^7\right]^2 + \cdots
\end{aligned}$$

$$= \frac{4}{K}\left[W^8 + 2W^9 + 3W^{10} + 4W^{11} + 3W^{12} + 2W^{13} + W^{14}\right] + \frac{24}{K}\left[W^{16} + 2W^{17} + W^{18}\right] + \cdots,$$

from which

$$P_b^{\mathrm{PCCC2}} \approx \frac{4}{K}[f(8) + 2f(9) + 3f(10) + 4f(11) + 3f(12) + 2f(13) + f(14)] + \frac{24}{K}[f(16) + 2f(17) + f(18)], \tag{8.17}$$

where $f(w) = Q\left(\sqrt{2wRE_b/N_0}\right)$ and $R = 1/3$.

We have plotted (8.16) and (8.17) in Figure 8.5 for $K = 100$, together with simulation results for a randomly selected interleaver (the same interleaver for both codes) and 20 decoding iterations. As predicted, PCCC1, for which $g^{(1)}(D)$ is primitive, has the superior performance. We also see the somewhat surprising close agreement between the ensemble performance bounds and the simulated performance curves for the two specific codes. Beside the fact that we are comparing an ensemble result with a specific code result, this plot is "surprising" because the bounds are estimates for maximum-likelihood decoders, whereas the simulations use the turbo decoder of the previous chapter. We remark that, in each case, increasing $K$ by a factor of $\alpha$ will lower the curves by a factor $\alpha$, as is evident from the error-rate expressions for $P_b^{\mathrm{PCCC1}}$ and $P_b^{\mathrm{PCCC2}}$, (8.16) and (8.17).

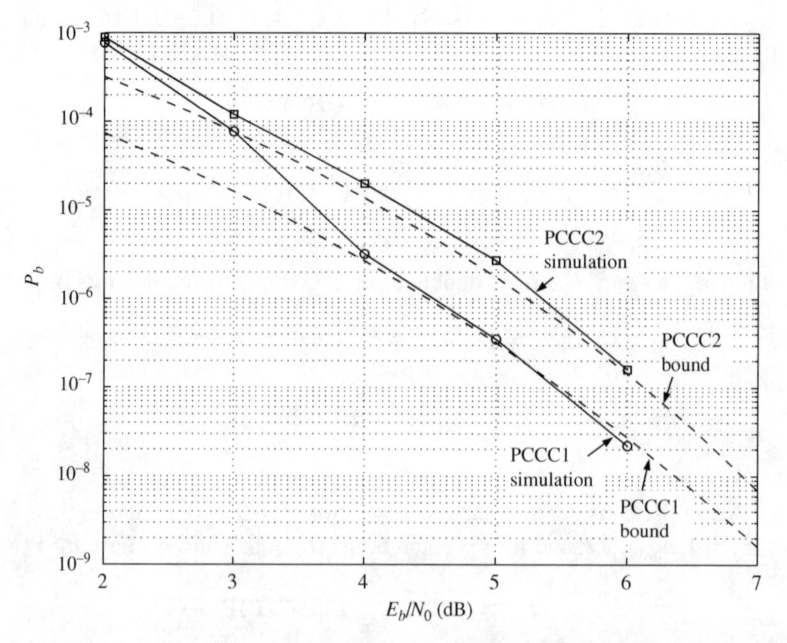


**Figure 8.5** A performance comparison between the PCCC1 and PCCC2 ensembles for $K = 100$. Simulation results are for one randomly selected instance in each ensemble, with 20 decoder iterations.

## 8.3 Ensemble Enumerators for Serial-Concatenated Codes

### 8.3.1 Preliminaries

As we did for the PCC case, we first examine how an enumerator for a serial-concatenated code (SCC) is related to that of its constituent codes, which may be either block or convolutional. The configuration assumed is pictured in Figure 8.6. The outer code, $C_1$, has parameters $(N_1, K_1)$ and code rate $R_1 = K_1/N_1$ and the inner code, $C_2$, has parameters $(N_2, K_2)$ and code rate $R_2 = K_2/N_2$. The overall SCC code rate is therefore $R = R_1R_2$. We shall assume that the size of the interleaver is $N_1$, although extension to the case $pN_1$, $p > 1$, is possible. As we shall see, in contrast with the PCC case for which the C-PWE $A_i(P)$ (or IP-WE $A(I,P)$) is appropriate (since parity weights are added), for SCCs the relevant WE is the C-OWE $A_i(W)$ (or IO-WE $A(I,W)$).

Consider an example for which $C_1$ is the $(4,3)$ SPC code and $C_2$ is the $(7,4)$ Hamming code, so that the interleaver size is 4. Consider also weight-1 SCC encoder inputs. For this code, $A_1^{C_1}(W) = 3W^2$ is obvious. Thus, a weight-1 SCC encoder input will yield a weight-2 $C_1$ output and, hence, a weight-2 $C_2$ input. Given that $A_2^{C_2}(W) = 3W^3 + 3W^4$ (from Section 8.1), we cannot know whether the $C_2$ (hence SCC) output has weight 3 or 4, since this depends on which of the $\binom{4}{2} = 6$ inputs occurs at the $C_2$ encoder input, which in turn depends on the interleaver chosen and its input. Thus, as was done for the PCC case, to avoid unmanageable combinatorics, we seek the average C-OWE, where the average is over all interleavers. This is easily done by letting the interleaver act as the uniform interleaver. Thus, in the present example, a given term in $A_1^{C_1}(W) = W^2 + W^2 + W^2$ will, in effect, select a term in $A_2^{C_2}(P) = W^3 + W^3 + W^3 + W^4 + W^4 + W^4$ with probability $1/\binom{4}{2} = 1/6$ in the computation of the C-OWE $A_1^{\text{SCC}}(W)$ for the SCC. That is, a given weight-2 $C_1$ output will select a given $C_2$ output with probability $1/6$. Thus, we have

$$A_{1,3}^{\text{SCC}} = \frac{A_{1,2}^{C_1} \cdot A_{2,3}^{C_2}}{\binom{4}{2}} = \frac{3 \cdot 3}{6} = 3/2.$$

As we will see in the next paragraph, in general, to obtain $A_{i,w}^{\text{SCC}}$ one would have to sum over all $C_1$ output weights possible for the given SCC input and output weights $i$ and $w$, but for this example there was only a single output weight: 2.

We now consider deriving $A_i^{\text{SCC}}(W)$ for the general SCC that employs the uniform interleaver. For constituent code $C_1$, the $\binom{K_1}{i}$ possible weight-$i$ inputs produce $A_{i,d}^{C_1}$ weight-$d$ outputs. For constituent code $C_2$, the $\binom{K_2}{d}$ weight-$d$ inputs produce

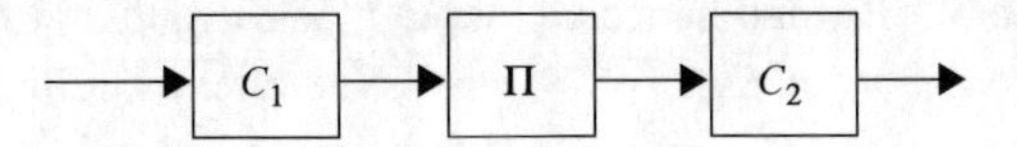


**Figure 8.6** The SCCC encoder: codes $C_1$ and $C_2$ separated by interleaver Π.

$A_{d,w}^{C_2}$ weight-$w$ outputs. Now consider concatenating $C_1$ with $C_2$ through a length-$K_2$ uniform interleaver. Each of the $A_{i,d}^{C_1}$ $C_1$ codewords will select one of the $A_{d,w}^{C_2}$ codewords of $C_2$ with probability $1/\binom{K_2}{d}$. Thus, the number of weight-$w$ codewords produced by weight-$i$ encoder inputs for the SCC with a length-$K_2$ uniform interleaver (equivalently, for the ensemble average) is given by the IO-WE coefficients $A_{i,w}^{\text{SCC}}$ computed as

$$A_{i,w}^{\text{SCC}} = \sum_{d=d_{\min,1}}^{K_2} \frac{A_{i,d}^{C_1} \cdot A_{d,w}^{C_2}}{\binom{K_2}{d}}, \tag{8.18}$$

where $d_{\min,1}$ is the minimum distance for the convolutional code $C_1$ considered as a block code. The (average) C-OWE for an SCC ensemble is thus

$$A_i^{\text{SCC}}(W) = \sum_{w=w_{\min}}^{N_2} A_{i,w}^{\text{SCC}} W^w = \sum_{d=d_{\min,1}}^{K_2} \frac{A_{i,d}^{C_1} \cdot A_d^{C_2}(W)}{\binom{K_2}{d}}, \tag{8.19}$$

where $A_d^{C_2}(W) = \sum_{w=w_{\min}}^{N_2} A_{d,w}^{C_2} W^w$ and $w_{\min}$ is the minimum SCC codeword weight. The IO-WE for an SCC ensemble is

$$A^{\text{SCC}}(I,W) = \sum_{i=i_{\min}}^{K_1} I^i A_i^{\text{SCC}}(W) = \sum_{d=d_{\min,1}}^{K_2} \frac{A_d^{C_1}(I) \cdot A_d^{C_2}(W)}{\binom{K_2}{d}}, \tag{8.20}$$

where $A_d^{C_1}(I) = \sum_{i=i_{\min}}^{K_1} A_{i,d}^{C_1} I^i$ enumerates the weights of the $C_1$ encoder inputs corresponding to weight-$d$ $C_1$ encoder outputs and $i_{\min}$ is the minimum input weight that creates a non-negligible error event in the SCC.

### 8.3.2 SCCC Ensemble Enumerators

We now consider SCCs whose constituent codes are convolutional codes, that is, serial-concatenated convolutional codes (SCCCs). SCCs with constituent block codes are possible, but, as we will see, those with constituent convolutional codes generally enjoy greater interleaver gain. Further, since convolutional-code trellises are less complex, their BCJR-based decoders are too. Our development follows that of [3].

The outer convolutional code $C_1$ has rate $R_1 = k/p = K_1/N_1$ and the inner convolutional code $C_2$ has rate $R_2 = p/m = K_2/N_2$, so that the SCCC rate is $R = k/m$. In these expressions, $K_1 = Sk$ is the SCCC (and hence $C_1$) input block-length and $K_2 = Sp$ is the $C_2$ input block-length, where $S = K_1/k = K_2/p$ is the number of trellis stages for both $C_1$and $C_2$. (One trellis stage in $C_1$ corresponds to $p$ output bits, which gives rise to one trellis stage in $C_2$, hence the equality.) The interleaver size is $K_2$ as before.

We now start with the upper bound on $P_b$, expressed as

$$P_b < \frac{1}{K_1} \sum_{i=i_{\min}}^{K_1} i A_i^{\mathrm{SCCC}}(W) \Bigg|_{W=\exp(-RE_b/N_0)}$$
$$= \frac{1}{K_1} \sum_{w=w_{\min}}^{K_2} \sum_{i=i_{\min}}^{K_1} i A_{i,w}^{\mathrm{SCCC}} \exp(-wRE_b/N_0), \tag{8.21}$$

where $A_{i,w}^{\mathrm{SCCC}}$ may be found for a given SCCC ensemble via (8.18). This implies that we will require knowledge of the IO-WE coefficients $A_{i,d}^{C_1}$ and $A_{d,w}^{C_2}$ for the constituent codes $C_1$ and $C_2$. Following the development of the PCCC case, which focused on parity weight, see (8.7), we can write for the equivalent block code of a constituent convolutional code within an SCCC with generic input length $K$,

$$A_i^{\mathrm{EBC}}(W) \lesssim \sum_{n=1}^{n_{\max}} \binom{S}{n} A_{i,n}(W). \tag{8.22}$$

We note that, in adapting (8.7) to obtain (8.22), $K$ is set to $S$ because the code rate assumed in (8.7) is 1/2. In (8.22),

$$A_{i,n}(W) = \sum_{\ell=1}^{K} A_{i,n,\ell}(W),$$

where $A_{i,n,\ell}(W)$ is the constituent convolutional-encoder output-weight enumerator conditioned on input weight $i$, which results in $n$ concatenated error events whose lengths total $\ell$ trellis stages. From (8.22), we may write for the $w$th coefficient of $A_i^{\mathrm{EBC}}(W)$,

$$A_{i,w}^{\mathrm{EBC}} \lesssim \sum_{n=1}^{n_{\max}} \binom{S}{n} A_{i,n,w}. \tag{8.23}$$

Substitution of (8.23) into (8.18) (twice, once for each constituent code) yields

$$A_{i,w}^{\mathrm{SCCC}} \lesssim \sum_{d=d_{\min,1}}^{K_2} \sum_{n_1=1}^{n_{1,\max}} \sum_{n_2=1}^{n_{2,\max}} \frac{\binom{S}{n_1}\binom{S}{n_2}}{\binom{K_2}{d}} A_{i,n_1,d} A_{d,n_2,w}. \tag{8.24}$$

As we did for the PCCC case, we can arrive at a simpler expression by further bounding $A_{i,w}^{\mathrm{SCCC}}$. Specifically, in the numerator we may (twice) use the bound

$$\binom{S}{n} < \frac{S^n}{n!}$$

and in the denominator we may use the bound

$$\binom{K_2}{d} > \frac{(K_2 - d + 1)^d}{d!} > \frac{(K_2)^d}{d^d d!},$$

where the second inequality follows from $(K_2 - d + 1) > K_2/d$ for all $d > 1$. Substitution of these bounds together with $S = K_2/p$ into (8.24) yields

$$A_{i,w}^{\text{SCCC}} \lesssim \sum_{d=d_{\min,1}}^{K_2} \sum_{n_1=1}^{n_{1,\max}} \sum_{n_2=1}^{n_{2,\max}} (K_2)^{n_1+n_2-d} \frac{d^d d!}{p^{n_1+n_2} \cdot n_1! n_2!} A_{i,n_1,d} A_{d,n_2,w}. \tag{8.25}$$

Finally, substitution of (8.25) into (8.21) yields

$$\begin{aligned} P_b < & \sum_{w=w_{\min}}^{K_2} \exp(-wRE_b/N_0) \\ & \times \sum_{i=i_{\min}}^{K_1} \sum_{d=d_{\min,1}}^{K_2} \sum_{n_1=1}^{n_{1,\max}} \sum_{n_2=1}^{n_{2,\max}} \frac{i}{k} (K_2)^{n_1+n_2-d-1} \\ & \times \frac{d^d d!}{p^{n_1+n_2-1} \cdot n_1! n_2!} A_{i,n_1,d} A_{d,n_2,w}, \end{aligned} \tag{8.26}$$

where we used $K_1 = (K_2/p)k$.

If we consider the quadruple sum in the second and third lines of (8.26) to be a coefficient of the exponential, then, for a given $w$, this coefficient is maximum (dominant) when the exponent of $K_2$, namely $n_1 + n_2 - d - 1$, is maximum. Noting that $n_1$ depends on the $C_1$ encoder input weight $i$ and $n_2$ depends on the $C_2$ encoder input weight $d$, we may write for this maximum exponent (which is a function of the SCCC codeword weight $w$)

$$\kappa_w = \max_{i,d} \{n_1 + n_2 - d - 1\}. \tag{8.27}$$

We now examine the worst-case exponent $\kappa_w$ for two important special cases.

### High-SNR Region: The Exponent $\kappa_w$ for $w = w_{\min}$

When $E_b/N_0$ is large, the $w = w_{\min}$ term in (8.26) can be expected to be dominant. Thus, we examine the worst-case exponent, $\kappa_w$, of $K_2$ for $w = w_{\min}$. Given $w = w_{\min}$, and recalling that $n_{2,\max}$ is the maximum number of error events possible for the given $w$, we may write

$$n_{2,\max} \leq \left\lfloor \frac{w_{\min}}{d_{\min,2}} \right\rfloor,$$

where $d_{\min,2}$ is the minimum weight for an error event of $C_2$. Similarly, for a given $C_1$ encoder output weight $d$,

$$n_{1,\max} \leq \left\lfloor \frac{d}{d_{\min,1}} \right\rfloor.$$

We may now rewrite (8.27) with $w = w_{\min}$ as

$$\begin{aligned} \kappa_{w_{\min}} & \leq \max_d \left\{ \left\lfloor \frac{d}{d_{\min,1}} \right\rfloor + \left\lfloor \frac{w_{\min}}{d_{\min,2}} \right\rfloor - d - 1 \right\} \\ & = \left\lfloor \frac{d(w_{\min})}{d_{\min,1}} \right\rfloor + \left\lfloor \frac{w_{\min}}{d_{\min,2}} \right\rfloor - d(w_{\min}) - 1, \end{aligned} \tag{8.28}$$

where $d(w_{\min})$ is the minimum weight $d$ of $C_1$ codewords that produces a weight-$w_{\min}$ $C_2$ codeword. It is generally the case that $d(w_{\min}) < 2d_{\min,1}$ and $w_{\min} < 2d_{\min,2}$, so that (8.28) becomes

$$\begin{aligned}\kappa_{w_{\min}} &\leq 1 - d(w_{\min}) \\ &\leq 1 - d_{\min,1},\end{aligned}$$

where the second inequality follows from the fact that $d(w_{\min}) \geq d_{\min,1}$ by virtue of the definition of $d(w_{\min})$.

Thus, focusing on the $w = w_{\min}$ term in (8.26), we have (for large values of $E_b/N_0$)

$$P_b \sim B_{\min} (K_2)^{1-d_{\min,1}} \exp(-w_{\min} R E_b/N_0), \tag{8.29}$$

where $B_{\min}$ is a constant corresponding to the four inner summations of (8.26) with $w = w_{\min}$. We see that, for relatively large values of $E_b/N_0$ (i.e., in the floor region), $P_b$ is dominated by the minimum distance of the SCCC, $w_{\min}$, and that interleaver gain is possible under the requirement that the outer code is non-trivial, that is, $d_{\min,1} > 1$. Further, the larger $w_{\min}$ and $d_{\min,1}$ are, the lower we can expect the floor to be. Note that an interleaver gain factor of $1/(K_2)^2$ is possible with only $d_{\min,1} = 3$, which is far better than is possible for PCCCs, which have an interleaver gain factor of $1/K$.

We remind the reader that (8.29) is a performance estimate for an SCCC *ensemble*. So an interleaver gain factor of $1/(K_2)^2$ in front of the exponential $\exp(-w_{\min} R E_b/N_0)$ means that, within the ensemble, a minimum distance of $w_{\min}$ for a given member of the ensemble is unlikely (since $1/(K_2)^2$ is so small) and a minimum distance greater than $w_{\min}$ is more common within the ensemble. That is, the performance for a *specific* SCCC will never go as $\left[B_{\min}/(K_2)^2\right]\exp(-w_{\min} R E_b/N_0)$; rather, taking the dominant ($w = w_{\min}$) term in (8.21), the performance of a specific SCCC will go as

$$[B_{\min}/K_1]\exp(-w_{\min} R E_b/N_0),$$

as for PCCCs. However, for specific codes, we can expect $w_{\min,\text{SCCC}}$ to be larger than $w_{\min,\text{PCCC}}$ because of the *serial* concatenation, resulting in lower error-rate floors for SCCCs. Lastly, we emphasize that these high-SNR results are true irrespective of the code type, namely block or convolutional, recursive or non-recursive.

### Low-SNR Region: The Maximum Exponent of $K_2$

The foregoing subsection examined $P_b$ for SCCC ensembles for large values of $E_b/N_0$, that is, for SCCC codeword weight fixed at $w = w_{\min}$. We now examine the low-SNR region, the so-called waterfall region that determines how closely a code's (or code ensemble's) performance approaches the capacity limit. In this case, the worst-case (over all $w$, $i$, and $d$) exponent $(n_1 + n_2 - d - 1)$ of $K_2$ in (8.26) comes into play. We denote this worst-case exponent by

$$\kappa^* = \max_{i,d,w}\{n_1 + n_2 - d - 1\}.$$

When the inner convolutional code is non-recursive, the maximum number of inner code error events, $n_{2,\max}$, is equal to $d$, so that

$$\kappa^* = n_{1,\max} - 1,$$

which is non-negative. Thus, low-SNR interleaver gain is not possible for non-recursive inner codes, so let us consider recursive inner convolutional codes.

For a recursive inner encoder input weight of $d$, the maximum number of error events is $n_{2,\max} = \lfloor d/2 \rfloor$, so that

$$\begin{aligned}\kappa^* &= \max_{i,d}\{n_{1,\max} + \lfloor d/2 \rfloor - d - 1\} \\ &= \max_{i,d}\left\{n_{1,\max} - \left\lfloor \frac{d+1}{2} \right\rfloor - 1\right\}.\end{aligned}$$

It can be shown [3] that $\kappa^*$ reduces to

$$\kappa^* = -\left\lfloor \frac{d_{\min,1}+1}{2} \right\rfloor,$$

so interleaver gain exists for a recursive inner code. Note that this result holds also for a rate-1 recursive inner code, such as an accumulator, as was studied in an example in the previous chapter.

We now summarize the SCCC design rules obtained from our ensemble performance analysis.

- At high SNR values, independently of the constituent encoder types, the ensemble interleaver gain factor is $(K_2)^{1-d_{\min,1}}$, so an outer code with a large value for $d_{\min,1}$ is mandatory if one seeks a code with a low error-rate floor.
- At low SNR values, no interleaver gain is realized when the inner code is non-recursive. When the inner code is recursive, the interleaver gain is $(K_2)^{-\lfloor (d_{\min,1}+1)/2 \rfloor}$, so we seek an outer code with a large $d_{\min,1}$ together with the inner recursive convolutional code.

## 8.4 Enumerators for Selected Accumulator-Based Codes

In this section we study weight enumerators for repeat–accumulate (RA) code ensembles and irregular repeat–accumulate (IRA) code ensembles. Both code types were introduced as a subclass of LDPC codes in Chapter 6, where it was shown that their parity-matrix structure is such that encoding may be accomplished using the parity-check matrix.

### 8.4.1 Enumerators for Repeat–Accumulate Codes

A repeat–accumulate code is a serially concatenated code whose outer constituent code is a rate-$1/q$ repetition code and whose inner constituent code is a rate-1 accumulator with transfer function $1/(1 + D)$. Thus, the encoder takes a length-$K$ data block and repeats it $q$ times before sending it through a length-$qK$ interleaver, whose output is then sent through the accumulator. Although the code may be made systematic by transmitting the length-$K$ data block together with the length-$qK$ accumulator output block, we consider only non-systematic $(qK, K)$ RA code ensembles. Application of (8.18) to RA code ensembles gives, for the IO-WE coefficients,

$$A_{i,w}^{\mathrm{RA}} = \sum_{d=q}^{qK} \frac{A_{i,d}^{\mathrm{rep}} \cdot A_{d,w}^{\mathrm{acc}}}{\binom{qK}{qi}}, \tag{8.30}$$

where $A_{i,d}^{\mathrm{rep}}$ is the number of weight-$d$ repetition-code codewords corresponding to weight-$i$ inputs and $A_{d,w}^{\mathrm{acc}}$ is the number of weight-$w$ accumulator codewords corresponding to weight-$d$ accumulator inputs. Note that, in comparison with (8.18), we have set $d_{\min,1} = q$ and $K_2 = qK$. The IO-WE coefficients for the repetition code with length-$K$ inputs are clearly given by

$$A_{i,d}^{\mathrm{rep}} = \begin{cases} \binom{K}{i}, & d = qi, \\ 0, & d \neq qi. \end{cases} \tag{8.31}$$

The IO-WE coefficients for the accumulator $1/(1 + D)$ with a length-$m$, weight-$d$ input and a length-$m$, weight-$w$ output are given [5] by

$$A_{d,w}^{\mathrm{acc}} = \binom{m-w}{\lfloor d/2 \rfloor}\binom{w-1}{\lceil d/2 \rceil - 1}. \tag{8.32}$$

We give a partial proof of this result as follows. First observe that error events are determined by all pairs of consecutive 1s at the accumulator input, and that the output due to two 1s separated by $h-1$ zeros is a string of $h$ consecutive 1s. Thus, $A_{d,w}^{\mathrm{acc}}$ should be determined in this light, that is, considering pairs of consecutive 1s among the $d$ 1s at the accumulator input (together with the "leftover" 1 when $d$ is odd). We consider a few values for $d$.

$\underline{d = 1}$: $A_{1,w}^{\mathrm{acc}} = 1 = \binom{m-w}{0}\binom{w-1}{0}$. To see that there is only a single weight-1 input word that yields a weight-$w$ output word, observe that the output from a single 1 followed by zeros is a string of 1s. This output string will have weight $w$ if and only if that single 1 at the accumulator input is followed by exactly $w - 1$ zeros that make up the end of the input block.

$\underline{d = 2}$: $A_{2,w}^{\mathrm{acc}} = m - w = \binom{m-w}{1}\binom{w-1}{0}$. To see that there are $m - w$ weight-2 input words that yield weight-$w$ output words, note that the pair of 1s at the accumulator input will yield $w$ 1s at the output only if they are separated by $w - 1$ zeros. Further, there are exactly $m - w$ ways in which a pair of 1s separated by $w - 1$ zeros can fit in a block of $m$ bits.

$\underline{d = 3}$: $A_{3,w}^{\mathrm{acc}} = (m - w)(w - 1) = \binom{m-w}{1}\binom{w-1}{1}$. In this case, the first two 1s create an error event and the third 1 yields at the accumulator output a string of 1s whose length $s$ equals $n_z + 1$, where $n_z$ is the number of zeros following the third 1. The length is $1 \leq s \leq w - 1$, for, if $s \geq w$, the weight of the output would exceed $w$. Thus, $0 \leq n_z \leq w - 2$, since $s = n_z + 1$. Consequently, for a given value of $n_z$, the output will have weight $w$, provided that the first two 1s are separated by $w - n_z - 1$ zeros. For each of the $w - 1$ values of $n_z$, the two 1s separated by $w - n_z - 1$ zeros can be in exactly $m - w$ places; hence, $A_{3,w}^{\mathrm{acc}} = (m - w)(w - 1)$.

The situation for $d = 4$ involves two error events, the situation for $d = 5$ involves two error events followed by a single 1, and so on for $d > 5$.

Substitution of (8.31) and (8.32) into (8.30) yields the following coefficients for the IO-WE for RA code ensembles:

$$A_{i,w}^{\mathrm{RA}} = \frac{\binom{K}{i}\binom{qK-w}{\lfloor qi/2 \rfloor}\binom{w-1}{\lceil qi/2 \rceil - 1}}{\binom{qK}{qi}}.$$

These results may be used to plot union-bound performance curves, which would show that RA codes perform very well. This is somewhat surprising because neither the repetition code alone nor the accumulator alone provides coding gain, but, when concatenated together through an interleaver, the resulting RA code provides substantial gain. For large values of $K$, the coding gain is within 1 dB of the capacity limit for $q \geq 4$ and within about 0.25 dB for $q \geq 8$. We will not examine RA code ensembles any further. Instead, we now examine the more practical IRA codes that allow more flexibility in code design and code-rate selection. The flexibility provided by irregularity leads to superior performance.

## 8.4.2 Enumerators for Irregular Repeat–Accumulate Codes

The techniques of the next section permit one to compute enumerators for IRA codes based on protographs, which are the most practical form of IRA code. However, it is illuminating to consider another subset of the IRA codes. This subset is characterized by the parameter triple $\langle N, K, w_c \rangle$, where $N$ is the codeword length, $K$ is the information word length, and $w_c$ is the column weight of $\mathbf{H}_u$, the systematic portion of the IRA code parity-check matrix, $\mathbf{H} = [\mathbf{H}_u \quad \mathbf{H}_p]$. We leave the row-weight distribution of $\mathbf{H}_u$ unspecified. To simplify the analysis, we assume that the $w_c$ 1s in each column of $\mathbf{H}_u$ are uniformly distributed and that the 1s in any two rows have at most one position in common, that is, there are no 4-cycles in the graph of $\mathbf{H}_u$.

It is convenient to use the following form of union bound on the bit-error rate for the IRA code ensemble,

$$P_b < \frac{1}{K} \sum_{w>0} \sum_{i:i+p=w} i A_{i,p} W^w \Bigg|_{W=e^{-RE_b/N_0}}, \tag{8.33}$$

where $A_{i,p}$ is the coefficient for the conditional parity-weight enumerator; that is, $A_{i,p}$ is the number of weight-$p$ parity words for weight-$i$ inputs. To facilitate the analysis, we use the encoder structure in Figure 8.7, which was introduced in Chapter 6, composed of the matrix $\mathbf{A} \triangleq \mathbf{H}_u^{\mathrm{T}} \mathbf{\Pi}^{-1}$, the uniform interleaver $\mathbf{\Pi}$, and the accumulator. If we treat the bottom branch of the encoder as a serial-concatenated code, then we may write

$$A_{i,p} = \sum_{\ell} \frac{A_{i,\ell}^{(\mathbf{A})} \cdot A_{\ell,p}^{(\mathrm{acc})}}{\binom{M}{\ell}}, \tag{8.34}$$

where $A_{i,\ell}^{(\mathbf{A})}$ is the IO-WE of the outer "code" corresponding to $\mathbf{A}$, $A_{\ell,p}^{(\mathrm{acc})}$ is the IO-WE of the inner accumulator, and $M = N - K$. From the previous section, the IO-WE of the accumulator is given by

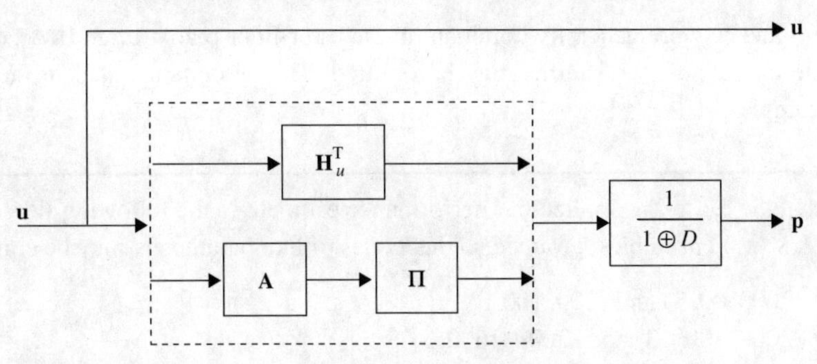


**Figure 8.7** An encoder for an irregular repeat–accumulate code.

$$A_{\ell,p}^{(\mathrm{acc})} = \binom{M-p}{\lfloor \ell/2 \rfloor}\binom{p-1}{\lceil \ell/2 \rceil - 1}. \tag{8.35}$$

The IO-WE $A_{i,\ell}^{(\mathbf{A})}$ of the outer "code" can be written as the number of weight-$i$ inputs, $\binom{K}{i}$, times the fraction, $F_{i,\ell}$, of those inputs that create weight $\ell$ outputs:

$$A_{i,\ell}^{(\mathbf{A})} = \binom{K}{i} F_{i,\ell}.$$

The computation of the fractions $F_{i,\ell}$ for all possible $i$ and $\ell$ is complex in general. However, in the high-SNR region $P_b$ is dominated by the low-weight codewords corresponding to low-weight inputs. Hence, we focus on $F_{i,\ell}$ for only small values of $i$. Now, when the input word $\mathbf{u}$ has weight $i = 1$, the output of "encoder" $\mathbf{A}$, given by $\mathbf{uA} = \mathbf{uH}_u^{\mathrm{T}}\mathbf{\Pi}^{-1}$, is simply a row of $\mathbf{H}_u^{\mathrm{T}}$ permuted by $\mathbf{\Pi}^{-1}$. But each row of $\mathbf{H}_u^{\mathrm{T}}$ has weight $w_c$ as per our assumption above. Thus, we have

$$F_{1,\ell} = \begin{cases} 1, & \text{for } \ell = w_c, \\ 0, & \text{for } \ell \neq w_c. \end{cases}$$

For weight-2 inputs ($i = 2$), the weight of the output of "encoder" $\mathbf{A}$ will be either $\ell = 2w_c$ or $\ell = 2w_c - 2$, since each weight-2 input adds two rows in $\mathbf{H}_u^{\mathrm{T}}$ and, by assumption, there is at most one overlap of 1s in the rows. Then we may write, for $i = 2$, $\ell = 2w_c$, and $\ell = 2w_c - 2$,

$$F_{2,\ell} = \frac{N_{2,\ell}}{N_{2,2w_c} + N_{2,2w_c-2}}, \tag{8.36}$$

where $N_{i,\ell}$ is the number of weight-$\ell$ outputs from "encoder" $\mathbf{A}$, given weight-$i$ inputs. The two terms $N_{i,\ell}$ that appear in (8.36) can be shown [7, 8] to be

$$N_{2,2w_c} = \frac{1}{2}\binom{M}{w_c}\binom{M-w_c}{w_c}$$

and

$$N_{2,2w_c-2} = \frac{1}{2}\binom{M}{w_c} w_c \binom{M-w_c}{w_c-1}.$$

$F_{1,\ell}$ and $F_{2,\ell}$ are generally dominant in the error-floor region of an IRA code's error-rate curve, so $i \geq 3$ terms may be omitted. This is demonstrated in the following example.

---

**Example 8.1** We applied our performance estimate to the following IRA code $\langle N, K, w_c \rangle$ ensembles, grouped so that codes of like parameters may be compared:

1. $\langle 200, 100, 3 \rangle$ and $\langle 200, 100, 5 \rangle$
2. $\langle 2000, 1000, 3 \rangle$ and $\langle 2000, 1600, 3 \rangle$
3. $\langle 200, 100, 3 \rangle$ and $\langle 2000, 1000, 3 \rangle$
4. $\langle 4161, 3431, 3 \rangle$, $\langle 4161, 3431, 4 \rangle$, and $\langle 4161, 3431, 5 \rangle$.

For each ensemble in the first three groups, two codes were randomly selected (designed) from the ensemble. For the last group, only one code is designed for each ensemble. The simulation results of the above four comparison groups are shown in Figure 8.8, together with the floor estimate (8.33). We observe that the floor estimates are consistent with the simulation results for the randomly selected codes. Moreover, we notice that increasing $w_c$ (to at least 5) tends to lower the floor dramatically.

---

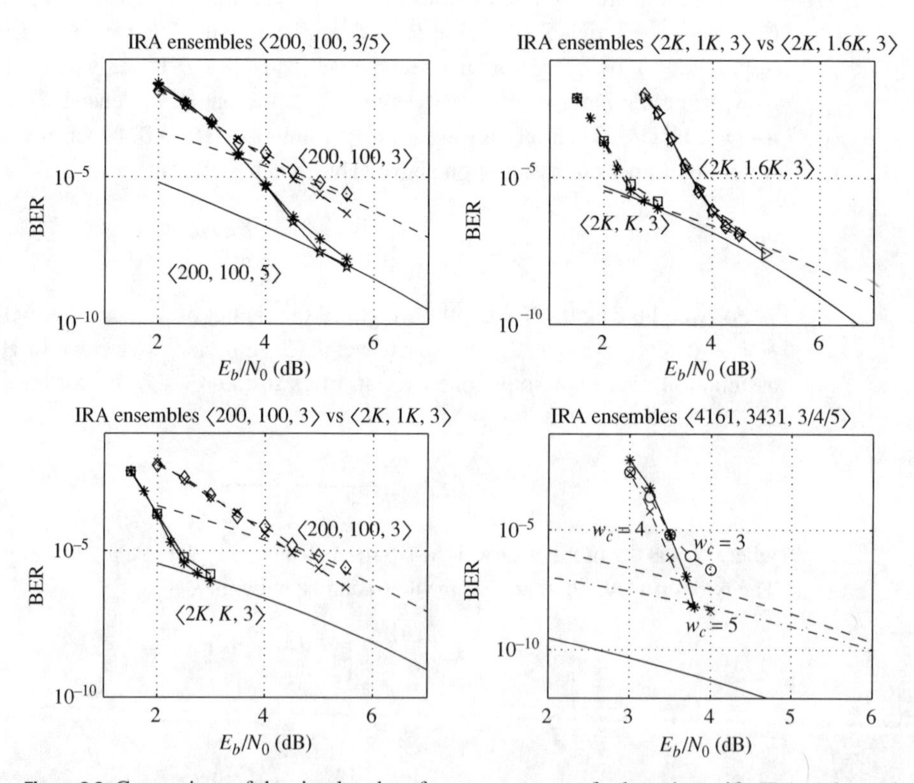


**Figure 8.8** Comparison of the simulated performance curves of selected specific IRA codes and their corresponding ensemble performance curves.

## 8.5 Enumerators for Protograph-Based LDPC Codes

The preceding sections in this chapter on enumerators first dealt with parallel turbo codes, then serial turbo codes, and then accumulator-based codes, which can be deemed as a hybrid of a turbo code and an LDPC code as discussed in Section 6.5.2. The next step in this progression is, naturally, LDPC codes. There exists a good number of results on enumerators for LDPC codes, but we will focus on LDPC codes based on protographs. The reason for this is that virtually all practical LDPC codes are quasi-cyclic and all quasi-cyclic codes are protograph codes. To see this, recall that, when the underlying permutation matrices involved in the protograph copy-and-permute procedure are circulant, the LDPC code is quasi-cyclic (see Chapter 6). Conversely, any quasi-cyclic code has an underlying protograph. Because essentially no extra work is involved, the results we present are for generalized LDPC (G-LDPC) codes, that is, codes whose Tanner graphs are sparse and contain constraint nodes beyond SPC nodes.

We first derive ensemble codeword weight enumerators for protograph-based G-LDPC codes for finite-length ensembles. We then use this result and extend it to infinite-length ensembles, and show how to determine when an ensemble has the property that its minimum distance grows linearly with code length. We then show how the codeword weight-enumerator approach may be adapted to obtain trapping-set enumerators and stopping-set enumerators. Our development closely follows [9–12]. For other important expositions on LDPC code ensemble enumerators, see [13–15].

A note of caution: as mentioned in the introduction to this chapter, if one finds an ensemble with excellent performance, then one can pick a "typical" code at random from the ensemble and expect excellent performance. However, code designers often choose a quasi-cyclic member from the ensemble. In the context of this section, this means that the various interleavers (see Figure 8.9) are chosen to be circulant permutation matrices. However, the fraction of quasi-cyclic codes goes to zero as the code length tends to infinity. That is, a quasi-cyclic code is not a typical member of the code ensemble and one cannot expect such a code to share the attractive properties seen in the ensemble average. (This was pointed out by D. Costello.) On the other hand, many excellent quasi-cyclic codes have been designed in the way suggested here [16–21].

### 8.5.1 Finite-Length Ensemble Weight Enumerators

Consider a G-LDPC protograph, $G = (V, C, E)$, where $V = \{v_1, v_2, \ldots, v_{n_v}\}$ is the set of $n_v$ variable nodes (VNs), $C = \{c_1, c_2, \ldots, c_{n_c}\}$ is the set of $n_c$ constraint nodes

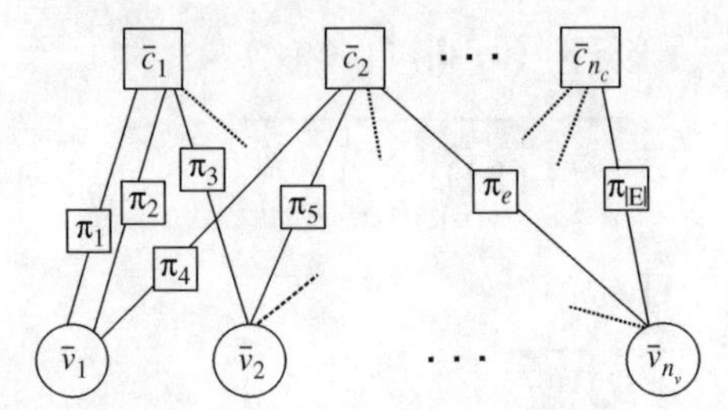

**Figure 8.9** A vectorized protograph, a shorthand for the copy-and-permute procedure.

(CNs), and $E$ is the set of edges. Denote by $q_{v_j}$ the degree of variable node $v_j$ and by $q_{c_i}$ the degree of constraint node $c_i$. Now consider the G-LDPC code constructed from a protograph $G$ by making $Q$ replicas of $G$ and using uniform interleavers, each of size $Q$, to permute the edges among the replicas of the protograph. As described in Chapter 6, this procedure is conveniently represented by an "expanded" protograph as in Figure 8.9. We refer to each of the $Q$ copies of protograph node $v_j$ as a "type $v_j$ node" and similarly for copies of protograph node $c_i$.

Recall that a length-$\ell$ uniform interleaver is a probabilistic device that maps each weight-$l$ input into the $\binom{\ell}{l}$ distinct permutations of it with equal probability, $1/\binom{\ell}{l}$. This approach allows ensemble-average weight enumerators to be derived for various types of concatenated codes. By exploiting the uniform-interleaver concept, the average weight enumerator for a protograph-based LDPC ensemble can be obtained. To do so, the VNs and CNs are treated as constituent codes in a serial-concatenated code (SCC) scheme, as explained further below.

In the case of two serially concatenated constituent codes, $C_1$ and $C_2$, separated by a uniform interleaver, the average number of weight-$w$ codewords created by weight-$l$ inputs in the SCC ensemble is given by (8.18). To apply this result to the G-LDPC code case, the group of $Q$ VNs of type $v_j$ is considered to be a constituent (repetition) code with a weight-$d_j$ input of length $Q$ and $q_{v_j}$ length-$Q$ outputs. Further, the group of $Q$ CNs of type $c_i$ is considered to be a constituent code with $q_{c_i}$ inputs, each of length $Q$, and a fictitious output of weight zero. Now let $A(\mathbf{w})$ be the average (over the ensemble) number of codewords having *weight vector* $\mathbf{w} = [w_1, w_2, \ldots, w_{n_v}]$ corresponding to the $n_v$ length-$Q$ inputs satisfying the protograph constraints. $A(\mathbf{w})$ is called the *weight-vector enumerator* for the ensemble of codes of length $Q \cdot n_v$ described by the protograph. Let us further define the following:

- $A^{v_j}(w_j, \mathbf{d}_j) = \binom{Q}{w_j}\delta_{w_j,d_{j,1}}\delta_{w_j,d_{j,2}} \cdots \delta_{w_j,d_{j,q_{v_j}}}$ is the weight-vector enumerator for the type-$v_j$ constituent code for a weight-$w_j$ input, where $\mathbf{d}_j = \left[d_{j,1}, d_{j,2}, \ldots, d_{j,q_{v_j}}\right]$ is a weight vector describing the constituent code's output (note that, while the elements of $\mathbf{w} = [w_1, w_2, \ldots, w_{n_v}]$ correspond to VN input weights, they may be deemed also as G-LDPC encoder output weights).
- $A^{c_i}(\mathbf{z}_i)$ is the weight-vector enumerator for the type-$c_i$ constituent code and $\mathbf{z}_i = \left[z_{i,1}, z_{i,2}, \ldots, z_{i,q_{c_i}}\right]$, where $z_{i,l} = d_{j,k}$ if the $l$th edge of CN $c_i$, is the $k$th edge of VN $v_j$.

A generalization of the development that led to (8.18) will yield for the G-LDPC protograph the average protograph weight-vector enumerator,

$$A(\mathbf{w}) = \sum_{w_{m,u}} \frac{\prod_{j=1}^{n_v} A^{v_j}(w_j, \mathbf{d}_j) \prod_{i=1}^{n_c} A^{c_i}(\mathbf{z}_i)}{\prod_{s=1}^{n_v} \prod_{r=1}^{q_{v_s}} \binom{Q}{d_{s,r}}}$$

$$= \frac{\prod_{i=1}^{n_c} A^{c_i}(\mathbf{w}_i)}{\prod_{j=1}^{n_v} \binom{Q}{w_j}^{q_{v_j}-1}}, \tag{8.37}$$

where the summation in the first line is over all weights $w_{m,u}$, where $w_{m,u}$ is the weight along the $u$th edge of VN $v_m$, where $m = 1, \ldots, n_v$ and $u = 1, \ldots, q_{v_m}$. The second line follows from the expression given above for $A^{v_j}(w_j, \mathbf{d}_j)$, a scaled product of Kronecker delta functions. The weight vector $\mathbf{w}_i = \left[w_{i_1}, w_{i_2}, \ldots, w_{i_{q_{c_i}}}\right]$ describes the weights of the $Q$-bit words on the edges connected to CN $c_i$, which are produced by the VNs neighboring $c_i$. The elements of $\mathbf{w}_i$ comprise a subset of the elements of $\mathbf{w}$.

Let $S_t$ be the set of transmitted VNs and let $S_p$ be the set of punctured VNs. Then the average number of codewords of weight $w$ in the ensemble, denoted by $A_w$, equals the sum of $A(\mathbf{w})$ over all $\mathbf{w}$ for which $\sum_{\{w_j : v_j \in S_t\}} w_j = w$. Notationally,

$$A_w = \sum_{\{w_j : v_j \in S_t\}} \sum_{\{w_k : v_k \in S_p\}} A(\mathbf{w}) \tag{8.38}$$

under the constraint $\sum_{\{w_j : v_j \in S_t\}} w_j = w$. Thus, to evaluate $A_w$ in (8.38), one first needs to compute the weight-vector enumerators, $A^{c_i}(\mathbf{w}_i)$, for the constraint nodes $c_i$, as seen in (8.37).

Consider the generic constituent $(\mu, \kappa)$ linear block code $\mathcal{C}$ in an expanded protograph. We need to find its weight-vector enumerator $A^{\mathcal{C}}(\mathbf{w})$, where $\mathbf{w} = [w_1, w_2, \ldots, w_\mu]$ is a constituent-code weight vector. The $\{A^{\mathcal{C}}(\mathbf{w})\}$ may easily be found as the coefficients of the multidimensional "W-transform" of $\{A^{\mathcal{C}}(\mathbf{w})\}$, as follows. Exploiting the uniform-interleaver property and the fact that the multi- dimensional W-transform of a single constraint node is $\sum_{\mathbf{x} \in \mathcal{C}} W_1^{x_1} W_2^{x_2} \ldots W_\mu^{x_\mu}$, the multidimensional W-transform for $Q$ copies of the protograph is

$$A^{\mathcal{C}}(W_1, W_2, \ldots, W_\mu) = \left(\sum_{\mathbf{x} \in \mathcal{C}} W_1^{x_1} W_2^{x_2} \ldots W_\mu^{x_\mu}\right)^Q, \tag{8.39}$$

where the $W_l$s are indeterminate bookkeeping variables and $\mathbf{x} = [x_1, x_2, \ldots, x_\mu]$, $x_l \in \{0, 1\}$, is a codeword in $\mathcal{C}$. Expanding the right-hand side of (8.39) will yield the form

$$A^{\mathcal{C}}(W_1, W_2, \ldots, W_\mu) = \sum_{\mathbf{w}} A^{\mathcal{C}}(\mathbf{w}) W_1^{w_1} W_2^{w_2} \ldots W_\mu^{w_\mu}, \tag{8.40}$$

from which we may obtain $A^{\mathcal{C}}(\mathbf{w})$. The direct application of the multinomial theorem on the right-hand side of (8.39) gives

$$A^{\mathcal{C}}(W_1, W_2, \ldots, W_\mu) = \sum_{\substack{n_1, n_2, \ldots, n_K \geq 0 \\ n_1 + n_2 + \cdots + n_K = Q}} C(Q; n_1, n_2, \ldots, n_K) \prod_{\mathbf{x} \in \mathcal{C}} \left(W_1^{x_1} W_2^{x_2} \ldots W_\mu^{x_\mu}\right)^{n_K}, \tag{8.41}$$

where $K = 2^\kappa$ is the number of codewords in $\mathcal{C}$, $n_K$ is the number of occurrences of the $K$th codeword, and $C\,(Q; n_1, n_2, \ldots, n_K)$ is the multinomial coefficient, given by

$$C(Q; n_1, n_2, \ldots, n_K) = \frac{Q!}{n_1! n_2! \ldots n_K!}. \tag{8.42}$$

Let $\mathbf{M}^{\mathcal{C}}$ be the $K \times \mu$ matrix with the codewords of $\mathcal{C}$ as its rows and $\mathbf{n} = [n_1, n_2, \ldots, n_K]$. Then (8.41) can be written as

$$A^{\mathcal{C}}(W_1, W_2, \ldots, W_\mu) = \sum_{\mathbf{w}} \sum_{\{\mathbf{n}\}} C\,(Q; n_1, n_2, \ldots, n_K)\, W_1^{w_1} W_2^{w_2} \ldots W_\mu^{w_\mu}, \qquad (8.43)$$

where $\{\mathbf{n}\}$ is the set of integer solutions to $\mathbf{w} = \mathbf{n} \cdot \mathbf{M}^{\mathcal{C}}$, under the constraints $n_1, n_2, \ldots, n_K \geq 0$ and $\sum_{k=1}^{K} n_k = Q$. To see the last step, note that the product in (8.41) can be manipulated as follows:

$$\prod_{\mathbf{x} \in \mathcal{C}} \left( W_1^{x_1} W_2^{x_2} \ldots W_\mu^{x_\mu} \right)^{n_k} = W_1^{w_1} W_2^{w_2} \ldots W_\mu^{w_\mu}, \qquad (8.44)$$

where $w_i = \sum_{\mathbf{x} \in \mathcal{C}} x_l n_k$, $l = \{1, 2, \ldots, \mu\}$. Also, if $\mathbf{w} = \mathbf{n} \cdot \mathbf{M}^{\mathcal{C}}$ has more than one solution of $\mathbf{n}$, the term $W_1^{x_1} W_2^{x_2} \ldots W_\mu^{x_\mu}$ will appear as a common factor in all of the terms that are associated with these solutions. This explains the presence of the second summation in (8.43). Finally, comparing (8.40) and (8.43) leads to the expression for the weight-vector enumerator,

$$A^{\mathcal{C}}(\mathbf{w}) = \sum_{\{\mathbf{n}\}} C(Q; n_1, n_2, \ldots, n_K), \qquad (8.45)$$

where $\{\mathbf{n}\}$ is the set of integer solutions to $\mathbf{w} = \mathbf{n} \cdot \mathbf{M}^{\mathcal{C}}$, with $n_1, n_2, \ldots, n_K \geq 0$ and $\sum_{k=1}^{K} n_k = Q$.

---

**Example 8.2** Consider the degree-4 SPC constraint node. The codeword set is $\mathcal{C} = \{0000, 1001, 0101, 1100, 0011, 1010, 0110, 1111\}$, so $K = 8$.
(a) Consider $Q = 3$ protograph copies and the weight vector $\mathbf{w} = [2, 2, 2, 2]$. From $\mathbf{w} = \mathbf{n} \cdot \mathbf{M}^{\mathcal{C}}$ and the associated constraints on $\mathbf{n}$, it is easy to see that $\{\mathbf{n}\} = \{[0, 0, 0, 1, 1, 0, 0, 1], [0, 0, 1, 0, 0, 1, 0, 1], [0, 1, 0, 0, 0, 0, 1, 1], [1, 0, 0, 0, 0, 0, 0, 2]\}$. From this, $A^{\mathcal{C}}(\mathbf{w}) = 21$ (via (8.45)).
(b) When $Q = 4$ and $\mathbf{w} = [4, 2, 2, 2]$, $\{\mathbf{n}\} = \{[0, 1, 0, 1, 0, 1, 0, 1]\}$ and so $A^{\mathcal{C}}(\mathbf{w}) = 24$.
(c) When $Q = 4$ and $\mathbf{w} = [3, 2, 2, 2]$, $\{\mathbf{n}\}$ is empty and so $A^{\mathcal{C}}(\mathbf{w}) = 0$.

---

**Example 8.3** Consider the protograph with a single (7, 4) Hamming constraint node and seven degree-1 VNs, all transmitted. Noting that the denominator in (8.37) is unity (since $q_{v_i} = 1$ for all $i$) and the numerator is $A^{c_1}(\mathbf{w}_1)$ (since $n_c = 1$), we will compute $A_w$ for $w = 0, 1, 2, 3$, assuming $Q = 4$ copies of the protograph. The Hamming code is generated by

$$\mathbf{G} = \begin{bmatrix} 1 & 0 & 0 & 0 & 0 & 1 & 1 \\ 0 & 1 & 0 & 0 & 1 & 1 & 0 \\ 0 & 0 & 1 & 0 & 1 & 1 & 1 \\ 0 & 0 & 0 & 1 & 1 & 0 & 1 \end{bmatrix},$$

from which the matrix $\mathbf{M}^{\mathcal{C}}$ may be obtained; we assume that the codewords are listed in the natural binary order with respect to the 16 input words. From (8.37) and (8.38), with $n_v = 7$ and $n_c = 1$, $A_w = \sum_{\mathbf{w}} A^{c_1}(\mathbf{w})$, where $\sum w_j = w$. Thus, $A_0 = A_{[0,0,0,0,0,0,0]} = C(4;4,0,\ldots,0) = 1$. $A_1 = \sum_{\mathbf{w}} A^{c_1}(\mathbf{w})$ with $\sum w_j = 1$, but any such $\mathbf{w}$ must result in $\{\mathbf{n}\}$ being empty. Consequently, $A_1 = 0$. Similarly, one finds $A_2 = 0$. With $A_3 = \sum_{\mathbf{w}} A^{c_1}(\mathbf{w})$ such that $\sum w_j = 3$, $\mathbf{w} = \{[1, 0, 0, 0, 0, 1, 1], [0, 1, 0, 0, 1, 1, 0], [1, 0, 1, 0, 1, 0, 0], [0, 1, 1, 0, 1, 0, 0], [0, 0, 0, 1, 1, 0, 1], [1, 1, 0, 1, 1, 0, 0, 0], [0, 0, 1, 1, 0, 1, 0]\}$. Each $\mathbf{w}$ yields only one solution $\mathbf{n}$ to the equation $\mathbf{w} = \mathbf{n} \cdot \mathbf{M}^{\mathcal{C}}$. Each solution has $n_1 = 3$ together with $n_k = 1$, where $k$ corresponds to a row in $\mathbf{M}^{\mathcal{C}}$ containing one of the seven weight-3 codewords. Note that the other $\mathbf{w}$s that achieve $\sum w_j = 3$ result in $\{\mathbf{n}\}$ being empty. Using (8.37), (8.38), and (8.45), $A_3 = \sum_{\mathbf{w}} A^{c_1}(\mathbf{w}) = \sum_{\{\mathbf{n}\}} C(4;3,0,\ldots,1,\ldots,0) = 28$.

---

## 8.5.2 Asymptotic Ensemble Weight Enumerators

For the asymptotic case, we define the normalized logarithmic asymptotic weight enumerator (we will simply call it the *asymptotic weight enumerator*) as

$$r(\delta) = \limsup_{n\to\infty} \frac{\ln A_w}{n}, \tag{8.46}$$

where $\delta \triangleq w/n$ (recall that $n$ is the number of transmitted variable nodes in the code). The matrix whose $2^k$ rows are the codewords of an $(n,k)$ linear code has equally probable ones and zeros in each of its $n$ columns. Thus, we can expect a fraction $\binom{n}{w}/2^n$ of these $2^k$ codewords to have weight $w$. From this, the weight enumerator for a rate-$R$, length-$n$ random linear code ensemble is

$$A_w = \binom{n}{w} 2^{-n(1-R)}, \tag{8.47}$$

so that the weight distribution is binomial for fixed $n$ and $R$. From this and *Stirling's approximation* of $n!$ for large $n$, namely, $n! \simeq \sqrt{2\pi n} n^n/e^n$, it can be shown that, for random linear codes,

$$r(\delta) = H(\delta) - (1-R)\ln(2),$$

where $H(\delta) = -(1-\delta)\ln(1-\delta) - \delta \ln \delta$.

It can be shown (with non-negligible effort) both for Gallager codes [22] and for protograph-based codes [23] that

$$A_w \le f(n) e^{nr(\delta)}, \tag{8.48}$$

where $f(n)$ is a sub-exponential function of $n$. (Compare this with a rearrangement of (8.46) with the lim sup operation removed: $A_w = e^{nr(\delta)}$.) The implication of (8.48) is that $A_w \simeq 0$ for large $n$ when $r(\delta) < 0$. This allows us to make statements about the minimum distance of a code ensemble, as follows. Observing that $r(0) = 0$, let $\delta_{\min}$ be the second zero-crossing of $r(\delta)$, if it exists. Assuming that $\delta_{\min}$ exists and $r(\delta) < 0$ for all $0 < \delta < \delta_{\min}$, $\delta_{\min}$ is called the *typical (normalized) minimum distance*. $\delta_{\min}$ is so called because, from its definition, we may conclude for large $n$ that $A_{\lfloor n\delta_{\min}\rfloor} \simeq 0$, so that the minimum distance of the code ensemble is lower bounded as $d_{\min} = w_{\min} \simeq n\delta_{\min}$. Observe from this that the ensemble $d_{\min}$ grows linearly with $n$, which is certainly a desirable property. For example, Gallager showed [22] that the typical minimum distance for regular LDPC codes with column weight 3 or greater increases linearly with code length $n$.

We now proceed to show how $\delta_{\min}$ may be numerically determined for protograph-based codes. Because the formulas in the previous section involve the number of copies $Q$ instead of the number of code bits $n$, it is convenient to define the function

$$\tilde{r}(\tilde{\delta}) = \lim_{Q\to\infty} \sup \frac{\ln A_w}{Q}, \tag{8.49}$$

where $\tilde{\delta} = w/Q$. Note that $n = |S_t| \cdot Q$ and so

$$r(\delta) = \frac{1}{|S_t|}\tilde{r}(|S_t| \cdot \delta).$$

We also define $\max^*(x, y) \triangleq \ln(e^x + e^y)$ and we similarly define $\max^*$ when more than two variables are involved. When $x$ and $y$ are large and distinct (so that $e^x$ and $e^y$ are vastly different), then $\max^*(x, y) \simeq \max(x, y)$. Similar comments apply for more than two variables.

From (8.38), we have

$$\begin{aligned}
\ln A_w &= \max^*_{\{w_l:v_l\in S_t\}} \left\{ \max^*_{\{w_k:v_k\in S_p\}} \{\ln A(\mathbf{w})\} \right\}, \\
&\approx \max_{\{w_l:v_l\in S_t\}} \left\{ \max_{\{w_k:v_k\in S_p\}} \{\ln A(\mathbf{w})\} \right\}, \\
&= \max_{\{w_l:v_l\in S_t\}} \max_{\{w_k:v_k\in S_p\}} \left\{ \sum_{i=1}^{n_c} \ln A^{c_i}(\mathbf{w}_i) - \sum_{j=1}^{n_v} (q_{v_j} - 1)\ln\binom{Q}{w_j} \right\},
\end{aligned}$$

under the constraint $\sum_{\{w_j:v_j\in S_t\}} w_j = w$. The second line holds when $Q$ is large and the third line is obtained by invoking (8.37). Taking the limit as $Q \to \infty$ and applying the result (from Stirling's formula)

$$\lim_{Q\to\infty} \sup \ln\binom{Q}{w_j}\Big/ Q = H(\tilde{\delta}_j) = -(1-\tilde{\delta}_j)\ln(1-\tilde{\delta}_j) - \tilde{\delta}_j \ln \tilde{\delta}_j,$$

where $\tilde{\delta}_j = w_j/Q$, we obtain

$$\tilde{r}(\tilde{\delta}) = \max_{\{\tilde{\delta}_l:v_l\in S_t\}} \max_{\{\tilde{\delta}_k:v_k\in S_p\}} \left\{ \sum_{i=1}^{n_c} a^{c_i}(\tilde{\delta}_i) - \sum_{j=1}^{n_v} (q_{v_j} - 1) H(\tilde{\delta}_j) \right\} \tag{8.50}$$

under the constraint $\sum_{\{\tilde{\delta}_j:v_j\in S_t\}} \tilde{\delta}_j = \tilde{\delta}$. In (8.50), $a^{c_i}(\tilde{\delta}_i)$ is the asymptotic weight-vector enumerator of the constraint node $c_i$, and $\tilde{\delta}_i = \mathbf{w}_i/Q$. For a generic constituent CN code $\mathcal{C}$, such an enumerator is defined as

$$a^{\mathcal{C}}(\omega) = \lim_{Q\to\infty} \sup \frac{\ln A^{\mathcal{C}}(\mathbf{w})}{Q}, \tag{8.51}$$

where $\omega = \mathbf{w}/Q$.

We may obtain a simple expression for $a^{\mathcal{C}}(\omega)$ using the method of types; see Chapter 12 of [24]. We define the type $P_\omega$ as the relative proportion of occurrences of each codeword of constituent CN code $\mathcal{C}$ in a sequence of $Q$ codewords. In other words, $P_\omega = [p_1, p_2, \ldots, p_K]$ is the empirical probability distribution of the codewords in $\mathcal{C}$ given a sequence of $Q$ such codewords, where $p_k = n_k/Q$ and $n_k$ is the number of occurrences of the $k$th codeword. Then the type class of $P_\omega$, $T(P_\omega)$, is the set of all length-$Q$ sequences of codewords in $\mathcal{C}$, each containing $n_k$ occurrences of the $k$th codeword in $\mathcal{C}$, for $k = 1, 2, \ldots, K$. Observe that $|T(P_\omega)| = C\,(Q; n_1, n_2, \ldots, n_K)$. From Theorem 12.1.3 in [24], $|T(P_\omega)| \to e^{Q \cdot H(P_\omega)}$ as $Q \to \infty$, where $H(P_\omega) = -\sum_{k=1}^{K} p_k \ln p_k$. Consequently, for $Q \to \infty$ we rewrite (8.45) as

$$\begin{aligned} A^{\mathcal{C}}(\mathbf{w}) &= \sum_{\{\mathbf{n}\}} C\,(Q; n_1, n_2, \ldots, n_K) \\ &= \sum_{\{P_\omega\}} |T(P_\omega)| \\ &\to \sum_{\{P_\omega\}} e^{Q \cdot H(P_\omega)}. \end{aligned} \tag{8.52}$$

It follows from (8.51) and (8.52) that

$$a^{\mathcal{C}}(\omega) = \max_{\{P_\omega\}} \{H(P_\omega)\}, \tag{8.53}$$

under the constraint that $\{P_\omega\}$ is the set of solutions to $\omega = P_\omega \cdot \mathbf{M}^{\mathcal{C}}$, where $p_1, p_2, \ldots, p_K \geq 0$ and $\sum_{k=1}^{K} p_k = 1$. These are the asymptotic equivalents of the constraints mentioned below (8.45).

---

**Example 8.4** In this example we evaluate the asymptotic weight enumerators for several related protograph-based LDPC code ensembles. In particular, we first consider the (6, 3) regular LDPC code ensemble, which has the protograph shown in Figure 8.10(a). (The simplest (6, 3) protograph has two degree-3 VNs connected to a single degree-6 CN, but we use the form in Figure 8.10(a) so that we may build upon it in the rest of Figure 8.10.) Its asymptotic weight-enumerator curve appears in

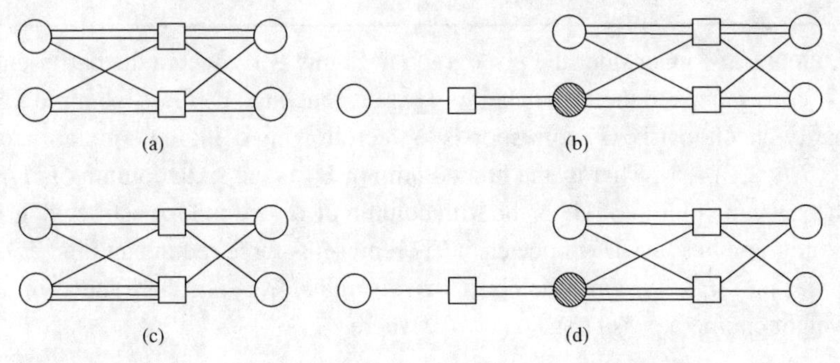


**Figure 8.10** Protographs for several rate-1/2 LDPC codes: (a) (6, 3) regular LDPC code, (b) precoded (6, 3) regular LDPC code, (c) RJA LDPC code, and (d) ARJA LDPC code. (The shaded VNs represent punctured nodes.)

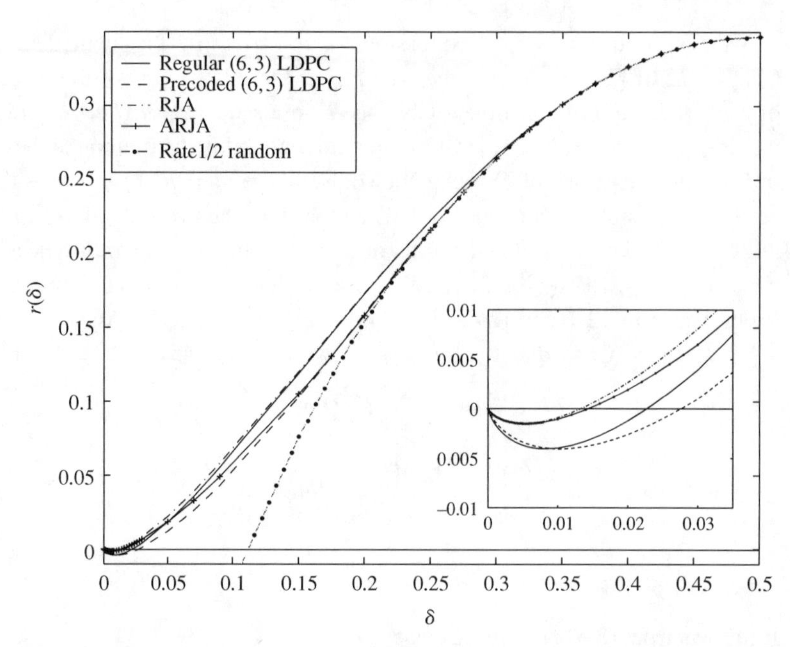


**Figure 8.11** Asymptotic weight enumerators for several protograph-based LDPC code ensembles [23].

Figure 8.11, which shows that this ensemble has $\delta_{\min} = 0.023$. (Again, $\delta_{\min}$ is the second zero-crossing of $r(\delta)$.) Consider also the asymptotic weight enumerator for a precoded version of the (6, 3) ensemble, where precoding is performed by an accumulator (see the discussion of ARA codes in Chapter 6). The protograph for this ensemble is shown in Figure 8.10(b), and the asymptotic enumerator results are given in Figure 8.11, which shows that precoding increases $\delta_{\min}$ from 0.023 to 0.028.

Next, we consider the RJA and the ARJA (precoded RJA) LDPC codes that have the protographs in Figures 8.10(c) and (d), respectively. The asymptotic weight enumerators for their ensembles appear in Figure 8.11. The latter figure shows that the RJA ensemble has $\delta_{\min} = 0.013$ and the ARJA has $\delta_{\min} = 0.015$.

---

**Example 8.5** Consider the protograph in Figure 8.12, where the parity-check matrix $\mathbf{H}_1$ corresponds to the (7, 4) Hamming code generated by $\mathbf{G}$ of Example 8.3. The parity-check matrix $\mathbf{H}_2$ corresponds to the following column permutation of $\mathbf{H}_1$: $(6, 7, 1, 2, 3, 4, 5)$. That is, the first column of $\mathbf{H}_2$ is the sixth column of $\mathbf{H}_1, \ldots,$ and the seventh column of $\mathbf{H}_2$ is the fifth column of $\mathbf{H}_1$. A code constructed as per this protograph has rate 1/7, since each CN represents three redundant bits. The variable nodes $v_1, v_2, \ldots, v_7$ have the normalized weights $\tilde{\delta}_1, \tilde{\delta}_2, \ldots, \tilde{\delta}_7$. The asymptotic weight enumerator is $r(\delta) = \tilde{r}(7\delta)/7$, where

$$\tilde{r}\left(\tilde{\delta}\right) = \max_{\tilde{\delta}_1, \ldots, \tilde{\delta}_7} \left\{ a^{\mathbf{H}_1}\left(\tilde{\delta}_1\right) + a^{\mathbf{H}_2}\left(\tilde{\delta}_2\right) - \sum_{j=1}^{7} H\left(\tilde{\delta}_j\right) \right\}, \tag{8.54}$$

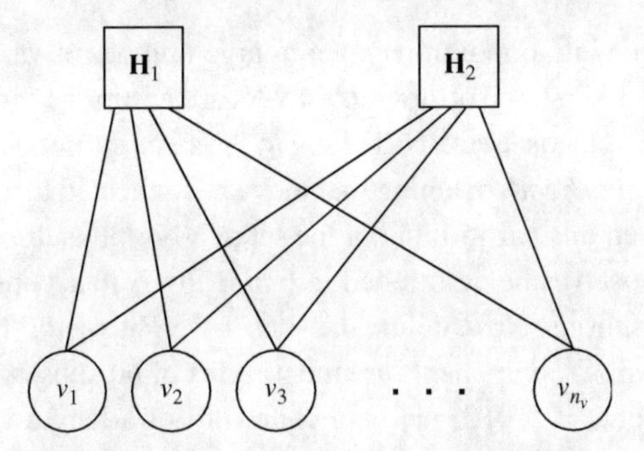


**Figure 8.12** A rate-1/7, $n_v = 7$ (or rate-7/15, $n_v = 15$) G-LDPC protograph.

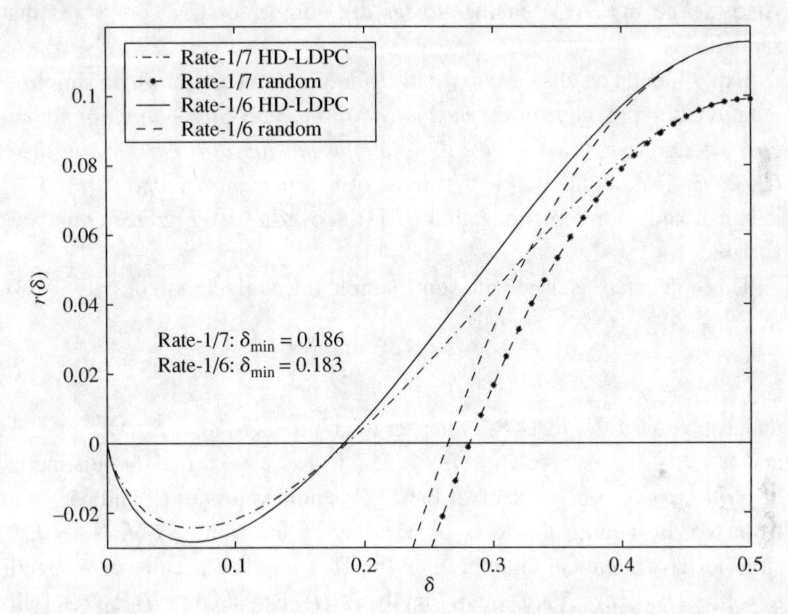


**Figure 8.13** Asymptotic weight enumerators for the rate-1/7 (1/6) G-LDPC code [23].

such that $\sum_{i=1}^{7} \tilde{\delta}_i = \tilde{\delta}$, and where $\tilde{\boldsymbol{\delta}}_1 = [\tilde{\delta}_1, \tilde{\delta}_2, \ldots, \tilde{\delta}_7]$ and $\tilde{\boldsymbol{\delta}}_2 = [\tilde{\delta}_6, \tilde{\delta}_7, \tilde{\delta}_1, \tilde{\delta}_2, \tilde{\delta}_3, \tilde{\delta}_4, \tilde{\delta}_5]$. We also evaluated the asymptotic weight enumerator for the rate-1/6 G-LDPC code that results from puncturing one of the VNs. The result is presented in Figure 8.13. Note that this ensemble has a large $\delta_{\min}$, relative to the earlier examples. Consequently, a long code based on this protograph has, with probability near unity, a good minimum distance.

## 8.5.3 On the Complexity of Computing Asymptotic Ensemble Enumerators

The drawback of the above method for evaluating the asymptotic enumerators can be seen from (8.53): the number of maximization arguments equals the number of codewords $K$ in the CN code $\mathcal{C}$. As an example, for the (15, 11) Hamming code, the maximization is over $K = 2^{11} = 2048$ variables.

To alleviate this issue of having to maximize over a large number of variables, the following steps are considered. First, the protograph's VNs are partitioned into subsets based on their neighborhoods. That is, two VNs belong to the same subset if and only if they have the same *type-neighborhood*, meaning that they are connected to an identical distribution of CN types. Given this partitioning of the set of VNs into subsets, the bits for a given CN code can themselves be partitioned into *bit subsets* in accordance with their membership in the VN subsets. Now, define the subset-weight vector (SWV) of a CN codeword as the vector whose components are the weights of bit subsets of the CN codeword. As an example, let $\bar{x} = [1010111]$ be a codeword of a length-7 CN code and let the CN code's bit subsets be $\{1,2,7\},\{3,4\},\{5,6\}$; then $\mathrm{SWV}(\bar{x}) = [2,1,2]$. Also, define the SWV enumerator as the number of CN codewords that have the same SWV.

Experimental results have led to the following conjecture, which simplifies the computational complexity: *in the maximization in (8.53), the optimal point occurs when codewords of equal SWV have the same proportion of occurrence in the constituent CN code*. The implication is that most of the elements of $P_{\omega} = [p_1, p_2, \ldots, p_K]$ are identical and so the maximization (8.53) is over a vastly reduced number of distinct variables.

This conjecture is used with some simple linear algebra to rewrite (8.53) as

$$a^{\mathcal{C}}(\omega) = \max_{\{\hat{P}\}} \left\{ H^*(\hat{P}) \right\}, \tag{8.55}$$

under the constraint that $\{\hat{P}\}$ is the set of solutions to $\omega = \hat{P} \cdot \hat{\mathbf{M}}^{\mathcal{C}}, p^{(1)}, p^{(2)}, \ldots \geq 0$ and $\Psi \cdot \hat{P}^{\mathrm{T}} = 1$, where $\hat{P} = [p^{(1)}, p^{(2)}, \ldots]$ is a vector of the distinct $p_k$s in $P_{\omega}$, $\Psi = [\psi_1, \psi_2, \ldots]$ is a vector of the SWV enumerators of $\mathcal{C}$, and $\hat{\mathbf{M}}^{\mathcal{C}}$ is constructed from $\mathbf{M}^{\mathcal{C}}$ by adding the rows of $\mathbf{M}^{\mathcal{C}}$ having the same SWV. Note that it is possible to have identical columns in $\hat{\mathbf{M}}^{\mathcal{C}}$. This implies that the corresponding $\omega_l$s in $\omega = [\omega_1, \omega_2, \ldots, \omega_\mu]$ are equal. Finally, $H^*(\hat{P})$ is related to $H(P_{\omega})$ as follows:

$$\begin{aligned} H(P_{\omega}) &= -\sum_{k=1}^{K} p_k \ln p_k \\ &= -\sum \left( p^{(l)} \ln p^{(l)} \right) \cdot \psi_l \\ &\triangleq H^*(\hat{P}). \end{aligned}$$

---

**Example 8.6** Consider again the protograph in Figure 8.12, but with (15, 11) Hamming codes for the constraints $\mathbf{H}_1$ and $\mathbf{H}_2$, where

$$\mathbf{H}_1 = [\mathbf{M}_1 \quad \mathbf{M}_2] = \begin{bmatrix} 1\,0\,1\,0\,1\,0\,1\,0 & 1\,0\,1\,0\,1\,0\,1 \\ 0\,1\,1\,0\,0\,1\,1\,0 & 0\,1\,1\,0\,0\,1\,1 \\ 0\,0\,0\,1\,1\,1\,1\,0 & 0\,0\,0\,1\,1\,1\,1 \\ 0\,0\,0\,0\,0\,0\,0\,1 & 1\,1\,1\,1\,1\,1\,1 \end{bmatrix},$$

$$\mathbf{H}_2 = [\mathbf{M}_2 \quad \mathbf{M}_1].$$

Note that there are $K = 2048$ codewords in each constituent code, so it would be difficult to evaluate (8.53) for this code. However, we may apply the conjecture because all of the VNs in this protograph have the same type-neighborhood. Also, after finding $\hat{\mathbf{M}}$, one finds that all of its columns are identical. Consequently, the VNs have the same normalized weight $\delta$.

The asymptotic weight enumerator is

$$r(\delta) = \frac{1}{15} \max_{\delta} \left\{ 2a^{\mathbf{H}_1}(\tilde{\boldsymbol{\delta}}) - 15H(\delta) \right\},$$

where $\tilde{\boldsymbol{\delta}} = [\delta, \delta, \ldots, \delta]$ (15 of them). Now, to find $a^{\mathbf{H}_1}(\tilde{\delta})$, define $p^{(\rho)}$ as the proportion of occurrence of a codeword of weight $\rho$ in the constituent CN code, so $\hat{P} = [p^{(0)}, p^{(3)}, p^{(4)}, p^{(5)}, p^{(6)}, p^{(7)}, p^{(8)}, p^{(9)}, p^{(10)}, p^{(11)}, p^{(12)}, p^{(15)}]$ and $\Psi = [1, 35, 105, 168, 280, 435, 435, 280, 168, 105, 35, 1]$. Consequently,

$$a^{\mathbf{H}_1}(\tilde{\delta}) = \max_{\{\hat{P}\}} \left\{ H^*(\hat{P}) \right\}, \tag{8.56}$$

under the constraint that $\{\hat{P}\}$ is the set of solutions to $\tilde{\delta} = \hat{P} \cdot \hat{\mathbf{M}}^{\mathcal{C}}$, $p^{(\rho)} \geq 0$, for all $p^{(\rho)}$ in $\hat{P}$, and $\Psi \cdot \hat{P}^{\mathrm{T}} = 1$. Clearly, under these assumptions, the computation of $a^{\mathbf{H}_1}(\delta)$, and hence of $r(\delta)$, is vastly simplified. The $r(\delta)$ result appears in Figure 8.14. Also included in Figure 8.14 is $r(\delta)$ for a rate-1/2 code obtained by puncturing one bit in the protograph. Note that, for both cases, $d_{\min}$ for the ensemble increases linearly with $n$.

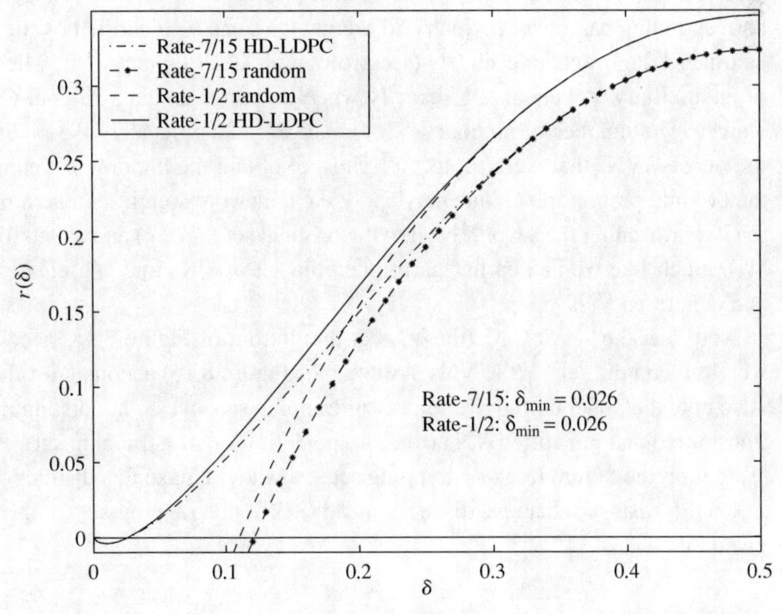


**Figure 8.14** Asymptotic weight enumerators for the rate-7/15 (1/2) G-LDPC code.

## 8.5.4 Ensemble Trapping-Set Enumerators

Classically, code design has involved finding code ensembles with large minimum distances. For modern codes with iterative decoders, additional work is necessary because iterative decoding gives rise to decoder pitfalls as a result of their distributed processing nature. The decoder pitfall associated with iterative decoders is a trapping set (see Chapter 5). An $(a, b)$ *general trapping set* [25], $\mathcal{T}_{a,b}$, is a set of VNs of size $a$, which induce a subgraph with exactly $b$ odd-degree check nodes (and an arbitrary number of even-degree check nodes). We use the qualifier "general" to make the distinction from elementary trapping sets, to be defined later. On the AWGN channel, an LDPC code will suffer from an error floor due to a small minimum distance or to trapping sets unless one selects from a "good" ensemble a code that has a large minimum distance (grows linearly with code length) and trapping sets with a negligible probability of occurrence.

### Finite-Size Trapping-Set Enumerators

The method for determining trapping-set enumerators for an LDPC code ensemble characterized by a given protograph involves first creating a modified protograph from the original protograph. Then the codeword weight enumerator for the modified protograph is determined, from which the trapping-set enumerator may be obtained. We describe the technique as follows.

Assume that we are interested in trapping sets of weight $a$ in the graph $G$ in Figure 8.9. The value of the companion parameter $b$ depends on which $a$ VNs are of interest, so we set to "1" the values of the $a$ VNs of interest and set to "0" the values of the remaining VNs. With the $a$ VNs so fixed, we are now interested in which CNs "see" an odd weight among their neighboring VNs, for the number of such CNs is the corresponding parameter $b$. Such odd-weight CNs can be identified by the addition of auxiliary "flag" VNs to each CN (see protograph $G'$ in Figure 8.15), where the value of an auxiliary VN equals "1" exactly when its corresponding original CN sees odd weight. The number of auxiliary VNs equal to "1" is precisely the parameter $b$ for the set of $a$ VNs that were set to "1." Thus, to obtain the trapping-set enumerator for the original protograph $G$, one may apply the codeword weight-enumerator technique to $G'$, partitioning the set of VNs into the original set $(S_t \cup S_p)$ and an auxiliary flag set $(S_f)$, much like we had earlier partitioned the set of VNs into subsets of transmitted and punctured VNs.

Note that the set $S_t \cup S_p$ (the VNs at the bottom of Figure 8.15) accounts for the weight $a$ and the set $S_f$ (the VNs at the top of Figure 8.15) accounts for the weight $b$. Also note that, in counting the number of trapping sets, we do not distinguish between transmitted and punctured VNs (refer to the definition of a trapping set). However, in evaluating the failure rate of a trapping set, one should make this distinction.

On the basis of the above discussion and (8.38), the trapping-set enumerator $A_{a,b}$ is given by

$$A_{a,b} = \sum_{\{w_j : v_j \in S_t \cup S_p\}} \sum_{\{w_k : v_k \in S_f\}} A(\mathbf{w}) \tag{8.57}$$

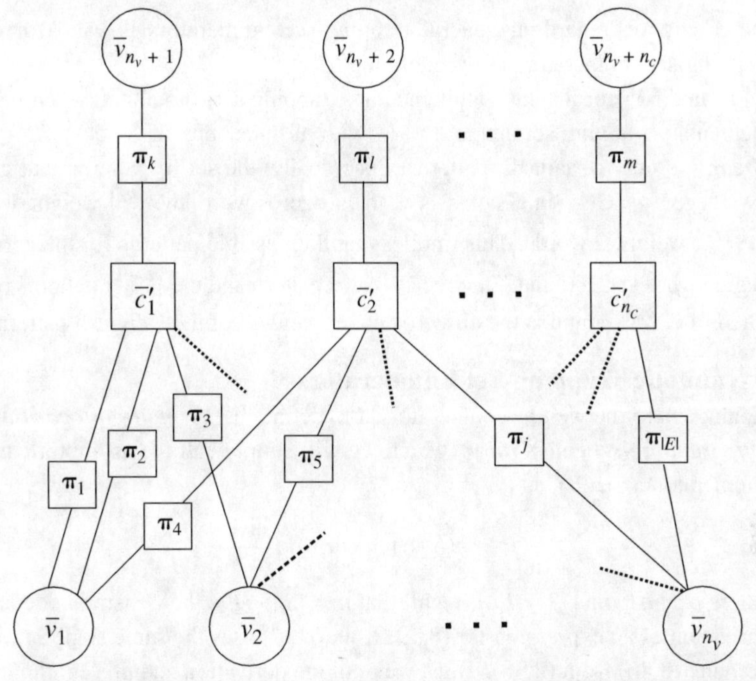


**Figure 8.15** A modified vectorized protograph $G'$ for trapping-set enumeration.

under the constraints $\sum_{\{w_j:v_j\in S_t\cup S_p\}} w_j = a$ and $\sum_{\{w_j:v_j\in S_f\}} w_j = b$, where

$$A(\mathbf{w}) = \frac{\prod_{i=1}^{n_c} A^{c'_i}(\mathbf{w}_i)}{\prod_{j=1}^{n_v} \binom{Q}{d_j}^{q_{v_j}-1}}. \tag{8.58}$$

Notice the use of $c'_i$ instead of $c_i$ in (8.58) to indicate that the weight-vector enumerators in (8.58) are the CNs in $G'$. Those weight-vector enumerators can be evaluated using (8.45).

## Elementary Trapping-Set Enumerators

An *elementary* $(a, b)$ *trapping set* $\mathcal{T}^{(e)}_{(a,b)}$ is a set of VNs that induces a subgraph with only degree-1 and degree-2 check nodes, and exactly $b$ of these CNs have degree 1. It has frequently been observed that the LDPC code error floors are due to elementary trapping sets. Some examples of elementary trapping sets are the $(12, 4)$ trapping set in the $(2640, 1320)$ Margulis code, and the $(4, 4)$ and $(5, 3)$ trapping sets in the $(1008, 504)$ and $(816, 408)$ MacKay codes. As a result, it is desired to compute the trapping-set enumerators just for the elementary trapping sets in the code ensembles.

To find elementary trapping-set enumerators, note that the only difference between an elementary and a general trapping set is the constraint on CN degrees in the induced subgraph of the former. This can be taken care of by choosing a proper matrix $\mathbf{M}^C$ in (8.45) and (8.53) when evaluating the weight-vector enumerators in (8.58) and the asymptotic weight-vector enumerators in (8.61), respectively. Consequently, the discussion regarding general trapping-set enumerators is valid for elementary trapping-set enumerators after redefining $\mathbf{M}^C$.

To find $\mathbf{M}^C$ for the case of elementary trapping sets, note that, when the bits of an elementary trapping set are set to "1" with all other bits in $G$ set to "0," the CNs in $G$ can see only certain bit patterns. Specifically, the set of patterns that can be seen by a degree-$q_c$ CN $c$ in $G$ consists of the all-zeros word, the $\binom{q_c}{1}$ weight-1 words, and the $\binom{q_c}{2}$ weight-2 words. This implies that the possible patterns for the corresponding degree-$(q_c+1)$ CN $c'$ in $G'$ are the all-zeros pattern and the $\binom{q_c+1}{2}$ patterns of weight 2. Therefore, $\mathbf{M}^C$ contains the all-zeros pattern and all of the weight-2 patterns.

### Asymptotic Trapping-Set Enumerators

Analogous to the weight-enumerator case, define the *normalized logarithmic asymptotic trapping-set enumerator* (which we will simply call the asymptotic trapping-set enumerator), $r(\alpha, \beta)$, as

$$r(\alpha, \beta) = \limsup_{n\to\infty} \frac{\ln A_{a,b}}{n}, \tag{8.59}$$

where $\alpha = a/n$ and $\beta = b/n$ (recall that $n = |S_t| \cdot Q$ is the transmit code length). The derivation of an expression for (8.59) from (8.57) uses the same steps as were used in deriving (8.46) from (8.38). Thus, we omit the derivation and present the final result:

$$r(\alpha, \beta) = \frac{1}{|S_t|}\tilde{r}(\alpha|S_t|, \beta|S_t|), \tag{8.60}$$

where

$$\tilde{r}(\tilde{\alpha}, \tilde{\beta}) = \max_{\{\tilde{\delta}_l : v_l \in S_t \cup S_p\}} \left\{ \max_{\{\tilde{\delta}_k : v_k \in S_f\}} \left\{ \sum_{i=1}^{n_c} a^{c'_i}(\tilde{\delta}_i) - \sum_{j=1}^{n_v} (q_{v_j} - 1) H(\tilde{\delta}_j) \right\} \right\}, \tag{8.61}$$

under the constraints $\sum_{\{\tilde{\delta}_j : v_j \in S_t \cup S_p\}} \tilde{\delta}_j = \tilde{\alpha}$ and $\sum_{\{\tilde{\delta}_j : v_j \in S_f\}} \tilde{\delta}_j = \tilde{\beta}$. The asymptotic weight-vector enumerator, $a^{c'_i}(\tilde{\delta}_i)$, can be evaluated using (8.53). (To avoid ambiguity, note the use of $a^c(\cdot)$ to refer to the asymptotic weight-vector enumerator and the use of $a$ to refer to the size of the $(a, b)$ trapping set.) It is possible to establish an analogy to the typical minimum distance in the weight-enumerator problem, but this is beyond our scope.

## 8.5.5 Ensemble Stopping-Set Enumerators

In the context of iterative decoding of LDPC codes on the BEC, a *stopping set* $S$ is a subset of the set of VNs whose neighboring SPC nodes of $S$ are connected to $S$ at least twice. The implication of this definition is that, if all of the bits in a stopping set are erased, the iterative erasure decoder will never resolve the values of those bits. For *single-CN-type* G-LDPC codes, a *generalized stopping set* $S^d$ is a subset of the set of VNs, such that all neighbors of $S^d$ are connected to $S^d$ at least $d$ times, where $d-1$ is the erasure capability of the neighboring CNs.

In this section, in addition to single-CN-type G-LDPC codes, we consider multi-CN-type G-LDPC codes, for which a mixture of CN types is permissible. We assume the use of a standard iterative erasure decoding tailored to the Tanner graph of the code, with bounded-distance algebraic decoders employed at the CNs. That is, CN $c_i$ can correct up to $d_{\min}^{(i)} - 1$ erasures. A set of erased VNs will fail to decode exactly when all of the neighboring CNs see more erasures than they are capable of resolving. Consequently, we introduce a new definition for the generalized stopping set as follows. A generalized stopping set $S^D$ of a G-LDPC code is a subset of the set of VNs, such that every neighboring CN $c_i$ of $S^D$ is connected to $S^D$ at least $d_{\min}^{(i)}$ times, for all $c_i$ in the neighborhood of $S^D$, where $D$ is the set of $d_{\min}^{(i)}$s for the neighboring CNs. Hereafter, we will use the term "stopping set" to refer to a generalized stopping set.

Much like in the trapping-set case, the method for finding weight enumerators for protograph-based G-LDPC code ensembles can be leveraged to obtain stopping-set enumerators for these ensembles. To see how, let us consider the mapping $\phi$ from the set of stopping sets $\{S^D\}$ to $\mathbb{F}_2^n$ defined as $\phi(S^D) = [x_0, x_1, \ldots, x_{n-1}]$, where $x_j = 1$, if and only if the corresponding VN $v_j$ in the Tanner graph is in the stopping set $S^D$. The set of all binary words $\{\phi(S^D)\}$ corresponding to a Tanner graph $G$ need not form a linear code, although the set is closed under bit-wise OR since the set of stopping sets $\{S^D\}$ is closed under unions. We call the (nonlinear) code $\{\phi(S^D)\}$ induced by the set of stopping sets of $G$ under the map $\phi$ a *stopping-set code* and we denote it by $C_{ss}(G) = \left\{\phi(S^D)\right\}$. Note that the weight of $\phi(S^D)$ equals the size of $S^D$, so the weight enumerator for the stopping-set code $C_{ss}(G)$ is identical to the stopping-set enumerator for $G$. For example, in a graph $G$ with a single CN $\mathcal{C}$, where $\mathcal{C}$ is a $(\mu, \kappa)$ linear block code of minimum distance $d_{\min}$, the stopping-set code is $C_{ss}(G) = \{\bar{x} \in \mathbb{F}_2^{\mu}\colon \text{weight}(\bar{x}) \geq d_{\min}\}$. Thus, the stopping-set enumerator for this simple graph is exactly the weight enumerator for $C_{ss}(G)$. This simple example also lets us speak of a stopping-set code for a single CN $c_i$, which is $C_{ss}(c_i) = \{\bar{x} \in \mathbb{F}_2^{\mu}\colon \text{weight}(\bar{x}) \geq d_{\min}^{(i)}\}$.

In the case of a nontrivial G-LDPC Tanner graph $G$, we would first find the stopping-set codes $C_{ss}(c_i)$ for each of the CNs $c_i$. We then form a new graph, $G'$, from $G$ by replacing CN $c_i$ in $G$ by $C_{ss}(c_i)$, for all $i$. The stopping-set enumerator for $G$ is then given by the weight enumerator for $C_{ss}(G')$, which is given by $C_{ss}(G') = \{\bar{x} \in \mathbb{F}_2^n\colon \bar{x} \text{ satisfies the constraints of } G'\}$. In summary, the ensemble *stopping-set enumerator* for a protograph-based G-LDPC code with graph $G$ is identically the ensemble *weight enumerator* for the graph $G'$ formed from $G$ by replacing the CNs in $G$ by their corresponding stopping-set code constraint nodes.

## Problems

**8.1** Consider the $(7, 4)$ Hamming code generated in systematic form by the generator $g(x) = 1 + x + x^3$. Plot the BER $P_b$ and FER $P_{cw}$ using (8.3) and (8.4) for $P_b$ and (8.1) and (8.2) for $P_{cw}$.

**8.2** Consider the $(8, 4)$ extended Hamming code generated in systematic form by the generator $g(x) = 1 + x^2 + x^3 + x^4$. Find the following enumerators for this code: $A(W)$, $A(I, P)$, $A_i(P)$ for $i = 1, 2, 3, 4$; $A(I, W)$, $A_i(W)$ for $i = 1, 2, 3, 4$; and $B(W)$, $B(I, P)$, $B_i(P)$ for $i = 1, 2, 3, 4$.

**8.3** Plot the $P_b$ ensemble bound for $\text{PCCC}_1$ whose constituent codes both have the RSC generator matrix

$$G_1(D) = \left[ 1 \quad \frac{1}{1+D} \right].$$

Compare this with the $P_b$ ensemble bound for PCCC$_2$ whose constituent codes both have the $\bar{\text{R}}$SC generator matrix

$$G_2(D) = [1 + D \quad 1].$$

Assume an information block of 200 bits in both cases.

**8.4** Plot the $P_b$ bound for an SCCC ensemble whose outer code has the $\bar{\text{R}}$SC generator matrix $G(D) = [1 + D \quad 1]$ and whose inner code is the rate-1 accumulator with transfer function $1/(1 + D)$. Assume an information block of 200 bits.

**8.5** Do the previous problem but now with outer code generator matrix $G(D) = \left[1 + D + D^2 \quad 1 + D\right]$.

**8.6** Plot the $P_b$ and $P_{cw}$ bounds for the (non-systematic) rate-1/3 RA code ensemble with codeword length 900.

**8.7** Find the input–output weight-enumerator coefficients $A_{i,w}$ for a rate-$1/(q + 1)$ *systematic* RA code.

**8.8** Plot the $P_b$ bounds for the rate-1/2 (2000, 1000) IRA code ensembles with column-weight parameter $w_c = 3$, 4, and 5.

**8.9** Consider the protograph-based LDPC codes considered in Section 8.5. For a given protograph LDPC code ensemble, show that the fraction of such codes that are quasi-cyclic goes to zero as the code length goes to infinity.

**8.10** Consider a (3, 2) SPC constraint node in a protograph. Find the vector weight enumerator $A(\mathbf{w})$ for $Q = 3$ copies of this code.

**8.11** Consider a (4, 3) SPC constraint node in a protograph. Find the vector weight enumerator $A(\mathbf{w})$ for $Q = 3$ copies of this code.

**8.12** Show that the weight enumerator for a rate-$R$, length-$n$ random linear code ensemble is

$$A_w = \binom{n}{w} 2^{-n(1-R)}.$$

From this and Stirling's approximation of $n!$ for large $n$, namely, $n! \simeq \sqrt{2\pi n} n^n / e^n$, show that, for random linear codes,

$$r(\delta) = H(\delta) - (1 - R)\ln(2),$$

where $H(\delta) = -(1 - \delta)\ln(1 - \delta) - \delta \ln \delta$ and $\delta = w/n$.

**8.13** (Project) Reproduce the asymptotic weight-enumerator curve $r(\delta)$ versus $\delta$ for the regular (6, 3) code ensemble that appears in Figure 8.11. The (6, 3) protograph has two degree-3 VNs connected to a single degree-6 CN. Your result should show that $\delta_{\min} = 0.023$.

**8.14** (With acknowledgment of D. Divsalar and S. Abu-Surra) (a) Show that the weight-vector enumerator for $Q$ copies of a $(4,3)$ SPC constraint node can be written as

$$A_Q^{(4,3)}(\mathbf{w}) = \sum_{l=1}^{Q} \frac{A_Q^{(3,2)}([w_1, w_2, l])A_Q^{(3,2)}([w_3, w_4, l])}{\binom{Q}{l}},$$

where $A_Q^{(k+1,k)}(\mathbf{w}) = A_Q^{(k+1,k)}([w_1, w_2, \ldots, w_{k+1}])$ is the weight-vector enumerator for $Q$ copies of a $(k+1,k)$ SPC constraint node. *Hint*: Consider a concatenation of constraint nodes. (b) Find an expression for the weight-vector enumerator for $Q$ copies of a $(6,5)$ SPC constraint node in terms of $A_Q^{(3,2)}(\mathbf{w})$.

**8.15** (With acknowledgment of D. Divsalar and S. Abu-Surra) (a) For the protograph below, argue that the optimizing values for $\tilde{\delta}_1$ and $\tilde{\delta}_2$ in (8.50) for VNs $v_1$ and $v_3$, respectively, must be equal. (b) Find $\hat{\mathbf{M}}$ and $\Psi$, as defined in (8.55) for the constraint node $c_2$. Use $\hat{\mathbf{M}}$ to explain your answer part (a).

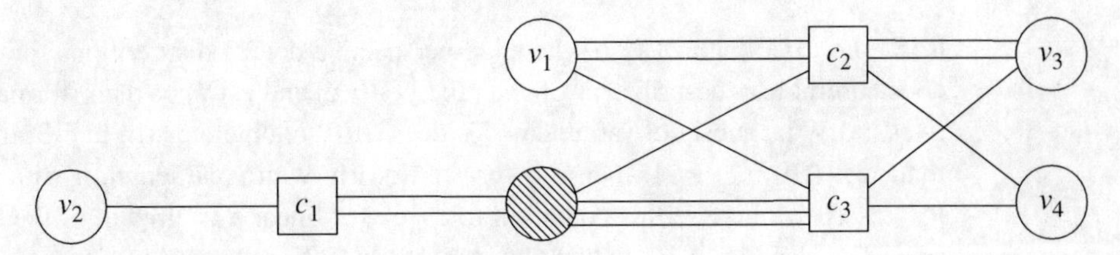


**8.16** (With acknowledgment of D. Divsalar and C. Jones) The figure below presents the protograph of a rate-1/6 RA code punctured to rate 1/2. (The shaded VNs are punctured.) Show that the punctured protograph can be reduced to an equivalent, simpler protograph with one (punctured) degree-6 VN, two degree-5 CNs, and two degree-2 VNs. Sketch this equivalent protograph.

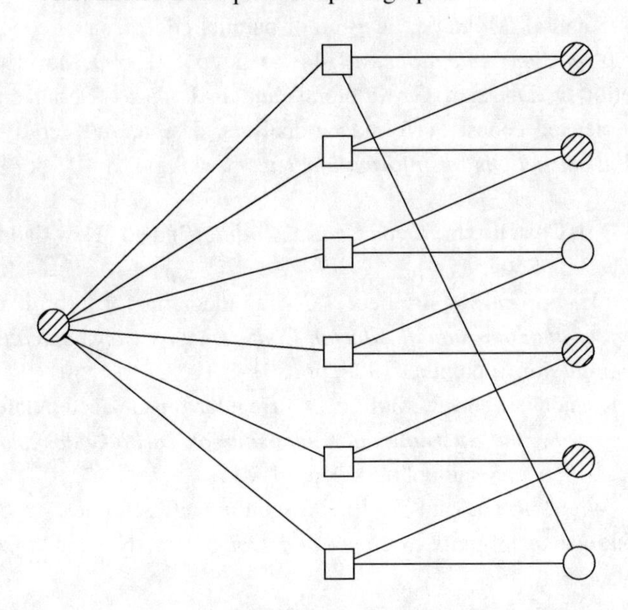

**8.17** (With acknowledgment of S. Abu-Surra) Argue that the minimum distance increases linearly with code length for the rate-2/5 ensemble corresponding to the protograph below, without the puncturing indicated. *Hint*: Using the ideas of the previous problem and Gallager's result for regular LDPC codes, show that the rate-2/3 ensemble corresponding to the punctured case is such that $d_{\min}$ grows linearly with code length. Then use the fact that the rate-2/3 ensemble's codewords are "sub-words" of the rate-2/5 ensemble's codewords.

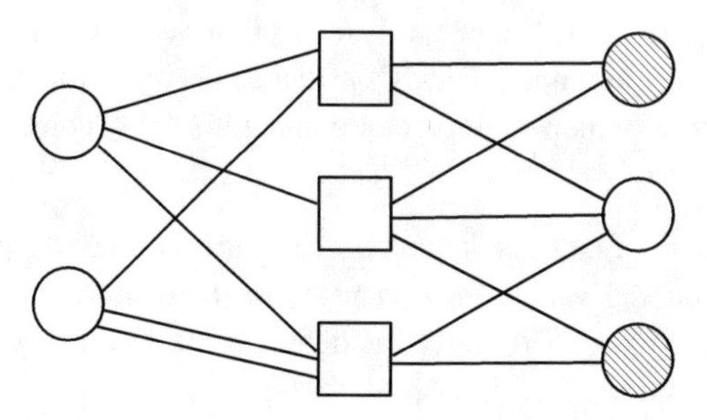

**8.18** Let $\lambda(X)$ and $\rho(X)$ be the edge-perspective degree distributions for an LDPC code ensemble as described in Chapter 5. Let $\lambda'(X)$ and $\rho'(X)$ be their (formal) derivatives. (a) What subset of variable nodes does $\lambda'(0)$ enumerate? (b) In [13] it is shown that, if $\lambda'(0)\rho'(1) < 1$, then $d_{\min}$ grows linearly with code length $n$ with probability $\sqrt{\lambda'(0)\rho'(1)}$. Use this condition to check for linear $d_{\min}$ growth of the following rate-1/2 ensembles presented in Chapter 6: RA, IRA, and ARJA.

## References


[1] S. Benedetto and G. Montorsi, "Unveiling turbo codes: Some results on parallel concatenated coding schemes," *IEEE Transactions on Information Theory*, vol. 42, no. 3, pp. 409–428, March 1996.

[2] S. Benedetto and G. Montorsi, "Design of parallel concatenated codes," *IEEE Transactions on Communications*, vol. 44, no. 5, pp. 591–600, May 1996.

[3] S. Benedetto, D. Divsalar, G. Montorsi, and F. Pollara, "Serial concatenation of interleaved codes: Performance analysis, design, and iterative decoding," *IEEE Transactions on Information Theory*, vol. 44, no. 5, pp. 909–926, May 1998.

[4] S. Lin and D. J. Costello, Jr., *Error Control Coding*, 2nd ed., New Saddle River, NJ, Prentice-Hall, 2004.

[5] D. Divsalar, H. Jin, and R. McEliece, "Coding theorems for turbo-like codes," *Proceedings of the 36th Annual Allerton Conference on Communication, Control, and Computing*, September 1998, pp. 201–210.

[6] H. Jin, A. Khandekar, and R. McEliece, "Irregular repeat–accumulate codes," *Proceedings of the 2nd International Symposium on Turbo Codes and Related Topics*, Brest, France, September 2000, pp. 1–8.

[7] Y. Zhang, W. E. Ryan, and Y. Li, "Structured eIRA codes," *38th Asilomar Conference on Signals, Systems and Computers*, November 2004, pp. 2005–2009.

[8] Y. Zhang and W. E. Ryan, "Structured IRA codes: Performance analysis and construction," *IEEE Transactions on Communications*, vol. 55, no. 5, pp. 837–844, May 2007.

[9] D. Divsalar, "Ensemble weight enumerators for protograph LDPC codes," *IEEE International Symposium on Information Theory*, July 2006, pp. 1554–1558.

[10] S. Abu-Surra, W. E. Ryan, and D. Divsalar, "Ensemble weight enumerators for protograph-based generalized LDPC codes," UCSD ITA Workshop, January 2007, pp. 342–348.

[11] S. Abu-Surra, W. E. Ryan, and D. Divsalar, "Ensemble trapping set enumerators for protograph-based LDPC codes," *Proceedings of the 45th Annual Allerton Conference on Communication, Control, and Computing*, October 2007.

[12] S. Abu-Surra, W. E. Ryan, and D. Divsalar, "Ensemble enumerators for protograph-based generalized LDPC codes," *2007 IEEE Global Telecommunications Conference*, November 2007, pp. 1492–1497.

[13] C. Di, T. Richardson, and R. Urbanke, "Weight distribution of low-density parity-check codes," *IEEE Transactions on Information Theory*, vol. 52, no. 11, pp. 4839–4855, November 2006.

[14] O. Milenkovic, E. Soljanin, and P. Whiting, "Asymptotic spectra of trapping sets in regular and irregular LDPC code ensembles," *IEEE Transactions on Information Theory*, vol. 53, no. 1, pp. 39–55, January 2007.

[15] T. Richardson and R. Urbanke, *Modern Coding Theory*, Cambridge, Cambridge University Press, 2008.

[16] D. Divsalar and C. Jones, "Protograph based low error floor LDPC coded modulation," *2005 IEEE MilCom Conference*, October 2005, pp. 378–385.

[17] D. Divsalar, C. Jones, S. Dolinar, and J. Thorpe, "Protograph based LDPC codes with minimum distance linearly growing with block size," *2005 IEEE GlobeCom Conference*, November–December 2006.

[18] D. Divsalar, S. Dolinar, and C. Jones, "Construction of protograph LDPC codes with linear minimum distance," *2006 International Symposium on Information Theory*, July 2006, pp. 664–668.

[19] D. Divsalar, S. Dolinar, and C. Jones, "Protograph LDPC codes over burst erasure channels," *2006 IEEE MilCom Conference*, October 2006, pp. 1–7.

[20] D. Divsalar and C. Jones, "Protograph LDPC codes with node degrees at least 3," *2006 IEEE GlobeCom Conference.*, November 2006, pp. 1–5.

[21] D. Divsalar, S. Dolinar, and C. Jones, "Short protograph-based LDPC codes," *2007 IEEE MilCom Conference*, October 2007, pp. 1–6.

[22] R. G. Gallager, *Low-Density Parity-Check Codes*, Cambridge, MA, MIT Press, 1963.

[23] S. Abu-Surra, D. Divsalar, and W. E. Ryan, "Ensemble enumerators for protograph-based generalized LDPC codes," *IEEE Transactions on Information Theory*, pp. 858–886, February 2011.

[24] T. Cover and J. Thomas, *Elements of Information Theory*, New York, Wiley, 1991.

[25] T. Richardson, "Error-floors of LDPC codes," *Proceedings of the 41st Annual Allerton Conference on Communication, Control, and Computing*, September 2003.

# 9 Ensemble Decoding Thresholds for LDPC and Turbo Codes

The previous chapter examined code properties responsible for the floor region (high-SNR region) of LDPC and turbo codes. Specifically, the emphasis was on the computation of various weight enumerators for LDPC and turbo code ensembles because a poor weight spectrum leads to poor performance in a code's floor region both for iterative decoders and for maximum-likelihood decoders. That chapter also introduced ensemble enumerators for trapping sets and stopping sets, both of which can lead to poor floor performance in iterative decoders. In this chapter we examine the iterative decoding performance of LDPC and turbo code ensembles in the complementary low-SNR region, the "waterfall" region. We show that the iterative decoding of long LDPC and turbo codes displays a threshold effect in which communication is reliable beyond this threshold and unreliable below it. The threshold is a function of code ensemble properties and the tools introduced in this chapter allow the designer to predict the decoding threshold and its gap from Shannon's limit. The ensemble properties for LDPC codes are the degree distributions that are typically the design targets for the code-design techniques presented in subsequent chapters. The ensemble properties for turbo codes are the selected constituent codes. Our development borrows heavily from the references listed at the end of the chapter and our focus is on the binary-input AWGN channel. The references and the problems consider other channels.

## 9.1 Density Evolution for Regular LDPC Codes

We first summarize the main results of [1] with regard to iterative (message-passing) decoding of long, regular LDPC codes.

1. A *concentration theorem* is proven. This asserts that virtually all codes in an ensemble have the same behavior, so that the performance prediction of a specific (long) code is possible via the ensemble average performance.
2. For long codes, this average performance is equal to the performance of cycle-free codes, which is computable via an algorithm called density evolution. *Density evolution* refers to the evolution of the probability density functions (pdfs) of the messages being passed around in the iterative decoder, where the messages are modeled as random variables (r.v.s). Such knowledge of the pdfs allows one to predict under which channel conditions (e.g., SNRs) the decoder bit-error probability will converge to zero.

3. Long codes possess a *decoding threshold*, which partitions the channel parameter (e.g., SNR) space into one region for which reliable communication is possible and another region for which it is not. The density-evolution algorithm allows us to determine the decoding threshold of an LDPC code ensemble.

In this section we present the computation, via density evolution, of the decoding thresholds for long, regular LDPC codes with variable-node (VN) degree $d_v$ and check-node (CN) degree $d_c$. The approach is quite general, and applies to a number of binary-input, symmetric-output channel models, although we focus primarily on the binary-input AWGN channel. The symmetric-output channel description for the binary-input AWGN channel means that the channel transition pdf satisfies $p(y|x = +1) = p(-y|x = -1)$. A similar relationship holds for the binary-symmetric channel (BSC) and other channels. Symmetry conditions are also required of the iterative decoder [1], all of which are satisfied by the sum–product algorithm, so we need not detail them here. It will be assumed throughout that the all-zeros codeword $\mathbf{c} = [0 \;\; 0 \;\; \ldots \;\; 0]$ is sent. Under the mapping $x = (-1)^c$, this means that the all-ones word $\mathbf{x} = [+1 \;\; +1 \;\; \ldots \;\; +1]$ is transmitted on the channel.

As mentioned above, the tool for determining the decoding threshold of a $(d_v, d_c)$-regular LDPC code is the density-evolution algorithm. Because $\mathbf{x} = [+1 \;\; +1 \;\; \ldots \;\; +1]$ is transmitted, an error will be made at the decoder after the maximum number of iterations if any of the signs of the cumulative variable-node LLRs, $L_j^{\text{total}}$, are negative. Let $p_\ell^{(v)}$ denote the pdf of a message $m^{(v)}$ to be passed from VN $v$ to some check node during the $\ell$th iteration.[1] Note that, under the above symmetry conditions for channel and decoder, the pdfs are identical for all such outgoing VN messages. Then, no decision error will occur after an infinite number of iterations if

$$\lim_{\ell \to \infty} \int_{-\infty}^{0} p_\ell^{(v)}(\tau)d\tau = 0, \tag{9.1}$$

because the probability that any $L_j^{\text{total}}$ is negative is zero when (9.1) holds. Note that $p_\ell^{(v)}(\tau)$ depends on the channel parameter, which we denote by $\alpha$. For example, $\alpha$ is the crossover probability $\varepsilon$ for the BSC and $\alpha$ is the standard deviation $\sigma$ of the channel noise for the AWGN channel. Then the decoding threshold $\alpha^*$ is given by

$$\alpha^* = \sup\left\{\alpha\colon \lim_{\ell \to \infty} \int_{-\infty}^{0} p_\ell^{(v)}(\tau)d\tau = 0\right\}$$

under the assumption that the codeword length $n \to \infty$. For the AWGN case, we will also say that the SNR, or $E_b/N_0$, value corresponding to $\alpha^* = \sigma^*$ is the decoding threshold. It is shown in [1] that if $\alpha > \alpha^*$ then Pr(error) will be bounded away from zero; otherwise, from the definition of $\alpha^*$, when $\alpha < \alpha^*$, Pr(error) $\to 0$ as $\ell \to \infty$.

[1] $m^{(v)}$ is one of the LLRs $L_{j \to i}$ from Chapter 5, but we avoid the latter notation in most of this chapter, since otherwise it would become overly cumbersome. Similarly, below we use $m^{(c)}$ for $L_{i \to j}$ to simplify the notation.

We now develop the density-evolution algorithm for computing $p_\ell^{(v)}(\tau)$. We start by recalling that an outgoing message from VN $v$ may be written as

$$m^{(v)} = m_0 + \sum_{j=1}^{d_v-1} m_j^{(c)} = \sum_{j=0}^{d_v-1} m_j^{(c)}, \tag{9.2}$$

where $m_0 \equiv m_0^{(c)}$ is the message from the channel and $m_1^{(c)}, \ldots, m_{d_v-1}^{(c)}$ are the messages received from the $d_v - 1$ neighboring CNs. Now let $p_j^{(c)}$ denote the pdfs for the $d_v$ incoming messages $m_{0,}^{(c)}, \ldots, m_{d_v-1}^{(c)}$. Then, ignoring the iteration-count parameter $\ell$, we have from (9.2)

$$\begin{aligned} p^{(v)} &= p_0^{(c)} * p_1^{(c)} * \cdots * p_{d_v-1}^{(c)} \\ &= p_0^{(c)} * \left[p^{(c)}\right]^{*(d_v-1)}, \end{aligned} \tag{9.3}$$

where independence among the messages is assumed and $*$ denotes convolution. The $(d_v - 1)$-fold convolution in the second line follows from the fact that the pdfs $p_j^{(c)}$ are identical because the code ensemble is regular, and we write $p^{(c)}$ for this common pdf. The computation of $p^{(v)}$ in (9.3) may be done via the fast Fourier transform according to

$$p^{(v)} = \mathcal{F}^{-1}\left\{\mathcal{F}\left\{p_0^{(c)}\right\}\left[\mathcal{F}\left\{p^{(c)}\right\}\right]^{d_v-1}\right\}. \tag{9.4}$$

At this point, we have a technique for computing the pdfs $p^{(v)}$ from the pdfs of the messages $m^{(c)}$ emanating from the CNs.

We need now to develop a technique to compute the pdf $p^{(c)}$ for a generic message $m^{(c)}$ emanating from a check node. From Chapter 5, we may write for $m^{(c)}$

$$m^{(c)} = \left(\prod_{i=1}^{d_c-1} \tilde{s}_i^{(v)}\right) \cdot \phi\left(\sum_{i=1}^{d_c-1} \phi\left(\left|m_i^{(v)}\right|\right)\right), \tag{9.5}$$

where $\tilde{s}_i^{(v)} = \text{sign}\left(m_i^{(v)}\right) \in \{+1, -1\}$ and $\phi(x) = -\ln \tanh(x/2)$. The messages $m_i^{(v)}$ are received by CN $c$ from $d_c - 1$ of its $d_c$ VN neighbors. We define $s_i^{(v)}$ informally as $s_i^{(v)} = \log_{-1}\left(\tilde{s}_i^{(v)}\right)$, so that $s_i^{(v)} = 1$ when $\tilde{s}_i^{(v)} = -1$ and $s_i^{(v)} = 0$ when $\tilde{s}_i^{(v)} = +1$. Then the first factor in (9.5) is equivalently

$$s^{(c)} = \sum_{i=1}^{d_c-1} s_i^{(v)} \pmod 2.$$

Note also that the second factor in (9.5) is non-negative, so that $s^{(c)}$ serves as the sign bit for the message $m^{(c)}$ and the second factor serves as its magnitude, $\left|m^{(c)}\right|$.

In view of this, to simplify the mathematics to follow, we represent the r.v. $m^{(c)}$ as the two-component random vector (r.$\bar{\text{v}}$.)

$$\bar{m}^{(c)} = \left[s^{(c)}, \left|m^{(c)}\right|\right].$$

Likewise, for the r.v.s $m_i^{(v)}$ we let

$$\bar{m}_i^{(v)} = \left[s_i^{(v)}, \left|m_i^{(v)}\right|\right].$$

Note that $\bar{m}^{(c)}, \bar{m}_i^{(v)} \in \{0,1\} \times [0,\infty)$. With the r.$\bar{\text{v}}$. representation, we rewrite (9.5) informally as

$$\bar{m}^{(c)} = \phi^{-1}\left(\sum_{i=1}^{d_c-1} \phi\left(\bar{m}_i^{(v)}\right)\right), \tag{9.6}$$

where we have replaced the outer $\phi$ in (9.5) by $\phi^{-1}$ for later use. (Recall that $\phi = \phi^{-1}$ for positive arguments.) We make use of (9.6) to derive the pdf of $\bar{m}^{(c)}$ (and hence of $m^{(c)}$) from $\{\bar{m}_i^{(v)}\}$ as follows.

1. Compute the pdf of the r.$\bar{\text{v}}$. $\bar{z}_i = \phi\left(\bar{m}_i^{(v)}\right)$. When $s_i^{(v)} = 0$ (equivalently, $m_i^{(v)} > 0$), we have $z_i = \phi\left(m_i^{(v)}\right) = -\ln\tanh\left(m_i^{(v)}/2\right)$, so that

$$\begin{aligned} p(0, z_i) &= \left.\left|\frac{dz_i}{dm_i^{(v)}}\right|^{-1} p^{(v)}\left(m_i^{(v)}\right)\right|_{m_i^{(v)}=\phi^{-1}(z_i)} \\ &= \frac{1}{\sinh(z_i)} p^{(v)}(\phi(z_i)), \end{aligned}$$

where we have used the fact that $\phi^{-1}(z_i) = \phi(z_i)$ because $z_i > 0$. Similarly, when $s_i^{(v)} = 1$ (equivalently, $m_i^{(v)} < 0$), $z_i = \phi\left(-m_i^{(v)}\right) = -\ln\tanh\left(-m_i^{(v)}/2\right)$, and from this it is easily shown that

$$p(1, z_i) = \frac{1}{\sinh(z_i)} p^{(v)}(-\phi(z_i)).$$

2. Compute the pdf of the r.$\bar{\text{v}}$. $\bar{w} = \sum_{i=1}^{d_c-1} \bar{z}_i = \sum_{i=1}^{d_c-1} \phi\left(\bar{m}_i^{(v)}\right)$. Under the assumption of independent messages $\bar{m}_i^{(v)}$, the r.$\bar{\text{v}}$.s $\bar{z}_i$ are independent, so we may write

$$\begin{aligned} p(\bar{w}) &= p(\bar{z}_1) * p(\bar{z}_2) * \cdots * p(\bar{z}_{d_c-1}) \\ &= p(\bar{z})^{*(d_c-1)}, \end{aligned} \tag{9.7}$$

where we write the second line, a $(d_c - 1)$-fold convolution of $p(\bar{z})$ with itself, because the pdfs $p(\bar{z}_i)$ are identical. These convolutions may be performed using a two-dimensional Fourier transform, where the first component of $\bar{z}_i$ takes values in $\{0, 1\}$ and the second component takes values in $[0, \infty)$. Note that the discrete Fourier transform $F_k$ of some function $f_n$: $\{0, 1\} \to R$, where $R$ is some range, is given by $F_0 = f_0 + f_1$ and $F_1 = f_0 - f_1$. In an analogous fashion, we write for the two-dimensional transform of $p(\bar{z})$

$$\begin{aligned} \mathcal{F}\{p(\bar{z})\}_{(0,\omega)} &= \mathcal{F}\{p(0,z)\}_\omega + \mathcal{F}\{p(1,z)\}_\omega, \\ \mathcal{F}\{p(\bar{z})\}_{(1,\omega)} &= \mathcal{F}\{p(0,z)\}_\omega - \mathcal{F}\{p(1,z)\}_\omega. \end{aligned}$$

Then, from (9.7), we may write

$$\mathcal{F}\{p(\bar{w})\}_{(0,\omega)} = \left[\mathcal{F}\{p(\bar{z})\}_{(0,\omega)}\right]^{d_c-1},$$

$$\mathcal{F}\{p(\bar{w})\}_{(1,\omega)} = \left[\mathcal{F}\{p(\bar{z})\}_{(1,\omega)}\right]^{d_c-1}.$$

$p(\bar{w})$ is then obtained by inverse transforming the previous expressions.

3. Compute the pdf $p^{(c)}$ of $\bar{m}^{(c)} = \phi^{-1}(\bar{w}) = \phi^{-1}\left(\sum_{i=1}^{d_c-1}\phi\left(\bar{m}_i^{(v)}\right)\right)$. The derivation is similar to that of the first step. The solution is

$$p\left(0, m^{(c)}\right) = \frac{1}{\sinh\left(m^{(c)}\right)} p(0,z)\bigg|_{z=\phi\left(m^{(c)}\right)}, \quad \text{when } s^{(c)} = 0, \tag{9.8}$$

$$p\left(1, m^{(c)}\right) = \frac{1}{\sinh\left(-m^{(c)}\right)} p(1,z)\bigg|_{z=\phi\left(-m^{(c)}\right)}, \quad \text{when } s^{(c)} = 1. \tag{9.9}$$

We are now equipped with the tools for computing $p^{(v)}$ and $p^{(c)}$, namely (9.4) and Steps 1–3 above. To perform density evolution to find the decoding threshold for a particular $(d_v,d_c)$-regular LDPC code ensemble, one mimics the iterative message-passing decoding algorithm, starting with the initial pdf $p_0^{(c)}$ for the channel LLRs. This is detailed in the algorithm description below.

**Algorithm 9.1** Density Evolution for Regular LDPC Codes

1. Set the channel parameter $\alpha$ to some nominal value expected to be less than the threshold $\alpha^*$. Set the iteration counter to $\ell = 0$.
2. Given $p_0^{(c)}$, obtain $p^{(v)}$ via (9.4) with $\mathcal{F}\left\{p^{(c)}\right\} = 1$ since initially $m^{(c)} = 0$.
3. Increment $\ell$ by 1. Given $p^{(v)}$, obtain $p^{(c)}$ using Steps 1–3 in the text above.
4. Given $p^{(c)}$ and $p_0^{(c)}$, obtain $p^{(v)}$ using (9.4).
5. If $\ell < \ell_{\max}$ and

$$\int_{-\infty}^{0} p^{(v)}(\tau)d\tau \leq p_e \tag{9.10}$$

for some prescribed error probability $p_e$ (e.g., $p_e = 10^{-6}$), increment the channel parameter $\alpha$ by some small amount and go to 2. If (9.10) does not hold and $\ell < \ell_{\max}$, then go back to 3. If (9.10) does not hold and $\ell = \ell_{\max}$, then the previous $\alpha$ is the decoding threshold $\alpha^*$.

**Example 9.1** Consider the binary symmetric channel with channel inputs $x \in \{+1,-1\}$, channel outputs $y \in \{+1,-1\}$, and error probability $\varepsilon$. We are interested in the decoding threshold $\varepsilon^*$ for regular LDPC codes (i.e., the channel parameter is $\alpha = \varepsilon$). A generic message $m_0^{(c)}$ from the channel is given by

$$m_0^{(c)} = \ln\left(\frac{P(y|x=+1)}{P(y|x=-1)}\right),$$

where $x$ is the channel input and $y$ is the channel output. It is shown in Problem 9.1 that, under the assumption that $x = +1$ is always transmitted, the pdf of $m_0^{(c)}$ is given by

$$p_0^{(c)}(\tau) = \varepsilon\delta\left(\tau - \ln\left(\frac{\varepsilon}{1-\varepsilon}\right)\right) + (1-\varepsilon)\delta\left(\tau - \ln\left(\frac{1-\varepsilon}{\varepsilon}\right)\right), \tag{9.11}$$

where $\delta(v) = 1$ when $v = 0$ and $\delta(v) = 0$ when $v \neq 0$. From the density-evolution algorithm above, we obtain the results in the table below [1], where $R = 1 - d_v/d_c$ is the code rate and $\varepsilon_{\text{cap}}$ is the solution to $R = 1 - \mathcal{H}(\varepsilon) = 1 + \varepsilon\log_2(\varepsilon) + (1-\varepsilon)\log_2(1-\varepsilon)$. Observe that decoding thresholds for these regular LDPC code ensembles are not very close to the capacity limits.

| $d_v$ | $d_c$ | $R$ | $\varepsilon^*$ | $\varepsilon_{\text{cap}}$ |
|---|---|---|---|---|
| 3 | 6 | 1/2 | 0.084 | 0.11 |
| 4 | 8 | 1/2 | 0.076 | 0.11 |
| 5 | 10 | 1/2 | 0.068 | 0.11 |
| 3 | 5 | 2/5 | 0.113 | 0.146 |
| 4 | 6 | 1/3 | 0.116 | 0.174 |
| 3 | 4 | 1/4 | 0.167 | 0.215 |

**Example 9.2** Consider the binary-input AWGN channel with channel inputs $x \in \{+1, -1\}$ and channel outputs $y = x + n$, where $n \sim \mathcal{N}\left(0, \sigma^2\right)$ is a white Gaussian noise sample. A generic message $m_0^{(c)}$ from the channel is given by

$$m_0^{(c)} = \ln\left(\frac{P(y \mid x = +1)}{P(y \mid x = -1)}\right) = \frac{2y}{\sigma^2},$$

which is clearly conditionally Gaussian, so we need only determine its mean and variance. Under the assumption that only $x = +1$ is sent, $y \sim \mathcal{N}\left(+1, \sigma^2\right)$, so that the mean value and variance of $m_0^{(c)}$ are

$$E\left(m_0^{(c)}\right) = \frac{2E(y)}{\sigma^2} = \frac{2}{\sigma^2},$$
$$\operatorname{var}\left(m_0^{(c)}\right) = \frac{4}{\sigma^4}\operatorname{var}(y) = \frac{4}{\sigma^2}.$$

From the density-evolution algorithm, we obtain the results in the table below [1], where the channel parameter in this case is $\alpha = \sigma$. In the table, $\sigma^*$ is the decoding threshold in terms of the noise standard deviation and $\sigma_{\text{cap}}$ is the standard deviation for which the channel capacity $C = C(\sigma)$ is equal to $R$. Again, for these regular LDPC code ensembles, the decoding thresholds are not very close to the capacity limits. Also included in the table are the $E_b/N_0$ values that correspond to $\sigma^*$ and $\sigma_{\text{cap}}$. (Observe that $1/\sigma = \sqrt{2RE_b/N_0}$, that is, $E_s = RE_b = 1$.)

| $d_v$ | $d_c$ | $R$ | $\sigma^*$ | $(E_b/N_0)^*$ (dB) | $\sigma_{\text{cap}}$ | $(E_b/N_0)_{\text{cap}}$ (dB) |
|---|---|---|---|---|---|---|
| 3 | 6 | 1/2 | 0.881 | 1.100 | 0.979 | 0.187 |
| 4 | 8 | 1/2 | 0.838 | 1.535 | 0.979 | 0.187 |
| 5 | 10 | 1/2 | 0.794 | 2.004 | 0.979 | 0.187 |
| 3 | 5 | 2/5 | 1.009 | 0.891 | 1.149 | −0.238 |
| 4 | 6 | 1/3 | 1.011 | 1.666 | 1.297 | −0.495 |
| 3 | 4 | 1/4 | 1.267 | 0.955 | 1.549 | −0.794 |

## 9.2 Density Evolution for Irregular LDPC Codes

We extend the results of the previous section to irregular LDPC code ensembles characterized by the degree-distribution pair $\lambda(X)$ and $\rho(X)$ first introduced in Chapter 5. Recall that

$$\lambda(X) = \sum_{d=1}^{d_v} \lambda_d X^{d-1}, \tag{9.12}$$

where $\lambda_d$ denotes the fraction of all edges connected to degree-$d$ VNs and $d_v$ denotes the maximum VN degree. Recall also that

$$\rho(X) = \sum_{d=1}^{d_c} \rho_d X^{d-1}, \tag{9.13}$$

where $\rho_d$ denotes the fraction of all edges connected to degree-$d$ CNs and $d_c$ denotes the maximum CN degree. Our goal is to incorporate these degree distributions into the density-evolution algorithm so that the decoding threshold of irregular LDPC code ensembles may be computed. We can leverage the results for regular LDPC ensembles if we first restate the density-evolution algorithm of the previous section in a more compact form. As before, we start with (9.3), but incorporate the iteration-count parameter $\ell$, so that

$$p_\ell^{(v)} = p_0^{(c)} * \left(p_{\ell-1}^{(c)}\right)^{*(d_v-1)}. \tag{9.14}$$

Next, let $\Gamma$ correspond to the "change of density" due to the transformation $\phi(\cdot)$ as occurs in the computation of $p^{(c)}$ in the previous section (Step 1). Similarly, let $\Gamma^{-1}$ correspond to the change of density due to the transformation $\phi^{-1}(\cdot)$ (Step 3). Then, in place of Steps 1–3 of the previous section, we may write the shorthand

$$p_\ell^{(c)} = \Gamma^{-1}\left[\left(\Gamma\left[p_\ell^{(v)}\right]\right)^{*(d_c-1)}\right]. \tag{9.15}$$

Substitution of (9.15) into (9.14) then gives

$$p_\ell^{(v)} = p_0^{(c)} * \left(\Gamma^{-1}\left[\left(\Gamma\left[p_{\ell-1}^{(v)}\right]\right)^{*(d_c-1)}\right]\right)^{*(d_v-1)}. \tag{9.16}$$

Equation (9.16) is a compact representation of (9.3) together with Steps 1–3 in the previous section. That is, it represents the entire density-evolution algorithm for regular LDPC ensembles.

For irregular ensembles, (9.14) must be modified to average over all possible VN degrees. This results in

$$\begin{aligned} p_\ell^{(v)} &= p_0^{(c)} * \sum_{d=1}^{d_v} \lambda_d \cdot \left(p_{\ell-1}^{(c)}\right)^{*(d-1)} \\ &= p_0^{(c)} * \lambda_*\left(p_{\ell-1}^{(c)}\right), \end{aligned} \tag{9.17}$$

where the notation $\lambda_*(X)$ in the second line is defined in the first line. Similarly, for irregular ensembles (9.15) becomes

$$\begin{aligned} p_\ell^{(c)} &= \Gamma^{-1}\left[\sum_{d=1}^{d_c} \rho_d \cdot \left(\Gamma\left[p_\ell^{(v)}\right]\right)^{*(d-1)}\right] \\ &= \Gamma^{-1}\left[\rho_*\left(\Gamma\left[p_\ell^{(v)}\right]\right)\right], \end{aligned} \tag{9.18}$$

where the notation $\rho_*(X)$ in the second line is defined in the first line. Substitution of (9.18) into (9.17) yields the irregular counterpart to (9.16),

$$p_\ell^{(v)} = p_0^{(c)} * \lambda_*\left(\Gamma^{-1}\left[\rho_*\left(\Gamma\left[p_{\ell-1}^{(v)}\right]\right)\right]\right). \tag{9.19}$$

Observe that (9.19) reduces to (9.16) when $\lambda(X) = X^{d_v-1}$ and $\rho(X) = X^{d_c-1}$.

Analogously to the regular ensemble case, the density-evolution recursion (9.19) is used to obtain $p_\ell^{(v)}$ as a function of the channel parameter $\alpha$, with $\lambda(X)$ and $\rho(X)$ fixed. That is, as specified in the algorithm presented in the previous section, $\alpha$ is incrementally increased until (9.10) fails; the value of $\alpha$ just prior to this failure is the decoding threshold $\alpha^*$.

Density evolution for irregular LDPC codes determines the decoding threshold for a given degree-distribution pair, $\lambda(X)$ and $\rho(X)$, but it does not by itself find the optimal degree distributions in the sense of the minimum threshold. To do the latter, one needs an "outer" global optimization algorithm that searches the space of polynomials, $\lambda(X)$ and $\rho(X)$, for the optimum threshold, assuming a fixed code rate. The density-evolution algorithm is, of course, the inner algorithm that determines the threshold for each trial polynomial pair. The global optimization algorithm suggested in [2] is the *differential-evolution* algorithm. This algorithm is in essence a combination of a hill-climbing algorithm and a genetic algorithm. Many researchers have used it successfully to determine optimal degree distributions. One observation that has reduced the search space for optimal $\lambda(X)$ and $\rho(X)$ is that only two or three (consecutive) nonzero coefficients of $\rho(X)$ are necessary and the nonzero coefficients of $\lambda(X)$ need only be $\lambda_2$, $\lambda_3$, $\lambda_{d_v}$, and a few intermediate coefficients. Of course, any global optimization algorithm will require the following constraints on $\lambda(X)$ and $\rho(X)$:

$$\lambda(1) = \rho(1) = 1, \tag{9.20}$$

$$\int_0^1 \rho(X)dX = (1-R)\int_0^1 \lambda(X)dX, \tag{9.21}$$

where $R$ is the design code rate.

**Example 9.3** The authors of [2] have determined the following rate-1/2 optimal degree distributions for the binary-input AWGN channel for maximum variable node degrees $d_v = 6$, 11, and 30. These are listed below together with their decoding thresholds. For comparison, the capacity limit for rate-1/2 coding on this channel is $(E_b/N_0)_{\text{cap}} = 0.187$ dB and the decoding threshold for a (3, 6) regular LDPC code is 1.11 dB. Figure 9.1 presents the AWGN performance of a $R(n, k) = 0.5(200\,012, 100\,283)$ LDPC code with degree distributions approximately equal to those given below for $d_v = 30$. Observe that at $P_b = 10^{-5}$ the simulated performance is about 0.38 dB from the decoding threshold and about 0.45 dB from the capacity limit. For comparison, the original 0.5(131 072, 65 536) turbo code performs about 0.51 dB from the capacity limit. As discussed in the example in the next section, it is possible to obtain performance extremely close to the capacity limit by allowing $d_v = 200$ with a codeword length of $10^7$ bits.

$\underline{d_v = 6}$

$$\lambda(X) = 0.332X + 0.247X^2 + 0.110X^3 + 0.311X^5,$$

$$\rho(X) = 0.766X^5 + 0.234X^6,$$

$$(E_b/N_0)^* = 0.627 \text{ dB}.$$

$\underline{d_v = 11}$

$$\lambda(X) = 0.239X + 0.295X^2 + 0.033X^3 + 0.433X^{10},$$

$$\rho(X) = 0.430X^6 + 0.570X^7,$$

$$(E_b/N_0)^* = 0.380 \text{ dB}.$$

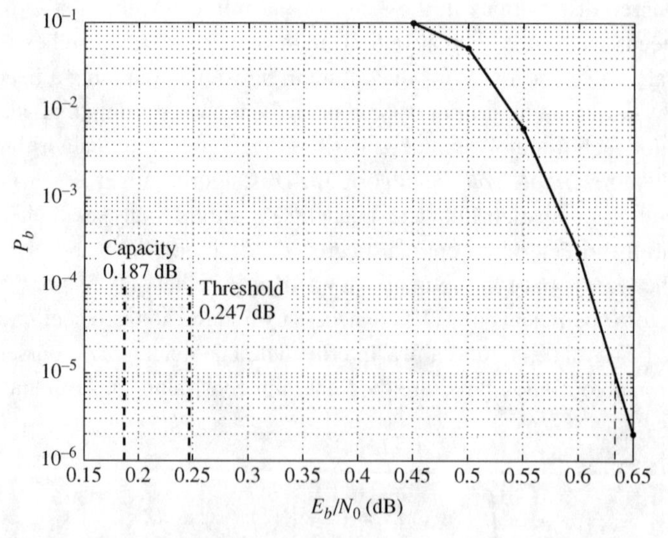


**Figure 9.1** Performance of 0.5(200 012, 100 283) LDPC code with approximately optimal degree distributions for the $d_v = 30$ case. ($d_v$ is the maximum variable-node degree.)

$$\begin{aligned}
&\underline{d_v = 30} \\
&\lambda(X) = 0.196X + 0.240X^2 + 0.002X^5 + 0.055X^6 + 0.166X^7 + 0.041X^8 \\
&\qquad\quad + 0.011X^9 + 0.002X^{27} + 0.287X^{29}, \\
&\rho(X) = 0.007X^7 + 0.991X^8 + 0.002X^9, \\
&(E_b/N_0)^* = 0.274\,\text{dB}.
\end{aligned}$$

---

**Example 9.4** It is important to remind the reader of the infinite-length, cycle-free assumptions underlying density evolution and the determination of optimal degree distributions. When optimal degree distributions are adopted in the design of short- or medium-length codes, the codes are generally susceptible to high error floors due to short cycles and trapping sets. In particular, the iterative decoder will suffer from the presence of cycles involving mostly, or only, degree-2 variable nodes.

Consider the design of an LDPC code with the parameters 0.82(4161, 3430) (to compete with other known codes with those parameters). Near-optimal degree distributions with $d_v = 8$ and $d_c = 20$ were found to be [3]

$$\begin{aligned}
\lambda(X) &= 0.2343X + 0.3406X^2 + 0.2967X^6 + 0.1284X^7, \\
\rho(X) &= 0.3X^{18} + 0.7X^{19}.
\end{aligned} \tag{9.22}$$

From $\lambda(X)$, the number of degree-2 variable nodes would be

$$4161 \cdot \tilde{\lambda}_2 = 4161 \cdot \left( \frac{\lambda_2/2}{\int_0^1 \lambda(X)dX} \right) = 1685.$$

In one of the Chapter 6 problems it is stated that, in a given Tanner graph (equivalently, $\mathbf{H}$ matrix), the maximum number of degree-2 variable nodes possible before a cycle involving only these degree-2 nodes is created is $n - k - 1$. Thus, a (4161, 3430) code with the above degree distributions will have many "degree-2 cycles" since 1685 is much greater than $n - k - 1 = 730$. Further, it can be shown (Problem 9.4) that the girth of this code can be no greater than 10, so these degree-2 cycles are likely somewhat short. Such graphical configurations give rise to an error floor when an iterative decoder is used.

Figure 9.2 presents simulations of the following four length-4161, rate-0.82 codes [3], where we observe that the irregular code with the near-optimal degree distribution in (9.22) does indeed appear to have the best decoding threshold, but it is achieved at the expense of a high error-rate floor due to its large number of degree-2 variable nodes and their associated cycles.

1. A (4161, 3430) irregular LDPC code with the degree distributions given in (9.22).
2. A (4161, 3431) (nearly) regular LDPC code due to MacKay having degree distributions

$$\lambda(X) = X^3, \quad \rho(X) = 0.2234X^{21} + 0.7766X^{22}.$$

   Note that there are no degree-2 nodes.

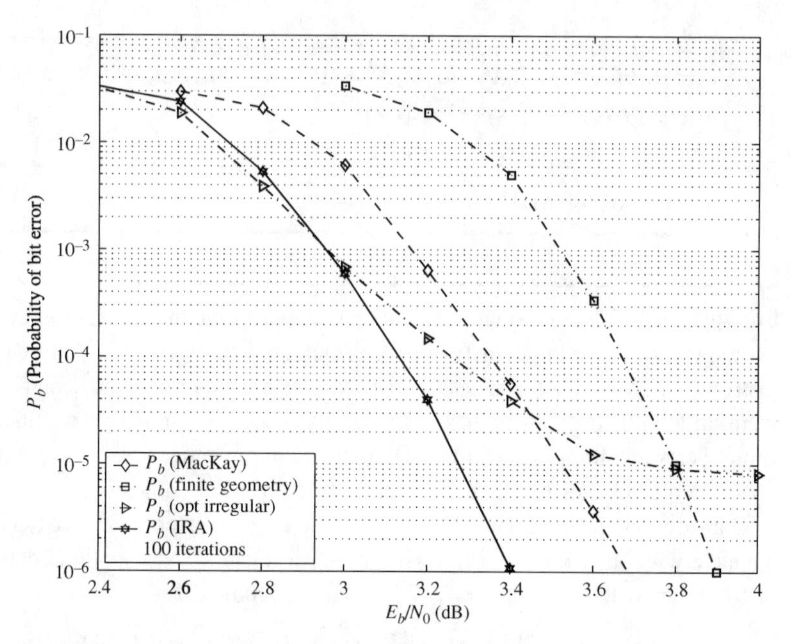


**Figure 9.2** Comparison of four length-4161, rate-0.82 LDPC codes, including a near-optimal ("opt" in the figure) irregular one that displays a high error-rate floor.

3. A (4161, 3430) regular finite-geometry-based LDPC due to Kou *et al.* [4] having degree distributions

$$\lambda(X) = X^{64}, \quad \rho(X) = X^{64}.$$

Note that there are no degree-2 nodes.

4. A (4161, 3430) IRA code with $4161 - 3430 - 1 = 730$ degree-2 nodes, with $d_v = 8$ and $d_c = 20$. The optimal IRA-constrained degree distributions (cf. Section 9.5) were found to be

$$\lambda(X) = 0.00007X^0 + 0.1014X + 0.5895X^2 + 0.1829X^6 + 0.1262X^7,$$
$$\rho(X) = 0.3037X^{18} + 0.6963X^{19}.$$

## 9.3 Quantized Density Evolution

Clearly the algorithms of the previous two sections involve large amounts of numerical computation that could easily become unstable unless care is taken to avoid this. One way to ensure stability is to quantize all of the quantities involved and design the density-evolution algorithm on this basis. Quantized density evolution has the added advantage that it corresponds to a quantized iterative decoder that would be employed in practice. This section describes the approach, following [5].

Let $\Delta$ be the quantization resolution and let $Q(m)$ be the quantized representation of the message $m$, a real number. Then

$$Q(m) = \begin{cases} \lfloor m/\Delta + 0.5 \rfloor \cdot \Delta, & \text{if } m \geq \Delta/2, \\ \lceil m/\Delta - 0.5 \rceil \cdot \Delta, & \text{if } m \leq -\Delta/2, \\ 0, & \text{otherwise,} \end{cases}$$

where $\lfloor x \rfloor$ is the largest integer not greater than $x$ and $\lceil x \rceil$ is the smallest integer not less than $x$. We will write $\breve{m}$ for $Q(m)$, so that the quantized version of (9.2) becomes

$$\breve{m}^{(v)} = \breve{m}_0 + \sum_{j=1}^{d_v-1} \breve{m}_j^{(c)} = \sum_{j=0}^{d_v-1} \breve{m}_j^{(c)}.$$

Because of the quantization, we speak of the probability mass function (pmf) of a random variable $\breve{m}$ rather than a pdf, and denote the pmf by $P_m[k] = \Pr(\breve{m} = k\Delta)$.

We first consider a $(d_v,d_c)$-regular LDPC code ensemble. Analogously to (9.3), we write for the quantized density evolution (actually, pmf evolution) of the variable nodes

$$\begin{aligned} P^{(v)} &= P_0^{(c)} * P_1^{(c)} * \cdots * P_{d_v-1}^{(c)} \\ &= P_0^{(c)} * \left[P^{(c)}\right]^{*(d_v-1)}, \end{aligned} \tag{9.23}$$

where $*$ now represents discrete convolution, $P^{(v)}$ is the pmf of $\breve{m}^{(v)}$, $P_j^{(c)}$ is the pmf of $\breve{m}_j^{(c)}$ for $j = 0, 1, \ldots, d_v - 1$, and $P_j^{(c)} = P^{(c)}$ for $j = 1, 2, \ldots, d_v - 1$. The computations in (9.23) can be efficiently performed using a fast Fourier transform.

As for the quantized density evolution for the check nodes, in lieu of the update equation in (9.5), we use the (quantized) box-plus ($\boxplus$) form of the computation,

$$\breve{m}^{(c)} = \boxplus_{i=1}^{d_c-1} \breve{m}_i^{(v)}, \tag{9.24}$$

where, for two quantized messages $\breve{m}_1$ and $\breve{m}_2$,

$$\breve{m}_1 \boxplus \breve{m}_2 = \breve{\mathcal{B}}(\breve{m}_1, \breve{m}_2) \triangleq Q\left(2\tanh^{-1}\left(\tanh\left(\breve{m}_1/2\right)\tanh\left(\breve{m}_2/2\right)\right)\right).$$

The box-plus operation $\breve{\mathcal{B}}(\cdot,\cdot)$ is implemented by use of a two-input look-up table. The pmf of $\breve{m} = \breve{m}_1 \boxplus \breve{m}_2$ is given by

$$P_m[k] = \sum_{(i,j)\,:\,k\Delta = \breve{\mathcal{B}}(i\Delta, j\Delta)} P_{m_1}[i] P_{m_2}[j].$$

To simplify the notation, we write this as

$$P_m = P_{m_1} \boxplus P_{m_2}.$$

This can be applied to (9.24) $d_c - 2$ times to obtain

$$\begin{aligned} P^{(c)} &= P_1^{(v)} \boxplus P_2^{(v)} \boxplus \cdots \boxplus P_{d_c-1}^{(v)} \\ &= \left(P^{(v)}\right)^{\boxplus(d_c-1)}, \end{aligned} \tag{9.25}$$

where on the second line we used the fact that $P_i^{(v)} = P^{(v)}$ for $i = 1, 2, \ldots, d_c - 1$.

In Section 9.2 we substituted (9.15) into (9.14) (and also (9.18) into (9.17)), that is, the expression for $p^{(c)}$ into the expression for $p^{(v)}$. Here, following [5], we go in the opposite direction and substitute (9.23) into (9.25) to obtain

$$P_\ell^{(c)} = \left(P_0^{(c)} * \left[P_{\ell-1}^{(c)}\right]^{*(d_v-1)}\right)^{\boxplus(d_c-1)}, \tag{9.26}$$

where we have added the iteration-count parameter $\ell$. Note that we need not have obtained the analogous expression to (9.16) because $\Pr(\breve{m}^{(c)} < 0) = 0$ exactly when $\Pr(\breve{m}^{(v)} < 0) = 0$ under the assumption that $+1$s are transmitted on the channel.

In summary, the quantized density-evolution algorithm for regular LDPC code ensembles follows the algorithm of Section 9.1, but uses (9.26) in place of (9.16) and, in place of (9.10), uses the following stopping criterion:

$$\sum_{k<0} P_\ell^{(c)}[k] \le p_e. \tag{9.27}$$

For irregular LDPC code ensembles with degree distributions $\lambda(X)$ and $\rho(X)$, analogously to (9.17), (9.23) becomes

$$P_\ell^{(v)} = P_0^{(c)} * \lambda_*\left(P_{\ell-1}^{(c)}\right). \tag{9.28}$$

Also, analogously to (9.18), (9.25) becomes

$$\begin{aligned} P_\ell^{(c)} &= \sum_{d=1}^{d_c} \rho_d \cdot \left(P_\ell^{(v)}\right)^{\boxplus(d-1)} \\ &= \rho_{\boxplus}\left(P_\ell^{(v)}\right), \end{aligned} \tag{9.29}$$

where the notation $\rho_{\boxplus}(X)$ is obviously defined. Finally, substitution of (9.28) into (9.29) yields

$$P_\ell^{(c)} = \rho_{\boxplus}\left(P_0^{(c)} * \lambda_*\left(P_{\ell-1}^{(c)}\right)\right). \tag{9.30}$$

At this point, the quantized density-evolution algorithm follows the same procedure as the others described earlier, with the recursion (9.30) at the core of the algorithm and (9.27) as the stopping criterion.

---

**Example 9.5** The authors of [5] have found optimal degree distributions for a rate-1/2 irregular LDPC code ensemble with $d_v = 200$. For 9 bits of quantization, they have found a decoding threshold of $\sigma^* = 0.975\,122$, corresponding to $(E_b/N_0)^* = 0.2188$ dB. For 14 bits of quantization, the threshold is $\sigma^* = 0.977\,041$, corresponding to $(E_b/N_0)^* = 0.2017$ dB, for an improvement of 0.0171 dB relative to 9 bits. Again, for comparison, $(E_b/N_0)_{\text{cap}} = 0.187$ dB. They have also designed and simulated a rate-1/2 LDPC code of length $10^7$ having $d_v = 200$ optimal degree distribution. Their code achieved a bit-error rate of $P_b = 10^{-6}$ at $E_b/N_0 = 0.225$ dB, only 0.038 dB from the 0.187 dB limit, and 0.0133 dB from the 0.2017 dB threshold.

---

## 9.4 The Gaussian Approximation

For the binary-input AWGN channel, an alternative to quantized density evolution that simplifies and stabilizes the numerical computations is approximate density evolution based on a *Gaussian approximation* (GA) [6]. The idea is to approximate the message pdfs by Gaussian densities (or Gaussian-mixture densities). Because Gaussian densities are fully specified by two parameters, the mean and variance, (approximate) density evolution entails only the evolution of these two parameters. A further simplification is possible under a consistency assumption, which allows the evolution of only the message means in order to determine approximate decoding thresholds.

A message $m$ satisfies the *consistency condition* if its pdf $p_m$ satisfies

$$p_m(\tau) = p_m(-\tau)e^{\tau}. \tag{9.31}$$

Under the assumption that all +1s are sent, from Example 9.2 the initial (channel) message for the AWGN channel is $m_0^{(c)} = 2y/\sigma^2$. This initial message $m_0^{(c)}$ has the normal pdf $\mathcal{N}\left(2/\sigma^2, 4/\sigma^2\right)$, from which

$$\begin{aligned} p_m(\tau) &= \frac{\sigma}{\sqrt{8\pi}} \exp\left[-\frac{\sigma^2}{8}\left(\tau - \frac{2}{\sigma^2}\right)^2\right] \\ &= \frac{\sigma}{\sqrt{8\pi}} \exp\left[-\frac{\sigma^2}{8}\left(-\tau - \frac{2}{\sigma^2}\right)^2 + \tau\right] \\ &= p_m(-\tau)e^{\tau}. \end{aligned}$$

Thus, the channel message satisfies the consistency condition, and the assumption is that other messages satisfy it at least approximately. Observe also that

$$\operatorname{var}\left(m_0^{(c)}\right) = 2E\left(m_0^{(c)}\right).$$

Such a relationship holds for all Gaussian pdfs that satisfy the consistency condition because

$$\frac{1}{\sqrt{2\pi}\sigma} \exp\left[-\frac{1}{2\sigma^2}(\tau - \mu)^2\right] = \frac{1}{\sqrt{2\pi}\sigma} \exp\left[-\frac{1}{2\sigma^2}(-\tau - \mu)^2\right] e^{\tau}$$

reduces to

$$\sigma^2 = 2\mu.$$

We call a normal density $\mathcal{N}(\mu, 2\mu)$ a *consistent normal density*. The implication is that one need only monitor the message means when performing density evolution using a Gaussian approximation with the consistency condition. It is interesting to note that, if one is concerned with the evolution of the SNR of messages instead of pdfs, a convenient SNR definition is

$$\text{SNR} = \frac{\mu^2}{\sigma^2} = \frac{\mu}{2}.$$

Consequently, also in the case of SNR evolution under the Gaussian/consistency assumptions, one need only propagate message means.

### 9.4.1 Gaussian Approximation for Regular LDPC Codes

Examining now the propagation of means for regular code ensembles, we first take the expected value of (9.2), the update equation for the VN-to-CN messages $m_\ell^{(v)}$, to obtain (after adding the iteration-count parameter $\ell$)

$$\begin{aligned}\mu_\ell^{(v)} &= \mu_0 + \sum_{j=1}^{d_v-1} \mu_{j,(\ell-1)}^{(c)} \\ &= \mu_0 + (d_v - 1)\,\mu_{\ell-1}^{(c)},\end{aligned} \tag{9.32}$$

where the second line follows from the fact that the code ensemble is regular so that all message means are identical during each iteration. For the CN-to-VN messages $m_\ell^{(c)}$, we rewrite (9.5) (or (9.24)) as

$$\tanh\left(\frac{m_\ell^{(c)}}{2}\right) = \prod_{i=1}^{d_c-1} \tanh\left(\frac{m_{i,\ell}^{(v)}}{2}\right). \tag{9.33}$$

Taking the expected value of this equation under an i.i.d. assumption for the messages $m_{i,\ell}^{(v)}$, we have

$$E\left\{\tanh\left(\frac{m_\ell^{(c)}}{2}\right)\right\} = \left[E\left\{\tanh\left(\frac{m_\ell^{(v)}}{2}\right)\right\}\right]^{d_c-1}, \tag{9.34}$$

where the expectations are taken with respect to a consistent normal density of the form $\mathcal{N}(\mu, 2\mu)$. We can therefore write (9.34) as

$$1 - \Phi\left(\mu_\ell^{(c)}\right) = \left[1 - \Phi\left(\mu_\ell^{(v)}\right)\right]^{d_c-1}, \tag{9.35}$$

where, for $\mu \geq 0$, we define

$$\Phi(\mu) \triangleq 1 - \frac{1}{\sqrt{4\pi\mu}} \int_{-\infty}^{\infty} \tanh(\tau/2)\exp\left[-(\tau-\mu)^2/(4\mu)\right]d\tau.$$

Note that $\Phi(\mu)$ need only be defined for $\mu \geq 0$ because we assume that only +1s are transmitted and hence all message means are positive. It can be shown that $\Phi(\mu)$ is continuous and decreasing for $\mu \geq 0$, with $\lim_{\mu\to 0}\Phi(\mu) = 1$ and $\lim_{\mu\to\infty}\Phi(\mu) = 0$. Equation (9.35) can be rearranged as

$$\mu_\ell^{(c)} = \Phi^{-1}\left(1 - \left[1 - \Phi\left(\mu_0 + (d_v - 1)\mu_{\ell-1}^{(c)}\right)\right]^{d_c-1}\right), \tag{9.36}$$

where we have also used (9.32) for $\mu_\ell^{(v)}$.

---

**Example 9.6** Using (9.32), (9.36), and the following approximation [6] for $\Phi(\mu)$,

$$\Phi(\mu) \simeq \begin{cases} \exp\left(-0.4527\mu^{0.86} + 0.0218\right), & \text{for } 0 < \mu < 10, \\ \sqrt{\dfrac{\pi}{\mu}}\exp(-\mu/4)\left(1 - \dfrac{10}{7\mu}\right), & \text{for } \mu > 10, \end{cases}$$

the authors of [6] have obtained the decoding thresholds ($\sigma_{GA}$ and $(E_b/N_0)_{GA}$) below for the binary-input AWGN channel using the Gaussian approximation. Also included in the table are the thresholds obtained using "exact" density evolution ($\sigma^*$ and $(E_b/N_0)^*$). Notice that the gaps between the "exact" and GA results are at most 0.1 dB. Figure 9.3 displays the $\mu_\ell^{(c)}$ versus $\ell$ trajectories for the $(d_v, d_c) = (3,6)$ ensemble for $E_b/N_0$ (dB) equal to 1.162, 1.163, 1.165, 1.170 (equivalently, $\sigma$ equal to 0.8748, 0.8747, 0.8745, 0.8740), where it is clearly seen that the threshold computed by the GA algorithm is $(E_b/N_0)_{GA} = 1.163$ dB (equivalently, $\sigma_{GA} = 0.8747$).

| $d_v$ | $d_c$ | $R$ | $\sigma^*$ | $(E_b/N_0)^*$ (dB) | $\sigma_{GA}$ | $(E_b/N_0)_{GA}$ (dB) |
|---|---|---|---|---|---|---|
| 3 | 6 | 1/2 | 0.8809 | 1.102 | 0.8747 | 1.163 |
| 4 | 8 | 1/2 | 0.8376 | 1.539 | 0.8323 | 1.594 |
| 5 | 10 | 1/2 | 0.7936 | 2.008 | 0.7910 | 2.037 |
| 3 | 5 | 2/5 | 1.009 | 0.8913 | 1.0003 | 0.9665 |
| 4 | 6 | 1/3 | 1.0109 | 1.667 | 1.0036 | 1.730 |
| 3 | 4 | 1/4 | 1.2667 | 0.9568 | 1.2517 | 1.060 |

## 9.4.2 Gaussian Approximation for Irregular LDPC Codes

For the irregular LDPC codes ensemble case, density evolution with the Gaussian approximation assumes that the outgoing messages both from the VNs and from the

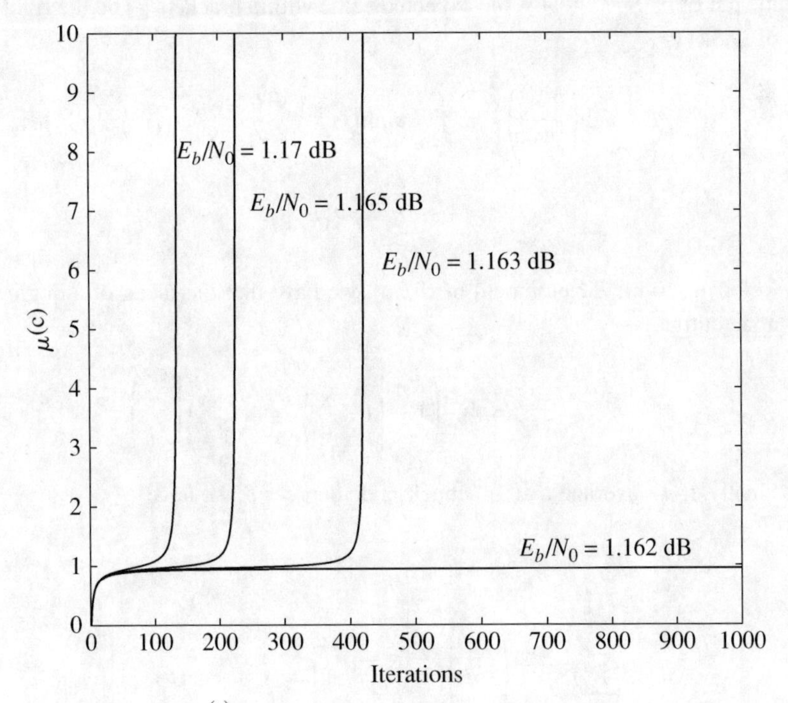


**Figure 9.3** Plots of $\mu_\ell^{(c)}$ versus $\ell$ trajectories for the $(d_v, d_c) = (3, 6)$ ensemble for $E_b/N_0$ (dB) = 1.162, 1.163, 1.165, 1.170 ($\sigma$ = 0.8748, 0.8747, 0.8745, 0.8740).

CNs are Gaussian (although they are not). Further, the assumption for both node types is that the incoming messages are Gaussian mixtures, that is, their pdfs are weighted sums of Gaussian pdfs.

Analogous to (9.32), the mean of an output message $m_\ell^{(v)}$ of a degree-$d$ VN is given by

$$\mu_{\ell,d}^{(v)} = \mu_0 + (d-1)\mu_{\ell-1}^{(c)}, \tag{9.37}$$

where $\mu_{\ell-1}^{(c)}$ is the mean of the message $m_{\ell-1}^{(c)}$, a Gaussian mixture (in general), and the $d-1$ incoming messages are assumed to be i.i.d. Since the output message $m_\ell^{(v)}$ is (assumed) Gaussian with mean $\mu_\ell^{(v)}$, its variance would be $2\mu_\ell^{(v)}$ under the consistency condition. Because a randomly chosen edge is connected to a degree-$d$ variable node with probability $\lambda_d$, averaging over all degrees $d$ yields the following Gaussian mixture pdf for the variable node messages:

$$p_{\ell,d}^{(v)}(\tau) = \sum_{d=1}^{d_v} \lambda_d \cdot \mathcal{N}_\tau\left(\mu_{\ell,d}^{(v)}, 2\mu_{\ell,d}^{(v)}\right).$$

In this expression, $\mathcal{N}_\tau(\mu, \sigma^2)$ represents the pdf for a Gaussian r.v. with mean $\mu$ and variance $\sigma^2$.

This pdf is important to us because, analogous to (9.34) through (9.36), we will need to take the expected value of a function of the form $\tanh(m/2)$ with respect to the pdf of $m$. Specifically, the expected value within brackets $[\cdot]$ on the right-hand side of (9.34) is

$$\begin{aligned} E\left\{\tanh\left(\frac{m_\ell^{(v)}}{2}\right)\right\} &= \int_{-\infty}^{\infty} \tanh\left(m_\ell^{(v)}\right) \sum_{d=1}^{d_v} \lambda_d \cdot \mathcal{N}_\tau\left(\mu_{\ell,d}^{(v)}, 2\mu_{\ell,d}^{(v)}\right) d\tau \\ &= 1 - \sum_{d=1}^{d_v} \lambda_d \Phi\left(\mu_{\ell,d}^{(v)}\right). \end{aligned}$$

Referring to the development of (9.36), we have that the mean of a degree-$\delta$ check-node output is

$$\mu_{\ell,\delta}^{(c)} = \Phi^{-1}\left(1 - \left[1 - \sum_{d=1}^{d_v} \lambda_d \Phi\left(\mu_{\ell,d}^{(v)}\right)\right]^{\delta-1}\right).$$

Finally, if we average over all check-node degrees $\delta$, we have

$$\begin{aligned} \mu_\ell^{(c)} &= \sum_{\delta=1}^{d_c} \rho_\delta \mu_{\ell,\delta}^{(c)} \\ &= \sum_{\delta=1}^{d_c} \rho_\delta \Phi^{-1}\left(1 - \left[1 - \sum_{d=1}^{d_v} \lambda_d \Phi\left(\mu_0 + (d-1)\mu_{\ell-1}^{(c)}\right)\right]^{\delta-1}\right), \end{aligned} \tag{9.38}$$

where we have also used (9.37).

Equation (9.38) is the Gaussian-approximation density-evolution recursion sought. Its notation can be simplified as follows. Let $s \in [0, \infty)$, $t \in [0, \infty)$, and define

$$f_\delta(s,t) = \Phi^{-1}\left(1 - \left[1 - \sum_{d=1}^{d_v} \lambda_d \Phi(s + (d-1)t)\right]^{\delta-1}\right),$$

$$f(s,t) = \sum_{\delta=1}^{d_c} \rho_\delta f_\delta(s,t).$$

Hence, the simplified recursion is

$$t_\ell = f(s, t_{\ell-1}),$$

where $s = \mu_0$, $t_\ell = \mu_\ell^{(c)}$, and $t_0 = 0$. The threshold sought is

$$s^* = \inf\left\{ s.s \in [0,\infty) \text{ and } \lim_{\ell\to\infty} t_\ell(s) = \infty \right\}.$$

Note that, since the mean of the initial message $m_0 = 2y/\sigma^2$ is $\mu_0 = 2/\sigma^2$, $s^* = \mu_0^*$ is of the form $s^* = 2/(\sigma^2)^*$, so that the noise threshold is

$$\sigma^* = \sqrt{2/s^*}.$$

## 9.5 On the Universality of LDPC Codes

A universal code is a code that may be used across a number of different channel types or conditions with little degradation relative to a good single-channel code. The explicit design of universal codes, which simultaneously seeks to solve a multitude of optimization problems, is a daunting task. This section shows that one channel may be used as a surrogate for an entire set of channels to produce good universal LDPC codes. This result suggests that sometimes a channel for which LDPC code design is simple (e.g., BEC) may be used as a surrogate for a channel for which LDPC code design is complex (e.g., Rayleigh). We examine in this section the universality of LDPC codes over the BEC, AWGN, and flat, independent (fully interleaved) Rayleigh-fading channels in terms of decoding threshold performance. Using excess mutual information (defined below) as a performance metric, we present design results supporting the contention that an LDPC code designed for a single channel can be universally good across the three channels. The section follows [7]. See also [8, 9, 10].

We will examine the SPA decoding thresholds of (non-IRA) LDPC codes and IRA codes of rate 1/4, 1/2, and 3/4 designed under various criteria. Quantized density evolution is used for the AWGN and Rayleigh channels and the result in Section 9.2 is used for the BEC. While the Gaussian approximation could be used for the AWGN channel, that is not the case for the Rayleigh channel. To see this, observe that the model for the Rayleigh channel is

$$y_k = a_k x_k + n_k,$$

where $a_k$ is a Rayleigh r.v., $x_k \in \{+1, -1\}$ represents a code bit, and $n_k \sim \mathcal{N}\left(0, \sigma^2\right)$ is an AWGN sample. The initial message is then

$$m_0 = 2\hat{a}_k y_k / \sigma^2,$$

where $\hat{a}_k$ is an estimate of $a_k$ (i.e., $\hat{a}_k$ is channel-state information, which we assume is perfect). Under the assumption that $x_k = +1$ (all-zeros codeword), the pdf of $m_0$ is the expectation, with respect to the Rayleigh pdf for $a_k$, of the $\mathcal{N}\left(2a/\sigma^2, 4a^2/\sigma^2\right)$ pdf conditioned on $a_k$. The resulting pdf is non-Gaussian.

Recall the constraints (9.20) and (9.21) for LDPC codes. The structure of IRA codes requires $n-k-1$ degree-2 nodes and a single degree-1 node, so that $\lambda_2 \cdot N_e = 2(n-k-1)$ and $\lambda_1 \cdot N_e = 1$, where $N_e$ is the number of edges. On combining these two equations with (9.20) and (9.21), we have

$$\frac{\lambda_2 R}{2(1-R) - 2/n} = \sum_{i=3}^{d_v} \frac{\lambda_i}{i},$$

$$\frac{\lambda_2 (1-R)}{2(1-R) - 2/n} = \sum_{j} \frac{\rho_j}{j},$$

$$\sum_{i=1}^{d_v} \lambda_i = 1,$$

$$\sum_{i=2}^{d_c} \rho_i = 1.$$

For $n > 200$, this is approximately equivalent to

$$\sum_{j} \frac{\rho_j}{j} = \frac{1-R}{R} \sum_{i=3}^{d_v} \frac{\lambda_i}{i},$$

$$\sum_{i=1}^{d_v} \lambda_i = 1,$$

$$\sum_{i=2}^{d_c} \rho_i = 1.$$

Thus, for IRA codes of practical length, the infinite-length assumption can still be applied.

In all density-evolution computations, the maximum number of decoder iterations is $\ell_{\max} = 1000$ and the stopping threshold is $p_e = 10^{-6}$. The ten code-design criteria are listed in Table 9.1. We consider both a large and a small maximum variable-node degree ($d_v = 50$ and $d_v = 8$), the former to approach theoretical limits and the latter to accommodate low-complexity encoding and decoding. For each design criterion and for each target channel, we compare the threshold-to-capacity gaps for each degree-distribution pair obtained. For the AWGN and Rayleigh channels these gaps are called "excess SNR" and for the BEC these gaps are called "excess $\delta_0$," where $\delta_0$ is the BEC erasure probability. With AWGN and Rayleigh as the target channels, Figure 9.4

**Table 9.1** Design criteria

| Entry | Type | Surrogate channel | $d_v$ |
|---|---|---|---|
| 1 | LDPC | BEC | 50 |
| 2 | LDPC | AWGN-GA | 50 |
| 3 | LDPC | AWGN | 50 |
| 4 | LDPC | Rayleigh | 50 |
| 5 | LDPC | BEC | 8 |
| 6 | LDPC | AWGN-GA | 8 |
| 7 | LDPC | AWGN | 8 |
| 8 | LDPC | Rayleigh | 8 |
| 9 | IRA | BEC | 8 |
| 10 | IRA | Rayleigh | 8 |

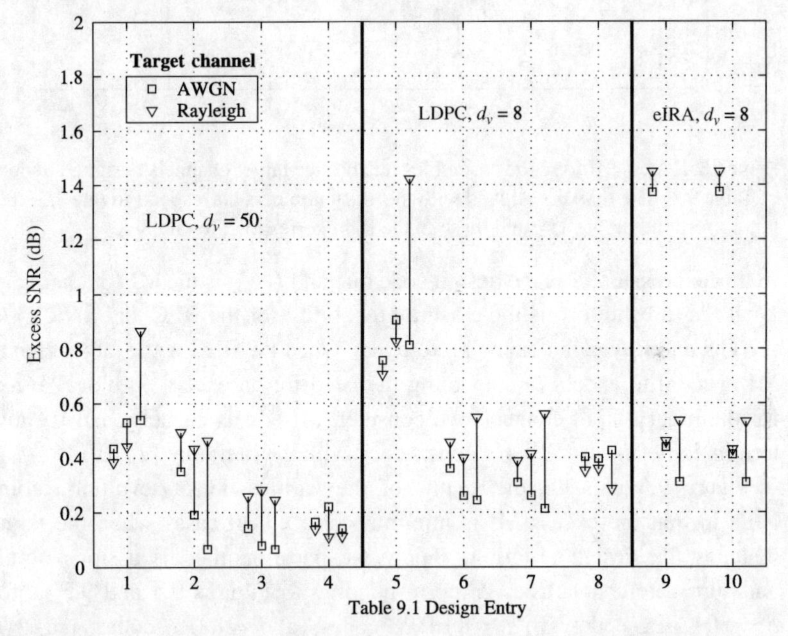


**Figure 9.4** Excess SNR ($E_b/N_0$) for codes designed under the criteria of Table 9.1. The markers aligned with the entry numbers correspond to rate 1/2, those to the left correspond to rate 1/4, and those to the right correspond to rate 3/4.

presents the excess-SNR results for the ten design criteria for all three code rates. This figure will be discussed shortly.

We can repeat this for the case where the BEC is the target channel and excess $\alpha$ is the performance metric, but we will find that such a plot would be redundant in view of a unifying performance metric we now consider. Specifically, it is convenient (in fact, proper) to present all of our results in a single plot using as the performance metric excess mutual information (MI) [7, 10], defined as

$$\text{excess MI} = I(\rho^*) - R.$$

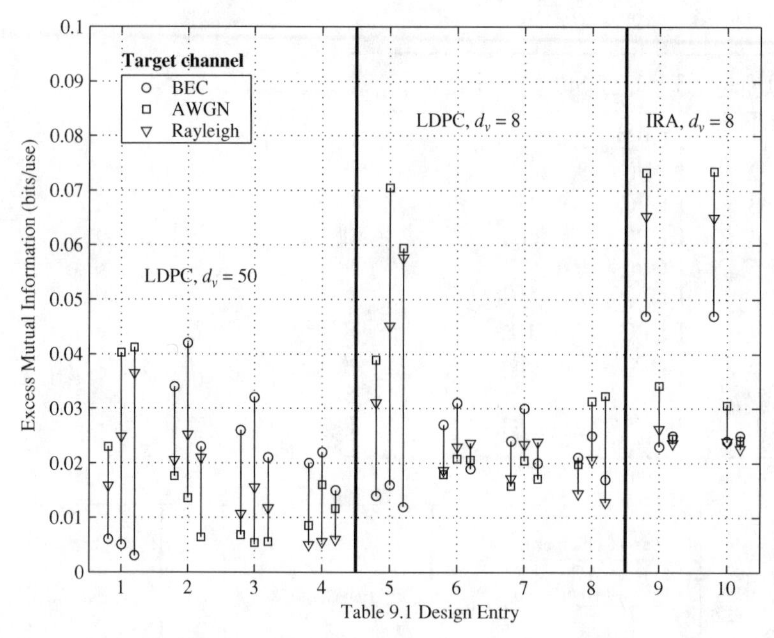


**Figure 9.5** Excess MI for code ensembles on all three target channels designed under the criteria of Table 9.1. The markers aligned with the entry numbers correspond to rate 1/2, those to the left correspond to rate 1/4, and those to the right correspond to rate 3/4.

In this expression, $R$ is the design code rate and $I(\rho^*)$ is the MI for channel parameter $\rho^*$ at the threshold. $\rho$ is the erasure probability for the BEC and SNR $E_b/N_0$ for the AWGN and Rayleigh channels. Note that, when the BEC is the target channel, excess MI is equal to excess $\delta_0$, obviating the need for an excess-$\delta_0$ plot. We remark that, for the binary-input channels we consider, MI equals capacity, but we maintain the terminology "excess MI" for consistency with the literature [7, 10].

Figure 9.5 presents the results of the various density-evolution computations. Note in that the excess MI is minimized for all 30 cases when the target channel matches the design criterion. Below we partition the discussions of universality and surrogate-channel design corresponding to Figures 9.4 and 9.5 as follows: (a) $d_v = 50$ LDPC codes, (b) $d_v = 8$ LDPC codes, (c) $d_v = 8$ IRA codes, and (d) design via surrogate channels.

(a) $d_v = 50$ LDPC codes. Starting with the excess-SNR metric in Figure 9.4 ($d_v = 50$), we observe that, for each code rate, the Rayleigh design criterion (entry 4) leads to codes that are universally good on the AWGN and Rayleigh channels. Specifically, the worst-case excess SNR is only 0.21 dB for rate -1/2 codes on the AWGN channel. At the other extreme, for a code rate of 3/4, the BEC design criterion (entry 1) leads to a worst-case excess SNR of 0.9 dB on the Rayleigh channel. While using the Rayleigh channel as a surrogate leads to the best universal codes, it is at the expense of much greater algorithm complexity. Using the AWGN channel as a surrogate (entry 3) might be preferred since it yields results that are nearly as good. In fact, using the AWGN-GA design criterion (entry 2) also appears to lead to universal codes that are quite good.

Similar comments can be made ($d_v = 50$) when using excess MI as the performance metric, as in Figure 9.5. We can add, however, that the BEC design criterion does not look quite as bad in this context. Consider that for the BEC surrogate channel (entry 1), the worst-case excess MI for rates 1/4, 1/2, and 3/4 is 0.023, 0.04, and 0.04, respectively. These correspond to worst-case throughput losses of $0.023/0.25 = 9.2\%$, $0.04/0.5 = 8\%$, and $0.04/0.75 = 5.3\%$, respectively. These are similar to the worst-case throughput losses of 8%, 4.4%, and 2% for the much more complex Rayleigh design criterion (entry 4).

(b) $d_v = 8$ LDPC codes. For $d_v = 8$, both in Figure 9.4 and in Figure 9.5, the BEC criterion (entry 5) leads to clearly inferior codes in terms of universality. For example, the worst-case excess SNR is 1.43 dB, which occurs for a rate-3/4 code on the Rayleigh channel. The corresponding excess MI value is 0.057, which corresponds to a throughput loss of 7.6%. On the other hand, the AWGN and Rayleigh criteria both lead to very good universal codes of nearly equal quality. The AWGN-GA criterion (entry 6) also results in codes that are good on all three channels.

(c) $d_v = 8$ IRA codes. Even though the structure of IRA codes forces additional constraints on the degree distributions of a code's Tanner graph, as shown in the next section on density evolution for IRA codes, the infinite-length assumption is still valid in the density-evolution process.

The empirical results in [11] indicate that the BEC design criterion may be used to design IRA codes for the Rayleigh-fading channel with negligible performance difference. Here, we reconsider this issue from the perspective of excess MI. As seen in Figure 9.5, there is negligible difference between the IRA codes designed using the BEC criterion (entry 9) and those designed using the Rayleigh criterion (entry 10). Thus, the BEC design technique should be used in the case of IRA codes for all three target channels. We note that, for rates 1/2 and 3/4, there are small excess MI losses (and occasionally gains) on going from $d_v = 8$ LDPC codes (entry 8) to $d_v = 8$ IRA codes (entry 9). However, the excess MI loss is substantial for rate 1/4. This is because IRA codes are constrained to $n - k - 1$ degree-2 variable nodes, which is substantially more than the number required for the optimal threshold for rate 1/4. For example, with $d_v = 8$, for rate 1/4, $\lambda_2 \approx 0.66$ for IRA codes, but $\lambda_2 \approx 0.43$ for an optimum LDPC code.

(d) Design via surrogate channels. The previous paragraph argued that the BEC may be used as a surrogate with negligible penalty when the designer is interested only in rate-1/2 and rate-3/4 IRA codes. For BEC-designed LDPC codes ($d_v = 50$) on the Rayleigh channel, the throughput loss compared with that for Rayleigh-designed codes is quite small for all three rates, with the worst-case loss occurring for rate 1/4: $0.012/0.25 = 4.8\%$ (entries 1 and 4). Additionally, the GA is a good "surrogate" when the target is AWGN, as is well known. As an example, the throughput loss compared with the AWGN criterion at $d_v = 50$ and rate 1/2 is $0.015/0.5 = 3\%$.

In summary, the main results of this section are that (1) an LDPC code can be designed to be universally good across all three channels; (2) the Rayleigh channel is a particularly good surrogate in the design of LDPC codes for the three channels, but the AWGN channel is typically an adequate surrogate; and (3) with the Rayleigh channel as the target, the BEC may be used as a faithful surrogate in the design of IRA codes with rates greater than or equal to 1/2, and there is a throughput loss (capacity reduction) of less than 6% if the BEC is used as a surrogate to design (non-IRA) LDPC codes.

## 9.6 EXIT Charts for LDPC Codes

As an alternative to density evolution, the extrinsic-information-transfer (EXIT) chart technique is a graphical tool for estimating the decoding thresholds of LDPC code and turbo code ensembles. The technique relies on the Gaussian approximation, but provides some intuition regarding the dynamics and convergence properties of an iteratively decoded code. The EXIT chart also possesses some additional properties, as covered in the literature [12–15] and briefly in Section 9.8. This section follows the development and the notation found in those publications.

The idea behind EXIT charts begins with the fact that the variable-node processors (VNPs) and check-node processors (CNPs) work cooperatively and iteratively to make bit decisions, with the metric of interest generally improving with each half-iteration. (Various metrics are mentioned below.) A transfer curve plotting the input metric versus the output metric can be obtained both for the VNPs and for the CNPs, where the transfer curve for the VNPs depends on the channel SNR. Further, since the output metric for one processor is the input metric for its companion processor, one can plot both transfer curves on the same axes, but with the abscissa and ordinate reversed for one processor. Such a chart aids in the prediction of the *decoding threshold* of the ensemble of codes characterized by given VN and CN degree distributions: the decoding threshold is the SNR at which the VNP transfer curve just touches the CNP curve, precluding convergence of the two processors. As with density evolution, decoding-threshold prediction via EXIT charts assumes a graph with no cycles, an infinite codeword length, and an infinite number of decoding iterations.

---

**Example 9.7** An EXIT chart example is depicted in Figure 9.6 for the ensemble of regular LDPC codes on the binary-input AWGN channel with $d_v = 3$ and $d_c = 6$. In Figure 9.6, the metric used for the transfer curves is extrinsic mutual information, hence the name extrinsic-information-transfer (EXIT) chart. (The notation used in the figure for the various information measures is given later in this chapter.) The top (solid) $I_{E,V}$ versus $I_{A,V}$ curve is an extrinsic-information curve corresponding to the VNPs. It plots the mutual information $I_{E,V}$ for the extrinsic information coming out of a VNP against the mutual information $I_{A,V}$ for the extrinsic (*a priori*) information going into the VNP. The bottom (dashed) $I_{A,C}$ versus $I_{E,C}$ curve is an extrinsic-information curve corresponding to the CNPs. It plots the mutual information $I_{A,C}$ for the extrinsic (*a priori*) information going into a CNP against the mutual information $I_{E,C}$ for the extrinsic information coming out of the CNP. (This curve is determined by computing $I_{E,C}$ as a function of $I_{A,C}$ and then, for the purposes

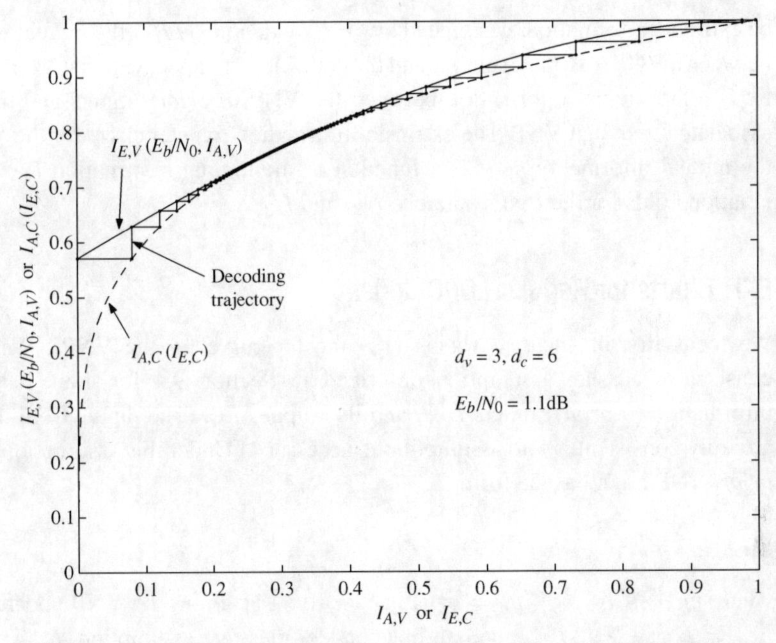


**Figure 9.6** An EXIT chart example for the $(d_v, d_c) = (3, 6)$ regular LDPC code ensemble.

of the EXIT chart, plotted on reversed axes.) Between these two curves is the decoding trajectory for an iterative SPA decoder. Note that it "bounces" between the two curves because the extrinsic information coming out of the VNPs (CNPs) is the *a priori* information going into the CNPs (VNPs). The trajectory starts at the $(0, 0)$ point (zero information) and eventually converges to the $(1, 1)$ point (one bit of information, zero errors). The trajectory allows one to visualize the amount of information (in bits) that is being exchanged between the VNPs and CNPs. As the amount of information exchanged approaches unity, the error rate approaches zero.

As the channel SNR increases, the top (VNP) curve shifts upward, increasing the "tunnel" between the two curves and thus the decoder convergence rate. The SNR for this figure is just above the decoding threshold for the $(d_v, d_c) = (3, 6)$ ensemble, that is, just above $(E_b/N_0)_{\text{EXIT}} = 1.1$ dB to allow a tunnel. If the SNR is below this value, then the tunnel will be closed, precluding the decoder trajectory from making it all the way through to the $(I_{E,V}, I_{E,C}) = (1, 1)$ point for which the error rate is zero.

Other metrics, such as the SNR or mean [15, 16] and the error probability [17] are possible, but mutual information generally gives the most accurate prediction of the decoding threshold [18, 19] and is a universally good metric across many channels [8–10, 14].

---

Following the work of ten Brink [18–13], we now consider the computation of EXIT transfer curves both for VNPs and for CNPs, first for regular LDPC codes and then for irregular codes. Except for the inputs from the channel, we consider VNP and CNP inputs to be *a priori* information, designated by "A," and their outputs to be extrinsic information, designated by "E." We denote by $I_{E,V}$ the mutual information between a VNP (extrinsic) output and the code bit associated with that VNP. We denote by $I_{A,V}$ the mutual information between the VNP (*a priori*) inputs and the code bit associated with that VNP. The extrinsic-information transfer curve for the VNPs plots the mutual information $I_{E,V}$ as a function of the mutual information $I_{A,V}$. A similar notation holds for the CNPs, namely, $I_{E,C}$ and $I_{A,C}$.

### 9.6.1 EXIT Charts for Regular LDPC Codes

We focus first on the $I_{E,V}$ versus $I_{A,V}$ transfer curve for the VNPs. We adopt the consistent-Gaussian assumption described in Section 9.4 for the VNP extrinsic-information (*a priori*) inputs $L_{i'\to j}$ and its output $L_{j\to i}$. (The inputs from the channel are truly consistent, so no assumption is necessary.) Under this assumption, such an *a priori* VNP input has the form

$$L_{i'\to j} = \mu_a \cdot x_j + n_j,$$

where $n_j \sim \mathcal{N}\left(0, \sigma_a^2\right)$, $\mu_a = \sigma_a^2/2$, and $x_j \in \{\pm 1\}$. From the VNP update equation, $L_{j\to i} = L_{ch,j} + \sum_{i'\neq i} L_{i'\to j}$, and an independent-message assumption, $L_{j\to i}$ is Gaussian with variance $\sigma^2 = \sigma_{ch}^2 + (d_v - 1)\sigma_A^2$ (and hence mean $\sigma^2/2$), where $\sigma_{ch}^2$ is the variance of the consistent-Gaussian input from the channel, $L_{ch,j}$. For simplicity, below we will write $L$ for the extrinsic LLR $L_{j\to i}$, $x$ for the code bit $x_j$, $X$ for the r.v. corresponding to the realization $x$, and $p_L(l|\pm)$ for $p_L(l|x = \pm 1)$. Then the mutual information between $X$ and $L$ is

$$\begin{aligned}
I_{E,V} &= H(X) - H(X|L) \\
&= 1 - E\left[\log_2\left(\frac{1}{p_{X|L}(x|l)}\right)\right] \\
&= 1 - \sum_{x=\pm 1} \frac{1}{2} \int_{-\infty}^{\infty} p_L(l|x) \cdot \log_2\left(\frac{p_L(l|+) + p_L(l|-)}{p_L(l|x)}\right) dl \\
&= 1 - \int_{-\infty}^{\infty} p_L(l|+) \log_2\left(1 + \frac{p_L(l|-)}{p_L(l|+)}\right) dl \\
&= 1 - \int_{-\infty}^{\infty} p_L(l|+) \log_2\left(1 + e^{-l}\right) dl,
\end{aligned}$$

where the last line follows from the consistency condition and because $p_L(l|x = -1) = p_L(-l|x = +1)$ for Gaussian densities.

Since $L_{j\to i} \sim \mathcal{N}\left(\sigma^2/2, \sigma^2\right)$ (when conditioned on $x_j = +1$), we have

$$I_{E,V} = 1 - \int_{-\infty}^{\infty} \frac{1}{\sqrt{2\pi}\sigma} e^{-(l-\sigma^2/2)^2/(2\sigma^2)} \log_2\left(1 + e^{-l}\right) dl. \tag{9.39}$$

For convenience, we write this as

$$I_{E,V} = J(\sigma) = J\left(\sqrt{(d_v - 1)\sigma_A^2 + \sigma_{ch}^2}\right). \tag{9.40}$$

To express $I_{E,V}$ as a function of $I_{A,V}$, we first exploit the consistent-Gaussian assumption for the inputs $L_{j \to i}$ to write

$$I_{A,V} = J(\sigma_A), \tag{9.41}$$

so that from (9.40) we have

$$I_{E,V} = J(\sigma) = J\left( \sqrt{(d_v - 1)\left[J^{-1}(I_{A,V})\right]^2 + \sigma_{ch}^2} \right). \tag{9.42}$$

The inverse function $J^{-1}(\cdot)$ exists since $J(\sigma_A)$ is monotonic in $\sigma_A$. Lastly, $I_{E,V}$ can be parameterized by $E_b/N_0$ for a given code rate $R$ since $\sigma_{ch}^2 = 4/\sigma_w^2 = 8R(E_b/N_0)$, where $\sigma_w^2$ is the white-noise variance. Approximations of the functions $J(\cdot)$ and $J^{-1}(\cdot)$ are given in [12], although numerical computations of these are fairly simple with modern mathematics programs.

To obtain the CNP EXIT curve, $I_{E,C}$ versus $I_{A,C}$, we can proceed as we did in the VNP case, that is, begin with the consistent-Gaussian assumption. However, this assumption is not sufficient because determining the mean and variance for a CNP output $L_{i \to j}$ is not straightforward, as is evident from the computation for CNPs in (9.5) or (9.24). Closed-form expressions have been derived for the check node EXIT curves [20, 21] and computer-based numerical techniques can also be used to obtain these curves. However, the simplest technique exploits the following duality relationship (which has been proven to be exact for the binary erasure channel [14]): the EXIT curves for a degree-$d_c$ check node (i.e., rate-$(d_c - 1)/d_c$ SPC code) and a degree-$d_c$ variable node (i.e., rate-$1/d_c$ repetition code (REP)) are related as

$$I_{E,\text{SPC}}(d_c, I_A) = 1 - I_{E,\text{REP}}(d_c, 1 - I_A).$$

This relationship was shown to be very accurate for the binary-input AWGN channel in [20, 21]. Thus,

$$\begin{aligned} I_{E,C} &= 1 - I_{E,V}(\sigma_{ch} = 0, d_v \leftarrow d_c, I_{A,V} \leftarrow 1 - I_{A,C}) \\ &= 1 - J\left( \sqrt{(d_c - 1)\left[J^{-1}(1 - I_{A,C})\right]^2} \right). \end{aligned} \tag{9.43}$$

Equations (9.42) and (9.43) were used to produce the plot in Figure 9.6.

### 9.6.2 EXIT Charts for Irregular LDPC Codes

For irregular LDPC codes, $I_{E,V}$ and $I_{E,C}$ are computed as weighted averages. The weighting is given by the coefficients of the "edge-perspective" degree-distribution polynomials $\lambda(X) = \sum_{d=1}^{d_v} \lambda_d X^{d-1}$ and $\rho(X) = \sum_{d=1}^{d_c} \rho_d X^{d-1}$, where $\lambda_d$ is the fraction of edges in the Tanner graph connected to degree-$d$ variable nodes, $\rho_d$ is the fraction of edges connected to degree-$d$ check nodes, and $\lambda(1) = \rho(1) = 1$. So, for irregular LDPC codes,

$$I_{E,V} = \sum_{d=1}^{d_v} \lambda_d I_{E,V}(d, I_{A,V}), \tag{9.44}$$

where $I_{E,V}(d)$ is given by (9.42) with $d_v$ replaced by $d$, and

$$I_{E,C} = \sum_{d=1}^{d_c} \rho_d I_{E,C}(d, I_{A,C}), \tag{9.45}$$

where $I_{E,C}(d)$ is given by (9.43) with $d_c$ replaced by $d$.

These equations allow one to compute the EXIT curves for the VNPs and the CNPs. For a given SNR and set of degree distributions, the latter curve is plotted on transposed axes together with the former curve, which is plotted in the standard way. If the curves intersect, the SNR must be increased; if they do not intersect, the SNR must be decreased; the SNR at which they just touch is the decoding threshold. After each determination of a decoding threshold, an outer optimization algorithm chooses the next set of degree distributions until an optimum threshold is attained. However, the threshold can be determined quickly and automatically without actually plotting the two EXIT curves, thus allowing the programmer to act as the outer optimizer.

It has been shown [14] that, to optimize the decoding threshold on the binary erasure channel, the shapes of the VNP and CNP transfer curves must be well matched in the sense that the CNP curve fits inside the VNP curve (an example will follow). This situation has also been observed on the binary-input AWGN channel [12]. This is, of course, intuitively clear, for if the shape of the VNP curve (nearly) exactly matches that of the CNP curve, then the channel parameter can be adjusted (e.g., SNR lowered) to its (nearly) optimum value so that the VNP curve lies just above the CNP curve. Further, to achieve a good match, the number of different VN degrees need be only about 3 or 4 and the number of different CN degrees need be only 1 or 2.

---

**Example 9.8** We consider the design of a rate-1/2 irregular LDPC code with four possible VN degrees and two possible CN degrees. Given that $\lambda(1) = \rho(1) = 1$ and $R = 1 - \int_0^1 \rho(X)dX / \int_0^1 \lambda(X)dX$, only two of the four coefficients for $\lambda(X)$ need be specified and only one of the two for $\rho(X)$ need be specified, after the six degrees have been specified. A non-exhaustive search yielded $\lambda(X) = 0.267X + 0.176X^2 + 0.127X^3 + 0.430X^9$ and $\rho(X) = 0.113X^4 + 0.887X^7$ with a decoding threshold of $(E_b/N_0)_{\text{EXIT}} = 0.414$ dB. The EXIT chart for $E_b/N_0 = 0.55$ dB is presented in Figure 9.7, where we see a narrow, but open, tunnel between the two transfer curves. Figure 9.7 also gives the "node-perspective" degree-distribution information. Thus, 17% of the CNs are degree-5 CNs and 85% are degree-8 CNs; 50% of the VNs are degree-2 VNs, 22% are degree-3 VNs, and so on.

---

The references contain additional information on EXIT charts, including EXIT charts for the Rayleigh channel, for higher-order modulation, and for multi-input/multi-output channels. The area property for EXIT charts and its significance is briefly reviewed in Section 9.8.

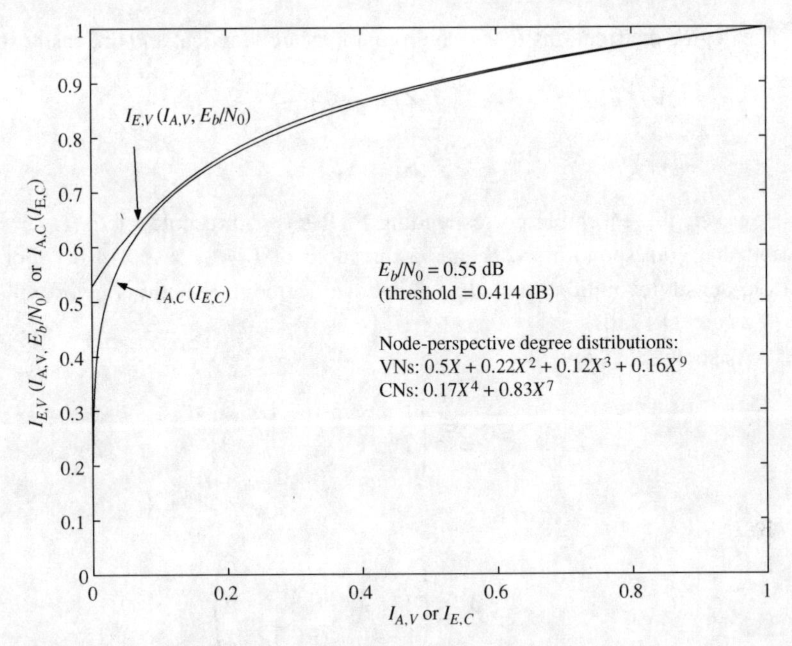


**Figure 9.7** An EXIT chart for a rate-1/2 irregular LDPC code ensemble.

## 9.6.3 EXIT Technique for Protograph-Based Codes

We present in this section an extension of the EXIT approach to codes defined by protographs, following [22, 23]. This extension is a multidimensional numerical technique and hence does not have a two-dimensional EXIT-chart representation of the iterative decoding procedure. Still, the technique yields decoding thresholds for LDPC code ensembles specified by protographs. This multidimensional technique is facilitated by the relatively small size of protographs and permits the analysis of protograph code ensembles characterized by the presence of *exceptional node types*, that is, node types that can lead to failed EXIT-based convergence. Examples of exceptional node types are degree-1 variable nodes and punctured variable nodes.

A code ensemble specified by a protograph is a refinement (sub-ensemble) of a code ensemble specified simply by the protograph's (and hence LDPC code's) degree distributions. To demonstrate this, we recall the base matrix $\mathbf{B} = \left[b_{ij}\right]$ for a protograph, where $b_{ij}$ is the number of edges between CN $i$ and VN $j$ in the protograph. As an example, consider the protographs with base matrices

$$\mathbf{B} = \begin{pmatrix} 2 & 1 & 1 \\ 1 & 1 & 1 \end{pmatrix}$$

and

$$\mathbf{B}' = \begin{pmatrix} 2 & 0 & 2 \\ 1 & 2 & 0 \end{pmatrix}.$$

The degree distributions for these protographs are identical and are easily seen to be

$$\lambda(X) = \frac{4}{7}X + \frac{3}{7}X^2,$$
$$\rho(X) = \frac{3}{7}X^2 + \frac{4}{7}X^3.$$

However, the ensemble corresponding to $\mathbf{B}$ has a threshold of $E_b/N_0 = 0.762$ dB and that corresponding to $\mathbf{B}'$ has a threshold of $E_b/N_0 = 0.814$ dB. For comparison, density evolution applied to the above degree distributions gives a threshold of $E_b/N_0 = 0.817$ dB.

As another example, let

$$\mathbf{B} = \begin{pmatrix} 1 & 2 & 1 & 1 & 0 \\ 2 & 1 & 1 & 1 & 0 \\ 1 & 2 & 0 & 0 & 1 \end{pmatrix}$$

and

$$\mathbf{B}' = \begin{pmatrix} 1 & 3 & 1 & 0 & 0 \\ 2 & 1 & 1 & 1 & 0 \\ 1 & 1 & 0 & 1 & 1 \end{pmatrix},$$

noting that they again have identical degree distributions. We also puncture the bits corresponding to the second column in each base matrix. Using the multidimensional EXIT algorithm described below, the thresholds for $\mathbf{B}$ and $\mathbf{B}'$ in this case were computed to be 0.480 dB and (about) 4.7 dB, respectively.

Thus, standard EXIT analysis based on degree distributions is inadequate for protograph-based LDPC code design. In fact, the presence of degree-1 variable nodes as in our second example implies that there is a term in the summation in (9.44) of the form

$$\lambda_1 I_{E,V}(1, I_{A,V}) = \lambda_1 J(\sigma_{ch}).$$

Since $J(\sigma_{ch})$ is always less than unity for $0 < \sigma_{ch} < \infty$ and since $\sum_{d=1}^{d_v} \lambda_d = 1$, the summation in (9.44), that is, $I_{E,V}$, will be strictly less than unity. Again, standard EXIT analysis implies failed convergence for codes with the same degree distributions as $\mathbf{B}$ and $\mathbf{B}'$. This is in contrast with the fact that codes in the $\mathbf{B}$ ensemble do converge when the SNR exceeds the threshold of 0.48 dB.

In the following, we present a multidimensional EXIT technique [22–24] that overcomes this issue and allows the determination of the decoding threshold for codes based on protographs (possibly with punctured nodes).

The algorithm presented in [22, 23] eliminates the average in (9.44) and considers the propagation of the messages on a decoding tree that is specified by the protograph of the ensemble. Let $\mathbf{B} = \left[b_{ij}\right]$ be the $M \times N$ base matrix for the protograph under analysis. Let $I_{E,V}^{j \to i}$ be the extrinsic mutual information between code bits associated with "type $j$" VNs and the LLRs $L_{j \to i}$ sent from these VNs to "type $i$" CNs. Similarly, let $I_{E,C}^{i \to j}$ be the extrinsic mutual information between code bits associated with "type $j$" VNs and the LLRs $L_{i \to j}$ sent from "type $i$" CNs to these VNs. Then, because $I_{E,C}^{i \to j}$

acts as *a priori* mutual information in the calculation of $I_{E,V}^{j\to i}$, following (9.42) we have (provided that there exists an edge between CN $i$ and VN $j$, that is, provided that $b_{ij} \neq 0$)

$$I_{E,V}^{j\to i} = J\left(\sqrt{\sum_{c=1}^{M}\left(b_{cj} - \delta_{ci}\right)\left(J^{-1}(I_{E,C}^{c\to j})\right)^2 + \sigma_{ch,j}^2}\right), \tag{9.46}$$

where $\delta_{ci} = 1$ when $c = i$ and $\delta_{ci} = 0$ when $c \neq i$. $\sigma_{ch,j}^2$ is set to zero if code bit $j$ is punctured. Similarly, because $I_{E,V}^{j\to i}$ acts as *a priori* mutual information in the calculation of $I_{E,C}^{i\to j}$, following (9.43) we have (when $b_{ij} \neq 0$)

$$I_{E,C}^{i\to j} = 1 - J\left(\sqrt{\sum_{v=1}^{N}\left(b_{iv} - \delta_{vj}\right)\left(J^{-1}(1 - I_{E,V}^{v\to i})\right)^2}\right). \tag{9.47}$$

The multidimensional EXIT algorithm can now be presented (see below). This algorithm converges only when the selected $E_b/N_0$ is above the threshold. Thus, the threshold is the lowest value of $E_b/N_0$ for which all $I_{CMI}^{j}$ converge to 1. As shown in [22, 23], the thresholds computed by this algorithm are typically within 0.05 dB of those computed by density evolution.

**Algorithm 9.2** Multidimensional EXIT Algorithm

1. **Initialization.** Select $E_b/N_0$. Initialize a vector $\boldsymbol{\sigma}_{ch} = (\sigma_{ch,0}, \ldots, \sigma_{ch,N-1})$ such that

$$\sigma_{ch,j}^2 = 8R(E_b/N_0)_j,$$

where $(E_b/N_0)_j$ equals zero when $x_j$ is punctured and equals the selected $E_b/N_0$ otherwise.
2. **VN to CN.** For $j = 0, \ldots, N-1$ and $i = 0, \ldots, M-1$, compute (9.46).
3. **CN to VN.** For $i = 0, \ldots, M-1$ and $j = 0, \ldots, N-1$, compute (9.47).
4. **Cumulative mutual information.** For $j = 0, \ldots, N-1$, compute

$$I_{CMI}^{j} = J\left(\sqrt{\sum_{c=1}^{M}\left(J^{-1}\left(I_{E,C}^{c\to j}\right)\right)^2 + \sigma_{ch,j}^2}\right).$$

5. **Stopping criterion.** If $I_{CMI}^{j} = 1$ (up to the desired precision) for all $j$, then stop; otherwise, go to Step 2.

## 9.7 EXIT Charts for Turbo Codes

The EXIT-chart technique for turbo codes is very similar to that for LDPC codes. For LDPC codes, there is one extrinsic-information transfer curve for the VNs and one for the CNs. The CNs are not connected to the channel, so only the VN transfer curve is affected by changes in $E_b/N_0$. The $E_b/N_0$ value at which the VN curve just touches the CN curve is the decoding threshold. For turbo codes, there is one extrinsic-information transfer curve for each constituent code (and we consider only turbo codes with two constituent codes). For parallel turbo codes, both codes are directly connected to the channel, so the transfer curves for both shift with changes in $E_b/N_0$. For serial turbo codes, only the inner code is generally connected to the channel, in which case only the transfer curve for that code is affected by changes in $E_b/N_0$. Our discussion in this section begins with parallel turbo codes [18] and then shows how the technique is straightforwardly extended to serial turbo codes.

Recall from Chapter 7 (e.g., Figure 7.3) that each constituent decoder receives as inputs $L^u_{ch} = 2y^u/\sigma^2$ (channel LLRs for systematic bits), $L^p_{ch} = 2y^p/\sigma^2$ (channel LLRs for parity bits), and $L_a$ (extrinsic information that was received from a counterpart decoder, used as *a priori* information). It produces as outputs $L^{\text{total}}$ and $L_e = L^{\text{total}} - L^u_{ch} - L_a$, where the extrinsic information $L_e$ is sent to the counterpart decoder, which uses it as *a priori* information. Let $I_E$ be the mutual information between the outputs $L_e$ and the systematic bits $u$ and let $I_A$ be the mutual information between the inputs $L_a$ and the systematic bits $u$. Then the transfer curve for each constituent decoder, is a plot of $I_E$ versus $I_A$. Each such curve is parameterized by $E_b/N_0$.

Let $p_E(l|U)$ be the pdf of $L_e$ conditioned on systematic input $U$. Then the mutual information between $L_e$ and the systematic bits $u$ is given by

$$I_E = 0.5 \sum_{u=\pm 1} \int_{-\infty}^{+\infty} p_E(l|U=u) \log_2\left(\frac{2p_E(l|U=u)}{p_E(l|U=+1)+p_E(l|U=-1)}\right) dl. \tag{9.48}$$

The conditional pdfs that appear in (9.48) are approximately Gaussian, especially for higher SNR values, but more accurate results are obtained if the pdfs are estimated via simulation of the encoding and decoding of the constituent code. During the simulation, two conditional histograms of $L_e = L^{\text{total}} - L^u_{ch} - L_a$ are accumulated, one corresponding to $p_E(l|U = +1)$ and one corresponding to $p_E(l|U = -1)$, and the respective pdfs are estimated from these histograms. $I_E$ is then computed numerically from (9.48) using these pdf estimates. Note that the inputs to this process are $E_b/N_0$ and $L_a$. In order to obtain the $I_E$ versus $I_A$ transfer curve mentioned in the previous paragraph, we need to obtain a relationship between the pdf of $L_a$ and $I_A$.

Recall that the input $L_a$ for one constituent decoder is the output $L_e$ for the counterpart decoder. As mentioned in the previous paragraph, $L_e$ (and hence $L_a$) is approximately conditionally Gaussian for both +1 and −1 inputs. For our purposes, it is (numerically) convenient to model $L_a$ as if it were precisely conditionally Gaussian; doing so results in negligible loss in accuracy. Further, we assume consistency so that

$$L_a = \mu_a \cdot u + n_a, \tag{9.49}$$

where $n_a \sim \mathcal{N}\left(0, \sigma_a^2\right)$ and $\mu_a = \sigma_a^2/2$. In this case, following (9.39), we have

$$I_A = 1 - \int_{-\infty}^{\infty} \frac{1}{\sqrt{2\pi}\sigma_a} \exp\left(-\left(l - \sigma_a^2/2\right)^2/(2\sigma_a^2)\right)\log_2\left(1+e^{-l}\right) dl. \tag{9.50}$$

Because $I_A$ is monotonic in $\sigma_a$, there is a one-to-one correspondence between $I_A$ and $\sigma_a$: given one, we may easily determine the other.

We are now prepared to enumerate the steps involved in obtaining the $I_E$ versus $I_A$ transfer curve for a constituent code. We assume a parallel turbo code with two constituent recursive systematic convolutional encoders.

1. Specify the turbo code rate $R$ and the SNR $E_b/N_0$ of interest. From these, determine the AWGN variance $N_0/2$. ($E_s = 1$ is assumed.)
2. Specify the mutual information $I_A$ of interest. From this determine $\sigma_a$ (e.g., from a computer program that implements (9.50)).
3. Run the RSC code simulator with the BCJR decoder under the model $y = x + n$, where $x \in \{\pm 1\}$ represent the transmitted code bits and $n \sim \mathcal{N}\left(0, \sigma^2\right)$, $\sigma^2 = N_0/2$. In addition to decoder inputs $L_{ch} = 2y/\sigma^2$ ($L_{ch}$ includes $L^u_{ch}$ and $L^p_{ch}$), the decoder has as inputs the *a priori* LLRs given by (9.49). The values for the systematic bits $u$ in (9.49) are exactly the systematic bits among the code bits $x$. The BCJR output is $L^{\text{total}}$, from which $L_e = L^{\text{total}} - L^u_{ch} - L_a$ may be computed for each systematic bit. For a long input block (100000 bits provides very stable results), histograms corresponding to $p_E(l|U = +1)$ and $p_E(l|U = -1)$ can be produced from the values collected for $L_e$ and used to estimate these conditional pdfs.
4. From (9.48) and the estimated pdfs, compute the mutual information $I_E$ that corresponds to the $I_A$ specified in Step 2.
5. If all of the $I_A$ values of interest have been exhausted, then stop; otherwise, go to Step 2.

---

**Example 9.9** We consider the computation of EXIT charts for a rate-1/2 parallel turbo code possessing two identical RSC constituent codes with generator polynomials $(7, 5)_8$. The component encoders are punctured to rate 2/3 to achieve the overall code rate of 1/2. In practice, only the first RSC code is terminated, but we can ignore termination issues because we choose 200000 as the encoder input length. The procedure above is used to obtain the $I_{E,1}$ versus $I_{A,1}$ extrinsic-information transfer curve for the first code. Because the two encoders are identical, this is also the $I_{E,2}$ versus $I_{A,2}$ transfer curve for the second code. Also, because the extrinsic output for the first code is the *a priori* input for the second code, and the extrinsic output for the second code is the *a priori* input for the first code, we can plot the $I_{E,2}$ versus $I_{A,2}$ transfer curve on the same plot as the $I_{E,1}$ versus $I_{A,1}$ transfer curve, but with opposite abscissa–ordinate conventions. In this way we can show the decoding trajectories between the two decoders using mutual information as the figure of merit. This is demonstrated for $E_b/N_0 = 0.6$ and 1 dB in Figure 9.8.

Note for both SNR values that the transfer curve for the second code is just that of the first code mirrored across the 45-degree line because the codes are identical. (Otherwise, a separate transfer curve for the second constituent code would have to be obtained.) Observe also that the decoding trajectory for the 0.6 dB case fails to make it through to the $(I_A, I_E) = (1, 1)$ convergence point because 0.6 dB is below the decoding threshold so that the two transfer curves intersect. The threshold, the value of $E_b/N_0$ at which the two transfer curves touch, is left to the reader in Problem 9.15. Note that, for SNR values just above the threshold, the tunnel between the two curves will be very narrow so that many decoding iterations would be required in order to reach the (1, 1) point (see, for example, Figure 9.6). Finally, these curves correspond to a very long turbo code, for which the information exchanged between the constituent decoders can be assumed to be independent. For turbo codes that are not quite so long, the independence assumption becomes false after some number of iterations, at which point the actual decoder trajectory does not reach the transfer-curve boundaries [18]. As a consequence, more decoder iterations become necessary.

---

The procedure for producing an EXIT chart for serial turbo codes is nearly identical to that for parallel turbo codes. We assume that only the inner code is connected to the channel. In this case, the transfer curve for that code is obtained in a fashion that is identical to that for a constituent code of a parallel turbo code. For the outer code, there are no inputs $L^u_{ch}$ and $L^p_{ch}$ from the channel, so its extrinsic output is computed as $L_e = L^{\text{total}} - L_a$ (i.e., not $L_e = L^{\text{total}} - L^u_{ch} - L_a$). Thus, to obtain the transfer curve for the outer code, one simulates the outer code decoder with inputs specified as in (9.49), from which estimates of the pdfs $p_E(l|U = +1)$ and $p_E(l|U = -1)$ are produced. These pdfs give $I_E$ through (9.48) as before. Specifying the parameter $\mu_a$ (equivalently, $\sigma_a^2$) in (9.49) pins down $I_A$ so each simulation of the model (9.49) with differing $\mu_a$ values gives different $(I_A, I_E)$ pairs from which the outer-code transfer curve is drawn.

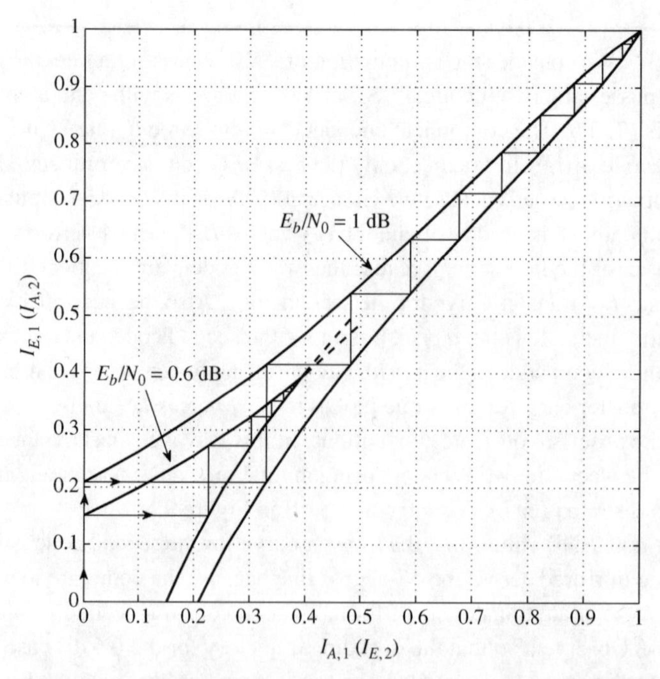


**Figure 9.8** EXIT charts at $E_b/N_0 = 0.6$ and 1 dB for a rate-1/2 parallel turbo code with RSC constituent codes having generators $g^{(1)}(D) = 1 + D + D^2$ and $g^{(2)}(D) = 1 + D^2$.

We remark that, unlike with LDPC codes, for which optimal degree distributions may be found via optimization techniques, for turbo codes, design involves a bit of trial and error. The reason is that there is a discrete search space for the turbo-code case, namely, the set of constituent-code generator polynomials.

## 9.8 The Area Property for EXIT Charts

When describing the design of iteratively decodable codes via EXIT charts, relying on common sense, it was remarked that the closer the shape of the CNP transfer curve matches the shape of the VNP transfer curve, the better the decoding threshold will be. The matched shapes allow one to lower the VNP curve (by lowering the SNR, for example) as much as possible without intersecting the CNP curve, thus improving (optimizing) the SNR decoding threshold. Information-theoretic support was given to this observation in [14, 25] for the BEC and later in [26, 27] for other binary-input memoryless channels. Following [14], we give a brief overview of this result in this section for serial-concatenated codes and LDPC codes, which have very similar descriptions. The results for parallel-concatenated codes are closely related, but require more subtlety [14]. The usual assumptions regarding independence, memorylessness, and long codes are in effect here.

### 9.8.1 Serial-Concatenated Codes

Consider a serial-concatenated code with EXIT characteristics $I_{E,1}(I_{A,1})$ for the outer code and $I_{E,2}(I_{A,2})$ for the inner code. For convenience, the dependence of $I_{E,2}$ on SNR is suppressed. For $j = 1, 2$, let

$$\mathcal{A}_j = \int_0^1 I_{E,j}(I_{A,j})dI_{A,j}$$

be the area under the $I_{E,j}(I_{A,j})$ curve. Then it can be shown that, for the outer code,

$$\mathcal{A}_1 = 1 - R_1, \tag{9.51}$$

where $R_1$ is the rate of the outer code. Note that this is the area *above* the $I_{E,1}(I_{A,1})$ curve when plotted on swapped axes. For the inner code, the result is

$$\mathcal{A}_2 = \frac{I(\underline{X};\underline{Y})/n}{R_2}, \tag{9.52}$$

where $R_2$ is the rate of the inner code, $n$ is the length of the inner code (and hence of the serial-concatenated code), and $I(\underline{X};\underline{Y})$ is the mutual information between the channel input vector (represented by $\underline{X}$) and the channel output vector (represented by $\underline{Y}$).

Equations (9.51) and (9.52) lead to profound results. First, we know that for successful iterative decoding the outer-code EXIT curve (plotted on swapped axes) must lie below the inner-code EXIT curve. This implies that $1 - \mathcal{A}_1 < \mathcal{A}_2$ or, from (9.51) and (9.52),

$$R_1 R_2 < I(\underline{X};\underline{Y})/n \leq C,$$

where $C$ is the channel capacity. This is, of course, in agreement with Shannon's result that the overall rate must be less than capacity, but we may draw additional conclusions. Because $1 - \mathcal{A}_1 < \mathcal{A}_2$, let $1 - \mathcal{A}_1 = \gamma \mathcal{A}_2$ for some $\gamma \in [0, 1)$. Then, from (9.51) and (9.52),

$$R_1 R_2 = \gamma I(\underline{X};\underline{Y})/n \leq \gamma C. \tag{9.53}$$

Because $\gamma$ is a measure of the mismatch between the two areas, $1 - \mathcal{A}_1$ and $\mathcal{A}_2$, (9.53) implies that the area-mismatch factor $\gamma$ between the two EXIT curves leads directly to a rate loss relative to $C$ by the same factor. Thus, the code-design problem in this sense leads to a curve-fitting problem: the shape of the outer-code curve must be identical to that of the inner code, otherwise a code rate loss will occur.

Note that, if $R_2 = 1$ (e.g., the inner code is an accumulator or an intersymbol-interference channel) and $I(\underline{X};\underline{Y})/n = C$, then $\mathcal{A}_2 = C$ (from (9.52)) and $\mathcal{A}_2 - (1 - \mathcal{A}_1) = C - R_1$; that is, the area between the two EXIT curves is precisely the code rate loss. Further, if $R_2 < 1$, then $I(\underline{X};\underline{Y})/n < C$, so that any inner code with $R_2 < 1$ creates an irrecoverable rate loss. One might consider a very strong inner code for which the loss is negligible, that is, $I(\underline{X};\underline{Y})/n \simeq C$, but the idea behind concatenated codes is to use weak, easily decodable component codes. Thus, rate-1 inner codes are highly recommended for serial-concatenated code design.

### 9.8.2 LDPC Codes

The LDPC code case is similar to that of the serial-concatenated code case in that the variable nodes act as the inner code and the check nodes act as the outer code (because the check nodes receive no channel information). Let $\mathcal{A}_C$ be the area under the $I_{E,C}(I_{A,C})$ curve and let $\mathcal{A}_V$ be the area under the $I_{E,V}(I_{A,V})$ curve. Then it can be shown that

$$\mathcal{A}_C = 1/\bar{d}_c, \tag{9.54}$$

$$\mathcal{A}_V = 1 - (1 - C)/\bar{d}_v, \tag{9.55}$$

where $C$ is the channel capacity as before and $\bar{d}_c$ ($\bar{d}_v$) is the average check (variable) node degree. $\bar{d}_c$ and $\bar{d}_v$ are easily derived from the degree distributions $\rho(X)$ and $\lambda(X)$ (or $\tilde{\rho}(X)$ and $\tilde{\lambda}(X)$). As we saw in our first description of EXIT charts in Section 9.6, the decoder will converge only if the CNP EXIT curve (plotted on swapped axes) lies below the VNP EXIT curve. This implies $1 - \mathcal{A}_C < \mathcal{A}_V$, or $1 - \mathcal{A}_V < \mathcal{A}_C$, which we may write as $1 - \mathcal{A}_V = \gamma \mathcal{A}_C$ for some $\gamma \in [0, 1)$. This equation, combined with (9.54) and (9.55) and some algebra, yields

$$R = \frac{C - (1 - \gamma)}{\gamma} < C, \tag{9.56}$$

where we have used the fact that $R = 1 - \bar{d}_v/\bar{d}_c$.

This equation is the LDPC-code counterpart to (9.53). Thus, any area difference between the CNP and VNP EXIT curves (quantified by the factor $\gamma$) corresponds to a rate loss relative to $C$. Consider, for example, the case $\gamma = 1$ in (9.56). Consequently, capacity can be approached by closely matching the two transfer curves, for example, by curve fitting the VNP EXIT curve to the CNP EXIT curve.

## 9.9 Reciprocal-Channel Approximation Algorithm

After covering several techniques for computing the decoding threshold of LDPC code ensembles, some of which are heavy in theory and/or computation, we have saved the best for last. The technique to be presented here is simple in principle and simple computationally. The idea is that we want to pass around a code ensemble's protograph – a parameter that serves as a proxy for a pdf, much like the Gaussian approximation approach.

The technique is called the (modified) *reciprocal-channel approximation* (RCA). It was devised by S. Y. Chung [28], popularized by Divsalar *et al.* [29], and modified by T.-Y. Chen *et al.* [30] for protographs with degree-1 variable nodes (VNs). We will present the modified RCA algorithm, but we will call it the RCA algorithm for short. It provides approximate but accurate thresholds for the binary-input AWGN channel, and it provides an exact threshold for the binary-erasure channel (BEC). We first develop the RCA algorithm for the BEC. The RCA algorithm for the binary-input AWGN channel will be modeled after the RCA algorithm for the BEC.

Following [29, 31], for reasons soon made clear, we put erasure probabilites in the form $e^{-s}$ and say edge $k$ passes erasure probability $e^{-s_k}$ from its VN to its CN. We call a BEC with erasure probability $e^{-r_k}$ a *reciprocal channel* to the BEC with erasure probability $e^{-s_k}$ if $e^{-s_k} + e^{-r_k} = 1$. As will be explained, it is these parameters, $s_k$ and $r_k$, that are passed around the protograph in the RCA algorithm that we now develop.

Consider a VN of degree $d_v + 1$, including the channel input. Because a VN represents a repetition code, for the BEC, an outgoing message from a VN is known (i.e., is not an erasure) if *any* of the inputs are *known*. In other words, the output is erased if and only if all of the inputs are erased, and this happens with probability

$$\prod_k e^{-s_k} = e^{-\sum_k s_k},$$

where the summation is over all $d_v+1$ edges connected to the VN. Because the parameters $\{s_k\}$ are added in the right-hand side of the above equation, for the RCA algorithm, the parameters $\{s_k\}$ are passed around and are added at the VNs; there is no need to pass around the erasure probabilities. Thus, the outgoing VN message along edge $k'$ in the RCA algorithm is the sum $\sum_{k \backslash k'} s_k$ of the parameters $\{s_k\}$ coming in from the *other* edges. This sum includes the edge coming from the channel, but we will write $s_{ch} + \sum_{k \backslash k'} s_k$ to isolate the channel parameter $s_{ch}$.

Now consider a CN of degree $d_c$. Because a CN represents a single parity-check code, for the BEC, an outgoing message is unknown (erased) if *any* of the inputs are *unknown* (erased). In other words, an outgoing CN message is known (unerased) if all of the inputs are known (unerased), and this happens with probability

$$\prod_k (1 - e^{-s_k}) = \prod_k e^{-r_k} = e^{-\sum_k r_k},$$

where each reciprocal channel parameter $r_k$ is chosen so that

$$e^{-s_k} + e^{-r_k} = 1. \tag{9.57}$$

Thus, while the parameters $s_k$ add at the VNs, the parameters $r_k$ add at the CNs.

Let $C(s) = 1 - e^{-s}$ represent the capacity of a BEC with parameter $s$ and erasure probability $e^{-s}$. Then, from (9.57), because $e^{-r}$ is the erasure probability of the reciprocal channel,

$$C(s) + C(r) = 1. \tag{9.58}$$

Because $C(\cdot)$ is monotonic in its argument, it is invertible and we may write

$$r = C^{-1}(1 - C(s)) \triangleq R(s) \tag{9.59}$$

and, similarly, $s = R(r)$.

We now have the necessary elements to describe the reciprocal channel technique for finding the decoding threshold for a protograph-based code on the BEC. First, the channel parameter $s_{ch}$ is fixed (to some positive number) and set as the channel inputs to all VNs in the protograph. Then, with all parameters $\{s_k\}$ and $\{r_k\}$ initialized to zero, for each VN and each edge $k'$ coming from a VN, the message $s_{k'} = s_{ch} + \sum_{k \setminus k'} s_k$ is sent to the neighboring CN. In turn, for each CN and each edge $k'$ coming from a CN, the message $r_{k'} = \sum_{k \setminus k'} R(s_k)$ is sent to the neighboring VN. The VNs take the messages $\{r_k\}$ as inputs and compute the messages $s_{k'} = s_{ch} + \sum_{k \setminus k'} R(r_k)$ for each edge $k'$, sending them back to the CNs, which again compute $r_{k'} = \sum_{k \setminus k'} R(s_k)$, and so on.

After some large number of iterations, the algorithm will have converged and the totals $S_{tot} = s_{ch} + \sum_k s_k$ for each VN are computed and compared to an empirically determined threshold, $S_{thres}$. The smallest value of $s_{ch}$, denoted by $s^*_{ch}$, for which the threshold is exceeded by $\min_v\{S_{tot}\}$ gives the decoding threshold for the protograph under test as $e^{-s^*_{ch}}$.

At this point we can present the algorithm a little more precisely. Further, we extend it to the binary-input AWGN channel, which means that the parameters $s_k$ and $r_k$ now represent signal-to-noise ratios of the form $2E_s/N_0$, where $E_s$ is the energy per channel bit and it equals the data bit energy $E_b$ scaled by the code rate, $E_s = RE_b$. The capacity for this channel was presented in Chapter 1 and is

$$C_{\text{BI-AWGN}}(s) = \int_{-\infty}^{+\infty} \frac{1}{\sqrt{2\pi}} e^{-z^2/2} \left(1 - \log_2\left(1 + e^{-2s + 2\sqrt{s}z}\right)\right) dz, \tag{9.60}$$

where $s = 2E_s/N_0$. We continue to have the reciprocal channel relationships $C(s) + C(r) = 1$, $r = R(s)$, and $s = R(r)$, with $R(s) = C^{-1}(1 - C(s))$ suitably defined for the binary-input AWGN channel. Lastly, we add a commonly used feature in the design of protograph-based LDPC codes, punctured VNs, whose SNR values from the channel are set to zero.

We emphasize that the reciprocal-channel algorithm is exact for the BEC, but is only approximate (but very accurate) for the binary-input AWGN channel, hence the name *reciprocal-channel approximation algorithm*. The implementation of the function $R(\cdot)$ in the algorithm below is by an interpolative look-up table. We define $E_c$ to be the set of edges connected to CN $c$ and $E_v$ to be the set of edges connected to VN $v$. The number of iterations, $N_{iter}$, is set to some large value such as 1000.

**The RCA Algorithm**

1. *Initialize*. Set $s_{ch} = 2E_s/N_0$ (initially some small value), set iteration counter $n = 0$, and set $N_{iter} = 1000$. For all VNs, if VN $v$ is punctured, set channel input $S_v = 0$; otherwise, set $S_v = s_{ch}$. For all edges $k$ connected to punctured nodes, set $s_k = 0$; for all other edges, set $s_k = s_{ch}$.
2. For all CNs $c$

   For all edges $k' \in E_c$

$$r_{k'} = \sum_{k \in E_c \backslash k'} R(s_k).$$

3. For all VNs $v$

   For all edges $k' \in E_v$

$$s_{k'} = S_v + \sum_{k \in E_v \backslash k'} R(r_k).$$

   If $++n \leq N_{iter}$, go to Step 2.
4. For all VNs $v$

$$S_{v,tot} = S_v + \sum_{k \in E_v} R(r_k).$$

5. If $\min_v\{S_{v,tot}\} < S_{thres}$, increase $2E_s/N_0$ and return to Step 1; else, stop and set the decoding threshold, $s^*_{ch} = (2E_s/N_0)^*$, to be the smallest value of $s_{ch} = 2E_s/N_0$ for which $\min_v\{S_{v,tot}\} > S_{thres}$.

---

**Example 9.10** We consider three protographs in this example, all specified by the base matrix for the protograph. The first is the protograph for a regular rate-1/2 $d_v = 3, d_c = 6$ LDPC code whose base matrix is given by

$$\mathbf{H}_{base} = \begin{bmatrix} 1 & 1 & 1 & 1 & 1 & 1 \\ 1 & 1 & 1 & 1 & 1 & 1 \\ 1 & 1 & 1 & 1 & 1 & 1 \end{bmatrix}.$$

Running the RCA algorithm for values of $s_{ch}$ in the range 0.1 to 2 dB gives the leftmost curve in Figure 9.9, where the vertical axis is the final $\min_v\{S_{v,tot}\}$ of the

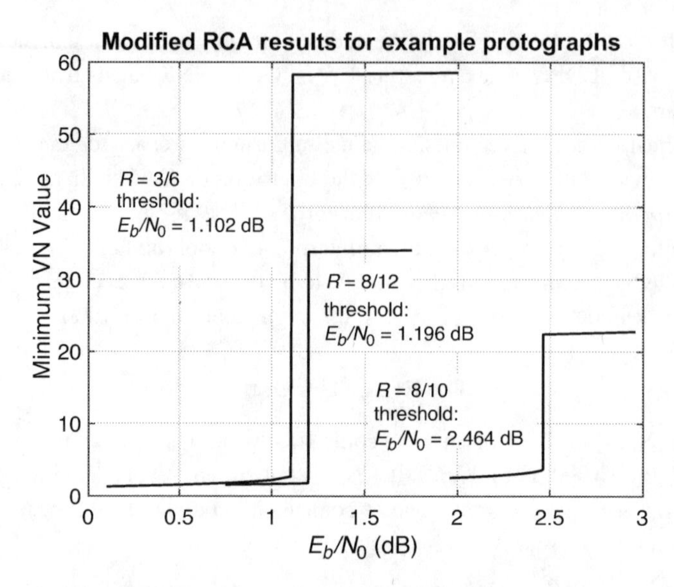


**Figure 9.9** RCA algorithm trajectories for the example protographs. $E_b/N_0 = s_{ch}/(2R)$. From the figure, $S_{thres} = 10, 15$, or 20 all appear reasonable.

algorithm. (Because $R = 1/2$, $E_b/N_0 = s_{ch}$ for this first example.) Observe that the threshold given by the RCA is $E_b/N_0 = 1.102$ dB, in agreement with Examples 9.2 and 9.6. (The levels of the plateau and the base of the trajectory are a function of the particular implementation of the RCA algorithm, specifically, the interpolative look-up table for the function $R(\cdot)$.)

The next example is the protograph for a rate-8/10 protograph-based Raptor-like (PBRL) LDPC code from [30]. (PBRL codes were discussed briefly in Chapter 6.) PBRL codes have base matrices of the form

$$\mathbf{H}_{base} = \begin{bmatrix} \mathbf{A} & \mathbf{0} \\ \mathbf{B} & \mathbf{I} \end{bmatrix},$$

where $\mathbf{0}$ and $\mathbf{I}$ are zero and identity matrices, respectively, of appropriate size. For this example,

$$\mathbf{A} = \begin{bmatrix} 3 & 1 & 3 & 1 & 3 & 1 & 3 & 1 & 2 & 1 \\ 1 & 3 & 1 & 3 & 1 & 3 & 1 & 2 & 1 & 2 \end{bmatrix}$$

and

$$\mathbf{B} = \begin{bmatrix} 2 & 0 & 0 & 0 & 0 & 0 & 0 & 0 & 0 & 0 \end{bmatrix}.$$

The base matrix $\mathbf{H}_{base}$ is $3 \times 11$. The first VN is punctured, giving a rate-8/10 code. From Figure 9.9, rightmost curve, the decoding threshold is at $E_b/N_0 = 2.464$ dB ($s_{ch}^* = 4.505$ dB), which is at a gap of 0.424 dB from the binary-input AWGN capacity limit for rate-0.8 codes. A PBRL code with a better decoding threshold is possible, but this code was designed for a short blocklength and a low error-rate floor.

The last example is also a PBRL code. Its base matrix is

$$\mathbf{H}_{base} = \begin{bmatrix} 3 & 2 & 1 & 1 & 1 & 1 & 1 & 1 & 0 & 0 & 1 & 0 & 0 \\ 1 & 1 & 2 & 2 & 2 & 2 & 2 & 2 & 2 & 1 & 1 & 0 & 0 \\ 2 & 0 & 0 & 0 & 0 & 0 & 0 & 0 & 1 & 2 & 0 & 0 & 0 \\ 2 & 1 & 2 & 1 & 0 & 0 & 0 & 0 & 1 & 0 & 0 & 1 & 0 \\ 1 & 2 & 1 & 0 & 1 & 0 & 0 & 0 & 0 & 1 & 0 & 0 & 1 \end{bmatrix}$$

and, upon puncturing the first VN, its rate becomes 8/12 and its decoding threshold from Figure 9.9 is $E_b/N_0 = 1.196$ dB ($s^*_{ch} = 2.445$ dB). This value is at a gap of 0.136 dB from the binary-input AWGN capacity limit for rate-2/3 codes.

## Problems

**9.1** Show that the pdf of the channel message $m_0^{(c)}$ for the BSC is given by (9.11) under the assumption that $x = +1$ is always transmitted.

**9.2** (Density evolution for the BEC [2]) Show that for the BEC with erasure probability $\delta_0$, the recursion (9.19) simplifies to

$$\delta_\ell = \delta_0 \lambda(1 - \rho(1 - \delta_{\ell-1})),$$

where $\delta_\ell$ is the probability of a bit erasure after the $\ell$th iteration. Note that convolution is replaced by multiplication. The decoding threshold $\delta_0^*$ in this case is the supremum of all erasure probabilities $\delta_0 \in (0, 1)$ for which $\delta_\ell \to 0$ as $\ell \to \infty$.

**9.3** (BEC stability condition [2]) Expand the recursion in the previous problem to show that

$$\delta_\ell = \delta_0 \lambda'(0) \rho'(1) \delta_{\ell-1} + O(\delta_{\ell-1}^2),$$

from which we can conclude that $\delta_\ell \to 0$ provided $\delta_0 < [\lambda'(0)\rho'(1)]^{-1}$, that is, $\delta_0^*$ is upper bounded by $[\lambda'(0)\rho'(1)]^{-1}$.

**9.4** Show that the girth of a (4161, 3430) code with the degree distributions in (9.22) is at most 10. To do this, draw a tree starting from a degree-2 variable node, at level 0. At level 1 of the tree there will be two check nodes, at level 3 there will be at least 36 variable nodes (since the check nodes have degree at least 19), and so on. Eventually, there will be a level $L$ with more than 4161 − 3430 = 731 check nodes in the tree, from which one can conclude that level $L$ has repetitions of the 731 check nodes from which the girth bound may be deduced.

**9.5** Reproduce Figure 9.3 of Example 9.6 using the Gaussian-approximation algorithm. While the expression given in the example for $\Phi(\mu)$, with $0 < \mu < 10$, is easily invertible, you will have to devise a way to implement $\Phi^{-1}(\cdot)$ in software for $\mu > 10$. One possibility is fzero in Matlab.

**9.6** Repeat the previous problem for the rate-1/3 regular-(4, 6) ensemble. You should find that $\sigma_{\mathrm{GA}} = 1.0036$, corresponding to $(E_b/N_0)_{\mathrm{GA}} = 1.730$ dB.

**9.7** Using density evolution, it was shown in Example 9.3 that the decoding threshold for the irregular $d_v = 11$ ensemble is $(E_b/N_0)^* = 0.380$ dB. Use the Gaussian-approximation algorithm of Section 9.4.2 to find the estimated threshold $(E_b/N_0)_{\mathrm{GA}}$ for the same degree-distribution pair.

**9.8** For quantized density evolution of *regular* LDPC ensembles, find a recursion analogous to (9.16). For quantized density evolution of *irregular* LDPC ensembles, find a recursion analogous to (9.19).

**9.9** Show that the consistency condition (9.31) is satisfied by the pdfs of the initial (channel) messages both from the BEC and from the BSC.

**9.10** Write an EXIT-chart computer program to reproduce Figure 9.6.

**9.11** Write an EXIT-chart computer program to reproduce Figure 9.7.

**9.12** (EXIT-like chart using SNR as convergence measure [15]) Suppose that, instead of mutual information, we use SNR as the convergence measure in an LDPC-code ensemble EXIT chart. Consider a $(d_v, d_c)$-regular LDPC code ensemble. Show that there is a straight-line equation for the variable nodes given by $\mathrm{SNR}_{\mathrm{out}} = (d_v - 1)\mathrm{SNR}_{\mathrm{in}} + 2RE_b/N_0$. Show also that there is a large-SNR asymptote for the check nodes given by $\mathrm{SNR}_{\mathrm{out}} = \mathrm{SNR}_{\mathrm{in}} - 2\ln(d_c - 1)$.

**9.13** Figure 9.7 allows four possible VN degrees and two possible CN degrees. Design degree distributions for a rate-1/2 ensemble with only one possible CN degree ($d_c$) and three possible VN degrees. Set $d_c = 8$ so that $\rho(X) = X^7$. Also, set the minimum and maximum VN degrees to be 2 and 18. You must choose (search for) a third VN degree. Since $\lambda(1) = 1$ and $R = 1/2 = 1 - \int_0^1 \rho(X)dX / \int_0^1 \lambda(X)dX$, only one coefficient of $\lambda(X)$ may be chosen freely, once all four degrees have been set. Try $\lambda_2$ in the vicinity of 0.5 and vary it by a small amount (in the range 0.49–0.51). You should be able to arrive at degree distributions with an EXIT-chart threshold close to $(E_b/N_0)_{\mathrm{EXIT}} = 0.5$ dB. Investigate the impact that the choice of $d_c$ has on the threshold and convergence speed.

**9.14** Use the multidimensional-EXIT-chart technique for protograph LDPC codes in Section 9.6.3 to confirm that the protographs with base matrices

$$\mathbf{B} = \begin{pmatrix} 2 & 1 & 1 \\ 1 & 1 & 1 \end{pmatrix}$$

and

$$\mathbf{B}' = \begin{pmatrix} 2 & 0 & 2 \\ 1 & 2 & 0 \end{pmatrix}$$

have decoding thresholds $E_b/N_0 = 0.762$ and 0.814 dB, respectively. As mentioned in Section 9.6.3, they have identical degree distributions (what are they?), for which density evolution yields a threshold of $E_b/N_0 = 0.817$ dB.

**9.15** Write a computer program to reproduce Figure 9.8. Also, use your program to find the decoding threshold for this turbo code, which is clearly between 0.6 and 1 dB from the figure.

**9.16** Produce an EXIT chart for a rate-1/2 serial-concatenated convolutional code with a rate-1/2 outer (non-recursive) convolutional code with generators $g^{(1)}(D) = 1+D+D^2$ and $g^{(2)}(D) = 1+D^2$ and a rate-1 inner code that is simply the accumulator $1/(1+D)$. Show that the decoding threshold is about 1 dB.

**9.17** Implement the RCA algorithm and reproduce Figure 9.9.

## References

[1] T. Richardson and R. Urbanke, "The capacity of low-density parity-check codes under message-passing decoding," *IEEE Transactions on Information Theory*, vol. 47, no. 2, pp. 599–618, February 2001.

[2] T. Richardson and R. Urbanke, "Design of capacity-approaching irregular LDPC codes," *IEEE Transactions on Information Theory*, vol. 47, no. 2, pp. 619–637, February 2001.

[3] M. Yang, W. E. Ryan, and Y. Li, "Design of efficiently encodable moderate-length high-rate irregular LDPC codes," *IEEE Transactions on Communications*, vol. 49, no. 4, pp. 564–571, April 2004.

[4] Y. Kou, S. Lin, and M. Fossorier, "Low-density parity-check codes based on finite geometries: A rediscovery and new results," *IEEE Transactions on Information Theory*, vol. 47, no. 11, pp. 2711–2736, November 2001.

[5] S.-Y. Chung, G. D. Forney, Jr., T. Richardson, and R. Urbanke, "On the design of low-density parity-check codes within 0.0045 dB of the Shannon limit," *IEEE Communications Letters*, vol. 4, no. 2, pp. 58–60, February 2001.

[6] S.-Y. Chung, T. Richardson, and R. Urbanke, "Analysis of sum–product decoding of LDPC codes using a Gaussian approximation," *IEEE Transactions on Information Theory*, vol. 47, no. 2, pp. 657–670, February 2001.

[7] F. Peng, W. E. Ryan, and R. Wesel, "Surrogate-channel design of universal LDPC codes," *IEEE Communications Letters*, vol. 10, no. 6, pp. 480–482, June 2006.

[8] C. Jones, A. Matache, T. Tian, J. Villasenor, and R. Wesel, "The universality of LDPC codes on wireless channels," *Proceedings of the Military Communications Conference (MILCOM)*, October 2003.

[9] M. Franceschini, G. Ferrari, and R. Raheli, "Does the performance of LDPC codes depend on the channel?" *Proceedings of the International Symposium on Information Theory and its Applications*, 2004.

[10] C. Jones, T. Tian, J. Villasenor, and R. Wesel, "The universal operation of LDPC codes over scalar fading channels," *IEEE Transactions on Communications*, vol. 55, no. 1, pp. 122–132, January 2007.

[11] F. Peng, M. Yang, and W. E. Ryan, "Simplified eIRA code design and performance analysis for correlated Rayleigh fading channels," *IEEE Transactions on Wireless Communications*, vol. 5, no. 4, pp. 720–725, March 2006.

[12] S. ten Brink, G. Kramer, and A. Ashikhmin, "Design of low-density parity-check codes for modulation and detection," *IEEE Transactions on Communications*, vol. 52, no. 4, pp. 670–678, April 2004.

[13] S. ten Brink and G. Kramer, "Design of repeat–accumulate codes for iterative detection and decoding," *IEEE Transactions on Signal Processing*, vol. 51, no. 11, pp. 2764–2772, November 2003.

[14] A. Ashikhmin, G. Kramer, and S. ten Brink, "Extrinsic information transfer functions: Model and erasure channel properties," *IEEE Transactions on Information Theory*, vol. 50, pp. 2657–2673, November 2004.

[15] D. Divsalar, S. Dolinar, and F. Pollara, "Iterative turbo decoder analysis based on density evolution," *IEEE Journal on Selected Areas in Communications*, vol. 19, no. 5, pp. 891–907, May 2001.

[16] H. El Gamal and A. R. Hammons, "Analyzing the turbo decoder using the Gaussian approximation," *IEEE Transactions on Information Theory*, vol. 47, no. 2, pp. 671–686, February 2001.

[17] M. Ardakani and F. R. Kschischang, "A more accurate one-dimensional analysis and design of LDPC codes," *IEEE Transactions on Communications*, vol. 52, no. 12, pp. 2106–2114, December 2004.

[18] S. ten Brink, "Convergence behavior of iteratively decoded parallel concatenated codes," *IEEE Transactions on Communications*, vol. 49, no. 10, pp. 1727–1737, October 2001.

[19] M. Tüchler, S. ten Brink, and J. Hagenauer, "Measures for tracing convergence of iterative decoding algorithms," *Proceedings of the 4th IEEE/ITG Conference on Source and Channel Coding*, Berlin, January 2002.

[20] E. Sharon, A. Ashikhmin, and S. Litsyn, "EXIT functions for the Gaussian channel," *Proceedings of the 40th Annual Allerton Conference on Communication, Control and Computers*, October 2003, pp. 972–981.

[21] E. Sharon, A. Ashikhmin, and S. Litsyn, "EXIT functions for binary input memoryless symmetric channels," *IEEE Transactions on Communications*, vol. 54, no. 7, pp. 1207–1214, July 2006.

[22] G. Liva, *Block Codes Based on Sparse Graphs for Wireless Communication Systems*, PhD thesis, Università degli Studi di Bologna, 2006.

[23] G. Liva and M. Chiani, "Protograph LDPC codes design based on EXIT analysis," *Proceedings of the 2007 IEEE GlobeCom Conference*, November 2007, pp. 3250–3254.

[24] G. Liva, S. Song, L. Lan, Y. Zhang, W. E. Ryan, and S. Lin, "Design of LDPC codes: A survey and new results," *Journal of Communications Software and Systems*, vol. 2, no. 9, pp. 191–211, September 2006.

[25] A. Ashikhmin, G. Kramer, and S. ten Brink, "Extrinsic information transfer functions: A model and two properties," *36th Annual Conference on Information Science Systems*, Princeton University, March 2002.

[26] C. Measson, A. Montanari, and R. Urbanke, "Why we cannot surpass capacity: The matching condition," *Proceedings of the 43rd Allerton Conference on Communications, Control, and Computing*, September 2005.

[27] K. Bhattad and K. Narayanan, "An MSE based transfer chart to analyze iterative decoding schemes," *IEEE Transactions on Information Theory*, vol. 53, no. 1, pp. 22–38, January 2007.

[28] S. Y. Chung, *On the Construction of Capacity-Approaching Coding Schemes*. PhD thesis, Department of Electrical Engineering and Computer Science, MIT, Cambridge, MA, 2000.

[29] D. Divsalar, S. Dolinar, C. Jones, and K. Andrews, "Capacity-approaching protograph codes," *IEEE Journal on Selected Areas in Communications*, vol. 27, no. 6 pp. 876–888, August 2009.

[30] T.-Y. Chen, K. Vakilinia, D. Divsalar, and R. D. Wesel, "Protograph-based Raptor-like LDPC codes," *IEEE Transactions on Communications*, vol. 63, no. 5 pp. 1522–1532, May 2015.

[31] H. Ochiai, K. Ikeya, and P. Mitran, "A new polar code design based on reciprocal channel approximation," *IEEE Transactions on Communications*, vol. 71, no. 2 pp. 631–643, February 2023.

# 10 Polar Codes

Polar codes, introduced by Arikan in 2009 [1], comprise another family of linear block codes, normally for binary code alphabets. In contrast with the BCH and RS constructions built on the algebra of finite fields, or with the graph-theoretic foundations of LDPC codes, polar codes are built around a simple idea, called *polarization*, that transforms $n$ uses of a symmetric binary memoryless channel, denoted by $W$, into a set of *bit channels* that are sequentially decided with the help of previous bit decisions. This transformation is provided by a linear matrix transformation and is called a *polar transform* on $n$ bits. Standard polar codes have blocklength $n = 2^m$, typically with $m \geq 7$.

As will be seen, the mutual information associated with a successively decided input bit and the entire set of $n$ channel outputs either approaches 1 (a perfect bit channel) or 0 (a useless channel) as $n$ becomes large. Moreover, the fraction of good channels approaches $I(W)$, the mutual information associated with a single channel use, with this fraction hardening as $n$ becomes large. Thus, Arikan was able to show formally that using these good bit channels to convey information, and inputting a fixed 0, say, to the polar transform at the remaining bad positions, allows reliable communication at information rates arbitrarily near Shannon's capacity of the underlying binary channel, without any further encoding beyond the polar transform process. This represents the first constructive block-coding technique that provably achieves the capacity limit for binary memoryless symmetric channels. In addition, both encoding and the successive cancellation decoding algorithm that produces this reliable communication have complexity $\mathcal{O}(n \log_2 n)$ computations per block, or $\mathcal{O}(\log_2 n/R)$ per message bit. Another helpful attribute is that the code rate $R$ can easily be tuned to suit an application.

Though capacity-achieving in a formal sense, and having attractive implementation complexity, short-to-moderate-blocklength polar codes using Arikan's original successive cancellation (SC) decoder are inferior to code designs seen earlier in this book (e.g., LDPC codes, turbo codes, etc). Subsequent research has shown how to improve this situation using list decoding combined with an outer error-detecting code, to the point that performance is quite close to the dispersion approximation [2] for performance of best finite-blocklength codes. Given these performance and complexity benefits, polar coding has attracted great attention in the coding world in the last decade, and has been adopted in standards, among them the coding method for short control-plane messages in cellular 5G NR (New Radio) [3].

The presentation here will highlight the important features of the polar coding idea, while leaving more technical details to further reading (e.g., [1, 4]). The chapter is organized as follows. In Section 10.1 we develop the key properties of the $2 \times 2$ case, then extend to the $n \times n$ polar transform, built on a recursive application of $2 \times 2$ binary mappings using a graph reminiscent of butterfly graphs for the FFT algorithm. This transformation takes $n$ input bits, some of which are known, or said to be *frozen*, and generates $n$ channel bits for transmission over the memoryless channel. This process is known as *channel combining*, and the mapping can be represented by a simple matrix that is the Kronecker product of the polarizing $2 \times 2$ binary matrix. The section then presents graphical and algebraic representations of larger blocklength designs, by recursive use of $2 \times 2$ transformations.

Section 10.2 develops the notion of polarization and takes up the design of polar codes. Once a blocklength $n$ is chosen, and a desired encoding rate $R = k/n$ is selected, the only design task is to designate the $k$-bit channels (those having greatest mutual information or smallest error probability under SC decoding) to convey information, while freezing the remainder at some agreed-upon value. Several good methods have been developed for bit-channel ranking, including the Bhattacharyya bounding method of Arikan and the Gaussian approximation of bit-channel mutual information used for the binary-input, additive-white-Gaussian-noise (BI-AWGN) channel. These design tools are used to illustrate the fundamental polarizing principle as $n$ becomes large.

Section 10.3 develops Arikan's SC decoding algorithm that uses all $n$ channel measurements for sequentially deciding input bits, then uses these previous decisions to support future decisions. This process produces a collection of *bit channels*, each mapping a single input bit to $n$ channel outputs. It is these bit channels that polarize into good and bad channels in proportion, matching the channel capacity. This polarization process is illustrated using the binary erasure channel (BEC) and BI-AWGN channel models.

An improvement to SC decoding performance introduced by Tal and Vardy [5] is successive cancellation list (SCL) decoding (Section 10.4), in which the binary decoding tree is explored in parallel with $L > 1$ SC decoders. The premise is that by keeping multiple paths in the tree alive at each level, we may avoid premature dismissal of a codeword eventually having greatest likelihood. A final decision outputs the sequence that has highest likelihood among the $L$ paths. An efficient data structure for list decoding can avoid excessive memory copying during the decoding cycle, and offers $\mathcal{O}(Ln \log_2 n)$ complexity (see, e.g., [6]).

List decoding alone improves performance, particularly at lower SNR, but SCL decoding gain quickly saturates in $L$. Tal and Vardy observed, however, that with modest $L$ the correct path is usually on the final list, but does not have best metric and is thus not selected by the SC decoder. They proposed a simple error-detecting code in the form of a CRC outer code to process list candidates, finally selecting the highest-metric path passing the CRC check; this provides dramatic improvement in performance for little overhead, beyond that of list decoding alone. Subsequently, others have explored the benefits of good CRC selection in improving the weight spectrum of the code. So-called *dynamic freezing*, which does a linear pretransformation on the message bits so that "frozen" positions also depend on the message bits of the code, and incorporating these constraints into the list decoding process, allows performance gains to accrue, to the point of closely approaching the dispersion (or normal) approximation discussed in Chapter 1. Arikan's polarization-adjusted convolutional (PAC) codes [7] represent an alternate precoding approach to stand alone polar constructions.

In Section 10.5, performance results for polar codes with $R = 1/2$ and $R = 3/4$, and blocklengths $n = 128$ and $n = 1024$, are presented for regular SC decoding and list decoding, with emphasis on block-error probability versus $E_b/N_0$ on the BI-AWGN channel. Comparisons are also shown relative to the dispersion bound for finite-length codes.

Though much of the theoretical interest in polar coding centers around the use of binary channels, or binary modulation, modern wireless applications seek greater spectral efficiency through use of higher-order (QAM) modulation, either with rectangular or PSK signal sets. Bit-interleaved coded modulation (BICM) is a simple and flexible way to couple binary polar coding with higher-order modulation, though multi-level coding is slightly more efficient in the information-theoretic sense. Both use a "demapping" operation on the received complex data to develop bit likelihoods for each of the modulator input bits. However, the quality of the recovered binary channel is not uniform across bits, and depends on constellation bit labeling. Moreover, bit LLRs attached to various QAM symbol positions are not independent. Section 10.6 discusses this polar coding extension, and illustrates polar coding with the wireless 5G NR standard for control channel messages needing low latency [3].

Finally, some topics for further reading are previewed in Section 10.7.

## 10.1 The Polarizing Transform

### 10.1.1 The 2 × 2 Case

To understand polarization and the polar transform, first consider a symmetric, binary-input memoryless channel designated by $W$, having an input variable $X \in \{0, 1\}$ and output $Y$. Examples might be a BSC, a BEC, or even a BI-AWGN model introduced in Chapter 1, but we'll confine the discussion to discrete-output models initially. Such a channel is described by a probability transition function $P(y|x)$ specified on all input–output combinations. We assume the input probabilities are equiprobable, and let $I(W)$ represent the mutual information between input and output variables in this case. (If the channel is not symmetric, $I(W)$ will be slightly less than $C$, the channel capacity.)

Now imagine two uses of this channel, typically provided by two sequential uses of the physical channel. Conceptually, we have a parallel channel model as shown in Figure 10.1. Here, the channel input vector is $\mathbf{x} = (x_0, x_1)$ with the two elements assumed equiprobable and independent.[1] The channel output vector in turn will be

[1] Row vectors are assumed throughout this chapter, with vectors and matrices denoted in bold font.

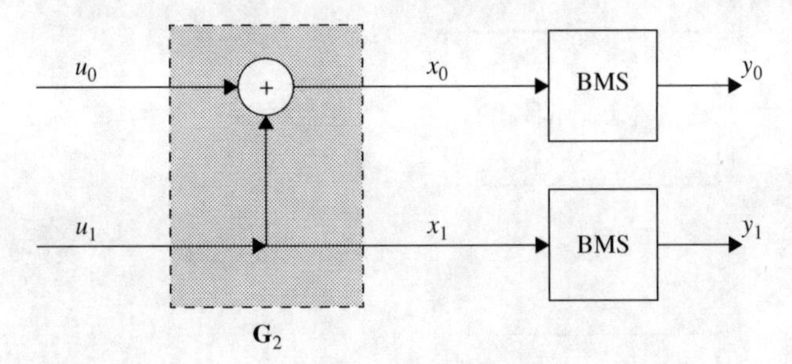


**Figure 10.1** Basic transformation on two message bits.

$\mathbf{y} = (y_0, y_1)$, and because of the memoryless channel assumption and independent inputs, we have

$$I(X_0, X_1; Y_0, Y_1) = 2I(W). \tag{10.1}$$

Now suppose we want to convey independent, equiprobable, binary message variables $(U_0, U_1)$ using this vector channel, and we prepare the channel input according to the simple precoding

$$\begin{aligned} x_0 &= u_0 + u_1, \\ x_1 &= u_1, \end{aligned} \tag{10.2}$$

where addition is modulo 2, or over GF(2). Figure 10.1 illustrates this *channel combining*, where BMS stands for a binary-input, memoryless symmetric channel.

By defining vectors $\mathbf{u} = (u_0, u_1)$ and $\mathbf{x} = (x_0, x_1)$, we have

$$\mathbf{x} = \mathbf{uF},$$

where

$$\mathbf{F} = \begin{bmatrix} 1 & 0 \\ 1 & 1 \end{bmatrix}$$

is called a *polarizing kernel*. We could also write $\mathbf{x} = \mathbf{uG}_2$, with $\mathbf{G}_2 = \mathbf{F}$ representing the generator matrix of a simple linear code on two binary bits.

Given the assumptions on the random variables $(U_0, U_1)$, we see that $X_0$ and $X_1$ are also equiprobable and independent. Also, since $\mathbf{F}$ is an invertible matrix, the mutual information $I(\mathbf{U};\mathbf{Y})$ is the same as $I(\mathbf{X};\mathbf{Y})$, namely $2I(W)$. In other words, vector mutual information is preserved under this precoding operation. So far, nothing interesting has been revealed.

Using the chain rule for mutual information (e.g., [8]), the information between input/output pairs may be decomposed as

$$I(U_0, U_1; Y_0, Y_1) = I(U_0; Y_0, Y_1) + I(U_1; Y_0, Y_1 | U_0). \tag{10.3}$$

The second mutual-information expression on the right can be written in terms of entropies as

$$\begin{aligned} I(U_1; Y_0, Y_1 | U_0) &= H(U_1 | U_0) - H(U_1 | U_0, Y_0, Y_1) \\ &= H(U_1) - H(U_1 | U_0, Y_0, Y_1) \\ &= I(U_1; Y_0, Y_1, U_0), \end{aligned}$$

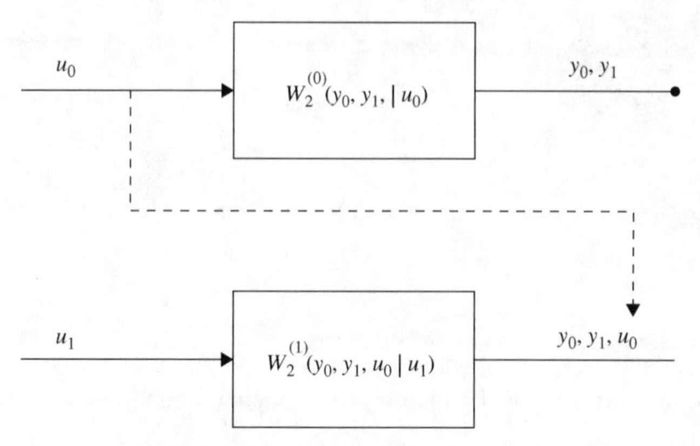


**Figure 10.2** Channel splitting into two bit channels.

where the second step follows from assumed independence of $U_0$ and $U_1$. Thus,

$$I(U_0, U_1;Y_0, Y_1) = 2I(W) = I(U_0;Y_0, Y_1) + I(U_1;Y_0, Y_1, U_0), \tag{10.4}$$

indicating that the same total information as in (10.1) can be obtained by a two-step process of first recovering $U_0$ from $(Y_0, Y_1)$, then obtaining $U_1$ from the triple $(Y_0, Y_1, U_0)$, that is, using the first input variable as side-information for deciding the second.[2] This suggests the decomposition shown in Figure 10.2, called a *channel-splitting* step, providing two so-called *bit channels* that eventually decode the same observation pair, where the second "virtual" channel has the benefit of knowing $U_0$.

Since adding extra variables cannot decrease information,

$$I(U_1;Y_0, Y_1, U_0) \geq I(U_1;Y_1) = I(W),$$

and because the sum in (10.4), is $2I(W)$ we have

$$I(U_0;Y_0, Y_1) \leq I(W) \leq I(U_1;Y_0, Y_1, U_0). \tag{10.5}$$

So, the first bit channel has smaller mutual information than the original single channel use provides, while the second bit channel's information is larger. This is the first indication of polarization – two equal channels can be transformed into two unequal channels, whose total information is unchanged.

The two bit channels are sometimes denoted by $W^-$ and $W^+$, respectively, alluding to their smaller and larger mutual information, but in the development here the two channels will be represented, as shown in Figure 10.2, by the mappings[3]

$$\begin{aligned} W_2^{(0)}&: u_0 \mapsto (y_0, y_1), \\ W_2^{(1)}&: u_1 \mapsto (y_0, y_1, u_0), \end{aligned} \tag{10.6}$$

where the second channel's notation indicates that the value of $u_0$ supplements the channel observations.

[2] We assume here that the inference about $U_0$ is correct, as indicated in Figure 10.2.

[3] The subscript will signify the polar transform size, here 2, and the superscript will denote the input index.

Decoding will require the likelihood formulation for these two bit channels, as follows:

$$\begin{aligned} W_2^{(0)}(y_0, y_1|u_0) &= \frac{1}{2}W(y_0, y_1|u_0, u_1 = 0) + \frac{1}{2}W(y_0, y_1|u_0, u_1 = 1) \\ &= \frac{1}{2}W(y_0, y_1|x_0 = u_0, x_1 = 0) + \frac{1}{2}W(y_0, y_1|x_0 = u_0 + 1, x_1 = 1) \\ &= \frac{1}{2}W(y_0|x_0 = u_0)W(y_1|x_1 = 0) + \frac{1}{2}W(y_0|x_0 = u_0 + 1)W(y_1|x_1 = 1) \end{aligned}$$

and

$$\begin{aligned} W_2^{(1)}(y_0, y_1, u_0|u_1) &= W(y_0, y_1|x_0 = u_0 + u_1, x_1 = u_1) \\ &= W(y_0|x_0 = u_0 + u_1)W(y_1|x_1 = u_1). \end{aligned}$$

These follow from the law of total probability and independence of the two underlying channels. Notice how $u_0$ modifies the conditional probability for $u_1$.

---

**Example 10.1** The BEC introduced in Chapter 1 provides a simple example for which exact results on bit-channel-error probability and mutual information can be obtained, and polarization is readily demonstrated. Let the output alphabet be $Y = \{0, 1, e\}$ and let $p$ denote the erasure probability of the BEC. Recall that the channel capacity for a single use is $C = 1 - p$ bits, and two uses of the channel supply mutual information $2(1 - p)$ bits.

The vector channel output $\mathbf{y} = (y_0, y_1)$ takes on one of nine combinations, and is $(u_0 + u_1, u_1)$ when no erasures occur. Either position in $\mathbf{y}$ is replaced by $e$ when a physical channel erases. The first channel, $W_2^{(0)}(\mathbf{y}|u_0)$, unambiguously conveys the value of $u_0$ only if *both* physical channels do not produce erasures. Equivalently, this synthetic channel is effectively a BEC with erasure probability $1 - (1 - p)^2 = 2p - p^2$, and its mutual information is thereby

$$I\left(W_2^{(0)}\right) = 1 - \left(2p - p^2\right) \text{ bits/vector.}$$

The synthetic channel $W_2^{(1)}(\mathbf{y}, u_0|u_1)$, on the other hand, has available to the inference process the value of $u_0$.[4] Now it can be seen that $u_1$ is correctly recovered unless *both* physical channels are erasure events. So this synthetic channel is also a BEC, but with erasure probability $p^2$ and corresponding mutual information

$$I\left(W_2^{(1)}\right) = 1 - p^2 \text{ bits/vector.}$$

For any $0 < p < 1/2$, the second synthetic channel has larger mutual information than the first, and the original channel $W$ as well, but the sum of the two remains $2(1 - p)$ bits/two channel uses.

How could we communicate with such a simple precoding? We might operate two separate channel encoder–decoder pairs, designed with code rates appropriate to the

[4] This side-information can be provided by knowledge of frozen input positions, or from previous decisions, which can, of course, be wrong. For now we again assume correct side-information.

respective mutual information, and with appeal to coding theorems assuming "error-free" operation as the blocklength grows. Such a strategy cannot increase the overall throughput, however, even for large coding complexity, as seen above. One simpler strategy is to freeze $u_0 = 0$ and make this *known* to the decoder. Then we could build an encoder for a stream of bits in position 1, and send at rate up to $(1-p^2)/2$ bits for this code (the factor of 2 accounting for the fact that two channel uses are required per $u_1$ bit). This is inferior, however, in terms of mutual information to simply using the raw channel to achieve $2(1-p)$ bits per two channel uses, but better possibilities await!

### 10.1.2 Decoding the General 2 × 2 Combined Channel

Next we describe successive cancellation (SC) decoding for the simple $2 \times 2$ case just presented. Optimal decisions about bits $u_0$ and $u_1$ are based on log-likelihood ratios (LLRs).[5] The upper input bit $u_0$ is decided based on $(y_0, y_1)$, then the decision $\widehat{u}_0$ is employed to aid the decision on $u_1$. First, we need

$$L_2^{(0)}(y_0, y_1) \triangleq \log\left(\frac{W_2^{(0)}(y_0, y_1|u_0 = 0)}{W_2^{(0)}(y_0, y_1|u_0 = 1)}\right) \tag{10.7}$$

with the decision $\widehat{u}_0$ in favor of 0 if this LLR is positive. The logarithm base is arbitrary, but we adopt base $e$ unless noted otherwise.

By use of Bayes' rule and assuming equiprobable input bits, we can recast this as

$$L_2^{(0)}(y_0, y_1) = \log\left(\frac{P(u_0 = 0|y_0, y_1)}{P(u_0 = 1|y_0, y_1)}\right). \tag{10.8}$$

The channel measurements supply LLRs about $x_0$ and $x_1$, denoted by $L_{x_0}$ and $L_{x_1}$:

$$L_{x_0} \triangleq \log\left(\frac{P(y_0|x_0 = 0)}{P(y_0|x_0 = 1)}\right) = \log\left(\frac{P(x_0 = 0|y_0)}{P(x_0 = 1|y_0)}\right) \tag{10.9}$$

and similarly for $L_{x_1}$. For the BI-AWGN model with channel outputs having mean values $\pm\sqrt{E_s}$ and noise variance $\sigma^2 = N_0/2$, the LLR (using base-$e$ logarithms) reduces to

$$L_{x_i} = \frac{4\sqrt{E_s}}{N_0} y_i,$$

incorporating proper scaling for channel parameters. In fact, LLR scaling is not important when the approximate LLR update rule described below is used (see Problem 10.22). For a BSC model we would use

$$L_{x_i} = \begin{cases} \log(1-\epsilon), & y_i = 0, \\ \log(\epsilon), & y_i = 1. \end{cases}$$

[5] Note that decisions based on LLRs match those of maximizing the log *a posteriori* ratio (LAPPR) assuming equiprobable message bits.

These can be translated by a common amount to give likelihoods of 0 and $\log(\epsilon/(1-\epsilon))$, respectively.

Exponentiating both sides of (10.9) and recalling that the sum of posterior probabilities is 1 leads to

$$P(x_0 = 0 \mid y_0) = \frac{e^{L_{x_0}}}{1+e^{L_{x_0}}}; \quad P(x_0 = 1 \mid y_0) = \frac{1}{1+e^{L_{x_0}}},$$
$$P(x_1 = 0 \mid y_1) = \frac{e^{L_{x_1}}}{1+e^{L_{x_1}}}; \quad P(x_1 = 1 \mid y_1) = \frac{1}{1+e^{L_{x_1}}}.$$

To convert these into likelihoods about $u_0$ and $u_1$, recall from the description of the channel combining in (10.2) that

$$\begin{aligned} u_0 &= x_0 + x_1, \\ u_1 &= u_0 + x_0. \end{aligned} \tag{10.10}$$

Then, from (10.10) and the memoryless channel assumption, we find

$$P(u_0 = 0 \mid y_0, y_1) = P(x_0 = 0 \mid y_0)P(x_1 = 0 \mid y_1) + P(x_0 = 1 \mid y_0)P(x_1 = 1 \mid y_1).$$

Substitution from above yields

$$\begin{aligned} P(u_0 = 0 \mid y_0, y_1) &= \frac{e^{L_{x_0}} e^{L_{x_1}}}{\left(1+e^{L_{x_0}}\right)\left(1+e^{L_{x_1}}\right)} + \frac{1}{\left(1+e^{L_{x_0}}\right)\left(1+e^{L_{x_1}}\right)} \\ &= \frac{1+e^{L_{x_0}+L_{x_1}}}{\left(1+e^{L_{x_0}}\right)\left(1+e^{L_{x_1}}\right)}. \end{aligned}$$

A similar argument for $P(u_0 = 1 \mid y_0, y_1)$ and substitution into the LLR definition for $u_0$ from (10.8) yields

$$L_2^{(0)}(y_0, y_1) = \log\left(\frac{1+e^{L_{x_0}+L_{x_1}}}{e^{L_{x_0}} e^{L_{x_1}}}\right). \tag{10.11}$$

This is the blending formula for forming the decision rule on the upper bit $u_0$ in Figure 10.1, given the two channel measurements, and is identical to that seen at the degree-3 check node involving $u_0, x_0$, and $x_1$ in LDPC message passing (Chapter 5), alternatively expressed as

$$L_2^{(0)}(y_0, y_1) = 2\tanh^{-1}\left(\tanh(L_{x_0}/2)\tanh(L_{x_1}/2)\right). \tag{10.12}$$

For shorthand this is often expressed as the "box-plus" operation, denoted by

$$L_2^{(0)}(y_0, y_1) = L_{x_0} \boxplus L_{x_1} \triangleq f(L_{x_0}, L_{x_1}), \tag{10.13}$$

and for the same complexity and numerical stability reasons discussed in Chapter 5 for LDPC decoding, this is commonly approximated by

$$L_2^{(0)}(y_0, y_1) \simeq \operatorname{sign}(L_{x_0})\operatorname{sign}(L_{x_1})\min(|L_{x_0}|, |L_{x_1}|). \tag{10.14}$$

Now, returning to the decision on $u_1$, we follow a more intuitive argument. Given that $u_0, y_0$ is either the result of the channel acting on $u_1 + 0$ or $u_1 + 1$, and $y_1$ is the response to $u_1$ in either case, a simple hypothesis test for $u_1$ then yields the LLR for the second bit as

$$L_2^{(1)}(y_0, y_1, \widehat{u}_0) \triangleq L_2(y_0, y_1, \widehat{u}_0 \mid u_1) = \begin{cases} L_{x_1} + L_{x_0}, & \widehat{u}_0 = 0 \\ L_{x_1} - L_{x_0}, & \widehat{u}_0 = 1 \end{cases}$$
$$= L_{x_1} + (1 - 2\widehat{u}_0)L_{x_0}$$
$$\triangleq g(L_{x_0}, L_{x_1}, \widehat{u}_0), \quad (10.15)$$

where $\widehat{u}_0$ is the decision regarding $u_0$. This follows because, from (10.10), if $u_0 = 0$, then $x_0 = x_1$ and so their LLRs should be added as in a repetition code. Similarly, if $u_0 = 1, x_1 = x_0^c$ and so $L_{x_0}$ should be subtracted from $L_{x_1}$.

The processing in (10.15) for $u_0$ is simpler than in (10.12) or (10.14) for $u_1$, involving only signed addition of channel LLRs, according to the previously decided information, and is basically the same as that performed at a variable node in an LDPC decoder. Clearly, if $\widehat{u}_0 \neq u_0$, the likelihood for $u_1$ is incorrect.

To summarize the processing, a decision on $u_0$ is formed using the LLR expression for the upper bit $u_0$ using (10.12) or (10.14), which we will designate using the function $f(L_{x_0}, L_{x_1})$, and this decision becomes side-information to the decision on the lower bit $u_1$ using (10.15), a function $g(L_{x_0}, L_{x_1}, \widehat{u}_0)$ of three variables – the two original LLRs and the new side-information. This two-step decoding process completes the definition of two virtual bit channels shown in Figure 10.2, and this depiction illustrates *channel splitting*. When extension is made to larger blocklength polar codes, this same $2 \times 2$ processing strategy is in force at each level of the polar transform graph (i.e., LLRs from previous stage are used to compute LLR's for following stages). Also observe that the only change when differing channel models are adopted is the initial computation of the LLRs; subsequent message passing is identical to that presented here.

A representation of the $2 \times 2$ processing as a simple binary tree leads to a helpful way to organize a general decoder. Figure 10.3 shows a one-level binary tree with leaf nodes $u_0$ and $u_1$. The root node at top gathers the channel observation pair $(y_0, y_1)$ and computes an LLR message for its left-child, using (10.12) or (10.14), (step 1), which makes a binary decision based on this LLR. This binary decision $\widehat{u}_0$ is returned to the root node (step 2), which uses the decision along with the original measurements to send an LLR to the right-child, using (10.15) (step 3). This LLR is thresholded to decide bit $u_1$. This binary tree fragment is embedded in decoding trees for larger codes, as described later.

### 10.1.3 Polar Transform for $n = 2^m$

The $2 \times 2$ transformation of binary pairs is interesting, but by itself has little practical value. To obtain powerful codes, we develop a transform on binary $n$-tuples, by co-mingling multiple instantiations of $2 \times 2$ building blocks, using a recursive structure that admits simple encoding and decoding. The final polar code is essentially a serial and parallel concatenation of these simple block codes, with interstage *deterministic*

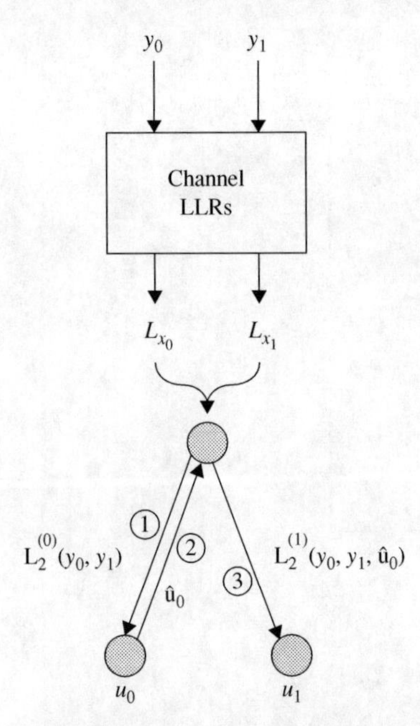


**Figure 10.3** Message-passing tree for $2 \times 2$ case. $f(L_{x_0}, L_{x_1})$ is used in step 1, $g(L_{x_0}, L_{x_1}, \widehat{u}_0)$ is used in step 3.

*interleaving*. This interleaving improves the distance properties of the code, and also, on the decoding side, maintains independence of likelihoods supplied by previous decodings. In this section we develop graphical models and algebraic descriptions for this polar transform.

There are several representations of polar codes that are equivalent, except for a possible bit-reversal reordering. We first illustrate the original Arikan formulation having $m = \log_2(n)$ columns of $n/2$ parallel $\mathbf{F}$ (or $\mathbf{G}_2$) processing units, interspersed with reverse-shuffle transformations. We begin with the $4 \times 4$ transformation as shown in Figure 10.4. Message vectors and code vectors are listed in natural order, top to bottom.

The input and output vectors are now binary 4-tuples, and notice pairs of $2 \times 2$ transformations on the input and output side. It is of no benefit to directly connect these modules together in parallel paths; instead, we interleave variables across the 4-tuples with the simple permutation shown, that is, swapping the order of bits $v_1$ and $v_2$ to produce the permuted vector $\tilde{\mathbf{v}} = [v_0, v_2, v_1, v_3]$. For four variables this permutation is defined by $\tilde{\mathbf{v}} = \mathbf{v}\mathbf{R}_4$, with the matrix

$$\mathbf{R}_4 = \begin{bmatrix} 1 & 0 & 0 & 0 \\ 0 & 0 & 1 & 0 \\ 0 & 1 & 0 & 0 \\ 0 & 0 & 0 & 1 \end{bmatrix}.$$

$\mathbf{R}_4$ is a special case of the *reverse shuffle tranformation* on $n$ variables that reforms the vector $\mathbf{v} = (v_0, v_1, v_2, ..., v_{n-2}, v_{n-1})$ as the vector $\tilde{\mathbf{v}} = (v_0, v_2, v_4, ..., v_1, v_3, ...., v_{n-1})$.

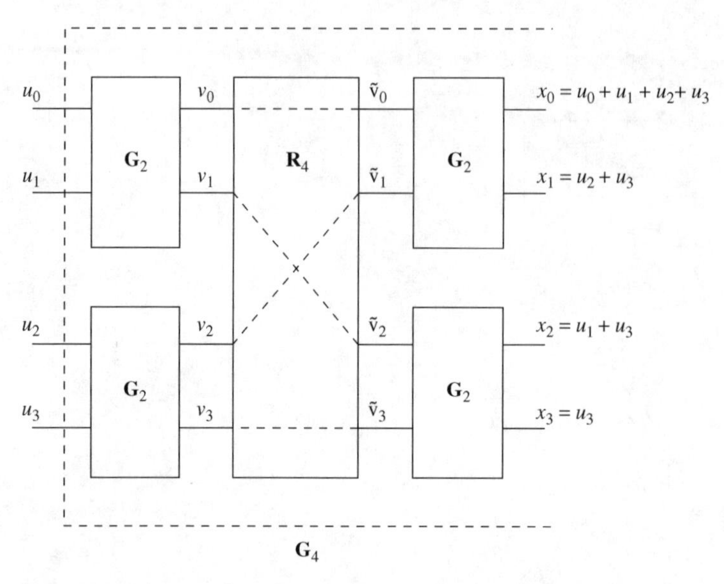


**Figure 10.4** Polar transform for $n = 4$.

Even-indexed variables move to the front, in order, while odd-indexed variables move to the rear. For $n$ variables, we again have

$$\tilde{\mathbf{v}} = \mathbf{v}\mathbf{R}_n$$

in the general case. Since $\mathbf{R}_n$ is a permutation matrix, $\mathbf{R}_n^{-1} = \mathbf{R}_n^{\mathrm{T}}$.

---

*Aside on Kronecker product of matrices*. The Kronecker product of matrices [9, 10] is a convenient way to mathematically represent the polar transform. Suppose $\mathbf{A}$ is an $L \times L$ matrix and $\mathbf{B}$ is $M \times M$. Then the Kronecker product of these, denoted by $\mathbf{A} \otimes \mathbf{B}$, is the $LM \times LM$ matrix defined as

$$\mathbf{A} \otimes \mathbf{B} = \begin{bmatrix} a_{11}\mathbf{B} & a_{12}\mathbf{B} & \dots & a_{1L}\mathbf{B} \\ \dots & \dots & \dots & \dots \\ a_{L1}\mathbf{B} & a_{L2}\mathbf{B} & \dots & a_{LL}\mathbf{B} \end{bmatrix}.$$

A very helpful identity for Kronecker product operations is the *mixed product identity*:

$$(\mathbf{A} \otimes \mathbf{B})(\mathbf{C} \otimes \mathbf{D}) = (\mathbf{AC} \otimes \mathbf{BD}), \tag{10.16}$$

provided the matrices $\mathbf{A}, \mathbf{B}, \mathbf{C}, \mathbf{D}$ are dimensionally consistent. As with normal matrix multiplication, the Kronecker product is non-commutative, but we do have

$$\mathbf{B} \otimes \mathbf{A} = \mathbf{R}^{\mathrm{T}}(\mathbf{A} \otimes \mathbf{B})\mathbf{R}, \tag{10.17}$$

where $\mathbf{R}$ is the reverse-shuffle permutation matrix of appropriate size defined above.

---

With $\mathbf{I}_2$ the $2\times2$ identity matrix, $\mathbf{I}_2 \otimes \mathbf{G}_2$ is simply the $4\times4$ transformation involving two parallel $\mathbf{F}$ (or $\mathbf{G}_2$) transformations on left and right in Figure 10.4. Thus, the $4 \times 4$ polar transformation specified above has a matrix representation

$$\mathbf{G}_4 = (\mathbf{I}_2 \otimes \mathbf{F})\mathbf{R}_4(\mathbf{I}_2 \otimes \mathbf{F}),$$

which amounts to two $2 \times 2$ operations followed by the reverse shuffle, followed by two more $2 \times 2$ operations.

Two redrawings of Figure 10.4 produce the schematics in Figure 10.5, previewing the general case. The corresponding algebraic mappings are $(\mathbf{F} \otimes \mathbf{I}_2)(\mathbf{I}_2 \otimes \mathbf{F})$ and $(\mathbf{I}_2 \otimes \mathbf{F})(\mathbf{F} \otimes \mathbf{I}_2)$, respectively. The Kronecker matrix product identity readily shows both equal $\mathbf{F} \otimes \mathbf{F} = \mathbf{F}^{\otimes 2}$. These avoid the explicit internal shuffle operation, but notice the input bits are reordered in Figure 10.5(a) to accommodate this. A bit-reversal operation, described below, at the input of Figure 10.5(a) can restore the natural order if necessary.

This $4 \times 4$ vector channel has a similar chain-rule decomposition for mutual information, as seen in the previous section, which leads to four synthetic bit channels, each mapping a single bit to *four* channel measurements along with an increasing amount of side-information from previous bit decisions. For example, the third of the four synthetic bit channels is a mapping

$$W_4^{(2)}: u_2 \mapsto (y_0, y_1, y_2, y_3, u_0, u_1).$$

This notation indicates side-information on $u_0$ and $u_1$ to infer $u_2$, with $u_3$ unknown.

At this point it is convenient to adopt shorthand notation for vectors, for example, $(y_0, y_1, \ldots, y_i)$ is represented as $y_0^i$. From the basic laws of probability, the bit channel just described has the conditional probability representation

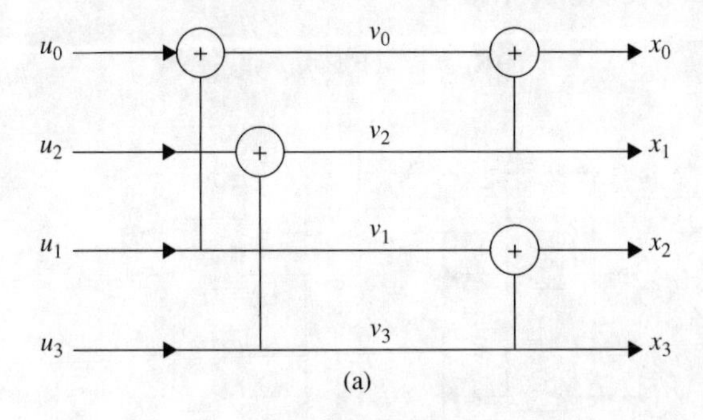


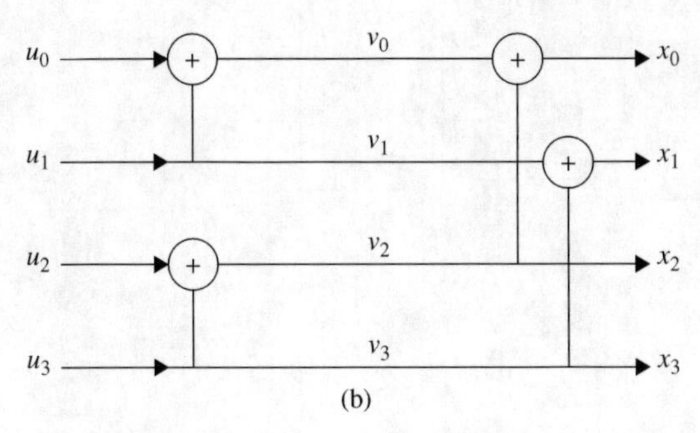


**Figure 10.5** Two versions of polar transform $\mathbf{G}_4$ without internal shuffle. Note the index orderings.

$$W_4^{(2)}\left(y_0^3, u_0^1 \mid u_2\right) = \frac{1}{2^3} \sum_{u_3 \in \{0,1\}} W_4\left(y_0^3 \mid u_0^3\right)$$

since $u_3$ is yet unknown.

The four mutual information values for these bit channels lie between 0 and 1, summing to $4I(W)$ bits/vector, with the first bit channel being the poorest, and the last bit channel being the best.

To construct longer polar codes, we proceed recursively, taking two 4-point transform blocks to transform eight input bits to eight output bits, as shown in Figure 10.6. This realization invokes two $\mathbf{G}_4$ modules defined above. A column of four $2 \times 2$ building blocks is followed by a mixing, or permutation block, here $\mathbf{R}_8$ (the reverse shuffle on eight variables), and this vector is processed by the two $\mathbf{G}_4$ blocks, as in Figure 10.6. The mapping has the matrix representation

$$\mathbf{G}_8 = (\mathbf{I}_4 \otimes \mathbf{F})\mathbf{R}_8(\mathbf{I}_2 \otimes \mathbf{G}_4). \tag{10.18}$$

In this diagram, input and output vectors are in natural order.

Generalizing this recursive construction gives the polar transform of order $n$ as

$$\mathbf{G}_n = (\mathbf{I}_{n/2} \otimes \mathbf{F})\mathbf{R}_n(\mathbf{I}_2 \otimes \mathbf{G}_{n/2}). \tag{10.19}$$

that is, $n/2$ $2 \times 2$ transformations on input pairs, followed by a shuffle, then by two separate $n/2$-point polar transformations.

It may be shown that (see Problem 10.7)

$$(\mathbf{I}_{n/2} \otimes \mathbf{F})\mathbf{R}_n = \mathbf{R}_n(\mathbf{F} \otimes \mathbf{I}_{n/2}). \tag{10.20}$$

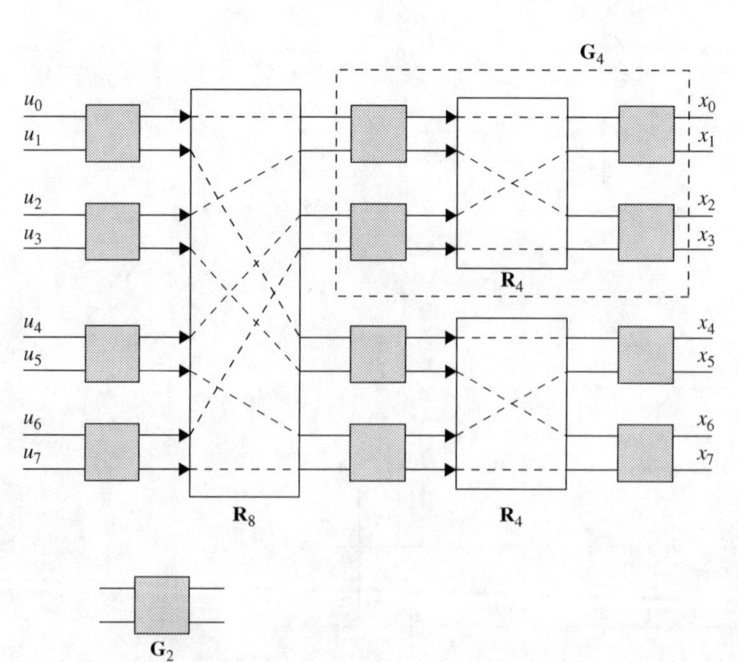


**Figure 10.6** Polar transform for $n = 8$.

so that another algebraic form is

$$\mathbf{G}_n = \mathbf{R}_n(\mathbf{F} \otimes \mathbf{I}_{n/2})(\mathbf{I}_2 \otimes \mathbf{G}_{n/2}). \tag{10.21}$$

Still another form of the transform is obtained by applying the matrix product identity to (10.21), producing

$$\mathbf{G}_n = \mathbf{R}_n(\mathbf{F} \otimes \mathbf{G}_{n/2}). \tag{10.22}$$

This can be unrolled to obtain

$$\mathbf{G}_n = \mathbf{R}_n(\mathbf{F} \otimes (\mathbf{R}_{n/2}(\mathbf{F} \otimes \mathbf{G}_{n/4}))). \tag{10.23}$$

Applying the Kronecker mixed-product identity with $\mathbf{A} = \mathbf{I}_2$, $\mathbf{C} = \mathbf{F}$, $\mathbf{B} = \mathbf{R}_{n/2}$, and $\mathbf{D} = \mathbf{F} \otimes \mathbf{G}_{n/4}$ gives

$$\mathbf{G}_n = \mathbf{R}_n(\mathbf{I}_2 \otimes \mathbf{R}_{n/2})\left(\mathbf{F}^{\otimes 2} \otimes \mathbf{G}_{n/4}\right), \tag{10.24}$$

where $\mathbf{F}^{\otimes 2} = \mathbf{F} \otimes \mathbf{F}$. Continuing this unrolling with use of the mixed product identity yields

$$\mathbf{G}_n = \mathbf{B}_n\mathbf{F}^{\otimes m} \quad (m = \log_2 n), \tag{10.25}$$

where $\mathbf{B}_n$ is the *bit-reversal permutation* on $n$-element vectors, recursively defined by

$$\mathbf{B}_n = \mathbf{R}_n(\mathbf{I}_2 \otimes \mathbf{B}_{n/2}) \tag{10.26}$$

with $\mathbf{B}_1 = 1$, and $\mathbf{F}^{\otimes m}$ is the $m$-fold Kronecker product of $\mathbf{F}$. $\mathbf{B}_n$ is another permutation matrix, like the shuffle matrix, which takes a variable with binary index $b_0b_1 \ldots b_{m-1}$ and places it in the location with reversed binary index $b_{m-1} \ldots b_1b_0$. The compact representation in (10.25) is sometimes taken as the definition for polar codes, but any of the representations can be turned into efficient software or hardware implementations.

Inclusion of $\mathbf{B}_n$, or not, in the definition in (10.25) does not affect the code performance in the end, provided frozen indices are chosen accordingly, and our discusion will thus adopt the simpler representation $\mathbf{x} = \mathbf{u}\mathbf{F}^{\otimes m}$, which is also the definition in much of the polar code literature, and for the 5G NR adoption of polar codes.

To summarize, several algebraic forms for the polar transform are

$$\mathbf{G}_n = (\mathbf{I}_{n/2} \otimes \mathbf{F})\mathbf{R}_n(\mathbf{I}_2 \otimes \mathbf{G}_{n/2}) \quad \text{recursive form;}$$
$$\mathbf{G}_n = \mathbf{R}_n(\mathbf{F} \otimes \mathbf{I}_{n/2})(\mathbf{I}_2 \otimes \mathbf{G}_{n/2}) \quad \text{alternate recursive form;}$$
$$\mathbf{G}_n = \mathbf{B}_n\mathbf{F}^{\otimes m} \; (\text{or } \mathbf{G}_n = \mathbf{F}^{\otimes m}) \quad \text{closed form.}$$

The last form is algebraically the most compact, but several realizations are possible. For example, the three-layer diagram of Figure 10.7 can be expressed as the matrix product

$$\mathbf{G}_8 = (\mathbf{I}_4 \otimes \mathbf{F})(\mathbf{I}_2 \otimes (\mathbf{F} \otimes \mathbf{I}_2))(\mathbf{F} \otimes \mathbf{I}_2), \tag{10.27}$$

which can be shown via Kronecker matrix identities to be precisely $\mathbf{F}^{\otimes 3}$. Moreover, it can be shown that the product of matrices in this expression has the commutative property, so that $3! = 6$ different layerings of the logic diagram are all equivalent to $\mathbf{G}_8 = \mathbf{F}^{\otimes 3}$. For clarity, an implementation with a regular progression as in Figure 10.7

makes the most sense. For future reference, we identify with bold dots on the graph of Figure 10.7 the location of necessary variables in encoding and decoding that store binary data (encoding) and log-likelihood plus binary information (decoding). Vertical levels are indexed left to right by $j$, for $j = 0, 1, \ldots, m$ and vertical position $i$, for $i = 0, 1, \ldots, n-1$.

Figure 10.8 depicts the binary matrix that represents the polar transform on 64 bits. Notice that upper rows of this matrix generally have the smallest Hamming weight, meaning an input bit in these positions contributes, by itself, less Hamming weight to the codeword.

The $\mathbf{G}_n$ transformation is a one-to-one mapping of input vectors $\mathbf{u}$ to code vectors $\mathbf{x}$, each an $n$-tuple. Though this mapping has memory across input bits, there is no redundancy, the other necessary ingredient for powerful channel codes. To achieve

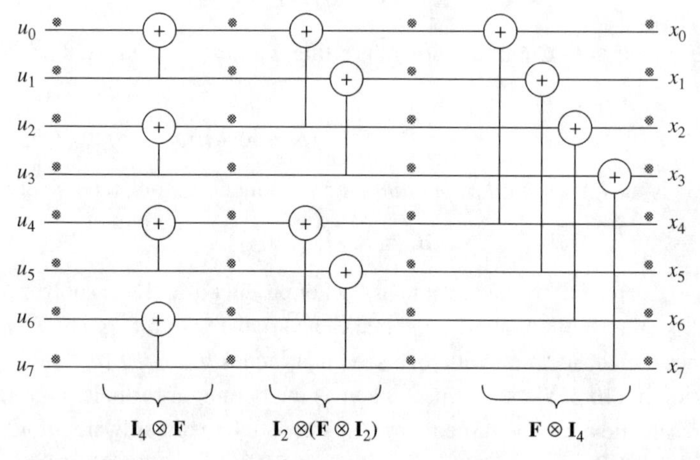


**Figure 10.7** Polar transform for $n = 8$, $\mathbf{F}^{\otimes 3} = (\mathbf{I}_4 \otimes \mathbf{F})(\mathbf{I}_2 \otimes (\mathbf{F} \otimes \mathbf{I}_2))(\mathbf{F} \otimes \mathbf{I}_4)$.

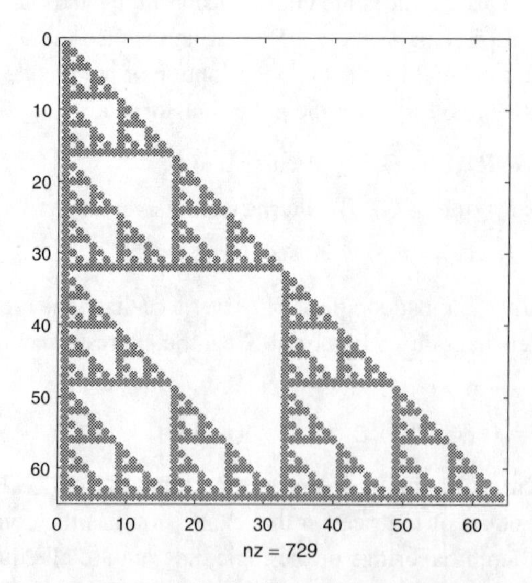


**Figure 10.8** Depiction of binary polar transform matrix for blocklength 64.

reliable communication according to information theory, the number of information-carrying bits must be reduced so that $R = k/n < I(W)$. The easiest way is to simply *freeze* $n - k$ positions to some fixed value, known to the decoder. The usual choice is to set these bits to 0, and this is equivalent to keeping only certain rows of $\mathbf{G}_n$ (the positions complementary to the frozen positions) as the generator matrix of an $(n, k)$ polar code. The obvious question is "which positions should be frozen (or non-frozen) for a rate $R$ polar code?" We address this by way of example.

---

**Example 10.2** The BEC model is especially convenient for analyzing polarization, as every layer of the graph corresponding to $\mathbf{F}^{\otimes m}$ turns a pair of BECs (with modified parameters) into a new pair of BECs. Consider the graph of Figure 10.7 for $n = 8$. The rightmost column of four $2 \times 2$ blocks changes two BECs each into a pair of new BECs with parameters $2p - p^2$ and $p^2$, as described earlier. (This is repeated four times in decoupled manner.) The eight pseudo-channels remain independent. In the middle layer, the $\mathbf{F}$ transformation is applied four times to these eight modified BECs, producing new BECs, and so on. Letting $h_0(x)$ and $h_1(x)$ denote the effective erasure probability seen at the upper and lower branches of a $2 \times 2$ combiner, where $x$ is the erasure probability of the channel "seen" at the output of the combiner, we have

$$h_0(x) = 2x - x^2,$$
$$h_1(x) = x^2.$$

At the middle layer the four input bits of each section present a BEC to the next stage to the left, with BEC parameters obtained by composition of these functions, that is

$$h_0(h_0(p)), \;\; h_1(h_0(p)), \;\; h_0(h_1(p)), \;\; h_1(h_1(p)). \tag{10.28}$$

Inspection of Figure 10.7 shows that the eight input bits "see" every permutation of the $m$-fold composition of the functions $h_i(x)$, and that the bit channel for position $i$, having binary expansion $b_{m-1} \cdots b_1 b_0$, will be equivalent, after combining and splitting with correct knowledge on previous bits, to a BEC with parameter

$$\widetilde{p}_i = h_{b_0}(h_{b_1}(\cdots(h_{b_{i-1}}(p)))).$$

For example, the binary variable $u_3$ having binary expansion 011 will experience a BEC with parameter $h_1(h_1(h_0(p)))$ under SC decoding. Indices chosen to carry information (non-frozen positions) are then based on finding the $k$ positions with smallest effective erasure probability.

---

### 10.1.4 Encoding

Before showing that this iterative transformation process is powerful as $n$ increases, and then tackling decoding, we describe polar encoding.

As with all channel-coding approaches, encoding is the easier task. Assuming the polar code design is given, including choice of $n, k, R$ and determination of message indices, the $k$ message positions are loaded into an $n$-vector along with $n - k$ frozen bits and this serves as input to the polar transform. Any of the graphical forms discussed before can be implemented in either software or hardware with comparable complexity; the only tricky part is memory addressing.

The representation in Figure 10.7 provides a simple recursive method, needing only an optional bit-reversal permutation and multiple instantiations of the $\mathbf{F}$ operation with modulo-2 adders. Observe that in each vertical section there are $n/2$ $2 \times 2$ operations each forming the $\mathbf{F}$ transformation, with the input and output indices of these operations dependent on the section number. These required indices have a very regular description, and there are $m$ such sections.

Pseudo-code for a software implementation is presented in Algorithm 10.1 following this structure. The algorithm computes the intermediate binary $n$-tuples at each of $m$ stages of the graph, beginning with the input vector $\mathbf{u}$, containing $k$ message bits and $n - k$ frozen bits in specified positions. The loop counter $j$ tracks the columns of the graph, left to right, and at each level the previous $\mathbf{v}$ vector is used to create a temporary $n$-tuple $\mathbf{t}$, which is then copied back into $\mathbf{v}$ for the next level. The variable

---

**Algorithm 10.1** Polar Encode

---

ASSUMPTIONS

1. Assumes the structure of Figure 10.7.
2. Zero-based indexing assumed.
3. Input $n$-tuple $\mathbf{u}$ with frozen entries.

ALGORITHM

```
Initialize. v = u. numsec = n/2. secsize = 2. step = 1.

for j = 1 to m do
    for r = 1 to numsec do
        first = 0
        for s = 1 to secsize do
            add1 = first
            add2 = first + step
            t(add1) = v(add1) + v(add2)          ▷ (mod-2 addition)
            t(add2) = v(add2)
        end for
        first = first + secsize
    end for
    v = t
    numsec = numsec/2
    secsize = secsize * 2
    step = step * 2
end for
x = v (output codeword)
```

---

*secsize* doubles at each level, and determines which bits of $\mathbf{v}$ are combined in the $2 \times 2$ modules. After $m$ cycles, the temporary $n$-tuple is output as the codeword $\mathbf{x}$.

An efficient hardware implementation can mimic this serial code, or a high-speed encoder can employ $n/2$ parallel $2\times2$ modules in a pipelined architecture, with proper level-dependent interleaving. Measured either in binary hardware operations or in software operations count, the encoding complexity per codeword is $\mathcal{O}(n \log_2 n)$.

As presented, this encoding produces non-systematic encoding. Any linear block code can be made equivalent to one in systematic form, but the question is whether the easy encoding and decoding is preserved. Arikan [11] has shown how to accomplish this if the systematic property is actually required.

As discussed later, polar codes are often precoded, either with a CRC code that generates $l_C$ check bits, or another means of linear precoding that also produces $k + l_C$ bits from the original $k$. In this case, the polar code design must reflect the need for $k + l_C$ positions to be unfrozen, assuming the overall code rate is to be maintained at $R = k/n$.

### 10.1.5 Relation to Reed–Muller Codes

Reed–Muller (RM) codes [12] have been known since the early days of channel coding as binary codes that are easy to encode and decode, at least with hard-decision decoding. Soft decoding using belief propagation has also been studied by Dumer [13].

Various descriptions exist (see, e.g., [14]), but one that connects here is that the $r$th-order RM code of length $n = 2^m$ is formed by a generator matrix consisting of the rows of $\mathbf{G}_n = \mathbf{F}^{\otimes m}$ having Hamming weight $2^{m-r}$ or greater. The number of such rows is

$$k = \sum_{i=0}^{r} \binom{m}{i}.$$

This subset of rows is typically slightly different than those produced by polar code designs (described next) with equal $k$, however.

The RM construction is less flexible in choice of $R = k/n$, but for $n = 2^p$ with $p$ odd, there is a $R = 1/2$ RM code with minimum Hamming distance $2^{(p+1)/2}$. For $n = 8$, polar and RM codes use the same subset of rows of $\mathbf{G}_n$, but for $n = 32$ and larger, the polar code constructions of Arikan begin to differ somewhat from RM design [15].

The rate-1/2 RM code with $n = 128$ will provide a useful comparison in our discussion of performance results. For this particular case, the polar code designed for the BI-AWGN channel at $E_b/N_0 = 2.5$ dB has $d_{\min} = 8$, with nearest-neighbor multiplicity $A_8 = 304$.[6] The $R = 1/2$ RM code, on the other hand, has $d_{\min} = 16$, but with much

[6] Weight spectra for small nonzero weights are found with use of a list decoder operating on a single code block at high SNR, with very large list size.

larger multiplicity, $A_{16} = 94\,488$. As will be seen in Section 10.5, the polar design wins under SC decoding, while the RM code emerges as better with list decoding for large $L$.

A hybrid design philosophy was proposed in [16], dubbed *RM-polar*. The procedure first designs a polar code, then locates the non-frozen indices whose row weights of $\mathbf{G}_n$ are smallest, and replaces these with indices having the next-best ranking in the polar design (and larger row weight). The polar transform matrix $\mathbf{G}_n = \mathbf{F}^{\otimes m}$ has the "visibility" property [17], meaning that the code defined as the span of any subset of rows $\mathbf{V}$ of $\mathbf{G}_n$ has minimum distance equivalent to the smallest row weight in the basis $\mathbf{V}$. Since the rows of $\mathbf{G}_n$ have weights that are powers of 2, this hybrid design process will double the minimum distance of the original code. A case that illustrates this is $n = 256, R = 1/2$, for which polar design at 2.5 dB produces $d_{\min} = 8$ still. There is only a single non-frozen index having this row weight, however, and $A_8 = 32$. Swapping out this row using the RM-polar idea bumps the minimum distance to 16, while giving $A_{16} = 54\,576$. Again, whether this yields better decoding efficiency depends on the decoding approach to be adopted.

This RM-polar modification does not always change the original polar construction. An illustrative case is a short $(64, 32)$ polar code designed at $E_b/N_0 = 2.5$ dB. The minimum distance of the polar code is 8, but all 15 weight-16 rows of $\mathbf{G}_{64}$ were originally chosen for the polar code, so replacing a weight-8 row with a previously unselected weight-16 row is not possible.

## 10.2 Polarization and Design of Polar Codes

In this section, we quantitatively illustrate polarization for increasing blocklength, and proceed to design polar $(n, k)$ codes, that is, identify the frozen positions given $n$ and $k$.

A useful connection to earlier discussions in this book is to note that specifying the $k$ message positions is equivalent to selecting $k$ rows from the matrix $\mathbf{G}_n$ to form a basis for a linear block code. Essentially one is forming an expurgated $(n, k)$ subcode of the $(n, n)$ "code" formed by $\mathbf{G}_n$. Choice of arbitrary frozen values in the remaining $n - k$ positions produces a *coset* of the code formed with all-zero frozen bits, and does not change the performance of the polar code. (It is crucial that the decoder knows the frozen bit choices, however.)

As discussed earlier, polarization is the phenomenon whereby, as blocklength grows, the synthetic binary channels associated with successive cancellation decoding separate into two classes, those with mutual information near 0 and those with mutual information near 1, and furthermore the fraction of bit channels with good mutual information is $C$, the channel capacity of $W$, in the large blocklength limit. For moderate blocklength this separation is less definitive, as will be seen, and in any case the good index positions must be identified.

The "best" positions actually have a subtle dependence on the decoding algorithm to be used. Arikan's idea was based on successive cancellation decoding as we have indicated, and minimizing the sum of bit-error probabilities over the $k$ message positions was suggested. Since exact computation of this metric is obviously difficult, a bounding procedure based on Bhattacharyya upper bounding was adopted [1].

### 10.2.1 Bhattacharyya Bound

Consider a binary hypothesis test, say for deciding whether $u_i$ is 0 or 1, in one of these synthetic decision problems. The probability of decision error for the SC decision-maker on message position $i$, conditioned on sending a 0 bit, is

$$P_{e_i|0} = \sum_{\mathbf{y} \in \mathcal{D}_1} P\left(\mathbf{y}, u_0^{i-1} \mid 0\right).$$

The sum is over observation vectors whose likelihood $P\left(\mathbf{y}, u_0^{i-1} \mid 1\right) / P\left(\mathbf{y}, u_0^{i-1} \mid 0\right)$ exceeds 1, that is, the decision region $\mathcal{D}_1$ in favor of $\widehat{u}_i = 1$. We can then upper bound $P_{e_i|0}$, for any $s \geq 0$, by

$$\begin{aligned} P_{e_i|0} &\leq \sum_{\mathbf{y} \in \mathcal{D}_1} P\left(\mathbf{y}, u_0^{i-1} \mid 0\right) \left[\frac{P\left(\mathbf{y}, u_0^{i-1} \mid 1\right)}{P\left(\mathbf{y}, u_0^{i-1} \mid 0\right)}\right]^s \\ &\leq \sum_{\text{all } \mathbf{y}} P\left(\mathbf{y}, u_0^{i-1} \mid 0\right) \left[\frac{P\left(\mathbf{y}, u_0^{i-1} \mid 1\right)}{P\left(\mathbf{y}, u_0^{i-1} \mid 0\right)}\right]^s. \end{aligned}$$

This follows since raising the likelihood ratio in the summand to a non-negative power cannot decrease the summand, and expanding the sum range in the second line only adds non-negative terms. A convenient choice for $s$ is $s = 1/2$, leaving

$$P_{e_i|0} \leq \sum_{\text{all } \mathbf{y}} \left[P\left(\mathbf{y}, u_0^{i-1} \mid 0\right) P\left(\mathbf{y}, u_0^{i-1} \mid 1\right)\right]^{1/2}. \tag{10.29}$$

This also bounds the error probability conditioned on 1, by symmetry, and is thus a bound on the unconditional error probability for $u_i$. The right-hand side is called the *Bhattacharyya bound* on error probability for our binary decision problem. Shortly, a recursive procedure will be described that computes (10.29) efficiently as blocklength grows.

A more traditional statement of the Bhattacharyya bound also holds for the raw channel $W$ and is

$$P_e \leq \sum_{\text{all } y} [P(y \mid x = 0) P(y \mid x = 1)]^{1/2},$$

whose right-hand side we designate by the real parameter $Z$. For the BSC, $Z = [4\epsilon(1 - \epsilon)]^{1/2}$, while for the BEC and BI-AWGN models we have $Z = p$ and $Z = e^{-E_s/N_0}$, respectively (see Problem 10.10).

This theory can be employed to bound the probability of decision error for each of the two inputs in Figure 10.1, as follows. Let $Z$ again represent the real-valued

Bhattacharyya parameter for the underlying channel $W$. Then it can be shown [1] that the two synthetic channels have Bhattacharyya parameters given by

$$Z^{(0)} \leq 2Z - Z^2,$$
$$Z^{(1)} = Z^2. \tag{10.30}$$

Because $Z \leq 1$, this expresses what we earlier found for the two virtual channels – the 0 channel is poorer (larger $Z$), while the 1 channel is better (smaller $Z$), in terms of bounds. This is also consistent with the statements about mutual information in the $2 \times 2$ case. For a general binary memoryless channel, the first expression above is an inequality as stated, but equality holds (in both) for the BEC case, as now demonstrated.

---

**Example 10.3** We now extend the earlier analysis for the BEC case to a $4 \times 4$ transform. Notice that bits designated $v_0$ and $v_1$ in either $4 \times 4$ transform shown in Figure 10.5 see independent BECs with effective erasure probability given by $2p - p^2$ and $p^2$ on the inputs to their $2 \times 2$ module. So the effective BEC seen at $u_0$ is $2(2p - p^2) - (2p - p^2)^2$, and that seen by $u_1$ is a BEC with effective erasure probability $(2p - p^2)^2$. Similar thinking gives the effective channels seen by $u_2$ and $u_3$ to be BECs with parameters $2p^2 - p^4$ and $p^4$, respectively.[7] This argument for the BEC shows that equality holds in the $Z$-parameter updates of (10.30) above. Also for the BEC, this leads to an exact recursion for bit-channel mutual information, since erasure probability is directly tied to mutual information.

We can extend (10.30) to larger transform sizes by again noting that computing the $n$ internal LLR variables at some level of the transform graph involves combining independent, statistically identical data supplied by the level to the right. Thus, whatever the error probability bounds are for a set of $2^p$ bits at one level, we can find bounds for $2^{p+1}$ bits at the next level to the left using a generalization of (10.30), except applied to vectors. In Matlab notation this becomes

```
Z = initial value    % channel dependent
for j = 1:m
    Z = [2*Z-Z.^2 , Z.^ 2 ];
end
```

The size of the $\mathbf{Z}$ vector doubles each step to a final size of $n = 2^m$.

Starting with $Z = 0.5$, corresponding to erasure probability 0.5, we have the following table for the first three steps of the recursion above, already exhibiting polarization.

[7] It is a key fact that the inputs on the right of any such $2 \times 2$ module are stochastically identical and independent.

| $n$ | $i = 0$ | 1 | 2 | 3 | 4 | 5 | 6 | 7 |
|---|---|---|---|---|---|---|---|---|
| 1 | 0.5000 | | | | | | | |
| 2 | 0.7500 | 0.2500 | | | | | | |
| 4 | 0.9375 | 0.4375 | 0.5625 | 0.0625 | | | | |
| 8 | 0.9961 | 0.4836 | 0.8086 | 0.1211 | 0.8789 | 0.1914 | 0.3164 | 0.0039 |

A toy design with $n = 8$ would then freeze indices 0, 1, 2, and 4 to produce an $(8, 4)$ code. This code is equivalent to an $(8, 4)$ extended Hamming code.

In Figure 10.9, the empirical cumulative distribution function (cdf) for bit-channel mutual information is provided for the BEC model. Blocklengths 128, 1024, and 8192 are presented. The BEC model has erasure probability $p = 0.5$, so the channel capacity is $C = 0.5$ bits/channel use. In this case the raw channel has $Z = 0.5$ and the above recursion on the $Z$-vector is exact, after which the vector of mutual information becomes $\mathbf{1} - \mathbf{Z}$. Notice that the cdf makes significant increase near $I = 0$ and $I = 1$, and that both 'jumps' have value 0.5. This is equivalent to saying that half the bit channels have mutual information near 0 and half have information near 1. Notice also the sharpening of the cdf transitions as $n$ increases, illustrating the fundamental polarization claim.

In Figure 10.10, we set $E_s/N_0 = 0.53 = -2.81$ dB on a BI-AWGN channel, again corresponding with $C = 0.5$. We employ the method using the Gaussian approximation

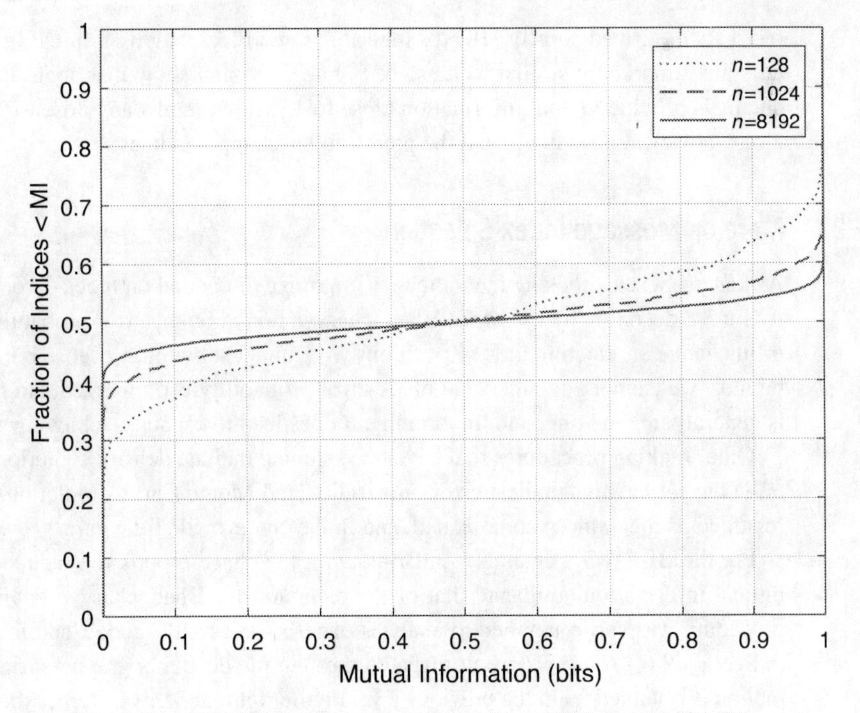


**Figure 10.9** The cdf for bit-channel mutual information, BEC channel, $p = 0.5$.

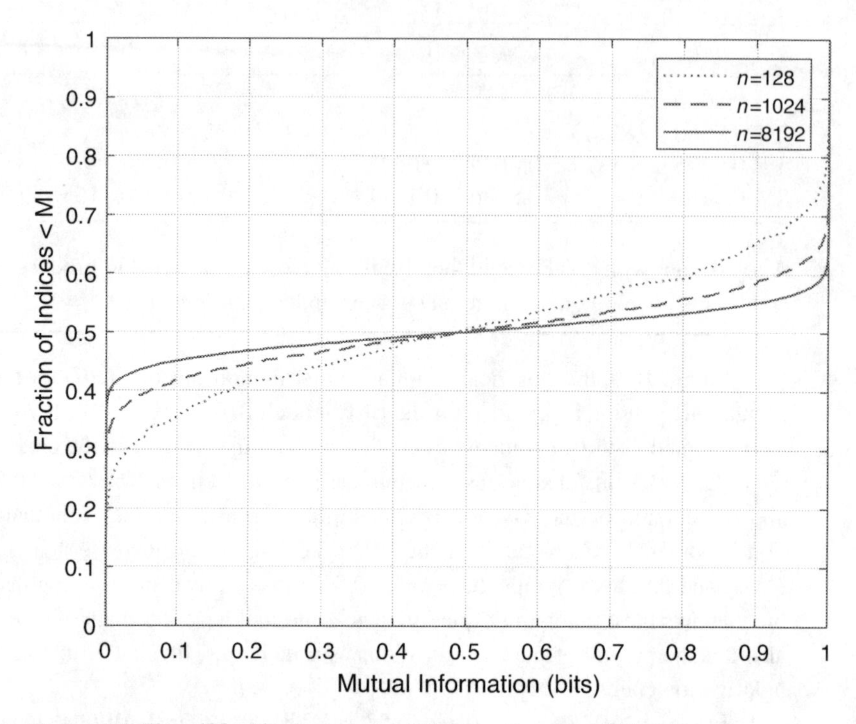


**Figure 10.10** The cdf for bit-channel mutual information, BI-AWGN channel, $E_s/N_0 = 0.53 \simeq -2.81$ dB, $J$-function method.

of LLRs described shortly [18] for the same three blocklengths. Mutual information cdfs are remarkably similar to those of Figure 10.9, and show that about half the bit channels provide mutual information close to 1, while the rest are close to 0. The cdf alone, however, does not reveal the preferred bit-channel indices.

### 10.2.2 More on Message Index Selection

Arikan proposed choosing the indices to minimize the bound on block-error probability for SC decoding by use of the Bhattacharyya bound on the error probability for the $n$-bit channels, selecting those $k$ positions with smallest bound. Because it is based on bounds, we cannot guarantee that block-error probability is minimized, but the design is generally a good one, and importantly, it does lead to capacity-achieving codes.

Other ranking procedures that have been studied include density evolution to determine the set having smallest error probability, and Monte Carlo simulation. Both are regarded as much more complicated, and in the end provide little gain.

For the BI-AWGN channel, ten Brink *et al.* [18] have proposed a recursion on the mutual information similar to that of the recursion for Bhattacharyya bounds. (This procedure was earlier applied to analysis of turbo and LDPC codes, and is presented in Section 9.6.) As LLRs are propagated through the decoding graph, extrinsic information is obtained from the previous level on the right, and this becomes the effective channel measurement for the next level of processing. Under this model, these LLRs are modeled as conditionally Gaussian at each point, so one need only track the means and variances, and the method is known as the Gaussian approximation (GA). These distributions can be mapped to a corresponding mutual information for a BI-AWGN model, as now described.

First, recall the mutual information between a binary channel input (±1 for now) and the output of a generic additive Gaussian channel, represented by

$$Y = \mu X + N, \tag{10.31}$$

where $\mu$ is the square root of the signal energy per symbol, and $N$ is zero-mean Gaussian with variance $\sigma^2$.

The mutual information between $X$ and $Y$ is, from Chapter 1, after simplification

$$I(X;Y) = \int_{-\infty}^{\infty} \frac{1}{(2\pi)^{1/2}} e^{-z^2/2} \left[1 - \log_2\left(1 + e^{-2\mu^2/\sigma^2 - 2\mu z/\sigma}\right)\right] dz \text{ bits.} \tag{10.32}$$

Observe that this is a function only of $\mu/\sigma$, which in common notation is $(2E_s/N_0)^{1/2}$.

The channel observation $Y$ is converted to the LLR about $X$ for decoding purposes, which is

$$L_x \triangleq \log_e \frac{f(y \mid +1)}{f(y \mid -1)} = \frac{2\mu y}{\sigma^2}. \tag{10.33}$$

This variable is also conditionally Gaussian, but with means $\mu_{ch} = \pm 2\mu^2/\sigma^2$ and variance $\sigma_{ch}^2 = 4\mu^2/\sigma^2$. Assuming +1 is transmitted, we could write

$$L_x \sim \mathcal{N}\left(\mu_{ch}, \sigma_{ch}^2\right) = \mathcal{N}\left(\frac{\sigma_{ch}^2}{2}, \sigma_{ch}^2\right), \tag{10.34}$$

which involves only a single parameter $\sigma_{ch}^2$. Substitution reveals that $\sigma_{ch}^2$ is actually an SNR measure, $\sigma_{ch}^2 = 8E_s/N_0$.

Scaling of $Y$ to produce the LLR does not change mutual information, so $I(X; L_x)$ is identical to $I(X; Y)$. Using the mutual information definition involving a binary input and Gaussian observation, we find

$$I(X; L_x) = \int_{-\infty}^{\infty} \frac{1}{(2\pi\sigma_{ch}^2)^{1/2}} e^{-(z-\mu_{ch})^2/2\sigma_{ch}^2} \left[1 - \log_2\left(1 + e^{2\mu_{ch}^2/\sigma_{ch}^2}\right)\right] dz \text{ bits.} \tag{10.35}$$

With a change of variable, this becomes

$$J(\sigma_{ch}) \triangleq I(X; L_x) = \int_{-\infty}^{\infty} \frac{1}{(2\pi)^{1/2}} e^{-z^2/2} \left[1 - \log_2\left(1 + e^{-2\mu_{ch}^2/\sigma_{ch}^2 - 2\mu_{ch} z/\sigma_{ch}}\right)\right] dz \text{ bits.} \tag{10.36}$$

Note that this is the same integral form as the expression for $I(X;Y)$ in (10.32), but (10.36) involves only the auxiliary parameter $\sigma_{ch}$. ten Brink *et al.* [18] define this latter expression as $J(\sigma_{ch})$ for converting back and forth between mutual information and the parameter $\sigma$ of an approximate Gaussian distribution. The $J$-function is monotonic, and its inverse exists, denoted by $J^{-1}(\cdot)$. The conversion is needed for performing transformations at each 2 × 2 module. Linkage between the two information-related measures in (10.32) and (10.36) is

$$I(E_s/N_0) = J\left([8E_s/N_0]^{1/2}\right). \tag{10.37}$$

Now consider a 2 × 2 module at the right edge of the polar transform graph. The channel measurements supply *a priori* information to the module, and we are interested in the extrinsic information attached to the input bits on the left. This then becomes prior information to a module further to the left. To propagate mutual information between intermediate binary variables and the $n$-tuple of channel observations, we adopt some design channel quality $E_s/N_0 = (k/n)E_b/N_0$, and first compute $J\left([8E_s/N_0]^{1/2}\right)$ to obtain mutual information for initialization. $J^{-1}(\cdot)$ acting upon this provides the effective channel parameter $\sigma_{ch}$ (not $\sigma_{ch}^2$).

Assuming correct knowledge on the bit modulating the $g$ function, the effective SNR parameter $\sigma_{ch}^2$ for the lower-bit's LLR is twice that of the input's effective SNR. So

$$I_{\text{lower,out}} = T_2(I_{\text{in}}) = J\left(\sqrt{2[J^{-1}(I_{\text{in}})]^2}\right). \tag{10.38}$$

Paralleling the discussion in Section 9.6 on EXIT methodology, and noting that the basic 2 × 2 module involves a degree-3 parity check between the upper input bit and the two output bits, the upper-bit update rule is given by (see (9.43))

$$I_{\text{upper,out}} = T_1(I_{\text{in}}) = 1 - J\left(\sqrt{2[J^{-1}(1 - I_{\text{in}})]^2}\right). \tag{10.39}$$

Each subsequent encounter with a 2 × 2 module transforms two input LLRs having the same distribution, approximated as Gaussian random variables with some associated mutual information and an effective SNR $\sigma^2$, to a new set of LLRs modeled by Gaussian variables and some new mutual information. The transformations for the upper and lower bits of a 2 × 2 block are given by

$$\begin{aligned} I_p^{(2i)} &= 1 - J\left(\sqrt{2\left[J^{-1}\left(1 - I_{p/2}^{(i)}\right)\right]^2}\right), \quad \text{even bits;} \\ I_p^{(2i+1)} &= J\left(\sqrt{2\left[J^{-1}\left(I_{p/2}^{(i)}\right)\right]^2}\right), \quad \text{odd bits.} \end{aligned} \tag{10.40}$$

Similar to the Bhattacharyya parameters, we can propagate (approximate) mutual information for a set of $p/2$ bits at one level of the polar transform graph to mutual information for $p$ bits at the next level to the left. Again, the recursion is initialized with $I_1^{(0)} = J\left(\sqrt{8E_s/N_0}\right)$.

Since the $J(\cdot)$ function involves a single-variable integral, polar code design benefits from simple approximations of the function. An accurate and efficient one is [19]

$$I = J(x) = \left(1 - 2^{-h_1 x^{2h_2}}\right)^{h_3} \tag{10.41}$$

with $h_1 = 0.3073$, $h_2 = 0.8935$, and $h_3 = 1.1064$. This also provides an approximation for the inverse function as

$$x = J^{-1}(I) = \left(-\frac{1}{h_1}\log_2\left(1 - I^{1/h_3}\right)\right)^{1/2h_2}. \tag{10.42}$$

This $J$-function methodology was employed to produce Figure 10.10.

Finally, it should be noted that polar code design is not universal, that is, for a given $(n, k)$, the chosen message indices, and thus the polar code, are dependent on the choice of channel quality in the design. For example, we may find the preferred $k = n/2$ message indices for a BI-AWGN $R = 1/2$ design at $E_b/N_0 = 2$ dB will differ slightly from those for $E_b/N_0 = 4$ dB. This design SNR should be chosen to match the expected channel quality, but if the channel differs a little, no major sacrifice in performance should result. Similar comments pertain to the design for a BSC, say.

Regarding choice of design SNR, it should normally be 1–3 dB greater than the theoretical minimum SNR according to the Shannon limit at code rate $R$, reflecting imperfect polarization at typical blocklengths. Finite-blocklength bounds can also be a guide to setting design SNR.

---

**Example 10.4** Let $\mathcal{F}$ and $\mathcal{F}^c$ respectively represent frozen index sets and message index sets. For a BI-AWGN channel model, an $R = 1/2$ $(64, 32)$ polar code designed using (10.40) with $E_b/N_0 = 2.5$ dB ($E_s/N_0 = -0.5$ dB) has message index set

$$\mathcal{F}^c = \{16, 24, 27, 28\text{–}32, 39, 40, 42\text{–}48, 50\text{–}64\}.$$

(Indexing here starts at 1, not 0.) Observe that the message indices tend to cluster in the higher positions of $(1, 2, \ldots, n)$, but this is not strictly true. These non-frozen positions also tend to index the higher-weight rows of $\mathbf{G}_{64}$. If the design SNR is changed to 5 dB, for example, only a minor change in the non-frozen set emerges; index position 27 is removed, and replaced by position 38.

With the same blocklength but $R = 3/4$, we simply add 16 *additional* message indices to the above set. With design SNR=2.5 dB, these are

$$\mathcal{F}^c_{add} = \{8, 12, 14, 15, 20, 21, 22, 23, 25, 26, 35\text{–}38, 41, 49\}. \tag{10.43}$$

These positions were lower-quality indices in the original $R = 1/2$ design.

For a BEC model with erasure probability 0.5 (a channel with roughly the same capacity in bits/channel use), the message index set for a $(64, 32)$ polar code is, based on selecting channels with smallest Bhattacharyya parameter,

$$\mathcal{F}^c = \{8, 12, 14, 15, 16, 20, 22, 23, 24, 26, 28, 30, 31, 32, 36, 38, 40, 44, 46, 47, 48, 52, 54, 55, 56, 58\text{–}64\}.$$

---

## 10.3 Successive Cancellation Decoding

Before describing the original decoding approach for polar codes, namely successive cancellation (SC), we recall the general decoding problem. The minimum probability of block error is achieved by maximizing the *a posteriori* probability (APP) $P(\mathbf{u}_j \mid \mathbf{y})$ over codeword index $j$ and, assuming messages are equiprobable, this is equivalent to ML decoding, that is, choosing the message $\mathbf{u}_j$ that maximizes $P(\mathbf{y} \mid \mathbf{u}_j) = P(\mathbf{y} \mid \mathbf{x}(\mathbf{u}_j))$.

Pursuing the APP, we have a chain rule representation

$$P(\mathbf{u} \mid \mathbf{y}) = \prod_{i=0}^{n-1} P\left(u_i \mid \mathbf{y}, u_0^{i-1}\right). \tag{10.44}$$

Each term in the product is the APP for bit $u_i$, given the observation vector $\mathbf{y}$ and lower-indexed bits. This could be a recipe for finding the optimal codeword choice, requiring calculation for all $2^k$ messages.

SC decoding instead sequentially determines the $u_i$ that maximize each term in the product above, using prior decisions on bit indices $\in \{0, 1, \ldots, i-1\}$ to decide bit $i$. That is, we make a sequence of binary decisions that maximize, for each $i$,

$$\widehat{u}_i = \underset{u_i}{\operatorname{argmax}}\ P\left(u_i \mid \mathbf{y}, \widehat{u}_0^{i-1}\right). \tag{10.45}$$

Equivalently, we could sequentially perform ML bit by bit:

$$\widehat{u}_i = \underset{u_i}{\operatorname{argmax}}\ P\left(\mathbf{y}, \widehat{u}_o^{i-1} \mid u_i\right),$$

which leads to the decoding procedure below.

The sequence obtained by these term-by-term maximizations is not in general the same as the global maximum APP (or ML) choice. But, lending support to SC decoding, provided previous decisions are correct, the next bit $\widehat{u}_{i+1}$ is chosen optimally. Any intermediate decision error, however, will contaminate a future bit decision.

To illustrate the two decision procedures, consider the elementary $2 \times 2$ scenario at the beginning of the chapter. The optimal message decision for the pair $(u_0, u_1)$ could be expressed as

$$\widehat{(u_0, u_1)} = \underset{u_0, u_1}{\operatorname{argmax}}\ P(u_0 \mid y_0, y_1) P(u_1 \mid y_0, y_1, u_0)$$

after computing $2^2 = 4$ vector APPs. Instead, SC decoding decides $u_0$ first, by maximizing the first term in the product above (actually maximizing $P(y_0, y_1 \mid u_0)$), then decides $u_1$ by maximizing the likelihood equivalent of the second term, under the assumption that the first decision is correct. The decision pairs of these two strategies are not always equal.

To progress to the general SC procedure, which propagates log-likelihood ratios, it is helpful to understand the case with $n = 4$, using Figure 10.5(b). Observe that the LLR for the bit designated $v_0$ is obtained from a $2 \times 2$ message-passing operation involving $y_0$ and $y_2$, namely the $f$ function

$$L_{v_0} = f\left(L_{x_0}, L_{x_2}\right) = L_{x_0} \boxplus L_{x_2},$$

and at the same time

$$L_{v_1} = f(L_{x_1}, L_{x_3}) = L_{x_1} \boxplus L_{x_3}.$$

(We cannot yet compute $L_{v_2}$ and $L_{v_3}$, awaiting decisions.) These $L_{v_i}$ constitute effective measurements about these internal binary variables, blending the original channel LLRs, and these are independent by virtue of the graph properties.

Then, proceeding to the left in the graph, we can compute the LLR for $u_0$:

$$L_{u_0} = L_{v_0} \boxplus L_{v_1}.$$

The decision for $u_0$ either compares this with 0 for a non-frozen position, or sets the decision to 0 for a frozen position, assuming frozen positions are set to 0, that is

$$\widehat{u}_0 = \begin{cases} 0, & L_{u_0} \geq 0, \\ 1, & L_{u_0} < 0, \\ 0, & u_0 \text{ is frozen bit.} \end{cases}$$

Next, this decision allows the computation of $L(u_1)$ using the $g$ function

$$L_{u_1} = g(L_{v_0}, L_{v_1}, \widehat{u}_0) \triangleq L_{v_1} + (1 - 2\widehat{u}_0)L_{v_0},$$

followed by a decision made on $u_1$.

Once this $2 \times 2$ subsection in the upper left of Figure 10.5(b) is completed, we can obtain estimates of both binary variables $v_0$ and $v_1$ as

$$\widehat{v}_0 = \widehat{u}_0 + \widehat{u}_1,$$
$$\widehat{v}_1 = \widehat{u}_1.$$

These in turn allow the calculation of *both* $L_{v_2}$ and $L_{v_3}$ based on the original LLR vector:

$$L_{v_2} = L_{x_0} \overset{\widehat{v}_0}{\pm} L_{x_2},$$
$$L_{v_3} = L_{x_1} \overset{\widehat{v}_1}{\pm} L_{x_3}.$$

Then, using these, the lower-left $2 \times 2$ section of the graph can be processed, as described above. Once all input bits have been decided, there is no need to propagate these decisions back to the right, since SC decoding is completed.

A general decoding algorithm follows, with the help of Figure 10.7, for $n = 8$. Define $L_j^{(i)}$ as the LLR for the $i$th bit at level $j \in \{0, 1, \ldots, m-1\}, i \in \{0, 1, \ldots, n-1\}$, and $\mathbf{L}_j$ as the $n$-tuple of these LLRs at level $j$. (For simplicity, we understand that these LLRs are implicit functions of the channel observation $\mathbf{y}$ and of a putative decision vector $\hat{\mathbf{u}}$.) Also, let $c_j^{(i)}$ be the binary intermediate bits at level $j$ of the same polar transform graph, which depend on $\mathbf{u}$ through the encoding rules of the polar transform. Observe that LLRs and the $c$-bits are combined with a counterpart at offset $\beta_j = 2^{j-1}$. Finally, define $d_j^{(2i)} = c_j^{(2i)} + c_j^{(2i+1+\beta_j)}$ as the control bit that modulates the $g$ functions at level $j$.

Noticing that LLRs for positions with even index and odd index are computed with the $f$ and $g$ functions, respectively, we separate the LLR recursion into even and odd bit updates. Then, supposing the state of bits $c_j^{(i)}$ is somehow known, we would compute the leftward recursion for $j = m, m-1, \ldots, 1$ using

$$L_{j-1}^{(2i)} = f\left(L_j^{(2i)}, L_j^{(2i+1+\beta_j)}\right),$$

$$L_{j-1}^{(2i+1)} = g\left(L_j^{(2i)}, L_j^{(2i+1+\beta_j)}, d_j^{(2i)}\right), \quad i = 0, 1, \ldots, \frac{n}{2} - 1.$$

The recursion from right to left in the graph is initialized with the $n$-tuple of channel LLRs $\mathbf{L}_m$. (Note the exact indexing progression, in particular $\beta_j$, will change with a different realization of the polar transform, discussed in the previous section.)

Of course, the decision vector $\hat{\mathbf{u}}$, and thus the states of $d_j^{(2i)}$, are unknown *a priori*, but the LLR for $u_0^{(0)}$ does not depend on any prior decision (or employs only the $f$ function). So SC decoding simply uses a scheduled, back-and-forth version of the recursion above to sequentially estimate $u(i)$ using available prior decisions, and each time a new decision is made, updating the $\mathbf{c}$ vectors in the graph.

### 10.3.1 Tree Viewpoint

Observe from Figure 10.7 that four likelihoods $L_2^{(0:3)}$ can be computed from the entire channel likelihood vector using a vector version of the $f$ function, and that these four are sufficient to compute $L_1^{(0:1)}$ and finally $L_0^{(0)}$, all using the $f$ operation. Moreover, once $u_0$ has been decided, $L_0^{(1)}$ can be evaluated using the $g$ function on $L_1^{(0:1)}$. Similar observations pertain elsewhere in the transform graph. Once both $f$ and $g$ operations have been applied at a node, a binary message $\mathbf{c}$-vector can be passed to the right in the graph to allow other sections of the graph to be processed.

This suggests bundling both LLRs and binary estimates into vectors attached to nodes, called supernodes, at level $j$ and horizontal position $t$, as seen in Figure 10.11, rotated by 90° relative to earlier transform diagrams. This data held at each supernode is sufficient to evaluate the complete subtree, and once this is complete, if desired, the subtree information can be overwritten and memory used for subsequent likelihood calculations.

The tree has nodes on $m+1$ levels, designated $0, 1, \ldots, m$, bottom to top. A root node sits at level $m$ and $n = 2^m$ leaf nodes reside at level 0, each corresponding to an input bit $u_i$. The total number of nodes is $2n-1$. We number the supernodes at level $j$ with an index $t \in \{0, 1, ..., 2^{m-j}\}$. It will also be convenient to assign a one-dimensional index *pos* to each node above level 0 in the set $\{1, ..., n-1\}$. We can refer to such a node by its two-dimensional index $(j, t)$, or by its position in a linear array at $pos = 2^{m-j} + t$.

Movement between parent and child nodes is facilitated by noticing that the two child nodes for a node at position $t$ at level $j$ reside at level $j-1$, with indices $2t$ and $2t+1$ for the left and right child.[8] Conversely, the parent node associated with a node at level $j$ having index $t$ is a node at level $j+1$ with index $\lfloor t/2 \rfloor$.

[8] Left and right refer to directions seen looking at the page.

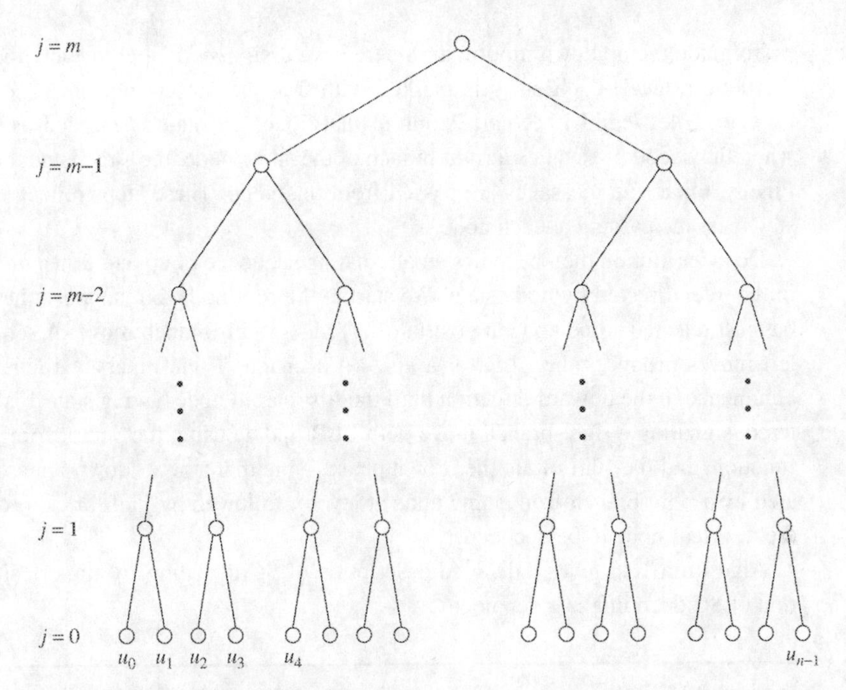


**Figure 10.11** General SC computation tree.

A data structure that admits the recursive nature of processing and maps directly to the tree uses a pair of two-dimensional arrays – an array **M** with size $m + 1 \times n$ that contains log-likelihood information compiled by the processor, and an array **C** with size $m + 1 \times n$ that stores all the binary messages entering and leaving a supernode.[9]

Traversal of the tree by the processor is depth-first, with bias to the left in the tree of supernodes. Here, however, the operation that generalizes the earlier sequence for processing an isolated $2 \times 2$ section does a vector version of earlier operations. The lifecycle of a supernode at level $j$ includes receiving $2^j$ LLRs from its parent node, first computing $2^{j-1}$ LLRs for the left-child supernode below, eventually receiving $2^{j-1}$ binary estimates from the left child below that influence the computation of LLRs to the child supernode to the right.[10] In other words, the processing is much like the three-step processing described for an isolated $2 \times 2$ section, with the difference being that the third step (applying the $g$ function to the vector LLR input data) must await the up-propagation from a leaf node through possibly several levels, when new LLRs can be calculated. Then a downward LLR recursion toward the next undecided leaf node is made, and a new message bit decision is made. Once the entire subtree of a node at level $j$ has been visited, binary messages in **C** are sufficient to send a binary $2^j$-tuple to the parent node above, enabling a right branch and another down phase.

[9] The present discussion borrows ideas from a tutorial by Professor A. Thangaraj, IIT-Madras.

[10] The root and leaf nodes have an abbreviated version of this cycle.

To guide the up/down motion in this tree, we assign *state*(*pos*) to each node above the bottom level. Each entry is marked with 0 at the outset, and once the decoder executes a left-child LLR calculation at that node, the state is marked as 1. Later, when the decoder executes a right branch at the same node, the state is denoted by 2. Finally, binary up messages are passed from this node. These steps reflect the three-step message passing at each node.

Decoder motion then becomes an alternating sequence of up and down movements in the tree, directed by node state. We start at the root node, but suppose the decoder has just reached some node at position $(j, t)$ via a right- branch move. A sequence of left moves produces the LLR for a new bit decision. Then, binary variable updates commence in the upward direction until the first parent node having state 1 is encountered, signaling a right-branch move with LLR update using the vector form of the $g$ function, and the start of another down phase. Note that a new "down" phase is initiated by a right branch from some node at level $j$, followed by $j$ left branches to reach the next leaf node to be decoded.

Algorithm 10.2 gives a detailed description of SC decoding. A numerical illustration of SC decoding is given next.

---

**Example 10.5** Suppose bits at indices 0, 1, 2, and 4 are frozen at 0, and the vector of LLRs from the raw channel is $\mathbf{llr} = [-0.70, 0.61, 0.81, 0.32, 1.11, 0.92, -0.41, 1.52]$. Hard decisions on this data would produce $\hat{\mathbf{x}} = [10000010]$, not a valid codeword.

To first clarify the tree search order, Figure 10.12 lists adjacent to each node the sequential order visits to that node, starting at 0 at the root node, and ending at 25 for the final decoding of $u_7$.

Table 10.1 compiles the *final* LLR values for the tree of the example. The top row is the channel LLR vector; from this the $f$ function supplies the first four values in row 2, then the first two values in row 3, etc. The last four values of row 2, on the other hand, are obtained with the $g$ function operating on row 1, using binary decisions propagated from below. (See Table 10.2.)

---

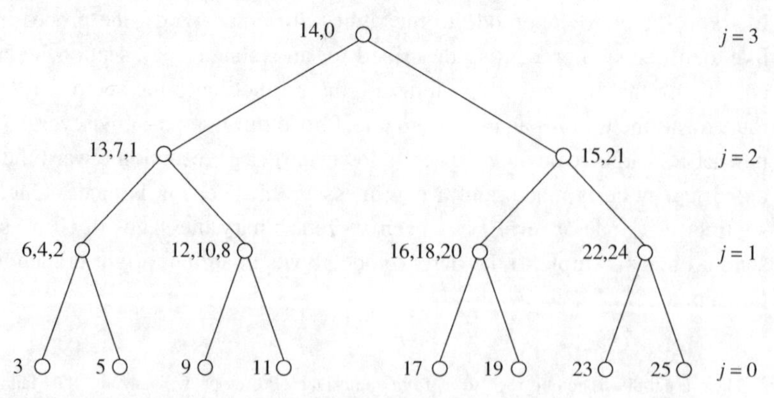


**Figure 10.12** Decoder evolution for (8, 4) example.

**Algorithm 10.2** SC Decoding

ASSUMPTIONS

1. Zero-based indexing
2. $f$ and $g$ functions invoked

ALGORITHM

**Initialize.** Input $n$, frozen bit indices, and **llr**. Set $m = \log_2 n$.

$\mathbf{M}(0, 0{:}n-1) = \mathbf{llr}$. $\mathbf{C}(0{:}m-1, 0{:}n-1) = 0$

$state(1{:}n-1) = 0$

$j = m, i = 0, t = 0, startdown = 1.$

**for** $i = 0$ to $n - 1$ step by 2 **do** ▹ bit index loop incrementing by 2

  $pos = startdown,\ t = rem(startdown/2^{m-j}), jump = 0;$

  **if** $state(1) = 1$ **then**

    $jump = n/2$ ▹ doing right sub tree

  **end if**

---

  **while** $j > 0$ **do** ▹ (Down movement; only left moves in tree)

    $len = 2^{j-1}$

    $off = t * len$

    $\mathbf{ml} = \mathbf{M}(m - j + 1, off + 1{:}off + len)$

    $\mathbf{mr} = \mathbf{M}(m - j + 1, off + len + 1{:}off + 2 * len)$

    $\mathbf{M}(m - j + 2, off + 1{:}off + len) = f(\mathbf{ml}, \mathbf{mr})$

    $pos = 2 * pos,\ t = 2 * t,\ state(pos) = 1;$

    $j = j - 1$ ▹ down one level

  **end while**

---

▹ (Decisions at level 0)

▹ Decide $\widehat{u}_i$ either from sign of LLR or frozen bit state

  $\mathbf{C}(0, i) = \widehat{u}_i$

  $\mathbf{M}(m + 1, i + 1) = \mathbf{mr} + (1 - 2 * \mathbf{C}(0, i)) * \mathbf{ml}$

▹ Decide $\widehat{u}_{i+1}$ either from sign of LLR or frozen bit state

  $\mathbf{C}(0, i + 1) = \widehat{u}_{i+1}$

---

▹ (Up movement)

  $cl = \mathbf{C}(m + 1, i + 1)$

  $cr = \mathbf{C}(m + 1, i + 2)$

  $a = cl + cr$, modulo-2; $b = cr$

  $\mathbf{C}(m, t + 1{:}t + 2) = [a, b]$ ▹ first level partial sums

  $j$=1

  $tup = rem(pos/2, 2^{m-j})$

  $startup = 2^{m-j} + tup$

  $pos = \lfloor startup/2 \rfloor$

  $t = \lfloor tup/2 \rfloor$

  $j = j + 1$ ▹ find turnaround in tree

```
while j ≤ m do
    if state(pos) = 1 then
        turnlev = j
        turnt = t
        state(pos) = 2
        break
    end if
    lenc = 2^(j-1)
    offc = t * lenc
    cl = C(m - j + 2, offc + 1:offc + lenc
    cr = C(m - j + 2, offc + 1 + lenc:offc + 2 * lenc)
    C(m - j + 1, 1:2 * lenc) = [mod(cl + cr, 2), cr]
    pos = ⌊pos/2⌋                                          ▹ move up a level
    t = ⌊t/2⌋
    j = j + 1
end while
                                              ▹ (prepare for next down cycle)
startdown = 2^(m-turnlev+1) + 2 * turnt + 1
t = rem(startdown, 2^(m-j+1))
len = 2^(j-1)
off = t * len
ml = M(m - j + 1, jump + 1:jump + len)
mr = M(m - j + 1, jump + len + 1:jump + 2 * len)
M(m-j+2, off+1:off+len) = mr+(1-2*C(m-j+2), jump+1:jump+len))*ml
j = turnlev - 1
end for                                                    ▹ end of bit loop
```

**Table 10.1** Final LLR table $\mathbf{M}$ for SC example

| $j$ | $t = 0$ | 1 | 2 | 3 | 4 | 5 | 6 | 7 |
|---|---|---|---|---|---|---|---|---|
| 3 | −0.70 | 0.61 | 0.81 | 0.32 | 1.11 | 0.92 | −0.41 | 1.52 |
| 2 | −0.70 | 0.61 | −0.41 | 0.32 | 1.81 | 0.31 | −1.22 | 1.20 |
| 1 | 0.41 | 0.32 | −1.11 | 0.93 | −1.22 | 0.31 | −3.03 | 0.89 |
| 0 | 0.32 | 0.73 | −0.93 | −0.18 | −0.31 | −0.91 | −0.89 | 3.92 |

In the example posed, the decision on the input vector is $\hat{\mathbf{u}}$ = (**000**1**0**110),[11] which has *sequence* log-likehood 3.92, as listed in the lower-right corner of $\mathbf{M}$. It might be noted that the decision on $u_3$ has the smallest LLR magnitude, 0.18. Had this bit been resolved in favor of 0, the final decoded vector is all zeros.

This example illustrates that SC decoding is not equivalent to ML decoding, as calculation will show that the sequence $\mathbf{u}$ = (**000**0**0**000) has higher metric, 4.18,

[11] Frozen decisions are denoted with bold entries.

**Table 10.2** Final binary variable table **C** for SC example. Bold entries correspond to frozen bits. Top row is actually unused

| $j$ | $t = 0$ | 1 | 2 | 3 | 4 | 5 | 6 | 7 |
|---|---|---|---|---|---|---|---|---|
| 3 | 1 | 0 | 0 | 1 | 0 | 1 | 1 | 0 |
| 2 | 1 | 1 | 1 | 1 | 0 | 1 | 1 | 0 |
| 1 | 0 | 0 | 1 | 1 | 1 | 1 | 1 | 0 |
| 0 | 0 | 0 | 0 | 1 | 0 | 1 | 1 | 0 |

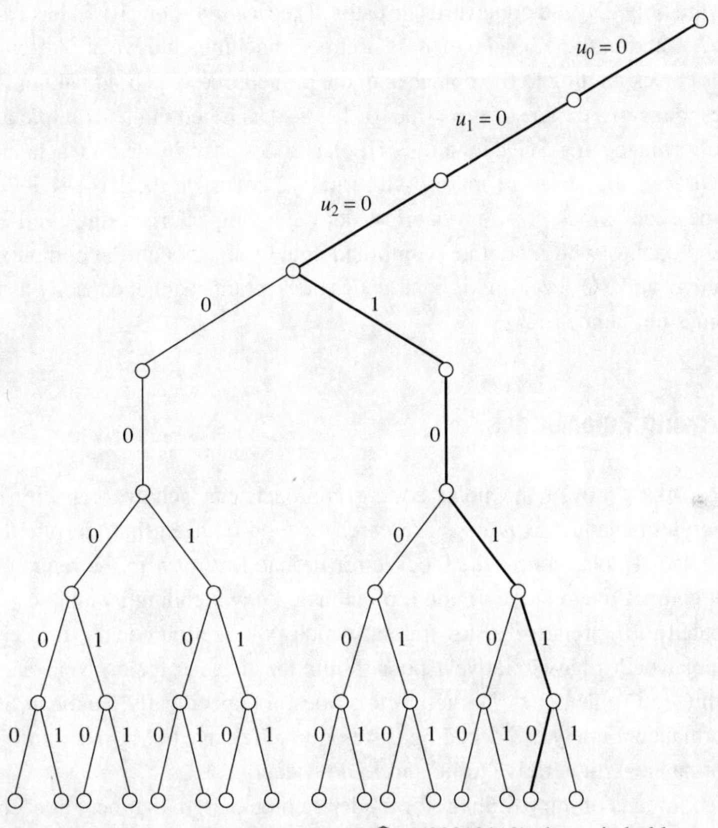


**Figure 10.13** SC decoding progress tree, $\widehat{\mathbf{u}}$ = (00010110) shown in bold.

which is also the sum of the eight LLRs. List decoding, discussed below, is a means of approaching ML performance for the code as list size grows.

A more efficient memory utilization (not presented above) is possible by realizing that, once all leaf nodes of a node at level $j$ have been decoded and **C** properly updated, the data in **M** and **C** will no longer be needed, and that memory can be reused for future tree updates. In fact, only $2n - 1$ LLR cells, including the raw input **llr**, are needed to perform SC decoding. Similarly, $2 \times (n - 1)$ binary cells are sufficient to manage the bit propagation in the tree. A more intricate control mechanism is needed to manage this data structure.

An alternative graphical visualization of decoding, useful in understanding list decoding, is shown in Figure 10.13 for $(n, k) = (8, 4)$, to continue the previous example. This tree is not a computational platform, but merely depicts decoder progress in a binary decision tree as the SC decoder commits to decisions $\widehat{u}_i, i = 0, 1, \ldots, n-1$. This tree is a pruned version of a full binary tree that has $n = 8$ levels, and represents the search problem for any decoding algorithm – the tree is explored in some manner to find a "good" choice for the transmitted message. The full tree for the general case will have $2^n$ leaf nodes and the various paths from root to terminal nodes correspond to the $2^n$ sequences $u_0^{n-1}$. Because $n - k$ of these bits are predefined (frozen), however, there are only $2^k$ valid codewords or paths. The tree in Figure 10.13 has frozen indices 0, 1, 2, 4 and at these levels, there is no tree branching, and we are left with 16 codewords corresponding to the branches in the pruned tree. The bold path in Figure 10.13 corresponds to a choice of $\mathbf{u} = (00010110)$ as developed in the example above.

Performance for SC decoding of polar codes is provided in a later section of this chapter, illustrated primarily with rate-1/2 codes on the BI-AWGN channel. As will be seen, block-error probability does not outperform other well-known coding approaches with moderate length and equal rate (though the complexity remains attractive, and SC decoding of polar codes does attain channel capacity as blocklength becomes unbounded.)

## 10.4 Improving Polar Codes

After Arikan proved the polar coding approach can achieve capacity of a binary memoryless channel as $n \to \infty$, researchers soon realized that, despite the attractive simplicity of polar codes, the block-error probability for $n$ in the range of $n < 1000$ is not competitive relative to good, previously known coding techniques. This can be attributed to the relatively slow rate at which polarization occurs; it is also apparent that polar codes have relatively poor minimum distance for a given rate and blocklength (see Problem 10.17), though this does not specifically handicap SC decoding performance. Finally, SC decoding is suboptimal relative to ML decoding (or near-ML performance with iterative turbo and LDPC decoding).

Two threads of improvement for moderate blocklength that benefit each other have emerged: (1) extending SC decoding to *list decoding*, and (2) *pre-coding* of the message prior to polar encoding. The latter is a concatenation approach designed to improve the distance properties of the code, by adaptively defining frozen positions in the inner polar code. SC decoding alone is incapable of exploiting this, however. Pre-coding, specifically the *polarization-adjusted convolutional* (PAC) extension to polar codes, will be discussed in a subsequent section.

### 10.4.1 List Decoding

SC decoding, though efficient to implement and capacity-approaching for very large blocklengths, suffers from occasional premature dismissal of the overall best codeword; in the context of the decision tree of Figure 10.13, an incorrect bit decision is made at some intermediate depth of the tree ($\hat{u}_3$ in the example), which cannot be later reversed. Tal and Vardy [5] proposed a decoder for the BI-AWGN channel that concurrently pursues $L > 1$ paths through this tree, and at the end produces the most likely path among these $L$ candidates (all valid codewords) by selecting the path with best final metric. The decoder is called a *successive cancellation list* (SCL) decoder, and the technique extends to any memoryless channel setting.

The simulation-based findings of Tal and Vardy, presented for $n = 2048, R = 1/2$, are twofold.

1. As $L$ increases, the transmitted message is increasingly likely to be on the final list, and sorting of the list produces the most likely codeword among these, meaning list decoding closes the performance gap between SC decoding and ML decoding for a given polar code. As intuitively expected, the required list size $L$ to reach the ML limit grows with blocklength $n$ and with decreasing SNR. Despite this gain, performance remains inferior for moderate blocklengths to other standard codes with the same parameters. This is attributable to the relative weakness of polar codes, specifically poor minimum distance.
2. However, though SCL decoding alone may produce a wrong codeword, as any ML decoder can do, it was noted that with moderate $L$, the correct vector was usually on the list (though was not the most likely codeword according to path metrics), and a simple CRC error-detecting code was suggested for scanning the list for this desired sequence. CRC-aided SCL decoding is usually dubbed CA-SCL. To illustrate performance, Tal and Vardy presented an $(n, k) = (2048, 1024)$ polar code with a 24-bit CRC, and showed that with $L = 32$, the performance closely approaches the performance of an ML decoder for this concatenated code at moderately large SNR, and shows gains of 1.5 dB at codeword error probability $10^{-3}$ over standard polar coding with SC decoding on the BI-AWGN channel.

SCL decoding essentially manages $L$ different SC decoders concurrently. When the SC decoder reaches a non-frozen leaf node, it normally makes a hard decision on $u_i$ according to the sign of the LLR. With list decoding, however, the $L$ current paths $\widehat{u}_0^{i-1}$ are extended with both this hard decision and its logical complement. Thus, each of $L$ former path candidates generate $2L$ candidate paths that are one bit longer. From these, the best $L$ paths, according to a path metric, are retained for the next cycle. The path metric is updated according to

$$\begin{aligned} P\left(\widehat{u}_0^i\right) &= P(\widehat{u}_0^{i-1}), \;\; \hat{u}_i = hard(LLR(u_i)), \\ P\left(\widehat{u}_0^i\right) &= P(\widehat{u}_0^{i-1}) + |LLR(u_i)|, \;\; \hat{u}_i, \neq hard(LLR(u_i)), \end{aligned} \tag{10.46}$$

where $hard(LLR(u_i))$ represents the normal decision based on the sign of $LLR(u_i)$. Essentially, the path metric is, aside from an overall bias that is of no importance, the sum of LLRs for bits along the candidate path, modulated by the sign of the LLR, together with a bias term. Path sorting of $2L$ paths retains those $L$ sequences having the *smallest* metric.

**Algorithm 10.3** SCL Decoding

ASSUMPTIONS

1. $L \geq 2$

ALGORITHM

1. Use SC decoder to build paths corresponding to $u_0 = 0$ and $u_0 = 1$. Set *numpaths* = 2. Maintain the associated data structure. Set $i = 0$.
2. Form both 0/1 extensions using SC algorithm, obtaining $2 * numpaths$ partial codewords. Set $i = i + 1$.
3. If $2 * numpaths > L$, sort partial codeword metrics for best $L$ choices. Set $numpaths = L$
4. While $i < n$, go to Step 2.
5. Do final sort on $L$ path metrics. Output **u** corresponding to best path.
6. (If CRC used) Perform CRC check on $L$ codeword candidates; the first passing CRC check is output by decoder.

Aside from an $L$-fold increase in computation and storage, list decoding imposes additional sorting overhead, and memory-management issues. Tal and Vardy showed how to avoid excessive sorting complexity and data copying using an efficient "lazy copy" or copy-on-write procedure to provide $\mathcal{O}(Ln\log_2 n)$ run-time complexity. Sorting of the best $L$ paths from $2L$ can be done efficiently for large $L$ by finding the median metric (with $\mathcal{O}(L)$ comparisons), then using up to $2L$ comparisons to find the best paths to extend. (For smaller $L$, simply bubble-sorting the list is probably faster, requiring $L^2/2$ comparisons.) Additionally, the memory complexity can be $\mathcal{O}(Ln)$. Both required computation and storage are thus $\mathcal{O}(L)$ times larger than the corresponding complexities for SC decoding, the essential penalty for list decoding. For a complete exposition and algorithm pseudo-code, readers should consult [5]. Balatsoukas-Stimming *et al.* [6] have put the algorithm into a numerically preferred log-domain version. Algorithm 10.3 gives pseudo-code for SCL decoding. Algorithm 10.3 gives pseudo-code for SCL decoding.

**Example 10.6** To continue the earlier toy example, Figure 10.14 represents an extension of Figure 10.13 introduced for SC decoding, and for simplicity we choose $L = 2$ for SCL decoding. Decoding starts with both paths initialized at the root node, and when unfrozen positions are encountered, both 0 and 1 extensions in the tree are evaluated. (This first occurs here for $u_3$.) The set of paths will soon exceed $L$, at which point we must choose the current "best" $L$ paths for further extension. This down-selection is based on sorting paths according to the path metrics in (10.46).

Figure 10.14 shows the tree fanout for $L = 2$ using the earlier example. At frozen poisitions, no tree expansion occurs. With SC decoding, the decision at index 3 is $\widehat{u}_3 = 1$, but now both choices $\widehat{u}_3 = 0$ and $\widehat{u}_3 = 1$ are kept in contention. No further fanout occurs at index 4, but at index 5, the two paths to that point spawn four paths, from which two are selected. Paths eliminated are denoted with an X. This repeats until level 7, where the two best paths are indicated with their metrics. With the same channel data as employed earlier for our SC decoding example, it is found that the best path among the two surviving paths is $\widehat{\mathbf{u}} = (\mathbf{0000}0000)$, with metric 4.18, better than that of $\widehat{\mathbf{u}} = (\mathbf{0001}\mathbf{0}110)$ which was the SC decoder selection, having final metric 3.92. The all-zeros codeword in fact is the best among all 16 possible polar codewords.

## CRC-Aided SCL

Now we elaborate on the CA-SCL modification introduced in [5]. The encoding procedure applies a systematic CRC code that appends $l_C$ bits to the $k$ message bits, prior to polar encoding, according to some CRC polynomial with degree $l_C$. For fair comparison, the overall code rate is maintained at $k/n$, necessitating that the polar code has $k_{\text{in}} = k + l_C$ positions as non-frozen, so its code rate is increased slightly, and the inner (polar) code's performance is thereby slightly degraded under SC decoding, but list decoding can overcome this. The CRC parity bits are normally installed in the high-indexed positions of the polar code input vector, and would be the last to be produced by the inner decoder.

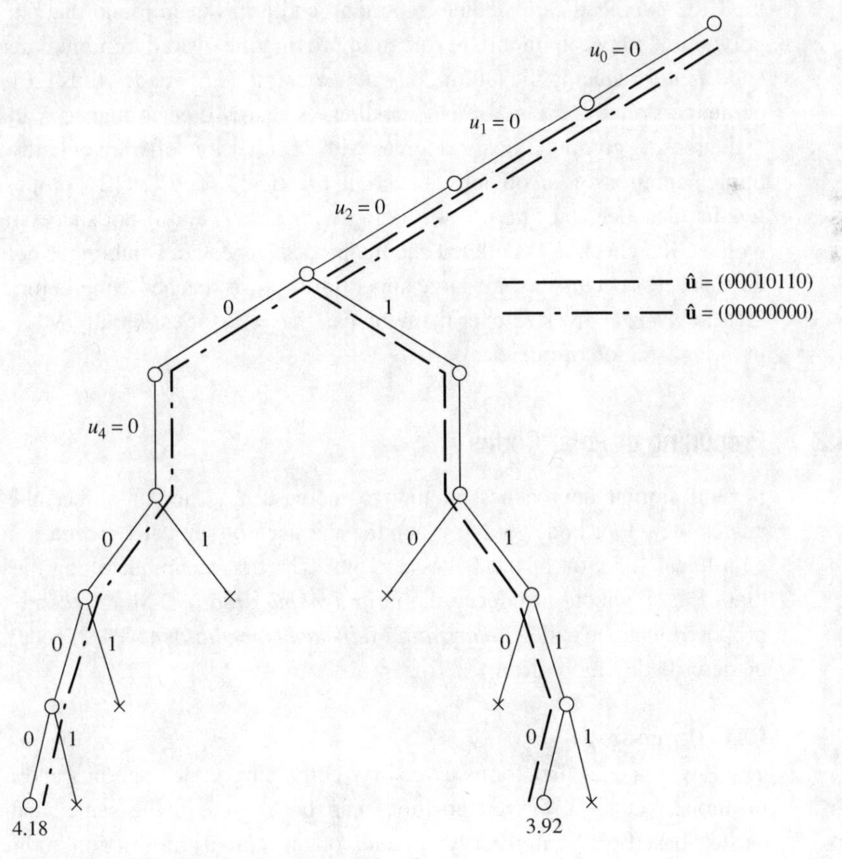


**Figure 10.14** SCL decoding tree, $n = 8, L = 2$.

To decode, we apply the list decoding mechanism, and for each path on the final list, test whether the CRC check is satisfied. (With a suitably long CRC code, the probability a random list member passes is negligibly small, however, and so if the correct vector is on the list, it is likely to be the only one passing the CRC check; if not, metric comparison can make the final decision on codeword.)

Two error mechanisms now exist: (1) the correct vector is not among the list of $L$ survivors (a *type I error*), and (2) the correct vector is on the list, but some other vector passes the CRC check and has overall higher likelihood than the correct codeword (a *type II error*). Notice that an ML decoder would also err in this second situation.

The codeword error probability can be written in terms of these event probabilities as

$$P_{cw} = P_I + (1 - P_I)P_{II}.$$

Depending on the code design, list size, and SNR, either term on the right may dominate, but as $L$ becomes large, $P_I$ diminishes toward 0 and $P_{cw} \to P_{II}$, and $P_{II}$ in turn approaches the maximum likelihood decoder error probability of the *concatenated code* as $L$ grows.

Tal and Vardy did not focus on choice of CRC code, but for shorter blocklengths, the CRC overhead can become important, and it has been found that choice of CRC polynomial plays an important role in improving the overall minimum distance of the code and/or reducing the multiplicity of nearest-neighbor codewords [20]. These code parameters are important for list decoding, as near-ML performance is attainable.

Increasing computational resources with $L$, either for software or hardware decoder implementations, is an obvious concern in practice. Li and Tse [21] proposed an *adaptive* list decoder that first tries decoding with $L = 1$, and if not successful as judged by the CRC check, $L$ is doubled and the process repeated. Doubling is continued until CRC success occurs or some large upper limit $L_{\max}$ is reached. Simulation results show that the *average* list size over many blocks is $L \simeq 2$ for essentially ML performance, in the $P_{cw}$ region of interest.

## 10.4.2 Precoding of Polar Codes

Several similar approaches attempt to increase $d_{\min}$ and/or reduce the number of nearest-neighbor codewords by making the usually statically frozen input positions be a linear function of the $k$ message bits. The first technique along these lines was that of [22], which introduced the term *dynamic freezing*. More recently, Arikan [7] proposed what he called *polarization-adjusted convolutional* (PAC) codes, which will be described in more detail.

### PAC Precoding

The PAC encoder first forms a vector $\mathbf{v}$ containing $k$ message bits and $n - k$ frozen positions, set to 0. Frozen positions can be chosen in the same manner as seen earlier, based on Bhattacharyya bounds or on mutual information, although Arikan preferred the choice based on maximizing minimum row weight, which he called RM rate profiling. In any case, this vector is convolved with a single impulse response $\mathbf{h} = (h_0, h_1, \ldots, h_\nu)$, with the convolution terminated after $n$ time steps. This new binary vector is denoted $\mathbf{u}$, and is also an $n$-tuple. The convolution is represented by

$$u_i = \sum_{j=0}^{\nu} h_j v_{i-j}, \; i = 0, 1, \ldots, n-1 \quad \text{(mod-2 addition).} \tag{10.47}$$

Note the convolutional encoder here does not add redundancy, so precoding has rate 1.

Equivalently we have

$$\mathbf{u} = \mathbf{vT}, \tag{10.48}$$

where $\mathbf{T}$ is the linear transformation defined by the upper-triangular matrix

$$\mathbf{T} = \begin{bmatrix} h_0 & h_1 & \cdots & h_\nu & 0 & 0 & 0 & 0 \\ 0 & h_0 & h_1 & \cdots & h_\nu & 0 & 0 & 0 \\ 0 & 0 & h_0 & h_1 & \cdots & h_\nu & 0 & 0 \\ \cdots & \cdots & \cdots & \cdots & \cdots & \cdots & \cdots & \cdots \\ 0 & 0 & 0 & 0 & 0 & 0 & h_0 & h_1 \\ 0 & 0 & 0 & 0 & 0 & 0 & 0 & h_0 \end{bmatrix}.$$

The vector $\mathbf{u}$ is then presented as input to the standard polar transform $\mathbf{G}_n$, and the transmitted codeword is

$$\mathbf{x} = \mathbf{vTG}_n,$$

representing a new linear mapping from message $k$-tuples to codeword $n$-tuples. None of these input bits to the polar transform are frozen, however, as in standard polar coding; the frozen positions reside in the original $\mathbf{v}$ vector. Using the polar coding mindset, we could say that formerly frozen bits are now dynamically determined by the message $\mathbf{v}$, that is, they are dynamically frozen.

The hope is that the minimum distance of this new concatenated mapping will increase over that of the stand alone polar code, or at least the nearest-neighbor multiplicity will drop substantially, while the attractive encoding and decoding properties of polar coding are essentially preserved. List decoding can exploit these features.

This modification of polar codes is illustrated for $n = 64$, $R = 1/2$ below. First, Figure 10.15 shows the $32 \times 64$ binary matrix of the polar code designed for a BI-AWGN channel with $E_b/N_0 = 2.5$ dB. This matrix comprises 32 rows of $\mathbf{G}_{64}$ depicted earlier. The minimum Hamming distance of this code is 8. Using $\mathbf{h} = (1011011)$ as an impulse response, from [7] we obtain the modified generator matrix shown in Figure 10.16. It appears that the code formed by this matrix might have larger distance, but in fact it remains at 8; the number of weight-8 words, however, is much smaller for the PAC construction.

### Decoding PAC Codes

The set of possible codewords in a PAC code can be represented by an irregular code tree [7]. For levels of the tree corresponding to frozen $v_i$, no branching occurs; for non-frozen levels the tree label of the $R = 1$ code is given by the convolution operation above.

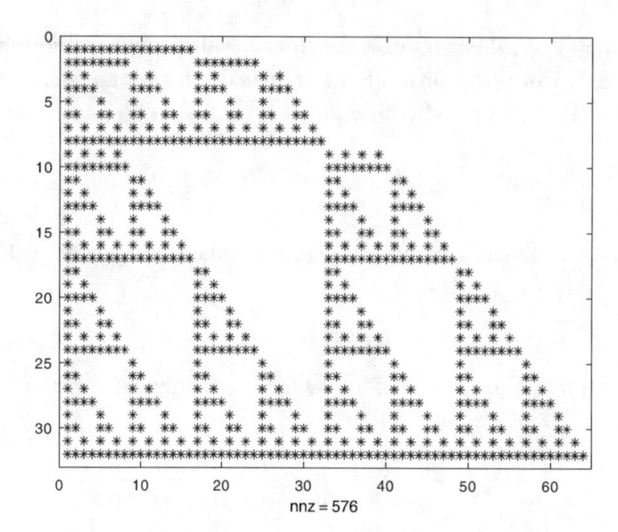


**Figure 10.15** Generator matrix for (64, 32) polar code.

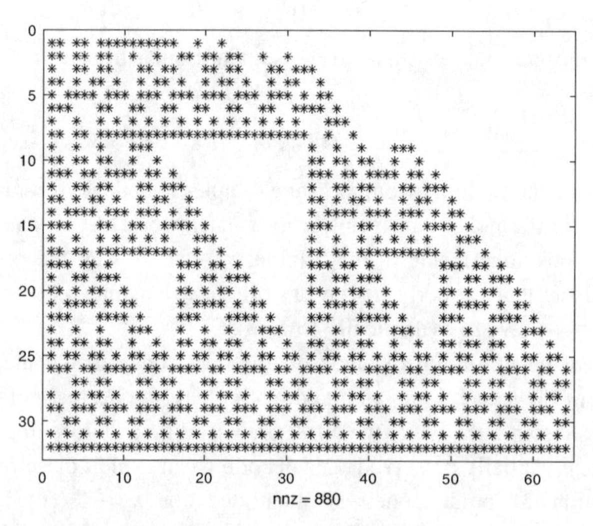


**Figure 10.16** Generator matrix for (64, 32) PAC code.

Decoding of this tree code by sequential decoding (a modified Fano algorithm) was suggested by Arikan as a means of approaching ML performance for the concatenated code. A list decoding alternative was described in [23], and for Arikan's same (128, 64) PAC code construction, a list size $L = 128$ gave essentially equal performance to sequential decoding. Both are quite close to the dispersion approximation for performance of finite-length code at this length and rate, as shown in Section 10.5.

The list decoder can be implemented similarly to a list decoder for normal polar codes, but must incorporate dynamic computation of what would have been frozen positions. As the tree search proceeds, $L$ paths are kept alive at each level of the decision tree, and shift-register structures contain the last $\nu$ bits of the inner decoder so that the convolution can be applied to these decisions. A detailed algorithm is presented in [23].

### CRC Precoding

Whereas PAC codes do not add redundancy to the input of the polar code, CRC outer codes do, by appending $l_C$ bits to the message and designing the polar code to accept this larger input. CRC coding of polar codes can also be viewed as a form of dynamic freezing, where $l_C$ formerly frozen positions are now linear functions of the message. In this case, however, decoding merely checks the integrity of the decoded message, and does not need to incorporate dynamic freezing into the decoding.

## 10.5 Performance Results

In this section, codeword error probability versus SNR on the BI-AWGN channel is presented for a range of polar code designs, with and without CRC-aided list decoding. We also briefly show the performance of PAC codes.

The results shown were obtained by traditional Monte Carlo simulation of decoding performance. Typically, for each SNR data point, the simulation was terminated when 200 000 codewords were decoded, or 100 block errors were encountered. Each block encoded a random $k$-bit message and independent Gaussian noise was added. To highlight the effect of list size, the random number generator producing messages and additive noise was reseeded to the same value at the beginning of each SNR step. Unsurprisingly, there is some simulation "unsmoothness" in the curves associated with Bernoulli trials, but the important trends are quite evident.

Our focus is on codeword error probability, denoted by $P_{cw}$, though data was compiled for bit-error probability, probability of "list misses" (type I errors), and probability that the correct vector is on the list but does not have the greatest metric (type II errors). When useful, these will be drawn upon in interpreting results.

### 10.5.1 List Decoding Performance without CRC

We begin with $n = 128, R = 1/2$ stand-alone polar coding designed at an SNR ($E_b/N_0$) of 2.5 dB, with list decoding (SC decoding corresponds with $L = 1$). As noted earlier, two types of error may occur with list decoding. Figure 10.17 shows the probability that the correct path is not on the list, $P_I$, for $L \in \{1, 4, 16, 64\}$. Clearly the miss probability diminishes with increasing list size, but the codeword error probability is eventually dominated by type II events, that is, some path on the list other than the correct path has the best metric. These would be error situations for an ML decoder as well.

In Figure 10.18, block-error probability is shown, and though increasing list size helps some, performance quickly saturates, with no benefit in using $L > 4$; performance is limited by occurrences where the transmitted codeword does not have the best metric. The minimum Hamming distance of the code is 8, with nearest-neighbor

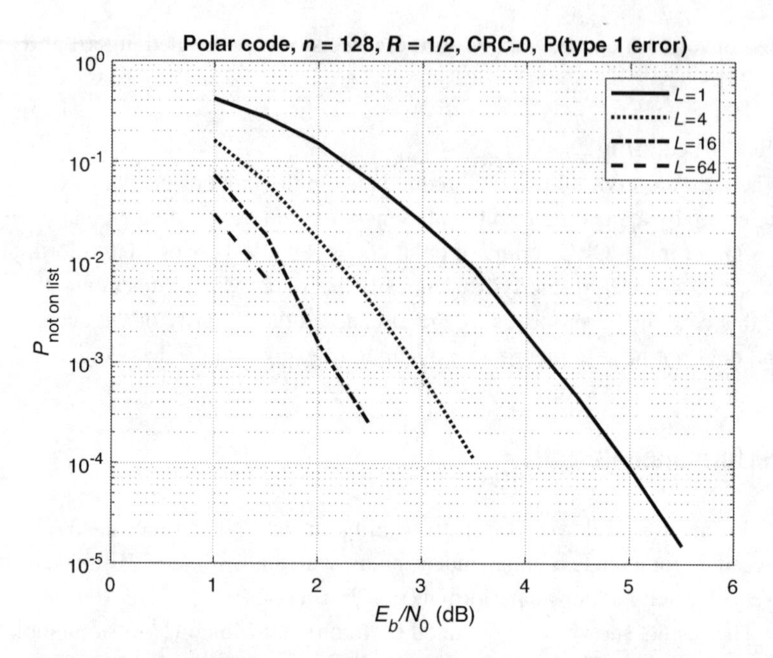


**Figure 10.17** Type I event error probability, polar code, $n = 128, R = 1/2$ list decoding.

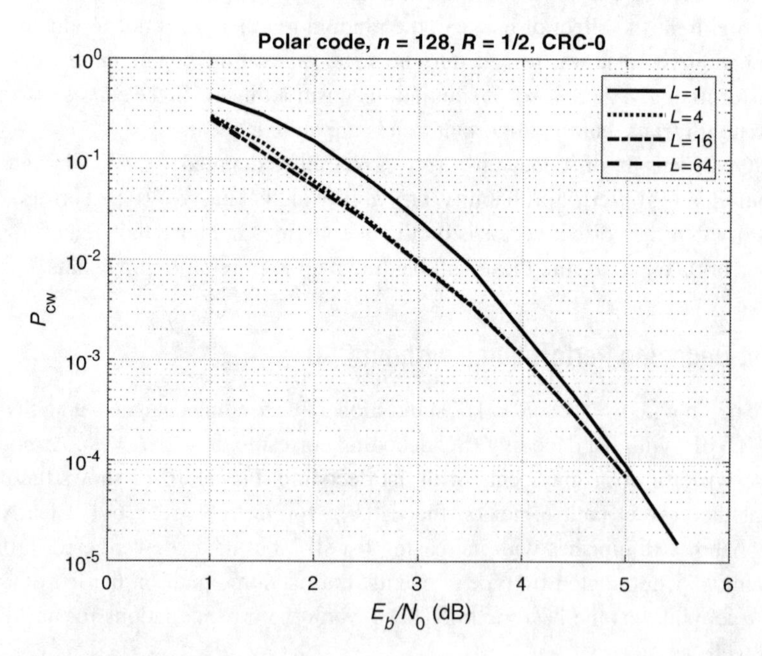


**Figure 10.18** Codeword error probability, polar code, $n = 128, R = 1/2$ list decoding with no CRC.

multiplicity 304, and the first term of the union bound on error probability for an ML decoder is [24]

$$P_{e,ML} \simeq 304 \cdot Q\left[\sqrt{(2E_b/N_0)(8/2)}\right].$$

Calculations show that is this matches simulated performance very closely as SNR increases.

Figure 10.19 illustrates similar behavior for a code with the same blocklength and $R = 3/4$. The design SNR is now 3.5 dB to adjust for the higher code rate. The curves shift to the right, as expected with higher-rate coding. Again, performance is only marginally improved with list decoding alone.

For additional comparison, Figure 10.20 shows the performance for a RM code with rate 1/2 and blocklength 128. Recall that this design merely selects a different subset of rows from $\mathbf{G}_{128}$, and in so doing, the minimum distance grows to 16. Despite this, performance under SC decoding ($L = 1$) is inferior to the polar design, as the polar design is "optimized" for SC processing. However, note that the performance gain with list decoding is more impressive, again approaching that of ML decoding for this higher-distance code.

Next, we present results for longer $R = 1/2$ polar codes, with $n = 1024$, still with no CRC (Figure 10.21). Now, performance curves are steeper and migrate toward the Shannon capacity limit, both expected with longer blocklength. But the benefit of larger list size is now more pronounced, especially at the lowest SNR. The $P_{cw}$ curve

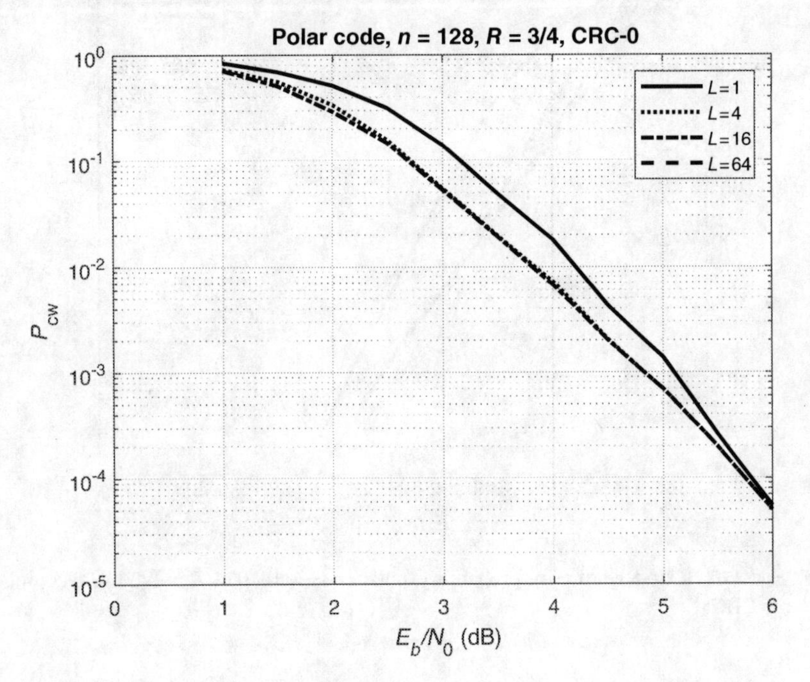


**Figure 10.19** Codeword error probability, polar code, $n = 128, R = 3/4$ list decoding with no CRC.

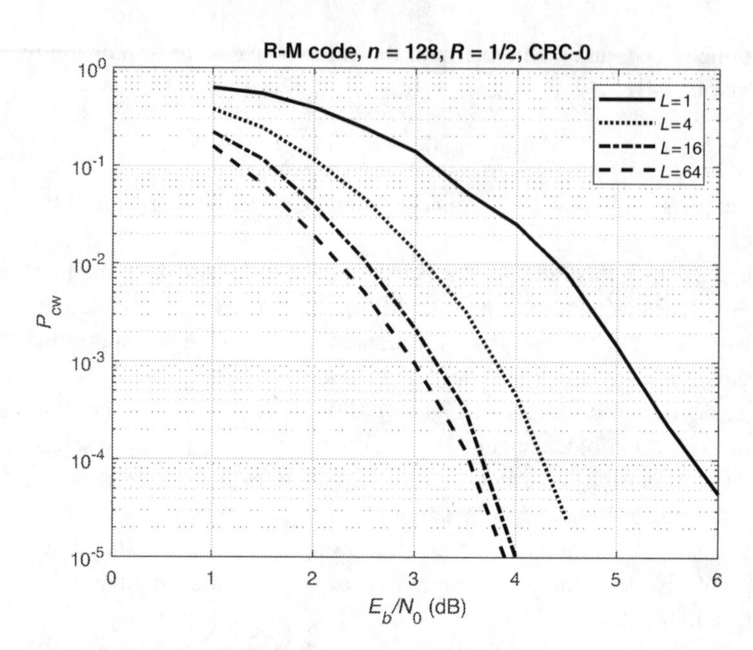


**Figure 10.20** Codeword error probability, RM code, $n = 128, R = 1/2$ list decoding with no CRC.

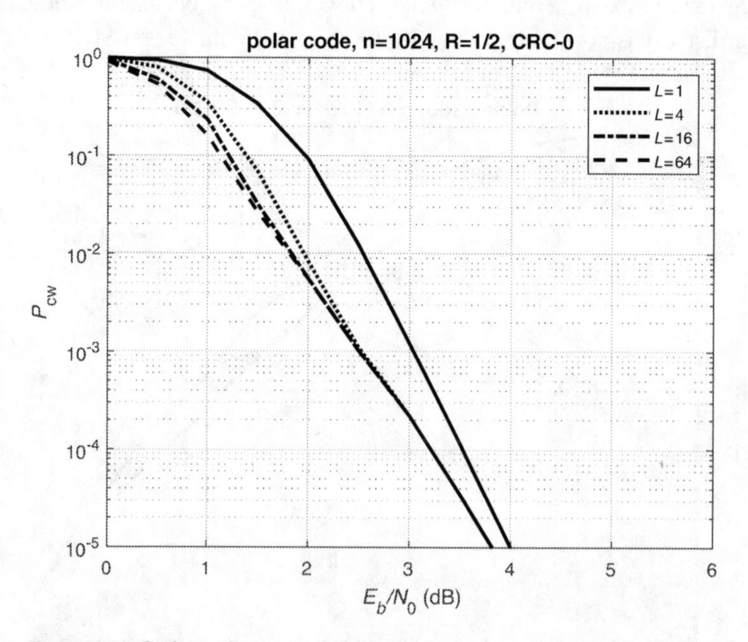


**Figure 10.21** Codeword error probability, polar code, $n = 1024, R = 1/2$ list decoding with no CRC.

is reminiscent of those for turbo codes, having a steep initial section that flares into a slower decay; the latter is determined by the minimum distance limitation for ML decoding.

### 10.5.2 Performance with CRC

List decoding with an outer CRC code has been shown to give impressive gains [5], at least for longer blocklength. To fairly accommodate the CRC overhead, we design the inner polar code to have higher code rate by a factor $(k + l_c)/k$, causing a small loss in performance for SC decoding, in exchange for the ability to find the correct codeword if it is on the output of a list decoder. Despite the greater inner-code rate, the outer precoding can actually increase the minimum distance of the overall code, as seen below, and list decoding can exploit this.

For choice of 8-bit CRC and $n = 128, R = 1/2$, Niu *et al.* [20, 25] find the best CRC polynomial is $D^8 + D^5 + D^4 + D + 1$, for which $d_{\min} = 12$. Simulation results are shown in Figure 10.22. With list size $L = 16$, performance is notably improved over SC decoding, and over $L = 16$ decoding without a CRC code (compare Figure 10.18).

Longer polar codes also benefit dramatically from use of an outer CRC code, as illustrated in Figure 10.23, where $n = 1024$. Here we employ the CCITT 16-bit CRC code, which has a rather small impact on inner code rate. Although not shown in the figure, the polar code with outer CRC and list decoding performs within about 0.6 dB of the normal approximation (see Chapter 1) at block-error probability $10^{-3}$.

### 10.5.3 PAC Code Performance

To illustrate performance with the PAC precoding idea, the list decoding results of [23] are presented for $R = 1/2$, $n = 128$ on the BI-AWGN channel from Figure 10.24.

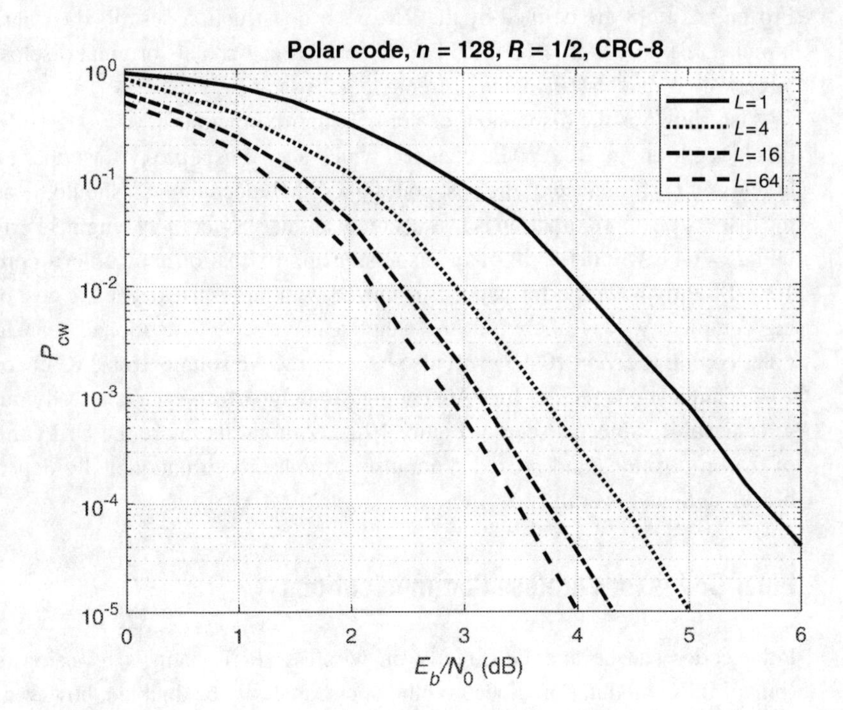


**Figure 10.22** Codeword error probability, polar code, $n = 128, R = 1/2$ list decoding with CRC-8.

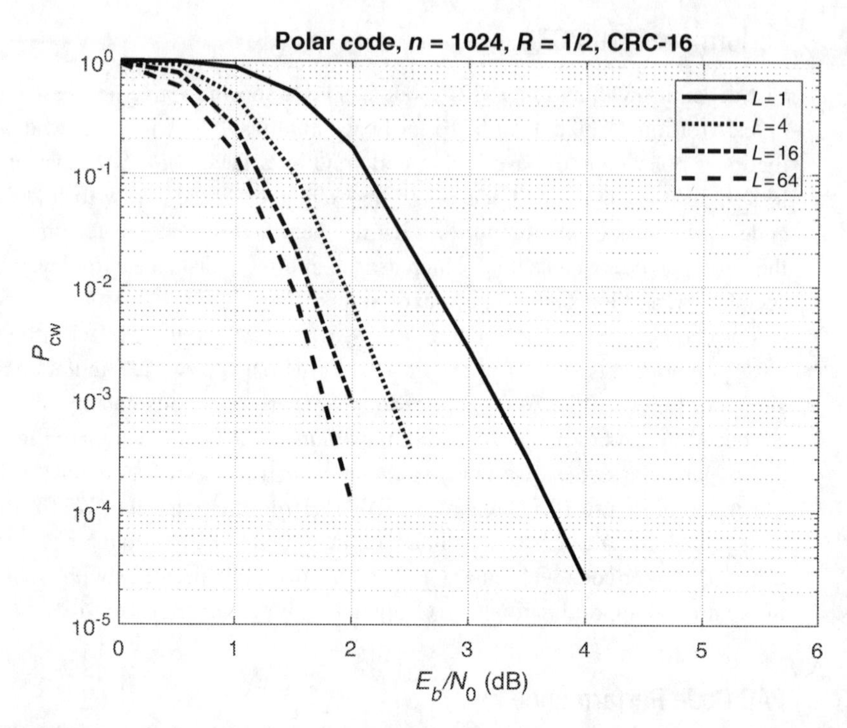


**Figure 10.23** Codeword error probability, polar code, $n = 1024, R = 1/2$ list decoding with CRC-16.

Frozen positions are defined by the RM-polar construction described earlier, and the impulse response is $\mathbf{h} = (1011011)$ as also used in Arikan's original disclosure. List sizes 1, 4, 16, and 64 are again presented.

Also shown is the dispersion, or normal, approximation for $R = 1/2$ codes of this blocklength on the BI-AWGN channel, which serves as a proxy for upper and lower bounds on performance of the best codes under ML decoding. (It should be noted that the dispersion approximation is less accurate at this shorter blocklength.) Performance with $L = 64$ is within 0.2 dB of the dispersion approximation at block-error probability $10^{-3}$. Slightly better performance is achieved with larger list size at the cost of decoding complexity. It is likely that the adaptive list decoder mentioned for CRC-aided polar coding (Section 10.4.1) will also be an attractive solution for PAC codes.

Open questions remain for PAC coding, though performance is already quite close to achievable limits, as seen in Figure 10.24. Among these are the best combination of frozen positions and precoder impulse response, both undoubtedly dependent on blocklength and overall code rate.

## 10.6 Polar Codes for Wireless Communications

Polar codes can be applied directly on wireless (RF) channels by using antipodal (binary PSK) modulation. Spectral efficiency can easily be doubled, however, without energy penalty, by using QPSK to modulate two code bits per modulator symbol. (This

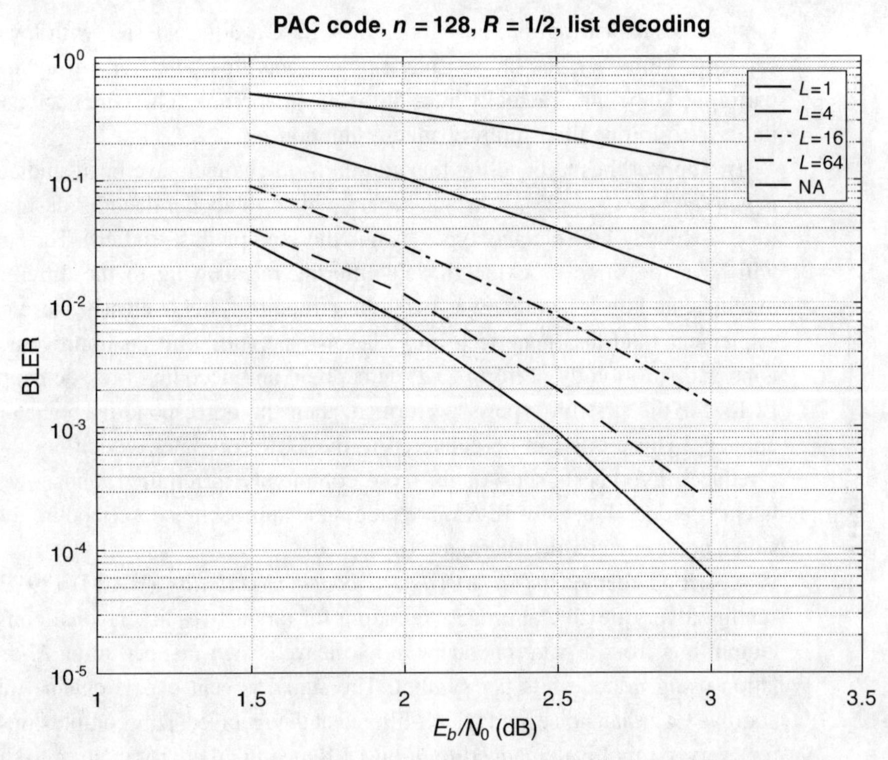


**Figure 10.24** Codeword error probability, PAC code, $n = 128, R = 1/2$, list decoding. Data taken from [23]. Note expanded SNR scale.

is the default modulation for control plane messaging in the 5G NR standard.) Many applications seek even better spectral efficiency, which for the Gaussian channel model can be obtained by modulating a higher-order QAM constellation, say 16-QAM or 64-QAM, to increase the minimum Euclidean distance between modulated sequences. Trellis-coded modulation [26] represents this approach. With increase in spectral efficiency in bits per second per hertz, the required $E_b/N_0$ will increase, as dictated by Shannon theory.

### 10.6.1 Polar-Coded QAM

On the surface, one can just take polar code bits, $m = \log_2 M$ at a time, where $M$ is the constellation size, and map these bits onto a constellation symbol according to some mapping rule, typically using Gray mapping. (This, however, does not guarantee good Euclidean distance between valid sequences.) Then at the decoder, bit LLRs can be produced using a log-MAP rule, or a simplified approximation to it, and these bit LLRs are sent to a traditional polar decoder.

However, the resulting binary channels seen between a code bit $x_i$ and its corresponding LLR $L_{x_i}$ vary with bit position in the $m$-bit symbol. For example, with 16-QAM and standard Gray mapping, the first two bits, of four, Gray map the real constellation value and the other two define the imaginary value. Bits 1 and 3 have the same bit LLR distribution, but bits 2 and 4 have a different one, with lower channel capacity. Moreover, the chosen mapping impacts this behavior, for example, a set-partitioned labeling is actually inferior to Gray mapping. Achieving good performance implies exploiting this "multi-channel" situation.

Two approaches to handling this non-uniform channel have been studied. The first is *multilevel coding* [27], in which separate binary encoder/decoder designs are fashioned for each of the channel types (two in the example just studied). The code designs will have the same blocklength, but differing rates owing to the different channel qualities, and each has its own frozen bit structure. This is clearly more complex to implement (and less generic to modulation type, etc). But an improvement on this approach is to jointly perform LLR demapping and decoding (i.e., demapping of the LLRs for the first bit type is performed, then the corresponding polar decoding is applied). These bits can be delivered to the LLR computer, which now works with a reduced hypothesis set. Again, more complexity is required, and now additional latency of decoding as well. A more pragmatic approach is described in [28], where a single polar encoder/decoder pair is designed.

A second method is *bit-interleaved coded modulation* (BICM) [29, 30], representing a very flexible approach to coding for larger two-dimensional constellations. Output bits from a polar encoder are interleaved, then mapped to an $M$-ary constellation using $m$ code bits per symbol. The signal is sent over a channel modeled as additive Gaussian noise, and match-filtered at the receiver. This complex measurement for every symbol is demapped to $m$-bit LLRs using an approximation to the bit LLR computation, then de-interleaved prior to standard polar decoding. De-interleaving turns the $m$-bit "multi-channel" with differing LLR distributions into a single mixed binary-input channel, whose capacity is surprisingly close, under Gray mapping, to the capacity of the standard QAM Gaussian channel [30]. Moreover, BICM is known to excel over fading channels, expressed as an increase in code diversity. The idea is very flexible, and easily accommodates all of the polar coding variations discussed earlier.

Figure 10.25 presents codeword error probability results for $R = 3/4$ coded 16-QAM, with polar blocklength $n = 1024$ and without CRC or other precoding. Four bits per modulator symbol implies 256 modulator symbols per codeword. Demapping uses the so-called max-log approximation. As expected, the performance benefits from list decoding, but is still well short of the channel capacity limit for 3 bits/symbol. This can be attributed to the relatively small blocklength in two-dimensional symbols.

### 10.6.2 Polar Codes in the 5G Standard

The vision of 5G wireless communication gives priority to low latency and machine-to-machine communication, and the attractiveness of polar codes has seen their adoption for encoding of control-plane information in both uplink and downlink directions. The desired flexibility in transmission rate and blocklength has introduced some simplifying design features that are found in [3], where many more details are found.

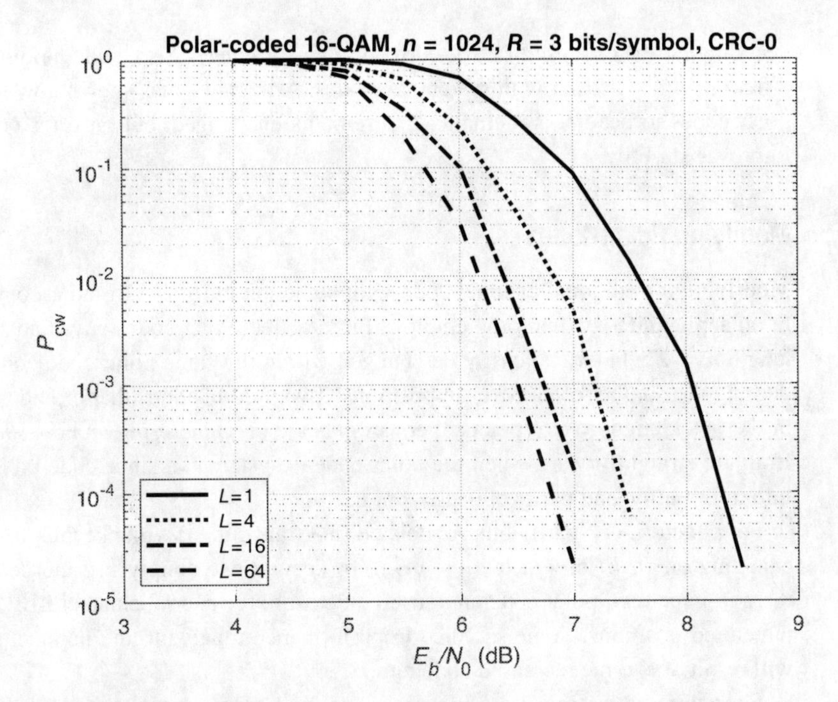


**Figure 10.25** Polar-coded 16-QAM, $n = 1024$, $R = 3$ bits/two-dimensional symbol.

The maximum blocklength defined is $n = 1024$, maintaining low latency. A range of code rates and shorter blocklengths is also defined. For blocklength 1024, a reliability sequence is defined that orders the 1024 positions of the polar transform in increasing reliability for the bit channels. For shorter blocklengths, subsets of this same reliability sequence are selected whose size is 512, 256, ..., 32. This avoids storage of many specific coding tables. The reliability sequence was chosen after extensive simulations for different coding needs; recall that the polar code design lacks universality, even for a fixed $(n, k)$. Other features described in this chapter included use of a CRC outer code, and list decoding with $L = 8$. (As typical in standards, the decoding aspect is not tightly specified, however.)

Other design aspects that provide flexibility and performance enhancements include interleaving of message and codword bits, and rate-matching to adjust to channel conditions. Readers are referred to [3] for further description.

## 10.7 Topics for Further Reading

### 10.7.1 Systematic Encoding

As described in this chapter, polar codes are non-systematic, that is, the $k$-bit message to be encoded does not appear explicitly in the codeword $\mathbf{x}$. For some applications, systematic encoding is useful, for example, the ability to extract the message without decoding, at the expense of error-correcting performance. Also, code modification to different $(n, k)$ is easier with systematic codes. Arikan [11] has shown how to modify polar codes to become systematic. $P_{cw}$ remains unchanged, but bit error probability improves slightly.

### 10.7.2 Modifying Polar Codes

Modifying a traditional channel code with some natural $(n, k)$ to more conveniently fit an application is commonly employed in practice. Polar codes have natural blocklength $n = 2^m$, but $k$ can easily be changed within the same polar code construct by altering the number of non-frozen positions. Two modifications that provide a change in the blocklength to suit practical constraints are (1) *puncturing* and (2) *shortening*. Both are straightforward when the polar code is systematic, since code bits separate into information and parity classes.

For puncturing, $p$ parity bits are deleted (not transmitted) while keeping $k$ constant, changing an $(n, k)$ design to $(n - p, k)$. To decode, we simply use the decoder for the non-punctured polar code, and insert an erasure (or neutral channel LLR) into the punctured positions. Some sacrifice in performance relative to the unpunctured case will result due to the increased code rate.

Shortening is the process of deleting $s$ of the $k$ message bits and not transmitting these. Encoding can be performed by freezing an additional $s$ message bits to 0 in a systematic code, then deleting these positions from the produced codeword. The original $(n, k)$ polar code becomes an $(n-s, k-s)$ code, still with $n-k$ frozen positions. Decoding again can employ the decoder for the mother code, inserting a large LLR in favor of 0 in the omitted message positions.

### 10.7.3 "Fast" decoding

Both SC and SCL decoding are inherently serial processes in exploring a tree of hypotheses, and not directly amenable to parallel processing. (The $L$ decoders in SCL can run in parallel, however, assuming access to common memory.) This can imply longer decoding latency than desired.

So-called fast decoders of polar codes exploit the fact that certain subtrees of the decoding tree correspond to:

(1) rate-0 subtrees, where all $u_i$ are frozen;
(2) rate-1 subtrees, where all leaf nodes are message bits;
(3) repetition code trees, where only the right most leaf node is a message bit;
(4) single-parity-check code trees, where the left most leaf is frozen and the rest are message bits.

Decoding algorithms tuned to these simple code structures allow a faster processing of a given subtree type, reducing decoding time. Sarkis *et al.* [31] develop this idea. Subsequently, Condo *et al.* [32] extended the subtree types that can be exploited. Typical speed-ups over a range of rates and blocklengths are in the range of 25%. It should be noted that exploiting such structure for a specific code makes the decoder less flexible to handle other polar codes.

## Problems

**10.1** Suppose the interstage permutations in Figures 10.4 or 10.6 are replaced by identity matrices in each case. Show that this reverts to just sending single unencoded bits over the raw physical channel, and thus offers no gain.

**10.2** Consider a $2 \times 2$ combined channel as in Figure 10.1, and suppose the channel outputs on a BI-AWGN channel are $(y_0, y_1) = (0.8, -0.2)$, where $y_i = \pm 1 + n_i$ and the noise variance is $\sigma^2 = 0.5$. First form the LLRs $L_{x_0}$ and $L_{x_1}$, then compute the LLRs for input bits $u_0$ and $u_1$ using SC decoding.

**10.3**

(a) List the 16 codewords for an $(8, 4)$ polar code defined by Figure 10.7, with positions 0, 1, 2, 4 frozen to 0. This code is equivalent to an $(8, 4)$ extended Hamming code, with $d_{\min} = 4$.
(b) With the above frozen positions and the message vector (1101), show algebraically that the codeword produced is $\mathbf{x} = (11000011)$.

**10.4** Develop an equivalent model to that of Example 10.1 for a BEC, where the physical channel model is a BSC with crossover probability $\epsilon < 1/2$. Using the laws of probability, determine $P(\mathbf{y} \mid u_0)$, $P(\mathbf{y}, u_0 \mid u_1)$, and so on, and compute the mutual information for the two bit channels. Confirm that the synthetic channel $W_2^{(0)}$ has smaller mutual information than $W_2^{(1)}$.

**10.5** The LLR blending formulas of (10.12) and (10.14) are generic to any channel model, provided the correct input LLRs $L_{x_0}$ and $L_{x_1}$ are computed. Derive expressions for the BSC LLR in terms of Hamming distance $d(x_i, y_i)$ and $\epsilon$. Repeat for the BEC with erasure parameter $p$.

**10.6** Consider the four-point polar transform $\mathbf{G}_4 = \mathbf{F}^{\otimes 2}$. Write out the four logical equations relating $x_i$ to the inputs $u_i$. With SC decoding, explain intuitively why the first bit channel, which attempts to infer $u_0$ from $(y_0, y_1, y_2, y_3)$, has poorest performance, while the last bit channel that seeks to recover $u_3$, given the same measurements and $(u_0, u_1, u_2)$, is the best.

**10.7** Verify the identity $(\mathbf{I}_{n/2} \otimes \mathbf{F})\mathbf{R}_n = \mathbf{R}_n(\mathbf{F} \otimes \mathbf{I}_{n/2})$. Use the facts that $\mathbf{R}^{-1} = \mathbf{R}^{\mathrm{T}}$ and the identity $\mathbf{B} \otimes \mathbf{A} = \mathbf{R}_n^{\mathrm{T}}(\mathbf{A} \otimes \mathbf{B})\mathbf{R}_n$. This leads to the algebraic form for the polar transform shown in Figure 10.7.

**10.8** Draw the polar transform graph corresponding to the recursive form in (10.22) for $n = 8$.

**10.9**

(a) Verify the matrix descriptions of each stage of the diagram in Figure 10.7, then show that the product of these $8 \times 8$ matrices is indeed $\mathbf{F}^{\otimes 3}$.

(b) Show that swapping the first and last matrices in this diagram gives exactly the same mapping. This diagram is an alternate representation seen in some presentations of polar codes.

(c) To the first graph, add the bit-reversal block $\mathbf{B}_8$ at the left, forming $\mathbf{G}_8 = \mathbf{B}_8\mathbf{F}^{\otimes 3}$, and encode the vector $\mathbf{u} = (00010111)$ from tracing results in the diagram. Verify that the algebraic route $\mathbf{x} = \mathbf{u}\mathbf{G}_8$ produces the same result.

**10.10** Using the definition of the Bhattacharyya bound in (10.30), verify the claims that $Z = p$ on the BEC, that $Z = \sqrt{4\epsilon(1-\epsilon)}$ for the BSC, and that $Z = e^{-E_s/N_0}$ for the BI-AWGN model. (Completing the square is helpful in the latter case.)

**10.11** Use the Matlab script below to propagate the Bhattacharyya bounds for bit channels, applying for different channel models and quality. The program steps through increasing blocklengths $n = 2^m$ and displays the histogram for bit-channel Bhattacharyya bounds.

```
%Bhattacharyya bounding for polar codes
e=.11; p=.5; esn0=10^(.189)/2);  %all channels have capacity 0.5
Z=sqrt(4*e*(1-e))  %or Z=1-p for BEC or Z=exp(-esn0) for BI-AWGN
x=[.025:.05:.975];  %histogram bin centers
for m=1:12
  p=[2*p-p.^2,p.^2];
  hist(p,x,1)
  pause(1)
end
```

**10.12** Use the bound

$$I(W) \geq \log_2\left(\frac{2}{1+Z(W)}\right) \tag{10.49}$$

and the script of the previous exercise to show that the fraction of bit channels having mutual information near 1 is approximately equal to the channel capacity as blocklength grows, that is, the histogram tends toward having equal probability near 0 and 1.

**10.13**

(a) Use the recursive procedure of Problem 10.11 for blocklength $n = 16$ on a BEC with erasure probability 0.5. Show that the eight indices with smallest Bhattacharyya bound are [9, 5, 3, 14, 13, 11, 7, 15]. These are the rankings for the mapping $\mathbf{F}^{\otimes 4}$ without additional bit reversal. Also recall, addressing is from 0 to 15.

(b) Repeat with erasure probability 0.05 to see if the same message indices emerge.

**10.14** The chosen indices for message bits in the previous small example are not in any natural or bit-reversed order. Decoding is simplified by processing bits in natural sequence, however. Argue that any such reordering does not affect the final block-error probability, and thus decoding order for message positions does not matter.

**10.15** Bhattacharyya bounds on error probability are linked to mutual information bounds through

$$I(W) \geq \log_2\left(\frac{2}{1+Z(W)}\right). \tag{10.50}$$

This holds true for a single binary channel and for virtual bit channels in polar codes, given the set of Bhattacharyya bounds.

(a) For a BEC, plot the exact $I(W) = 1 - p$ as well as the bound given above, versus erasure probability $p \in [0, 0.5]$.
(b) Repeat for a BSC channel as a function of crossover probability $\epsilon$.

**10.16** Perform SC decoding for a $4 \times 4$ polar code, using Figure 10.5(b), with $\mathbf{y} = (0.8, -0.2, 0.5, 1.2)$. List the intermediate LLRs on a copy of the diagram.

**10.17** Polar codes do not have notably good minimum distance versus blocklength $n$, and do not qualify as "good" codes in the asymptotic minimum distance sense, since it can be shown that codes exist with rate 1/2 having minimum distance growing as $d_{\min} \simeq 0.11n$ [14, 24]. To confirm this, the following small table shows minimum distance versus $n$ for polar codes, both for design SNR 2.5 dB and 4 dB on a BI-AWGN channel. Also tabulated is the minimum distance for extended BCH codes of the same (approximate) rate.

| $n$ | $d_{\min}$(2.5 dB) | $d_{\min}$(4 dB) | $d_{\text{eBCH}}$ |
|---|---|---|---|
| 16 | 4 | 4 | 6 |
| 32 | 4 | 8 | 8 |
| 64 | 8 | 8 | 12 ($k = 36$) |
| 128 | 8 | 8 | 22 |
| 256 | 8 | 16 | 38 ($k = 131$) |
| 512 | 16 | 16 | 62 ($k = 259$) |
| 1024 | 16 | 16 | 116 ($k = 513$) |
| 2048 | 16 | 32 | 214 |

(a) Fit an exponential function to the 4 dB polar code data to determine $\beta$ in $d_{\min} \simeq n^{\beta}$.
(b) Observe the much faster growth of minimum distance for corresponding BCH codes, indicating that these binary codes are much better packings of points on an $n$-dimensional sphere. Even so, BCH codes are known to be "asymptotically bad." (They are also much more difficult to decode in ML manner.)

Despite this apparent deficiency, polar codes with SC decoding are capacity-achieving, and this calls into question the emphasis on minimum distance alone.

**10.18**

(a) Consider the $4 \times 4$ polar transform with the first bit frozen to 0, so we have a $(4, 3)$ block code. The generator matrix is the bottom three rows of $\mathbf{F}^{\otimes 2}$. Use this construction to list the eight codewords, and observe that the code is equivalent to a single-parity check code. Each codeword has six neighbors at Hamming distance 2 and one neighbor at distance 4. The optimal (ML) decoding error probability is thus union bounded by

$$P_e \leq 6 \cdot Q\left(\sqrt{\frac{2E_b}{N_0}\frac{6}{4}}\right) + 1 \cdot Q\left(\sqrt{\frac{2E_b}{N_0}\frac{6}{2}}\right).$$

Use the code on the BI-AWGN channel with noise variance 1. Build a small Monte Carlo simulation of *optimal* decoding using the vector LLR to decide the three input bits, and compute the error probability. *Hint*: An optimal decoder for this code does hard decisions on each measurement, and if the parity check fails, the decoder flips the bit with poorest likelihood. Compare with the union bound above.

(b) Repeat for SC decoding to sequentially decode the input bits for the *same* channel observation sequence. Input bit $u_0$ is frozen at 0. Use the standard simple approximation to the $\boxplus$ computation. Compare the results.

**10.19** Explain the difference in the error probability curves for polar and RM-polar designs under SC and SCL decoding with $n = 128$ presented in Section 10.5, referring to the decomposition into type I and type II errors.

**10.20** Suppose a polar decoder has been built for a rate-1/2 code with blocklength $n = 256$ and $L = 8$ that requires $M$ bytes of memory and can execute one decoding block in 100 $\mu$s. Use "big-O" scaling laws to estimate the required memory and execution time for $n = 2048$ and $L = 32$. Assume a serial processor.

**10.21** Compare the LLR update rules of (10.12) and (10.14) by plotting level contours of $L_2^{(0)}$ versus the two-dimensional input $(L_{x_0}, L_{x_1})$ over the range of $(-5, 5)$ for both input variables. The level contours should be similar, supporting the claim that (10.14) is a good approximation to the exact expression.

**10.22** Argue that LLR scaling is not an issue when the "box-plus" approximation is used in decoding. Final LLRs will scale accordingly, but the bit decisions remain unchanged.

**10.23** Verify the table contents of Tables 10.1 and 10.2, by hand or by program.

**10.24**

(a) Create an SC decoder in your favorite language for general $(n, k)$, following the discussion of Section 10.4.

(b) Create a more memory-efficient version that uses about $n$ locations to store the LLR updates, and about $2n$ locations for updating the stored bits in the array $\mathbf{C}$ in the discussion of SC decoding. New auxiliary pointers are likely required.

## References

[1] E. Arikan, "Channel polarization: A method for constructing capacity-achieving codes for symmetric binary-input memoryless channels," *IEEE Transactions on Information Theory*, vol. 55, no. 7, pp. 3051–3073, 2009.

[2] Y. Polyanskiy, H. V. Poor, and S. Verdu, "Channel coding rate in the finite blocklength regime," *IEEE Transactions on Information Theory*, vol. 55, no. 5, pp. 2307–2593, 2010.

[3] V. Bioglio, C. Condo, and I. Land, "Design of polar codes in 5G New Radio," *IEEE Communications Surveys and Tutorials*, vol. 23, no. 1, pp. 29–40, 2021.

[4] T. K. Moon, *Error Correction Coding*, 2nd ed., Chichester, Wiley, 2021.

[5] I. Tal and A. Vardy, "List decoding of polar codes," *IEEE Transactions on Information Theory*, vol. 61, no. 5, pp. 2213–2226, 2015.

[6] A. Balatsoukas-Stimming, M. B. Parizi, and A. Berg, "LLR-based successive cancellation list decoding of polar codes," *IEEE Transactions on Signal Processing*, vol. 63, no. 10, pp. 6562–6582, 2015.

[7] E. Arikan, "From sequential decoding to channel-polarization and back again," arXiv, abs/1908.09594, 2019.

[8] T. M. Cover and J. A. Thomas, *Principles of Information Theory*, 2nd ed., New York, Wiley, 2006.

[9] G. H. Golub and C. F. Van Loan, *Matrix Computations*, 3rd ed., Baltimore, MD, Johns Hopkins University Press, 1996.

[10] Unknown, Kronecker product, Wikimedia Foundation, 2022.

[11] E. Arikan, "Systematic polar coding," *IEEE Communication Letters*, vol. 15, no. 8, pp. 860–862, 2011.

[12] I. S. Reed, "A class of multiple-error-correcting codes and the decoding scheme," *Transactions of the IRE Group on Information Theory*, vol. 4, no. 4, pp. 38–49, 1954.

[13] I. Dumer and K. Shuunov, "Soft-decision decoding of Reed–Muller codes: A simplified algorithm," *IEEE Transactions on Information Theory*, vol. 52, no. 10, pp. 954–963, 2006.

[14] J. MacWilliams and N. J. A. Sloane, *The Theory of Error-Correcting Codes*, Amsterdam, North Holland, 1983.

[15] E. Arikan, "A performance comparison of polar codes and Reed–Muller codes," *IEEE Communications Letters*, vol. 12, no. 6, pp. 447–449, 2008.

[16] B. Li and D. Tse, "RM-polar codes," *IEE Letters*, 2015.

[17] H. N. Ward, "Visible codes," *Archiv der Mathematik*, vol. 54, pp. 307–312, 1990.

[18] S. ten Brink, G. Kramer, and A. Ashikhmin, "Design of low-density parity-check codes for modulation and detection," *IEEE Transactions on Communications*, vol. 52, no. 4, pp. 670–678, 2004.

[19] L. K. Rasmussen, F. Brannstrom, and A. J. Grant, "Convergence analysis and optimal scheduling for multiple concatenated codes," *IEEE Transactions on Information Theory*, vol. 51, no. 9, pp. 3354–3364, 2005.

[20] K. Niu and K. Chen, "CRC-aided decoding of polar codes," *IEEE Communications Letters*, vol. 16, no. 10, pp. 1668–1671, 2012.

[21] B. Li, H. Shen, and D. Tse, "An adaptive successive cancellation list decoding of polar codes," *IEEE Communication Letters*, vol. 51, no. 9, pp. 3354–3364, 2005.

[22] P. Trifonov and V. Miloslavskaya, "Polar subcodes," *IEEE Journal on Selected Areas in Communications*, vol. 34, no. 2, pp. 254–266, 2016.

[23] H. Yao, A. Fazeli, and A. Vardy, "List decoding of Arikan's PAC codes," *Entropy*, vol. 23, June 2021.

[24] S. G. Wilson, *Digital Modulation and Coding*, Upper Saddle River, NJ, Prentice-Hall, 1996.

[25] J. Piao, K. Niu, J. Dai, and C. Dong, "Sphere constraint based enumeration methods to analyze the minimum weight distribution of polar codes," *IEEE Transactions on Vehicular Technology*, vol. 69, no. 10, pp. 11557–11569, 2020.

[26] G. Ungerboeck, "Channel coding with multilevel/phase signals," *IEEE Transactions on Information Theory*, vol. 28, no. 1, pp. 55–67, 1982.

[27] M. Seidl, A. Schenk, C. Stierstorfer, and J. B. Huber, "Polar coded modulation," *IEEE Transactions on Communications*, vol. 61, no. 10, pp. 4108–4119, 2013.

[28] H. Mahdavifar, M. El-Khamy, J. Lee, and I. Kang, "Polar coding for bit-interleaved coded modulation," *IEEE Transactions on Vehicular Technology*, vol. 65, no. 5, pp. 3115–3127, 2016.

[29] X. Li and J. Ritcey, "Bit-interleaved coded modulation with iterative decoding," *IEEE Communications Letters*, vol. 1, no. 6, pp. 169–171, 1997.

[30] G. Caire, G. Taricco, and E. Biglieri, "Bit-interleaved coded modulation," *IEEE Transactions on Information Theory*, vol. 44, no. 3, pp. 927–946, 1998.

[31] G. Sarkis, P. Giard, A. Vardy, C. Thibeault, and W. Gross, "Fast polar decoders: Algorithm and implementation," *IEEE Journal on Selected Areas in Communications*, vol. 32, no. 5, pp. 946–957, 2014.

[32] C. Condo, V. Bioglio, and I. Land, "Generalized fast decoding of polar codes," arXiv, abs/1804.09508v3, 2020.

# 11 Finite-Geometry LDPC Codes

Finite geometries, such as Euclidean and projective geometries, are powerful mathematical tools for constructing error-control codes. In the 1960s and 1970s, finite geometries were successfully used to construct many classes of easily implementable majority-logic decodable codes. In 2000, Kou, Lin, and Fossorier [1–3] showed that finite geometries can also be used to construct LDPC codes that perform well and close to the Shannon theoretical limit with iterative decoding based on belief propagation. These codes are called *finite-geometry* (*FG*)-LDPC codes. FG-LDPC codes form the first class of LDPC codes that are constructed algebraically. Since 2000, there have been many major developments in the construction of LDPC codes based on various structural properties of finite geometries [4–18]. In this chapter, we put together all the major constructions of LDPC codes based on finite geometries under a unified frame. We begin with code constructions based on Euclidean geometries and then go on to discuss those based on projective and partial geometries.

## 11.1 Construction of LDPC Codes Based on Lines of Euclidean Geometries

This section presents a class of cyclic LDPC codes and a class of quasi-cyclic (QC) LDPC codes constructed using lines of Euclidean geometries. Before we present the constructions of these two classes of Euclidean-geometry (EG)-LDPC codes, we recall some fundamental structural properties of a Euclidean geometry that have been discussed in Chapter 2.

Consider the $m$-dimensional Euclidean geometry EG$(m, q)$ over the Galois field GF$(q)$. This geometry consists of $q^m$ points and

$$J \triangleq J_{\text{EG}}(m, 1) = q^{m-1}(q^m - 1)/(q - 1) \tag{11.1}$$

lines. A point in EG$(m, q)$ is simply an $m$-tuple over GF$(q)$. A line in EG$(m, q)$ is simply a one-dimensional subspace or its coset of the vector space $\mathbf{V}$ of all the $m$-tuples over GF$(q)$. Each line consists of $q$ points. Two lines either do not have any point in common or they have one and only one point in common. Let $\mathcal{L}$ be a line in EG$(m, q)$ and $\mathbf{p}$ a point on $\mathcal{L}$. We say that $\mathcal{L}$ passes through the point $\mathbf{p}$. If two lines have a common point $\mathbf{p}$, we say that they intersect at $\mathbf{p}$. For any point $\mathbf{p}$ in EG$(m, q)$, there are

$$g \triangleq g_{\text{EG}}(m, 1, 0) = (q^m - 1)/(q - 1) \tag{11.2}$$

lines intersecting at $\mathbf{p}$ (or passing through $\mathbf{p}$). As described in Chapter 2, the Galois field GF($q^m$) as an extension field of GF($q$) is a realization of EG($m, q$). Let $\alpha$ be a primitive element of GF($q^m$). Then the powers of $\alpha$,

$$\alpha^{-\infty} \triangleq 0, \alpha^0 = 1, \alpha, \alpha^2, \ldots, \alpha^{q^m-2},$$

represent the $q^m$ points of EG($m, q$), where $\alpha^{-\infty}$ represents the origin point of EG($m, q$).

Let EG*($m, q$) be a *subgeometry* of EG($m, q$) obtained by removing the origin $\alpha^{-\infty}$ and all the lines passing through the origin from EG($m, q$). This subgeometry consists of $q^m - 1$ non-origin points and

$$J_0 \triangleq J_{0,\mathrm{EG}}(m, 1) = (q^{m-1} - 1)(q^m - 1)/(q - 1) \tag{11.3}$$

lines not passing through the origin. Let $\mathcal{L}$ be a line in EG*($m, q$) consisting of the points $\alpha^{j_1}, \alpha^{j_2}, \ldots, \alpha^{j_q}$, that is

$$\mathcal{L} = \{\alpha^{j_1}, \alpha^{j_2}, \ldots, \alpha^{j_q}\}.$$

For $0 \le t < q^m - 1$,

$$\alpha^t\mathcal{L} = \{\alpha^{j_1+t}, \alpha^{j_2+t}, \ldots, \alpha^{j_q+t}\} \tag{11.4}$$

is also a line in EG*($m, q$), where the powers of $\alpha$ in $\alpha^t\mathcal{L}$ are taken modulo $q^m - 1$. In Chapter 2, it has been shown that the $q^m - 1$ lines $\mathcal{L}, \alpha\mathcal{L}, \alpha^2\mathcal{L}, \ldots, \alpha^{q^m-2}\mathcal{L}$ are all different. Since $\alpha^{q^m-1} = 1$, $\alpha^{q^m-1}\mathcal{L} = \mathcal{L}$. These $q^m - 1$ lines form a *cyclic class*. The $J_0$ lines in EG*($m, q$) can be partitioned into

$$K_c \triangleq K_{c,\mathrm{EG}}(m,1) = (q^{m-1} - 1)/(q - 1) \tag{11.5}$$

cyclic classes, denoted by $\mathcal{S}_1, \mathcal{S}_2, \ldots, \mathcal{S}_{K_c}$. The above structure of the lines in EG*($m, q$) is called the *cyclic structure*.

For any line $\mathcal{L}$ in EG*($m, q$) not passing through the origin, we define the following ($q^m - 1$)-tuple over GF(2):

$$\mathbf{v}_{\mathcal{L}} = (v_0, v_1, \ldots, v_{q^m-2}), \tag{11.6}$$

whose components correspond to the $q^m - 1$ non-origin points $\alpha^0, \alpha, \ldots, \alpha^{q^m-2}$ in EG*($m, q$), where $v_i = 1$ if $\alpha^i$ is a point on $\mathcal{L}$, otherwise $v_i = 0$. If $\mathcal{L} = \{\alpha^{j_1}, \alpha^{j_2}, \ldots, \alpha^{j_q}\}$, then $v_{j_1} = v_{j_2} = \cdots = v_{j_q} = 1$. The ($q^m - 1$)-tuple $\mathbf{v}_{\mathcal{L}}$ over GF(2) is called the *type-1 incidence vector* of $\mathcal{L}$. The weight of $\mathbf{v}_{\mathcal{L}}$ is $q$. From (11.4), we can readily see that, for $0 \le t < q^m - 1$, the incidence vector $\mathbf{v}_{\alpha^{t+1}\mathcal{L}}$ of the line $\alpha^{t+1}\mathcal{L}$ can be obtained by cyclically shifting all the components of the incidence vector $\mathbf{v}_{\alpha^t\mathcal{L}}$ of the line $\alpha^t\mathcal{L}$ one place to the right. We call $\mathbf{v}_{\alpha^{t+1}\mathcal{L}}$ the *right cyclic shift* of $\mathbf{v}_{\alpha^t\mathcal{L}}$.

### 11.1.1 A Class of Cyclic EG-LDPC Codes

For each cyclic class $\mathcal{S}_i$ of lines in EG*($m, q$) with $1 \le i \le K_c$, we form a $(q^m - 1) \times (q^m - 1)$ matrix $\mathbf{H}_{c,i}$ over GF(2) with the incidence vectors $\mathbf{v}_{\mathcal{L}}, \mathbf{v}_{\alpha\mathcal{L}}, \ldots, \mathbf{v}_{\alpha^{q^m-2}\mathcal{L}}$ of the lines $\mathcal{L}, \alpha\mathcal{L}, \ldots, \alpha^{q^m-2}\mathcal{L}$ in $\mathcal{S}_i$ as rows arranged in cyclic order. Then $\mathbf{H}_{c,i}$ is a

$(q^m - 1) \times (q^m - 1)$ circulant over GF(2) with both column and row weight $q$. For $1 \leq k \leq K_c$, we form the following $k(q^m - 1) \times (q^m - 1)$ matrix over GF(2):

$$\mathbf{H}^{(1)}_{\text{EG},c,k} = \begin{bmatrix} \mathbf{H}_{c,1} \\ \mathbf{H}_{c,2} \\ \vdots \\ \mathbf{H}_{c,k} \end{bmatrix}, \tag{11.7}$$

which has column and row weights $kq$ and $q$, respectively. The subscript "$c$" of $\mathbf{H}^{(1)}_{\text{EG},c,k}$ stands for "cyclic." Since the rows of $\mathbf{H}^{(1)}_{\text{EG},c,k}$ correspond to lines in EG*(EG) (or EG$(m, q)$) and no two lines have more than one point in common, it follows that no two rows (or two columns) in $\mathbf{H}^{(1)}_{\text{EG},c,k}$ have more than one 1-component in common. Hence $\mathbf{H}^{(1)}_{\text{EG},c,k}$ satisfies the RC constraint. The null space of $\mathbf{H}^{(1)}_{\text{EG},c,k}$ gives a cyclic $(kq, q)$-regular LDPC code $\mathcal{C}_{\text{EG},c,k}$ of length $q^m - 1$ with minimum distance at least $kq + 1$, whose Tanner graph has a girth of at least 6. The above construction gives a class of cyclic EG-LDPC codes. Cyclic EG-LDPC codes with $k = K_c$ were actually discovered in the late 1960s [19, 20, 21] and also shown to form a subclass of polynomial codes [22], but were not recognized to form a class of LDPC codes until 2000 [4].

Since $\mathcal{C}_{\text{EG},c,k}$ is cyclic, it is uniquely specified by its generator polynomial $\mathbf{g}(X)$ (see Chapter 3). To find the generator polynomial of $\mathcal{C}_{\text{EG},c,k}$, we first express each row of $\mathbf{H}^{(1)}_{\text{EG},c,k}$ as a polynomial of degree $q^m - 2$ or less with the leftmost entry of the row as the constant term and the rightmost entry as the coefficient of $X^{q^m-2}$. Let $\mathbf{h}(X)$ be the *greatest common divisor* of the row polynomials of $\mathbf{H}^{(1)}_{\text{EG},c,k}$. Let $\mathbf{h}^*(X)$ be the *reciprocal polynomial* of $\mathbf{h}(X)$. Then

$$\mathbf{g}(X) = (X^{q^m-1} - 1)/\mathbf{h}^*(X). \tag{11.8}$$

For $k = K_c$, the roots of $\mathbf{g}(X)$ can be completely determined [4, 22, 23–21]. Of particular interest is the cyclic code constructed from the subgeometry EG*$(2, q)$ of the two-dimensional Euclidean geometry EG$(2, q)$. The subgeometry EG*$(2, q)$ of EG$(2, q)$ consists of $q^2 - 1$ non-origin points and $q^2 - 1$ lines not passing through the origin. The $q^2 - 1$ lines of EG*$(2, q)$ form a single cyclic class $\mathcal{S}$. Using the incidence vectors of the lines in this single cyclic class $\mathcal{S}$, we can form a $(q^2 - 1) \times (q^2 - 1)$ circulant matrix $\mathbf{H}^{(1)}_{\text{EG},c,1}$ over GF(2) with both column and row weight $q$. The null space of $\mathbf{H}^{(1)}_{\text{EG},c,1}$ gives a cyclic EG-LDPC code $\mathcal{C}_{\text{EG},c,1}$ of length $q^2 - 1$ with minimum distance at least $q + 1$. If $q$ is a power of 2, say $2^s$, $\mathcal{C}_{\text{EG},c,1}$ has the following parameters [4, 23]:

| | |
|---|---|
| Length | $2^{2s} - 1$ |
| Number of parity-check bits | $3^s - 1$ |
| Minimum distance | $2^s + 1$ |

Let $\mathbf{g}(X)$ be the generator polynomial of $\mathcal{C}_{\text{EG},c,1}$. The roots of $\mathbf{g}(X)$ in GF$(2^s)$ can be completely determined [22, 23]. For any integer $h$ such that $0 \leq h \leq 2^{2s} - 1$, it can be expressed in *radix*-$2^s$ form as follows:

$$h = c_0 + c_1 2^s, \tag{11.9}$$

where $0 \leq c_0$ and $c_1 < 2^s$. The sum $W_{2^s}(h) = c_0 + c_1$ of the coefficients of $h$ in the radix-$2^s$ form is called the $2^s$-*weight* of $h$. For any non-negative integer $l$, let $h^{(l)}$ be the remainder resulting from dividing $2^l h$ by $2^{2s} - 1$. Then $0 \leq h^{(l)} < 2^{2s} - 1$. The radix-$2^s$ form and $2^s$-weight of $h^{(l)}$ are $h^{(l)} = c_0^{(l)} + c_1^{(l)} 2^s$ and $W_{2^s}(h^{(l)}) = c_0^{(l)} + c_1^{(l)}$, respectively. Let $\alpha$ be a primitive element of GF($2^{2s}$). Then $\alpha^h$ is a root of $\mathbf{g}_{\mathrm{EG},2,c}(X)$ if and only if

$$0 < \max_{0 \leq l < s} W_{2^s}(h^{(l)}) < 2^s. \tag{11.10}$$

Knowing its roots in GF($2^{2s}$), $\mathbf{g}(X)$ can easily be determined (see the discussion of BCH codes in Chapter 3). The smallest integer that does not satisfy the condition given by (11.10) is $2^s + 1$. Hence, $\mathbf{g}(X)$ has as roots the following consecutive powers of $\alpha$:

$$\alpha, \alpha^2, \ldots, \alpha^{2^s}.$$

Then it follows from the BCH bound [23, 24, 25] (see Section 3.3) that the minimum distance of $\mathcal{C}_{\mathrm{EG},2,c}$ is at least $2^s + 1$, which is exactly equal to the column weight $2^s$ of the parity-check matrix $\mathbf{H}_{\mathrm{EG},c,1}$ of $\mathcal{C}_{\mathrm{EG},c,1}$ plus 1.

In fact, the minimum distance is exactly equal to $2^s + 1$. To show this, we consider the polynomial $X^{2^{2s}-1} - 1$ over GF(2), which can be factored as follows:

$$X^{2^{2s}-1} - 1 = \left(X^{2^s-1} - 1\right)\left(1 + X^{2^s-1} + X^{2(2^s-1)} + \cdots + X^{2^s(2^s-1)}\right). \tag{11.11}$$

First, we note that $X^{2^{2s}-1} - 1$ has all the nonzero elements of GF($2^{2s}$) as roots and the factor $X^{2^s-1} - 1$ has $\alpha^0, \alpha^{2^s+1}, \alpha^{2(2^s+1)}, \alpha^{3(2^s+1)}, \ldots, \alpha^{(2^s-2)(2^s+1)}$ as roots. It can easily be checked that powers of the roots of $X^{2^s-1} - 1$ (i.e., $0, 2^s + 1, \ldots, (2^s - 2)(2^s + 1)$) do not satisfy the condition given by (11.10). Hence, the polynomial $1 + X^{2^s-1} + \cdots + X^{2^s(2^s-1)}$ over GF(2) contains the roots $\alpha^h$ whose powers satisfy the condition given by (11.10). Consequently, the polynomial $1 + X^{2^s-1} + \cdots + X^{2^s(2^s-1)}$ is divisible by $\mathbf{g}(X)$ and is a code polynomial (or codeword) in $\mathcal{C}_{\mathrm{EG},c,1}$. This code polynomial has weight $2^s + 1$. Since the minimum distance of $\mathcal{C}_{\mathrm{EG},c,1}$ is lower bounded by $2^s + 1$, the minimum distance of $\mathcal{C}_{\mathrm{EG},c,1}$ is exactly $2^s + 1$. It has been proved [14] that $\mathcal{C}_{\mathrm{EG},c,1}$ has no pseudo-codeword with weight less than its minimum weight $2^s + 1$, that is, the Tanner graph of $\mathcal{C}_{\mathrm{EG},c,1}$ has no trapping set with size less than $2^s + 1$.

In the following, we use an example to illustrate the above construction of cyclic EG-LDPC codes. In this and other examples in this chapter, we decode an LDPC code with iterative decoding using the sum–product algorithm (SPA). The maximum number of decoding iterations is set to 100 (or 50). We also assume the binary-input AWGN channel, a model for BPSK transmission.

---

**Example 11.1** Consider the two-dimensional Euclidean geometry EG($2, 2^6$). The subgeometry EG*($2, 2^6$) of EG($2, 2^6$) consists of 4095 non-origin points and 4095 lines not passing through the origin of EG($2, 2^6$). Using the type-1 incidence vectors of the lines in EG*($2, 2^6$), we can construct a single $4095 \times 4095$ circulant matrix over GF(2) with both column and row weight 64. The null space of this circulant gives a

(4095, 3367) cyclic EG-LDPC code with rate 0.822 and minimum distance exactly equal to 65. The performance of this code decoded with iterative decoding using the SPA (100 iterations) is shown in Figure 11.1. At a BER of $10^{-6}$, the code performs 1.65 dB from the Shannon limit. Since it has a very large minimum distance and a minimum trapping set of the same size, it has a very low error floor. Furthermore, decoding of this code converges very fast, as shown in Figure 11.2. We see that the performance curves of 10 and 100 iterations are basically on top of each other.

### 11.1.2 A Class of Quasi-cyclic EG-LDPC Codes

Consider a cyclic class $\mathcal{S}_j$ of lines in the subgeometry EG$^*(m, q)$ of the $m$-dimensional Euclidean geometry EG$(m, q)$ with $1 \le j \le K_c$. We can form a $(q^m - 1) \times (q^m - 1)$ circulant $\mathbf{G}_{c,j}$ over GF(2) by taking the type-1 incidence vector of a line in $\mathcal{S}_j$ as the first column and its $q^m - 2$ *downward* cyclic shifts as the other columns. In this case, the columns of $\mathbf{G}_{c,j}$ are the type-1 incidence vectors of the lines in $\mathcal{S}_j$. For $1 \le k \le K_c$, form the following $(q^m - 1) \times k(q^m - 1)$ matrix over GF(2):

$$\mathbf{H}^{(2)}_{\text{EG},qc,k} = \begin{bmatrix} \mathbf{G}_{c,1} & \mathbf{G}_{c,2} & \cdots & \mathbf{G}_{c,k} \end{bmatrix}, \tag{11.12}$$

which consists of a row of $k$ circulants over GF(2) of size $(q^m - 1) \times (q^m - 1)$. $\mathbf{H}^{(2)}_{\text{EG},qc,k}$ has column and row weights $q$ and $kq$, respectively. Actually, $\mathbf{H}^{(2)}_{\text{EG},qc,k}$ is the transpose

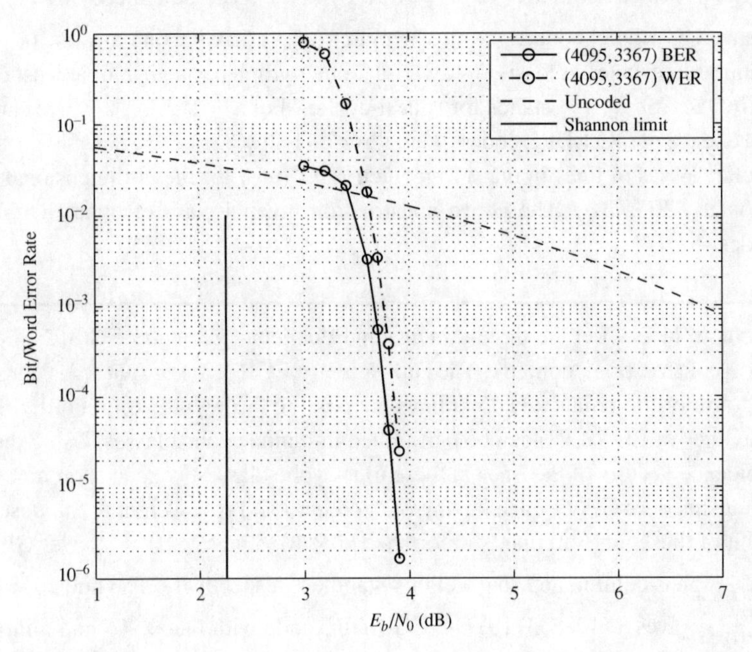


**Figure 11.1** Error performance of the cyclic (4095, 3367) EG-LDPC code given in Example 11.1.

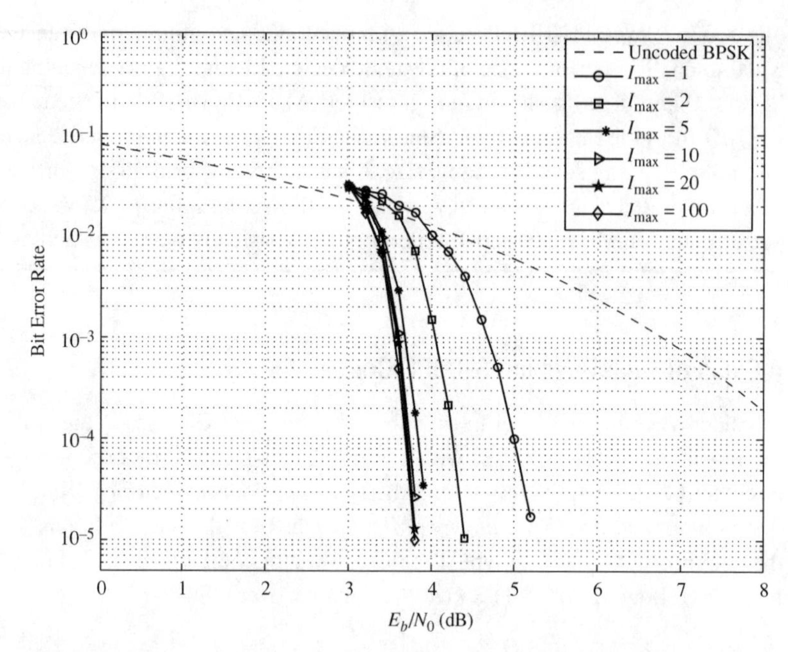


**Figure 11.2** The convergence rate of decoding of the cyclic (4095, 3367) EG-LDPC code given in Example 11.1.

of the parity-check matrix $\mathbf{H}^{(1)}_{\mathrm{EG},c,k}$ given by (11.7). The null space of $\mathbf{H}^{(2)}_{\mathrm{EG},qc,k}$ gives a binary QC-LDPC code $\mathcal{C}^{(2)}_{\mathrm{EG},qc,k}$ of length $k(q^m - 1)$ with rate at least $(k - 1)/k$ and minimum distance at least $q + 1$, whose Tanner graph has a girth of at least 6. The subscript "$qc$" of $\mathcal{C}^{(2)}_{\mathrm{EG},qc,k}$ stands for "quasi-cyclic." For $k = 1, 2, \ldots, K_c$, we can construct a sequence of QC-LDPC codes of lengths $q^m - 1, 2(q^m - 1), \ldots, K_c(q^m - 1)$ based on the cyclic classes of lines in the sub-geometry $\mathrm{EG}^*(m, q)$ of the $m$-dimensional Euclidean geometry $\mathrm{EG}(m, q)$. The above construction gives a class of binary QC-EG-LDPC codes.

---

**Example 11.2** Let the three-dimensional Euclidean geometry $\mathrm{EG}(3, 2^3)$ be the code-construction geometry. The subgeometry $\mathrm{EG}^*(3, 2^3)$ of $\mathrm{EG}(3, 2^3)$ consists of 511 non-origin points and 4599 lines not passing through the origin of $\mathrm{EG}(3, 2^3)$. The lines of $\mathrm{EG}^*(3, 2^3)$ can be partitioned into nine cyclic classes. Using the incidence vectors of the lines in these nine cyclic classes, we can form nine $511 \times 511$ circulants over GF(2), each having both column and row weight 8. Suppose we take eight of these nine circulants to form a $511 \times 4088$ matrix $\mathbf{H}^{(2)}_{\mathrm{EG},qc,8}$ over GF(2). $\mathbf{H}^{(2)}_{\mathrm{EG},qc,8}$ has column and row weights 8 and 64, respectively. The null space of $\mathbf{H}^{(2)}_{\mathrm{EG},qc,8}$ gives a (4088, 3716) QC-EG-LDPC code with rate 0.909 and minimum distance at least 9. The error performance of this code decoded with iterative decoding using the SPA (50 iterations) is shown in Figure 11.3. At a BER of $10^{-6}$, the code performs 1.2 dB from the Shannon limit.

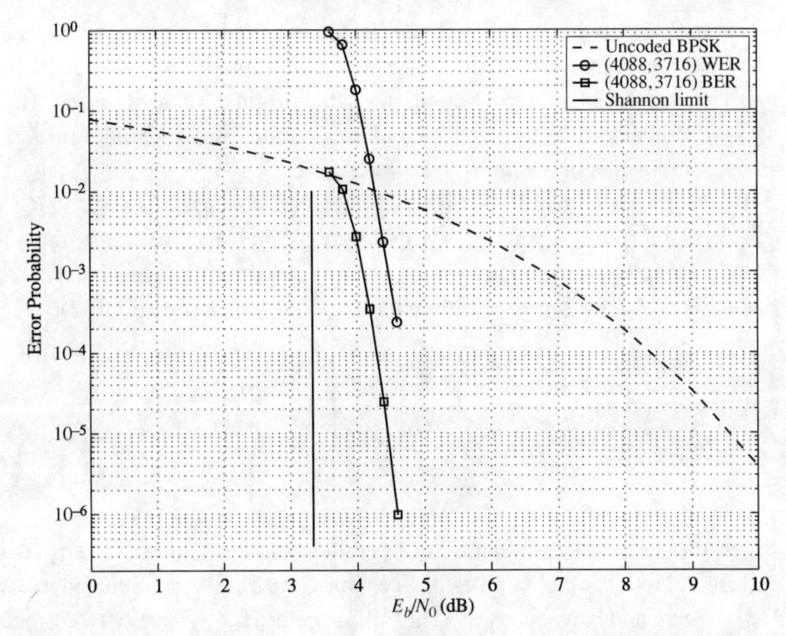


**Figure 11.3** The error performance of the (4088, 3716) QC-EG-LDPC code given in Example 11.2.

Since a cyclic EG-LDPC code has a large minimum distance, in general it has a very low error floor.

## 11.2 Construction of LDPC Codes Based on the Parallel Bundles of Lines in Euclidean Geometries

In the previous section, we presented two classes of structured LDPC codes constructed by exploiting the cyclic structure of lines in Euclidean geometries. Another important structure of lines in Euclidean geometries is the parallel structure as described in Chapter 2. On the basis of this parallel structure of lines, another class of EG-LDPC codes can be constructed.

Consider the $m$-dimensional Euclidean geometry EG$(m, q)$. As described in Chapter 2, the $J \triangleq J_{\text{EG}}(m, 1) = q^{m-1}(q^m - 1)/(q - 1)$ lines in EG$(m, q)$ can be partitioned into $K_p = (q^m - 1)/(q - 1)$ parallel bundles, $\mathcal{P}_1(m, 1), \mathcal{P}_2(m, 1), \ldots, \mathcal{P}_{K_p}(m, 1)$, each consisting of $q^{m-1}$ parallel lines. The lines in a parallel bundle contain all $q^m$ points of EG$(m, q)$, each point appearing on one and only one line. Each parallel bundle contains a line passing through the origin. For a line $\mathcal{L}$ in EG$(m, q)$ passing or not passing through the origin of EG$(m, q)$, we define a $q^m$-tuple over GF(2),

$$\mathbf{v}_{\mathcal{L}} = (v_{-\infty}, v_0, v_1, \ldots, v_{q^m-2}), \tag{11.13}$$

whose components correspond to all the points in EG$(m, q)$ represented by the elements of GF$(q^m)$, $\alpha^{-\infty} = 0, \alpha^0, \alpha, \ldots, \alpha^{q^m-2}$, where $v_j = 1$ if and only if the point

$\alpha^j$ is on $\mathcal{L}$, otherwise $v_j = 0$. This $q^m$-tuple is called the *type-2 incidence vector* of line $\mathcal{L}$. Note that the type-2 incidence vector of a line in EG$(m, q)$ contains a coordinate (or component) $v_{-\infty}$ corresponding to the origin $\alpha^{-\infty}$ of EG$(m, q)$. The weight of $\mathbf{v}_{\mathcal{L}}$ is $q$. It is clear that the type-2 incidence vectors of two parallel lines do not have any 1-components in common.

For each parallel bundle $\mathcal{P}_i(m, 1)$ of lines in EG$(m, q)$ with $1 \leq i \leq K_p$, we form a $q^{m-1} \times q^m$ matrix $\mathbf{H}_{p,i}$ over GF(2) with the type-2 incidence vectors of the $q^{m-1}$ parallel lines in $\mathcal{P}_i(m, 1)$ as rows. The column and row weights of $\mathbf{H}_{p,i}$ are 1 and $q$, respectively. For $1 \leq k \leq K_p$, we form the following $kq^{m-1} \times q^m$ matrix over GF(2):

$$\mathbf{H}_{\text{EG},p,k} = \begin{bmatrix} \mathbf{H}_{p,1} \\ \mathbf{H}_{p,2} \\ \vdots \\ \mathbf{H}_{p,k} \end{bmatrix}, \tag{11.14}$$

where the subscript "$p$" stands for "parallel bundle of lines." The rows of $\mathbf{H}_{\text{EG},p,k}$ correspond to $k$ parallel bundles of lines in EG$(m, q)$. The column and row weights of $\mathbf{H}_{\text{EG},p,k}$ are $k$ and $q$, respectively. The null space of $\mathbf{H}_{\text{EG},p,k}$ gives a $(k, q)$-regular LDPC code $\mathcal{C}_{\text{EG},p,k}$ of length $q^m$ with minimum distance at least $k + 1$. For $k = 1, 2, \ldots, K_p$, we can construct a family of regular LDPC codes of the same length $q^m$ with different rates and minimum distances. The above construction gives a class of binary regular LDPC codes.

---

**Example 11.3** Consider the three-dimensional Euclidean geometry EG$(3, 2^3)$ over GF$(2^3)$. This geometry consists of 512 points and 4672 lines. Each line consists of eight points. The 4672 lines can be grouped into 73 parallel bundles, each consisting of 64 parallel lines. Suppose we take six parallel bundles of lines and form a $384 \times 512$ matrix $\mathbf{H}_{\text{EG},p,6}$ with column and row weights 6 and 8, respectively. The null space of $\mathbf{H}_{\text{EG},p,6}$ gives a (6, 8)-regular (512, 256) EG-LDPC code with rate 0.5 and minimum distance at least 7. The error performance of this code decoded with iterative decoding using the SPA (100 iterations) is shown in Figure 11.4.

---

**Example 11.4** Consider the two-dimensional Euclidean geometry EG$(2, 2^5)$. This geometry consists of 1024 points and 1056 lines. The 1056 lines can be partitioned into 33 parallel bundles, each consisting of 32 parallel lines. Suppose we take eight parallel bundles and form a $256 \times 1024$ matrix $\mathbf{H}_{\text{EG},p,8}$ over GF(2) with column and row weights 8 and 32. The null space of $\mathbf{H}_{\text{EG},p,8}$ gives an (8, 32)-regular (1024, 845) LDPC code with rate 0.825 and minimum distance at least 9. The error performance of this code over the binary-input AWGN channel decoded with iterative decoding using the SPA (100 iterations) is shown in Figure 11.5. At a BER of $10^{-6}$, the code performs 1.9 dB from the Shannon limit.

---

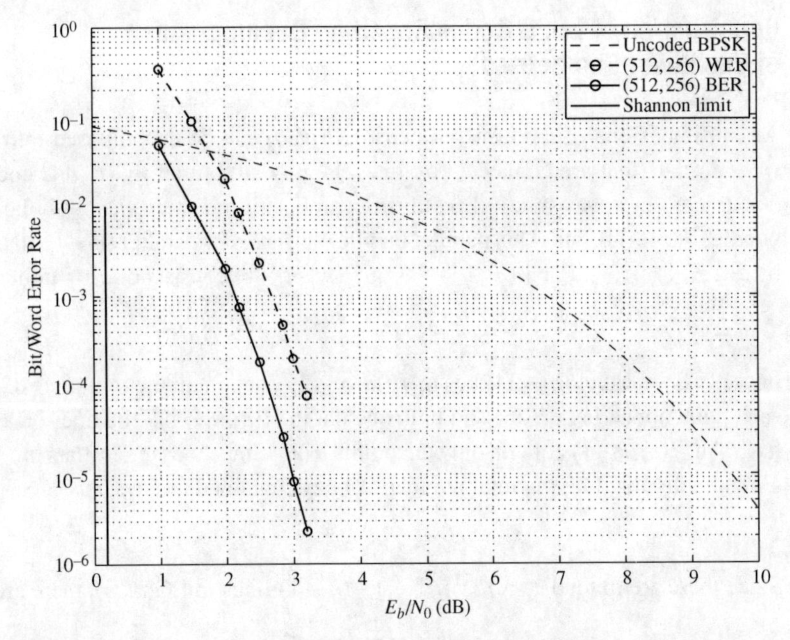


**Figure 11.4** The error performance of the (512, 256) EG-LDPC code given in Example 11.3.

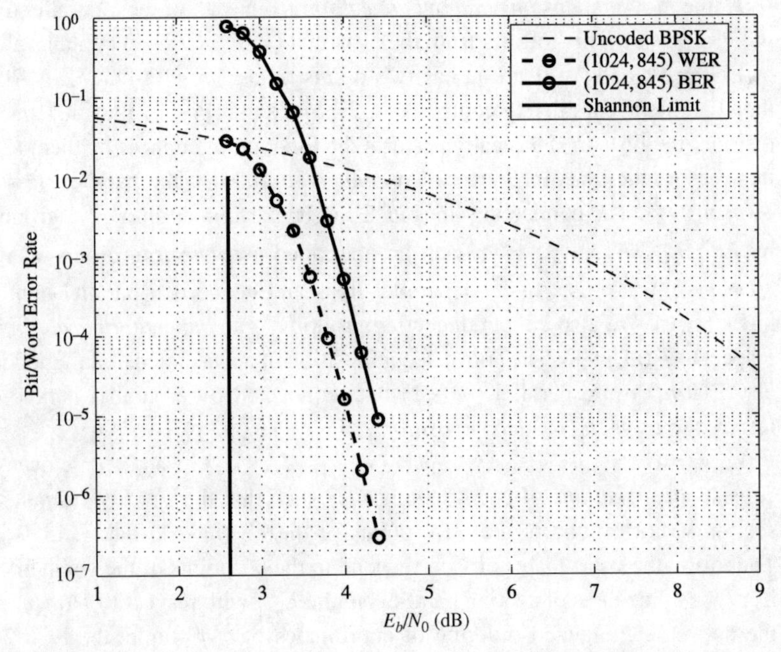


**Figure 11.5** The error performance of the (1024, 845) EG-LDPC code given in Example 11.4.

## 11.3 Construction of LDPC Codes Based on Decomposition of Euclidean Geometries

An $m$-dimensional Euclidean geometry EG$(m, q)$ can be decomposed into $q$ parallel $(m-1)$-flats that are connected by lines. Highly structured EG-LDPC codes can be constructed in a way that utilizes this decomposition. Let $\mathcal{P}(m, m-1)$ be a parallel bundle of $(m-1)$-flats. This parallel bundle consists of $q$ parallel $(m-1)$-flats, denoted by $\mathcal{F}_1, \mathcal{F}_1, \ldots, \mathcal{F}_q$. Each $(m-1)$-flat in $\mathcal{P}(m, m-1)$ consists of $q^{m-1}$ points and

$$J_1 \triangleq J_{\mathrm{EG}}(m-1, 1) = q^{m-2}(q^{m-1}-1)/(q-1)$$

lines. The $q$ parallel $(m-1)$-flats in $\mathcal{P}(m, m-1)$ contain all the $q^m$ points of EG$(m, q)$, each point appearing on one and only one flat in $\mathcal{P}(m, m-1)$. The lines on an $(m-1)$-flat $\mathcal{F}_i$ in $\mathcal{P}(m, m-1)$ contain only the points in $\mathcal{F}_i$. Since the $(m-1)$-flats in $\mathcal{P}(m, m-1)$ are mutually disjoint, the total number of lines contained in $\mathcal{P}(m, m-1)$ is

$$J_2 = qJ_1 = q^{m-1}(q^{m-1}-1)/(q-1).$$

Since there are in total $J = q^{m-1}(q^m - 1)/(q-1)$ lines in EG$(m, q)$, there are

$$J_3 = J - J_2 = q^{2(m-1)}$$

lines in EG$(m, q)$ that are not contained in $\mathcal{P}(m, m-1)$. We denote this set of lines by $E$.

A line in $E$ contains *one and only one point* from each of the $q$ parallel $(m-1)$-flats in $\mathcal{P}(m, m-1)$. This follows from the facts that (1) $\mathcal{P}(m, m-1)$ contains all the points of EG$(m, q)$; and (2) if a line has two points on an $(m-1)$-flat $\mathcal{F}_i$ in $\mathcal{P}(m, m-1)$, then the line is completely contained in $\mathcal{F}_i$. Since the $(m-1)$-flats in $\mathcal{P}(m, m-1)$ are disjoint, the lines in $E$ basically perform the function of connecting them together. For this reason, the lines in $E$ are called *connecting lines* of the parallel $(m-1)$-flats in $\mathcal{P}(m, m-1)$. The connecting lines in $E$ can be partitioned into $q^{m-1}$ groups, denoted by $E_{m,1}^{(0)}, E_{m,1}^{(1)}, \ldots, E_{m,1}^{(q^{m-1}-1)}$. Each group, called a *connecting group* with respect to $\mathcal{P}(m, m-1)$, consists of $q^{m-1}$ parallel lines and hence is a parallel bundle of lines in EG$(m, q)$. We also call a connecting group $E_{m,1}^{(i)}$ a *connecting parallel bundle* with respect to $\mathcal{P}(m, m-1)$. The above decomposition of an $m$-dimensional Euclidean geometry EG$(m, q)$ into parallel $(m-1)$-flats connected by a parallel bundle of lines is referred to as *geometry decomposition* [15].

Let $\mathbf{v} = (v_{-\infty}, v_0, v_1, \ldots, v_{q^m-2}) = (\mathbf{v}_0, \mathbf{v}_1, \ldots, \mathbf{v}_{q-1})$ be a $q^m$-tuple over GF(2) that consists of $q$ *sections* of equal length $q^{m-1}$ such that the $q^{m-1}$ coordinates in the $i$th section $\mathbf{v}_i$ correspond to the $q^{m-1}$ points of the $i$th $(m-1)$-flat $\mathcal{F}_i$ in $\mathcal{P}(m, m-1)$. Therefore, the coordinates of $\mathbf{v}$ correspond to the $q^m$ points of the geometry EG$(m, q)$. Let $\mathcal{L}$ be a line in a connecting parallel bundle $E_{m,1}^{(i)}$ with respect to $\mathcal{P}(m, m-1)$. Using the above sectionalized ordering of coordinates of a $q^m$-tuple, the type-2 incidence vector $\mathbf{v}_{\mathcal{L}} = (\mathbf{v}_0, \mathbf{v}_1, \ldots, \mathbf{v}_{q-1})$ of $\mathcal{L}$ consists of $q$ sections of equal length $q^{m-1}$ such that the $i$th section of $\mathbf{v}_i$ has *one and only one* 1-component corresponding to the point in the $(m-1)$-flat $\mathcal{F}_i$ where the line $\mathcal{L}$ passes through. Therefore, $\mathbf{v}_{\mathcal{L}}$ has one and only one 1-component in each section. This type-2 incidence vector is called a *sectionalized type-2 incidence vector*.

For each connecting parallel bundle $E_{m,1}^{(i)}$ of lines with $0 \le i < q^{m-1}$, we form a $q^{m-1} \times q^m$ matrix $\mathbf{H}_{d,i}$ over GF(2) using the sectionalized type-2 incidence vectors of the parallel lines in $E_{m,1}^{(i)}$ as rows. The subscript "$d$" of $\mathbf{H}_{d,i}$ stands for "decomposition." It follows from the sectionalized structure of the type-2 incidence vector of a connecting line in $E_{m,1}^{(i)}$ that $\mathbf{H}_{d,i}$ consists of a row of $q$ permutation matrices of size $q^{m-1} \times q^{m-1}$,

$$\mathbf{H}_{d,i} = \begin{bmatrix} \mathbf{A}_{i,0} & \mathbf{A}_{i,1} & \cdots & \mathbf{A}_{i,q-1} \end{bmatrix}. \tag{11.15}$$

$\mathbf{H}_{d,i}$ is a matrix with column and row weights 1 and $q$, respectively. $\mathbf{H}_{d,i}$ is called the *incidence matrix* of the connecting parallel bundle $E_{m,1}^{(i)}$ of lines with respect to $\mathcal{P}(m, m-1)$. Form the following $q^{m-1} \times q$ array of $q^{m-1} \times q^{m-1}$ permutation matrices over GF(2):

$$\mathbf{H}_{\text{EG},d}^{(1)} = \begin{bmatrix} \mathbf{H}_{d,0} \\ \mathbf{H}_{d,1} \\ \vdots \\ \mathbf{H}_{d,q^{m-1}-1} \end{bmatrix} = \begin{bmatrix} \mathbf{A}_{0,0} & \mathbf{A}_{0,1} & \cdots & \mathbf{A}_{0,q-1} \\ \mathbf{A}_{1,0} & \mathbf{A}_{1,1} & \cdots & \mathbf{A}_{1,q-1} \\ \vdots & \vdots & \ddots & \vdots \\ \mathbf{A}_{q^{m-1}-1,0} & \mathbf{A}_{q^{m-1}-1,1} & \cdots & \mathbf{A}_{q^{m-1}-1,q-1} \end{bmatrix}. \tag{11.16}$$

$\mathbf{H}_{\text{EG},d}^{(1)}$ is a $q^{2(m-1)} \times q^m$ matrix over GF(2) with column and row weights $q^{m-1}$ and $q$, respectively. Since the rows of $\mathbf{H}_{\text{EG},d}^{(1)}$ correspond to $q^{2(m-1)}$ lines in EG$(m, q)$, no two rows (or two columns) of $\mathbf{H}_{\text{EG},d}^{(1)}$ have more than one 1-component in common. Hence, $\mathbf{H}_{\text{EG},d}^{(1)}$ satisfies the RC constraint.

For a pair $(g, r)$ of integers with $1 \le g < q^{m-1}$ and $1 \le r \le q$, let $\mathbf{H}_{\text{EG},d}^{(1)}(g, r)$ be a $g \times r$ subarray of $\mathbf{H}_{\text{EG},d}^{(1)}$. Then $\mathbf{H}_{\text{EG},d}^{(1)}(g, r)$ is a $gq^{m-1} \times rq^{m-1}$ matrix over GF(2) with column and row weights $g$ and $r$, respectively. The null space of $\mathbf{H}_{\text{EG},d}^{(1)}(g, r)$ gives a $(g, r)$-regular EG-LDPC code $\mathcal{C}_{\text{EG},d}^{(1)}$ of length $rq^{m-1}$ with minimum distance at least $g+1$. $\mathbf{H}_{\text{EG},d}^{(1)}(g, r)$ consists of $r$ columns of permutation matrices, each column consisting of $g$ permutation matrices. If we add $l$ columns (or column-vectors) of $\mathbf{H}_{\text{EG},d}^{(1)}(g, r)$, we obtain a column vector $\mathbf{h}$ with $g$ sections of $q^{m-1}$ bits each. For this column vector $\mathbf{h}$ to be zero, each section of $\mathbf{h}$ must be zero. For each section of $\mathbf{h}$ to be zero, $l$ must be even. This implies that the minimum distance of $\mathcal{C}_{\text{EG},d}^{(1)}$ must be even. Since the minimum distance of $\mathcal{C}_{\text{EG},d}^{(1)}$ is at least $g + 1$, the minimum distance of $\mathcal{C}_{\text{EG},d}^{(1)}$ is at least $g + 2$ for even $g$ and at least $g + 1$ for odd $g$. The above construction gives a class of regular EG-LDPC codes.

---

**Example 11.5** Let the two-dimensional Euclidean geometry EG$(2, 2^6)$ be the code-construction geometry. This geometry consists of 4096 points and 4160 lines. Each line consists of 64 points. The lines in EG$(2, 2^6)$ can be partitioned into 65

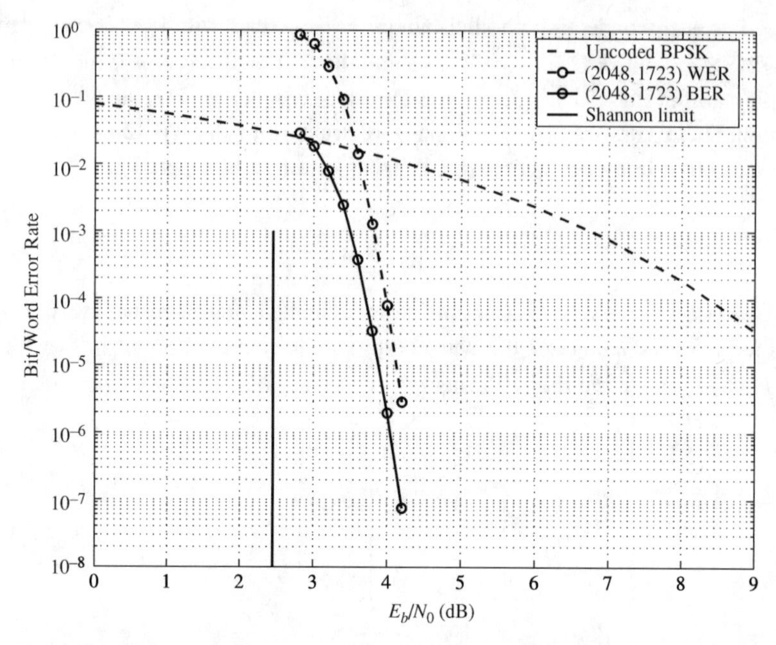


**Figure 11.6** The error performance of the (2048, 1723) LDPC code given in Example 11.5.

parallel bundles, each consisting of 64 lines. Suppose we decompose this geometry in accordance with a chosen parallel bundle $\mathcal{P}(2, 1)$ of lines. Then there are 64 connecting parallel bundles of lines, $E_{2,1}^{(0)}, E_{2,1}^{(1)}, \ldots, E_{2,1}^{(63)}$. For $0 \leq i < 64$, the incidence matrix $\mathbf{H}_{d,i} = [\mathbf{A}_{i,0} \ \ \mathbf{A}_{i,1} \ \ \cdots \ \ \mathbf{A}_{i,63}]$ of the connecting parallel bundle $E_{2,1}^{(i)}$ consists of a row of 64 permutation matrices of size $64 \times 64$. Based on (11.16), we can form a $64 \times 64$ array $\mathbf{H}_{\text{EG},d}^{(1)}$ of $64 \times 64$ permutation matrices. From this array, we can construct a family of regular EG-LDPC codes with various lengths, rates, and minimum distances. Suppose we take the $6 \times 32$ subarray $\mathbf{H}_{\text{EG},d}^{(1)}(6, 32)$ at the upper-left corner of $\mathbf{H}_{\text{EG},d}^{(1)}$ as the parity-check matrix for an EG-LDPC code. $\mathbf{H}_{\text{EG},d}^{(1)}(6, 32)$ is a $384 \times 2048$ matrix over GF(2) with column and row weights 6 and 32, respectively. The null space of this matrix gives a (6, 32)-regular (2048, 1723) EG-LDPC code with rate 0.841 and minimum distance at least 8. The error performance of this code over the binary-input AWGN channel decoded with iterative decoding using the SPA (100 iterations) is shown in Figure 11.6. At a BER of $10^{-6}$, the code performs 1.55 dB from the Shannon limit and achieves a 6-dB coding gain over uncoded BPSK. This code is equivalent to the IEEE 802.3 standard (2048, 1723) LDPC code for the 10-G Base-T Ethernet, which has no error floor down to a BER of $10^{-12}$.

---

**Example 11.6** We continue Example 11.5. Suppose we take an $8 \times 64$ subarray $\mathbf{H}_{\text{EG},d}^{(1)}(8, 64)$ from the $64 \times 64$ array $\mathbf{H}_{\text{EG},d}^{(1)}$ given in Example 11.5, say the top eight rows of permutation matrices of $\mathbf{H}_{\text{EG},d}^{(1)}$. $\mathbf{H}_{\text{EG},d}^{(1)}(8, 64)$ is a $512 \times 4096$ matrix over

GF(2) with column and row weights 8 and 64, respectively. The null space of this matrix gives an (8, 64)-regular (4096, 3687) EG-LDPC code with rate 0.9 and minimum distance at least 10. The error performance, error floor, and convergence rate of decoding of this code are shown in Figures 11.7 to 11.9, respectively. At a

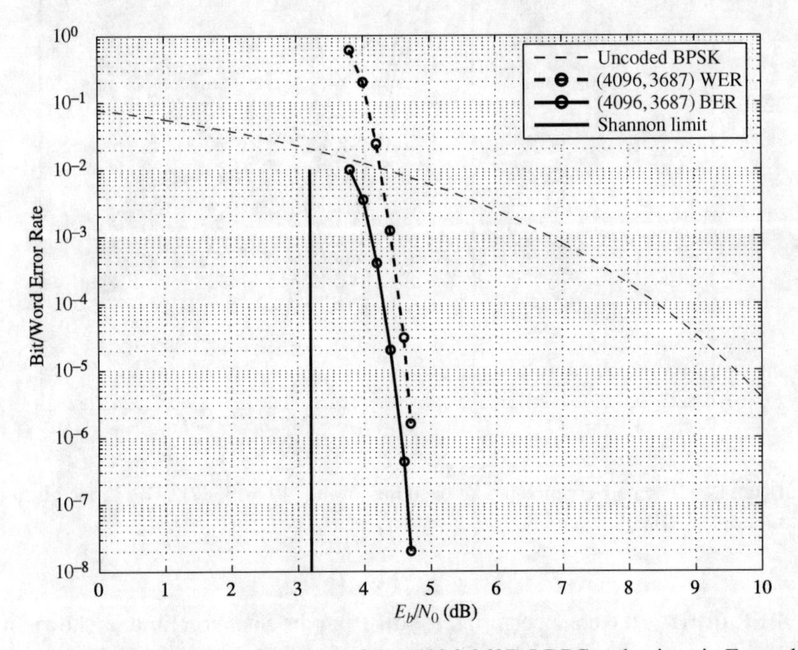


**Figure 11.7** The error performance of the (4096, 3687) LDPC code given in Example 11.6.

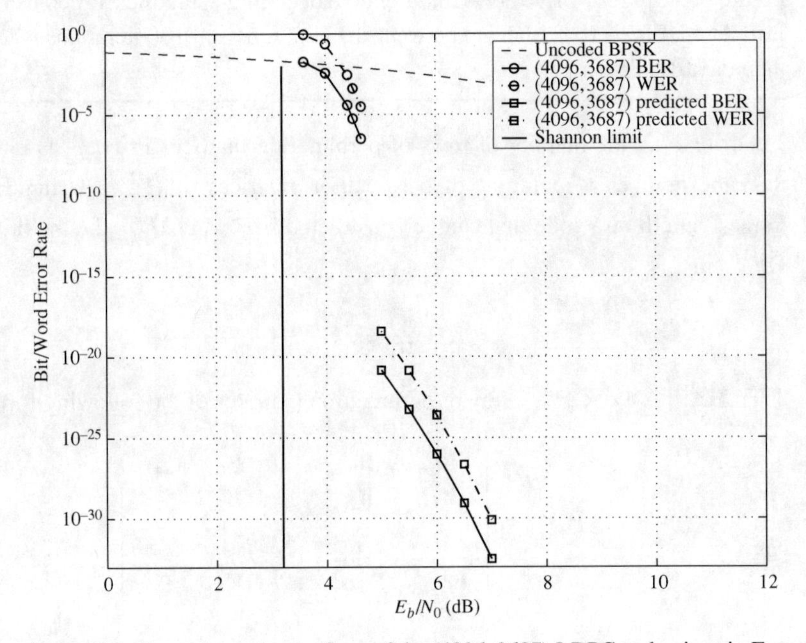


**Figure 11.8** The estimated error floor of the (4096, 3687) LDPC code given in Example 11.6.

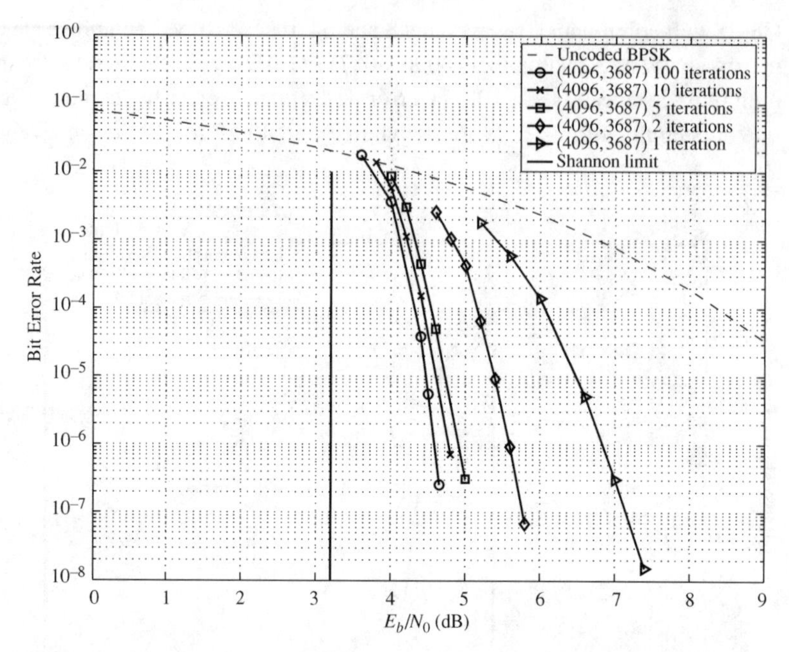


**Figure 11.9** The convergence rate of decoding of the (4096, 3687) LDPC code given in Example 11.6.

BER of $10^{-6}$, the code performs 1.3 dB from the Shannon limit as shown in Figure 11.7 and its error floor for BER is estimated to be below $10^{-21}$ as shown in Figure 11.8. Figure 11.9 shows that iterative decoding using the SPA converges very fast. At a BER of $10^{-6}$, the gap between 10 iterations and 100 iterations is less than 0.2 dB.

---

For $m > 2$, the number of rows of permutation matrices in $\mathbf{H}_{\text{EG},d}^{(1)}$ is much larger than the number of columns of permutation matrices in $\mathbf{H}_{\text{EG},d}^{(1)}$. Using $\mathbf{H}_{\text{EG},d}^{(1)}$, the longest length of a code that can be constructed is $q^m$. Let $\mathbf{H}_{\text{EG},d}^{(2)}$ be the transpose of $\mathbf{H}_{\text{EG},d}^{(1)}$, that is

$$\mathbf{H}_{\text{EG},d}^{(2)} = \left[\mathbf{H}_{\text{EG},d}^{(1)}\right]^{\text{T}}. \tag{11.17}$$

Then $\mathbf{H}_{\text{EG},d}^{(2)}$ is a $q \times q^{m-1}$ array of permutation matrices of the following form:

$$\mathbf{H}_{\text{EG},d}^{(2)} = \begin{bmatrix} \mathbf{B}_{0,0} & \mathbf{B}_{0,1} & \cdots & \mathbf{B}_{0,q^{m-1}-1} \\ \mathbf{B}_{1,0} & \mathbf{B}_{1,1} & \cdots & \mathbf{B}_{1,q^{m-1}-1} \\ \vdots & \vdots & \ddots & \vdots \\ \mathbf{B}_{q-1,0} & \mathbf{B}_{q-1,1} & \cdots & \mathbf{B}_{q-1,q^{m-1}-1} \end{bmatrix}, \tag{11.18}$$

where $\mathbf{B}_{i,j} = \mathbf{A}^{\mathrm{T}}_{j,i}$ for $0 \le i < q$ and $0 \le j < q^{m-1}$. Using $\mathbf{H}^{(2)}_{\mathrm{EG},d}$, we can construct longer codes with higher rates. However, the largest minimum distance of an EG-LDPC code constructed from $\mathbf{H}^{(2)}_{\mathrm{EG},d}$ is lower bounded by $q + 1$. The advantage of using the array $\mathbf{H}^{(1)}_{\mathrm{EG},d}$ for code construction is that the codes constructed have a wider range of minimum distances.

---

**Example 11.7** Suppose we use the three-dimensional Euclidean geometry EG(3, 13) over GF(13) for code construction. Suppose we decompose this geometry in terms of a parallel bundle $\mathcal{P}(3, 2)$ of 2-flats. The decomposition results in a $169 \times 13$ array $\mathbf{H}^{(1)}_{\mathrm{EG},d}$ of $169 \times 169$ permutation matrices. Using this array, the longest length of a code that can be constructed is 2197. Suppose we take the transpose $\mathbf{H}^{(2)}_{\mathrm{EG},d}$ of $\mathbf{H}^{(1)}_{\mathrm{EG},d}$. Then $\mathbf{H}^{(2)}_{\mathrm{EG},d}$ is a $13 \times 169$ array of $169 \times 169$ permutation matrices. The longest length of a code that can be constructed from $\mathbf{H}^{(2)}_{\mathrm{EG},d}$ is 28 561. Suppose we take a $6 \times 48$ subarray $\mathbf{H}^{(2)}_{\mathrm{EG},d}(6, 48)$ from $\mathbf{H}^{(2)}_{\mathrm{EG},d}$. $\mathbf{H}^{(2)}_{\mathrm{EG},d}(6, 48)$ is a $1014 \times 8112$ matrix over GF(2) with column and row weights 6 and 48, respectively. The null space of this matrix gives a (6, 48)-regular (8112, 7103) EG-LDPC code with rate 0.8756 and minimum distance at least 8. The error performance of this code with iterative decoding using the SPA (100 iterations) is shown in Figure 11.10. At a BER of $10^{-10}$, the code performs 1.5 dB from the Shannon limit.

---

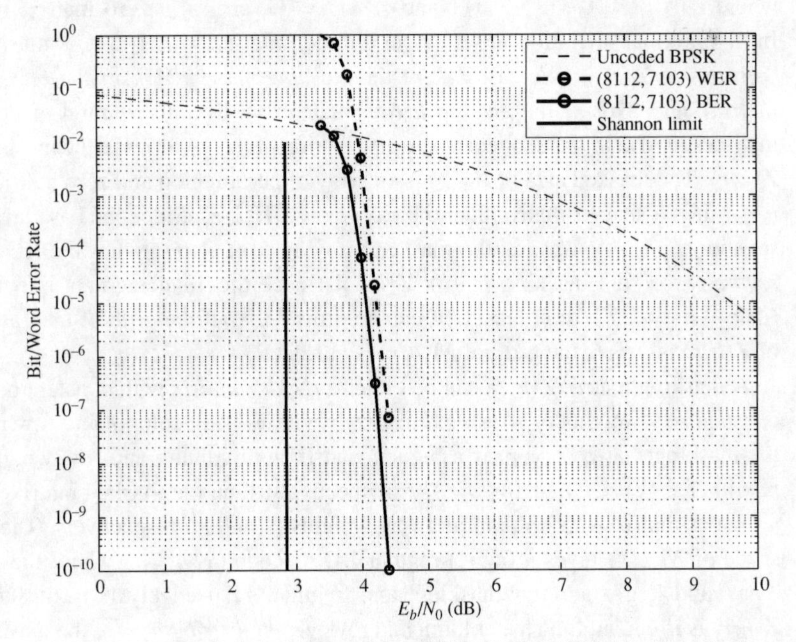


**Figure 11.10** The error performance of the (8112, 7103) LDPC code given in Example 11.7.

## 11.4 Construction of EG-LDPC Codes by Masking

In the previous section, we presented two RC-constrained arrays of permutation matrices for constructing regular EG-LDPC codes. Although these arrays are highly structured, their constituent permutation matrices are densely packed. The density of such an array or its subarrays can be reduced by replacing a set of permutation matrices by zero matrices. This replacement of permutation matrices by zero matrices is referred to as *masking* [15, 23]. Masking an array of permutation matrices results in an array of permutation and zero matrices whose Tanner graph has fewer edges and hence has fewer short cycles and possibly larger girth. Masking subarrays of arrays $\mathbf{H}_{\mathrm{EG},d}^{(1)}$ and $\mathbf{H}_{\mathrm{EG},d}^{(2)}$ given by (11.16) and (11.18) results in new EG-LDPC codes, including irregular codes.

### 11.4.1 Masking

The masking operation can be modeled mathematically as a *special matrix product*. Let $\mathbf{H}(k,r) = \left[\mathbf{A}_{i,j}\right]$ be an RC-constrained $k \times r$ array of $n \times n$ permutation matrices over GF(2). $\mathbf{H}(k,r)$ may be a subarray of the array $\mathbf{H}_{\mathrm{EG},d}^{(1)}$ (or $\mathbf{H}_{\mathrm{EG},d}^{(2)}$) given by (11.16) (or (11.18)). Let $\mathbf{Z}(k,r) = \left[z_{i,j}\right]$ be a $k \times r$ matrix over GF(2). Define the following (Kronecker) product of $\mathbf{Z}(k,r)$ and $\mathbf{H}(k,r)$:

$$\mathbf{M}(k,r) = \mathbf{Z}(k,r) \circledast \mathbf{H}(k,r) = \left[z_{i,j}\mathbf{A}_{i,j}\right], \tag{11.19}$$

where $z_{i,j}\mathbf{A}_{i,j} = \mathbf{A}_{i,j}$ if $z_{i,j} = 1$ and $z_{i,j}\mathbf{A}_{i,j} = \mathbf{O}$, an $n \times n$ zero matrix, if $z_{i,j} = 0$. In this product operation, a set of permutation matrices in $\mathbf{H}(k,r)$ is masked by the 0-entries in $\mathbf{Z}(k,r)$. We call $\mathbf{Z}(k,r)$ the *masking matrix*, $\mathbf{H}(k,r)$ the *base array* (or matrix), and $\mathbf{M}(k,r)$ the *masked array* (or matrix). The distribution of permutation matrices in $\mathbf{M}(k,r)$ is identical to the distribution of the 1-entries in the masking matrix $\mathbf{Z}(k,r)$. The masked array $\mathbf{M}(k,r)$ is an array of permutation and zero matrices of size $n \times n$. It is a sparser matrix than the base matrix $\mathbf{H}(k,r)$. Since the base array $\mathbf{H}(k,r)$ satisfies the RC constraint, the masked array $\mathbf{M}(k,r)$ also satisfies the RC constraint *regardless of* the masking matrix $\mathbf{Z}(k,r)$. Hence the Tanner graph associated with $\mathbf{M}(k,r)$ has a girth of at least 6. It can be proved that, if the girth of the Tanner graph of $\mathbf{Z}(k,r)$ is $\lambda > 6$, the girth of $\mathbf{M}(k,r)$ is at least $\lambda$.

A masking matrix $\mathbf{Z}(k,r)$ can be either a *regular matrix* with constant column and constant row weights or an *irregular matrix* with *varying* column and row weights. If the masking matrix $\mathbf{Z}(k,r)$ is a regular matrix with column and row weights $k_z$ and $r_z$ with $1 \le k_z \le k$ and $1 \le r_z \le r$, respectively, then the masked matrix $\mathbf{M}(k,r)$ is a regular matrix with column and row weights $k_z$ and $r_z$, respectively. Then the null space of $\mathbf{M}(k,r)$ gives a $(k_z, r_z)$-regular LDPC code $C_{\mathrm{EG,mas,reg}}$ where the subscripts "mas" and "reg" stand for "masking" and "regular," respectively. If the masking matrix $\mathbf{Z}(k,r)$ is irregular, then the column and row *weight distributions* of the masked matrix $\mathbf{M}(k,r)$ are identical to the column and row weight distributions of $\mathbf{Z}(k,r)$. The null space of $\mathbf{M}(k,r)$ gives an irregular EG-LDPC code $C_{\mathrm{EG,mas,irreg}}$, where the subscript "irreg" of $C_{\mathrm{EG,mas,irreg}}$ stands for "irregular."

### 11.4.2 Regular Masking

Regular masking matrices can be constructed algebraically. There are several construction methods [15]. One is presented in this section and the others will be presented in the next section and Chapter 12. Let $r = lk$, where $l$ and $k$ are two positive integers. Suppose we want to construct a $k \times r$ masking matrix $\mathbf{Z}(k, r)$ with column weight $k_z$ and row weight $r_z = lk_z$. A $k$-tuple $\mathbf{g} = (g_0, g_1, \ldots, g_{k-1})$ over GF(2) is said to be *primitive* if executing $k$ right cyclic shifts of $\mathbf{g}$ gives $k$ different $k$-tuples. For example, (1011000) is a primitive 7-tuple. Two primitive $k$-tuples are said to be *nonequivalent* if one cannot be obtained from the other by cyclic shifting. To construct $\mathbf{Z}(k, r)$, we choose $l$ primitive nonequivalent $k$-tuples of weight $k_z$, denoted by $\mathbf{g}_0, \mathbf{g}_1, \ldots, \mathbf{g}_{l-1}$. For each primitive $k$-tuple $\mathbf{g}_i$, we form a $k \times k$ circulant $\mathbf{G}_i$ with $\mathbf{g}_i$ as its top row and all the other $k-1$ right cyclic shifts as the other $k-1$ rows. Then

$$\mathbf{Z}(k, r) = \begin{bmatrix} \mathbf{G}_0 & \mathbf{G}_1 & \cdots & \mathbf{G}_{l-1} \end{bmatrix}, \tag{11.20}$$

which consists of a row of $l$ circulants of size $k \times k$. It is a $k \times lk$ matrix over GF(2) with column and row weights $k_z$ and $lk_z$, respectively. The $k$-tuples, $\mathbf{g}_0, \mathbf{g}_1, \ldots, \mathbf{g}_{l-1}$, are called the *generators* of the circulants $\mathbf{G}_0, \mathbf{G}_1, \ldots, \mathbf{G}_{l-1}$, or the masking-matrix generators. These masking-matrix generators should be designed to achieve two objectives: (1) maximizing the girth of the Tanner graph of the masking matrix $\mathbf{Z}(k, r)$; and (2) preserving the minimum distance of the code given by the null space of base array $\mathbf{H}(k, r)$ after masking, that is, keeping the minimum distance of the code given by the null space of the masked array $\mathbf{M}(k, r) = \mathbf{Z}(k, r) \circledast \mathbf{H}(k, r)$ the same as that of the code given by the null space of the base array $\mathbf{H}(k, r)$. *How to achieve these two objectives is unknown and is a very challenging research problem.*

A masking matrix that consists of a row of circulants can also be constructed by decomposing RC-constrained circulants constructed from finite geometries, for example, the circulants constructed from the cyclic classes of lines of a Euclidean geometry as described in Section 11.1.1. This construction will be presented in the next section.

---

**Example 11.8** Consider the two-dimensional Euclidean geometry EG$(2, 2^7)$ over GF$(2^7)$. On decomposing this geometry with respect to a chosen parallel bundle $\mathcal{P}(2, 1)$ of lines, we obtain 128 connecting parallel bundles of lines with respect to $\mathcal{P}(2, 1)$. From these 128 connecting parallel bundles of lines with respect to $\mathcal{P}(2, 1)$, we can construct a $128 \times 128$ array $\mathbf{H}^{(1)}_{\text{EG},d}$ of $128 \times 128$ permutation matrices. Take a $32 \times 64$ subarray $\mathbf{H}^{(1)}_{\text{EG},d}(32, 64)$ from $\mathbf{H}^{(1)}_{\text{EG},d}$ as the base array for masking. Construct a $32 \times 64$ masking matrix $\mathbf{Z}(32, 64) = [\mathbf{G}_0 \mathbf{G}_1]$ over GF(2) that consists of a row of two $32 \times 32$ circulants, $\mathbf{G}_0$ and $\mathbf{G}_1$, whose generators are two primitive nonequivalent 32-tuples over GF(2) with weight 3:

$$\mathbf{g}_0 = (10100100000000000000000000000000)$$

and

$$\mathbf{g}_1 = (10000010000000100000000000000000).$$

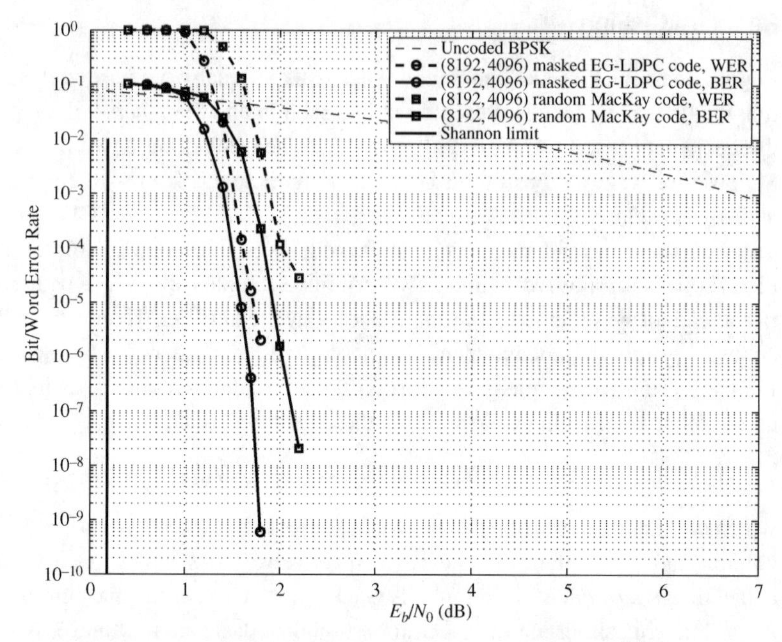


**Figure 11.11** Error performances of the (8192, 4096) masked EG-LDPC code and a (8192, 4096) random MacKay code given in Example 11.8.

These two masking-matrix generators are designed such that the Tanner graph of the masking matrix $\mathbf{Z}(k, r) = [\mathbf{G}_0\mathbf{G}_1]$ is free of cycles of length 4 and has a small number of cycles of length 6. By masking the base array $\mathbf{H}^{(1)}_{\mathrm{EG},d}(32, 64)$ with $\mathbf{Z}(32, 64)$, we obtain a masked array $\mathbf{M}(32, 64)$ with 192 permutation and 1856 zero matrices of size $128 \times 128$. However, the base array $\mathbf{H}^{(1)}_{\mathrm{EG},d}(32, 64)$ consists of 2048 permutation matrices. Therefore, $\mathbf{M}(32, 64)$ is a much sparser matrix than $\mathbf{H}^{(1)}_{\mathrm{EG},d}(32, 64)$. $\mathbf{M}(32, 64)$ is a $4096 \times 8192$ matrix over GF(2) with column and row weights 3 and 6, respectively. The null space of $\mathbf{M}(32, 64)$ gives a rate-1/2 (3, 6)-regular (8192, 4096) EG-LDPC code. The error performance of this code with iterative decoding using the SPA (100 iterations) is shown in Figure 11.11. At a BER of $10^{-9}$, it performs 1.6 dB from the Shannon limit. It has no error floor down to a BER of $5 \times 10^{-10}$. Also included in Figure 11.11 is the error performance of a rate-1/2 (3, 6)-regular (8192, 4096) random code constructed by computer. We see that the EG-LDPC code outperforms the random code.

### 11.4.3 Irregular Masking

As defined in Chapter 5, an irregular LDPC code is given by the null space of a sparse matrix with varying column and/or row weights. Consequently, its Tanner graph $\mathcal{T}$ has varying variable-node degrees and/or varying check-node degrees. As shown in Chapter 5, the degree distributions of these two types of nodes are expressed in terms of two polynomials, $\tilde{\lambda}(X) = \sum_{i=1}^{d_{\tilde{\lambda}}} \tilde{\lambda}_i X^{i-1}$ and $\tilde{\rho}(X) = \sum_{i=1}^{d_{\tilde{\rho}}} \tilde{\rho}_i X^{i-1}$, where $\tilde{\lambda}_i$ and $\tilde{\rho}_i$ denote the fractions of variable and check nodes in $\mathcal{T}$ with degree $i$, respectively, and $d_{\tilde{\lambda}}$ and $d_{\tilde{\rho}}$ denote the maximum variable- and check-node degrees, respectively. Since the variable and check nodes of $\mathcal{T}$ correspond to the columns and rows of the adjacency matrix $\mathbf{H}$ of $\mathcal{T}$, $\tilde{\lambda}(X)$ and $\tilde{\rho}(X)$ also give the column and row weight distributions of $\mathbf{H}$. It has been shown that the error performance of an irregular LDPC code depends on the variable- and check-node degree distributions of its Tanner graph $\mathcal{T}$ [26] and Shannon-limit-approaching LDPC codes can be designed by optimizing the two degree distributions by utilizing the evolution of the probability densities (called *density evolution*, Chapter 9) of the messages passed between the two types of nodes in a belief-propagation decoder. In code construction, once the degree distributions $\tilde{\lambda}(X)$ and $\tilde{\rho}(X)$ have been derived, a Tanner graph is constructed by connecting the variable nodes and check nodes with edges based on these two degree distributions. Since the selection of edges in the construction of a Tanner graph is not unique, edge selection is carried out in a *random manner* by computer search. During the edge-selection process, an effort must be made to ensure that the resultant Tanner graph does not contain short cycles, especially cycles of length 4. Once the code's Tanner graph $\mathcal{T}$ has been constructed, the incidence matrix $\mathbf{H}$ of $\mathcal{T}$ is derived from the edges that connect the variable and check nodes of $\mathcal{T}$ and is used as the parity-check matrix of an irregular LDPC code. The null space of $\mathbf{H}$ gives a *random-like* irregular LDPC code.

Geometry decomposition (presented in Section 11.3) and array masking (presented earlier) can be used for constructing irregular LDPC codes based on the degree distributions of variable and check nodes of their Tanner graphs derived from density evolution. First, we choose an appropriate Euclidean geometry with which to construct an RC-constrained array $\mathbf{H}$ of permutation matrices $\left(\text{either } \mathbf{H}_{\text{EG},d}^{(1)} \text{ or } \mathbf{H}_{\text{EG},d}^{(2)}\right)$ using geometry decomposition. Take a $k \times r$ subarray $\mathbf{H}(k,r)$ from $\mathbf{H}$ such that the null space of $\mathbf{H}(k,r)$ gives an LDPC code with length and rate equal to or close to the desired length $n$ and rate $R$. Let $\tilde{\lambda}(X)$ and $\tilde{\rho}(X)$ be the designed degree distributions of the variable and check nodes of the Tanner graph of a desired LDPC code of rate $R$. Construct a $k \times r$ masking matrix $\mathbf{Z}(k,r)$ over GF(2) with column and row weight distributions equal to or close to $\tilde{\lambda}(X)$ and $\tilde{\rho}(X)$, respectively. Then the masked matrix $\mathbf{M}(k,r) = \mathbf{Z}(k,r) \circledast \mathbf{H}(k,r)$ has column and row weight distributions *identical* to or *close* to $\tilde{\lambda}(X)$ and $\tilde{\rho}(X)$. The null space of $\mathbf{M}(k,r)$ gives an irregular LDPC code whose Tanner graph has variable- and check-node degree distributions identical to or close to $\tilde{\lambda}(X)$ and $\tilde{\rho}(X)$.

The above construction of irregular LDPC codes by masking and geometry decomposition is basically algebraic. It avoids the effort needed to construct a large random graph by computer search without cycles of length 4. Since the size of the masking matrix $\mathbf{Z}(k,r)$ is relatively small compared with the size of the code's Tanner graph, it is very easy to construct a matrix with column and row weight distributions identical to or close to the degree distributions $\tilde{\lambda}(X)$ and $\tilde{\rho}(X)$ derived from density evolution. Since the base array $\mathbf{H}(k,r)$ satisfies the RC constraint, the masked matrix $\mathbf{M}(k,r)$ also satisfies the RC constraint regardless of whether the masking matrix $\mathbf{Z}(k,r)$ satisfies

**Table 11.1** Column and row weight distributions of the masking matrix $\mathbf{Z}(12, 63)$ of Example 11.9

| Column weight distribution | | Row weight distribution | |
|---|---|---|---|
| Column weight | No. of columns | Row weight | No. of rows |
| 3 | 25 | 23 | 4 |
| 4 | 25 | 24 | 8 |
| 8 | 9 | | |
| 9 | 4 | | |

the RC constraint. Hence the irregular LDPC code given by the null space of $\mathbf{M}(k, r)$ contains no cycles of length 4 in its Tanner graph.

In [26], the degree distributions of a code graph that optimize the code performance over the AWGN channel for a given code rate are derived under the assumptions of infinite code length, a cycle-free Tanner graph, and an infinite number of decoding iterations. These degree distributions are no longer optimal when used for constructing codes of finite length and, in general, result in high error floors, mostly due to the large number of low-degree variable nodes, especially degree-2 variable nodes. Hence, they must be adjusted for constructing codes of finite length. Adjustment of degree distributions for finite-length LDPC codes will be discussed in Chapter 12. In the following, we simply give an example to illustrate how to construct an irregular code by masking an array of permutation matrices using degree distributions of the two types of nodes in a Tanner graph.

**Example 11.9** The following degree distributions of variable and check nodes of a Tanner graph are designed for an irregular LDPC code of length around 4000 and rate 0.82: $\tilde{\lambda}(X) = 0.4052X^2 + 0.3927X^3 + 0.1466X^7 + 0.0555X^8$ and $\tilde{\rho}(X) = 0.3109X^{22} + 0.6891X^{23}$. From this pair of degree distributions, we construct a $12 \times 63$ masking matrix $\mathbf{Z}(12, 63)$ with column and row weight distributions given in Table 11.1.

Note that the column and row weight distributions of the constructed masking matrix $\mathbf{Z}(12, 63)$ are not exactly identical to the designed degree distributions of the variable and check nodes given above. The average column and row weights of the masking matrix $\mathbf{Z}(12, 63)$ are 4.19 and 23.66. Next, we construct a $64 \times 64$ array $\mathbf{H}^{(1)}_{\mathrm{EG},d}$ of $64 \times 64$ permutation matrices over GF(2) based on the two-dimensional Euclidean geometry EG$(2, 2^6)$. Take a $12 \times 63$ subarray $\mathbf{H}^{(1)}_{\mathrm{EG},d}(12, 63)$ from $\mathbf{H}^{(1)}_{\mathrm{EG},d}$. $\mathbf{H}^{(1)}_{\mathrm{EG},d}(12, 63)$ is a $768 \times 4032$ matrix over GF(2) with column and row weights 12 and 63, respectively. By masking $\mathbf{H}^{(1)}_{\mathrm{EG},d}(12, 63)$ with $\mathbf{Z}(12, 63)$, we obtain a masked $768 \times 4032$ matrix $\mathbf{M}(12, 63) = \mathbf{Z}(12, 63) \circledast \mathbf{H}^{(1)}_{\mathrm{EG},d}(12, 63)$ with the column and row weight distributions given in Table 11.2. The null space of $\mathbf{M}(12, 63)$ gives a $(4032, 3264)$ irregular EG-LDPC code with rate 0.8113 whose Tanner graph has a girth of at least 6. The performance of this code over the binary-input AWGN channel

**Table 11.2** The weight distribution of the masked matrix $\mathbf{M}(12, 63)$ of Example 11.9

| Column weight distribution | | Row weight distribution | |
|---|---|---|---|
| Column weight | No. of columns | Row weight | No. of rows |
| 3 | 1536 | 23 | 256 |
| 4 | 1664 | 24 | 512 |
| 8 | 576 | | |
| 9 | 256 | | |

decoded with the iterative decoding using the SPA with 100 iterations is shown in Figure 11.12. At a BER of $10^{-6}$, it performs 1.2 dB from the Shannon limit.

## 11.5 Construction of QC-EG-LDPC Codes by Circulant Decomposition

In Section 11.1, we have shown that RC-constrained circulants can be constructed from type-1 incidence vectors of lines in a Euclidean geometry not passing through the origin. These circulants can be used to construct either cyclic EG-LDPC codes as shown in Section 11.1.1 or QC-EG-LDPC codes as shown in Section 11.1.2. In this section, we show that QC-EG-LDPC codes can be constructed by decomposing these circulants into arrays of circulants using *column and row decompositions* [27].

Consider an $n \times n$ circulant $\mathbf{G}$ over GF(2) with both column and row weight $w$. Since the column and row weights of circulant $\mathbf{G}$ are both $w$, we say that $\mathbf{G}$ has weight $w$. Label the rows and columns of $\mathbf{G}$ from 0 to $n-1$. $\mathbf{G}$ can be decomposed into a row of $n \times n$ circulants by *column splitting*. For $1 \leq t \leq w$, let $w_0, w_1, \ldots, w_{t-1}$ be a set of positive integers such that $w = w_0 + w_1 + \cdots + w_{t-1}$. Let $\mathbf{g}$ be the first column of $\mathbf{G}$. Partition the locations of the $w$ 1-components of $\mathbf{g}$ into $t$ disjoint sets, $R_0, R_1, \ldots, R_{t-1}$. The $j$th location set $R_j$ consists of $w_j$ locations of $\mathbf{g}$ where the components are 1-components. Split $\mathbf{g}$ into $t$ columns of the same length $n$, $\mathbf{g}_0, \mathbf{g}_1, \ldots, \mathbf{g}_{t-1}$, with the $w$ 1-components of $\mathbf{g}$ distributed among these new columns. For $0 \leq j < t$, we put the $w_j$ 1-components of $\mathbf{g}$ at the locations in $R_j$ into the $j$th new column $\mathbf{g}_j$ at the locations in $R_j$ of $\mathbf{g}_j$ and set all the other components of $\mathbf{g}_j$ to zero. For each new column $\mathbf{g}_j$, we form an $n \times n$ circulant $\mathbf{G}_j$ with $\mathbf{g}_j$ as the first column and its $n-1$ downward cyclic shifts as the other $n-1$ columns. This results in $t$ circulants of size $n \times n$, $\mathbf{G}_0, \mathbf{G}_1, \ldots, \mathbf{G}_{t-1}$, with weights $w_0, w_1, \ldots, w_{t-1}$, respectively. These circulants are called the *column descendants* of $\mathbf{G}$. The above process of column splitting decomposes the circulant $\mathbf{G}$ into a row of $t$ $n \times n$ circulants,

$$\mathbf{G}_{\text{col,decom}} = \begin{bmatrix} \mathbf{G}_0 & \mathbf{G}_1 & \cdots & \mathbf{G}_{t-1} \end{bmatrix}. \tag{11.21}$$

$\mathbf{G}_{\text{col,decom}}$ is called a *column decomposition* of $\mathbf{G}$. The subscript "col, decom" stands for "column decomposition." Since $\mathbf{G}$ satisfies the RC constraint, it is clear that $\mathbf{G}_{\text{col,decom}}$ and each circulant $\mathbf{G}_j$ in $\mathbf{G}_{\text{col,decom}}$ also satisfy the RC-constraint, that is, column decomposition preserves the RC-constraint structure.

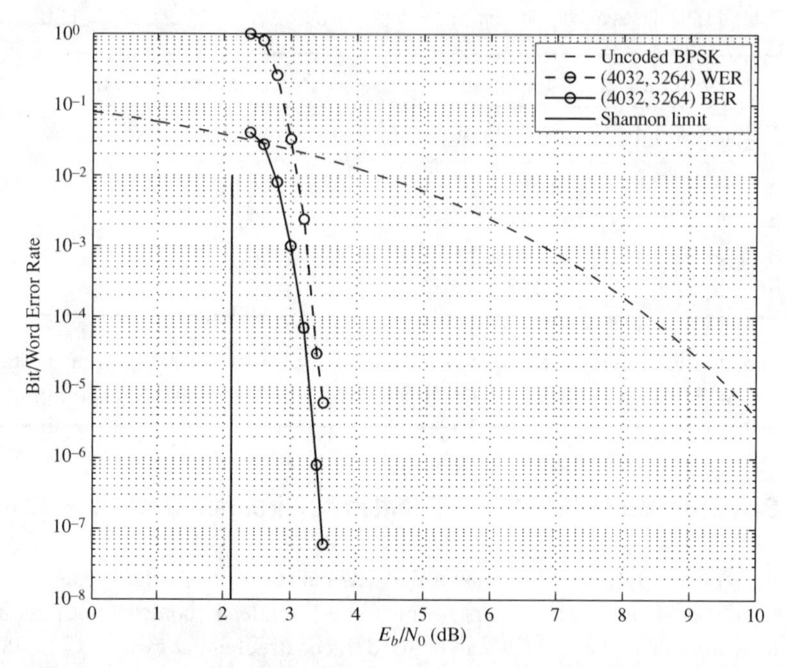


**Figure 11.12** The error performance of the (4032, 3264) irregular EG-LDPC code given in Example 11.9.

Let $c$ be a positive integer such that $1 \leq c \leq \max\{w_j: 0 \leq j < t\}$. For $0 \leq j < t$, let $w_{0,j}, w_{1,j}, \ldots, w_{c-1,j}$ be a set of non-negative integers such that $w_{0,j} + w_{1,j} + \cdots + w_{c-1,j} = w_j$. Each circulant $\mathbf{G}_j$ in $\mathbf{G}_{\text{col,decom}}$ can be decomposed into a column of $c$ circulants of size $n \times n$ with weights $w_{0,j}, w_{1,j}, \ldots, w_{c-1,j}$, respectively. This is accomplished by splitting the first row $\mathbf{r}_j$ of $\mathbf{G}_j$ into $c$ rows of the same length $n$, denoted by $\mathbf{r}_{0,j}, \mathbf{r}_{1,j}, \ldots, \mathbf{r}_{c-1,j}$, with $w_j$ 1-components of $\mathbf{r}_j$ distributed among these $c$ new rows, where the $i$th row $\mathbf{r}_{i,j}$ contains $w_{i,j}$ 1-components of $\mathbf{r}_j$. The row-splitting process is exactly the same as the column-splitting process described above. For $0 \leq j < t$, partition the locations of the $w_j$ 1-components of $\mathbf{r}_j$ into $c$ disjoint sets, $R_{0,j}, R_{1,j}, \ldots, R_{c-1,j}$. For $0 \leq i < c$, the $i$th location set $R_{i,j}$ consists of $w_{i,j}$ locations of $\mathbf{r}_j$ where the components are 1-components. For $0 \leq i < c$, we put the $w_{i,j}$ 1-components of $\mathbf{r}_j$ at the locations in $R_{i,j}$ into the $i$th new row $\mathbf{r}_{i,j}$ at the locations in $R_{i,j}$ of $\mathbf{r}_{i,j}$ and set all the other components of $\mathbf{r}_{i,j}$ to zeros. For each new row $\mathbf{r}_{i,j}$ with $0 \leq i < c$, we form an $n \times n$ circulant $\mathbf{G}_{i,j}$ with $\mathbf{r}_{i,j}$ as the first row and its $n-1$ right cyclic shifts as the other $n-1$ rows. The above row decomposition of $\mathbf{G}_j$ results in a column of $c$ circulants of size $n \times n$,

$$\mathbf{G}_{\text{row,decom},j} = \begin{bmatrix} \mathbf{G}_{0,j} \\ \mathbf{G}_{1,j} \\ \vdots \\ \mathbf{G}_{c-1,j} \end{bmatrix}, \tag{11.22}$$

where the $i$th circulant $\mathbf{G}_{i,j}$ has weight $w_{i,j}$ with $0 \leq i < c$. The circulants $\mathbf{G}_{0,j}, \mathbf{G}_{1,j}, \ldots, \mathbf{G}_{c-1,j}$ are called the *row descendants* of $\mathbf{G}_j$ and $\mathbf{G}_{\text{row,decom},j}$ is called a *row decomposition* of $\mathbf{G}_j$. In row splitting, we allow $w_{i,j} = 0$. If $w_{i,j} = 0$, then $\mathbf{G}_{i,j}$ is an $n \times n$ zero matrix. Since each circulant $\mathbf{G}_j$ in $\mathbf{G}_{\text{col,decom}}$ satisfies the RC constraint, the row decomposition $\mathbf{G}_{\text{row,decom},j}$ of $\mathbf{G}_j$ must also satisfy the RC constraint.

On replacing each $n \times n$ circulant $\mathbf{G}_j$ in $\mathbf{G}_{\text{col,decom}}$ given by (11.21) by its row decomposition $\mathbf{G}_{\text{row,decom},j}$, we obtain the following RC-constrained $c \times t$ array of $n \times n$ circulants over GF(2):

$$\mathbf{H}_{\text{array,decom}} = \begin{bmatrix} \mathbf{G}_{0,0} & \mathbf{G}_{0,1} & \cdots & \mathbf{G}_{0,t-1} \\ \mathbf{G}_{1,0} & \mathbf{G}_{1,1} & \cdots & \mathbf{G}_{1,t-1} \\ \vdots & \vdots & \ddots & \vdots \\ \mathbf{G}_{c-1,0} & \mathbf{G}_{c-1,1} & \cdots & \mathbf{G}_{c-1,t-1} \end{bmatrix}, \tag{11.23}$$

where the constituent circulant $\mathbf{G}_{i,j}$ has weight $w_{i,j}$ for $0 \leq i < c$ and $0 \leq j < t$. $\mathbf{H}_{\text{array,decom}}$ is called a $c \times t$ *array decomposition* of circulant $\mathbf{G}$. If all the constituent circulants in $\mathbf{H}_{\text{array,decom}}$ have the same weight $\sigma$, then $\mathbf{H}_{\text{array,decom}}$ is an RC-constrained $cn \times tn$ matrix with constant column and row weights $c\sigma$ and $t\sigma$, respectively. If all the constituent circulants of $\mathbf{H}_{\text{array,decom}}$ have weights equal to 1, then $\mathbf{H}_{\text{array,decom}}$ is a $c \times t$ array of $n \times n$ circulant permutation matrices (CPMs).

For any pair $(k, r)$ of integers with $1 \leq k \leq c$ and $1 \leq r \leq t$, let $\mathbf{H}_{\text{array,decom}}(k, r)$ be a $k \times r$ subarray of $\mathbf{H}_{\text{array,decom}}$. The null space of $\mathbf{H}_{\text{array,decom}}(k, r)$ gives a QC-EG-LDPC code $\mathcal{C}_{\text{EG},qc,\text{decom}}$ over GF(2) of length $rn$, whose Tanner graph has a girth of at least 6. The above construction gives a class of QC-EG-LDPC codes.

---

**Example 11.10** Consider the three-dimensional Euclidean geometry EG$(3, 2^3)$. As shown in Example 11.2, using the type-1 incidence vectors of the lines in this geometry not passing through the origin, we can construct nine circulants of size $511 \times 511$, $\mathbf{G}_0, \mathbf{G}_1, \ldots, \mathbf{G}_8$, each with weight 8. Suppose we take eight of these circulants and arrange them in a row $\mathbf{G} = [\mathbf{G}_0\mathbf{G}_1 \cdots \mathbf{G}_7]$. For $0 \leq j < 8$, decompose each constituent circulant $\mathbf{G}_j$ in $\mathbf{G}$ into a $2 \times 2$ array of $511 \times 511$ circulants over GF(2),

$$\mathbf{M}_j = \begin{bmatrix} \mathbf{G}_{0,0}^{(j)} & \mathbf{G}_{0,1}^{(j)} \\ \mathbf{G}_{1,0}^{(j)} & \mathbf{G}_{1,1}^{(j)} \end{bmatrix},$$

where each constituent circulant in $\mathbf{M}_j$ has weight 2. $\mathbf{M}_j$ is a $1022 \times 1022$ matrix over GF(2) with both column and row weight 4. On replacing each constituent circulant $\mathbf{G}_j$ in $\mathbf{G}$ by its $2 \times 2$ array decomposition $\mathbf{M}_j$, we obtain a $2 \times 16$ array $\mathbf{H}_{\text{array,decom}}$ of $511 \times 511$ circulants, each with weight 2. $\mathbf{H}_{\text{array,decom}}$ is a $2044 \times 8176$ matrix with column and row weights 4 and 32, respectively. The null space of this matrix gives a $(4, 32)$-regular $(8176, 7156)$ QC-EG-LDPC code with rate 0.8752, whose Tanner graph has a girth of at least 6. The error performance of this code decoded with iterative decoding using the SPA (50 iterations) is shown in Figures 11.13 and 11.14. At a BER of $10^{-6}$, the code performs 1 dB from the Shannon limit. The error floor of this code for the bit-error-rate performance is estimated to be below $10^{-15}$. This code is used in the NASA Landsat Data Continuation, which was launched in February

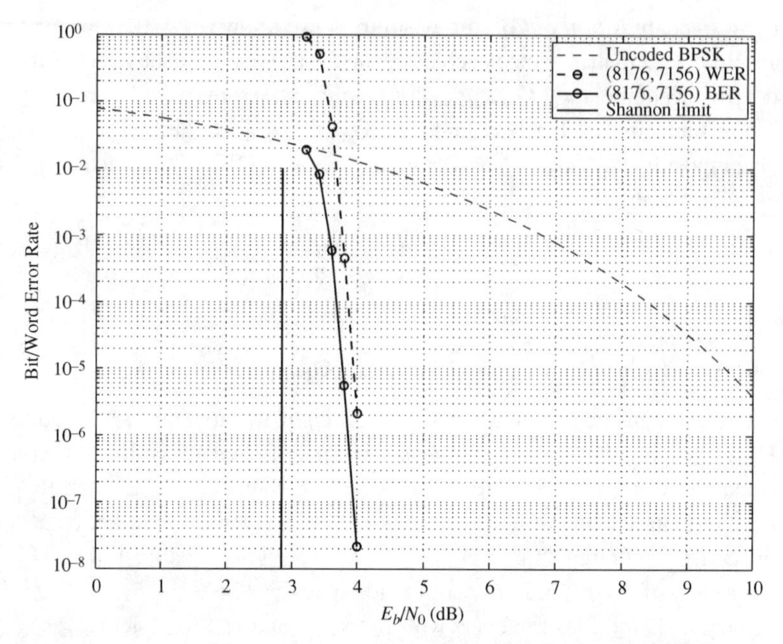


**Figure 11.13** The error performance of the (8176, 7156) QC-EG-LDPC code given in Example 11.10.

2013. Figure 11.14 shows the rate of decoding convergence of the code. This code is also being used in NASA's Cruise Exploration Shuttle Mission.

A VLSI decoder for this code has been built. Using this decoder, bit-error performance down to $10^{-14}$ can be simulated as shown in Figure 11.15. We see that there is no error floor even down to this low bit error rate.

---

Note that the matrix $\mathbf{G}_{col,decom}$ given by (11.21) consists of a row of $t$ $n{\times}n$ circulants. For $1 \leq k \leq t$, any $k$ circulants in $\mathbf{G}_{col,decom}$ can be used to form a masking matrix for masking an $n \times kn$ array of permutation matrices constructed as in Section 11.3 for code construction. For example, consider the two-dimensional Euclidean geometry EG$(2, 2^3)$ over GF$(2^3)$. Using the type-1 incidence vectors of lines in EG$(2, 2^3)$ not passing through the origin, we can construct a single $63 \times 63$ circulant $\mathbf{G}$ with both column and row weight 8. By column decomposition, we can decompose $\mathbf{G}$ into two $63 \times 63$ column-descendant circulants, $\mathbf{G}_0$ and $\mathbf{G}_1$, each having both column and row weight 4. Then $[\mathbf{G}_0\mathbf{G}_1]$ is a $63 \times 126$ matrix over GF(2) with column and row weights 4 and 8, respectively, and the matrix $[\mathbf{G}_0\mathbf{G}_1]$ can be used as a $(4, 8)$-regular masking matrix. We can also decompose $\mathbf{G}$ into three $63 \times 63$ column-descendant circulants, $\mathbf{G}_0^*$, $\mathbf{G}_1^*$, and $\mathbf{G}_2^*$, with weights 3, 3, and 2, respectively, by column decomposition. We can use $\mathbf{G}_0^*$ and $\mathbf{G}_1^*$ to form a $63 \times 126$ $(3, 6)$-regular masking matrix $[\mathbf{G}_0^*\mathbf{G}_1^*]$ or we can use $\mathbf{G}_0^*$, $\mathbf{G}_1^*$, and $\mathbf{G}_2^*$ to form an irregular masking matrix $[\mathbf{G}_0^*\mathbf{G}_1^*\mathbf{G}_2^*]$ with two different column weights.

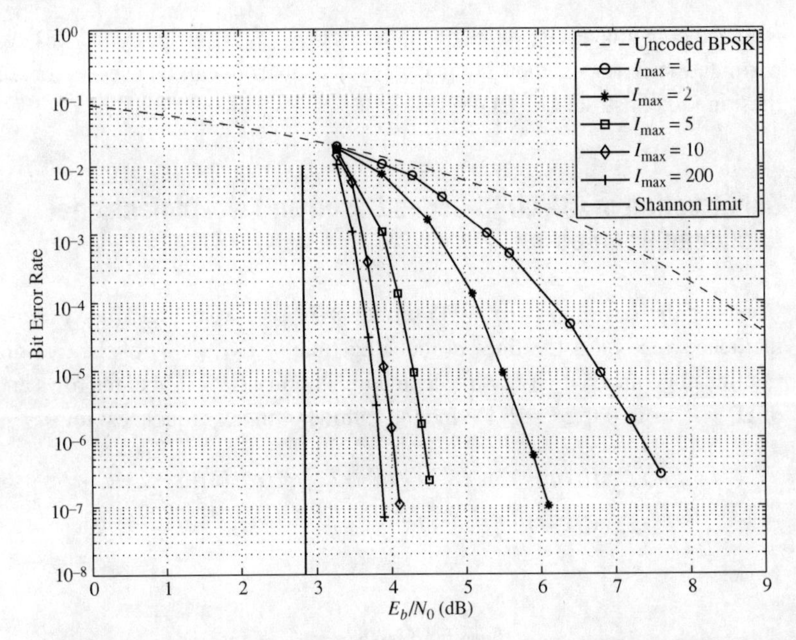


**Figure 11.14** The convergence rate of decoding of the (8176, 7156) QC-EG-LDPC code given in Example 11.10.

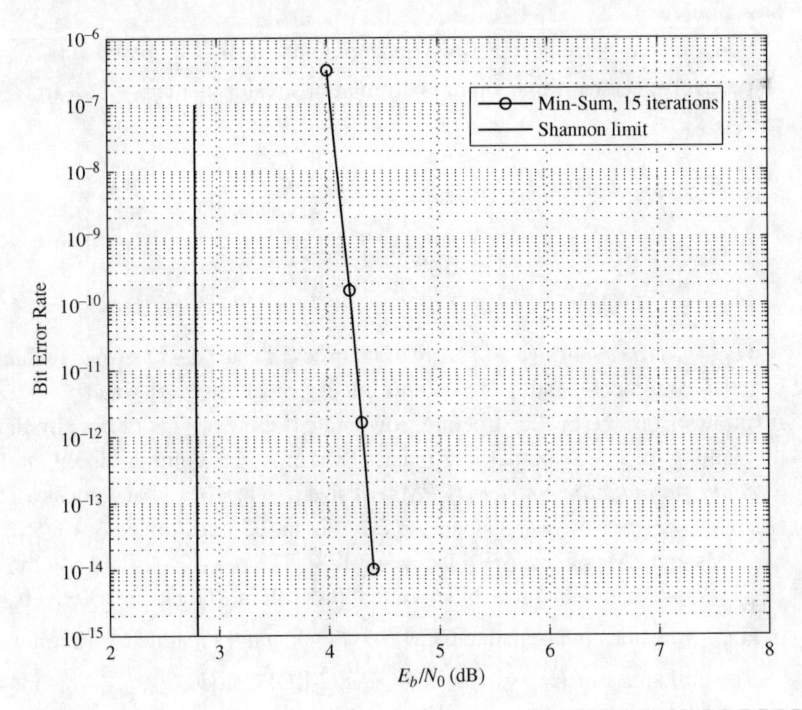


**Figure 11.15** The VLSI-simulated error performance of the (8176, 7156) QC-EG-LDPC code given in Example 11.10.

If the constituent circulants of the array $\mathbf{H}_{\text{array,decom}}$ given by (11.23) are circulant permutation matrices, then $\mathbf{H}_{\text{array,decom}}$ or its subarray can be used as a base array for masking to construct QC-LDPC codes.

## 11.6 Construction of QC-LDPC Codes Based on Two-Dimensional Euclidean Geometries

Consider the $(q^2-1) \times (q^2-1)$ circulant $\mathbf{H}^{(1)}_{\text{EG},c,1}$ constructed by using the type-1 incidence vectors of the lines in the subgeometry EG*(2, $q$) of the two-dimensional Euclidean geometry EG(2, $q$) presented in Section 11.1.1. Label the rows and columns of $\mathbf{H}^{(1)}_{\text{EG},c,1}$ from 0 to $q^2-2$. Define the following index sequences: for $0 \le i, j \le q$,

$$\pi_{row,i} = [i, (q-1)+i, 2(q-1)+i, \ldots, (q-2)(q-1)+i], \tag{11.24}$$

$$\pi_{col,j} = [j, (q-1)+j, 2(q-1)+j, \ldots, (q-2)(q-1)+j]. \tag{11.25}$$

Let

$$\pi_{row} = [\pi_{row,0}, \pi_{row,1}, \ldots, \pi_{row,q}], \tag{11.26}$$

$$\pi_{col} = [\pi_{col,0}, \pi_{col,1}, \ldots, \pi_{col,q}]. \tag{11.27}$$

Then $\pi_{row}$ and $\pi_{col}$ define reorderings of the column indices and row indices, respectively.

Suppose we permute the rows and columns of $\mathbf{H}^{(1)}_{\text{EG},c,1}$ based on $\pi_{row}$ and $\pi_{col}$, respectively. The row and column permutations result in a $(q^2-1) \times (q^2-1)$ matrix over GF(2):

$$\mathbf{H}^{(1)}_{\text{EG},qc,1} = [\mathbf{A}_{i,j}]_{0 \le i,j \le q} = \begin{bmatrix} \mathbf{A}_{0,0} & \mathbf{A}_{0,1} & \cdots & \mathbf{A}_{0,q} \\ \mathbf{A}_{1,0} & \mathbf{A}_{0,1} & \cdots & \mathbf{A}_{1,q} \\ \vdots & \vdots & & \vdots \\ \mathbf{A}_{q,0} & \mathbf{A}_{q,1} & \cdots & \mathbf{A}_{q,q} \end{bmatrix}. \tag{11.28}$$

$\mathbf{H}^{(1)}_{\text{EG},qc,1}$ is a $(q+1)\times(q+1)$ array of matrices of size $(q-1)\times(q-1)$ which consists of $q+1$ row-blocks and $q+1$ column-blocks of matrices of size $(q-1) \times (q-1)$. It follows from the cyclic structure, row and column weights of the circulant $\mathbf{H}^{(1)}_{\text{EG},c,1}$ that among the $q+1$ submatrices in each row-block (or column-block), $q$ of them are *circulant permutation matrices* (CPMs) of size $(q-1)\times(q-1)$ and one is a zero matrix (ZM) of size $(q-1)\times(q-1)$. Hence, $\mathbf{H}^{(1)}_{\text{EG},qc,1}$ is an RC-constrained $(q+1)\times(q+1)$ array of CPMs and ZMs of size $(q-1)\times(q-1)$ [28]. The row-blocks and column-blocks of $\mathbf{H}^{(1)}_{\text{EG},qc,1}$ are called CPM row-blocks and CPM column-blocks, respectively. Each ZM in $\mathbf{H}^{(1)}_{\text{EG},qc,1}$ resides in a separate CPM row-block and a separate CPM column-block.

The null space of $\mathbf{H}^{(1)}_{\text{EG},qc,1}$ gives a QC-EG-LDPC code $C_{\text{EG},qc,1}$ which is combinatorically equivalent to the cyclic EG-LDPC code $C_{\text{EG},c,1}$. Let $g$ and $r$ be two positive integers such that $1 \le g \le r \le q+1$. Take a $g \times r$ subarray $\mathbf{H}^{(1)}_{\text{EG},qc,1}(g,r)$ from $\mathbf{H}^{(1)}_{\text{EG},qc,1}$.

The null space over GF(2) of $\mathbf{H}^{(1)}_{\text{EG},qc,1}(g,r)$ gives a QC-EG-LDPC code $\mathcal{C}_{\text{EG},qc,1}(g,r)$ of length $r(q-1)$ with rate at least $(r-g)/r$. If $\mathbf{H}^{(1)}_{\text{EG},qc,1}(g,r)$ does not contain ZMs, $\mathcal{C}_{\text{EG},qc,1}(g,r)$ is a $(g,r)$-regular QC-EG-LDPC code. With various choices of the $(g,r)$ pairs, we can construct a family of QC-EG-LDPC codes of various lengths and rates for a given two-dimensional Euclidean geometry EG(2, $q$) over GF($q$).

Since any two rows in a CPM row-block of $\mathbf{H}^{(1)}_{\text{EG},qc,1}(g,r)$ do not have any 1-entry in common (in the same location), the $q-1$ rows in each CPM row-block of $\mathbf{H}^{(1)}_{\text{EG},qc,1}(g,r)$ correspond to $q-1$ parallel lines in a parallel bundle of EG*(2, $q$). The array $\mathbf{H}^{(1)}_{\text{EG},qc,1}(g,r)$ has another interesting and useful structural property. Each CPM row-block of $\mathbf{H}_{\text{EG},qc,1}$ is the cyclic shift of the CPM row-block above it, but when the rightmost CPM, denoted by CPM$_r$, is shifted around to the leftmost position, and all the rows in CPM$_r$ are simultaneously cyclically shifted one position to the right within the $q-1$ column positions of CPM$_r$, which gives another CPM, denoted by CPM$_l$, that is, the generator of CPM$_l$ is the cyclic shift of the generator of CPM$_r$ one place to the right. The top CPM row-block of $\mathbf{H}^{(1)}_{\text{EG},qc,1}(g,r)$ is the block cyclic shift of its last CPM row-block. This property is referred to as the *CPM row-block cyclic structure* of $\mathbf{H}^{(1)}_{\text{EG},qc,1}(g,r)$.

With the CPM row-block cyclic structure, the array $\mathbf{H}^{(1)}_{\text{EG},qc,1}(g,r)$ can be formed by using its top CPM row-block $\mathbf{A}_0 = [\mathbf{A}_{0,0}\mathbf{A}_{0,1}\ldots\mathbf{A}_{0,q}]$. The top CPM row-block $\mathbf{A}_0$ of $\mathbf{H}^{(1)}_{\text{EG},qc,1}$ can be formed by taking the $q-1$ rows, labeled by $0, q+1, 2(q+1), \ldots, (q-2)(q+1)$, from the circulant $\mathbf{H}^{(1)}_{\text{EG},c,1}$ to form a $(q-1)\times(q^2-1)$ matrix $\mathbf{H}_0$. Then permute the $(q^2-1)$ columns of $\mathbf{H}_0$ with the permutation $\pi_{col}$ defined by (11.25) and (11.27). The column permutation of $\mathbf{H}_0$ results in the top CPM row-block $\mathbf{A}_0$ of $\mathbf{H}^{(1)}_{\text{EG},qc,1}$. In fact, the matrix $\mathbf{H}_0$ can be formed by taking the top row $\mathbf{h}_0$ of the circulant $\mathbf{H}^{(1)}_{\text{EG},c,1}$ and cyclically shifting it to the right $q+1, 2(q+1), \ldots, (q-2)(q+1)$ positions to obtain the other $q-2$ rows of $\mathbf{H}_0$. Hence, to form the array $\mathbf{H}^{(1)}_{\text{EG},qc,1}$, we only need the top row of $\mathbf{H}^{(1)}_{\text{EG},c,1}$ (or the incidence vector of any line in the subgeometry EG*(2, $q$) of EG(2, $q$)). This enables us to construct of $\mathbf{H}^{(1)}_{\text{EG},qc,1}$ without going through the construction of the circulant $\mathbf{H}^{(1)}_{\text{EG},c,1}$ and permuting its rows and columns.

The above developments show that both the circulant $\mathbf{H}^{(1)}_{\text{EG},c,1}$ and the array $\mathbf{H}^{(1)}_{\text{EG},qc,1}$ of CPMs for generating the cyclic EG-LDPC code $\mathcal{C}_{\text{EG},c,1}$ and the QC-EG-LDPC code $\mathcal{C}_{\text{EG},qc,1}$ can be constructed based on the first row $\mathbf{h}_0$ of $\mathbf{H}^{(1)}_{\text{EG},c,1}$. Note that the first row $\mathbf{h}_0$ of $\mathbf{H}^{(1)}_{\text{EG},c,1}$ can be the type-1 incidence vectors of *any line* in the subgeometry EG*(2, $q$) of the two-dimensional Euclidean geometry EG(2, $q$) over GF($q$). In the following, we give a more concise representation of $\mathbf{H}^{(1)}_{\text{EG},qc,1}$ or EG*(2, $q$) which facilitates design and construction of QC-EG-LDPC codes.

Applying the permutation $\pi_{col} = [\pi_{col,0}, \pi_{col,1}, \ldots, \pi_{col,q}]$ defined by (11.27) to the $q^2-1$ coordinates of the first row $\mathbf{h}_0$ of $\mathbf{H}^{(1)}_{\text{EG},c,1}$, we obtain a permuted $(q^2-1)$-tuple $\mathbf{h}^*0 = (\mathbf{h}^*_{0,0}, \mathbf{h}^*_{0,1}, \ldots, \mathbf{h}^*_{0,q})$ which consists of $q+1$ sections, each consisting of $q-1$ components of $\mathbf{h}_0$. Following the construction of $\mathbf{H}^{(1)}_{\text{EG},qc,1}$ given above, we readily see that there are $q$ sections in $\mathbf{h}^*_0$, each consisting of a *single* 1-component, and one section which consists of only 0-components. The permuted $(q^2 - 1)$-tuple $\mathbf{h}_0^*$ is simply the top row of $\mathbf{H}_{\mathrm{EG},qc,1}^{(1)}$. The nonzero sections are the generators (top rows) of the CPMs in the top CPM row-block $\mathbf{A}_0$ of $\mathbf{H}_{\mathrm{EG},qc,1}^{(1)}$.

Let $\beta$ be a primitive element of GF($q$), a subfield of GF($q^2$). Then $\beta^{-\infty} = 0$, $\beta^0 = 1, \beta, \beta^2, \ldots, \beta^{q-2}$ form all the elements of GF($q$). Since GF($q$) is a subfield of GF($q^2$), $\beta = \alpha^{q+1}$ where $\alpha$ is a primitive element of GF($q^2$). Let $\{-\infty, 0, 1, \ldots, q-2\}$ be the set of powers of $\beta^{-\infty} = 0, \beta^0 = 1, \beta, \beta^2, \ldots, \beta^{q-2}$. For $0 \leq j \leq q$ and $l_j \in \{-\infty, 0, 1, \ldots, q-2\}$, let $l_j$ be the location of the single 1-component in the $j$th section $\mathbf{h}_{0,j}^*$ of $\mathbf{h}_0^*$. If $l_j = -\infty$, the $j$th section of $\mathbf{h}_{0,j}^*$ is the zero-section of $\mathbf{h}_0^*$. We represent the $j$th section $\mathbf{h}_{0,j}^*$ of $\mathbf{h}_0^*$ by the element $\beta^{l_j}$ in GF($q$). If the $j$th section $\mathbf{h}_{0,j}^*$ is the zero-section, it is represented by the zero element $\beta^{-\infty} = 0$ of GF($q$). Note that the representation is unique, and the mapping $\mathbf{h}_{0,j}^* \longleftrightarrow \beta^{l_j}$ is *one-to-one*. Since $\mathbf{h}_{0,j}^*$ is the generator of the $j$th CPM $\mathbf{A}_{0,j}$ of the top CPM row-block $\mathbf{A}_0$ of $\mathbf{H}_{\mathrm{EG},qc,1}^{(1)}$, it uniquely specifies $\mathbf{A}_{0,j}$. Since $\mathbf{h}_{0,j}^* \longleftrightarrow \beta^{l_j}$ is one-to-one, the element $\beta^{l_j}$ uniquely specifies the CPM $\mathbf{A}_{0,j}$. Hence, we can represent $\mathbf{A}_{0,j}$ by $\beta^{l_j}$ and the mapping $\beta^{l_j} \longleftrightarrow \mathbf{A}_{0,j}$ is one-to-one. We call $\mathbf{A}_{0,j}$ the *CPM-dispersion* of $\beta^{l_j}$ with respect to GF($q$), denoted by CPM($\beta^{l_j}$). Note that if $\mathbf{A}_{0,j}$ is the zero matrix, it is represented by the 0 element of GF($q$).

The mapping $\mathbf{h}_{0,j}^* \longleftrightarrow \beta^{l_j}, 0 \leq j \leq q$, maps $\mathbf{h}_0^* = (\mathbf{h}_{0,0}^*, \mathbf{h}_{0,1}^*, \ldots, \mathbf{h}_{0,q}^*)$ into a $(q+1)$-tuple $(\beta^{l_0}, \beta^{l_1}, \ldots, \beta^{l_q})$ over GF($q$). The CPM dispersion of each component in $(\beta^{l_0}, \beta^{l_1}, \ldots, \beta^{l_q})$ with respect to GF($q$) gives the top CPM-row-block $\mathbf{A}_0$ of the array $\mathbf{H}_{\mathrm{EG},qc,1}^{(1)}$. Cyclically shifting $\mathbf{A}_0$ in the block $q$ times gives the other $q$ CPM row-blocks of $\mathbf{H}_{\mathrm{EG},qc,1}^{(1)}$.

Based on the above developments, the array $\mathbf{H}_{\mathrm{EG},qc,1}^{(1)}$ can be uniquely represented by the following $(q+1) \times (q+1)$ matrix over GF($q$) with $(\beta^{l_0}, \beta^{l_1}, \ldots, \beta^{l_q})$ as the top row [28]:

$$\mathbf{B}_{\mathrm{EG},1} = \begin{bmatrix} \beta^{l_0} & \beta^{l_1} & \beta^{l_2} & \cdots & \beta^{l_{q-1}} & \beta^{l_q} \\ \beta^{l_q+1} & \beta^{l_0} & \beta^{l_1} & \cdots & \beta^{l_{q-2}} & \beta^{l_{q-1}} \\ \beta^{l_{q-1}+1} & \beta^{l_q+1} & \beta^{l_0} & \cdots & \beta^{l_{q-3}} & \beta^{l_{q-2}} \\ \vdots & \vdots & & & \vdots & \vdots \\ \beta^{l_2+1} & \beta^{l_3+1} & \beta^{l_4+1} & \cdots & \beta^{l_0} & \beta^{l_1} \\ \beta^{l_1+1} & \beta^{l_2+1} & \beta^{l_3+1} & \cdots & \beta^{l_q+1} & \beta^{l_0} \end{bmatrix}. \tag{11.29}$$

From (11.29), we see that each row $\mathbf{B}_{\mathrm{EG},1}$ is a cyclic shift of the row above it one position to the right, but when the rightmost component is shifted around to the leftmost position, it is multiplied $\beta$. This cyclic shift is called the $\beta$-multiplied cyclic shift. Note that the CPM obtained by simultaneously cyclically shifting all the rows of the CPM dispersion of $\beta^{l_j}$ one place to the right is the CPM-dispersion of $\beta^{l_j+1}$. Then it follows from the cyclic structure of $\mathbf{H}_{\mathrm{EG},qc,1}^{(1)}$ and $\mathbf{B}_{\mathrm{EG},1}$ that $\mathbf{H}_{\mathrm{EG},qc,1}^{(1)}$ is the CPM dispersion CPM($\mathbf{B}_{\mathrm{EG},1}$) of $\mathbf{B}_{\mathrm{EG},1}$ (the CPM dispersion of every entry in $\mathbf{B}_{\mathrm{EG},1}$).

Since the CPM dispersion of $\mathbf{B}_{\text{EG},1}$ gives the array $\mathbf{H}^{(1)}_{\text{EG},qc,1}$ and the array $\mathbf{H}^{(1)}_{\text{EG},qc,1}$ is the incidence matrix (in permuted form) of subgeometry $\text{EG}^*(2,q)$ of $\text{EG}(2,q)$, we call $\mathbf{B}_{\text{EG},1}$ the base matrix of $\text{EG}^*(2,q)$ (or $\mathbf{H}^{(1)}_{\text{EG},qc,1}$). Note that every row (column) of $\mathbf{B}_{\text{EG},1}$ contains a single 0 element. In constructing QC-EG-LDPC codes, we can use submatrices of $\mathbf{B}_{\text{EG},1}$ in conjunction with CPM-dispersion. This will make code design and construction much simpler.

Let $g$ and $r$ be two positive integers such that $1 \leq g \leq r \leq q+1$. Take a $g \times r$ submatrix $\mathbf{B}_{\text{EG},1}(g,r)$ from $\mathbf{B}_{\text{EG},1}$. The CPM dispersion of $\mathbf{B}_{\text{EG},1}(g,r)$ with respect to the multiplicative group of GF($q$) gives a $g \times r$ array $\mathbf{H}^{(1)}_{\text{EG},qc,1}(g,r)$ of CPMs (or ZMs) of size $(q-1)\times(q-1)$, which is a $g(q-1)\times r(q-1)$ matrix over GF(2). The null space over GF(2) of $\mathbf{H}^{(1)}_{\text{EG},qc,1}(g,r)$ gives a QC-EG-LDPC code $\mathcal{C}_{\text{EG},qc,1}(g,r)$ of length $r(q-1)$ with rate at least $(r-g)/r$. If $\mathbf{B}_{\text{EG},1}(g,r)$ does not contain zeros, the column and row weights of $\mathbf{H}^{(1)}_{\text{EG},qc,1}(g,r)$ are $g$ and $r$, respectively. In this case, $\mathcal{C}_{\text{EG},qc,1}(g,r)$ is a $(g,r)$-regular QC-EG-LDPC code. If $\mathbf{B}_{\text{EG},1}(g,r)$ contains zeros, $\mathbf{H}^{(1)}_{\text{EG},qc,1}(g,r)$ has two column weights, $g-1$ and $g$, and two row weights, $r-1$ and $r$. In this case, $\mathcal{C}_{\text{EG},qc,1}(g,r)$ is an irregular QC-EG-LDPC code.

---

**Example 11.11** Consider the subgeometry $\text{EG}^*(2,2^6)$ of the two-dimensional Euclidean geometry $\text{EG}(2,2^6)$ over $\text{GF}(2^6)$ constructed in Example 11.1. Take a line $\mathcal{L}$ in $\text{EG}^*(2,2^6)$ and form its type-1 incidence vector $\mathbf{v} = (v_0, v_1, \ldots, v_{4094})$, which is a 4095-tuple over GF(2). Permuting the coordinates with the permutation $\pi_{col}$ defined by (11.27) with $q = 2^6$, we obtain a permuted vector $\mathbf{v}^* = (\mathbf{v}_0, \mathbf{v}_1, \ldots, \mathbf{v}_{64})$, which consists of 65 sections, each with 63 components of $\mathbf{v}$. There are 64 sections, each consisting of a single 1-component, and one zero-section.

Let $\alpha$ be a primitive element of $\text{GF}(2^{12})$ and $\beta = \alpha^{65}$. The order of $\beta$ is 63. The set $\{\beta^{-\infty} = 0, \beta^0 = 1, \beta, \beta^2, \ldots, \beta^{62}\}$ forms a subfield of $\text{GF}(2^{12})$. Next, we represent each section $\mathbf{v}_j$ in $\mathbf{v}^*$ by an element $\beta^{l_j}$ in accordance with the location $l_j$ of its single-1 component. This results in a 65-tuple $(\beta^{l_0}, \beta^{l_1}, \ldots, \beta^{l_{64}})$ over $\text{GF}(2^6)$ (see Problem 11.20).

Using this 65-tuple and its 64 $\beta$-multiplied cyclic shifts, we form the base matrix $\mathbf{B}_{\text{EG},1}$ of $\text{EG}^*(2,2^6)$, which is a $65\times 65$ matrix over $\text{GF}(2^6)$. The CPM-dispersion of $\mathbf{B}_{\text{EG},1}$ with respect to the multiplicative group of $\text{GF}(2^6)$ gives a $65\times 65$ array $\mathbf{H}^{(1)}_{\text{EG},qc,1}$ of CPMs and ZMs of size $63\times 63$, which is a $4095\times 4095$ matrix with column and row weights 64. $\mathbf{H}^{(1)}_{\text{EG},qc,1}$ is the incidence matrix of $\text{EG}^*(2,2^6)$ in block-cyclic form.

The null space over GF(2) of $\mathbf{H}^{(1)}_{\text{EG},qc,1}$ gives a $(4095, 3367)$ QC-EG-LDPC code, which is combinatorically equivalent to the $(4095, 3367)$ cyclic EG-LDPC code constructed in Example 11.1. Both codes have the same error performance (see Figure 11.1).

---

So far, the SPA has been used to decode all the EG-LDPC codes constructed in preceding examples. Decoding a long LDPC code with the SPA requires a very large computational complexity, which may exclude it from practical applications. It was shown in Chapter 5 that by decoding an LDPC code with the min-sum algorithm (MSA), computational complexity can be significantly reduced. With proper scaling, the MSA can achieve performance very close to the SPA with a small or negligible degradation. In the following examples, we use the MSA to decode long QC-EG-LDPC codes.

---

**Example 11.12** Consider the $65 \times 65$ array $\mathbf{H}^{(1)}_{\mathrm{EG},qc,1}$ of CPMs and ZMs of size $63 \times 63$ constructed in Example 11.11.

Let $g = 5$ and $r = 50$. Take a $5 \times 50$ submatrix $\mathbf{B}_{\mathrm{EG},1}(5, 50)$ from $\mathbf{B}_{\mathrm{EG},1}$ (avoiding 0-entries). The CPM dispersions of the entries in $\mathbf{B}_{\mathrm{EG},1}(5, 50)$ with respect to the multiplicative group of GF($2^6$) give a $5 \times 50$ array $\mathbf{H}^{(1)}_{\mathrm{EG},qc,1}(5, 50)$ of CPMs of size $63 \times 63$, which is a $315 \times 3150$ matrix over GF(2) with column and row weights 5 and 50, respectively.

The null space of $\mathbf{H}^{(1)}_{\mathrm{EG},qc,1}(5, 50)$ gives a $(5, 50)$-regular $(3150, 2827)$ QC-EG-LDPC code $\mathcal{C}_{\mathrm{EG},qc,1}(5, 50)$ of length 3150 with rate 0.8974. Decode the code $\mathcal{C}_{\mathrm{EG},qc,1}(5, 50)$ with the scaled MSA. The BER and BLER performances of the code over the AWGN channel using the BPSK signaling decoded with 5, 10, and 50 iterations of the scaled MSA are shown in Figure 11.16.

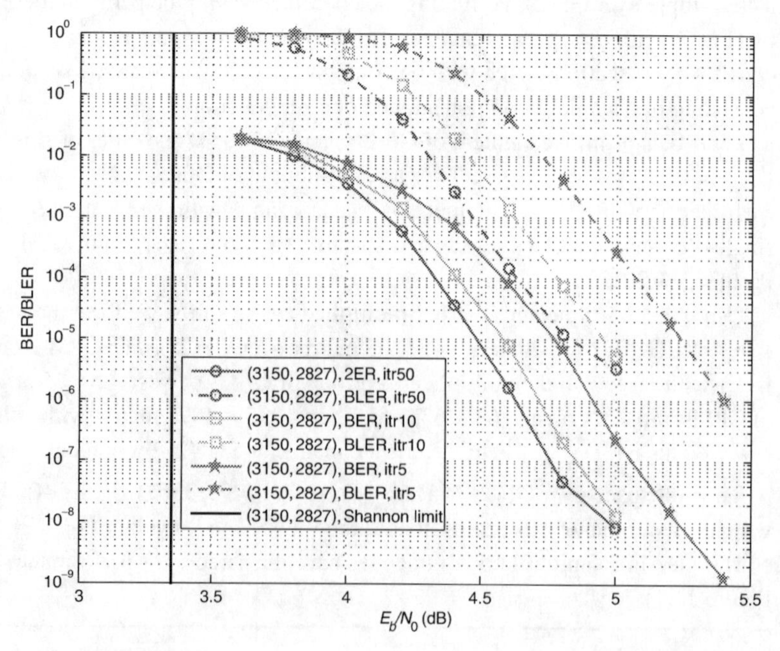


**Figure 11.16** Simulated error performance of the $(5, 50)$-regular $(3150, 2827)$ QC-EG-LDPC code $\mathcal{C}_{\mathrm{EG},qc,1}(5, 50)$ given in Example 11.12.

---

**Example 11.13** In this example, we construct a relatively long QC-EG-LDPC code based on the subgeometry $\mathrm{EG}^*(2,2^7)$ of the two-dimensional Euclidean geometry $\mathrm{EG}(2,2^7)$ over $\mathrm{GF}(2^7)$. Using a single line in $\mathrm{EG}^*(2,2^7)$, we construct the base matrix $\mathbf{B}_{\mathrm{EG},1}$ of $\mathrm{EG}^*(2,2^7)$, which is a $129 \times 129$ matrix over $\mathrm{GF}(2^7)$. The CPM dispersion of $\mathbf{B}_{\mathrm{EG},1}$ with respect to the multiplicative group of $\mathrm{GF}(2^7)$ gives a $129 \times 129$ array $\mathbf{H}^{(1)}_{\mathrm{EG},qc,1}$ of CPMs and ZMs of size $127 \times 127$.

Let $g = 6$ and $r = 80$. Take a $6 \times 80$ submatrix $\mathbf{B}_{\mathrm{EG},1}(6,80)$ from $\mathbf{B}_{\mathrm{EG},1}$ (avoiding 0-entries), say, deleting 49 columns from the top six rows of $\mathbf{B}_{\mathrm{EG},1}$. The CPM dispersion of $\mathbf{B}_{\mathrm{EG},1}(6,80)$ with respect to $\mathrm{GF}(2^7)$ gives a $6 \times 80$ array $\mathbf{H}^{(1)}_{\mathrm{EG},qc,1}(6,80)$ of CPMs of size $127 \times 127$, which is a $762 \times 10\,160$ matrix over GF(2). The null space over GF(2) of $\mathbf{H}^{(1)}_{\mathrm{EG},qc,1}(6,80)$ is a (6, 80)-regular (10 160, 9473) QC-EG-LDPC code $\mathcal{C}_{\mathrm{EG},qc,1}(6,80)$ of length 10 160 with rate 0.9323, a long
high-rate code.

The BER and BLER performances of the code $\mathcal{C}_{\mathrm{EG},qc,1}(6,80)$ decoded with 5, 10, and 50 iterations of the scaled MSA are shown in Figure 11.17. At a BER of $10^{-8}$ with 10 and 50 decoding iterations, the code performs 1.26 and 1.2 dB from the Shannon limit, respectively.

Next, we construct a short low-rate code by setting $g = 4$ and $r = 8$. Take a $4 \times 8$ submatrix $\mathbf{B}_{\mathrm{EG},1}(4,8)$ from $\mathbf{B}_{\mathrm{EG},1}$ (avoiding 0-entries), say, eight columns from the top four rows of $\mathbf{B}_{\mathrm{EG},1}$. Suppose we mask $\mathbf{B}_{\mathrm{EG},1}(4,8)$ with the following $4 \times 8$ matrix over GF(2):

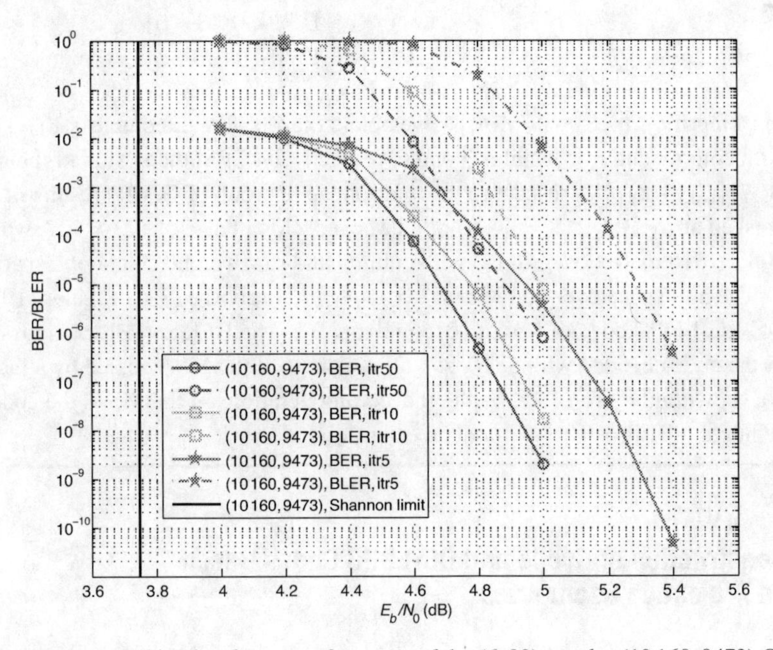


**Figure 11.17** Simulated error performance of the (6, 80)-regular (10 160, 9473) QC-EG-LDPC code $\mathcal{C}_{\mathrm{EG},qc,1}(6,80)$ given in Example 11.13.

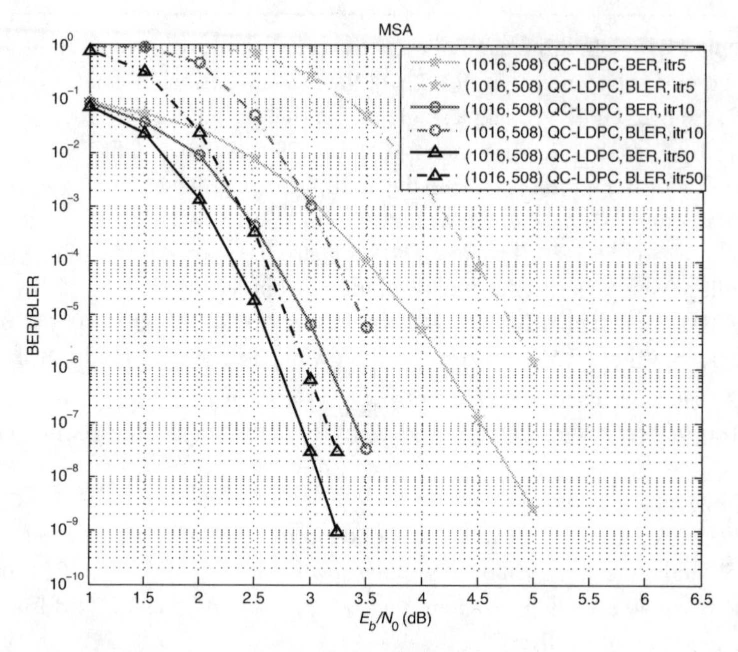


**Figure 11.18** Simulated error performance of the (3, 6)-regular (1016, 508) QC-EG-LDPC code given in Example 11.13.

$$\mathbf{Z}(4,8) = \begin{bmatrix} 1 & 0 & 1 & 0 & 1 & 1 & 1 & 1 \\ 0 & 1 & 0 & 1 & 1 & 1 & 1 & 1 \\ 1 & 1 & 1 & 1 & 1 & 0 & 1 & 0 \\ 1 & 1 & 1 & 1 & 0 & 1 & 0 & 1 \end{bmatrix}. \quad (11.30)$$

Masking $\mathbf{B}_{\text{EG},1}(4,8)$ with $\mathbf{Z}(4,8)$, we obtain the $4 \times 8$ masked matrix $\mathbf{B}_{\text{EG},1,\text{mask}}(4,8)$ with column and row weights 3 and 6, respectively. CPM dispersion of the matrix $\mathbf{B}_{\text{EG},1,\text{mask}}(4,8)$ with respect to the multiplicative group of GF($2^7$) results in a $4 \times 8$ masked array $\mathbf{H}^{(1)}_{\text{EG},qc,1,\text{mask}}(4,8)$ of CPMs and ZMs of size of $127 \times 127$, which is a $508 \times 1016$ matrix over GF(2) with column and row weights 3 and 6, respectively.

The null space over GF(2) of $\mathbf{H}^{(1)}_{\text{EG},qc,1,\text{mask}}(4,8)$ gives a (3, 6)-regular (1016, 508) QC-EG-LDPC code of length 1016 and rate 0.5. The BER and BLER performances of the code decoded with 5, 10, and 50 iterations of the MSA scaled by a factor 0.5 are shown in Figure 11.18. With 50 decoding iterations of the MSA, the code achieves a BER of $10^{-9}$ without an error floor.

## 11.7 Construction of Cyclic and QC-LDPC Codes Based on Projective Geometries

In the last six sections, various methods have been presented for constructing cyclic, quasi-cyclic, and regular LDPC codes based on the cyclic and parallel structures of lines in Euclidean geometries. In this section, we will show that lines of projective geometries can also be used for constructing LDPC codes. However, the lines in a projective geometry have the cyclic structure but not the parallel structure. As a result, only the construction methods based on the cyclic structure of lines presented in Sections 11.1, 11.4, and 11.5 can be applied to construct cyclic and quasi-cyclic LDPC codes based on the lines of projective geometries. Projective geometries and their structural properties have been discussed in Chapter 2. In the following, we give a brief review of the structural properties of lines in these geometries for code construction.

### 11.7.1 Cyclic PG-LDPC Codes

Consider the $m$-dimensional projective geometry PG$(m,q)$ over GF$(q)$ with $m \geq 2$. This geometry consists of

$$n = (q^{m+1} - 1)/(q-1) \tag{11.31}$$

points and

$$J_4 \triangleq J_{\text{PG}}(m,1) = \frac{(q^m - 1)(q^{m+1} - 1)}{(q^2 - 1)(q - 1)} \tag{11.32}$$

lines. Each line consists of $q+1$ points. Two lines in PG$(m,q)$ either do not have any point in common or they intersect at one and only one point. For any given point, there are

$$g \triangleq g_{\text{PG}}(m,1) = (q^m - 1)/(q-1) \tag{11.33}$$

lines that intersect at this point. Let GF$(q^{m+1})$ be the extension field of GF$(q)$. Let $\alpha$ be a primitive element in GF$(q^{m+1})$. The $n = (q^{m+1}-1)/(q-1)$ points of PG$(m,q)$ can be represented by the $n$ elements $\alpha^0, \alpha, \alpha^2, \ldots, \alpha^{n-1}$ of GF$(q^{m+1})$ (see Chapter 2).

Let $\mathcal{L}$ be a line in PG$(m,q)$. The incidence vector of $\mathcal{L}$ is defined as an $n$-tuple, $\mathbf{v}_{\mathcal{L}} = (v_0, v_1, \ldots, v_{n-1})$, whose components correspond to the $n$ points of PG$(m,q)$, where $v_j = 1$ if and only if $\alpha^j$ is a point on $\mathcal{L}$, otherwise $v_j = 0$. The weight of the incidence vector of a line in PG$(m,q)$ is $q+1$. The (right or left) cyclic shift of $\mathbf{v}_{\mathcal{L}}$ is the incidence vector of another line in PG$(m,q)$. Form a $J_4 \times n$ matrix $\mathbf{H}_{\text{PG},c}$ over GF(2) with the incidence vectors of all the $J_4$ lines in PG$(m,q)$ as rows. $\mathbf{H}_{\text{PG},c}$ has column and row weights $(q^m-1)/(q-1)$ and $q+1$, respectively. Since no two lines in PG$(m,q)$ have more than one point in common, no two rows (or two columns) of $\mathbf{H}_{\text{PG},c}$ have more than one 1-component in common. Hence, $\mathbf{H}_{\text{PG},c}$ satisfies the RC constraint. The null space of $\mathbf{H}_{\text{PG},c}$ gives a cyclic LDPC code $\mathcal{C}_{\text{PG},c}$ of length $n = (q^{m+1} - 1)/(q-1)$ with minimum distance at least $(q^m - 1)/(q-1) + 1$.

The above construction gives a class of cyclic PG-LDPC codes. The generator polynomial of $\mathcal{C}_{\text{PG},c}$ can be determined in exactly the same way as was given for a cyclic EG-LDPC code in Section 11.1.1. Cyclic PG-LDPC codes were also discovered in the late 1960s [19, 20, 29–31] and shown to form a subclass of polynomial codes [32, 33], again not being recognized as LDPC codes until 2000 [1].

An interesting case is that in which the code-construction geometry is a two-dimensional projective geometry PG$(2, q)$ over GF$(q)$. For this case, the geometry consists of $q^2 + q + 1$ points and $q^2 + q + 1$ lines. The parity-check matrix $\mathbf{H}_{\mathrm{PG},c}$ of the cyclic code $\mathcal{C}_{\mathrm{PG},c}$ constructed from the incidence vectors of the lines in PG$(2, q)$ consists of a single $(q^2+q+1)\times(q^2+q+1)$ circulant over GF(2) with both column and row weight $q + 1$. If $q$ is a power of 2, say $2^s$, then $\mathcal{C}_{\mathrm{PG},c}$ has the following parameters [4, 23]:

| | |
|---|---|
| Length | $2^{2s} + 2^s + 1$ |
| Number of parity-check bits | $3^s + 1$ |
| Minimum distance (lower bound) | $2^s + 2$ |

Let $\mathbf{g}(X)$ be the generator polynomial of $\mathcal{C}_{\mathrm{PG},c}$ and let $\alpha$ be a primitive element of GF$(2^{3s})$. Let $h$ be a non-negative integer less than $2^{3s} - 1$. Then $\alpha^h$ is a root of $\mathbf{g}(X)$ if and only if $h$ is divisible by $2^s - 1$ and

$$0 \leq \max_{0\leq l<s} W_{2^s}(h^{(l)}) = 2^s - 1. \tag{11.34}$$

This result was proved in [22].

---

**Example 11.14** Consider the two-dimensional projective geometry PG$(2, 2^5)$ over GF$(2^5)$. This geometry consists of 1057 points and 1057 lines. Each line consists of 33 points. Using the incidence vectors of the 1057 lines in PG$(2, 2^5)$, we can form a single $1057 \times 1057$ circulant $\mathbf{H}_{\mathrm{PG},c}$ with both column and row weight 33. The null space of this circulant gives a (33, 33)-regular (1057, 813) cyclic PG-LDPC code with rate 0.7691 and minimum distance at least 34, whose Tanner graph has a girth of at least 6. The error performance of this code with iterative decoding using the SPA with 100 iterations is shown in Figure 11.19. At a BER of $10^{-6}$, it performs 1.85 dB from the Shannon limit. Since the code has a very large minimum distance, its error performance should have a very low error floor.

---

In the above construction of a cyclic PG-LDPC code, we use the incidence vectors of all the lines in a projective geometry to form the parity-check matrix. A larger class of cyclic PG-LDPC codes can be constructed by grouping the incidence vectors of the lines in a projective geometry. For a Euclidean geometry EG$(m, q)$, the type-1 incidence vector of any line is primitive, that is, the incidence vector $\mathbf{v}_{\mathcal{L}}$ of a line $\mathcal{L}$ in EG$(m, q)$ not passing through the origin and its $q^m - 2$ cyclic shifts are all different $(q^m - 1)$-tuples over GF(2). However, this is not the case for a projective geometry PG$(m, q)$. Whether the incidence vector of a line in PG$(m, q)$ is primitive or not depends on whether the dimension $m$ of the geometry is even or odd [4, 11, 12].

For even $m$, the incidence vector of each line in PG$(m, q)$ is primitive. For this case, the incidence vectors of the lines in PG$(m, q)$ can be partitioned into

$$K_{c,\mathrm{even}} = K^{(\mathrm{e})}_{c,\mathrm{PG}}(m, 1) = (q^m - 1)/(q^2 - 1) \tag{11.35}$$

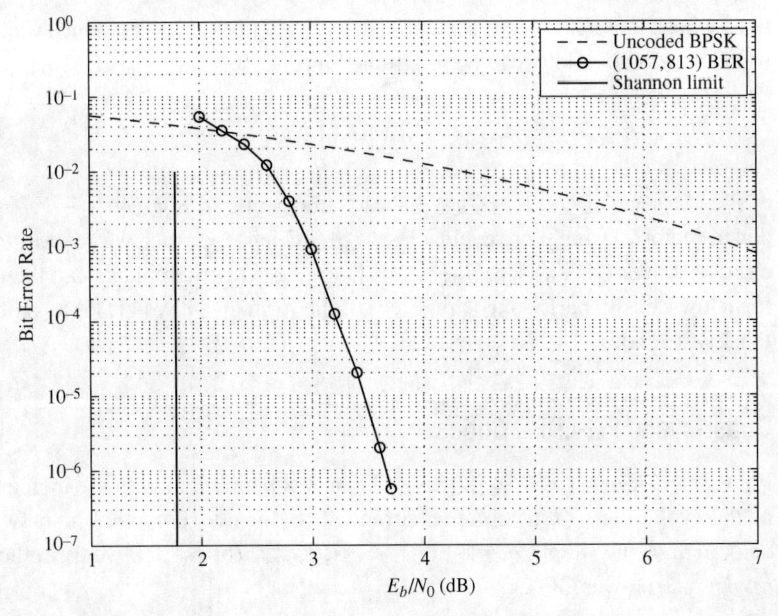


**Figure 11.19** The error performance of the (1057, 813) cyclic PG-LDPC code given in Example 11.11.

cyclic classes, $Q_0, Q_1, \ldots, Q_{K_{c,\text{even}}-1}$, each consisting of $n = (q^{m+1} - 1)/(q - 1)$ incidence vectors. For $0 \leq i < K_{c,\text{even}}$, the incidence vectors of cyclic class $Q_i$ can be obtained by cyclically shifting any incidence vector in $Q_i$ $n$ times. For each cyclic class $Q_i$, we can form an $n \times n$ circulant $\mathbf{H}_{c,i}$ using the incidence vectors in $Q_i$ as rows arranged in cyclic order. Both the column weight and the row weight of $\mathbf{H}_{c,i}$ are $q + 1$. For $1 \leq k \leq K_{c,\text{even}}$, we form the following $kn \times n$ matrix:

$$\mathbf{H}_{\text{PG},c,k}^{(1)} = \begin{bmatrix} \mathbf{H}_{c,0} \\ \mathbf{H}_{c,1} \\ \vdots \\ \mathbf{H}_{c,k-1} \end{bmatrix}, \tag{11.36}$$

which consists of a column of $k$ circulants of size $n \times n$. $\mathbf{H}_{\text{PG},c,k}^{(1)}$ is a $kn \times n$ matrix over GF(2) with column and row weights $k(q + 1)$ and $q + 1$, respectively, and it satisfies the RC constraint. The null space of $\mathbf{H}_{\text{PG},c,k}^{(1)}$ gives a cyclic PG-LDPC code of length $n$ with minimum distance at least $k(q + 1) + 1$, whose Tanner graph has a girth of at least 6.

For odd $m \geq 3$, there are $l_0 = (q^{m+1} - 1)/(q^2 - 1)$ lines in PG$(m, q)$ whose incidence vectors are nonprimitive and there are

$$J_5 = \frac{q(q^{m+1} - 1)(q^{m-1} - 1)}{(q^2 - 1)(q - 1)} \tag{11.37}$$

lines whose incidence vectors are primitive. The $J_5$ primitive incidence vectors of lines in PG$(m, q)$ with odd $m$ can be partitioned into

$$K_{c,\text{odd}} = K_{c,\text{PG}}^{(o)}(m, 1) = \frac{q(q^{m-1} - 1)}{q^2 - 1} \tag{11.38}$$

cyclic classes, $Q_0, Q_1, \ldots, Q_{K_{c,\text{odd}}-1}$, each consisting of $n = (q^{m+1} - 1)/(q - 1)$ incidence vectors. Using these cyclic classes of incidence vectors, we can form $K_{c,\text{odd}}$ $n \times n$ circulants over GF(2), $\mathbf{H}_{c,0}, \mathbf{H}_{c,1}, \ldots, \mathbf{H}_{c,K_{c,\text{odd}}-1}$. These circulants can then be used to form low-density parity-check matrices of the form given by (11.36) to generate cyclic PG-LDPC codes.

### 11.7.2 Quasi-cyclic PG-LDPC Codes

For $0 \leq i < K_{c,\text{even}}$ (or $K_{c,\text{odd}}$), let $\mathbf{G}_{c,i}$ be an $n \times n$ circulant with the incidence vectors of the cyclic class $Q_i$ as columns arranged in downward cyclic order. $\mathbf{G}_{c,i}$ has both column and row weight $q + 1$. For $1 \leq k < K_{c,\text{even}}$ (or $K_{c,\text{odd}}$), we form the following $n \times kn$ matrix over GF(2):

$$\mathbf{H}_{\text{PG},qc,k}^{(2)} = \left[ \begin{array}{cccc} \mathbf{G}_{c,0} & \mathbf{G}_{c,1} & \cdots & \mathbf{G}_{c,k-1} \end{array} \right], \tag{11.39}$$

which consists of a row of $k$ $n \times n$ circulants. The column and row weights of $\mathbf{H}_{\text{PG},qc,k}^{(2)}$ are $q + 1$ and $k(q + 1)$, respectively. The null space of $\mathbf{H}_{\text{PG},qc,k}^{(2)}$ gives a QC-PG-LDPC code of length $kn$ with minimum distance at least $q + 2$, whose Tanner graph is at least 6. For a given projective geometry PG$(m, q)$ with $1 \leq k \leq K_{c,\text{even}}$ (or $K_{c,\text{odd}}$), we can construct a sequence of QC-PG-LDPC codes with different lengths and rates. The above construction gives a class of QC-PG-LDPC codes.

---

**Example 11.15** Consider the three-dimensional projective geometry PG$(3, 2^3)$ over GF$(2^3)$. This geometry consists of 585 points and 4745 lines. Each line consists of nine points. The incidence vector of a line in PG$(3, 2^3)$ is a 585-tuple over GF(2) with weight 9. There are 65 lines in PG$(3, 2^3)$ whose incidence vectors are nonprimitive. The incidence vectors of the other 4680 lines in PG$(3, 2^3)$ are primitive. These primitive incidence vectors can be partitioned into eight cyclic classes, $Q_0, Q_1, \ldots, Q_7$, each consisting of 585 incidence vectors. For each cyclic class $Q_j$ with $0 \leq j \leq 7$, we form a $585 \times 585$ circulant $\mathbf{G}_{c,j}$ with the 585 incidence vectors of $Q_i$ as columns arranged in downward cyclic order. The column weight and row weight of $\mathbf{G}_{c,j}$ are both 9. Suppose we set $k = 6$ and form the following $585 \times 3510$ matrix over GF(2): $\mathbf{H}_{\text{PG},qc,6}^{(2)} = [\mathbf{H}_{c,0} \ \ \mathbf{H}_{c,1} \ \ \cdots \ \ \mathbf{H}_{c,5}]$, which has column and row weights 9 and 54, respectively. The null space of this matrix gives a $(9, 54)$-regular $(3510, 3109)$ QC-PG-LDPC code with rate 0.8858 and minimum distance at least 10. The error performance of this code with iterative decoding using the SPA with 100 iterations is shown in Figure 11.20. At a BER of $10^{-6}$, it performs 1.3 dB from the Shannon limit.

---

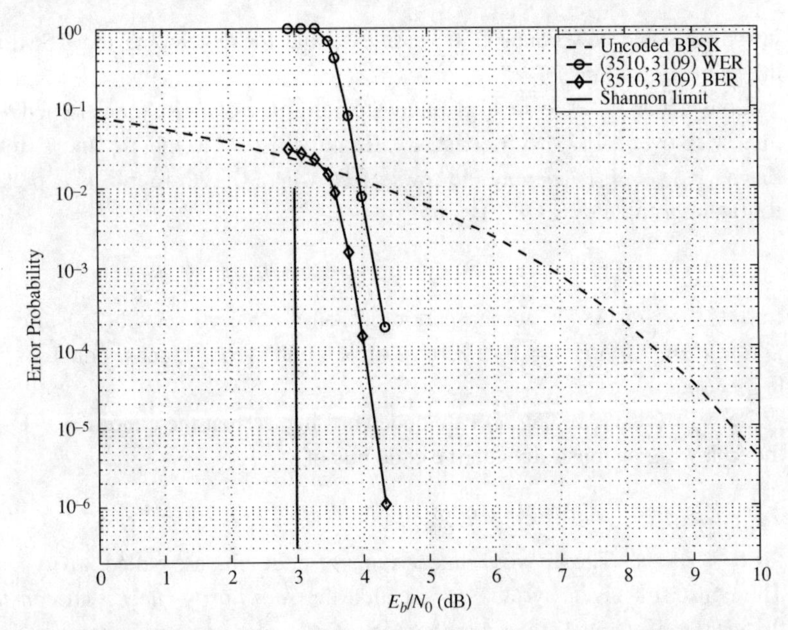


**Figure 11.20** The error performance of the (3510, 3109) QC-PG-LDPC code given in Example 11.12.

If we decompose each $n \times n$ circulant $\mathbf{G}_{c,j}$ in $\mathbf{H}^{(2)}_{\mathrm{PG},c,k}$ into a $c \times t$ array of $n \times n$ circulants as described in Section 11.5, we obtain a $c \times kt$ array $\mathbf{H}_{\mathrm{array,decom}}$ of $n \times n$ circulants. The null space of $\mathbf{H}_{\mathrm{array,decom}}$ gives a QC-PG-LDPC code of length $nkt$.

## 11.8 One-Step Majority-Logic and Bit-Flipping Decoding Algorithms for FG-LDPC Codes

As shown in earlier sections of this chapter, FG-LDPC codes perform very well over the binary-input AWGN channel with iterative decoding using the SPA. In this and the next two sections, we show that FG-LDPC codes can also be decoded with other decoding algorithms that require much less decoding (or computational) complexity to provide a wide range of effective trade-offs among error performance, decoding complexity, and decoding speed. These decoding algorithms include the *one-step majority-logic* (OSMLG) decoding algorithm, the *bit-flipping* (BF) decoding algorithm, and various *weighted BF (WBF) decoding algorithms*. The OSMLG and BF decoding algorithms are *hard-decision* decoding algorithms that can easily be implemented. The WBF decoding algorithms are *reliability-based* decoding algorithms, which require larger computational complexity than the hard-decision OSMLG and BF decoding algorithms, but provide better performance.

In this section, we first present the hard-decision OSMLG and BF decoding algorithms for LDPC codes. Before we introduce these decoding algorithms, we give a brief review of some basic concepts of linear block codes presented in Chapter 3 and introduce some new concepts that are essential for developing the OSMLG and BF decoding algorithms.

Consider an LDPC code $C$ given by the null space of an RC-constrained $m \times n$ sparse matrix $\mathbf{H}$ over GF(2) with constant column and row weights $g$ and $r$, respectively. Let $\mathbf{h}_0, \mathbf{h}_1, \ldots, \mathbf{h}_{m-1}$ denote the rows of $\mathbf{H}$, where the $i$th row $\mathbf{h}_i$ is expressed as the following $n$-tuple over GF(2):

$$\mathbf{h}_i = (h_{i,0}, h_{i,1}, \ldots, h_{i,n-1}),$$

for $0 \leq i < m$. As shown in Section 3.2, an $n$-tuple $\mathbf{v}$ over GF(2) is a codeword in $C$ *if and only if* $\mathbf{v}\mathbf{H}^{\mathrm{T}} = \mathbf{0}$ (a zero $m$-tuple), that is, the inner product $\mathbf{v} \cdot \mathbf{h}_i = 0$ for $0 \leq i < m$.

Let $\mathbf{v} = (v_0, v_1, \ldots, v_{n-1})$ be a codeword in $C$. Then the condition $\mathbf{v}\mathbf{H}^{\mathrm{T}} = \mathbf{0}$ gives the following $m$ constraints on the code bits of $\mathbf{v}$:

$$\mathbf{v} \cdot \mathbf{h}_i = v_0 h_{i,0} + v_1 h_{i,1} + \cdots + v_{n-1} h_{i,n-1} = 0, \tag{11.40}$$

for $0 \leq i < m$. The above $m$ linear sums of code bits are called *parity-check sums*. The constraint given by (11.33) is called the *zero-parity-check-sum constraint*. For $0 \leq j < n$, if $h_{i,j} = 1$, then the $j$th code bit $v_j$ participates (or is contained) in the $i$th parity-check sum given by (11.40). In this case, we say that the $i$th row $\mathbf{h}_i$ of $\mathbf{H}$ *checks* on the $j$th code bit $v_j$ of $\mathbf{v}$ (or the $j$th code bit $v_j$ of $\mathbf{v}$ is *checked* by the $i$th row $\mathbf{h}_i$ of $\mathbf{H}$). Since $\mathbf{H}$ has constant column weight $g$, there are $g$ rows of $\mathbf{H}$ that check on $v_j$ for $0 \leq j < n$. Hence, there are $g$ zero-parity-check sums that contain (or check on) the code bit $v_j$.

Suppose a codeword $\mathbf{v} = (v_0, v_1, \ldots, v_{n-1})$ in $C$ is transmitted over the BSC (a binary-input AWGN channel with hard-decision output). Let $\mathbf{z} = (z_0, z_1, \ldots, z_{n-1})$ be the *hard-decision received* vector (an $n$-tuple over GF(2)). To decode $\mathbf{z}$ with any decoding method for a linear block code, the first step is to compute its $m$-tuple *syndrome*,

$$\mathbf{s} = (s_0, s_1, \ldots, s_{m-1}) = \mathbf{z}\mathbf{H}^{\mathrm{T}}, \tag{11.41}$$

where

$$s_i = \mathbf{z} \cdot \mathbf{h}_i = z_0 h_{i,0} + z_1 h_{i,1} + \cdots + z_{n-1} h_{i,n-1}, \tag{11.42}$$

for $0 \leq i < m$, which is called a *syndrome sum* of $\mathbf{z}$. If $\mathbf{s} = \mathbf{0}$, then the received bits in $\mathbf{z}$ satisfy all the $m$ zero-parity-check-sum constraints given by (11.40) and hence $\mathbf{z}$ is a codeword. In this case, the receiver assumes that $\mathbf{z}$ is the transmitted codeword and accepts it as the decoded codeword. If $\mathbf{s} \neq \mathbf{0}$, then the received bits in $\mathbf{z}$ do not satisfy all the $m$ zero-parity-check-sum constraints given by (11.33) and $\mathbf{z}$ is not a codeword. In this case, we say that errors in $\mathbf{z}$ are *detected*. Then an error-correction process is initiated.

A syndrome sum $s_i$ of the received sequence $\mathbf{z}$ that is not equal to zero is referred to as a *parity-check failure*, that is, the received bits that participate in the syndrome sum $s_i$ of $\mathbf{z}$ fail to satisfy the zero-parity-check-sum constraint given by (11.40). It is clear that the number of parity-check failures of $\mathbf{z}$ is equal to the number of 1-components in the syndrome $\mathbf{s}$. A syndrome sum $s_i$ that contains a received bit $z_j$ is said to *check on* $z_j$. Consider the syndrome sums of $\mathbf{z}$ that check on the received bit $z_j$ but fail to satisfy the zero-parity-check-sum constraints given by (11.40). We call these syndrome sums the *parity-check failures* on $z_j$. The number of parity-check failures on a received bit $z_j$ gives a *measure of the reliability* of the received bit $z_j$. The larger the number of parity-check failures on $z_j$, the less reliable $z_j$ is. Conversely, the smaller the number of parity-check failures on $z_j$, the more reliable $z_j$ is.

In the following, we will use the above-defined reliability measures of the bits of a hard-decision received sequence $\mathbf{z}$ to devise the OSMLG and BF decoding algorithms.

### 11.8.1 The OSMLG Decoding Algorithm for LDPC Codes over the BSC

Consider an LDPC code given by the null space of an RC-constrained $m \times n$ parity-check matrix $\mathbf{H}$ with constant column and row weights $g$ and $r$, respectively. Express the $m$ rows of $\mathbf{H}$ as follows:

$$\mathbf{h}_i = (h_{i,0}, h_{i,1}, \ldots, h_{i,n-1}),$$

for $1 \leq i < m$. For $0 \leq i < m$ and $0 \leq j < n$, we define the following two index sets:

$$N_i = \{j: 0 \leq j < n, h_{i,j} = 1\}, \tag{11.43}$$

$$M_j = \{i: 0 \leq i < m, h_{i,j} = 1\}. \tag{11.44}$$

The indices in $N_i$ are simply the locations of the 1-components in the $i$th row $\mathbf{h}_i$ of $\mathbf{H}$. $N_i$ is called the *support* of $\mathbf{h}_i$. The indices in $M_j$ give the rows of $\mathbf{H}$ whose $j$th components are equal to 1. Since $\mathbf{H}$ satisfies the RC constraint, it is clear that (1) for $0 < i, i' < m$, and, $i \neq i'$, $N_i$ and $N_{i'}$ have *at most one* index in common; and (2) for $0 \leq j, j' < n$, and $j \neq j'$, $M_j$ and $M_{j'}$ have at most one index in common. Since $\mathbf{H}$ has constant column weight $g$ and constant row weight $r$, $|M_j| = g$ for $0 \leq j < n$ and $|N_i| = r$ for $0 \leq i < m$.

For $0 \leq j < n$, define the following set of rows of $\mathbf{H}$:

$$\mathcal{A}_j = \{\mathbf{h}_i: 0 \leq i < m, i \in M_j\}. \tag{11.45}$$

Then it follows from the RC constraint on the rows of $\mathbf{H}$ that $\mathcal{A}_j$ has the following structural properties: (1) every row in $\mathcal{A}_j$ checks on the code bit $v_j$; and (2) any code bit other than $v_j$ is checked by at most one row in $\mathcal{A}_j$. The rows in $\mathcal{A}_j$ are said to be *orthogonal* on the code bit $v_i$ (or the $i$th bit position of $\mathbf{v}$).

It follows from the definition of $N_i$ that the $i$th syndrome sum $s_i$ of the syndrome $\mathbf{s}$ of a hard-decision received sequence $\mathbf{z}$ given by (11.35) can be expressed in terms of the indices in $N_i$ as follows:

$$s_i = \sum_{j \in N_i} z_j h_{i,j}, \tag{11.46}$$

for $0 \leq i < m$. The expression (11.39) simply says that only the received bits of $\mathbf{z}$ with indices in $N_i$ participate in the $i$th syndrome sum $s_i$ (or are checked by $s_i$). For $0 \leq j < n$, define the following subset of syndrome sums of $\mathbf{z}$:

$$S_j = \{s_i = \mathbf{z} \cdot \mathbf{h}_i: \mathbf{h}_i \in \mathcal{A}_j, i \in M_j\}. \tag{11.47}$$

Then the $j$th received bit $z_j$ of $\mathbf{z}$ is contained in every syndrome sum in $S_j$ and any received bit other than $z_j$ is contained in (or is checked by) at most one syndrome sum in $S_j$. The $g$ syndrome sums in $S_j$ are said to be *orthogonal* on the $j$th received bit $z_j$ or are called the *orthogonal syndrome sums* on $z_j$.

Let $\mathbf{e} = (e_0, e_1, \ldots, e_{n-1})$ be the error pattern caused by the channel noise during the transmission of the codeword $\mathbf{v}$ over the BSC. Then $\mathbf{z} = \mathbf{v} + \mathbf{e}$. Since $\mathbf{v} \cdot \mathbf{h}_i = \mathbf{0}$ for $1 \leq i < m$,

$$s_i = \mathbf{e} \cdot \mathbf{h}_i. \tag{11.48}$$

For $0 \leq j < n$, it follows from (11.41) and the orthogonality property of $S_j$ that, for $i \in M_j$, we have the following $g$ relationships between the $g$ syndrome sums in $S_j$ and the error digits $e_0, e_1, \ldots, e_{n-1}$ in $\mathbf{z}$:

$$s_i = e_j + \sum_{l \in M_j, l \neq j} e_l h_{i,l}. \tag{11.49}$$

We see that all the above $g$ syndrome sums contain (or check on) the error digit $e_j$. Furthermore, due to the orthogonal property of $S_j$ (or $\mathcal{A}_j$), any error digit other than $e_j$ is checked by (or appears in) at most one of the $g$ syndrome sums given by (11.42). These syndrome sums are said to be *orthogonal on* the error digit $e_j$ and are called *orthogonal syndrome sums* on $e_j$.

Next we show that the $g$ orthogonal syndrome sums given by (11.42) can be used to decode the error digits of an error pattern $\mathbf{e}$. Suppose that there are $\lfloor g/2 \rfloor$ or fewer errors in the error pattern $\mathbf{e}$. The error digit $e_j$ at the $j$th position of $\mathbf{e}$ has two possible values, 1 or 0. If $e_j = 1$, the other nonzero error digits in $\mathbf{e}$ can distribute among at most $\lfloor g/2 \rfloor - 1$ syndrome sums orthogonal on $e_j$. Hence, at least $g - \lfloor g/2 \rfloor + 1$ (*more than half*) of the orthogonal syndrome sums on $e_j$ are equal to $e_j = 1$. If $e_j = 0$, the nonzero error digits in $\mathbf{e}$ can distribute among at most $\lfloor g/2 \rfloor$ syndrome sums orthogonal on $e_j$. Hence, at least $g - \lfloor g/2 \rfloor$ (*half or more than half*) of the orthogonal syndrome sums on $e_j$ are equal to $e_j = 0$. Let $w(\mathbf{e})$ be the weight of the error pattern $\mathbf{e}$ (or the number of errors in $\mathbf{e}$). For $w(\mathbf{e}) \leq \lfloor g/2 \rfloor$, the above analysis simply says that (1) the value of the $j$th error digit $e_j$ of $\mathbf{e}$ is equal to the value assumed by a *majority* of the orthogonal syndrome sums on $e_j$; and (2) if no value, 0 or 1, is assumed by a clear majority of the orthogonal syndrome sums on $e_j$ (i.e., *half of them are equal to 0 and half of them are equal to 1*), then the value of the $j$th error digit $e_j$ is equal to 0. Given the above facts, a simple algorithm for decoding the error digits in an error pattern $\mathbf{e}$ can be devised as follows.

> *For $0 \leq j \leq n - 1$, the error digit $e_j$ at the jth location of* $\mathbf{e}$ *is decoded as "1" if a majority of the orthogonal syndrome sums on $e_j$ assumes the value "1"; otherwise, $e_j$ is decoded as "0."*

The above method for decoding the error pattern $\mathbf{e}$ contained in a hard-decision received sequence $\mathbf{z}$ is referred to as *OSMLG decoding*. Correct decoding of an error pattern $\mathbf{e}$ is *guaranteed* if the number of errors in $\mathbf{e}$ is $\lfloor g/2 \rfloor$ or less. However, correct decoding is *not guaranteed* if an error pattern $\mathbf{e}$ contains more than $\lfloor g/2 \rfloor$ errors (see Problem 11.13). Therefore, an LDPC code $C$ given by the null space of an

RC-constrained parity-check matrix **H** with constant column weight $g$ is capable of correcting *any combination* of $\lfloor g/2 \rfloor$ or fewer errors with OSMLG decoding. The *parameter* $\lfloor g/2 \rfloor$ is called the *OSMLG error-correction capability* of the code $C$. Even though OSMLG decoding guarantees to correct all the error patterns with just $\lfloor g/2 \rfloor$ or fewer errors, it actually can correct a *very large fraction* of error patterns with more than $\lfloor g/2 \rfloor$ errors. Codes that can be decoded with OSMLG decoding are said to be *OSMLG-decodable* and are called *OSMLG-decodable codes*.

The above OSMLG decoding is based on the number of parity-check failures in the set of syndrome sums orthogonal on each received bit. The larger the number of parity-check failures in the set of syndrome sums orthogonal on a received bit, the less reliable the received bit is. Conversely, the smaller the number of parity-check failures in the set of syndrome sums orthogonal on a received bit, the more reliable the received bit is. Therefore, the number of parity-check failures orthogonal on a received bit can be used as a *measure of the reliability* of the received bit.

For an LDPC code $C$ given by the null space of an RC-constrained parity-check matrix **H** with constant column weight $g$, a set of $g$ syndrome sums orthogonal on each received bit can be formed. Consider the $j$th received bit $z_j$ in the received sequence **z**. As shown earlier, the set $S_j$ of $g$ syndrome sums orthogonal on $z_j$ is given by (11.40). For $0 \leq j < n$, define the following *integer sum*:

$$\sum_{i \in M_j} s_i. \tag{11.50}$$

The range of this sum consists of $g + 1$ non-negative integers from 0 to $g$, denoted by $[0, g]$. For OSMLG decoding, $z_j$ is decoded as "1" if

$$\sum_{i \in M_j} s_i > \lfloor g/2 \rfloor, \tag{11.51}$$

otherwise, $z_j$ is decoded as "0." The above inequality can be put into the following form:

$$\sum_{i \in M_j} (2s_i - 1) > 0. \tag{11.52}$$

Let

$$\Lambda_j = \sum_{i \in M_j} (2s_i - 1). \tag{11.53}$$

The range of $\Lambda_j$ consists of $2g + 1$ integers from $-g$ to $g$, denoted by $[-g, g]$. From (11.46), we see that the greater the number of syndrome sums orthogonal on the received bit $z_j$ that satisfy the zero-parity-check-sum constraint given by (11.33), the more negative $\Lambda_j$ is and hence the more reliable the received bit $z_j$ is. Conversely, the greater the number of syndrome sums orthogonal on $z_j$ that fail the zero-parity-check-sum constraint given by (11.33), the more positive $\Lambda_j$ is and hence the less reliable $z_j$ is. Therefore, $\Lambda_j$ also gives a measure of the reliability of the received bit $z_j$.

With the above measure of reliability of a received bit, the OSMLG decoding algorithm can be formulated as follows.

**Algorithm 11.1** The OSMLG Decoding Algorithm

1. Compute the syndrome $\mathbf{s} = (s_0, s_1, \ldots, s_{m-1})$ of the received sequence $\mathbf{z} = (z_0, z_1, \ldots, z_{n-1})$. If $\mathbf{s} = \mathbf{0}$, stop decoding; otherwise, go to Step 2.
2. For each received bit $z_j$ of $\mathbf{z}$ with $0 \le j < n$, compute its reliability measure $\Lambda_j$. Go to Step 3.
3. For $0 \le j < n$, if $\Lambda_j > 0$, decode the error digit $e_j$ as "1"; otherwise, decode $e_j$ as "0." Form the estimated error pattern $\mathbf{e}^* = (e_0, e_1, \ldots, e_{n-1})$. Go to Step 4.
4. Decode $\mathbf{z}$ into $\mathbf{v}^* = \mathbf{z} + \mathbf{e}^*$. Compute $\mathbf{s}^* = \mathbf{v}^*\mathbf{H}^T$. If $\mathbf{s}^* = \mathbf{0}$, the decoding is successful; otherwise, declare a decoding failure.

Since OSMLG decoding involves only logical operations, an OSMLG decoder can be implemented with a combinational logic circuit. For high-speed decoding, all the error digits can be decoded in parallel. For a cyclic OSMLG-decodable code, the OSMLG decoding can be carried out in *serial manner*, decoding one error digit at a time using the same decoding circuit. Serial decoding further reduces decoding complexity. The OSMLG decoding is effective only when the column weight $g$ of the RC-constrained parity-check matrix $\mathbf{H}$ of an LDPC code is relatively large. For more on majority-logic decoding, readers are referred to [23].

The classes of FG-LDPC codes presented in Sections 11.1–11.3, 11.6, and 11.7 are *particularly effective* with OSMLG decoding. For a code in each of these classes, a large number of syndrome sums orthogonal on every error digit can be formed for decoding. Consider the classes of cyclic EG-LDPC codes constructed on the basis of Euclidean geometries over finite fields given in Section 11.1.1. Let $C^{(1)}_{\mathrm{EG},c,k}$ be the cyclic EG-LDPC code of length $n = q^m - 1$ constructed on the basis of the $m$-dimensional Euclidean geometry EG$(m, q)$ over GF$(q)$, where $1 \le k \le K_c$, with $K_c = (q^{m-1} - 1)/(q - 1)$. The parity-check matrix $\mathbf{H}^{(1)}_{\mathrm{EG},c,k}$ of this cyclic EG-LDPC code is given by (11.7) and consists of a column of $k$ circulants of size $(q^m - 1) \times (q^m - 1)$, each having both column and row weight $q$. The parity-check matrix $\mathbf{H}^{(1)}_{\mathrm{EG},c,k}$ has constant column and row weights $kq$ and $q$, respectively. Since $\mathbf{H}^{(1)}_{\mathrm{EG},c,k}$ satisfies the RC constraint, $kq$ syndrome sums orthogonal on each received bit can be formed. Hence, this code $C^{(1)}_{\mathrm{EG},c,k}$ given by the null space of $\mathbf{H}^{(1)}_{\mathrm{EG},c,k}$ is capable of correcting any combination of $\lfloor kq/2 \rfloor$ or fewer errors. Therefore, for $k = 1, 2, \ldots, K_c$, a sequence of OSMLG-decodable cyclic EG-LDPC codes can be constructed on the basis of the $m$-dimensional Euclidean geometry EG$(m, q)$ over GF$(q)$.

**Example 11.16** Consider the $(4095, 3367)$ cyclic EG-LDPC code $C^{(1)}_{\mathrm{EG},c,1}$ given in Example 11.1. This code is constructed using the two-dimensional Euclidean geometry EG$(2, 2^6)$ over GF$(2^6)$. The parity-check matrix $\mathbf{H}^{(1)}_{\mathrm{EG},c,1}$ of this code consists of a single $4095 \times 4095$ circulant with both column and row weight 64, which is formed by the type-1 incidence vectors of the lines of EG$(2, 2^6)$ not passing through the origin. Since the parity-check matrix of this code satisfies the RC constraint, 64 syndrome sums orthogonal on any error-digit position can be formed.

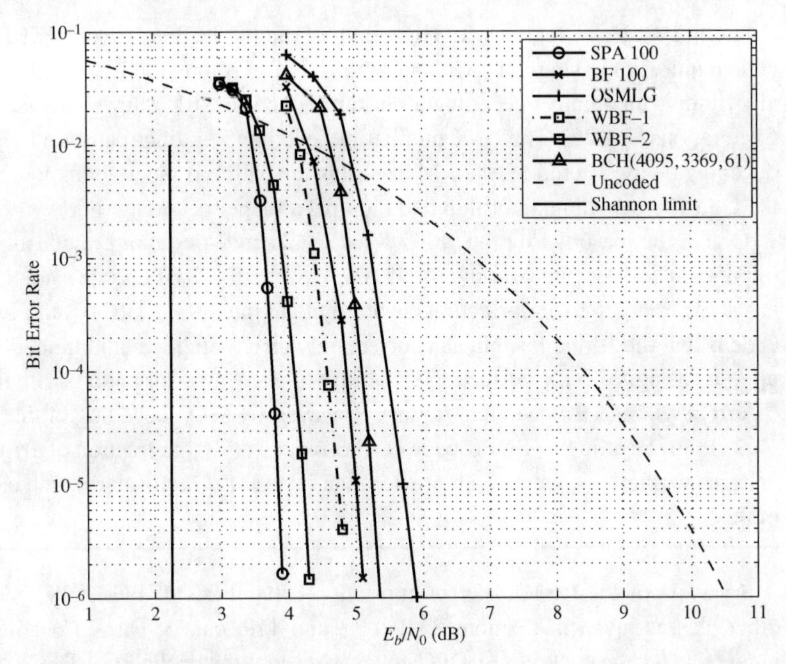


**Figure 11.21** Error performances of the (4095, 3367) cyclic EG-LDPC code given in Example 11.13 with various decoding methods.

Hence, this cyclic EG-LDPC code is capable of correcting any combination of 32 or fewer errors with the OSMLG decoding. The minimum distance of this code is exactly 65 (1 greater than the column weight of the parity-check matrix of the code).

The error performance of this code with OSMLG decoding over the hard-decision AWGN channel is shown in Figure 11.18. At a BER of $10^{-6}$, it achieves a coding gain of 4.7 dB over the uncoded BPSK system. Figure 11.21 also includes the error performance of the (4095, 3367) cyclic EG-LDPC code over the binary-input AWGN channel decoded using the SPA with 100 iterations. We see that OSMLG decoding of this code loses about 2 dB in SNR compared with the SPA decoding of the code at the BER of $10^{-6}$ and below. However, OSMLG decoding requires *much lower* computational complexity than SPA decoding.

Also included in Figure 11.21 is the error performance of a (4095, 3367) BCH code with *designed minimum distance* 123, which is about twice that of the (4095, 3367) cyclic EG-LDPC code. This BCH code is capable of correcting 61 (*designed error-correction capability*) or fewer errors with the (*hard-decision*) Berlekamp–Massey (BM) algorithm [23, 25, 34]. In terms of error-correction capability, the BCH code is twice as powerful as the cyclic EG-LDPC code. However, from Figure 11.21, we see that at a BER of $10^{-6}$, the (4095, 3367) BCH code decoded with the BM algorithm has a coding gain of less than 0.5 dB over the (4095, 3367) cyclic EG-LDPC code decoded with OSMLG decoding. Decoding the (4095, 3367) BCH code with BM algorithm requires computations over the Galois field GF($2^{12}$). However, OSMLG decoding of the (4095, 3367) cyclic EG-LDPC code requires only simple logical operations. Furthermore, the BM decoding algorithm is an iterative decoding algorithm. To correct 61 or fewer errors, it requires 61 iterations [23, 25]. The large number of decoding iterations causes a long decoding delay. Therefore, the BM decoding algorithm is much more complex than the OSMLG decoding algorithm and requires a longer decoding time.

One of the reasons why the (4095, 3367) BCH code does not provide an impressive coding gain over the (4095,3367) cyclic EG-LDPC code even though it has a much larger error-correction capability than the (4095, 3367) cyclic EG-LDPC code is that the BM algorithm can correct only error patterns with numbers of errors up to its designed error-correction capability of 61. Any error pattern with more than 61 errors will cause a decoding failure. However, OSMLG decoding of the (4095, 3367) cyclic EG-LDPC code can correct a very large fraction of error patterns with error counts much greater than its OSMLG error-correction capability of 32.

---

Analysis of the OSMLG error-correction capabilities of FG-LDPC codes in the other classes given in Sections 11.1–11.3 and 11.6 can be carried out in the same manner as for the cyclic EG-LDPC codes. Consider the cyclic PG-LDPC code $C^{(1)}_{\mathrm{PG},c,k}$ of length $n = (q^{m+1} - 1)/(q - 1)$ given by the null space of the parity-check matrix $\mathbf{H}^{(1)}_{\mathrm{PG},c,k}$ of (11.36), which consists of a column of $k$ circulants of size $n \times n$ (see Section 11.6.1). The $k$ circulants of $\mathbf{H}^{(1)}_{\mathrm{PG},c,k}$ are constructed from $k$ cyclic classes of lines in the $m$-dimensional projective geometry PG$(m, q)$ with $2 \leq m$ and $1 \leq k \leq K_{c,\mathrm{even}}$ (or $K_{c,\mathrm{odd}}$), where $K_{c,\mathrm{even}}$ and $K_{c,\mathrm{odd}}$ are given by (11.35) and (11.38), respectively. Each circulant of $H^{(1)}_{\mathrm{PG},c,k}$ has both column and row weight $q + 1$. The column and row weights of $H^{(1)}_{\mathrm{PG},c,k}$ are $k(q + 1)$ and $q + 1$, respectively. Since $\mathbf{H}^{(1)}_{\mathrm{PG},c,k}$ satisfies the RC constraint, $k(q + 1)$ syndrome sums orthogonal on every code bit position can be formed for OSMLG decoding. Hence, the OSMLG error-correction capability of the cyclic PG-LDPC code $C^{(1)}_{\mathrm{PG},c,k}$ is $\lfloor k(q + 1)/2) \rfloor$.

---

**Example 11.17** For $m = 2$, let the two-dimensional projective geometry PG$(2, 2^6)$ over GF$(2^6)$ be the code-construction geometry. This geometry consists of 4161 lines. Since $m = 2$ is even, $K_{c,\mathrm{even}} = 1$. Using (11.36), we can form a parity-check matrix $\mathbf{H}^{(1)}_{\mathrm{PG},c,1}$ with a single circulant constructed from the incidence vectors of the 4161 lines of PG$(2, 2^6)$. The column weight and row weight of $\mathbf{H}_{\mathrm{PG},c,1}$ are 65. The null space of $\mathbf{H}^{(1)}_{\mathrm{PG},c,1}$ gives a (4161, 3431) cyclic PG-LDPC code with rate 0.823 and minimum distance at least 66. For this code, 65 syndrome sums orthogonal on each code bit position can be formed. Hence, the OSMLG error-correction capability of this cyclic PG-LDPC code is 32, the same as the OSMLG error-correction capability of the (4095, 3367) cyclic EG-LDPC code given in Example 11.16. The error performances of this code with various decoding algorithms are shown in Figure 11.22.

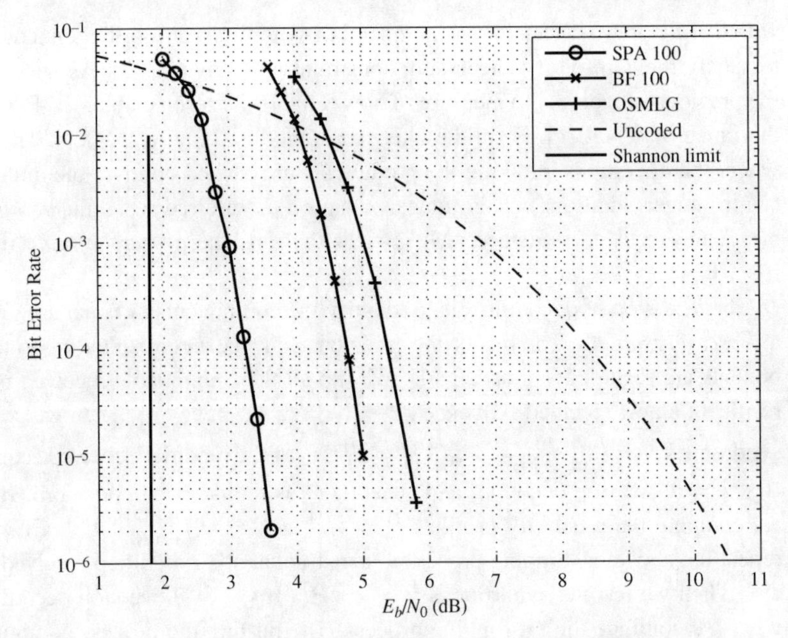


**Figure 11.22** Error performances of the (4161, 3431) cyclic PG-LDPC code given in Example 11.14 with various decoding methods.

---

Consider the regular LDPC codes constructed on the basis of decomposition of Euclidean geometries given in Section 11.3. From the $m$-dimensional Euclidean geometry EG($m, q$) over GF($q$), an RC-constrained $q^{m-1} \times q$ array $\mathbf{H}_{\mathrm{EG},d}^{(1)}$ of $q^{m-1} \times q^{m-1}$ permutation matrices can be formed. Let $1 \leq g \leq q^{m-1}$ and $1 \leq r \leq q$. The null space of any $g \times r$ subarray $\mathbf{H}_{\mathrm{EG},d}^{(1)}(g, r)$ of $\mathbf{H}_{\mathrm{EG},d}^{(1)}$ gives a $(g, r)$-regular LDPC code $C_{\mathrm{EG},d}^{(1)}$ of length $n = rq^{m-1}$ that is OSMLG decodable. For this code, $g$ syndrome sums orthogonal on any code bit position can be formed for OSMLG decoding. Hence, the OSMLG error-correction capability is $\lfloor g/2 \rfloor$. Furthermore, any subarray of the $q \times q^{m-1}$ array $\mathbf{H}_{\mathrm{EG},d}^{(2)} = [\mathbf{H}_{\mathrm{EG},d}^{(1)}]^{\mathrm{T}}$ also gives an OSMLG-decodable EG-LDPC code. Decomposition of Euclidean geometries gives a large class of OSMLG-decodable EG-LDPC codes.

### 11.8.2 The BF Algorithm for Decoding LDPC Codes Over the BSC

The BF algorithm for decoding LDPC codes over the BSC was discussed in Section 5.7.4. In this section, we show that bit-flipping decoding is very effective for decoding FG-LDPC codes in terms of the trade-off between error performance and decoding complexity. Before we do so, we give a brief review of BF decoding that is based on the concept of a reliability measure of received bits developed in the previous section.

Again we consider an LDPC code $C$ given by the null space of an RC-constrained $m \times n$ parity-check matrix $\mathbf{H}$ with column and row weights $g$ and $r$, respectively. Let $\mathbf{z} = (z_0, z_1, \ldots, z_{n-1})$ be the hard-decision received sequence. The first step of decoding $\mathbf{z}$ is to compute its syndrome $\mathbf{s} = (s_0, s_1, \ldots, s_{m-1}) = \mathbf{z}\mathbf{H}^{\mathrm{T}}$. For each received bit $z_j$

with $0 \leq j < n$, we determine the number $f_j$ of syndrome sums orthogonal on $z_i$ that fail to satisfy the zero-parity-check-sum constraint given by (11.33). As we described in the previous section, $f_j$ is a measure of how reliable the received bit $z_j$ is. For $0 \leq j < n$, the range of $f_j$ is $[0, g]$. Form the integer $n$-tuple $\mathbf{f} = (f_0, f_1, \ldots, f_{n-1})$. Then $\mathbf{f} = \mathbf{sH}$, where the operations in taking the product $\mathbf{sH}$ are carried out over the integer system (with integer additions). This integer $n$-tuple $\mathbf{f}$ is referred to as the *reliability profile* of the received sequence $\mathbf{z}$. It is clear that $\mathbf{f}$ is the *all-zero n-tuple* if and only if $\mathbf{s} = \mathbf{0}$.

The next step of decoding $\mathbf{z}$ is to identify the components of $\mathbf{f}$ that are greater than a *preset threshold* $\delta$. The bits of the received sequence $\mathbf{z}$ corresponding to these components are regarded as not reliable. We flip all these unreliable received bits, which results in a new received sequence $\mathbf{z}^{(1)} = \left(z_0^{(1)}, z_1^{(1)}, \ldots, z_{n-1}^{(1)}\right)$. Then we compute the syndrome $\mathbf{s}^{(1)} = \left(s_0^{(1)}, s_1^{(1)}, \ldots, s_{n-1}^{(1)}\right) = \mathbf{z}^{(1)}\mathbf{H}^{\mathrm{T}}$ of the modified received sequence $\mathbf{z}^{(1)}$. If $\mathbf{s}^{(1)} = \mathbf{0}$, we stop decoding and accept $\mathbf{z}^{(1)}$ as the decoded codeword. If $\mathbf{s}^{(1)} \neq \mathbf{0}$, we compute the reliability profile $\mathbf{f}^{(1)} = \left(f_0^{(1)}, f_1^{(1)}, \ldots, f_{n-1}^{(1)}\right)$ of $\mathbf{z}^{(1)}$. Given $\mathbf{f}^{(1)}$, we repeat the above bit-flipping process to construct another modified received sequence $\mathbf{z}^{(2)}$. Then we test the syndrome $\mathbf{s}^{(2)} = \mathbf{z}^{(2)}\mathbf{H}^{\mathrm{T}}$. If $\mathbf{s}^{(2)} = \mathbf{0}$, we stop decoding; otherwise, we continue the bit-flipping process. The bit-flipping process continues until a zero syndrome is obtained or a preset maximum number of iterations is reached. If the syndrome is equal to zero, decoding is successful; otherwise, a decoding failure is declared.

The above BF decoding process is an iterative decoding algorithm. The threshold $\delta$ is a design parameter that should be chosen in such a way as to optimize the error performance while minimizing the computations of parity-check sums. The computations of parity-check sums are binary operations; however, the computations of the reliability profiles and comparisons with the threshold are integer operations. Hence, the computational complexity of the above BF decoding is larger than that of OSMLG decoding. The value of the threshold depends on the code parameters, $g$ and $r$, and the SNR. The optimum threshold for bit-flipping has been derived by Gallager [32], but we will not discuss it here. Instead, we present a simple bit-flipping mechanism that leads to a very simple BF decoding algorithm displayed as Algorithm 11.2 below. Since, at each iteration, we flip the bits with the largest parity-check failures (LPCFs), we call the above simple BF decoding the LPCF-BF decoding algorithm.

---

**Algorithm 11.2** A Simple BF Decoding Algorithm

---

**Initialization**. Set $k = 0$, $\mathbf{z}^{(0)} = \mathbf{z}$, and the maximum number of iterations to $k_{\max}$.

1. Compute the syndrome $\mathbf{s}^{(k)} = \mathbf{z}^{(k)}\mathbf{H}^{\mathrm{T}}$ of $\mathbf{z}^{(k)}$. If $\mathbf{s}^{(k)} = \mathbf{0}$, stop decoding and output $\mathbf{z}^{(k)}$ as the decoded codeword; otherwise, go to Step 2.
2. Compute the reliability profile $\mathbf{f}^{(k)}$ of $\mathbf{z}^{(k)}$.
3. Identify the set $F_k$ of bits in $\mathbf{z}^{(k)}$ that have the largest parity-check failures.
4. Flip the bits in $F_k$ to obtain an $n$-tuple $\mathbf{z}^{(k+1)}$ over GF(2).
5. $k \leftarrow k + 1$. If $k > k_{\max}$, declare a decoding failure and stop the decoding process; otherwise, go to Step 1.

---

The LPCF-BF decoding algorithm is very effective for decoding FG-LDPC codes since it allows the reliability measure of each received bit to vary over a large range due to the large column weights of the parity-check matrices of these codes. To demonstrate this, we again use the (4095, 3367) cyclic EG-LDPC code given in Example 11.1 (or Example 11.13). The parity-check matrix of this code has column weight 64. Hence, the reliability measure $f_j$ of each received bit $z_j$ can vary between 0 and 64. The error performance of this code over the BSC decoded using the LPCF-BF decoding presented above with 100 iterations is shown in Figure 11.18. We see that, at a BER of $10^{-6}$, LPCF-BF decoding outperforms OSMLG decoding by about 0.7 dB. Of course, this coding gain is achieved at the expense of a larger computational complexity. We also see that the (4095, 3367) cyclic EG-LDPC code decoded with LPCF-BF decoding outperforms the (4095, 3369) BCH code decoded by BM decoding, with less computational complexity.

## 11.9 Weighted BF Decoding: Algorithm 1

The performance of the simple hard-decision LPCF-BF decoding algorithm presented in Section 11.7.2 can be improved by *weighting* its bit-flipping decision function with some type of *soft* reliability information about the received symbols. In this and the following sections, we present three weighted BF decoding algorithms with increasing performance but also increasing decoding complexity. We will show that these weighted BF decoding algorithms are very effective for decoding FG-LDPC codes.

Let $C$ be an LDPC code of length $n$ given by the null space of an $m \times n$ RC-constrained parity-check matrix $\mathbf{H}$ with constant column and row weights $g$ and $r$, respectively. Suppose a codeword $\mathbf{v} = (v_0, v_1, \ldots, v_{n-1})$ in $C$ is transmitted over the binary-input AWGN channel with two-sided power spectral density $N_0/2$. Assuming transmission using BPSK signaling with unit energy per signal, the transmitted codeword $\mathbf{v}$ is mapped into a sequence of BPSK signals that is represented by a bipolar code sequence

$$(2v_0 - 1, 2v_1 - 1, \ldots, 2v_{n-1} - 1),$$

where the $j$th component $2v_j - 1 = +1$ for $v_j = 1$ and $2v_j - 1 = -1$ for $v_j = 0$. Let $\mathbf{y} = (y_0, y_1, \ldots, y_{n-1})$ be the sequence of *samples* (or *symbols*) at the output of the channel-receiver sampler. This sequence is commonly called a *soft-decision* received sequence. The samples of $\mathbf{y}$ are real numbers with $y_j = (2v_j - 1) + x_j$ for $0 \le j < n$, where $x_j$ is a Gaussian random variable with zero mean and variance $N_0/2$. For $0 \le j < n$, suppose each sample $y_j$ of $\mathbf{y}$ is decoded *independently* in accord with the following hard-decision rule:

$$z_j = \begin{cases} 0, & \text{for } y_j \le 0, \\ 1, & \text{for } y_j > 0. \end{cases} \tag{11.54}$$

Then we obtain a binary sequence $\mathbf{z} = (z_0, z_1, \ldots, z_{n-1})$, the hard-decision received sequence. The $j$th bit $z_j$ of $\mathbf{z}$ is simply an estimate of the $j$th code bit $v_j$ of the transmitted codeword $\mathbf{v}$. If $z_j = v_j$ for $0 \leq j < n$, then $\mathbf{z} = \mathbf{v}$; otherwise, $\mathbf{z}$ contains transmission errors. Therefore, $\mathbf{z}$ is an estimate of the transmitted codeword $\mathbf{v}$ *prior* to decoding. The above hard-decision rule is optimal in the sense of minimizing the estimation error probability of a code bit as described in Chapter 1.

With the hard-decision rule given by (11.54), the magnitude $|y_j|$ of the $j$th sample $y_j$ of $\mathbf{y}$ can be used as a reliability measure of the hard-decision decoded bit $z_j$, since the magnitude of the log-likelihood ratio

$$\log\left[\frac{\Pr(y_j|v_j = 1)}{\Pr(y_j|v_j = 0)}\right]$$

associated with the hard decision given by (11.54) is proportional to $|y_j|$. The larger the magnitude $|y_j|$ of $y_j$, the more reliable the hard-decision estimation $z_j$ of $v_j$ is.

For $0 \leq i < m$, we define

$$\phi_i = \min_{j \in N_i} |y_j|, \tag{11.55}$$

which is the *minimum magnitude* of the samples of $\mathbf{y}$ whose hard-decision bits participate in the $i$th syndrome sum $s_i$ given by (11.46). The value of $\phi_i$ gives an indication of how much *confidence* we have in the $i$th syndrome sum $s_i$ satisfying the zero-parity-check-sum constraint given by (11.40). The larger $\phi_i$, the more reliable the hard-decision bits that participate in the $i$th syndrome sum $s_i$ and the higher the probability that $s_i$ satisfies the zero-parity-check-sum constraint given by (11.40). Therefore, $\phi_i$ may be taken as a reliability measure of the syndrome sum $s_i$.

Suppose we weight each term $2s_i - 1$ in the summation of (11.53) by $\phi_i$ for $i \in M_j$. We obtain the following weighted sum:

$$E_j = \sum_{i \in M_j} (2s_i - 1)\phi_i, \tag{11.56}$$

which is simply the sum of weighted syndrome sums orthogonal on the $j$th hard-decision bit $z_j$ of $\mathbf{z}$. $E_j$ is a real number with a range from $-|M_j|\phi_i$ to $+|M_j|\phi_i$. Since $|M_j| = g$, the range of $E_j$ is the real-number interval $[-g\phi_i, +g\phi_i]$. For hard-decision OSMLG decoding, $\Lambda_j = \sum_{i \in M_j}(2s_i - 1)$ gives a measure of the reliability of $z_j$ that is used as the decoding function to determine whether the $j$th received bit $z_j$ is error-free or erroneous. Then $E_j$ gives a *weighted reliability measure* of the hard-decision received bit $z_j$. From (11.56), we see that the more negative the reliability measure $E_j$, the more syndrome sums orthogonal on $z_j$ satisfy the zero-parity-check-sum constraint given by (11.40), and hence the more reliable $z_j$ is. In this case, $z_j$ is less likely to be different from the transmitted code bit $v_j$ and hence is most likely to be error-free. Conversely, the more positive the reliability measure $E_j$, the more syndrome sums orthogonal on $z_j$ fail to satisfy the zero-parity-check-sum constraint given by (11.40), and hence the less reliable $z_j$ is. In this case, $z_j$ is most likely erroneous and should be flipped. Form over the real-number system the $n$-tuple

$$\mathbf{E} = (E_0, E_1, \ldots, E_{n-1}), \tag{11.57}$$

whose components give the reliability measures of the bits in the hard-decision received sequence $\mathbf{z} = (z_0, z_1, \ldots, z_{n-1})$. This real $n$-tuple $\mathbf{E}$ is called the *weighted reliability profile* of the hard-decision received sequence $\mathbf{z}$. The received bit in $\mathbf{z}$, say $z_j$, that has the *largest* $E_j$ in the weighted reliability profile $\mathbf{E}$ of $\mathbf{z}$ is the *most unreliable* bit in $\mathbf{z}$ and should be flipped in a BF decoding algorithm.

With the concepts developed above and the weighted reliability measure of a hard-decision received bit defined by (11.56), an iterative weighted BF decoding algorithm can be formulated. For $0 \le k \le k_{\max}$, let $\mathbf{z}^{(k)} = \left(z_0^{(k)}, z_1^{(k)}, \ldots, z_{n-1}^{(k)}\right)$ be the modified received sequence available at the beginning of the $k$th iteration of the BF decoding process and let $\mathbf{E}^{(k)} = \left(E_0^{(k)}, E_1^{(k)}, \ldots, E_{n-1}^{(k)}\right)$ be the weighted reliability profile of $\mathbf{z}^{(k)}$ computed via (11.56) and (11.57).

**Algorithm 11.3** Weighted BF Decoding Algorithm 1

**Initialization**. Set $k = 0$, $\mathbf{z}^{(0)} = \mathbf{z}$, and the maximum number of iterations to $k_{\max}$. Compute and store $\phi_i$ for $0 \le i < m$.

1. Compute the syndrome $\mathbf{s}^{(k)} = \mathbf{z}^{(k)}\mathbf{H}^{\mathrm{T}}$ of $\mathbf{z}^{(k)}$. If $\mathbf{s}^{(k)} = \mathbf{0}$, stop decoding and output $\mathbf{z}^{(k)}$ as the decoded codeword; otherwise, go to Step 2.
2. Compute the reliability profile $\mathbf{E}^{(k)}$ of $\mathbf{z}^{(k)}$ on the basis of (11.56). Go to Step 3.
3. Identify the bit position $j$ for which $E_j$ is largest. Go to Step 4. *Remark*: If $j$ is not unique, then choose a $j$ at random.
4. Flip the $j$th received bit $z_j^{(k)}$ of $\mathbf{z}^{(k)}$ to obtain a modified received sequence $\mathbf{z}^{(k+1)}$. Go to Step 5.
5. $k \leftarrow k + 1$. If $k > k_{\max}$, declare a decoding failure and stop the decoding process; otherwise, go to Step 1.

Weighted BF Decoding Algorithm 1 was first proposed in [4]. The five decoding steps of the algorithm are the same as those of the hard-decision LPCF-BF decoding algorithm. However, the weighted BF decoding algorithm is more complex in computation than the LPCF-BF decoding algorithm, since it requires real-number additions and comparisons in order to carry out Steps 2 and 3, rather than the integer comparisons and additions used in the same two steps in the LPCF-BF decoding algorithm. Furthermore, additional memory is needed, to store $\phi_i$ for $0 \le i < m$.

To show the performance improvement of Weighted BF Decoding Algorithm 1 over the hard-decision LPCF-BF decoding algorithm, we again consider the (4095, 3367) cyclic EG-LDPC code $C_{\mathrm{EG},c,1}^{(1)}$ constructed from the two-dimensional Euclidean geometry EG$(2, 2^6)$ over GF$(2^6)$ given in Example 11.16. The error performance of this code over the binary-input AWGN channel decoded using Weighted BF Decoding Algorithm 1 with a maximum of 100 iterations is shown in Figure 11.21. We see that, at a BER of $10^{-6}$, Weighted BF Decoding Algorithm 1 achieves a 0.6 dB coding gain over the hard-decision LPCF-BF-decoding algorithm. Of course, this coding gain is achieved at the expense of a larger computational complexity.

Steps 3 and 4 of Weighted BF Decoding Algorithm 1 can be modified to allow flipping of multiple bits at a time in Step 4 (see Problem 11.18).

## 11.10 Weighted BF Decoding: Algorithms 2 and 3

The performance of Weighted BF Decoding Algorithm 1 can be further enhanced by improving the reliability measure of a received bit given by (11.49) and preventing the possibility of the decoding being trapped in a *loop*. Of course, this performance enhancement comes with a cost in terms of additional computational complexity and memory requirement. In this section, we present two enhancements of Weighted BF Decoding Algorithm 1. These enhancements are based on the work on weighted BF decoding given in [17, 35–38].

For $0 \leq i < m$ and $j \in N_i$, we define a new reliability measure of the syndrome sum $s_i$ that checks on the $j$th received bit $z_j$ as follows:

$$\phi_{i,j} = \min_{j' \in N_i \setminus j} |y_{j'}|. \tag{11.58}$$

We notice that the above reliability measure $\phi_{i,j}$ of the syndrome sum $s_i$ checking on the $j$th received bit $z_j$ is different from the reliability measure $\phi_i$ of the same syndrome sum that is defined by (11.55). The reliability measure $\phi_i$ of the syndrome sum $s_i$ is defined by considering the reliabilities of *all the received bits* participating in $s_i$, while the reliability measure $\phi_{i,j}$ of $s_i$ is defined by *excluding* the reliability measure $|y_j|$ of the $j$th received bit $z_j$ that is checked by $s_i$. Therefore, $\phi_{i,j}$ is a function both of the row index $i$ and of the column index $j$ of the parity-check matrix $\mathbf{H}$ of the code to be decoded, while $\phi_i$ is a function of the row index $i$ alone.

Now define a new weight reliability measure of a hard-decision received bit $z_j$ as follows:

$$E_{j,\epsilon} = \sum_{i \in M_j} (2s_i - 1)\phi_{i,j} - \epsilon|y_j|, \tag{11.59}$$

which consists of two parts. The first part of $E_{j,\epsilon}$ is the sum part of (11.52), which contains the reliability information coming from all the syndrome sums orthogonal on the $j$th received bit $z_j$ but not including the reliability information of $z_j$. This part indicates to *what extent* the received bit $z_j$ should be flipped. The second part $\epsilon|y_j|$ of $E_{j,\epsilon}$ gives reliability information on the received bit $z_j$. It basically indicates to what extent the received bit $z_j$ should maintain its value unchanged. The parameter $\epsilon$ of the second part of $E_{j,\epsilon}$ is a positive real number, a design parameter, which is called the *confidence coefficient* of received bit $z_j$. For a given LDPC code, the confidence coefficient $\epsilon$ should be chosen so as to optimize the reliability measure $E_j$ and hence to minimize the error rate of the code with BF decoding. The optimum value of $\epsilon$ is hard to derive analytically and it is usually found by computer simulation [17, 36, 38]. Experimental results show that the optimum choice of $\epsilon$ varies slightly with the SNR [36]. For simplicity, it is kept constant during the decoding process.

Weighted BF Decoding Algorithm 1 can be modified by using the new reliability measures of a syndrome sum and a received bit defined by (11.51) and (11.52), respectively. For $0 \leq k < k_{\max}$, let $\mathbf{z}^{(k)} = \left(z_0^{(k)}, z_1^{(k)}, \ldots, z_{n-1}^{(k)}\right)$ be the hard-decision received sequence generated in the $k$th decoding iteration and let $\mathbf{E}_\epsilon^{(k)} = \left(E_{0,\epsilon}^{(k)}, E_{1,\epsilon}^{(k)}, \ldots, E_{n-1,\epsilon}^{(k)}\right)$

be the weighted reliability profile of $\mathbf{z}^{(k)}$, where $E_{j,\epsilon}^{(k)}$ is the weighted reliability measure of the $j$th bit of $\mathbf{z}^{(k)}$, computed using (11.52) and the syndrome $\mathbf{s}^{(k)}$ of $\mathbf{z}^{(k)}$.

---

**Algorithm 11.4** Weighted BF Decoding Algorithm 2

---

**Initialization**. Set $k = 0$, $\mathbf{z}^{(0)} = \mathbf{z}$, and the maximum number of iterations to $k_{\max}$. Store $\epsilon|y_j|$ for $0 \le j < n$. Compute and store $\phi_{i,j}$ for $0 \le i < m$ and $j \in N_i$.

1. Compute the syndrome $\mathbf{s}^{(k)} = \mathbf{z}^{(k)}\mathbf{H}^{\mathrm{T}}$ of $\mathbf{z}^{(k)}$. If $\mathbf{s}^{(k)} = \mathbf{0}$, stop decoding and output $\mathbf{z}^{(k)}$ as the decoded codeword; otherwise, go to Step 2.
2. Compute the weighted reliability profile $\mathbf{E}_{\epsilon}^{(k)}$ on the basis of (11.58) and (11.58). Go to Step 3.
3. Identify the bit position $j$ for which $E_{j,\epsilon}^{(k)}$ is largest. Go to Step 4. *Remark*: If $j$ is not unique, then choose a $j$ at random.
4. Flip the $j$th received bit of $\mathbf{z}^{(k)}$ to obtain a modified received sequence $\mathbf{z}^{(k+1)}$. Go to Step 5.
5. $k \leftarrow k+1$. If $k > k_{\max}$, stop decoding; otherwise, go to Step 1.

---

On comparing (11.58) with (11.56), we see that Weighted BF Decoding Algorithm 2 requires more real-number additions than does Weighted BF Decoding Algorithm 1. Furthermore, it requires more storage than does Algorithm 1 because it has to store $\epsilon|y_j|$ for $0 \le j < n$ in addition to $\phi_{i,j}$ for $0 \le i < m$ and $j \in N_i$.

Again, we consider the (4095, 3367) cyclic EG-LDPC code constructed in Example 11.16. The bit-error performance of this code over the binary-input AWGN channel decoded using Weighted BF Decoding Algorithm 2 with a maximum of 100 iterations is shown in Figure 11.21. The confidence coefficient $\epsilon$ is chosen to be 2.64 by computer search. At a BER of $10^{-6}$, Weighted BF Decoding Algorithm 2 outperforms Weighted BF Decoding Algorithm 1 by 0.5 dB and it performs only 0.5 dB from the SPA with a maximum of 50 iterations.

During the BF decoding process, it may happen that the generated received sequence at some iteration, say the $k$th iteration, is a *repetition* of the received sequence generated at an earlier iteration, say the $k_0$th iteration, with $1 \le k_0 < k$, that is, $\mathbf{z}^{(k)} = \mathbf{z}^{(k_0)}$. In this case, $\mathbf{s}^{(k)} = \mathbf{s}^{(k_0)}$ and $\mathbf{E}_{\epsilon}^{(k)} = \mathbf{E}_{\epsilon}^{(k_0)}$. Consequently, the decoder may enter into a *loop* and generate the same set of received sequences over and over again until a preset maximum number of iterations is reached. This loop is referred to as a *trapping loop*. When a BF decoder enters into a trapping loop, decoding will never converge.

A trapping loop can be detected and avoided by carefully choosing the bit to be flipped such that all the generated received sequences $\mathbf{z}^{(k)}, 0 \le k < k_{\max}$, are different. The basic idea is very straightforward. We list all the received sequences, say $\mathbf{z}^{(0)}$ to $\mathbf{z}^{(k-1)}$, that have been generated and compare the newly generated received sequence $\mathbf{z}^{(k)}$ with them. Suppose $\mathbf{z}^{(k)}$ is obtained by flipping the $j$th bit $z_j^{(k-1)}$ of $\mathbf{z}^{(k-1)}$, which has largest value $E_{j,\epsilon}^{(k-1)}$ in the reliability profile $\mathbf{E}_{\epsilon}^{(k-1)}$ of $\mathbf{z}^{(k-1)}$. If the newly generated sequence $\mathbf{z}^{(k)}$ is different from the sequences on the list, we add $\mathbf{z}^{(k)}$ to the list and continue the decoding process. If the newly generated sequence $\mathbf{z}^{(k)}$ has already been generated before, we discard it and flip the bit of $\mathbf{z}^{(k-1)}$, say the $l$th bit, $z_l^{(k-1)}$, with the *next-largest component* in $\mathbf{E}_\varepsilon^{(k-1)}$. We keep flipping the next unreliable bits of $\mathbf{z}^{(k-1)}$, one at a time, until we generate a sequence that has never been generated before. Then we add this new sequence to the list and continue the decoding process. This approach to detecting and avoiding a trapping loop in weighted BF decoding was first proposed in [37].

It seems that the above straightforward approach to detecting and avoiding a trapping loop requires a large memory to store all the generated received sequences and a large number of vector comparisons. Fortunately, it can be accomplished with a simple mechanism that does not require either a large memory to store all the generated received sequences or a large number of vector comparisons. For $0 \leq l < n$, define the following $n$-tuple over GF(2):

$$\mathbf{u}_l = (0, 0, \ldots, 0, 1, 0, \ldots, 0),$$

whose $l$th component is equal to 1, with all the other components being equal to 0. This $n$-tuple over GF(2) is called a *unit vector*. Let $j_k$ be the bit position of $\mathbf{z}^{(k-1)}$ chosen to be flipped at in the $k$th decoding iteration. Then

$$\mathbf{z}^{(k)} = \mathbf{z}^{(k-1)} + \mathbf{u}_{j_k}, \tag{11.60}$$

$$\mathbf{z}^{(k)} = \mathbf{z}^{(k_0)} + \mathbf{u}_{j_{k_0+1}} + \cdots + \mathbf{u}_{j_k}, \tag{11.61}$$

for $0 \leq k_0 < k$. For $0 \leq l \leq k$, define the following sum of unit vectors:

$$\mathbf{U}_l^{(k)} = \sum_{t=l}^{k} \mathbf{u}_{j_t}. \tag{11.62}$$

It follows from (11.61) and (11.62) that we have

$$\mathbf{z}^{(k)} = \mathbf{z}^{(k_0)} + \mathbf{U}_{k_0+1}^{(k)}. \tag{11.63}$$

It is clear that $\mathbf{z}^{(k)} = \mathbf{z}^{(k_0)}$ if and only if $\mathbf{U}_{k_0+1}^{(k)} = \mathbf{0}$. This implies that, to avoid a possible trapping loop, the vector sums $\mathbf{U}_1^{(k)}, \mathbf{U}_2^{(k)}, \ldots, \mathbf{U}_{k-1}^{(k)}$ must all be *nonzero* vectors. Note that all these vector sums depend only on the bit positions $j_1, j_2, \ldots, j_k$ at which bits were flipped from the first iteration to the $k$th iteration. Using our knowledge of these bit positions, we don't have to store all the generated received sequences.

Note that $\mathbf{U}_l^{(k)}$ can be constructed recursively:

$$\mathbf{U}_l^{(k)} = \begin{cases} \mathbf{u}_{j_k}, & \text{for } l = k, \\ \mathbf{U}_{l+1}^{(k)} + \mathbf{u}_{j_l}, & \text{for } 1 \leq l < k. \end{cases} \tag{11.64}$$

To check whether $\mathbf{U}_l^{(k)}$ is zero, we define $w_l$ as the Hamming weight of $\mathbf{U}_l^{(k)}$. Then $w_l$ can be computed recursively as follows, without counting the number of 1s in $\mathbf{U}_l^{(k)}$:

$$w_l = \begin{cases} 1, & \text{if } l = k, \\ w_{l+1} + 1, & \text{if the } l\text{th bit of } \mathbf{U}_{l+1}^{(k)} = 0, \\ w_{l+1} - 1, & \text{if the } l\text{th bit of } \mathbf{U}_{l+1}^{(k)} = 1. \end{cases} \tag{11.65}$$

Obviously, $\mathbf{U}_l^{(k)} = \mathbf{0}$ if and only if $w_l = 0$.

Let $B$ denote a list containing the bit positions that would cause trapping loops during decoding. This list of the bit positions is called a *loop-exclusion list*, which will be built up as the decoding iterations go on. With the above developments, Weighted BF Decoding Algorithm 2 can be modified with a simple loop-detection mechanism as follows.

**Algorithm 11.5** Weighted BF Decoding Algorithm 3 with Loop Detection

**Initialization**. Set $k = 0$, $\mathbf{z}^{(0)} = \mathbf{z}$, the maximum number of iterations to $k_{\max}$, and the loop-exclusion list $B = \varnothing$ (empty set). Compute and store $\phi_{i,j}$ for $0 \leq i < m$ and $j \in N_i$. Store $\epsilon|y_j|$ for $0 \leq j < n$.

1. Compute the syndrome $\mathbf{s}^{(k)} = \mathbf{z}^{(k)}\mathbf{H}^{\mathrm{T}}$ of $\mathbf{z}^{(k)}$. If $\mathbf{s}^{(k)} = \mathbf{0}$, stop decoding and output $\mathbf{z}^{(k)}$ as the decoded codeword; otherwise, go to Step 2.
2. $k \leftarrow k + 1$. If $k > k_{\max}$, declare a decoding failure and stop the decoding process; otherwise, go to Step 3.
3. Compute the weighted reliability profile $\mathbf{E}_{\epsilon}^{(k)}$ of $\mathbf{z}^{(k)}$. Go to Step 4.
4. Choose the bit position
$$j_k = \arg \max_{0 \leq j < n, j \notin B} E_{j,\epsilon}^{(k)}.$$
Go to Step 5.
5. Compute $\mathbf{U}_l^{(k)}$ and $w_l$ on the basis of (11.64) and (11.65) for $1 \leq l < k$. If $w_l = 0$ for any $l$, then $B \leftarrow B \cup \{j_k\}$ and go to Step 4. If all $w_1, w_2, \ldots, w_{k-1}$ are nonzero, go to Step 6.
6. Compute $\mathbf{z}^{(k)} = \mathbf{z}^{(k-1)} + \mathbf{u}_{j_k}$. Go to Step 1.

Compared with Weighted BF Decoding Algorithm 2, Weighted BF Decoding Algorithm 3 with loop detection requires additional memory to store a loop-exclusion list and more computations to carry out Steps 4 and 5. Avoiding trapping loops in flipping during decoding would often make the decoding of many LDPC codes, be they algebraically or randomly constructed, converge faster and would lower their error floors. However, it might not make much difference for the finite-geometry codes constructed in Sections 11.1–11.3, 11.6, and 11.7 because they already perform well with the weighted BF decoding algorithms without loop detection in terms of decoding convergence and the error floor. Clearly, loop detection can be included in the LPCF-BF and weighted BF decoding algorithms (see Problem 11.19).

## 11.11 Quasi-cyclic Codes Based on Partial Geometries

Besides Euclidean and projective geometries, there is another category of finite geometries, called *partial geometries* (PaGs) [39–44]. Partial geometries are geometries of two dimensions, which consists of only points and lines. Two-dimensional Euclidean and projective geometries form two subclasses of partial geometries. Partial geometries were first introduced by Bose [39] in 1963. Since then, they have become an interesting branch in combinatorial mathematics, and lately they have been shown to be effective in construction of LDPC codes with distinct geometric and algebraic structures [45, 46]. PaG-based LDPC codes perform well over AWGN channels with various iterative decoding algorithms using assorted iterative decoding algorithms based on belief propagation. The graphical structures of PaG-based LDPC codes are like those of the LDPC codes constructed based on the two-dimensional EG- and PG-LDPC codes [47].

In this section, we first introduce the basic concepts and structures of a partial geometry. Then we present a very special class of partial geometries, constructed based on prime fields. More on partial geometries and their associated LDPC codes can be found in [45–46]. In Chapter 12, we will show that certain classes of QC-LDPC codes constructed based on finite fields are PaG-LDPC codes.

### 11.11.1 Basic Concepts and Structures of a Partial Geometry

Consider a system composed of a set $\mathbf{N}$ of $n$ points and a set $\mathbf{M}$ of $m$ lines where each line is a set of points. If a line $\mathcal{L}$ contains a point $\mathbf{p}$, we say that $\mathbf{p}$ is on $\mathcal{L}$ and that $\mathcal{L}$ passes through $\mathbf{p}$. If two points are on a line, then we say that the two points are adjacent. If two lines pass through the same point, then we say that the two lines intersect, otherwise they are parallel. The system composed of the sets $\mathbf{N}$ and $\mathbf{M}$ is a partial geometry if the following conditions are satisfied for some fixed integers $\gamma \geq 2$, $\rho \geq 2$, and $\delta \geq 1$. (1) Any two points are on at most one line. (2) Each point is on $\gamma$ lines. (3) Each line passes through $\rho$ points. (4) If a point $\mathbf{p}$ is not on a line $\mathcal{L}$, then there are exactly $\delta$ lines, each passing through $\mathbf{p}$ and a point on $\mathcal{L}$. Such a partial geometry will be denoted by PaG$(\gamma, \rho, \delta)$ and $\gamma, \rho,$ and $\delta$ are called the parameters of the partial geometry. The parameter $\delta$ is called the *connection number* of the geometry. A simple counting argument [42] shows that the partial geometry PaG$(\gamma, \rho, \delta)$ has exactly

$$n = \rho[(\rho - 1)(\gamma - 1) + \delta]/\delta \tag{11.66}$$

points and

$$m = \gamma[(\rho - 1)(\gamma - 1) + \delta]/\delta = \gamma n/\rho \tag{11.67}$$

lines.

If $\delta = \gamma - 1$, then the partial geometry PaG$(\gamma, \rho, \gamma - 1)$ consists of $n = \rho^2$ points and $m = \gamma\rho$ lines. In this case, a point $\mathbf{p}$ not on a line $\mathcal{L}$ in PaG$(\gamma, \rho, \gamma - 1)$ is on a unique line that is parallel to $\mathcal{L}$. The $\gamma\rho$ lines in PaG$(\gamma, \rho, \gamma - 1)$ can be partitioned into $\gamma$ parallel bundles, each consisting of $\rho$ parallel lines. A partial geometry PaG$(\gamma, \rho, \gamma - 1)$ with parameters $\gamma, \rho, \gamma - 1$ is called a *net*. A special type of net is $\rho = \gamma$ and $\delta = \gamma - 1$, that is, a partial geometry PaG$(\gamma, \gamma, \gamma - 1)$ with parameters $\gamma, \gamma, \gamma - 1$. In this case, $n = m = \rho^2 = \gamma^2$.

Examples of partial geometries are two-dimensional Euclidean and projective geometries over GF$(q)$. The two-dimensional Euclidean geometry EG$(2, q)$ over GF$(q)$ is a partial geometry PaG$(q + 1, q, q)$ with parameters $\gamma = q + 1$, $\rho = q$, and

$\delta = q$, which has $q^2$ points and $q(q+1)$ lines, each line consisting of $q$ points. Since $\delta = \gamma - 1$, EG(2, $q$) is a net. The two-dimensional projective geometry PG(2, $q$) over GF($q$) is a partial geometry PaG($q+1, q+1, q$) with parameters $\gamma = \rho = q + 1$ and $\delta = q$, which consists of $q^2 + q + 1$ points and $q^2 + q + 1$ lines, each line consisting of $q + 1$ points, and is also a net.

Let $\mathbf{H}_{\mathrm{PaG}}$ be the line–point incidence matrix of the partial geometry PaG($\gamma, \rho, \delta$), that is, the rows of $\mathbf{H}_{\mathrm{PaG}}$ are the incidence vectors of the lines in PaG($\gamma, \rho, \delta$). $\mathbf{H}_{\mathrm{PaG}}$ is an $m \times n$ matrix over GF(2). Since two lines in PaG($\gamma, \rho, \delta$) are either parallel or intersect at one and only one point, no two rows (columns) in $\mathbf{H}_{\mathrm{PaG}}$ have more than one location where they both have 1-components. Hence, $\mathbf{H}_{\mathrm{PaG}}$ satisfies the RC constraint.

The null space over GF(2) of a submatrix of $\mathbf{H}_{\mathrm{PaG}}$ gives a PaG-LDPC code whose Tanner graph has girth at least 6. If $\mathbf{H}_{\mathrm{PaG}}$ is an array of CPMs, then QC-PaG-LDPC codes of various lengths and rates can be constructed. The construction is like the construction of QC-EG-LDPC codes using two-dimensional Euclidean geometry EG(2, $q$) over GF(2).

### 11.11.2 A Class of Partial Geometries Based on Prime Fields

In this section, we present a class of partial geometries that are nets. The construction is based on a theorem proved in [47]. We restate the theorem without proof.

**Theorem 11.1** *Let* $\mathbf{H}_{\mathrm{PaG}}$ *be an* $m \times n$ *RC-constrained matrix with column and row weights* $\gamma$ *and* $\rho$*, respectively, which is a* $\gamma \times \rho$ *array of CPMs of size* $\gamma \times \gamma$*, where* $m = \gamma^2$ *and* $n = \gamma\rho$*. Then* $\mathbf{H}_{\mathrm{PaG}}$ *is the line–point incidence matrix of a partial geometry PaG(*$\gamma, \rho, \rho - 1$*) that has* $n$ *points corresponding to the columns of* $\mathbf{H}_{\mathrm{PaG}}$ *and* $m$ *lines corresponding to the rows of* $\mathbf{H}_{\mathrm{PaG}}$*.*

Let $p$ be a prime. Then the set of integers $\{0, 1, 2, \ldots, p-1\}$ forms a field GF($p$) under modulo-$p$ addition and multiplication. Such a field is called a prime field (see Chapter 2). For each element $i$ in GF($p$), we form a $p$-tuple over GF(2),

$$\mathbf{z}(i) = (z_0, z_1, \ldots, z_{p-1}), \tag{11.68}$$

whose components correspond to all the elements in GF($p$), labeled from 0 to $p - 1$, where the $i$th component $z_i = 1$ and all the other components are zeros. This $p$-tuple $\mathbf{z}(i)$ is called the *location vector* of the element $i$.

The location vector of the 0 element of GF($p$) is $z(0) = (1, 0, 0, \ldots, 0)$. The 1-components of the location vectors of two different elements in GF($p$) are at two different positions. For a given element $k$ in GF($p$), $k + 0, k + 1, k + 2, \ldots, k + (p - 1)$ (modulo-$p$) are different integers and they form all the $p$ elements of GF($p$). Form a $p \times p$ matrix $\mathbf{A}$ with the location vectors $\mathbf{z}(k), \mathbf{z}(k + 1), \mathbf{z}(k + 2), \ldots, \mathbf{z}(k + (p - 1))$ of $k + 0, k + 1, k + 2, \ldots, k + (p - 1)$ as rows (top-down). $\mathbf{A}$ is a CPM of size $p \times p$ with $\mathbf{z}(k)$ (the top row of $\mathbf{A}$) as the generator. As defined in Section 11.6, $\mathbf{A}$ is the CPM dispersion CPM($k$) of the element of $k$ in GF($p$). The CPM dispersions of the $p$ elements in GF($p$) are all different.

Form the following $p \times p$ matrix over GF($p$) with columns and rows labeled from 0 to $p-1$:

$$\mathbf{B}_{\text{PaG}} = \begin{bmatrix} \mathbf{b}_0 \\ \mathbf{b}_1 \\ \vdots \\ \mathbf{b}_{p-1} \end{bmatrix} = \begin{bmatrix} 0 \cdot 0 & 0 \cdot 1 & \cdots & 0 \cdot (p-1) \\ 1 \cdot 0 & 1 \cdot 1 & \cdots & 1 \cdot (p-1) \\ 2 \cdot 0 & 2 \cdot 1 & \cdots & 2 \cdot (p-1) \\ \vdots & \vdots & & \vdots \\ (p-1) \cdot 0 & (p-1) \cdot 1 & \cdots & (p-1) \cdot (p-1) \end{bmatrix}, \tag{11.69}$$

where the multiplication $i \cdot j$ of two elements $i$ and $j$ in GF($p$) is carried out modulo $p$. Label the rows and columns from 0 to $p-1$. Matrix $\mathbf{B}_{\text{PaG}}$ has the following structural properties: (1) all the entries in the 0th (or top) row are zeros; (2) all the entries in any row other than the top row are different and they form all the elements of GF($p$); (3) all the entries in the 0th (or leftmost) column are zeros; (4) all the entries in any column other than the leftmost column are different and they form all the elements of GF($p$); and (5) any two different rows (or different columns) have the 0 element of GF($p$) in common at the leftmost column (or top row) and differ in all the other $p-1$ positions.

From the structural properties of $\mathbf{B}_{\text{PaG}}$, we can prove that the rows of $\mathbf{B}_{\text{PaG}}$ satisfy the constraints given by the following two lemmas (see Problem 11.21).

**Lemma 11.2** *Let* $\mathbf{b}_i = (i \cdot 0, i \cdot 1, \ldots, i \cdot (p-1))$, $0 \le i < p$, *be the ith row of* $\mathbf{B}_{\text{PaG}}$. *For different elements k and l in GF(p), the two p-tuples* $(i \cdot 0 + k, i \cdot 1 + k, \ldots, i \cdot (p-1) + k)$ *and* $(i \cdot 0 + l, i \cdot 1 + l, \ldots, i \cdot (p-1) + l)$ *differ in all p positions.*

**Lemma 11.3** *For* $0 \le i, j < p$ *and* $i \ne j$, *let* $\mathbf{b}_i = (i \cdot 0, i \cdot 1, \ldots, i \cdot (p-1))$ *and* $\mathbf{b}_j = (j \cdot 0, j \cdot 1, \ldots, j \cdot (p-1))$ *be two different rows of* $\mathbf{B}_{\text{PaG}}$. *For any two elements k and l in GF(p) with* $0 \le k, l < p$, *the two p-tuples over GF(p),* $(i \cdot 0 + k, i \cdot 1 + k, \ldots, i \cdot (p-1) + k)$ *and* $(j \cdot 0 + l, j \cdot 1 + l, \ldots, j \cdot (p-1) + l)$ *differ in at least* $p-1$ *positions.*

The two constraints on the rows of $\mathbf{B}_{\text{PaG}}$ given by Lemmas 11.2 and 11.3 are referred to as row constraints 1 and 2.

Now, we represent each entry $i \cdot j$ in $\mathbf{B}_{\text{PaG}}$ by a CPM, denoted by CPM($i \cdot j$), with its location vector $\mathbf{z}(i \cdot j)$ as the generator. CPM($i \cdot j$) is referred to as the CPM dispersion of $i \cdot j$ with respect to the field GF($p$) as defined in Section 11.6. The CPM dispersions of the entries in $\mathbf{B}_{\text{PaG}}$ result in a $p \times p$ array $\mathbf{H}_{\text{PaG}}$ of CPMs of size $p \times p$, which is a $p^2 \times p^2$ matrix over GF(2) with both column and row weight $p$. The array $\mathbf{H}_{\text{PaG}}$ consists of $p$ CPM row-blocks and $p$ CPM column-blocks. $\mathbf{H}_{\text{PaG}}$ is referred to as the CPM dispersion of $\mathbf{B}_{\text{PaG}}$.

It follows from the structural properties of $\mathbf{B}_{\text{PaG}}$ that $\mathbf{H}_{\text{PaG}}$ has the following structural properties: (1) the top CPM row-block consists of $p$ identity matrices; (2) the leftmost CPM column-block consists of $p$ identity matrices; (3) except for the top CPM row-block, the $p$ CPMs of each of the other $\gamma - 1$ CPM row-blocks are different; (4) except for the leftmost CPM column-block, the $p$ CPMs of each of the other $\gamma - 1$ CPM column-blocks are different; and (5) any two different CPM-row-blocks (or different CPM column-blocks) have the identity matrix CPM(0) in common at the leftmost CPM column-block (or top CPM row-block) and differ in all the other $p - 1$ positions.

It follows from Lemmas 11.2 and 11.3 that $\mathbf{H}_{\text{PaG}}$, as a matrix, satisfies the RC constraint. Hence, it follows from Theorem 11.1 that $\mathbf{H}_{\text{PaG}}$ is the incidence matrix of a partial geometry PaG$(p, p, p - 1)$ that consists of $p^2$ points and $p^2$ lines. Each line consists of $p$ points and each point is on $p$ lines. The $p$ rows of a CPM row-block are the incidence vectors of $p$ parallel lines in PaG$(p, p, p - 1)$. Therefore, the $p^2$ lines can be partitioned into $p$ parallel bundles. Since each column of $\mathbf{H}_{\text{PaG}}$ has weight $p$, each point in PaG$(p, p, p - 1)$ is on $p$ lines.

PaG$(p, p, p - 1)$ is a net. The matrix $\mathbf{B}_{\text{PaG}}$ over GF$(p)$ is referred to as the *base matrix* for the construction of the partial geometry PaG$(p, p, p - 1)$ (or $\mathbf{H}_{\text{PaG}}$).

Like the construction of QC-EG-LDPC codes presented in Section 11.6, submatrices of $\mathbf{B}_{\text{PaG}}$ in conjunction with CPM dispersion can be used to construct QC-PaG-LDPC codes. Let $g$ and $r$ be two positive integers such that $1 \leq g \leq r \leq p$. Take a $g \times r$ submatrix $\mathbf{B}_{\text{PaG}}(g, r)$ from $\mathbf{B}_{\text{PaG}}$. The CPM dispersion of $\mathbf{B}_{\text{PaG}}$ with respect to GF$(p)$ gives a $g \times r$ array $\mathbf{H}_{\text{PaG}}(g, r)$ of CPMs of size $p \times p$, which is a $gp \times rp$ matrix over GF(2) with column and row weights $g$ and $r$, respectively. The null space over GF(2) gives a $(g, r)$-regular QC-PaG-LDPC code $\mathcal{C}_{\text{PaG}}(g, r)$ of length $rp$ with rate at least $(r - g)/r$. The Tanner graph of $\mathcal{C}_{\text{PaG}}(g, r)$ has girth at least 6. With various choices of $(g, r)$ pairs, we can construct a family of QC-PaG-LDPC codes of different lengths and rates.

---

**Example 11.18** Suppose we use the prime field GF(73) to construct a base matrix $\mathbf{B}_{\text{PaG}}$ in the form of (11.69). The CPM dispersion with respect to GF(73) of $\mathbf{B}_{\text{PaG}}$ gives a $73 \times 73$ array $\mathbf{H}_{\text{PaG}}$ of CPMs of size $73 \times 73$, which is a $5329 \times 5329$ matrix over GF(2) with both column and row weight 73. $\mathbf{H}_{\text{PaG}}$ is the incidence matrix of a partial geometry PaG$(73, 73, 72)$, which consists of 5329 points and 5329 lines. Each line in PaG$(73, 73, 72)$ consists of 73 points and each point is on 73 lines. The lines in PaG$(73, 73, 72)$ can be partitioned into 73 parallel bundles, each consisting of 73 parallel lines.

Choose $g = 6$ and $r = 72$. Take a $6 \times 72$ submatrix $\mathbf{B}_{\text{PaG}}(6, 72)$ from $\mathbf{B}_{\text{PaG}}$. The CPM dispersion of $\mathbf{B}_{\text{PaG}}(6, 72)$ gives a $6 \times 72$ array $\mathbf{H}_{\text{PaG}}(6, 72)$ of CPMs of size $73 \times 73$, which is a subarray of $\mathbf{H}_{\text{PaG}}$. $\mathbf{H}_{\text{PaG}}(6, 72)$ is a $438 \times 5256$ RC-constrained matrix with column and row weights 6 and 72, respectively. The rank of $\mathbf{H}_{\text{PaG}}(6, 72)$

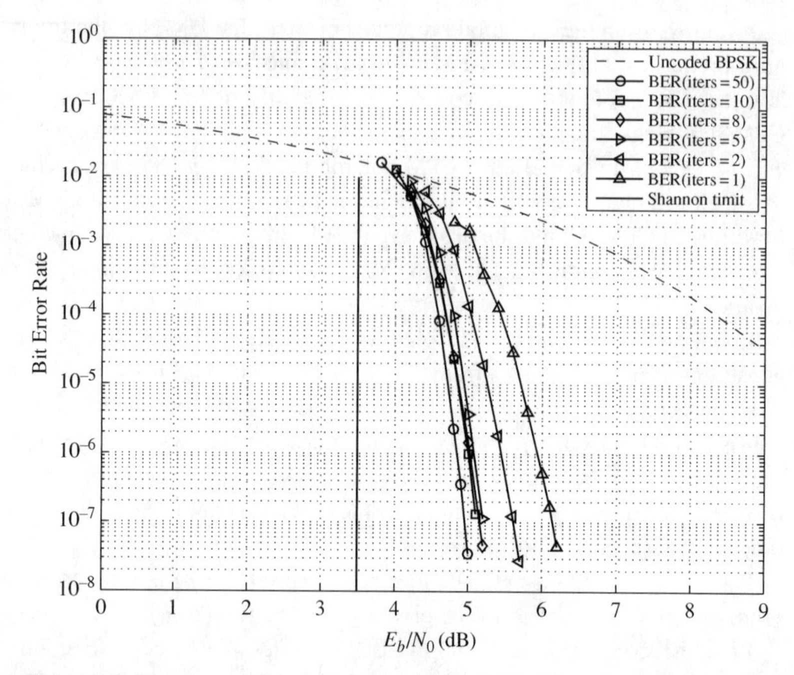


**Figure 11.23** The error performance of the (5256,4823) QC-PaG-LDPC code given in Example 11.18.

is 433. The null space of $\mathbf{H}_{\text{PaG}}(6, 72)$ is a (5256, 4823) QC-PaG-LDPC code $\mathcal{C}_{\text{PaG}}(6, 72)$ with rate 0.9176. The estimated minimum distance of CPaG(6, 72) is 20.

The bit-error performances of this code decoded with iterative decoding using the SPA with various numbers of decoding iterations are shown in Figure 11.23. At a BER of $10^{-6}$ with 50 decoding iterations, the code performs 1.3 dB from the Shannon limit. We also see that the decoding of this code converges very fast. At a BER of $10^{-6}$, the performance gap between 5 and 50 iterations is about 0.2 dB, while the gap between 10 and 50 iterations is about 0.1 dB. The code also has a very low error floor as shown in Figure 11.24. With 50 iterations of the SPA, the estimated error floor of the code is below $10^{-25}$ for bit-error rate and below $10^{-22}$ for word-error rate.

An FPGA decoder for this code has been built. Using this decoder and 15 iterations of the SPA, the performance of this code can be simulated down to a BER of $10^{-12}$ and a WER of almost $10^{-10}$, as shown in Figure 11.25. From Figures 11.23 and 11.25 we see that, to achieve a BER of $10^{-7}$, both 15 and 50 iterations of the SPA require 5 dB.

The (5256, 4823) QC-PaG-LDPC code constructed above can be shortened or punctured to varying degrees to obtain codes with assorted lengths or rates for different applications to communication and data storage systems.

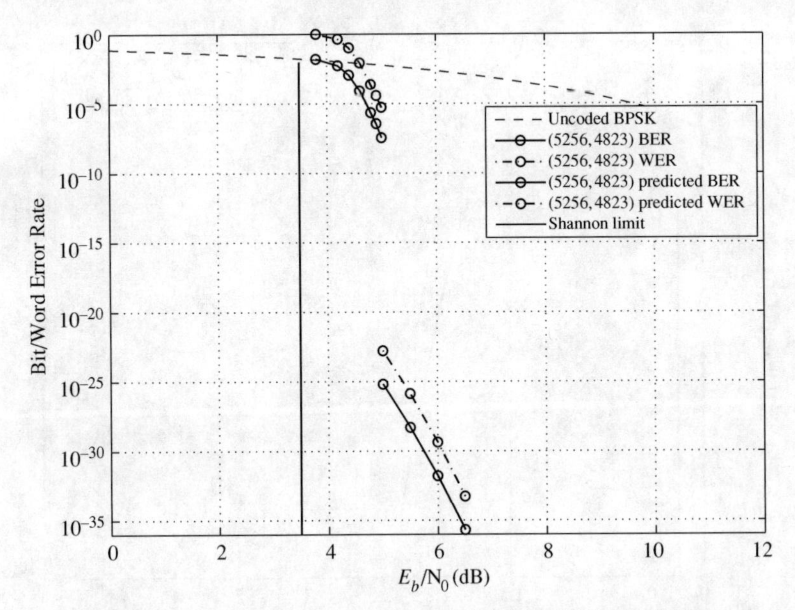


**Figure 11.24** The estimated error floor of the (5256, 4823) QC-PaG-LDPC code given in Example 11.18.

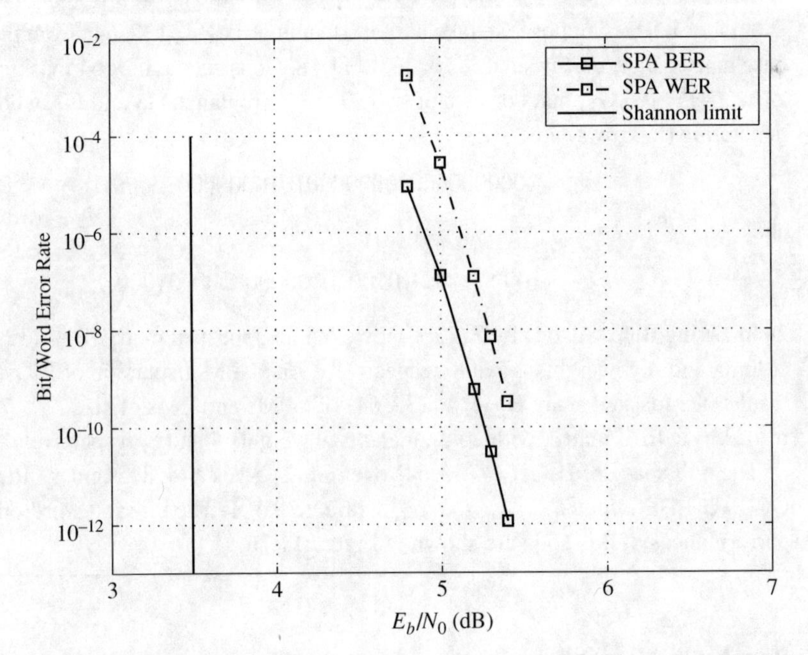


**Figure 11.25** Bit- and word-error performances of the (5256, 4823) QC-PaG-LDPC code given in Example 11.18 simulated by an FPGA decoder using 15 iterations of the SPA.

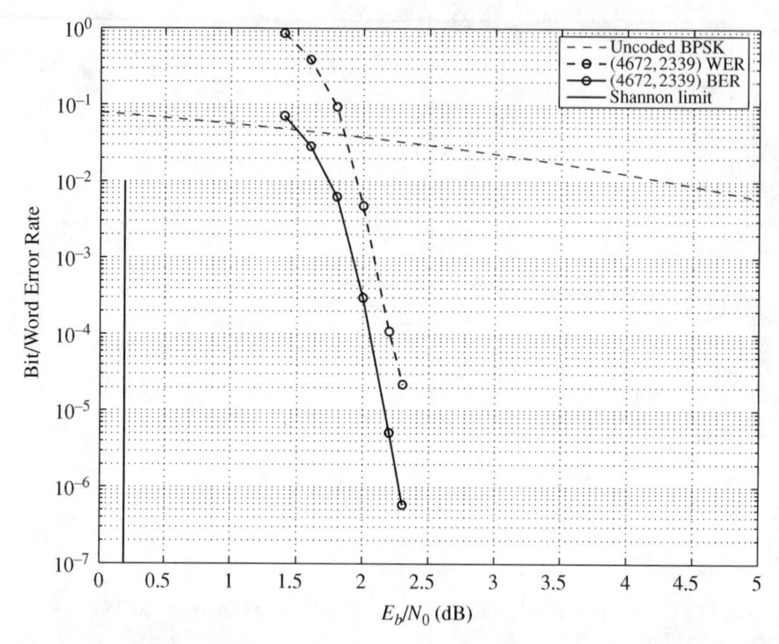


**Figure 11.26** Error performances of the (4672, 2339) QC-PaG-LDPC codes given in Example 11.19.

---

**Example 11.19** Suppose we take a $32 \times 64$ submatrix $\mathbf{B}_{\text{PaG}}(32, 64)$ from the $73 \times 73$ base matrix $\mathbf{B}_{\text{PaG}}$ constructed in Example 11.18. Construct a $32 \times 64$ masking matrix $\mathbf{Z}(32, 64) = [\mathbf{G}_0\mathbf{G}_1]$ that consists of two $32 \times 32$ circulants, $\mathbf{G}_0$ and $\mathbf{G}_1$, whose generators (top rows) are

$$\mathbf{g}_0 = (00000000010000000101000000010000)$$

and

$$\mathbf{g}_1 = (00000000010100100000000000001000).$$

On masking $\mathbf{B}_{\text{PaG}}(32, 64)$ by $\mathbf{Z}(32, 64)$, we obtain a masked matrix $\mathbf{B}_{\text{PaG}}(32, 64)$ with column and row weights 4 and 8, respectively. The CPM dispersion of $\mathbf{B}_{\text{PaG}}(32, 64)$ results in a masked array $\mathbf{H}_{\text{PaG,mask}}(32, 64)$ of CPMs and ZMs of size $73 \times 73$, which is a $2336 \times 4673$ matrix with column and row weights 4 and 8, respectively.

The null space of $\mathbf{H}_{\text{PaG,mask}}(32, 64)$ over GF(2) gives a (4, 8)-regular (4672, 2339) QC-PaG-LDPC code $\mathcal{C}_{\text{PaG,mask}}(32, 64)$ with rate 0.501. The bit- and word-error performances of this code are shown in Figure 11.26.

---

## 11.12 Concluding Remarks

Before we conclude this chapter, we must emphatically point out that all the hard-decision and reliability-based decoding algorithms presented in Sections 11.8 and

11.10 can be applied to decode any LDPC code whose parity-check matrix satisfies the RC constraint, not just the FG-LDPC codes constructed in Sections 11.1, 11.3, 11.6, 11.7, and 11.11. In fact, they are very effective in decoding the codes constructed based on finite fields to be presented in the next chapter. These decoding algorithms, together with the iterative decoding algorithms based on belief propagation presented in Chapter 5, provide a wide spectrum of trade-offs among error performance, decoding complexity, and decoding speed. See also [33, 54].

In Section 11.11, prime fields are used to construct partial geometries and their associated QC-LDPC codes. In fact, cyclic subgroups of any finite field can be used to construct partial geometries. This will be discussed in Chapters 12 and 13.

## Problems

**11.1** Construct a cyclic EG-LDPC code based on the lines of the two-dimensional Euclidean geometry $\mathrm{EG}(2,13)$ over $\mathrm{GF}(13)$ not passing through the origin of the geometry. Determine the length, dimension, minimum distance, and generator polynomials of the code. Compute the bit- and word-error performances of the code over the binary-input AWGN channel decoded using the SPA with a maximum of 50 iterations.

**11.2** Construct a cyclic EG-LDPC code based on the lines of the two-dimensional Euclidean geometry $\mathrm{EG}(2,2^4)$ over $\mathrm{GF}(2^4)$ not passing through the origin of the geometry. Determine its generator polynomial and compute its bit- and word-error performances over the binary-input AWGN channel using the SPA with a maximum of 50 iterations.

**11.3** Consider the three-dimensional Euclidean geometry $\mathrm{EG}(3,7)$ over $\mathrm{GF}(7)$. How many cyclic classes of lines not passing through the origin of this geometry can be formed? Construct a QC-EG-LDPC code of the longest possible length based on these cyclic classes of lines. Compute the bit- and word-error performances of this code over the binary-input AWGN channel using the SPA with a maximum of 50 iterations.

**11.4** Consider the two-dimensional Euclidean geometry $\mathrm{EG}(2,2^6)$ over $\mathrm{GF}(2^6)$. How many parallel bundles of lines can be formed? Suppose we choose six parallel bundles of lines in this geometry to construct a regular LDPC code.

(a) Determine the length and dimension of the code.
(b) Compute the bit- and word-error performances of the code over the binary-input AWGN channel decoded using the SPA with a maximum of 50 iterations.
(c) Display the rate of decoding convergence of the constructed code with 1, 3, 5, 10, 20, and 50 iterations.

**11.5** The incidence vectors of the lines in the two-dimensional Euclidean geometry $\mathrm{EG}(2,2^5)$ over $\mathrm{GF}(2^5)$ not passing through the origin of the geometry form a single $1023 \times 1023$ circulant with both column and row weight 32.

(a) Using column splitting, decompose this circulant into a row $\mathbf{G}_{\mathrm{col,decom}} = [\mathbf{G}_0 \;\; \mathbf{G}_1 \;\; \cdots \;\; \mathbf{G}_{15}]$ of 16 $1023 \times 1023$ circulants, each having column weight 4 and row weight 4. Take the first eight circulants of $\mathbf{G}_{\text{col,decom}}$ to form a parity-check matrix $\mathbf{H}_{\text{col,decom}}(1,8) = [\mathbf{G}_0 \ \ \mathbf{G}_1 \ \ \cdots \ \ \mathbf{G}_7]$. Determine the QC-LDPC code given by the null space of $\mathbf{H}_{\text{col,decom}}(1,8)$. Compute its bit and word-error performances over the binary-input AWGN channel using the SPA with 50 iterations.

(b) Using row splitting, decompose each circulant of $\mathbf{H}_{\text{col,decom}}(1,8)$ into a column of two $1023 \times 1023$ circulants of weight 2. This decomposition results in a $2 \times 8$ array $\mathbf{H}_{\text{array,decom}}(2,8)$ of $1023 \times 1023$ circulants, each with weight 2. Determine the QC-LDPC code given by the array $\mathbf{H}_{\text{array,decom}}(2,8)$ and compute its bit- and word-error performances over the binary-input AWGN channel using the SPA with a maximum 50 iterations.

**11.6** This problem is a continuation of Problem 11.5.

(a) Using row splitting, decompose each circulant of $\mathbf{H}_{\text{col,decom}}(1,8)$ into a column of four $1023 \times 1023$ circulant permutation matrices. This decomposition results in a $4 \times 8$ array $\mathbf{H}_{\text{array,decom}}(4,8)$ of $1023 \times 1023$ circulant permutation matrices. Determine the QC-LDPC code given by the null space of $\mathbf{H}_{\text{array,decom}}(4,8)$ and compute its bit- and word-error performances over the binary-input AWGN channel using the SPA with 50 iterations.

(b) Replace a circulant permutation matrix in each column of the $4 \times 8$ array $\mathbf{H}_{\text{array,decom}}(4,8)$ that you have constructed by a zero matrix such that the resultant array $\mathbf{H}^*_{\text{array,decom}}(4,8)$ has three circulant permutation matrices and one zero matrix in each column, and six circulant permutation matrices and two zero matrices in each row. Determine the QC-LDPC code given by the null space of $\mathbf{H}^*_{\text{array,decom}}(4,8)$ and compute its bit- and word-error performances over the binary-input AWGN channel using the SPA with a maximum of 50 iterations.

**11.7** Decompose the two-dimensional Euclidean geometry EG$(2,2^6)$ based on a chosen parallel bundle $\mathcal{P}(2,1)$ of lines (see Example 11.5). From this decomposition, a $64 \times 64$ array $\mathbf{H}^{(1)}_{\text{EG},d}$ of $64 \times 64$ permutation matrices over GF(2) can be constructed. Take the first four rows of $\mathbf{H}^{(1)}_{\text{EG},d}$ to form a $4 \times 64$ subarray $\mathbf{H}^{(1)}_{\text{EG},d}(4,64)$, which is a $256 \times 4096$ matrix over GF(2) with column and row weights 4 and 64, respectively. Determine the $(4,64)$-regular LDPC code given by the null space of $\mathbf{H}^{(1)}_{\text{EG},d}(4,64)$ and compute its bit- and word-error performances of the code over the binary-input AWGN channel using the SPA with a maximum of 50 iterations.

**11.8** This problem is a continuation of Problem 11.7. Suppose we take a $32 \times 64$ subarray $\mathbf{H}^{(1)}_{\text{EG},d}(32,64)$ from the $64 \times 64$ array $\mathbf{H}^{(1)}_{\text{EG},d}$ of $64 \times 64$ permutation matrices over GF(2) constructed in Problem 11.7, say the first 32 rows of $\mathbf{H}^{(1)}_{\text{EG},d}$. Use the subarray $\mathbf{H}^{(1)}_{\text{EG},d}(32,64)$ as a base array for masking. Construct a masking matrix $\mathbf{Z}(32,64)$ over GF(2) that consists of two $32 \times 32$ circulants in a row, each with column weight 3 and row weight 3. Then $\mathbf{Z}(32,64)$ is a $32 \times 64$ matrix over GF(2) with column and row weights 3 and 6, respectively. Masking $\mathbf{H}^{(1)}_{\text{EG},d}(32,64)$ with $\mathbf{Z}(32,64)$ results in a $32 \times 64$ masked array $\mathbf{M}^{(1)}_{\text{EG},d}(32,64)$ that is a $4096 \times 2048$ matrix over GF(2) with column and row weights 3 and 6, respectively. Determine the code given by the null space of $\mathbf{M}_{\text{EG},d}^{(1)}(32, 64)$ and compute its bit- and word-error performance over the binary-input AWGN channel using the SPA with a maximum of 50 iterations.

**11.9** Prove that, for even $m$, the incidence vector of any line in the $m$-dimensional projective geometry PG$(m, q)$ over GF$(q)$ is primitive.

**11.10** Prove that, for odd $m$, there are $(q^{m+1} - 1)/(q^2 - 1)$ lines in the $m$-dimensional projective geometry PG$(m, q)$ over GF$(q)$ whose incidence vectors are not primitive.

**11.11** Consider the cyclic PG-LDPC code whose parity-check matrix is formed by using the incidence vectors of the lines of the two-dimensional projective geometry PG$(2, 2^4)$ over GF$(2^4)$ as rows. Determine the code and compute its bit- and word-error performances over the binary-input AWGN channel using the SPA with a maximum of 50 iterations. Also display the rate of decoding convergence of this code for iterations 1, 3, 5, 10, and 50.

**11.12** Consider the three-dimensional Euclidean geometry EG$(3, 2^3)$ over GF$(2^3)$. The Galois field GF$(2^9)$ as an extension field of GF$(2^3)$ is a realization of EG$(3, 2^3)$. Let $\alpha$ be a primitive element of GF$(2^9)$. Let $\mathcal{F}$ be a 2-flat in EG$(3, 2^3)$ not passing through the origin of the geometry. Define the incidence vector of $\mathcal{F}$ as a 511-tuple over GF(2),

$$\mathbf{v}_{\mathcal{F}} = (v_0, v_1, \ldots, v_{510}),$$

whose components correspond to the non-origin points $\alpha^0, \alpha, \ldots, \alpha^{511}$ of EG$(3, 2^3)$, where $v_i = 1$ if and only if $\alpha^i$ is a point on $\mathcal{F}$, otherwise $v_i = 0$. Form a matrix $\mathbf{H}$ with the incidence vectors of the 2-flats of EG$(3, 2^3)$ not passing through the origin of the geometry as rows. Prove that the null space of $\mathbf{H}$ gives a cyclic code. Does the Tanner graph of this code contain cycles of length 4? If it does, how many cycles of length 4 are there? Decode this code using the SPA with a maximum of 50 iterations and compute its bit- and word-error performances over the binary-input AWGN channel.

**11.13** Consider an OSMLG-decodable code $C$ with OSMLG error-correction capability $\lfloor g/2 \rfloor$. Show that there is at least one error pattern with $\lfloor g/2 \rfloor + 1$ errors that is not correctable with OSMLG decoding. Also show that there are some error patterns with more than $\lfloor g/2 \rfloor$ errors that are correctable with OSMLG decoding.

**11.14** Let the two-dimensional Euclidean geometry EG$(2, 2^5)$ over GF$(2^5)$ be the code-construction geometry. A cyclic EG-LDPC code of length 1023 can be constructed by utilizing the type-1 incidence vectors of the lines in EG$(2, 2^5)$ not passing through the origin.

(a) Determine the dimension of this code and its OSMLG error-correction capability.
(b) Compute its error performance over the BSC with OSMLG decoding.
(c) Compute its error performance over the BSC with LPCF-BF decoding using 10 and 50 iterations, respectively.
(d) Compute its error performances over the binary-input AWGN channel using weighted BF decoding algorithms 1 and 2 with 50 iterations.

(e) Compute its error performance over the binary-input AWGN channel decoded with the SPA using 50 iterations.

**11.15** A $64 \times 64$ array $H^{(1)}_{EG,d}$ of $64 \times 64$ permutation matrices over GF(2) can be constructed by utilizing decomposition of the two-dimensional Euclidean geometry EG$(2, 2^6)$ over GF$(2^6)$ (see Example 11.5). Take a $16 \times 64$ subarray $H^{(1)}_{EG,d}(16, 64)$ from $H^{(1)}_{EG,d}$.

(a) Determine the EG-LDPC code given by the null space of $H^{(1)}_{EG,d}(16, 64)$.
(b) Compute the error performance of the code over the BSC decoded with the LPCF-BF decoding algorithm using 100 iterations.
(c) Compute the error performances of the code over the binary-input AWGN channel decoded with weighted BF decoding algorithms 1 and 2 with 100 iterations.
(d) Compute the error performance of the code over the binary-input AWGN channel decoded with the SPA with 100 iterations.

**11.16** Decode the $(6, 32)$-regular $(2048, 1723)$ LDPC code constructed in Example 11.5 by utilizing decomposition of the Euclidean geometry EG$(2, 2^6)$ over GF$(2^6)$ with the Weighted BF Decoding Algorithm 2 using 50 and 100 iterations, and compute its error performances over the binary-input AWGN channel.

**11.17** Decode the $(3510, 3109)$ QC-PG-LDPC code given in Example 11.12 with Weighted BF Decoding Algorithm 2 using 50 and 100 iterations, and compute its error performances over the binary-input AWGN channel.

**11.18** Modify Weighted BF Decoding algorithm 1 to allow flipping of multiple bits at Step 4.

**11.19** Devise an LPCF-BF decoding Algorithm with loop detection. With your algorithm, decode the $(4095, 3367)$ cyclic EG-LDPC code constructed in Example 11.1 (or Example 11.13).

**11.20** Construct the $65 \times 65$ base matrix $\mathbf{B}_{EG,1}$ of the subgeometry EG$^*(2, 2^6)$ of the two-dimensional Euclidean geometry EG$(2, 2^6)$ over GF$(2^6)$. Take a $5 \times 30$ submatrix of $\mathbf{B}_{EG,1}$ and construct a QC-EG-LDPC code of length 1890. Decode the code with 5, 10, and 50 iterations of the MSA using a chosen scale factor. Plot its bit- and block-error performance.

**11.21** Prove the base matrix $\mathbf{B}_{PaG}$ of the partial geometry PaG$(p, p, p-1)$ satisfies the constraints stated in Lemmas 11.2 and 11.3.

## References


[1] Y. Kou, S. Lin, and M. Fossorier, "Low density parity check codes based on finite geometries: A rediscovery," *Proceedings of the IEEE International Symposium on Information Theory*, Sorrento, June 25–30, 2000, p. 200.

[2] Y. Kou, S. Lin, and M. Fossorier, "Construction of low density parity check codes: A geometric approach," *Proceedings of the 2nd International*

*Symposium on Turbo Codes and Related Topics*, Brest, September 2000, pp. 137–140.

[3] Y. Kou, S. Lin, and M. Fossorier, "Low density parity-check codes: Construction based on finite geometries," *Proceedings of IEEE Globecom*, San Francisco, CA, November–December, 2000, pp. 825–829.

[4] Y. Kou, S. Lin, and M. Fossorier, "Low-density parity-check codes based on finite geometries: A rediscovery and new results," *IEEE Transactions on Information Theory*, vol. 47, no. 7, pp. 2711–2736, November 2001.

[5] Y. Kou, J. Xu, H. Tang, S. Lin, and K. Abdel-Ghaffar, "On circulant low density parity check codes," *Proceedings of the IEEE International Symposium on Information Theory*, Lausanne, June–July 2002, p. 200.

[6] S. Lin, H. Tang, and Y. Kou, "On a class of finite geometry low density parity check codes," *Proceedings of the IEEE International Symposium on Information Theory*, Washington, DC, June 2001, p. 24.

[7] S. Lin, H. Tang, Y. Kou, and K. Abdel-Ghaffar, "Codes on finite geometries," *Proceedings of the IEEE Information Theory Workshop*, Cairns, Australia, September 2–7, 2001, pp. 14–16.

[8] H. Tang, *Codes on Finite Geometries*, PhD thesis, University of California, Davis, CA, 2002.

[9] H. Tang, Y. Kou, J. Xu, S. Lin, and K. Abdel-Ghaffar, "Codes on finite geometries: Old, new, majority-logic and iterative decoding," *Proceedings of the 6th International Symposium on Communication Theory and Applications*, Ambleside, UK, July 2001, pp. 381–386.

[10] H. Tang, J. Xu, Y. Kou, S. Lin, and K. Abdel-Ghaffar, "On algebraic construction of Gallager low density parity-check codes," *Proceedings of the IEEE International Symposium on Information Theory*, Lausanne, June–July, 2002, p. 482.

[11] H. Tang, J. Xu, Y. Kou, S. Lin, and K. Abdel-Ghaffar, "On algebraic construction of Gallager and circulant low density parity check codes," *IEEE Transactions on Information Theory*, vol. 50, no. 6, pp. 1269–1279, June 2004.

[12] H. Tang, J. Xu, S. Lin, and K. Abdel-Ghaffar, "Codes on finite geometries," *IEEE Transactions on Information Theory*, vol. 51, no. 2, pp. 572–596, 2005.

[13] Y. Y. Tai, L. Lan, L.-Q. Zeng, S. Lin, and K. Abdel Ghaffar, "Algebraic construction of quasi-cyclic LDPC codes for AWGN and erasure channels," *IEEE Transactions on Communications*, vol. 54, no. 10, pp. 1765–1774, October 2006.

[14] P. O. Vontobel and R. Koetter, "Graph-cover decoding and finite-length analysis of message-passing iterative decoding of LDPC codes," unpublished work at https://arxiv.org/pdf/cs/0512078.

[15] J. Xu, L. Chen, I. Djurdjevic, S. Lin, and K. Abdel-Ghaffar, "Construction of regular and irregular LDPC codes: Geometry decomposition and masking," *IEEE Transactions on Information Theory*, vol. 53, no. 1, pp. 121–134, January 2007.

[16] L.-Q. Zeng, L. Lan, Y. Y. Tai, B. Zhou, S. Lin, and K. Abdel-Ghaffar, "Construction of nonbinary quasi-cyclic LDPC codes: A finite geometry approach," *IEEE Transactions on Communications*, vol. 56, no. 3, pp. 378–387, March 2008.

[17] J. Zhang and M. P. C. Fossorier, "A modified weighted bit-flipping decoding for low-density parity-check codes," *IEEE Communication Letters*, vol. 8, pp. 165–167, March 2004.

[18] B. Zhou, J.-Y. Kang, Y. Y. Tai, S. Lin, and Z. Ding, "High performance nonbinary quasi-cyclic LDPC codes on Euclidean geometries," *IEEE Transactions on Communications*, vol. 57. no. 5, pp. 1298–1311, May 2009.

[19] L. D. Rudolph, *Geometric Configuration and Majority Logic Decodable Codes*, MEE thesis, University of Oklahoma, Norman, OK, 1964.

[20] L. D. Rudolph, "A class of majority logic decodable codes," *IEEE Transactions on Information Theory*, vol. 13. pp. 305–307, April 1967.

[21] E. J. Weldon, Jr., "Euclidean geometry cyclic codes," *Proceedings of the Symposium on Combinatorial Mathematics*, University of North Carolina, Chapel Hill, NC, April 1967.

[22] T. Kasami, S. Lin, and W. W. Peterson, "Polynomial codes," *IEEE Transactions on Information Theory*, vol. 14, no. 6, pp. 807–814, November 1968.

[23] S. Lin and D. J. Costello, Jr., *Error Control Coding: Fundamentals and Applications*, 2nd ed., Upper Saddle River, NJ, Prentice-Hall, 2004.

[24] W. W. Peterson and E. J. Weldon, *Error-Correcting Codes*, 2nd ed., Cambridge, MA, MIT Press, 1972.

[25] E. R. Berlekamp, *Algebraic Coding Theory*, New York, McGraw-Hill, 1968. (Revised edition, Laguna Hills, NY, Aegean Park Press, 1984).

[26] T. Richardson, A. Shokrollahi, and R. Urbanke, "Design of capacity approaching irregular codes," *IEEE Transactions on Information Theory*, vol. 47, no. 2, pp. 619–637, February 2001.

[27] L. Chen, J. Xu, I. Djurdjevic, and S. Lin, "Near Shannon limit quasi-cyclic low-density parity-check codes," *IEEE Transactions on Communications*, vol. 52, no. 7, pp. 1038–1042, July 2004.

[28] Q. Huang, Q. Diao, S. Lin, and K. Abdel-Ghaffar, "Cyclic and quasi-cyclic LDPC codes on constrained parity-check matrices and their trapping sets," *IEEE Transactions on Information Theory*, vol. 58, no. 5, pp. 2648–2671, May 2012.

[29] J. M. Goethals and P. Delsarte, "On a class of majority-logic decodable codes," *IEEE Transactions on Information Theory*, vol. 14, no. 2, pp. 182–189, March 1968.

[30] K. J. C. Smith, *Majority Decodable Codes Derived from Finite Geometries*, Institute of Statistics Memo Series No. 561, University of North Carolina, Chapel Hill, NC, 1967.

[31] E. J. Weldon, Jr., "New generalization of the Reed–Muller codes, part II: Nonprimitive codes," *IEEE Transactions on Information Theory*, vol. 14, no. 2, pp. 199–205, March 1968.

[32] R. G. Gallager, "Low-density parity-check codes," *IRE Transactions on Information Theory*, vol. IT-8, no. 1, pp. 21–28, January 1962.

[33] T. Kasami and S. Lin, "On majority-logic decoding for duals of primitive polynomial codes," *IEEE Transactions on Information Theory*, vol. 17, no. 3, pp. 322–331, May 1971.

[34] J. L. Massey, "Shift-register synthesis and BCH decoding," *IEEE Transactions on Information Theory*, vol. IT-15, no. 1, pp. 122–127, January 1969.

[35] J. Chen and M. P. C. Fossorier, "Near optimum universal belief propagation based decoding of low-density parity check codes," *IEEE Transactions on Communications*, vol. 50, no. 3, pp. 406–614, March 2002.

[36] M. Jiang, C. Zhao, Z. Shi, and Y. Chen, "An improvement on the modified weighted bit-flipping decoding algorithm for LDPC codes," *IEEE Communications Letters*, vol. 9, no. 7, pp. 814–816, September 2005.

[37] Z. Liu and D. A. Pados, "A decoding algorithm for finite-geometry LDPC codes," *IEEE Transactions on Communications*, vol. 53, no. 3, pp. 415–421, March 2005.

[38] M. Shan, C. Zhao, and M. Jian, "An improved weighted bit-flipping algorithm for decoding LDPC codes," *IEE Proceedings Communications*, vol. 152, pp. 1350–2425, 2005.

[39] R. C. Bose, "Strongly regular graphs, partial geometries and partially balanced designs," *Pacific Journal of Mathematics*, vol. 13, no. 2, pp. 389–419, April 1963.

[40] J. H. van Lint, "Partial geometries," *Proceedings of the International Congress of Mathematicians*, Warszawa, August 16–24, 1983, pp. 1579–1589.

[41] P. J. Cameron and J. H. van Lint, *Designs, Graphs, Codes, and their Links*, Cambridge, Cambridge University Press, 1991.

[42] L. M. Batten, *Combinatorics of Finite Geometries*, 2nd ed., Cambridge, Cambridge University Press, 1997.

[43] E. J. Kamischke, *Benson's Theorem for Partial Geometries*, MSc thesis, Michigan Technological University, 2013.

[44] S. J. Johnson and S. R. Weller, "Codes for iterative decoding from partial geometries," *IEEE Transactions on Communications*, vol. 52, no. 2, pp. 236–243, February 2004.

[45] Q. Diao, Y. Y. Tai, S. Lin, and K. Abdel-Ghaffar, "LDPC codes on partial geometries: Construction, trapping sets structure, and puncturing," *IEEE Transactions on Information Theory*, vol. 59, no. 12, pp. 7898–7914, September 2013.

[46] S. Lin and J. Li, *Fundamentals of Classical and Modern Error-Correcting Codes*, Cambridge, Cambridge University Press, 2022.

[47] S. Lin, Q. Diao, and I. F. Blake, "Error floors and finite geometries," *Proceedings of the International Symposium*, Germany, August 18–22, 2014.

[48] Q. Diao, J. Li, S. Lin, and I. F. Blake, "New classes of partial geometries and their associated LDPC codes," *IEEE Transactions on Information*, vol. 62, no. 6, pp. 2947–2965.

[49] J. Li, K. Liu, S. Lin, and K. Abdel-Ghaffar, "Construction of partial geometries and LDPC codes based on Reed–Solomon codes," *Proceedings of the International Symposium on Information Theory*, July 7–12, 2019, Paris.

[50] S. Lin, "On the number of information symbols in polynomial codes," *IEEE Transactions on Information Theory*, vol. 18, no. 6, pp. 785–794, November 1972.

[51] N. Miladinovic and M. P. Fossorier, "Improved bit-flipping decoding of low-density parity-check codes," *IEEE Transactions on Information Theory*, vol. 51, no. 4, pp. 1594–1606, April 2005.

[52] N. Mobini, A. H. Banihashemi, and S. Hemati, "A differential binary-passing LDPC decoder," *Proceedings of IEEE Globecom*, Washington, DC, November 2007, pp. 1561–1565.

[53] T. J. Richardson, "Error floors of LDPC codes," *Proceedings of the 41st Allerton Conference on Communications, Control and Computing*, Monticello, IL, October 2003, pp. 1426–1435.

[54] E. J. Weldon, Jr., "Some results on majority-logic decoding," in H. Mann (ed.), *Error-Correcting Codes*, New York, Wiley, 1968.

# 12 Constructions of LDPC Codes Based on Finite Fields

In the late 1950s and early 1960s, finite fields were successfully used to construct linear block codes, especially cyclic codes, with large minimum distances for hard-decision algebraic decoding, such as BCH codes [1, 2] and RS codes [3]. There have recently been major developments in using finite fields to construct LDPC codes [4–11]. LDPC code constructions based on finite fields perform well over the binary-input AWGN channel with iterative decoding based on belief propagation. Most importantly, these finite-field LDPC codes were shown to have low error floors, which is important for communication and data-storage systems, for which very low bit and/or word-error rates is required. Furthermore, most of the LDPC code construction based on finite fields is quasi-cyclic and hence can be efficiently encoded using simple shift registers with linear complexity [12, 13] (see Chapter 3). This chapter is devoted to constructions of LDPC codes based on finite fields.

## 12.1 Matrix Dispersions of Elements of a Finite Field

Consider the Galois field GF($q$), where $q$ is a power of a prime. Let $\alpha$ be a primitive element of GF($q$). Then the $q$ powers of $\alpha$,

$$\alpha^{-\infty} \triangleq 0, \alpha^0 = 1, \alpha, \ldots, \alpha^{q-2},$$

form all the $q$ elements of GF($q$) and $\alpha^{q-1} = 1$. GF($q$) consists of two groups, the *additive* and *multiplicative* groups. The $q$ elements of GF($q$) under addition form the additive group and the $q-1$ nonzero elements of GF($q$) under multiplication form the multiplicative group of GF($q$).

For each nonzero element $\alpha^i$ with $0 \leq i < q-1$, we define a $(q-1)$-tuple over GF(2),

$$\mathbf{z}(\alpha^i) = (z_0, z_1, \ldots, z_{q-2}), \tag{12.1}$$

whose components correspond to the nonzero elements $\alpha^0, \alpha, \ldots, \alpha^{q-2}$ of GF($q$), where the $i$th component $z_i = 1$ and all the other $q-2$ components are set to zero. This $(q-1)$-tuple over GF(2) with a single 1-component is referred to as the *location vector* of element $\alpha^i$ with respect to the multiplicative group of GF($q$). We call $\mathbf{z}(\alpha^i)$ the $\mathcal{M}$*-location vector* of $\alpha^i$, where "$\mathcal{M}$" stands for "multiplicative group." Clearly, the 1-components of the $\mathcal{M}$-location vectors of two different nonzero elements of GF($q$) reside at two different positions. The $\mathcal{M}$-location vector $\mathbf{z}(0)$ of the 0 element of GF($q$) is defined as the all-zero $(q-1)$-tuple, $(0, 0, \ldots, 0)$.

Let $\delta$ be a nonzero element of GF($q$). For $1 \leq i < q-1$, the $\mathcal{M}$-location vector $\mathbf{z}(\alpha^i\delta)$ of element $\alpha^i\delta$ is the cyclic shift (one place to the right) of the $\mathcal{M}$-location vector $\mathbf{z}(\alpha^{i-1}\delta)$ of the element $\alpha^{i-1}\delta$. Since $\alpha^{q-1} = 1$, $\mathbf{z}(\alpha^{q-1}\delta) = \mathbf{z}(\delta)$. This is to say that $\mathbf{z}(\delta)$ is the cyclic shift of $\mathbf{z}(\alpha^{q-2}\delta)$. Form a $(q-1)\times(q-1)$ matrix $\mathbf{A}$ over GF(2) with the $\mathcal{M}$-location vectors of

$$\delta,\ \alpha\delta,\ \ldots,\ \alpha^{q-2}\delta$$

as rows. Then $\mathbf{A}$ is a $(q-1)\times(q-1)$ *circulant permutation matrix (CPM)*, that is, $\mathbf{A}$ is a permutation matrix for which each row is the right cyclic shift of the row above it and the first row is the right cyclic shift of the last row. $\mathbf{A}$ is referred to as the $(q-1)$-*fold matrix dispersion* (or *expansion*) of the field element $\delta$ over GF(2). Clearly, the $(q-1)$-fold matrix dispersions of two different nonzero elements of GF($q$) are two different CPMs over GF(2). As defined in Section 11.6, $\mathbf{A}$ is referred to as the CPM dispersion of $\delta$, denoted by CPM($\delta$).

## 12.2 A General Construction of QC-LDPC Codes Based on Finite Fields

This section presents a general method for constructing QC-LDPC codes based on finite fields [8, 9]. Let $\alpha$ be a primitive element of GF($q$). Construction begins with an $m \times n$ matrix over GF($q$),

$$\mathbf{W} = \begin{bmatrix} \mathbf{w}_0 \\ \mathbf{w}_1 \\ \vdots \\ \mathbf{w}_{m-1} \end{bmatrix} = \begin{bmatrix} w_{0,0} & w_{0,1} & \cdots & w_{0,n-1} \\ w_{1,0} & w_{1,1} & \cdots & w_{1,n-1} \\ \vdots & \vdots & \ddots & \vdots \\ w_{m-1,0} & w_{m-1,1} & \cdots & w_{m-1,n-1} \end{bmatrix}, \tag{12.2}$$

whose rows satisfy the following two constraints: (1) for $0 \leq i < m$ and $0 \leq k, l < q-1$ and $k \neq l$, $\alpha^k\mathbf{w}_i$ and $\alpha^l\mathbf{w}_i$ have at most one position where both have the same symbol from GF($q$) (i.e., they differ in at least $n-1$ positions); and (2) for $0 \leq i, j < m$, $i \neq j$, and $0 \leq k, l < q-1$, $\alpha^k\mathbf{w}_i$ and $\alpha^l\mathbf{w}_j$ differ in at least $n-1$ positions. The above two constraints on the rows of matrix $\mathbf{W}$ are referred to as *$\alpha$-multiplied row constraints* 1 and 2, respectively. The first $\alpha$-multiplied row constraint implies that each row has at most one 0-component and the second $\alpha$-multiplied row constraint implies that any two rows of $\mathbf{W}$ differ in at least $n-1$ positions. Matrices over GF($q$) that satisfy $\alpha$-multiplied row constraints 1 and 2 are referred to as *$\alpha$-multiplied row-constrained* matrices.

On replacing each entry of $\mathbf{W}$ by its $(q-1)$-fold matrix dispersion over GF(2), we obtain the following $m \times n$ array of $(q-1)\times(q-1)$ square submatrices:

$$\mathbf{H}_{\text{disp}} = \begin{bmatrix} \mathbf{A}_{0,0} & \mathbf{A}_{0,1} & \cdots & \mathbf{A}_{0,n-1} \\ \mathbf{A}_{1,0} & \mathbf{A}_{1,1} & \cdots & \mathbf{A}_{1,n-1} \\ \vdots & \vdots & \ddots & \vdots \\ \mathbf{A}_{m-1,0} & \mathbf{A}_{m-1,1} & \cdots & \mathbf{A}_{m-1,n-1} \end{bmatrix}, \tag{12.3}$$

where $\mathbf{A}_{i,j}$ is a $(q-1) \times (q-1)$ CPM if the entry $w_{i,j}$ is a nonzero element of GF($q$) and a $(q-1) \times (q-1)$ zero matrix if $w_{i,j} = 0$. $\mathbf{H}_{\text{disp}}$ is an $m(q-1) \times n(q-1)$ matrix over GF(2). It follows from $\alpha$-multiplied row constraint 1 that each row of the array $\mathbf{H}_{\text{disp}}$ has at most one $(q-1) \times (q-1)$ zero matrix. It follows from $\alpha$-multiplied row constraint 2 that any two rows of $\mathbf{H}_{\text{disp}}$ can have at most one place where both have the same CPM, that is, except at this place, the two CPMs at any other place in these two rows are different. This implies that $\mathbf{H}_{\text{disp}}$, as an $m(q-1) \times n(q-1)$ matrix over GF(2), satisfies the RC constraint. $\mathbf{H}_{\text{disp}}$ is called the *multiplicative* $(q-1)$*-fold array dispersion* (or simply array dispersion) of $\mathbf{W}$. The subscript "disp" of $\mathbf{H}_{\text{disp}}$ stands for "dispersion." $\mathbf{W}$ is referred to as the base matrix of $\mathbf{H}_{\text{disp}}$.

For any pair $(g, r)$ of integers with $1 \leq g \leq m$ and $1 \leq r \leq n$, let $\mathbf{H}_{\text{disp}}(g, r)$ be a $g \times r$ subarray of $\mathbf{H}_{\text{disp}}$. $\mathbf{H}_{\text{disp}}(g, r)$ is a $g(q-1) \times r(q-1)$ matrix over GF(2). Since it is a submatrix of $\mathbf{H}_{\text{disp}}$, it also satisfies the RC constraint. Since $\mathbf{H}_{\text{disp}}(g, r)$ is an array of circulant permutation and/or zero matrices of the same size, the null space of $\mathbf{H}_{\text{disp}}(g, r)$ gives a QC-LDPC code $\mathcal{C}_{qc,\text{disp}}$ of length $r(q-1)$ with rate at least $(r-g)/r$, whose Tanner graph has a girth of at least 6. If $\mathbf{H}_{\text{disp}}(g, r)$ does not contain zero matrices, it has constant column and row weights $g$ and $r$, respectively. Then $\mathcal{C}_{qc,\text{disp}}$ is a $(g, r)$-regular QC-LDPC code with minimum distance at least $g + 2$ for even $g$ and $g + 1$ for odd $g$. The proof is the same as that given in Section 11.3 for regular LDPC codes constructed from arrays of permutation matrices. If $\mathbf{H}_{\text{disp}}(g, r)$ contains zero matrices, it might not have constant column or row weights. In this case, the null space of $\mathbf{H}_{\text{disp}}(g, r)$ gives an irregular QC-LDPC code.

The above $(q-1)$-fold array dispersion of a matrix $\mathbf{W}$ over a finite field GF($q$) that satisfies two $\alpha$-multiplied row constraints gives a method for constructing QC-LDPC codes. For any given finite field GF($q$), a family of QC-LDPC codes with various lengths, rates, and minimum distances can be constructed. Now the question is how to construct $\alpha$-multiplied row-constrained matrices over finite fields for array dispersion. In the next eight sections, various methods for constructing $\alpha$-multiplied row-constrained matrices will be presented.

## 12.3 Construction of QC-LDPC Codes Based on the Minimum-Weight Codewords of an RS Code with Two Information Symbols

Consider a cyclic $(q-1, 2, q-2)$ RS code $\mathcal{C}_b$ over GF($q$) of length $q-1$ with two information symbols whose generator polynomial $\mathbf{g}(X)$ has $\alpha, \alpha^2, \ldots, \alpha^{q-3}$ as roots (see Section 3.4). The minimum distance (or minimum weight) of this RS code is $q-2$. It consists of one codeword of weight 0, $(q-1)^2$ codewords of weight $q-2$, and $2(q-1)$ codewords of weight $q-1$ [13]. Since the two polynomials over GF($q$), namely

$$\mathbf{v}_1(X) = 1 + X + X^2 + \cdots + X^{q-2}$$

and

$$\mathbf{v}_2(X) = 1 + \alpha X + \alpha^2 X + \cdots + \alpha^{q-2} X^{q-2},$$

have $\alpha, \alpha^2, \ldots, \alpha^{q-2}$ as roots, they are divisible by $\mathbf{g}(X)$ and hence are code polynomials of $\mathcal{C}_b$. Their corresponding codewords are $\mathbf{v}_1 = (1, 1, \ldots, 1)$ and $\mathbf{v}_2 = (1, \alpha, \ldots, \alpha^{q-2})$, respectively. They both have weight $q-1$. Let

$$\mathbf{w}_0 = \mathbf{v}_2 - \mathbf{v}_1 = (0, \alpha - 1, \alpha^2 - 1, \ldots, \alpha^{q-2} - 1). \tag{12.4}$$

Then $\mathbf{w}_0$ is a codeword in $\mathcal{C}_b$ with weight $q-2$ (minimum weight). The $q-1$ components of $\mathbf{w}_0$ are different elements of GF($q$). There is one and only one 0 element in $\mathbf{w}_0$.

Form the following $(q-1) \times (q-1)$ matrix over GF($q$) with $\mathbf{w}_0$ and its $q-2$ right cyclic shifts, $\mathbf{w}_1, \mathbf{w}_2, \ldots, \mathbf{w}_{q-2}$, as rows:

$$\mathbf{W}^{(1)} = \begin{bmatrix} \mathbf{w}_0 \\ \mathbf{w}_1 \\ \vdots \\ \mathbf{w}_{q-2} \end{bmatrix} = \begin{bmatrix} 0 & \alpha - 1 & \alpha^2 - 1 & \cdots & \alpha^{q-2} - 1 \\ \alpha^{q-2} - 1 & 0 & \alpha - 1 & \cdots & \alpha^{q-3} - 1 \\ \vdots & \vdots & \vdots & \ddots & \vdots \\ \alpha - 1 & \alpha^2 - 1 & \alpha^3 - 1 & \cdots & 0 \end{bmatrix}. \tag{12.5}$$

Each row (or each column) of $\mathbf{W}^{(1)}$ contains one and only one 0 entry and the $q-1$ 0 entries of $\mathbf{W}^{(1)}$ lie on the main diagonal of $\mathbf{W}^{(1)}$. Since $\mathcal{C}_b$ is cyclic and $\mathbf{w}_0$ is a minimum-weight codeword in $\mathcal{C}_b$, all the rows of $\mathbf{W}^{(1)}$ are minimum-weight codewords of $\mathcal{C}_b$. Any two rows of $\mathbf{W}^{(1)}$ differ in all $q-1$ positions. For $0 \le i < q-1$ and $0 \le k, l < q-1$ and $k \ne l$, $\alpha^k \mathbf{w}_i$ and $\alpha^l \mathbf{w}_i$ are two different minimum-weight codewords in $\mathcal{C}_b$ and they both have a 0 element at position $i$. Hence, they differ in all the other $q-2$ positions. Therefore, the rows of $\mathbf{W}^{(1)}$ satisfy $\alpha$-multiplied row constraint 1. For $0 \le i, j < q-1$, $i \ne j$, and $0 \le k, l < q-1$, $\alpha^k \mathbf{w}_i$ and $\alpha^l \mathbf{w}_j$ are two different minimum-weight codewords in $\mathcal{C}_b$, and they differ in at least $q-2$ places. Consequently, the rows of $\mathbf{W}^{(1)}$ satisfy $\alpha$-multiplied row constraint 2.

On replacing each entry in $\mathbf{W}^{(1)}$ by its multiplicative $(q-1)$-fold matrix dispersion, we obtain the following $(q-1) \times (q-1)$ array of $(q-1) \times (q-1)$ circulant permutation and zero matrices:

$$\mathbf{H}_{qc,\text{disp}}^{(1)} = \begin{bmatrix} \mathbf{O} & \mathbf{A}_{0,1} & \cdots & \mathbf{A}_{0,q-2} \\ \mathbf{A}_{0,q-2} & \mathbf{O} & \cdots & \mathbf{A}_{0,q-3} \\ \vdots & \vdots & \ddots & \vdots \\ \mathbf{A}_{0,1} & \mathbf{A}_{0,2} & \cdots & \mathbf{O} \end{bmatrix}. \tag{12.6}$$

The array $\mathbf{H}_{qc,\text{disp}}^{(1)}$ has the following structural properties: (1) each row (or column) has one and only one zero matrix; (2) all the zero matrices lie on the main diagonal of the array; (3) each row (or column) is a right (or downward) cyclic shift of the row above it (or the column on its left) and the first row (or first column) is the right (or downward) cyclic shift of the last row (or last column); and (4) all the $q-2$ CPMs in each row (or each column) are different from each other. $\mathbf{H}^{(1)}_{qc,\text{disp}}$ is a $(q-1)^2 \times (q-1)^2$ matrix over GF(2) that satisfies the RC constraint and has both column and row weight $q-2$. The $(q-1)^2$ rows of $\mathbf{H}^{(1)}_{qc,\text{disp}}$ correspond to the $(q-1)^2$ minimum-weight codewords of $\mathcal{C}_b$. A row in $\mathbf{H}^{(1)}_{qc,\text{disp}}$ is obtained by representing the $q-1$ components of a minimum-weight codeword in $\mathcal{C}_b$ by their respective $\mathcal{M}$-location vectors.

For any pair $(g, r)$ of integers with $1 \leq g, r < q$, let $\mathbf{H}^{(1)}_{qc,\text{disp}}(g, r)$ be a $g \times r$ subarray of $\mathbf{H}^{(1)}_{qc,\text{disp}}$. Then $\mathbf{H}^{(1)}_{qc,\text{disp}}(g, r)$ is a $g(q-1) \times r(q-1)$ matrix over GF(2) that satisfies the RC constraint. The null space of $\mathbf{H}^{(1)}_{qc,\text{disp}}(g, r)$ gives a QC-LDPC code $\mathcal{C}^{(1)}_{qc,\text{disp}}$ of length $r(q-1)$ with rate at least $(r-g)/r$, whose Tanner graph has a girth of at least 6. If $\mathbf{H}^{(1)}_{qc,\text{disp}}(g, r)$ lies either above or below the main diagonal of $\mathbf{H}^{(1)}_{qc,\text{disp}}$, it contains no zero submatrices of $\mathbf{H}^{(1)}_{qc,\text{disp}}$ and hence has constant column and row weights $g$ and $r$, respectively. Then $\mathcal{C}^{(1)}_{qc,\text{disp}}$ is a $(g, r)$-regular QC-LDPC code with minimum distance at least $g+2$ for even $g$ and $g+1$ for odd $g$. If $\mathbf{H}^{(1)}_{qc,\text{disp}}(g, r)$ contains some zero submatrices of $\mathbf{H}^{(1)}_{qc,\text{disp}}$, it may have two different column weights $g-1$ and $g$, and/or two different row weights $r-1$ and $r$. In this case, $\mathcal{C}^{(1)}_{qc,\text{disp}}$ is a near-regular QC-LDPC code with minimum distance at least $g$ for even $g$ and $g+1$ for odd $g$. For a given field GF($q$), we can construct a family of QC-LDPC codes of various lengths, rates, and minimum distances.

The above construction gives a class of RS-based QC-LDPC codes. If we mask subarrays of the array $\mathbf{H}^{(1)}_{qc,\text{disp}}$ given by (12.6), we can construct many more QC-LDPC codes, regular or irregular. Codes in this class are called QC-RS-LDPC codes.

In the following, we give several examples to illustrate the above code construction of QC-RS-LDPC with or without masking. Again, we assume transmission by the binary-input AWGN channel.

---

**Example 12.1** Consider the $(n, k, d_{\min}) = (256, 2, 255)$ RS code over the prime field GF(257). Based on this RS code and (12.4)–(12.6), we can construct a $256 \times 256$ array $\mathbf{H}^{(1)}_{qc,\text{disp}}$ of circulant permutation and zero matrices over GF(2) of size $256 \times 256$ with the 256 zero matrices on the main diagonal of $\mathbf{H}^{(1)}_{qc,\text{disp}}$. For any pair $(g, r)$ of integers with $1 \leq g, r < 257$, the null space of a $k \times r$ subarray $\mathbf{H}^{(1)}_{qc,\text{disp}}(g, r)$ of $\mathbf{H}^{(1)}_{qc,\text{disp}}$ gives a QC-LDPC code whose Tanner graph has a girth of at least 6. If we choose $r = 2^i$ with $2 \leq i \leq 8$ and various $g$s, we can construct a family of QC-LDPC codes of lengths that are powers of 2. For example, we set $g = 4$ and $r = 32$, taking a $4 \times 32$ subarray $\mathbf{H}^{(1)}_{qc,\text{disp}}(4, 32)$ from $\mathbf{H}^{(1)}_{qc,\text{disp}}$ above or below the main diagonal of $\mathbf{H}^{(1)}_{qc,\text{disp}}$. $\mathbf{H}^{(1)}_{qc,\text{disp}}(4, 32)$ is a $1024 \times 8192$ matrix over GF(2) with column and row weights 4 and 32, respectively. The null space of this matrix gives a $(4, 32)$-regular $(8192, 7171)$ QC-RS-LDPC code with rate 0.8753. The error performance of this code over the binary-input AWGN channel with iterative decoding using the SPA (100 iterations) is shown in Figure 12.1. At a BER of $10^{-6}$, it performs 1 dB from the Shannon limit.

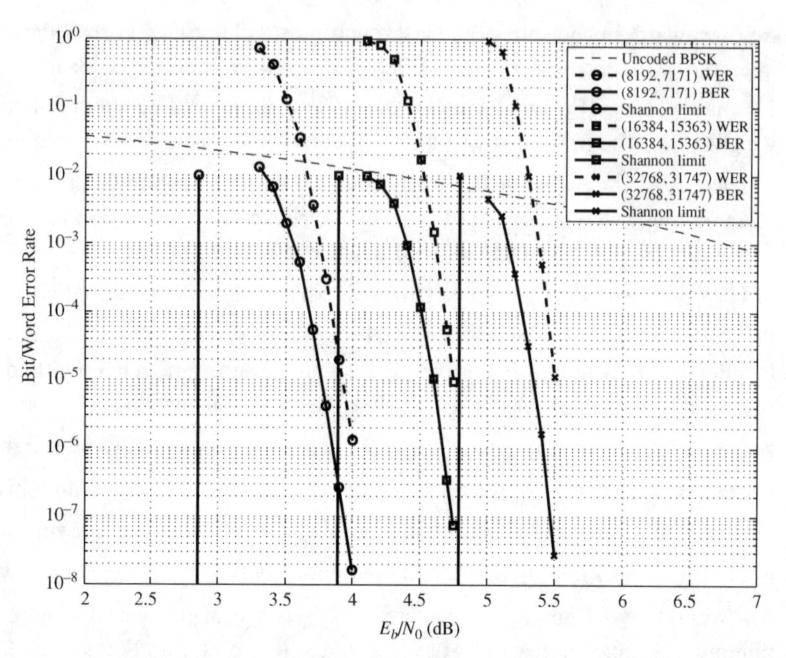


**Figure 12.1** Error performances of the (8192, 7171), (16384, 15363), and (32786, 31747) QC-RS-LDPC codes given in Example 12.1.

Suppose we set $g = 4$ and $r = 64$. Take a $4 \times 64$ subarray $\mathbf{H}^{(1)}_{qc,\mathrm{disp}}(4, 64)$ from $\mathbf{H}^{(1)}_{qc,\mathrm{disp}}$, avoiding the zero submatrices on the main diagonal of $\mathbf{H}^{(1)}_{qc,\mathrm{disp}}$. $\mathbf{H}^{(1)}_{qc,\mathrm{disp}}(4, 64)$ is a $1024 \times 16\,384$ matrix over GF(2) with column and row weights 4 and 64, respectively. The null space of this matrix gives a (4, 64)-regular (16 384, 15 363) QC-RC-LDPC code with rate 0.9376. Its error performance with iterative decoding using the SPA (100 iterations) is also shown in Figure 12.1. At a BER of $10^{-6}$, it performs 0.75 dB from the Shannon limit. If we set $g = 4$ and $r = 128$, we can construct a (4, 128)-regular RS-based (32 786, 31 747) QC-RS-LDPC code with rate 0.9688. Its error performance decoded with iterative decoding using the SPA (100 iterations) is also shown in Figure 12.1. At a BER of $10^{-6}$, it performs 0.6 dB from the Shannon limit.

---

**Example 12.2** A subarray of the array given by (12.6) can be masked to form the parity-check matrix of a QC-RS-LDPC code. Consider the (72, 2, 71) RS code over the prime field GF(73). From this RS code and (12.4)–(12.6), we can construct a $72 \times 72$ array $\mathbf{H}^{(1)}_{qc,\mathrm{disp}}$ of circulant permutation and zero matrices of size $72 \times 72$. Set $g = 32$ and $r = 64$. Take a $32 \times 64$ subarray $\mathbf{H}^{(1)}_{qc,\mathrm{disp}}(32, 64)$ from $\mathbf{H}^{(1)}_{qc,\mathrm{disp}}$ as the base array for masking, avoiding zero submatrices on the main diagonal of $\mathbf{H}_{qc,\text{disp}}^{(1)}$. Construct a $32 \times 64$ masking matrix $\mathbf{Z}(32, 64)$ that consists of a row of two $32 \times 32$ circulants. The generators of the two circulants in $\mathbf{Z}(32, 64)$ are two primitive and nonequivalent 32-tuples with weight 3:

$$\mathbf{g}_1 = (10100100000000000000000000000000)$$

and

$$\mathbf{g}_2 = (10000010000000100000000000000000).$$

On masking the base array $\mathbf{H}_{qc,\text{disp}}^{(1)}(32, 64)$ with $\mathbf{Z}(32, 64)$, we obtain a $32 \times 64$ masked array $\mathbf{M}(32, 64) = \mathbf{Z}(32, 64) \circledast \mathbf{H}_{qc,\text{disp}}^{(1)}(32, 64)$. The masked array $\mathbf{M}(32, 64)$ consists of 192 CPMs and 1856 zero matrices of size $72 \times 72$. It is a $2304 \times 4608$ matrix with column and row weights 3 and 6, respectively. The null space of $\mathbf{M}(32, 64)$ gives a (3, 6)-regular (4608, 2304) QC-RS-LDPC code with rate 0.5 whose Tanner graph has a girth of at least 6. The bit- and word-error performances of this code decoded with iterative decoding using the SPA with 50 iterations are shown in Figure 12.2. At a BER of $10^{-6}$, it performs 1.7 dB from the Shannon limit. At a BER of $10^{-9}$, it performs 1.9 dB from the Shannon limit and has no error floor down to a BER of $3 \times 10^{-10}$. It has a beautiful waterfall performance. It is a very good LDPC code.

Suppose we construct a $32 \times 64$ irregular masking matrix $\mathbf{Z}^*(32, 64)$ by computer in accordance with the following degree distributions of variable and check nodes of a Tanner graph, designed for a code of length 4608 with rate 0.5:

$$\tilde{\lambda}(X) = 0.25X + 0.625X^2 + 0.125X^8$$

and

$$\tilde{\rho}(X) = X^6.$$

The column and row weight distributions of $\mathbf{Z}^*(32, 64)$ are given in Table 12.1.

On masking the base array $\mathbf{H}_{qc,\text{disp}}^{(1)}(32, 64)$ given above with the irregular masking matrix $\mathbf{Z}^*(32, 64)$, we obtain a $32 \times 64$ irregular masked matrix $\mathbf{M}^*(32, 64)$. The average column and row weights of $\mathbf{M}^*(32, 64)$ are 3.5 and 7. The null space of $\mathbf{M}^*(32, 64)$ gives a (4608, 2304) irregular QC-RS-LDPC code. The bit and word-error performance of this code with iterative decoding using the SPA with 50 iterations is also shown in Figure 12.2. At a BER of $10^{-6}$, it performs 1.45 dB from the Shannon limit. We see that, in the waterfall region, the irregular (4608, 2304) QC-RS-LDPC code outperforms the (3, 6)-regular (4608, 2304) QC-RS-LDPC code by 0.25 dB. However, the irregular code requires a larger computational complexity per iteration.

**Table 12.1** Column and row distributions of the masking matrix $\mathbf{Z}^*(32, 64)$ of Example 12.2

| Column weight distribution | | Row weight distribution | |
|---|---|---|---|
| Column weight | No. of columns | Row weight | No. of rows |
| 2 | 16 | 7 | 32 |
| 3 | 40 | | |
| 9 | 8 | | |

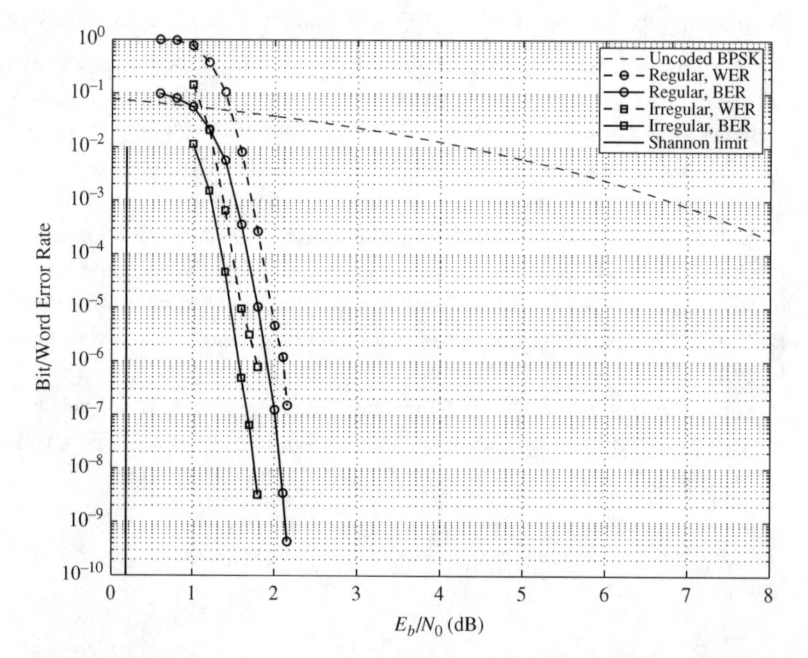


**Figure 12.2** Error performances of the (4608, 2304) regular and irregular QC-RS-LDPC codes given in Example 12.2.

**Example 12.3** In this example, we construct another rate-1/2 irregular QC-LDPC code by masking a subarray of the $256 \times 256$ array $\mathbf{H}^{(1)}_{qc,\text{disp}}$ constructed from the $(n, k, d_{\min}) = (256, 2, 255)$ RS code over the prime field GF(257) given in Example 12.1. Take a $16 \times 32$ subarray $\mathbf{H}^{(1)}_{qc,\text{disp}}(16, 32)$ from $\mathbf{H}^{(1)}_{qc,\text{disp}}$ as the base array for masking, avoiding zero submatrices on the main diagonal of $\mathbf{H}^{(1)}_{qc,\text{disp}}$. Construct a $16 \times 32$ masking matrix $\mathbf{Z}(16, 32)$ over GF(2) based on the following variable- and check-node degree distributions of a Tanner graph designed for a rate-1/2 irregular code of length 8196:

$$\tilde{\lambda}(X) = 0.25X + 0.625X^2 + 0.125X^8$$

and

$$\tilde{\rho}(X) = X^6.$$

The column and weight distributions of $\mathbf{Z}(16, 32)$ are given in Table 12.2. The row weight and average column weight are 7 and 3.5, respectively.

Masking the base array $\mathbf{H}^{(1)}_{qc,\text{disp}}(16, 32)$ with $\mathbf{Z}(16, 32)$ results in a $16 \times 32$ masked array $\mathbf{M}(16, 32) = \mathbf{Z}(16, 32) \circledast \mathbf{H}^{(1)}_{qc,\text{disp}}(16, 32)$ with 112 circulant permutation and 400 zero matrices of size $256 \times 256$. $\mathbf{M}(16, 32)$ is an irregular $4096 \times 8192$ matrix over GF(2) with column and row weight distributions given in Table 12.3.

The null space of $\mathbf{M}(16, 32)$ gives an (8192, 4096) irregular QC-LDPC code with rate 0.5. The error performances of this code over the binary-input AWGN channel with iterative decoding using the SPA with 20, 50, and 100 iterations are shown in Figure 12.3. With 100 iterations, the code performs 1.2 dB from the Shannon limit at a BER of $10^{-6}$ and a little less than 1.4 dB from the Shannon limit at a BER of $10^{-9}$. Decoding of this code converges relatively fast. At a BER of $10^{-6}$, the gap between 20 and 100 iterations is less than 0.25 dB, while the gap between 50 and 100 iterations is less than 0.05 dB.

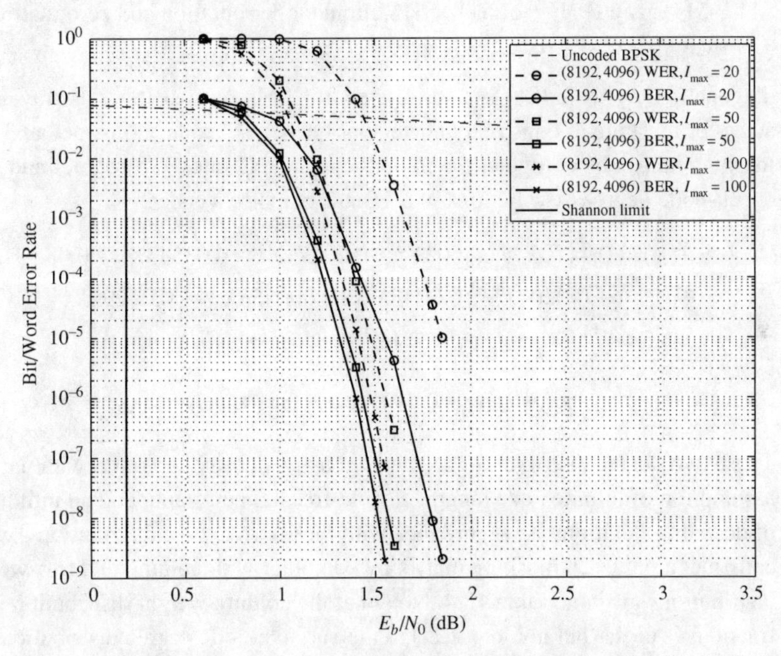


**Figure 12.3** The error performance of the (8192, 4096) irregular QC-RS-LDPC of Example 12.3.

**Table 12.2** Column and row weight distributions of the masking matrix $\mathbf{Z}(16, 32)$ of Example 12.3

| Column weight distribution | | Row weight distribution | |
|---|---|---|---|
| Column weight | No. of columns | Row weight | No. of rows |
| 2 | 8 | 7 | 16 |
| 3 | 20 | | |
| 9 | 4 | | |

**Table 12.3** Column and row weight distributions of the masked matrix $\mathbf{M}(16, 32)$ of Example 12.3

| Column weight distribution | | Row weight distribution | |
|---|---|---|---|
| Column weight | No. of columns | Row weight | No. of rows |
| 2 | 2048 | 7 | 4096 |
| 3 | 5120 | | |
| 9 | 1024 | | |

**Example 12.4** The objective of this example is to demonstrate that a long algebraic LDPC code can perform very close to the Shannon limit. Consider the $(511, 2, 510)$ RS code over GF($2^9$). From this RS code and (12.4)–(12.6), we can construct a $511 \times 511$ array $\mathbf{H}_{qc,\text{disp}}^{(1)}$ of $511 \times 511$ circulant permutation and zero matrices over GF(2). Take a $63 \times 126$ subarray $\mathbf{H}_{qc,\text{disp}}^{(1)}(63, 126)$ from $\mathbf{H}_{qc,\text{disp}}^{(1)}$ above its main diagonal. $\mathbf{H}_{qc,\text{disp}}^{(1)}(63, 126)$ is a solid array of CPMs. We use this subarray as the base array for masking to construct an irregular QC-LDPC code. Construct a $63 \times 126$ masking matrix by computer based on the following rate-1/2 variable- and check-node degree distributions of a Tanner graph of a code:

$$\tilde{\lambda}(X) = 0.4410X + 0.3603X^2 + 0.00171X^5 + 0.03543X^6 + 0.093\,31X^7 + 0.0204X^8 + 0.0048X^9 + 0.000353X^{27} + 0.04292X^{29}$$

and

$$\tilde{\rho}(X) = 0.00842X^7 + 0.99023X^8 + 0.00135X^9.$$

These two degree distributions are derived using density evolution with the assumption of infinite code length, a cycle-free Tanner graph, and an infinite number of decoding iterations. In accordance with these two degree distributions, we construct a $63 \times 126$ masking matrix by computer with column and row weight distributions given in Table 12.4. Note that the column weight distribution in terms of fractions is close (but not identical) to the variable-node degree distribution $\tilde{\lambda}(X)$ given above and the row weight distribution in terms of fractions is quite different from the check-node degree distribution $\tilde{\rho}(X)$ given above.

On masking the base array $\mathbf{H}_{qc,\text{disp}}^{(1)}$ with $\mathbf{Z}(63, 126)$, we obtain a $63 \times 126$ masked array $\mathbf{M}(63, 126) = \mathbf{Z}(63, 126) \circledast \mathbf{H}_{qc,\text{disp}}^{(1)}(63, 126)$. It is a $32193 \times 64386$ matrix over GF(2). The null space of $\mathbf{M}(63, 126)$ gives a (64 386, 32 193) irregular QC-LDPC code whose bit- and word-error performances with iterative decoding using the SPA (100 iterations) is shown in Figure 12.4. At a BER of $10^{-6}$, it performs only 0.48 dB from the Shannon limit.

The RS-based QC-LDPC codes presented in this section can be effectively decoded with the OSMLG, BF, and WBF decoding algorithms presented in Sections 11.7–11.9.

**Table 12.4** Column and row weight distributions of the masking matrix $\mathbf{Z}(63, 126)$ of Example 12.4

| Column weight distribution | | Row weight distribution | |
|---|---|---|---|
| Column weight | No. of columns | Row weight | No. of rows |
| 2 | 55 | 5 | 1 |
| 3 | 45 | 6 | 1 |
| 7 | 4 | 7 | 2 |
| 8 | 12 | 8 | 5 |
| 9 | 3 | 9 | 19 |
| 10 | 1 | 10 | 35 |
| 30 | 6 | | |

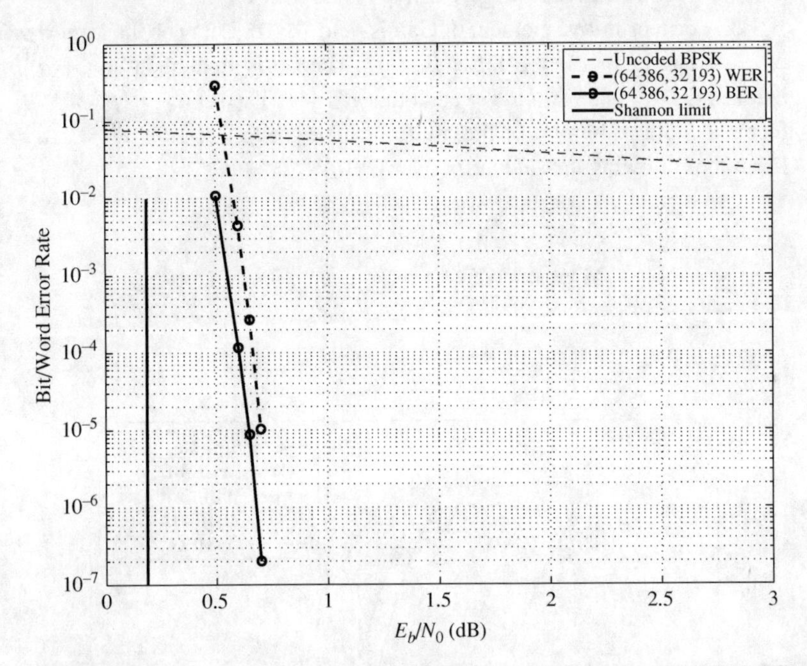


**Figure 12.4** The error performance of the (64 386, 32 193) irregular QC-RS-LDPC code of Example 12.4.

---

**Example 12.5** Consider the $(31, 2, 30)$ RS code $C_b$ over GF($2^5$). From this RS code, we can construct a $31 \times 31$ array $\mathbf{H}^{(1)}_{qc,\mathrm{disp}}$ of $31 \times 31$ circulant permutation and zero matrices over GF(2). This array is a $961 \times 961$ matrix over GF(2) with both column and row weight 30. The null space of this array gives a $(961, 721)$ QC-RS-LDPC code over GF(2) with rate 0.750 and minimum distance at least 32. The OSMLG error-correction capability of this code is 15. The error performances of this code decoded with the OSMLG, LPCF-BF, WBF-2, and SPA decoding algorithms are shown in Figure 12.5. Also included in this figure is the performance of a $(1023, 768)$ BCH code decoded with the BM algorithm. This BCH code is constructed using the

Galois field GF($2^5$) and has rate 0.751 and minimum distance 53. We see that the (961, 721) QC-LDPC code decoded with the simple LPCF-BF decoding algorithm outperforms the (1023, 768) BCH code decoded with the BM algorithm which requires computations over GF($2^5$).

## 12.4 Construction of QC-LDPC Codes Based on the Universal Parity-Check Matrices of a Special Subclass of RS Codes

In this section, we show that an RC-constrained array of CPMs can be constructed from the universal parity-check matrix for a family of RS codes over a given field GF($q$). This universal parity-check matrix is a matrix for which the null space of any number of consecutive rows gives an RS code over GF($q$).

Let $\alpha$ be a primitive element of Galois field GF($q$). Let $m$ be the *largest prime factor* of $q - 1$ and let $q - 1 = cm$. Let $\beta = \alpha^c$. Then $\beta$ is an element in GF($q$) of order $m$, that is, $m$ is the *smallest integer* such that $\beta^m = 1$. The set $\mathcal{G}_m = \{\beta^0 = 1, \beta, \ldots, \beta^{m-1}\}$ forms a cyclic subgroup of the multiplicative group $\mathcal{G}_{q-1} = \{\alpha^0 = 1, \alpha, \ldots, \alpha^{q-2}\}$. Form the following $m \times m$ matrix over GF($q$):

$$\mathbf{W}^{(2)} = \begin{bmatrix} \mathbf{w}_0 \\ \mathbf{w}_1 \\ \mathbf{w}_2 \\ \vdots \\ \mathbf{w}_{m-1} \end{bmatrix}$$

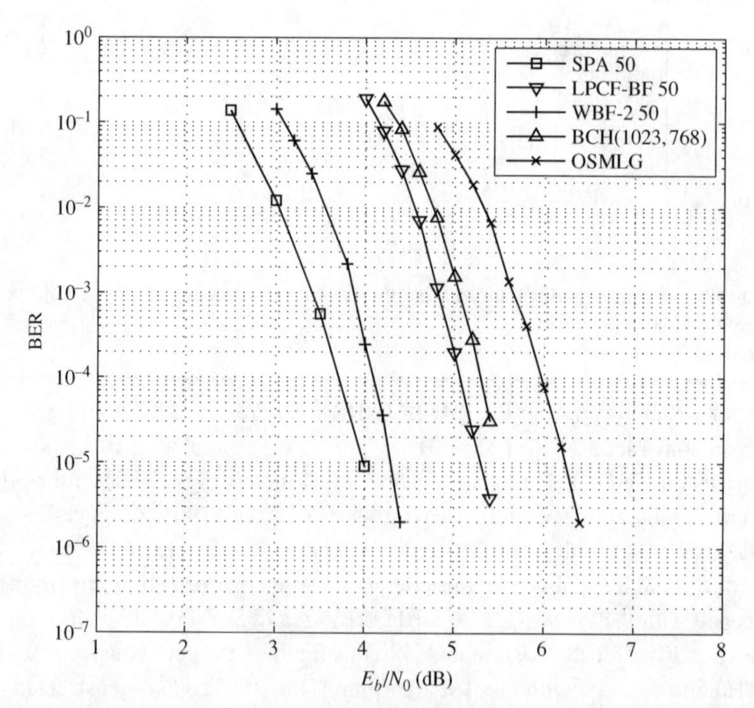


**Figure 12.5** Error performances of the (961, 721) QC-RS-LDPC code given in Example 12.5 decoded with various decoding algorithms.

$$= \begin{bmatrix} 1 & 1 & 1 & \cdots & 1 \\ 1 & \beta & \beta^2 & \cdots & \beta^{m-1} \\ 1 & \beta^2 & (\beta^2)^2 & \cdots & (\beta^2)^{m-1} \\ \vdots & \vdots & \vdots & \ddots & \vdots \\ 1 & \beta^{m-1} & (\beta^{m-1})^2 & \cdots & (\beta^{m-1})^{m-1} \end{bmatrix}, \qquad (12.7)$$

where the powers of $\beta$ are taken modulo $m$. For $1 \leq t \leq m$, any $t$ consecutive rows of $\mathbf{W}^{(2)}$ form a parity-check matrix of a cyclic $(m, m-t, t+1)$ RS code over GF($q$), including the *end-around* case (e.g., the last $l$ rows plus the first $t - l$ rows with $0 \leq l \leq t$). Its generator polynomial has $t$ consecutive powers of $\beta$ as roots. For this reason, we call $\mathbf{W}^{(2)}$ a *universal parity-check matrix* for a family of RS codes.

Since $m$ is a prime, we can easily prove that $\mathbf{W}^{(2)}$ has the following structural properties: (1) except for the first row, all the entries in a row are different and they form all the $m$ elements of the cyclic subgroup $\mathcal{G}_m$ of the multiplicative group of GF($q$); (2) except for the first column, all the entries in a column are different and they form all the $m$ elements of $\mathcal{G}_m$; (3) any two rows have only the first entries identical (equal to 1) and are different in all the other $m - 1$ positions; and (4) any two columns have only the first entries identical (equal to 1) and are different in all the other $m - 1$ positions.

$\mathbf{W}^{(2)}$ satisfies $\alpha$-multiplied row constraints 1 and 2. The proof of this is given in the following two lemmas.

**Lemma 12.1** *Let $\mathbf{w}_i$ be the ith row of $\mathbf{W}^{(2)}$ with $0 \leq i < m$. For $0 \leq k, l < m$, and $k \neq l$, $\alpha^k \mathbf{w}_i$ and $\alpha^l \mathbf{w}_i$ differ in at least $m - 1$ positions.*

*Proof* This lemma is a direct consequence of the fact that all the components of $\mathbf{w}_i$ are nonzero. □

**Lemma 12.2** *Let $\mathbf{w}_i$ and $\mathbf{w}_j$ be two rows of $\mathbf{W}^{(2)}$ with $0 \leq i, j < m$, and $i \neq j$. For $0 \leq k, l < q - 1$, the two m-tuples, $\alpha^k \mathbf{w}_i$ and $\alpha^l \mathbf{w}_j$, over GF($q$) can have at most one position where they have identical components, that is, they differ in at least $q - 1$ positions.*

*Proof* Suppose $\alpha^k \mathbf{w}_i$ and $\alpha^l \mathbf{w}_j$ have two positions, say $s$ and $t$ with $t > s$, where they have identical components. Then $\alpha^k(\beta^i)^s = \alpha^l(\beta^j)^s$ and $\alpha^k(\beta^i)^t = \alpha^l(\beta^j)^t$. We assume that $i < j$. Then the above two equalities imply that $\beta^{(j-i)(t-s)} = 1$. Since the order of $\beta$ is $m$, $(j - i)(t - s)$ must be a multiple of $m$. However, since $m$ is a prime and both $j - i$ and $t - s$ are smaller than $m$, we have that $j - i$ and $t - s$ must be relatively prime to $m$. Consequently, $(j - i)(t - s)$ cannot be a multiple of $m$ and hence the above hypothesis that $\alpha^k \mathbf{w}_i$ and $\alpha^l \mathbf{w}_j$ have two positions where they have identical components is invalid. This proves the lemma. □

It follows from Lemmas 12.1 and 12.2 that the matrix $\mathbf{W}^{(2)}$ given by (12.7) satisfies $\alpha$-multiplied row constraints 1 and 2. Hence, $\mathbf{W}^{(2)}$ can be used as the base matrix for array dispersion. On replacing each entry of $\mathbf{W}^{(2)}$ by its multiplicative $(q - 1)$-fold matrix dispersion, we obtain the following RC-constrained $m \times m$ array of $(q - 1) \times (q - 1)$ CPMs:

$$\mathbf{H}_{qc,\text{disp}}^{(2)} = \begin{bmatrix} \mathbf{A}_{0,0} & \mathbf{A}_{0,1} & \cdots & \mathbf{A}_{0,m-1} \\ \mathbf{A}_{1,0} & \mathbf{A}_{1,1} & \cdots & \mathbf{A}_{1,m-1} \\ \vdots & \vdots & \ddots & \vdots \\ \mathbf{A}_{m-1,0} & \mathbf{A}_{m-1,1} & \cdots & \mathbf{A}_{m-1,m-1} \end{bmatrix}. \tag{12.8}$$

Since all the entries of $\mathbf{W}^{(2)}$ are nonzero, $\mathbf{H}_{qc,\text{disp}}^{(2)}$ contains no zero matrix. $\mathbf{H}_{qc,\text{disp}}^{(2)}$ is an $m(q-1) \times m(q-1)$ matrix over GF(2) with both column and row weight $m$.

For any pair $(g, r)$ of integers with $1 \leq g, r \leq m$, let $\mathbf{H}_{qc,\text{disp}}^{(2)}(g, r)$ be a $g \times r$ subarray of $\mathbf{H}_{qc,\text{disp}}^{(2)}$. $\mathbf{H}_{qc,\text{disp}}^{(2)}(g, r)$ is a $g(q-1) \times r(q-1)$ matrix over GF(2) with column and row weights $g$ and $r$, respectively. The null space of $\mathbf{H}_{qc,\text{disp}}^{(2)}(g, r)$ gives a $(g, r)$-regular QC-LDPC code $\mathcal{C}_{qc,\text{disp}}^{(2)}$ of length $r(q-1)$ with rate at least $(r-g)/r$ and minimum distance $g+2$ for even $g$ and $g+1$ for odd $g$. The above construction gives a class of QC-LDPC codes.

---

**Example 12.6** Let GF($2^7$) be the field for code construction. Since $2^7 - 1 = 127$ is a prime, the largest prime factor of 127 is itself. Using (12.7) and (12.8), we can construct a $127 \times 127$ array $\mathbf{H}_{qc,\text{disp}}^{(2)}$ of $127 \times 127$ CPMs over GF(2). Take a $4 \times 40$ subarray $\mathbf{H}_{qc,\text{disp}}^{(2)}(4, 40)$ from $\mathbf{H}_{qc,\text{disp}}^{(2)}$, say the $4 \times 40$ subarray at the upper-left corner of $\mathbf{H}_{qc,\text{disp}}^{(2)}$. $\mathbf{H}_{qc,\text{disp}}^{(2)}(4, 40)$ is a $508 \times 5080$ matrix over GF(2) with column and row weights 4 and 40, respectively. The null space of this matrix gives a $(4, 40)$-regular $(5080, 4575)$ QC-RS-LDPC code with rate 0.9006. The error performance of this code over the binary-input AWGN channel decoded with iterative decoding using the SPA (100 iterations) is shown in Figure 12.6. At a BER of $10^{-6}$, it performs 1.1 dB from the Shannon limit. Also included in Figure 12.6 is a $(5080, 4575)$ random MacKay code constructed by computer search, whose parity-check matrix has column weight 4 and average row weight 40. We see that the error performance curves of the two codes are almost on top of each other, but the quasi-cyclic code performs slightly better than the corresponding random MacKay code.

---

The above array dispersion of $\mathbf{W}^{(2)}$ is based on the multiplicative group $\mathcal{G}_{q-1} = \{\alpha^0 = 1, \alpha, \ldots, \alpha^{q-2}\}$ of GF($q$). Every nonzero entry in $\mathbf{W}^{(2)}$ is replaced by a $(q-1) \times (q-1)$ CPM. Since the entries in $\mathbf{W}^{(2)}$ are elements in the cyclic subgroup $\mathcal{G}_m$ of $\mathcal{G}_{q-1}$, $\mathbf{W}^{(2)}$ can be dispersed using a cyclic subgroup of $\mathcal{G}_{q-1}$ that contains $\mathcal{G}_m$ as a subgroup.

Suppose that $q-1$ can be factored as $q-1 = ltm$, where $m$ again is the largest prime factor of $q-1$. Let $\beta = \alpha^{lt}$ and $\delta = \alpha^l$. The orders of $\beta$ and $\delta$ are $m$ and $tm$, respectively. Then the two sets of elements $\mathcal{G}_m = \{\beta^0 = 1, \beta, \ldots, \beta^{m-1}\}$ and $\mathcal{G}_{tm} = \{\delta^0 = 1, \delta, \ldots, \delta^{tm-1}\}$ form two cyclic subgroups of $\mathcal{G}_{q-1}$, and $\mathcal{G}_{tm}$ contains $\mathcal{G}_m$ as a subgroup. $\mathcal{G}_{tm}$ is a supergroup of $\mathcal{G}_m$. Next we define the location vector of an element $\delta^i$ in $\mathcal{G}_{tm}$ as a $tm$-tuple over GF(2),

$$\mathbf{z}(\delta^i) = (z_0, z_1, \ldots, z_{tm-1}),$$

whose components correspond to the $tm$ elements of $\mathcal{G}_{km}$, where $z_i = 1$ and all the other components are set to zero. We refer to this location vector as the location vector of $\delta^i$ *with respect to the group* $\mathcal{G}_{tm}$ (or call it the $\mathcal{G}_{tm}$ location vector of $\delta^i$). Let $\sigma$

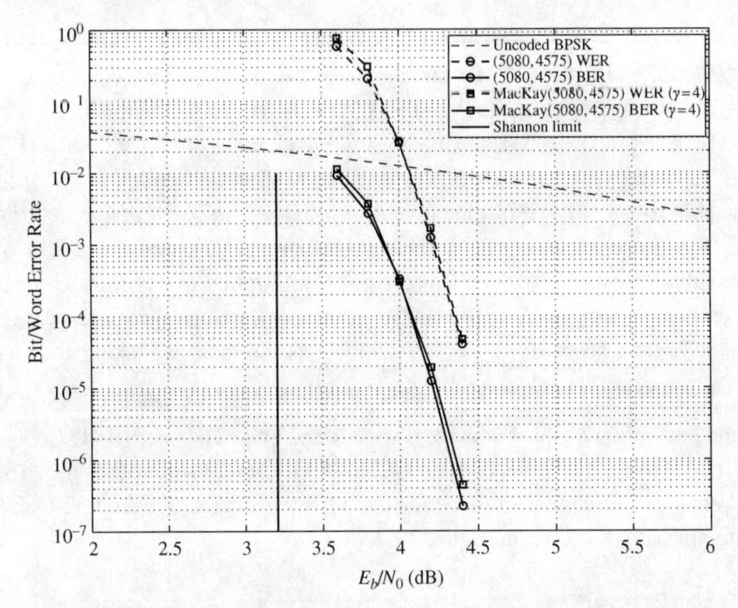


**Figure 12.6** Error performances of the (5080, 4575) QC-RS-LDPC code of Example 12.6 and a (5080, 4565) random MacKay code over the binary-input AWGN channel.

be an element of $\mathcal{G}_{tm}$. Then $\sigma, \delta\sigma, \ldots, \delta^{tm-1}\sigma$ are distinct and they form all the $tm$ elements of $\mathcal{G}_{tm}$. Form a $tm \times tm$ matrix $\mathbf{A}^*$ over $\mathcal{G}_{tm}$ with the $\mathcal{G}_{tm}$location vectors of $\sigma, \delta\sigma, \ldots, \delta^{tm-1}\sigma$ as rows. Then $\mathbf{A}^*$ is a $tm \times tm$ CPM over GF(2). $\mathbf{A}^*$ is called the $tm$-fold matrix dispersion of $\sigma$ with respect to the cyclic group $\mathcal{G}_{tm}$.

Since $\mathcal{G}_m$ is a cyclic subgroup of $\mathcal{G}_{tm}$, $\mathbf{W}^{(2)}$ satisfies $\delta$-multiplied row constraints 1 and 2. Now, on replacing each entry in $\mathbf{W}^{(2)}$ by its $tm$-fold matrix dispersion, we obtain the following RC-constrained $m \times m$ array of $tm \times tm$ CPMs over GF(2):

$$\mathbf{H}^{(3)}_{qc,\text{disp}} = \begin{bmatrix} \mathbf{A}^*_{0,0} & \mathbf{A}^*_{0,1} & \cdots & \mathbf{A}^*_{0,m-1} \\ \mathbf{A}^*_{1,0} & \mathbf{A}^*_{1,1} & \cdots & \mathbf{A}^*_{1,m-1} \\ \vdots & \vdots & \ddots & \vdots \\ \mathbf{A}^*_{m-1,0} & \mathbf{A}^*_{m-1,1} & \cdots & \mathbf{A}^*_{m-1,m-1} \end{bmatrix}. \tag{12.9}$$

$\mathbf{H}^{(3)}_{qc,\text{disp}}$ is a $tm^2 \times tm^2$ matrix over GF(2) with both column and row weight $m$. For any pair $(g, r)$ of integers with $1 \le g, r \le m$, let $\mathbf{H}^{(3)}_{qc,\text{disp}}(g, r)$ be a $g \times r$ subarray of $\mathbf{H}^{(3)}_{qc,\text{disp}}$. $\mathbf{H}^{(3)}_{qc,\text{disp}}(g, r)$ is a $gtm \times rtm$ matrix over GF(2) with column and row weights $g$ and $r$, respectively. The null space of $\mathbf{H}^{(3)}_{qc,\text{disp}}(g, r)$ gives a $(g, r)$-regular QC-RS-LDPC code of length $rtm$, whose Tanner graph has a girth of at least 6.

If the array dispersion of $\mathbf{W}^{(2)}$ is carried out with respect to the cyclic subgroup $\mathcal{G}_m$ of the multiplicative group $\mathcal{G}_{q-1}$ of GF($q$), we obtain an $m \times m$ array $\mathbf{H}^{(3)}_{qc,\text{disp}}$ of $m \times m$ CPMs. From the subarrays of this array, we can construct QC-LDPC codes. The above construction based on various supergroups of $\mathcal{G}_m$ gives various sizes of CPMs in $\mathbf{H}^{(3)}_{qc,\text{disp}}$. As a result, we obtain various families of QC-LDPC codes.

**Example 12.7** Let GF($2^{10}$) be the code-construction field. We can factor $2^{10} - 1 = 1023$ as the product $3 \cdot 11 \cdot 31$. The largest prime factor of 1023 is 31. Let $m = 31$, $l = 11$, and $t = 3$. Let $\alpha$ be a primitive element of GF($2^{10}$), $\beta = \alpha^{3\times 11}$, and $\delta = \alpha^{11}$. Then $\mathcal{G}_{31} = \{\beta^0 = 1, \beta, \ldots, \beta^{30}\}$ and $\mathcal{G}_{93} = \{\delta^0 = 1, \delta, \ldots, \delta^{92}\}$ form two cyclic subgroups of the multiplicative group $\mathcal{G}_{1023} = \{\alpha^0 = 1, \alpha, \ldots, \alpha^{1022}\}$ of GF($2^{10}$). $\mathcal{G}_{93}$ contains $\mathcal{G}_{31}$ as a subgroup. Utilizing (12.7), we form a $31 \times 31$ matrix $\mathbf{W}^{(2)}$ with entries from $\mathcal{G}_{31}$, which satisfies $\delta$-multiplied row constraints 1 and 2. On replacing each entry in $\mathbf{W}^{(2)}$ by its 93-fold matrix dispersion with respect to $\mathcal{G}_{93}$, we obtain a $31 \times 31$ array $\mathbf{H}^{(3)}_{qc,\text{disp}}$ of $93 \times 93$ CPMs. Suppose we take a $4 \times 16$ subarray $\mathbf{H}^{(3)}_{qc,\text{disp}}(4, 16)$ from $\mathbf{H}^{(3)}_{qc,\text{disp}}$. $\mathbf{H}^{(3)}_{qc,\text{disp}}(4, 16)$ is a $372 \times 1488$ matrix over GF(2) with column and row weights 4 and 16, respectively. The null space of $\mathbf{H}^{(3)}_{qc,\text{disp}}(4, 16)$ gives a (4, 16)-regular (1488, 1125) QC-RS-LDPC code with rate 0.756. The bit- and word-error performances of this code decoded with iterative decoding using the SPA (50 iterations) are shown in Figure 12.7.

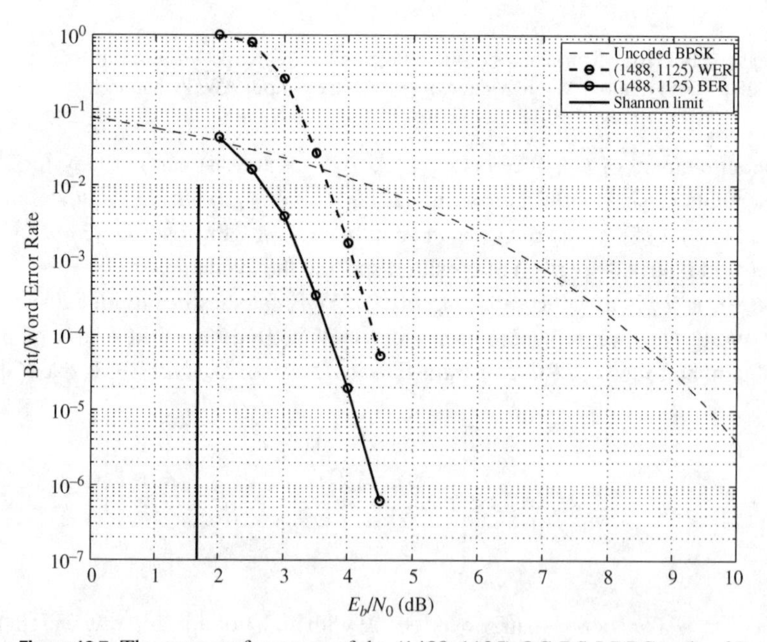


**Figure 12.7** The error performance of the (1488, 1125) QC-RS-LDPC code of Example 12.7.

**Example 12.8** Suppose we take an $8 \times 16$ subarray $\mathbf{H}^{(3)}_{qc,\text{disp}}(8, 16)$ from the array $\mathbf{H}^{(3)}_{qc,\text{disp}}$ of $93 \times 93$ CPMs constructed in Example 12.7. We use this subarray as the base array for masking. Construct an $8 \times 16$ masking matrix $\mathbf{Z}(8, 16)$ that consists of a row of two $8 \times 8$ circulants with generators $\mathbf{g}_1 = (01011001)$ and $\mathbf{g}_2 = (10010110)$. The masking matrix $\mathbf{Z}(8, 16)$ has column and row weights 4 and 8, respectively. Masking $\mathbf{H}^{(3)}_{qc,\text{disp}}(8, 16)$ with $\mathbf{Z}(8, 16)$, we obtain an $8 \times 16$ masked array $\mathbf{M}_{mask}(8, 16)$,

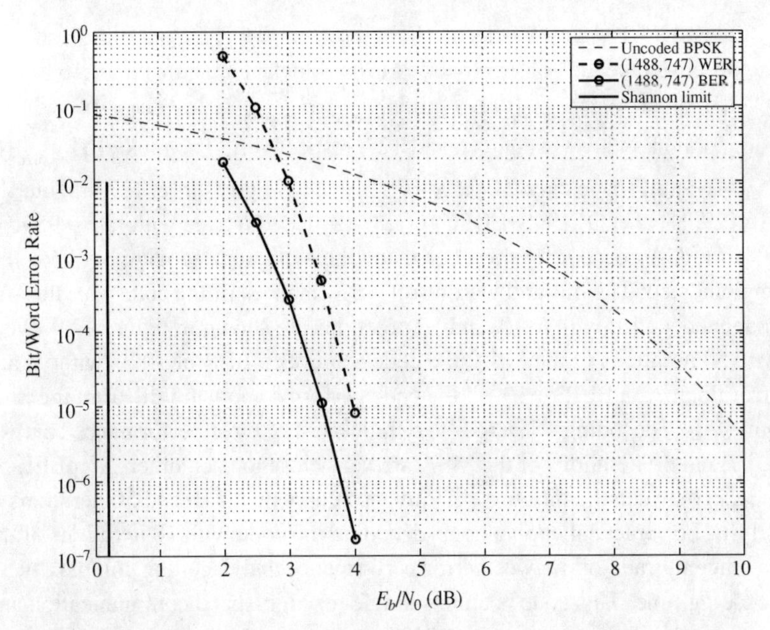


**Figure 12.8** The error performance of the (1488, 747) QC-RS-LDPC code of Example 12.8.

which is a $744 \times 1488$ matrix with column and row weights 4 and 8, respectively. The null space of $\mathbf{M}_{mask}(8, 16)$ gives a (4, 8)-regular (1488, 747) QC-RS-LDPC code with rate 0.502. The bit- and word-error performances of this code with iterative decoding using the SPA with 50 iterations are shown in Figure 12.8.

---

In the above construction, if we set $t = 1$ and take $m$-fold array dispersion of $\mathbf{W}^{(2)}$ with respect to a cyclic subgroup $G_m$ of order $m$ of the multiplicative group of GF($q$), we obtain an $m \times m$ array $\mathbf{H}_{qc,\text{disp}}^{(3)}$ of CPMs of size $m \times m$, which is an $m^2 \times m^2$ matrix with both column and row weight $m$. Then it follows from Theorem 11.1 that $\mathbf{H}_{qc,\text{disp}}^{(3)}$ is the line–point incidence matrix of a partial geometry PaG$(m, m, m-1)$ with $m^2$ points and $m^2$ lines. Each line in PaG$(m, m, m-1)$ consists of $m$ points and each point is on $m$ lines. In this case, QC-RS-LDPC codes constructed based on subarrays of $\mathbf{H}_{qc,\text{disp}}^{(3)}$ (or the base matrix $\mathbf{W}^{(2)}$ over $G_m$) are also partial geometry QC-LDPC codes.

---

**Example 12.9** Consider the array $\mathbf{H}_{qc,\text{disp}}^{(2)}$ constructed using the field GF($2^7$) given in Example 12.6. It is a $127 \times 127$ array of CPMs of size $127 \times 127$, which is a $16\,129 \times 16\,129$ matrix with both column and row weight 127. It follows from Theorem 11.1 that $\mathbf{H}_{qc,\text{disp}}^{(2)}$ is the line–point incidence matrix of a partial geometry PaG(127, 127, 126) with 16 129 points and 16 129 lines. Each line in PaG(127, 127, 126) consists of 127 points and each point is on 127 lines. Therefore, the code constructed in Example 12.6 is also a partial geometry QC-LDPC code. Using subarrays of $\mathbf{H}_{qc,\text{disp}}^{(2)}$ of various sizes, we can construct QC-PaG-LDPC codes (from a geometric point of view) of various lengths and rates.

Label the CPM row-blocks of $\mathbf{H}^{(2)}_{qc,\text{disp}}$ from 0 to 126. Set $g = 6$ and $r = 127$. Suppose we take the consecutive CPM row-blocks labeled with 1 to 6 to form a $6 \times 127$ subarray $\mathbf{H}^{(2)}_{qc,\text{disp}}(6, 127)$ of $\mathbf{H}^{(2)}_{qc,\text{disp}}$. $\mathbf{H}^{(2)}_{qc,\text{disp}}(6, 127)$ is a $762 \times 16\,129$ matrix with column and row weights 6 and 127, respectively. The rank of $\mathbf{H}^{(2)}_{qc,\text{disp}}(6, 127)$ is 757. The null space over GF(2) of $\mathbf{H}^{(2)}_{qc,\text{disp}}(6, 127)$ gives a (6, 127)-regular (16 129, 15 372) QC-PaG-LDPC (or QC-RS-LDPC) code with rate 0.953.

An FPGA decoder for this code using the MSA for decoding has been implemented. The bit- and word-error performances of the code over the AWGN channel decoded with 5, 10, and 50 iterations are shown in Figure 12.9. We see that, with 50 iterations of the MSA, the code achieves a BER of $10^{-15}$ without a visible error floor. At a BER of $10^{-15}$, the code performs about 1.5 dB from the Shannon limit. Also, decoding converges very quickly. The error performance curves with 5, 10, and 50 iterations of the MSA are very close to each other. At a BER of $10^{-12}$, the gap between the bit-error performance curves with 10 and 50 iterations is within 0.1 dB. For practical applications, decoding the code with 10 iterations of the MSA is an efficient trade-off between error performance and decoding complexity and decoding time. This code is quite suitable for high-speed communication and high-density data storage systems that require very low error rates, such as optical communication systems and flash memories.

---

In the construction of the base matrix $\mathbf{W}^{(2)}$ given by (12.7), we used the largest prime factor $m$ of $q - 1$. In fact, any prime factor $p$ of $q - 1$, not necessarily the largest prime factor, can be used in the construction of $\mathbf{W}^{(2)}$. In this case, the entries of $\mathbf{W}^{(2)}$ are elements in an order-$p$ cyclic subgroup $G_p$ of the multiplicative group of GF($q$). The $p$-fold array dispersion of $\mathbf{W}^{(2)}$ with respect to $G_p$ gives a $p \times p$ array $\mathbf{H}^{(3)}_{qc,\text{disp}}$ of CPMs of size $p \times p$, which is a $p^2 \times p^2$ matrix over GF(2) with column and row weights both equal to $p$. It follows from Theorem 11.1 that $\mathbf{H}^{(3)}_{qc,\text{disp}}$ is the line–point incidence matrix of a partial geometry PaG($p, p, p - 1$) with $p$ points and $p$ lines. Each line consists of $p$ points, and a point is on $p$ lines.

With the above construction of base matrices using cyclic subgroups of finite fields, a large class of partial geometries can be constructed and hence a large class of QC-PaG-LDPC (or QC-RS-LDPC) codes can be constructed.

---

**Example 12.10** Consider the field GF($2^{11}$). Note that $2^{11} - 1 = 2047$ can be factored as the product of two primes 23 and 89. The multiplicative group of GF($2^{11}$) contains two cyclic subgroups $G_{23}$ and $G_{89}$ of orders 23 and 89, respectively.

Using $G_{89}$, we construct an $89 \times 89$ base matrix $\mathbf{W}^{(2)}_{89}$ over $G_{89}$ in the form of (12.7) with $t = 1$. Dispersing each entry in $\mathbf{W}^{(2)}_{89}$ into an $89 \times 89$ CPM with respect to $G_{89}$, we obtain an $89 \times 89$ array $\mathbf{H}^{(3)}_{89,qc,\text{disp}}$ of CPMs of size $89 \times 89$, which is a $7921 \times 7921$ matrix over GF(2) with both column and row weight 89. $\mathbf{H}^{(3)}_{89,qc,\text{disp}}$ is the line–point incidence matrix of a partial geometry PaG(89, 89, 88) with 7921 points

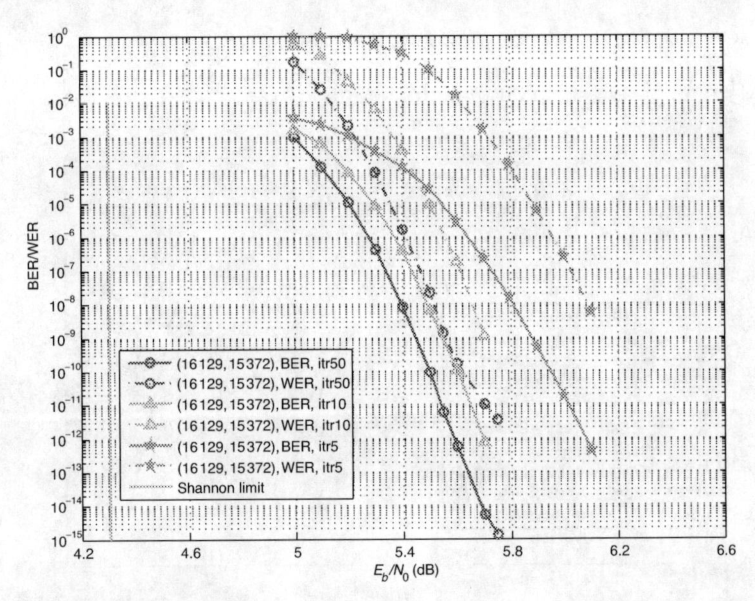


**Figure 12.9** Bit- and word-error performances of the (16 129, 15 372) QC-PaG-LDPC code given in Example 12.9.

and 7921 lines. Each line in PaG(89, 89, 88) consists of 89 points and each point is on 89 lines.

Suppose we use the factor 23 of 2047 and the cyclic subgroup $G_{23}$ to construct a $23 \times 23$ base matrix $\mathbf{W}_{23}^{(2)}$ over $G_{23}$ in the form of (12.7). The 23-fold array dispersion of $\mathbf{W}_{23}^{(2)}$ with respect to $G_{23}$ gives a $23 \times 23$ array $\mathbf{H}_{23,qc,\text{disp}}^{(3)}$ of CPMs of size $23 \times 23$, which is a $529 \times 529$ matrix over GF(2) with both column and row weight 23. $\mathbf{H}_{23,qc,\text{disp}}^{(3)}$ is the line–point incidence matrix of a partial geometry PaG(23, 23, 22) with 529 points and 529 lines. Each line consists of 23 points and each point is on 23 lines.

Using the submatrices of $\mathbf{W}_{23}^{(2)}$ and $\mathbf{W}_{89}^{(2)}$ in conjunction with CPM dispersion, we can construct partial-geometry-based (or RS-based) QC-LDPC codes of various lengths and rates.

Suppose we set $g = 4$ and $r = 89$. Label the rows and columns $\mathbf{W}_{89}^{(2)}$ from 0 to 88. Take four consecutive rows labeled from 1 to 4, from $\mathbf{W}_{89}^{(2)}$. This gives a $4 \times 89$ submatrix $\mathbf{W}_{89}^{(2)}(4, 89)$ of $\mathbf{W}_{89}^{(2)}$. The 89-fold array dispersion of $\mathbf{W}_{89}^{(2)}(5, 89)$ with respect to $G_{89}$ gives a $4 \times 89$ array $\mathbf{H}_{89,qc,\text{disp}}^{(3)}(4, 89)$ of CPMs of size $89 \times 89$, which is a $356 \times 7921$ matrix with column and row weights 4 and 89, respectively. The null space over GF(2) of $\mathbf{H}_{89,qc,\text{disp}}^{(3)}(4, 89)$ gives a (4, 89)-regular (7921, 7566) QC-PaG-LDPC (or QC-RS-LDPC) code of length 7921 with rate 0.955. The bit and word-error rates of the code over the AWGN channel using BPSK signaling decoded with 5, 10, and 50 iterations of the MSA are shown in Figure 12.10.

Set $g = 4$ and $r = 8$. Take a $4 \times 8$ submatrix $\mathbf{W}_{89}^{(2)}(4, 8)$ from $\mathbf{W}_{89}^{(2)}$ (excluding the top row of $\mathbf{W}_{89}^{(2)}$). Masking $\mathbf{W}_{89}^{(2)}(4, 8)$ with the $4 \times 8$ masking matrix,

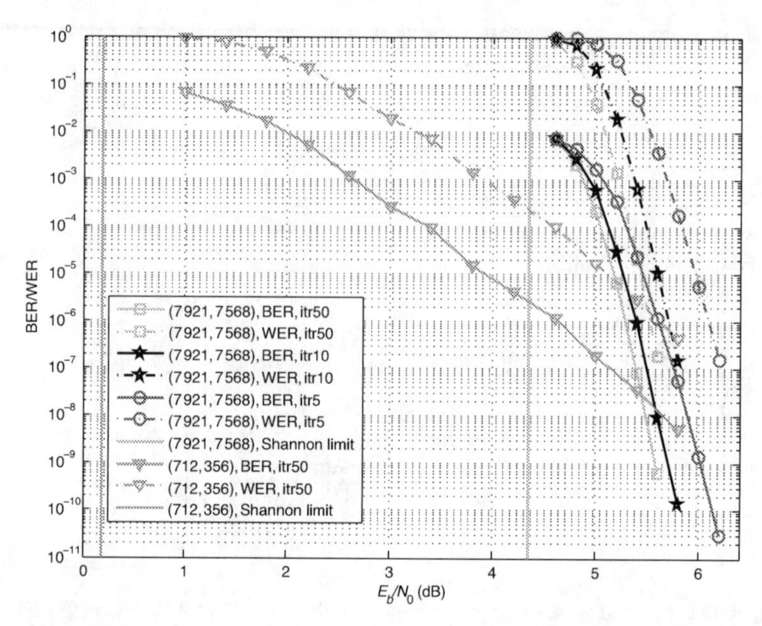


**Figure 12.10** The error performances of the two codes given in Example 12.10.

$$\mathbf{Z}(4,8) = \begin{bmatrix} 1 & 0 & 1 & 0 & 1 & 1 & 1 & 1 \\ 0 & 1 & 0 & 1 & 1 & 1 & 1 & 1 \\ 1 & 1 & 1 & 1 & 1 & 0 & 1 & 0 \\ 1 & 1 & 1 & 1 & 0 & 1 & 0 & 1 \end{bmatrix},$$

we obtain a masked $4 \times 6$ matrix $\mathbf{W}^{(2)}_{89,\text{mask}}(4,8)$ with column and row weights 3 and 6, respectively. The 89-fold array dispersion $\mathbf{H}^{(3)}_{89,qc,\text{disp}}(4,8)$ of $\mathbf{W}^{(2)}_{89,\text{mask}}(4,8)$ is a $4 \times 8$ array of CPMs and zero matrices of size $89 \times 89$, which is a $356 \times 712$ matrix over GF(2) with column and row weights 3 and 6, respectively. The null space over GF(2) of $\mathbf{H}^{(3)}_{89,qc,\text{disp}}(4,8)$ gives a (3, 6)-regular (712, 356) code of length 712 with rate 0.5. The bit- and word-error performances decoded with 50 iterations of the MSA are also shown in Figure 12.10.

## 12.5 Construction of QC-LDPC Codes Based on Subgroups of a Finite Field

Subgroups of the additive or multiplicative groups of a finite field can be used to construct RC-constrained arrays of CPMs. QC-LDPC codes can be constructed from these arrays [10].

### 12.5.1 Construction of QC-LDPC Codes Based on Subgroups of the Additive Group of a Finite Field

Let $q = p^m$, where $p$ is a prime and $m$ is a positive integer. Let GF($q$) be an extension field of the prime field GF($p$). Let $\alpha$ be a primitive element of GF($q$). Then

$\alpha^0, \alpha, \ldots, \alpha^{m-1}$ are linearly independent over GF($q$). They form a basis of GF($q$), called the polynomial basis. Any element $\alpha^i$ of GF($q$) can be expressed as a linear combination of $\alpha^0, \alpha, \ldots, \alpha^{m-1}$ as follows:

$$\alpha^i = c_{i,0}\alpha^0 + c_{i,1}\alpha + \cdots + c_{i,m-1}\alpha^{m-1},$$

with $c_{i,j} \in$ GF($p$). We say that the $m$ independent elements $\alpha^0, \alpha, \ldots, \alpha^{m-1}$ over GF($p$) span the field GF($q$). For $1 \le t < m$, let $\mathcal{G}_1 = \{\beta_0 = 0, \beta_1, \ldots, \beta_{p^t-1}\}$ be the additive subgroup of GF($q$) generated by the linear combinations of $\alpha^0, \alpha, \ldots, \alpha^{t-1}$, that is

$$\beta_i = c_{i,0}\alpha^0 + c_{i,1}\alpha + \cdots + c_{i,t-1}\alpha^{t-1}.$$

Let $\mathcal{G}_2 = \{\delta_0 = 0, \delta_1, \ldots, \delta_{p^{m-t}-1}\}$ be the additive subgroup of GF($q$) generated by the linear combinations of $\alpha^t, \alpha^{t+1}, \ldots, \alpha^{m-1}$, that is

$$\delta_i = c_{i,t}\alpha^t + c_{i,t+1}\alpha^{t+1} + \cdots + c_{i,m-1}\alpha^{m-1}.$$

It is clear that $\mathcal{G}_1 \cap \mathcal{G}_2 = \{0\}$. For $0 \le i < p^{m-t}$, define the following set of elements:

$$\delta_i + \mathcal{G}_1 = \{\delta_i, \delta_i + \beta_1, \ldots, \delta_i + \beta_{p^t-1}\}.$$

This set is simply a coset of $\mathcal{G}_1$ with coset leader $\delta_i$. There are $p^{m-t}$ cosets of $\mathcal{G}_1$, including $\mathcal{G}_1$ itself. These $p^{m-t}$ cosets of $\mathcal{G}_1$ form a *partition* of the $q$ elements of the field GF($q$). Two cosets of $\mathcal{G}_1$ are mutually disjoint.

Form the following $p^{m-t} \times p^t$ matrix over GF($q$):

$$\mathbf{W}^{(4)} = \begin{bmatrix} \mathbf{w}_0 \\ \mathbf{w}_1 \\ \vdots \\ \mathbf{w}_{p^{m-t}-1} \end{bmatrix} = \begin{bmatrix} 0 & \beta_1 & \cdots & \beta_{p^t-1} \\ \delta_1 & \delta_1 + \beta_1 & \cdots & \delta_1 + \beta_{p^t-1} \\ \vdots & \vdots & \ddots & \vdots \\ \delta_{p^{m-t}-1} & \delta_{p^{m-t}-1} + \beta_1 & \cdots & \delta_{p^{m-t}-1} + \beta_{p^t-1} \end{bmatrix}, \quad (12.10)$$

where the $i$th row consists of the elements of the $i$th coset of $\mathcal{G}_i$, with $0 \le i < p^{m-t}$. Every element of GF($q$) appears once and only once in $\mathbf{W}^{(4)}$. Except for the first row, every row $\mathbf{w}_i$ of $\mathbf{W}^{(4)}$ consists of only nonzero elements of GF($q$). The first row of $\mathbf{W}^{(4)}$ contains the 0 element of GF($q$) at the first position. Except for the first column, every column of $\mathbf{W}^{(4)}$ consists of only nonzero elements of GF($q$). The first column of $\mathbf{W}^{(4)}$ contains the 0 element of GF($q$) at the first position. Since two cosets of $\mathcal{G}_1$ are disjoint, two different rows $\mathbf{w}_i$ and $\mathbf{w}_j$ of $\mathbf{W}^{(4)}$ differ in all positions. We also note that the elements of each column form a coset of $\mathcal{G}_2$. As a result, two different columns of $\mathbf{W}^{(4)}$ differ in all positions. From the above structural properties of $\mathbf{W}^{(4)}$, we can easily prove that $\mathbf{W}^{(4)}$ satisfies $\alpha$-multiplied row constraints 1 and 2.

On replacing each entry of $\mathbf{W}^{(4)}$ by its multiplicative $(q-1)$-fold matrix dispersion, we obtain the following RC-constrained $p^{m-t} \times p^t$ array $\mathbf{H}^{(4)}_{qc,\text{disp}}$ of $(q-1) \times (q-1)$ CPMs:

$$\mathbf{H}^{(4)}_{qc,\text{disp}} = \begin{bmatrix} \mathbf{A}_{0,0} & \mathbf{A}_{0,1} & \cdots & \mathbf{A}_{0,p^t-1} \\ \mathbf{A}_{1,0} & \mathbf{A}_{1,1} & \cdots & \mathbf{A}_{1,p^t-1} \\ \vdots & \vdots & \ddots & \vdots \\ \mathbf{A}_{p^{m-t}-1,0} & \mathbf{A}_{p^{m-t}-1,1} & \cdots & \mathbf{A}_{p^{m-t}-1,p^t-1} \end{bmatrix}, \quad (12.11)$$

where $\mathbf{A}_{0,0}$ is a $(q-1) \times (q-1)$ zero matrix, the only zero submatrix in the array. Since all the entries in $\mathbf{W}^{(4)}$ are different, all the permutation matrices in $\mathbf{H}^{(4)}_{qc,\text{disp}}$ are different.

For any pair $(g, r)$ of integers, with $1 \leq g \leq p^{m-t}$ and $1 \leq r \leq p^t$, let $\mathbf{H}^{(4)}_{qc,\text{disp}}(g, r)$ be a $g \times r$ subarray of $\mathbf{H}^{(4)}_{qc,\text{disp}}$, which does not contain the single zero matrix of $\mathbf{H}^{(4)}_{qc,\text{disp}}$. $\mathbf{H}^{(4)}_{qc,\text{disp}}(g, r)$ is a $g(q-1) \times r(q-1)$ matrix over GF(2) with column and row weights $g$ and $r$, respectively. The null space of $\mathbf{H}^{(4)}_{qc,\text{disp}}(g, r)$ gives a $(g, r)$-regular QC-LDPC code $C^{(4)}_{qc,\text{disp}}$ of length $r(q-1)$, whose Tanner graph has a girth of at least 6. The above construction gives a class of QC-LDPC codes.

---

**Example 12.11** In this example, we choose $m = 8$ and use GF($2^8$) as the code-construction field. Let $\alpha$ be a primitive element of GF($2^8$). Set $t = 5$. Then $m - t = 3$. Let $\mathcal{G}_1$ and $\mathcal{G}_2$ be two subgroups of the additive group of GF($2^8$) with orders 32 and 8 spanned by the elements in $\{\alpha^0, \alpha, \alpha^2, \alpha^3, \alpha^4\}$ and the elements in $\{\alpha^5, \alpha^6, \alpha^7\}$, respectively. Via these two groups, we can form an $8 \times 32$ array $\mathbf{H}^{(4)}_{qc,\text{disp}}$ of 255 CPMs of size $255 \times 255$ and a single $255 \times 255$ zero matrix. Choose $g = 4$ and $r = 32$. Take a $4 \times 32$ subarray $\mathbf{H}^{(4)}_{qc,\text{disp}}(4, 32)$ from $\mathbf{H}^{(4)}_{qc,\text{disp}}$, avoiding the single zero matrix. $\mathbf{H}^{(4)}_{qc,\text{disp}}(4, 32)$ is a $1020 \times 8160$ matrix over GF(2) with column and row weights 4 and 32, respectively. The null space of this matrix gives a (4, 32)-regular (8160, 7159) QC-LDPC code with rate 0.8773. The bit- and word-error performances of this code decoded with iterative decoding using the SPA with 100 iterations are

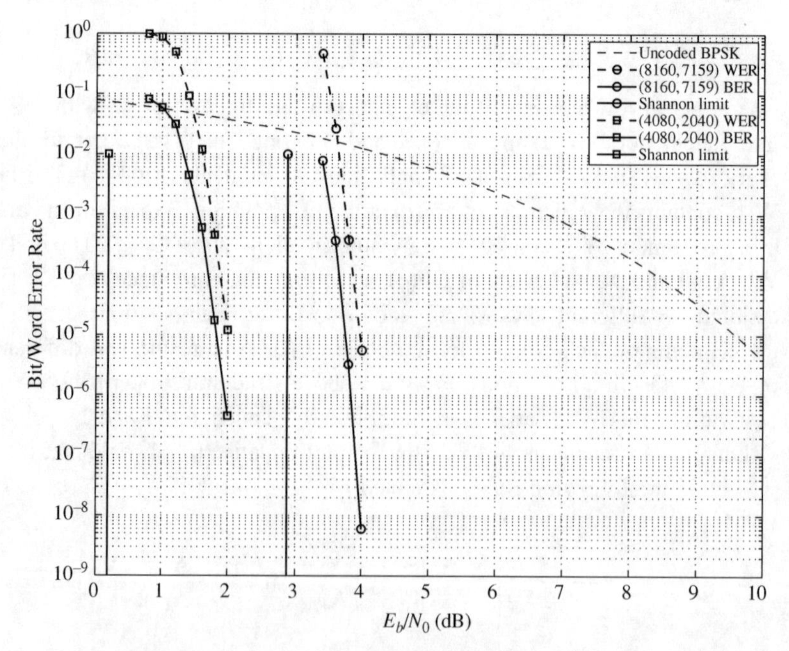


**Figure 12.11** Error performances of the (8160, 7159) QC-LDPC code and the (4080, 2040) QC-LDPC code given in Example 12.11.

shown in Figure 12.11. At a BER of $10^{-6}$, the code performs 0.98 dB from the Shannon limit.

Suppose we take the $8 \times 16$ subarray $\mathbf{H}^{(4)}_{qc,\text{disp}}(8,16)$ from $\mathbf{H}^{(4)}_{qc,\text{disp}}$ and use it as a base array for masking. Construct an $8 \times 16$ masking matrix $\mathbf{Z}(8,16) = [\mathbf{G}_0\mathbf{G}_1]$ over GF(2) that consists of two $8 \times 8$ circulants, $\mathbf{G}_0$ and $\mathbf{G}_1$. The two circulants in $\mathbf{Z}(8,16)$ are generated by $\mathbf{g}_0 = (01011000)$ and $\mathbf{g}_1 = (00101010)$, respectively. Then $\mathbf{Z}(8,16)$ is a (3, 6)-regular masking matrix with column and row weights 3 and 6, respectively. On masking $\mathbf{H}^{(4)}_{qc,\text{disp}}$ with $\mathbf{Z}(8,16)$, we obtain an $8 \times 16$ masked array $\mathbf{M}^{(4)}(8,16) = \mathbf{Z}(8,16) \circledast \mathbf{H}^{(4)}_{qc,\text{disp}}$ that is a $2040 \times 4080$ matrix over GF(2) with column and row weights 3 and 6, respectively. The null space of $\mathbf{M}^{(4)}(8,16)$ gives a (3, 6)-regular (4080, 2040) QC-LDPC code with rate 1/2, whose bit- and word-error performances with iterative decoding using SPA (50 iterations) are also shown in Figure 12.11. We see that, at a BER of $10^{-6}$, it performs 1.77 dB from the Shannon limit.

### 12.5.2 Construction of QC-LDPC Codes Based on Subgroups of the Multiplicative Group of a Finite Field

Again, we use GF($q$) for code construction. Let $\alpha$ be a primitive element of GF($q$). Suppose $q-1$ is not prime. We can factor $q-1$ as a product of two relatively prime factors $c$ and $m$ such that $q-1 = cm$. Let $\beta = \alpha^c$ and $\delta = \alpha^m$. Then $\mathcal{G}_{c,1} = \{\beta^0 = 1, \beta, \ldots, \beta^{m-1}\}$ and $\mathcal{G}_{c,2} = \{\delta^0 = 1, \delta, \ldots, \delta^{c-1}\}$ form two cyclic subgroups of the multiplicative group of GF($q$) and $\mathcal{G}_{c,1} \cap \mathcal{G}_{c,2} = \{1\}$. For $0 \le i < c$, the set

$$\delta^i \mathcal{G}_{c,1} = \{\delta^i, \delta^i\beta, \ldots, \delta^i\beta^{m-1}\}$$

forms a multiplicative coset of $\mathcal{G}_{c,1}$. The cyclic subgroup $\mathcal{G}_{c,1}$ has $c$ multiplicative cosets, including itself. For $0 \le j < m$, the set

$$\beta^j \mathcal{G}_{c,2} = \{\beta^j, \beta^j\delta, \ldots, \beta^j\delta^{c-1}\}$$

forms a multiplicative coset of $\mathcal{G}_{c,2}$. The cyclic subgroup $\mathcal{G}_{c,2}$ has $m$ multiplicative cosets, including itself.

Form the following $c \times m$ matrix over GF($q$):

$$\mathbf{W}^{(5)} = \begin{bmatrix} \mathbf{w}_0 \\ \mathbf{w}_1 \\ \vdots \\ \mathbf{w}_{c-1} \end{bmatrix} = \begin{bmatrix} \beta^0 - 1 & \beta - 1 & \cdots & \beta^{m-1} - 1 \\ \delta\beta^0 - 1 & \delta\beta - 1 & \cdots & \delta\beta^{m-1} - 1 \\ \vdots & \vdots & \ddots & \vdots \\ \delta^{c-1}\beta^0 - 1 & \delta^{c-1}\beta - 1 & \cdots & \delta^{c-1}\beta^{m-1} - 1 \end{bmatrix}, \tag{12.12}$$

where (1) the components of the $i$th row are obtained by subtracting the 1 element of GF($q$) from each element in the $i$th multiplicative coset of $\mathcal{G}_{c,1}$ and (2) the components of the $j$th column are obtained by subtracting the 1 element from each element in the $j$th multiplicative coset of $\mathcal{G}_{c,2}$. $\mathbf{W}^{(5)}$ has the following structural properties: (1) there is one and only one 0 entry, which is located in the upper-left corner, and all the other $cm-1$ entries are nonzero elements in GF($q$); (2) all the entries are different elements in GF($q$); (3) any two rows differ in all the $m$ positions; and (4) any two columns differ in all $c$ positions. Given all the above structural properties, we can easily prove that $\mathbf{W}^{(5)}$ satisfies $\alpha$-multiplied row constraints 1 and 2.

By dispersing each nonzero entry of $\mathbf{W}^{(5)}$ into a $(q-1)\times(q-1)$ CPM and the single 0 entry into a $(q-1)\times(q-1)$ zero matrix, we obtain the following RC-constrained $c \times m$ array of $cm-1$ CPMs of size $(q-1)\times(q-1)$ and a single $(q-1)\times(q-1)$ zero matrix:

$$\mathbf{H}^{(5)}_{qc,\text{disp}} = \begin{bmatrix} \mathbf{O} & \mathbf{A}_{0,1} & \cdots & \mathbf{A}_{0,m-1} \\ \mathbf{A}_{1,0} & \mathbf{A}_{1,1} & \cdots & \mathbf{A}_{1,m-1} \\ \vdots & \vdots & \ddots & \vdots \\ \mathbf{A}_{c-1,0} & \mathbf{A}_{c-1,1} & \cdots & \mathbf{A}_{c-1,m-1} \end{bmatrix}. \tag{12.13}$$

Since all the nonzero entries of $\mathbf{W}^{(5)}$ are different, all the CPMs in $\mathbf{H}^{(5)}_{qc,\text{disp}}$ are different. The null space of any subarray of $\mathbf{H}^{(5)}_{qc,\text{disp}}$ gives a QC-LDPC code whose Tanner graph has a girth of at least 6. The above construction gives a class of QC-LDPC codes.

Of course, masking subarrays of the arrays given by (12.11) and (12.13) will result in many more QC-LDPC codes.

---

**Example 12.12** Let GF($2^8$) be the code-construction field. We find that $2^8-1=255$ can be factored as the product of 5 and 51, which are relatively prime. Let $\beta=\alpha^5$ and $\delta=\alpha^{51}$. The sets $\mathcal{G}_{c,1}=\{\beta^0,\beta,\ldots,\beta^{50}\}$ and $\mathcal{G}_{c,2}=\{\delta^0,\delta,\delta^2,\delta^3,\delta^4\}$ form two cyclic groups of the multiplicative group of GF($2^8$) and $\mathcal{G}_{c,1}\cap\mathcal{G}_{c,2}=\{1\}$. Given these two cyclic subgroups of GF($2^8$), (12.12) and (12.13), we can construct a $5\times 51$ array $\mathbf{H}^{(5)}_{qc,\text{disp}}$ of 254 CPMs of size $255\times 255$ and a single $255\times 255$ zero matrix. Take a $4\times 16$ subarray $\mathbf{H}^{(5)}_{qc,\text{disp}}(4,16)$ from $\mathbf{H}^{(5)}_{qc,\text{disp}}$, avoiding the single zero matrix. $\mathbf{H}^{(5)}_{qc,\text{disp}}(4,16)$ is a $1020\times 4080$ matrix with column and row weights 4 and 16, respectively. The null space of this matrix gives a (4, 16)-regular (4080, 3093) QC-LDPC code with rate 0.758. The error performance of this code with iterative decoding using the SPA with 50 iterations is shown in Figure 12.12. At a BER of $10^{-6}$, it performs 1.3 dB from the Shannon limit.

Suppose we remove the first column of $\mathbf{H}^{(5)}_{qc,\text{disp}}$. This removal results in a $5\times 50$ subarray $\mathbf{H}^{(5)}_{qc,\text{disp}}(5,50)$, which is a $1275\times 12\,750$ matrix with column and row weights 5 and 50, respectively. The null space of $\mathbf{H}^{(5)}_{qc,\text{disp}}(5,50)$ gives a (5, 50)-regular (12 750, 11 553) QC-LDPC code with rate 0.9061. The error performance of this code is also shown in Figure 12.12. At a BER of $10^{-6}$, it performs 0.9 dB from the Shannon limit.

Suppose we use $\mathbf{H}^{(5)}_{qc,\text{disp}}(5,50)$ as the base array for masking. Construct a $5\times 50$ masking matrix $\mathbf{Z}(5,50)=[\mathbf{G}_0\mathbf{G}_1\cdots\mathbf{G}_9]$ over GF(2), where, for $0\le i<9$,

$$\mathbf{G}_i = \begin{bmatrix} 1 & 1 & 1 & 0 & 0 \\ 0 & 1 & 1 & 1 & 0 \\ 0 & 0 & 1 & 1 & 1 \\ 1 & 0 & 0 & 1 & 1 \\ 1 & 1 & 0 & 0 & 1 \end{bmatrix}.$$

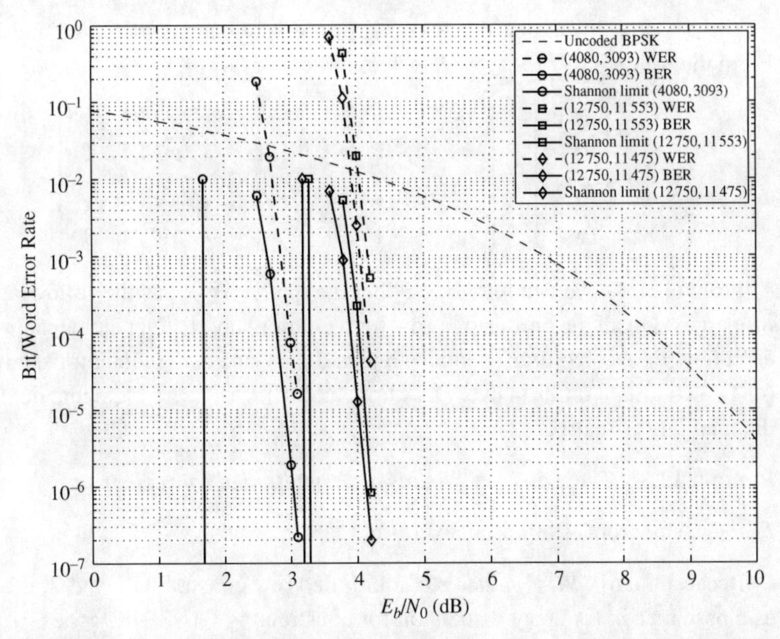


**Figure 12.12** Error performances of the (4080, 3093) and (12 750, 11 553) QC-LDPC codes given in Example 12.12.

$\mathbf{Z}(5, 50)$ has column and row weights 3 and 30, respectively. On masking $\mathbf{H}^{(5)}_{qc,\text{disp}}(5, 50)$ with $\mathbf{Z}(5, 50)$, we obtain a $5 \times 50$ masked array $\mathbf{M}^{(5)}(5, 50) = \mathbf{Z}(5, 10) \circledast \mathbf{H}^{(5)}(5, 50)$, which is a $1275 \times 12750$ matrix over GF(2) with column and row weights 3 and 30, respectively. The null space of this matrix gives a (3, 30)-regular (12 750, 11 475) QC-LDPC code with rate 0.9, whose bit- and word-error performances with iterative decoding using the SPA (50 iterations) are also shown in Figure 12.12.

## 12.6 Construction of QC-LDPC Codes Based on Primitive Elements of a Field

This section gives a method of constructing QC-LDPC codes based on primitive elements of a finite field. Consider the Galois field GF($q$), where $q$ is a power of a prime. The number of primitive elements in GF($q$) can be enumerated with Euler's formula as given by (2.43). First, we factor $q - 1$ as a product of powers of primes,

$$q - 1 = p_1^{k_1} p_2^{k_2} \cdots p_t^{k_t},$$

where $p_i$ is a prime, $1 \leq i \leq t$. Then the number of primitive elements in GF($q$) is given by (Euler's totient formula)

$$K = (q - 1) \prod_{i=1}^{t} (1 - 1/p_i).$$

Let $\alpha^{j_0}, \alpha^{j_1}, \ldots, \alpha^{j_K}$ be the set of $K$ primitive elements of GF($q$) and let $j_0 = 0$. Form the following $(K+1) \times (K+1)$ matrix over GF($q$):

$$\mathbf{W}^{(6)} = \begin{bmatrix} \mathbf{w}_0 \\ \mathbf{w}_1 \\ \vdots \\ \mathbf{w}_K \end{bmatrix} = \begin{bmatrix} \alpha^{j_0-j_0}-1 & \alpha^{j_1-j_0}-1 & \cdots & \alpha^{j_K-j_0}-1 \\ \alpha^{j_0-j_1}-1 & \alpha^{j_1-j_1}-1 & \cdots & \alpha^{j_K-j_1}-1 \\ \vdots & \vdots & \ddots & \vdots \\ \alpha^{j_0-j_K}-1 & \alpha^{j_1-j_K}-1 & \cdots & \alpha^{j_K-j_K}-1 \end{bmatrix}. \tag{12.14}$$

From (12.14), we can readily see that the matrix $\mathbf{W}^{(6)}$ has the following structural properties: (1) all the entries of a row (or a column) are distinct elements in GF($q$); (2) each row (or each column) contains one and only one zero element; (3) any two rows (or two columns) differ in every position; and (4) $K+1$ zero elements lie on the main diagonal of the matrix.

**Lemma 12.3** *The matrix* $\mathbf{W}^{(6)}$ *satisfies* $\alpha$*-multiplied row constraints 1 and 2.*

*Proof* The proof is left as an exercise. □

Because matrix $\mathbf{W}^{(6)}$ satisfies $\alpha$-multiplied row constraints 1 and 2, it can be used as a base matrix for array dispersion for constructing QC-LDPC codes. If we replace each entry by its multiplicative $(q-1)$-fold matrix dispersion, we obtain the following RC-constrained $(K+1) \times (K+1)$ array $\mathbf{H}^{(6)}_{qc,\text{disp}}$ of CPMs and zero matrices, with the zero matrices on the main diagonal of the array:

$$\mathbf{H}^{(6)}_{qc,\text{disp}} = \begin{bmatrix} \mathbf{A}_{1,0} & \mathbf{A}_{1,1} & \cdots & \mathbf{A}_{1,K} \\ \mathbf{A}_{2,0} & \mathbf{A}_{2,1} & \cdots & \mathbf{A}_{2,K} \\ \vdots & \vdots & \ddots & \vdots \\ \mathbf{A}_{K,0} & \mathbf{A}_{K,1} & \cdots & \mathbf{A}_{K,K} \end{bmatrix}. \tag{12.15}$$

For any pair $(g, r)$ of integers satisfying $1 \le g, r \le K$, let $\mathbf{H}^{(6)}_{qc,\text{disp}}(g, r)$ be a $g \times r$ subarray of $\mathbf{H}^{(6)}_{qc,\text{disp}}$. The null space of $\mathbf{H}^{(6)}(g, r)$ gives a QC-LDPC code of length $r(q-1)$ whose Tanner graph is free of length-4 cycles. The construction based on the subarrays of $\mathbf{H}^{(6)}_{qc,\text{disp}}$ gives a family of QC-LDPC codes. Of course, masking subarrays of $\mathbf{H}^{(6)}_{qc,\text{disp}}$ gives even more QC-LDPC codes.

---

**Example 12.13** Let GF($2^6$) be the code-construction field. Note that $2^6 - 1 = 63$ can be factored as $3^2 \times 7$. Using Euler's formula, we find that GF($2^6$) has $K = 63 \times (1 - 1/3)(1 - 1/7) = 36$ primitive elements. Using (12.14) and (12.15), we can construct a $37 \times 37$ array $\mathbf{H}^{(6)}_{qc,\text{disp}}$ of $63 \times 63$ circulant permutation and zero matrices. The zero matrices lie on the main diagonal of $\mathbf{H}^{(6)}_{qc,\text{disp}}$. Choose $g = 6$ and $r = 37$. Take the first six rows of $\mathbf{H}^{(6)}_{qc,\text{disp}}$ to form a $6 \times 37$ subarray $\mathbf{H}^{(6)}_{qc,\text{disp}}(6, 37)$, which is a $378 \times 2331$ matrix over GF(2) with row weight 36 and two column weights 5 and 6. Its null space gives a (2331, 2007) QC-LDPC code with rate 0.861. The bit- and word-error performances of this code decoded using the SPA with 50 iterations are shown in Figure 12.13. At a BER of $10^{-6}$, it performs 1.5 dB from the

Shannon limit. The error floor of this code is estimated to lie below the bit-error rate $10^{-16}$ and below the word-error rate $10^{-14}$, as shown in Figure 12.14. The decoding of this code also converges fast, as shown in Figure 12.15. At a BER of $10^{-6}$, the performance gap between 5 and 50 iterations is less than 0.4 dB.

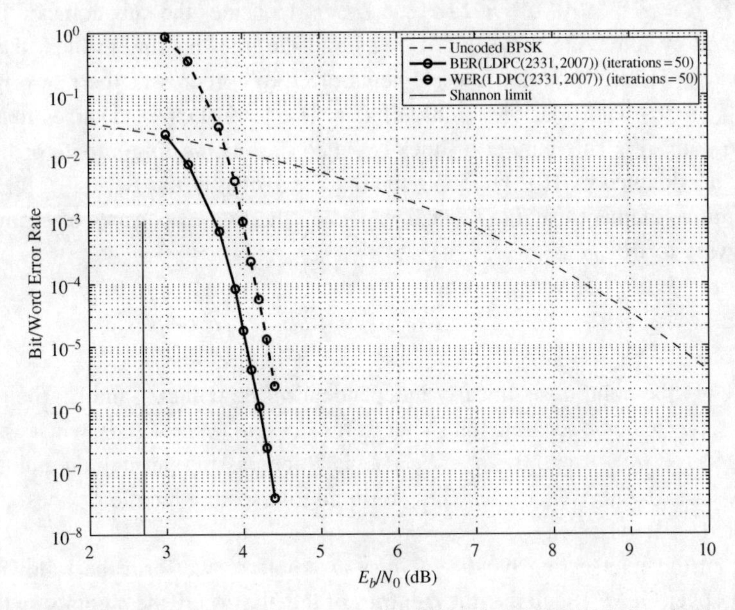


**Figure 12.13** The error performance of the (2331, 2007) QC-LDPC code given in Example 12.13.

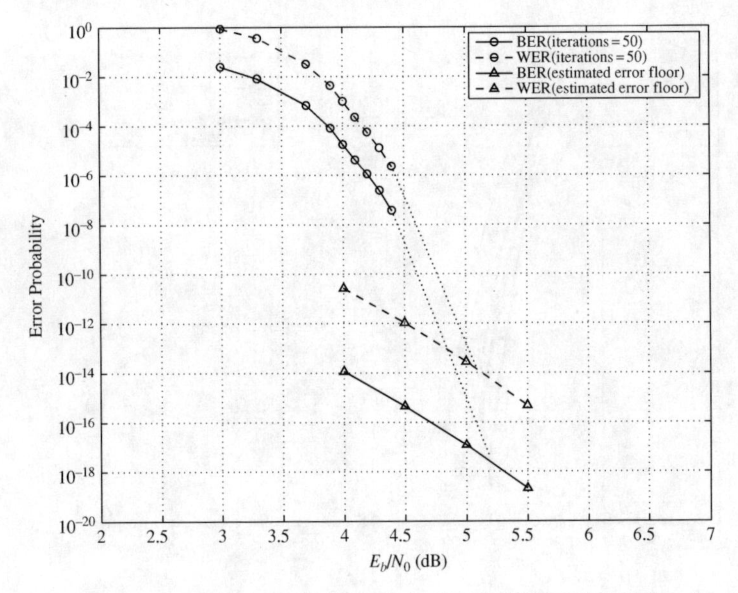


**Figure 12.14** The estimated error floor of the (2331, 2007) QC-LDPC code given in Example 12.13.

## 12.7 Construction QC-LDPC Codes Based on the Intersecting Bundles of Lines of Euclidean Geometries

Consider the $m$-dimensional Euclidean geometry EG$(m, q)$ over GF$(q)$. Let $\alpha$ be a primitive element of GF$(q^m)$. Represent the $q^m$ points of EG$(m, q)$ by the $q^m$ elements of GF$(q^m)$, $\alpha^{-\infty} = 0$, $\alpha^0 = 1$, $\alpha, \ldots, \alpha^{q^m-2}$. Consider the subgeometry EG$^*(m, q)$ obtained by removing the origin $\alpha^{-\infty}$ and all the lines passing through the origin in EG$(m, q)$. Then all the nonzero elements of GF$(q^m)$ represent the non-origin points of EG$^*(m, q)$. Consider the bundle of $K = q(q^{m-1} - 1)/(q - 1)$ lines that intersect at the point $\alpha^0$. This bundle of lines is called the *intersecting bundle* of lines at the point $a^0$, denoted by $\mathbf{I}(\alpha^0)$. Denote the lines in $\mathbf{I}(\alpha^0)$ by $\mathcal{L}_0, \mathcal{L}_1, \ldots, \mathcal{L}_{K-1}$. Let $\beta$ be a primitive element of GF$(q)$. For $0 \leq i < K$, the line $\mathcal{L}_i$ consists of $q$ points of the following form:

$$\mathcal{L}_i = \{\alpha^0, \alpha^0 + \beta^0\alpha^{j_i}, \alpha^0 + \beta\alpha^{j_i}, \ldots, \alpha^0 + \beta^{q-2}\alpha^{j_i}\}, \tag{12.16}$$

where (1) the point $\alpha^{j_i}$ is linearly independent of the point $\alpha^0$ and (2) the points $\alpha^{j_0}$, $\alpha^{j_1}, \ldots, \alpha^{j_{K-1}}$ lie on separate lines, that is, for $k \neq i$, $\alpha^{j_k} \neq \alpha^0 + \beta^l\alpha^{j_i}$ with $0 \leq l < q-1$. For $0 \leq t < q^m - 1$ and $0 \leq i < K$, $\alpha^t\mathcal{L}_i$ is a line passing through the point $\alpha^t$. Then the $K$ lines $\alpha^t\mathcal{L}_0$, $\alpha^t\mathcal{L}_1$, ..., $\alpha^t\mathcal{L}_{K-1}$ form an intersecting bundle of lines at point $\alpha^t$, denoted by $\mathbf{I}(\alpha^t)$.

From the intersecting bundle of lines at point $\alpha^0$, we form the following $K \times q$ matrix over GF$(q^m)$ such that the $q$ entries of the $i$th row are the $q$ points on the $i$th line $\mathcal{L}_i$ of $\mathbf{I}(\alpha^0)$:

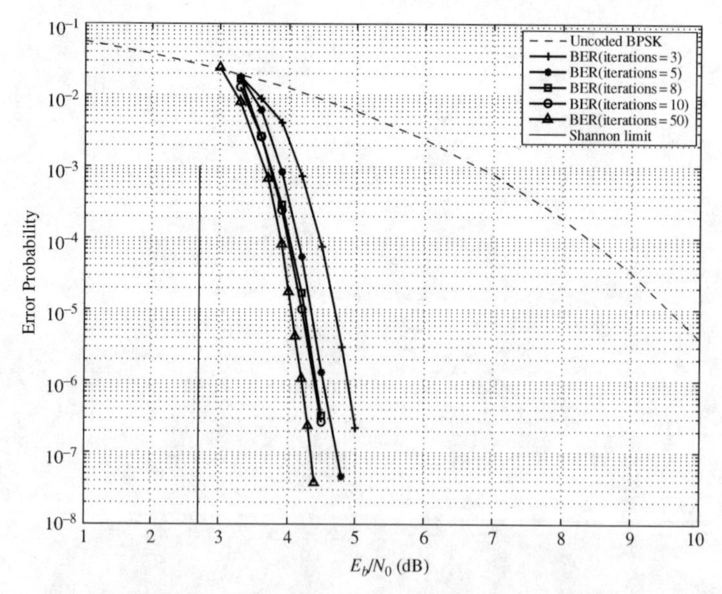


**Figure 12.15** The rate of decoding convergence of the (2331, 2007) QC-LDPC code given in Example 12.13.

$$\mathbf{W}^{(7)}_{qc,\text{disp}} = \begin{bmatrix} \mathcal{L}_0 \\ \mathcal{L}_1 \\ \vdots \\ \mathcal{L}_{K-1} \end{bmatrix}$$
$$= \begin{bmatrix} \alpha^0 & \alpha^0 + \beta^0\alpha^{j_0} & \alpha^0 + \beta\alpha^{j_0} & \cdots & \alpha^0 + \beta^{q-2}\alpha^{j_0} \\ \alpha^0 & \alpha^0 + \beta^0\alpha^{j_1} & \alpha^0 + \beta\alpha^{j_1} & \cdots & \alpha^0 + \beta^{q-2}\alpha^{j_1} \\ \vdots & \vdots & \vdots & \cdots & \vdots \\ \alpha^0 & \alpha^0 + \beta^0\alpha^{j_{K-1}} & \alpha^0 + \beta\alpha^{j_{K-1}} & \cdots & \alpha^0 + \beta^{q-2}\alpha^{j_{K-1}} \end{bmatrix}. \tag{12.17}$$

Label the rows and columns of $\mathbf{W}^{(7)}$ from 0 to $K-1$ and $-\infty, 0, 1, \ldots, q-2$, respectively. $\mathbf{W}^{(7)}$ has the following structural properties: (1) all the entries of each row of $\mathbf{W}^{(7)}$ are nonzero elements of GF$(q^m)$; (2) except for the leftmost column, all the entries in each column are distinct nonzero elements in GF$(q^m)$; (3) any two rows have identical entries at the leftmost position and differ in all the other $q-1$ positions; (4) any two columns differ in all the $K$ positions; and (5) for $0 \le i, j < K$, $i \ne j$, $0 \le k$, and $l < q^m - 1$, $\alpha^k\mathcal{L}_i$ and $\alpha^l\mathcal{L}_j$ are different lines and they differ in at least $q-1$ positions. Properties (1) and (5) imply that $\mathbf{W}^{(7)}$ satisfies $\alpha$-multiplied row constraints 1 and 2. Hence, $\mathbf{W}^{(7)}$ can be used as the base matrix for array dispersion to construct an RC-constrained array of CPMs.

By dispersing each entry of $\mathbf{W}^{(7)}$ into a $(q^m-1)\times(q^m-1)$ CPM over GF(2), we obtain an RC-constrained $K \times q$ array $\mathbf{H}^{(7)}_{qc,\text{EG},1}$ of $(q^m-1)\times(q^m-1)$ CPMs. It is a $K(q^m-1)\times q(q^m-1)$ matrix over GF(2) with column and row weights $K$ and $q$, respectively. For $m > 2$, $K$ is much larger than $q$. In this case, there are more rows than columns in $\mathbf{H}^{(7)}_{qc,\text{EG},1}$. Let $\mathbf{H}^{(7)}_{qc,\text{EG},2}$ be the transpose of $\mathbf{H}^{(7)}_{qc,\text{EG},1}$, that is

$$\mathbf{H}^{(7)}_{qc,\text{EG},2} = \left[\mathbf{H}^{(7)}_{qc,\text{EG},1}\right]^{\text{T}}. \tag{12.18}$$

Then $\mathbf{H}^{(7)}_{qc,\text{EG},2}$ is an RC-constrained $q \times K$ array of $(q^m-1)\times(q^m-1)$ CPMs over GF(2). It is a $q(q^m-1)\times K(q^m-1)$ matrix over GF(2) with column and row weights $q$ and $K$, respectively. For $m > 2$, $\mathbf{H}^{(7)}_{qc,\text{EG},2}$ has more columns than rows. Both arrays $\mathbf{H}^{(7)}_{qc,\text{EG},1}$ and $\mathbf{H}^{(7)}_{qc,\text{EG},2}$ can be used for constructing QC-LDPC codes.

For $1 \le g \le K$ and $1 \le r \le q$, let $\mathbf{H}^{(7)}_{qc,\text{EG},1}(g,r)$ be a $g \times r$ subarray of $\mathbf{H}^{(7)}_{qc,\text{EG},1}$. The null space of $\mathbf{H}^{(7)}_{qc,\text{EG},1}(g,r)$ gives a binary QC-LDPC code $\mathcal{C}^{(7)}_{qc,\text{EG},1}$, whose Tanner graph has a girth of at least 6. For $1 \le g \le q$ and $1 \le r \le K$, let $\mathbf{H}^{(7)}_{qc,\text{EG},2}(g,r)$ be a $g \times r$ subarray of $\mathbf{H}^{(7)}_{qc,\text{EG},2}$. Then the null space of $\mathbf{H}^{(7)}_{qc,\text{EG},2}(g,r)$ gives a QC-LDPC code $\mathcal{C}^{(7)}_{qc,\text{EG},2}$ whose Tanner graph has a girth of at least 6. For $m > 2$, using $\mathbf{H}^{(7)}_{qc,\text{EG},2}$ allows us to construct longer and higher-rate codes. The above construction results in another class of Euclidean-geometry QC-LDPC codes.

---

**Example 12.14** From the three-dimensional Euclidean EG$(3, 2^3)$ over GF$(2^3)$, we can construct an $8 \times 72$ array $\mathbf{H}^{(7)}_{qc,\text{EG},2}$ of $511 \times 511$ CPMs. For the pairs of integers (3, 6), (4, 16), (4, 20), and (4, 32), we take the corresponding subarrays from

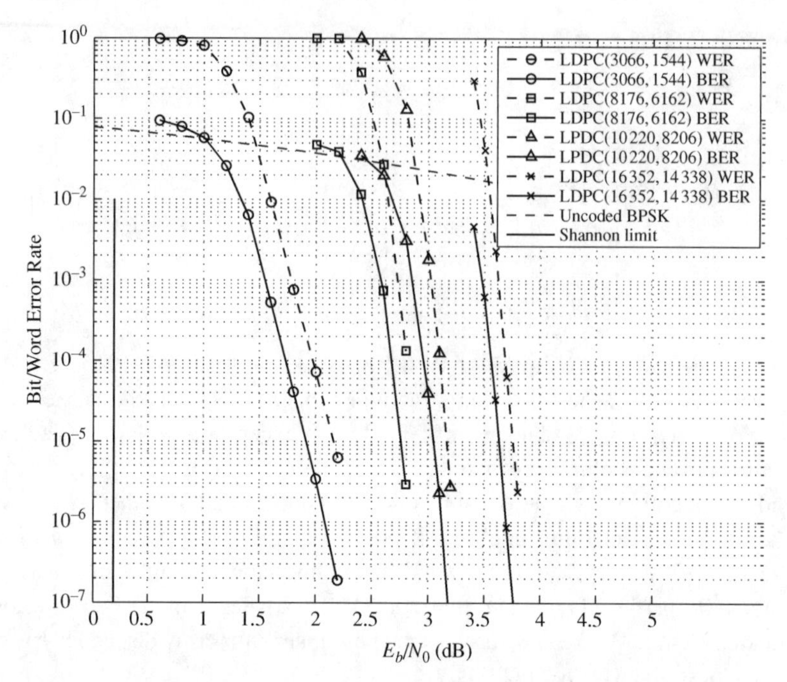


**Figure 12.16** Error performances of the QC-LDPC codes given in Example 12.14.

$\mathbf{H}^{(7)}_{qc,\text{EG},2}$, $\mathbf{H}^{(7)}_{qc,\text{EG},2}(3,6)$, $\mathbf{H}^{(7)}_{qc,\text{EG},2}(4,16)$, $\mathbf{H}^{(7)}_{qc,\text{EG},2}(4,20)$, and $\mathbf{H}^{(7)}_{qc,\text{EG},2}(4,32)$. These subarrays are $1533 \times 3066$, $2044 \times 8176$, $2044 \times 10\,220$, and $2044 \times 16\,352$ matrices with column- and row-weight pairs (3, 6), (4, 16), (4, 20), and (4, 32). The null spaces of these four matrices give (3066, 1544), (8176, 6162), (10 220, 8206), and (16 352, 14 338) EG-QC-LDPC codes with rates 0.5036, 0.7537, 0.8029, and 0.8768, respectively. The performances of these codes over the binary-input AWGN channel decoded with the SPA using 100 iterations are shown in Figure 12.16. Consider the (8176, 6162) code. At a BER of $10^{-6}$, it performs 1.2 dB from the Shannon limit for rate 0.7537. For the (16 352, 14 338) code, it performs 0.8 dB from the Shannon limit for rate 0.8768 at a BER of $10^{-6}$.

---

In forming the base matrix $\mathbf{W}^{(7)}$, we use the bundle of lines intersecting at the point $\alpha^0$. However, we can use the bundle of lines intersecting at any point $\alpha^j$ in EG$(m, q)$. Of course, we can mask subarrays of $\mathbf{H}^{(7)}_{qc,\text{EG},1}$ and $\mathbf{H}^{(7)}_{qc,\text{EG},2}$ to construct both regular and irregular QC-LDPC codes. We can also construct RC-constrained arrays of CPMs and QC-LDPC codes based on bundles of intersecting lines in projective geometries in a similar way.

## 12.8 A Class of Structured RS-Based LDPC Codes

Section 12.5.1 presented a method for constructing QC-LDPC codes based on the additive group of a prime field using additive matrix dispersions of the field elements.

This method can be generalized to construct structured LDPC codes from any field. However, the codes constructed are not quasi-cyclic unless the field is a prime field. In this section, we apply the method presented in Section 12.5.1 to construct a class of LDPC codes based on extended RS codes with two information symbols. The codes constructed perform very well over the binary-input AWGN channel with iterative decoding using the SPA. One such code has been chosen as the IEEE 802.3 standard code for the 10G BASE-T Ethernet.

Consider the Galois field GF($q$), where $q$ is a power of a prime $p$. Let $\alpha$ be a primitive element of GF($q$). Then $\alpha^{-\infty} = 0, \alpha^0 = 1, \alpha, \alpha^2, \ldots, \alpha^{q-2}$ give all the elements of GF($q$) and they form the additive group of GF($q$) under the addition operation of GF($q$). For each element $\alpha^i$ with $i = -\infty, 0, 1, \ldots, q-2$, define a $q$-tuple over GF(2),

$$\mathbf{z}(\alpha^i) = (z_{-\infty}, z_0, z_1, \ldots, z_{q-2}), \tag{12.19}$$

whose components correspond to the $q$ elements $\alpha^{-\infty} = 0, \alpha^0 = 1, \alpha, \alpha^2, \ldots, \alpha^{q-2}$ of GF($q$), where the $i$th component $z_i = 1$ and all the other $q-1$ components are set to zero. This $q$-tuple $\mathbf{z}(\alpha^i)$ over GF(2) is called the location vector of $\alpha^i$ with respect to the additive group of GF($q$). For simplicity, we call $\mathbf{z}(\alpha^i)$ the $A$-location vector of $\alpha^i$ (note that this is just a generalization of the $\mathcal{A}$-location vector of an element of a prime field GF($p$) defined in Section 12.5.1). For $i = -\infty$, the $A$-location vector of $\alpha^{-\infty}$ is $\mathbf{z}(\alpha^{-\infty}) = (1, 0, 0, \ldots, 0)$.

For any element $\delta$ in GF($q$), the sums $\delta + \alpha^{-\infty}, \delta + \alpha^0, \delta + \alpha, \ldots, \delta + \alpha^{q-2}$ are distinct and they give all the $q$ elements of GF($q$). Form the following $q \times q$ matrix over GF(2) with the $A$-location vectors of $\delta + \alpha^{-\infty}, \delta + \alpha^0, \delta + \alpha, \ldots, \delta + \alpha^{q-2}$ as the rows:

$$A = \begin{bmatrix} \mathbf{z}(\delta + \alpha^{-\infty}) \\ \mathbf{z}(\delta + \alpha^0) \\ \mathbf{z}(\delta + \alpha) \\ \vdots \\ \mathbf{z}(\delta + \alpha^{q-2}) \end{bmatrix}. \tag{12.20}$$

Then $A$ is a $q \times q$ permutation matrix (PM). This PM is called the $q$-fold matrix dispersion of $\delta$ with respect to the additive group of GF($q$). The $q$-fold matrix dispersion of the 0 element of GF($q$) is simply a $q \times q$ identity matrix.

Consider the $(q-1, 2, q-2)$ cyclic RS code $\mathcal{C}$ over GF($q$). As shown in Section 12.3, the two $(q-1)$-tuples over GF($q$)

$$\mathbf{b} = (1, \alpha, \alpha^2, \ldots, \alpha^{q-2}) \tag{12.21}$$

and

$$\mathbf{c} = (1, 1, \ldots, 1) \tag{12.22}$$

are two codewords in $\mathcal{C}$ with weight $q-1$. Note that the $q-1$ components of $\mathbf{b}$ are the $q-1$ nonzero elements of GF($q$). For $0 \leq i < q-1$, the $(q-1)$-tuple

$$\alpha^i \mathbf{b} = (\alpha^i, \alpha^{i+1}, \ldots, \alpha^{q-2+i}) \tag{12.23}$$

is also a codeword in $\mathcal{C}$, where the power of each component in $\alpha^i\mathbf{b}$ is reduced by modulo-$(q-1)$. The codeword $\alpha^i\mathbf{b}$ can be obtained by cyclically shifting every component of $\mathbf{b}$ $i$ places to the left. It is clear that, for $i = 0$, $\alpha^0\mathbf{b} = \mathbf{b}$. It is also clear that the $q-1$ components of $\alpha^i\mathbf{b}$ are still distinct and that they give the $q-1$ nonzero elements of GF($q$). If we extend the codeword $\alpha^i\mathbf{b}$ by adding an overall parity-check symbol to its left, we obtain a codeword in the extended $(q, 2, q-1)$ RS code $\mathcal{C}_e$ over GF($q$) (see Section 3.4). Since the sum of the $q-1$ nonzero elements of GF($q$) is equal to the 0 element of GF($q$), the overall parity-check symbol of $\alpha^i\mathbf{b}$ is 0. Hence, for $0 \le i < q-1$,

$$\mathbf{w}_i = (0, \alpha^i\mathbf{b}) = (0, \alpha^i, \alpha^{i+1}, \ldots, \alpha^{q-2+i}) \tag{12.24}$$

is a codeword in the extended $(q, 2, q-1)$ RS code over GF($q$) with weight $q-1$. Let $\mathbf{w}_{-\infty} = (0, 0, \ldots, 0)$ be the all-zero $q$-tuple over GF($q$). This all-zero $q$-tuple is the zero codeword of $\mathcal{C}_e$.

Since $q = 0$ modulo-$q$, the overall parity-check symbol of the codeword $\mathbf{c} = (1, 1, \ldots, 1)$ in $\mathcal{C}$ is 1. On extending $\mathbf{c}$ by adding its overall parity-check symbol, we obtain a $q$-tuple $\mathbf{c}_e = (1, 1, \ldots, 1)$, which consists of $q$ 1-components. This $q$-tuple $\mathbf{c}_e$ is a codeword of weight $q$ in the extended $(q, 2, q-1)$ RS code $\mathcal{C}_e$ over GF($q$). For $k = -\infty, 0, 1, \ldots, q-2$,

$$\alpha^k\mathbf{c}_e = (\alpha^k, \alpha^k, \ldots, \alpha^k) \tag{12.25}$$

is also a codeword in the extended $(q, 2, q-1)$ RS code $\mathcal{C}_e$ over GF($q$). For $k = -\infty$, $\alpha^{-\infty}\mathbf{c}_e = (0, 0, \ldots, 0)$ is the zero codeword in $\mathcal{C}_e$. Note that $\mathbf{w}_{-\infty} = \alpha^{-\infty}\mathbf{c}_e$. For $k \ne -\infty$, $\alpha^k\mathbf{c}_e$ is a codeword in $\mathcal{C}_e$ with weight $q$. For $i, k = -\infty, 0, 1, \ldots, q-2$, the sum

$$\mathbf{w}_i + \alpha^k\mathbf{c}_e \tag{12.26}$$

is a codeword in $\mathcal{C}_e$. From the structures of $\mathbf{w}_i$ and $\alpha^k\mathbf{c}_e$ given by (12.24) and (12.25), we readily see that, for $i, k \ne -\infty$, the weight of the codeword $\mathbf{w}_i + \alpha^k\mathbf{c}_e$ is $q-1$. From the above analysis, we see that the extended $(q, 2, q-1)$ RS code $\mathcal{C}_e$ has $q-1$ codewords with weight $q$, $(q-1)q$ codewords with weight $q-1$, and one codeword with weight zero.

Form a $q \times q$ matrix over GF($q$) with the codewords $\mathbf{w}_{-\infty}, \mathbf{w}_0, \mathbf{w}_1, \ldots, \mathbf{w}_{q-2}$ of $\mathcal{C}_e$ as the rows:

$$\mathbf{W}^{(8)} = \begin{bmatrix} \mathbf{w}_{-\infty} \\ \mathbf{w}_0 \\ \mathbf{w}_1 \\ \vdots \\ \mathbf{w}_{q-2} \end{bmatrix} = \begin{bmatrix} 0 & 0 & 0 & \cdots & 0 \\ 0 & 1 & \alpha & \cdots & \alpha^{(q-2)} \\ 0 & \alpha & \alpha^2 & \cdots & \alpha^{(q-1)} \\ \vdots & \vdots & \vdots & & \vdots \\ 0 & \alpha^{(q-2)} & \alpha^{(q-1)} & \cdots & \alpha^{2(q-2)} \end{bmatrix}, \tag{12.27}$$

where the power of each nonzero entry is reduced by modulo-$(q-1)$. We readily see that $[\mathbf{W}^{(8)}]^{\mathrm{T}} = \mathbf{W}^{(8)}$. The matrix $\mathbf{W}^{(8)}$ has the following structural properties: (1) except for the first row, each row consists of the $q$ elements of GF($q$); (2) except for the first column, every column consists of the $q$ elements of GF($q$); (3) any two rows differ in exactly $q-1$ positions; and (4) any two columns differ in exactly $q-1$ positions.

Given the facts that the minimum distance of the extended $(q, 2, q-1)$ RS code $\mathcal{C}_e$ is $q-1$ and $\mathbf{w}_i + \alpha^k \mathbf{c}_e$ for $i, k = -\infty, 0, 1, \ldots, q-2$, is a codeword in $\mathcal{C}_e$, we can easily see that the following two lemmas hold.

**Lemma 12.4** *For any row $\mathbf{w}_i$ in $\mathbf{W}^{(8)}$, $\mathbf{w}_i + \alpha^k \mathbf{c}_e \neq \mathbf{w}_i + \alpha^l \mathbf{c}_e$ for $k \neq l$.*

**Lemma 12.5** *For two different rows $\mathbf{w}_i$ and $\mathbf{w}_j$ of $\mathbf{W}^{(8)}$, $\mathbf{w}_i + \alpha^k \mathbf{c}_e \neq \mathbf{w}_j + \alpha^l \mathbf{c}_e$ differ in at least $q-1$ positions.*

Lemmas 12.4 and 12.5 are simply generalizations of Lemmas 11.2 and 11.3 from a prime field GF($p$) to a field GF($q$), where $q$ is a power of $p$. The conditions on the rows of $\mathbf{W}^{(8)}$ given by Lemmas 12.4 and 12.5 are referred to as additive row constraints 1 and 2.

On replacing each entry of $\mathbf{W}^{(8)}$ by its additive $q$-fold matrix dispersion as given by (12.20), we obtain the following $q \times q$ array of PMs over GF(2):

$$\mathbf{H}^{(8)}_{rs,\text{disp}} = \begin{bmatrix} \mathbf{A}_{-\infty,-\infty} & \mathbf{A}_{-\infty,0} & \mathbf{A}_{-\infty,1} & \cdots & \mathbf{A}_{-\infty,q-2} \\ \mathbf{A}_{0,-\infty} & \mathbf{A}_{0,0} & \mathbf{A}_{0,1} & \cdots & \mathbf{A}_{0,q-2} \\ \mathbf{A}_{1,-\infty} & \mathbf{A}_{1,0} & \mathbf{A}_{1,1} & \cdots & \mathbf{A}_{0,q-2} \\ \vdots & \vdots & \vdots & & \vdots \\ \mathbf{A}_{q-2,-\infty} & \mathbf{A}_{q-2,0} & \mathbf{A}_{q-2,1} & \cdots & \mathbf{A}_{q-2,q-2} \end{bmatrix}, \tag{12.28}$$

where the PMs in the top row and leftmost column are $q \times q$ identity matrices. For $0 \leq i < q-1$, since the $q$ entries in the $i$th row $\mathbf{w}_i$ of $\mathbf{W}^{(8)}$ are $q$ different elements of GF($q$), the $q$ PMs in the $i$th row of the array $\mathbf{H}^{(8)}_{rs,\text{disp}}$ are distinct. Also, for $j \neq -\infty$, the $q$ PMs in the $j$th column of $\mathbf{H}^{(8)}_{rs,\text{disp}}$ are distinct. Since any two rows (or two columns) of $\mathbf{W}^{(8)}$ differ in $q-1$ places, any two row-blocks (or two column-blocks) of $\mathbf{H}^{(8)}_{rs,\text{disp}}$ have one and only one position where they have identical PMs. $\mathbf{H}^{(8)}_{rs,\text{disp}}$ is called the additive $q$-fold array dispersion of $\mathbf{W}^{(8)}$.

Array $\mathbf{H}^{(8)}_{rs,\text{disp}}$ is a $q^2 \times q^2$ matrix over GF(2) with both column and row weight $q$. It follows from additive row constraints 1 and 2 on the base matrix $\mathbf{W}^{(8)}$ that $\mathbf{H}^{(8)}_{rs,\text{disp}}$, being a $q^2 \times q^2$ matrix over GF(2), satisfies the RC constraint. The $i$th row $(\mathbf{A}_{i,-\infty}\ \mathbf{A}_{i,0}\ \mathbf{A}_{i,1}\ \ldots\ \mathbf{A}_{i,q-2})$ of $\mathbf{H}^{(8)}_{rs,\text{disp}}$ is a $q \times q^2$ matrix GF(2), which is obtained by replacing each entry of the following $q \times q$ matrix over GF($q$) by its $A$-location vector:

$$\mathbf{R}_i = \begin{bmatrix} \mathbf{w}_i + \alpha^{-\infty}\mathbf{c}_e \\ \mathbf{w}_i + \alpha^{0}\mathbf{c}_e \\ \mathbf{w}_i + \alpha^{1}\mathbf{c}_e \\ \vdots \\ \mathbf{w}_i + \alpha^{q-2}\mathbf{c}_e \end{bmatrix}. \tag{12.29}$$

Since $\alpha^{-\infty}\mathbf{c}_e, \alpha^0\mathbf{c}_e, \alpha^1\mathbf{c}_e, \ldots, \alpha^{q-2}\mathbf{c}_e$ form a $(q, 1, q)$ subcode $\mathcal{C}_{e,\text{sub}}$ of the extended $(q, 2, q-1)$ RS code $\mathcal{C}_e$, the rows of $\mathbf{R}_i$ are simply a coset of $\mathcal{C}_{e,\text{sub}}$ in $\mathcal{C}_e$. $\mathbf{H}^{(8)}_{rs,\text{disp}}$ is the same array of PMs as was constructed in [6].

For any pair $(g, r)$ of integers with $1 \le g, r \le q$, take a $g \times r$ subarray $\mathbf{H}^{(8)}_{rs,\text{disp}}(g, r)$ from $\mathbf{H}^{(8)}_{rs,\text{disp}}$. $\mathbf{H}^{(8)}_{rs,\text{disp}}(g, r)$ is a $gq \times rq$ matrix over GF(2) with column and row weights $g$ and $r$, respectively. Since $\mathbf{H}^{(8)}_{rs,\text{disp}}$ satisfies the RC constraint, so does $\mathbf{H}^{(8)}_{rs,\text{disp}}(g, r)$. The null space of $\mathbf{H}^{(8)}_{rs,\text{disp}}(g, r)$ gives a $(g, r)$-regular LDPC code $\mathcal{C}^{(8)}_{rs,\text{disp}}$ whose Tanner graph has a girth of at least 6. The minimum distance of this code is at least $g + 2$ for even $g$ and $g + 1$ for odd $g$. For a given finite field, the above construction gives a family of structured LDPC codes. Of course, subarrays of $\mathbf{H}^{(8)}_{rs,\text{disp}}$ can be masked to construct many more LDPC codes, regular or irregular. If $q$ is a power of 2, LDPC codes with lengths that are powers of 2 or multiples of an 8-bit byte can be constructed. In many practical applications in communication and storage systems, codes with lengths that are powers of 2 or multiples of an 8-bit byte are preferred due to the data-frame requirements.

---

**Example 12.15** Let GF($2^6$) be the code-construction field. From the extended (64, 2, 63) RS code over GF($2^6$), we can construct a $64 \times 64$ array $\mathbf{H}^{(8)}_{rs,\text{disp}}$ of $64 \times 64$ PMs. Choose $g = 6$ and $r = 32$. Take the $6 \times 32$ subarray $\mathbf{H}^{(8)}_{rs,\text{disp}}(6, 32)$ at the upper-left corner of $\mathbf{H}^{(8)}_{rs,\text{disp}}$. This subarray is a $384 \times 2048$ matrix over GF(2) with column and row weights 6 and 32, respectively. The null space of this matrix gives a (6, 32)-regular (2048, 1723) LDPC code with minimum distance at least 8. This code has been chosen as the IEEE 802.3 standard code for the 10G BASE-T Ethernet. The bit-error performances of this code over the binary-input AWGN channel decoded using the SPA with 1, 5, 10, 50, and 100 iterations are shown in Figure 12.17. We see that decoding of this code converges very fast. The performance curves of this code with 50 and 100 iterations basically overlap each other. The error floor of this code is below the BER of $10^{-12}$.

---

## 12.9 Construction of QC-LDPC Codes Based on Two Arbitrary Subsets of a Finite Field

In Sections 12.3 to 12.7, several methods were presented for constructing QC-LDPC codes from finite fields. The parity-check matrices of these QC-LDPC codes are RC-constrained arrays of CPMs, which are constructed based on CPM dispersions of base matrices that satisfy $\alpha$-multiplied row constraints.

In this section, another approach to construct base matrices whose CPM dispersions produce RC-constrained arrays of CPMs is presented. This approach was proposed in [14, 15] and is rephrased in the following theorem without proof.

**Theorem 12.6** *Let* $\mathbf{W}$ *be an* $m \times n$ *matrix over GF(q) where q is a power of a prime. Let* $\mathbf{H}$ *be the* $m \times n$ *array of CPMs/or zero matrices of size* $(q-1) \times (q-1)$ *obtained by* $(q-1)$*-fold CPM array dispersion of* $\mathbf{W}$*. The matrix* $\mathbf{H}$ *satisfies the RC constraint if and only if every* $2 \times 2$ *submatrix of* $\mathbf{W}$ *contains at least one zero entry or is nonsingular.*

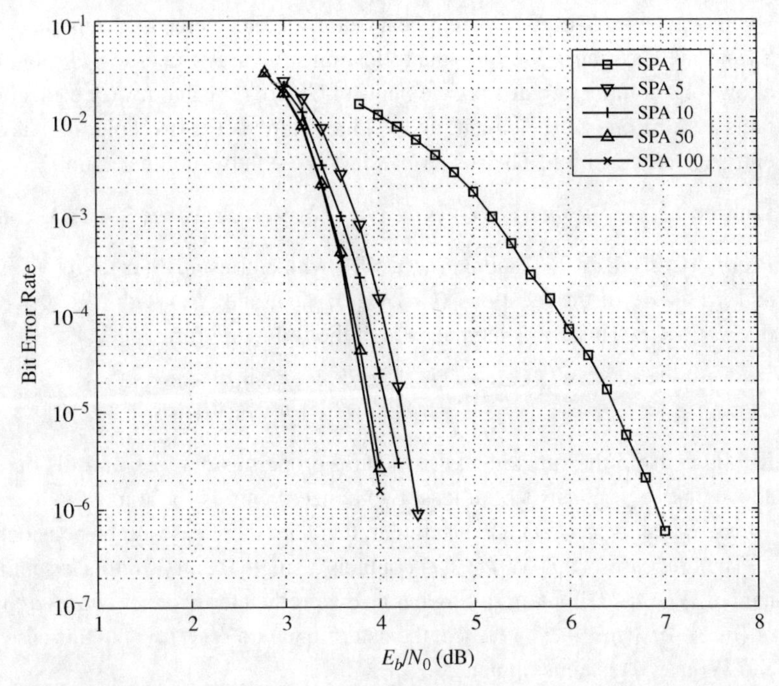


**Figure 12.17** The rate of decoding convergence of the (2048, 1723) LDPC code given in Example 12.15.

For convenience, the necessary and sufficient condition on **W** given in Theorem 12.6 is referred to as the $2 \times 2$ *submatrix (SM) constraint*, or $2 \times 2$ SM constraint for short. It can be proved that the base matrices presented in Sections 12.3 to 12.7 satisfy the $2 \times 2$ SM constraint (Problem 12.11).

In the following, a class of $2 \times 2$ SM-constrained base matrices over GF($q$) is constructed based on two arbitrary subsets with GF($q$) [16–18]. Using matrices in this class in conjunction with CPM dispersion, QC-LDPC codes of various lengths and rates can be constructed.

Let $\alpha$ be a primitive element of the field GF($q$). For $1 \le m, n \le q$, let

$$\mathbf{S}_0 = \{\alpha^{i_0}, \alpha^{i_1}, \ldots, \alpha^{i_{m-1}}\},$$
$$\mathbf{S}_1 = \{\alpha^{j_0}, \alpha^{j_1}, \ldots, \alpha^{j_{n-1}}\}$$

be two arbitrary subsets of elements in GF($q$) with $i_k$ and $j_l$ in $L = \{-\infty, 0, 1, 2, \ldots, q-2\}$, $i_0 < i_1 < \cdots < i_{m-1}$ and $j_0 < j_1 < \cdots < j_{n-1}$. Let $\eta$ be any nonzero element in GF($q$). We then form the following $m \times n$ matrix over GF($q$):

$$\mathbf{W}_s^{(9)} = \begin{bmatrix} \eta\alpha^{i_0} + \alpha^{j_0} & \eta\alpha^{i_0} + \alpha^{j_1} & \cdots & \eta\alpha^{i_0} + \alpha^{j_{n-1}} \\ \eta\alpha^{i_1} + \alpha^{j_0} & \eta\alpha^{i_1} + \alpha^{j_1} & \cdots & \eta\alpha^{i_1} + \alpha^{j_{n-1}} \\ \vdots & \vdots & \cdots & \vdots \\ \eta\alpha^{i_{m-1}} + \alpha^{j_0} & \eta\alpha^{i_{m-1}} + \alpha^{j_1} & \cdots & \eta\alpha^{i_{m-1}} + \alpha^{j_{n-1}} \end{bmatrix}, \quad (12.30)$$

where the entries in $\mathbf{W}_s^{(9)}$ are linear sums of the elements from the two subsets $\mathbf{S}_0$ and $\mathbf{S}_1$. The matrix $\mathbf{W}_s^{(9)}$ has the following structural properties: (1) all the entries in a row (or a column) are distinct elements of GF($q$); (2) each row (or each column) contains at most one zero element; and (3) no two rows (or two columns) have identical entries in any position. The subscript $s$ in $\mathbf{W}_s^{(9)}$ stands for "linear sum."

**Lemma 12.7** *The matrix* $\mathbf{W}_s^{(9)}$ *given by (12.30) satisfies the* $2 \times 2$ *SM constraint.*

*Proof* Consider a 2×2 submatrix in $\mathbf{W}_s^{(9)}$ with entries $\eta\alpha^{i_k}+\alpha^{j_s}$, $\eta\alpha^{i_k}+\alpha^{j_t}$, $\eta\alpha^{i_l}+\alpha^{j_s}$, and $\eta\alpha^{i_l}+\alpha^{j_t}$ at the locations $(k,s)$, $(k,t)$, $(l,s)$, and $(l,t)$ with $0 \leq k < l < m$ and $0 \leq s < t < n$. This $2 \times 2$ submatrix is of the following form:

$$\mathbf{W}(k,l;s,t) = \begin{bmatrix} \eta\alpha^{i_k} + \alpha^{j_s} & \eta\alpha^{i_k} + \alpha^{j_t} \\ \eta\alpha^{i_l} + \alpha^{j_s} & \eta\alpha^{i_l} + \alpha^{j_t} \end{bmatrix}.$$

It follows from the fundamental structural properties (1), (2), and (3) of $\mathbf{W}_s^{(9)}$ given above that $\mathbf{W}_s^{(9)}$ consists of at least two nonzero entries (or at most two zero entries). To prove that $\mathbf{W}(k,l;s,t)$ is nonsingular, there are three cases to be considered.

The first case is that $\mathbf{W}(k,l;s,t)$ contains a single 0 entry. In this case, the determinant of $\mathbf{W}(k,l;s,t)$ contains a single nonzero product term, either $(\eta\alpha^{i_k}+\alpha^{j_s})(\eta\alpha^{i_l}+\alpha^{j_t})$ or $(\eta\alpha^{i_k}+\alpha^{j_t})(\eta\alpha^{i_l}+\alpha^{j_s})$. Hence, the determinant of $\mathbf{W}(k,l;s,t)$ in this case is nonzero and $\mathbf{W}(k,l;s,t)$ is nonsingular.

The second case is that $\mathbf{W}(k,l;s,t)$ contains two 0 entries. It follows from the structural properties (2) and (3) of $\mathbf{W}_s^{(9)}$ that these two 0 entries must lie either on the main diagonal or the anti-diagonal of $\mathbf{W}(k,l;s,t)$. In this case, the determinant of $\mathbf{W}(k,l;s,t)$ also contains a single nonzero product term, either $(\eta\alpha^{i_k}+\alpha^{j_s})(\eta\alpha^{i_l}+\alpha^{j_t})$ or $(\eta\alpha^{i_k}+\alpha^{j_t})(\eta\alpha^{i_l}+\alpha^{j_s})$, and hence $\mathbf{W}(k,l;s,t)$ is nonsingular.

The third case is that all the four entries in $\mathbf{W}(k,l;s,t)$ are nonzero. In this case, the determinant of $\mathbf{W}(k,l;s,t)$ is $(\eta\alpha^{i_k}+\alpha^{j_s})(\eta\alpha^{i_l}+\alpha^{j_t}) - (\eta\alpha^{i_k}+\alpha^{j_t})(\eta\alpha^{i_l}+\alpha^{j_s})$. Suppose $\mathbf{W}(k,l;s,t)$ is singular. Then the following equality must hold: $(\eta\alpha^{i_k}+\alpha^{j_s})(\eta\alpha^{i_l}+\alpha^{j_t}) - (\eta\alpha^{i_k}+\alpha^{j_t})(\eta\alpha^{i_l}+\alpha^{j_s}) = 0$. With some algebraic manipulations of the above equality, we have $\alpha^{i_k} = \alpha^{i_l}$ and $\alpha^{j_s} = \alpha^{j_t}$, which contradicts the fact that all the elements in $\mathbf{S}_0$ and $\mathbf{S}_1$ are distinct. Hence, $\mathbf{W}(k,l;s,t)$ cannot be singular and must be nonsingular.

With the above results, we conclude that $\mathbf{W}(k,l;s,t)$ is nonsingular. Hence $\mathbf{W}_s^{(9)}$ satisfies the $2 \times 2$ SM constraint. □

It follows from Theorem 12.6 and Lemma 12.7 that the $(q-1)$-fold CPM dispersion of $\mathbf{W}_s^{(9)}$ gives an $m \times n$ RC-constrained array $\mathbf{H}_{s,qc,\text{disp}}^{(9)}$ of CPMs and/or zero matrices of size $(q-1) \times (q-1)$.

The null space over GF(2) of $\mathbf{H}_{s,qc,\text{disp}}^{(9)}$ gives a QC-LDPC code, denoted by $\mathcal{C}_{s,qc}(m,n)$, of length $n(q-1)$ with rate at least $(n-m)/n$. The Tanner graph of $\mathcal{C}_{s,qc}(m,n)$ has a girth of at least 6. With different choices of $m$, $n$, $\mathbf{S}_0$, $\mathbf{S}_2$, and the multiplier $\eta$ in GF($q$), we can construct a family of 2×2 SM-constrained base matrices over GF($q$) in the sum form given by (12.30) with various sizes. Using these base matrices and their $(q-1)$-fold CPM dispersions as LDPC matrices, we can construct a family of QC-LDPC codes with various lengths and rates whose Tanner graphs have girth at least 6.

There are two special cases of the base matrix $\mathbf{W}_s^{(9)}$ given in (12.30). The first case is that $\mathbf{S}_0 = \mathbf{S}_1 = \text{GF}(q) = \{0, 1, \alpha, \alpha^2, \ldots, \alpha^{q-2}\}$. In this case, $\mathbf{W}_s^{(9)}$ is a $q \times q$ matrix over GF($q$) in which each row and each column contains every element of GF($q$) exactly once. The matrix $\mathbf{W}_s^{(9)}$ is a Latin square over GF($q$) [19]. The construction of QC-LDPC codes based on Latin squares was proposed in [20].

The second case is that $\mathbf{S}_0$ and $\mathbf{S}_1$ are two additive subgroups of the additive group of GF($q$) with orders $m$ and $n$, respectively, such that $m + n \leq q$ and $\mathbf{S}_0 \cap \mathbf{S}_1 = \{0\}$. Let $\alpha^{i_0} = \alpha^{j_0} = 0$. In this case, the first row of matrix $\mathbf{W}_s^{(9)}$ contains the elements of the subgroup $\mathbf{S}_1$ and all the other rows of $\mathbf{B}_s$ are cosets of $\mathbf{S}_1$ with the elements in the set $\eta\mathbf{S}_0 = \{\eta\alpha^{i_0}, \eta\alpha^{i_1}, \ldots, \eta\alpha^{i_{m-1}}\}$ as coset leaders. The construction of QC-LDPC codes based on two additive subgroups of the additive group of GF($q$) with orders $m$ and $n = q - m$, respectively, and $\eta = 1$, was presented in Section 12.5.1. Hence, $\mathbf{W}_s^{(9)}$ is identical to $\mathbf{W}^{(4)}$ given by (12.10).

The construction of a $2 \times 2$ SM-constrained base matrix based on the two subsets $\mathbf{S}_0$ and $\mathbf{S}_1$ of GF($q$) can also be put in *product form* as follows [16–18]:

$$\mathbf{W}_p^{(9)} = \begin{bmatrix} \alpha^{i_0}\alpha^{j_0} - \eta & \alpha^{i_0}\alpha^{j_1} - \eta & \cdots & \alpha^{i_0}\alpha^{j_{n-1}} - \eta \\ \alpha^{i_1}\alpha^{j_0} - \eta & \alpha^{i_1}\alpha^{j_1} - \eta & \cdots & \alpha^{i_1}\alpha^{j_{n-1}} - \eta \\ \vdots & \vdots & \cdots & \vdots \\ \alpha^{i_{m-1}}\alpha^{j_0} - \eta & \alpha^{i_{m-1}}\alpha^{j_1} - \eta & \cdots & \alpha^{i_{m-1}}\alpha^{j_{n-1}} - \eta \end{bmatrix}. \tag{12.31}$$

Then the null space of the CPM dispersion of $\mathbf{W}_p^{(9)}$, denoted by $\mathbf{H}_{p,\text{disp}}^{(9)}$, gives a QC-LDPC code $\mathcal{C}_{p,qc}(m, n)$. The subscript $p$ of $\mathbf{W}_p^{(9)}$, $\mathbf{H}_{p,\text{disp}}^{(9)}$, and $\mathcal{C}_{p,qc}(m, n)$ stands for the product form of the base matrix $\mathbf{W}_p^{(9)}$ given by (12.31).

There are four special cases of the base matrix $\mathbf{W}_p^{(9)}$ in the product form given in (12.31). The first case is that $\mathbf{S}_0$ and $\mathbf{S}_1$ are two cyclic subgroups of the multiplicative group of GF($q$) with order $m$ and $n$, respectively, such that $\mathbf{S}_0 \cap \mathbf{S}_1 = \{1\}$, and $m$ and $n$ are relatively prime factors of $q - 1$. Let $\alpha^{i_0} = \alpha^{j_0} = 1$ and $\eta = 1$. Then $\mathbf{W}_p^{(9)}$ constructed based on $\mathbf{S}_0$ and $\mathbf{S}_1$ is identical to the base matrix $\mathbf{W}^{(5)}$ given by (12.12) constructed based on two cyclic subgroups of the multiplicative groups of GF($q$) presented in Section 12.5.2. The second case is that $\mathbf{S}_0 = \mathbf{S}_1 = \text{GF}(q)\backslash\{0\} = \{\alpha^0, \alpha, \alpha^2, \ldots, \alpha^{(q-3)}, \alpha^{(q-2)}\}$ and $\eta = 1$. In this case, the base matrix $\mathbf{W}_p^{(9)}$ is identical to the base matrix $\mathbf{W}^{(1)}$ given by (12.5) in Section 12.3.

The third case is that

$$\mathbf{S}_0 = \{\alpha^{-j_1}, \alpha^{-j_2}, \ldots, \alpha^{-j_k}\},$$
$$\mathbf{S}_1 = \{\alpha^{j_1}, \alpha^{j_2}, \ldots, \alpha^{j_k}\},$$

and $\eta = 1$, where $\{\alpha^{j_1}, \alpha^{j_2}, \ldots, \alpha^{j_k}\}$ are the primitive elements of GF($q$). Euler's totient function gives the number of primitive elements of GF($q$):

$$k = (q-1)\prod_{i=1}^{t}(1 - 1/p_i),$$

in which $p_1, p_2, \ldots, p_t$ are the prime factors of $q - 1$. Using these two sets of elements of GF($q$), the base matrix $\mathbf{W}_p^{(9)}$ is identical to the base matrix $\mathbf{W}^{(6)}$ given by (12.14) in Section 12.6.

For the fourth case, let GF($q^2$) be an extension field of GF($q$). Let $\alpha$ and $\beta$ be two primitive elements of GF($q$) and GF($q^2$), respectively. Then we have GF($q$) $= \{0, 1, \alpha, \alpha^2, \ldots, \alpha^{q-2}\}$ and GF($q^2$) $= \{0, 1, \beta, \beta^2, \ldots, \beta^{q^2-2}\}$. There exist $q$ distinct elements $\beta^{j_0}, \beta^{j_1}, \ldots, \beta^{j_{q-1}}$ in GF($q^2$) with $j_0, j_1, \ldots, j_{q-1} \in \{0, 1, 2, \ldots, q^2-2\}$, which are independent of $\beta^0$, and, for $i \neq k$, $\beta^{j_k} \neq 1 + \alpha^l \beta^{j_i}$ with $0 \leq l < q-1$. Let $\mathbf{S}_0 = \{\beta^{j_0}, \beta^{j_1}, \ldots, \beta^{j_{q-1}}\}$, $\mathbf{S}_1 = \mathrm{GF}(q) = \{0, 1, \alpha, \alpha^2, \ldots, \alpha^{(q-2)}\}$, and $\eta = -\beta^0$. In this case, the matrix $\mathbf{W}_p^{(9)}$ constructed based on these two sets $\mathbf{S}_0$ and $\mathbf{S}_1$ taken from GF($q^2$) becomes the base matrix $\mathbf{W}^{(7)}$ given by (12.17), constructed based on the two-dimensional Euclidean geometry EG($2, q$) presented in Section 12.7.

---

**Example 12.16** Let $\alpha$ be a primitive element in the prime field GF(101). Suppose we form a $4 \times 100$ base matrix $\mathbf{W}_s^{(9)}$ over GF(101) in the form of (12.30) using two subsets $\mathbf{S}_0 = \{0, 1, \alpha, \alpha^2\}$ and $\mathbf{S}_1 = \{1, \alpha, \alpha^2, \ldots, \alpha^{99}\}$ of GF(101). The CPM dispersion of $\mathbf{W}_s^{(9)}$ gives a $4 \times 100$ array $\mathbf{H}_{s,qc,\mathrm{disp}}^{(9)}$ of CPMs of size $100 \times 100$. The array $\mathbf{H}_{s,qc,\mathrm{disp}}^{(9)}$ is a $400 \times 10\,000$ matrix over GF(2) with average column and row weights 3.97 and 99.25, respectively, which satisfies the RC constraint. The null space of $\mathbf{H}_{s,qc,\mathrm{disp}}^{(9)}$ over GF(2) gives a binary nearly (4, 100)-regular (10 000, 9600) QC-LDPC code $\mathcal{C}_{s,qc}(4, 100)$ with rate 0.96 (a very high-rate code).

The BER performance of this code over the AWGN channel decoded with 5, 10, and 50 iterations of the MSA scaled by a factor of 0.75 is shown in Figure 12.18 [21]. We see that the code performs well. At a BER of $10^{-8}$, it performs 0.9 dB from the Shannon limit (which is 4.48 dB) with 50 iterations of the scaled MSA.

If we delete various numbers of columns from $\mathbf{W}_s^{(9)}$, we can construct QC-LDPC codes of various lengths (from long to short) and rates (high to low). Suppose we take

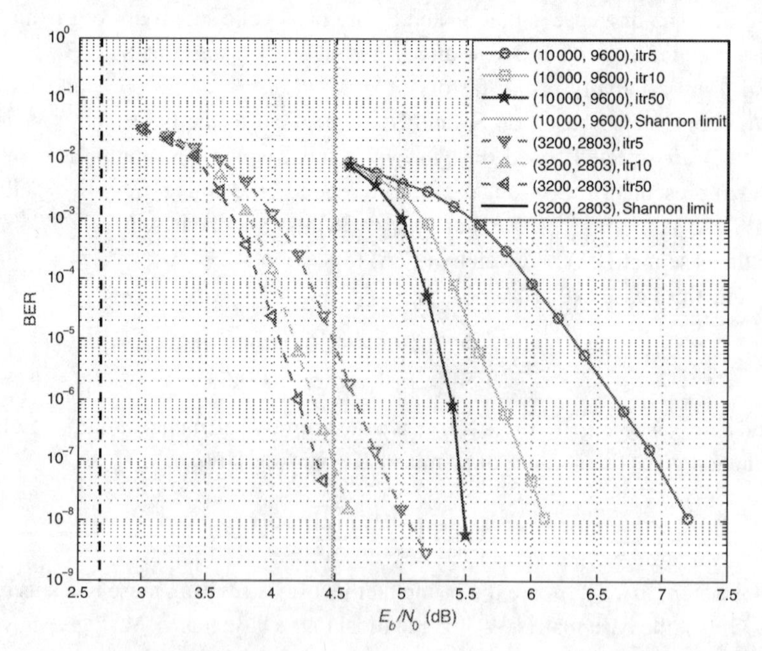


**Figure 12.18** The BER performances of the (10 000, 9600) and the (3200, 2803) QC-LDPC codes given in Example 12.16.

the first 32 columns of $\mathbf{W}_s^{(9)}$ to form a $4 \times 32$ base matrix $\mathbf{W}_s^{(9)}(4, 32)$. The null space over GF(2) of the CPM-dispersion $\mathbf{H}_{s,qc,\text{disp}}^{(9)}(4, 32)$ of $\mathbf{W}_s^{(9)}(4, 32)$ gives a $(4, 32)$-regular $(3200, 2803)$ QC-LDPC code $\mathcal{C}_{s,qc}(4, 32)$ with rate 0.8759. The BER performance of $\mathcal{C}_{s,qc}(4, 32)$ decoded with 5, 10, and 50 iterations of the MSA scaled by a factor 0.75 is also shown in Figure 12.18. At a BER of $10^{-8}$, the code performs 1.3 dB from the Shannon limit (2.65 dB).

---

**Example 12.17** Suppose we take the first four columns from $\mathbf{W}_s^{(9)}$ constructed in Example 12.16 to form a $4 \times 8$ submatrix $\mathbf{W}_s^{(9)}(4, 8)$. The 100-fold CPM dispersion of $\mathbf{W}_s^{(9)}(4, 8)$ gives a $4 \times 8$ array $\mathbf{W}_{s,qc,\text{disp}}^{(9)}(4, 8)$. Next, we mask $\mathbf{W}_s^{(9)}(4, 8)$ with the $\mathbf{Z}(4, 8)$ masking matrix

$$\mathbf{Z}(4, 8) = \begin{bmatrix} 1 & 0 & 1 & 0 & 1 & 1 & 1 & 1 \\ 0 & 1 & 0 & 1 & 1 & 1 & 1 & 1 \\ 1 & 1 & 1 & 1 & 1 & 0 & 1 & 0 \\ 1 & 1 & 1 & 1 & 0 & 1 & 0 & 1 \end{bmatrix}.$$

Masking results in a $4 \times 8$ masked matrix $\mathbf{W}_{s,\text{mask}}^{(9)}(4, 8)$. The 100-fold CPM dispersion of $\mathbf{W}_{s,\text{mask}}^{(9)}(4, 8)$ gives a $4 \times 8$ array $\mathbf{H}_{s,qc,\text{disp},\text{mask}}^{(9)}(4, 8)$ of CPMs and zero matrices of size $100 \times 100$, which is a $400 \times 800$ matrix with column and row weights 3 and 6, respectively.

The null space over GF(2) of $\mathbf{H}_{s,qc,\text{disp},\text{mask}}^{(9)}(4, 8)$ generates a $(3, 6)$-regular $(800, 400)$ QC-LDPC code $\mathcal{C}_{s,qc,\text{mask}}(4, 8)$ of length 800 with rate 0.5. The bit and word-error performances of the code decoded with 5, 10, and 50 iterations are shown in Figure 12.19.

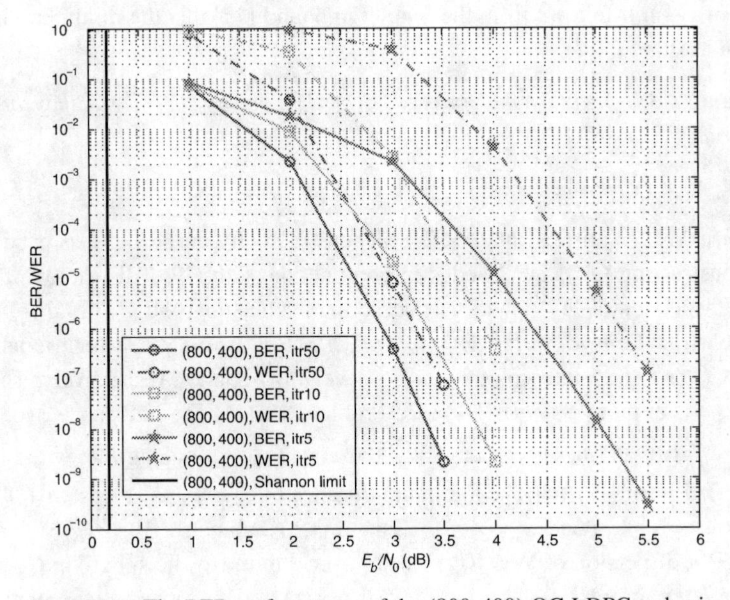


**Figure 12.19** The BER performances of the (800, 400) QC-LDPC code given in Example 12.17.

## 12.10 QC-LDPC Codes Constructed Based on Cyclic Codes of Prime Lengths

This section presents a method for constructing QC-LDPC codes based on the roots of the generator polynomial of a cyclic code of prime length. This construction produces another class of QC-LDPC codes.

Let $C$ be an $(n, n-m)$ cyclic code of prime length $n$ and dimension $n-m$ over the finite field GF($q$) of characteristic 2, say $q = 2^\kappa$ for some positive integer $\kappa$. The cyclic code $C$ is uniquely specified by a generator polynomial $g(X) = \prod_{i=0}^{m-1}(X - \beta^{l_i})$ over GF($q$) with $m$ distinct roots $\{\beta^{l_0}, \beta^{l_1}, \beta^{l_2}, \ldots, \beta^{l_{m-1}}\}$ where $0 \leq l_0 < l_1 < l_2, \ldots, < l_{m-1} < n$ and $\beta$ is an element in a field GF($q^\tau$) (an extension field of GF($q$)) of order $n$ that divides $q^\tau - 1$ for some positive integer $\tau$. A parity-check matrix of the cyclic code $C$ in terms of its roots has the following form (see Chapter 3):

$$\mathbf{W}_c^{(10)} = \begin{bmatrix} 1 & \beta^{l_0} & \beta^{2l_0} & \cdots & \beta^{(n-1)l_0} \\ 1 & \beta^{l_1} & \beta^{2l_1} & \cdots & \beta^{(n-1)l_1} \\ \vdots & \vdots & \vdots & \cdots & \vdots \\ 1 & \beta^{l_{m-1}} & \beta^{2l_{m-1}} & \cdots & \beta^{(n-1)l_{m-1}} \end{bmatrix}. \quad (12.32)$$

Note that the entries of $\mathbf{W}_c^{(10)}$ are elements in the cyclic subgroup $G_n = \{\beta^0 = 1, \beta, \beta^2, \ldots, \beta^{(n-1)}\}$ of GF($q^\tau$) generated with $\beta$ as the generator. The subscript $c$ stands for "cyclic code."

A vector $\mathbf{v} = (v_0, v_1, \ldots, v_{n-1})$ of length $n$ over GF($q$) is a codeword in $C$ if and only if it is orthogonal to every row in $\mathbf{W}_c^{(10)}$, that is, $\mathbf{v} \cdot [\mathbf{W}_c^{(10)}]^\mathrm{T} = \mathbf{0}$. In polynomial form, the vector $\mathbf{v}$ is represented by a polynomial $v(X) = v_0 + v_1X + \cdots + v_{n-1}X^{n-1}$ over GF($q$) with degree $n-1$ or less. Then $v(X)$ is a code polynomial if and only if $v(X)$ is divisible by the generator polynomial $g(X)$ of $C$, that is, if only if $\beta^{l_0}, \beta^{l_1}, \beta^{l_2}, \ldots, \beta^{l_{m-1}}$ are roots of $v(X)$. It follows from the Singleton bound [22] that the minimum distance of $C$ is at most $m + 1$.

**Lemma 12.8** *The $m \times n$ matrix $\mathbf{W}_c^{(10)}$ over $G_n$ given by (12.32) satisfies the $2 \times 2$ SM constraint.*

*Proof* See Problem 12.12. □

Since $\mathbf{W}_c^{(10)}$ satisfies the $2 \times 2$ SM constraint, it can be used as a base matrix for constructing QC-LDPC codes in conjunction with CPM dispersion. The $n$-fold CPM dispersion of $\mathbf{W}_c^{(10)}$ with respect to the cyclic subgroup $G_n$ of GF($q^\tau$) gives an RC-constrained $m \times n$ array $\mathbf{H}_{c,qc,\text{disp}}^{(10)}$ of CPMs of size $n \times n$, which is an $mn \times n^2$ matrix over GF(2) with column and row weights $m$ and $n$, respectively. The rank of $\mathbf{H}_{c,qc,\text{disp}}^{(10)}$ is at most $1 + m(n-1)$ (see Problem 12.13). The null space over GF(2) of $\mathbf{H}_{c,qc,\text{disp}}^{(10)}$ gives an $(m, n)$-regular QC-LDPC code $C_{c,qc,\text{disp}}$ of length $n^2$ with rate at least $(n-m)/n + (m-1)/n^2$. The minimum distance of $C_{c,qc,\text{disp}}$ is at least $m + 1$.

Let $1 \leq g \leq m$ and $g \leq r \leq n$. Take a $g \times r$ submatrix $\mathbf{W}_c^{(10)}(g, r)$ from $\mathbf{W}_c^{(10)}$. The CPM dispersion of $\mathbf{W}_c^{(10)}(g, r)$ with respect to the cyclic subgroup $G_n$ of GF($q^\tau$) gives a $g \times r$ array $\mathbf{H}_{c,qc,\text{disp}}^{(10)}(g, r)$ of CPMs of size $n \times n$, which is a $gn \times rn$ matrix over GF(2) with column and row weights $g$ and $r$, respectively. The null space over GF(2) of $\mathbf{H}^{(10)}_{c,qc,\text{disp}}(g,r)$ generates a $(g,r)$-regular QC-LDPC code $\mathcal{C}_{c,qc,\text{disp}}(g,r)$ of length $rn$ with rate at least $(r-g)/r$. Masking can be applied to $\mathbf{W}^{(10)}_c$ (or $\mathbf{W}^{(10)}_c(g,r)$) to construct irregular QC-LDPC codes based on designed VN and CN degree distributions (or column and row weight distributions of the masking matrix). The above construction of QC-LDPC codes based on cyclic codes of prime length was proposed in [23–25].

Suppose we set $\kappa = 1$ and $\tau \geq 3$. Let $\beta$ be an element in GF($2^\tau$) of order $n$, which is a prime factor of $2^\tau - 1$. Choose a set $\{\beta^{l_0}, \beta^{l_1}, \beta^{l_2}, \ldots, \beta^{l_{m-1}}\}$ of $m$ elements in GF($2^\tau$), which consists of $2t$ consecutive powers of $\beta$, say, $\beta, \beta^2, \ldots, \beta^{2t}$ and their conjugates, with $1 \leq 2t \leq m < n$. Then the cyclic code given by the null space over GF(2) of the base matrix $\mathbf{W}^{(10)}_c = [\beta^{jl_i}]_{0\leq i<m, 0\leq j<n}$ over GF($2^\tau$) in the form of (12.32) is a binary $(n, n-m)$ BCH code, denoted by $\mathcal{C}_{\text{BCH}}$, with rate $(n-m)/n$ and minimum distance at least $2t+1$ (see Section 3.3). In this case, $\mathbf{W}^{(10)}_c$ is the parity-check matrix of a $t$-error-correcting BCH code, which can be decoded with the Berlekamp–Massey algorithm. The null space over GF(2) of the CPM dispersion $\mathbf{H}^{(10)}_{c,qc,\text{disp}}$ of $\mathbf{W}^{(10)}_c$ gives a BCH-associated QC-LDPC-code, called the QC-BCH-LDPC code $\mathcal{C}_{\text{BCH},qc,\text{disp}}$, with rate $(n-m)/n + (m-1)/n^2$ and minimum distance at least $m+1$. Note that the rate of $\mathcal{C}_{\text{BCH},qc,\text{disp}}$ is slightly higher than its associated BCH code and the minimum distance $m+1$ of $\mathcal{C}_{\text{BCH},qc,\text{disp}}$ may be greater (or much greater) than the minimum distance of $\mathcal{C}_{\text{BCH}}$.

Suppose we set $\tau = 1$. Let $\beta$ be an element in GF($2^\kappa$) of order $n$, which is a prime factor of $2^\kappa - 1$, and $l_0, l_1, l_2, \ldots, l_{m-1}$ be $m$ consecutive integers modulo $n$. Then the null space over GF($2^\kappa$) of $\mathbf{W}^{(10)}_c$ is a $2^\kappa$-ary $(n, n-m, m+1)$ cyclic RS code $\mathcal{C}_{RS}$ of length $n$ and dimension $n-m$ with minimum distance $m+1$. Commonly, $l_0, l_1, l_2, \ldots, l_{m-1}$ are taken to be $1, 2, \ldots, m$. In this case, $\mathbf{W}^{(10)}_c$ is identical to the RS base matrix $\mathbf{W}^{(2)}$ given by (12.7), with $t = 1$. Let $\mathbf{H}^{(10)}_{c,qc,\text{disp}}$ be the CPM dispersion of $\mathbf{W}^{(10)}_c$ with respect to the cyclic subgroup $G_n = \{\beta^0 = 1, \beta, \beta^2, \ldots, \beta^{(n-1)}\}$ of GF($2^\kappa$). The null space over GF(2) of the array $\mathbf{H}^{(10)}_{c,qc,\text{disp}}$ gives a QC-RS-LDPC code presented in Section 12.5 (see Examples 12.6 and 12.7).

With different choices of $\kappa, \tau, m, n$, and the set $\{\beta^{l_0}, \beta^{l_1}, \beta^{l_2}, \ldots, \beta^{l_{m-1}}\}$ of roots, we can construct base matrices $\mathbf{W}^{(10)}_c$ of various sizes and characteristics. Using these base matrices in conjunction with CPM dispersion, we can construct a large class of QC-LDPC codes associated with cyclic codes of all types, besides BCH and RS codes, say cyclic Reed–Muller (RM) codes [26, 27], quadratic residue (QR) codes [28], and polynomial codes [29].

---

**Example 12.18** Set $\kappa = 1$ and $\tau = 7$. Let $\beta$ be a primitive element of GF($2^7$). The order of $\beta$ is 127, which is a prime. Set $n = 127$ and $t = 2$. Consider the $(127, 113)$ double-error-correcting BCH code $\mathcal{C}_{\text{BCH}}$ whose generator polynomial $g_{\text{BCH}}(X)$ has as roots $\{\beta, \beta^2, \beta^3, \beta^4\}$ and their 10 conjugates $\beta^6, \beta^8, \beta^{12}, \beta^{16}, \beta^{24}, \beta^{32}, \beta^{48}, \beta^{64}, \beta^{65}$, and $\beta^{96}$. Based on the 14 roots of $g_{\text{BCH}}(X)$, we construct a $14 \times 127$ matrix $\mathbf{W}^{(10)}_{\text{BCH}}$ over GF($2^7$) in the form given by (12.32). $\mathbf{W}^{(10)}_{\text{BCH}}$ is a parity-check matrix of $\mathcal{C}_{\text{BCH}}$.

The null space over GF(2) of $\mathbf{W}_{\text{BCH}}^{(10)}$ gives the (127, 113) BCH code $\mathcal{C}_{\text{BCH}}$, which has rate 0.8897 and minimum distance 5.

The 127-fold CPM dispersion of $\mathbf{W}_{\text{BCH}}^{(10)}$ with respect to the multiplicative group of GF($2^7$) is a $14 \times 127$ array $\mathbf{H}_{\text{BCH},qc,\text{disp}}^{(10)}$ of CPMs of size $127 \times 127$, which is a $1778 \times 16\,129$ matrix with column and row weights 14 and 127, respectively. The rank of $\mathbf{H}_{\text{BCH},qc,\text{disp}}^{(10)}$ is 1765. The null space over GF(2) of $\mathbf{H}_{\text{BCH},qc,\text{disp}}^{(10)}$ gives a (16 129, 14 364) QC-LDPC code $\mathcal{C}_{\text{BCH},qc,\text{disp}}$ with rate 0.8905 and minimum distance at least 15. Note that the rate of $\mathcal{C}_{\text{BCH},qc,\text{disp}}$ is slightly higher than the rate of its associated BCH code $\mathcal{C}_{\text{BCH}}$ and the minimum distance of $\mathcal{C}_{\text{BCH},qc,\text{disp}}$ is at least three times larger than the minimum distance 5 of $\mathcal{C}_{\text{BCH}}$. Hence, the array dispersion increases the code rate and enlarges the minimum distance of the base code.

The bit- and word-error performances of $\mathcal{C}_{\text{BCH},qc,\text{disp}}$ over the AWGN channel decoded with 5, 10, and 50 iterations of the MSA scaled by a factor 0.5 are shown in Figure 12.20. We see that the code has a beautiful waterfall error performance and decoding with the MSA converges quickly.

The 14 roots of the generator polynomial $g_{\text{BCH}}(X)$ of the BCH code $\mathcal{C}_{\text{BCH}}$ are from two conjugate sets, $\{\beta, \beta^2, \beta^4, \beta^8, \beta^{16}, \beta^{32}, \beta^{64}\}$ and $\{\beta^3, \beta^6, \beta^{12}, \beta^{24}, \beta^{48}, \beta^{65}, \beta^{96}\}$. Suppose we take seven rows from $\mathbf{W}_{\text{BCH}}^{(10)}$, which correspond to the seven conjugate roots in the set $\{\beta, \beta^2, \beta^4, \beta^8, \beta^{16}, \beta^{32}, \beta^{64}\}$. These seven rows form a $7 \times 127$ submatrix $\mathbf{W}_{\text{Hamm}}^{(10)}(7, 127)$ of $\mathbf{W}_{\text{BCH}}^{(10)}$. The null space over GF(2) of $\mathbf{W}_{\text{Hamm}}^{(10)}(7, 127)$ gives the (127, 120) Hamming code with rate 0.9449 and minimum distance 3. The subscript, "Hamm" in $\mathbf{W}_{\text{Hamm}}^{(10)}(7, 127)$ stands for "Hamming."

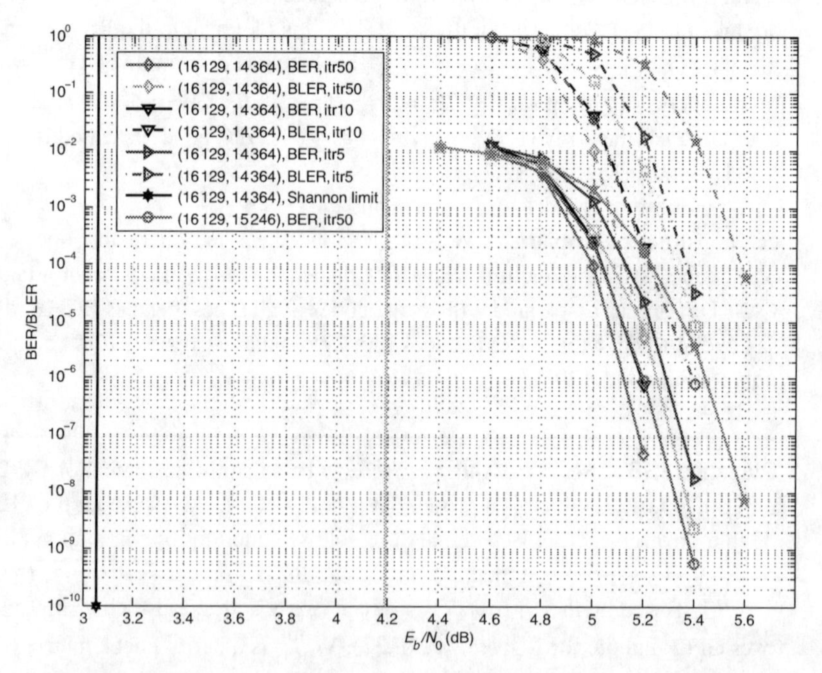


**Figure 12.20** The BER performances of the QC-LDPC codes given in Example 12.18.

The 127-fold CPM dispersion of $\mathbf{W}_{\text{Hamm}}^{(10)}(7, 127)$ with respect to the multiplicative group of GF($2^7$) gives a $7 \times 127$ array $\mathbf{W}_{\text{Hamm},qc,\text{disp}}^{(10)}(7, 127)$ of CPMs of size $127 \times 127$, which is an $889 \times 16\,129$ matrix with column and row weights 7 and 127, respectively, The rank of $\mathbf{W}_{\text{Hamm},qc,\text{disp}}^{(10)}(7, 127)$ is 883. The null space over GF(2) of $\mathbf{H}_{\text{BCH},qc,\text{disp}}^{(10)}(7, 127)$ gives a $(16\,129, 15\,246)$ QC-LDPC code $\mathcal{C}_{\text{Hamm},qc,\text{disp}}(7, 127)$ with rate 0.9452 and minimum distance at least 8.

The bit- and word-error performances of $\mathcal{C}_{\text{Hamm},qc,\text{disp}}(7, 127)$ over the AWGN channel decoded with 5, 10, and 50 iterations of the MSA scaled by a factor 0.5 are also shown in Figure 12.20. As shown in the figure, $\mathcal{C}_{\text{Hamm},qc,\text{disp}}(7, 127)$ performs very well.

## Problems

**12.1** Consider the $(63, 2, 62)$ RS code over GF($2^6$). From this code and (12.5), a $63 \times 63$ array $\mathbf{H}^{(1)}$ of $63 \times 63$ CPMs can be constructed. Take a $4 \times 28$ subarray $\mathbf{H}^{(1)}(4, 28)$ from $\mathbf{H}^{(1)}$, avoiding zero matrices of $\mathbf{H}^{(1)}$. Construct a QC-LDPC code using $\mathbf{H}^{(1)}(4, 8)$ as the parity-check matrix and compute its error performance over the binary-input AWGN channel using the SPA with 10 and 50 iterations.

**12.2** Take a $15 \times 30$ subarray $\mathbf{H}^{(1)}(15, 30)$ from the $63 \times 63$ array $\mathbf{H}^{(1)}$ of $63 \times 63$ CPMs constructed in Problem 12.1. Design a $15 \times 30$ matrix $\mathbf{Z}(15, 30)$ over GF(2) with column and row weights 3 and 6, respectively. Using $\mathbf{H}^{(1)}(15, 30)$ as the base array and $\mathbf{Z}(15, 30)$ as the masking matrix, construct a QC-LDPC code and compute its error performance using the SPA with 50 iterations.

**12.3** Show that an RC-constrained $q \times q$ array of $(q - 1) \times (q - 1)$ CPMs can be constructed using the minimum-weight codewords of the extended $(q, 2, q - 1)$ RS code over GF($q$) (see the discussion of extended RS codes in Chapter 3).

**12.4** Let GF($2^7$) be the code-construction field. Using (12.7) and (12.8), a $127 \times 127$ array $\mathbf{H}_{qc,\text{disp}}^{(2)}$ of $127 \times 127$ CPMs can be constructed. Take an $8 \times 16$ subarray $\mathbf{H}_{8,16}^{(2)}$ from $\mathbf{H}^{(2)}$. Construct an $8 \times 16$ masking matrix $\mathbf{Z}(8, 16) = [\mathbf{G}_0\mathbf{G}_1]$ that consists of two $8 \times 8$ circulants $\mathbf{G}_0$ and $\mathbf{G}_1$ with generators $\mathbf{g}_0 = (1011000)$ and $\mathbf{g}_1 = (01010100)$, respectively. Masking $\mathbf{H}_{8,16}^{(2)}$ with $\mathbf{Z}_{8,16}$ results in a masked array $\mathbf{M}_{8,16}^{(2)}$ that is a $1016 \times 2032$ matrix over GF(2) with column and row weights 3 and 6, respectively. Determine the QC-LDPC code given by the null space of $\mathbf{M}_{8,16}^{(2)}$ and compute its bit- and word-error performances over the binary-input AWGN channel using BPSK transmission decoded with 50 iterations of the SPA.

**12.5** Prove Lemmas 12.3 and 12.4.

**12.6** Prove that the array $\mathbf{H}_{qc,\text{disp}}^{(6)}$ given by (12.16) satisfies the RC constraint.

**12.7** Use the prime field GF(53) and the method presented in Section 12.6 to construct a $53 \times 53$ array $\mathbf{H}$ of $53 \times 53$ CPMs. Take a $6 \times 48$ subarray $\mathbf{H}(6, 48)$ from $\mathbf{H}_{qc,\text{disp}}^{(6)}$. Determine the QC-LDPC code with $\mathbf{H}_{qc,\text{disp}}^{(6)}(6, 48)$ as the parity-check matrix and compute its error performance over the binary-input AWGN channel using the SPA with 50 iterations.

**12.8** This is a continuation of Problem 12.7. Take a $26 \times 52$ subarray $\mathbf{H}^{(6)}_{qc,\text{disp}}(26, 52)$ from the array $\mathbf{H}^{(6)}_{qc,\text{disp}}$ constructed in Problem 12.7. Construct an irregular masking matrix $\mathbf{Z}(26, 52)$ based on the variable- and check-node degree distributions given in Example 12.2. Masking $\mathbf{H}^{(6)}(26, 52)$ with $\mathbf{Z}(26, 52)$ results in a masked array $\mathbf{M}^{(6)}(26, 52)$. Determine the column and row weight distributions of the masking matrix $\mathbf{Z}(26, 52)$. Determine the irregular QC-LDPC code given by the null space of $\mathbf{M}^{(6)}(26, 52)$ and compute its bit- and word-error performances over the binary-input AWGN channel using BPSK transmission decoded with 5, 10, and 50 iterations of the SPA, respectively.

**12.9** Prove Lemma 12.5.

**12.10** Show that QC-LDPC codes can also be constructed from an intersecting bundle of lines in a projective geometry using a method similar to that given in Section 12.8. Choose a projective geometry, construct a code, and compute the bit- and word-error performances of the code constructed.

**12.11** Prove Theorem 12.6.

**12.12** Prove Lemma 12.8.

**12.13** Prove that the rank of the $n \times n$ CPM dispersion $\mathbf{H}^{(10)}_{c,qc,\text{disp}}$ of the matrix $\mathbf{W}^{(10)}_c$ defined by (12.32) is upper bounded by $1 + m(n - 1)$.

## References

[1] R. C. Bose and D. K. Ray-Chaudhuri, "On a class of error correcting binary group codes," *Information and Control*, vol. 3, no. 3, pp. 68–79, March 1960.

[2] A. Hocquenghem, "Codes correcteurs d'erreurs," *Chiffres*, vol. 2, pp. 147–156, 1959.

[3] I. S. Reed and G. Solomon, "Polynomial codes over certain finite fields," *Journal of the Society for Industrial and Applied Mathematics*, vol. 8, pp. 300–304, June 1960.

[4] L. Chen, J. Xu, I. Djurdjevic, and S. Lin, "Near-Shannon limit quasi-cyclic low-density parity-check codes," *IEEE Transactions on Communications*, vol. 52, no. 7, pp. 1038–1042, July 2004.

[5] L. Chen, L. Lan, I. Djurdjevic, S. Lin, and K. Abdel-Ghaffar, "An algebraic method for constructing quasi-cyclic LDPC codes," *Proceedings of the International Symposium on Information Theory and Its Applications (ISITA)*, Parma, October 2004, pp. 535–539.

[6] I. Djurdjevic, J. Xu, K. Abdel-Ghaffar, and S. Lin, "A class of low-density parity-check codes constructed based on Reed–Solomon codes with two information symbols," *IEEE Communications Letters*, vol. 7, no. 7, pp. 317–319, July 2003.

[7] J. L. Fan, "Array codes as low-density parity-check codes," *Proceedings of the 2nd International Symposium on Turbo Codes and Related Topics*, Brest, September 2000, pp. 543–546.

[8] L. Lan, L.-Q. Zeng, Y. Y. Tai, S. Lin, and K. Abdel-Ghaffar, "Constructions of quasi-cyclic LDPC codes for AWGN and binary erasure channels based on finite fields and affine permutations," *Proceedings of the IEEE International Symposium on Information Theory*, Adelaide, September 2005, pp. 2285–2289.

[9] L. Lan, L.-Q. Zeng, Y. Y. Tai, L. Chen, S. Lin, and K. Abdel-Ghaffar, "Construction of quasi-cyclic LDPC codes for AWGN and binary erasure channels: A finite field approach," *IEEE Transactions on Information Theory*, vol. 53, no. 7, pp. 2429–2458, July 2007.

[10] S. M. Song, B. Zhou, S. Lin, and K. Abdel-Ghaffar, "A unified approach to the construction of binary and nonbinary quasi-cyclic LDPC codes based on finite fields," *IEEE Transactions on Communications*, vol. 57, no. 1, pp. 84–93, January 2009.

[11] Y. Y. Tai, L. Lan, L.-Q. Zheng, S. Lin, and K. Abdel-Ghaffar, "Algebraic construction of quasi-cyclic LDPC codes for the AWGN and erasure channels," *IEEE Transactions on Communications*, vol. 54, no. 10, pp. 1765–1774, October 2006.

[12] Z.-W. Li, L. Chen, L.-Q. Zeng, S. Lin, and W. Fong, "Efficient encoding of quasi-cyclic low-density parity-check codes," *IEEE Transactions on Communications*, vol. 54, no. 1, pp. 71–81, January 2006.

[13] S. Lin and D. J. Costello, Jr., *Error Control Coding: Fundamentals and Applications*, 2nd ed., Upper Saddle River, NJ, Prentice-Hall, 2004.

[14] Q. Diao, Q. Huang, S. Lin, and K. Abdel-Ghaffar, "A transform approach for analyzing and constructing quasi-cyclic low-density parity-check codes," *Proceedings of the Information Theory Applications Workshop*, San Diego, CA, February 6–11, 2011, pp. 1–8.

[15] Q. Diao, Q. Huang, S. Lin, and K. Abdel-Ghaffar, "A matrix-theoretic approach for analyzing quasi-cyclic low-density parity-check codes," *IEEE Transactions on Information Theory*, vol. 58, no. 6, pp. 4030–4048, June 2012.

[16] J. Li, K. Liu, S. Lin, and K. Abdel-Ghaffar, "Quasi-cyclic LDPC codes on two arbitrary sets of a finite field," *Proceedings of the International Symposium on Information Theory (ISIT)*, Honolulu, HI, June 29 – July 4, 2014, pp. 2454–2458.

[17] J. Li, K. Liu, S. Lin, and K. Abdel-Ghaffar, "Algebraic quasi-cyclic LDPC codes: Construction, low error-floor, large girth and a reduced-complexity decoding scheme," *IEEE Transactions on Communications*, vol. 62, no. 8, pp. 2626–2637, August 2014.

[18] J. Li, S. Lin, K. Abdel-Ghaffar, W. E. Ryan, and D. J. Costello, Jr., *LDPC Code Designs, Constructions, and Unification*, Cambridge, Cambridge University Press, 2017.

[19] R. Lidl and H. Niederreiter, *Introduction to Finite Fields and their Applications*, revised ed., Cambridge, Cambridge University Press, 1994.

[20] L. Zhang, Q. Huang, S. Lin, K. Abdel-Ghaffar, and I. Blake, "Quasi-cyclic LDPC codes: An algebraic construction, rank analysis, and codes on Latin squares," *IEEE Transactions on Communications*, vol. 58, no. 11, pp. 3126–3139, November 2010.

[21] M. Nasseri, X. Xiao, S. Lin, and B Vasic, "Globally coupled finite geometry and finite field LDPC coding," *IEEE Transactions on Vehicular Technology*, vol. 70, no. 9, pp. 9207–9216, September 2021.

[22] R. C. Singleton, "Maximum distance $q$-ary codes," *IEEE Transactions on Information Theory*, vol. 10, no. 2, pp. 116–118, April 1964.

[23] S. Lin, K. Abdel-Ghaffar, J. Li, and K. Liu, "A novel coding scheme for encoding and iterative soft-decision decoding of binary BCH codes of prime lengths," *Proceedings of Information Theory and Applications (ITA)*, San Diego, CA, February 11–16, 2018.

[24] S. Lin, K. Abdel-Ghaffar, J. Li, and K. Liu, "Collective encoding and iterative soft-decision decoding of cyclic codes of prime lengths in Galois–Fourier transform domain," *Proceedings of the 10th International Symposium on Turbo Codes and Iterative Information Processing (ISTC)*, December 3–7, 2018, pp. 1–8.

[25] S. Lin, K. Abdel-Ghaffar, J. Li, and K. Liu, "A scheme for collective encoding and iterative soft-decision decoding of cyclic codes of prime lengths: Applications to Reed–Solomon, BCH, and quadratic residue codes," *IEEE Transactions on Information Theory*, vol. 66, no. 9, September 2020, pp. 5358–5378.

[26] S. Lin, "Some codes which are invariant under a transitive permutation group and their connection with balanced incomplete block designs," in R. C. Bose and T. A. Dowling (eds), *Proceedings of Combinational Mathematics and its Applications*, University of Carolina Press, Chapel Hill, NC, pp. 388–401, 1967.

[27] T. Kasami, S. Lin, and W. W. Peterson, "New generalizations of the Reed–Muller codes, Pt. I: Primitive codes," *IEEE Transactions on Information Theory*, vol. IT-14, no. 2, pp. 189–199, March 1968.

[28] I. F. Blake and R. C. Mullin, *The Mathematical Theory of Coding*, Cambridge, Cambridge University Press, 1975.

[29] T. Kasami, S. Lin, and W. W. Peterson, "Polynomial codes," *IEEE Transactions on Information Theory*, vol. 14, pp. 807–814, 1968.

# 13 LDPC Codes Based on Combinatorial Designs, Graphs, and Superposition

Combinatorial designs [1–8] form an important branch in combinatorial mathematics. In the late 1950s and during the 1960s, special classes of combinatorial designs, such as *balanced incomplete block designs*, were used to construct error-correcting codes, especially majority-logic-decodable codes. More recently, combinatorial designs were successfully used to construct structured LDPC codes [9–12]. LDPC codes of practical lengths constructed from several classes of combinatorial designs were shown to perform very well over the binary-input AWGN channel with iterative decoding.

Graphs form another important branch in combinatorial mathematics. They were used to construct error-correcting codes in the early 1960s, but not very successfully. Only a few small classes of majority-logic-decodable codes were constructed. However, since the rediscovery of LDPC codes in the middle of the 1990s, graphs have become an important tool for constructing LDPC codes. One example is to use protographs for constructing iteratively decodable codes, as described in Chapters 6 and 8.

This chapter presents several methods for constructing LDPC codes based on special types of combinatorial designs and graphs.

## 13.1 Balanced Incomplete Block Designs and LDPC Codes

Balanced incomplete block designs (BIBDs) form an important class of combinatorial designs. A special subclass of BIBDs can be used to construct RC-constrained matrices or arrays of CPMs from which LDPC codes can be constructed. This section gives a brief description of BIBDs. For an in-depth understanding of this subject, readers are referred to [1–8].

Let $\mathcal{X} = \{x_1, x_2, \ldots, x_m\}$ be a set of $m$ objects. A BIBD $\mathcal{B}$ of $\mathcal{X}$ is a collection of $n$ $g$-subsets of $\mathcal{X}$, denoted by $B_1, B_2, \ldots, B_n$, called *blocks*, which have the following structural properties: (1) each object $x_i$ appears in exactly $r$ of the $n$ blocks; and (2) every two objects appear together in exactly $\lambda$ of the $n$ blocks. Since a BIBD is characterized by five parameters, $m$, $n$, $g$, $r$, and $\lambda$, it is also called an $(m, n, g, r, \lambda)$-BIBD. For the special case with $\lambda = 1$, each pair of objects in $\mathcal{X}$ appears in *one and only one block*. Consequently, any two blocks have *exactly one* object in common. BIBDs of this special type will be used for constructing LDPC codes whose Tanner graphs have girths of at least 6.

Instead of listing all the blocks, an $(m, n, g, r, \lambda)$-BIBD $\mathcal{B}$ of $\mathcal{X}$ can be efficiently described by an $m \times n$ matrix $\mathbf{H}_{\mathrm{BIBD}} = \left[h_{i,j}\right]$ over GF(2) as follows: (1) the rows of $\mathbf{H}_{\mathrm{BIBD}}$ correspond to the $m$ objects of $\mathcal{X}$; (2) the columns of $\mathbf{H}_{\mathrm{BIBD}}$ correspond to the $n$ blocks of the design $\mathcal{B}$; and (3) the entry $h_{i,j}$ at the $i$th row and $j$th column is set to "1" if and only if the $i$th object $x_i$ of $\mathcal{X}$ is contained in the $j$th block $B_j$ of the design $\mathcal{B}$; otherwise, it is set to "0." This matrix over GF(2) is called the *incidence matrix* of the design $\mathcal{B}$. It follows from the properties of an $(m, n, g, r, \lambda)$-BIBD $\mathcal{B}$ that the incidence matrix $\mathbf{H}_{\mathrm{BIBD}}$ of $\mathcal{B}$ has the following structural properties: (1) every column has weight $g$; (2) every row has weight $r$; and (3) any two columns (or two rows) have exactly $\lambda$ 1-components in common.

For the special case with $\lambda = 1$, the incidence matrix $\mathbf{H}_{\mathrm{BIBD}}$ of an $(m, n, g, r, 1)$-BIBD $\mathcal{B}$ satisfies the RC constraint and hence its null space gives an LDPC code whose Tanner graph has a girth of at least 6.

As an example, consider a set $\mathcal{X} = \{x_1, x_2, x_3, x_4, x_5, x_6, x_7\}$ of seven objects. The blocks

$$\begin{aligned}
&B_1 = \{x_1, x_2, x_4\}, \quad && B_2 = \{x_2, x_3, x_5\}, \quad && B_3 = \{x_3, x_4, x_6\}, \\
&B_4 = \{x_4, x_5, x_7\}, \quad && B_5 = \{x_5, x_6, x_1\}, \quad && B_6 = \{x_6, x_7, x_2\}, \\
&B_7 = \{x_7, x_1, x_3\}
\end{aligned}$$

form a (7, 7, 3, 3, 1)-BIBD $\mathcal{B}$ of $\mathcal{X}$. Every block consists of three objects, each object appears in three blocks, and every pair of objects appear together in one and only one block. The incidence matrix of this (7, 7, 3, 3, 1)-BIBD is a $7 \times 7$ matrix over GF(2), given as follows:

$$\mathbf{H}_{\mathrm{BIBD}} = \begin{bmatrix} 1 & 0 & 0 & 0 & 1 & 0 & 1 \\ 1 & 1 & 0 & 0 & 0 & 1 & 0 \\ 0 & 1 & 1 & 0 & 0 & 0 & 1 \\ 1 & 0 & 1 & 1 & 0 & 0 & 0 \\ 0 & 1 & 0 & 1 & 1 & 0 & 0 \\ 0 & 0 & 1 & 0 & 1 & 1 & 0 \\ 0 & 0 & 0 & 1 & 0 & 1 & 1 \end{bmatrix}.$$

Note that $\mathbf{H}_{\mathrm{BIBD}}$ is also a circulant. Its null space gives a $(3, 3)$-regular $(7, 3)$ cyclic LDPC code with minimum distance at least 4. However, since the vector sum of columns 1, 2, 4, and 5 of $\mathbf{H}_{\mathrm{BIBD}}$ gives a zero column vector, the minimum distance of the code is then exactly 4.

## 13.2 Class-I Bose BIBDs and QC-LDPC Codes

Combinatorial design is an old and rich subject in combinatorial mathematics. Over the years, many classes of BIBDs have been constructed using various methods. Extensive coverage of these designs can be found in [4]. In this section and the next, two classes of BIBDs with $\lambda = 1$ are presented. These classes of BIBDs with $\lambda = 1$ were constructed by Bose [1] from finite fields and can be used to construct QC-LDPC codes. In the following, we present the construction of these designs without providing the proofs. For proofs, readers are referred to [1].

### 13.2.1 Class-I Bose BIBDs

Let $t$ be a positive integer such that $12t + 1$ is a prime. Then there exists a prime field $\mathrm{GF}(12t + 1) = \{0, 1, \ldots, 12t\}$. Let the elements of GF($12t + 1$) represent a set $\mathcal{X}$ with $12t + 1$ objects for which a BIBD is to be constructed. Suppose GF($12t + 1$) has a primitive element $\alpha$ such that the condition

$$\alpha^{4t} - 1 = \alpha^{c} \tag{13.1}$$

holds, where $c$ is an odd non-negative integer less than $12t + 1$. Under such a condition on GF($12t + 1$), Bose [1] showed that there exists an $(m, n, g, r, 1)$-BIBD with $m = 12t + 1$, $n = t(12t + 1)$, $g = 4$, $r = 4t$, and $\lambda = 1$. This BIBD is referred to as a *class-I Bose BIBD*.

Since $\alpha$ is a primitive element of GF($12t+1$), $\alpha^{-\infty} = 0, \alpha^0 = 1, \alpha, \ldots, \alpha^{12t-1}$ form the $12t + 1$ elements of GF($12t + 1$) and $\alpha^{12t} = 1$. Note that the $12t + 1$ powers of $\alpha$ simply give the $12t + 1$ integral elements, $0, 1, \ldots, 12t + 1$, of GF($12t + 1$). To form a class-I Bose BIBD, denoted by $\mathcal{B}^{(1)}$, we first form $t$ *base blocks* [1], which are given as follows: for $0 \le i < t$,

$$B_{i,0} = \{\alpha^{-\infty}, \alpha^{2i}, \alpha^{2i+4t}, \alpha^{2i+8t}\}. \tag{13.2}$$

For each base block $B_{i,0}$, we form $12t + 1$ blocks $B_{i,0}, B_{i,1}, \ldots, B_{i,12t}$ by adding each element of GF($12t + 1$) in turn to the elements in $B_{i,0}$. Then, for $0 \le j \le 12t$,

$$B_{i,j} = \{j + \alpha^{-\infty}, j + \alpha^{2i}, j + \alpha^{2i+4t}, j + \alpha^{2i+8t}\}, \tag{13.3}$$

with modulo-($12t + 1$) addition. The $12t + 1$ blocks $B_{i,0}, B_{i,1}, \ldots, B_{i,12t}$ are called *co-blocks* of the base block $B_{i,0}$ and they form a *translate class*, denoted by $T_i$. The $t(12t + 1)$ blocks in the $t$ translate classes $T_0, T_1, \ldots, T_{t-1}$ form a class-I ($12t + 1$, $t(12t + 1)$, 4, $4t$, 1) Bose BIBD $\mathcal{B}^{(1)}$. Table 13.1 gives a list of $t$s such that $12t + 1$ is a prime and the prime field GF($12t + 1$) has a primitive element $\alpha$ that satisfies the condition given by (13.1).

### 13.2.2 Type-I Class-I Bose BIBD-LDPC Codes

For the $j$th block $B_{i,j}$ in the $i$th translate class $T_i$ of a class-I Bose BIBD with $0 \le i < t$ and $0 \le j \le 12t$, we define a ($12t + 1$)-tuple over GF(2),

$$\mathbf{v}_{i,j} = (v_{i,j,0}, v_{i,j,1}, \ldots, v_{i,j,12t}), \tag{13.4}$$

whose components correspond to the $12t + 1$ elements of GF($12t + 1$), where the $k$th component $v_{i,j,k} = 1$ if $k$ is an element in $B_{i,j}$, otherwise $v_{i,j,k} = 0$. This ($12t + 1$)-tuple $\mathbf{v}_{i,j}$ is simply the *incidence vector of the block* $B_{i,j}$ and it has weight 4. For $0 \le j \le 12t$, it follows from (13.3) that the incidence vector $\mathbf{v}_{i,j+1}$ of the ($j + 1$)th block $B_{i,j+1}$ in $T_i$

**Table 13.1** A list of $t$s for which $12t + 1$ is a prime and the condition given by (13.1) holds

| $t$ | Field | $(\alpha, c)$ |
|---|---|---|
| 1 | GF(13) | (2,1) |
| 6 | GF(73) | (5,33) |
| 8 | GF(97) | (5,27) |
| 9 | GF(109) | (6,71) |
| 15 | GF(181) | (2,13) |
| 19 | GF(229) | (6,199) |
| 20 | GF(241) | (7,191) |
| 23 | GF(277) | (5,209) |
| 28 | GF(337) | (10,129) |
| 34 | GF(409) | (21,9) |
| 35 | GF(421) | (2,167) |
| 38 | GF(457) | (13,387) |
| 45 | GF(541) | (2,7) |
| 59 | GF(709) | (2,381) |
| 61 | GF(733) | (6,145) |

is the right cyclic shift of the incidence vector $\mathbf{v}_{i,j}$ of the $j$th block $B_{i,j}$ in $T_i$. Note that $\mathbf{v}_{i,12t+1} = \mathbf{v}_{i,0}$. For $0 \le i < t$, form a $(12t+1)\times(12t+1)$ circulant $\mathbf{G}_i$ with the incidence vector $\mathbf{v}_{i,0}$ of the base block $B_{i,0}$ of $T_i$ as the first column and its $12t$ downward cyclic shifts as the other $12t$ columns. The $12t + 1$ columns of $\mathbf{G}_i$ are simply the *transposes* of the incidence vectors of the $12t + 1$ blocks in the $i$th translate class $T_i$ of the class-I $(12t + 1, t(12t + 1), 4, 4t, 1)$ Bose BIBD $\mathcal{B}^{(1)}$. The column and row weights of $\mathbf{G}_i$ are both 4. Form the following $(12t + 1) \times t(12t + 1)$ matrix over GF(2):

$$\mathbf{H}^{(1)}_{\text{BIBD}} = \begin{bmatrix} \mathbf{G}_0 & \mathbf{G}_1 & \cdots & \mathbf{G}_{t-1} \end{bmatrix}, \tag{13.5}$$

which consists of a row of $t$ circulants. $\mathbf{H}^{(1)}_{\text{BIBD}}$ is the incidence matrix of the class-I $(12t+1, t(12t+1), 4, 4t, 1)$ Bose BIBD $\mathcal{B}^{(1)}$ given above. Since $\lambda = 1$, $\mathbf{H}^{(1)}_{\text{BIBD}}$ satisfies the RC constraint. It has column and row weights 4 and $4t$, respectively.

For $1 \le k \le t$, let $\mathbf{H}^{(1)}_{\text{BIBD}}(k)$ be a subarray of $\mathbf{H}^{(1)}_{\text{BIBD}}$ that consists of $k$ circulants of $\mathbf{H}^{(1)}_{\text{BIBD}}$. $\mathbf{H}^{(1)}_{\text{BIBD}}(k)$ is a $(12t+1)\times k(12t+1)$ matrix with column and row weights 4 and $4k$, respectively. The null space of $\mathbf{H}^{(1)}_{\text{BIBD}}(k)$ gives a $(4, 4k)$-regular QC-LDPC code of length $k(12t + 1)$ with rate at least $(k - 1)/k$. The above construction gives a class of *type-I QC-BIBD-LDPC codes* with various lengths and rates.

---

**Example 13.1** For $t = 15$, $12t + 1 = 181$ is a prime. Then there exists a prime field $\text{GF}(181) = \{0, 1, \ldots, 180\}$. This field has a primitive element $\alpha = 2$ that satisfies the condition (13.1) with $c = 13$ (see Table 13.1). From this field, we can construct a class-I $(181, 2715, 4, 60, 1)$ Bose BIBD. The incidence matrix of this BIBD is

$$\mathbf{H}^{(1)}_{\text{BIBD}} = \begin{bmatrix} \mathbf{G}_0 & \mathbf{G}_1 & \cdots & \mathbf{G}_{14} \end{bmatrix},$$

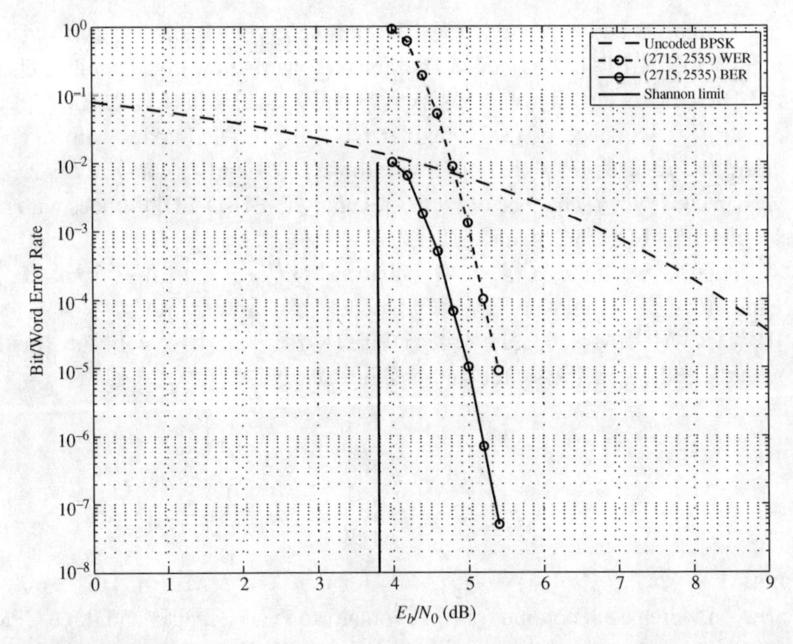


**Figure 13.1** The error performance of the (2715, 2535) type-I QC-BIBD-LDPC code given in Example 13.1.

which consists of 15 circulants of size $181 \times 181$, each with both column and row weight 4. $\mathbf{H}^{(1)}_{\text{BIBD}}$ is a $181 \times 2715$ matrix with column and row weights 4 and 60, respectively. Suppose we choose $\mathbf{H}^{(1)}_{\text{BIBD}}$ as the parity-check matrix for code generation. Then the null space of $\mathbf{H}^{(1)}_{\text{BIBD}}$ gives a (4, 60)-regular (2715, 2535) type-I QC-BIBD-LDPC code with rate 0.934, whose Tanner graph has a girth of at least 6.

Assume BPSK transmission over the binary-input AWGN channel. The error performance of this code with iterative decoding using the SPA with 100 iterations is shown in Figure 13.1. At a BER of $10^{-6}$, it performs 1.3 dB from the Shannon limit.

## 13.2.3 Type-II Class-I Bose BIBD-LDPC Codes

For $0 \leq i < t$, if we decompose each circulant $\mathbf{G}_i$ in $\mathbf{H}^{(1)}_{\text{BIBD}}$ given by (13.5) into a column of four $(12t + 1) \times (12t + 1)$ CPMs using the row decomposition presented in Section 11.5 (see (11.22)), we obtain the following $4 \times t$ array of $(12t + 1) \times (12 + 1)$ CPMs:

$$\mathbf{H}^{(1)}_{\text{BIBD,decom}} = \begin{bmatrix} \mathbf{A}_{0,0} & \mathbf{A}_{0,1} & \cdots & \mathbf{A}_{0,t-1} \\ \mathbf{A}_{1,0} & \mathbf{A}_{1,1} & \cdots & \mathbf{A}_{1,t-1} \\ \mathbf{A}_{2,0} & \mathbf{A}_{2,1} & \cdots & \mathbf{A}_{2,t-1} \\ \mathbf{A}_{3,0} & \mathbf{A}_{3,1} & \cdots & \mathbf{A}_{3,t-1} \end{bmatrix}. \tag{13.6}$$

For $3 \leq k \leq 4$ and $4 \leq r \leq t$, let $\mathbf{H}^{(1)}_{\text{BIBD,decom}}(k, r)$ be a $k \times r$ subarray of $\mathbf{H}^{(1)}_{\text{BIBD,decom}}$. $\mathbf{H}^{(1)}_{\text{BIBD,decom}}(k, r)$ is a $k(12t + 1) \times r(12t + 1)$ matrix over GF(2) with column and row weights $k$ and $r$, respectively. The null space of $\mathbf{H}^{(1)}_{\text{BIBD,decom}}(k,r)$ gives a $(k,r)$-regular QC-BIBD-LDPC code. The above construction gives another class of QC-BIBD-LDPC codes. If we choose $k = 3$ and $r = 3l$ with $l = 2, 3, 4, 5, \ldots$, we can construct a sequence of $(3, 3l)$-regular QC-BIBD-LDPC codes with rates equal (or close) to $1/2, 2/3, 3/4, 4/5, \ldots$. If we choose $k = 4$ and $r = 4l$ with $l = 2, 3, 4, 5, \ldots$, we can construct a sequence of $(4, 4l)$-regular QC-BIBD-LDPC codes with rates equal (or close) to 1/2,2/3,3/4, and 4/5, ....

Suppose we take a $4 \times 4l$ subarray $\mathbf{H}^{(1)}_{\text{BIBD,decom}}(4, 4l)$ from $\mathbf{H}^{(1)}_{\text{BIBD,DECOM}}$ with $4l \leq t$ and $1 \leq l$. Divide this array into $l$ $4 \times 4$ subarrays, $\mathbf{H}^{(1)}_0(4,4), \mathbf{H}^{(1)}_1(4,4), \ldots, \mathbf{H}^{(1)}_{l-1}(4,4)$. For $0 \leq i < l$, mask the $4{\times}4$ subarray $\mathbf{H}^{(1)}_i(4,4)$ with the following $4 \times 4$ circulant masking matrix:

$$\mathbf{Z}_i(4,4) = \begin{bmatrix} 1 & 0 & 1 & 1 \\ 1 & 1 & 0 & 1 \\ 1 & 1 & 1 & 0 \\ 0 & 1 & 1 & 1 \end{bmatrix}. \tag{13.7}$$

Let $\mathbf{M}^{(1)}_i(4,4) = \mathbf{Z}_i(4,4) \circledast \mathbf{H}^{(1)}_i(4,4)$ for $0 \leq i < l$. $\mathbf{M}^{(1)}_i(4,4)$ is a masked $4 \times 4$ array in which each column (or row) contains one zero matrix and three CPMs of size $(12t+1) \times (12t+1)$. Form the following $4 \times 4l$ masked array of circulant permutation and zero matrices of size $(12t + 1) \times (12t + 1)$:

$$\mathbf{M}^{(1)}_{\text{BIBD,decom}}(4, 4l) = \left[\mathbf{M}^{(1)}_0(4,4) \quad \mathbf{M}^{(1)}_1(4,4) \quad \cdots \quad \mathbf{M}^{(1)}_{l-1}(4,4)\right]. \tag{13.8}$$

This array is a $4(12t + 1) \times 4l(12t + 1)$ matrix over GF(2) with column and row weights 3 and $3l$. The null space of this masked matrix gives a $(3, 3l)$-regular QC-BIBD-LDPC code. Since every $4 \times 4$ subarray $\mathbf{H}^{(1)}_i(4,4)$ is masked with the same masking matrix, the above masking is referred to as *uniform circulant masking*. Of course, the subarrays $\mathbf{H}^{(1)}_0(4,4), \mathbf{H}^{(1)}_1(4,4), \ldots, \mathbf{H}^{(1)}_{l-1}(4,4)$ can be masked with different masking circulant matrices. With non-uniform circulant masking of the subarrays, we obtain a $4 \times 4l$ masked array $\mathbf{M}^{(1)}_{\text{BIBD,decom}}(4, 4l)$ with multiple column weights but constant row weight. Then the null space of this irregular array gives an irregular QC-BIBD-LDPC code.

---

**Example 13.2** Consider the array $\mathbf{H}^{(1)}_{\text{BIBD}}$ constructed in Example 13.1. Decompose each circulant $\mathbf{G}_i$ in this array into a column of four $181 \times 181$ CPMs. The decomposition results in a $4 \times 15$ array $\mathbf{H}^{(1)}_{\text{BIBD,decom}}$ of $181 \times 181$ CPMs. Suppose we take the $4 \times 12$ subarray $\mathbf{H}^{(1)}_{\text{BIBD,decom}}(4, 12)$ from $\mathbf{H}^{(1)}_{\text{BIBD,decom}}$, say the first 12 columns of $\mathbf{H}^{(1)}_{\text{BIBD,decom}}$. $\mathbf{H}^{(1)}_{\text{BIBD,decom}}(4, 12)$ is a $724 \times 2172$ matrix over GF(2) with column and row weights 4 and 12, respectively. Divide $\mathbf{H}^{(1)}_{\text{BIBD,decom}}(4, 12)$ into three $4 \times 4$ subarrays and then mask each of these $4 \times 4$ subarrays with the circulant matrix given by (13.7). This results in a regular $4 \times 12$ masked array $\mathbf{M}^{(1)}_{\text{BIBD,decom}}(4, 12)$ of $181 \times 181$ circulant permutation and zero matrices. $\mathbf{M}^{(1)}_{\text{BIBD,decom}}(4, 12)$ is a $724 \times 2172$ matrix with column and row weights 3 and 9, respectively. The null spaces of $\mathbf{H}^{(1)}_{\text{BIBD,decom}}(4, 12)$ and $\mathbf{M}^{(1)}_{\text{BIBD,decom}}(4, 12)$ give (2172, 1451) and

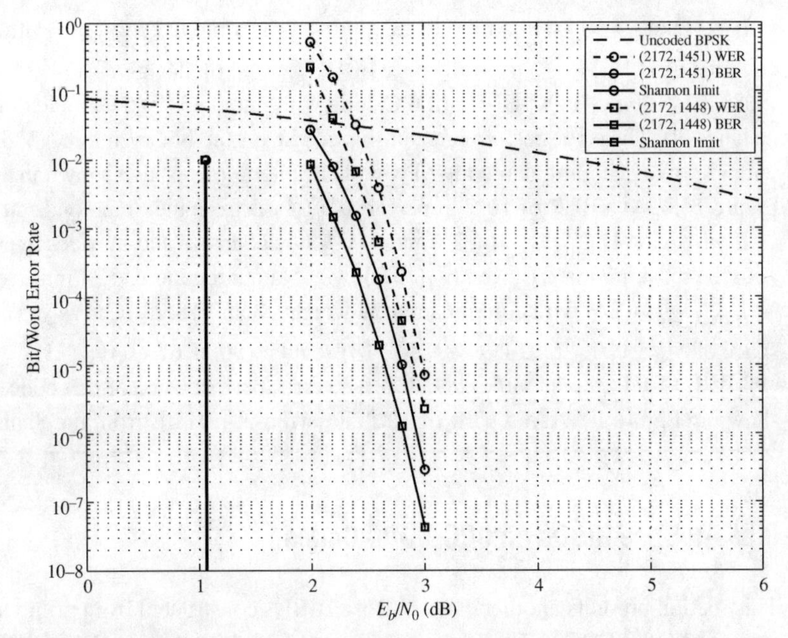


**Figure 13.2** The error performances of the (2172, 1451) and (2172, 1448) codes given in Example 13.2.

(2172, 1448) QC-BIBD-LDPC codes with rates 0.668 and 0.667, respectively. Their error performances with iterative decoding using the SPA with 100 iterations are shown in Figure 13.2.

---

**Example 13.3** Let $t = 28$. Then $12t + 1 = 337$ is a prime and there is a prime field GF(337). From Table 13.1, we see that this prime field satisfies the condition given by (13.1). Using this prime field, we can construct a class-I $(m, n, g, r, 1)$ Bose BIBD with $m = 337$, $n = 9436$, $g = 4$, $r = 112$, and $\lambda = 1$. This BIBD consists of 28 translate classes of blocks. Using the incidence vectors of the blocks in these translate classes, we can form the following row of 28 $337 \times 337$ circulants, each having column weight and row weight 4: $\mathbf{H}^{(1)}_{\text{BIBD}} = \begin{bmatrix} \mathbf{G}_0 & \mathbf{G}_1 & \cdots & \mathbf{G}_{27} \end{bmatrix}$. Decompose each circulant $\mathbf{G}_i$ in $\mathbf{H}^{(1)}_{\text{BIBD}}$ into a column of four $337 \times 337$ CPMs by row decomposition. The decomposition results in a $4 \times 28$ array $\mathbf{H}^{(1)}_{\text{BIBD,decom}}$ of $337 \times 337$ CPMs. Suppose we take a $3 \times 6$ subarray $\mathbf{H}^{(1)}_{\text{BIBD,decom}}(3, 6)$ from $\mathbf{H}^{(1)}_{\text{BIBD,decom}}$, say the first three rows of the first six columns of $\mathbf{H}^{(1)}_{\text{BIBD,decom}}$. $\mathbf{H}^{(1)}_{\text{BIBD,decom}}(3, 6)$ is a $1011 \times 2022$ matrix over GF(2) with column and row weights 3 and 6, respectively. The null space of this matrix gives a (3, 6)-regular (2022, 1013) type-II QC-BIBD-LDPC code with rate 0.501. The error performance of this code over the binary-input AWGN channel with iterative decoding using the SPA with 100 iterations is shown in Figure 13.3. At a BER of $10^{-6}$, it performs 2 dB from the Shannon limit. It has no error floor down to a BER of $10^{-8}$.

If we take a $4 \times 24$ subarray $\mathbf{H}^{(1)}_{\text{BIBD,decom}}(4,24)$ from $\mathbf{H}^{(1)}_{\text{BIBD,decom}}$, say the first 24 columns of $\mathbf{H}^{(1)}_{\text{BIBD,decom}}$, the null space of $\mathbf{H}^{(1)}_{\text{BIBD,decom}}(4,24)$ gives a (4, 24)-regular (8088, 6743) type-II QC-BIBD-LDPC code with rate 0.834 and estimated minimum distance 30. The error performance of this code over the binary-input AWGN channel with iterative decoding using the SPA with 100 iterations is also shown in Figure 13.3. At a BER of $10^{-6}$, it performs 1.04 dB from the Shannon limit.

If we divide $\mathbf{H}^{(1)}_{\text{BIBD,decom}}(4,24)$ into six $4 \times 4$ subarrays and mask each subarray with the masking matrix given by (13.7), we obtain a regular $4 \times 24$ masked array $\mathbf{M}^{(1)}_{\text{BIBD,decom}}(4,24)$ with column and row weights 3 and 18, respectively. The null space of $\mathbf{M}^{(1)}_{\text{BIBD,decom}}(4,24)$ gives a (3, 18)-regular (8088, 6740) type-II QC-BIBD-LDPC code with rate 0.833. The error performance of this code is also shown in Figure 13.3. At a BER of $10^{-6}$, it performs 1.04 dB from the Shannon limit.

## 13.3 Class-II Bose BIBDs and QC-LDPC Codes

This section presents another class of Bose BIBDs constructed from prime fields. Two classes of QC-LDPC codes can be constructed from this class of Bose BIBDs.

### 13.3.1 Class-II Bose BIBDs

Let $t$ be a positive integer such that $20t + 1$ is a prime. Then there exists a prime field $\text{GF}(20t + 1) = \{0, 1, \ldots, 20t\}$ under modulo-$(20t + 1)$ addition and multiplication.

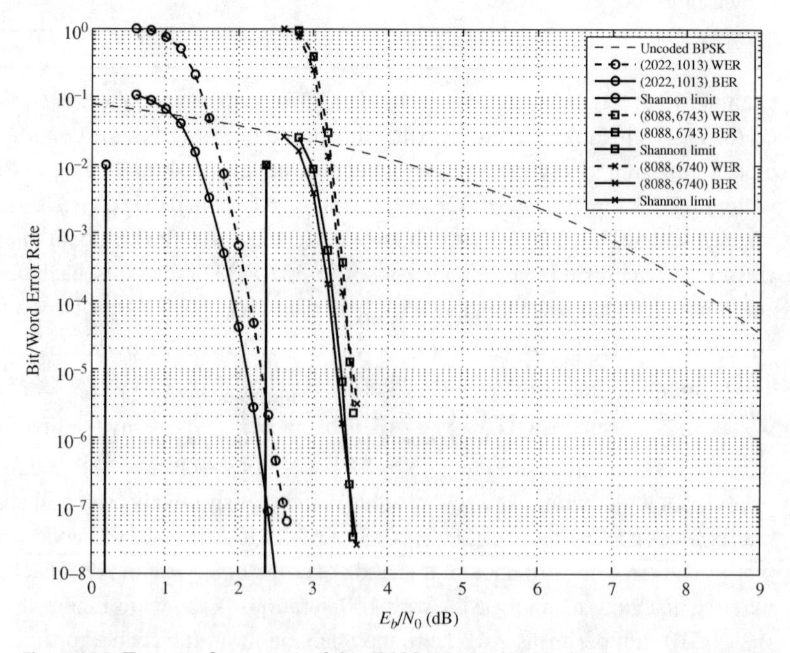


**Figure 13.3** Error performances of the (2022, 1013), (8088, 6743), and (8088, 6740) codes given in Example 13.3.

Let the elements of GF($20t + 1$) represent a set $\mathcal{X}$ of $20t + 1$ objects. Suppose there exists a primitive element $\alpha$ for which the condition

$$\alpha^{4t} + 1 = \alpha^c \tag{13.9}$$

holds, where $c$ is a positive odd integer less than $20t + 1$. Bose [1] showed that under this condition there exists an $(m, n, g, r, 1)$-BIBD with $m = 20t + 1$, $n = t(20t + 1)$, $g = 5$, $r = 5t$, and $\lambda = 1$. For this BIBD, there are $t$ base blocks, which are

$$\mathbf{B}_{i,0} = \left\{\alpha^{2i}, \alpha^{2i+4t}, \alpha^{2i+8t}, \alpha^{2i+12t}, \alpha^{2i+16t}\right\}, \tag{13.10}$$

with $0 \leq i < t$. For each base block $\mathbf{B}_{i,0}$, we form $20t+1$ co-blocks $\mathbf{B}_{i,0}, \mathbf{B}_{i,1}, \ldots, \mathbf{B}_{i,20t}$ by adding the $20t + 1$ elements of GF($2t + 1$) in turn to the elements in $\mathbf{B}_{i,0}$. Then the $j$th co-block of $\mathbf{B}_{i,0}$ is given by

$$\mathbf{B}_{i,j} = \left\{j + \alpha^{2i}, j + \alpha^{2i+4t}, j + \alpha^{2i+8t}, j + \alpha^{2i+12t}, j + \alpha^{2i+16t}\right\}, \tag{13.11}$$

with $j \in$ GF($20t + 1$). The $20t + 1$ co-blocks of a base block $\mathbf{B}_{i,0}$ form a translate class $T_i$ of the design. The $t(20t + 1)$ blocks in the $t$ translate classes $T_0, T_1, \ldots, T_{t-1}$ form a type-II $(20t + 1, t(20t + 1), 5, 5t, 1)$ Bose BIBD design, denoted by $\mathcal{B}^{(2)}$. Table 13.2 gives a list of $t$s that satisfy the condition given by (13.9).

### 13.3.2 Type-I Class-II Bose BIBD-LDPC Codes

The incidence matrix of a class-II $(20t + 1, t(20t + 1), 5, 5t, 1)$ Bose BIBD, $\mathcal{B}^{(2)}$, can be arranged as a row of $t$ circulants of size $(20t + 1) \times (20t + 1)$ as follows:

$$\mathbf{H}^{(2)}_{\text{BIBD}} = \begin{bmatrix}\mathbf{G}_0 & \mathbf{G}_1 & \cdots & \mathbf{G}_{t-1}\end{bmatrix}, \tag{13.12}$$

where the $i$th circulant $\mathbf{G}_i$ is formed by the incidence vectors of the blocks in the $i$th translate class $T_i$ of $\mathcal{B}^{(2)}$ arranged as columns in downward cyclic order. Each circulant $\mathbf{G}_i$ in $\mathbf{H}^{(2)}_{\text{BIBD}}$ has both column and row weight 5. Therefore, $\mathbf{H}^{(2)}_{\text{BIBD}}$ is an RC-constrained $(20t + 1) \times t(20t + 1)$ matrix over GF(2) with column and row weights 5 and $5t$, respectively.

For $1 \leq k \leq t$, let $\mathbf{H}^{(2)}_{\text{BIBD}}(k)$ be a subarray of $\mathbf{H}^{(2)}_{\text{BIBD}}$ that consists of $k$ circulants of $\mathbf{H}^{(2)}_{\text{BIBD}}$. $\mathbf{H}^{(2)}_{\text{BIBD}}(k)$ is a $(20t + 1) \times k(20t + 1)$ matrix over GF(2) with column and row weights 5 and $5k$, respectively. The null space of $\mathbf{H}^{(2)}_{\text{BIBD}}(k)$ gives a $(5, 5k)$-regular *type-I class-II QC Bose BIBD-LDPC code* of length $k(20t + 1)$ with rate at least $(k - 1)/k$. The above construction gives a class of QC-LDPC codes.

---

**Example 13.4** Let $t = 21$. From Table 13.2, we see that a type-II (421, 8841, 5, 105, 1) Bose BIBD, $\mathcal{B}^{(2)}$, can be constructed from the prime field GF(421). $\mathcal{B}^{(2)}$ consists of 21 translate classes, each having 421 co-blocks. Using the incidence vectors of the blocks in these translate classes, we can form the incidence matrix $\mathbf{H}^{(2)}_{\text{BIBD}} = \begin{bmatrix}\mathbf{G}_0 & \mathbf{G}_1 & \cdots & \mathbf{G}_{20}\end{bmatrix}$ of $\mathcal{B}^{(2)}$ that consists of 21 $421 \times 421$ circulants, each with both column and row weight 5. Set $k = 10$. Take the first ten circulants of $\mathbf{H}_{\text{BIBD}}^{(2)}$ to form a subarray $\mathbf{H}_{\text{BIBD}}^{(2)}(10)$ of $\mathbf{H}_{\text{BIBD}}^{(2)}$. $\mathbf{H}_{\text{BIBD}}^{(2)}(10)$ is a $421 \times 4210$ matrix with column and row weights 5 and 50, respectively. The null space of $\mathbf{H}_{\text{BIBD}}^{(2)}(10)$ gives a $(5, 50)$-regular $(4210, 3789)$ QC-BIBD-LDPC code with rate 0.9. The error performance of this code with iterative decoding using the SPA with 100 iterations is shown in Figure 13.4.

**Table 13.2** A list of $t$s for which $20t + 1$ is a prime and the condition $\alpha^{4t} + 1 = \alpha^c$ holds

| $t$ | Field | $(\alpha, c)$ |
|---|---|---|
| 2 | GF(41) | (6,3) |
| 3 | GF(61) | (2,23) |
| 12 | GF(241) | (7,197) |
| 14 | GF(281) | (3,173) |
| 21 | GF(421) | (2,227) |
| 30 | GF(601) | (7,79) |
| 32 | GF(641) | (3,631) |
| 33 | GF(661) | (2,657) |
| 35 | GF(701) | (2,533) |
| 41 | GF(821) | (2,713) |

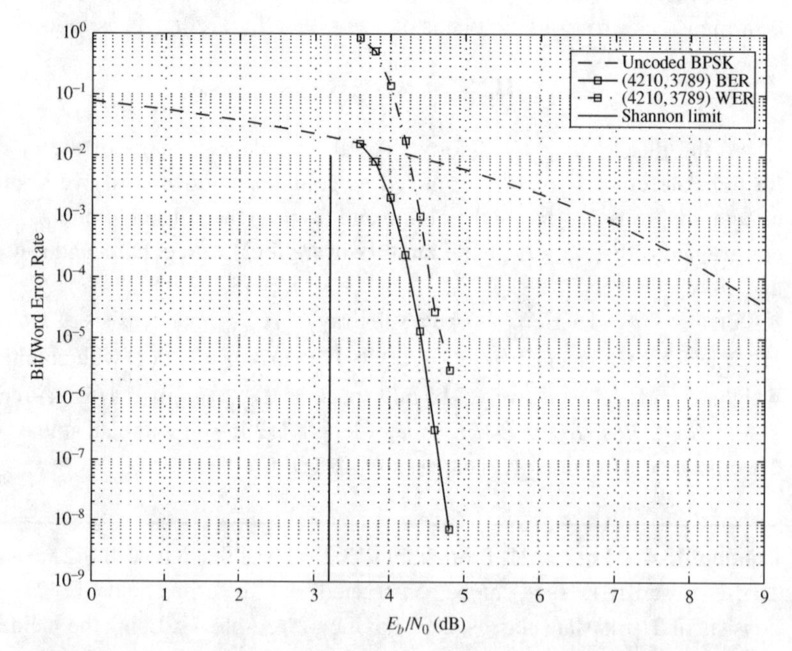


**Figure 13.4** The error performance of the (4210, 3789) QC-BIBD-LDPC code given in Example 13.4.

### 13.3.3 Type-II Class-II QC-BIBD-LDPC Codes

If we decompose each circulant $\mathbf{G}_i$ in $\mathbf{H}_{\text{BIBD}}^{(2)}$ given by (13.12) into a column of five $(20t+1) \times (20t+1)$ CPMs with row decomposition, we obtain an RC-constrained $5 \times t$ array of $(20t+1) \times (20t+1)$ CPMs as follows:

$$\mathbf{H}_{\text{BIBD,decom}}^{(2)} = \begin{bmatrix} \mathbf{A}_{0,0} & \mathbf{A}_{0,1} & \cdots & \mathbf{A}_{0,t-1} \\ \mathbf{A}_{1,0} & \mathbf{A}_{1,1} & \cdots & \mathbf{A}_{1,t-1} \\ \mathbf{A}_{2,0} & \mathbf{A}_{2,1} & \cdots & \mathbf{A}_{2,t-1} \\ \mathbf{A}_{3,0} & \mathbf{A}_{3,1} & \cdots & \mathbf{A}_{3,t-1} \\ \mathbf{A}_{4,0} & \mathbf{A}_{4,1} & \cdots & \mathbf{A}_{4,t-1} \end{bmatrix}. \tag{13.13}$$

For $3 \leq k \leq 5$ and $3 \leq r \leq t$, let $\mathbf{H}_{\text{BIBD,decom}}^{(2)}(k,r)$ be a $k \times r$ subarray of $\mathbf{H}_{\text{BIBD,decom}}^{(2)}$. $\mathbf{H}_{\text{BIBD,decom}}^{(2)}(k,r)$ is a $k(20t+1) \times r(20t+1)$ matrix over GF(2) with column and row weights $k$ and $r$, respectively. The null space of $\mathbf{H}_{\text{BIBD,decom}}^{(2)}(k,r)$ gives a $(k,r)$-regular QC-LDPC code of length $r(20t+1)$ with rate at least $(r-k)/r$. The above construction gives another class of QC-BIBD-LDPC codes.

---

**Example 13.5** For $t = 21$, consider the class-II (421, 8841, 5, 105, 1) Bose BIBD constructed from the prime field GF(421) given in Example 13.4. From the incidence matrix of this Bose BIBD and row decompositions of its constituent circulants, we obtain a $5 \times 21$ array $\mathbf{H}_{\text{BIBD,decom}}^{(2)}$ of $421 \times 421$ CPMs. Take a $4 \times 20$ subarray $\mathbf{H}_{\text{BIBD,decom}}^{(2)}(4,20)$ from $\mathbf{H}_{\text{BIBD,decom}}^{(2)}$. This subarray is a $1684 \times 8420$ matrix over GF(2) with column and row weights 4 and 20, respectively. The null space of this matrix gives a $(4,20)$-regular $(8420,6739)$ QC-BIBD-LDPC code with rate 0.8004. The error performance of this code with iterative decoding using the SPA with 100 iterations is shown in Figure 13.5. At a BER of $10^{-6}$, it performs 1.1 dB from the Shannon limit.

---

Let $r = 5k$ such that $5k \leq t$. Take a $5 \times 5k$ subarray $\mathbf{H}_{\text{BIBD,decom}}^{(2)}(5,5k)$ from $\mathbf{H}_{\text{BIBD,decom}}^{(2)}$ and divide it into $k$ $5 \times 5$ subarrays $\mathbf{H}_0^{(2)}(5,5), \mathbf{H}_1^{(2)}(5,5), \ldots, \mathbf{H}_{k-1}^{(2)}(5,5)$. Then

$$\mathbf{H}_{\text{BIBD,decom}}^{(2)}(5,5k) = \left[\mathbf{H}_0^{(2)}(5,5) \quad \mathbf{H}_1^{(2)}(5,5) \quad \cdots \quad \mathbf{H}_{k-1}^{(2)}(5,5)\right]. \tag{13.14}$$

For $0 \leq i < k$, we can mask each constituent subarray $\mathbf{H}_i^{(2)}(5,5)$ of $\mathbf{H}_{\text{BIBD,decom}}^{(2)}(5,5k)$ with either of the following two circulant masking matrices:

$$\mathbf{Z}_1(5,5) = \begin{bmatrix} 1 & 0 & 0 & 1 & 1 \\ 1 & 1 & 0 & 0 & 1 \\ 1 & 1 & 1 & 0 & 0 \\ 0 & 1 & 1 & 1 & 0 \\ 0 & 0 & 1 & 1 & 1 \end{bmatrix}, \tag{13.15}$$

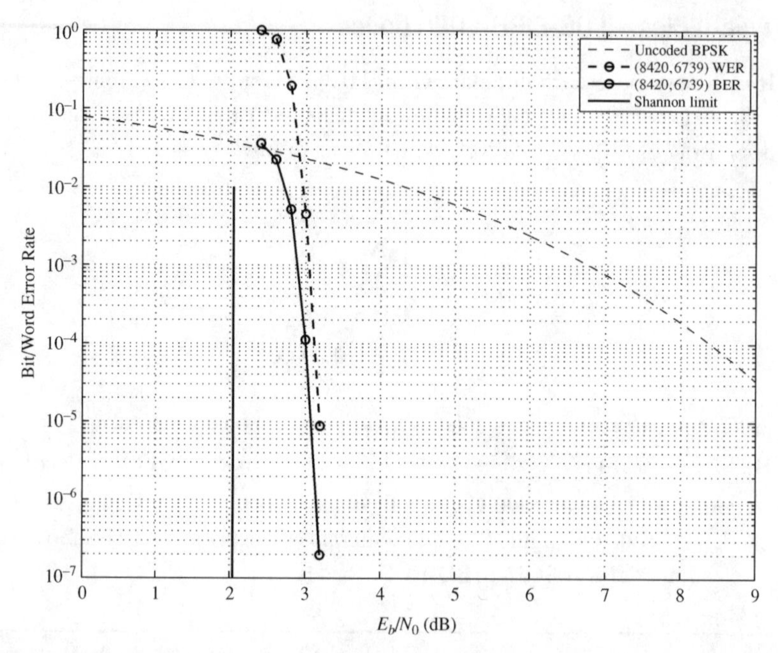


**Figure 13.5** The error performance of the (8420, 6739) QC-BIBD-LDPC code given in Example 13.5.

$$\mathbf{Z}_2(5,5) = \begin{bmatrix} 1 & 0 & 1 & 1 & 1 \\ 1 & 1 & 0 & 1 & 1 \\ 1 & 1 & 1 & 0 & 1 \\ 1 & 1 & 1 & 1 & 0 \\ 0 & 1 & 1 & 1 & 1 \end{bmatrix}. \tag{13.16}$$

If we mask each constituent $5 \times 5$ subarray $\mathbf{H}_i^{(2)}(5,5)$ of $\mathbf{H}_{\text{BIBD,decom}}^{(2)}(5,5k)$ with $\mathbf{Z}_1(5,5)$ given by (13.15), we obtain a $5 \times 5k$ uniformly masked array of circulant permutation and zero matrices, for $0 \leq i < k$.

$$\mathbf{M}_{\text{BIBD,decom,1}}^{(2)}(5,5k) = \left[\mathbf{M}_{0,1}^{(2)}(5,5) \quad \mathbf{M}_{1,1}^{(2)}(5,5) \quad \cdots \quad \mathbf{M}_{k-1,1}^{(2)}(5,5)\right], \tag{13.17}$$

where, for $0 \leq i < k$, $\mathbf{M}_{i,1}^{(2)}(5,5) = \mathbf{Z}_1(5,5) \circledast \mathbf{H}_i^{(2)}(5,5)$. $\mathbf{M}_{\text{BIBD,decom,1}}^{(2)}(5,5k)$ is a $5(20t+1) \times 5k(20t+1)$ matrix over GF(2) with column and row weights 3 and $3k$, respectively. The null space of $\mathbf{M}_{\text{BIBD,decom}}^{(2)}(5,5k)$ gives a $(3,3k)$-regular QC-BIBD-LDPC code.

For $0 \leq i < k$, if we mask each constituent $5 \times 5$ subarray of $\mathbf{H}_{\text{BIBD,decom}}^{(2)}(5,5k)$ with $\mathbf{Z}_2(5,5)$ given by (13.16), we obtain a $5 \times 5k$ uniformly masked array of circulant permutation and zero matrices,

$$\mathbf{M}_{\text{BIBD,decom,2}}^{(2)}(5,5k) = \left[\mathbf{M}_{0,2}^{(2)}(5,5) \quad \mathbf{M}_{1,2}^{(2)}(5,5) \quad \cdots \quad \mathbf{M}_{k-1,2}^{(2)}(5,5)\right], \tag{13.18}$$

where, for $0 \leq i < k$, $\mathbf{M}_{i,2}^{(2)}(5,5) = \mathbf{Z}_2(5,5) \circledast \mathbf{H}_i^{(2)}(5,5)$. $\mathbf{M}_{\text{BIBD,decom,2}}^{(2)}(5,5k)$ is a $5(20t+1) \times 5k(20t+1)$ matrix over GF(2) with column and row weights 4 and $4k$, respectively. The null space of this matrix gives a $(4,4k)$-regular QC-BIBD-LDPC code.

The above maskings of the base array $\mathbf{H}^{(2)}_{\text{BIBD,decom}}(5,5k)$ are uniform maskings. Each constituent $5 \times 5$ subarray $\mathbf{H}^{(2)}_i(5,5)$ is masked by the same masking matrix. However, the constituent $5 \times 5$ subarrays of the base array $\mathbf{H}^{(2)}_{\text{BIBD,decom}}(5,5k)$ can be masked with different $5 \times 5$ circulant masking matrices with different column weights. This non-uniform masking results in a masked array $\mathbf{M}^{(2)}_{\text{BIBD,decom}}(5,5k)$ with multiple column weights. Besides the masking matrices given by (13.15) and (13.16), we define the following circulant masking matrix with column weight 2:

$$\mathbf{Z}_0(5,5) = \begin{bmatrix} 1 & 0 & 0 & 0 & 1 \\ 1 & 1 & 0 & 0 & 0 \\ 0 & 1 & 1 & 0 & 0 \\ 0 & 0 & 1 & 1 & 0 \\ 0 & 0 & 0 & 1 & 1 \end{bmatrix}. \tag{13.19}$$

Let $k_0$, $k_1$, $k_2$, and $k_3$ be four non-negative integers such that $k_0 + k_1 + k_2 + k_3 = k$. Suppose we mask the base array $\mathbf{H}^{(2)}_{\text{BIBD,decom}}(5,5k)$ as follows: (1) the first $k_0$ constituent $5 \times 5$ subarrays are masked with $\mathbf{Z}_0(5,5)$; (2) the next $k_1$ constituent $5 \times 5$ subarrays are masked with $\mathbf{Z}_1(5,5)$; (3) the next $k_2$ constituent $5 \times 5$ subarrays are masked with $\mathbf{Z}_2(5,5)$; and (4) the last $k_3$ constituent $5 \times 5$ subarrays are unmasked. This non-uniform masking results in a $5 \times 5k$ masked array $\mathbf{M}^{(2)}_{\text{BIBD,decom,3}}(5,5k)$ with multiple column weights but constant row weight. The average column weight of the masked array is $(2k_0+3k_1+4k_2+5k_3)/k$. The masked array has a constant row weight, which is $2k_0 + 3k_1 + 4k_2 + 5k_3$. The null space of the non-uniformly masked array $\mathbf{M}^{(2)}_{\text{BIBD,decom,3}}(5,5k)$ gives an irregular QC-BIBD-LDPC code. The error performance of this irregular code over the binary-input AWGN channel with iterative decoding depends on the choice of the parameters $k_0$, $k_1$, $k_2$, and $k_3$.

---

**Example 13.6** Consider the $5 \times 21$ array $\mathbf{H}^{(2)}_{\text{BIBD,decom}}$ of $421 \times 421$ CPMs constructed from the class-II (421, 8841, 5, 105, 1) Bose BIBD using the prime field GF(421) with $t = 21$ given in Example 13.5. Take a $5 \times 20$ subarray $\mathbf{H}^{(2)}_{\text{BIBD,decom}}(5,20)$ from $\mathbf{H}^{(2)}_{\text{BIBD,decom}}$ and divide this subarray into four $5 \times 5$ subarrays $\mathbf{H}^{(2)}_0(5,5)$, $\mathbf{H}^{(2)}_1(5,5)$, $\mathbf{H}^{(2)}_2(5,5)$, and $\mathbf{H}^{(2)}_3(5,5)$. Choose $k_0 = k_1 = k_2 = k_3 = 1$. We mask the first three $5 \times 5$ subarrays with $\mathbf{Z}_0(5,5)$, $\mathbf{Z}_1(5,5)$, and $\mathbf{Z}_2(5,5)$, respectively, and leave the fourth subarray $\mathbf{H}^{(2)}_3(5,5)$ unmasked. The masking results in a $5 \times 20$ masked array $\mathbf{M}^{(2)}_{\text{BIBD,decom,3}}(5,20)$ of circulant permutation and zero matrices. This masked array has average column weight 3.5 and constant row weight 14. The null space of this masked array gives an irregular (8240, 6315) QC-BIBD-LDPC code with rate 0.7664. The error performance of this code with iterative decoding using the SPA with 100 iterations is shown in Figure 13.6. At a BER of $10^{-6}$, the code performs 1 dB from the Shannon limit.

---

*Remark:* Before we conclude this section, we remark that any $k \times kl$ sub- array $\mathbf{H}^{(e)}_{qc,\text{disp}}(k,kl)$ of the array $\mathbf{H}^{(e)}_{qc,\text{disp}}$ of CPMs with $1 \le e \le 6$ constructed in Chapter 12 can be divided into $l$ subarrays of size $k \times k$. Each constituent $k \times k$ subarray of $\mathbf{H}^{(e)}_{qc,\text{disp}}(k,k)$ can be masked with a $k \times k$ circulant masking matrix with weight less

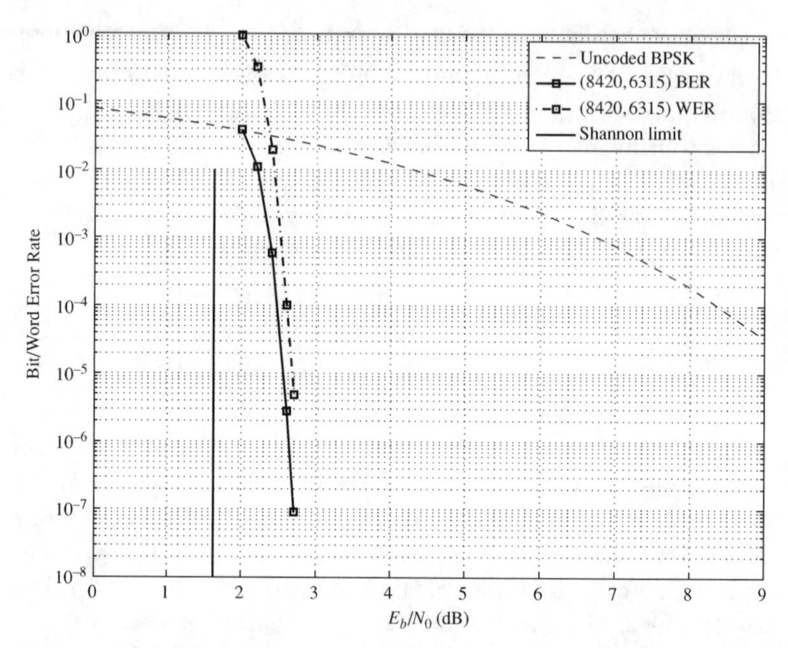


**Figure 13.6** The error performance of the (8240, 6315) QC-BIBD-LDPC code given in Example 13.6.

than or equal to $k$. Then the masking results in a $k \times kl$ masked array $\mathbf{M}^{(e)}_{qc,\text{disp}}(k, kl)$. The null space of this masked array gives a regular QC-LDPC code if masking is uniform; otherwise, it is an irregular QC-LDPC code with non-uniform masking.

## 13.4 Construction of Type-II Bose BIBD-LDPC Codes by Dispersion

Since the class-I and -II Bose BIBDs given in Sections 13.2.1 and 13.3.1 are constructed using prime fields, RC-constrained arrays of CPMs in the forms given by (13.6) and (13.13) can also be obtained by the additive-dispersion technique presented in Section 12.6. For the explanation of array and code construction, we use the class-II Bose BIBDs given in Section 13.3.

Let $t$ be a positive integer such that $20t + 1$ is a prime and the condition of (13.9) holds. Then there exists a class-II $(20t + 1, t(20t + 1), 5, 5t, 1)$ Bose BIBD constructed from the prime field GF$(20t + 1)$. Using this Bose BIBD, we form the following $t \times 5$ matrix over GF$(20t + 1)$ with the elements of the base blocks $B_{0,0}, B_{1,0}, \ldots, B_{t-1,0}$ (given by (13.10)) as rows:

$$\mathbf{W}_{\text{BIBD}} = \begin{bmatrix} \alpha^0 & \alpha^{4t} & \alpha^{8t} & \alpha^{12t} & \alpha^{16t} \\ \alpha^2 & \alpha^{2+4t} & \alpha^{2+8t} & \alpha^{2+12t} & \alpha^{2+16t} \\ \vdots & \vdots & \vdots & \vdots & \vdots \\ \alpha^{2(t-1)} & \alpha^{2(t-1)+4t} & \alpha^{2(t-1)+8t} & \alpha^{2(t-1)+12t} & \alpha^{2(t-1)+16t} \end{bmatrix}. \quad (13.20)$$

It follows from the structural property of an $(m, n, g, r, 1)$-BIBD that $\mathbf{W}_{\text{BIBD}}$ satisfies additive row constraints 1 and 2 (given by Lemmas 12.3 and 12.4). Hence, $\mathbf{W}_{\text{BIBD}}$ can be used as a base matrix for dispersion.

On replacing each entry of $\mathbf{W}_{\text{BIBD}}$ by its additive $(20t + 1)$-fold matrix dispersion, we obtain the following RC-constrained $t \times 5$ array of $(20t + 1) \times (20t + 1)$ CPMs over GF(2):

$$\mathbf{M}_{\text{BIBD,disp}} = \begin{bmatrix} \mathbf{M}_{0,0} & \mathbf{M}_{0,1} & \mathbf{M}_{0,2} & \mathbf{M}_{0,3} & \mathbf{M}_{0,4} \\ \mathbf{M}_{1,0} & \mathbf{M}_{1,1} & \mathbf{M}_{1,2} & \mathbf{M}_{1,3} & \mathbf{M}_{1,4} \\ \vdots & \vdots & \vdots & \vdots & \vdots \\ \mathbf{M}_{t-1,0} & \mathbf{M}_{t-1,1} & \mathbf{M}_{t-1,2} & \mathbf{M}_{t-1,3} & \mathbf{M}_{t-1,4} \end{bmatrix}. \tag{13.21}$$

On taking the transpose of $\mathbf{M}_{\text{BIBD}}$, we obtain the following RC-constrained $5 \times t$ array of CPMs:

$$\mathbf{H}^{(3)}_{\text{BIBD,disp}} = \mathbf{M}^{\text{T}}_{\text{BIBD,disp}} = \begin{bmatrix} \mathbf{A}_{0,0} & \mathbf{A}_{0,1} & \cdots & \mathbf{A}_{0,t-1} \\ \mathbf{A}_{1,0} & \mathbf{A}_{1,1} & \cdots & \mathbf{A}_{1,t-1} \\ \vdots & \vdots & \vdots & \vdots \\ \mathbf{A}_{5,0} & \mathbf{A}_{5,1} & \cdots & \mathbf{A}_{5,t-1} \end{bmatrix}, \tag{13.22}$$

where $\mathbf{A}_{i,j} = [\mathbf{M}_{j,i}]^{\text{T}}$ with $0 \leq i < 5$ and $0 \leq j < t$. $\mathbf{H}^{(3)}_{\text{BIBD,disp}}$ and $\mathbf{H}^{(2)}_{\text{BIBD,decom}}$ are structurally the same.

The null space of any subarray of $\mathbf{H}^{(3)}_{\text{BIBD,disp}}$ gives a QC-BIBD-LDPC code. Masking a subarray of $\mathbf{H}^{(3)}_{\text{BIBD,disp}}$ also gives a QC-LDPC code.

Similarly, using additive dispersion, we can construct an RC-constrained $4 \times t$ array of $(12t+1) \times (12t+1)$ CPMs based on a class-I $(12t+1, t(12t+1), 4, 4t, 1)$ Bose BIBD constructed from the field GF$(12t + 1)$, provided that $12t + 1$ is a prime and condition (13.1) holds.

## 13.5 A Trellis-Based Construction of LDPC Codes

Graphs are not only useful for interpretation of iterative types of decoding as described in Chapter 5, but also form an important combinatorial tool for constructing iteratively decodable codes. A class of iteratively decodable codes constructed from a special type of graphs, called *protographs*, has been presented in Chapter 6. Constructions of LDPC codes with girth 6 or larger based on graphs can be found in [13–15]. In this section, we present a trellis-based method to construct LDPC codes with large girth progressively from a bipartite graph with a small girth.

### 13.5.1 A Trellis-Based Method for Removing Short Cycles from a Bipartite Graph

Consider a *simple connected* bipartite graph $\mathcal{G}_0 = (V_1, V_2)$ with two disjoint sets of nodes, $V_1 = \{v_0, v_1, \ldots, v_{n-1}\}$ and $V_2 = \{c_0, c_1, \ldots, c_{m-1}\}$. Any edge in $\mathcal{G}_0$ connects a node in $V_1$ and a node in $V_2$. Since it is simple, there are no multiple edges between a

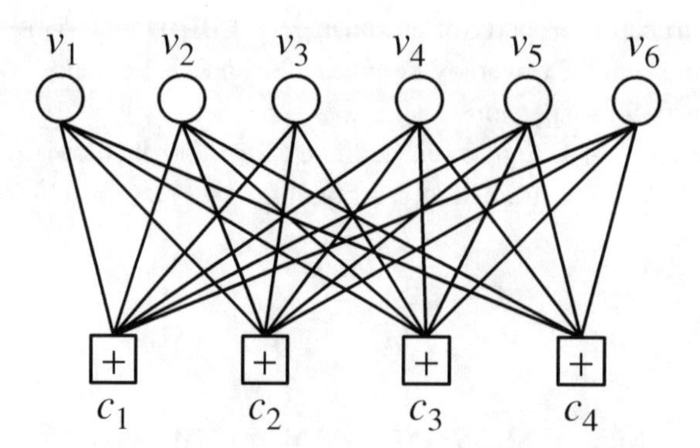


**Figure 13.7** A simple bipartite graph with girth 4.

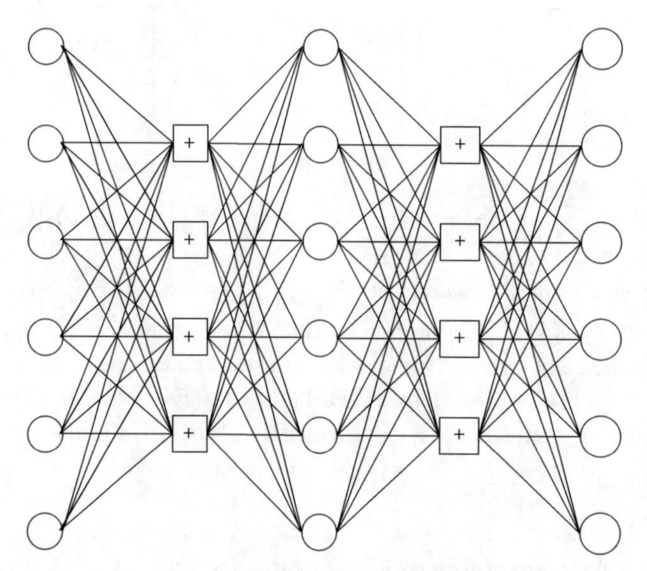

**Figure 13.8** A four-section trellis constructed from the bipartite graph shown in Figure 13.7.

node in $V_1$ and a node in $V_2$. Let $\lambda_0$ be the girth of $\mathcal{G}_0$. Form a trellis $\mathcal{T}$ of $\lambda_0$ sections with $\lambda_0+1$ levels of nodes, labeled from 0 to $\lambda_0$ [16]. Recall that the girth of a bipartite graph is even. For $0 \leq l \leq \lambda_0/2$, the nodes at the $2l$th level of $\mathcal{T}$ are the nodes in $V_1$, and the nodes at the $(2l+1)$th level of $\mathcal{T}$ are the nodes in $V_2$. Two nodes $v_i$ and $c_j$ at two consecutive levels are connected by a *branch* if and only if $(v_i, c_j)$ is an edge in $\mathcal{G}_0$. Every section of $\mathcal{T}$ is simply a representation of the bipartite graph $\mathcal{G}_0$. Therefore, $\mathcal{T}$ is simply a repetition of $\mathcal{G}_0$ $\lambda_0$ times, and every section is the *mirror image* of the preceding section. For example, consider the bipartite graph shown in Figure 13.7. A four-section trellis constructed from this bipartite graph is shown in Figure 13.8.

The nodes at the 0th level of $\mathcal{T}$ are called the *initial nodes* and the nodes at the $\lambda_0$th level are called the *terminal nodes*. The initial nodes and terminal nodes of $\mathcal{T}$ are identical and they are nodes in $V_1$. For $0 \leq i < n$, an *elementary* $(i, i)$-path of $\mathcal{T}$

is defined as a sequence of $\lambda_0$ connected branches starting from the initial node $v_i$ at the 0th level of $\mathcal{T}$ and ending at the terminal node $v_i$ at the $\lambda_0$th level of $\mathcal{T}$ such that the following constraints are satisfied: (1) if $(v_j, c_k)$ is a branch in the sequence, then $(c_k, v_j)$ cannot be a branch in the sequence and vice versa; and (2) every branch in the sequence appears once and only once. Then an elementary $(i, i)$-path represents a cycle of length $\lambda_0$ in $\mathcal{G}_0$ starting and ending at node $v_i$; conversely, a cycle of length $\lambda_0$ in $\mathcal{G}_0$ starting and ending at node $v_i$ is represented by an elementary $(i, i)$-path in $\mathcal{T}$. For an elementary $(i, i)$-path in $\mathcal{T}$, if we remove the last branch that connects a node $c_k$ at the $(\lambda_0 - 1)$th level of $\mathcal{T}$ to the terminal node $v_i$ at the $\lambda_0$th level of $\mathcal{T}$, we break all those cycles of length $\lambda_0$ in $\mathcal{G}_0$ that contain $(c_k, v_i)$ as an edge. On this basis, we can develop a procedure to process $\mathcal{T}$ and to identify all the elementary paths in $\mathcal{T}$. Once all the elementary paths in $\mathcal{T}$ have been identified, we remove their last branches systematically to break all the cycles of length $\lambda_0$ in $\mathcal{G}_0$. Since branches in the last section of $\mathcal{T}$ are removed to break the cycles of length $\lambda_0$ in $\mathcal{G}_0$, the last section of $\mathcal{T}$ at the end of the branch-removal process gives a new bipartite graph $\mathcal{G}_1$ with girth $\lambda_0 + 2$.

To identify the elementary $(i, i)$-path for all $i$, we process the trellis $\mathcal{T}$ level by level [14]. Suppose we have processed the trellis $\mathcal{T}$ up to level $l$ with $0 \leq l < \lambda_0$. For every node $v_j$ (or $c_j$) at the $l$th level, we list all the *partial elementary paths* (PEPs) of length $l$ that originate from the initial node $v_i$ at the 0th level and terminate at $v_j$ (or $c_j$), denoted by PEP$(i, j, l)$. We extend each partial elementary path on the list PEP$(i, j, l)$ to the $(l + 1)$th level of $\mathcal{T}$ through every branch diverging from $v_j$ (or $c_j$) that does not appear in any partial elementary path on the list PEP$(i, j, l)$ before $v_j$ (or $c_j$). For each node $v_j$ (or $c_j$) at the $(l + 1)$th level of $\mathcal{T}$, we again form a list PEP$(i, j, l + 1)$ of all the partial elementary paths of length $l + 1$ that originate from the initial node $v_i$ and terminate at $v_j$ (or $c_j$). At the $(2m - 1)$th level of $\mathcal{T}$ with $0 < m < \lambda_0/2$, a partial elementary path originating from the initial node $v_i$ at the 0th level of $\mathcal{T}$ and ending at a node $c_j$ in the $(2m - 1)$th level of $\mathcal{T}$ cannot be extended to the node $v_i$ at the $2m$th level through the branch $(c_j, v_i)$, because this would create a cycle of length less than $\lambda_0$ that does not exist in $\mathcal{G}_0$ (since $\lambda_0$ is the shortest length of a cycle in $\mathcal{G}_0$). Continue the above *extend-and-list* (EAL) process until the $(\lambda_0 - 1)$th level of $\mathcal{T}$ is reached. For each note $c_j$ at the $(\lambda_0 - 1)$th level of $\mathcal{T}$, all the partial elementary paths that originate from $v_i$ and terminate at $c_j$ are extended through the branch $(c_j, v_i)$, if it exists, to the terminal node $v_i$ at the $\lambda_0$th level of $\mathcal{T}$. This extension gives the list PEP$(i, i, \lambda_0)$ of all the elementary $(i, i)$-paths in $\mathcal{T}$. The union of all the PEP$(i, i, \lambda_0)$ lists at the $\lambda_0$th level of $\mathcal{T}$ forms a table with all the elementary $(i, i, \lambda_0)$-paths in $\mathcal{T}$, which is denoted by EPT$(\mathcal{T})$ and called the *elementary path table* of $\mathcal{T}$.

The elementary paths of $\mathcal{T}$ give all the cycles of length $\lambda_0$ in $\mathcal{G}_0$. The next step is to break all the cycles of length $\lambda_0$ in $\mathcal{G}_0$ by removing the last branches of some or all of the elementary $(i, i)$-paths in EPT$(\mathcal{T})$. Let $c_j$ be the node at the $(\lambda_0 - 1)$th level of $\mathcal{T}$ that has the largest degree and is connected to the terminal node $v_i$ of the largest degree at the $\lambda_0$th level of $\mathcal{T}$. Remove the branch $(c_j, v_i)$ from the last section of $\mathcal{T}$. Removal of $(c_j, v_i)$ breaks all the elementary $(i, i)$-paths in $\mathcal{T}$ with $(c_j, v_i)$ as the last branch and hence breaks all the cycles of length $\lambda_0$ in $\mathcal{G}_0$ with $v_i$ as the starting and ending nodes that contain $(c_j, v_i)$ as an edge. The degrees of $v_i$ and $c_j$ are reduced by 1. Removing the branch $(c_j, v_i)$ from $\mathcal{T}$ also breaks all the other elementary $(k, k)$-paths in $\mathcal{T}$ that have $(c_j, v_i)$ as a branch. These elementary $(i, i)$- and $(k, k)$-paths are then removed from EPT($\mathcal{T}$). Removal of the branch $(c_j, v_i)$ results in a new last section of $\mathcal{T}$. We repeat the above branch-removal process until all the elementary paths have been removed from EPT($\mathcal{T}$). As a result, we break all the cycles of length $\lambda_0$ in the bipartite graph $\mathcal{G}_0$. Then the last section of the new trellis gives a pruned bipartite graph $\mathcal{G}_1$ with girth at least $\lambda_1 = \lambda_0 + 2$.

By applying the above cycle-removal process to the new bipartite graph $\mathcal{G}_1$, we can construct another new bipartite graph $\mathcal{G}_2$ with girth $\lambda_2 = \lambda_0 + 4$. We continue this process until we have obtained a bipartite graph with the desired girth or the degrees of some nodes in $V$ (or $S$) have become too small.

### 13.5.2 Code Construction

The trellis-based cycle-removal process can be used either to construct LDPC codes or to improve the error performance of existing codes by increasing their girth [14]. We begin with a bipartite graph $\mathcal{G}_0$ with girth $\lambda_0$, which is either the Tanner graph of an existing LDPC code $\mathcal{C}_0$ or a chosen bipartite graph. We apply the cycle-removal process repeatedly. For $1 \leq k$ , at the end of the $k$th application of the cycle-removal process, we obtain a new bipartite graph $\mathcal{G}_k = \left(V_1^{(k)}, V_2^{(k)}\right)$ with girth $\lambda_k = \lambda_0 + 2k$. Then we construct the incidence matrix $\mathbf{H}_k = \left[h_{i,j}\right]$ of $\mathcal{G}_k$ whose rows correspond to the nodes in $V_2^{(k)}$ and whose columns correspond to the nodes in $V_1^{(k)}$, where $h_{i,j} = 1$ if and only if the node $c_i$ in $V_2^{(k)}$ and the node $v_j$ in $V_1^{(k)}$ are connected by an edge in $\mathcal{G}_k$. Then the null space of $\mathbf{H}_k$ gives an LDPC code $\mathcal{C}_k$. Next, we compute the error performance of $\mathcal{C}_k$ with iterative decoding using the SPA. We compare the error performance of $\mathcal{C}_k$ with that of $\mathcal{C}_{k-1}$. If $\mathcal{C}_k$ performs better than $\mathcal{C}_{k-1}$, we continue the cycle-removal process; otherwise, we stop the cycle-removal process and $\mathcal{C}_{k-1}$ becomes the end code.

As the cycle-removal process continues, the degrees of variable nodes (also check nodes) of the resultant Tanner graph become smaller. At a certain point, when there are too many nodes of small degree, especially nodes of degree 2, the error performance of the resultant LDPC code starts to degrade and a high error floor appears. A general guide is that one should stop the cycle-removal process when the number of variable nodes with degree 2 becomes greater than the number of check nodes.

---

**Example 13.7** Consider the three-dimensional Euclidean geometry EG$(3, 2^3)$. A 2-flat in EG$(3, 2^3)$ consists of 64 points (see Chapter 2). There are 511 2-flats in EG$(3, 2^3)$ not containing the origin of the geometry. The incidence vectors of these 511 2-flats not containing the origin form a single $511 \times 511$ circulant $\mathbf{G}$ with both column and row weight 64. The Tanner graph of $\mathbf{G}$ contains 3 605 616 cycles of length 4. We can decompose $\mathbf{G}$ into a $4 \times 8$ array $\mathbf{H}_0$ of $511 \times 511$ circulants with column and row decompositions (see Section 11.5), each with column weight and row weight 2. $\mathbf{H}_0$ is a matrix over GF(2) with column and row weights 8 and 16,

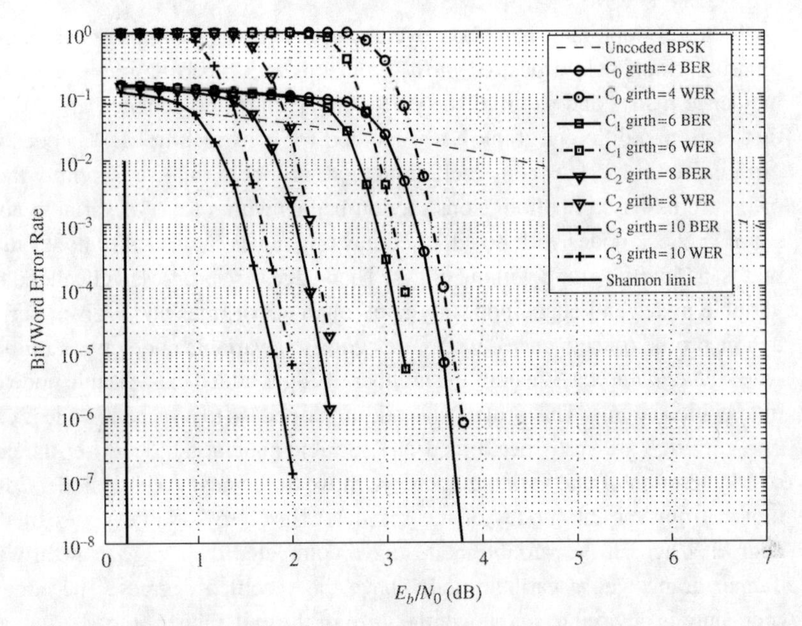


**Figure 13.9** The error performances of the LDPC codes given in Example 13.7.

respectively. The Tanner graph $\mathcal{G}_0$ of $\mathbf{H}_0$ has a girth of 4 and contains 13 286 cycles of length 4. The null space of $\mathbf{H}_0$ gives a (4088, 2046) code $\mathcal{C}_0$ with rate 0.5009 whose error performance is shown in Figure 13.9. At a BER of $10^{-6}$, it performs 3.4 dB from the Shannon limit. On starting from $\mathcal{G}_0$ and applying the cycle-removal process repeatedly, we obtain three bipartite graphs $\mathcal{G}_1$, $\mathcal{G}_2$, and $\mathcal{G}_3$ with girths 6, 8, and 10, respectively. The null spaces of the incidence matrices $\mathbf{H}_1$, $\mathbf{H}_2$, and $\mathbf{H}_3$ of $\mathcal{G}_1$, $\mathcal{G}_2$, and $\mathcal{G}_3$ give three rate-1/2 (4088, 2044) LDPC codes $\mathcal{C}_1$, $\mathcal{C}_2$, and $\mathcal{C}_3$, respectively. The error performances of these three LDPC codes are also shown in Figure 13.9. We see that the error performances of the codes are improved as the girth increases from 4 to 10. However, as the girth is increased from 10 to 12, the error performance of code $\mathcal{C}_4$ becomes poor and has a high error floor. So, $\mathcal{C}_3$ is the end code in our code-construction process. At a BER of $10^{-6}$, $\mathcal{C}_3$ performs only 1.6 dB from the Shannon limit.

---

The above trellis-based method for constructing LDPC codes of moderate lengths with large girth is quite effective. However, it becomes ineffective when the bipartite graph to be processed becomes too big.

## 13.6 Construction of LDPC Codes Based on Progressive Edge-Growth Tanner Graphs

In the previous section, a trellis-based method for constructing LDPC codes with large girth was presented. The construction begins with a given simple connected bipartite (or Tanner) graph and then short cycles are progressively removed to obtain an end bipartite graph with a desired girth. From this end bipartite graph, we construct its incidence matrix and use it as the parity-check matrix to generate an LDPC code. In this section, a different graph-based method for constructing LDPC codes with large girth is presented. This construction method, proposed in [13], is simply the opposite of the trellis-based method. Construction begins with a set of $n$ variable nodes and a set of $m$ check nodes with no edges connecting nodes in one set to nodes in the other, that is, a bipartite graph without edges. Then edges are progressively added to connect variable nodes and check nodes by applying a set of rules and a given variable-node degree profile (or sequence). Edges are added to a variable node one at a time using an *edge-selection procedure* until the number of edges added to a variable node is equal to its specified degree. This progressive addition of edges to a variable node is called *edge growth*. Edge growth is performed one variable node at a time. After the completion of edge growth of one variable node, we move to the next variable node. Edge growth moves from variable nodes of the smallest degree to variable nodes of the largest degree. When all the variable nodes have completed their edge growth, we obtain a Tanner graph whose variable nodes have the specified degrees. The edge-selection procedure is devised to maximize the girth of the end Tanner graph.

Before we present the edge-growth procedure used to construct Tanner graphs with large girth, we introduce some concepts. Consider a bipartite graph $\mathcal{G} = (V_1, V_2)$ with variable-node set $V_1 = \{v_0, v_1, \ldots, v_{n-1}\}$ and check-node set $V_2 = \{c_0, c_1, \ldots, c_{m-1}\}$. Consider the variable node $v_i$. Let $\lambda_i$ be the length of the shortest cycle that passes through (or contains) $v_i$. This length $\lambda_i$ of the shortest cycle passing through $v_i$ is called the *local girth* of $v_i$. Then the girth $\lambda$ of $\mathcal{G}$ is given by $\lambda = \min\{\lambda_i : 0 \leq i < n\}$. The edge-growth procedure for constructing a Tanner graph is devised to maximize the local girth of every variable node (a greedy algorithm).

For a given variable node $v_i$ in a Tanner graph $\mathcal{G}$, let $N_i^{(l)}$ denote the set of check nodes in $\mathcal{G}$ that are connected to $v_i$ within a distance $2l + 1$. The distance between two nodes in a graph is defined as the length of a *shortest path* between the two nodes (see Chapter 2). The shortest paths that connect $v_i$ to the check nodes in $N_i^{(l)}$ can be represented by a tree $\mathcal{R}_i$ with $v_i$ as the root, as shown in Figure 13.10. This tree consists of $l + 1$ levels and each level consists of two layers of nodes, a layer of variable nodes and a layer of check nodes. We label the levels from 0 to $l$. Each path in $\mathcal{R}_i$ consists of a sequence of alternate variable and check nodes; each path in $\mathcal{R}_i$ terminates at a check node in $N_i^{(l)}$. Every node on a path in $\mathcal{R}_i$ appears once and only once. This path tree $\mathcal{R}_i$ with variable node $v_i$ as its root can be constructed progressively. Starting from the variable node $v_i$, we transverse all the edges $(v_i, c_{i_1}), (v_i, c_{i_2}), \ldots, (v_i, c_{i_{d_{v_i}}})$ that connect $v_i$ to its $d_{v_i}$ *nearest-neighbor* check nodes $c_{i_1}, c_{i_2}, \ldots, c_{i_{d_{v_i}}}$. This results in the 0th level of the *path* tree $\mathcal{R}_i$. Next, we transverse all the edges that connect the check nodes in the 0th level to their respective nearest-neighbor variable nodes in the first level. Then we transverse all the edges that connect the variable nodes at the first layer of the first level of the tree $\mathcal{R}_i$ to their respective nearest-neighbor check nodes in the second layer of the first level. This completes the first level of the tree $\mathcal{R}_i$. The

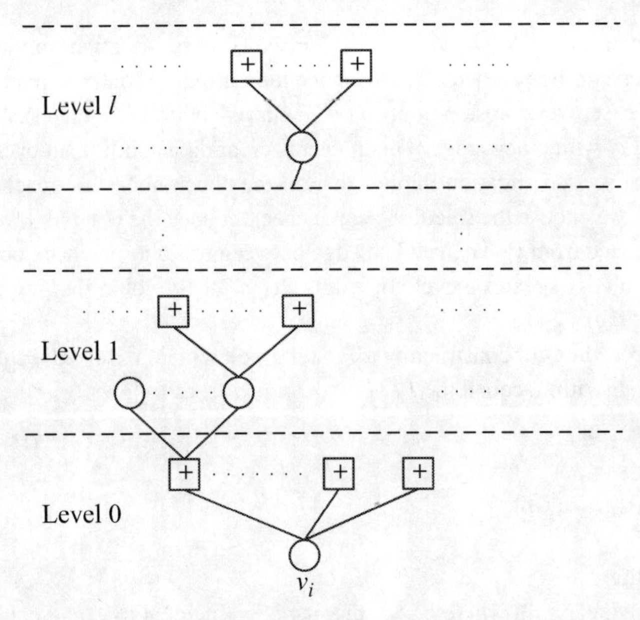


**Figure 13.10** Tree representation of the neighborhood within depth $l$ of a variable node.

transversing process continues level by level until the $l$th level, or a level from which the tree cannot grow further, has been reached. It is clear that the distance between $v_i$ and a variable node in the $l$th level is $2l$ and the distance between $v_i$ and a check node in the $l$th level is $2l + 1$. If there is an edge in $\mathcal{G}$ that connects a check node in the $l$th level of the tree $\mathcal{R}_i$ to the root variable node $v_i$, then adding this edge between the check node and $v_i$ creates a cycle of length $2(l + 1)$ in $\mathcal{G}$ that passes through $v_i$. The set $N_i^{(l)}$ is referred to as the *neighborhood within depth* $l$ of $v_i$. Let $\bar{N}_i^{(l)}$ be the *complementary set* of $N_i^{(l)}$, that is, $\bar{N}_i^{(l)} = V_2 \backslash N_i^{(l)}$.

Using the concepts of the local girth of a variable node $v_i$, the neighborhood $N_i^{(l)}$ within depth $l$ of $v_i$, and the tree representation of the paths that connect $v_i$ to the check nodes in $N_i^{(l)}$, a progressive edge growth (PEG) procedure was devised by Hu *et al.* [13] to construct Tanner graphs with relatively large girth. Suppose that, for the given number $n$ of variable nodes, the given number $m$ of check nodes, and the degree profile $D_v = (d_0, d_1, \ldots, d_{n-1})$ of variable nodes, we have completed the edge-growth of the first $i$ variable nodes $v_0, v_1, \ldots, v_{i-1}$ with $1 \leq i \leq n$, using the PEG procedure. At this point, we have a *partially connected* Tanner graph. Next, we grow the edges incident from the $i$th variable node $v_i$ to $d_{v_i}$ check nodes. The growth is done one edge at a time. Suppose $k - 1$ edges have been added to $v_i$ with $1 \leq k \leq d_{v_i}$. Before the $k$th edge is added to $v_i$, we construct a path tree $\mathcal{R}_i$ with $v_i$ as the root, as shown in Figure 13.10, based on the current partially connected Tanner graph, denoted by $\mathcal{G}_{i,k-1}$. We keep growing the tree until it reaches a level, say the $l$th level, such that one of the following two situations occurs: (1) the tree cannot grow further but the cardinality $\left|N_i^{(l)}\right|$ of $N_i^{(l)}$

is smaller than $m$; or (2) $\bar{N}_i^{(l)} \neq \emptyset$ but $\bar{N}_i^{(l+1)} = \emptyset$. The first situation implies that not all check nodes can be reached from $v_i$ under the current partially connected graph $\mathcal{G}_{i,k-1}$. In this case, we choose a check node $c_j$ in $\bar{N}_i^{(l)}$ with the smallest degree and connect $v_i$ and $c_i$ with a new edge. This prevents creating an additional cycle passing through $v_i$. The second situation implies that all the check nodes are reachable from $v_i$. In this case, we choose the smallest-degree check node at the $(l+1)$th level that has the largest distance from $v_i$. Then add an edge between this chosen check node and $v_i$. Adding such an edge creates a cycle of length $2(l+2)$. By doing this, we maximize the local girth $\lambda_i$ of $v_i$.

The PEG procedure for constructing a Tanner graph with maximized local girth can be put into an algorithm, called the *PEG algorithm* [13], as follows:

**for** $i = 0$ to $n-1$, **do**
**begin**
  **for** $k = 0$ to $d_{v_i} - 1$, **do**
  **begin**
    **if** $k = 0$, **then**
      $\mathcal{E}_i^{(0)} \leftarrow \text{edge}(v_i, c_j)$, where $\mathcal{E}_i^{(0)}$ is the first edge incident to $v_i$ and $c_j$ is a check node that has the lowest degree in the current partially connected Tanner graph
    **else**
      grow a path tree $\mathcal{R}_i$ with $v_i$ as the root to a level, say the $l$th level, based on the current partially connected Tanner graph such that either the cardinality of $N_i^{(l)}$ stops increasing (i.e., the path tree cannot grow further) but is less than $m$, or $\bar{N}_i^{(l)} \neq \emptyset$ but $\bar{N}_i^{(l+1)} = \emptyset$, then $\mathcal{E}_i^{(k)} \leftarrow (v_i, c_j)$, where $\mathcal{E}_i^{(k)}$ is the $k$th edge incident to $v_i$ and $c_j$ is a check node chosen from $\bar{N}_i^{(l)}$ that has the lowest degree
    **end**
**end**

Unlike the trellis-based method for constructing Tanner graphs, the PEG algorithm cannot predict the girth of the end Tanner graph, so it is known only at the end of the implementation of the algorithm.

---

**Example 13.8** The following degree distributions of variable and check nodes of a Tanner graph are designed for a rate-1/2 LDPC code of length 4088:

$$\tilde{\lambda}(X) = 0.451X + 0.3931X^2 + 0.1558X^9,$$
$$\tilde{\rho}(X) = 0.7206X^6 + 0.2994X^7.$$

From these degree distributions, using the PEG algorithm we constructed a rate-1/2 (4088, 2044) irregular LDPC code whose Tanner graph has a girth of 8. The error performance of this code is shown in Figure 13.11.

---

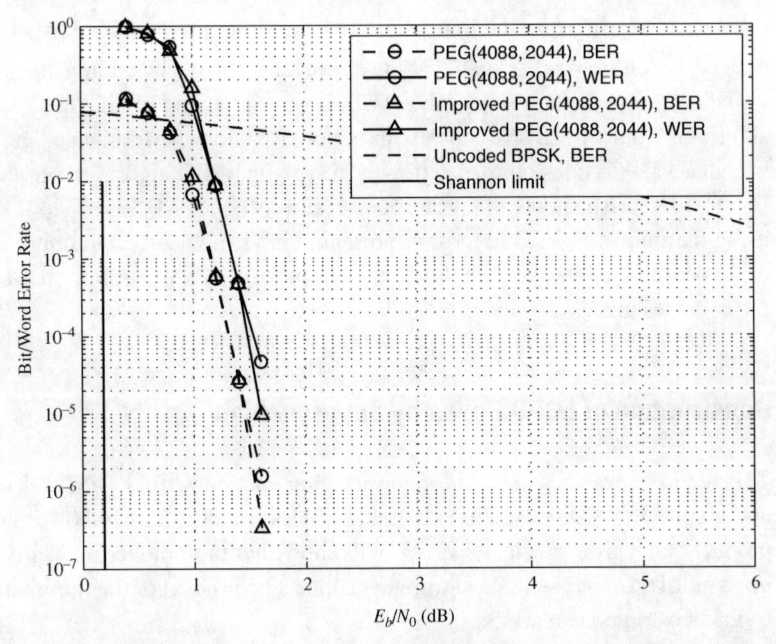


**Figure 13.11** The error performance of the (4088, 2044) irregular LDPC code given in Example 13.8.

The PEG algorithm given above can be improved to construct irregular LDPC codes with better performance in the high-SNR region without performance degradation in the low-SNR region. An improved PEG algorithm was presented in [15]. To present this improved PEG algorithm, we first introduce a new concept. A cycle $\mathcal{C}$ of length $2t$ in a Tanner graph $\mathcal{G}$ consists of $t$ variable nodes and $t$ check nodes. Let

$$\epsilon = \sum_{i=1}^{t}(d_i - 2), \tag{13.23}$$

where $d_i$ is the degree of the $i$th variable node on $\mathcal{C}$. The sum $\epsilon$ is simply the number of edges that connect the cycle $\mathcal{C}$ to the rest of the Tanner graph, outside the cycle. This sum $\epsilon$ is a measure of the *connectivity* of cycle $\mathcal{C}$ to the rest of the Tanner graph. The larger this connectivity, the more messages from outside of the cycle $\mathcal{C}$ are available for the variable nodes on $\mathcal{C}$. For this reason, $\epsilon$ is called the *approximate cycle extrinsic message degree* (ACE) [17, 18].

The improved PEG algorithm presented in [15] is identical to the PEG algorithm except for the case in which $k \geq 1$, $\bar{N}_i^{(l)} \neq \emptyset$, and $\bar{N}_i^{(l+1)} = \emptyset$, when building the path tree $\mathcal{R}_i$ with $v_i$ as the root. In this case, there may be more than one candidate check node with the smallest degree. Let $\mathcal{V}_{2,i}^{(l,k)}$ denote the set of candidate check nodes in $\bar{N}_i^{(l)}$ with the smallest degree. For any check node $c_j \in \mathcal{V}_{2,i}^{(l,k)}$, there is at least one path of length $2(l+1)+1$ between $v_i$ and $c_j$, but no shorter path between them. Hence, the placement of edge $\mathcal{E}_i^{(k)}$ between $v_i$ and $c_j$ will create at least one new cycle of length $2(l+2)$, but no shorter cycles. In the PEG algorithm, a check node is chosen randomly from $\mathcal{V}_{2,i}^{(l,k)}$. However, in the improved PEG algorithm presented in [15], a check node $c_{\max}$ is chosen from $\mathcal{V}_{2,i}^{(l,k)}$ such that the new cycles created by adding an edge between $v_i$ and $c_{\max}$ have the largest possible ACE. It was shown in [15] that this modified PEG algorithm results in LDPC codes with better error-floor performance than that of the irregular LDPC codes constructed with the original PEG algorithm presented above. Figure 13.11 also shows the performance of a (4088, 2044) LDPC code constructed using the improved PEG algorithm presented in [15]. Clearly, the improvement is at the expense of additional computational complexity, but this applies once only, during the code design.

## 13.7 Construction of LDPC Codes by Superposition

This section presents a method for constructing long powerful LDPC codes from short and simple LDPC codes. This method includes the classic method for constructing product codes as a special case. We will show that the product of two LDPC codes gives an LDPC code with its minimum distance the product of the minimum distances of the two component codes.

### 13.7.1 A General Superposition Construction of LDPC Codes

Let $\mathbf{B} = \left[b_{i,j}\right]$ be a sparse $c \times t$ matrix over GF(2) that satisfies the RC constraint. The null space of $\mathbf{B}$ gives an LDPC code whose Tanner graph has a girth of at least 6. Let $\mathcal{Q} = \{\mathbf{Q}_1, \mathbf{Q}_2, \ldots, \mathbf{Q}_m\}$ be a class of sparse $k \times n$ matrices over GF(2) that has the following structural properties: (1) each member matrix in $\mathcal{Q}$ satisfies the RC constraint; and (2) a matrix formed by any two member matrices in $\mathcal{Q}$ arranged either in a row or in a column satisfies the RC constraint, which is called the *pair-wise RC constraint*. The pair-wise RC constraint implies that, for $1 \leq l \leq m$, a matrix formed by taking $l$ member matrices in $\mathcal{Q}$ arranged either in a row or in a column also satisfies the RC constraint. Since each member matrix in $\mathcal{Q}$ satisfies the RC constraint, its null space gives an LDPC code.

If we replace each 1-entry in $\mathbf{B}$ by a member matrix in $\mathcal{Q}$ and a 0-entry by a $k \times n$ zero matrix, we obtain a $ck \times tn$ matrix $\mathbf{H}_{\text{sup}}$ over GF(2). For $\mathbf{H}_{\text{sup}}$ to satisfy the RC constraint, the replacement of the 1-entries in $\mathbf{B}$ by the member matrices in $\mathcal{Q}$ is carried out under the rule that all the 1-entries in a column or in a row must be replaced by distinct member matrices in $\mathcal{Q}$. This replacement rule is called the *replacement constraint*. Since $\mathbf{B}$ and the member matrices in $\mathcal{Q}$ are sparse, $\mathbf{H}_{\text{sup}}$ is also a sparse matrix. $\mathbf{H}_{\text{sup}}$ is simply a $c \times t$ array of $k \times n$ submatrices, each either a member matrix in $\mathcal{Q}$ or a $k \times n$ zero matrix. Since $\mathbf{B}$ satisfies the RC constraint, there are no four 1-entries at the four corners of a rectangle in $\mathbf{B}$. This implies that there are no four member matrices in $\mathcal{Q}$ at the four corners of a rectangle in $\mathbf{H}_{\text{sup}}$, viewed as a $c \times t$ array of $k \times n$ submatrices. Then it follows from the pair-wise RC constraint on the member matrices in $\mathcal{Q}$ and the constraint on the replacement of the 1-entries in $\mathbf{B}$ by the member matrices in $\mathcal{Q}$ that $\mathbf{H}_{\text{sup}}$ satisfies the RC constraint. Hence, the null space of $\mathbf{H}_{\text{sup}}$ gives an LDPC code $\mathcal{C}_{\text{sup}}$ of length $tn$ with rate at least $(tn - ck)/(tn)$.

The above construction of LDPC codes is referred to as the *superposition construction* [19, 20]. The parity-check matrix $\mathbf{H}_{\text{sup}}$ is obtained by superimposing the member matrices in $\mathcal{Q}$ onto the matrix $\mathbf{B}$. The subscript "sup" of $\mathbf{H}_{\text{sup}}$ stands for "superposition." The $\mathbf{B}$ matrix is referred to as the *base matrix* for superposition, and the member matrices in $\mathcal{Q}$ are called the *constituent matrices*. If the base matrix $\mathbf{B}$ has constant column and row weights $w_{b,c}$ and $w_{b,r}$, respectively, and each constituent matrix in $\mathcal{Q}$ has the same constant column and row weights $w_{\text{con},c}$ and $w_{\text{con},r}$, respectively, then $\mathbf{H}_{\text{sup}}$ has constant column and row weights $w_{b,c}w_{\text{con},c}$ and $w_{b,r}w_{\text{con},r}$, respectively. In this case, the null space of $\mathbf{H}_{\text{sup}}$ gives a regular LDPC code with minimum distance at least $w_{b,c}w_{\text{con},c} + 1$. The subscripts "$b$," "$c$," "$r$," and "con" of $w_{b,c}$, $w_{b,r}$, $w_{\text{con},c}$, and $w_{\text{con},r}$ stand for "base," "column," "row," and "constituent," respectively.

If all the constituent matrices in $\mathcal{Q}$ are arrays of CPMs of the same size, then $\mathbf{H}_{\text{sup}}$ is an array of circulant permutation and zero matrices. In this case, the null space of $\mathbf{H}_{\text{sup}}$ gives a QC-LDPC code. Any array $\mathbf{H}_{qc,\text{disp}}^{(e)}$ with $1 \leq e \leq 6$ constructed in Chapter 12 can be divided into subarrays to form constituent matrices in $\mathcal{Q}$.

It is clear that the number of constituent matrices in $\mathcal{Q}$ must be large enough that the parity-check matrix $\mathbf{H}_{\text{sup}}$ can be constructed in such a way as to satisfy the replacement constraint. The assignment of constituent matrices in $\mathcal{Q}$ to the 1-entries in the base matrix $\mathbf{B}$ to satisfy the replacement constraint is equivalent to *coloring the edges* of the Tanner graph of $\mathbf{B}$ (a bipartite graph) such that no two adjacent edges have the same color [21, 22]. It is known in graph theory that, for any bipartite graph, the minimum number of colors that is sufficient to achieve the coloring constraint is equal to the maximum node degree of the graph [22]. Suppose $\mathbf{B}$ has constant column and row weights $w_{b,c}$ and $w_{b,r}$, respectively. Let $w_{b,\max} = \max\{w_{b,c}, w_{b,r}\}$. Then $w_{b,\max}$ constituent matrices in $\mathcal{Q}$ will suffice to satisfy the replacement constraint.

If $\mathbf{B}$ is a $t \times t$ circulant with row weight $w_{b,r}$, then the replacement of the 1-entries in $\mathbf{B}$ can be carried out in a cyclic manner. First, we replace the $w_{b,r}$ 1-entries in the top row of $\mathbf{B}$ by $w_{b,r}$ distinct constituent matrices in $\mathcal{Q}$ and the $t - w_{b,r}$ 0-entries by $t - w_{b,r}$ zero matrices of size $k \times n$. This results in a row of $t$ submatrices of size $k \times n$. Then this row and its $t - 1$ right cyclic shifts (with each $k \times n$ submatrix as a shifting unit) give a $tk \times tn$ superimposed matrix $\mathbf{H}_{\text{sup}}$ that satisfies the replacement constraint. This replacement of 1-entries in $\mathbf{B}$ by the constituent matrices in $\mathcal{Q}$ is called *cyclic replacement*. If $\mathbf{B}$ is an array of circulants, then cyclic replacement is applied to each circulant in $\mathbf{B}$.

If each constituent matrix in $\mathcal{Q}$ is a row of permutation (or circulant permutation) matrices then, in replacing the 1-entries in $\mathbf{B}$ by constituent matrices in $\mathcal{Q}$, only the requirement that all the 1-entries in a column of $\mathbf{B}$ be replaced by distinct constituent matrices in $\mathcal{Q}$ is sufficient to guarantee that $\mathbf{H}_{\text{sup}}$ satisfies the RC constraint. This is due to the fact that a row of permutation matrices will satisfy the RC constraint, no matter whether the permutation matrices in the row are all distinct or not.

If each constituent matrix in $\mathcal{Q}$ is a permutation matrix, then it is not necessary to follow replacement rules at all. In this case, if the base matrix **B** satisfies the RC constraint, the superimposed matrix $\mathbf{H}_{\text{sup}}$ also satisfies the RC constraint.

### 13.7.2 Construction of Base and Constituent Matrices

RC-constrained base matrices and pair-wise RC-constrained constituent matrices can be constructed using finite geometries, finite fields, or BIBDs.

Consider the $m$-dimensional Euclidean geometry EG($m$, $q$) over GF($q$). As shown in Section 11.1 (or Section 2.7.1), $K_c = q^{m-1}$ circulants of size $(q^m - 1) \times (q^m - 1)$ can be constructed from the type-1 incidence vectors of the lines in EG($m$, $q$) not passing through the origin of the geometry. Each of these circulants has weight $q$ (i.e., both the column weight and the row weight are $q$). These circulants satisfy the pair-wise RC constraints. One or a group of these circulants arranged in a row (see (11.12)) can be used as a base matrix. If the weight $q$ of a circulant is too large, it can be decomposed into a row of column descendant circulants of the same size with smaller weights by column splitting as shown in Section 11.5. These column-descendant circulants also satisfy the pair-wise RC constraint. One or a group of these column-descendant circulants arranged in a row can be used as a base matrix.

The circulants constructed from EG($m$, $q$) can be decomposed into a class of column (or row)-descendant circulants by column (or row) splitting. These column (or row) descendants can be grouped to form a class of pair-wise RC-constraint constituent matrices for superposition. This is best explained by an example.

---

**Example 13.9** Consider the two-dimensional Euclidean geometry EG(2, 3) over GF(3). The type-1 incidence vectors of the lines in EG(2, 3) not passing through the origin form a single $8 \times 8$ circulant matrix **B** with both column and row weight 3 that satisfies the RC constraint. We use this circulant as the base matrix for superposition code construction. Next we consider the three-dimensional Euclidean geometry EG($3, 2^3$) over GF($2^3$). Nine $511 \times 511$ circulants $\mathbf{G}_1, \mathbf{G}_2, \ldots, \mathbf{G}_9$ can be constructed via the type-1 incidence vectors of the lines in EG($3, 2^3$) not passing through the origin (see Example 11.2). Each of these circulants has both column and row weight 8. For $0 \leq i \leq 9$, we decompose $\mathbf{G}_i$ into eight $511 \times 511$ CPMs $\mathbf{G}_{i,1}, \mathbf{G}_{i,2}, \ldots, \mathbf{G}_{i,8}$ by column decomposition. Using these eight CPMs, we form four $511 \times 1002$ matrices $\mathbf{Q}_{i,1} = [\mathbf{G}_{i,1}\mathbf{G}_{i,2}]$, $\mathbf{Q}_{i,2} = [\mathbf{G}_{i,3}\mathbf{G}_{i,4}]$, $\mathbf{Q}_{i,3} = [\mathbf{G}_{i,5}\mathbf{G}_{i,6}]$, and $\mathbf{Q}_{i,4} = [\mathbf{G}_{i,7}\mathbf{G}_{i,8}]$. Each $\mathbf{Q}_{i,j}$, $1 \leq j \leq 4$, has column and row weights 1 and 2, respectively. Then $\mathcal{Q} = \{\mathbf{Q}_{i,1}, \mathbf{Q}_{i,2}, \mathbf{Q}_{i,3}, \mathbf{Q}_{i,4}: 1 \leq i \leq 9\}$ forms a class of constituent matrices for superposition code construction. To construct the superimposed parity-check matrix $\mathbf{H}_{\text{sup}}$, for $1 \leq i \leq 8$, we replace the three 1-entries of the $i$th row by three constituent matrices from the group $\{\mathbf{Q}_{i,1}, \mathbf{Q}_{i,2}, \mathbf{Q}_{i,3}, \mathbf{Q}_{i,4}\}$. In the replacement, the three 1-entries in a column must be replaced by three constituent matrices in $\mathcal{Q}$ with three different first indices. For $0 \leq i \leq 8$, the three 1-entries in the $i$th row of **B** are replaced by three constituent matrices with the same first indices $i$. The replacement results in an

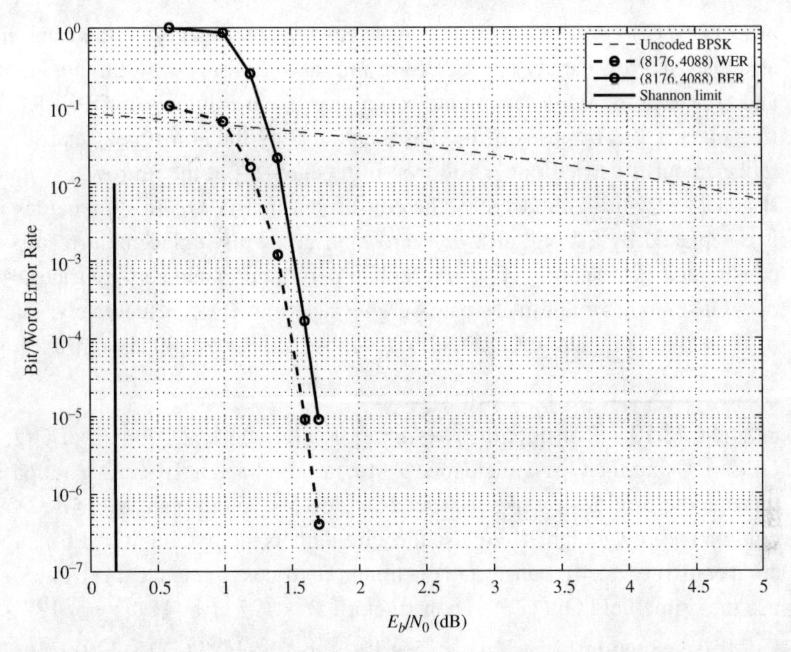


**Figure 13.12** The error performance of the (8176, 4088) QC-LDPC code given in Example 13.9.

RC-constrained $4088 \times 8176$ superimposed matrix $\mathbf{H}_{\text{sup}}$ with column and row weights 3 and 6, respectively. It is an $8 \times 16$ array of 48 $511 \times 511$ CPMs and 80 $511 \times 511$ zero matrices. The null space of $\mathbf{H}_{\text{sup}}$ gives a (3, 6)-regular (8176, 4088) QC-LDPC code with rate 1/2, whose Tanner graph has a girth of at least 6. The error performance of this code with iterative decoding using the SPA with 100 iterations is shown in Figure 13.12. At a BER of $10^{-6}$, it performs 1.5 dB from the Shannon limit.

---

For $1 \leq e \leq 6$, any RC-constrained array $\mathbf{H}^{(e)}_{qc,\text{disp}}$ of CPMs constructed as in Sections 12.3–12.7 can be partitioned into subarrays of the same size to form a class of constituent matrices for superposition code construction. Suppose the base matrix $\mathbf{B} = \left[b_{i,j}\right]$ is an RC-constrained $c \times t$ matrix over GF(2) with column and row weights $w_{b,c}$ and $w_{b,r}$, respectively. Choose two positive integers $k$ and $n$ such that $ck$ and $tn$ are smaller than the number of rows and the number of columns of $\mathbf{H}^{(e)}_{qc,\text{disp}}$, respectively. Take a $ck \times tn$ subarray $\mathbf{H}^{(e)}_{qc,\text{disp}}(ck, tn)$ from $\mathbf{H}^{(e)}_{qc,\text{disp}}$. Divide the $\mathbf{H}^{(e)}_{qc,\text{disp}}(ck, tn)$ horizontally into $c$ subarrays $\mathbf{Q}_1, \mathbf{Q}_2, \ldots, \mathbf{Q}_c$, where $\mathbf{Q}_i$ consists of the $i$th group of $k$ consecutive rows of CPMs of $\mathbf{H}^{(e)}_{qc,\text{disp}}(ck, tn)$. For $1 \leq i \leq c$, divide $\mathbf{Q}_i$ vertically into $t$ subarrays $\mathbf{Q}_{i,1}, \mathbf{Q}_{i,2}, \ldots, \mathbf{Q}_{i,t}$ of the same size $k \times n$. Then the set of $k \times n$ subarrays of $\mathbf{H}^{(e)}_{qc,\text{disp}}(ck, tn)$ given by

$$\mathcal{Q} = \{\mathbf{Q}_{i,j}: 1 \leq i \leq c, \quad 1 \leq j \leq t\} \tag{13.24}$$

can be used as constituent matrices for superposition code construction with a $c \times t$ base matrix $\mathbf{B}$. Note that the member matrices in $\mathcal{Q}$ do not exactly satisfy the pair-wise RC constraint. However, the member matrices in $\mathcal{Q}$ have the following RC-constraint properties: (1) for $j_1 \neq j_2$, any two matrices $\mathbf{Q}_{i,j_1}$ and $\mathbf{Q}_{i,j_2}$ with the same first index $i$ arranged in a row satisfy the RC constraint; and (2) for $i_1 \neq i_2$, any two matrices $\mathbf{Q}_{i_1,j}$ and $\mathbf{Q}_{i_2,j}$ with the same second index $j$ arranged in a column satisfy the RC constraint. In replacing the 1-entries of the base matrix $\mathbf{B}$ by the constituent matrices in $\mathcal{Q}$, the replacement is carried out as follows: if the entry $b_{i,j}$ at the $i$th row and $j$th column of $\mathbf{B}$ is a 1-entry, it is replaced by the constituent matrix $\mathbf{Q}_{i,j}$ in $\mathcal{Q}$, whereas if $b_{i,j} = 0$, it is replaced by a $k \times n$ array of zero matrices. This replacement results in an RC-constrained $ck \times tn$ array $\mathbf{H}_{\text{sup}}$ of circulant permutation and zero matrices. As a matrix over GF(2), it has column and row weights $kw_{b,c}$ and $nw_{b,r}$, respectively. The null space of $\mathbf{H}_{\text{sup}}$ gives a regular QC-LDPC code whose Tanner graph has a girth of at least 6.

---

**Example 13.10** Consider the two-dimensional Euclidean geometry EG$(2, 2^2)$ over GF$(2^2)$. Using the type-1 incidence vectors of the lines in EG$(2, 2^2)$ not passing through the origin, we can construct a single $15 \times 15$ circulant over GF(2) with both column and row weight 4. We use this circulant as the base matrix $\mathbf{B}$ for superposition code construction. To construct constituent matrices to replace the 1-entries in $\mathbf{B}$, we use the prime field GF(127) to construct a $127 \times 127$ array $\mathbf{H}^{(6)}_{qc,\text{disp}}$ of $127 \times 127$ CPMs (see Section 12.6). Take a $15 \times 120$ subarray $\mathbf{H}^{(6)}_{qc,\text{disp}}(15, 120)$ from $\mathbf{H}^{(6)}_{qc,\text{disp}}$. Choose $k = 1$ and $n = 8$. Using the method described above, we partition $\mathbf{H}^{(6)}_{qc,\text{disp}}(15, 120)$ into the following class of $1 \times 8$ subarrays of $127 \times 127$ CPMs: $\mathcal{Q} = \{\mathbf{Q}_{i,j}: 1 \leq i \leq 15,\ 1 \leq j \leq 15\}$. We replace the 1-entries in $\mathbf{B}$ by the constituent arrays in $\mathcal{Q}$ using the replacement rule as described above. The replacement results in a $15 \times 120$ superimposed array $\mathbf{H}_{\text{sup}}$ of circulant permutation and zero matrices of size $127 \times 127$. $\mathbf{H}_{\text{sup}}$ is a $1905 \times 15\,240$ matrix over GF(2) with column and row weights 4 and 32, respectively. The null space of this matrix gives a (4, 32)-regular (15 240, 13 342) QC-LDPC code with rate 0.875 whose Tanner graph has a girth of at least 6. The error performance of this code with iterative decoding using the SPA with 50 iterations is shown in Figure 13.13. At a BER of $10^{-6}$, the code performs 0.86 dB from the Shannon limit. It has a beautiful waterfall error performance.

---

The circulants constructed using class-I or class-II Bose BIBDs given in Section 13.2 can also be used to construct base matrices for the superposition construction of LDPC codes. For example, let $t = 1$. There is a class-I (13, 13, 4, 4, 1) Bose BIBD whose incidence matrix consists of a single $13 \times 13$ circulant $\mathbf{G}$ over GF(2) with both column and row weight 4. This circulant can be used as a base matrix for superposition to construct LDPC codes. If we decompose $\mathbf{G}$ into two column descendants $\mathbf{G}_1$ and $\mathbf{G}_2$, with column splitting such that $\mathbf{G}_1$ is a circulant with both column and row weight 3 and $\mathbf{G}_2$ is a CPM, the circulant $\mathbf{G}_1$ can be used as a base matrix for superposition construction of LDPC codes. Suppose we construct a class $\mathcal{Q}$ of pair-wise RC-constrained constituent matrices for which each member constituent matrix consists of a row of $l$ CPMs of size $n \times n$. On replacing each 1-entry of $\mathbf{G}_1$ by a constituent matrix in $\mathcal{Q}$ under the replacement constraint, we obtain a $13 \times 13l$ RC-constrained array $\mathbf{H}_{\text{sup}}$ of $n \times n$ circulant permutation and zero matrices. This array $\mathbf{H}_{\text{sup}}$ is a $13n \times 13ln$ matrix

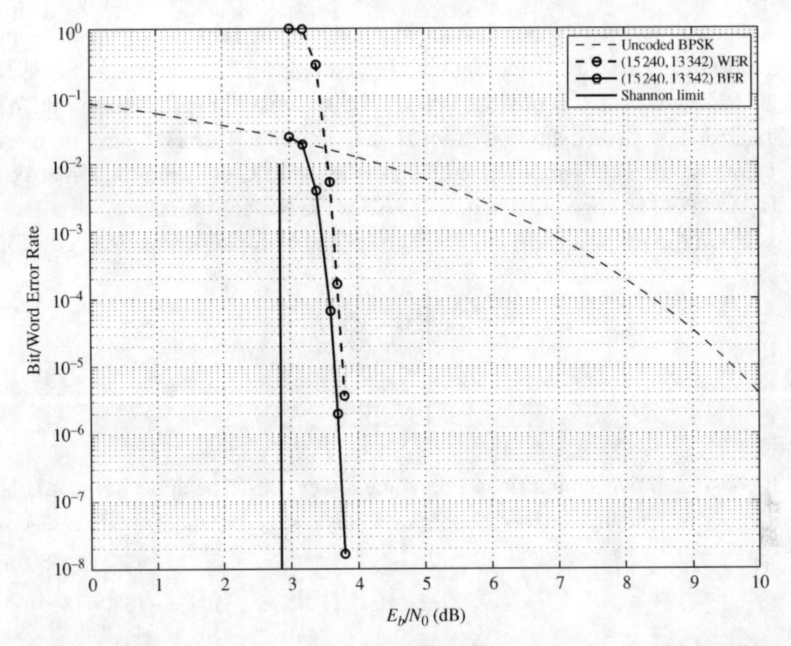


**Figure 13.13** The error performance of the (15 240, 13 342) QC-LDPC code given in Example 13.10.

over GF(2) with column and row weights 3 and $3l$, respectively. The null space of $\mathbf{H}_{\text{sup}}$ gives a $(3, 3l)$-regular QC-LDPC code of length $13ln$.

### 13.7.3 Superposition Construction of Product LDPC Codes

For $i = 1$ and 2, let $\mathcal{C}_i$ be an $(n_i, k_i)$ LDPC code with minimum distance $d_i$ that is given by the null space of an $m_i \times n_i$ RC-constrained parity-check matrix $\mathbf{H}_i$ over GF(2). For $i = 2$, express $\mathbf{H}_2$ in terms of its columns,

$$\mathbf{H}_2 = \left[\mathbf{h}_1^{(2)} \quad \mathbf{h}_2^{(2)} \quad \cdots \quad \mathbf{h}_{n_2}^{(2)}\right].$$

For $1 \leq j \leq n_2$, we form the following $m_2n_1 \times n_1$ matrix over GF(2) by shifting the $j$th column $\mathbf{h}_j^{(2)}$ downward $n_1$ times:

$$\mathbf{H}_j^{(2)} = \begin{bmatrix} \mathbf{h}_j^{(2)} & \mathbf{0} & \cdots & \mathbf{0} \\ \mathbf{0} & \mathbf{h}_j^{(2)} & \cdots & \mathbf{0} \\ \vdots & \vdots & \ddots & \vdots \\ \mathbf{0} & \mathbf{0} & \cdots & \mathbf{h}_j^{(2)} \end{bmatrix}, \tag{13.25}$$

where $\mathbf{0}$ is a column vector of length $m_2$. Each row of $\mathbf{H}_j^{(2)}$ has at most one 1-component and no two columns have any 1-component in common. It is clear that $\mathbf{H}_j^{(2)}$ satisfies the RC constraint. It follows from the structure of $\mathbf{H}_j^{(2)}$ with $1 \leq j \leq n_2$ that the $m_2n_1 \times n_2n_1$ matrix

$$\mathbf{H}_{2,\text{int}} = \left[\mathbf{H}_1^{(2)} \quad \mathbf{H}_2^{(2)} \quad \cdots \quad \mathbf{H}_{n_2}^{(2)}\right]$$

is simply obtained by interleaving the columns of $\mathbf{H}^{(2)}$ with a span (or interleaving depth) of $n_1$. The subscript "int" of $\mathbf{H}_{2,\text{int}}$ stands for "interleaving." Since $\mathbf{H}_2$ satisfies the RC constraint, it is obvious that $\mathbf{H}_{2,\text{int}}$ also satisfies the RC constraint. We also note that, for $1 \leq j \leq n_2$, any row from $\mathbf{H}_1$ and any row from $\mathbf{H}_j^{(2)}$ have no more than one 1-component in common. Since $\mathbf{H}_1$ satisfies the RC constraint, the matrix formed by $\mathbf{H}_1$ and $\mathbf{H}_j^{(2)}$ arranged in a column satisfies the RC constraint.

Form the following $(n_2 + 1) \times n_2$ base matrix for superposition code construction:

$$\mathbf{B} = \left[\begin{array}{ccccc} 1 & 0 & 0 & \cdots & 0 \\ 0 & 1 & 0 & \cdots & 0 \\ \vdots & \vdots & \vdots & \ddots & \vdots \\ 0 & 0 & 0 & \cdots & 1 \\ \hline 1 & 1 & 1 & \cdots & 1 \end{array}\right]. \tag{13.26}$$

$\mathbf{B}$ consists of two submatrices, upper and lower ones. The upper submatrix is an $n_2 \times n_2$ identity matrix and the lower submatrix is a $1 \times n_2$ row matrix with $n_2$ 1-components. Let

$$\mathcal{Q} = \{\mathbf{H}_1, \mathbf{H}_1^{(2)}, \mathbf{H}_2^{(2)}, \ldots, \mathbf{H}_{n_2}^{(2)}\}$$

be the class of constituent matrices for superposition code construction.

On replacing each 1-entry in the upper $n_2 \times n_2$ identity matrix of $\mathbf{B}$ by $\mathbf{H}_1$ and the $j$th 1-entry in the lower submatrix of $\mathbf{B}$ by the $\mathbf{H}_j^{(2)}$ in $\mathcal{Q}$ for $1 \leq j \leq n_2$, we obtain the following $(m_1n_2 + m_2n_1) \times n_1n_2$ matrix over GF(2):

$$\mathbf{H}_{\text{sup},p} = \left[\begin{array}{ccccc} \mathbf{H}_1 & 0 & 0 & \cdots & 0 \\ 0 & \mathbf{H}_1 & 0 & \cdots & 0 \\ \vdots & \vdots & \vdots & \ddots & \vdots \\ 0 & 0 & 0 & \cdots & \mathbf{H}_1 \\ \hline \mathbf{H}_1^{(2)} & \mathbf{H}_2^{(2)} & \mathbf{H}_3^{(2)} & \cdots & \mathbf{H}_{n_2}^{(2)} \end{array}\right]. \tag{13.27}$$

The matrix $\mathbf{H}_{\text{sup},p}$ consists of two parts. The upper part of $\mathbf{H}_{\text{sup},p}$ consists of an $n_2 \times n_2$ array of $m_1 \times n_1$ submatrices with $\mathbf{H}_1$ on its main diagonal and zero matrices elsewhere. The lower part of $\mathbf{H}_{\text{sup},p}$ is simply the matrix $\mathbf{H}_{2,\text{int}}$. It follows from the RC constraint structure of $\mathbf{H}_1$, $\mathbf{H}_2$, $\mathbf{H}_j^{(2)}$ with $1 \leq j \leq n_2$ and $\mathbf{H}_{2,\text{int}}$ that $\mathbf{H}_{\text{sup},p}$ satisfies the RC constraint.

The null space of $\mathbf{H}_{\text{sup},p}$ gives an LDPC code $\mathcal{C}_{\text{sup},p}$ of length $n_1n_2$ whose Tanner graph has a girth of at least 6. From the structure of $\mathbf{H}_{\text{sup},p}$, we can readily prove that $\mathbf{H}_{\text{sup},p}$ is a parity-check matrix of the direct product of $\mathcal{C}_1$ and $\mathcal{C}_2$, with $\mathcal{C}_1$ and $\mathcal{C}_2$ as the row and column codes, respectively. Therefore, $\mathcal{C}_{\text{sup},p} = \mathcal{C}_1 \times \mathcal{C}_2$ and the minimum distance of $\mathcal{C}_{\text{sup},p}$ is $d_{\text{sup},p} = d_1 \times d_2$, the product of the minimum distances of $\mathcal{C}_1$ and $\mathcal{C}_2$. The subscript "$p$" of $\mathbf{H}_{\text{sup},p}$ and $\mathcal{C}_{\text{sup},p}$ stands for "product." If $\mathcal{C}_1$ and $\mathcal{C}_2$ are both QC-LDPC codes, the encoding of $\mathcal{C}_{\text{sup},p}$ can be done with simple shift registers by encoding the row code and the column code separately. Taking the product of more than two codes can be carried out recursively.

Three methods can be used to decode a product LDPC code $\mathcal{C}_{\text{sup},p}$ with two component codes $\mathcal{C}_1$ and $\mathcal{C}_2$. The first method decodes the product LDPC code $\mathcal{C}_{\text{sup},p}$ with iterative decoding based on the parity-check matrix $\mathbf{H}_{\text{sup},p}$ of the code given by (13.27). The second method decodes the row code $\mathcal{C}_1$ and then the column code $\mathcal{C}_2$ iteratively, like turbo decoding. The third method is a hybrid decoding. First, we perform turbo decoding based on the two component codes. After a preset number of iterations of turbo decoding, we switch to iterative decoding of the product code $\mathcal{C}_{\text{sup},p}$ based on the parity-check matrix $\mathbf{H}_{\text{sup},p}$ given by (13.27).

---

**Example 13.11** Let $\mathcal{C}_1$ be the (1023, 781) cyclic EG-LDPC code constructed from the two-dimensional Euclidean geometry EG$(2, 2^5)$ over GF$(2^5)$. The parity-check matrix $\mathbf{H}_1$ of $\mathcal{C}_1$ is a $1023 \times 1023$ circulant with both column and row weight 32 (see Section 11.1.1 for its construction). The minimum distance of $\mathcal{C}_1$ is exactly 33. Let $\mathcal{C}_2$ be the (32, 31) single parity-check code with minimum distance 2. The parity-check matrix $\mathbf{H}_2$ of $\mathcal{C}_2$ is simply a row of 32 1-components. From (13.25), we see that, for $1 \leq j \leq 32$, $\mathbf{H}_j^{(2)}$ is a $1023 \times 1023$ identity matrix. It follows from (13.27) that we can form the parity-check matrix $\mathbf{H}_{\text{sup},p}$ of the product LDPC code $\mathcal{C}_{\text{sup},p} = \mathcal{C}_1 \times \mathcal{C}_2$. The null space of $\mathbf{H}_{\text{sup},p}$ gives a (32 736, 24 211) product LDPC code with rate 0.74 with minimum distance 66. The error performance of this product LDPC code with iterative decoding based on its parity-check matrix $\mathbf{H}_{\text{sup},p}$ using the SPA with 100 iterations is shown in Figure 13.14. We see that the code has a beautiful straight-down waterfall error performance. Since the code has a very large minimum distance, it should have a very low error floor.

---

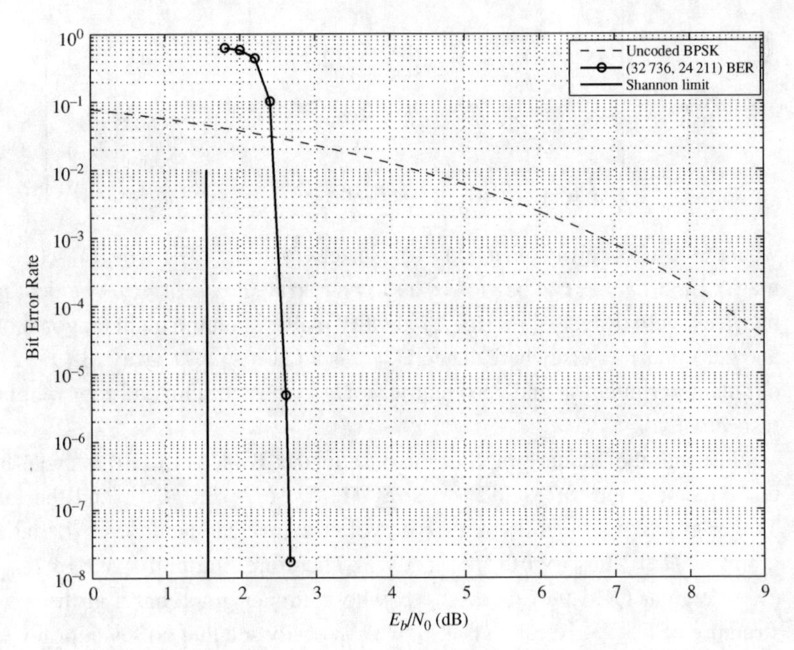


**Figure 13.14** The error performance of the (32 736, 24 211) product LDPC code given in Example 13.11.

## 13.8 Two Classes of LDPC Codes with Girth 8

For $1 \leq e \leq 6$, consider an RC-constrained array $\mathbf{H}^{(e)}_{qc,\text{disp}}$ of $(q-1) \times (q-1)$ CPMs constructed using the Galois field GF($q$) (see Sections 12.3–12.7). Let $\mathbf{H}^{(e)}_{qc,\text{disp}}(2, r)$ be a $2 \times r$ subarray of $\mathbf{H}^{(e)}_{qc,\text{disp}}$. The associated Tanner graph of this $2 \times r$ subarray is not only free of cycles of length 4 but also free of cycles of length 6, since forming a cycle of length 6 requires at least three rows of CPMs. This cycle structure can be used to construct a class of $(3, r)$-regular QC-LDPC codes whose Tanner graphs have a girth of 8, using the superposition code-construction method.

Take a $2k \times r$ subarray $\mathbf{H}^{(e)}_{qc,\text{disp}}(2k, r)$ from $\mathbf{H}^{(e)}_{qc,\text{disp}}$ with $2k$ and $r$ smaller than the numbers of rows and columns of $\mathbf{H}^{(e)}_{qc,\text{disp}}$, respectively. Assume that $\mathbf{H}^{(e)}_{qc,\text{disp}}(2k, r)$ contains no zero submatrix. Slice $\mathbf{H}^{(e)}_{qc,\text{disp}}(2k, r)$ horizontally into $k$ subarrays of size $2 \times r$, $\mathbf{H}^{(e)}_{1,qc,\text{disp}}(2, r), \mathbf{H}^{(e)}_{2,qc,\text{disp}}(2, r), \ldots, \mathbf{H}^{(e)}_{k,qc,\text{disp}}(2, r)$. Let

$$\mathcal{Q} = \{\mathbf{H}^{(e)}_{1,qc,\text{disp}}(2, r), \mathbf{H}^{(e)}_{2,qc,\text{disp}}(2, r), \ldots, \mathbf{H}^{(e)}_{k,qc,\text{disp}}(2, r)\} \tag{13.28}$$

be the class of constituent matrices for superposition code construction. Form a $(k+1) \times k$ base matrix $\mathbf{B}$ of the form given by (13.26). Next, we replace each 1-entry of the upper $k \times k$ identity submatrix of $\mathbf{B}$ by a constituent matrix in $\mathcal{Q}$ and each 1-entry of the lower submatrix of $\mathbf{B}$ by an $r(q-1) \times r(q-1)$ identity matrix. This replacement gives the following $(2k + r)(q-1) \times kr(q-1)$ matrix over GF(2):

$$\mathbf{H}_{\text{sup},p} = \left[\begin{array}{cccc} \mathbf{H}^{(e)}_{1,qc,\text{disp}}(2, r) & \mathbf{O} & \cdots & \mathbf{O} \\ \mathbf{O} & \mathbf{H}^{(e)}_{2,qc,\text{disp}}(2, r) & \cdots & \mathbf{O} \\ \vdots & \vdots & \ddots & \vdots \\ \mathbf{O} & \mathbf{O} & \cdots & \mathbf{H}^{(e)}_{k,qc,\text{disp}}(2, r) \\ \hline \mathbf{I}_{r(q-1)\times r(q-1)} & \mathbf{I}_{r(q-1)\times r(q-1)} & \cdots & \mathbf{I}_{r(q-1)\times r(q-1)} \end{array}\right], \tag{13.29}$$

where $\mathbf{O}$ is a $2 \times r$ array of $(q-1) \times (q-1)$ zero matrices. The $r(q-1) \times r(q-1)$ identity matrix $\mathbf{I}_{r(q-1)\times r(q-1)}$ can be viewed as an $r \times r$ array of $(q-1) \times (q-1)$ identity and zero matrices, with the $(q-1) \times (q-1)$ identity matrices on the main diagonal of the array and zero matrices elsewhere. Then $\mathbf{H}_{\text{sup},p}$ is a $(2k + r) \times kr$ array of $(q-1) \times (q-1)$ circulant permutation and zero matrices. $\mathbf{H}_{\text{sup},p}$ has column and row weights 3 and $r$, respectively.

Note that the Tanner graph of the base matrix $\mathbf{B}$ (see (13.26)) is cycle-free. Given the cycle structure of each $2 \times r$ array $\mathbf{H}^{(e)}_{i,qc,\text{disp}}(2, r)$ in $\mathbf{H}_{\text{sup},p}$ and the fact that the Tanner graph of an identity matrix is cycle-free, we can readily prove that the Tanner graph of $\mathbf{H}_{\text{sup},p}$ has a girth of exactly 8. Therefore, the null space of $\mathbf{H}_{\text{sup},p}$ gives a $(3, r)$-regular QC-LDPC code $\mathcal{C}_{\text{sup},p}$, whose Tanner graph has a girth of 8. From the structure of $\mathbf{H}_{\text{sup},p}$ given by (13.29), we can easily see that no seven or fewer columns of $\mathbf{H}_{\text{sup},p}$ can be added to a zero column vector. Hence, the minimum distance of $\mathcal{C}_{\text{sup},p}$ is at least 8. The above superposition code construction gives a class of QC-LDPC codes. Actually, $C_{\text{sup},p}$ is a generalized product code with $k$ row codes and one column code. The $i$th row code is given by the null space of the $2 \times r$ array $\mathbf{H}^{(e)}_{i,qc,\text{disp}}(2,r)$ of $(q-1)\times(q-1)$ CPMs and the column code is simply the $(k, k-1)$ *single-parity-check (SPC)* code whose parity-check matrix $\mathbf{H}_{\text{spc}} = [1 \quad 1 \quad \cdots \quad 1]$ consists of a row of $k$ 1-components.

---

**Example 13.12** Consider the $127 \times 127$ array $\mathbf{H}^{(1)}_{qc,\text{disp}}$ of $127 \times 127$ circulant permutation and zero matrices constructed from the field GF($2^7$) using the method given in Section 12.3, where the zero matrices lie on the main diagonal of $\mathbf{H}^{(1)}_{qc,\text{disp}}$. Set $k = r = 6$. Take a $12 \times 6$ subarray $\mathbf{H}^{(1)}_{qc,\text{disp}}(12,6)$ from $\mathbf{H}^{(1)}_{qc,\text{disp}}$, avoiding the zero submatrices on the main diagonal of $\mathbf{H}^{(1)}_{qc,\text{disp}}$. Slice $\mathbf{H}^{(1)}_{qc,\text{disp}}(12,6)$ into six $2 \times 6$ subarrays, $\mathbf{H}^{(1)}_{1,qc,\text{disp}}(2,6), \mathbf{H}^{(1)}_{1,qc,\text{disp}}(2,6), \ldots, \mathbf{H}^{(1)}_{6,qc,\text{disp}}(2,6)$. We use these $2 \times 6$ subarrays for superposition code construction. Construct a $7 \times 6$ base matrix $\mathbf{B}$ of the form given by (13.26). Using the above superposition construction method, we construct an $18 \times 36$ array $\mathbf{H}_{\text{sup},p}$ of $127 \times 127$ circulant permutation and zero matrices. $\mathbf{H}_{\text{sup},p}$ is a $2286 \times 4572$ matrix with column and row weights 3 and 6, respectively. The null space of $\mathbf{H}_{\text{sup},p}$ gives a (3, 6)-regular (4572, 2307) QC-LDPC code with rate 0.5045, whose Tanner graph has a girth of 8. The error performance of this code with iterative decoding using the SPA with 100 iterations is shown in Figure 13.15. At a BER of $10^{-6}$, it performs 1.6 dB from the Shannon limit.

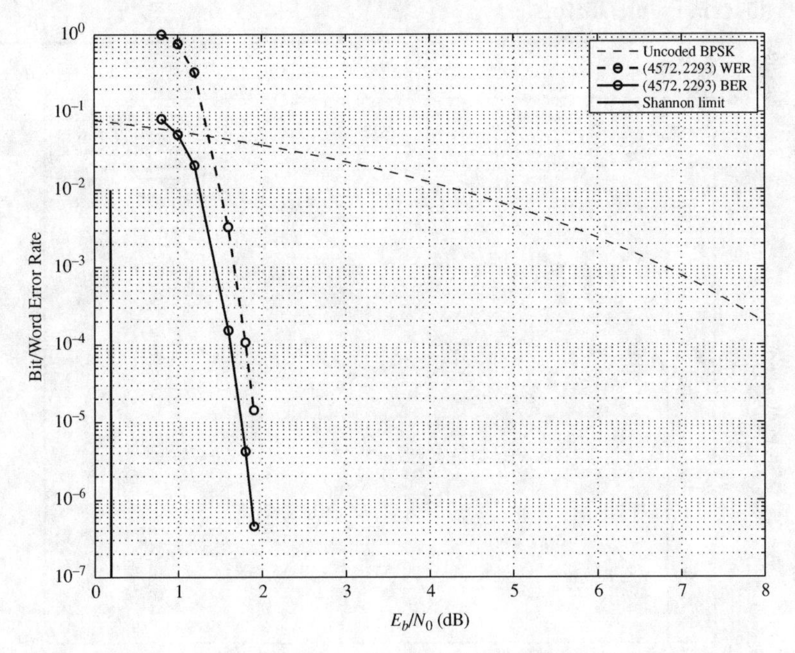


**Figure 13.15** The error performance of the (4572, 2307) QC-LDPC code given in Example 13.12.

---

The superposition construction given by (13.27) can be applied repeatedly to construct a large sparse matrix from a single small and simple matrix. A very simple case is to take the product of a short SPC code with itself repeatedly. Let $\mathcal{C}_{\text{spc}}$ be an $(n, n-1)$ SPC code with parity-check matrix $\mathbf{H}_{\text{spc}} = [1 \quad 1 \quad \cdots \quad 1]$ that consists of a row of $n$ 1-components. First, we use $\mathbf{H}_{\text{spc}}$ as both $\mathbf{H}_1$ and $\mathbf{H}_2$ to construct a superimposed matrix $\mathbf{H}_{\text{sup},p}$ based on (13.27). $\mathbf{H}_{\text{sup},p}$ is called the *second power* of $\mathbf{H}_{\text{spc}}$, denoted $\mathbf{H}^2_{\text{spc}}$. $\mathbf{H}^2_{\text{spc}}$ has column and row weights 2 and $n$, respectively. We can easily check that the girth of the Tanner graph of $\mathbf{H}^2_{\text{spc}}$ is 8. Next, we use $\mathbf{H}^2_{\text{spc}}$ as $\mathbf{H}_1$ and $\mathbf{H}_{\text{spc}}$ as $\mathbf{H}_2$ to construct the third power $\mathbf{H}^3_{\text{spc}}$ of $\mathbf{H}_{\text{spc}}$ using (13.27). We can easily prove that the girth of the Tanner graph of $\mathbf{H}^3_{\text{spc}}$ is again 8. $\mathbf{H}^3_{\text{spc}}$ has column and row weights 3 and $n$, respectively. Repeating the above recursive construction forms the $k$th power $\mathbf{H}^k_{\text{spc}}$ of $\mathbf{H}_{\text{spc}}$, which has column and row weights $k$ and $n$, respectively. The girth of $\mathbf{H}^k_{\text{spc}}$ is again 8. $\mathbf{H}^k_{\text{spc}}$ can be used either as a base matrix for superposition construction of an LDPC code, or as the parity-check matrix to generate an LDPC code $\mathcal{C}^k_{\text{spc}}$ of length $n^k$ with rate $((n-1)/n)^k$, girth 8, and minimum distance $2^k$.

---

**Example 13.13** Suppose we use the (6, 5) SPC code for product LDPC code construction. The parity-check matrix of this SPC code is $\mathbf{H}_{\text{spc}} = [1 \quad 1 \quad 1 \quad 1 \quad 1 \quad 1]$. Using (13.27) twice, we form the third power $\mathbf{H}^3_{\text{spc}}$ of $\mathbf{H}_{\text{spc}}$, which is a $128 \times 216$ matrix over GF(2) with column and row weights 3 and 6, respectively. The null space of $\mathbf{H}^3_{\text{spc}}$ gives a (216, 125) product LDPC code with rate 0.579 and minimum distance 8, whose error performance with iterative decoding using the SPA (100 iterations) is shown in Figure 13.16.

---

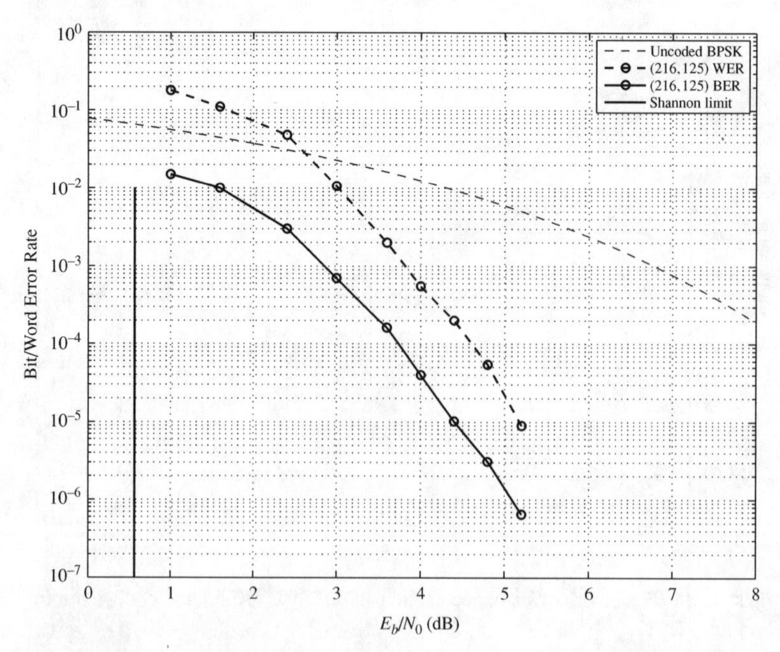


**Figure 13.16** The error performance of the (216, 125) product LDPC code given in Example 13.13.

## 13.9 Globally Coupled LDPC Codes

This section presents a special type of LDPC code, called a globally coupled (GC) LDPC code.

### 13.9.1 Basic Characteristics of a Globally Coupled LDPC Code

The Tanner graph $G_{global}$ of a GC-LDPC code $C_{global}$ consists of a set of disjoint subgraphs, denoted by $G_{local}$, of the same size (i.e., the same number of VNs and the same number of CNs), called *local graphs*. The VNs of the local graphs are globally connected by a group of CNs of another subgraph $G_{coupling}$ of $G_{global}$, called a *global coupling graph*, as shown in Figure 13.17. The graph $G_{global}$ is called a *globally coupled graph* (global graph). The VNs and CNs of the local graphs are called local VNs and local CNs of the global graph $G_{global}$. The CNs of the global coupling graph $G_{coupling}$ are called *global CNs* of $G_{global}$. For complete global coupling, each local VN is directly connected to at least one global CN. A GC-LDPC code is referred to as a *CN-based GC-LDPC code*.

Each local graph in $G_{global}$ specifies a local LDPC code $C_{local}$ and the global coupling graph $G_{coupling}$ of $G_{global}$ specifies a global coupling code $C_{coupling}$. Suppose $G_{global}$ consists of $c$, $c > 1$, local graphs, denoted by $G_{local,0}, G_{local,1}, \ldots, G_{local,c-1}$. Let $\mathbf{H}_{local,0}, \mathbf{H}_{local,1}, \ldots, \mathbf{H}_{local,c-1}$ be the parity-check matrices of local codes $C_{local,0}, C_{local,1}, \ldots, C_{local,c-1}$ specified by the local graphs $G_{local,0}, G_{local,1}, \ldots, G_{local,c-1}$ of $G_{global}$. Let $\mathbf{H}_{coupling}$ be the parity-check matrix of the global coupling code $C_{coupling}$ specified by the global coupling graph $G_{coupling}$. Let $\mathbf{H}_{global}$ be the parity-check matrix of the CN-based globally coupled LDPC code $C_{global}$ specified by $G_{global}$. The number of columns in the global coupling parity-check matrix $\mathbf{H}_{coupling}$ is at least equal to the sum of columns of the $c$ local parity-check matrices.

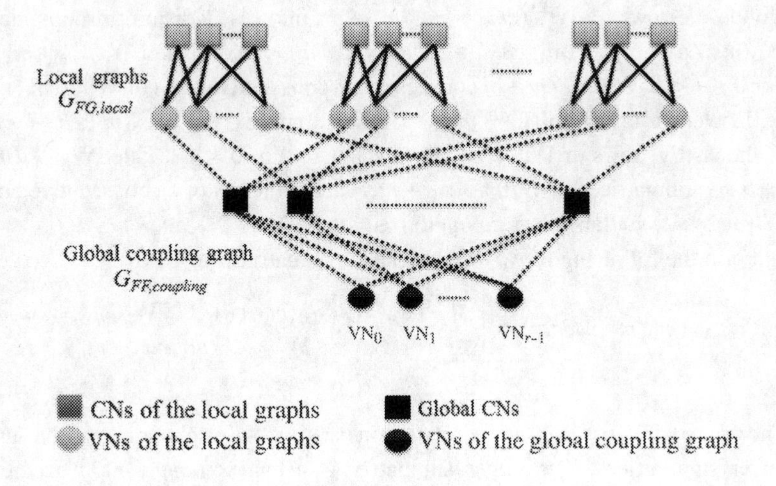


**Figure 13.17** The global coupling structure of the Tanner graph of a CN-based GC-LDPC code.

It follows from the above characteristics of a CN-based GC-LDPC code $C_{global}$ that its parity-check matrix must be in the following form:

$$\mathbf{H}_{global} = \begin{bmatrix} \mathbf{0} & \text{diag}(\mathbf{H}_{local,0}, \mathbf{H}_{local,1}, \ldots, \mathbf{H}_{local,c-1}) \\ & \mathbf{H}_{coupling} \end{bmatrix}, \tag{13.30}$$

where $\text{diag}(\mathbf{H}_{local,0}, \mathbf{H}_{local,1}, \ldots, \mathbf{H}_{local,c-1})$ is a $c \times c$ array with $c$ local parity-check matrices lying on its main diagonal and $\mathbf{0}$ is a zero matrix of appropriate size. If the number of columns in $\mathbf{H}_{coupling}$ is equal to the number of columns in $\text{diag}(\mathbf{H}_{local,0}, \mathbf{H}_{local,1}, \ldots, \mathbf{H}_{local,c-1})$, then the zero matrix $\mathbf{0}$ is a null matrix (i.e., it does not exist).

If $\mathbf{H}_{global}$ satisfies the RC constraint, its associated Tanner graph $G_{global}$ has a girth of at least 6. If $\mathbf{H}_{global}$ is an array of CPMs and/or zero matrices of the same size, the null space over GF(2) of $\mathbf{H}_{global}$ gives a CN-based QC-GC-LDPC code $C_{global}$. The decoding of $C_{global}$ can be carried out in one phase based on the global parity-check matrix $\mathbf{H}_{global}$ or in two phases with one phase based on the local parity-check matrices and the other phase based on the global coupling matrix.

If $\mathbf{H}_{global}$ does not satisfy the RC constraint, but the local parity-check matrices and the global coupling matrix satisfy the RC constraint separately, then the globally coupled code $C_{global}$ needs to be decoded in two phases to achieve good error-rate performance.

### 13.9.2 A Construction of CN-Based QC-GC-LDPC Codes

Let $m_0, m_1, n_0, c$, and $r$ be five positive integers with $m_0 < n_0$. Construct a $(cm_0 + m_1) \times (cn_0 + r)$ base matrix $\mathbf{W}_s^{(9)}(cm_0 + m_1, cn_0 + r)$ over GF($q$) in the form of (13.30) using two subsets, $\mathbf{S}_0 = \{\alpha^{i_0}, \alpha^{i_1}, \ldots, \alpha^{i_{m-1}}\}$ and $\mathbf{S}_1 = \{\alpha^{j_0}, \alpha^{j_1}, \ldots, \alpha^{j_{n-1}}\}$, of GF($q$) with sizes $m = cm_0 + m_1$ and $n = cn_0 + r$, respectively, and $n \leq q$. The matrix $\mathbf{W}_s^{(9)}(cm_0 + m_1, cn_0 + r)$ satisfies the $2 \times 2$ SM constraint. Label the rows and columns of $\mathbf{W}_s^{(9)}(cm_0 + m_1, cn_0 + r)$ from 0 to $cm_0 + m_1 - 1$ and 0 to $cn_0 + r - 1$, respectively. Divide the rows of $\mathbf{W}_s^{(9)}(cm_0 + m_1, cn_0 + r)$ into $c + 1$ disjoint groups, denoted by $\mathbf{W}_0(m_0, cn_0 + r), \mathbf{W}_1(m_1, cn_0 + r), \ldots, \mathbf{W}_{c-1}(m_{c-1}, cn_0 + r)$ and $\mathbf{W}_{coupling}(m_1, cn_0 + r)$. For $0 \leq l < c$, $\mathbf{W}_l(m_0, cn_0 + r)$ consists of $m_0$ consecutive rows of $\mathbf{W}_s^{(9)}(cm_0 + m_1, cn_0 + r)$, labeled from $lm_0$ to $(l+1)m_0 - 1$. The submatrix $\mathbf{W}_{coupling}(m_1, cn_0 + r)$ consists of the last $m_1$ rows of $\mathbf{W}_s^{(9)}(cm_0 + m_1, cn_0 + r)$. For $0 \leq l < c$, let $\mathbf{W}_{local,l}(m_0, n_0)$ be $m_0 \times n_0$ submatrices of $\mathbf{W}_l(m_0, cn_0 + r)$, which consists of $n_0$ consecutive columns of $\mathbf{W}_l(m_0, cn_0 + r)$, labeled from $ln_0$ to $(l+1)n_0 - 1$.

Form the following $(cm_0 + m_1) \times (cn_0 + r)$ matrix over GF($q$):

$$\mathbf{W}_{global}(m_0, m_1, n_0, c, r) = \begin{bmatrix} \mathbf{0} & \text{diag}(\mathbf{W}_{local,0}(m_0, n_0), \ldots, \mathbf{W}_{local,c-1}(m_0, n_0)) \\ & \mathbf{W}_{coupling}(m_1, cn_0 + r) \end{bmatrix}. \tag{13.31}$$

The matrix $\mathbf{W}_{global}(m_0, m_1, n_0, c, r)$ consists of two submatrices, the upper and lower submatrices. The upper submatrix is a $cm_0 \times (cn_0 + r)$ matrix consisting of two parts, where the first part $\mathbf{0}$ is a $cm_0 \times r$ zero matrix and the second part, $\text{diag}(\mathbf{W}_{local,0}(m_0, n_0), \ldots, \mathbf{W}_{local,c-1}(m_0, n_0))$ is a $c \times c$ diagonal array with the $c$ submatrices $\mathbf{W}_{local,0}(m_0, n_0), \ldots, \mathbf{W}_{local,c-1}(m_0, n_0)$ of size $m_0 \times n_0$ lying on its main diagonal and zeros elsewhere. The lower submatrix $\mathbf{W}_{coupling}(m_1, cn_0 + r)$ of $\mathbf{W}_{global}(m_0, m_1, n_0, c, r)$ is an $m_1 \times (cn_0 + r)$ matrix, which globally connects the $c$ disjoint local base matrices, $\mathbf{W}_{local,0}(m_0, n_0), \ldots, \mathbf{W}_{local,c-1}(m_0, n_0)$. Hence, $\mathbf{W}_{global}(m_0, m_1, n_0, c, r)$ is a globally coupled matrix over GF($q$). The matrix $\mathbf{W}_{global}(m_0, m_1, n_0, c, r)$ is a submatrix of the base matrix $\mathbf{W}_s^{(9)}(cm_0+m_1, cn_0+r)$. Since $\mathbf{W}_s^{(9)}(cm_0+m_1, cn_0+r)$ satisfies the $2 \times 2$ SM constraint, $\mathbf{W}_{global}(m_0, m_1, n_0, c, r)$ also satisfies this constraint.

Dispersing each nonzero entry in $\mathbf{W}_{global}(m_0, m_1, n_0, c, r)$ into a $(q-1) \times (q-1)$ CPM and each zero entry into a $(q-1) \times (q-1)$ zero matrix (ZM) with respect to GF($q$), we obtain the following $(cm_0 + m_1) \times (cn_0 + r)$ array of CPMs and ZMs of size $(q-1) \times (q-1)$:

$$\mathbf{H}_{global}(m_0, m_1, n_0, c, r) = \begin{bmatrix} \mathbf{0} \quad \text{diag}(\mathbf{H}_{local,0}(m_0, n_0), \ldots, \mathbf{H}_{local,c-1}(m_0, n_0)) \\ \mathbf{H}_{coupling}(m_1, cn_0 + r) \end{bmatrix}, \tag{13.32}$$

where for $0 \le l < c$, $\mathbf{H}_{local,0}(m_0, n_0) = \text{CPM}(\mathbf{W}_{local,l}(m_0, n_0))$, $\mathbf{0}$ is a $cm_0(q-1) \times r(q-1)$ zero matrix, and $\mathbf{H}_{coupling}(m_1, cn_0 + r) = \text{CPM}(\mathbf{W}_{coupling}(m_1, cn_0 + r))$. The array $\mathbf{H}_{global}(m_0, m_1, n_0, c, r)$ is a $(cm_0 + m_1)(q-1) \times (cn_0 + r)(q-1)$ matrix over GF(2), in which the submatrix $\mathbf{H}_{coupling}(m_1, cn_0 + r)$ globally couples $c$ disjoint local submatrices $\mathbf{H}_{local,l}(m_0, n_0)$, $0 \le l < c$. $\mathbf{H}_{global}(m_0, m_1, n_0, c, r)$ satisfies the RC constraint.

The null space over GF(2) of each local parity-check matrix $\mathbf{H}_{local,l}(m_0, n_0)$ gives a local QC-LDPC code $C_{local,l}(m_0, n_0)$ of length $n_0$. The null space over GF(2) of the coupling matrix $\mathbf{H}_{coupling}(m_1, cn_0 + r)$ gives a global coupling QC-LDPC code of length $cn_0 + r$. The null space over GF(2) of $\mathbf{H}_{global}(m_0, m_1, n_0, c, r)$ gives a CN-based QC-GC-LDPC code $C_{global}(m_0, m_1, n_0, c, r)$ of length $(cn_0+r)(q-1)$, which is composed of $c$ local QC-LDPC codes $C_{local,l}(m_0, n_0)$, $0 \le l < c$, of length $n_0$ and a global coupling QC-LDPC code $C_{coupling}(m_1, cn_0 + r)$ of length $(cn_0 + r)(q-1)$. The Tanner graph of $C_{global}(m_0, m_1, n_0, c, r)$ has a girth of at least 6. Clearly, the Tanner graph $G_{local,l}(m_0, n_0)$ of each local code $C_{local,l}(m_0, n_0)$ has a girth of at least 6 and the Tanner graph $G_{coupling}(m_1, cn_0+r)$ of the global coupling code $C_{coupling}(m_1, cn_0+r)$ has a girth of at least 6. The global coupling between the local graphs and the connecting graph in $G_{global}(m_0, m_1, n_0, c, r)$ is shown in Figure 13.17.

### 13.9.3 Encoding of a CN-based QC-GC-LDPC Code

Let $k_0$ be the dimension of each local code $C_{local,l}(m_0, n_0)$. Hence, each local code is an $(n_0, k_0)$ code. Let $\mathbf{u}$ be a sequence of $ck_0$ binary information symbols. Encoding of this information sequence consists of two stages, local and global coupling encodings. First, the information sequence $\mathbf{u}$ is divided into $c$ subsequences, denoted by $\mathbf{u}_0, \mathbf{u}_1, \ldots, \mathbf{u}_{c-1}$, each consisting of $k_0$ information symbols and called a message. For $0 \le l < c$, the $l$th message $\mathbf{u}_l$ is encoded into a codeword $\mathbf{v}_l$ of $n_0$ code symbols in the

$l$th local code $C_{local,l}(m_0, n_0)$. Encoding results in $c$ local codewords, $\mathbf{v}_0, \mathbf{v}_1, \ldots, \mathbf{v}_{c-1}$, each in a different local code. Cascading these $c$ local codewords, we obtain a sequence $\mathbf{v} = (\mathbf{v}_0, \mathbf{v}_1, \ldots, \mathbf{v}_{c-1})$ of $cn_0$ code symbols, called a *cascaded codeword*.

This completes the first stage of encoding, referred to as the *local encoding*. There are $2^{ck_0}$ such cascaded codewords, which form a $(cn_0, ck_0)$ linear code of length $cn_0$, denoted by $C_{casc}(c)$. The parity-check matrix of $C_{casc}(c)$ is the $c \times c$ diagonal array $\text{diag}(\mathbf{H}_{local,0}(m_0, n_0), \ldots, \mathbf{H}_{local,c-1}(m_0, n_0))$. The code $C_{casc}(c)$ is called a *cascaded code* of the $c$ local codes. Note that for multi-user communications, the $c$ messages may come from $c$ sources (users). In this case, interleaving is used to combine the $c$ different local codewords.

In the second stage of encoding, a cascaded codeword $\mathbf{v}$ in $C_{casc}(c)$ is encoded into a codeword $\mathbf{w} = (\mathbf{p}, \mathbf{v})$ of $cn_0 + r$ code symbols in the global coupling LDPC code $C_{coupling}(m_1, cn_0 + r)$. Assume systematic encoding. The codeword $\mathbf{w}$ consists of two parts, $\mathbf{v}$ and $\mathbf{p}$. The first part $\mathbf{v}$ is a cascaded codeword in $C_{casc}(c)$ and the second part $\mathbf{p}$ consists of $r$ parity-check symbols, which are formed based on the parity-check matrix $\mathbf{H}_{coupling}(m_1, cn_0 + r)$ of $C_{coupling}(m_1, cn_0 + r)$. These $r$ parity symbols globally connect the $c$ local codewords $\mathbf{v}_0, \mathbf{v}_1, \ldots, \mathbf{v}_{c-1}$ in $\mathbf{v}$. The encoding performed at the second stage is referred to as *global coupling encoding*. Local encoding, cascading, and global coupling encoding result in $2^{ck_0}$ codewords in $C_{coupling}(m_1, cn_0 + r)$. It is clear that $\mathbf{w} \cdot \mathbf{H}^{\mathrm{T}}_{global}(m_0, m1, n_0, c, r) = \mathbf{0}$. Hence, $\mathbf{w}$ is a codeword in the CN-based QC-GC-LDPC code $C_{global}(m_0, m_1, n_0, c, r)$.

### 13.9.4 Two-Phase Iterative Decoding of a CN-Based QC-GC-LDPC Code

Suppose a codeword $\mathbf{w} = (\mathbf{p}, \mathbf{v}_0, \mathbf{v}_1, \ldots, \mathbf{v}_{c-1})$ in $C_{global}(m_0, m_1, n_0, c, r)$ is transmitted. Let $\mathbf{r} = (\mathbf{r}_p, \mathbf{r}_0, \mathbf{r}_1, \ldots, \mathbf{r}_{c-1})$ be the received word. Since the global parity-check matrix $\mathbf{H}_{global}(m_0, m_1, n_0, c, r)$ satisfies the RC constraint, the received word $\mathbf{r}$ can be decoded based on $\mathbf{H}_{global}(m_0, m_1, n_0, c, r)$ in one phase iteratively, using either the SPA or the MSA. However, based on the global coupling structure of $C_{global}(m_0, m_1, n_0, c, r)$, $\mathbf{r}$ can be decoded in two phases iteratively. In the following, two two-phase iterative decoding schemes, namely *global/local* and *local/global*, are presented.

With the global/local iterative decoding, the received word $\mathbf{r}$ is first decoded based on the global coupling parity-check matrix $\mathbf{H}_{coupling}(m_1, cn_0+r)$ of the global coupling code $C_{coupling}(m_1, cn_0 + r)$. Set the maximum number of global coupling decoding iterations to $I_{coupling,max}$. Decode $\mathbf{r}$ based on a chosen iterative decoding algorithm. At the end of each global coupling decoding iteration, we check the syndrome $\mathbf{S}_{coupling}$ of the decoded vector $\mathbf{y}$ based on $\mathbf{H}_{coupling}(m_1, cn_0 + r)$. If $\mathbf{S}_{coupling} = \mathbf{0}$, then $\mathbf{y}$ is a codeword in $C_{coupling}(m_1, cn_0 + r)$. If $\mathbf{S}_{coupling} \neq \mathbf{0}$, we continue the global coupling decoding until either the decoded vector $\mathbf{y}$ is a codeword in $C_{coupling}(m_1, cn_0 + r)$ or the preset maximum number $I_{coupling,max}$ of decoding iterations is reached.

If a codeword $\mathbf{w}$ in $C_{coupling}(m_1, cn_0 + r)$ is obtained during the global coupling decoding phase, we remove all of the $r$ parity-check symbols from $\mathbf{w}$. This gives $c$ decoded words $\mathbf{v}^*_0, \mathbf{v}^*_1, \ldots, \mathbf{v}^*_{c-1}$ for the $c$ transmitted local codewords. For each decoded vector $\mathbf{v}_l^*$, $0 \le l < c$, we compute its syndrome $\mathbf{S}_{local,l}$ based on the local parity-check matrix $\mathbf{H}_{local,l}(m_0, n_0)$ of the local code $C_{local,l}(m_0, n_0)$. If the syndromes of the $c$ decoded words are all zero, then $\mathbf{v}_0^*, \mathbf{v}_1^*, \ldots, \mathbf{v}_{c-1}^*$ are codewords in the local codes. In this case, the decoding process is complete. Then all the parity-check symbols from the $c$ decoded local codewords are removed and the $ck_0$ decoded information symbols are delivered to the user (or users in multi-user communications). If the syndrome of any of the $c$ decoded vectors is not zero, the corresponding local code decoder is activated to perform decoding on the failed local word.

For the decoding of each failed local word at the output of the global coupling decoder, we set the maximum number of local decoding iterations to $I_{local,\max}$. If the decoding of all the failed local words is successful, all the parity-check symbols from all the decoded local codewords are removed and the $ck_0$ decoded information symbols are delivered to the user(s). If the decoding of any failed local word is unsuccessful after $I_{local,\max}$ iterations, decoding is switched back to the global coupling decoding phase with the decoded information and the channel information as input to decode the received vector $\mathbf{r}$ again.

The global coupling decoding and the local phase decoding form a decoding loop, referred to as a global/local decoding loop. The global/local decoding process continues until either the entire decoding is successful, or a present maximum number $I_{\max}$ of global/local decoding iterations is reached. With the global/local decoding of the QC-GC-LDPC code, the local decoder is used to correct the *local errors* that the global coupling decoder fails to correct.

Note that if the global coupling decoding is successful in an iteration loop, the entire decoding process can be terminated without checking the syndromes of the decoded local codewords to determine whether the local decoding should be carried out. This reduces computational complexity and decoding delay.

Opposite to global/local decoding as presented above, the received vector $\mathbf{r}$ can be decoded in reverse order, that is, perform local decoding before global coupling decoding in each global decoding loop. Recall that each globally coupled codeword $\mathbf{w}$ in systematic form consists of $c$ local codewords in cascade followed by $r$ parity-check symbols of the global coupling code $C_{coupling}(m_1, cn_0 + r)$. Local decoding is carried out as soon as a local received word $\mathbf{r}_l$ , $0 \le l < c$, of $n_0$ symbols corresponding to a transmitted local codeword $\mathbf{v}_l$ is received. The decoding of the $l$th local received word $\mathbf{r}_l$ is based on the $l$ th local parity-check matrix $\mathbf{H}_{local,l}(m_0, n_0)$ using a chosen iterative decoding algorithm. If the local phase decoding is successful and produces $c$ codewords in $c$ local codes, the decoding process stops and the decoded messages are delivered to the user(s). If the local phase decoding fails to decode $c$ received local words into $c$ codewords in the $c$ local codes, the global coupling iterative decoding phase is then activated. The rest of the decoding process is the same as the global/local iterative decoding. The local/global iterative decoding is effective for correcting local random errors.

Since every codeword in $C_{global}(m_0, m_1, n_0, c, r)$ consists of $c$ local codewords, the global coupling decoding is performed on a collection of $c$ received local words jointly. During the decoding process, the reliability information of each decoded local word is shared by the others to enhance the overall reliability of all decoded local codewords. This joint decoding and information sharing at the global coupling decoding phase reduces the probability of performing the local decoding phase and hence makes the decoding converge more quickly. The joint decoding and information sharing is a key feature of two-phase decoding. For more on decoding of globally coupled LDPC codes, readers are referred to [23, 24].

---

**Example 13.14** In the following, the prime field GF(101) (see Example 13.16) is used to construct a CN-based QC-GC-LDPC code. Set $m_0 = m_1 = 4, n_0 = 32, c = 3$, and $r = 4$. Let $\alpha$ be a primitive element in GF(101). Form a $16 \times 100$ base matrix $\mathbf{W}_s^{(9)}(16, 100)$ over GF(101) in the form (13.30) based on two subsets $\mathbf{S}_0 = \{0, 1, \alpha, \alpha^2, \ldots, \alpha^{15}\}$ and $\mathbf{S}_1 = \{1, \alpha, \alpha^2, \ldots, \alpha^{99}\}$ of GF(101) with $\eta = 1$. Label the rows and columns of $\mathbf{W}_s^{(9)}(16, 100)$ as from 0 to 15 and 0 to 99, respectively. Divide the 16 rows of $\mathbf{W}_s^{(9)}(16, 100)$ into four disjoint groups $\mathbf{W}_0(4, 100), \mathbf{W}_1(4, 100), \mathbf{W}_2(4, 100)$, and $\mathbf{W}_{coupling}(4, 100)$. For $0 \leq l < 3$, $\mathbf{W}_l(4, 100)$ consists of four consecutive rows of $\mathbf{W}_s^{(9)}(16, 100)$, labeled from $4l$ to $4(l + 1) - 1$. The submatrix $\mathbf{W}_{coupling}(4, 100)$ consists of the last four rows of $\mathbf{W}_s^{(9)}(16, 100)$. For $0 \leq l < 3$, let $\mathbf{W}_{local,l}(4, 32)$ be a $4 \times 32$ submatrix of $\mathbf{W}_l(4, 100)$, which consists of 32 consecutive columns of $\mathbf{W}_l(4, 100)$, labeled from $32l$ to $32(l + 1) - 1$. Form the following $16 \times 100$ matrix over GF(101):

$$\mathbf{W}_{global}(4, 4, 32, 3, 4) = \begin{bmatrix} \mathbf{0} & \text{diag}(\mathbf{W}_{local,0}(4, 32), \mathbf{W}_{local,1}(4, 32), \mathbf{W}_{local,2}(4, 32)) \\ & \mathbf{W}_{coupling}(4, 100) \end{bmatrix}, \tag{13.33}$$

where $\mathbf{0}$ is a $4 \times 4$ zero matrix.

The $100 \times 100$ CPM dispersion of $\mathbf{W}_{global}(4, 4, 32, 3, 4)$ with respect to the multiplicative group of GF(101) gives the following $16 \times 100$ array of CPMs and ZMs of size $100 \times 100$:

$$\mathbf{H}_{global}(4, 4, 32, 3, 4) = \begin{bmatrix} \mathbf{0} & \text{diag}(\mathbf{H}_{local,0}(4, 32), \mathbf{H}_{local,1}(4, 32), \mathbf{H}_{local,2}(4, 32)) \\ & \mathbf{H}_{coupling}(4, 100) \end{bmatrix}, \tag{13.34}$$

which is a $1600 \times 10\,000$ matrix over GF(2) with two constant column weights, 4 and 8, and two row weights, 32 and 100.

The null space of $\mathbf{H}_{global}(4, 4, 32, 3, 4)$ over GF(2) gives a (10 000, 8400) CN-based QC-GC-LDPC code $C_{global}(4, 4, 32, 3, 4)$ of length 10 000 and rate 0.84, which is composed of three local QC-LDPC codes of length 3200 and a global coupling QC-LDPC code of length 10 000. The code can be decoded either in one phase based on the global parity-check matrix $\mathbf{H}_{global}(4, 4, 32, 3, 4)$ or in two phases based on local and global coupling parity-check matrices, respectively.

The BER performance of $C_{global}(4, 4, 32, 3, 4)$ over an AWGN channel using BPSK signaling decoded in one phase based on the global matrix $\mathbf{H}_{global}(4, 4, 32, 3, 4)$ with 20 and 100 iterations of the MSA, scaled by a factor of

0.275, is shown in Figure 13.18. We see that the performance gap between 20 and 100 iterations is very small, about 0.1 dB.

The BER performance of $C_{GC,FF,FF}(4, 4, 32, 3, 4)$ decoded using the global/local two-phase decoding of the MSA with $I_{\text{out,max}} = I_{\text{in,max}} = I_{\text{max}} = 5$ is also shown in Figure 13.18.

Figure 13.18 shows that the one-phase decoding and the two-phase decoding of the code give almost the same error performance in the simulation range. Simulation results show that one-phase decoding converges slightly faster than two-phase decoding. However, implementation of two-phase decoding is simpler.

The code described above was constructed in [24].

## 13.9.5 Remarks

If the number of columns in the global coupling parity-check matrix is greater than the sum of columns in the local parity-check matrices (i.e., $r > 0$), a CN-based globally coupled LDPC code is a type of concatenated LDPC code with local codes as *outer codes* and the global coupling code as the *inner code*. Such a concatenated code has multiple outer codes. If the number of columns in the global coupling parity-check matrix is equal to the sum of columns in the local parity-check matrices (i.e., $r = 0$), then the CN-based globally coupled LDPC codes presented in this section form a subclass of globally coupled LDPC codes presented in [23, 24].

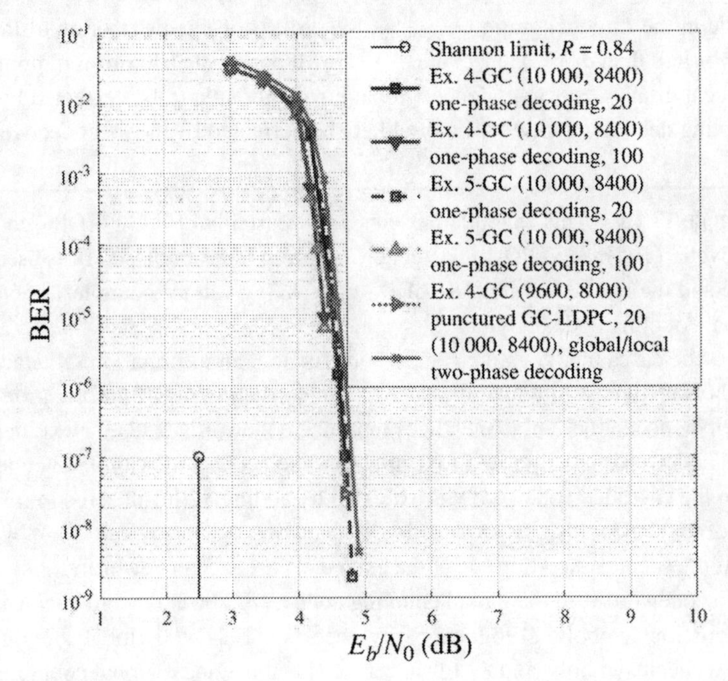


**Figure 13.18** The BER performance of the CN-based QC-GC-LDPC code given in Example 13.14.

The base matrix in the form given by (13.31) can be constructed using any of the methods presented in Chapter 12 besides the construction based on two subsets of a finite field.

If we don't require the globally coupled parity-check matrix $\mathbf{H}_{global}$ to satisfy the RC constraint, then the local parity-check matrices can be made the same. Also, the local parity-check matrices and the global coupling matrix can be constructed from two different fields, or different cyclic subgroups of a finite field.

Furthermore, the local codes and the global coupling codes can be two different types of LCPC codes. One such case is that all the local codes are copies of a finite geometry (FG) LDPC code and the global coupling code is a finite-field (FF) code [24]. Since an FG-LDPC code has a large minimum distance and its Tanner graph contains no small harmful trapping sets and has a large connectivity (see Chapter 12), it can achieve a very low error rate without a visible error floor and its decoding can converge rapidly with a small number of decoding iterations, say 5 to 10.

The global coupling of multiple copies of an FG-LDPC code and an FF-LDPC code, called CN-based *FG/FF-GC-LDPC coupling*, combines the distinct features of both FG- and FF-LDPC codes to achieve low error rates at a rapid decoding convergence and an error performance close to the Shannon limit. Since each local graph $G_{FG,local}$ in the global graph $G_{FG,FF,global}$ is the Tanner graph of the local FG-LDPC code with minimum distance $d$, it contains no harmful trapping sets of size smaller than $d - 1$ [25]. For $\kappa < d - 1$, if the global coupling graph $G_{FF}$ contains a harmful trapping set $T$ of size $\kappa$, then the $\kappa$ VNs in $T$ either reside in one local graph or distribute among the $c$ local graphs. For $0 \leq i < c$, let $\kappa_i$ be the number of VNs in $T$ that reside in the $i$th local graph $G_{FG,local,i}$. Regardless of the distribution of the $\kappa$ VNs in $T$, $\kappa_i$ is less than $d - 1$. Hence, the $\kappa_i$ VNs will not create a harmful trapping set in the $i$th local graph $G_{FG,local,i}$. Hence, global errors not able to be corrected by the global coupling decoding phase will more likely be corrected by the local decoding phase.

---

**Example 13.15** This example presents a CN-based FG/FF-GC-LDPC code in which the cyclic (1057, 813) PG-LDPC code $C_{PG}$ given in Example 12.14 is used as local code and the (10 000, 9600) QC-LDPC code $C_{FF}$ given in Example 13.16 is used as global coupling code.

Set the cascading degree $c = 9$. In construction, an information sequence of 7317 symbols is divided into nine messages, each consisting of 813 information symbols. We first encode these nine messages into nine codewords in $C_{PG}$. Next, these nine local codewords are cascaded to form a cascaded codeword $\mathbf{v}$ of 9513 code symbols. Then the cascaded codeword $\mathbf{v}$ is extended by adding 87 fill-in zeros to form a word $\mathbf{v}_{ext}$ of 9600 symbols.

At the second encoding stage, the extended cascaded codeword $\mathbf{v}_{ext}$ is encoded into a codeword $\mathbf{w}$ in the global coupling code $C_{FF}$. The codeword $\mathbf{w}$ consists of 7317 information symbols, 2196 local parity-check symbols, 400 global coupling parity-check symbols, and 87 fill-in zeros. The above local/global coupling encoding results in a (10 000, 7317) PG/FF-GC-LDPC code $C_{PG,FF,global}$ with rate 0.7317.

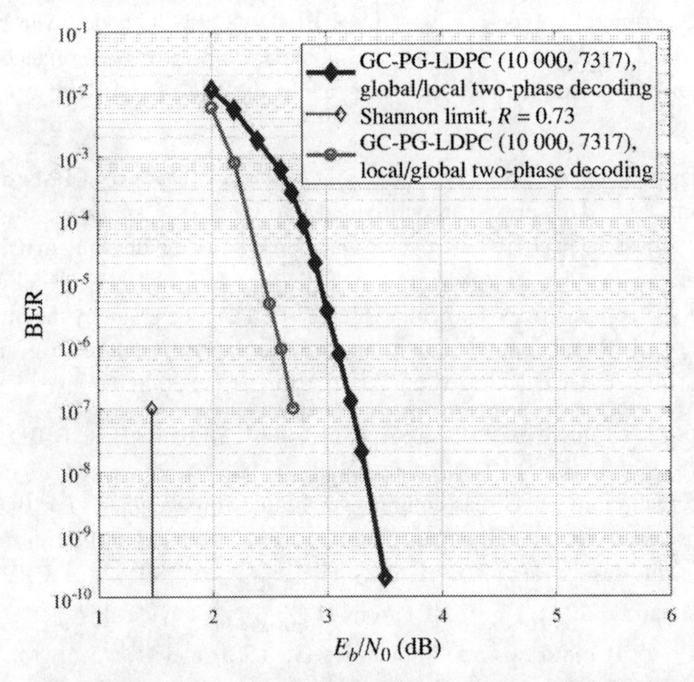


**Figure 13.19** The BER performance of the PG/FF-GC-LDPC code given in Example 13.15.

To decode $C_{PG,FF,global}$, we use local/global two-phase iterative decoding. The BER performance of the code over an AWGN channel decoded using the MSA for both local and global coupling codes with $I_{local}$, $I_{local,max}$, and $I_{max}$ set to 5, 5, and 5, respectively, is shown in Figure 13.19. The scaling factors for decoding the local and global codes are 0.275 and 0.625, respectively. When decoding is switched from a global coupling decoding phase to a local decoding phase, the 87 fill-in zeros are removed; conversely, when decoding is switched from a local decoding phase to a global coupling decoding phase, the 87 fill-in zeros are added back.

The above globally coupled FG/FF-LDPC code was constructed in [24]. The product QC-LDPC codes presented in Sections 13.7 and 13.8 can be viewed as globally coupled LDPC codes.

## Problems

**13.1** For $t = 9$, there exists a class-I $(109, 981, 4, 36, 1)$ Bose BIBD. Utilizing this BIBD, a row of nine $109 \times 109$ circulants $\mathbf{H}^{(1)}_{\text{BIBD}} = [\mathbf{G}_0 \quad \mathbf{G}_1 \quad \cdots \quad \mathbf{G}_8]$ can be constructed. Each circulant $\mathbf{G}_i$ has column weight and row weight 4. The null space of $\mathbf{H}^{(1)}_{\text{BIBD}}$ gives a type-I, class-I Bose BIBD-QC-LDPC code of length 981. Determine the code and compute its bit- and word-error performances over the AWGN channel with iterative decoding using the SPA for 10 and 50 iterations, respectively.

**13.2** Consider the class-I $(109, 981, 4, 36, 1)$ Bose BIBD constructed in Problem 12.1. Set $k = 4$ and $r = 8$. Construct a $4 \times 8$ array $\mathbf{H}^{(1)}_{\text{BIBD,decom}}(4, 8)$ of $109 \times 109$

CPMs. Determine the type-II class-1 Bose BIBD-QC-LDPC code given by the null space of $\mathbf{H}^{(1)}_{\text{BIBD,decom}}(4,8)$. Compute its bit- and word-error performances over the AWGN channel with iterative decoding using the SPA for 10 and 50 iterations, respectively.

**13.3** Consider the $4 \times 8$ array $\mathbf{H}^{(1)}_{\text{BIBD,decom}}(4,8)$ of $109 \times 109$ CPMs constructed in Problem 13.2. This array can be divided into two $4 \times 4$ subarrays of $109 \times 109$ CPMs. By masking each of these two $4\times4$ subarray with the masking matrix given by (13.7), we obtain a $4 \times 8$ masked array $\mathbf{M}^{(1)}_{\text{BIBD,decom}}(4,8)$ of $109 \times 109$ circulant permutation and zero matrices. Determine the QC-LDPC code given by the null space of $\mathbf{M}^{(1)}_{\text{BIBD,decom}}(4,8)$ and compute its bit- and word-error performances over the AWGN channel with iterative decoding using the SPA for 10 and 50 iterations.

**13.4** For $t = 41$, there exists a class-II $(821, 33661, 5, 205, 1)$ Bose BIBD. From this BIBD, a row of 41 $821 \times 821$ circulants $\mathbf{H}^{(2)}_{\text{BIBD}} = [\mathbf{G}_0 \quad \mathbf{G}_2 \quad \cdots \quad \mathbf{G}_{40}]$ can be constructed. Each circulant $\mathbf{G}_i$ has column weight and row weight 5. For $0 \le i \le 40$, decompose each circulant $\mathbf{G}_i$ into a column of five $821 \times 821$ CPMs with row decomposition. This results in a $5 \times 41$ array $\mathbf{H}^{(2)}_{\text{BIBD,decom}}$ of $821 \times 821$ CPMs. Take a $5 \times 10$ subarray $\mathbf{H}^{(2)}_{\text{BIBD,decom}}(5,10)$ from $\mathbf{H}^{(2)}_{\text{BIBD,decom}}$. Divide this $5 \times 10$ subarray $\mathbf{H}^{(2)}_{\text{BIBD,decom}}(5,10)$ into two $5 \times 5$ subarrays $\mathbf{H}^{(2)}_1(5,5)$ and $\mathbf{H}^{(2)}_2(5,5)$. Mask each of these two $5 \times 5$ subarrays with the $5 \times 5$ masking matrix $\mathbf{Z}_1(5,5)$ given by (13.15). This results in a $5 \times 10$ masked array $\mathbf{M}^{(2)}_{\text{BIBD,decom}}(5,10)$. This masked array is an RC-constrained $4105 \times 8210$ matrix over GF(2) with column and row weights 3 and 6, respectively. Determine the QC-LDPC code given by the null space of $\mathbf{M}^{(2)}_{\text{BIBD,decom}}(5,10)$ and compute its bit- and word-error performances over the AWGN channel using the SPA with 10 and 50 iterations.

**13.5** Consider the $q^m$ points of the $m$-dimensional Euclidean geometry EG($m$, $q$) over GF($q$) as a collection of $q^m$ objects. Given the structural properties of the lines in EG($m$, $q$), show that there exists a BIBD with $\lambda = 1$ for a collection of $q^m$ objects. Give the parameters of this BIBD.

**13.6** For the $m$-dimensional projective geometry PG($m$, $q$) over GF($q$), show that there exists a BIBD with $\lambda = 1$ for a collection of $(q^{m+1} - 1)/(q - 1)$ objects.

**13.7** Using the lines of the two-dimensional Euclidean geometry EG$(2, 2^4)$ over GF$(2^4)$ that do not pass the origin of EG$(2, 2^5)$, a $255 \times 255$ circulant $\mathbf{G}$ with both column and row weight 16 can be constructed. The Tanner graph $\mathcal{G}$ of $\mathbf{G}$ has a girth of 6 (it is free of cycles of length 4). Decompose $\mathbf{G}$ with column decomposition into a $255 \times 510$ matrix $\mathbf{H}_0$ that consists of two circulants $\mathbf{G}_0$ and $\mathbf{G}_1$, each having both column and row weight 8.

(a) Determine the number of cycles of length 6 in the Tanner graph $\mathbf{G}_0$ of $\mathbf{H}_0$. Compute the bit- and word-error performances of the code $\mathcal{C}_0$ given by the null space of $\mathbf{H}_0$.
(b) Use the trellis-based cycle-removal technique given in Section 13.5 to remove the cycles of length 6 in the Tanner graph $\mathcal{G}_0$ of $\mathbf{H}_0$. The cycle-removal process

results in a Tanner graph $\mathcal{G}_1$ that is free of cycles of length 6. From this Tanner graph, construct an LDPC code $\mathcal{C}_1$ and compute its bit- and word-error performances.

(c) Remove the cycles of length 8 from the Tanner graph $\mathcal{G}_1$. This results in a new Tanner graph $\mathcal{G}_2$ that has a girth of at least 10. From $\mathcal{G}_2$, construct a new LDPC code $\mathcal{C}_2$ and compute its bit- and word-error performances.

**13.8** Use the PEG algorithm to construct a rate-1/2 (2044, 1022) irregular LDPC code based on the degree distributions given in Example 13.8. Compute the bit- and word-error performances of the code constructed. Also construct a rate-1/2 (2044, 1022) irregular LDPC code with the improved PEG algorithm using ACE and compute its bit- and word-error performances.

**13.9** Consider the three-dimensional Euclidean geometry EG(3, 3) over GF(3). From the lines of EG(3, 3) not passing through the origin, four $26\times26$ circulants over GF(2) can be constructed, each having both column and row weight 3. Take two of these circulants and arrange them in a row to form a $26 \times 52$ RC-constrained matrix $\mathbf{B}$ with column and row weights 3 and 6, respectively. $\mathbf{B}$ can be used as a base matrix for superposition construction of LDPC codes. Suppose we replace each 1-entry of $\mathbf{B}$ by a distinct $126\times126$ circulant permutation matrix and each 0-entry by a $126\times126$ zero matrix. This replacement results in an RC-constrained $26\times52$ array $\mathbf{H}_{\text{sup}}$ of $126\times126$ circulant permutation and zero matrices. $\mathbf{H}_{\text{sup}}$ is a $4056 \times 8112$ matrix over GF(2) with column and row weights 3 and 6, respectively. Determine the code given by the null space of $\mathbf{H}_{\text{sup}}$ and compute its bit- and word-error performances over the AWGN channel using the SPA with 10 and 50 iterations.

**13.10** Let $\mathcal{C}_1$ be the (63, 37) cyclic EG-LDPC code constructed using the two-dimensional Euclidean geometry EG(2, $2^3$) over GF($2^3$). The parity-check matrix $\mathbf{H}_1$ of $\mathcal{C}_1$ is a $63\times63$ circulant with both column and row weight 8. The minimum distance is exactly 9. Let $\mathcal{C}_2$ be the (8, 7) single parity-check code whose parity-check matrix $\mathbf{H}_2 = [1\ \ 1\ \ 1\ \ 1\ \ 1\ \ 1\ \ 1\ \ 1]$ is a row of eight 1s. The parity-check matrix $\mathbf{H}_{\text{sup},p}$ of the product of $\mathcal{C}_1$ and $\mathcal{C}_2$ is of the form given by (13.27). The product code $\mathcal{C}_{\text{sup},p}$ of $\mathcal{C}_1$ and $\mathcal{C}_2$ is a (504, 259) LDPC code with minimum distance 18. Compute the bit- and word-error performances of $\mathcal{C}_{\text{sup},p}$ over the AWGN channel using the SPA with 50 iterations.

## References


[1] R. C. Bose, "On the construction of balanced incomplete block designs," *Annals of Eugenics*, vol. 9, pp. 353–399, 1939.

[2] I. F. Blake and R. C. Mullin, *The Mathematical Theory of Coding*, New York, Academic Press, 1975.

[3] R. D. Carmichael, *Introduction to Theory of Groups of Finite Orders*, Boston, MA, Gin & Co., 1937.

[4] C. J. Colbourn and J. H. Dintz (eds.), *The Handbook of Combinatorial Designs*, Boca Raton, FL, CRC Press, 1996.

[5] D. J. Finney, *An Introduction to the Theory of Experimental Design*, Chicago, IL, University of Chicago Press, 1960.

[6] M. Hall, Jr., *Combinatorial Theory*, 2nd ed., New York, Wiley, 1986.

[7] H. B. Mann, *Analysis and Design of Experiments*, New York, Dover Publications, 1949.

[8] H. J. Ryser, *Combinatorial Mathematics*, New York, Wiley, 1963.

[9] B. Ammar, B. Honary, Y. Kou, J. Xu, and S. Lin, "Construction of low density parity-check codes based on balanced incomplete designs," *IEEE Transactions on Information Theory*, vol. 50, no. 6, pp. 1257–1268, June 2004.

[10] S. Johnson and S. R. Weller, "Regular low-density parity-check codes form combinatorial designs," *Proceedings of the 2001 IEEE Information Theory Workshop*, Cairns, Australia, September 2001, pp. 90–92.

[11] L. Lan, Y. Y. Tai, S. Lin, B. Memari, and B. Honary, "New constructions of quasi-cyclic LDPC codes based on special classes of BIBDs for the AWGN and binary erasure channels," *IEEE Transactions on Communications*, vol. 56, no. 1, pp. 39–48, January 2008.

[12] B. Vasic and O. Milenkovic, "Combinatorial construction of low density parity-check codes for iterative decoding," *IEEE Transactions on Information Theory*, vol. 50, no. 6, pp. 1156–1176, June 2004.

[13] X.-Y. Hu, E. Eleftheriou, and D. M. Arnold, "Regular and irregular progressive edge-growth Tanner graphs," *IEEE Transactions on Information Theory*, vol. 51, no. 1, pp. 386–398, January 2005.

[14] L. Lan, Y. Y. Tai, L. Chen, S. Lin, and K. Abdel-Ghaffar, "A trellis-based method for removal of cycles from bipartite graphs and construction of low density parity check codes," *IEEE Communications Letters*, vol. 8, no. 7, pp. 443–445, July 2004.

[15] H. Xiao and A. H. Banihashemi, "Improved progressive-edge-growth (PEG) construction of irregular LDPC codes," *IEEE Communications Letters*, vol. 8, no. 12, pp. 715–717, December 2004.

[16] S. Lin and D. J. Costello, Jr., *Error Control Coding: Fundamentals and Applications*, Upper Saddle River, NJ, Prentice-Hall, 2004.

[17] T. Tian, C. Jones, J. D. Villasenor, and R. D. Wesel, "Construction of irregular LDPC codes with low error floors," *Proceedings of the IEEE International Conference on Communications*, vol. 5, Anchorage, AK, May 2003, pp. 3125–3129.

[18] T. Tian, C. Jones, J. D. Villasenor, and R. D. Wesel, "Selective avoidance of cycles in irregular LDPC code construction," *IEEE Transactions on Communications*, vol. 52, no. 8, pp. 1242–1247, August 2004.

[19] J. Xu and S. Lin, "A combinatoric superposition method for constructing low-density parity-check codes," *Proceedings of the International Symposium on Information Theory*, vol. 30, Yokohama, June–July 2003.

[20] J. Xu, L. Chen, L.-Q. Zeng, L. Lan, and S. Lin, "Construction of low-density parity-check codes by superposition," *IEEE Transactions on Communications*, vol. 53, no. 2, pp. 243–251, February 2005.

[21] N. Deo, *Graph Theory and Applications to Engineering and Computer Engineering*, Englewood Cliffs, NJ, Prentice-Hall, 1974.

[22] D. B. West, *Introduction to Graph Theory*, 2nd ed., Upper Saddle River, NJ, Prentice-Hall, 2001.

[23] J. Li, S. Lin, K. Abdel-Ghaffar, W. Ryan, and D. J. Costello, *LDPC Code Design, Constructions and Unifications*, Cambridge, Cambridge University Press, 2017.

[24] M. Nasseri, X. Xiao, B. Vasic, and S. Lin, "Globally coupled finite-geometry and finite-field LDPC coding schemes." *IEEE Transactions on Vehicular Technology*, vol. 70, no. 9, September 2021, pp. 9207–9216.

[25] Q. Diao, Y. Y. Tai, S. Lin, and K. Abdel-Ghaffar, "LDPC codes on partial geometries: Construction, trapping sets structure, and puncturing," *IEEE Transactions on Information Theory*, vol. 59, no. 12, pp. 7898–7914, September 2013.

# 14 LDPC Codes for Binary Erasure Channels

Many channels, such as wireless, magnetic recording, and jammed channels, tend to suffer from time intervals during which their reliability deteriorates significantly, to a degree that compromises data integrity. In some scenarios, receivers are able to detect the presence of these time intervals and may choose, accordingly, to "erase" some (or all of the) symbols received during such intervals. This technique causes symbol losses at known locations. This chapter is devoted to LDPC codes for correcting (or recovering) transmitted symbols that have been erased, called *erasures*. The simplest channel model with erasures is the *binary erasure channel* over which a transmitted bit is either correctly received or erased. There are two basic types of binary erasure channel, *random* and *burst*. Over a random *binary erasure channel* (BEC), erasures occur at random locations, each with the same probability of occurrence, whereas over a *binary burst erasure channel* (BBEC), erasures cluster into bursts. In this chapter, we first show that the LDPC codes constructed in Chapters 11 to 13, besides performing well over the AWGN channel, also perform well over the BEC. Then we construct LDPC codes for correcting bursts of erasures.

## 14.1 Iterative Decoding of LDPC Codes for the BEC

For transmission over the BEC, a transmitted symbol 0 or 1 is either correctly received with probability $1 - p$ or erased with probability $p$, called the *erasure probability*, as shown in Figure 14.1. The channel output alphabet consists of three symbols, namely 0, 1, and "?," where the symbol "?" denotes a transmitted symbol that has been erased, called an *erasure*.

Consider an LDPC code $\mathcal{C}$ of length $n$ given by the null space of an $m \times n$ sparse matrix $\mathbf{H}$ over GF(2). Let $\mathbf{v} = (v_0, v_1, \ldots, v_{n-1})$ be an $n$-tuple over GF(2). Then $\mathbf{v}$ is a codeword in $\mathcal{C}$ if and only if $\mathbf{v}\mathbf{H}^{\mathrm{T}} = \mathbf{0}$. Suppose a codeword $\mathbf{v} = (v_0, v_1, \ldots, v_{n-1})$ in $\mathcal{C}$ is transmitted and $\mathbf{r} = (r_0, r_1, \ldots, r_{n-1})$ is the corresponding received sequence. Let $\mathcal{E} = \{j_1, j_2, \ldots, j_t\}$ be the set of locations in $\mathbf{r}$ with $0 \leq j_0 < j_1 < \cdots < j_t < n$, where the transmitted symbols are erased, that is, $r_{j_1}, r_{j_2}, \ldots, r_{j_t} = ?$. The set $\mathcal{E}$ displays the pattern of the erased symbols (or erasure locations) in $\mathbf{r}$ and hence is called an *erasure pattern*. Let $\{0, 1, \ldots, n-1\}$ be the index set of the components of a codeword in $\mathcal{C}$. Define $\bar{\mathcal{E}} \triangleq \{0, 1, \ldots, n-1\} \backslash \mathcal{E}$. Then $\bar{\mathcal{E}}$ is the set of positions in $\mathbf{r}$ where the transmitted symbols are correctly received, that is, $r_i = v_i$ for $i \in \bar{\mathcal{E}}$. Decoding $\mathbf{r}$ involves determining the value of $v_{j_l}$ (or $r_{j_l}$) for each $j_l \in \mathcal{E}$. An erasure pattern $\mathcal{E}$ is

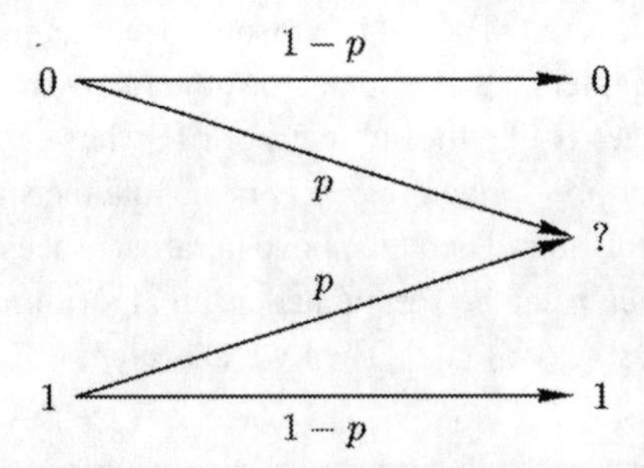


**Figure 14.1** A model of the BEC.

said to be *recoverable* (or *correctable*) if the value of each erased transmitted symbol $v_{j_l}$ with $j_l \in \mathcal{E}$ can be correctly determined.

In the following, we describe a simple iterative decoding method for correcting erasures. Label the rows of the parity-check matrix $\mathbf{H}$ of $\mathcal{C}$ from 0 to $m-1$. For $0 \leq i < m$, let

$$\mathbf{h}_i = (h_{i,0}, h_{i,1}, \ldots, h_{i,n-1}) \tag{14.1}$$

denote the $i$th row of $\mathbf{H}$. For a received sequence $\mathbf{r} = (r_0, r_1, \ldots, r_{n-1})$ to be a codeword, it must satisfy the parity-check constraint

$$s_i = r_0 h_{i,0} + r_1 h_{i,1} + \cdots + r_{n-1} h_{i,n-1} = 0, \tag{14.2}$$

for $i = 0, 1, \ldots, m-1$, which is called a *check sum*. The received symbol $r_j$ in the above check sum is said to be *checked* by the row $\mathbf{h}_i$, if $h_{i,j} = 1$. In this case, if $r_j = ?$ and all the other received symbols checked by $\mathbf{h}_i$ are not erasures (i.e., they are correctly received), then (14.2) contains only one unknown, $r_j$. Consequently, $r_j$ can be determined from

$$r_j = \sum_{k=0, k \neq j}^{n-1} r_k h_{i,k}. \tag{14.3}$$

For each erased position $j_l$ in an erasure pattern $\mathcal{E} = \{j_1, j_2, \ldots, j_t\}$, if there exists a row $\mathbf{h}_i$ in $\mathbf{H}$ that checks only the erased symbol $r_{j_l}$ and not any of the other $t-1$ erased symbols with locations in $\mathcal{E}$, then it follows from (14.3) that the value of the erased symbol $j_l \in \mathcal{E}$ can be correctly determined from the correctly received symbols that are checked by $\mathbf{h}_i$. Such an erasure pattern is said to be *recoverable in one step* (or *one iteration*). However, there are erasure patterns that are not recoverable in one step, but recoverable in *multiple steps iteratively*. Given an erasure pattern $\mathcal{E}$, we first determine the values of those erased symbols that can be recovered in one step by using (14.3). Then we remove the recovered erased symbols from $\mathcal{E}$. This results in a new erasure pattern $\mathcal{E}_1$ of smaller size. Next, we determine the values of erased symbols in $\mathcal{E}_1$ that are recoverable using (14.3). On removing the recovered erased symbols from $\mathcal{E}_1$, we obtain an erasure pattern $\mathcal{E}_2$ of size smaller than that of $\mathcal{E}_1$. We repeat the above process until either all the erased symbols in $\mathcal{E}$ have been recovered, or an erasure pattern $\mathcal{E}_m$ is obtained such that no erasure in $\mathcal{E}_m$ can be recovered using (14.3). In the latter case, some erasures in $\mathcal{E}$ cannot be recovered. The set of erasure locations in $\mathcal{E}_m$ is said to form a *stopping set* [1, 2] (see Chapters 5 and 8) that stops the recovery process.

For a given erasure pattern $\mathcal{E}$, let $\mathbf{H}_{\mathcal{E}}$ be the submatrix of $\mathbf{H}$ whose columns correspond to the locations of erased symbols given in $\mathcal{E}$. Let $\mathbf{r}_{\mathcal{E}}$ denote the subsequence of $\mathbf{r}$ that consists of the erased symbols in $\mathbf{r}$. Then the above iterative process for recovering erasures in a received sequence $\mathbf{r}$ can be formulated as an algorithm. To initialize the recovery process, we set $k = 0$ and $\mathcal{E}_0 = \mathcal{E}$. Then we execute the following steps iteratively.

---

1. Determine $\mathcal{E}_k$ and form $\mathbf{r}_{\mathcal{E}}$. If $\mathcal{E}_k$ is empty, stop decoding; otherwise, go to Step 2.
2. Form $\mathbf{H}_{\mathcal{E}_k}$ on the basis of $\mathbf{r}_{\mathcal{E}}$.
3. Find the set $\mathcal{R}_k$ of rows in $\mathbf{H}_{\mathcal{E}}$ such that each row in $\mathcal{R}_k$ contains a single 1-component. If $\mathcal{R}_k \neq \emptyset$, determine the erasures in $\mathcal{E}_k$ that are checked by the rows in $\mathcal{R}_k$. Determine values of these erasures by application of (14.3) based on the rows of $\mathbf{H}$ that correspond to the rows in $\mathcal{R}_k$. Then go to Step 4. If $\mathcal{R}_k = \emptyset$, stop decoding.
4. Remove the locations of the erasures recovered at Step 3 from $\mathcal{E}_k$. Set $k = k + 1$ and go to Step 1.

---

If the decoding stops at Step 1, either there is no erasure in $\mathbf{r}$ or all the erasures in the erasure pattern $\mathcal{E}$ have been recovered and decoding is successful. If decoding stops at Step 3, some erasures in $\mathcal{E}$ cannot be recovered.

The above decoding algorithm for recovering erased symbols was first proposed in [3] and then in [1]. It is equivalent to the iterative erasure-filling algorithm presented in Chapter 5. Since the parity-check matrix $\mathbf{H}$ of an LDPC code is a sparse matrix, the weight of each row is relatively small compared with the length of the code and hence the number of terms in the sum given by (14.3) is small. Therefore, hardware implementations of the above iterative decoding algorithm can be very simple.

## 14.2 Random-Erasure-Correction Capability

The performance of an LDPC code over the BEC is determined by the stopping sets and the degree distributions of the nodes of its Tanner graph $\mathcal{G}$. Let $\mathcal{V}$ be a set of variable nodes in $\mathcal{G}$, and $\mathcal{S}$ be a set of check nodes in $\mathcal{G}$ that are adjacent to the nodes in $\mathcal{V}$, that is, each check node in $\mathcal{S}$ is connected to *at least one* variable node in $\mathcal{V}$. The nodes in $\mathcal{S}$ are called the *neighbors* of the nodes in $\mathcal{V}$. A set $\mathcal{V}$ of variable nodes in $\mathcal{G}$ is called a *stopping set* of $\mathcal{G}$, if each check node in the neighbor set $\mathcal{S}$ of $\mathcal{V}$ is connected to *at least two variable nodes* in $\mathcal{V}$. If an erasure pattern $\mathcal{E}$ corresponds to a stopping set in the Tanner graph $\mathcal{G}$ of an LDPC code, then a row of the parity-check matrix $\mathbf{H}$ of the code that checks an erasure in $\mathcal{E}$ also checks at least one other erasure in $\mathcal{E}$. In this case, the sum given by (14.2) contains at least two unknowns (two erasures). As a result, no erasure in $\mathcal{E}$ can be determined with (14.3).

A set $\mathcal{V}$ of variable nodes in $\mathcal{G}$ may contain more than one stopping set. The union of two stopping sets is also a stopping set and the union of all stopping sets contained in $\mathcal{V}$ gives the *maximum stopping set* of $\mathcal{V}$. A set $\mathcal{V}_{\text{ssf}}$ of variable nodes in $\mathcal{G}$ is said to be *stopping-set-free* (SSF), if it does not contain any stopping set. It is clear that, for a given erasure pattern $\mathcal{E}$, the erasures that are contained in the maximum stopping set of $\mathcal{E}$ cannot be recovered. It is also clear that any erasure pattern $\mathcal{E}$ is recoverable if it is SSF. Let $\mathcal{V}_{\min}$ be a stopping set of *minimum size* in the Tanner graph $\mathcal{G}$ of an LDPC code, called a *minimum stopping set*. If the code symbols corresponding to the variable nodes in $\mathcal{V}_{\min}$ are erased during the transmission, then $\mathcal{V}_{\min}$ forms an irrecoverable erasure pattern of minimum size. Therefore, for random erasure correction (REC) with the above iterative decoding algorithm, it is desired to construct codes with the largest possible minimum stopping sets in their Tanner graphs. The error-floor performance of an LDPC code with the above iterative decoding depends on the size and number of minimum stopping sets in the Tanner graph $\mathcal{G}$ of the code, or more precisely, on the stopping-set distribution of $\mathcal{G}$. A good LDPC code for REC must have no or very few small stopping sets in its Tanner graph.

Consider a $(g, r)$-regular LDPC code given by an RC-constrained parity-check matrix $\mathbf{H}$ with column and row weights $g$ and $r$, respectively. The size of a minimum possible stopping set in its Tanner graph is $g + 1$. To prove this, let $\mathcal{E} = \{j_1, j_2, \ldots, j_t\}$ be an erasure pattern that consists of $t$ erasures with $0 \leq t \leq g$. Consider the erasure at a location $j_l$ in $\mathcal{E}$. Since the column weight of $\mathbf{H}$ is $g$, there exists a set of $g$ row $\Lambda_l = \{\mathbf{h}_{i_1}, \mathbf{h}_{i_2}, \ldots, \mathbf{h}_{i_g}\}$, such that all the rows in $\Lambda_l$ check the erased symbol $r_{j_l}$. The other $t - 1$ erasures in $\mathcal{E}$ can be checked by at most $t - 1$ rows in $\Lambda_l$. Therefore, there is at least one row in $\Lambda_l$ that checks only the erased symbol $r_{j_l}$ and $r - 1$ other correctly received symbols. Since $\mathbf{H}$ satisfies the RC constraint, no two erasures in $\mathcal{E}$ can be checked simultaneously by two rows in $\Lambda_l$. The above two facts guarantee that every erasure in $\mathcal{E}$ can be recovered. Therefore, the size of a minimum possible stopping set of the Tanner graph of an LDPC code given by the null space of an RC-constrained $(g, r)$-regular parity-check matrix $\mathbf{H}$ is at least $g + 1$. Now suppose that an erasure pattern $\mathcal{E} = \{j_1, j_2, \ldots, j_{g+1}\}$ with $g + 1$ erasures occurs. Consider the $g$ rows in $\Lambda_l$ that check the erased symbol $r_{j_l}$. If, in the worst case, each of the other $g$ erasures is checked separately by the $g$ rows in $\Lambda_l$, then each row in $\Lambda_l$ checks two erasures. In this case, $\mathcal{E}$ corresponds to a stopping set of size $g + 1$. Therefore, the size of a minimum possible stopping set of the Tanner graph of a $(g, r)$-regular LDPC code given by the null space of an RC-constrained parity-check matrix $\mathbf{H}$ is $g + 1$.

For $(g, r)$-regular LDPC codes, it has been proved in [2, 4] that the sizes of minimum stopping sets of their Tanner graphs with girths 4, 6, and 8 are 2, $g + 1$, and $2g$, respectively. Hence, for correcting random erasures with an LDPC code using iterative decoding, the most critical cycles in the code's Tanner graph are cycles of length 4. Therefore, in construction of LDPC codes for the BEC, cycles of length 4 must be avoided in their Tanner graphs. It is proved in [5] that a code with minimum distance $d_{\min}$ contains stopping sets of size $d_{\min}$. In fact, the *support* of a codeword (defined as the locations of nonzero components of a codeword) forms a stopping set of size equal to the weight of the codeword [5]. Therefore, in the construction of an LDPC code for correcting random erasures, we need to keep its minimum distance large.

## 14.3 Good LDPC Codes for the BEC

The cyclic EG-LDPC codes in Section 11.1.1 based on Euclidean geometries have minimum stopping sets of large sizes in their Tanner graphs. Consider the cyclic EG-LDPC code $\mathcal{C}_{\text{EG},c,k}$ of length $q^m - 1$ produced by the null space of the parity-check matrix $\mathbf{H}^{(1)}_{\text{EG},c,k}$ given by (11.7) constructed on the basis of the $m$-dimensional Euclidean geometry EG$(m, q)$ over GF$(q)$. $\mathbf{H}^{(1)}_{\text{EG},c,k}$ consists of a column of $k$ circulants of size $(q^m-1)\times(q^m-1)$ over GF(2), each having both column weight and row weight $q$. Since $\mathbf{H}^{(1)}_{\text{EG},c,k}$ satisfies the RC constraint and has column weight $g = kq$, it follows that the size of a minimum stopping set of the Tanner graph of $\mathcal{C}_{\text{EG},c,k}$ is $kq + 1$, with $1 \leq k < q^{m-1}$. Using the iterative decoding algorithm presented in Section 14.1, the code is capable of correcting any erasure pattern with $kq$ or fewer random erasures. Consider the special case for which $m = 2$ and $q = 2^s$. The cyclic EG-LDPC code $\mathcal{C}_{\text{EG},c,1}$ based on the two-dimensional Euclidean geometry EG$(2, 2^s)$ over GF$(2^s)$ has length $2^{2s} - 1$ and minimum distance $2^s + 1$. The parity-check matrix $\mathbf{H}^{(1)}_{\text{EG},c,1}$ of this code consists of a single $(2^{2s} - 1) \times (2^{2s} - 1)$ circulant with both column and row weight $2^s$. Hence, the size of a minimum stopping set in the Tanner graph of this code is $2^s + 1$. Since cyclic EG-LDPC codes have minimum stopping sets of large size in their Tanner graphs, they should perform well over the BEC.

Many experimental results have shown that the structured LDPC codes constructed in Chapters 11 to 13 perform well not only over the AWGN channel but also over the BEC with the iterative decoding presented in Section 14.1. In the following, we will use structured LDPC codes from different classes to demonstrate this phenomenon. For each code, we give its performance and compare it with the Shannon limit in terms of the *unresolved* (or *unrecovered*) *erasure bit rate* (UEBR). For transmitting information at a rate $R$ (information bits per channel usage) over the BEC, the Shannon limit is $1 - R$. The implication of this Shannon limit is that for erasure probability $p$ smaller than $1 - R$, information can be transmitted reliably over the BEC using a sufficiently long code with rate $R$ and, conversely, reliable transmission is not possible if the erasure probability $p$ is larger than the Shannon limit.

---

**Example 14.1** Consider the (4095, 3367) cyclic EG-LDPC code of rate 0.822 with minimum distance 65 based on the two-dimensional Euclidean geometry EG$(2, 2^6)$ over GF$(2^6)$ given in Example 11.1. The parity-check matrix of this code is a $4095 \times 4095$ circulant over GF(2) with both column and row weight 64. Since the parity-check matrix of this code satisfies the RC constraint, the size of a minimum stopping set in its Tanner graph is 65. Therefore, any erasure pattern with 64 or fewer random erasures is recoverable with the iterative decoding algorithm given in Section 14.1. The Shannon limit for rate 0.822 is $1 - 0.822 = 0.178$. The error performance

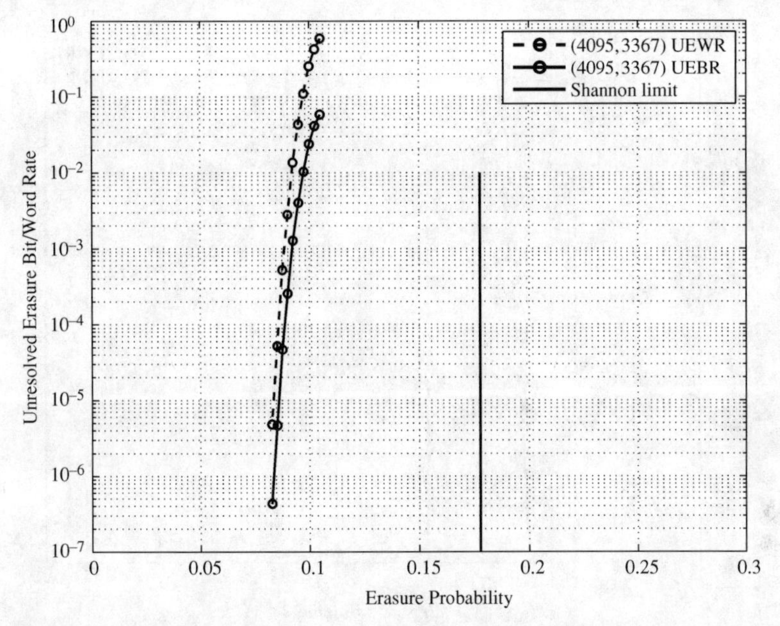


**Figure 14.2** The error performance of the (4095, 3367) cyclic EG-LDPC code over the BEC (UEWR stands for unresolved erasure word rate).

of this code over the BEC with iterative decoding is shown in Figure 14.2. At a UEBR of $10^{-6}$, the code performs 0.095 from the Shannon limit. As shown in Figure 11.1, this code also performs very well over the AWGN channel. So, this code performs well over both the BEC and AWGN channels.

---

**Example 14.2** Consider the (8192, 7171) QC-LDPC code based on the (256, 2, 255) RS code over the prime field GF(257) given in Example 12.1. The code has rate $R = 0.8753$. Figure 12.1 shows that the code performs very well over the binary-input AWGN channel. At a BER of $10^{-6}$, it performs only 1 dB from the Shannon limit. The error performance of this code over the BEC with iterative decoding is shown in Figure 14.3. At a UEBR of $10^{-6}$, the code performs 0.040 from the Shannon limit, 0.1247. From Figures 12.1 and 14.3, we see that the code performs well over both the AWGN and binary random erasure channels.

---

**Example 14.3** Consider the (5256, 4823) QC-LDPC code constructed using the additive group of the prime field GF(73) given in Example 12.11. This code has rate 0.918. Figure 14.4 shows its error performance over the BEC with iterative decoding. We see that, at a UEBR of $10^{-6}$, the gap between the code performance and the Shannon limit, 0.0824, is less than 0.03. Also from Figure 14.4, we see that the code performs smoothly down to a UEBR of $10^{-8}$ without showing any error floor. As

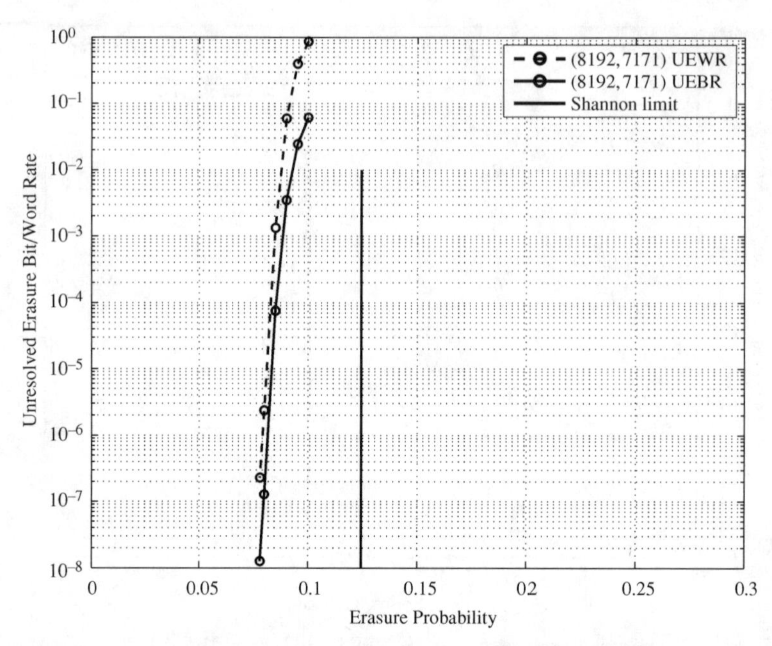


**Figure 14.3** The error performance of the (8192, 7171) QC-LDPC code over the BEC.

shown in Figure 12.11, the code also performs well over the binary-input AWGN channel, and iterative decoding of this code converges very fast. Its estimated error floor for the AWGN channel is below $10^{-25}$, as shown in Figure 12.12.

---

**Example 14.4** Consider the $256 \times 256$ array $\mathbf{H}^{(1)}_{qc,\text{disp}}$ of circulant permutation and zero matrices over GF(2) based on the (256, 2, 255) RS code over GF(257) given in Example 12.1. Take an $8 \times 128$ subarray $\mathbf{H}^{(1)}_{qc,\text{disp}}(8, 128)$ from $\mathbf{H}^{(1)}_{qc,\text{disp}}$, avoiding the zero matrices on the main diagonal of $\mathbf{H}^{(1)}_{qc,\text{disp}}$. We use this subarray as the base array for masking. Design an $8 \times 128$ masking matrix $\mathbf{Z}(8, 128)$ that consists of a row of 16 $8 \times 8$ circulants $[\mathbf{G}_0 \quad \mathbf{G}_1 \quad \cdots \quad \mathbf{G}_{15}]$ whose generators are given in Table 14.1. Since each generator has weight 4, the masking matrix $\mathbf{Z}(8, 128)$ has column and row weights 4 and 64, respectively. By masking $\mathbf{H}^{(1)}_{qc,\text{disp}}(8, 128)$ with $\mathbf{Z}(8, 128)$, we obtain an $8 \times 128$ masked array $\mathbf{M}^{(1)}(8, 128) = \mathbf{Z}(8, 128) \circledast \mathbf{H}^{(1)}_{qc,\text{disp}}(8, 128)$ of circulant permutation and zero matrices. $\mathbf{M}^{(1)}(8, 128)$ is a $2048 \times 32\,768$ matrix over GF(2) with column and row weights 4 and 64, respectively. The null space of $\mathbf{M}^{(1)}(8, 128)$ gives a (32 768, 30 721) QC-LDPC code with rate 0.9375. The performances of this code over the AWGN channel and the BEC are shown in Figures 14.5 and 14.6. From Figure 14.5, we see that, at a BER of $10^{-6}$, the code performs only 0.65 dB from the Shannon limit for the binary-input AWGN channel. From Figure 14.6, we see that, at a UEBR of $10^{-6}$, the code performs 0.017 from the Shannon limit, 0.0625, for the BEC. Again this long QC-LDPC code performs well over both the AWGN and binary random erasure channels.

**Table 14.1** The generators of the masking matrix $\mathbf{Z}(8, 128)$ of Example 14.4

| | |
|---|---|
| $\mathbf{g}_0 = (10100101)$ | $\mathbf{g}_1 = (01101010)$ |
| $\mathbf{g}_2 = (10101100)$ | $\mathbf{g}_3 = (10100110)$ |
| $\mathbf{g}_4 = (01011100)$ | $\mathbf{g}_5 = (10111000)$ |
| $\mathbf{g}_6 = (01010110)$ | $\mathbf{g}_7 = (01110010)$ |
| $\mathbf{g}_8 = (10010110)$ | $\mathbf{g}_9 = (01011010)$ |
| $\mathbf{g}_{10} = (00011110)$ | $\mathbf{g}_{11} = (11000110)$ |
| $\mathbf{g}_{12} = (00111010)$ | $\mathbf{g}_{13} = (01011010)$ |
| $\mathbf{g}_{14} = (00111010)$ | $\mathbf{g}_{15} = (11001100)$ |

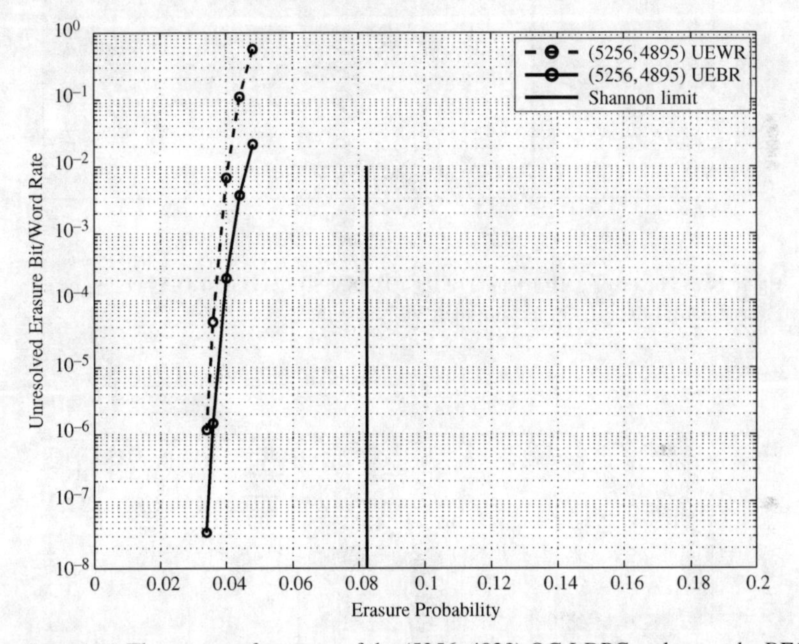


**Figure 14.4** The error performance of the (5256, 4823) QC-LDPC code over the BEC.

---

**Example 14.5** In this example, we consider the (4, 24)-regular (8088, 6743) type-II QC-BIBD-LDPC code with rate 0.834 constructed in Example 13.3. The error performance of this code over the binary-input AWGN channel, decoded with the SPA, is shown in Figure 13.3. At a BER of $10^{-6}$, it performs 1.05 dB from the Shannon limit for the AWGN channel. The error performance of this code over the BEC, decoded with the iterative decoding algorithm given in Section 14.1, is shown in Figure 14.7. For code rate 0.834, the Shannon limit for the BEC is 0.166. From Figure 14.7, we see that, at a UEBR of $10^{-6}$, the code performs 0.055 from the Shannon limit, 0.166, for the BEC.

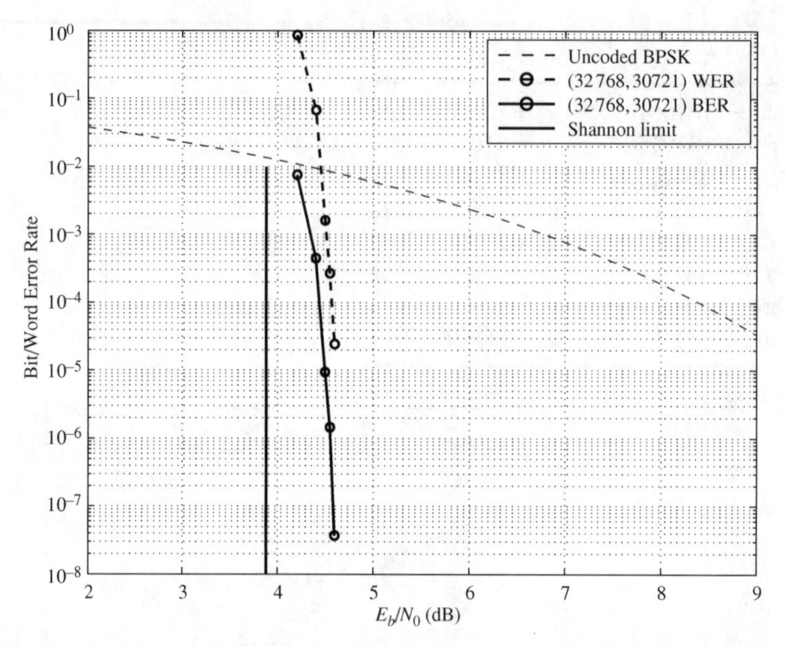


**Figure 14.5** The error performance of the (32 786, 30 721) QC-LDPC code given in Example 14.4 over the binary-input AWGN channel.

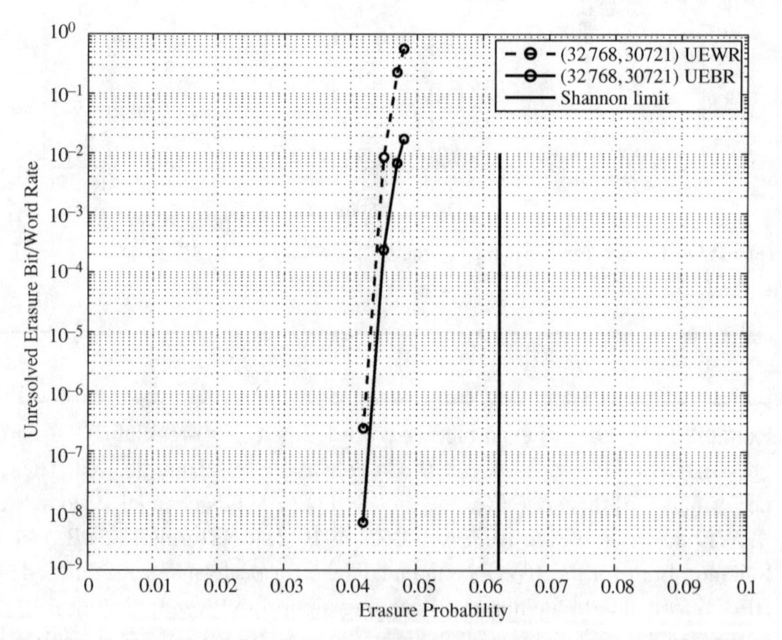


**Figure 14.6** The error performance of the (32 786, 30721) QC-LDPC code given in Example 14.4 over the BEC.

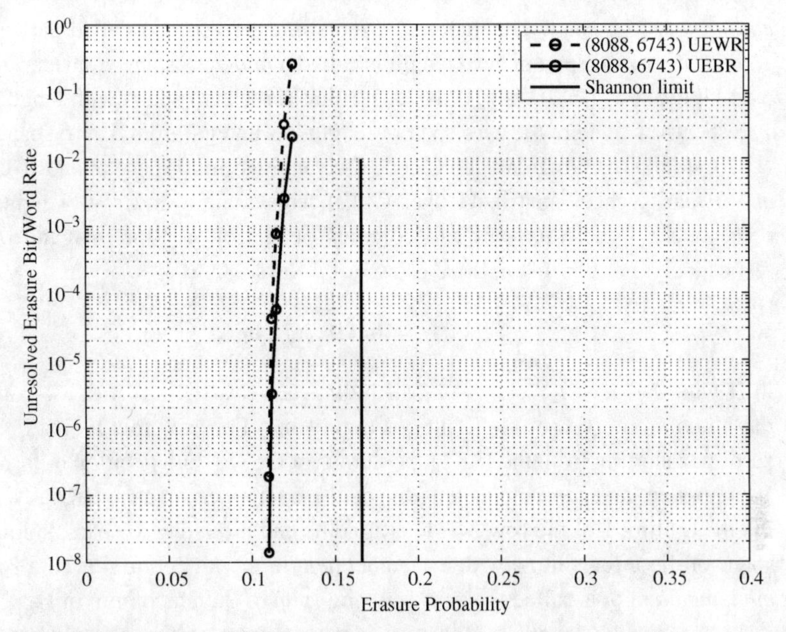


**Figure 14.7** The error performance of the (8088, 6743) QCBIBD-LDPC code given in Example 14.5 over the BEC.

---

In [6–9], many other structured LDPC codes constructed from finite geometries, finite fields, and combinatorial designs have been shown to perform well over both the AWGN and binary random erasure channels.

## 14.4 Correction of Erasure Bursts

An erasure pattern $\mathcal{E}$ is called an *erasure burst* of length $b$ if the erasures in $\mathcal{E}$ are confined to *b consecutive locations*, the first and last of which are erasures. Erasure bursts occur often in recording, jammed, and some fading channels. This section is concerned with the correction of erasure bursts with LDPC codes using iterative decoding. LDPC codes for correcting erasure bursts over the BBEC were first investigated in [10–13] and later in [6–9, 14] using a different approach that leads to construction of good erasure-burst-correction LDPC codes algebraically. The erasure-burst-correction algorithm, code designs, and constructions presented in this and the next two sections follow the approach given in [7, 9].

Let $\mathbf{v} = (v_0, v_1, \ldots, v_{n-1})$ be a nonzero $n$-tuple over GF(2). The first (leftmost) 1-component of $\mathbf{v}$ is called the *leading* 1 of $\mathbf{v}$ and the last (rightmost) 1-component of $\mathbf{v}$ is called the *trailing* 1 of $\mathbf{v}$. If $\mathbf{v}$ has only a single 1-component, then the leading 1 and trailing 1 of $\mathbf{v}$ are the same. A *zero span* of $\mathbf{v}$ is defined as *a sequence of consecutive zeros* between two 1-components. The zeros to the right of the trailing 1 of $\mathbf{v}$ together with the zeros to the left of the leading 1 of $\mathbf{v}$ also form a zero span, called the *end-around zero span*. The number of zeros in a zero span is called the *length* of the zero span. A zero span of zero length is called a *null* zero span. Consider the 16-tuple over GF(2), $\mathbf{v} = (0001000010010000)$. It has three zero spans with lengths 4, 2, and 7, respectively, where the zero span of length 7 is an end-around zero span.

Consider a $(g, r)$-regular LDPC code $\mathcal{C}$ given by the null space of an RC-constrained $m \times n$ parity-check matrix $\mathbf{H}$ over GF(2) with column and row weights $g$ and $r$, respectively. Label the rows and columns of $\mathbf{H}$ from 0 to $m - 1$ and 0 to $n - 1$, respectively. For $0 \leq j < n$, there is a set of $g$ rows,

$$\Lambda_j = \{\mathbf{h}_{i_1}, \mathbf{h}_{i_2}, \ldots, \mathbf{h}_{i_g}\},$$

in $\mathbf{H}$ with $0 \leq i_1 < i_2, \ldots, i_g < m$, such that each row $\mathbf{h}_{i_l}$ in $\Lambda_j$ has a 1-component in the $j$th column of $\mathbf{H}$ (or at the $j$th position of $\mathbf{h}_{i_l}$). For each row $\mathbf{h}_{i_l}$ in $\Lambda_j$, we find its zero span starting from the $(j+1)$th column (or the $(j+1)$th position of $\mathbf{h}_{i_l}$) to the next 1-component and compute its length. If the 1-component of a row $\mathbf{h}_{i_l}$ in $\Lambda_j$ at position $j$ is the trailing 1 of the row, we determine its end around zero span. Among the zero spans of the $g$ rows in $\Lambda_j$ with a 1-component in the $j$th column of $\mathbf{H}$, a zero span of *maximum length* is called the *zero covering span* of the $j$th column of $\mathbf{H}$.

Let $\sigma_j$ be the length of the zero covering span of the $j$th column of $\mathbf{H}$. The $n$-sequence of integers

$$(\sigma_0, \sigma_1, \ldots, \sigma_{n-1}) \tag{14.4}$$

is called the *profile* of column zero covering spans of the parity-check matrix $\mathbf{H}$. The column zero covering span of *minimum length* is called the *zero covering span of the parity-check matrix* $\mathbf{H}$. The length of the zero covering span of $\mathbf{H}$ is given by

$$\sigma = \min\{\sigma_j \colon 0 \leq j < n\}. \tag{14.5}$$

By employing the concept of zero covering spans of columns of the parity-check matrix $\mathbf{H}$ of an LDPC code $\mathcal{C}$, a simple erasure-burst decoding algorithm can be devised. The erasure-burst-correction capability of code $\mathcal{C}$ is determined by the length $\sigma$ of the zero covering span of its parity-check matrix $\mathbf{H}$.

---

**Example 14.6** Consider the (3, 3)-regular cyclic (7, 3) LDPC code given by the null space of the following $7 \times 7$ circulant over GF(2):

$$\mathbf{H} = \begin{bmatrix} \mathbf{h}_0 \\ \mathbf{h}_1 \\ \mathbf{h}_2 \\ \mathbf{h}_3 \\ \mathbf{h}_4 \\ \mathbf{h}_5 \\ \mathbf{h}_6 \end{bmatrix} = \begin{bmatrix} 1 & 0 & 0 & 0 & 1 & 1 & 0 \\ 0 & 1 & 0 & 0 & 0 & 1 & 1 \\ 1 & 0 & 1 & 0 & 0 & 0 & 1 \\ 1 & 1 & 0 & 1 & 0 & 0 & 0 \\ 0 & 1 & 1 & 0 & 1 & 0 & 0 \\ 0 & 0 & 1 & 1 & 0 & 1 & 0 \\ 0 & 0 & 0 & 1 & 1 & 0 & 1 \end{bmatrix}. \tag{14.6}$$

We can easily check that the length of each column zero covering span is 3. Hence, the profile of the column zero covering spans of **H** is $(3, 3, 3, 3, 3, 3, 3)$ and the length of the zero covering span of the parity-check matrix **H** is $\sigma = 3$.

---

Consider a $(g, r)$-regular LDPC code $\mathcal{C}$ over GF(2) given by the null space of an $m \times n$ parity-check matrix **H** with column and row weights $g$ and $r$, respectively, for which the profile of the column zero covering span is $(\sigma_0, \sigma_1, \ldots, \sigma_{n-1})$. Suppose this code is used for correcting erasure bursts. Let $\mathbf{v} = (v_0, v_1, \ldots, v_{n-1})$ be the transmitted codeword and $\mathbf{r} = (r_0, r_1, \ldots, r_{n-1})$ the corresponding received sequence. For $0 \leq j < n$, if $j$ is the starting position of an erasure-burst pattern $\mathcal{E}$ of length at most $\sigma_j + 1$ that occurs during the transmission of **v**, then there is a row $\mathbf{h}_i = (h_{i,0}, h_{i,1}, \ldots, h_{i,n-1})$ in **H** for which the $j$th component $h_{i,j}$ is "1" and that is followed by a zero span of length $\sigma_j$. Since the burst starts at position $j$ and has length at most $\sigma_j + 1$, all the received symbols checked by $\mathbf{h}_i$, other than the $j$th received symbol $r_j$, are known. Consequently, the value of the $j$th erased symbol $r_j$ (or the transmitted symbol $v_j$) can be determined from (14.3). Replacing the erased symbol at the $j$th position by its value results in a shorter erasure-burst pattern $\mathcal{E}\backslash\{j\}$. The erasure-recovery procedure can be repeated until all erased symbols in $\mathcal{E}$ have been determined.

The above iterative algorithm for correcting erasure bursts that is based on the profile of column zero covering spans of the parity-check matrix of an LDPC code can be formulated as follows.

---

1. Check the received sequence **r**. If there are erasures, determine the starting position of the erasure burst in **r**, say $j$, and go to Step 2. If there is no erasure in **r**, stop decoding.
2. Determine the length of the erasure burst, say $b$. If $b \leq \sigma_j + 1$, go to Step 3; otherwise, stop decoding.
3. Determine the value of the erasure at position $j$ using (14.3) and go to Step 1.

---

It follows from (14.5) that any erasure burst of length $\sigma + 1$ or less is guaranteed to be recoverable regardless of its starting position. Therefore, $\sigma + 1$ is a lower bound on the erasure-burst-correction capability of an LDPC code using the iterative decoding algorithm given above. The simple $(7, 3)$ cyclic LDPC code given in Example 14.1 is capable of correcting any erasure burst of length 4 or less.

From the definition of the zero covering span of the parity-check matrix **H** of an LDPC code and (14.5), the Tanner graph of the code has the following structural property: for an LDPC code whose parity-check matrix **H** has a zero covering span of length $\sigma$, no $\sigma + 1$ or fewer consecutive variable nodes in its Tanner graph form a stopping set and hence any $\sigma + 1$ consecutive variable nodes form a *stopping-set-free zone*.

The erasure-burst-correction capability $l_b$ of an $(n, k)$ LDPC code (or any $(n, k)$ linear block code) of length $n$ and dimension $k$ is upper bounded by $n - k$, the number of parity-check symbols of the code. This can easily be proved as follows. Suppose the code is capable of correcting all the erasure bursts of length $n - k + 1$ with the iterative decoding algorithm presented above. Then it must have a parity-check matrix with a zero covering span of length $n - k$. This implies that the parity-check matrix of the code has at least $n - k + 1$ linearly independent rows. Consequently, the rank of the parity-check matrix is at least $n - k + 1$, which contradicts the fact that the rank of any parity-check matrix of an $(n, k)$ linear block code is $n - k$. To measure the efficiency of erasure burst correction of an $(n, k)$ LDPC code with the iterative decoding algorithm presented above, we define its *erasure-burst-correction efficiency* as

$$\eta = \frac{l_b}{n - k}. \tag{14.7}$$

Clearly, $\eta$ is upper bounded by 1. An LDPC code is said to be *optimal* for erasure burst correction if its erasure-burst efficiency is equal to 1.

The concepts and iterative decoding algorithm developed for a regular LDPC code can be applied to irregular LDPC codes and, in fact, can be applied to any linear block codes. For an LDPC code, the same parity-check matrix $\mathbf{H}$ can be used for iterative decoding over both the AWGN and binary erasure channels.

## 14.5 Erasure-Burst-Correction Capabilities of Cyclic Finite-Geometry and Superposition LDPC Codes

### 14.5.1 Erasure Burst Correction with Cyclic Finite-Geometry LDPC Codes

The erasure-burst-correction capabilities of cyclic LDPC codes based on two-dimensional Euclidean (or projective) geometries can easily be determined. Consider a cyclic EG-LDPC code $\mathcal{C}_{\text{EG},c,1}$ based on the two-dimensional Euclidean geometry EG$(2, q)$ over GF$(q)$ (see Section 11.1). Take a line $\mathcal{L}$ in EG$(2, q)$ not passing through the origin, and form its incidence vector $\mathbf{v}_{\mathcal{L}}$, a $(q^2 - 1)$-tuple over GF(2). A parity-check matrix $\mathbf{H}^{(1)}_{\text{EG},c,1}$ of $\mathcal{C}_{\text{EG},c,1}$ is a $(q^2 - 1) \times (q^2 - 1)$ circulant, which is formed with $\mathbf{v}_{\mathcal{L}}$ as the first (or top) row and its $q^2 - 2$ cyclic shifts as the other rows. Since the incidence vector $\mathbf{v}_{\mathcal{L}}$ has $q$ 1-components, it has $q$ zero spans. Determine the maximum zero span of $\mathbf{v}_{\mathcal{L}}$ and its length. Note that the maximum zero span of $\mathbf{v}_{\mathcal{L}}$ is unique, otherwise $\mathbf{H}^{(1)}_{\text{EG},c,1}$ would not satisfy the RC constraint. Since all the other rows of $\mathbf{H}^{(1)}_{\text{EG},c,1}$ are cyclic shifts of $\mathbf{v}_{\mathcal{L}}$, their maximum zero spans have the same length as that of the maximum zero span of $\mathbf{v}_{\mathcal{L}}$. The starting positions of the maximum zero spans of $\mathbf{v}_{\mathcal{L}}$ and its $q^2-2$ cyclic shifts are all different and they spread from 0 to $n-1$. Consequently, the lengths of the zero covering spans of the $q^2 - 1$ columns of the parity-check matrix $\mathbf{H}^{(1)}_{\text{EG},c,1}$ of $\mathcal{C}_{\text{EG},c,1}$ are all equal to the length of the maximum zero span of $\mathbf{v}_{\mathcal{L}}$. Hence, the length $\sigma$ of the zero covering span of the parity-check matrix $\mathbf{H}^{(1)}_{\text{EG},c,1}$ of $\mathcal{C}_{\text{EG},c,1}$ is equal to the length of the maximum zero span of $\mathbf{v}_{\mathcal{L}}$. Table 14.2 gives the lengths of zero covering spans $\sigma$ and actual erasure-burst-correction capabilities $l_b$ of a list of cyclic EG-LDPC codes.

Computing the length of the zero covering span of the parity-check matrix of a cyclic PG-LDPC code constructed from the two-dimensional projective geometry PG$(2, q)$ given in Section 11.6 can be done in the same manner. The lengths of the zero covering spans $\sigma$ of the parity-check matrices and the actual erasure-burst-correction capabilities $l_b$ of some cyclic PG-LDPC codes are also given in Table 14.2.

**Table 14.2** Erasure-burst-correction capabilities of some finite-geometry cyclic LDPC codes ($l_b$ is the actual erasure-burst-correction capability)

| Geometry | Codes | Size of minimal stopping set | $\sigma$ | $l_b$ |
|---|---|---|---|---|
| EG$(2, 2^4)$ | (255, 175) | 17 | 54 | 70 |
| EG$(2, 2^5)$ | (1023, 781) | 33 | 121 | 157 |
| EG$(2, 2^6)$ | (4095, 3367) | 65 | 309 | 367 |
| EG$(2, 2^7)$ | (16 383, 14 197) | 129 | 561 | 799 |
| PG$(2, 2^4)$ | (273, 191) | 18 | 61 | 75 |
| PG$(2, 2^5)$ | (1057, 813) | 34 | 90 | 124 |
| PG$(2, 2^6)$ | (4161, 3431) | 66 | 184 | 303 |
| PG$(2, 2^7)$ | (16 513, 14 326) | 130 | 1077 | 1179 |

### 14.5.2 Erasure Burst Correction with Superposition LDPC Codes

Consider the superposition product $C_{\text{sup},p} = C_1 \times C_2$ of an $(n_1, k_1)$ LDPC code $C_1$ and an $(n_2, k_2)$ LDPC code $C_2$ given by the null spaces of an $m_1 \times n_1$ and an $m_2 \times n_2$ low-density parity-check matrix $\mathbf{H}_1$ and $\mathbf{H}_2$, respectively. Let $\sigma_1$ and $\sigma_2$ be the lengths of the zero covering spans of $\mathbf{H}_1$ and $\mathbf{H}_2$, respectively. Consider the parity-check matrix $\mathbf{H}_{\text{sup},p}$ of the superposition product code $C_{\text{sup},p}$ given by (13.27). On examining the structures of $\mathbf{H}_{\text{sup},p}$ and $\mathbf{H}_j^{(2)}$ given by (13.25), we can readily see that, for each column of $\mathbf{H}_{\text{sup},p}$, there is a 1-component in the lower part of $\mathbf{H}_{\text{sup},p}$ that is followed by a span of at least $\sigma_2(n_1 - 1)$ consecutive zeros. This implies that the length $\sigma_{\text{sup},p}$ of the zero covering span of $\mathbf{H}_{\text{sup},p}$ is at least $\sigma_2(n_1 - 1)$. Consequently, using $\mathbf{H}_{\text{sup},p}$ for decoding, the superposition product LDPC code $C_{\text{sup},p}$ is capable of correcting any erasure burst of length up to at least $\sigma_2(n_1 - 1) + 1$ with the erasure-burst decoding algorithm presented in Section 14.4. The superposition product LDPC code $C_{\text{sup},p}$ generated by the parity-check matrix $\mathbf{H}_{\text{sup},p}$ given by (13.27) has $C_1$ as the row code and $C_2$ as the column code. If we switch the roles of $C_1$ and $C_2$, with $C_2$ as the row code and $C_1$ as the column code, then the length of the zero covering span of the parity-check matrix of the resultant product code $C^*_{\text{sup},p} = C_2 \times C_1$ is at least $\sigma_1(n_2 - 1)$. In this case, the superposition product LDPC code $C^*_{\text{sup},p}$ is capable of correcting any erasure burst of length up to at least $\sigma_1(n_2 - 1) + 1$. Summarizing the above results, we can conclude that the erasure-burst-correction capability of the superposition product of two LDPC codes $C_1$ and $C_2$ is lower bounded as follows:

$$b_{\text{sup},p} = \max(\sigma_1(n_2 - 1) + 1, \sigma_2(n_1 - 1) + 1). \tag{14.8}$$

Consider the parity-check matrix $\mathbf{H}_{\text{sup},p}$ of a product QC-LDPC code $C_{\text{sup},p}$ given by (13.29), constructed by superposition. The lower part of the matrix consists of a row of $k$ identity matrices of size $r(q - 1) \times r(q - 1)$. On examining the structure of $\mathbf{H}_{\text{sup},p}$ we find that, for each column of $\mathbf{H}_{\text{sup},p}$, there is a 1-component in the lower part of $\mathbf{H}_{\text{sup},p}$ that is followed by a span of $r(q - 1) - 1$ zeros and then a 1-component. This zero span is actually the span of the $r(q - 1) - 1$ zeros between a 1-component in a $r(q - 1) \times r(q - 1)$ identity matrix and the corresponding 1-component in the next $r(q-1) \times r(q-1)$ identity matrix of the lower part of $\mathbf{H}_{\text{sup},p}$, including the end-around case. Consequently, the length of the zero covering span of $\mathbf{H}_{\text{sup},p}$ is at least $r(q-1)-1$ (actually it is the exact length of the zero covering span of the parity-check matrix $\mathbf{H}_{\text{sup},p}$). Therefore, the product QC-LDPC code $\mathcal{C}_{\text{sup},p}$ given by the null space of $\mathbf{H}_{\text{sup},p}$ given by (13.29) is capable of correcting any erasure burst of length up to at least $r(q-1)$ with the iterative decoding algorithm presented in the previous section.

---

**Example 14.7** Consider the (3, 6)-regular (4572, 2307) product QC-LDPC code with rate 0.5015 given in Example 13.12 that was constructed using the superposition method presented in Section 13.7.4. The length of the zero covering span of the parity-check matrix $\mathbf{H}_{\text{sup},p}$ of this code is $(6 \times 127) - 1 = 761$. Then 762 is a lower bound on the erasure-burst-correction capability of the code. However, by computer search, we find that the actual erasure-burst-correction capability is $l_b = 1015$. In Figure 13.15, we showed that this code performs well over the AWGN channel with iterative decoding using the SPA. The performance of this code over the BEC with iterative decoding presented in Section 14.1 is shown in Figure 14.8. At a UEBR of $10^{-6}$, we see that the code performs 0.11 from the Shannon limit, 0.4985. So, the code performs universally well over all three types of channel: AWGN, binary random, and erasure-burst channels.

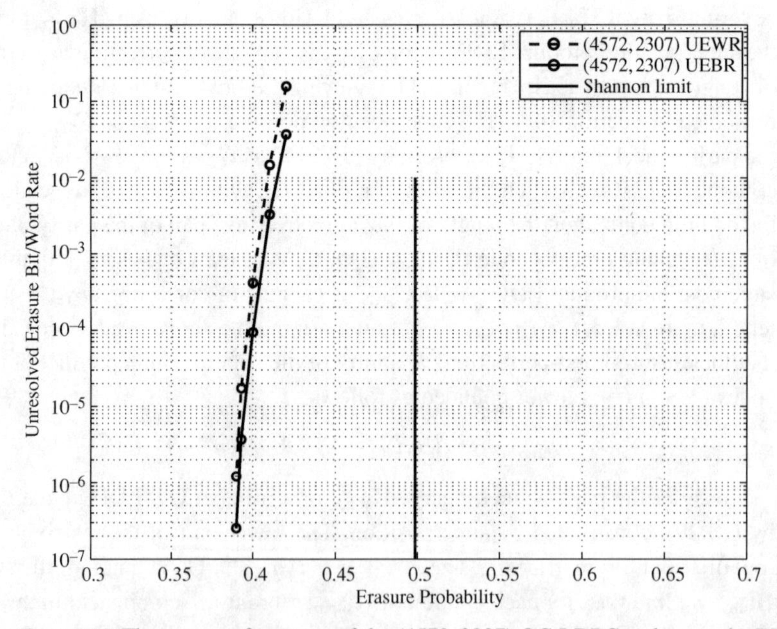


**Figure 14.8** The error performance of the (4572, 2307) QC-LDPC code over the BEC channel.

---

## 14.6 Asymptotically Optimal Erasure-Burst-Correction QC-LDPC Codes

Asymptotically optimal QC-LDPC codes can be constructed by masking the arrays of CPMs constructed in Chapter 12 using a special class of masking matrices.

Let $k$, $l$, and $s$ be three positive integers such that $2 \le k, l$ and $1 \le s < k$. Form $l$ $k$-tuples over GF(2), $\mathbf{u}_0, \mathbf{u}_1, \ldots, \mathbf{u}_{l-1}$, where (1) $\mathbf{u}_0 = (1, 0, 0, \ldots, 0)$ consists of a single 1-component at the leftmost position followed by $k - 1$ consecutive zeros; (2) $\mathbf{u}_{l-1} = (0, \ldots, 0, 1, \ldots, 1)$ consists of a sequence of $s$ consecutive zeros followed by $k - s$ consecutive 1s; and (3) the other $k$-tuples, $\mathbf{u}_1$ to $\mathbf{u}_{l-2}$, are zero $k$-tuples, that is, $\mathbf{u}_1 = \mathbf{u}_2 = \cdots = \mathbf{u}_{l-2} = (0, 0, \ldots, 0)$. Define the following $kl$-tuple over GF(2):

$$\mathbf{u} = (\mathbf{u}_0, \mathbf{u}_1, \ldots, \mathbf{u}_{l-1}). \tag{14.9}$$

Then the $kl$-tuple $\mathbf{u}$ has a single 1-component at the left end and $k - s$ 1-components at the right end. It has weight $k - s + 1$. This $kl$-tuple has one and only one zero span of length $k(l - 1) + s - 1$.

Form a $kl \times kl$ circulant $\mathbf{G}$ with $\mathbf{u}$ as the generator. Then $\mathbf{G}$ has the following structural properties: (1) its column and row weights are both $k - s + 1$; and (2) each row (or column) has a unique zero span of length $k(l-1) + s - 1$, including the end-around zero span. For $t \ge 1$, we form the following $kl \times klt$ matrix over GF(2):

$$\mathbf{Z}(kl, klt) = [\mathbf{G} \;\; \mathbf{G} \;\; \cdots \;\; \mathbf{G}], \tag{14.10}$$

which consists of a row of $t$ $\mathbf{G}$ matrices. This matrix $\mathbf{Z}(kl, klt)$ has constant column and row weights $k - s + 1$ and $t(k - s + 1)$, respectively. It has the following structural properties: (1) each column is a downward cyclic shift of the column on its left (including the transition across the boundary of two neighboring $\mathbf{G}$ matrices) and the first column is the downward cyclic shift of the last column; (2) each row is the right cyclic shift of the row above it and the first row is the right cyclic shift of the last row; (3) each row consists of $t$ zero spans (including the end-around zero span), each of length $k(l - 1) + s - 1$; and (4) each column has a unique zero span of length $k(l - 1) + s - 1$. It follows from the above structural properties of $\mathbf{Z}(kl, klt)$ that the length of the zero covering span of each column of $\mathbf{Z}(kl, klt)$ is $k(l - 1) + s - 1$. Note that there are zero-spans in some rows of $\mathbf{Z}(kl, klt)$ that run across the boundary of two neighboring $\mathbf{G}$ matrices. Consequently, the length of the zero covering span of $\mathbf{Z}(kl, klt)$ is $\sigma = k(l - 1) + s - 1$.

---

**Example 14.8** Let $k = 4$, $l = 3$, $t = 2$, and $s = 2$. Construct three 4-tuples over GF(2), $\mathbf{u}_0 = (1000)$, $\mathbf{u}_1 = (0000)$, and $\mathbf{u}_2 = (0011)$. Form the following 12-tuple over GF(2):

$$\mathbf{u} = (\mathbf{u}_0, \mathbf{u}_1, \mathbf{u}_2) = (100000000011).$$

Use $\mathbf{u}$ as the generator to form a $12 \times 12$ circulant $\mathbf{G}$ with both column and row weight 3. Then use two $\mathbf{G}$ matrices to form the following matrix $\mathbf{Z}(12, 24)$ with column and row weights 3 and 6, respectively:

$$\mathbf{Z}(12,24)$$
$$= \begin{bmatrix}
1&0&0&0&0&0&0&0&0&0&1&1&1&0&0&0&0&0&0&0&0&0&1&1\\
1&1&0&0&0&0&0&0&0&0&0&1&1&1&0&0&0&0&0&0&0&0&0&1\\
1&1&1&0&0&0&0&0&0&0&0&0&1&1&1&0&0&0&0&0&0&0&0&0\\
0&1&1&1&0&0&0&0&0&0&0&0&0&1&1&1&0&0&0&0&0&0&0&0\\
0&0&1&1&1&0&0&0&0&0&0&0&0&0&1&1&1&0&0&0&0&0&0&0\\
0&0&0&1&1&1&0&0&0&0&0&0&0&0&0&1&1&1&0&0&0&0&0&0\\
0&0&0&0&1&1&1&0&0&0&0&0&0&0&0&0&1&1&1&0&0&0&0&0\\
0&0&0&0&0&1&1&1&0&0&0&0&0&0&0&0&0&1&1&1&0&0&0&0\\
0&0&0&0&0&0&1&1&1&0&0&0&0&0&0&0&0&0&1&1&1&0&0&0\\
0&0&0&0&0&0&0&1&1&1&0&0&0&0&0&0&0&0&0&1&1&1&0&0\\
0&0&0&0&0&0&0&0&1&1&1&0&0&0&0&0&0&0&0&0&1&1&1&0\\
0&0&0&0&0&0&0&0&0&1&1&1&0&0&0&0&0&0&0&0&0&1&1&1
\end{bmatrix}. \tag{14.11}$$

We can readily see that each row of $\mathbf{Z}(12,24)$ has two zero spans of length 9. The profile of the column zero covering span is $(9,9,9,9,9,9,9,9,9,9,9,9)$. The length of the zero covering span of $\mathbf{Z}(12,24)$ is $\sigma = 9$.

---

Take a $kl \times klt$ subarray $\mathbf{H}(kl, klt)$ from any RC-constrained array $\mathbf{H}$ of $(q-1)\times(q-1)$ CPMs constructed in Chapter 12 as the base array for masking. If $\mathbf{H}$ contains zero matrices (such as $\mathbf{H}^{(1)}_{qc,\text{disp}}$), $\mathbf{H}(kl, klt)$ is taken avoiding the zero matrices in $\mathbf{H}$. On masking $\mathbf{H}(kl, klt)$ with $\mathbf{Z}(kl, klt)$, we obtain an RC-constrained $kl \times klt$ masked array $\mathbf{M}(kl, klt) = \mathbf{Z}(kl, klt) \circledast \mathbf{H}(kl, klt)$ of $(q-1)\times(q-1)$ circulant permutation and zero matrices. In each column of $\mathbf{M}(kl, klt)$ (as an array), there is a circulant permutation matrix in a row followed by $k(l-1)+s-1$ consecutive $(q-1)\times(q-1)$ zero matrices and ending with a CPM. $\mathbf{M}(kl, klt)$ is an RC-constrained $kl(q-1) \times klt(q-1)$ matrix over GF(2) with column and row weights $k-s+1$ and $t(k-s+1)$, respectively. The length of the zero covering span of each column of $\mathbf{M}(kl, klt)$ is at least $[k(l-1)+s-1](q-1)$. Consequently, the length of the zero-covering-span of $\mathbf{M}(kl, klt)$ is at least $[k(l-1)+s-1](q-1)$. The null space of $\mathbf{M}(kl, klt)$ gives a $(k-s+1, t(k-s+1))$-regular QC-LDPC code $\mathcal{C}_{qc,\text{mas}}$ of length $klt(q-1)$, rate $(t-1)/t$, and minimum distance at least $k-s+2$, whose Tanner graph has a girth of at least 6. The code is capable of correcting any erasure-burst of length up to

$$[k(l-1)+s-1](q-1)+1. \tag{14.12}$$

The number of parity-check bits of $\mathcal{C}_{qc,\text{mas}}$ is at most $kl(q-1)$. The erasure-burst-correction efficiency is

$$\eta \geq \frac{[k(l-1)+s-1](q-1)+1}{kl(q-1)}. \tag{14.13}$$

For large $q$, $k$, $l$ and $k-s=2$ or 3 (or $k-s$ small relative to $k$), the above lower bound on $\eta$ is close to 1. For $k-s=2$, $\mathcal{C}_{qc,\text{mas}}$ is a $(3,3t)$-regular QC-LDPC code, and for $k-s=3$, $\mathcal{C}_{qc,\text{mas}}$ is a $(4,4t)$-regular QC-LDPC code.

If the base array $\mathbf{H}(kl, klt)$ contains zero matrices, the above construction simply gives a near-regular code. The erasure burst correction of the code is still lower bounded by (14.12). If the base array $\mathbf{H}(kl, klt)$ is taken from the array $\mathbf{H}^{(6)}_{qc,\text{disp}}$ constructed from a prime field GF($p$) (see Section 11.6), then $\mathbf{H}(kl, klt)$ is a $kl \times klt$ array of $p \times p$ circulant permutation matrices. In this case, the code $C_{qc,\text{disp}}$ given by the null space of $\mathbf{M}(kl, klt)$ is capable of correcting any erasure burst of length up to $[k(l-1)+s-1]p+1$.

Also, subarrays of the arrays of permutation matrices constructed on the basis of the decomposition of finite Euclidean geometries given in Section 11.3 (see (11.16)) can be used as base arrays for masking to construct asymptotically optimal LDPC codes for correcting erasure bursts.

---

**Example 14.9** Consider the RC-constrained $72 \times 72$ array $\mathbf{H}^{(1)}_{qc,\text{disp}}$ of $72 \times 72$ circulant permutation and zero matrices based on the $(72, 2, 71)$ RS code over the prime field GF(73) (see Example 12.2). Set $k = 4$, $l = 2$, $t = 8$, and $s = 1$. Then $kl = 8$ and $klt = 64$. Take the first eight rows of $\mathbf{H}^{(1)}_{qc,\text{disp}}$ and remove the first and last seven columns. This results in an $8 \times 64$ subarray $\mathbf{H}^{(1)}_{qc,\text{disp}}(8, 64)$ of CPMs (no zero matrices). This subarray $\mathbf{H}^{(1)}_{qc,\text{disp}}(8, 64)$ will be used as the base array for masking to construct a QC-LDPC code for correcting erasure-bursts over the BEC. To construct an $8 \times 64$ masking matrix $\mathbf{Z}(8, 64)$, we first form two 4-tuples over GF(2), $\mathbf{u}_0 = (1000)$ and $\mathbf{u}_1 = (0111)$. Concatenate these two 4-tuples to form an 8-tuple over GF(2), $\mathbf{u} = (\mathbf{u}_0\mathbf{u}_1) = (10000111)$. Use $\mathbf{u}$ as the generator to form the following $8 \times 8$ circulant over GF(2):

$$\mathbf{G} = \begin{bmatrix} 1 & 0 & 0 & 0 & 0 & 1 & 1 & 1 \\ 1 & 1 & 0 & 0 & 0 & 0 & 1 & 1 \\ 1 & 1 & 1 & 0 & 0 & 0 & 0 & 1 \\ 1 & 1 & 1 & 1 & 0 & 0 & 0 & 0 \\ 0 & 1 & 1 & 1 & 1 & 0 & 0 & 0 \\ 0 & 0 & 1 & 1 & 1 & 1 & 0 & 0 \\ 0 & 0 & 0 & 1 & 1 & 1 & 1 & 0 \\ 0 & 0 & 0 & 0 & 1 & 1 & 1 & 1 \end{bmatrix}.$$

On the basis of (14.10), we construct an $8 \times 64$ masking matrix $\mathbf{Z}(8, 64)$ that consists of a row of eight $\mathbf{G}$ matrices, that is, $\mathbf{Z}(8, 64) = [\mathbf{GGGGGGGG}]$. The length of the zero covering span of $\mathbf{Z}(8, 64)$ is 4. The column and row weights of $\mathbf{Z}(8, 64)$ are 4 and 32, respectively. By masking $\mathbf{H}^{(1)}_{qc,\text{disp}}(8, 64)$ with $\mathbf{Z}(8, 64)$, we obtain an RC-constrained $8 \times 64$ masked array $\mathbf{M}^{(1)}(8, 64) = \mathbf{Z}(8, 64) \circledast \mathbf{H}^{(1)}_{qc,\text{disp}}(8, 64)$. This masked array is a $576 \times 4608$ matrix over GF(2) with column and row weights 4 and 32, respectively. The length $\sigma$ of the zero covering span of $\mathbf{M}^{(1)}(8, 64)$ is $4 \times 72 = 288$. The null space of $\mathbf{M}^{(1)}(8, 64)$ gives a $(4608, 4035)$ QC-LDPC code with rate 0.8757 that is capable of correcting any erasure burst of length up to $\sigma + 1 = 289$. By computer search, we find that the code can actually correct any erasure-burst of length up to 375. Thus, the erasure-burst-correction efficiency is

0.654. This (4608, 4035) QC-LDPC code also performs well over the AWGN channel and the BEC as shown in Figures 14.9 and 14.10. For the AWGN, it performs 1.15 dB from the Shannon limit at a BER of $10^{-8}$. For the BEC, it performs 0.045 from the Shannon limit, 0.1243.

---

**Example 14.10** In this example, we construct a long QC-LDPC code with high erasure-burst-correction efficiency, which also performs well over both the AWGN and BEC channels. The array for code construction is the $256 \times 256$ array $\mathbf{H}^{(1)}_{qc,\text{disp}}$ of $256 \times 256$ circulant permutation and zero matrices constructed on the basis of the (256, 2, 255) RS code over GF(257) given in Example 12.1. Set $k = 4$, $l = 8$, $t = 8$, and $s = 1$. Then $kl = 32$ and $klm = 256$. Take a $32 \times 256$ subarray $\mathbf{H}^{(1)}_{qc,\text{disp}}(32, 256)$ from $\mathbf{H}^{(1)}_{qc,\text{disp}}$, say the first 32 rows of $\mathbf{H}^{(1)}_{qc,\text{disp}}$, and use it as the base array for masking. Each of the first 32 columns of $\mathbf{H}^{(1)}_{qc,\text{disp}}(32, 256)$ contains a single $256 \times 256$ zero matrix. Next, we construct a 32-tuple $\mathbf{u} = (\mathbf{u}_0, \mathbf{u}_1, \ldots, \mathbf{u}_7)$, where $\mathbf{u}_0 = (1000)$, $\mathbf{u}_7 = (0111)$, and $\mathbf{u}_1 = \cdots = \mathbf{u}_6 = (0000)$. Construct a $32 \times 32$ circulant $\mathbf{G}$ with $\mathbf{u}$ as the generator. Both the column and row weight of $\mathbf{G}$ are 4. Then form a $32 \times 256$ masking matrix $\mathbf{Z}(32, 256)$ with eight $\mathbf{G}$ matrices arranged in a row, that is, $\mathbf{Z}(32, 258) = [\mathbf{GGGGGGGG}]$, which has column and row weights 4 and 32, respectively. By masking $\mathbf{H}^{(1)}_{qc,\text{disp}}(32, 256)$ with $\mathbf{Z}(32, 256)$, we obtain a $32 \times 256$ masked matrix $\mathbf{M}^{(1)}(32, 256) = \mathbf{Z}(32, 256) \circledast \mathbf{H}^{(1)}_{qc,\text{disp}}(32, 256)$, which is an

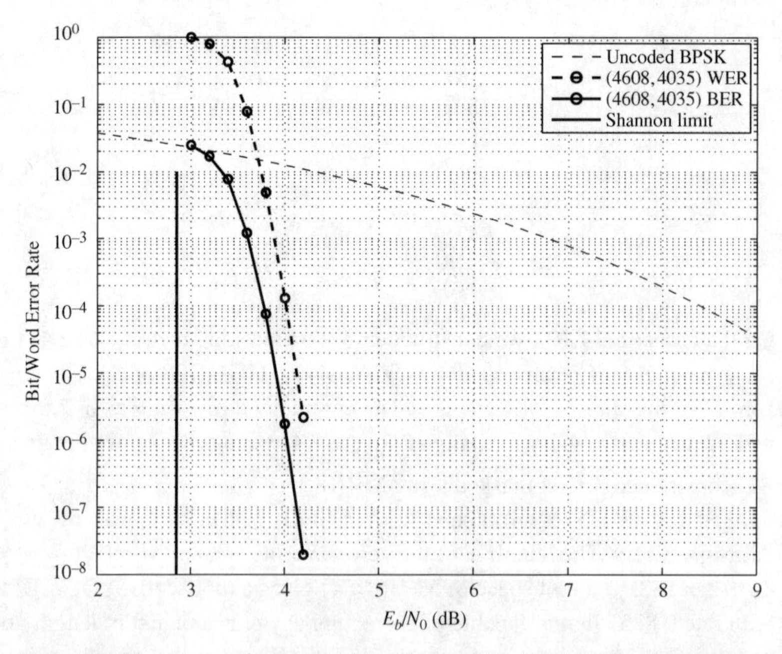


**Figure 14.9** The error performance of the (4608, 4035) QC-LDPC code over the binary-input AWGN channel given in Example 14.9.

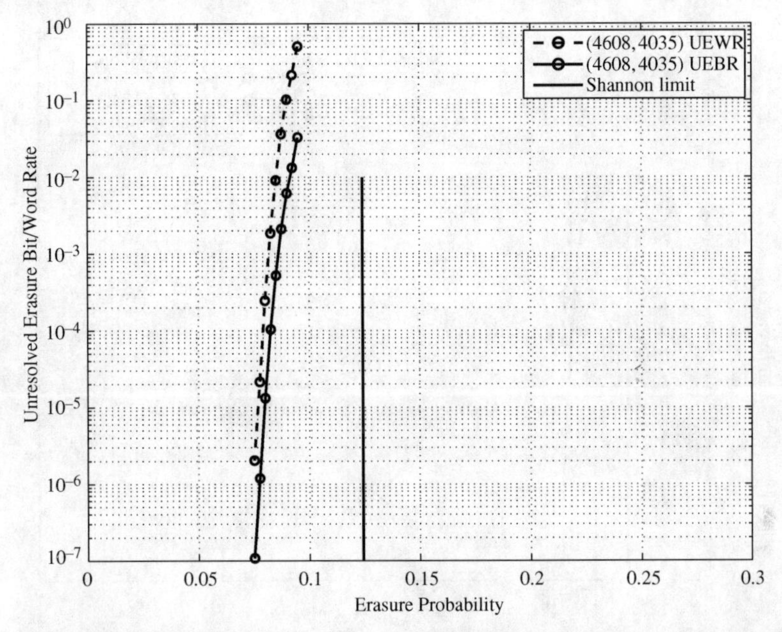


**Figure 14.10** The error performance of the (4608, 4035) QC-LDPC code over the BEC given in Example 14.9.

$8192 \times 65536$ matrix with two different column weights, 3 and 4, and two different row weights, 31 and 32. The length of the zero covering span $\sigma$ of $\mathbf{M}^{(1)}(32, 256)$ is $[k(l-1)+s-1](q-1) = [4(8-1)+1-1] \times 256 = 7168$. The null space of $\mathbf{M}^{(1)}(32, 256)$ gives a (65 536, 57 345) near-regular QC-LDPC code with rate 0.875. The code is capable of correcting any erasure burst of length at least up to 7169. Its erasure-burst-correction efficiency is lower bounded by 0.8752. The error performances of this code over the binary-input AWGN channel and BEC are shown in Figures 14.11 and 14.12, respectively. For the AWGN channel, the code performs 0.6 dB from the Shannon limit at a BER of $10^{-6}$. For the BEC, it performs 0.03 from the Shannon limit, 0.125.

## 14.7 Construction of QC-LDPC Codes by Array Dispersion

A subarray of the array $\mathbf{H}$ of circulant permutation matrices constructed in Chapters 12 and 13 can be dispersed into a larger array with a lower density to construct new QC-LDPC codes. In this section, we present a *dispersion technique* [9] by which to construct a large class of QC-LDPC codes. Codes constructed by this dispersion technique not only have good *erasure-burst-correction* capabilities but also perform well AWGN and binary random erasure channels.

Let $\mathbf{H}$ be an $c \times n$ RC-constrained array of $(q-1) \times (q-1)$ CPMs as constructed in either Chapter 12 or Chapter 13. The values of $c$ and $n$ depend on the method that is

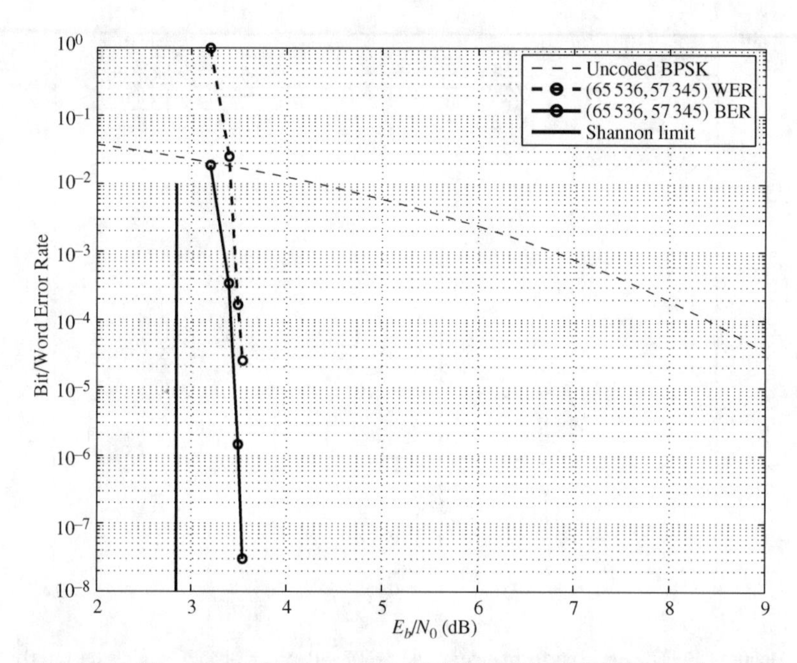


**Figure 14.11** The error performance of the (65536, 57345) QC-LDPC code over the AWGN channel given in Example 14.10.

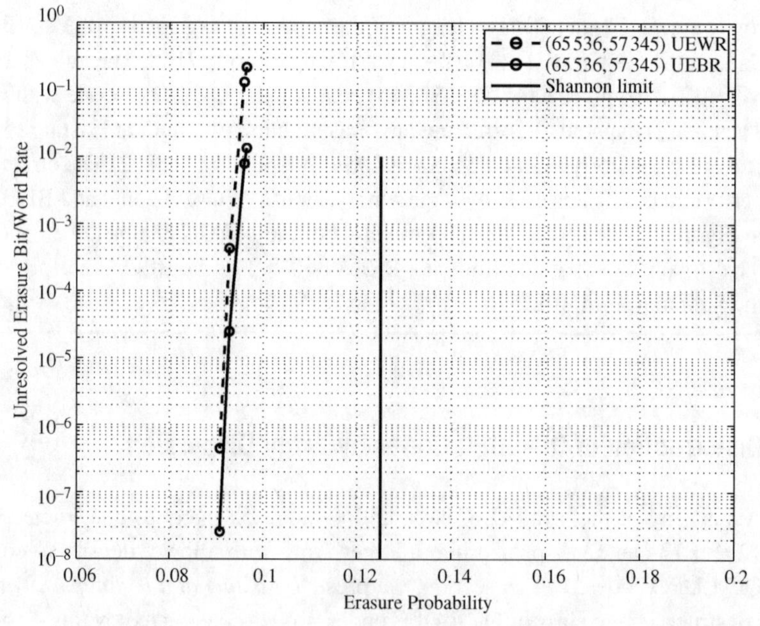


**Figure 14.12** The error performance of the (65536, 57345) QC-LDPC code over the BEC given in Example 14.10.

used to construct $\mathbf{H}$. For example, if $\mathbf{H}$ is the array $\mathbf{H}^{(1)}_{qc,\text{disp}}$ given by (12.6) constructed from the $(q-1, 2, q-2)$ RS code over GF($q$), then $c = n = q-1$. Let $k$ and $t$ be two positive integers such that $1 < k$, $2 < t \le c$, and $kt \le n$, and let $\mathbf{H}(t, kt)$ be a $t \times kt$ subarray of $\mathbf{H}$. We assume that $\mathbf{H}(t, kt)$ does not contain any zero submatrix of $\mathbf{H}$ (if any exists). Divide $\mathbf{H}(t, kt)$ into $k$ $t \times t$ subarrays, $\mathbf{H}_1(t,t), \mathbf{H}_2(t,t), \ldots, \mathbf{H}_k(t,t)$, such that

$$\mathbf{H}(t, kt) = [\mathbf{H}_1(t,t) \;\; \mathbf{H}_2(t,t) \;\; \ldots \;\; \mathbf{H}_k(t,t)], \tag{14.14}$$

where, for $1 \le j \le k$, the $j$th $t \times t$ subarray $\mathbf{H}_j(t,t)$ is expressed in the following form:

$$\mathbf{H}_j(t,t) = \begin{bmatrix} \mathbf{A}^{(j)}_{0,0} & \mathbf{A}^{(j)}_{0,1} & \cdots & \mathbf{A}^{(j)}_{0,t-1} \\ \mathbf{A}^{(j)}_{1,0} & \mathbf{A}^{(j)}_{1,1} & \cdots & \mathbf{A}^{(j)}_{1,t-1} \\ \vdots & \vdots & \ddots & \vdots \\ \mathbf{A}^{(j)}_{t-1,0} & \mathbf{A}^{(j)}_{t-1,1} & \cdots & \mathbf{A}^{(j)}_{t-1,t-1} \end{bmatrix}, \tag{14.15}$$

where each $\mathbf{A}^{(j)}_{i,l}$ with $0 \le i, l < t$ is a $(q-1) \times (q-1)$ CPM over GF(2). Since $\mathbf{H}(t, kt)$ satisfies the RC constraint, each subarray $\mathbf{H}_j(t,t)$ also satisfies the RC constraint.

Cut $\mathbf{H}_j(t,t)$ into two *triangles*, namely *upper* and *lower triangles*, along its main diagonal, where the lower triangle contains the CPMs on the main diagonal of $\mathbf{H}_j(t,t)$. Form two $t \times t$ arrays of circulant permutation and zero matrices as follows:

$$\mathbf{H}_{j,U}(t,t) = \begin{bmatrix} \mathbf{O} & \mathbf{A}^{(j)}_{0,1} & \mathbf{A}^{(j)}_{0,2} & \cdots & \mathbf{A}^{(j)}_{0,t-1} \\ \mathbf{O} & \mathbf{O} & \mathbf{A}^{(j)}_{1,2} & \cdots & \mathbf{A}^{(j)}_{1,t-1} \\ \vdots & \vdots & \vdots & \ddots & \vdots \\ \mathbf{O} & \mathbf{O} & \mathbf{O} & \cdots & \mathbf{A}^{(j)}_{t-2,t-1} \\ \mathbf{O} & \mathbf{O} & \mathbf{O} & \cdots & \mathbf{O} \end{bmatrix} \tag{14.16}$$

and

$$\mathbf{H}_{j,L}(t,t) = \begin{bmatrix} \mathbf{A}^{(j)}_{0,0} & \mathbf{O} & \mathbf{O} & \cdots & \mathbf{O} \\ \mathbf{A}^{(j)}_{1,0} & \mathbf{A}^{(j)}_{1,1} & \mathbf{O} & \cdots & \mathbf{O} \\ \vdots & \vdots & \vdots & \ddots & \vdots \\ \mathbf{A}^{(j)}_{t-2,0} & \mathbf{A}^{(j)}_{t-2,1} & \mathbf{A}^{(j)}_{t-2,2} & \cdots & \mathbf{O} \\ \mathbf{A}^{(j)}_{t-1,0} & \mathbf{A}^{(j)}_{t-1,1} & \mathbf{A}^{(j)}_{t-1,2} & \cdots & \mathbf{A}^{(j)}_{t-1,t-1} \end{bmatrix}, \tag{14.17}$$

where $\mathbf{O}$ is a $(q-1) \times (q-1)$ zero matrix. From (14.16), we see that the upper triangle of the $t \times t$ array $\mathbf{H}_{j,U}(t,t)$ above the main diagonal is identical to the upper triangle of $\mathbf{H}_j(t,t)$ above the main diagonal, and the rest of the submatrices in $\mathbf{H}_{j,U}(t,t)$ are zero matrices. From (14.17), we see that the lower triangle of $\mathbf{H}_{j,L}(t,t)$ including the submatrices on the main diagonal is identical to that of $\mathbf{H}_j(t,t)$, and the submatrices above the main diagonal are zero matrices. Since $\mathbf{H}_j(t,t)$ satisfies the RC constraint, it is clear that $\mathbf{H}_{j,U}(t,t)$ and $\mathbf{H}_{j,L}(t,t)$ also satisfy the RC constraint.

For $1 \leq j \leq k$ and $2 \leq l$, we form the following $l \times l$ array of $t \times t$ subarrays:

$$\mathbf{H}_{j,l\text{-f,disp}}(lt, lt) = \begin{bmatrix} \mathbf{H}_{j,L}(t,t) & \mathbf{O} & \mathbf{O} & \cdots & \mathbf{O} & \mathbf{H}_{j,U}(t,t) \\ \mathbf{H}_{j,U}(t,t) & \mathbf{H}_{j,L}(t,t) & \mathbf{O} & \cdots & \mathbf{O} & \mathbf{O} \\ \mathbf{O} & \mathbf{H}_{j,U}(t,t) & \mathbf{H}_{j,L}(t,t) & \cdots & \mathbf{O} & \mathbf{O} \\ \vdots & \vdots & \vdots & \ddots & \vdots & \vdots \\ \mathbf{O} & \mathbf{O} & \mathbf{O} & \cdots & \mathbf{H}_{j,U}(t,t) & \mathbf{H}_{j,L}(t,t) \end{bmatrix}, \tag{14.18}$$

where $\mathbf{O}$ is a $t \times t$ array of $(q-1) \times (q-1)$ zero matrices. From (14.18), we see that each row of $\mathbf{H}_{j,l\text{-f,disp}}(lt, lt)$ is a right cyclic shift of the row above it and the first row is the right cyclic shift of the last row. Also, the $t \times t$ subarrays $\mathbf{H}_{j,L}(t,t)$ and $\mathbf{H}_{j,U}(t,t)$ in $\mathbf{H}_{j,l\text{-f,disp}}(lt, lt)$ are separated by a *span* of $l-2$ $t \times t$ zero subarrays, including the *end-around* case with $\mathbf{H}_{j,L}(t,t)$ as the *starting* subarray and $\mathbf{H}_{j,U}(t,t)$ as the *ending* subarray. From (14.16)–(14.18), we readily see that the CPMs in each row (or each column) of $\mathbf{H}_{j,l\text{-f,disp}}(lt, lt)$ together form the $j$th subarray $\mathbf{H}_j(t,t)$ of the $t \times kt$ array $\mathbf{H}(t, kt)$ given by (14.14). $\mathbf{H}_{j,l\text{-f,disp}}(lt, lt)$ is called the *$l$-fold dispersion* of $\mathbf{H}_j(t,t)$, where the subscripts "$l$-f" and "disp" of $\mathbf{H}_{j,l\text{-f,disp}}(lt, lt)$ stand for "$l$-fold" and "dispersion," respectively. $\mathbf{H}_{j,l\text{-f,disp}}(lt, lt)$ is an $lt(q-1) \times lt(q-1)$ matrix over GF(2) with both column and row weight $t$. Since $\mathbf{H}_j(t,t)$, as a $t(q-1) \times t(q-1)$ matrix, satisfies the RC constraint, it follows readily from (14.18) that the $l$-fold dispersion of $\mathbf{H}_j(t,t)$, as an $lt(q-1) \times lt(q-1)$ matrix, also satisfies the RC constraint. Since any two subarrays in $\mathbf{H}(t, kt)$ given by (14.14) jointly satisfy the RC constraint, their $l$-fold dispersions jointly satisfy the RC constraint.

Now we view $\mathbf{H}_{j,l\text{-f,disp}}(lt, lt)$ as an $lt \times lt$ array of $(q-1) \times (q-1)$ circulant permutation and zero matrices. From the structures of $\mathbf{H}_{j,U}(t,t)$, $\mathbf{H}_{j,L}(t,t)$, and $\mathbf{H}_{j,l\text{-f,disp}}(lt, lt)$ given by (14.16), (14.17), and (14.18), respectively, we readily see that each row (or each column) of $\mathbf{H}_{j,l\text{-f,disp}}(lt, lt)$ contains a single span of $(l-1)t$ zero matrices of size $(q-1) \times (q-1)$ between two CPMs, including the end-around case. For $0 \leq s < t$, by replacing the $s$ CPMs right after the single span of zero matrices by $s$ $(q-1) \times (q-1)$ zero matrices, we obtain a new $lt \times lt$ array $\mathbf{H}_{j,l\text{-f,disp},s}(lt, lt)$ of circulant permutation and zero matrices. $\mathbf{H}_{j,l\text{-f,disp},s}(lt, lt)$ is called the *$s$-masked* and *$l$-fold dispersion* of $\mathbf{H}_j(t,t)$. Each row of $\mathbf{H}_{j,l\text{-f,disp},s}(lt, lt)$ contains a single span of $(l-1)t + s$ zero matrices of size $(q-1) \times (q-1)$. The number $s$ is called the *masking parameter*.

On replacing each $t \times t$ subarray in the $t \times kt$ array $\mathbf{H}(t, kt)$ of (14.14) by its $s$-masked $l$-fold dispersion, we obtain the following $lt \times klt$ array of $(q-1) \times (q-1)$ circulant permutation and zero matrices over GF(2):

$$\mathbf{H}_{l\text{-f,disp},s}(lt, lkt) = \left[\mathbf{H}_{1,l\text{-f,disp},s}(lt, lt) \;\; \mathbf{H}_{2,l\text{-f,disp},s}(lt, lt) \;\; \cdots \;\; \mathbf{H}_{k,l\text{-f,disp},s}(lt, lt)\right]. \tag{14.19}$$

$\mathbf{H}_{l\text{-f,disp},s}(lt, klt)$ is referred to as the *$s$-masked* and *$l$-fold dispersion* of the array $\mathbf{H}(t, kt)$ given by (14.14). As an $lt \times klt$ array of circulant permutation and zero matrices, each row of $\mathbf{H}_{l\text{-f,disp},s}(lt, klt)$ contains $k$ spans of zero matrices, each consisting of $(l-1)t + s$ zero matrices of size $(q-1) \times (q-1)$, including the end-around case. $\mathbf{H}_{l\text{-f,disp},s}(lt, klt)$

is an $lt(q-1) \times klt(q-1)$ matrix over GF(2) with column and row weights $t-s$ and $k(t-s)$, respectively. It satisfies the RC constraint.

The null space over GF(2) of $\mathbf{H}_{l\text{-f,disp},s}(lt, klt)$ gives a $(t-s, k(t-s))$-regular QC-LDPC code $C_{l\text{-f,disp},s}$ of length $klt(q-1)$ with rate at least $(k-1)/k$, whose Tanner graph has a girth of at least 6. The above construction by multi-fold array dispersion gives a large class of QC-LDPC codes. This multi-fold array dispersion allows us to construct long codes of various rates. There are five degrees of freedom in the code construction, namely, $q$, $k$, $l$, $s$, and $t$. The parameters $k$ and $t$ are limited by $n$, that is, $kt \leq n$. To avoid having a column weight of $\mathbf{H}_{l\text{-f,disp},s}(kt, klt)$ less than 3, we need to choose $s$ and $t$ such that $t-s \geq 3$. However, there is no limitation on $l$. Therefore, for given $q$, $k$, $s$, and $t$, we can construct very long codes over the same field GF(2) by varying $l$. A special subclass of QC-LDPC codes is obtained by choosing $s$ and $t$ such that $t-s=3$. This gives a class of $(3,3k)$-regular QC-LDPC codes. On setting $k=2,3,4,5,\ldots$, we obtain a sequence of codes with rational rates equal to (or at least equal to) $1/2, 2/3, 3/4, 4/5, \ldots$. If we choose $t$ and $s$ such that $t-s=4$, then we obtain a class of $(4,4k)$-regular QC-LDPC codes.

Next, we show that the QC-LDPC codes constructed by array dispersion given above are effective for correcting erasure bursts. Consider the $s$-masked and $l$-fold dispersed $lt \times klt$ array $\mathbf{H}_{l\text{-f,disp},s}(lt, klt)$ of $(q-1)\times(q-1)$ circulant permutation and zero matrices over GF(2) given by (14.19). For $1 \leq j \leq k$, the $j$th subarray $\mathbf{H}_{j,l\text{-f,disp},s}(lt, lt)$ of $\mathbf{H}_{l\text{-f,disp},s}(lt, klt)$ is an $lt \times lt$ array of $(q-1)\times(q-1)$ circulant permutation and zero matrices. From the structure of the arrays, $\mathbf{H}_{j,U}(t,t)$, $\mathbf{H}_{j,L}(t,t)$, and $\mathbf{H}_{j,l\text{-f,disp}}(lt, lt)$ given by (14.16), (14.17), and (14.18), respectively, we readily see that, for each column of $\mathbf{H}_{j,l\text{-f,disp},s}(lt, lt)$, there is a row with a CPM in that column, which is followed horizontally along the row by a span of $(l-1)t+s$ consecutive $(q-1)\times(q-1)$ zero matrices before it is ended with another CPM in the same row, including the end-around case. Now we consider $\mathbf{H}_{j,l\text{-f,disp},s}(lt, lt)$ as an $lt(q-1)\times lt(q-1)$ matrix over GF(2). Then, for each column of $\mathbf{H}_{j,l\text{-f,disp},s}(lt, lt)$, there is a row with a nonzero entry in that column, which is followed horizontally along the row by a span of at least $[(l-1)t+s](q-1)$ zero entries before it is ended with another nonzero entry in the same row, including the end-around case.

Since all the subarrays

$$\mathbf{H}_{1,l\text{-f,disp},s}(lt, lt)\mathbf{H}_{2,l\text{-f,disp},s}(lt, lt), \ldots, \mathbf{H}_{k,l\text{-f,disp},s}(lt, lt)$$

of the $s$-masked $l$-fold dispersed array $\mathbf{H}_{l\text{-f,disp},s}(lt, klt)$ have identical structure, for each column of the array $\mathbf{H}_{l\text{-f,disp},s}$ there is a row with a CPM in that column, which is followed horizontally along the row across all the boundaries of neighboring subarrays by a span of $(l-1)t+s$ consecutive $(q-1)\times(q-1)$ zero matrices, including the end-around case. Now view $\mathbf{H}_{l\text{-f,disp},s}(lt, klt)$ as an $lt(q-1)\times klt(q-1)$ matrix over GF(2). Then, for each column in $\mathbf{H}_{l\text{-f,disp},s}(lt, klt)$, there is a row with a nonzero entry in that column, which is followed horizontally along the row by a span of at least $[(l-1)t+s](q-1)$ zero entries before it is ended with another nonzero entry in the same row, including the end-around case. Therefore, the length of the zero covering span of each column of $\mathbf{H}_{l\text{-f,disp},s}(lt, klt)$ is at least $[(l-1)t+s](q-1)$ and hence the length of the zero covering span of the $s$-masked and $l$-fold dispersed $lt(q-1)\times klt(q-1)$ matrix $\mathbf{H}_{l\text{-f,disp},s}(lt, klt)$ over GF(2) is lower bounded by $[(l-1)t+s)](q-1)$.

Consider the QC-LDPC code $\mathcal{C}_{l\text{-f,disp},s}$ given by the null space of the $lt(q-1)\times klt(q-1)$ matrix $\mathbf{H}_{l\text{-f,disp},s}(lt, klt)$ over GF(2). Then, with the simple iterative decoding algorithm presented in Section 14.4, the code $\mathcal{C}_{l\text{-f,disp},s}$ is capable of correcting any erasure burst of length up to $[(l-1)t+s](q-1)+1$. Since the row rank of $\mathbf{H}_{l\text{-f,disp},s}(lt, klt)$ is at most $lt(q-1)$, the erasure-burst-correction efficiency $\eta$ is at least

$$\frac{[(l-1)t+s](q-1)+1}{lt(q-1)} \approx \frac{(l-1)t+s}{lt}. \tag{14.20}$$

For large $l$, the erasure-burst-correction efficiency of $\mathcal{C}_{l\text{-f,disp},s}$ approaches unity. Therefore, the class of QC-LDPC codes constructed by array dispersion is asymptotically optimal for correcting bursts of erasures.

---

**Example 14.11** Suppose we construct a $63\times 63$ array $\mathbf{H}^{(1)}_{qc,\text{disp}}$ of $63\times 63$ circulant permutation and zero matrices based on the (63, 2, 62) RS code over GF($2^6$) using the construction method presented in Section 12.3 (see (12.6)). Set $k=2$, $l=8$, $s=5$, and $t=8$. Take an $8\times 16$ subarray $\mathbf{H}(8,16)$ from $\mathbf{H}^{(1)}_{qc,\text{disp}}$, avoiding zero matrices on the main diagonal of $\mathbf{H}^{(1)}_{qc,\text{disp}}$. The 5-masked and 8-fold dispersion of $\mathbf{H}(8,16)$ gives a $64\times 128$ array $\mathbf{H}_{8\text{-f,disp},s}(64,128)$ of $63\times 63$ circulant permutation and zero matrices. It is a $4032\times 8064$ matrix over GF(2) with column and row weights 3 and 6, respectively. The null space of this matrix gives an (8064, 4032) QC-LDPC code with rate 0.5. This code is capable of correcting any erasure burst of length up to at least 3844. Hence, the erasure-burst-correction efficiency of this code is at least 0.9533.

If we keep $k=2$, $t=8$, and $s=5$, and let $l=16$, 32, and 64, we obtain three long codes with rates 0.5. They are (16 128, 8064), (32 256, 16 128), and (64 512, 32 256) QC-LDPC codes. They can correct erasure bursts of length up to at least 7876, 15 939, and 32 067, respectively. Their erasure-burst-correction efficiencies are 0.9767, 0.9883, and 0.9941.

---

For small $l$, say $l=2$, QC-LDPC codes constructed by array dispersion also perform well over the AWGN channel. This is illustrated in the next example.

---

**Example 14.12** Utilizing the four-dimensional Euclidean geometry EG(4, $2^2$) over GF($2^2$), (12.19), (12.20), and (12.21), we can form a $4\times 84$ $\mathbf{H}^{(8)}_{qc,\text{EG},2}$ of $255\times 255$ CPMs. Set $k=5$, $l=2$, $s=0$, and $t=4$. Take a $4\times 20$ subarray $\mathbf{H}(4,20)$ from $\mathbf{H}^{(8)}_{qc,\text{EG},2}$. The 0-masked and 2-fold dispersion of $\mathbf{H}(4,20)$ gives an $8\times 40$ array $\mathbf{H}_{2\text{-f,disp},0}(8,40)$ of $255\times 255$ circulant permutation and zero matrices. It is a $2040\times 10\,200$ matrix over GF(2) with column and row weights 4 and 20, respectively. The null space of $\mathbf{H}_{2\text{-f,disp},0}$ gives a (10 200, 8191) QC-LDPC code with rate 0.803; the lower bound on its erasure-burst-correction capability is 1021. However, by computer search, it is found that the code can actually correct any erasure burst of length up to 1366. The performances of this code over the AWGN and binary random erasure channels decoded with iterative decoding are shown in Figures 14.13 and 14.14, respectively. We see that the (10 200, 8191) QC-LDPC code performs well over both channels. For the AWGN channel, the code performs 1 dB from the Shannon limit at a BER of $10^{-6}$. For the BEC, the code performs 0.055 from the Shannon limit, 0.197, at a UEBR of $10^{-6}$.

## 14.8 Cyclic Codes for Correcting Bursts of Erasures

Except for the cyclic EG- and PG-LDPC codes presented in Sections 11.1.1 and 11.6, no other well-known cyclic codes perform well with iterative decoding over either the AWGN or random erasure channels. However, cyclic codes are very effective for correcting bursts of erasures with the simple iterative decoding algorithm presented in Section 14.4.

Consider an $(n, k)$ cyclic code $\mathcal{C}$ over GF($q$) with generator polynomial

$$\mathbf{g}(X) = g_0 + g_1X + \cdots + g_{n-k-1}X^{n-k-1} + X^{n-k},$$

where $g_0 \neq 0$ and $g_i \in \mathrm{GF}(q)$ (see (3.48)). Its parity-check polynomial is given by

$$\mathbf{h}(X) = h_0 + h_1X + \cdots + h_{k-1}X^{k-1} + X^k = \frac{X^n - 1}{g_0^{-1}X^{n-k}\mathbf{g}(X^{-1})}, \tag{14.21}$$

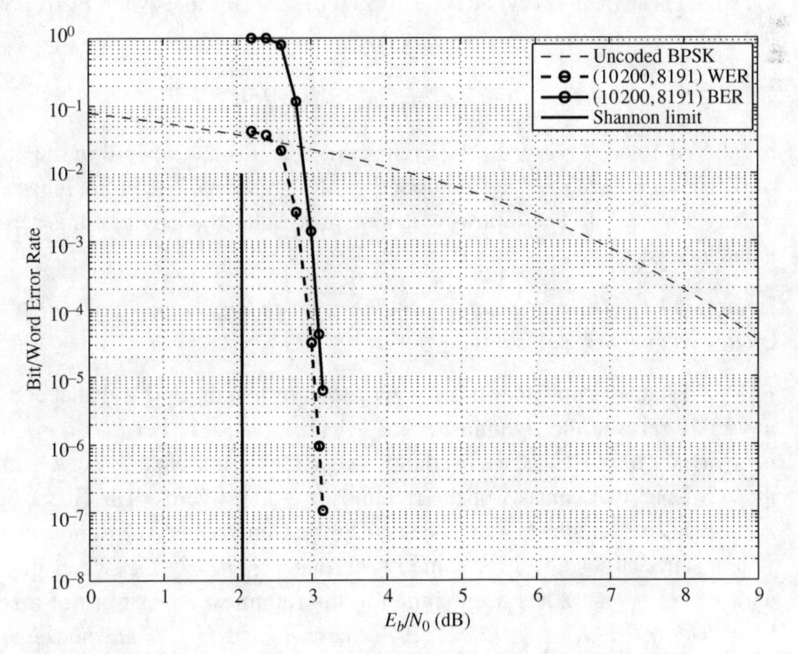


**Figure 14.13** The error performance of the (10 200, 8191) QC-LDPC code over the binary-input AWGN channel given in Example 14.12.

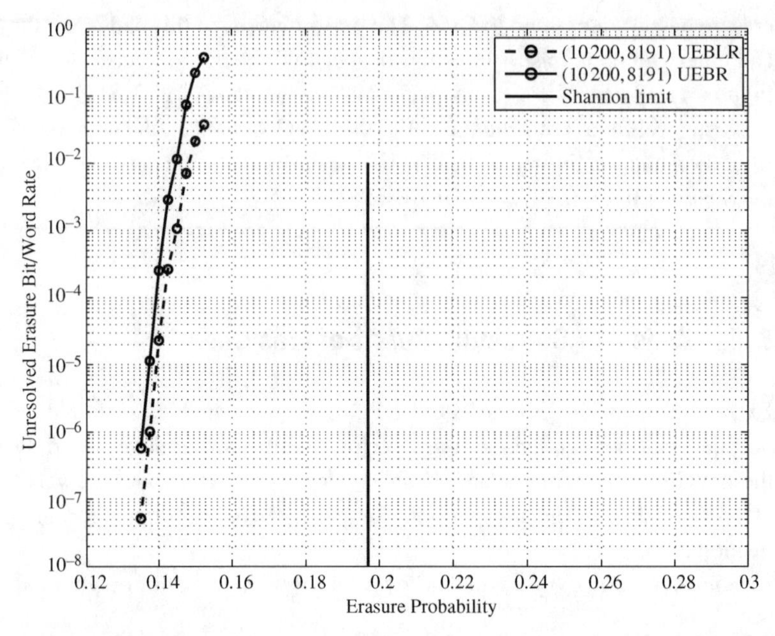


**Figure 14.14** The error performance of the (10 200, 8191) QC-LDPC code over the BEC given in Example 14.12.

where $h_0 \neq 0, h_i \in \text{GF}(q)$, and $g_0^{-1}$ is the multiplicative inverse of $g_0$. Let $\mathcal{C}_d$ be the dual code of $\mathcal{C}$. Then the parity-check polynomial $\mathbf{h}(X)$ of $\mathcal{C}$ is the generator polynomial of $\mathcal{C}_d$, which is an $(n, n-k)$ cyclic code over GF($q$). The $n$-tuple over GF($q$) corresponding to $\mathbf{h}(X)$,

$$\mathbf{h} = (h_0, h_1, \ldots, h_{k-1}, 1, 0, 0, \ldots, 0), \tag{14.22}$$

is called the *parity vector* of the cyclic code $\mathcal{C}$ and is a codeword in the dual code $\mathcal{C}_d$ of $\mathcal{C}$. The rightmost $n-k-1$ components of $\mathbf{h}$ are zeros. Therefore, $\mathbf{h}$ has a zero span of length $n-k-1$. The following lemma proves that the length of this zero span of $\mathbf{h}$ is the longest.

**Lemma 14.1** *The maximum length of a zero span in the parity-vector* $\mathbf{h}$ *of an* $(n, k)$ *cyclic code over GF($q$) is* $n-k-1$.

*Proof* First we know that $\mathbf{h}$ has a zero span of length $n-k-1$ that consists of the $n-k-1$ zeros at the rightmost $n-k-1$ positions of $\mathbf{h}$. For $k < n-k$, the lemma is obviously true. Hence, we need only prove the lemma for $k \geq n-k$. Let $\mathbf{z}_0$ denote the zero span that consists of the rightmost $n-k-1$ zeros of $\mathbf{h}$. Suppose there is a zero-span $\mathbf{z}$ of length $\lambda \geq n-k$ in $\mathbf{h}$. Since $h_0$ and $h_k = 1$ are nonzero, this zero span $\mathbf{z}$ starts at position $i$ for some $i$ such that $1 \leq i \leq k-\lambda$. We cyclically shift $\mathbf{h}$ until the zero span $\mathbf{z}$ has been shifted to the rightmost $\lambda$ positions of a new $n$-tuple $\mathbf{h}^* = (h_0^*, h_1^*, \ldots, h_{n-\lambda-1}^*, 0, 0, \ldots, 0)$, where $h_0^*$ and $h_{n-\lambda-1}^*$ are nonzero. This new $n$-tuple $\mathbf{h}^*$ and its $\lambda$ cyclic shifts form a set of $\lambda + 1 > n-k$ linearly independent vectors and they are codewords in the dual code $\mathcal{C}_d$ of $\mathcal{C}$. However, this contradicts the fact that $\mathcal{C}_d$, of dimension $n-k$, has at most $n-k$ linearly independent codewords. This proves the lemma. □

Form an $n \times n$ matrix over GF($q$) with the parity vector $\mathbf{h}$ and its $n-1$ cyclic shifts as rows, as follows:

$$\mathbf{H}_n = \begin{array}{c} n-k\left\{\begin{array}{c} \\ \\ \\ \\ \\ \\ \\ \end{array}\right. \\ k\left\{\begin{array}{c} \\ \\ \\ \\ \\ \\ \end{array}\right. \end{array} \left( \begin{array}{cccccccccc} h_0 & h_1 & h_2 & \cdots & h_{k-1} & 1 & 0 & 0 & \cdots & 0 \\ 0 & h_0 & h_1 & h_2 & \cdots & h_{k-1} & 1 & 0 & \cdots & 0 \\ 0 & 0 & h_0 & h_1 & h_2 & \cdots & h_{k-1} & 1 & \cdots & 0 \\ \cdots & \cdots & \cdots & \cdots & \cdots & \cdots & \cdots & \cdots & \cdots & \cdots \\ \cdots & \cdots & \cdots & \cdots & \cdots & \cdots & \cdots & \cdots & \cdots & \cdots \\ \cdots & \cdots & \cdots & \cdots & \cdots & \cdots & \cdots & \cdots & \cdots & \cdots \\ 0 & 0 & \cdots & 0 & h_0 & h_1 & h_2 & \cdots & h_{k-1} & 1 \\ 1 & 0 & 0 & \cdots & 0 & h_0 & h_1 & h_2 & \cdots & h_{k-1} \\ h_{k-1} & 1 & 0 & 0 & \cdots & 0h_0 & h_1 & h_2 & \cdots & h_{k-2} \\ \cdots & \cdots & \cdots & \cdots & \cdots & \cdots & \cdots & \cdots & \cdots & \cdots \\ \cdots & \cdots & \cdots & \cdots & \cdots & \cdots & \cdots & \cdots & \cdots & \cdots \\ \cdots & \cdots & \cdots & \cdots & \cdots & \cdots & \cdots & \cdots & \cdots & \cdots \\ h_1 & h_0 & \cdots & h_{k-1} & 1 & 0 & 0 & \cdots & 0 & h_0 \end{array} \right). \tag{14.23}$$

Since $\mathcal{C}_d$, the dual code of $\mathcal{C}$, is cyclic and the first row $(h_0, h_1, \ldots, h_{k-1}, 1, 0, 0, \ldots, 0)$ of $\mathbf{H}_n$ is a codeword in $\mathcal{C}_d$, all the rows of $\mathbf{H}_n$ are codewords in $\mathcal{C}_d$. Furthermore, the first $n-k$ rows of $\mathbf{H}_n$ are linearly independent and they form a full-rank parity-check matrix $\mathbf{H}_{n-k}$, which is often used to define $\mathcal{C}$ (see (3.30)), that is, $\mathcal{C}$ is the null space of $\mathbf{H}_{n-k}$. The matrix $\mathbf{H}_n$ is simply a *redundant expansion* of the parity-check matrix $\mathbf{H}_{n-k}$. Therefore, $\mathbf{H}_n$ is also a parity-check matrix of $\mathcal{C}$.

From $\mathbf{H}_n$, we see that each row has a zero span of length $n-k-1$ between $h_k = 1$ and $h_0$ (including the end-around case), each starting at a different position. This implies that every column of $\mathbf{H}_n$ has a zero covering span of length $n-k-1$ and hence the length of the zero covering span of $\mathbf{H}_n$ is $n-k-1$. Suppose a codeword $\mathbf{v}$ in $\mathcal{C}$ is transmitted over a $q$-ary erasure-burst channel. Let $\mathbf{r} = (r_0, r_1, \ldots, r_{n-1})$ be the corresponding received sequence with an erasure-burst pattern $\mathcal{E}$ of length $n-k$ or less. Suppose the starting position of the erasure burst $\mathcal{E}$ is $j$ with $0 \le j < n$. Then there exists a row $\mathbf{h}_i = (h_{i,0}, h_{i,1}, \ldots, h_{i,n-1})$ in $\mathbf{H}_n$ for which the $j$th component $h_{i,j}$ is equal to the 1 element of GF($q$) (see $\mathbf{H}_n$ given by (14.23)) and followed by a zero span of length $n-k-1$. Therefore, the row checks only the erasure at the $j$th position but not other erasures in $\mathcal{E}$. On setting the inner product $\langle \mathbf{r}, \mathbf{h}_i \rangle = 0$, we have the following equation:

$$h_{i,0}r_0 + h_{i,1}r_1 + \cdots + h_{i,n-1}r_{n-1} = 0,$$

which contains only one unknown $r_j$, the erased symbol at the $j$th position. From this equation, we can determine the value of the $j$th transmitted symbol as follows:

$$v_j = r_j = -h_{i,j}^{-1} \sum_{l=0, l \neq j}^{n-1} r_l h_{i,l}. \tag{14.24}$$

Once the symbol $v_j$ has been recovered, the index $j$ is removed from the erasure pattern $\mathcal{E}$. The procedure can be repeated to recover all the erased symbols in $\mathcal{E}$ iteratively using the decoding algorithm presented in Section 14.4.

Since the length of the zero covering span of $\mathbf{H}_n$ is $n-k-1$, $C$ is capable of correcting any erasure burst of length up to $n-k$, which is the limit of the erasure-burst-correction capability of any $(n, k)$ linear block code, binary or nonbinary. Therefore, using the expanded parity-check matrix $\mathbf{H}_n$ and the simple iterative decoding algorithm presented in Section 14.4, any $(n, k)$ cyclic code is optimal in terms of correcting a single erasure burst over a span of $n$ code symbols. RS codes are effective not only at correcting random erasures but also at correcting bursts of erasures.

See also [15–23] for further information on LDPC codes for binary erasure channels.

## Problems

**14.1** Compute the error performance of the $(8176, 7156)$ QC-EG-LDPC code given in Example 11.10 over the binary random erasure channel with the iterative decoding given in Section 14.1. How far from the Shannon limit does the code perform at a UEBR of $10^{-6}$?

**14.2** Consider the three-dimensional Euclidean geometry EG$(3, 3)$ over GF$(3)$. From the lines in EG$(3, 3)$ not passing through the origin, four $26\times 26$ circulants over GF$(2)$ can be constructed. Each of these circulants has both column weight and row weight 3. Determine the length of the zero covering span of each of these four circulants. Do all these four circulants have the same length of zero covering span?

**14.3** Using one of the four $26 \times 26$ circulants, denoted $\mathbf{G}$, constructed in Problem 14.2, form a $26\times 52$ matrix $\mathbf{Z}(26, 52) = [\mathbf{GG}]$ over GF$(2)$. Construct a $127\times 127$ array $\mathbf{H}^{(6)}_{qc,\text{disp}}$ of $127 \times 127$ circulant permutation matrices based on the prime field GF$(127)$ and the additive 127-fold matrix dispersion technique presented in Section 12.6. Take a $26 \times 52$ subarray $\mathbf{H}^{(6)}_{qc,\text{disp}}(26, 52)$ from the array $\mathbf{H}^{(6)}_{qc,\text{disp}}$. Masking $\mathbf{H}^{(6)}_{qc,\text{disp}}(26, 52)$ with $\mathbf{Z}(26, 52)$ results in a $26\times 52$ masked array $\mathbf{M}^{(6)}(26, 52)$ of circulant permutation and zero matrices. $\mathbf{M}^{(6)}(26, 52)$ is a $3302 \times 6604$ matrix over GF$(2)$ with column and row weights 3 and 6, respectively.

(a) Determine the QC-LDPC code given by the null space of $\mathbf{M}^{(6)}(26, 52)$.
(b) Compute the bit- and block-error performance of the code given in (a) over the binary-input AWGN channel using the SPA with 50 iterations.
(c) Compute the error performance of the code given in (a) over the binary random erasure channel.
(d) Determine the length of the zero covering span of $\mathbf{M}^{(6)}(26, 52)$ and the erasure burst correction capability.

**14.4** Consider the $127 \times 127$ array $\mathbf{H}^{(6)}_{qc,\text{disp}}$ of $127 \times 127$ CPMs constructed using GF$(127)$ given in Problem 14.3. Set $t = 5$, $k = 4$, $l = 3$, and $s = 2$. Take a $5\times 20$ subarray $\mathbf{H}^{(6)}_{qc,\text{disp}}(5, 20)$ from $\mathbf{H}^{(6)}_{qc,\text{disp}}$. Using the array dispersion technique given in Section 14.7, construct a 2-masked and 3-fold dispersion $\mathbf{H}^{(6)}_{3\text{-f,disp},2}(15, 60)$ of $\mathbf{H}^{(6)}_{qc,\text{disp}}(5, 20)$, which is a $15 \times 60$ array of a $127 \times 127$ circulant permutation and zero matrix.

(a) Determine the QC-LDPC code given by the null space of the array $\mathbf{H}^{(6)}_{3\text{-f,disp},2}(15, 60)$. What is its erasure-burst-correction capability and efficiency?
(b) Compute the bit- and word-error performance of the code given in (a) over the binary-input AWGN channel using the SPA with 50 iterations.
(c) Compute the error performance of the code given in (a) over the binary random erasure channel.

**14.5** Prove that the maximum zero span of the parity-vector **h** of an $(n, k)$ cyclic code over GF($q$) is unique.

## References

[1] C. Di, D. Proietti. I. E. Teletar, T. J. Richardson, and R. L. Urbanke, "Finite length analysis of low-density parity-check codes on the binary erasure channels," *IEEE Transactions on Information Theory*, vol. 48, no. 6, pp. 1576–1579, June 2002.

[2] A. Orlitsky, R. Urbanke, K. Viswanathan, and J. Zhang, "Stopping sets and the girth of Tanner graphs," *Proceedings of the IEEE International Symposium on Information Theory*, Lausanne, June 2002, p. 2.

[3] M. G. Luby, M. Mitzenmacker, M. A. Sokrollahi, and D. A. Spilman, "Efficient erasure correcting codes," *IEEE Transactions on Information Theory*, vol. 47, no. 2, pp. 569–584, February 2001.

[4] A. Orlitsky, K. Viswanathan, and J. Zhang, "Stopping sets distribution of LDPC code ensemble," *IEEE Transactions on Information Theory*, vol. 51, no. 3, pp. 929–953, March 2005.

[5] T. Tian, C. Jones, J. D. Villasenor, and R. D. Wesel, "Construction of irregular LDPC code with low error floors," *Proceedings of the IEEE International Conference on Communications*, Anchorage, AK, May 2003, pp. 3125–3129.

[6] L. Lan, Y. Y. Tai, S. Lin, B. Memari, and B. Honary, "New construction of quasi-cyclic LDPC codes based on special classes of BIBDs for the AWGN and binary erasure channels," *IEEE Transactions on Communications*, vol. 56, no. 1, pp. 39–48, January 2008.

[7] L. Lan, L.-Q. Zeng, Y. Y. Tai, L. Chen, S. Lin, and K. Abdel-Ghaffar, "Construction of quasi-cyclic LDPC codes for AWGN and binary erasure channels: A finite field approach," *IEEE Transactions on Information Theory*, vol. 53, no. 7, pp. 2429–2458, July 2007.

[8] S. Song, S. Lin, and K. Addel-Ghaffar, "Burst-correction decoding of cyclic LDPC codes," *Proceedings of the IEEE International Symposium on Information Theory*, Seattle, WA, July 9–14, 2006, pp. 1718–1722.

[9] Y. Y. Tai, L. Lan, L.-Q. Zeng, S. Lin, and K. Abdel-Ghaffar, "Algebraic construction of quasi-cyclic LDPC codes for the AWGN and erasure channels," *IEEE Transactions on Communications*, vol. 54, no. 10, pp. 1765–1774, October 2006.

[10] J. Ha and S. W. McLaughlin, "Low-density parity-check codes over Gaussian channels with erasures," *IEEE Transactions on Information Theory*, vol. 49, no. 7, pp. 1801–1809, July 2003.

[11] F. Peng, M. Yang, and W. E. Ryan, "Design and analysis of eIRA codes on correlated fading channels," *Proceedings of the IEEE Global Telecommunications Conference*, Dallas, TX, November–December 2004, pp. 503–508.
[12] M. Yang and W. E. Ryan, "Design of LDPC codes for two-state fading channel models," *Proceedings of the 5th International Symposium on Wireless Personal Multimedia Communications*, Honolulu, HI, October 2002, pp. 503–508.
[13] M. Yang and W. E. Ryan, "Performance of efficiently encodable low-density parity-check codes in noise bursts on the EPR4 channel," *IEEE Transactions on Magnetics*, vol. 40, no. 2, pp. 507–512, March 2004.
[14] S. Song, S. Lin, K. Abdel-Ghaffar, and W. Fong, "Erasure-burst and error-burst decoding of linear codes," *Proceedings of the IEEE Information Theory Workshop*, Lake Tahoe, CA, September 2–6, 2007, pp. 132–137.
[15] D. Burshtein and B. Miller, "An efficient maximum-likelihood decoding of LDPC codes over the binary erasure channel," *IEEE Transactions on Information Theory*, vol. 50, no. 11, pp. 2837–2844, November 2004.
[16] S. Lin and D. J. Costello, Jr., *Error Control Coding: Fundamentals and Applications*, Upper Saddle River, NJ, Prentice-Hall, 2004.
[17] P. Oswald and A. Shokrollahi, "Capacity-achieving sequences for erasure channels," *IEEE Transactions on Information Theory*, vol. 48, no. 12, pp. 3017–3028, December 2002.
[18] H. D. Pfister, I. Sason, and R. L. Urbanke, "Capacity-approaching ensembles for the binary erasure channel with bounded complexity," *IEEE Transactions on Information Theory*, vol. 51, no. 7, pp. 2352–2379, July 2005.
[19] H. Pishro-Nik and F. Fekri, "On decoding of low-density parity-check codes over the binary erasure channel," *IEEE Transactions on Information Theory*, vol. 50, vol 3, pp. 439–454, March 2004.
[20] T. J. Richardson and R. L. Urbanke, *Modern Coding Theory*, Cambridge, Cambridge University Press, 2008.
[21] M. Rashidpour, A. Shokrollahi, and S. H. Jamali, "Optimal regular LDPC codes for the binary erasure channel," *IEEE Communications Letters,* vol. 9, no. 6, pp. 546–548, June 2005.
[22] H. Saeedi and A. H. Banihashimi, "Deterministic design of low-density parity-check codes for binary erasure channel," *Proceedings of IEEE Globecom*, San Francisco, CA, November 2006, pp. 1566–1570.
[23] B. N. Vellambi and F. Fekri, "Results on the improved decoding algorithm for low-density parity-check codes over the binary erasure channels," *IEEE Transactions on Information Theory*, vol. 53, no. 4, pp. 1510–1520, April 2007.

# 15 Nonbinary LDPC Codes

Although a great deal of research effort has been expended on the design, construction, encoding, decoding, performance analysis, and applications of binary LDPC codes in communication and storage systems, very little has been done on nonbinary LDPC codes in these respects. The first study of nonbinary LDPC codes was conducted by Davey and MacKay in 1998 [1]. In their paper, they generalized the SPA for decoding binary LDPC codes to decode $q$-ary LDPC codes, called *QSPA*. Later, in 2000, MacKay and Davey introduced a *fast-Fourier-transform* (FFT)-based QSPA to reduce the decoding computational complexity of QSPA [2]. This decoding algorithm is referred to as FFT-QSPA. MacKay and Davey's work on FFT-QSPA was further improved by Barnault and Declercq in 2003 [3] and Declercq and Fossorier in 2007 [4]. Significant works on the design, construction, and analysis of nonbinary LDPC codes didn't appear until the middle of 2000. Results in these works are very encouraging. They show that nonbinary LDPC codes have a great potential to replace the widely used RS codes in some applications in communication and storage systems. This chapter is devoted to nonbinary LDPC codes.

Just like binary LDPC codes, nonbinary LDPC codes can be classified into two major categories: (1) random-like nonbinary codes constructed by computer under certain design criteria or rules; and (2) structured nonbinary codes constructed using algebraic or combinatorial tools, such as finite fields and finite geometries. In this chapter, we focus on algebraic constructions of nonbinary LDPC codes. The design and construction of *random-like* nonbinary LDPC codes can be found in [1, 5, 6, 7, 8] and that of *structured* nonbinary LDPC codes can be found in [9, 10, 11, 12–21, 22].

## 15.1 Definitions

Fundamental concepts, structural properties, and methods of construction, encoding, and decoding developed for binary LDPC codes in the previous chapters can be generalized to LDPC codes with symbols from nonbinary fields.

Let GF($q$) be a Galois field with $q$ elements, where $q$ is a power of a prime. A $q$-ary *regular* LDPC code $\mathcal{C}$ of length $n$ is given by the null space over GF($q$) of a sparse parity-check matrix $\mathbf{H}$ over GF($q$) that has the following structural properties: (1) each row has weight $r$; and (2) each column has weight $g$, where $r$ and $g$ are small compared with the length of the code. Such a $q$-ary LDPC code is said to be $(g, r)$-regular. If the columns and/or rows of the parity-check matrix $\mathbf{H}$ have *varying* (*multiple*) weights, then the null space over GF($q$) of $\mathbf{H}$ gives a $q$-ary *irregular* LDPC code. If $\mathbf{H}$ is an array of sparse circulants of the same size over GF($q$), then the null space over GF($q$) of $\mathbf{H}$ gives a $q$-ary quasi-cyclic (QC) LDPC code. If $\mathbf{H}$ consists of a single sparse circulant or a column of sparse circulants, then the null space over GF($q$) of $\mathbf{H}$ gives a $q$-ary cyclic LDPC code. Encoding of $q$-ary cyclic and QC-LDPC codes can be implemented with shift registers just like encoding of binary cyclic and QC codes using the circuits as presented in Section 3.2 and 3.6 with some modifications.

The Tanner graph of a $q$-ary LDPC code $\mathcal{C}$ given by the null space of a sparse $m \times n$ parity-check matrix $\mathbf{H} = [h_{i,j}]$ over GF($q$) is constructed in the same way as that for a binary LDPC code. The graph has $n$ variable nodes that correspond to the $n$ code symbols of a codeword in $\mathcal{C}$ and $m$ check nodes that correspond to $m$ check-sum constraints on the code symbols. The $j$th variable node $v_j$ is connected to the $i$th check node $c_i$ with an edge if and only if the $j$th code symbol $v_j$ is contained in the $i$th check-sum $c_i$, that is, if and only if the entry $h_{i,j}$ at the intersection of the $i$th row and the $j$th column of $\mathbf{H}$ is a nonzero element of GF($q$). To ensure that the Tanner graph of the $q$-ary LDPC code $\mathcal{C}$ is free of cycles of length 4 (or has a girth of at least 6), we further impose the following constraint on the rows and columns of $\mathbf{H}$: no two rows (or two columns) of $\mathbf{H}$ have more than one position where they both have nonzero components. This constraint is referred to as the row–column (RC) constraint. This RC constraint was imposed on the parity-check matrices of all the binary LDPC codes presented in previous chapters, regardless of their methods of construction. For a $(g, r)$-regular $q$-ary LDPC code $\mathcal{C}$, the RC constraint on $\mathbf{H}$ also ensures that the minimum distance of the $q$-ary LDPC code $\mathcal{C}$ is at least $g + 1$, where $g$ is the column weight of $\mathbf{H}$.

## 15.2 Decoding of Nonbinary LDPC Codes

This section will present the most popular decoding algorithm for LDPC codes over GF($q$), where $q = 2^s$, $s > 1$. This algorithm, which relies on the fast Hadamard transform, was first presented in [1, 2].

We will assume that each $q$-ary symbol is transmitted as a group of $s$ bits over the binary-input AWGN channel, although the results can be extended to $q$-ary-input channels. We will let $n$ and $m$ denote the number of columns and rows, respectively, of the code's (low-density) parity-check matrix over GF($q$), $\mathbf{H} = [h_{i,j}]_{0 \le i < m, 0 \le j < n}$ with $h_{i,j} \in$ GF($q$).

Recall that, for binary LDPC codes, the decoder iteratively updates estimates on $P_j[0]$ and $P_j[1]$, the probabilities that code bit $v_j$ is 0 and 1, respectively, for $j = 0, 1, \ldots, n-1$. For mathematical convenience and reduced implementational complexity, the decoder estimates these probabilities indirectly by instead iteratively estimating the log-likelihood ratio (LLR) for $v_j$, given by

$$L_j = \log\left[\frac{P_j[0 \mid \mathbf{y}]}{P_j[1 \mid \mathbf{y}]}\right], \tag{15.1}$$

where the dependency of these probabilities on the channel output vector $\mathbf{y}$ is made explicit in the *a posteriori* probabilities $P_j[0 \mid \mathbf{y}]$ and $P_j[1 \mid \mathbf{y}]$ for code bit $v_j$. Additional conditioning is implied; namely, the code constraints as encapsulated in the code's parity-check matrix $\mathbf{H}$.

For $q$-ary LDPC codes, a pair of probabilities or a single LLR for each code symbol $v_j$ is not appropriate because $v_j$ can take on one of $q$ values and so $q$ probabilities must be estimated. These $q$ probabilities form a probability mass function (pmf). Thus, the $q$-ary LDPC code decoder, which consists of VNs and CNs connected by edges labeled with elements from GF($q$), must propagate pmfs instead of LLRs along the edges of the code's Tanner graph representation. Once the decoder discontinues its iterative computations, its decision for $v_j$ is that value of $v \in \mathrm{GF}(q)$ which maximizes the *a posteriori* probability $P_j[v \mid \mathbf{y}]$.

### 15.2.1 Algorithm Derivation

As with the binary LDPC code decoding algorithm, the algorithm for $q$-ary LDPC codes receives information from the channel and sends that information up to the CN processors. The CN processors then send their processed outputs to the VN processors, which take these inputs along with the channel information to produce information that is sent back to the CN processors. The VN/CN processing iterations continue until a codeword is found or the preset maximum number of iterations is reached.

For the AWGN channel with two-sided power spectral density $N_0/2$, the appropriate bit-wise information computed from the channel output $y$ for candidate binary input $b \in \{\pm 1\}$ is

$$\Pr(b \mid y) = p(y \mid b)\Pr(b)/p(y) = \frac{1}{1 + \exp\left(-4yb/N_0\right)}. \tag{15.2}$$

These bit-wise probabilities are converted by a preprocessor to symbol-wise probabilities by computing appropriate products of the former. For example, for $s = 4$, suppose $\alpha^2 \in \mathrm{GF}(2^4)$ has the binary representation [0 1 0 0], where $\alpha$ is a primitive element of GF($2^4$). Then $\Pr(\alpha^2 \mid \overline{y}) = \Pr(0 \mid y_3)\Pr(1 \mid y_2)\Pr(0 \mid y_1)\Pr(0 \mid y_0)$, where $\overline{y} = [y_3\ y_2\ y_1\ y_0]$ is the group of channel outputs corresponding to $s = 4$ consecutive binary inputs. Each VN processor receives $q$ such symbol-wise probabilities from the decoder preprocessor; that is, each VN processor receives a (conditional) pmf on the elements of GF($q$). The preprocessor pmf for VN $j$ will be denoted by $P_j$, for $j = 0, 1, \ldots, n-1$. The $q$ elements of $P_j$ will be denoted by $P_j(0), P_j(1), P_j(\alpha), P_j(\alpha^2), \ldots, P_j(\alpha^{q-2})$, where

$$P_j(\beta) = \Pr(v_j = \beta \mid \overline{y}) \tag{15.3}$$

for all $\beta \in \mathrm{GF}(q)$, with $P_j(\beta)$ given by a product of $s$ probabilities $\Pr(b \mid y)$.

Given the channel outputs and computed pmfs, we now need to develop the VN and CN update equations involved in the iterative decoding. To aid the discussion, we define the *CN neighborhood* of VN $j$, for $0 \le j < n$, to be

$$M_j = \left\{ i{:}h_{i,j} \neq 0 \right\}. \tag{15.4}$$

Similarly, we define the *VN neighborhood* of CN $i$, for $0 \leq i < m$, to be

$$N_i = \left\{ j{:}h_{i,j} \neq 0 \right\}. \tag{15.5}$$

## 15.2.2 VN Update

As is customary, we regard each VN to be a repetition code so that all of the edges leaving a VN carry the same value in GF($q$). (These values are later multiplied by the edge labels on their way up to the CNs.) Moreover, we assume that the messages from the CNs to the VNs are independent. Because the optimum repetition code decoder adds LLRs, or multiplies probabilities, it follows that, for each $\beta \in \text{GF}(q)$, the (extrinsic) message $m_{j\to i}(\beta)$ to be sent from VN $j$ to CN $i$ is given by

$$m_{j\to i}(\beta) = P_j(\beta) \prod_{k \in M_j \backslash i} m_{k\to j}(\beta), \tag{15.6}$$

where $m_{k\to j}(\beta)$ is the message sent from CN $k$ to VN $j$ about the probability that code symbol $v_j$ is equal to $\beta$. Note that the independence assumption on the incoming messages at VN $j$ actually holds only for the first $g/2$ iterations, where $g$ is the girth of the code's Tanner graph. Note also, because of the exclusion $M_j \backslash i$ in the product, that the message $m_{j\to i}(\beta)$ does not send CN $i$ information that it already has.

We emphasize that, while a single computation of the form (15.6) is performed in the binary case, $q$ such computations are performed for $q$-ary LDPC codes, one for each value of $\beta \in \text{GF}(q)$.

The computation of the CN-to-VN messages is much more involved, as is its development, which we will do in stages.

## 15.2.3 CN Update: Complex Version

The CN update equation takes more time to develop, but the concepts are well known to readers of this book. First, note that a CN is considered to be a $q$-ary single-parity check (SPC) code. Thus, the $i$th CN, representing the $i$th row $[h_{i,0}\ h_{i,1}\ \ldots\ h_{i,n-1}]$ of the parity-check matrix $\mathbf{H}$, corresponds to the parity-check equation

$$\sum_{j \in N_i} v_j h_{i,j} = 0. \tag{15.7}$$

In this equation, $v_j \in \text{GF}(q)$ is the code symbol corresponding to VN $j$, $\{h_{i,j}\} \subset \text{GF}(q)$, with $j = 0, 1, \ldots, n-1$, are the nonzero elements of the $i$th row of $\mathbf{H}$, and addition and multiplication are over GF($q$).

Letting $v'_j = v_j h_{i,j}$, the above sum can be rewritten as

$$\sum_{j \in N_i} v'_j = 0. \tag{15.8}$$

The previous equation leads to the CN update equation given CN inputs from neighbouring VNs. In particular, consider the $i$th CN, its neighbourhood of VNs, $N_i$, and their pmfs. From these pmfs (messages), we are to compute the pmf to be sent from CN $i$ to VN $j$, for all $j$s in $N_i$. The pdf to be sent to VN $j$ is the pmf of

$$v'_j = \sum_{\ell \in N_i \backslash j} v'_\ell. \tag{15.9}$$

When a discrete-valued random variable $v'_j$ is the sum of independent discrete-valued random variables $v'_\ell$, the pmf of $v'_j$ is given by the *cyclic* convolution of the pmfs of the $v'_\ell$s.

It is important to highlight here an important detail regarding the convolution of pmfs of elements of GF($q$). Consider two random variables, $X$ and $Y$, taking values in GF($q$). Let their corresponding pmfs be denoted by $\mathbf{p}_X$ and $\mathbf{p}_Y$. We are interested in determining the pmf of their sum, $Z = X + Y$, noting that the sum is performed via the modulo-2 addition of the binary $s$-tuple representations of $X$ and $Y$. Thus, it is convenient to represent the elements of GF($q$) by the binary $s$-tuples or their decimal equivalents. We have then that

$$\mathbf{p}_Z(z) = \sum_{x,y:\, x+y=z} \mathbf{p}_X(x)\mathbf{p}_Y(y) = \sum_x \mathbf{p}_X(x)\mathbf{p}_Y(z-x), \tag{15.10}$$

or, using the cyclic convolution operator shorthand,

$$\mathbf{p}_Z = \mathbf{p}_X \circledast \mathbf{p}_X. \tag{15.11}$$

This generalizes to a sum of more than two random variables. To apply this result to (15.9), we let $\mathbf{m}_{i\to j} = [m_{i\to j}(0)\ m_{i\to j}(1)\ \ldots\ m_{i\to j}(\alpha^{q-2})]$ be the conditional pmf of $v'_j$ and $\mathbf{m}_{\ell\to i} = [m_{\ell\to i}(0)\ m_{\ell\to i}(1)\ \ldots\ m_{\ell\to i}(\alpha^{q-2})]$ the conditional pmf of $v'_\ell$. It follows from the development above that the conditional pdf of $v'_j$ is

$$\mathbf{m}_{i\to j} = \circledast_{\ell \in N_i \backslash j} \mathbf{m}_{\ell\to i}. \tag{15.12}$$

## 15.2.4 CN Update: Fast Hadamard Transform Version

The convolution of multiple pmfs to obtain the pmf of $v'_j$ is obviously quite complex. The computational complexity may be vastly reduced by using fast Hadamard transform (FHT) techniques to perform the convolutions. To introduce the FHT-based algorithm, some preliminary information on the Hadamard transform itself is necessary. We will present the fast Hadamard transform algorithm later.

The Hadamard transform of a length-$q$, real-valued (row) vector $\mathbf{p} = [p_0\ p_1\ \ldots\ p_{q-1}]$, considered to be a pmf here, is given by

$$\mathbf{P} = \mathcal{H}(\mathbf{p}) = \mathbf{p}\mathbf{H}_q, \tag{15.13}$$

where $\mathbf{H}_q$ is recursively defined as

$$\mathbf{H}_q = \frac{1}{\sqrt{2}} \begin{bmatrix} \mathbf{H}_{q/2} & \mathbf{H}_{q/2} \\ \mathbf{H}_{q/2} & -\mathbf{H}_{q/2} \end{bmatrix}, \tag{15.14}$$

with the initial condition

$$\mathbf{H}_2 = \frac{1}{\sqrt{2}} \begin{bmatrix} 1 & 1 \\ 1 & -1 \end{bmatrix}. \tag{15.15}$$

Letting $\boldsymbol{\phi}_x$ represent the $x$th row of $\mathbf{H}_q$, the Hadamard transform of $\mathbf{p}$ may be written as

$$\mathbf{P} = \sum_{x=0}^{q-1} p_x \boldsymbol{\phi}_x. \tag{15.16}$$

Because $\mathbf{H}_q$ is recursively defined, it is not difficult to prove by induction the following properties of the Hadamard transform.

1. $\mathbf{H}_q^{\mathrm{T}} = \mathbf{H}_q$ (where superscript T denotes matrix transpose).
2. $\mathbf{H}_q\mathbf{H}_q^{\mathrm{T}} = \mathbf{H}_q^{\mathrm{T}}\mathbf{H}_q = \mathbf{I}_q$ (where $\mathbf{I}_q$ is the $q \times q$ identity matrix). In particular, the inverse Hadamard transform is performed by $\mathbf{H}_q^{\mathrm{T}} = \mathbf{H}_q^{-1}$, that is, $\mathcal{H}^{-1}(\mathbf{P}) = \mathbf{P}\mathbf{H}_q^{\mathrm{T}} = \mathbf{p}\mathbf{H}_q\mathbf{H}_q^{\mathrm{T}} = \mathbf{p}$. A corollary to this property is that the rows of $\mathbf{H}_q$ are orthonormal, $\boldsymbol{\phi}_x\boldsymbol{\phi}_y^{\mathrm{T}} = \boldsymbol{\delta}_{x-y}$, where $\boldsymbol{\delta}_n$ is the Kronecker delta function.
3. The component-wise multiplication, denoted by $\odot$, of the $x$th and $y$th rows of $\mathbf{H}_q$ is $\boldsymbol{\phi}_x \odot \boldsymbol{\phi}_y = \frac{1}{\sqrt{2}}\boldsymbol{\phi}_{x\oplus y}$, where $x \oplus y$ is component-wise mod-2 addition of the binary representations of $x$ and $y$. For example, $13 \oplus 10 = 7$.

Observe that the component-wise mod-2 addition mentioned in Property 3 is identical to addition of two elements in GF($2^s$), although, following convention, we have used $\oplus$ for the former and + for the latter. Property 3 is critical to the proof of the following cyclic convolution theorem for Hadamard transforms that is central to the low-complexity computation of the cyclic convolution of pmfs of random variables that take values in GF($q$). It will be seen in the proof that the indices to the rows of $\mathbf{H}_q$ are put into a one-to-one correspondence with the elements ($s$-tuples) of the additive group in GF($q$).

**Theorem 15.1** *Consider two independent random variables, $X$ and $Y$, defined on GF($q$) with pmfs $\mathbf{p}_X$ and $\mathbf{p}_Y$. Then the pmf $\mathbf{p}_Z$ of their sum in GF($q$), $Z = X + Y$, may be computed from $\sqrt{2}\,\mathcal{H}^{-1}[\mathcal{H}(\mathbf{p}_X) \odot \mathcal{H}(\mathbf{p}_Y)]$.*

*Proof* For simplicity of notation, instead of $p_X(x)$ and $p_Y(y)$, $p(x)$ and $p(y)$ will be used for the individual components of $\mathbf{p}_X$ and $\mathbf{p}_Y$. All summations below are from 0 to $q-1$, or their binary $s$-tuple equivalents.

$$\begin{aligned}
\sqrt{2}\,\mathcal{H}^{-1}[\mathcal{H}(\mathbf{p}_X) \odot \mathcal{H}(\mathbf{p}_Y)] &= \sqrt{2}\,\mathcal{H}^{-1}\left[\sum_x p(x)\boldsymbol{\phi}_x \odot \sum_y p(y)\boldsymbol{\phi}_y\right] \\
&= \sqrt{2}\,\mathcal{H}^{-1}\left[\sum_x\sum_y p(x)\boldsymbol{\phi}_x \odot p(y)\boldsymbol{\phi}_y\right] \\
&= \sqrt{2}\,\mathcal{H}^{-1}\left[\sum_x\sum_y p(x)p(y)\frac{1}{\sqrt{2}}\boldsymbol{\phi}_{x\oplus y}\right]
\end{aligned}$$

$$= \sum_x \sum_y \sum_z p(x)p(y)\boldsymbol{\phi}_{x\oplus y}\boldsymbol{\phi}_z^{\mathrm{T}} \tag{15.17}$$

$$= \sum_x \sum_y \sum_z p(x)p(y)\boldsymbol{\delta}_{(x\oplus y)-z}$$

$$= \sum_{x,y:x\oplus y=z} p(x)p(y)$$

$$= \sum_x p(x)p(z-x)$$

$$= \mathbf{p}_X \circledast \mathbf{p}_Y.$$

The first two lines are obvious, the third follows from Property 3, the fourth from Properties 1 and 2, the fifth from the corollary to Property 2, and the last three are obvious. □

---

**Example 15.1** Although the development so far has focused on $q$-ary LDPC codes, the results are applicable to binary LDPC codes as well, that is, the case $q = 2$. Substitution of $\mathbf{p}_X = [p_X(0)\ \ p_X(1)]$ and $\mathbf{p}_Y = [p_Y(0)\ \ p_Y(1)]$ into $\sqrt{2}\ \mathcal{H}^{-1}\left[\mathcal{H}(\mathbf{p}_X) \odot \mathcal{H}(\mathbf{p}_Y)\right]$ yields $\mathbf{p}_X \circledast \mathbf{p}_Y = [p_X(0)p_Y(0) + p_X(1)p_Y(1)$ $p_X(0)p_Y(1) + p_X(1)p_Y(0)]$. Clearly, the first component of this vector is equal to $\Pr[X + Y = 0]$ and the second component is equal to $\Pr[X + Y = 1]$. The logarithm of the ratio of the second component to the first appears in the proof of Lemma 5.2, which shows how to compute the LLR of a sum (over GF(2)) of two independent binary random variables, given their individual LLRs.

---

## The Fast Hadamard Transform

Efficient computation of a Hadamard transform, via the FHT algorithm (also called fast Fourier transform (FFT) in [1, 2]) is possible because of the recursive nature of Hadamard matrices. Further, from Properties 1 and 2, the Hadamard transform and the inverse Hadamard transform are identical, so we need only discuss one fast algorithm.

Presentation of the FHT is easiest by way of example. Let us consider the 16-ary case so that $q = 16$, $\mathbf{p} = [p_0\, p_1\, \ldots\, p_{15}]$, and $\mathbf{P} = \mathcal{H}(\mathbf{p}) = \mathbf{p}\mathbf{H}_{16}$. Observe that

$$\mathbf{H}_{16} = \frac{1}{\sqrt{2}}\begin{bmatrix} \mathbf{H}_8 & \mathbf{H}_8 \\ \mathbf{H}_8 & -\mathbf{H}_8 \end{bmatrix} = \frac{1}{\sqrt{2}}\begin{bmatrix} \mathbf{H}_8 & \mathbf{0} \\ \mathbf{0} & \mathbf{H}_8 \end{bmatrix}\begin{bmatrix} \mathbf{I}_8 & \mathbf{I}_8 \\ \mathbf{I}_8 & -\mathbf{I}_8 \end{bmatrix}. \tag{15.18}$$

Figure 15.1 illustrates the implementation of this equation, where $\mathbf{p}_0^7$ is the subvector $[p_0\ p_1\ \ldots\ p_7]$ of $\mathbf{p}$ and similarly for the other subvectors. In the figure, quantities along edges that diverge are identical and quantities along edges that merge are added.

This implementation does not look advantageous until we observe that $\mathbf{H}_8$ decomposes as

$$\mathbf{H}_8 = \frac{1}{\sqrt{2}}\begin{bmatrix} \mathbf{H}_4 & \mathbf{0} \\ \mathbf{0} & \mathbf{H}_4 \end{bmatrix}\begin{bmatrix} \mathbf{I}_4 & \mathbf{I}_4 \\ \mathbf{I}_4 & -\mathbf{I}_4 \end{bmatrix}, \tag{15.19}$$

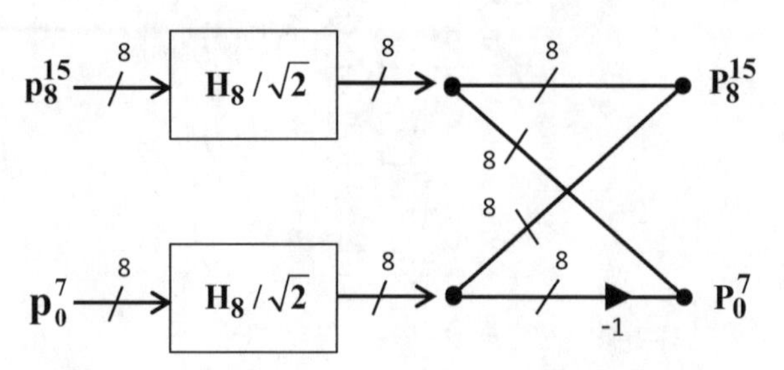


**Figure 15.1** Diagram of implementation of $\mathbf{P} = \mathbf{p}\mathbf{H}_{16}$.

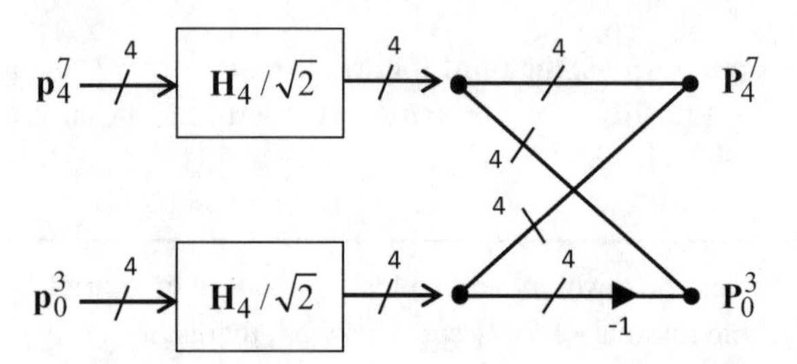


**Figure 15.2** Diagram of implementation of $\mathbf{P} = \mathbf{p}_0^7\mathbf{H}_8$.

which is implemented as in Figure 15.2 for the input $\mathbf{p}_0^7$. The implementation for the input $\mathbf{p}_8^{15}$ is essentially identical. The setup in Figure 15.2 and the analogous one for $\mathbf{p}_8^{15}$ replace the two $\mathbf{H}_8$ blocks in Figure 15.1. Next, each of the $\mathbf{H}_4$ blocks in the implementation of $\mathbf{H}_8$ are replaced by the implementation of

$$\mathbf{H}_4 = \frac{1}{\sqrt{2}}\begin{bmatrix} \mathbf{H}_2 & \mathbf{0} \\ \mathbf{0} & \mathbf{H}_2 \end{bmatrix}\begin{bmatrix} \mathbf{I}_2 & \mathbf{I}_2 \\ \mathbf{I}_2 & -\mathbf{I}_2 \end{bmatrix}, \tag{15.20}$$

resulting in eight $\mathbf{H}_2$ blocks.

The culmination of all these steps is Figure 15.3, which is a concatenation of $\log_2(q) = 4$ levels of "butterflies," with eight butterflies in the first level, four in the second, two in the third, and one in the last. The scale factor of 1/4 at the output accounts for the four factors of $1/\sqrt{2}$ that would occur in each of the four stages, but combined near the output.

### 15.2.5 The $q$-ary LDPC Decoding Algorithm

Below we summarize the $q$-ary LDPC code decoding algorithm, which is also called the fast Fourier transform $q$-ary sum–product algorithm (FFT-QSPA). The messages $m_{j\to i}(\beta)$ computed in Step 2 correspond to $v_j$, whereas the convolutions in (15.12) correspond to $v'_j$, so the translations in Steps 3 and 5 are necessary. Each of the steps is to be applied to all $\beta \in \mathrm{GF}(q)$ and all $i$ and $j$ for which $h_{i,j} \neq 0$, with $0 \leq i < m$ and $0 \leq j < n$.

1. Initialize the probabilities $P_j(\beta)$ according to (15.2) and (15.3) and the discussion in between. Initialize also the CN messages $m_{i\to j}(\beta) = 1$.

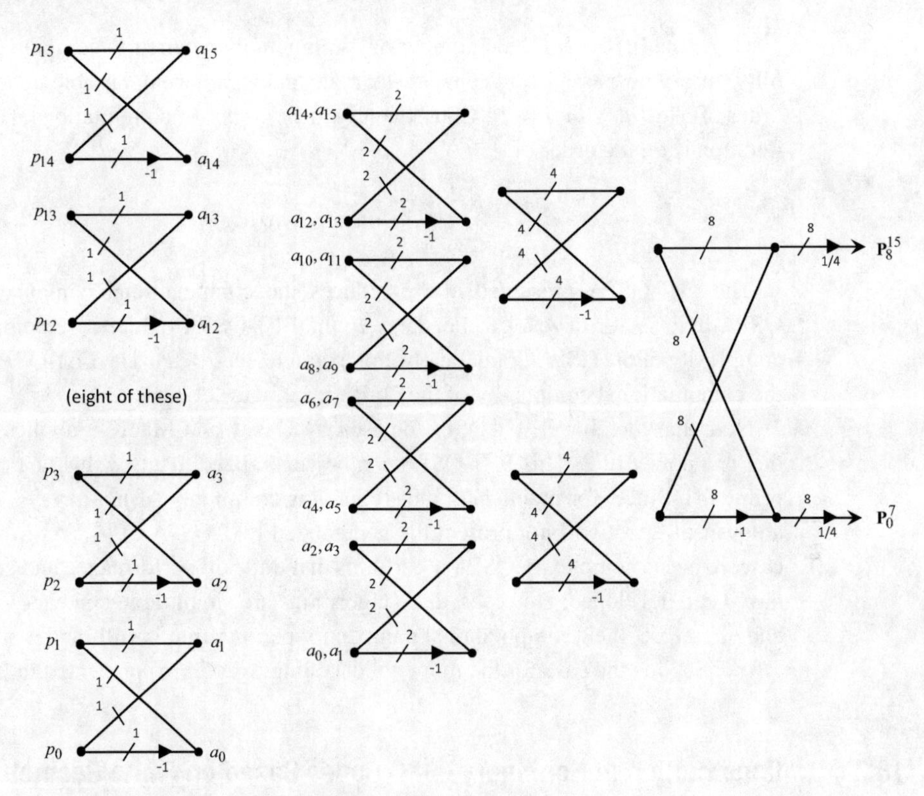


**Figure 15.3** Diagram of the fast Hadamard transform.

2. Compute the VN messages $m_{j\to i}(\beta)$ according to (15.6).
3. Translate the messages $m_{j\to i}(\beta)$ to the messages $m_{j\to i}(\beta')$ according to $m_{j\to i}(\beta') = m_{j\to i}(\beta)$, where $\beta' = h_{i,j}\beta$.
4. Using the Hadamard transform convolution theorem and the FHT algorithm, compute the CN messages

$$\mathbf{m}_{i\to j} = \mathcal{H}^{-1}\left\{ \prod_{\ell\in N_i\backslash j} \mathcal{H}\{\mathbf{m}_{\ell\to i}\} \right\}. \tag{15.21}$$

5. Translate the messages $m_{i\to j}(\beta')$ to the messages $m_{i\to j}(\beta)$ according to $m_{i\to j}(\beta) = m_{i\to j}(\beta')$, where $\beta' = h_{i,j}\beta$.
6. Make symbol decisions using

$$\hat{v}_j = \text{argmax}_\beta\ m_j(\beta), \tag{15.22}$$

where

$$m_j(\beta) = P_j(\beta) \prod_{k\in M_j} m_{k\to j}(\beta). \tag{15.23}$$

If found codeword (if $\hat{\mathbf{v}}\mathbf{H}^{\mathrm{T}} = \mathbf{0}$) or if reached the maximum number of iterations, then stop; otherwise, go to Step 2.

Using the FFT (FHT), the number of computations required to compute the probability messages passed between a check node and an adjacent variable node is on the order of $q \log q$, with $q = 2^s$. Consequently, the number of computations required per iteration is on the order of

$$\sum_{i=0}^{m-1} |N_i| q \log q = D_H q \log q. \tag{15.24}$$

The FFT-QSPA presented above reduces the computational complexity of the QSPA by a factor of $q/\log q$. For large $q$, the FFT-QSPA reduces the computational complexity of the QSPA dramatically. For example, let $q = 2^8$. The FFT-QSPA reduces the computational complexity of the QSPA by a factor of 32.

Note that the above FFT-QSPA devised by Davey and MacKay applies only to $q$ that is a power of 2. This FFT-QSPA can be generalized to any $q$ that is a power of a prime [4]. Since, for practical applications, $q$ is commonly (if not always) chosen as a power of 2, we will not present the generalized FFT-QSPA of [4]. As in the binary case, reduced-complexity QSPA algorithms that trade off performance and complexity have been developed. However, these algorithms are simplified versions of the QSPA and, therefore, their computational complexity per iteration is still on the order of $q^2$, $O(q^2)$. Such is the case for the min-sum decoding over GF($q$) presented in [23].

## 15.3 Construction of Nonbinary LDPC Codes Based on Finite Geometries

All the algebraic methods and techniques presented in Chapters 11 to 14 can be applied, with some modifications, for constructing LDPC codes over nonbinary fields. In this section, we start with constructions of nonbinary LDPC codes based on the lines and flats of finite geometries. In presenting the constructions, we will follow the fundamental concepts and notations presented in Chapter 11.

### 15.3.1 A Class of $q^m$-ary Cyclic EG-LDPC Codes

Consider the $m$-dimensional Euclidean geometry EG($m, q$) over GF($q$). Recall that the Galois field GF($q^m$) is a realization of the $m$-dimensional Euclidean geometry EG($m, q$) (see Chapters 2 and 11). Let $\alpha$ be a primitive element of GF($q^m$). Then the powers of $\alpha$, $\alpha^{-\infty} = 0, \alpha^0 = 1, \alpha, \ldots, \alpha^{q^m-2}$, represent the $q^m$ points of EG($m, q$) and $\alpha^{-\infty} = 0$ represents the origin of EG($m, q$). Let EG$^*(m, q)$ denote the subgeometry obtained from EG($m, q$) by removing the origin and all the lines passing through the origin of EG($m, q$). Then EG$^*(m, q)$ contains $n = q^m - 1$ non-origin points and

$$J_{0,\mathrm{EG}}(m, 1) = (q^{m-1} - 1)(q^m - 1)/(q - 1)$$

lines (see (2.54)) not passing through the origin of EG($m, q$).

Let $\mathcal{L} = \{\alpha^{j_1}, \alpha^{j_2}, \ldots, \alpha^{j_q}\}$ be a line in EG$^*(m, q)$ that consists of the points $\alpha^{j_1}, \alpha^{j_2}, \ldots, \alpha^{j_q}$, with $0 \leq j_1, j_2, \ldots, j_q < q^m - 1$. Define the following $(q^m - 1)$-tuple over GF($q^m$) based on the points of $\mathcal{L}$:

$$\mathbf{v}_{\mathcal{L}} = (v_0, v_1, \ldots, v_{q^m-2}), \tag{15.25}$$

whose $q^m - 1$ components $v_0, v_1, \ldots, v_{q^m-2}$ correspond to the $q^m - 1$ non-origin points $\alpha^0, \alpha, \ldots, \alpha^{q^m-2}$ of EG*$(m, q)$, where the $j_1$th, $j_2$th, ..., $j_q$th components are $v_{j_1} = \alpha^{j_1}$, $v_{j_2} = \alpha^{j_2}, \ldots, v_{j_q} = \alpha^{j_q}$, and other components are equal to the 0 element of GF($q^m$). This $(q^m - 1)$-tuple $\mathbf{v}_{\mathcal{L}}$ over GF($q^m$) is called the *type-1 $q^m$-ary incidence vector* of line $\mathcal{L}$, in contrast to the type-1 binary incidence vector of a line in EG*$(m, q)$ defined by (11.6) in Section 11.1. This vector displays the $q$ points on line $\mathcal{L}$, with not only their locations but also their values represented by nonzero elements of GF($q^m$). Consider the line $\alpha\mathcal{L} = \{\alpha^{j_1+1}, \alpha^{j_2+1}, \ldots, \alpha^{j_q+1}\}$. The type-1 $q^m$-ary incidence vector $\mathbf{v}_{\alpha\mathcal{L}}$ of the line $\alpha\mathcal{L}$ is the right cyclic shift of the type-1 $q^m$-ary incidence vector $\mathbf{v}_{\mathcal{L}}$ of the line $\mathcal{L}$ multiplied by $\alpha$.

Recall that, for any line $\mathcal{L}$ in EG*$(m, q)$, the $q^m - 1$ lines $\mathcal{L}, \alpha\mathcal{L}, \ldots, \alpha^{q^m-2}\mathcal{L}$ form a cyclic class (see Sections 2.7.1 and 11.1) and the $J_{0,\text{EG}}(m, 1)$ lines in EG*$(m, q)$ can be partitioned into $K_c = (q^{m-1} - 1)/(q - 1)$ cyclic classes, $S_1, S_2, \ldots, S_{K_c}$. For each cyclic class $S_i$ of lines in EG*$(m, q)$, we form a $(q^m - 1) \times (q^m - 1)$ matrix $\mathbf{H}_{q^m,c,i}$ over GF($q^m$) with the type-1 $q^m$-ary incidence vectors $\mathbf{v}_{\mathcal{L}}, \mathbf{v}_{\alpha\mathcal{L}}, \ldots, \mathbf{v}_{\alpha^{q^m-2}\mathcal{L}}$ of the lines $\mathcal{L}, \alpha\mathcal{L}, \ldots, \alpha^{q^m-2}\mathcal{L}$ in $S_i$ as rows arranged in cyclic order. The matrix $\mathbf{H}_{q^m,c,i}$ is a *special type of circulant* over GF($q^m$) in which each row is the right cyclic shift of the row above it multiplied by $\alpha$, and the first row is the right cyclic shift of the last row multiplied by $\alpha$. Such a circulant is called an *$\alpha$-multiplied circulant* over GF($q^m$). Both the column and row weight of $\mathbf{H}_{q^m,c,i}$ are $q$. This $\alpha$-multiplied circulant $\mathbf{H}_{q^m,c,i}$ over GF($q^m$) is simply the $q^m$-ary counterpart of the binary circulant $\mathbf{H}_{c,i}$ over GF(2) constructed from type-1 binary incidence vectors of the lines in the cyclic class $S_i$ of EG*$(m, q)$ that was defined in Section 11.1.1.

For $1 \le k \le K_c$, if we replace each $(q^m - 1) \times (q^m - 1)$ circulant $\mathbf{H}_{c,i}$ over GF(2) in the matrix $\mathbf{H}^{(1)}_{\text{EG},c,k}$ over GF(2) given by (11.7) by the corresponding $\alpha$-multiplied $(q^m - 1) \times (q^m - 1)$ circulant over GF($q^m$), we obtain the following $k(q^m - 1) \times (q^m - 1)$ matrix over GF($q^m$):

$$\mathbf{H}^{(1)}_{q^m,\text{EG},c,k} = \begin{bmatrix} \mathbf{H}_{q^m,c,1} \\ \mathbf{H}_{q^m,c,2} \\ \vdots \\ \mathbf{H}_{q^m,c,k} \end{bmatrix}, \tag{15.26}$$

which consists of a column of $k$ $\alpha$-multiplied $(q^m - 1) \times (q^m - 1)$ circulants over GF($q^m$). This matrix has column and row weights $kq$ and $q$, respectively. From the definition of the type-1 $q^m$-ary incidence vector of a line in EG*$(m, q)$, it is clear that each nonzero entry of $\mathbf{H}^{(1)}_{q^m,\text{EG},c,k}$ is a nonzero point of EG*$(m, q)$, represented by a nonzero element of GF($q^m$). Since the rows of $\mathbf{H}^{(1)}_{q^m,\text{EG},c,k}$ correspond to different lines in EG*$(m, q)$ and two lines have at most one point in common, no two rows (or two columns) of $\mathbf{H}^{(1)}_{q^m,\text{EG},c,k}$ have more than one position where they have identical nonzero components. Consequently, $\mathbf{H}^{(1)}_{q^m,\text{EG},c,k}$ satisfies the RC constraint.

The null space of $\mathbf{H}^{(1)}_{q^m,\text{EG},c,k}$ gives a $q^m$-ary cyclic EG-LDPC code $C_{q^m,\text{EG},c,k}$ over GF($q^m$) of length $q^m - 1$ with minimum distance at least $kq + 1$, whose Tanner graph has a girth of at least 6. This $q^m$-ary cyclic EG-LDPC code is the $q^m$-ary counterpart of the binary cyclic EG-LDPC code $\mathcal{C}_{\mathrm{EG},c,k}$ given by the null space of the binary matrix $\mathbf{H}^{(1)}_{\mathrm{EG},c,k}$ given by (11.7). The generator polynomial $\mathbf{g}_{q^m}(X)$ of $\mathcal{C}_{q^m,\mathrm{EG},c,k}$ can be determined in exactly the same way as that for its binary counterpart $\mathcal{C}_{\mathrm{EG},c,k}$ (see (11.8)). The most interesting case is that $m = 2$. In this case, $\mathbf{H}^{(1)}_{q^2,\mathrm{EG},c,1}$ consists of a single $\alpha$-multiplied $(q^2-1)\times(q^2-1)$ circulant over GF$(q^2)$ that is constructed from the $q^2-1$ lines in the subgeometry EG$^*(2,q)$ of the two-dimensional Euclidean geometry EG$(2,q)$ over GF$(q)$. The null space of $\mathbf{H}^{(1)}_{q^2,\mathrm{EG},c,1}$ gives a $q^2$-ary cyclic EG-LDPC code over GF$(q^2)$ of length $q^2-1$ with minimum distance at least $q+1$.

In the following, two examples are given to illustrate the above construction of $q^m$-ary cyclic LDPC codes. In these and subsequent examples in the rest of this chapter, we set $q$ as a power of 2, say $q = 2^s$. In decoding, we use the FFT-QSPA with 50 iterations. For each constructed code, we compute its error performance over a binary-input AWGN channel using BPSK signaling and compare its word-error performance with that of an RS code of the same length, rate, and symbol alphabet decoded with the hard-decision (HD) Berlekamp–Massey (BM) algorithm [24, 25] (or the Euclidean algorithm) (see Sections 3.3 and 3.4) and the algebraic soft-decision (ASD) Koetter–Vardy (KV) algorithm [26] (the most well-known soft-decision decoding algorithm for RS codes). The ASD-KV algorithm for decoding an RS code consists of three steps: *multiplicity assignment*, *interpolation*, and *factorization*. The major part of the computational complexity (70%) involved in application of the ASD-KV algorithm comes from the interpolation step and is on the order of $\lfloor\lambda\rfloor^4N^2$ [27, 28], denoted $\mathcal{O}(\lfloor\lambda\rfloor^4N^2)$, where $N$ is the length of the code and $\lambda$ is a complexity parameter that is determined by the *interpolation cost* of the *multiplicity matrix* constructed at the multiplicity-assignment step. As $\lambda$ increases, the performance of the ASD-KV algorithm improves, but the computational complexity increases drastically. As $\lambda$ approaches $\infty$, the performance of the ASD-KV algorithm reaches its limit. The parameter is called the *interpolation complexity coefficient*.

---

**Example 15.2** Let the two-dimensional Euclidean geometry EG$(2,2^3)$ over GF$(2^3)$ be the code-construction geometry. The subgeometry EG$^*(2,2^3)$ of EG$(2,2^3)$ consists of 63 lines not passing through the origin of EG$(2,2^3)$. These lines form a single cyclic class. Let $\alpha$ be a primitive element of GF$(2^6)$. Form an $\alpha$-multiplied $63\times 63$ circulant over GF$(2^6)$ with the 64-ary incidence vectors of the 63 lines in EG$^*(2,2^3)$ as rows arranged in cyclic order. Both the column and row weight of this circulant are 8. Use this circulant as the parity-check matrix $\mathbf{H}_{2^6,\mathrm{EG},c,1}$ of a cyclic EG-LDPC code. The null space over GF$(2^6)$ of this parity-check matrix gives a 64-ary $(63,37)$ cyclic EG-LDPC code $\mathcal{C}_{2^6,\mathrm{EG},c,1}$ over GF$(2^6)$ with minimum distance at least 9, whose Tanner graph has a girth of at least 6. The generator polynomial of this cyclic LDPC code is

$$\begin{aligned}\mathbf{g}_{2^6}(X) = {} & \alpha^{26}+\alpha^{24}X^2+\alpha^{20}X^6+\alpha^{16}X^{10}+\alpha^{14}X^{12}\\ & +\alpha^{13}X^{13}+\alpha^{12}X^{14}+\alpha^{11}X^{15}+\alpha^2X^{24}+X^{26}.\end{aligned}$$

It has eight consecutive powers of $\alpha$ as roots, from $\alpha^2$ to $\alpha^9$. The BCH lower bound on the minimum distance (see Section 3.3) of this 64-ary cyclic LDPC code is 9, which is the same as the column weight of $\mathbf{H}^{(1)}_{2^6,\mathrm{EG},c,1}$ plus 1.

The symbol- and word-error performances of this code decoded with the FFT-QSPA using 50 iterations over the binary-input AWGN channel with BPSK transmission are shown in Figure 15.4, which also includes the word performances of the (63, 37, 27) RS code over GF($2^6$) decoded with the HD-BM and ASD-KV algorithms, respectively. At a WER of $10^{-6}$, the 64-ary (63, 37) cyclic LDPC code achieves a coding gain of 2.6 dB over the (63, 37, 27) RS code over GF($2^6$) decoded with the HD-BM algorithm, while achieving coding gains of 1.8 and 1.2 dB over the RS code decoded using the ASD-KV algorithm with interpolation complexity coefficients 4.99 and $\infty$, respectively. The FFT-QSPA decoding of the 64-ary (63, 37) cyclic LDPC code also converges very fast, as shown in Figure 15.5. At a WER of $10^{-6}$, the performance gap between 3 and 50 iterations is only 0.1 dB, while the performance gap between 2 and 50 iterations is 0.6 dB. We see that, even with three iterations of the FFT-QSPA, the 64-ary (63, 37) QC-LDPC code still achieves a coding gain of 1.7 dB over the (63, 37, 27) RS code decoded with the ASD-KV algorithm with interpolation-complexity coefficient 4.99.

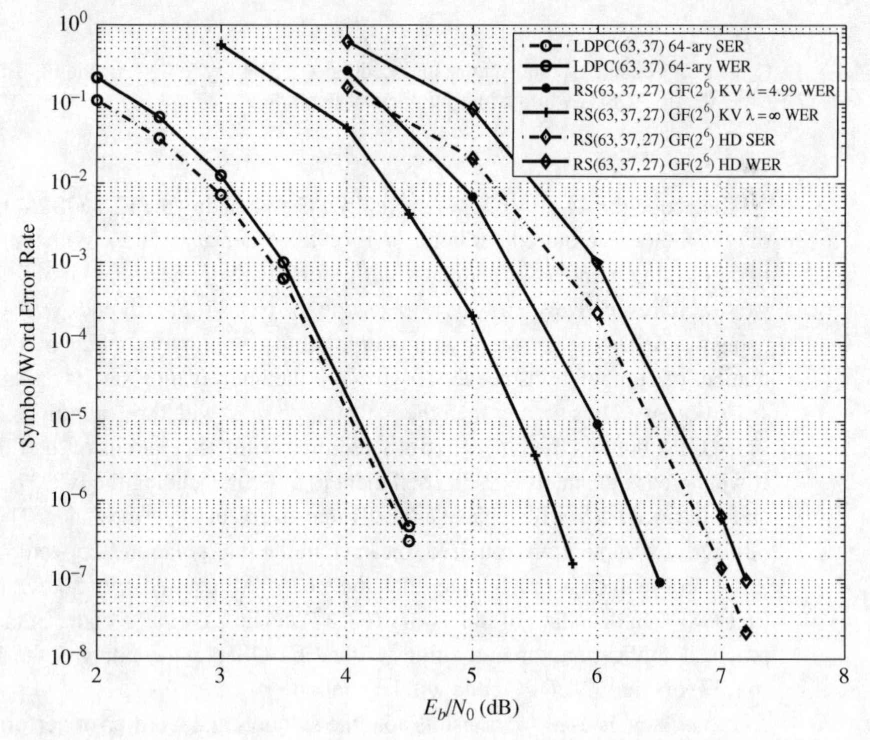


**Figure 15.4** Symbol- and word-error performances of the 64-ary (63, 37) cyclic EG-LDPC code decoded with 50 iterations of the FFT-QSPA and the word-error performances of the (63, 37, 27) RS code over GF($2^6$) decoded with the ASD-KV algorithm with interpolation-complexity coefficients 4.99 and $\infty$.

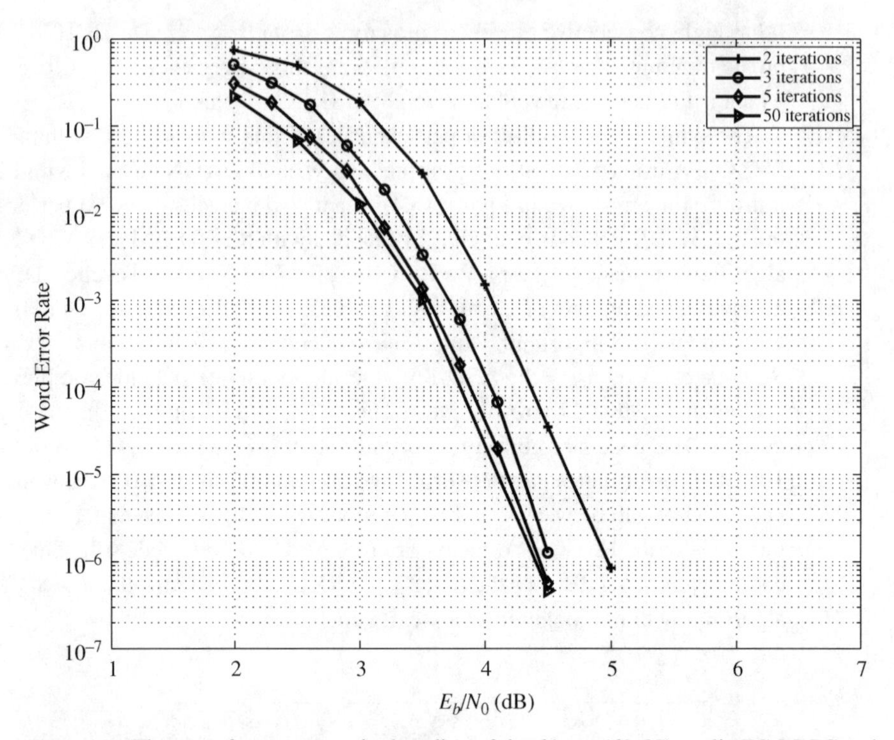


**Figure 15.5** The rate of convergence in decoding of the 64-ary (63, 37) cyclic EG-LDPC code using the FFT-QSPA with various numbers of iterations.

To decode the 64-ary (63, 37) cyclic LDPC code using the FFT-QSPA, the number of computations required per iteration is on the order of 193 536. With 3 and 50 iterations, the numbers of computations required are on the orders of 580 608 and 9 676 800, respectively. However, to decode the (63, 37, 27) RS code over GF($2^6$) using the ASD-KV algorithm with interpolation-complexity coefficient 4.99, the number of computations needed to carry out the interpolation step is on the order of 1 016 064, which is greater than the 580 608 required with three iterations of the FFT-QSPA but less than the 9 676 800 required with 50 iterations of the FFT-QSPA. If we increase the interpolation coefficient to 9.99 in decoding the (63, 37, 27) RS code, the ASD-KV will achieve a 0.2 dB improvement in performance. Then the number of computations required to carry out the interpolation step would be 26 040 609, which is much larger than the 9 676 000 required with 50 iterations of the FFT-QSPA in decoding of the 64-ary (63, 37) cyclic EG-LDPC code. In fact, for practical application, three iterations of the FFT-QSPA in decoding of the 64-ary (63, 37) cyclic EG-LDPC code will be enough.

If 64-QAM is used for transmission, the symbol- and word-error performances of the 64-ary (63, 37) cyclic EG-LDPC code are as shown in Figure 15.6. At a WER of $10^{-5}$, the 64-ary (63, 37) cyclic EG-LDPC code achieves a coding gain of 3.3 dB over the (63, 37, 27) RS code decoded with the HD-BM algorithm.

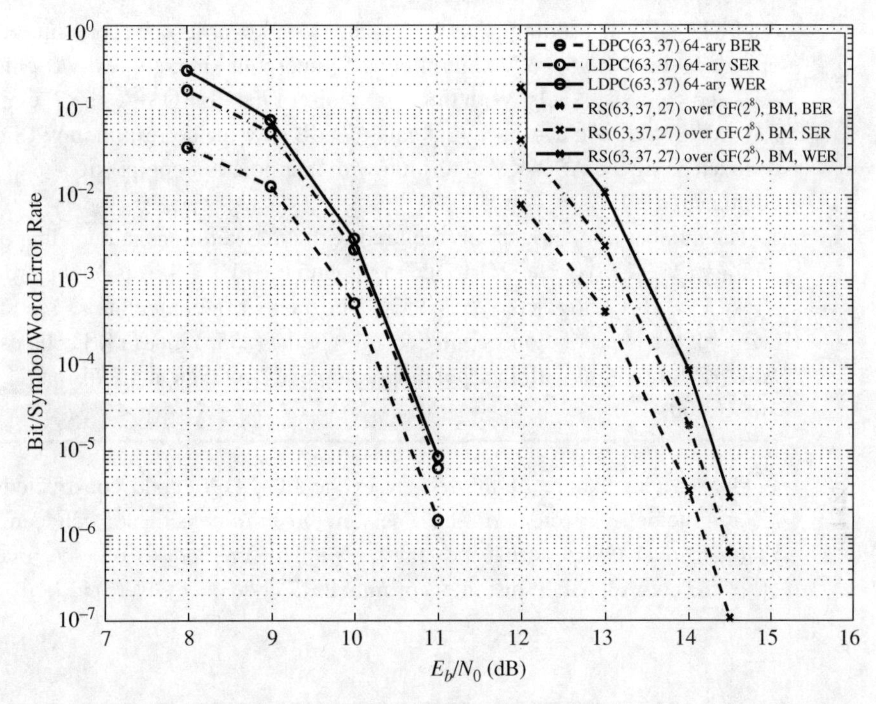


**Figure 15.6** Symbol- and word-error performances of the 64-ary (63, 37) cyclic EG-LDPC code over an AWGN channel with 64-QAM signaling.

---

**Example 15.3** Let the two-dimensional Euclidean geometry EG(2, $2^4$) over GF($2^4$) be the code-construction geometry. Let $\alpha$ be a primitive element of GF($2^8$). From the 256-ary incidence vectors of the 255 lines in EG(2, $2^4$) not passing through the origin, we can form an $\alpha$-multiplied $255 \times 255$ circulant over GF($2^8$) whose column and row weight are both 16. The null space over GF($2^8$) of this circulant gives a 256-ary (255, 175) cyclic EG-LDPC code over GF($2^6$) with minimum distance at least 17 whose Tanner graph has a girth of at least 6. The symbol and word-error performances of this cyclic EG-LDPC code over a binary-input AWGN channel decoded using the FFT-QSPA with 50 iterations are shown in Figure 15.7, which also includes the symbol- and word-error performances of the (255, 175, 81) RS code over GF($2^8$) decoded with the HD-BM algorithm. At a WER of $10^{-5}$, the 256-ary (255, 175) cyclic EG-LDPC code decoded with 50 iterations of the FFT-QSPA achieves a coding gain of 1.5 dB over the (255, 175, 81) RS code decoded with the HD-BM algorithm.

Figure 15.8 shows the word-error performances of the 256-ary (255, 175) cyclic EG-LDPC code decoded with 3 and 50 iterations of the FFT-QSPA and the word-error performances of the (255, 175, 81) RS code decoded with the ASD-KV algorithm using interpolation-complexity coefficients 4.99 and $\infty$, respectively. We see that at a WER of $10^{-5}$, the 256-ary cyclic EG-LDPC code with 50 iterations of the FFT-QSPA has coding gains of 1.1 and 0.7 dB over its corresponding RS code decoded with the ASD-KV algorithm using interpolation-complexity coefficients 4.99 and $\infty$, respectively. With three iterations of the FFT-QSPA, the 256-ary cyclic EG-LDPC code achieves a coding gain of 1 dB over the corresponding RS code decoded using the ASD-KV algorithm with interpolation-complexity coefficient 4.99.

The number of computations required with three iterations in decoding of the 256-ary (255, 175) cyclic EG-LDPC code with the FFT-QSPA is on the order of $3 \times (255 \times 16 \times 256 \times 8) = 25\,067\,520$. The number of computations required to carry out the interpolation step in decoding of the (255, 175, 81) RS code using the ASD-KV algorithm with interpolation-complexity coefficient 4.99 is on the order of 16 646 400.

---

One special feature of the $q^m$-ary cyclic EG-LDPC code constructed from the $q^m$-ary incidence vectors of the lines in the $m$-dimensional Euclidean geometry EG$(m, q)$ over GF$(q)$ is that the length $q^m - 1$ of the code is one less than the size $q^m$ of the code alphabet, just like a primitive RS code over GF$(q^m)$ [29].

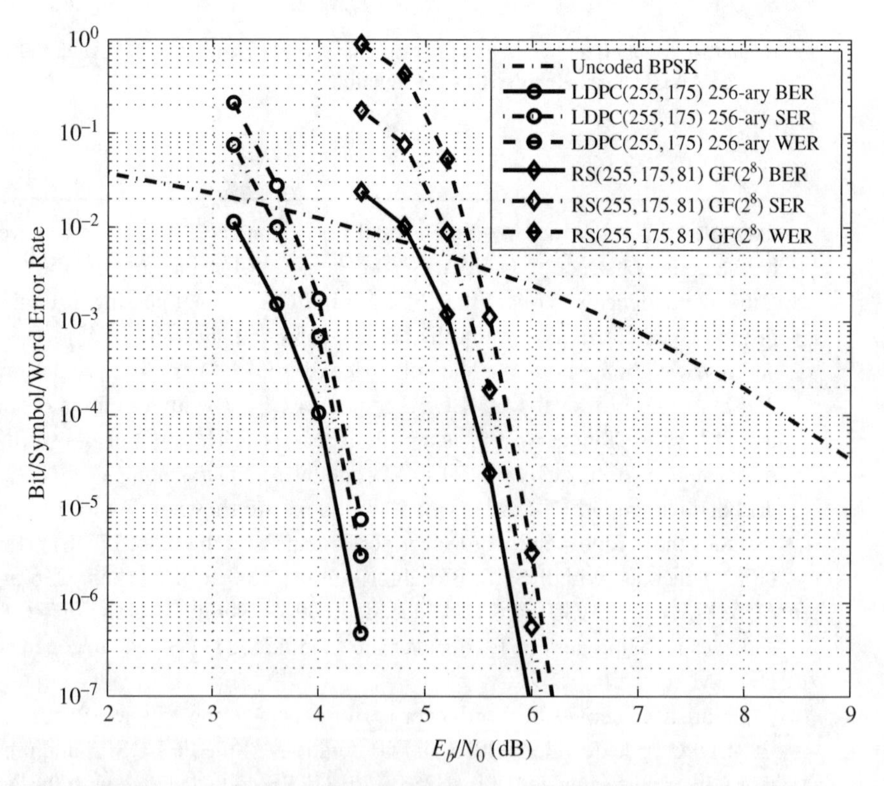


**Figure 15.7** Bit-, symbol-, and word-error performances of the 256-ary (255, 175) cyclic EG-LDPC code decoded with the FFT-QSPA using 50 iterations and the (255, 175, 81) RS code decoded with the HD-BM algorithm.

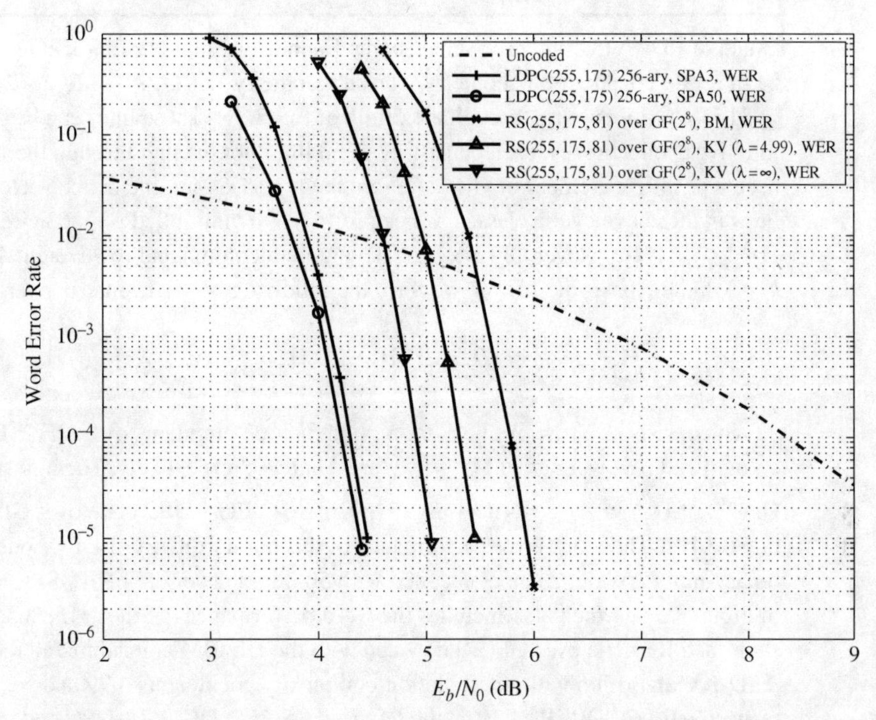


**Figure 15.8** Word-error performances of the 256-ary (255, 175) cyclic EG-LDPC code decoded with 3 and 50 iterations of the FFT-QSPA and the word-error performance of the (255, 175, 81) RS code decoded using the ASD-KV algorithm with interpolation-complexity coefficients 4.99 and ∞.

## 15.3.2 A Class of Nonbinary Quasi-cyclic EG-LDPC Codes

For $1 \leq k \leq K_c$, let $\mathbf{H}^{(2)}_{q^m,\mathrm{EG},qc,k}$ be the transpose of the matrix $\mathbf{H}^{(1)}_{q^m,\mathrm{EG},c,k}$ given by (14.29), that is

$$\begin{aligned}\mathbf{H}^{(2)}_{q^m,\mathrm{EG},c,k} &= \left[\mathbf{H}^{(1)}_{q^m,\mathrm{EG},qc,k}\right]^{\mathrm{T}} \\ &= \left[\mathbf{H}^{\mathrm{T}}_{q^m,c,1} \quad \mathbf{H}^{\mathrm{T}}_{q^m,c,2} \quad \cdots \quad \mathbf{H}^{\mathrm{T}}_{q^m,c,k}\right],\end{aligned} \tag{15.27}$$

where, for $1 \leq i < k$, $\mathbf{H}^{\mathrm{T}}_{q^m,c,i}$ is the transpose of the submatrix $\mathbf{H}_{q^m,c,i}$ in $\mathbf{H}^{(1)}_{q^m,\mathrm{EG},c,k}$. $\mathbf{H}^{(2)}_{q^m,\mathrm{EG},qc,k}$ consists of a row of $k$ $\alpha$-multiplied $(q^m - 1) \times (q^m - 1)$ circulants that is a $(q^m - 1) \times k(q^m - 1)$ matrix over GF($q^m$) with column and row weights $q$ and $kq$, respectively. Since $\mathbf{H}^{(1)}_{q^m,\mathrm{EG},c,k}$ satisfies the RC constraint, $\mathbf{H}^{(2)}_{q^m,\mathrm{EG},qc,k}$ also satisfies the RC constraint. The null space of $\mathbf{H}^{(2)}_{q^m,\mathrm{EG},qc,k}$ gives a $q^m$-ary QC-EG-LDPC code of length $k(q^m - 1)$ with minimum distance at least $q + 1$. For $m > 2$, $K_c$ is greater than unity. As a result, for $k > 1$, the nonbinary QC-EG-LDPC code given by the null space of $\mathbf{H}^{(2)}_{q^m,\mathrm{EG},qc,k}$ has a length $k(q^m - 1)$ that is longer than the size $q^m$ of the code alphabet GF($q^m$). The above construction allows us to construct longer nonbinary LDPC codes from a smaller nonbinary field.

**Example 15.4** Let the three-dimensional Euclidean geometry $\mathrm{EG}(3, 2^2)$ over $\mathrm{GF}(2^2)$ be the code-construction geometry. The subgeometry $\mathrm{EG}^*(3, 2^2)$ of $\mathrm{EG}(3, 2^2)$ consists of 315 lines not passing through the origin of $\mathrm{EG}(3,2^2)$. These lines can be partitioned into five cyclic classes, each consisting of 63 lines not passing through the origin. Let $\alpha$ be a primitive element of $\mathrm{GF}(2^6)$. From the type-1 64-ary incidence vectors of the lines in these five cyclic classes, we can form five $\alpha$-multiplied $63 \times 63$ circulants $\mathbf{H}_{2^6,c,1}, \ldots, \mathbf{H}_{2^6,c,5}$ over $\mathrm{GF}(2^6)$, each having both column and row weight 4. Choose $k = 5$. On the basis of (15.27), we form the following $63 \times 315$ matrix over $\mathrm{GF}(2^6)$:

$$\mathbf{H}^{(2)}_{2^6,\mathrm{EG},qc,5} = \left[\mathbf{H}^{\mathrm{T}}_{2^6,c,1} \quad \mathbf{H}^{\mathrm{T}}_{2^6,c,2} \quad \mathbf{H}^{\mathrm{T}}_{2^6,c,3} \quad \mathbf{H}^{\mathrm{T}}_{2^6,c,4} \quad \mathbf{H}^{\mathrm{T}}_{2^6,c,5}\right],$$

which consists of a row of five $\alpha$-multiplied $63 \times 63$ circulants over $\mathrm{GF}(2^6)$. The column and row weights of $\mathbf{H}^{(2)}_{2^6,\mathrm{EG},qc,5}$ are 4 and 20, respectively. The null space over $\mathrm{GF}(2^6)$ of $\mathbf{H}^{(2)}_{2^6,\mathrm{EG},qc,5}$ gives a 64-ary (315, 265) QC-EG-LDPC code over $\mathrm{GF}(2^6)$ with rate 0.8412. The word-error performance of this 64-ary QC-EG-LDPC code over a binary-input AWGN channel decoded with 50 iterations of the FFT-QSPA is shown in Figure 15.9, which also includes the word performances of the (315, 265, 51) shortened RS code over $\mathrm{GF}(2^9)$ decoded with the HD-BM algorithm and the ASD-KV algorithm with interpolation-complexity coefficients 4.99 and $\infty$, respectively. At a WER of $10^{-5}$, the 64-ary (315, 265) QC-EG-LDPC code achieves a coding gain of 2.1 dB over the (315, 265, 51) shortened RS code decoded with the HD-BM algorithm, while achieving coding gains of 1.8 and 1.5 dB over the shortened RS code decoded using the ASD-KV algorithm with interpolation-complexity coefficients 4.99 and $\infty$, respectively. As shown in Figure 15.9, even with five iterations of the FFT-QSPA, the 64-ary (315, 265) QC-EG-LDPC code achieves a coding gain of 1.5 dB over the (315, 256, 51) shortened RS code decoded using the ASD-KV algorithm with interpolation-complexity coefficient 4.99.

To decode the 64-ary (315, 265) QC-EG-LDPC code with 5 and 50 iterations of the FFT-QSPA, the numbers of computations required are on the order of 2 381 400 and 23 814 000, respectively. However, the number of computations required to carry out the interpolation step in decoding of the (315, 256, 51) shortened RS code using the ASD-KV algorithm with interpolation-complexity coefficient 4.99 is on the order of 25 401 600, which is larger than both 2 381 400 and 23 814 000, the numbers of computations required in order to decode the 64-ary (315, 265) QC-EG-LDPC code with 5 and 50 iterations of the FFT-QSPA, respectively.

### 15.3.3 A Class of Nonbinary Regular EG-LDPC Codes

The construction of binary regular LDPC codes based on the parallel bundles of lines in Euclidean geometries presented in Section 11.2 can be generalized to construct nonbinary regular LDPC codes in a straightforward manner.

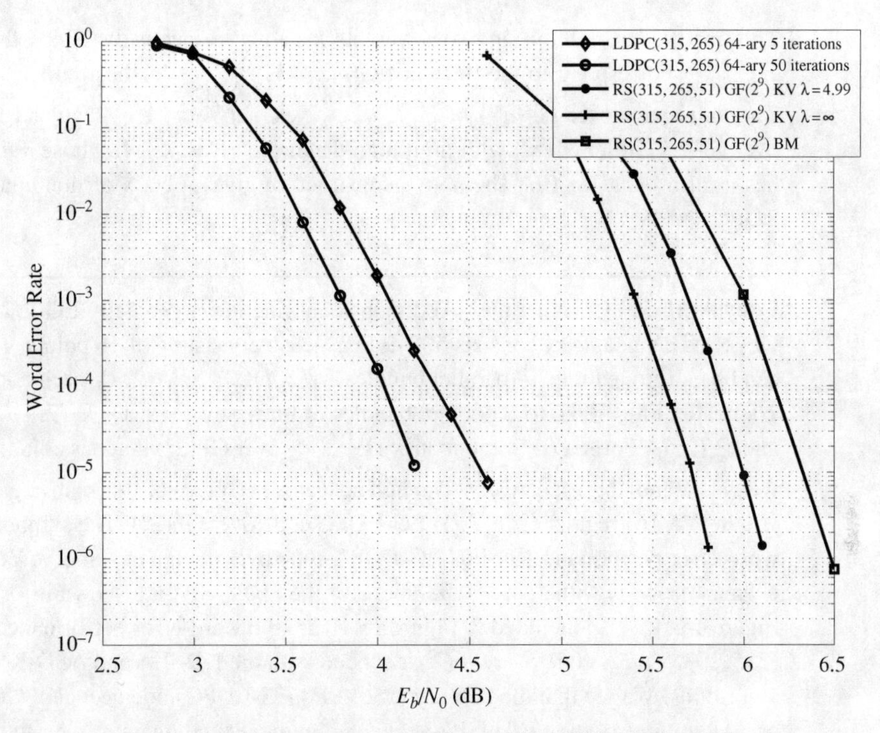


**Figure 15.9** Word-error performances of the 64-ary (315, 265) QC-EG-LDPC code over a binary-input AWGN channel decoded with 5 and 50 iterations of the FFT-QSPA.

Consider the $m$-dimensional Euclidean geometry EG$(m, q)$ over GF$(q)$. Let $\mathcal{L}$ be a line in EG$(m, q)$, passing or not passing through the origin of EG$(m, q)$. Let $\alpha$ be a primitive element of GF$(q^m)$. Define the following $q^m$-tuple over GF$(q^m)$ based on the points on $\mathcal{L}$:

$$\mathbf{v}_{\mathcal{L}} = (v_{-\infty}, v_0, \ldots, v_{q^m-2}),$$

where $v_i = \alpha^i$ if $\alpha^i$ is a point on $\mathcal{L}$, otherwise $v_i = 0$. This $q^m$-tuple over GF$(q^m)$ is referred to as the *type-2* $q^m$*-ary incidence vector* of $\mathcal{L}$, in contrast to the binary type-2 incidence vector of $\mathcal{L}$ defined in Section 11.2 (see (11.13)).

As presented in Section 2.7.1 (and also in Section 11.2), the lines in EG$(m, q)$ can be partitioned into $K_p = (q^m - 1)/(q - 1)$ parallel bundles, denoted $\mathcal{P}_1(m, 1)$, $\mathcal{P}_2(m, 1), \ldots, \mathcal{P}_{K_p}(m, 1)$, each consisting of $q^{m-1}$ parallel lines. For $1 \le i \le K_p$, form a $q^{m-1} \times q^m$ matrix $\mathbf{H}_{q^m,p,i}$ over GF$(q^m)$ with the type-2 $q^m$-ary incidence vectors of the $q^{m-1}$ parallel lines in the parallel bundle $\mathcal{P}_i(m, 1)$ as rows. It is clear that $\mathbf{H}_{q^m,p,i}$ has column and row weights 1 and $q$, respectively. Choose a positive integer $k$ such that $1 \le k \le K_p$. Form a $kq^{m-1} \times q^m$ matrix over GF$(q^m)$ as follows:

$$\mathbf{H}^{(3)}_{q^m,\mathrm{EG},p,k} = \begin{bmatrix} \mathbf{H}_{q^m,p,1} \\ \mathbf{H}_{q^m,p,2} \\ \vdots \\ \mathbf{H}_{q^m,p,k} \end{bmatrix}. \tag{15.28}$$

This matrix has column and row weights $k$ and $q$, respectively. Since the rows of $\mathbf{H}^{(3)}_{q^m,\mathrm{EG},p,k}$ correspond to the lines in EG$(m, q)$, $\mathbf{H}^{(3)}_{q^m,\mathrm{EG},p,k}$ satisfies the RC constraint. The null space over GF$(q^m)$ of $\mathbf{H}^{(3)}_{q^m,\mathrm{EG},p,k}$ gives a $q^m$-ary $(k, q)$-regular LDPC code over GF$(q^m)$ of length $q^m$ with minimum distance at least $k + 1$, whose Tanner graph has a girth of at least 6. The above construction gives a class of nonbinary regular LDPC codes.

---

**Example 15.5** Consider the two-dimensional Euclidean geometry EG$(2, 2^4)$ over GF$(2^4)$. This geometry consists of 272 lines, each consisting of 16 points. These lines can be partitioned into 17 parallel bundles $\mathcal{P}_1(2, 1), \ldots, \mathcal{P}_{17}(2, 1)$, each consisting of 16 parallel lines. Take four parallel bundles of lines, say $\mathcal{P}_1(2, 1)$, $\mathcal{P}_2(2, 1)$, $\mathcal{P}_3(2, 1)$, and $\mathcal{P}_4(2, 1)$. Form a $64 \times 256$ matrix $\mathbf{H}^{(3)}_{2^8,\mathrm{EG},p,4}$ over GF$(2^8)$ that has column and row weights 4 and 16, respectively. The null space over GF$(2^8)$ of this matrix gives a 256-ary (4, 16)-regular (256, 203) EG-LDPC code over GF$(2^8)$. The symbol- and word-error performances of this EG-LDPC code over the binary-input AWGN channel decoded with 5 and 50 iterations of the FFT-QSPA are shown in Figure 15.10. Also included in Figure 15.10 are the word-error performances of the (255, 203, 53) RS code over GF$(2^8)$ decoded with the HD-BM and ASD-KV algorithms. We see that the 256-ary (256, 203) EG-LDPC code decoded with either 5 or 50 iterations of the FFT-QSPA achieves significant coding gains over the (255, 203, 53) RS code decoded with either the HD-BM algorithm or the ASD-KV algorithm.

---

### 15.3.4 Nonbinary LDPC Code Constructions Based on Projective Geometries

The two methods for constructing binary PG-LDPC codes presented in Section 11.6 can be generalized to construct nonbinary PG-LDPC codes in a straightforward manner.

Consider the $m$-dimensional projective geometry PG$(m, q)$ over GF$(q)$ with $m \geq 2$. This geometry consists of

$$n = (q^{m+1} - 1)/(q - 1)$$

points and

$$J_4 = J_{\mathrm{G}}(m, 1) = \frac{(q^m - 1)(q^{m+1} - 1)}{(q^2 - 1)(q - 1)}$$

lines (see (11.24) and (11.25)), each line consisting of $q + 1$ points.

Let $\alpha$ be a primitive element of GF$(q^{m+1})$. The $n$ points of GF$(q^m)$ can be represented by the elements $\alpha^0, \alpha, \ldots, \alpha^{n-1}$ of GF$(q^{m+1})$ (see Sections 2.7.2 and 11.6.1). Let $\mathcal{L}$ be a line in PG$(m, q)$. The $q^{m+1}$-ary incidence vector of $\mathcal{L}$ is defined by the following $n$-tuple over GF$(q^{m+1})$:

$$\mathbf{v}_{\mathcal{L}} = (v_0, v_1, \ldots, v_{n-1}),$$

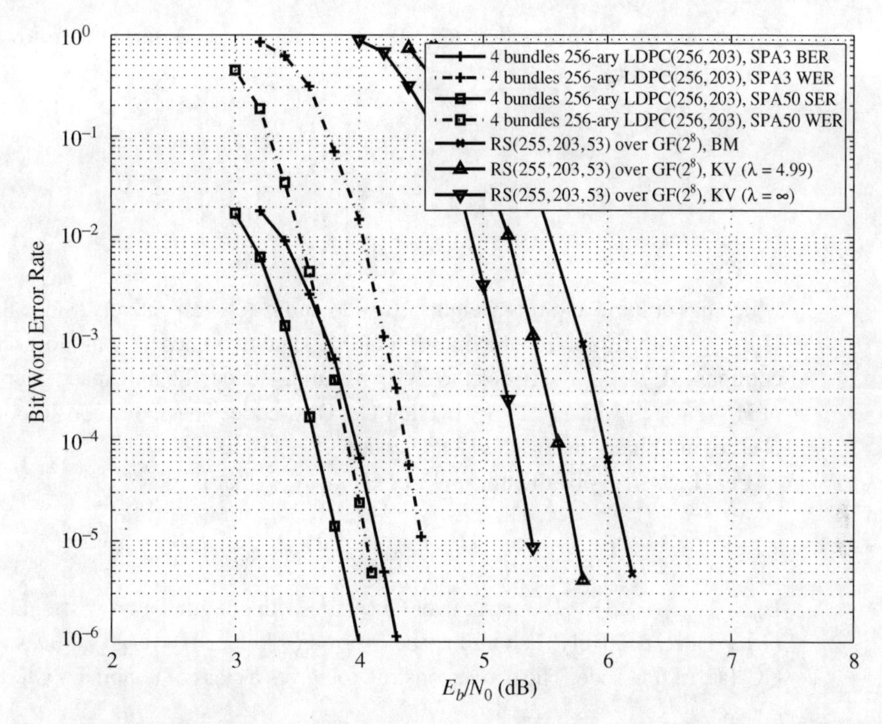


**Figure 15.10** Symbol- and word-error performances of the 256-ary (256, 203) EG-LDPC code given in Example 15.5.

whose components correspond to the $n$ points $\alpha^0, \alpha, \ldots, \alpha^{n-1}$ in PG$(m, q)$, where for $0 \leq i < n$, $v_i = \alpha^i$ if and only if $\alpha^i$ is a point on $\mathcal{L}$, otherwise $v_i = 0$. The weight of $\mathbf{v}_{\mathcal{L}}$ is $q + 1$. The $n$-tuple over GF$(q^{m+1})$ obtained by cyclic shifting $\mathbf{v}_{\mathcal{L}}$ one place to the right and multiplying every component of $\mathbf{v}_{\mathcal{L}}$ by $\alpha$ is also a $q^{m+1}$-ary incidence vector of a line in PG$(m, q)$ (the power of $\alpha^{i+1}$ is taken modulo $n$).

As described in Section 2.6, for even $m$, the lines in PG$(m, q)$ can be partitioned into

$$K^{(e)}_{c,\mathrm{PG}}(m, 1) = \frac{q^m - 1}{q^2 - 1}$$

cyclic classes of size $n$ (see (2.72)). For odd $m$, the lines in PG$(m, q)$ can be partitioned into

$$K^{(o)}_{c,\mathrm{PG}}(m, 1) = \frac{q(q^{m-1} - 1)}{q^2 - 1}$$

cyclic classes of size $n$ (see (2.73)) and a single cyclic class of size

$$l_0 = (q^{m+1} - 1)/(q^2 - 1).$$

For each cyclic class $S_i$ with $0 \leq i < K^{(e)}_{c,\mathrm{PG}}(m, 1)$ (or $K^{(o)}_{c,\mathrm{PG}}(m, 1)$), we form an $n \times n$ $\alpha$-multiplied circulant $\mathbf{H}_{q^{m+1},c,i}$ over GF$(q^{m+1})$ with the $q^{m+1}$-ary incidence vectors of the $n$ lines in $S_i$ as rows arranged in cyclic order. Both the column and row weight of

$\mathbf{H}_{q^{m+1},c,i}$ are $q + 1$. For $1 \le k \le K^{(e)}_{c,\text{PG}}(m, 1)$ (or $K^{(o)}_{c,\text{PG}}(m, 1)$), form the following $kn \times n$ matrix over GF($q^{m+1}$):

$$\mathbf{H}_{q^{m+1},\text{PG},c,k} = \begin{bmatrix} \mathbf{H}_{q^{m+1},c,0} \\ \mathbf{H}_{q^{m+1},c,1} \\ \vdots \\ \mathbf{H}_{q^{m+1},c,k-1} \end{bmatrix}, \tag{15.29}$$

which has column and row weights $k(q + 1)$ and $q + 1$, respectively. Since the rows of $\mathbf{H}_{q^{m+1},\text{PG},c,k}$ correspond to the lines in PG($m, q$) and two lines have at most one point in common, $\mathbf{H}_{q^{m+1},\text{PG},c,k}$ satisfies the RC constraint. Hence, the null space over GF($q^{m+1}$) of $\mathbf{H}_{q^{m+1},\text{PG},c,k}$ gives a $q^{m+1}$-ary cyclic PG-LDPC code. The above construction gives a class of nonbinary cyclic PG-LDPC codes.

Let $\mathbf{H}_{q^{m+1},\text{PG},qc,k}$ be the transpose of $\mathbf{H}_{q^{m+1},\text{PG},c,k}$, that is,

$$\mathbf{H}_{q^{m+1},\text{PG},qc,k} = \mathbf{H}^{\text{T}}_{q^{m+1},PG,c,k}. \tag{15.30}$$

$\mathbf{H}_{q^{m+1},\text{PG},qc,k}$ is an $n \times kn$ matrix over GF($q^{m+1}$) with column and row weights $q + 1$ and $k(q + 1)$, respectively. The null space over GF($q^{m+1}$) of $\mathbf{H}_{q^{m+1},\text{PG},qc,k}$ gives a $q^{m+1}$-ary QC-PG-LDPC code. The above construction gives a class of nonbinary QC-PG-LDPC codes.

---

**Example 15.6** Let the two-dimensional projective geometry PG($2, 2^3$) over GF($2^3$) be the code-construction geometry. This geometry has 73 points and 73 lines, each line consisting of nine points. The 73 lines of PG($2, 2^3$) form a single cyclic class. Let $\alpha$ be a primitive element of GF($2^9$). Using the $2^9$-ary incidence vectors of the lines in PG($2, 2^3$) as rows, we can form an $\alpha$-multiplied $73 \times 73$ circulant over GF($2^9$) with both column and row weight 9. The null space over GF($2^9$) of this $\alpha$-multiplied circulant gives a $2^9$-ary (73, 45) cyclic PG-LDPC code over GF($2^9$) with minimum distance at least 10. The word-error performances of this code over the binary-input AWGN channel with BPSK transmission on decoding this code with 5 and 50 iterations of the FFT-QSPA are shown in Figure 15.11. Also included in Figure 15.11 are the word-error performances of the (73, 45, 39) shortened RS code over GF($2^9$) decoded with the HD-BM and ASD-KV algorithms. At a WER of $10^{-5}$, the $2^9$-ary (73, 45) cyclic PG-LDPC code achieves a coding gain of 2.6 dB over the (73, 45, 39) shortened RS code decoded with the HD-BM algorithm, while achieving coding gains of 2.0 and 1.4 dB over the RS code decoded using the ASD-KV algorithm with interpolation complexity coefficients 4.99 and $\infty$, respectively. Figure 15.12 shows that the FFT-QSPA decoding of the $2^9$-ary (73, 45) cyclic PG-LDPC code converges very fast. The word-error performance of the code with five iterations is almost the same as that with 50 iterations.

The numbers of computations required in decoding of the $2^9$-ary (73, 45) cyclic PG-LDPC code with 5 and 50 iterations are on the order of 15 107 715 and 151 077 150, respectively. However, the number of computations required to carry out the interpolation step in decoding the (73, 45, 39) shortened RS code with the ASD-KV algorithm with interpolation-complexity coefficient 4.99 is on the order of 1 360 000.

## 15.4 Constructions of Nonbinary QC-LDPC Codes Based on Finite Fields

In Chapter 12, methods for constructing binary QC-LDPC codes based on finite fields have been presented. These methods can be generalized to construct nonbinary QC-LDPC codes. The generalization is based on dispersing the elements of a nonbinary finite field GF($q$) into *special circulant permutation matrices* over GF($q$) in a way very similar to the binary matrix dispersions of elements of a finite field presented in Section 12.1.

### 15.4.1 Dispersion of Field Elements into Nonbinary Circulant Permutation Matrices

Consider the Galois field GF($q$). Let $\alpha$ be a primitive element of GF($q$). Then the $q$ powers of $\alpha$, $\alpha^{-\infty}, \alpha^0, \alpha, \ldots, \alpha^{q-2}$, form all the elements of GF($q$). For each non

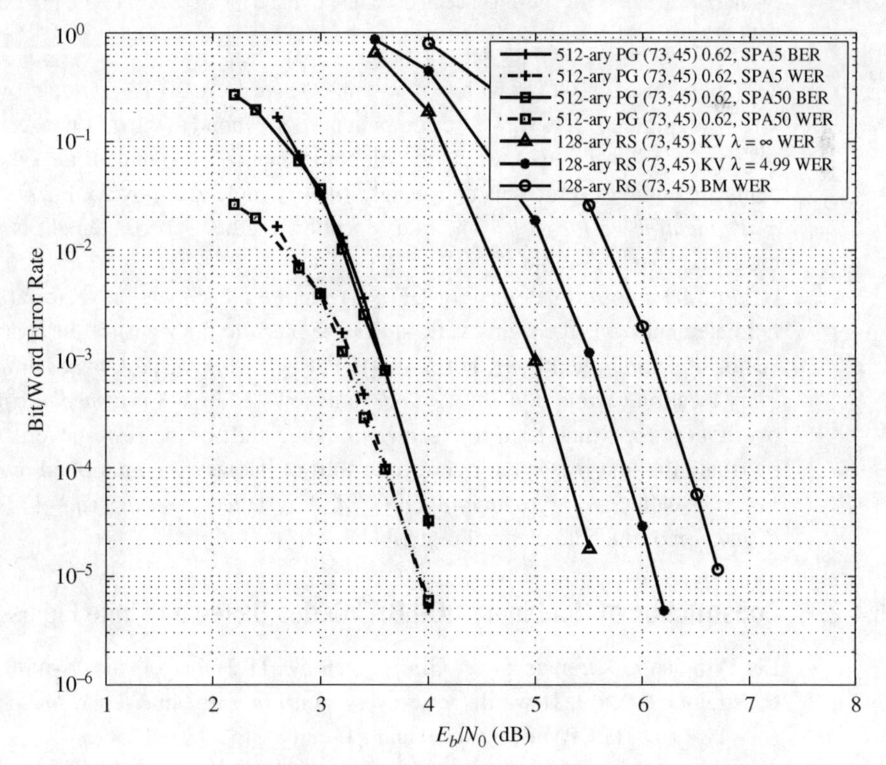


**Figure 15.11** Word-error performances of the $2^9$-ary (73, 45) cyclic PG-LDPC code over a binary-input AWGN channel decoded with 5 and 50 iterations of the FFT-QSPA.

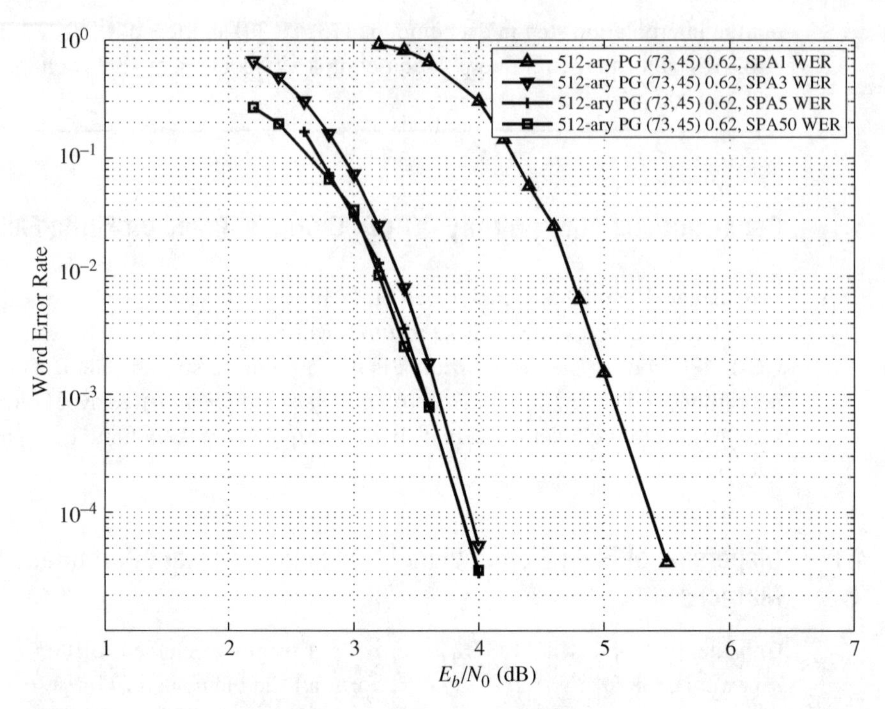


**Figure 15.12** The rate of convergence in decoding of the $2^9$-ary (73, 45) cyclic PG-LDPC code with the FFT-QSPA.

zero element $\alpha^i$ in GF($q$) with $0 \leq i < q-1$, we form a $(q-1)$-tuple over GF($q$), $\mathbf{z}(\alpha^i) = (z_0, z_1, \ldots, z_{q-2})$, whose components correspond to the $q-1$ nonzero elements $\alpha^0, \alpha, \ldots, \alpha^{q-2}$ of GF($q$), where the $i$th component $z_i = \alpha^i$ and all the other components are equal to zero. This unit-weight $(q-1)$-tuple $\mathbf{z}(\alpha^i)$ over GF($q$) is called the *q-ary location vector* of $\alpha^i$. The $q$-ary location vector of the 0 element is defined as the all-zero $(q-1)$-tuple.

Let $\delta$ be a nonzero element of GF($q$). Then the $q$-ary location vector $\mathbf{z}(\alpha\delta)$ of the field element $\alpha\delta$ is the right cyclic shift (one place to the right) of the $q$-ary location vector of $\delta$ multiplied by $\alpha$. Form a $(q-1) \times (q-1)$ matrix $\mathbf{A}$ over GF($q$) with the $q$-ary location vectors of $\delta, \alpha\delta, \ldots, \alpha^{q-1}\delta$ as rows. Matrix $\mathbf{A}$ is *a special type of CPM* over GF($q$) for which each row is a right cyclic shift of the row above it multiplied by $\alpha$ and the first row is the right cyclic shift of the last row multiplied by $\alpha$. Such a matrix is called a *q-ary $\alpha$-multiplied CPM*. $\mathbf{A}$ is called the *q-ary $(q-1)$-fold matrix dispersion* of the field element $\delta$ of GF($q$).

## 15.4.2 Construction of Nonbinary QC-LDPC Codes Based on Finite Fields

Let $\mathbf{W}$ be an $m \times n$ matrix over GF($q$) given by (11.2) that satisfies $\alpha$-multiplied row constraints 1 and 2. If we disperse every nonzero entry in $\mathbf{W}$ into an $\alpha$-multiplied $(q-1) \times (q-1)$ CPM over GF($q$) and a 0-entry into a $(q-1) \times (q-1)$ zero matrix, we obtain an $m \times n$ array of $\alpha$-multiplied $(q-1) \times (q-1)$ circulant permutation and/or zero matrices over GF($q$),

$$\mathbf{H}_{q,\mathrm{disp}} = \left[\mathbf{A}_{i,j}\right]_{0\le i<m}^{0\le j<n}, \tag{15.31}$$

where $\mathbf{A}_{i,j}$ is either an $\alpha$-multiplied $(q-1)\times(q-1)$ CPM or a $(q-1)\times(q-1)$ zero matrix. It can be proved that $\mathbf{H}_{q,\mathrm{disp}}$ satisfies the RC constraint. Just like in the binary case, we call $\mathbf{W}$ the base matrix for dispersion and $\mathbf{H}_{q,\mathrm{disp}}$ the $q$-ary dispersion of $\mathbf{W}$.

For $1 \le g \le m$ and $1 \le r \le n$, let $\mathbf{H}_{q,\mathrm{disp}}(g, r)$ be a $g \times r$ subarray of $\mathbf{H}_{q,\mathrm{disp}}$. Then $\mathbf{H}_{q,\mathrm{disp}}(g, r)$ is a $g(q-1)\times r(q-1)$ matrix over GF($q$). The null space of $\mathbf{H}_{q,\mathrm{disp}}(g, r)$ over GF($q$) gives a $q$-ary QC-LDPC code over GF($q$).

Any of the ten base matrices $\mathbf{W}^{(1)}$ to $\mathbf{W}^{(10)}$ given in Chapter 12 can be dispersed into an RC-constrained array of $\alpha$-multiplied $(q-1)\times(q-1)$ CPMs over GF($q$). For $1 \le i \le 10$, let $\mathbf{H}^{(i)}_{q,\mathrm{disp}}$ be the $q$-ary dispersion of the base matrix $\mathbf{W}^{(i)}$. Utilizing the $q$-ary arrays $\mathbf{H}^{(1)}_{q,\mathrm{disp}}$ to $\mathbf{H}^{(10)}_{q,\mathrm{disp}}$, we can construct ten classes of $q$-ary QC-LDPC codes over GF($q$). Furthermore, the masking technique presented in Section 11.4, the superposition technique presented in Section 13.7, and the array dispersion technique presented in Section 14.7 can also be used to construct $q$-ary QC-LDPC codes. In the following, we give several examples to illustrate various constructions of nonbinary QC-LDPC codes based on finite fields.

---

**Example 15.7** From the $(31, 2, 30)$ RS code over GF($2^5$) and (12.5), we can construct a $31 \times 31$ base matrix $\mathbf{W}^{(1)}$ over GF($2^5$) that satisfies $\alpha$-multiplied row constraints 1 and 2, where $\alpha$ is a primitive element of GF($2^5$). The dispersion of $\mathbf{W}^{(1)}$ gives a $31 \times 31$ array $\mathbf{H}^{(1)}_{2^5,\mathrm{disp}}$ of $\alpha$-multiplied $31 \times 31$ circulant permutation and zero matrices over GF($2^5$) with the zero matrices on the main diagonal of $\mathbf{H}^{(1)}_{2^5,\mathrm{disp}}$. Take the first four rows of $\mathbf{H}^{(1)}_{2^5,\mathrm{disp}}$ and remove the first three columns. This results in a $4 \times 28$ subarray $\mathbf{H}^{(1)}_{2^5,\mathrm{disp}}(4, 28)$ of $\mathbf{H}^{(1)}_{2^5,\mathrm{disp}}$. The last submatrix of the first column of $\mathbf{H}^{(1)}_{2^5,\mathrm{disp}}$ is a $31 \times 31$ zero matrix. $\mathbf{H}^{(1)}_{2^5,\mathrm{disp}}(4, 28)$ is a $124 \times 868$ matrix over GF($2^5$). The null space over GF($2^5$) of $\mathbf{H}^{(1)}_{2^5,\mathrm{disp}}(4, 28)$ gives a 32-ary near-regular $(868, 756)$ QC-LDPC code over GF($2^5$) with rate 0.871. The symbol- and word-error performances of this code are shown in Figure 15.13. Also included in Figure 15.13 are the word-error performances of the $(868, 756, 113)$ shortened RS code over GF($2^{10}$) decoded with the HD-BM and ASD-KV algorithms with interpolation-complexity coefficients 4.99 and $\infty$, respectively. We see that, at a WER of $10^{-5}$, the 32-ary $(868, 756)$ QC-LDPC code decoded with 50 iterations of the FFT-QSPA achieves a coding gain of 1.9 dB over the $(868, 756, 113)$ shortened RS code decoded with the HD-BM algorithm, while achieving coding gains of 1.6 and 1.4 dB over the RS code decoded with the ASD-KV algorithm with interpolation-complexity coefficients 4.99 and $\infty$, respectively. The number of computations required to decode the 32-ary $(868, 756)$ QC-LDPC code with 50 iterations of the FFT-QSPA is on the order of 27 776 000; however, the number of computations required to carry out the interpolation step of the ASD-KV algorithm is on the order of 192 876 544.

---

### 15.4.3 Construction of Nonbinary QC-LDPC Codes by Masking

As shown in Chapters 11 to 14, masking can be applied to arrays of binary (circulant) permutation matrices to produce masked arrays with a low density of (circulant) permutation matrices. The null spaces of these masked arrays give binary LDPC codes whose Tanner graphs have fewer edges and short cycles. As a result, these binary LDPC codes may have very good error performances with reduced decoding complexity. Masking can also be applied to arrays of $\alpha$-multiplied CPMs to produce high-performance nonbinary QC-LDPC codes. This is illustrated in the next example.

---

**Example 15.8** Let $\alpha$ be a primitive element of GF($2^6$). Utilizing the $(63, 2, 62)$ RS code over GF($2^6$) and (12. 5), we can construct a $63 \times 63$ matrix $\mathbf{W}^{(1)}$ over GF($2^6$) that satisfies $\alpha$-multiplied row constraints 1 and 2. On dispersing each nonzero entry in $\mathbf{W}^{(1)}$ into an $\alpha$-multiplied $63 \times 63$ CPM over GF($2^6$) and a 0-entry into a $63 \times 63$ zero matrix, we obtain a $63 \times 63$ RC-constrained array $\mathbf{H}^{(1)}_{2^6,\text{disp}}$ of $\alpha$-multiplied circulant permutation and zero matrices, with the zero matrices lying on the main

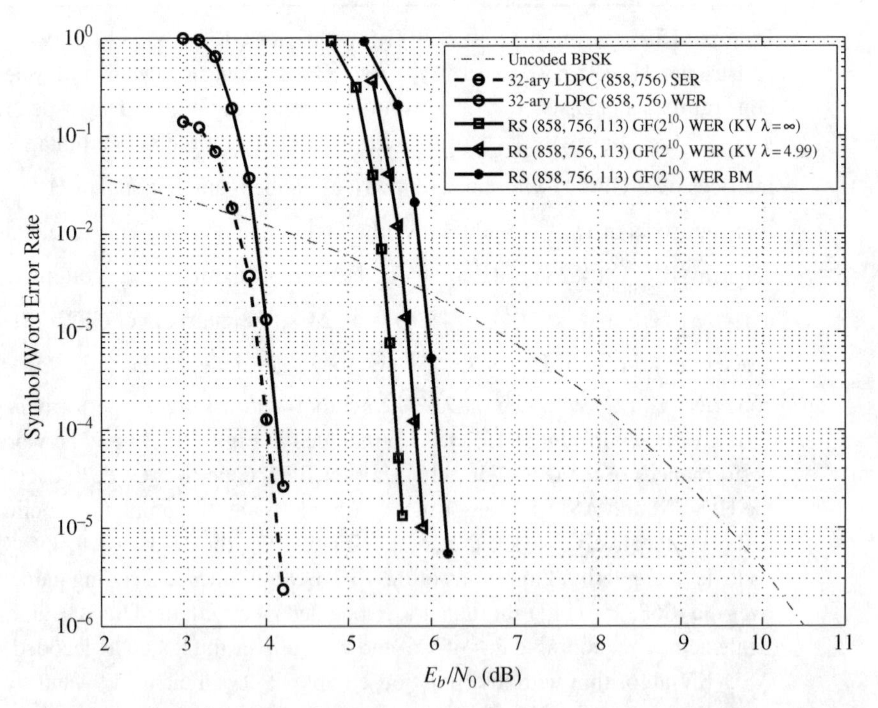


**Figure 15.13** Symbol- and word-error performances of the 32-ary $(868, 756)$ QC-LDPC code given in Example 15.7 over a binary-input AWGN channel decoded with 50 iterations of the FFT-QSPA and the word-error performances of the $(868, 756, 113)$ shortened RS code over GF($2^{10}$) decoded with the HD-BM algorithm and the ASD-KV algorithm.

diagonal of $\mathbf{H}^{(1)}_{2^6,\mathrm{disp}}$. Take an $8 \times 16$ subarray $\mathbf{H}^{(1)}_{2^6,\mathrm{disp}}(8,16)$ from $\mathbf{H}^{(1)}_{2^6,\mathrm{disp}}$, avoiding zero matrices. We use $\mathbf{H}^{(1)}_{2^6,\mathrm{disp}}(8,16)$ as the base array for masking. Construct an $8 \times 16$ masking matrix $\mathbf{Z}(8,16)$ over GF(2) that consists of two $8 \times 8$ circulants, $\mathbf{G}_1$ and $\mathbf{G}_2$, in a row, each having both column and row weight 3. By computer search, we find two primitive generators (top rows) of the two circulants in $\mathbf{Z}(8,16)$, which are $\mathbf{g}_1 = (01011000)$ and $\mathbf{g}_2 = (00101010)$. By masking the base array $\mathbf{H}^{(1)}_{2^6,\mathrm{disp}}(8,16)$ with $\mathbf{Z}(8,16)$, we obtain an $8 \times 16$ masked array $\mathbf{M}^{(1)}_{2^6}(8,16) = \mathbf{Z}(8,16) \otimes \mathbf{H}^{(1)}_{2^6,\mathrm{disp}}(8,16)$, which is a $504 \times 1008$ matrix over GF($2^6$) with column and row weights 3 and 6, respectively. The null space over GF($2^6$) of $\mathbf{M}^{(1)}_{2^6}(8,16)$ gives a 64-ary (1008, 504) QC-LDPC code over GF($2^6$) with rate 0.5. The bit-, symbol-, and word-error performances of this code decoded with 5 and 50 iterations of the FFT-QSPA are shown in Figure 15.14, which also includes the word-error performances of the (1008, 504, 505) shortened RS code over GF($2^6$) decoded with the HD-BM algorithm and the ASD-KV algorithm with interpolation-complexity coefficients 4.99 and $\infty$, respectively. At a WER of $10^{-5}$, the 64-ary (1008, 504) QC-LDPC code achieves a coding gain of 4 dB over the (1008, 504, 505) RS code over GF($2^{10}$) decoded with the HD-BM algorithm, while achieving coding gains of 3.3 and 2.8 dB over the shortened RS code decoded using the ASD-KV algorithm with interpolation-complexity coefficients 4.99 and $\infty$, respectively. Even with five iterations of the FFT-QSPA, the 64-ary (1008, 504) QC-LDPC code has a coding gain of 2.3 dB over the RS code decoded using the ASD-KV algorithm with interpolation-complexity coefficient 4.99.

The numbers of computations required to decode the 64-ary (1008, 504) QC-LDPC code with 5 and 50 iterations of the FFT-QSPA are on the order of 5 806 080 and 58 060 800, respectively. However, the number of computations required to carry out the interpolation step of the ASD-KV algorithm with interpolation-complexity coefficient 4.99 in decoding of the (1008, 504, 505) RS code is on the order of 260 112 384.

### 15.4.4 Construction of Nonbinary QC-LDPC Codes by Array Dispersion

In Section 14.7, a technique called *array dispersion* was presented for constructing binary QC-LDPC codes both for the AWGN and erasure channels using arrays of binary CPMs. This technique can be generalized to construct nonbinary QC-LDPC codes. We begin with an array $\mathbf{H}_{q,\mathrm{disp}}$ of $\alpha$-multiplied CPMs over GF($q$). Then take a subarray from $\mathbf{H}_{q,\mathrm{disp}}$ and disperse it as described in Section 14.7. The null space of the dispersed $q$-ary array gives a $q$-ary QC-LDPC code. We illustrate this construction by two examples.

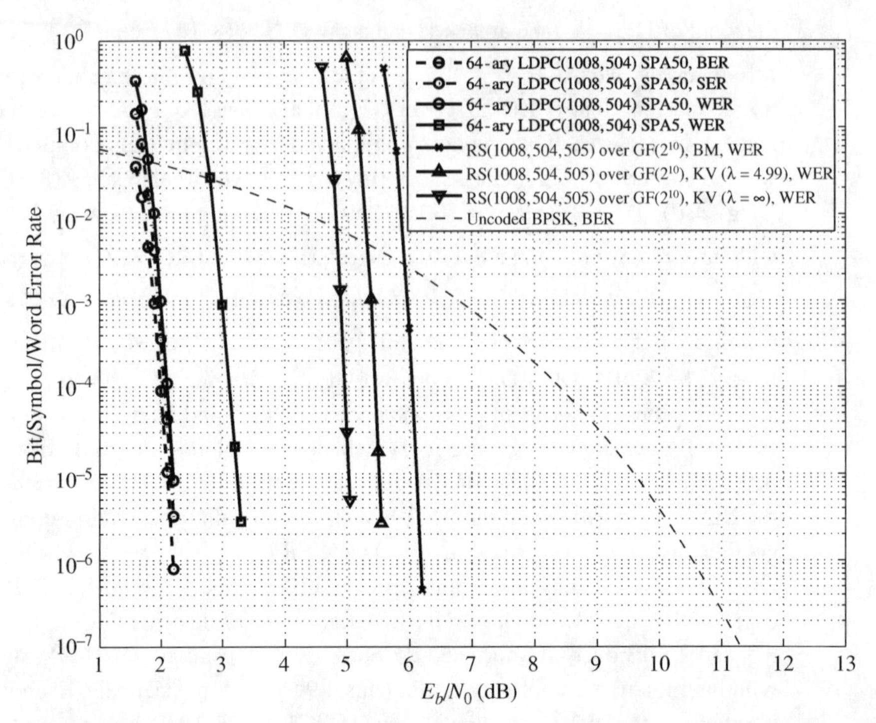


**Figure 15.14** Error performances of the 64-ary $(1008, 504)$ QC-LDPC code given in Example 15.8 over a binary-input AWGN channel decoded with 5 and 50 iterations of the FFT-QSPA.

---

**Example 15.9** Consider the $31 \times 31$ array $\mathbf{H}^{(1)}_{2^5,\text{disp}}$ of $\alpha$-multiplied $31 \times 31$ CPMs constructed in Example 15.7 using the $(31, 2, 30)$ RS code over GF$(2^5)$. Set $k = 2$, $l = 2$, $s = 2$, and $t = 5$. Take a $5 \times 10$ subarray $\mathbf{H}^{(1)}_{2^5,\text{disp}}(5, 10)$ from $\mathbf{H}^{(1)}_{2^5,\text{disp}}$, avoiding the zero matrices on the main diagonal of $\mathbf{H}^{(1)}_{2^5,\text{disp}}$. The 2-masked and 2-fold dispersion of $\mathbf{H}^{(1)}_{2^5,\text{disp}}(5, 10)$ gives a $10 \times 20$ array $\mathbf{H}^{(1)}_{2^5,\text{2-f.disp},2}(10, 20)$ of $\alpha$-multiplied $31 \times 31$ circulant permutation and zero matrices over GF$(2^5)$. $\mathbf{H}^{(1)}_{2^5,\text{2-f,disp},2}(10, 20)$ is a $310 \times 620$ matrix over GF$(2^5)$ with column and row weights 3 and 6, respectively. The null space of this matrix gives a 32-ary $(3, 6)$-regular $(620, 310)$ QC-LDPC code over GF$(2^5)$. The bit-, symbol-, and word-error performances of this code over a binary-input AWGN channel decoded with 50 iterations of the FFT-QSPA are shown in Figure 15.15, which also includes the word-error performances of the $(620, 310, 311)$ shortened RS code over GF$(2^{10})$ decoded using the HD-BM algorithm and the ASD-KV algorithm with interpolation-complexity coefficients 4.99 and $\infty$, respectively. At a WER of $10^{-5}$, the 32-ary $(620, 310)$ QC-LDPC code has a coding gain of 4 dB over the $(620, 310, 311)$ RS code decoded with the HD-BM algorithm, while achieving coding gains of 3.2 and 2.8 dB over the RS code decoded using the ASD-KV with interpolation-complexity coefficients 4.99 and $\infty$, respectively.

The number of computations required in decoding the 32-ary $(620, 310)$ QC-LDPC code with 50 iterations of the FFT-QSPA is on the order of 14 880 000. However, the number of computations required to carry out the interpolation step of

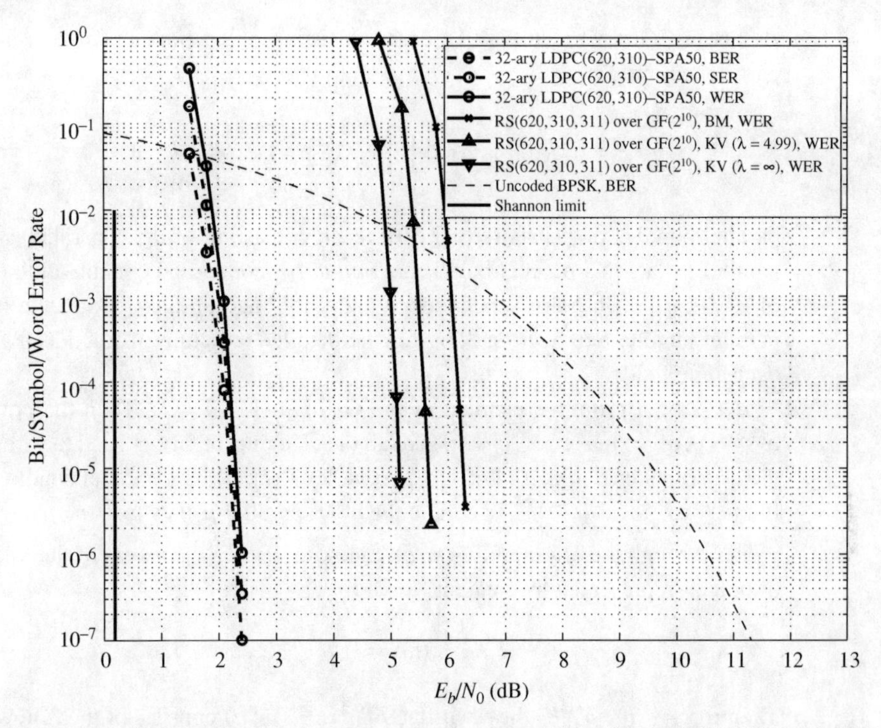


**Figure 15.15** Bit-, symbol-, and word-error performances of the 32-ary (620, 310) QC-LDPC code over a binary-input AWGN channel decoded with 50 iterations of the FFT-QSPA.

the ASD-KV algorithm with interpolation-complexity coefficient 4.99 in decoding of the (620, 310, 311) RS code is on the order of 101 606 400.

The 32-ary (620, 310) QC-LDPC code over GF($2^5$) is also capable of correcting any erasure burst of length up to at least 218 symbols. The erasure-burst-correction efficiency of this code is at least 0.703.

## 15.5 Construction of QC-EG-LDPC Codes Based on Parallel Flats in Euclidean Geometries and Matrix Dispersion

In Section 15.3, methods for constructing nonbinary EG-LDPC codes based on two types of nonbinary incidence vectors of lines in Euclidean geometries were presented. In this section, we show that matrices based on the points on parallel bundles of $\mu$-flats in Euclidean geometries that satisfy $\alpha$-multiplied row constraints 1 and 2 can be constructed. By dispersing the entries (represented by elements of finite fields) of these matrices into either $\alpha$-multiplied CPMs or zero matrices, we obtain RC-constrained arrays of $\alpha$-multiplied circulant permutation and/or zero matrices. From these arrays, we can construct nonbinary QC-EG-LDPC codes.

Consider the $m$-dimensional Euclidean geometry EG($m, q$) over GF($q$). As shown in Section 2.7.1, for $0 \leq \mu \leq m$, there are

$$J_{\rm EG}(m, \mu) = q^{m-u} \prod_{i=1}^{\mu} (q^{m-i+1} - 1)/(q^{\mu-i+1} - 1)$$

$\mu$-flats in EG$(m, q)$ (see (2.49)), each consisting of $q^\mu$ points. These $\mu$-flats can be partitioned into

$$K_{\text{EG}}(m, \mu) = \prod_{i=1}^{\mu} (q^{m-i+1} - 1)/(q^{\mu-i+1} - 1)$$

parallel bundles of $\mu$-flats (see (2.50)), each consisting of $q^{m-\mu}$ parallel $\mu$-flats. The parallel $\mu$-flats in a parallel bundle are *mutually disjoint* and contain all the $q^m$ points of EG$(m, q)$; each point appears once and only once on only one of the flats in the parallel bundle. See Section 2.7.1 for the detailed structure of Euclidean geometries over finite fields.

Let $\alpha$ be a primitive element of the Galois field GF$(q^m)$. Then the powers of $\alpha$, $\alpha^{-\infty} = 0, \alpha^0 = 1, \alpha^2, \ldots, \alpha^{q^m-2}$, give all the elements of GF$(q^m)$ and they represent all the $q^m$ points of EG$(m, q)$. For $1 \le \mu < m$, let $\mathcal{P}(m, \mu)$ be a parallel bundle of $\mu$-flats. Let $\mathcal{F}_0^{(\mu)}, \mathcal{F}_1^{(\mu)}, \ldots, \mathcal{F}_{q^{m-\mu}-1}^{(\mu)}$ denote the $q^{m-\mu}$ parallel $\mu$-flats in $\mathcal{P}(m, \mu)$, where $\mathcal{F}_0^{(\mu)}$ passes through the origin of EG$(m, q)$. Suppose that $\mathcal{F}_0^{(\mu)}$ consists of the following set of points (represented by elements in GF$(q^m)$):

$$\mathcal{F}_0^{(\mu)} = \{\alpha^{j_0} = 0, \alpha^{j_1}, \ldots, \alpha^{j_{q^\mu-1}}\}. \tag{15.32}$$

Then for $1 \le i < q^{m-\mu}$, the $i$th $\mu$-flat $\mathcal{F}_i^{(\mu)}$ in $\mathcal{P}(m, \mu)$ consists of the following set of points:

$$\begin{aligned} \mathcal{F}_i^{(\mu)} &= \delta_i + \mathcal{F}_0^{(\mu)} \\ &= \{\delta_i, \delta_i + \alpha^{j_1}, \ldots, \delta_i + \alpha^{j_{q^\mu-1}}\}, \end{aligned} \tag{15.33}$$

where $\delta_i$ is a point *not in any other* $\mu$-flat in $\mathcal{P}(m, \mu)$.

Let $n_1 = q^\mu$ and $n_2 = q^{m-\mu}$. Utilizing the parallel bundle $\mathcal{P}(m, \mu)$ of $\mu$-flats in EG$(m, q)$, we form the following $n_2 \times n_1$ matrix over GF$(q^m)$:

$$\mathbf{W}_{q^m,\text{EG},p,\mu} = \begin{bmatrix} 0 & \alpha^{j_1} & \cdots & \alpha^{j_{n_1-1}} \\ \delta_1 & \delta_1 + \alpha^{j_1} & \cdots & \delta_1 + \alpha^{j_{n_1-1}} \\ \vdots & \vdots & \ddots & \vdots \\ \delta_{n_2-1} & \delta_{n_2-1} + \alpha^{j_1} & \cdots & \delta_{n_2-1} + \alpha^{j_{n_1-1}} \end{bmatrix}, \tag{15.34}$$

where the entries of the $i$th row are the points of the $i$th $\mu$-flat of $\mathcal{P}(m, \mu)$. $\mathbf{W}_{q^m,\text{EG},p,\mu}$ has the following structural properties: (1) all the $q^m$ entries are distinct and they form all the elements of GF$(q^m)$, that is, every element of GF$(q^m)$ appears once and only once in the matrix; (2) any two rows (two columns) differ in all positions; and (3) the 0-entry appears at the upper-left corner of the matrix. From these structural properties, it can easily be proved that $\mathbf{W}_{q^m,\text{EG},p,\mu}$ satisfies $\alpha$-multiplied row constraints 1 and 2.

By dispersing every nonzero entry in $\mathbf{W}_{q^m,\text{EG},p,\mu}$ into an $\alpha$-multiplied $(q^m - 1) \times (q^m - 1)$ CPM and the single 0-entry into a $(q^m - 1) \times (q^m - 1)$ zero matrix, we obtain the following $n_2 \times n_1$ array $\mathbf{H}_{q^m,\text{EG},p,\mu}$ of $q^m - 1$ $\alpha$-multiplied CPMs and a single zero matrix:

$$\mathbf{H}_{q^m,\mathrm{EG},p,\mu} = \begin{bmatrix} \mathbf{A}_{0,0} & \mathbf{A}_{0,1} & \cdots & \mathbf{A}_{0,n_1-1} \\ \mathbf{A}_{1,0} & \mathbf{A}_{1,1} & \cdots & \mathbf{A}_{1,n_1-1} \\ \vdots & \vdots & \ddots & \vdots \\ \mathbf{A}_{n_2-1,0} & \mathbf{A}_{n_2-1,1} & \cdots & \mathbf{A}_{n_2-1,n_1-1} \end{bmatrix}, \tag{15.35}$$

where $\mathbf{A}_{0,0} = \mathbf{O}$ (a $(q^m-1)\times(q^m-1)$ zero matrix). Since all the entries in $\mathbf{W}_{q^m,\mathrm{EG},p,\mu}$ are distinct, all the $\alpha$-multiplied CPMs in $\mathbf{H}_{q^m,\mathrm{EG},p,\mu}$ are distinct. $\mathbf{H}_{q^m,\mathrm{EG},p,\mu}$ is an $n_2(q^m - 1) \times n_1(q^m - 1)$ matrix over GF($q^m$) that satisfies the RC constraint.

For any pair $(g, r)$ of integers with $1 \le g \le n_2$ and $1 \le r \le n_1$, let $\mathbf{H}_{q^m,\mathrm{EG},p,\mu}(g,r)$ be a $g \times r$ subarray of $\mathbf{H}_{q^m,\mathrm{EG},p,\mu}$. Then $\mathbf{H}_{q^m,\mathrm{EG},p,\mu}(g,r)$ is a $g(q^m - 1) \times r(q^m - 1)$ matrix over GF($q^m$) that also satisfies the RC constraint. If $\mathbf{H}_{q^m,\mathrm{EG},p,\mu}(g,r)$ does not contain the single zero matrix of $\mathbf{H}_{q^m,\mathrm{EG},p,\mu}$, it has column and row weights $g$ and $r$, respectively. The null space over GF($q^m$) of $\mathbf{H}_{q^m,\mathrm{EG},p,\mu}(g,r)$ gives a $(g,r)$-regular $q^m$-ary QC-EG-LDPC code of length $rn_1$ with rate at least $(r - g)/r$, whose Tanner graph has a girth of at least 6. If $\mathbf{H}_{q^m,\mathrm{EG},p,\mu}(g,r)$ contains the single zero matrix of $\mathbf{H}_{q^m,\mathrm{EG},p,\mu}$, it has two different column weights, $g-1$ and $g$, and two row weights, $r-1$ and $r$. Then the null space of $\mathbf{H}_{q^m,\mathrm{EG},p,\mu}(g,r)$ gives a *near-regular* $q^m$-ary QC-EG-LDPC code. The above construction gives a class of $q^m$-ary QC-EG-LDPC codes. For $q = 2^s$, we obtain a subclass of $2^{ms}$-ary QC-LDPC codes. In this subclass, a special case is that of $s = 1$. For this case, the geometry for code construction is EG($m$, 2) over GF(2), the simplest Euclidean geometry.

In the following, we use an example to illustrate the construction of $q^m$ QC-EG-LDPC codes based on a parallel bundle of $\mu$-flats of EG($m, q$).

---

**Example 15.10** Consider the five-dimensional Euclidean geometry EG(5, 2) over GF(2) ($m = 5$ and $q = 2$). Let $\mu = 3$. For this geometry, each 3-flat consists of eight points. Each parallel bundle of 3-flats consists of four 3-flats. Take a parallel bundle $P(5, 3)$ of 3-flats for code construction. From this parallel bundle, (15.34) and (15.35), we can construct a $4 \times 8$ array $\mathbf{H}_{2^5,\mathrm{EG},p,3}$ of 31 $\alpha$-multiplied $31 \times 31$ CPMs and a single zero matrix over GF($2^5$). Set $g = 4$ and $r = 8$. Then $\mathbf{H}_{2^5,\mathrm{EG},p,3}(4, 8)$ is the entire array $\mathbf{H}_{2^5,\mathrm{EG},p,3}$. It is a $124 \times 248$ matrix over GF($2^5$) with two different column weights, 3 (31 columns) and 4 (217 columns), and two different row weights, 7 (31 rows) and 8 (93 rows). The null space over GF($2^5$) of this matrix gives a near-regular 32-ary (248, 136) QC-LDPC code with rate 0.5484, whose Tanner graph has a girth of at least 6. For BPSK transmission over the binary-input AWGN channel, the bit-, symbol-, and word-error performances of this code decoded with iterative decoding using FFT-QSPA with 50 iterations are shown in Figure 15.16, which also includes the bit-, symbol-, and word-error performances of the (248, 136, 113) shortened RS code over GF($2^8$) decoded with the HD-BM algorithm. At a BER of $10^{-6}$ (or a WER of $10^{-5}$), the 32-ary (248, 136) QC-LDPC code achieves a coding gain of 2.6 dB over the 256-ary (248, 136, 113) RS code. Figure 15.17 shows the word-error performances of the (248, 136, 113) shortened RS code over GF($2^8$) decoded with the ASD-KV algorithm with interpolation-complexity coefficients equal to 4.99 and $\infty$, respectively. We see that, at a WER of $10^{-5}$, the 32-ary (248, 136) QC-EG-LDPC code has coding gains of 2.102 and 1.574 dB over the (248, 136, 113) shortened RS

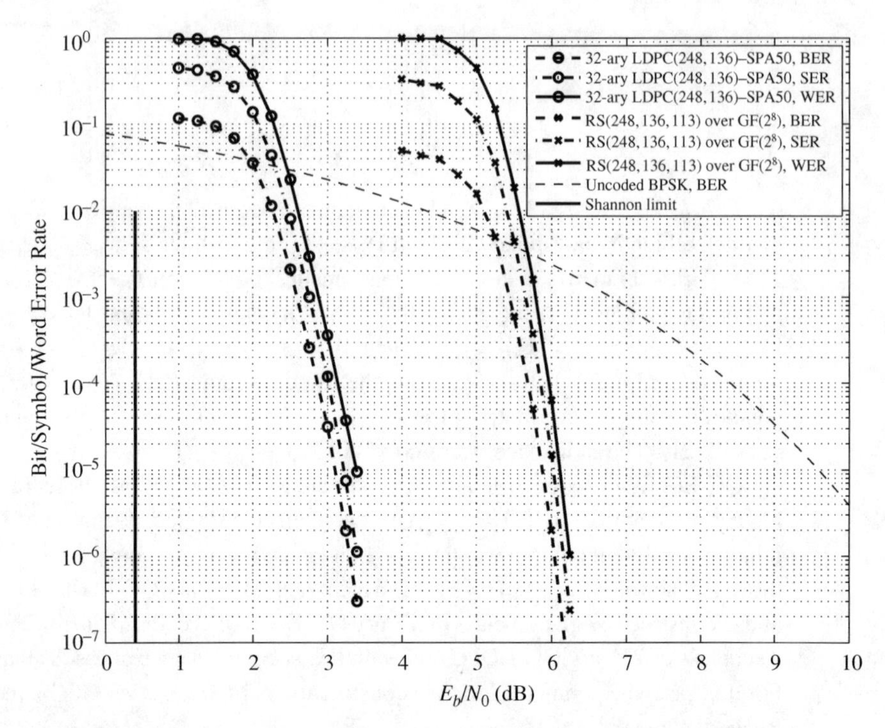


**Figure 15.16** Bit-, symbol-, and word-error performances of the 32-ary (248,136) QC-EG-LDPC code given in Example 15.10.

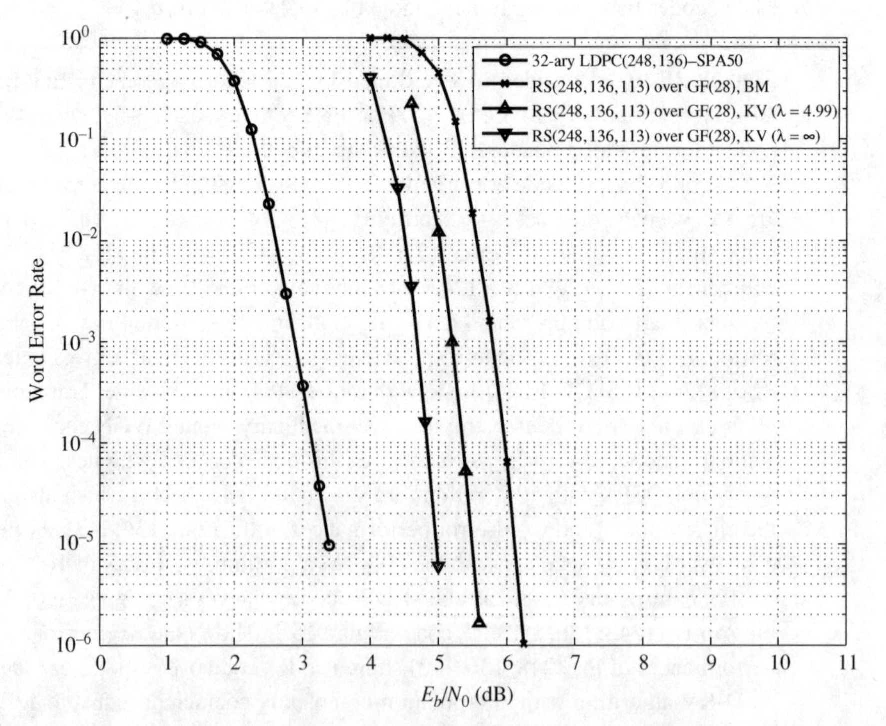


**Figure 15.17** Word performances of the 32-ary (248, 136) QC-EG-LDPC code decoded with the FFT-QSPA and the (248, 136, 113) shortened RS code over GF($2^9$) decoded with the HD-BM and the ASD-KV algorithms.

code over GF($2^8$) decoded using the ASD-KV algorithm with interpolation-complexity coefficients 4.99 and $\infty$, respectively.

The number of computations required to decode the 32-ary (248, 136) QC-EG-LDPC code with 50 iterations of the FFT-QSPA is on the order of 7 936 000; however, the number of computations required to carry out the interpolation step of the ASD-KV algorithm with interpolation-complexity coefficient 4.99 is on the order of 15 745 024. That is to say, at a WER of $10^{-5}$, the 32-ary (248, 136) QC-EG-LDPC code decoded with 50 iterations of the FFT-QSPA achieves a coding gain of 2.102 dB over the (248, 136, 113) shortened RS code over GF($2^8$) using the ASD-KV algorithm with fewer computations. Furthermore, the FFT-QSPA decoding of the (248, 136) QC-LDPC code converges very fast, as shown in Figure 15.18. At a WER of $10^{-5}$, even with three iterations of the FFT-QSPA, the 32-ary (248, 136) QC-LDPC code provides coding gains of 1 and 0.5 dB over the (248, 136, 113) shortened RS code over GF($2^8$) decoded using the ASD-KV algorithm with interpolation-complexity coefficients 4.99 and $\infty$, respectively. With five iterations of the FFT-QSPA, the coding gains are 1.1 and 1.5 dB, respectively.

## 15.6 Construction of Nonbinary QC-EG-LDPC Codes Based on Intersecting Flats in Euclidean Geometries and Matrix Dispersion

In this section, we show that a matrix that satisfies $\alpha$-multiplied row constraints 1 and 2 can be constructed utilizing the points of a bundle of intersecting flats

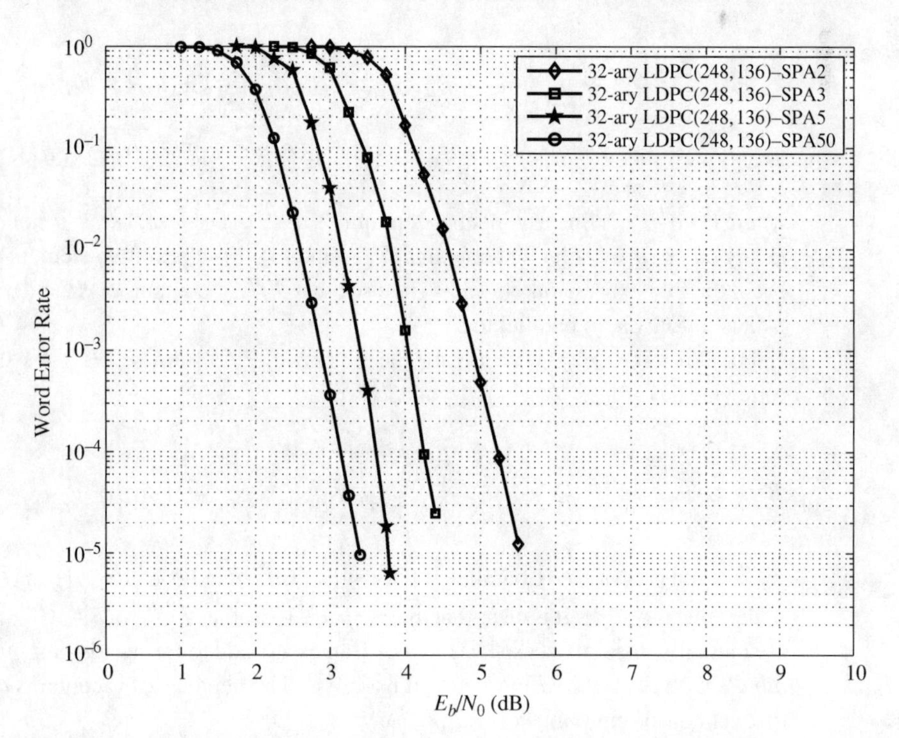


**Figure 15.18** The rate of decoding convergence of the FFT-QSPA for the (248, 136) QC-EG-LDPC code given in Example 15.10.

of a Euclidean geometry. On dispersing this matrix, we obtain an RC-constrained array of $\alpha$-multiplied CPMs. From this array, nonbinary QC-EG-LDPC codes can be constructed.

For $0 \leq \mu < m - 1$, two $(\mu + 1)$-flats in EG$(m, q)$ are either disjoint or intersect on a flat with smaller dimension. The largest flat that two $(\mu + 1)$-flats can intersect on is a $\mu$-flat. The number of $(\mu + 1)$-flats in EG$(m, q)$ that intersect on a given $\mu$-flat is $(q^{m-\mu} - 1)/(q - 1)$ (see (2.51)). Let EG$^*(m, q)$ be the subgeometry obtained from EG$(m, q)$ by removing the origin and all the flats in EG$(m, q)$ that pass through the origin. Then, for a given $\mu$-flat $\mathcal{F}^{(\mu)}$ in EG$^*(m, q)$, there are $n_3 \triangleq q(q^{m-\mu-1} - 1)/(q - 1)$ $(\mu + 1)$-flats in EG$^*(m, q)$ that intersect on $\mathcal{F}^{(\mu)}$. Let $\mathcal{F}_0^{(\mu+1)}, \mathcal{F}_1^{(\mu+1)}, \ldots, \mathcal{F}_{n_3-1}^{(\mu+1)}$ denote the $n_3$ $(\mu+1)$-flats that intersect on the given $\mu$-flat $\mathcal{F}^{(\mu)}$. These $n_3$ $(\mu+1)$-flats form an *intersecting bundle* with respect to $\mathcal{F}^{(\mu)}$. Since $\mathcal{F}^{(\mu)}$ corresponds to a coset of a $\mu$-dimensional subspace of the vector space $V$ of all the $m$-tuples over GF$(q)$, it contains $q^\mu$ points (represented by nonzero elements of GF$(q^m)$) of the following form:

$$\mathcal{F}^{(\mu)} = \left\{ \alpha^{j_0} + \sum_{k=1}^{\mu} \beta_k \alpha^{j_k} \colon \beta_k \in \mathrm{GF}(q), 1 \leq k \leq \mu \right\}, \tag{15.36}$$

where $\alpha$ is a primitive element of GF$(q^m)$, and $\alpha^{j_0}, \alpha^{j_1}, \ldots, \alpha^{j_\mu}$ represent $\mu+1$ linearly independent points in EG$(m, q)$. We denote the points on $\mathcal{F}^{(\mu)}$ by $\delta_0, \delta_1, \ldots, \delta_{q^\mu-1}$, where, for $0 \leq j < q^\mu$, $\delta_j$ is represented by a power of $\alpha$. Then, for $0 \leq i < n_3$, the $i$th $(\mu + 1)$-flat $\mathcal{F}_i^{(\mu+1)}$ that contains $\mathcal{F}^{(\mu)}$ consists of $q^{\mu+1}$ points of the following form:

$$\begin{aligned}\mathcal{F}_i^{(\mu+1)} &= \left\{ \alpha^{j_0} + \sum_{k=1}^{\mu} \beta_k \alpha^{j_k} + \beta_{\mu+1}\alpha^{j_{\mu+1,i}} \colon \beta_k, \beta_{\mu+1} \in \mathrm{GF}(q), 1 \leq k \leq \mu \right\} \\ &= \{\delta_j, \delta_j + \beta^0 \alpha^{j_{\mu+1,i}}, \ldots, \delta_j + \beta^{q-2}\alpha^{j_{\mu+1,i}} \colon \delta_j \in \mathcal{F}^{(\mu)}, 0 \leq j < q^\mu\},\end{aligned} \tag{15.37}$$

where (1) $\alpha^{j_{\mu+1,i}}$ is linearly independent of $\alpha^{j_0}, \alpha^{j_1}, \ldots, \alpha^{j_\mu}$; (2) $\alpha^{j_{\mu+1,i}}$ is not contained in any other $(\mu + 1)$-flat that contains $\mathcal{F}^{(\mu)}$; and (3) $\beta$ is a primitive element of GF$(q)$. From the first expression of (15.37), we see that $\mathcal{F}_i^{(\mu+1)}$ contains $\mathcal{F}^{(\mu)}$ and $q - 1$ other $\mu$-flats in EG$^*(m, q)$ parallel to $\mathcal{F}^{(\mu)}$.

For $0 \leq i < n_3$, on removing the points in $\mathcal{F}^{(\mu)}$ from $\mathcal{F}_i^{(\mu+1)}$, we obtain the following set of $n_4 \triangleq (q - 1)q^\mu$ points:

$$\begin{aligned}\bar{\mathcal{F}}_i^{(\mu+1)} &= \mathcal{F}_i^{(\mu+1)} \backslash \mathcal{F}^{(\mu)} \\ &= \{\delta_j + \beta^0 \alpha^{j_{\mu+1,i}}, \ldots, \delta_j + \beta^{q-2}\alpha^{j_{\mu+1,i}} \colon \delta_j \in \mathcal{F}^{(\mu)}, 0 \leq j < q^\mu\}.\end{aligned} \tag{15.38}$$

$\bar{\mathcal{F}}_i^{(\mu+1)}$ consists of $q - 1$ parallel $\mu$-flats in EG$^*(m, q)$ and is called a $(\mu, q - 1)$-frame with respect to $\mathcal{F}^{(\mu)}$. It is clear that the $n_3$ $(\mu, q-1)$-frames $\bar{\mathcal{F}}_0^{(\mu+1)}, \bar{\mathcal{F}}_1^{(\mu+1)}, \ldots, \bar{\mathcal{F}}_{n_3-1}^{(\mu+1)}$ are mutually disjoint. These $n_3$ $(\mu, q - 1)$-frames are said to form a *disjoint intersecting bundle* with respect to $\mathcal{F}^{(\mu)}$, denoted by $\mathbb{I}^{(\mu+1)}$. The bundle $\mathbb{I}^{(\mu+1)}$ contains $q^m - q^{\mu+1}$ distinct non-origin points of EG$^*(m, q)$.

Using the disjoint intersecting bundle $\mathbb{I}^{(\mu+1)}$ with respect to a given $\mu$-flat $\mathcal{F}^{(\mu)} = \{\delta_0, \delta_1, \ldots, \delta_{q^\mu-1}\}$, we form the following $n_3 \times n_4$ matrix over GF($q^m$):

$$\mathbf{W}_{q^m,\mathrm{EG,int},\mu} = \begin{bmatrix} \mathbf{B}_0 & \mathbf{B}_1 & \cdots & \mathbf{B}_{q^\mu-1} \end{bmatrix}, \tag{15.39}$$

where, for $0 \le j < q^\mu$, $\mathbf{B}_j$ is an $n_3 \times (q-1)$ matrix over GF($q^m$) defined as

$$\mathbf{B}_j = \begin{bmatrix} \delta_j + \beta^0\alpha^{j_{\mu+1,0}} & \delta_j + \beta^1\alpha^{j_{\mu+1,0}} & \cdots & \delta_j + \beta^{q-2}\alpha^{j_{\mu+1,0}} \\ \delta_j + \beta^0\alpha^{j_{\mu+1,1}} & \delta_j + \beta^1\alpha^{j_{\mu+1,1}} & \cdots & \delta_j + \beta^{q-2}\alpha^{j_{\mu+1,1}} \\ \vdots & \vdots & \ddots & \vdots \\ \delta_j + \beta^0\alpha^{j_{\mu+1,n_3-1}} & \delta_j + \beta^1\alpha^{j_{\mu+1,n_3-1}} & \cdots & \delta_j + \beta^{q-2}\alpha^{j_{\mu+1,n_3-1}} \end{bmatrix}. \tag{15.40}$$

The subscript "int" in $\mathbf{W}_{q^m,\mathrm{EG,int},\mu}$ stands for "intersecting." From (15.38) and (15.40), we readily see that, for $0 \le i < n_3$, the entries of the $i$th row of $\mathbf{W}_{q^m,\mathrm{EG,int},\mu}$ are simply the points on the $i$th $(\mu, q-1)$-frame $\bar{\mathcal{F}}_i^{(\mu+1)}$ in $\mathbb{I}^{(\mu+1)}$. Since the $(\mu, q-1)$-frames in $\mathbb{I}^{(\mu+1)}$ are mutually disjoint, any two rows (or two columns) of $\mathbf{W}_{q^m,\mathrm{EG,int},\mu}$ differ in all positions. Furthermore, all the entries in each row of $\mathbf{W}_{q^m,\mathrm{EG,int},\mu}$ are nonzero and distinct. From the structural properties of $\mathbb{I}^{(\mu+1)}$ and $\mathbf{W}_{q^m,\mathrm{EG,int},\mu}$, it can easily be proved that $\mathbf{W}_{q^m,\mathrm{EG,int},\mu}$ satisfies $\alpha$-multiplied row constraints 1 and 2.

On dispersing each entry in $\mathbf{W}_{q^m,\mathrm{EG,int},\mu}$ into an $\alpha$-multiplied $(q^m-1) \times (q^m-1)$ circulant permutation matrix over GF($q^m$), we obtain an $n_3 \times n_4$ array $\mathbf{H}_{q^m,\mathrm{EG,int},\mu}$ of $\alpha$-multiplied $(q^m-1) \times (q^m-1)$ CPMs,

$$\mathbf{H}_{q^m,\mathrm{EG,int},\mu} = \begin{bmatrix} \mathbf{A}_{0,0} & \mathbf{A}_{0,1} & \cdots & \mathbf{A}_{0,n_4-1} \\ \mathbf{A}_{1,0} & \mathbf{A}_{1,1} & \cdots & \mathbf{A}_{1,n_4-1} \\ \vdots & \vdots & \ddots & \vdots \\ \mathbf{A}_{n_3-1,0} & \mathbf{A}_{n_3-1,1} & \cdots & \mathbf{A}_{n_3-1,n_4-1} \end{bmatrix}. \tag{15.41}$$

Since all the entries of $\mathbf{W}_{q^m,\mathrm{EG,int},\mu}$ are nonzero and distinct, all the $\alpha$-multiplied CPMs in $\mathbf{H}_{q^m,\mathrm{EG,int},\mu}$ are distinct. The array $\mathbf{H}_{q^m,\mathrm{EG,int},\mu}$ is an $n_3(q^m-1) \times n_4(q^m-1)$ matrix over GF($q^m$) with column and row weights $n_3$ and $n_4$, respectively. Since $\mathbf{W}_{q^m,\mathrm{EG,int},\mu}$ satisfies $\alpha$-multiplied row constraints 1 and 2, $\mathbf{W}_{q^m,\mathrm{EG,int},\mu}$, as a matrix, satisfies the RC constraint.

For any pair $(g, r)$ of integers with $1 \le g \le n_3$ and $1 \le r \le n_4$, let $\mathbf{H}_{q^m,\mathrm{EG,int},\mu}(g, r)$ be a $g \times r$ subarray of $\mathbf{H}_{q^m,\mathrm{EG,int},\mu}$. $\mathbf{H}_{q^m,\mathrm{EG,int},\mu}(g, r)$ is a $g(q^m-1) \times r(q^m-1)$ matrix over GF($q^m$) with column and row weights $g$ and $r$, respectively. The null space over GF($q^m$) of $\mathbf{H}_{q^m,\mathrm{EG,int},\mu}(g, r)$ gives a $(g, r)$-regular $q^m$-ary QC-LDPC code of length $r(q^m-1)$ with rate at least $(r-g)/r$, whose Tanner graph has a girth of at least 6. The above construction based on the intersecting structure of flats of a Euclidean geometry over a finite field gives a family of $q^m$-ary QC-EG-LDPC codes.

A special case of the above construction is the case for which $\mu = 0$. For $\mu = 0$, a 0-flat is simply a non-origin point of EG$^*(m, q)$. Then the rows of $\mathbf{W}_{q^m,\mathrm{EG,int},0}$ are the points of the lines in EG$^*(m, q)$ that intersect at a given non-origin point $\alpha^j$ in EG$^*(m, q)$ with the point $\alpha^j$ excluded. The number of lines in EG$^*(m, q)$ that intersect at a point in EG$^*(m, q)$ is $n_3 \triangleq q(q^{m-1}-1)/(q-1)$. A $(0, q-1)$-frame with respect to a non-origin point in EG$^*(m, q)$ has $n_4 = q-1$ points. Based on the $n_3$ $(0, q-1)$-frames with respect to a non-origin point in $\mathrm{EG}^*(m,q)$, (15.39)–(15.41), we can construct an $n_3 \times (q-1)$ array $\mathbf{H}_{q^m,\mathrm{EG,int},\mu}$ of $\alpha$-multiplied $(q^m-1)\times(q^m-1)$ CPMs over $\mathrm{GF}(q^m)$. The null space of any subarray of $\mathbf{H}_{q^m,\mathrm{EG,int},0}$ gives a $q^m$-ary QC-LDPC code.

For $m \geq 2$, there are more rows in $\mathbf{H}_{q^m,\mathrm{EG,int},\mu}$ than there are columns. Let $\mathbf{H}^{\mathrm{T}}_{q^m,\mathrm{EG,int},\mu}$ be the transpose of $\mathbf{H}_{q^m,\mathrm{EG,int},\mu}$. Using the subarrays of $\mathbf{H}^{\mathrm{T}}_{q^m,\mathrm{EG,int},\mu}$, we can construct longer codes than we can using the subarrays of $\mathbf{H}_{q^m,\mathrm{EG,int},\mu}$.

---

**Example 15.11** Let the three-dimensional Euclidean geometry $\mathrm{EG}(3,2^2)$ over $\mathrm{GF}(2^2)$ be the code-construction geometry. The geometry contains 64 points. For any given 1-flat $\mathcal{F}^{(1)}$ (a line) in the subgeometry $\mathrm{EG}^*(3,2^2)$ of $\mathrm{EG}(3,2^2)$, there are four 2-flats intersecting on it. From these four intersecting 2-flats on $\mathcal{F}^{(1)}$, we can construct four $(1,3)$-frames with respect to $\mathcal{F}^{(1)}$. From these four $(1,3)$-frames, (15.39)–(15.41), we can construct a $4\times 12$ array $\mathbf{H}_{2^6,\mathrm{EG,int},1}$ of $\alpha$-multiplied $63\times 63$ CPMs over $\mathrm{GF}(2^6)$. Choose $k=4$ and $r=16$. The array $\mathbf{H}_{2^6,\mathrm{EG,int},1}(4,16)$ is simply the entire array $\mathbf{H}_{2^6,\mathrm{EG,int},1}$, which is a $252\times 756$ matrix over $\mathrm{GF}(2^6)$ with column and row weights 4 and 16, respectively. The null space over $\mathrm{GF}(2^6)$ of this matrix gives a $(4,16)$-regular 64-ary $(756,522)$ QC-EG-LDPC code with rate 0.69. The bit-, symbol-, and word-error performances of this code decoded using the FFT-QSPA with 50 iterations are shown in Figure 15.19, which also includes the word-error

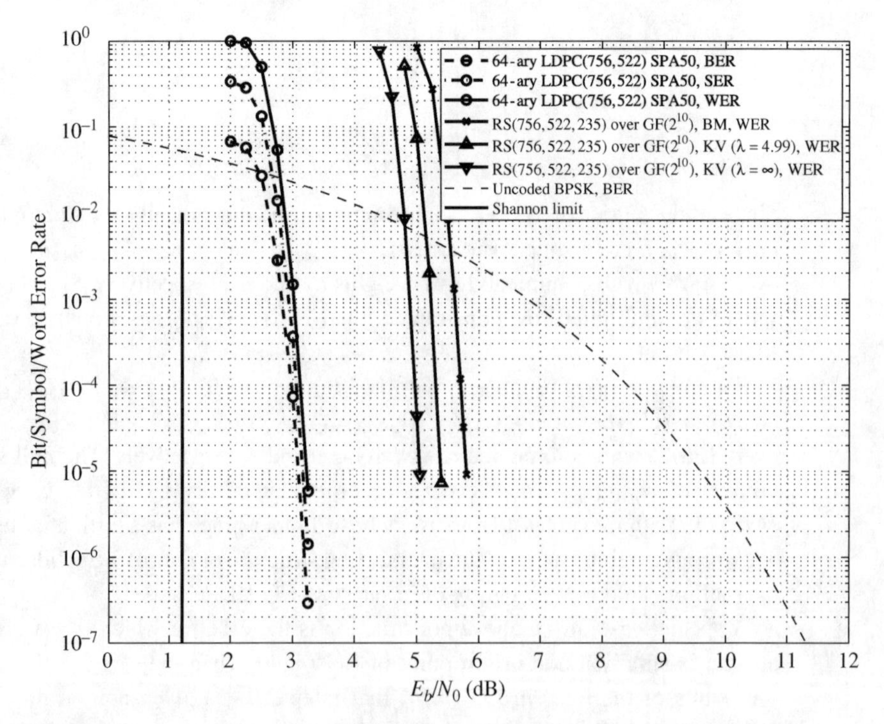


**Figure 15.19** Error performances of the 64-ary $(756,522)$ QC-EG-LDPC codes and the $(756,522,235)$ shortened RS code over $\mathrm{GF}(2^{10})$ given in Example 15.11.

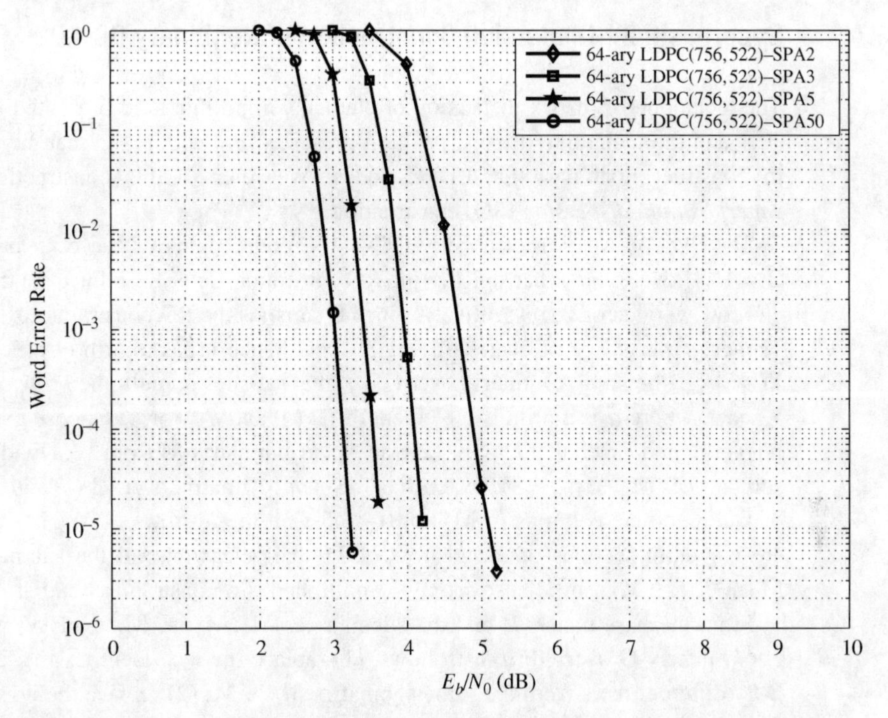


**Figure 15.20** The rate of decoding convergence of the FFT-QSPA for the 64-ary (756, 522) QC-EG-LDPC code given in Example 15.11.

performances of the (756, 522, 235) shortened RS code over GF($2^{10}$) decoded using the HD-BM algorithm and the ASD-KV algorithm with interpolation-complexity coefficients 4.99 and $\infty$, respectively. At a WER of $10^{-5}$, the 64-ary (756, 522) QC-LDPC code achieves a coding gain of 2.555 dB over the (756, 522, 235) shortened RS code over GF($2^{10}$) decoded with the HD-BM algorithm, while achieving coding gains of 2.15 and 1.8 dB over the RS code decoded using the ASD-KV algorithm with interpolation-complexity coefficients 4.99 and $\infty$, respectively. The iterative decoding of the (756, 522) QC-LDPC code also converges fast, as shown in Figure 15.20.

The number of computations required to decode the (756, 522) QC-LDPC code with 50 iterations of the FFT-QSPA is on the order of 77 414 400, while the number of computations required in order to carry out the interpolation step of the ASD-KV algorithm for decoding the (756, 522, 235) shortened RS code over GF($2^{10}$) with interpolation-complexity coefficient 4.99 is on the order of 146 313 216. So, the (756, 522) QC-LDPC code decoded with 50 iterations of the FFT-QSPA achieves a coding gain of 2.15 dB over the (756, 522, 235) shortened RS code decoded using the ASD-KV algorithm with interpolation-complexity coefficient 4.99 with fewer computations.

## 15.7 Superposition–Dispersion Construction of Nonbinary QC-LDPC Codes

The technique of matrix dispersion of elements in a finite field presented in Section 12.1 and the technique of superposition presented in Section 13.7 can be combined to construct nonbinary QC-LDPC codes. We refer to this construction as the *superposition–dispersion* (SD) construction.

The SD construction of a $q$-ary QC-LDPC code begins with a $c \times t$ base matrix $\mathbf{B} = \left[b_{i,j}\right]$ over GF(2) that satisfies the RC constraint. Let $w_{b,c}$ and $w_{b,r}$ be the column and row weights of $\mathbf{B}$, respectively. Since $\mathbf{B}$ satisfies the RC constraint, its associated Tanner graph is free of cycles of length 4 and hence it has a girth of at least 6. Let $\mathbf{W} = \left[w_{i,j}\right]$ be an $m \times n$ matrix over GF($q$) that satisfies $\alpha$-multiplied row constraints 1 and 2, where $\alpha$ is a primitive element of GF($q$) and $m \geqslant w_{b,c}$. Choose two positive integers $k$ and $l$ such that $kl \leq n$. Take an $m \times kl$ submatrix $\mathbf{M}$ from $\mathbf{W}$. Divide $\mathbf{M}$ into $l$ submatrices $\mathbf{M}_0, \mathbf{M}_1, \ldots, \mathbf{M}_{l-1}$, each being an $m \times k$ matrix over GF($q$) and consisting of $k$ consecutive columns of $\mathbf{M}$. For $0 \leq i < m$, the $i$th rows of $\mathbf{M}_0, \mathbf{M}_1, \ldots, \mathbf{M}_{l-1}$ are $k$-symbol sections of the $i$th row of $\mathbf{M}$. It is clear that all the submatrices $\mathbf{M}$, $\mathbf{M}_0, \mathbf{M}_1, \ldots, \mathbf{M}_{l-1}$ of $\mathbf{W}$ also satisfy $\alpha$-multiplied row constraints 1 and 2.

From the base matrix $\mathbf{B}$ and the submatrices $\mathbf{M}_0, \mathbf{M}_1, \ldots \mathbf{M}_{l-1}$, of $\mathbf{W}$, we form a $c \times kt$ matrix $\mathbf{Q}$ over GF($q$) as follows: (1) replace the $w_{b,c}$ 1-entries in a column of $\mathbf{B}$ by different rows from the same submatrix $\mathbf{M}_i$ of $\mathbf{M}$; (2) replace the $w_{b,r}$ 1-entries in a row of $\mathbf{B}$ by the rows of $\mathbf{M}_0, \mathbf{M}_1, \ldots, \mathbf{M}_{l-1}$ that are sections from the same row of matrix $\mathbf{M}$; and (3) replace each 0-entry of $\mathbf{B}$ by a row of $k$ zeros. The constraint on the replacement of 1-entries in a column of the base matrix $\mathbf{B}$ by the rows of a submatrix $\mathbf{M}_i$ of $\mathbf{M}$ is called the *column-replacement constraint*. The constraint on the replacement of 1-entries in a row of $\mathbf{B}$ by sections of a row in matrix $\mathbf{M}$ is called the *row-replacement constraint*. The RC constraint on the rows and columns of $\mathbf{B}$, the structural properties of $\mathbf{W}$ (see Section 12.2), and the constraints on the replacement of the 1-entries in the base matrix $\mathbf{B}$ by the rows of $\mathbf{M}_0, \mathbf{M}_1, \ldots, \mathbf{M}_{l-1}$ ensure that no two rows (or two columns) of $\mathbf{Q}$ have more than one position where they both have the same nonzero symbol in GF($q$), that is, two rows (or two columns) of $\mathbf{Q}$ have at most one position where they have identical nonzero symbols in GF($q$).

Next, we disperse each nonzero entry of $\mathbf{Q}$ into an $\alpha$-multiplied $(q-1) \times (q-1)$ CPM over GF($q$) and each zero entry into a $(q-1) \times (q-1)$ zero matrix. The matrix dispersions of the entries of $\mathbf{Q}$ result in a $c \times kt$ array $\mathbf{H}_{\text{sup,disp}}(c, kt)$ of $\alpha$-multiplied $(q-1)\times(q-1)$ circulant permutation and zero matrices, where "sup" and "disp" stand for superposition and dispersion, respectively. $\mathbf{H}_{\text{sup,disp}}(c, kt)$ is a $c(q-1) \times kt(q-1)$ matrix over GF($q$) that satisfies the RC constraint. If $\mathbf{M}$ does not contain zero entries, then the column and row weights of $\mathbf{H}_{\text{sup,disp}}(c, kt)$ are $w_{b,c}$ and $kw_{b,r}$, respectively. The null space over GF($q$) of $\mathbf{H}_{\text{sup,disp}}(c, kt)$ gives a $q$-ary ($w_{b,c}$,$w_{b,r}$)-regular QC-LDPC code. Any of the matrices $\mathbf{W}^{(1)}$ to $\mathbf{W}^{(8)}$ given in Chapter 12 can be used as $\mathbf{W}$ for SD construction of $q$-ary QC-LDPC codes.

---

**Example 15.12** Let $\mathbf{B}$ be the $8 \times 8$ circulant over GF(2) constructed from the type-1 binary incidence vectors of the eight lines in the two-dimensional Euclidean geometry EG(2, 3) over GF(3) not passing through the origin of EG(2, 3). The column and row weights of **B** are both 3. The matrix **B** is used as the base matrix for the SD construction of a $q$-ary QC-LDPC code. Let $\mathbf{W}^{(1)}$ be the $31 \times 31$ matrix over GF($2^5$) constructed from the (31, 2, 30) RS code over GF($2^5$) given in Example 14.6. We use $\mathbf{W}^{(1)}$ as the matrix **W** for the SD construction of a 32-ary QC-LDPC code. Choose $k = 5$ and $l = 6$. Take the first 30 columns of $\mathbf{W}^{(1)}$ to form the matrix **M** over GF($2^5$). Divide **M** into six $31 \times 5$ submatrices, $\mathbf{M}_0, \mathbf{M}_1, \ldots, \mathbf{M}_5$. Using the matrices $\mathbf{B}, \mathbf{M}_0, \mathbf{M}_1, \ldots, \mathbf{M}_5$ and the constraints on the replacement of the 1-entries in **B** by rows from $\mathbf{M}_0, \mathbf{M}_1, \ldots, \mathbf{M}_5$, we construct an $8 \times 40$ array $\mathbf{H}_{\text{sup,disp}}(8, 40)$ of $\alpha$-multiplied $31 \times 31$ circulant permutation and zero matrices over GF($2^5$). This array is a $248 \times 1240$ matrix over GF($2^5$) with column and row weights 3 and 15, respectively. The null space over GF($2^5$) gives a 32-ary (1240, 992) QC-LDPC code over GF($2^5$) with rate 0.8. The symbol- and word-error performances of this code over the binary-input AWGN channel decoded with 50 iterations of the FFT-QSPA are shown in Figure 15.21, which also includes the word performances of the (1240, 992, 249) shortened RS code over GF($2^{11}$) decoded with the HD-BM algorithm and the ASD-KV algorithm for comparison. We see that, at a WER of $10^{-5}$, the 32-ary (1240, 992) QC-LDPC code achieves a coding gain of 2.5 dB over the RS code decoded with the HD-BM algorithm, while achieving coding gains of 2.15 and 1.85 dB over the RS code decoded using the ASD-KV algorithm with interpolation-complexity coefficients 4.99 and $\infty$, respectively.

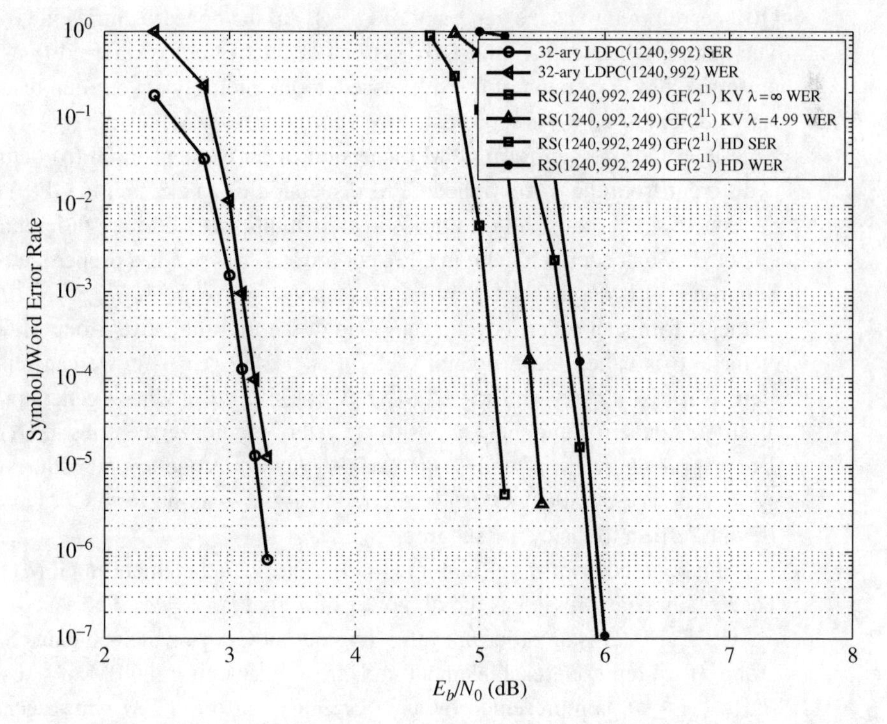


**Figure 15.21** Symbol- and word-error performances of the 32-ary (1240, 992) QC-LDPC code over the AWGN channel decoded with 50 iterations of the FFT-QSPA.

The number of computations required to decode the 32-ary (1240, 992) QC-LDPC code with 50 iterations of the FFT-QSPA is on the order of 29 760 000. However, the number of computations required to carry out the interpolation step of the ASD-KV algorithm with interpolation-complexity coefficient 4.99 in decoding of the (1240, 992, 249) shortened RS code is on the order of 393 625 600. See also [29–36].

## 15.8 Construction of Nonbinary LDPC Codes Based on RS Codes

In this section, two methods are presented for constructing nonbinary QC-LDPC codes based on the conventional parity-check matrices of RS codes in conjunction with nonbinary uniform CPM dispersions of elements of a finite field.

### 15.8.1 Nonbinary Uniform CPM Dispersions of Elements of a Finite Field

Let GF($q$) be a finite field with $q$ elements and let $\alpha$ be a primitive element of GF($q$). The powers of $\alpha$, namely $\alpha^{-\infty} = 0$, $\alpha^0 = 1$, $\alpha$, $\alpha^2, \ldots, \alpha^{q-2}$, give all the elements of GF($q$) and $\alpha^{q-1} = 1$. For $0 \leq l < q-1$, we represent the element $\alpha^l$ by a $q$-ary CPM, denoted by $\mathbf{Q}(\alpha^l)$, of size $(q-1) \times (q-1)$ (with rows and columns labeled from 0 to $q-2$, respectively), whose top row has $\alpha^l$ as its single nonzero component at position $l$ and all the other rows of $\mathbf{Q}(\alpha^l)$ are cyclic shifts of the top row. The top row of $\mathbf{Q}(\alpha^l)$ is referred to as the generator of $\mathbf{Q}(\alpha^l)$. All the nonzero entries in $\mathbf{Q}(\alpha^l)$ are $\alpha^l$. The mapping between $\alpha^l$ and $\mathbf{Q}(\alpha^l)$ is one-to-one. $\mathbf{Q}(\alpha^l)$ is referred to as the $q$-ary uniform CPM dispersion of $\alpha^l$ with respect to the multiplicative group of GF($q$). The zero element of GF($q$) is represented by a zero matrix (ZM) of size $(q-1) \times (q-1)$. Note that the $q$-ary uniform CPM dispersion of an element in GF($q$) defined above is different from the $\alpha$-multiplied CPM dispersion of an element in GF($q$) defined in Section 15.4.1, because all the entries in an $\alpha$-multiplied CPM are different.

Let $\beta$ be a nonzero element in GF($q$) of order $n$, where $n$ is a proper factor of $q-1$. Suppose $q-1 = cn$. Then $\beta = \alpha^c$. The set $\mathbf{S}_n = \{\beta^0 = 1, \beta^1, \beta^2, \ldots, \beta^{n-1}\}$ of $n$ elements forms an order-$n$ cyclic subgroup of the multiplicative group of GF($q$). The element $\beta$ is called the generator of $\mathbf{S}_n$. For $0 \leq j \leq n-1$, we can represent the element $\beta^j$ by a $q$-ary uniform CPM $\mathbf{Q}(\beta^j)$ of size $n \times n$ whose generator has $\beta^j$ as its single nonzero component at position $j$. All of the nonzero entries in $\mathbf{Q}(\beta^j)$ are $\beta^j$. The mapping between $\beta^j$ and $\mathbf{Q}(\beta^j)$ for $0 \leq j \leq n-1$ is one-to-one with respect to the cyclic group $\mathbf{S}_n$ of GF($q$). $\mathbf{Q}(\beta^j)$ is referred to as the $q$-ary uniform CPM dispersion of $\beta^j$ with respect to the cyclic subgroup $\mathbf{S}_n$.

Let $r$ be a factor of $q-1$ and let $r = ln$. Let $\delta$ be an element in GF($q$) of order $r$. Then $\beta$ can be expressed as the $l$th power of $\delta$, that is, $\beta = \delta^l$. The set $\mathbf{S}_r = \{\delta^0 = 1, \delta^1, \delta^2, \ldots, \delta^{ln-1}\}$ is an order-$ln$ cyclic subgroup of GF($q$), which contains $\mathbf{S}_n$ as a subgroup. If we represent each element in $\mathbf{S}_r$ by a $q$-ary uniform CPM of size $r \times r$, then $\beta^i$, $0 \leq i < n$, is represented by an $r \times r$ $q$-ary uniform CPM whose generator has $\beta^i = \delta^i l$ as its single nonzero component at the location $il$. In this case, every element in $\mathbf{S}_n$ is dispersed into a $q$-ary uniform CPM of size $r \times r$, and the dispersion is referred to as uniform CPM dispersion with respect to the cyclic group $\mathbf{S}_r$, a supergroup of $\mathbf{S}_n$.

Suppose GF($q$) is a subfield of GF($q'$) with $q' > q$. Let $\zeta$ be a primitive element in GF($q'$). Then a nonzero element $\omega$ in GF($q$) can be expressed as a power of $\zeta$, say $\omega = \zeta^l$, with $0 \leq l \leq q' - 2$. With respect to the larger field GF($q'$), we can disperse $\omega$ into a $q'$-ary uniform CPM $\mathbf{Q}(\omega)$ of size $(q' - 1) \times (q' - 1)$ whose generator has $\omega$ as its single nonzero component at location $l$. Again, the mapping between $\omega$ and $\mathbf{Q}(\omega)$ is one-to-one. In this case, an element of GF($q$) is dispersed into a nonbinary uniform CPM over a larger field.

The above shows that an element in a field GF($q$) can be one-to-one dispersed (or mapped) into a $q$-ary uniform CPM of size equal to, smaller, or larger than $q - 1$. This one-to-one $q$-ary uniform CPM dispersion of a field element in GF($q$) will be used to construct $q$-ary QC-LDPC codes based on the conventional parity-check matrices of RS codes over GF($q$) in the next two subsections. These codes are called $q$-ary QC-RS-LDPC codes. In the constructions, the fields used are of characteristic 2.

### 15.8.2 Construction of $q$-ary QC-RS-LDPC Codes: Method 1

Let $\alpha$ be a primitive element in the field GF($2^s$) and $n$ be a factor of $2^s - 1$ such that $2^s - 1 = cn$, with $n > 1$. Let $p_s$ be the smallest prime factor of $n$ and let $\beta = \alpha^c$. The order of $\beta$ is $n$ and the set $\mathbf{S}_n = \{1, \beta, \ldots, \beta^{n-1}\}$ forms a cyclic subgroup of the multiplicative group of GF($2^s$). Let $m$ be a positive integer less than or equal to $p_s$ that is, $1 \leq m \leq p_s$. Consider the $(n, k, d_{\min}) = (n, n - m, m + 1)$ RS code $C_{RS}$ over GF($2^s$) of length $n$ with minimum distance $m + 1$, whose generator polynomial $g(X)$ has $\beta, \beta^2, \ldots, \beta^m$ as roots. The conventional parity-check matrix of this RS code in terms of its roots is given as follows (see Chapter 3):

$$\mathbf{W}_{RS}(m, n) = \begin{bmatrix} 1 & \beta & \beta^2 & \cdots & \beta^{n-1} \\ 1 & \beta^2 & (\beta^2)^2 & \cdots & (\beta^2)^{n-1} \\ \vdots & \vdots & \vdots & & \vdots \\ 1 & \beta^m & (\beta^m)^2 & \cdots & (\beta^m)^{n-1} \end{bmatrix}. \tag{15.42}$$

If $n = 2^s - 1$, the RS code generated by $\mathbf{W}_{RS}(m, n)$ is a *primitive* RS code of length $2^s - 1$, otherwise, it is a *non-primitive* RS code.

The matrix $\mathbf{W}_{RS}(m, n)$ satisfies the $2 \times 2$ SM constraint. Since $\mathbf{W}_{RS}(m, n)$ satisfies the $2 \times 2$ SM constraint, it can be used as the base matrix to construct a $2^s$-ary QC-LDPC code using the $2^s$-ary uniform CPM dispersion with respect to the cyclic group $\mathbf{S}_n$ (or a supergroup of $\mathbf{S}_n$). The $2^s$-ary uniform CPM dispersion of $\mathbf{W}_{RS}(m, n)$ gives an $m \times n$ array $\mathbf{H}_{2^s,RS,\text{disp}}(m, n)$ of $2^s$-ary uniform CPMs of size $n \times n$. The null space over GF($2^s$) of $\mathbf{H}_{2^s,RS,disp}(m, n)$ gives an $(m, n)$-regular $2^s$-ary QC-RS-LDPC code $C_{2^s,RS,\text{disp}}(m, n)$ whose Tanner graph has girth at least 6.

Let $d$ and $r$ be two positive integers such than $1 < d < m$ and $d < r \leq n$. Let $\mathbf{W}_{RS}(d, r)$ be a $d \times r$ submatrix of $\mathbf{W}_{RS}(d, r)$. $\mathbf{W}_{RS}(d, r)$ also satisfies the $2 \times 2$ SM constraint. Hence, it can be used as a base matrix to construct a $2^s$-ary QC-RS-LDPC using $2^s$-ary uniform CPM dispersion.

**Example 15.13** Let GF($2^6$) be the field for code construction and let $\alpha$ be a primitive element of GF($2^6$). Set $n = 2^6 - 1 = 63$, the order of $\alpha$. In this case, $\beta = \alpha$. The smallest prime factor $p_s$ of 63 is 3. Set $m = 3$. Then the parity-check matrix $\mathbf{W}_{RS}(3, 63)$ of the (63, 60, 4) RS code over GF($2^6$) is a $3 \times 63$ matrix over GF($2^6$), which satisfies the $2 \times 2$ SM constraint. The 64-ary uniform CPM dispersion of $\mathbf{W}_{RS}(3, 63)$ with respect to the multiplicative group of GF($2^6$) gives a $3 \times 63$ array $\mathbf{H}_{2^6,RS,\text{disp}}(3, 63)$ of 64-ary uniform CPMs of size $63 \times 63$, which is a $189 \times 3969$ matrix over GF($2^6$) with column and row weights 3 and 63, respectively. The null space over GF($2^6$) of $\mathbf{H}_{2^6,RS,\text{disp}}(3, 63)$ gives a 64-ary (3, 60)-regular (3969, 3780) QC-RS-LDPC code $C_{2^6,RS,qc}(3, 63)$ with rate 0.9524. The symbol- and word-error performances of the $C_{2^6,RS,qc}(3, 63)$ code over the AWGN channel decoded with 50 iterations of the FFT-QSPA are shown in Figure 15.22.

Note that $2^6 - 1 = 63$ has a factor 21. Set $n = 21$ and let $\beta = \alpha^3$. Then the order of $\beta$ is 21 and the set $\mathbf{S}_{21} = \{1, \beta, \beta^2, \ldots, \beta^{20}\}$ generated by $\beta$ forms an order-21 cyclic subgroup of the multiplicative group of GF($2^6$). The smallest prime factor $p_s$ of 21 is 3. Set $m = 3$. Consider the (21, 18, 4) RS code of length 21 whose generator $g(X)$ has $\beta$, $\beta^2$, and $\beta^3$ as roots. The parity-check matrix $\mathbf{W}_{RS}(3, 21)$ of the RS code is a $3 \times 21$, matrix over the cyclic group $\mathbf{S}_{21}$.

The $2^6$-ary uniform CPM dispersion of $\mathbf{W}_{RS}(3, 21)$ with respect to $\mathbf{S}_{21}$ gives a $3 \times 21$ array $\mathbf{H}_{2^6,RS,\text{disp}}(3, 21)$ of $2^6$-ary uniform CPMs of size $21 \times 21$, which is a $63 \times 441$ matrix over $\mathbf{S}_{21}$. The null space over GF($2^6$) of $\mathbf{H}_{2^6,RS,\text{disp}}(3, 21)$ gives a 64-ary (3, 21)-regular (441, 378) QC-RS-LDPC code $C_{2^6,RS,qc}(3, 63)$ of length 441 with rate 0.875. The symbol- and word-error performances of $C_{2^6,RS,qc}(3, 63)$ over the AWGN channel decoded with 50 iterations of the FFT-QSPA are also shown in Figure 15.22.

**Example 15.14** In this example, we use GF($2^8$) as the field for code construction. Note that $2^8 - 1 = 255$ can be factored as the product of three prime factors, 3, 5, and 17, that is, $255 = 3 \times 5 \times 17$. Set $c = 3$ and $n = 85 = 5 \times 17$. The smallest prime factor $p_s$ of 85 is 5. Let $\alpha$ be a primitive element of GF($2^8$) and let $\beta = \alpha^3$. The order of $\beta$ is then 85. Consider the (85, 80, 6) RS code over GF($2^8$) whose generator polynomial has $\beta, \beta^2, \beta^3, \beta^4$, and $\beta^5$ as roots. The parity-check matrix $\mathbf{W}_{RS}(5, 85)$ of this code is a $5 \times 85$ matrix over GF($2^8$) that satisfies the $2 \times 2$ SM constraint. All the entries in $\mathbf{W}_{RS}(5, 85)$ are elements in the cyclic subgroup $\mathbf{S}_{85}$ of GF($2^8$) generated by $\beta$.

Dispersing each entry of $\mathbf{W}_{RS}(5, 85)$ into a $2^8$-ary uniform CPM of size $85 \times 85$, we obtain a $5 \times 85$ array $\mathbf{H}_{2^8,RS,\text{disp}}(5, 85)$ of $2^8$-ary uniform CPMs of size $85 \times 85$, which is a $255 \times 7225$ matrix over GF($2^8$) with constant column and row weights 5 and 85, respectively. The null space over GF($2^8$) of $\mathbf{H}_{2^8,RS,\text{disp}}(5, 85)$ gives a $2^8$-ary (5, 85)-regular (7225, 6804) QC-RS-LDPC code $C_{2^8,RS,qc}(5, 85)$ of length 7225 with rate 0.9417. The symbol- and word-error performances of $C_{2^8,RS,qc}(5, 85)$ over the AWGN channel decoded with 50 iterations of the FFT-QSPA are shown in Figure 15.23.

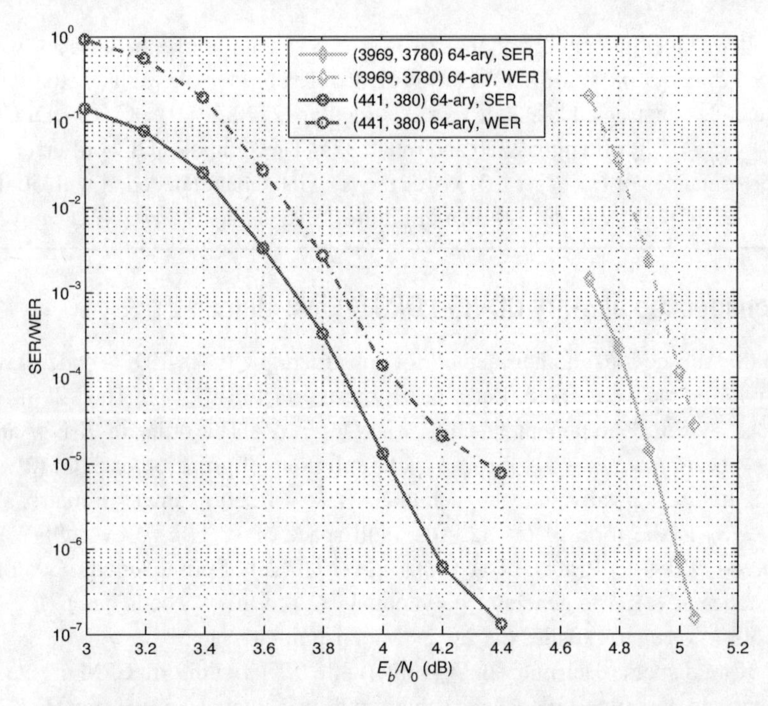


**Figure 15.22** The error performances of the two 64-ary QC-RS-LDPC codes given in Example 15.13.

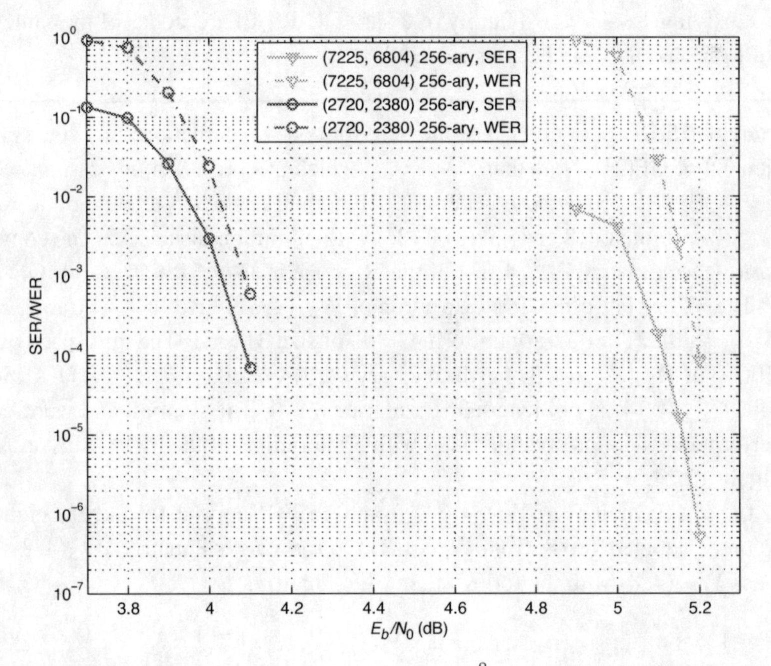


**Figure 15.23** The error performances of the two $2^8$-ary QC-RS-LDPC codes given in Example 15.14.

If we take the first 32 columns of $\mathbf{W}_{RS}(5, 85)$ and remove the last row, we obtain a $4 \times 32$ submatrix $\mathbf{W}_{RS}(4, 32)$ of $\mathbf{W}_{RS}(5, 85)$. Using $\mathbf{W}_{RS}(4, 32)$ as the base matrix, the code constructed is a $2^8$-ary $(4, 32)$-regular $(2720, 2380)$ QC-RS-LDPC code $C_{2^8,RS,qc}(4, 32)$ of length 2720 and rate 0.875. The symbol- and word-error performances of $C_{2^8,RS,qc}(4, 32)$ over the AWGN channel decoded with 50 iterations of the FFT-QSPA are also shown in Figure 15.23.

### 15.8.3 Construction of $q$-ary QC-RS-LDPC Codes: Method 2

In this subsection, another method for constructing QC-RS-LDPC codes based on the field GF($2^s$) is presented. Let $n$ be any prime factor of $2^s-1$. Then $2^s-1 = cn$. Let $\alpha$ be a primitive element of GF($2^s$) and $\beta = \alpha^c$. The order of $\beta$ is $n$ and the set $\mathbf{S}_n = \{1, \beta, \beta^2, \ldots, \beta^{n-1}\}$ forms a cyclic subgroup of prime order of GF($2^s$).

Let $m$ be a positive integer such that $1 \leq m < n$. Form an $m \times n$ matrix $\mathbf{W}_{RS}(m, n)$ over $\mathbf{S}_n$ in the form of (15.42). The null space of $\mathbf{W}_{RS}(m, n)$ over GF($2^s$) gives an $(n, n-m, m+1)$ RS code $C_{RS,qc}(m, n)$ over GF($2^s$) of prime length $n$ with minimum distance $m + 1$. The generator polynomial $g(X)$ of this RS code has $\beta, \beta^2, \ldots, \beta^m$ as roots. $\mathbf{W}_{RS}(m, n)$ satisfies the $2 \times 2$ SM constraint.

If we disperse each entry in $\mathbf{W}_{RS}(m, n)$ into a $2^s$-ary uniform CPM of size $n \times n$ with respect to the prime-order cyclic subgroup $\mathbf{S}_n$, we obtain an $m \times n$ array $\mathbf{H}_{2^s,RS,\mathrm{disp}}(m, n)$ of $2^s$-ary uniform CPMs of size $n \times n$. The null space of $\mathbf{H}_{2^s,RS,\mathrm{disp}}(m, n)$ over GF($2^s$) gives a $2^s$-ary QC-RS-LDPC code $C_{2^s,RS,qc}(m, n)$ of length $n^2$. For various choices of $m$ satisfying $1 < m < n$, a family of $2^s$-ary QC-RS-LDPC codes of the same length $n^2$ with various rates can be constructed.

**Example 15.15** Let GF($2^5$) be the field for code construction and $\alpha$ be a primitive element of GF($2^5$). Note that $2^5-1 = 31$, which is a prime. In this case, $n = 31$. Set $m = 4$. The RS code whose generator polynomial has $\alpha, \alpha^2, \alpha^3$, and $\alpha^4$ as roots is a $(31, 27)$ primitive RS code over GF($2^5$) with minimum distance 5. The conventional parity-check matrix $\mathbf{W}_{RS}(4, 31)$ of this RS code is a $4 \times 31$ matrix over GF($2^5$).

The 32-ary uniform CPM dispersion of $\mathbf{W}_{RS}(4, 31)$ gives a $4 \times 31$ array $\mathbf{H}_{2^5,RS,\mathrm{disp}}(4, 31)$ of 32-ary uniform CPMs of size $31 \times 31$. The null space over GF($2^5$) of $\mathbf{H}_{2^5,RS,\mathrm{disp}}(4, 31)$ gives a $2^5$-ary $(4, 31)$-regular $(961, 840)$ QC-RS-LDPC code $C_{2^5,RS,qc}(4, 31)$ of length 961 with rate 0.874. The symbol- and word-error performances of this code decoded with 50 iterations of the FFT-QSPA are shown in Figure 15.24.

Label the columns of $\mathbf{W}_{RS}(4, 31)$ from 0 to 30. Suppose we take the eight columns, labeled 1 to 8, from $\mathbf{W}_{RS}(4, 31)$ to form a $4 \times 8$ matrix $\mathbf{W}_{RS}(4, 8)$. Next, we mask $\mathbf{W}_{RS}(4, 8)$ with the following masking matrix:

$$\mathbf{Z}(4, 8) = \begin{bmatrix} 1 & 0 & 1 & 0 & 1 & 1 & 1 & 1 \\ 0 & 1 & 0 & 1 & 1 & 1 & 1 & 1 \\ 1 & 1 & 1 & 1 & 1 & 0 & 1 & 0 \\ 1 & 1 & 1 & 1 & 0 & 1 & 0 & 1 \end{bmatrix}. \tag{15.43}$$

Masking $\mathbf{W}_{RS}(4,8)$ with $\mathbf{Z}(4,8)$ results in a $4 \times 8$ masked matrix $\mathbf{W}_{RS,\text{mask}}(4,8)$ with column and row weights 3 and 6, respectively. The $2^5$-ary uniform CPM dispersion of $\mathbf{W}_{RS,\text{mask}}(4,8)$ with respect to the multiplicative group of GF($2^5$) gives a $4 \times 8$ array $\mathbf{H}_{2^5,RS,\text{disp,mask}}(4,8)$ of $2^5$-ary uniform CPMs and zero matrices of size $31 \times 31$, which is a $124 \times 248$ matrix over GF($2^5$) with column and row weights 3 and 6, respectively.

The null space over GF($2^5$) of $\mathbf{H}_{2^5,RS,\text{disp,mask}}(4,8)$ gives a 32-ary (3, 6)-regular (248, 124) QC-RS-LDPC code $C_{2^5,S,qc,\text{mask}}(4,8)$ with rate 0.5. The symbol- and word-error performances of this code decoded with 50 iterations of the FFT-QSPA are also shown in Figure 15.24.

---

**Example 15.16** In this example, the field GF($2^8$) is used to construct $2^8$-ary QC-RS-LDPC codes. Note that $2^8 - 1 = 255$ can be factored as the product of 17 and 15 with 17 a prime. Set $n = 17, m = 4$, and $\beta = \alpha^{15}$, where $\alpha$ is a primitive element of GF($2^8$). With these settings, we can construct the $4 \times 17$ parity-check matrix $\mathbf{W}_{RS}(4,17)$ of a (17, 13, 5) RS code $C_{RS,qc}(4,17)$ over GF($2^8$) with entries from the order-17 cyclic subgroup $\mathbf{S}_{17}$ of GF($2^8$) generated by $\beta$, which satisfies the RC constraint.

The $2^8$-ary uniform CPM dispersion of $\mathbf{W}_{RS}(4,17)$ with respect to the cyclic subgroup $\mathbf{S}_{17}$ of GF($2^8$) gives a $4 \times 17$ array $\mathbf{H}_{2^8,RS,\text{disp}}(4,17)$ of $2^8$-ary uniform CPMs of size $17 \times 17$, which is a $68 \times 289$ matrix over GF($2^8$) with column and row weights 4 and 17, respectively. The null space over GF($2^8$) of $\mathbf{H}_{2^8,RS,\text{disp}}(4,17)$ gives a $2^8$-ary (4, 17)-regular (289, 221) QC-RS-LDPC code $C_{2^8,RS,qc}(4,17)$ of length 289

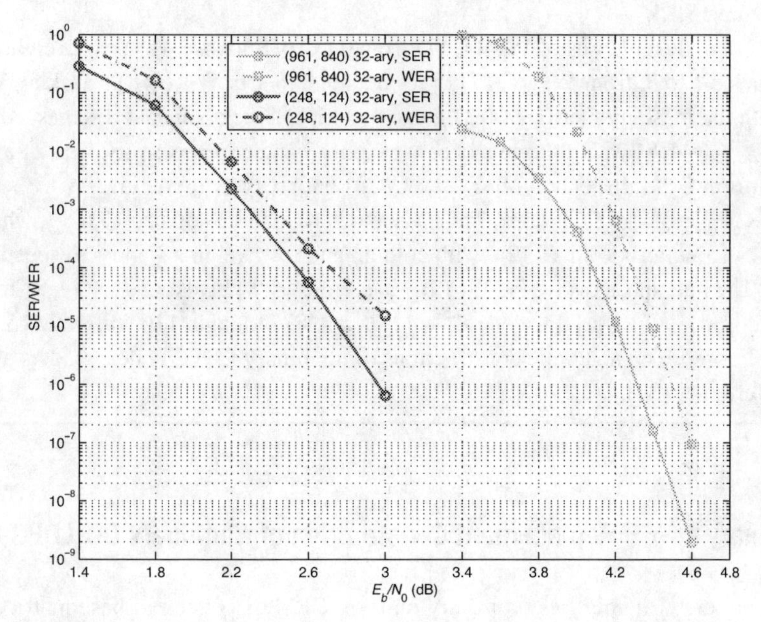


**Figure 15.24** The error performances of the two 32-ary QC-RS-LDPC codes given in Example 15.15.

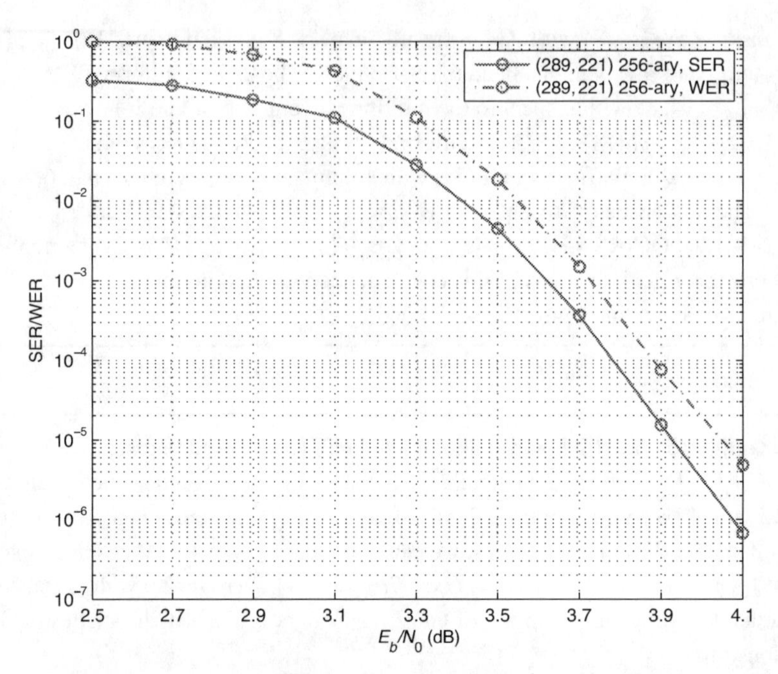


**Figure 15.25** The error performance of the (289, 221) $2^8$-ary QC-RS-LDPC codes given in Example 15.16.

with rate 0.7647. The symbol- and word-error performances of $C_{2^8,RS,qc}(4,17)$ over the AWGN channel decoded with 50 iterations of the FFT-QSPA are shown in Figure 15.25.

Label the columns of $\mathbf{W}_{RS}(4,17)$ from 0 to 16. Suppose we take the eight columns, labeled 1 to 8, from $\mathbf{W}_{RS}(4,17)$ to from a $4 \times 8$ matrix $\mathbf{W}_{RS}(4,8)$. Masking $\mathbf{W}_{RS}(4,8)$ with the masking matrix $\mathbf{Z}(4,8)$ given by (15.43), we obtain a $4 \times 8$ masked matrix $\mathbf{W}_{RS,\text{mask}}(4,8)$ with column and row weights 3 and 6, respectively. The $2^8$-ary uniform CPM dispersion of $\mathbf{W}_{RS,\text{mask}}(4,8)$ with respect to $\mathbf{S}_{17}$ gives a $4 \times 8$ array $\mathbf{H}_{2^8,RS,\text{disp,mask}}(4,8)$ of uniform CPMs and zero matrices of size $17 \times 17$, which is a $68 \times 136$ matrix over GF($2^8$) with column and row weights 3 and 6, respectively.

The null space over GF($2^8$) of $\mathbf{H}_{2^8,RS,\text{disp,mask}}(4,8)$ gives a $2^8$-ary $(4,8)$-regular $(136,68)$ QC-RS-LDPC code $C_{2^8,RS,qc,\text{mask}}(4,8)$ of length 136 with rate 0.5.

For more on algebraic constructions of nonbinary LDPC codes, readers are referred to [37].

## 15.9 Binary-to-NB Replacement Construction of Nonbinary QC-LDPC Codes

Using either $\alpha$-multiplied or $q$-ary uniform CPM dispersion of base matrices to construct nonbinary QC-LDPC codes, if the sizes of the code symbol field and CPMs are too big, the decoding computational complexity will be enormous. To mitigate this problem, we can combine the binary CPM dispersion with a *binary-to-NB (B-to-NB) replacement* method to construct a nonbinary QC-LDPC code whose parity-check array either consists of CPMs of large size over a small symbol field or consists of CPMs of small size over a large symbol field.

First, a chosen $m \times n$ RS base matrix $\mathbf{W}(m,n) = [b_{i,j}]_{0 \le i < m, 0 \le j < n}$ over GF($q$) is dispersed into an $m \times n$ array $\mathbf{H}_{b,\text{disp}}(m,n) = [\mathbf{A}_{i,j}]_{0 \le i < m, 0 \le j < n}$ of binary CPMs and/or ZMs of size, say, $n \times n$. Then, for $0 \le i < m$ and $0 \le j < n$, we replace all the 1-entries in the constituent binary CPM $\mathbf{A}_{i,j}$ of $\mathbf{H}_{b,\text{disp}}(m,n)$ by the same nonzero element from a chosen code symbol field GF($q^*$). The field GF($q^*$) for code symbols can be smaller than or larger than the base field GF($q$), or the same. This binary to $q^*$-ary replacement operation transforms the binary CPM $\mathbf{A}_{i,j}$ into a $q^*$-ary uniform CPM $\mathbf{Q}_{i,j}$. Applying this binary-to-$q^*$-ary replacement to each constituent binary CPM in the binary array $\mathbf{H}_{b,\text{disp}}(m,n)$, we obtain an $m \times n$ array $\mathbf{H}_{q^*,\text{disp}}(m,n) = [\mathbf{Q}_{i,j}]_{0 \le i < m, 0 \le j < n}$ of $q^*$-ary uniform CPMs of size $n \times n$. The null space over GF($q^*$) of $\mathbf{H}_{q^*,\text{disp}}(m,n)$ gives a $q^*$-ary QC-LDPC code $C_{q^*,qc}$. The above construction of a nonbinary QC-LDPC code is referred to as binary-to-nonbinary (B-to-NB) replacement construction.

Using the B-to-NB replacement construction, we can construct an array of $q^*$-ary uniform CPMs of small size over a large code symbol field, denoted by $\mathbf{H}_{q^*,\text{disp}}(m,n)$. The null space over GF($q^*$) of $\mathbf{H}_{q^*,\text{disp}}(m,n)$ gives a nonbinary QC-LDPC code over a large field. Or we can construct an array $\mathbf{H}_{q^*,\text{disp}}(m,n)$ of $q^*$-ary uniform CPMs of large size over a small code symbol field. Then the null space of $\mathbf{H}_{q^*,\text{disp}}(m,n)$ gives a nonbinary QC-LDPC code over a small field.

The EG-base matrix given by (11.29), the base matrices constructed based on finite fields presented in Chapter 12, and the base matrices constructed based on the conventional parity-check matrices of RS codes presented in the last section can be used for binary CPM dispersion.

In the following, several examples are given to illustrate the B-to-NB replacement construction of nonbinary codes. In these examples, fields of characteristic 2 are used as symbol fields and nonbinary uniform CPMs are used to construct parity-check arrays.

---

**Example 15.17** Suppose we construct a $(7,4)$ RS code over GF($2^3$) whose generator polynomial has $\alpha$, $\alpha^2$, and $\alpha^3$ as roots, where $\alpha$ is a primitive element of GF($2^3$). Its parity-check matrix $\mathbf{W}_{RS}(3,7)$ is a $3 \times 7$ matrix over GF($2^3$), which satisfies the $2 \times 2$ constraint. Dispersing $\mathbf{W}_{RS}(3,7)$ into a $3 \times 7$ array $\mathbf{H}_{b,RS,\text{disp}}(3,7)$ of binary CPMs of size $7 \times 7$, next, we transform the binary array $\mathbf{H}_{b,RS,disp}(3,7)$ into an array over a larger field, say GF($2^8$), using the B-to-NB replacement (random replacement). The replacement results in a $3 \times 7$ array $\mathbf{H}_{2^8,RS,\text{disp}}(3,7)$ of $2^8$-ary uniform CPMs of size $7 \times 7$ over GF($2^8$). The null space over GF($2^8$) of $\mathbf{H}_{b,RS,\text{disp}}(3,7)$ gives a $2^8$-ary $(3,7)$-regular $(49,28)$ QC-RS-LDPC $C_{2^8,RS,qc}$ of length 49 with rate 0.5714. It is a short code over a large field. The symbol- and word-error performances of the code are shown in Figure 15.26.

---

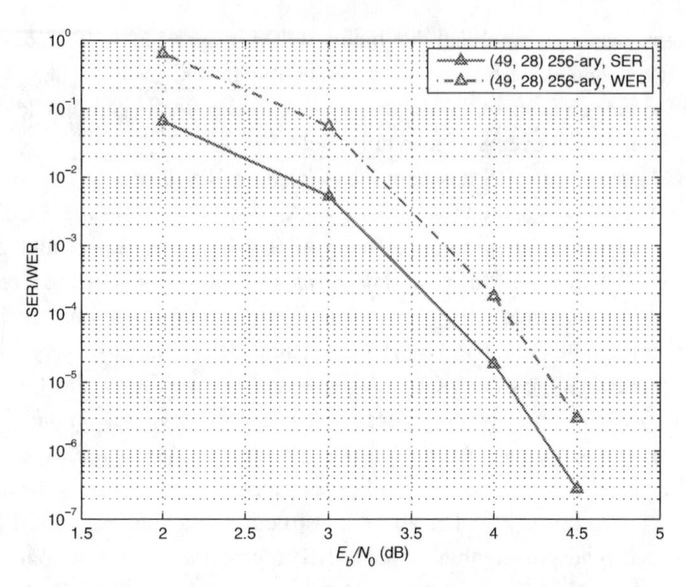


**Figure 15.26** The error performance of the $2^8$-ary QC-RS-LDPC code given in Example 15.17.

---

**Example 15.18** In this example, we use the RS-matrix $\mathbf{W}_{RS}(5, 85)$ over GF($2^8$) constructed in Example 15.14 as the base matrix to construct a nonbinary QC-RS-LDPC code with the B-to-NB replacement method.

We now disperse $\mathbf{W}_{RS}(5, 85)$ into a $5 \times 85$ array $\mathbf{H}_{b,RS,\text{disp}}(5, 85)$ of binary CPMs of size $85 \times 85$. Let GF($2^4$) be the field for the B-to-NB replacement. We replace each binary CPM in $\mathbf{H}_{b,RS,\text{disp}}(5, 85)$ by a 16-ary uniform CPM (random replacement). The replacement results in a $5 \times 85$ array $\mathbf{H}_{2^4,RS,\text{disp}}(5, 85)$ of 16-ary uniform CPMs of size $85 \times 85$ over GF($2^4$). The null space over GF($2^4$) of $\mathbf{H}_{2^4,RS,\text{disp}}(5, 85)$ gives a 16-ary (5, 85)-regular (7225, 6804) QC-RS-LDPC code $C_{2^4,RS,qc}(5, 85)$ of length 7225 with rate 0.941. It is a long code over a small field. The symbol- and word-performances of this code over the AWGN channel decoded with 50 iterations of the FFT-QSPA are shown in Figure 15.27.

Suppose we use the $4 \times 8$ masked matrix $\mathbf{W}_{RS,\text{mask}}(4, 8)$ over GF($2^8$) constructed in Example 15.16 as the base matrix for B-to-NB replacement construction with GF($2^4$) as the symbol field. The code constructed is a $2^4$-ary (4, 8)-regular (136, 68) QC-RS-LDPC code $C_{2^4,RS,qc,\text{mask}}(4, 8)$ of length 136 with rate 0.5.

---

---

**Example 15.19** In this example, the method presented in Section 12.9 is used to construct a $2 \times 2$ SM-constrained base matrix $\mathbf{W}_s^{(9)}$ for B-to-NB replacement construction of two nonbinary QC-LDPC codes. The prime field GF(17) is used to construct the base matrix and GF($2^4$) is used as the symbol field.

Let $\alpha$ be a primitive element in GF(17). Form a $4 \times 16$ base matrix $\mathbf{W}_s^{(9)}(4, 16)$ over GF(17) in the form of (12.30) using two subsets $\mathbf{S}_0 = \{1, \alpha, \alpha^2, \alpha^3\}$ and

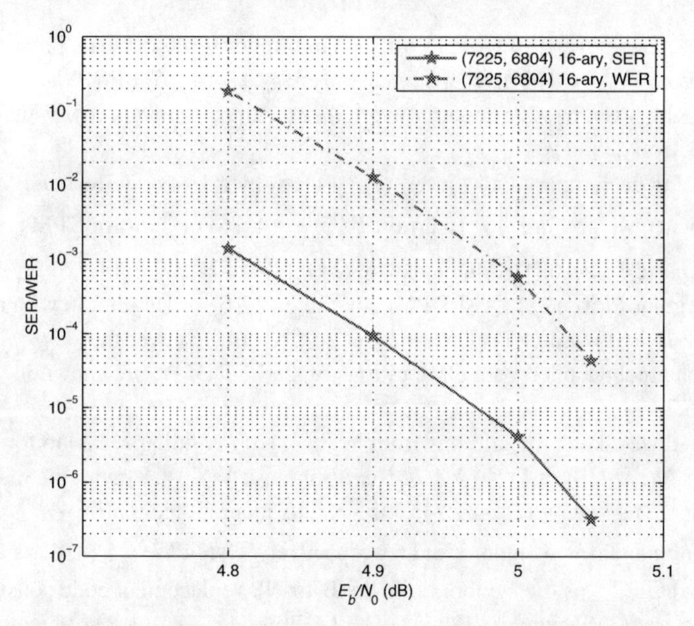


**Figure 15.27** The error performances of the two $2^4$-ary QC-RS-LDPC codes given in Example 15.18.

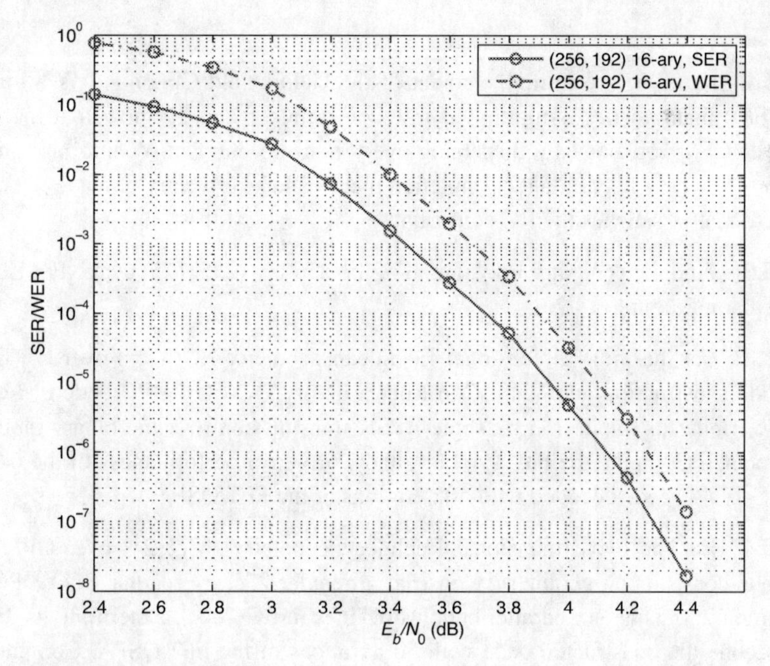


**Figure 15.28** The error performance of the (256, 192) $2^4$-ary QC-RS-LDPC code given in Example 15.19.

$\mathbf{S}_1 = \{1, \alpha, \alpha^2, \ldots, \alpha^{15}\}$ of GF(17) with the multiplier $\eta$ set to 1. The binary CPM dispersion of $\mathbf{W}_s^{(9)}(4, 16)$ gives a $4 \times 16$ array $\mathbf{H}_{b,qc,\text{disp}}(4, 16)$ of CPMs of size $16 \times 16$, which is a $64 \times 256$ matrix over GF(2) with column and row weights 4 and 16, respectively.$^{(9)}$

Replacing each binary CPM in $\mathbf{H}_{b,qc,\text{disp}}^{(9)}(4, 16)$ by a 16-ary uniform CPM (random replacement), we obtain a $4 \times 16$ array $\mathbf{H}_{2^4,qc,\text{disp}}^{(9)}(4, 16)$ of uniform CPMs over GF($2^4$) of size $16 \times 16$. The null space over GF($2^4$) of $\mathbf{H}_{2^4,qc,disp}^{(9)}(4, 16)$ gives a 16-ary (4, 16)-regular (256, 192) QC-LDPC code $C_{2^4,qc}(4, 16)$ of length 256 with rate 0.75. The symbol- and word-error performances of $C_{2^4,qc}(4, 16)$ over the AWGN channel decoded with 50 iterations of the FFT-QSPA are shown in Figure 15.28.

Suppose we take eight columns from $\mathbf{W}_s^{(9)}(4, 16)$ (avoiding zeros) to form a $4 \times 8$ submatrix $\mathbf{W}_s^{(9)}(4, 8)$ of $\mathbf{W}_s^{(9)}(4, 16)$. Next, we mask $\mathbf{W}_s^{(9)}(4, 8)$ with the masking matrix $\mathbf{Z}(4, 8)$ given by (15.43). Masking results in a masked matrix $\mathbf{W}_{s,\text{mask}}^{(9)}(4, 8)$ with column and row weights 3 and 6, respectively. Using $\mathbf{W}_{s,\text{mask}}^{(9)}(4, 8)$ as the base matrix and GF($2^4$) as the symbol field, the B-to-NB replacement code construction gives a 16-ary (3, 6)-regular (128, 64) QC-LDPC code $C_{2^4,qc}(4, 8)$ of length 64 with rate 0.5.

---

## Problems

**15.1** Let $\alpha$ be a primitive element of the Galois field GF($q$) and $\mathbf{W}$ a matrix over GF($q$) that satisfies $\alpha$-multiplied row constraints 1 and 2. Show that the matrix $\mathbf{H}$ over GF($q$) obtained by dispersing every nonzero entry of $\mathbf{W}$ into an $\alpha$-multiplied $(q-1) \times (q-1)$ circulant permutation matrix and a zero entry into a $(q-1) \times (q-1)$ zero matrix satisfies the RC constraint.

**15.2** Find the generator polynomial of the 256-ary (255, 175) cyclic EG-LDPC code given in Example 14.2.

**15.3** Let the three-dimensional Euclidean geometry EG($3, 2^2$) over GF($2^2$) be the code-construction geometry. Construct a 64-ary QC-EG-LDPC code of length 126. Determine its dimension. Assume BPSK transmission over the binary-input AWGN channel. Compute the bit-, symbol-, and word-error performances of the 64-ary QC-EG-LDPC code decoded with 50 iterations of the FFT-QSPA.

**15.4** Let the two-dimensional Euclidean geometry EG($2, 2^4$) over GF($2^4$) be the code-construction geometry. Construct a regular 255-ary regular EG-LDPC code of length 256 using six parallel bundles of lines in EG($2, 2^4$). Determine its dimension. Decode the constructed code with 50 iterations of the FFT-QSPA. Compute its bit-, symbol-, and block-error performances over the binary-input AWGN channel with BPSK signaling.

**15.5** Construct a $15 \times 15$ array $\mathbf{H}_{2^4,\text{disp}}^{(1)}$ of $\alpha$-multiplied $15 \times 15$ circulant permutation and zero matrices over GF($2^4$) based on the (15, 2, 14) RS code over GF($2^4$).

(a) Determine the 16-ary QC-LDPC code given by the null space over GF($2^4$) of the $15 \times 15$ array $\mathbf{H}^{(1)}_{2^4,\text{disp}}$.
(b) Decode the code constructed in (a) with 50 iterations of the FFT-QSPA. Compute its bit-, symbol-, and word-error performances over the binary-input AWGN channel with BPSK signaling.

**15.6** Starting from the $(31, 2, 30)$ RS code over GF($2^5$), construct a $31 \times 31$ array $\mathbf{H}^{(1)}_{2^5,\text{disp}}$ of $\alpha$-multiplied $31 \times 31$ circulant permutation and zero matrices over GF($2^5$). Take a $4 \times 16$ subarray $\mathbf{H}^{(1)}_{2^5,\text{disp}}(4, 16)$ from $\mathbf{H}^{(1)}_{2^5,\text{disp}}$, avoiding zero matrices in $\mathbf{H}^{(1)}_{2^5,\text{disp}}$.
(a) Determine the 32-ary QC-LDPC code given by the null space over GF($2^5$) of $\mathbf{H}^{(1)}_{2^5,\text{disp}}(4, 16)$.
(b) Assume BPSK transmission over the binary-input AWGN channel. Decode the code constructed in (a) using the FFT-QSPA with 50 iterations. Compute its bit-, symbol-, and word-error performances. Compare the word-error performance of the code with that of a shortened RS code of the same length and rate decoded with the HD-BM algorithm.
(c) Compute the word-error performances of the code constructed in (a) with 3, 5, 10, and 50 iterations of the FFT-QSPA.

**15.7** Consider the $31 \times 31$ array $\mathbf{H}^{(1)}_{2^5,\text{disp}}$ constructed in Problem 15.6. Take an $8 \times 16$ subarray $\mathbf{H}^{(1)}_{2^5,\text{disp}}(8, 16)$ from $\mathbf{H}^{(1)}_{2^5,\text{disp}}$. Use $\mathbf{H}^{(1)}_{2^5,\text{disp}}(8, 16)$ as the base array for masking. Construct an $8 \times 16$ masking matrix $\mathbf{Z}(8, 16)$ with column and row weights 3 and 6, respectively. Mask $\mathbf{H}^{(1)}_{2^5,\text{disp}}(8, 16)$ with $\mathbf{Z}(8, 16)$ to obtain an $8 \times 16$ masked array $\mathbf{M}^{(1)}_{2^5,\text{disp}}(8, 16)$ of $\alpha$-multiplied $31 \times 31$ circulant permutation and zero matrices over GF($2^5$).
(a) Determine the 32-ary QC-LDPC code given by the null space over GF($2^5$) of the masked array $\mathbf{M}^{(1)}_{2^5,\text{disp}}(8, 16)$.
(b) Decode the code constructed in (a) with 50 iterations of the FFT-QSPA. Compute its bit-, symbol-, and word-error performances of the code over the binary-input AWGN channel with BPSK signaling.

**15.8** Let GF($2^6$) be the field for code construction and let $\alpha$ be a primitive element of GF($2^6$). Let $\mathcal{G}_1$ and $\mathcal{G}_2$ be two additive subgroups of GF($2^6$) spanned by the elements in $\{\alpha^0, \alpha, \alpha^2, \alpha^3\}$ and $\{\alpha^4, \alpha^5\}$, respectively. From these two groups, construct a $4 \times 16$ matrix $\mathbf{W}^{(4)}$ over GF($2^6$) using (12.10) that satisfies $\alpha$-multiplied row constraints 1 and 2. Using $\mathbf{W}^{(4)}$ as the base matrix for dispersion, construct a $4 \times 16$ array $\mathbf{H}^{(4)}_{2^6,qc,\text{disp}}$ of 63 $\alpha$-multiplied $63 \times 63$ CPMs and a single zero matrix over GF($2^6$). Take a $4 \times 12$ subarray $\mathbf{H}^{(4)}_{2^6,qc,\text{disp}}(4, 12)$ from $\mathbf{H}^{(4)}_{2^6,qc,\text{disp}}$, avoiding the zero matrix at the top of the first column of $\mathbf{H}^{(4)}_{2^6,qc,\text{disp}}$.
(a) Determine the 64-ary QC-LDPC code given by the null space over GF($2^6$) of $\mathbf{H}^{(4)}_{2^6,qc,\text{disp}}(4, 12)$.
(b) Assume BPSK transmission over the AWGN channel. Decode the code constructed in (a) with 5 and 50 iterations of the FFT-QSPA. Compute the bit-, symbol-, and word-performances of the code.

**15.9** Consider the $15\times15$ array $\mathbf{H}^{(1)}_{2^4,\text{disp}}$ of $\alpha$-multiplied $15\times15$ circulant permutation and zero matrices over GF($2^4$) constructed using the (15, 2, 30) RS code over GF($2^4$). Choose $k = 2$, $l = 6$, $s = 0$, and $t = 4$. Take a $4 \times 8$ subarray $\mathbf{H}^{(1)}_{2^4,\text{disp}}(4,8)$ (avoiding zero matrices on the main diagonal of the array $\mathbf{H}^{(1)}_{2^4,\text{disp}}$). Use $\mathbf{H}^{(1)}_{2^4,\text{disp}}(4,8)$ as the base array for dispersion as described in Sections 13.7 and 14.4.4. The 0-masked 6-fold dispersion of $\mathbf{H}^{(1)}_{2^4,\text{disp}}(4,8)$ gives a $24 \times 48$ array $\mathbf{H}^{(1)}_{2^4,\text{6-f,disp},0}(24,48)$ of $\alpha$-multiplied $15 \times 15$ circulant permutation and zero matrices over GF($2^4$).

(a) Determine the 16-ary QC-LDPC code given by the null space over GF($2^4$) of $\mathbf{H}^{(1)}_{2^4,\text{6-f,disp},0}(24,48)$.
(b) Assume BPSK transmission over the AWGN channel. Decode the code constructed in (a) with 50 iterations of the FFT-QSPA. Compute the bit-, symbol-, and word-error performances of the code.
(c) What are the erasure-burst-correction capability and efficiency of the code?

**15.10** In Problem 15.9, we now set $s = 1$ and all the other parameters remain the same. Determine the QC-LDPC code given by the null space over GF($2^4$) of the 1-masked 6-fold dispersed array $\mathbf{H}^{(1)}_{2^4,\text{6-f,disp},1}(24,48)$. Compute the bit-, symbol-, and word-error performances of the code decoded with 50 iterations of the FFT-QSPA over the AWGN channel with BPSK signaling.

**15.11** Assume 16-QAM transmission over the AWGN channel. Compute the symbol- and word-error performances of the code constructed in Problem 15.9 with 50 iterations of the FFT-QSPA.

## References

[1] M. C. Davey and D. J. C. MacKay, "Low-density parity check codes over GF($q$)," *IEEE Communications Letters*, vol. 2, no. 6, pp. 165–167, June 1998.
[2] D. J. C. MacKay and M. C. Davey, "Evaluation of Gallager codes of short block length and high rate applications," *Proceedings of the IMA International Conference on Mathematics and Its Applications: Codes, Systems and Graphical Models*, New York, Springer-Verlag, 2000, pp. 113–130.
[3] L. Barnault and D. Declercq, "Fast decoding algorithm for LDPC over GF($2^q$)," *Proceedings of the 2003 IEEE Information Theory Workshop*, Paris, March 31–April 4, 2003, pp. 70–73.
[4] D. Declercq and M. Fossorier, "Decoding algorithms for nonbinary LDPC codes over GF($q$)," *IEEE Transactions on Communications*, vol. 55, no. 4, pp. 633–643, April 2007.
[5] J. I. Boutros, A. Ghaith, and Yi Yun-Wu, "Non-binary adaptive LDPC codes for frequency selective channels: Code construction and iterative decoding," *Proceedings of the IEEE Information Theory Workshop*, Chengdu, October 2007, pp. 184–188.
[6] C. Poulliat, M. Fossorier, and D. Declercq, "Optimization of non binary LDPC codes using their binary images," *Proceedings of the International Turbo Code Symposium*, Munich, April 2006.

[7] C. Poulliat, M. Fossorier, and D. Declercq, "Design of non binary LDPC codes using their binary images: Algebraic properties," *Proceedings of the IEEE International Symposium on Information Theory*, Seattle, WA, July 2006.

[8] A. Voicila, D. Declercq, F. Verdier, M. Fossorier, and P. Urard, "Split non-binary LDPC codes," *Proceedings of the IEEE International Symposium on Information Theory*, Toronto, Ontario, July 6–11, 2008.

[9] S. Lin, S. Song, B. Zhou, J. Kang, Y. Y. Tai, and Q. Huang, "Algebraic constructions of nonbinary quasi-cyclic LDPC codes: Array masking and dispersion," *Proceedings of the 2nd Workshop for the Center of Information Theory and its Applications*, UCSD Division of Calit2 and the Jacobs School of Engineering, San Diego, CA, January 29–February 2, 2007.

[10] V. Rathi and R. Urbanke, "Density evolution, thresholds and the stability condition for non-binary LDPC codes," *IEE Proceedings on Communications*, vol. 152, no. 6, pp. 1069–1074, December 2005.

[11] S. Song, L. Zeng, S. Lin, and K. Abdel-Ghaffar, "Algebraic constructions of nonbinary quasi-cyclic LDPC codes," *Proceedings of the IEEE International Symposium on Information Theory*, Seattle, WA, July 9–14, 2006, pp. 83–87.

[12] L. Zeng, L. Lan, Y. Y. Tai, and S. Lin, "Dispersed Reed–Solomon codes for iterative decoding and construction of $q$-ary LDPC codes," in *Proceedings of the IEEE Global Telecommunications Conference*, St. Louis, MO, November 28–December 2, 2005, pp. 1193–1198.

[13] L. Zeng, L. Lan, Y. Y. Tai, B. Zhou, S. Lin, and K. A. S. Abdel-Ghaffar, "Construction of nonbinary cyclic, quasi-cyclic and regular LDPC codes: A finite geometry approach," *IEEE Transactions on Communications*, vol. 56, no. 3, pp. 378–387, March 2008.

[14] L. Zeng, L. Lan, Y. Y. Tai, S. Song, S. Lin, and K. Abdel-Ghaffar, "Constructions of nonbinary quasi-cyclic LDPC codes: A finite field approach," *IEEE Transactions on Communications*, vol. 56, no. 4, pp. 545–554 , April 2008.

[15] B. Zhou, J. Kang, Y. Y. Tai, S. Lin, and Z. Ding, "High performance non-binary quasi-cyclic LDPC codes on Euclidean geometries," *IEEE Transactions on Communications*, vol. 57, no. 4, pp. 545–554, April 2009.

[16] B. Zhou, J. Kang, S. Song, S. Lin, K. Abdel-Ghaffar, and M. Xu, "Construction of non-binary quasi-cyclic LDPC codes by arrays and array dispersions," *IEEE Transactions on Communications*, vol. 57, no. 6, pp. 1652–1662, June 2009.

[17] B. Zhou, J. Kang, Y. Y. Tai, Q. Huang, and S. Lin, "High performance nonbinary quasi-cyclic LDPC codes on Euclidean geometries," *Proceedings of the 2007 Military Communications Conference (MILCOM 2007)*, Orlando, FL, October 29–31, 2007.

[18] B. Zhou, Y. Y. Tai, L. Lan, S. Song, L. Zeng, and S. Lin, "Construction of high performance and efficiently encodable nonbinary quasi-cyclic LDPC codes," *Proceedings of the IEEE Global Telecommunications Conference*, San Francisco, CA, November 27–December 1, 2006.

[19] B. Zhou, L. Zhang, J. Kang, Q. Huang, Y. Y. Tai, and S. Lin, "Non-binary LDPC codes vs. Reed–Solomon codes," *Proceedings of the 3rd Workshop for the*

*Center of Information Theory and its Applications*, UCSD Division of Calit2 and the Jacobs School of Engineering, San Diego, CA, January 27–February 1, 2008.

[20] B. Zhou, L. Zhang, Q. Huang, S. Lin, and M. Xu, "Constructions of high performance non-binary quasi-cyclic LDPC codes," *Proceedings of the IEEE Information Theory Workshop (ITW)*, pp. 71–75, Porto, May 5–9, 2008.

[21] B. Zhou, L. Zhang, J. Y. Kang, Q. Huang, S. Lin and K. Abdel-Ghaffar, "Array dispersions of matrices and constructions of quasi-cyclic LDPC codes over non-binary fields," *Proceedings of the IEEE International Symposium on Information Theory*, Toronto, Ontario, July 6–11, 2008.

[22] W. J. Gross, F. R. Kschischang, R. Kötter, and P. G. Gulak, "Applications of algebraic soft-decision decoding of Reed–Solomon codes," *IEEE Transactions on Communications*, vol. 54, no. 7, pp. 1224–1234, July 2006.

[23] H. Wymeersch, H. Steendam, and M. Moeneclaey, "Log-domain decoding of LDPC codes over GF($q$)," *Proceedings of the IEEE International Conference on Communications*, Paris, June 2004, pp. 772–776.

[24] E. R. Berlekamp, *Algebraic Coding Theory*, New York, McGraw-Hill, 1968 (revised edition Laguna Hill, CA, Aegean Park Press, 1984).

[25] J. L. Massey, "Shift-register synthesis and BCH decoding," *IEEE Transactions on Information Theory*, vol. 15, no. 1, pp. 122–127, January 1969.

[26] R. Kötter and A. Vardy, "Algebraic soft-decision decoding of Reed–Solomon codes," *IEEE Transactions on Information Theory*, vol. 49, no. 11, pp. 2809–2825, November 2003.

[27] M. El-Khamy and R. McEliece, "Iterative algebraic soft-decision list decoding of RS codes," *IEEE Journal on Selected Areas in Communications*, vol. 24, no. 3, pp. 481–490, March 2006.

[28] S. Lin and D. J. Costello, Jr., *Error Control Coding: Fundamentals and Applications*, 2nd ed., Upper Saddle River, NJ, Prentice-Hall, 2004.

[29] A. Bennatan and D. Burshtein, "Design and analysis of nonbinary LDPC codes for arbitrary discrete-memoryless channels," *IEEE Transactions on Information Theory*, vol. 52, no. 2, pp. 549–583, February 2006.

[30] D. J. C. MacKay, "Good error-correcting codes based on very sparse matrices," *IEEE Transactions on Information Theory*, vol. 45, no. 2, pp. 399–432, March 1999.

[31] A. Goupil, M. Colas, G. Gelle, and D. Declercq, "On BP decoding of LDPC codes over groups," *Proceedings of the International Symposium on Turbo Codes*, Munich, April 2006, CD-ROM.

[32] A. Goupil, M. Colas, G. Gelle, and D. Declercq, "FFT-based BP decoding of general LDPC codes over Abelian groups," *IEEE Transactions on Communications*, vol. 55, no. 4, pp. 644–649, April 2007.

[33] Y. Kou, S. Lin, and M. Fossorier, "Low-density parity-check codes based on finite geometries: A rediscovery and new results," *IEEE Transactions on Information Theory*, vol. 47, no. 7, pp. 2711–2736, November 2001.

[34] Z. Li, L. Chen, L. Zeng, S. Lin, and W. Fong, “Efficient encoding of quasi-cyclic low-density parity-check codes,” *IEEE Transactions on Communications*, vol. 54, no. 1, pp. 71–81, January 2006.

[35] X. Li and M. R. Soleymani, “A proof of the Hadamard transform decoding of the belief propagation decoding for LDPCC over GF($q$),” *Proceedings of the Fall Vehicle Technology Conference*, vol. 4, September 2004, pp. 2518–2519.

[36] H. Song and J. R. Cruz, “Reduced-complexity decoding of $Q$-ary LDPC codes for magnetic recording,” *IEEE Transactions on Magnetics*, vol. 39, no. 2, pp. 1081–1087, March 2003.

[37] J. Li, S. Lin, K. Abdel-Ghaffar, W. E. Ryan, and D. J. Costello, *LDPC Code Designs, Constructions, and Unification*, Cambridge, Cambridge University Press, 2017.

# Index

Printed in the United States
by Baker & Taylor Publisher Services